29th Symposium of Malaysian Chemical Engineers (SOMChE 2016)

IOP Conference Series: Materials Science and Engineering
Volume 206

Miri, Malaysia
1 – 3 December 2016

ISBN: 978-1-5108-4325-7
ISSN: 1757-8981

Printed from e-media with permission by:

Curran Associates, Inc.
57 Morehouse Lane
Red Hook, NY 12571

Some format issues inherent in the e-media version may also appear in this print version.

Printed by Curran Associates, Inc. (2017)

Institute of Physics
Dirac House, Temple Back
Bristol BS1 6BE UK

Phone: 44 1 17 929 7481
Fax: 44 1 17 920 0979

techtracking@iop.org

Additional copies of this publication are available from:

Curran Associates, Inc.
57 Morehouse Lane
Red Hook, NY 12571 USA
Phone: 845-758-0400
Fax: 845-758-2633
Email: curran@proceedings.com
Web: www.proceedings.com

Table of contents

Volume 206

29th Symposium of Malaysian Chemical Engineers (SOMChE) 2016

1–3 December 2016, Miri, Sarawak, Malaysia

Accepted papers received: 26 May 2017
Published online: 21 June 2017

Preface

Carbon dioxide capture technology

Engineering education and other relevant topics

Energy

Food Process Engineering

Material Science and Engineering

Process System Engineering

Waste and Water Treatment & Management

IOP Conf. Series: Materials Science and Engineering **206** (2017) 011001 doi:10.1088/1757-899X/206/1/011001

EDITORIAL NOTE

29th Symposium of Malaysian Chemical Engineers (SOMChE) 2016 was held in Miri Marriott Resort & Spa, Miri, Sarawak, Malaysia from 1st to 3rd December 2016. This symposium was jointly co-organised by Department of Chemical Engineering, Curtin University Malaysia and Institution of Chemical Engineers (IChemE), Malaysia. The theme of SOMChE 2016 is "Engineering Solutions for Sustainable Development". Population growth and industrial development place increasing pressures on the available resources contributing to challenging provision of food, water and energy supplies, environmental issues and climate change. These grand challenges for humanity require immediate solutions in sustainable ways without harming the environment. Therefore, SOMChE 2016 brought together a community of chemical and process engineers and other related engineers from Malaysia and South East Asia region and beyond to discuss the latest advances and best practices of engineering solutions in industry and academic for sustainable food, water, energy and environment. More than 150 delegates, including international participants from Indonesia, Vietnam, Philippines, Brunei, Singapore, India, Iran, Australia and Japan, attended and presented their papers at the symposium. This proceeding is compilation of the selected, accepted papers and covers 9 subject areas: Carbon Dioxide Capture Technology (CA), Engineering Education and Other Relevant Topics (ED), Energy (EN), Food Process Engineering (FD), Material Science and Engineering (MA), Process System Engineering (PS), Reaction Engineering (RE), Separation Processes (SE) and Waste and Water Treatment & Management (WT). Each paper had undergone double peer review to ensure its quality. We believe that the proceedings will serve as a book of knowledge and an essential literature, which will lead to not only scientific and engineering progress but also other novel processes and new products. SOMChE 2016 might not be successful without the supports from various parties. Therefore, we would like to thank the participants, organising committee, sponsors, and all those who have contributed directly and indirectly in making this symposium a big success.

Editors: Dr. Agus Saptoro; Dr. Wee Siaw Khur; Dr. Lau Shiew Wei; Dr. Wendy Pei Qin Ng; Dr. Mahmood Anwar; Ms. Christine Yeo; Dr. Khor Ee Huey
Curtin University Malaysia

IOP Conf. Series: Materials Science and Engineering **206** (2017) 011001 doi:10.1088/1757-899X/206/1/011001

ORGANIZING COMMITTEE

Chairman Dr. Agus Saptoro

Deputy Chairman Assoc. Prof. Chua Han Bing

Secretary and Registration Committee Dr. Lau Shiew Wei, Dr. Wendy Pei Qin Ng, Dr. Wee Siaw Khur, Ms. Christine Yeo, Dr. Khor Ee Huey, Dr. Mahmood Anwar

Treasurer Dr. Zeinab Abbas Jawad, Ms. Hii Kiew Ling, Dr. Stephanie Chan

Program & Publication Committee Dr. Agus Saptoro, Dr. Wee Siaw Khur, Dr. Wendy Pei Qin Ng, Dr. Hendra Gunawan Harno, Dr. Mahmood Anwar, Assoc. Prof. Mubarak Mujawar, Prof. Michael Kobina Danquah, Assoc. Prof. Hari Babu Vuthaluru, Assoc. Prof. Perumal Kumar, Assoc. Prof. Jobrun Nandong

Sponsorship Committee Dr. Bridgid Chin, Ms. Hii Kiew Ling, Ms. Loh Wan Ning, Mr. Freddie Panau

Logistic & Exhibition Committee Dr. John Lau, Dr. Jibrail Kansedo, Dr. Henry Foo Chee Yew, Dr. Tan Inn Shi, Dr. Faisal Amri Tanjung

Publicity Committee Mr. Tay Chen Lim, Dr. Yalun Arifin

SCIENTIFIC COMMITTEE

1. Prof. Brent Young (The University of Auckland, New Zealand)
2. Prof. Manabu Kano (Kyoto University, Japan)
3. Associate Prof. Yusuke Asakuma (University of Hyogo, Japan)
4. Prof. Xiao Dong Chen (Soochow University, China)
5. Prof. Rahmat Sotudeh-Gharebagh (University of Tehran, Iran)
6. Prof. Chun-Zhu Li (Curtin University, Australia)
7. Dr. Sharif Hussein Sharif Zein (University of Hull, UK)
8. Dr. Aditya Putranto (Monash University, Australia)
9. Dr. Ujjal Kumar Ghosh (Qatar University, Qatar)
10. Dr. Jens Kromer (The University of Queensland, Australia)
11. Prof. Bassim H. Hameed (USM, Malaysia)

IOP Conf. Series: Materials Science and Engineering **206** (2017) 011001 doi:10.1088/1757-899X/206/1/011001

12. Associate Prof. Lim Jit Kang (USM, Malaysia)
13. Dr. Low Siew Chun (USM, Malaysia)
14. Associate Prof. Suzana Yusup (UTP, Malaysia)
15. Prof. Denny Ng Kok Sum (University of Nottingham, Malaysia)
16. Associate Prof. Dr Lam Hon Loong (University of Nottingham, Malaysia)
17. Dr. Chong Chien Hwa (Taylor University, Malaysia)
18. Dr. Irene Chew Mei Leng (Monash University, Malaysia)
19. Associate Prof. Farouq Twaiq (Swinburne University, Malaysia Campus)
20. Prof. Awang Bono (UMS, Malaysia)
21. Associate Prof. Christopher Chu (UMS, Malaysia)
22. Prof. T. Murugesan (UTP, Malaysia)
23. Dr. Yeong Yin Fong (UTP, Malaysia)
24. Prof. Duduku Krishniah (UMS, Malaysia)
25. Dr. Lam Man Kee (UTP, Malaysia)
26. Associate Prof. Dr. Syamsul Rizal Abd Shukor (USM, Malaysia)
27. Prof. Siti Kartom Kamarudin (UKM, Malaysia)
28. Associate Prof. Chan Eng Seng (Monash University, Malaysia)
29. Prof. Mohamed Kheireddine Bin Taieb Aroua (UM, Malaysia)
30. Associate Prof. Chong Meng Nan (Monash University, Malaysia)
31. Prof. Siti Rozaimah Sheikh Abdullah (UKM, Malaysia)
32. Associate Prof. Rozita Yusoff (UM, Malaysia)
33. Prof. Zainuddin Abdul Manan (UTM, Malaysia)
34. Dr Teoh Yi Peng (UNIMAP, Malaysia)
35. Dr Mohamad Amran Mohd Salleh (UPM, Malaysia)
36. Prof. Mohamed Azlan Hussain (UM, Malaysia)
37. Prof. Robiah Yunus (UPM, Malaysia)
38. Prof Luqman Chuah Bin Abdullah (UPM, Malaysia)
39. Dr Rebecca Lim Xiao Yien (Taylors University, Malaysia)
40. Associate Prof. Haslenda Hashim (UTM, Malaysia)
41. Prof. Thomas Choong Shean Yaw (UPM, Malaysia)
42. Prof. Abdul Aziz Abdul Raman (UM, Malaysia)
43. Dr Farizul Hafiz Bin Kasim (UNIMAP, Malaysia)
44. Dr Muhammad Syarhabil Bin Ahmad (UNIMAP, Malaysia)
45. Prof. Mohd Sobri Takriff (UKM, Malaysia)
46. Prof. Azman Bin Hassan (UTM, Malaysia)
47. Dr Suryadi Ismadji (Widya Mandala Catholic University, Indonesia)
48. Professor Rochmadi (Gadjah Mada University, Indonesia)
49. Prof. Clem Kuek (Curtin University Malaysia)
50. Prof. Michael Kobina Danquah (Curtin University Malaysia)

IOP Conf. Series: Materials Science and Engineering **206** (2017) 011001 doi:10.1088/1757-899X/206/1/011001

SPONSORS

Sarawak Convention Bureau

Sarawak Shell Berhad

Solution Engineering

MMI Engineering

CREST NanoSolutions

Young Scientists Network Malaysia

INSTITUTE OF MATERIALS, MALAYSIA

SARAWAK TOURISM BOARD

ACS Chemistry for Life® MALAYSIA CHAPTER

IOP Conf. Series: Materials Science and Engineering **206** (2017) 011002 doi:10.1088/1757-899X/206/1/011002

Peer review statement

All papers published in this volume of *IOP Conference Series: Materials Science and Engineering* have been peer reviewed through processes administered by the proceedings Editors. Reviews were conducted by expert referees to the professional and scientific standards expected of a proceedings journal published by IOP Publishing.

Carbon Dioxide (CO_2) Adsorption by Activated Carbon Functionalized with Deep Eutectic Solvent (DES)

N Z Zulkurnai[1], U F Md. Ali[1*], N Ibrahim[1], N S Abdul Manan[2]
[1]School of Environmental Engineering, Universiti Malaysia Perlis, Kompleks Pusat Pengajian Jejawi 3, 02600 Arau, Perlis, Malaysia
[2]Department of Chemistry, Faculty of Science, University of Malaya, 50603 Kuala Lumpur, Malaysia

Email: umifazara@unimap.edu.my

Abstract. In recent years, carbon dioxide (CO_2) emission has become a major concern as the amount of the emitted gas significantly increases annually. Consequently, this phenomenon contributes to global warming. Several CO_2 capture methods, including chemical adsorption by activated carbon, have been proposed. In this study, activated carbon was prepared from sea mango (*Cerbera odollam*), which was functionalized with deep eutectic solvent (DES) composed of choline chloride and glycerol to increase the efficiency of CO_2 capture. The samples underwent pre-carbonization and carbonization processes at 200 °C and 500 °C, respectively, with nitrogen gas and flowing several gases, namely, CO_2 and steam, and then followed by impregnation with 50% phosphoric acid (H_3PO_4) at 1:2 precursor-to-activant ratio. The prepared activated carbon was impregnated with DES at 1:2 precursor-to-activant ratio. The optimum CO_2 adsorption capacity of the activated carbon was obtained by using CO_2 gas treatment method (9.851 mgCO_2/g_{sol}), followed by the absence of gases (9.685 mgCO_2/g_{sol}), steam (9.636 mgCO_2/g_{sol}), and N_2 (9.536 mgCO_2/g_{sol}).

1. Introduction

Current atmospheric carbon dioxide (CO_2) levels are considered alarming; such levels have contributed to global environmental issues. Santos [1] reported that 70% of CO_2 emission is attributed to different forms of energy generated and used in various processes, such as burning of fossil fuels, including natural gas, oil, or coal, and industrial processes, such as cement production, oil refinery, power plant processes, and iron and steel manufacturing. Furthermore, CO_2 emission not only adversely affects climate change but also threatens human health.

Several effective methods of carbon capture and storage (CCS), such as adsorption [2], membrane separation [3], and cryogenic separation, have been proposed to reduce the amount of emitted CO_2 in the atmosphere [4]. Lee [5] demonstrated that adsorption is a promising method to remove CO_2 because it is cost effective.

Adsorption processes are generally performed using activated carbon [6], activated [7], zeolite [8], and polymeric adsorbents [9]. A few researchers [5, 10, 11] also revealed that activated carbon can capture CO_2 because it consists of a large surface area per unit volume and submicroscopic pores, in which contaminant adsorption occurs. Moreover, activated carbon is stable under acidic and basic conditions. It is also cost effective because it can be regenerated and thus suitable for organic compound removal.

Considering cost effectiveness in activated carbon production, researchers developed different precursors from abundant waste materials, such as palm shells, sea mango, cocoa pod shells, and rice

IOP Conf. Series: Materials Science and Engineering **206** (2017) 012001 doi:10.1088/1757-899X/206/1/012001

husks. For instance, Siti Noraishah [12] successfully produced activated carbon from rice husks and utilized it to remove CO_2 generated from industrial activities.

The integration of chemicals into activated carbon has shown an excellent performance in CO_2 uptake. Saad [13] and Karousos [14] demonstrated that the impregnation of activated carbon with chemicals, such as amine and ionic liquids (ILs), has increased the amount of captured CO_2. Although the immobilization of these chemicals in activated carbon assists CO_2 capture, such chemicals are difficult to prepare because they are toxic and relatively expensive [15]. Abbott [16] developed a novel green solvent known as Deep Eutectic Solvent (DES) to overcome the limitation of amine and ILs [17].

DES is used in numerous possible applications, including metal electro deposition, metal electropolishing, and synthesis applications, such as biodiesel purification and manufacturing, biotransformation, and CO_2 capture [17]. DES for CO_2 capture has been extensively investigated because of its easy preparation, biodegradability, and nontoxicity; DES can also yield results similar to amine, which is rigidly known for good CO_2 capture [18].

DES can be classified into four types: Type 1, quaternary ammonium cation + metal halide; Type 2, quaternary ammonium cation + hydrated metal halide; Type 3, quaternary ammonium cation + hydrogen bond donor; and Type 4, metal halide + hydrogen bond donor. Among these types, Type 3 is the most favorable because it is easily produced, mostly unreactive in water, biodegradable, and inexpensive [17].

In the current research, the prepared activated carbon from sea mango (*Cerbera odollam*) was modified to improve the CO_2 adsorption capacity by promoting DES on the surface of the activated carbon through chemical impregnation. Thus far, very few studies have focused on CO_2 capture with DES. Hence, the present work focused on the characterization and performance of CO_2 capture by using DES-functionalized activated carbon.

2. Procedure

2.1. Preparation of activated carbon

Sea mango (*Cerbera odollam*) was collected around Perlis and Kedah. The skin and seeds of the raw material were removed to obtain the fibrous shell. The samples were then washed and dried in an oven for 24 h at 105 °C to remove the moisture content. Afterward, the samples were crushed and sieved to obtain fine particles measuring 1 mm to 2 mm. The raw material was then subjected to semi-carbonization in a tube furnace for 30 min at 200 °C by flowing with nitrogen gas. Char was subsequently impregnated with 50 wt% H_3PO_4 at 1:2 precursor-to-activant ratio for 4 h and then dehydrated in the oven at 110 °C until dryness.

The char was weighed and placed in a tube furnace with three different gases, namely, nitrogen, carbon dioxide, and steam, for activation. One tube furnace was not filled with gases. The furnace was heated until the desired temperature of 500 °C was reached. Activation was conducted for 2 h. The samples were cooled to room temperature (25°C) and washed few times with hot distilled water to remove excess H_3PO_4.

Activated carbon was then filtered and dried in the oven until it is completely dried. The activated carbon samples for gas activation of nitrogen, carbon dioxide, steam, and absence of any gases were denoted as AC-N_2, AC-CO_2, AC-S, and AC-AAG, respectively.

2.2. Preparation of DES

Choline chloride and glycerol (1:2) were added at 80 °C under stirring. The mixture was stirred for approximately 1 h, and the resulting solution was clear and homogeneous.

2.3. Functionalization of activated carbon with DES

Each type of activated carbon was impregnated by using DES with a 1:2 ratio of activated carbon to DES for 4 h by using the vacuum impregnation method with a vacuum pressure of −0.5 bar. The activated carbon was completely dried at 150 °C. The impregnated activated carbon samples for AC-N_2, AC-CO_2, AC-S, and AC-AAG were denoted as AC-N_2-D, AC-CO_2-D, AC-S-D, and AC-AAG-D, respectively.

2.4. Carbon dioxide adsorption

The experimental set-up employed in this study is as shown in figure 1. The adsorption column was made of glass with 1 cm inner diameter and 30 cm long, enclosed with an insulator to retain a constant

IOP Conf. Series: Materials Science and Engineering **206** (2017) 012001 doi:10.1088/1757-899X/206/1/012001

temperature throughout the whole experiments. The glass column was vertically installed and implanted within a glass water jacket with inlet and outlet connections. The activated carbon (3 g) was inserted into this column with a piece of cotton wool at the top and bottom for breakthrough study. During the experiments, the gas flow was in an up-flow manner at 5ml/min.

A Guardian NG CO_2 analyzer with measuring a range of 0-30% CO_2 (Edinburgh Instruments Ltd, Chorley, Lancashire, England) was used to measure the concentration of CO_2 exiting the adsorption bed. The CO_2 analyzer displays a value when the adsorption column is saturated with CO_2 and thus causes the gas to break through out of the column. The breakthrough time was recorded for every sample.

The gases flow from the gas cylinders was regulated using gas regulators Weldmark provided by (Weldmarks's Corporate, Indianapolis, Indiana, United State). A set of gas flow controller (Brooks Instrument, PA, USA) model SLA 5850 and model 0254 Secondary Electronics were used to accurately measure and control the gas flow rate. The gases, purified N_2 and 15 % CO_2 (balance N_2) were obtained from The Linde Group (Munich, Germany).

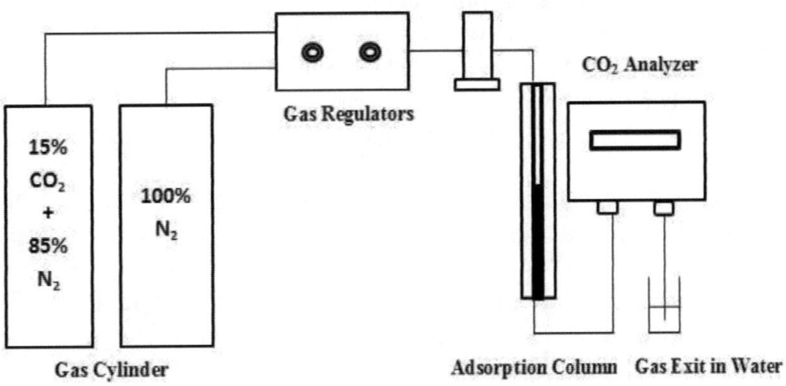

Figure 1. Experimental Setup.

2.5. Characterization of activated carbon

The surface area of the activated carbon was determined using a BELSORP-Mini machine. Fourier transform infrared (FTIR-KBR) with the obtained spectra between 450 and 4000 cm^{-1} was used to determine the surface properties and functional groups. Surface morphology and element composition analyses were performed using a field emission scanning electron microscope (FESEM), and the total ash content of the adsorbent was determined in accordance with ASTM standard D2866 – 11[19].

3. Results and Discussion

3.1. Characterization results

3.1.1. BET surface area and pore characteristics. Table 1 shows the BET surface area and characteristics for all the samples. Different gas treatments showed varying surface areas. The highest surface area of the non-functionalized activated carbon was obtained from the absence of any gas treatment at 882.71 m^2/g, whereas the lowest surface area from steam treatment was 842.25 m^2/g.

After impregnation with DES was conducted, the surface area of all of the gas treatments reduced almost to half at 435.63, 426.40, 423.99, and 455.18 m^2/g for AC-AAG-D, AC-S-D, AC-N_2-D, and AC-CO_2-D, respectively. This result is attributed to the deposition of DES on the surface and inside the pore, which reduced the surface area of the activated carbon [20].

Table 1. Total surface area and pore volume of different types of activated carbon.

Sample	Total surface area [$m^2\ g^{-1}$]	Vmicropore [$cm^3\ g^{-1}$]	Vmesopore [$cm^3\ g^{-1}$]	Total pore volume [$cm^3\ g^{-1}$]	Main type of pore
AC-AAG	882.71	0.4539	0.1520	0.4696	Micropore
AC-S	842.25	0.4556	0.1543	0.4696	Micropore
AC-N_2	852.06	0.4318	0.1475	0.4445	Micropore
AC-CO_2	880.39	0.4438	0.1516	0.4603	Micropore
AC-AAG-D	435.63	0.2197	0.0733	0.2276	Micropore
AC-S-D	426.40	0.2129	0.0693	0.2191	Micropore
AC-N_2-D	423.99	0.2125	0.0713	0.2195	Micropore
AC-CO_2-D	455.18	0.2263	0.0722	0.2334	Micropore

3.1.2. FTIR Spectra. FTIR is an adaptable technique that can be applied to determine surface chemistry. In particular, chemical bonding was observed in the sample. The following chemical bonds (table 2) were observed in all of the samples: OH-stretch, H-bonded, and C-H stretch interactions. These bonds were possibly ascribed to the cerebrin structure of sea mango. The N-H bonds from the primary amine group only existed in the DES-functionalized activated carbon because the presence of choline chloride in the activated carbon contributed to the N-H bond. Figure 2 shows the suggested attachment of DES on the surface of the activated carbon. The elemental N which is believed to assist in CO_2 capture into the substrate was successfully attached on the surface of the activated carbon as per illustrated in figure 3 [21]. The C-C stretch from the aromatic group was present in the raw and non-functionalized activated carbon; however, this bond was absent in the DES-functionalized activated carbon because the reaction of H from DES with C from cerebrin promoted a new band, which is a C-H bond from the alkane group at peaks 1477.89 cm^{-1} to 1478.35 cm^{-1}.

The C-O stretch existed in the non-functionalized activated carbon at peaks 1055.02 cm^{-1} to 1150.09 cm^{-1}. However, this bond was absent in the DES-functionalized activated carbon. The C-N stretch bond from the aliphatic amine group existed only in the DES-functionalized activated carbon because it is attributed to the choline chloride of DES.

The C-H bond from the alkane group was present at peaks 955.94 cm^{-1} to 956.58 cm^{-1} for the DES-functionalized activated carbon but was absent in the raw and non-functionalized activated carbon. The C-H bond was likely attributed to glycerol in DES. The C-Cl stretch from the alkyl halide group was present only in the DES-functionalized activated carbon because of the presence of Cl element in the choline chloride. As such, the functionalization process added several functional groups to the surface of the activated carbon. The presence of DES increases the number of active sites on the surface and enhances the CO_2 adsorption capacity [22].

Table 2. Fourier transforms infrared analysis results.

Bond	Functional group	Raw (cm^{-1})	AC-AAG	AC-CO_2	AC-N_2	AC-S	AC-AAG-D	AC-CO_2-D	AC-N_2-D	AC-S-D
O–H stretch, H–bonded	alcohols, phenols	3411.10	3429.82	3430.09	3415.09	3429.17	3401.20	3399.77	3399.93	3400.00
C–H stretch	alkanes	2923.78	2043.6	2051.88	2053.3	2081.40	2942.90	2941.62	2944.48	2937.06
N–H bend	primary amine	-	-	-	-	-	1635.98	1638.00	1638.51	1636.21
C–C stretch (in–ring)	aromatics	1622.97	1617.60	1554.40	1542.23	1570.20	-	-	-	-
C–H bend	alkanes	1452.05					1478.35	1478.02	1477.98	1477.89
C–O stretch	alcohols, carboxylic acids, esters, ethers	1055.02	1146.76	1146.78	1144.05	1150.09	-	-	-	-
C–N stretch	aliphatic amines	-	-	-	-	-	-	1205.81	1206.90	-
		-	-	-	-	-	1083.92	1084.17	1084.19	1084.09
		-	-	-	-	-	1048.33	1046.35	1048.45	1046.89
C–H bend	alkanes	-	-	-	-	-	955.95	956.19	956.58	955.94
C–Cl stretch	alkyl halides	-	-	-	-	-	-	864.41	885.70	864.92
		-	-	-	-	-	561.70	570.22	553.18	565.59

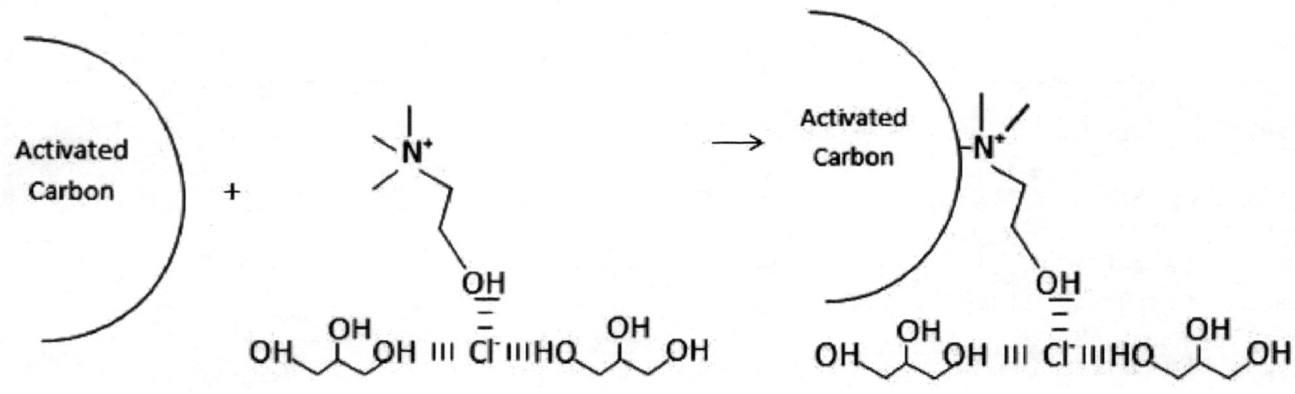

Figure 2. Reaction of DES on the surface of the activated carbon.

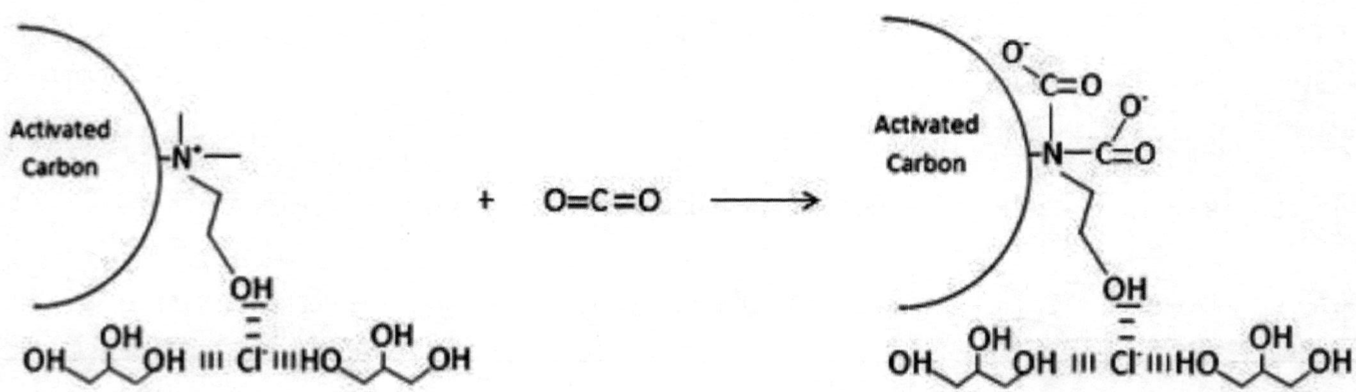

Figure 3. Mechanism of CO_2 interaction with DES.

IOP Conf. Series: Materials Science and Engineering **206** (2017) 012001 doi:10.1088/1757-899X/206/1/012001

3.1.3. Surface Morphology. Surface morphology represents the physical structure of each sample as shown in figure 4(a)-(i).

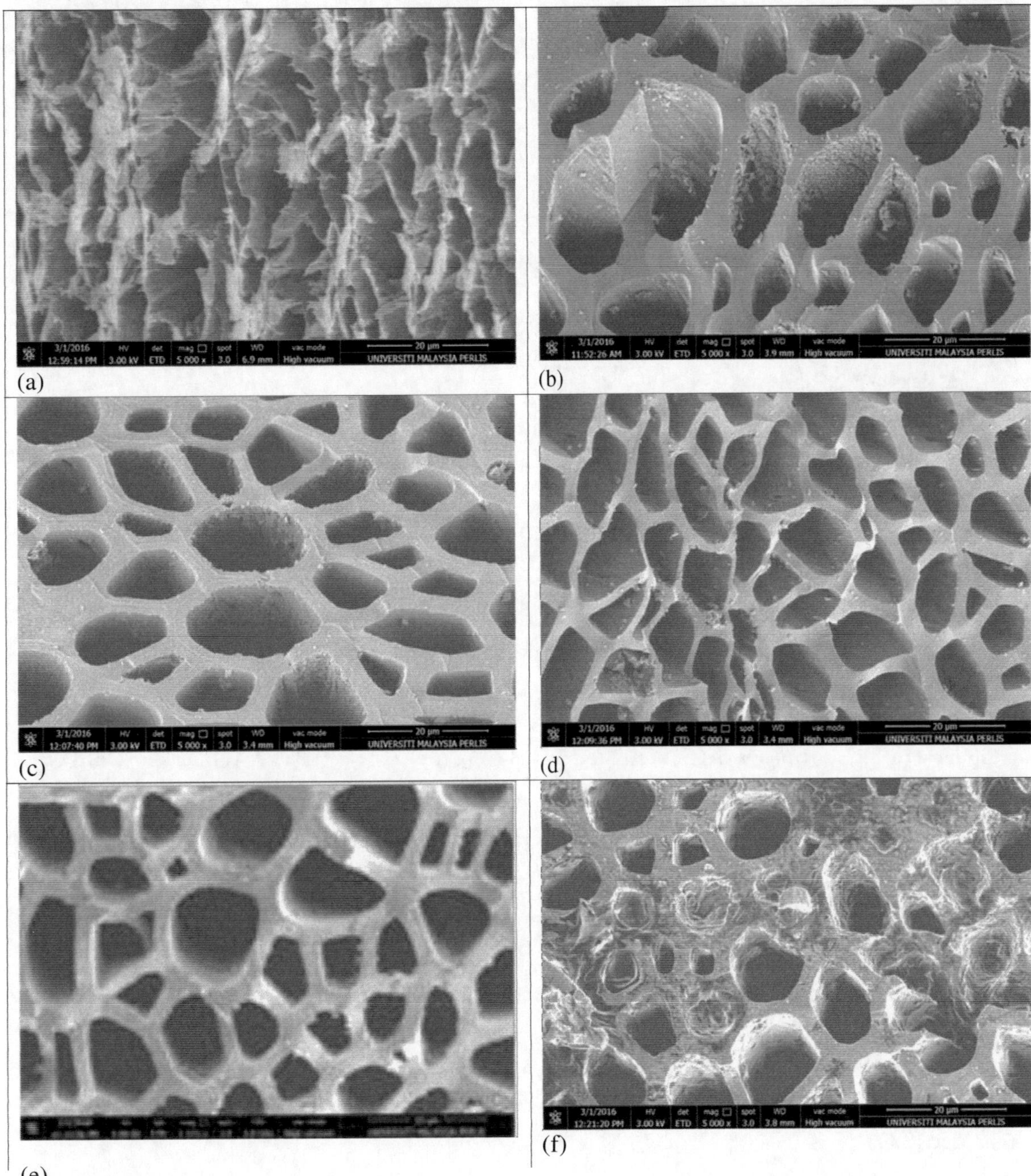

IOP Conf. Series: Materials Science and Engineering **206** (2017) 012001 doi:10.1088/1757-899X/206/1/012001

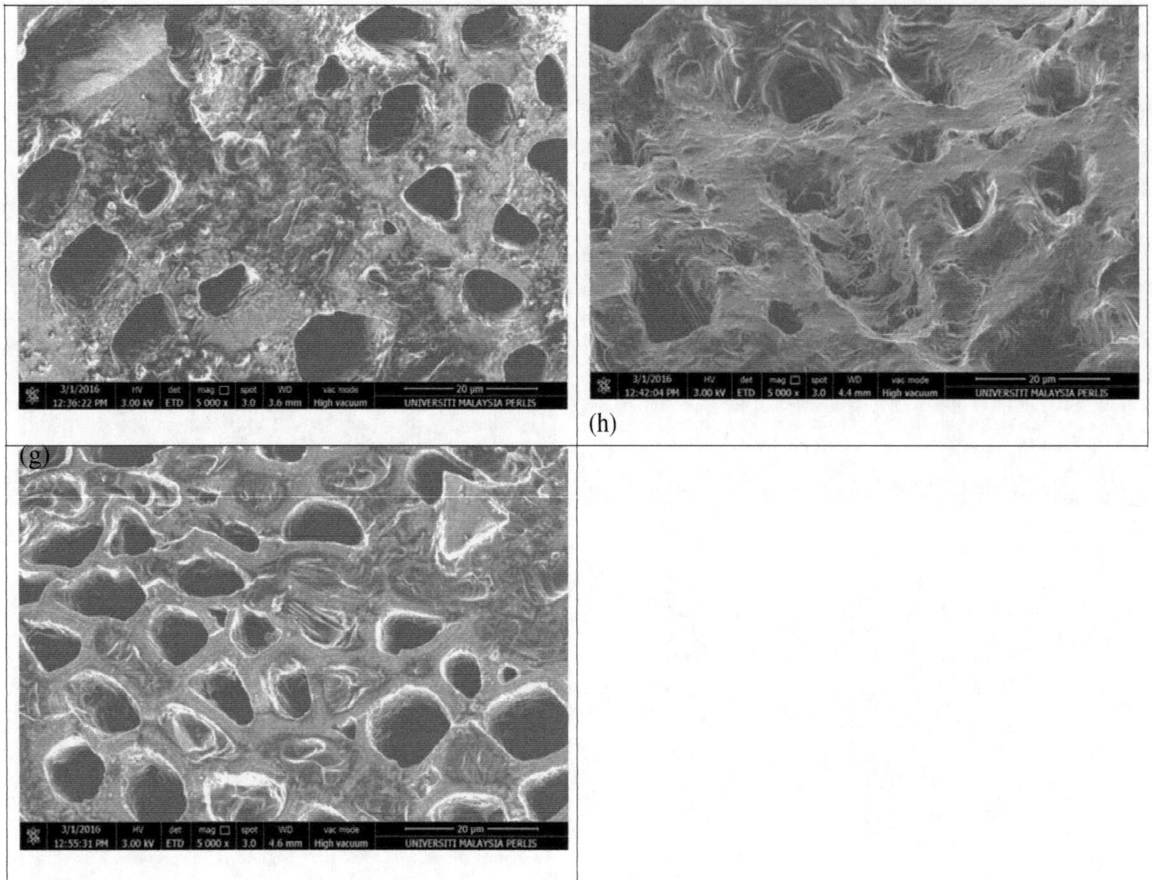

Figure 4 (a). Raw, (b) AC-AAG, (c) AC-S, (d) AC-CO_2, (e) AC-N_2, (f) AC-AAG-D, (g) AC-S-D, (h) AC-CO_2-D, (i) AC-N_2-D.

Figure 4(a) shows the physical structure of a raw sample. Evidently, the pore volume was very limited and not well developed. After carbonization and impregnation with H_3PO_4 were conducted, a honeycomb-like pore structure with a high pore volume was clearly observed, as confirmed by the BET results [figures 4(b) to 4(e)]. The pore structures were developed during physiochemical activation with the help of H_3PO_4, which assists the development of pores [23]. The well-developed pores resulted in the large surface area and pore structure, which allowed a good surface for CO_2 adsorption. These images also confirmed that no chemical was attached to the surface of the activated carbon. Thus, CO_2 adsorption occurred between the activated carbon particle surface and CO_2 molecules, and this process is known as physisorption.

After impregnation with DES was completed, few pores were found clogged. Consequently, the pore volume decreased. CO_2 adsorption occurred primarily through chemisorption as the DES on the surface of activated carbon assisted CO_2 adsorption; thus, CO_2 adsorption in the activated carbon was enhanced [22].

3.1.4. Energy-dispersive X-Ray (EDX) analysis. EDX is an X-ray technique used to identify the elemental composition of materials. Elemental analysis revealed that the raw, non-functionalized, and DES-functionalized activated carbon contained C, O, P, N, and Cl (table 3). Approximately 58.00% to 88.00% carbon content was found in all the samples related to the original content in sea mango (Cerbera odollam). After the raw material was activated, the composition of C increased significantly from 66.00% to 88.00%. This finding showed that the activation process helped increase C composition because heat supply during activation initiates thermal degradation; thus, volatile matter is removed and a stable carbon element is retained [24].

P was present in the non-functionalized activated carbon and DES-functionalized activated carbon, and this finding indicated a reaction with H_3PO_4 during the impregnation process. N and Cl were also detected in the DES-functionalized activated carbon possibly because of the presence of choline chloride of DES. This finding confirmed that N was successfully attached to the surface of the activated carbon; as a result, more active sites become available for CO_2 adsorption, and CO_2 capture is enhanced [22].

Table 3. Elemental analyses of the samples.

Sample	Element (wt%)				
	C	O	P	N	Cl
RAW	58.5	36.67	0	0	0
AC-AAG	86.93	5.51	1.83	0	0
AC-S	87.99	5.92	0.99	0	0
AC-CO_2	88.6	6.87	0.77	0	0
AC-N_2	83.53	9.49	2.8	0	0
AC-AAG-D	68.05	14.98	0.94	5.66	8.53
AC-S-D	66.07	16.75	0.53	5.16	9.01
AC-CO_2-D	66.13	17.05	1.15	5.97	8.93
AC-N_2-D	70.14	16.87	0.06	3.79	7.75

3.2. Breakthrough time of CO_2 adsorption

The CO_2 adsorption test was conducted to determine the breakthrough time of each sample. Figure 5 illustrates the breakthrough time curve for the non-functionalized sample by using different types of flowing gas treatment.

Table 4. Breakthrough time and CO_2 adsorption capacity of the samples.

Sample	Breakthrough time (seconds)	CO_2 adsorption capacity (mgCO2/gsol)
AC-AAG	1152	9.585
AC-S	1140	9.485
AC-CO_2	1008	8.412
AC-N_2	1116	9.286
AC-AAG-D	1164	9.685
AC-S-D	1158	9.636
AC-CO_2-D	1184	9.851
AC-N_2-D	1146	9.536

In figure 5(a) and table 4, the highest CO_2 adsorption capacity and the longest breakthrough time of the non-functionalized activated carbon were observed in the AC-AAG sample at 9.585 mg_{CO2}/g_{sol}, followed by AC-S at 9.485mg_{CO2}/g_{sol}, AC-N_2 at 9.286 mg_{CO2}/g_{sol}, and AC-CO_2 at 8.412 mg_{CO2}/g_{sol}. The highest adsorption capacity was detected in AC-AAG because it exhibited the highest surface area, which allows more CO_2 molecules to bind to the active sites of the activated carbon [5].

IOP Conf. Series: Materials Science and Engineering **206** (2017) 012001 doi:10.1088/1757-899X/206/1/012001

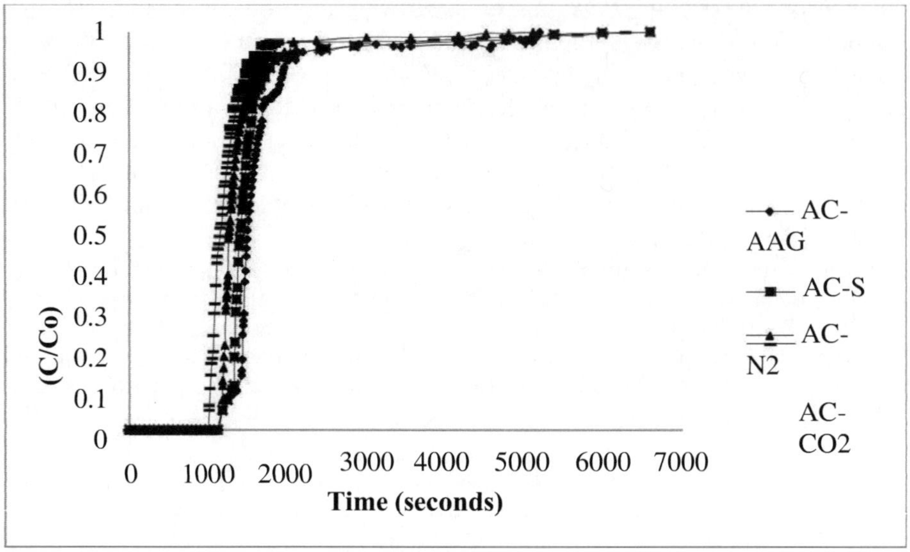

Figure 5(a). Relative concentration against time of non-functionalized activated carbon.

For the DES-functionalized activated carbon [figure 5(b) and table 4], the highest CO_2 adsorption capacity was found in AC-CO_2-D at 9.851 mg_{CO2}/g_{sol}, followed by AC-AAG-D at 9.685 mg_{CO2}/g_{sol}, AC-S-D at 9.636 mg_{CO2}/g_{sol}, and AC-N_2-D at 9.536 mg_{CO2}/g_{sol}. The highest CO_2 adsorption capacity was observed in AC-CO_2-D because of its surface area and available elemental N on the surface, which significantly accelerated CO_2 adsorption [21]. Hence, the DES-functionalized activated carbon demonstrated a more efficient performance than the non-functionalized activated carbon did.

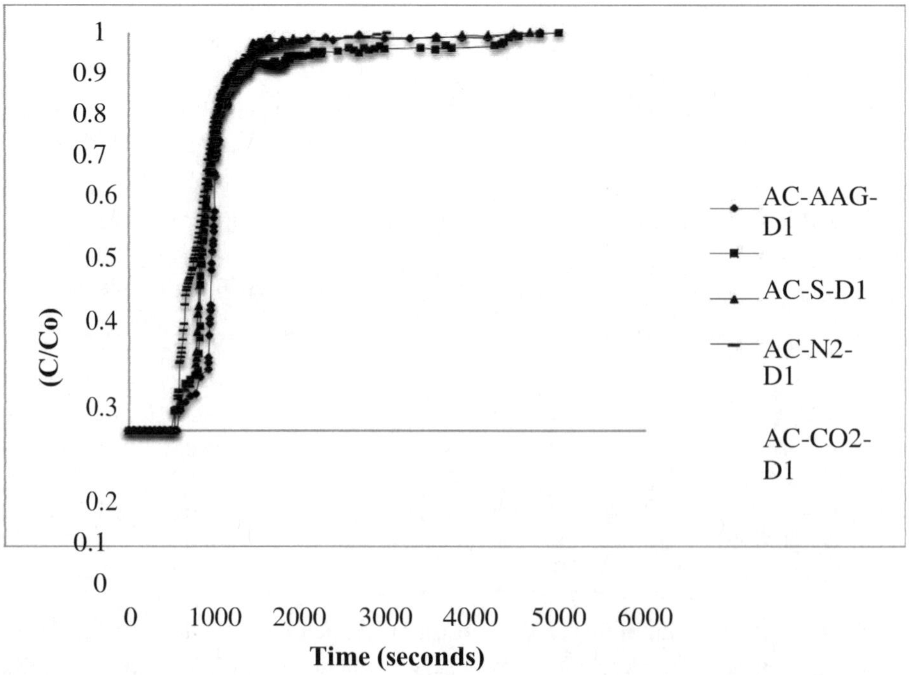

Figure 5(b). Relative concentration against time of DES-functionalized activated carbon

4. Conclusion

The DES-functionalized activated carbon yielded a higher CO_2 adsorption capacity than the non-functionalized activated carbon did because the elemental N on the surface of the activated carbon increased the number of active sites for CO_2 capture. Although the surface area of the DES-functionalized activated carbon was reduced to almost half of the non-functionalized activated carbon, the adsorption capacity of the DES-functionalized activated carbon was improved. Therefore, the DES-functionalized activated carbon underwent chemisorption in its active site. Elemental N from DES played a vital role in CO_2 adsorption.

References

[1] Santos S C G, Garrido Pedrosa A M, Souza M J B, Cecilia J A and Rodríguez-Castellón E 2015 Carbon dioxide adsorption on micro-mesoporous composite materials of ZSM-12/MCM-48 type: The role of the contents of zeolite and functionalized amine *Materials Research Bulletin* **70** 663–672

[2] Oliveres-Marin M, Sanz-Perez E S, Wong M S and Maroto- Valer M M 2011 Development of regenerable sorbents from abundant wastes for capture of CO_2 *Energy Procedia* **4** 1118-1124

[3] Brunetti A, Scura F, Barbieri G and Drioli E 2010 Membrane Technology for CO_2 separation *J. of Membrane Sci.* **359** 115-125

[4] Tuinier M J, Van Sint Annaland M, Kramer G J and Kuipers J A M 2010 Cryogenic capture using dynamically operated packed beds *Chemical Eng. Sci.* **65** 114-119

[5] Lee C S, Ong Y L, Aroua M K and Daud W M A W 2013 Impregnation of palm shell-based activated carbon with sterically hindered amines for CO_2 adsorption *Chemical Eng. J.* **219** 558–564

[6] Mohammad Saleh S, Wan Mohd Ashri W D, Amirhossein H and Ahmad Shamiri 2010 A review on surface modification of activated carbon for carbon dioxide adsorption *J. of Analytical and Applied Pyrolysis* **89** 143–151

[7] Tony S S and Pant K K 2004 Equilibrium, kinetics and thermodynamic studies for adsorption of As(III) on activated alumina *Separation and Purification Tech.* **36** 139–147

[8] Chung K L, Shin S L, Lain C J, Cheng C W, Kuen S L and Meng D L 2007 Application of MCM-41 for dyes removal from wastewater *J. of Hazardous Materials* **147** 997–1005

[9] Jones D A, Lelyveld T P, Mavrofidis S D, Kingman S W and Miles N J 2002 Microwave heating applications in environmental engineering—a review *Res., Conservation and Recycling* **34** 75–90

[10] Duan X, Srinivasakannan C and Liang J S 2014 Process optimization of thermal regeneration of spent coal based activated carbon using steam and application to methylene blue dye adsorption *J. of the Taiwan Inst. of Chemical Engineers* **45** 1618–1627

[11] El-Shafey E I, Syeda N F A, Saleh A, Haider A J and Al-Lawati 2016 Preparation and characterization of surface functionalized activated carbons from date palm leaflets and application for methylene blue removal *J. of Environ. Chemical Eng.* **4** 2713–2724

[12] Siti Nuraishah M R, Saidatul Shima J, Noor Asma F A S and Nur Aminatulmimi I 2016 Investigation of Impregnated Commercial Rice Husks, Activated Carbon with Alkanolamines as Carbon Dioxide (CO_2) Adsorbent *ARPN J. of Eng. and Applied Sci.* **11** 4

[13] Saad H K, Aroua M K and Daud W M A W 2012 Study on the improvement of the capacity of amine- impregnated commercial activated carbon beds for CO_2 adsorbing *Chemical Eng. J.* **183** 15-20

IOP Conf. Series: Materials Science and Engineering **206** (2017) 012001 doi:10.1088/1757-899X/206/1/012001

[14] Karousos D S, Vangeli O C, Athanasekou C P, Sapalidis A A, Kouvelos E P, Romanos G E and Kanellopoulos N K 2016 Physically bound and chemically grafted activated carbon supported 1-hexyl-3-methylimidazolium bis(trifluoromethylsulfonyl) imide and 1-ethyl-3-methylimidazolium acetate ionic liquid absorbents for SO_2/CO_2 gas separation *Chemical Eng. J.* **306** 146–154

[15] García G, Santiago A, Ruh U and Mert A 2015 Deep Eutectic Solvents: Physicochemical Properties and Gas Separation Applications *Energy Fuel* **29** 2616-2644

[16] Abbot A P, Barron J C, Ryder K S and Wilson D 2007 Eutectic-based ionic liquids with metal-containing anions and cations *Chemical Review* **13** 6945-6501

[17] Emma L S, Andrew P A and Karl S R 2014 Deep Eutectic Solvents (DESs) and Their Application *Chemical Review* **114** 11060-11082

[18] Ling G, Huang W, Shaokun T, Songjiang T and Xiangwen Z 2015 A novel deep eutectic solvent for biodiesel preparation using a homogeneous base catalyst *Chemical Eng. J.* **259** 647–652

[19] ASTM D2866-11 2011 Standard Test Method for Total Ash Content of Activated Carbon ASTM International West Conshohocken PA

[20] Gray M L, Champagne K J, Fauth D, Baltrus J P and Henry P 2008 Performance of immobilized tertiary amine solid sorbents for the capture of carbon dioxide *Inter. J. of Green House Gas Control* **2** 3–8

[21] Yew Liang O 2014 *Impregnation of palm shell-based activated carbon with sterically hindered amines for carbon dioxide adsorption in packed coloumn* Unpublished Master's Thesis University of Malaya, Kuala Lumpur, Malaysia

[22] Khalid M A ,Abdul hakim M A, Hayyana M,Shatirah A and Mohd Ali H 2016 Functionalization of CNTs surface with phosphonuim based deep eutectic solvents for arsenic removal from water *Applied Surface Sci.* **389** 216–226

[23] Budinova T, Ekinci E, Yardim F, Grimm A, Björnbom E, Minkova V and Goranova M 2006 Characterization and application of activated carbon produced by H_3PO_4 and water vapor activation *Fuel Processing Tech.* **87** 899–905

[24] Olugbenga S B, Mohd Azmier A and Tan T S 2011 Utilization of Cocoa Pod Husk for the Removal of Remazol Black B. Reactive Dye from Aqueous Solutions: Kinetic, Equilibrium and Thermodynamic Studies *Trends in Applied Sci. Res.* **6(8)** 794-812

CO_2 storage in heterogeneous aquifer: A study on the effect of temperature and mineral precipitation

A Raza[1*], R Gholami[1], R Rezaee[2], C H Bing[1], R Nagarajan[1], M A Hamid[1]

[1] Faculty of Engineering and Science, Curtin University, Miri, Sarawak, Malaysia
[2] Faculty of Science and Engineering, Curtin University, Bentley WA 6102, Australia

Email: arshad.raza@postgrad.curtin.edu.my

Abstract. CO_2 storage in suitable geologic media has been recognized as a major strategy taken to have a carbon free environment. This practice can be done in depleted reservoirs as well as brine aquifers where sufficient storage capacity is available to hold carbon dioxide for thousands of years. Storage in an aquifer is often achieved through four trapping mechanisms, among which capillary trapping is a rapid and effective phenomenon. Although, there have been studies pointing out the relationships of different storage related factors with capillary trapping, more studies are still required to recognize other parameters linked to this effective trapping mechanism. The aim of this paper is to evaluate the effect of temperature and mineral precipitation on trapping mechanisms of heterogeneous aquifer. A dynamic numerical simulation was run by the commercial reservoir simulator Eclipse300 to simulate 30 years of CO_2 injection. A synthetic but realistic model of a geologic formation was considered to evaluate the efficiency of trapping mechanisms under different temperature and mineral precipitation conditions. The results obtained indicated that trapping mechanisms are affected by both temperature and mineral precipitation in a short and long terms - temperature is indirectly affecting the trapping ability regardless of the precipitation effect. However, precipitation have a severe impact on injectivity as well as trapping mechanisms in the long term. Although some practical conclusions were drawn, the results obtained and presented in this study may need experimental verification before taking into serious consideration.

1. Introduction

CO_2 storage in suitable geologic media has been recognized as a major strategy taken in recent years to reduce the amount of greenhouse gas released into the atmosphere [1, 2]. The storage capacity and characteristics of saline aquifers make them a suitable candidate for CO_2 storage. A saline aquifer is often composed of sandstone and carbonates and located at the depth of 800 to 3000 m. They offer natural immobilization of CO_2 due to having favourable chemistry, porosity and subsurface conditions [3]. The subsurface temperature and pressure conditions at these depths are within 25°C to 200°C and 10 bar to 300 bar, respectively [4], which ensures that CO_2 appears in its supercritical form and occupies a lesser space for a higher volume injection [5].

Four trapping mechanisms are often generated due to injection of CO_2 in geologic structures [6]: i) structural trapping, ii) capillary trapping, iii) solubility trapping, and, iv) mineral trapping [7, 8]. These trapping mechanisms are significantly affected by subsurface pressure and temperature conditions as well as rocks and fluids characteristics [9]. Comparatively, capillary trapping is arguably recognized as a rapid, effective and safe mechanism to immobilize CO_2 in subsurface formations [10-13]. Moreover,

IOP Conf. Series: Materials Science and Engineering **206** (2017) 012002 doi:10.1088/1757-899X/206/1/012002

it may offer more CO_2 entrapment compared to other trappings such as solubility and mineral trappings [14].

During the injection period, rapid and significant changes in hydrodynamic, chemical, thermal, and mechanical aspects of subsurface formations are observed [9]. The chemical activity of CO_2 with the resident fluid and minerals creates precipitation which affects different aspects of storage sites [15, 16]. For instance, injection of CO_2 causes mineral precipitations near the wellbore region which ultimately impacts the injectivity [17,18]. Saeedi [19] briefly explained the geochemical reactions taking place during the process of mineral precipitation. These reactions might be sensitive to pressure and temperature and are often faster in carbonates compared to siliciclastic rocks [15]. Tables 1 summarises studies carried out in recent years to evaluate the effect of dissolution/precipitation on rocks properties.

The aim of this paper is to evaluate the effect of temperature and mineral precipitation on structural (free), capillary (residual) and solubility (dissolved) trapping mechanisms of heterogeneous aquifers. A numerical simulation was run for this purpose and trapping mechanisms were evaluated in terms of temperature and mineral precipitation.

Table 1. Effect of dissolution/precipitation on rocks properties [15]

Contributors	Remarks/Approach
Bacci et al., 2011	Dissolution/precipitation dependent on pressure and temperature have the significant effect on injectivity/ Experimental & Simulation
Liu et al., 2013	Precipitation process result permeability change/ Experimental
Peysson et al., 2014	Drying and mineral precipitation due to dry gas injection reduce the rock permeability/ Experimental
André et al., 2014	Dissolution/precipitation result change in permeability and injectivity caused by the interplay of capillary forces and the salinity of the initial brine / Experimental & Simulation
Zheng et al., 2015	Deformation of quartz–feldspar–detrital sandstone occurs by dissolution effect due to NaCl solution and CO_2–NaCl solution / Experimental
Ott et al., 2015	Drying and mineral precipitation during dry CO_2 injection affect the permeability in unimodal sandstone / Experimental

2. Simulation approach

CO2STORE dynamic numerical modeling [20, 21], which is a part of Schlumberger Eclipse300 Simulator, was used for the purpose of this study to estimate CO_2 trapping during injection. To make a realistic and static geologic reservoir sophisticated for a storage job, grid geometry, porosity and permeability properties were borrowed from Juanes Research Group [22].

The geometry of the 3D model was consisted of a dome in the center and five layers of fluvial sands and shales. The average reservoir thickness was 15 m and the model was discretized into 19×28×5 grid blocks, of which 1761 blocks were active. The X and Y dimension of each block was 180 meter and the permeability anisotropy ratio was about 3. The injection well was assumed to be located in the middle of the model. Two cases with and without precipitation were considered having an average porosity and permeability of 20% and 100 mD, respectively. The porosity distribution map obtained is given in figure 1. For the sake of simplicity and to ensure that CO_2 appears under supercritical condition, top depth of the reservoir was set to be 840m. The rock compressibility used in this study was 5×10^{-4} bar^{-1}.

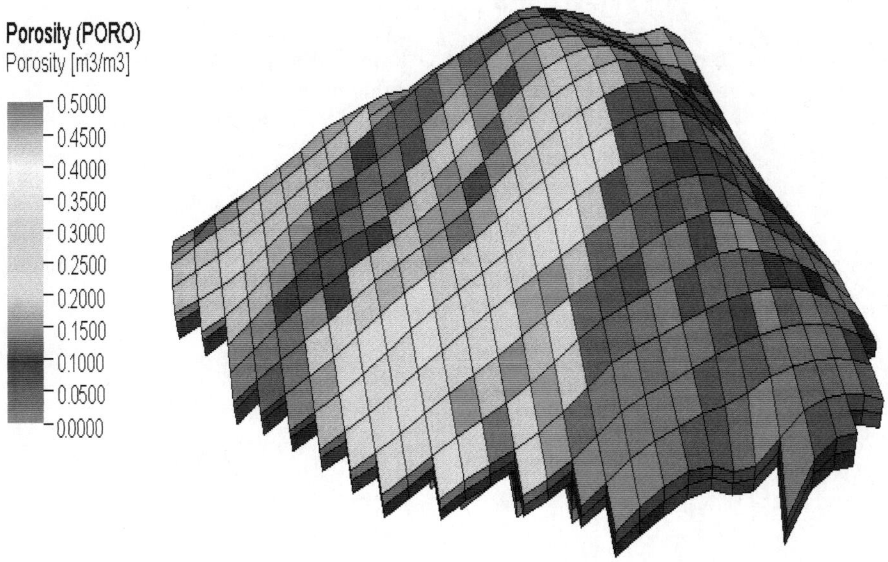

Figure 1. Plot of porosity distribution in CO2STORE Model

The liquid phase in both cases was having H_2O (0.9258 mole fraction), CO_2 (0 mole fraction), NaCl (0.0741 mole fraction) and $CaCl_2$ (0.0001 mole fraction). NaCl and $CaCl_2$ were present in the aqueous and the solid phases according to following reactions [20, 21]:

$$H_2O \rightleftharpoons H^+ + OH^- \quad (1)$$

$$CO_2 + H_2O \rightleftharpoons HCO_3^- + H^+ \quad (2)$$

$$HCO_3^- \rightleftharpoons CO_3^{--} + H^+ \quad (3)$$

$$CaCl^+ \rightleftharpoons Ca^{2+} + Cl^- \quad (4)$$

$$NaCl_{(s)} \rightleftharpoons Na^+ + Cl^- \quad (5)$$

$$CaCl_2(s) \rightleftharpoons CaCl^+ + Cl^- \quad (6)$$

Before injection, the initial pressure was set to be 278 bar (4032 psia) at the top of the structure while the temperature was varied from 49 to 148°C, ensuring that CO_2 appears under supercritical state. Three phases of CO_2 rich (gas), H_2O rich (water phase) and a solid were taken into consideration. The diffusion function was also introduced to cover the effect of CO_2 interaction with formation water. In particular, this diffusion results in transportation of CO_2 in each grid block causing more CO_2 dissolution within the medium. It should be noticed that the simulator generally calculates the mutual solubility of CO_2 and H_2O to match it with experimental data under a typical storage temperature and pressure condition. The effect of mineral and CO_2 was considered using the Ezrokhi's method (Eq. (7)) to calculate the water density [20, 21].

$$\log 10(\rho) = \log 10(\rho_o(P,T)) + \sum_i A_i(T) w_i \quad (7)$$

In Eq. (7) w_i is the weight fraction of the non-water component i, ρ_o (P, T) is the density of pure water, A_i (T) is the temperature coefficients determined by Eq. (8) in °C.

$$A_I(T) = a_{0.i} + a_{1.i}T + a_2.iT^2 \tag{8}$$

The functions to consider the effect of solid concentration on mobility in case of precipitation effect near the wellbore and overburden pressure effect were introduced.

The relative permeability and capillary pressure curves were generated using the Corey and van Genuchten correlations (i.e., Eq. (9-11)) [23].

$$k_{rw} = \left(\frac{S_w - S_{wr}}{1 - S_{wr}}\right)^4 \tag{9}$$

$$k_{rg} = k_{rg-\max}\left(1 - \frac{S_w - S_{wr}}{1 - S_{wr} - S_{gr}}\right)^2 \tag{10}$$

$$P_c = P_o\left[\left(\frac{S_w - S_{wr}}{1 - S_{wr}}\right)^{-\frac{1}{\lambda}-1} - 1\right]^{1-\lambda} \tag{11}$$

where k_{rw} and k_{rg} are the water and gas relative permeability, respectively, S_w is the water saturation, S_{wr} is the residual water saturation, S_{gr} is the residual gas saturation, k_{rg-max} is the maximum gas relative permeability, P_c is the capillary pressure, P_o is the capillary entry pressure and λ is the capillary pressure exponent.

In both cases, supercritical CO_2 was injected into the four layers of the storage formation at 5.66 million sm^3/day (standard cubic meter per day) equivalent to 200 million standard cubic feet per day for 30 years. The CO_2 was injected at the pressure below the facture pressure of the seal (i.e., 500 bar) and maximum bottom-hole pressure (i.e., 483 bar). The values of the parameters used in this studies are reported in table 2.

Table 2. Scenarios adopted in simulations cases

	CASE A	CASE B
	Temperature effect	
Case. No.	*Without precipitation*	*With precipitation*
1	49°C (Base Case)	49°C
2	60 °C	60 °C
3	71 °C	71 °C
4	82 °C	82 °C
5	93 °C	93 °C
6	104 °C	104 °C
7	115 °C	115 °C
8	126 °C	126 °C
9	137 °C	137 °C
10	148°C	148°C

Out of two cases, a base case (Case A) was developed without considering the effect of precipitation. This case was simulated at the temperature of 49°C and the pressure of 278 bar with a 5.66 million sm^3/day injection rate. Nine subcases were then generated by changing the temperature from 49°C to 148°C. In the second scenario (Case B), similar subcases were simulated under the same conditions by taking the precipitation effect into consideration. At the end, the results obtained from this scenario were compared with those of the Case A to evaluate the importance of precipitation and temperature on structural, capillary and solubility trappings.

3. Results Analysis

The results obtained from running the simulation for both cases indicated the effect of the temperature on the injection rate and pressure buildup as seen in figures 2. In the Case A, the injection rate stabilizes for 20 years and then declines rapidly. In fact, the temperature revealed to have an indirect relationship with the injection rate after 20 years. The reservoir pressure was increasing because of the continuous injection of CO_2 and ultimately affecting the trend of the injection rate. It was also found that temperature is changing by the variation of the reservoir pressure. This might be due to changes in the density of CO_2. A very same trends was observed for the injection rate and pressure build-up of the Case B. It should be noticed that the injection rate begins to decline after 11 to 15 years in different cases depending on the temperature.

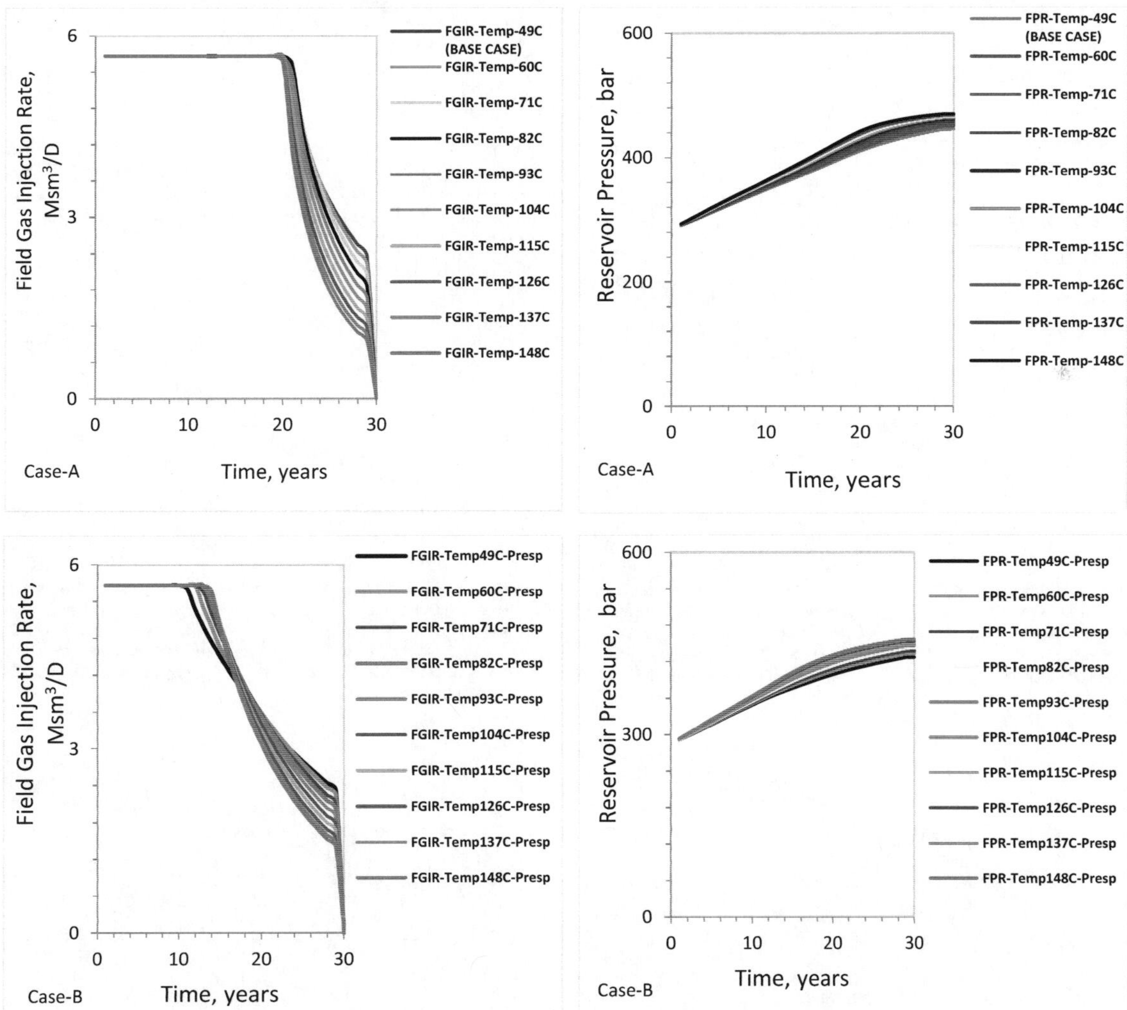

Figure 2. Effect of temperature on injection rate (FGIR) and on pressure buildup (FPR) for Case A and Case B.

IOP Conf. Series: Materials Science and Engineering **206** (2017) 012002 doi:10.1088/1757-899X/206/1/012002

Considering the relationship of temperature with the free, residual and dissolved CO_2 saturations in the absence and presence of precipitation, it can be concluded that a medium with a higher temperature would trap lesser amount of CO_2 compared to a low-temperature medium (See figure 3a, b). In fact, temperature is showing an inverse effect on capillary and solubility trappings. This effect becomes significant after five years of injection and has not been observed in any of the earlier studies. However, precipitation reduces the trapping ability in later stages.

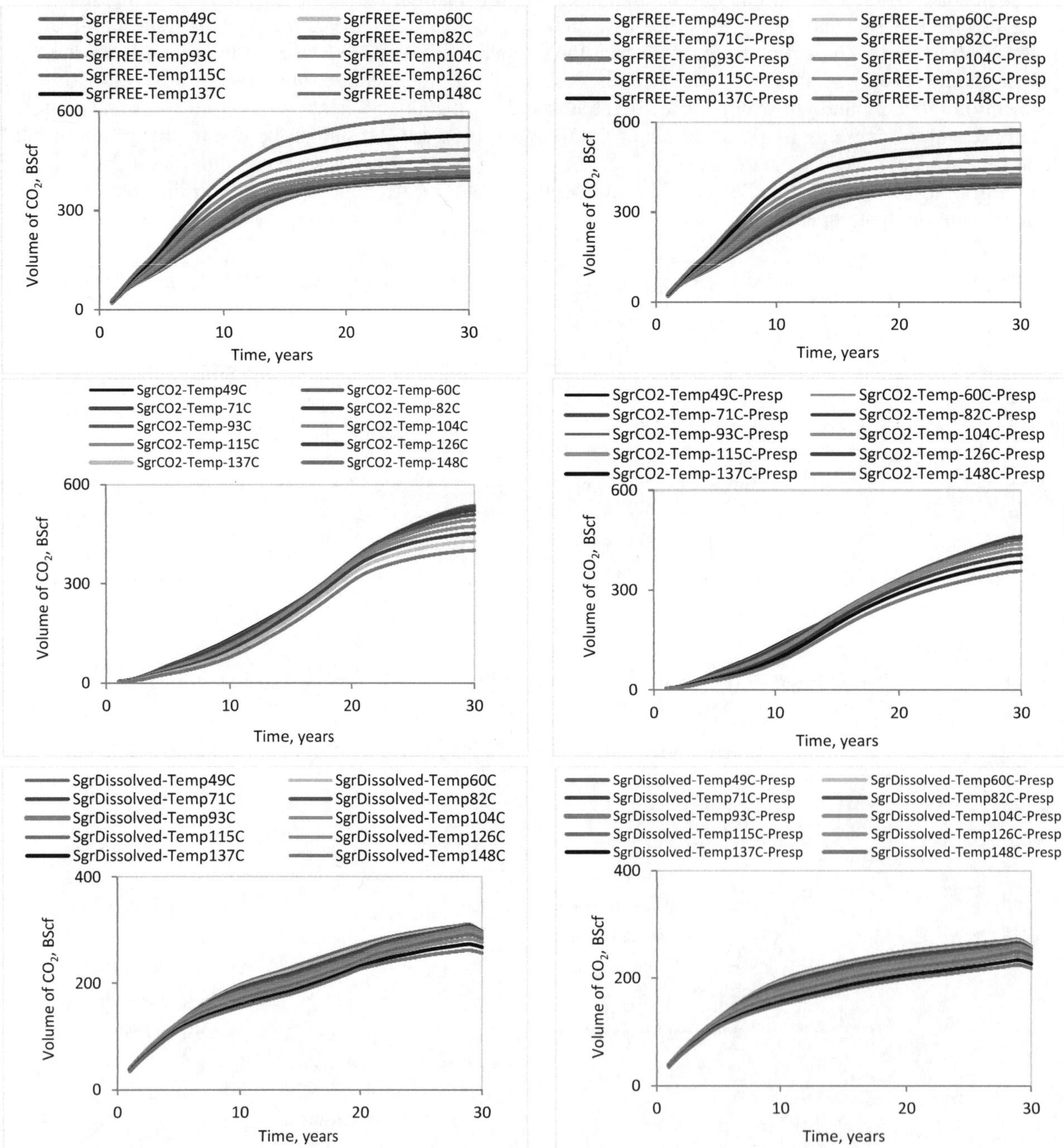

Figure 3. a) Effect of temperature on trapping mechanisms considering no precipitation effect (Case-A-left); b) Effect of temperature on trapping mechanisms considering precipitation effect (Case-B-right).

IOP Conf. Series: Materials Science and Engineering **206** (2017) 012002 doi:10.1088/1757-899X/206/1/012002

This was aligned with the statement of Metz et al. [24] where it was indicated that low-temperature gradient basins are more favourable than the high-temperature ones due to achieving a high density CO_2. It was also found that the ganglia of CO_2 is left behind in the pore space and surrounded by water during the capillary trapping. It might be possible that the temperature variation is affecting the water movement and lowering the capillary trapping ability causing the free gas saturation to be enhanced. It could also be due to the inverse effect of temperature on CO_2 solubility in brine which causes CO_2 plume to flow at a high rate with a lesser solubility trapping [25].

Comparing the results obtained from the Case A at two different temperatures (49°C and 148°C) with those of the Case B revealed a quantitative impact of precipitations on the capillary trapping including structural and solubility trappings as shown in figure 4. It was also revealed that the capillary and solubility trappings would be in their maximum and minimum range at low temperatures while the structural trapping is more active at high temperatures (See figure 5). However, it was found that the residual and dissolved gas saturations of CO_2 significantly reduce due to mineral precipitation after 50% of injection. A minor impact was observed on the free gas saturation as well. On the other hand, it was concluded that precipitation may have impacts on the near wellbore injectivity due to the variation of permeability and capillary pressure forces [17, 18]. It was also noticed that precipitations may change capillary, structural (free) and solubility trappings of the storage site in the long term.

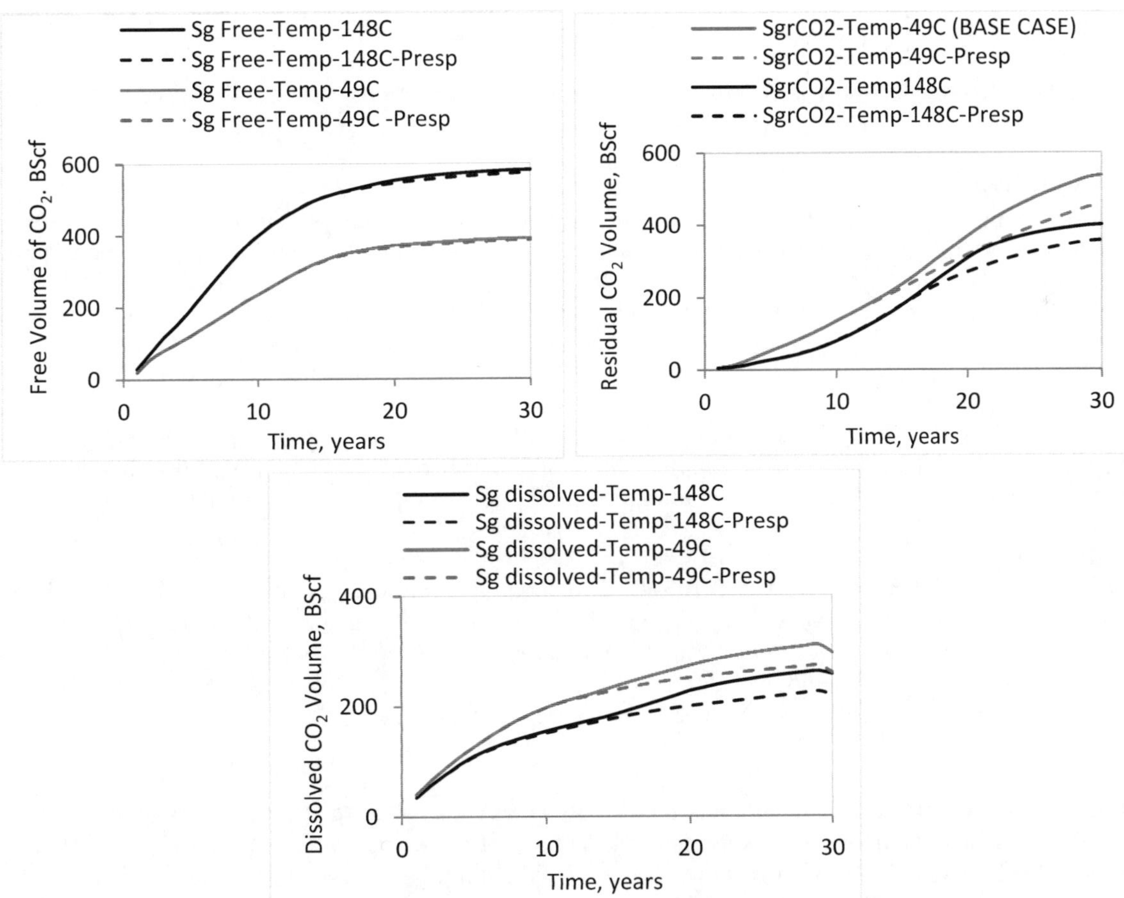

Figure 4. Comparison of trapping mechanisms at 49°C and 148°C with and without precipitation effect.

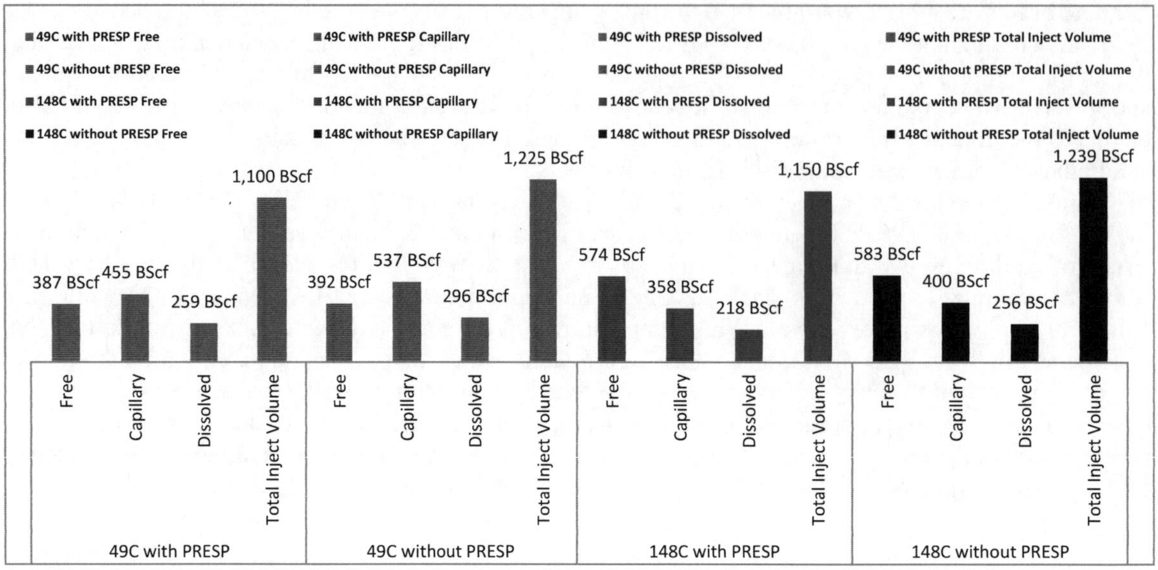

Figure 5. Comparison of CO_2 volume in total injected and different phases considering temperature and precipitation effects

4. Conclusions

This study investigated the effect of temperature and precipitation on trappings of CO_2 during a storage practice in aquifers. The results obtained indicated that temperature is indirectly affecting the trapping ability regardless of the precipitation effect. Precipitation, however, may have a severe impact on rock properties, injectivity, as well as trapping mechanisms in the long term. It is suggested to have the precipitation effect included in the dynamic modelling to mimic the real scenario of CO_2 injection.

Acknowledgments

This study was funded by Curtin Sarawak University, Malaysia through the Curtin Sarawak Research Institute (CSRI) Flagship scheme under the grant number CSRI-6015. We also acknowledge Schlumberger Malaysia for providing Petrel tool and Eclipse compositional simulator.

References

[1] Ketzer J M, Iglesias R and Einloft S 2012 Reducing greenhouse gas emissions with co$_2$ capture and geological storage. In: W.-Y. Chen, J. Seiner, T. Suzuki and M. Lackner (Editors) *Handbook of Climate Change Mitigation. Springer* US, pp. 1405-1440.

[2] Surdam R C 2013 Geological CO_2 storage characterization: The key to deploying clean fossil energy technology *Springer Sci. & Business Media*. 1-100.

[3] Riaz A and Cinar Y 2014 Carbon dioxide sequestration in saline formations: Part I—Review of the modeling of solubility trapping *J. of Petroleum Sci. and Eng.*, **124**: 367-380.

[4] Bachu S 2003 Screening and ranking of sedimentary basins for sequestration of co$_2$ in geological media in response to climate change *Environmental Geology*, **44(3)**: 277-289.

[5] Van Der Meer B 2005 Carbon dioxide storage in natural gas reservoir *Oil & Gas Sci. and Tech.* - Rev. IFP, **60(3)**: 527-536.

[6] Sarmadivaleh M, Al-Yaseri A Z and Iglauer S 2015 Influence of temperature and pressure on quartz–water– CO_2 contact angle and CO_2–water interfacial tension *J. of Colloid and Interface Sci.*, **441(0)**: 59-64.

[7] Benson S M and Cole D R 2008 CO_2 sequestration in deep sedimentary formations *Elements*,

4(5): 325-331.

[8] Shamshiri H and Jafarpour B 2012 Controlled CO_2 injection into heterogeneous geologic formations for improved solubility and residual trapping *Water Res. Research*, **48(2)**. 1-15.

[9] Han W, Kim K Y, Esser R, Park E and McPherson B 2011 Sensitivity Study of Simulation Parameters Controlling CO_2 trapping mechanisms in saline formations *Transp Porous Med*, **90(3)**: 807-829.

[10] Burnside N M and Naylor M 2014 Review and implications of relative permeability of CO_2/brine systems and residual trapping of CO_2 *Inter. J. of Greenhouse Gas Control*, **23(0)**: 1-11.

[11] Herring AL, Andersson L, Schlüter S, Sheppard A and Wildenschild D 2015 Efficiently engineering pore-scale processes: The role of force dominance and topology during nonwetting phase trapping in porous media *Advances in Water Res.*, **79(0)**: 91-102.

[12] Pentland CH, El-Maghraby R, Iglauer S and Blunt M J 2011 Measurements of the capillary trapping of super-critical carbon dioxide in berea sandstone *Geophysical Research Letters*, **38(6)**.

[13] Qi R LaForce, T C and Blunt M J 2009 Design of carbon dioxide storage in aquifers *Inter. J. of Greenhouse Gas Control*, **3(2)**: 195-205.

[14] Kimbrel E H, Herring A L, Armstrong R T, Lunati I, Bay B K and Wildenschild D 2015 Experimental characterization of nonwetting phase trapping and implications for geologic CO_2 sequestration *Inter. J. of Greenhouse Gas Control*, **42**: 1-15.

[15] Raza A, Rezaee R, Gholami R, Rasouli V, Bing C H, Nagarajan R and Hamid M A 2015a Injectivity and quantification of capillary trapping for CO_2 storage: A review of influencing parameters *J. of Natural Gas Sci. and Eng.*, **26**: 510-517.

[16] Raza A, Rezaee R, Bing C H, Gholami R, Hamid M A and Nagarajan R 2015b Carbon dioxide storage in subsurface geologic medium: A review on capillary trapping mechanism *Egyptian J. of Petroleum (EGYJP)*-Elsevier, **25**(3): 367-373.

[17] Kim K Y, Han W S, Oh J, Kim T and Kim J C 2012 Characteristics of salt-precipitation and the associated pressure build-up during CO_2 storage in saline aquifers *Transport in Porous Media*, **92**(2): 397-418.

[18] Zeidouni M, Pooladi-Darvish M and Keith D 2009 Analytical solution to evaluate salt precipitation during CO_2 injection in saline aquifers *Inter. J. of Greenhouse Gas Control*, 3(5): 600-611.

[19] Saeedi A 2012 Experimental study of multiphase flow in porous media during CO_2 Geo-Sequestration Processes. *Springer Sci. & Business Media*: 1-120.

[20] Schlumberger 2014a Reference manual. Vol 2014.2.

[21] Schlumberger 2014b Technical description. Vol 2014.2.

[22] Juanes Group 2016 Modell description. *In: J.R. Group (Editor), Massachusetts Institute of Technology· Department of Civil and Environmental Engineering.*

[23] Hussain F, Michael K and Cinar Y 2015 A numerical study of the effect of brine displaced from CO_2 storage in a saline formation on groundwater *Greenhouse Gases: Sci. and Tech.*. 1-17.

[24] Metz B, Davidson O, De Coninck, H, Loos M and Meyer L 2005 *IPCC special report on carbon dioxide capture and storage. Prepared by working group III of the intergovernmental panel on climate change.* Cambridge, United Kingdom and New York, NY, USA: 1-200.

[25] Raza A, Rezaee R, Gholami R, Bing CH, Nagarajan R and Hamid M A 2016 A screening criterion for selection of suitable CO_2 storage sites *J. of Natural Gas Sci. and Eng.*, **28**: 317-327.

IOP Conf. Series: Materials Science and Engineering **206** (2017) 012003 doi:10.1088/1757-899X/206/1/012003

Physical Activation of Oil Palm Empty Fruit Bunch via CO_2 Activation Gas for CO_2 Adsorption

C G Joseph[1*], K S Quek[1], W M A W Daud[2], P Y Moh[1]

[1] Faculty of Science and Natural Resources, Universiti Malaysia Sabah, 88400 Kota Kinabalu, Sabah, Malaysia
[2] Department of Chemical Engineering, Faculty of Engineering, University of Malaya, 50603 Kuala Lumpur, Malaysia

*Email: collin@ums.edu.my (Collin G. Joseph)

Abstract. In this study, different parameters for the preparation of activated carbon were investigated for their yield and CO_2 capture capabilities. The activated carbon was prepared from Oil Palm Empty Fruit Bunch (OPEFB) via a 2-step physical activation process. The OPEFB was pyrolyzed under inert conditions at 500 °C and activated via CO_2. A 2-factorial design was employed and the effects of activation temperature, activation dwell time and gas flow rate on yield and CO_2 capture capabilities were compared and studied. The yield obtained ranged from between 20 – 26%, whereby the temperature was determined to be the most significant factor in influencing CO_2 uptake. The CO_2 capture capacity was determined using Temperature Programmed Desorption (TPD) technique. The CO_2 uptake of EFB activated carbon achieved was between 1.85 – 2.09 mmol/g. TPD analysis has shown that the surface of AC were of basic nature. AC was found to be able to withhold the CO_2 up to 663°C before maximum desorption occurs. The surface area and pore size of OPEFB obtained from BET analysis is 2.17 $m^2\ g^{-1}$ and 0.01 $cm^3\ g^{-1}$. After activation, both surface area and pore size increased with a maximum observed surface area and pore size of 548.07 $m^2\ g^{-1}$ and 0.26 $cm^3\ g^{-1}$. Surface morphology, functional groups, pore size and surface area were analyzed using SEM, FT-IR, TPD and BET.

1. Introduction

In recent years, researchers are concern about the environment due to global warming, which has resulted in the change in climate and weather patterns. This has in turn driven researchers to focus on developing improved and efficient methods and technologies to reduce the environmental impact from human activities. Many reports have shown that the global temperature has risen with measurable consistency to the increase in atmospheric CO_2. In the mid 1960's, the global temperature has risen by 0.2°C and by the 1980's, the global temperature has again risen by 0.4°C [1]. This is further supported by a mathematical model done by Specht et al., 2016 and a continued increase in temperature is predicted in the near future [2].

Among the many contributors of global warming are the anthropogenic emissions of CO_2 gases [3]. It has been reported that CO_2 has the highest negative impact on the observed greenhouse effect, which causes up to approximately 55% of the global warming [4]. This is mainly attributed to CO_2 causing greater adsorption of thermal radiation from the Earth's surface and thus, increases the overall air temperature within the troposphere [5].

Many alternative solutions are being proposed to tackle this situation, Carbon Capture and Storage (CCS) technologies is currently the most viable short term option to reduce the anthropogenic CO_2 emissions from industrial sources (Metz et al., 2005; Yang et al., 2008). One of the promising technologies of CCS is to employ carbonaceous materials for the adsorption of CO_2. Carbonaceous materials which are converted to activated carbon (AC) has been proven to be exceptional in capturing post-combustion CO_2 [6]. Commercially, AC is one of the most common adsorbents used for the removal of organic compounds from both air and water sources. This is because the raw materials required to produce AC are cheap and abundant.

Production of activated carbons is carried out by pyrolysis from a precursor; mainly from an organic substance or synthetic polymers into char. The char produced from pyrolysis yields varying degree of surface area and porosity, which depends on the material used and also pyrolysis conditions. However, the surface area and porosity may be insufficient for industrial applications [7]. Therefore, further activation, either by physical or by chemical process, is required. Agricultural by-products have been widely used as the raw material due to its high carbon content, low inorganics and availability at low prices or sometimes freely available as it is considered as waste materials [8].

Oil palm is one of the most economical perennial oil crops for its valuable oil-producing fruits in tropical regions such as West Africa and Southeast Asia [9]. Malaysia is currently the world's largest exporter of palm oil. Malaysia contributes to about 47% of the world's supply of palm oil. Malaysia also accounts the highest percentage of global vegetable oils and fats trade in year 2005 [10]. Therefore, in this study, the agricultural waste, which is the oil palm empty fruit bunch (OPEFB), is studied for its suitability as a precursor to preparing AC for CO_2 capture.).

2. Materials and Method

2.1. Materials

The OPEFB is collected from a local oil palm plantation, in Kota Kinabalu, Sabah, Malaysia. The OPEFB collected were thoroughly washed to remove unwanted impurities such as dirt, pesticide residue and other inorganic compounds. Then, it is thoroughly dried and cut into small pieces, approximately 2-3 cm and further washed with hot water to remove excess oil. These small pieces were placed in the oven overnight at 130 ºC, for drying. The drying process will to reduce the moisture content in the empty fruit bunch. After that, the empty fruit bunches were grounded and sieved, using a mechanical sieve, to a size of 2-3 mm. Finally, it is stored in an air-tight container, for further analysis and processing.

Two-step physical activation was chosen to prepare AC from OPEFB. Carbonization of the raw OPEFB were carried out in a horizontal tube furnace (LT furnace model STF50-1200), at 500°C under N_2 flow for 2 hours. Then, for the activation process, N_2 was stopped and gasification by CO_2 was carried out at different activating temperatures, dwell time and CO_2 flowrate. The experimental parameters, the percentage of yield and resulting CO_2 uptake is shown in table 1.

2.2 Characterization

The constituents of OPEFB were determined using Thermogravimetric analysis (TGA). The sample was heated in an inert condition, from 30°C to 700°C at a ramping rate of 5°C per minute. Fourier Transform Infrared spectroscopy (FTIR) was used to characterize the surface functional groups of both the OPEFB and the prepared AC surface. Scanning electron microscopy (Hitachi S-3400 SEM) and energy dispersive X–ray spectrometry (EDX) were used to observe the morphology and to conduct the elemental mapping of the samples. The acid/base character of the surface groups of the carbon samples were determined by Temperature Programmed Desorption (ThermoFinnigan TPDRO 1100 Series).

IOP Conf. Series: Materials Science and Engineering **206** (2017) 012003 doi:10.1088/1757-899X/206/1/012003

2.3 CO_2 Adsorption

CO_2 adsorption capacity of the AC were determined by temperature-programmed desorption coupled with spectrometry (TPD–MS). Each AC sample weighing approximately 5 mg was placed in a quartz tube in a furnace. The AC was pretreated by flowing Nitrogen gas into the chamber before the analysis to produce an inert environment. Then, CO_2 gas was flowed into the chamber for an hour with simultaneous heating causing the AC pores to become saturated with CO_2. After that, Nitrogen gas was passed through for 5 minutes under low temperature to remove excess CO_2. Finally, Helium gas was flowed into the chamber with a linear heating rate of 5 °C /min in vacuum of up to 900°C. The mass spectrometer was set to detect CO_2. The total amount of CO_2 desorbed is the total uptake capacity of CO_2 by the AC.

3. Results and Discussion

Table 1. Activation conditions, percentage of yield and CO_2 uptake capacity

	Preparation parameters				CO_2 adsorption capacity
AC Sample	Temp (°C)	Time (hr)	Flow (mL/min)	Yield (%)	(mmol/g)
1	500	1	30	25.57	2.09
2	700	1	30	20.55	1.93
3	500	3	30	25.16	2.01
4	700	3	30	20.42	1.93
5	500	1	50	24.48	2.08
6	700	1	50	20.76	1.86
7	500	3	50	25.06	2.01
8	700	3	50	20.32	1.86

3.1. Characterization

3.1.1. Thermogravetric Analysis. OPEFB is a lignocellulosic biomass, which is mainly composed of cellulose, hemicellulose and lignin, as well as other minor components [11]. A study was made by Vamvuka *et al.*, which shows the TG analysis of olive kernel and cotton residues. In this study, the main thermal decomposition of lignocellulosic materials was found to occur over the temperature range of 200 – 400°C. Lignin was the first component to decompose at a low temperature which continues on until approximately 900°C. Hemicellulose is a light fraction component which also decomposes at the low temperature regions, ranging from 160 and 360°C. Lastly, cellulose was determined to be the last component to decompose at the high temperatures, ranging from 240 – 390°C [12].

The thermogravimetric (TG) data and its first derivative (DTG) data of oil palm OPEFB were plotted. The OPEFB sample was pyrolyzed at 5°C/min from 50°C to 700°C. From the graph in figure 1 it can be seen that pyrolysis of oil palm EFB starts above 250°C. From the DTG curve, the maximum rate of decomposition was observed at approximately 300° C, with a smaller shoulder on the lower temperature side.

This indicates that there is at least one major reaction scheme occurring during the pyrolysis process. The maximum peak observed on the DTG curve is probably contributed by the decomposition of the lighter fraction (i.e. cellulose) and the small shoulder corresponds to the decomposition of the heavier component (i.e. hemicellulose).

The graph and result obtained is similar to the results obtained by Luangkiattikhun *et al.*, who studied the thermal decomposition of palm shell, fibre and kernel [13], and Vamvuka *et al.*, who studied the thermal decomposition of olive kernel and cotton residues [12].

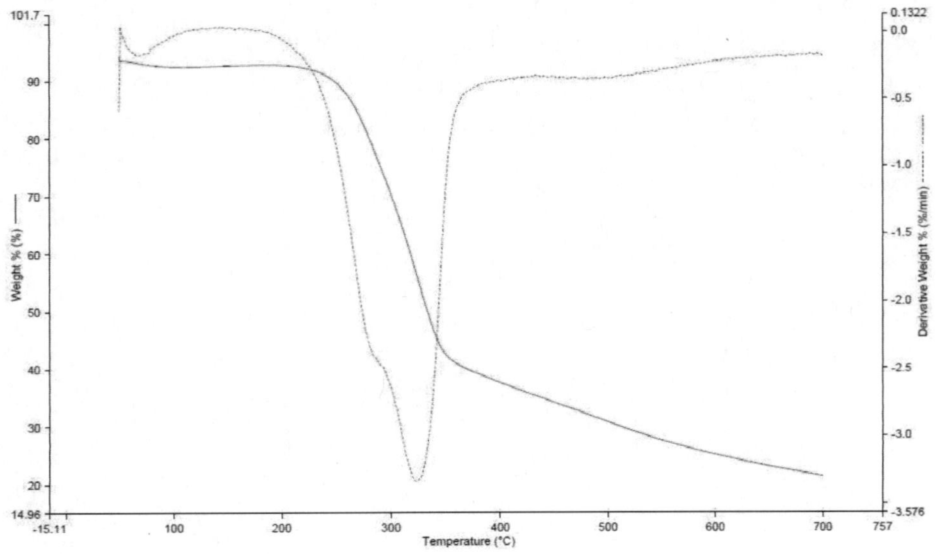

Figure 1. TG and DTG data of Oil Palm EFB

3.1.2. Fourier Transformed Infrared. The FTIR spectra of the raw OPEFB and AC were presented in figure 2. Each peak found in the OPEFB were labeled and assigned accordingly in table 2. Based on the figure, a broad adsorption peak for raw OPEFB is observed at 3345 cm^{-1} which is attributed to O-H stretching functional group and this indicates the presence of bonded hydroxide in the raw EFB. The EFB sample also shows two other prominent adsorption peaks at 1252 cm^{-1} and 1041 cm^{-1} which refers to C-O stretching functional group.

For activated carbon OPEFB, during the carbonization and activation process most of the adsorption peaks of functional groups have disappeared. This is because the functional groups from the raw material spectrum were vaporized as volatile materials when heat was supplied to the sample [14]. This shows that the activation process has taken place successfully.

IOP Conf. Series: Materials Science and Engineering **206** (2017) 012003 doi:10.1088/1757-899X/206/1/012003

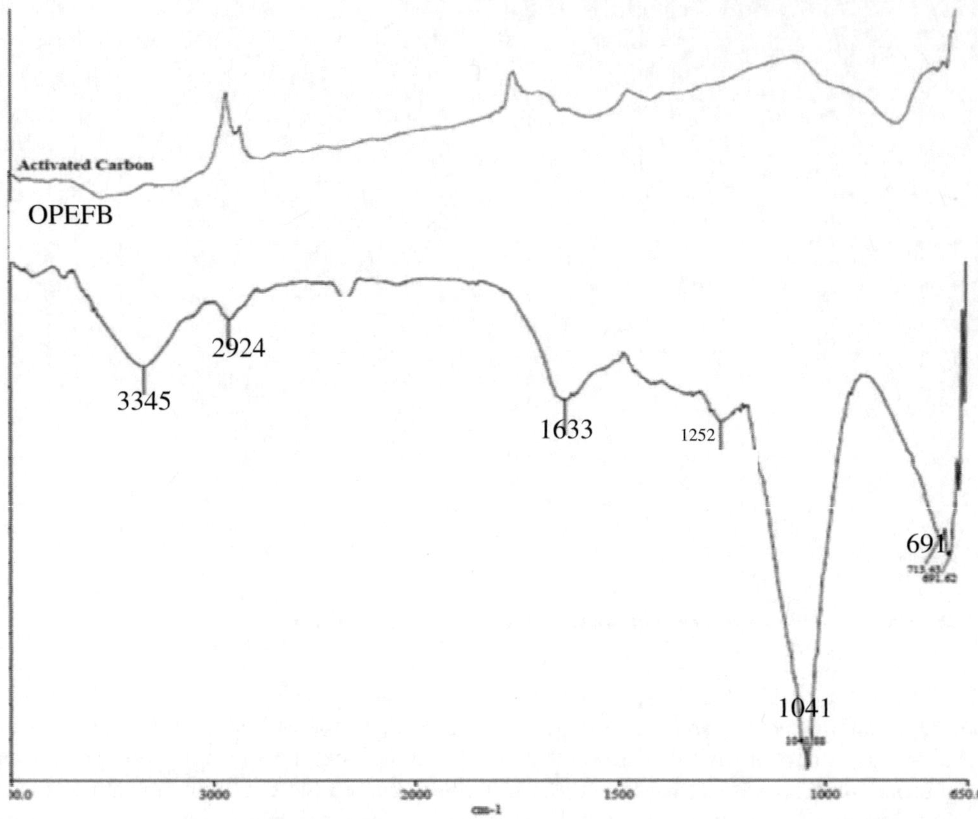

Figure 2. FTIR spectra of Raw EFB and AC EFB

Table 2. FTIR peaks and its corresponding assignments

Wave number (cm^{-1})	Assignments	EFB	AC
3500 – 3200	O-H Stretching	✓	
2930 – 2900	C-H Asym. Stretching	✓	
1610 – 1640	C=C Aromatic skeletal stretching	✓	
1240 – 1280	C-O Asym. stretching of aromatic ethers, esters and phenols	✓	
1260 – 1000	C-O Carboxylic acids, alcohols, phenols and esters or the P=O bond in phosphate esters	✓	
700 – 400	C-C stretching	✓	✓

IOP Conf. Series: Materials Science and Engineering **206** (2017) 012003 doi:10.1088/1757-899X/206/1/012003

3.1.3. Temperature Programmed Desorption. Temperature programmed desorption (TPD-CO_2) analysis was carried out on selected AC samples, prepared at pyrolysis temperature 500°C and 700°C. From figure 3, the TPD-CO_2 profile shows the basic sites density and basic strength of the AC and AC-ZIF. Both AC shows two desorption peaks. The desorption peaks at T_{max} 269°C and 445°C can be assigned to weak and medium basic sites whereas T_{max} > 600°C can be assigned to super basic sites. These basic sites can be attributed with the resonating π-electrons of the carbon aromatic rings which attract protons and also the basic surface functional groups such as nitrogen containing groups [15]. These basic sites may also be caused by the high pyrolysis temperature used. Studies have shown that a there is a decomposition of oxygen functional groups at elevated temperatures, which in turn increases the basicity of carbon surface [16]. It was also stated in a research by Shafeeyan and his team of researchers than carbon with a more basic surface area is more favorable for CO_2 capture [15].

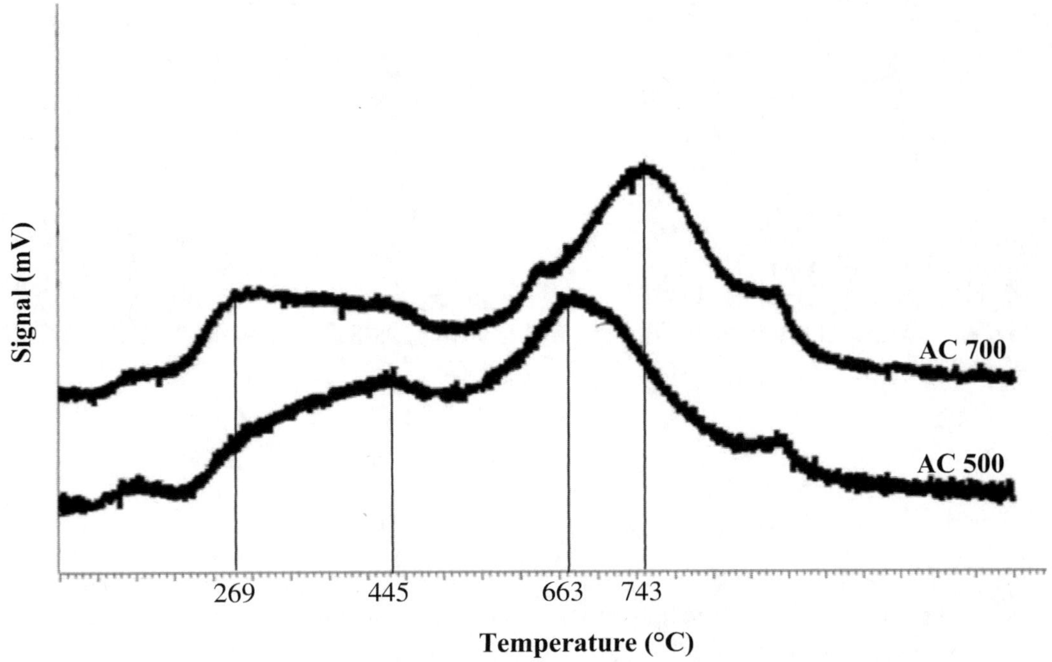

Figure 3. TPD-CO_2 Profile for AC and AC-ZIF

3.1.4. Scanning Electron Microscopy (SEM). The SEM images of the activated carbons are shown in figure 4. Figure 4(a) and 4(b) shows the SEM images of OPEFB before activation, whereas 4(c) and 4(d) shows the OPEFB after activation. By comparing figure 4(a) and 4 (c), the pyrolyzed OPEFB showed a much denser ridges and rougher surface compared to the raw OPEFB. This roughness is increased during the CO_2 activation which resulted in largely uniformed pores. The comparison of pore can be seen in figure 4(b) and 4(c).

IOP Conf. Series: Materials Science and Engineering **206** (2017) 012003 doi:10.1088/1757-899X/206/1/012003

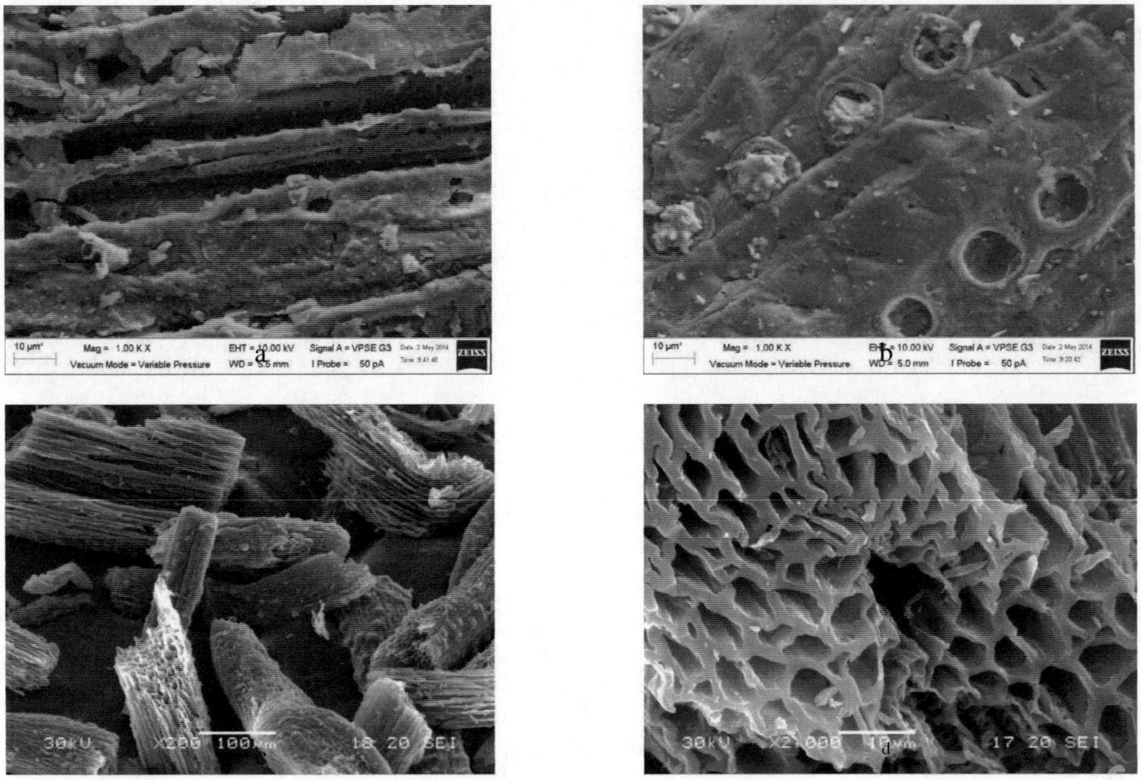

Figure 4. SEM images of Activated carbon

3.2. Yield of AC prepared using different parameters

Table 1 summarizes the parameters of pyrolysis or activation used in this experiment and also the yield of each set of parameters and is represented in a 3d plot in figure 5. From the figure, it can be seen that the most prominent parameter affecting the yield is the activation temperature. As temperature increases, yield decreases. This is primarily due to the moisture evaporation and also large amount of volatiles released during the pyrolysis stage. The large amount of volatiles released, known as the devolatilisation step, can be mainly attributed to the decomposition of cellulose, hemicelluloses and lignin found in the OPEFB [17]. The duration of dwell time shows very little effect on the yield, whereby a longer dwell time decreases the yield of activated carbon. This is also attributed to devolatilisation, as a longer dwell time allows for higher loss of volatile matter [18]. The different flow rate of CO_2 however, shows no identifiable trend and effects towards the yield of activated carbon.

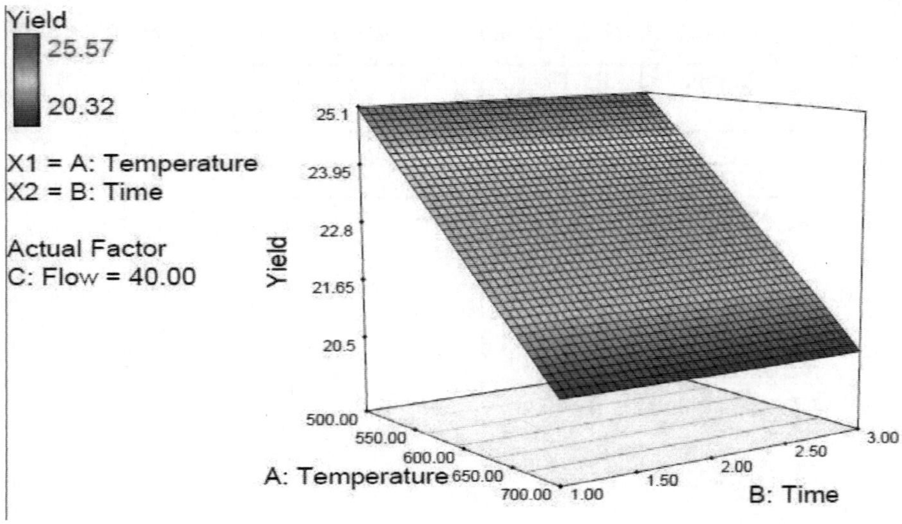

Figure 5. 3D plot of Activation Parameters against Yield

3.3. CO_2 uptake capacity

From the table 1 and figure 6, it can be seen that the adsorption capacity of CO_2 decreases when using AC prepared at higher activating temperature and longer dwell time. The highest amount of CO_2 adsorbed was 2090 µmol/g, achieved by the AC produced at activating temperature of 500 °C and 1 hr dwell time. A higher activating temperature and longer dwell time may increase surface functional group which contains oxygen. This may lead to acidic oxidation which in turn leads to pore blocking [17].

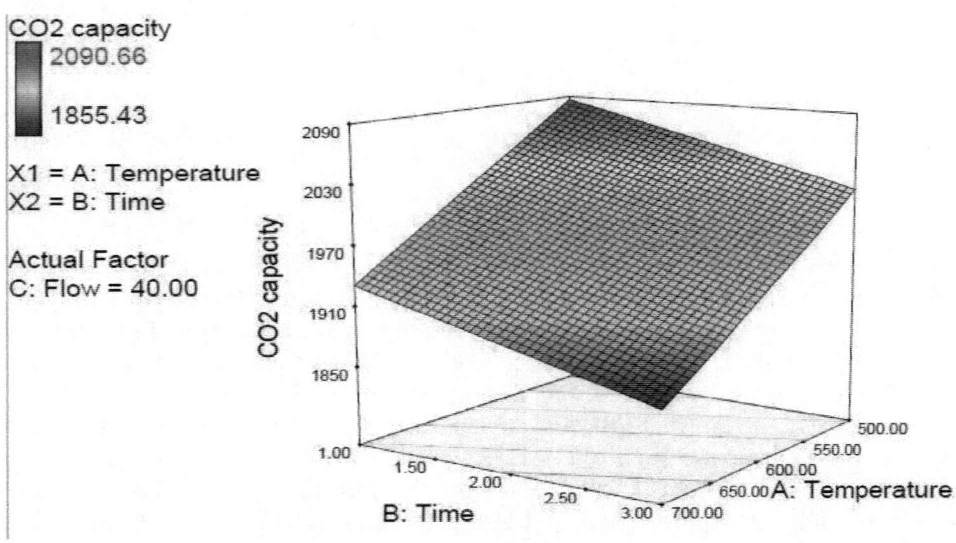

Figure 6. 3D plot of Activation Parameters against CO_2 uptake capacity

3.4. BET Analysis

Table 3. BET analysis of selected samples

Sample	BET surface area (m^2 g^{-1})	total pore volume (cm^3 g^{-1})
Raw EFB	2.17	0.0100
AC 700	14.32	0.0123
AC 500	548.07	0.2552

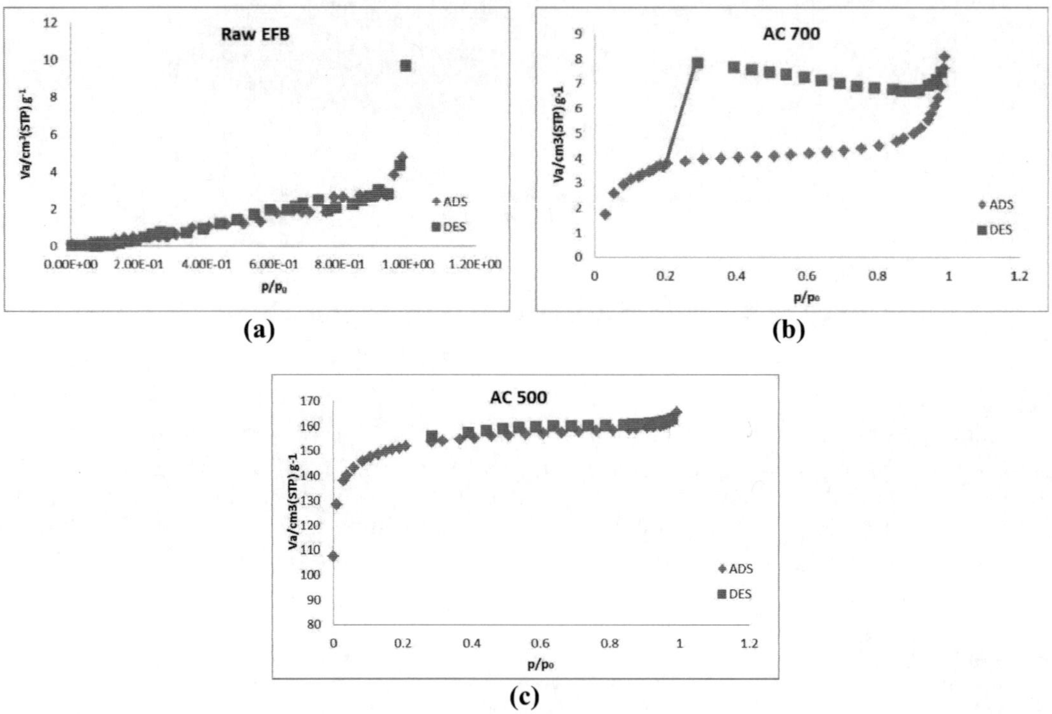

Figure 7. N_2 isotherms of (a) OPEFB, (b) EFB-AC activated at 700 °C and (c) EFB-AC activated at 500 °C

From the BET surface shown in table 3, higher surface area is seen after activation. All the samples after activation have shown an increase in surface area. AC prepared at 500 °C yields higher surface area than AC prepared at 700 °C. This is consistent with the uptake capacity of CO_2 shown in tale 1. Figure shows graph of volume (Va/cm^3 (STP) g^{-1}) against relative pressure (P/P_o) for OPEFB, AC activated at 700 °C and AC activated at 500 °C respectively. From figure 7(a), OPEFB shows type III isotherm [20], whereas figure 7(b) EFB-AC activated at 500 °C shows type I isotherm, which is assigned to microporous materials [21]. EFB-AC activated at 700 °C (figure 7(c)) however, shows a type IV isotherm with H2 hysteresis loop. A review by Sing & Williams, 2004, relates the H2 hysteresis loop with ink bottle pore system and also pore blocking effects [22]. Nguyen et al., 2013 came to the same conclusion whereby H2 hysteresis loop represents either a cavitation or pore-blocking mechanism. Nguyen and his team further explain that this is caused when the cavity is stretched beyond its tensile strength [23]. This result in a lower adsorption capacity and pore size of the EFB-AC activated higher temperatures.

IOP Conf. Series: Materials Science and Engineering **206** (2017) 012003 doi:10.1088/1757-899X/206/1/012003

4. Conclusions

Porous activated carbon with high specific surface area, large pore volume, and wide pore-size distribution has been successfully produced from oil palm empty fruit bunch via 2 step CO_2 physical activation. The prepared carbon has shown to be able to adsorbed and retain CO_2. While its CO_2 capture abilities are still limited, the OPEFB has shown to be a promising adsorbent for CCS technology. Due to its abundance, it also provides a waste to wealth opportunity.

References

[1] Hansen J, Johnson D, Lacis a, Lebedeff S, Lee P, Rind D and Russell G 1981 Climate impact of increasing atmospheric carbon dioxide. *Science* **213** 957–66

[2] Specht E, Redemann T and Lorenz N 2016 Simplified mathematical model for calculating global warming through anthropogenic CO2 *Int. J. Therm. Sci.* **102** 1–8

[3] Samanta A, Zhao A, Shimizu G K H, Sarkar P and Gupta R 2011 Post-combustion CO2 capture using solid sorbents: A review *Ind. Eng. Chem. Res.* **51** 1438–63

[4] Romeo L M, Bolea I and Escosa J M 2008 Integration of power plant and amine scrubbing to reduce CO2 capture costs *Appl. Therm. Eng.* **28** 1039–46

[5] Charney J G, Arakawa A, Baker J D, Bolin B, Dickinson R E, Goody R M, Leith C E, Stommel H M and Wunsch C I 1979 Carbon Dioxide and Climate: A Scientific Assessment

[6] Rashidi N A and Yusup S 2016 An overview of activated carbons utilization for the post-combustion carbon dioxide capture *J. CO2 Util.* **13** 1–16

[7] Nowicki P, Pietrzak R and Wachowska H 2008 Comparison of physicochemical properties of nitrogen-enriched activated carbons prepared by physical and chemical activation of brown coal
Energy & Fuels **22** 4133–8

[8] Cuhadaroglu D and Uygun O A 2008 Production and characterization of activated carbon from a bituminous coal by chemical activation *African J. Biotechnol.* **7** 3703–10

[9] Law K-N, Daud W R W and Ghazali A 2007 Morphological And Chemical Nature Of Fiber Strands Of Oil Palm Empty-Fruit-Bunch (OPEFB) *BioResources* **2** 351–62

[10] Sumathi S, Chai S P and Mohamed A R 2008 Utilization of oil palm as a source of renewable energy in Malaysia *Renew. Sustain. Energy Rev.* **12** 2404–21

[11] Harmsen P, Huijgen W, Bermudez L and Bakker R 2010 *Literature Review of Physical and Chemical Pretreatment Processes for Lignocellulosic Biomass*

[12] Vamvuka D, Kakaras E, Kastanaki E and Grammelis P 2003 Pyrolysis characteristics and kinetics of biomass residuals mixtures with lignite *Fuel* **82** 1949–60

[13] Luangkiattikhun P, Tangsathitkulchai C and Tangsathitkulchai M 2008 Non-isothermal thermogravimetric analysis of oil-palm solid wastes *Bioresour. Technol.* **99** 986–97

[14] Hidayu a. R, Mohamad N F, Matali S and Sharifah a. S a. K 2013 Characterization of Activated Carbon Prepared from Oil Palm Empty Fruit Bunch Using BET and FT-IR Techniques
Procedia Eng. **68** 379–84

[15] Shafeeyan M S, Mohd W, Wan A, Houshmand A and Shamiri A 2010 Journal of Analytical and Applied Pyrolysis A review on surface modification of activated carbon for carbon dioxide adsorption *J. Anal. Appl. Pyrolysis* **89** 143–51

[16] Otake Y and Jenkins R G 1993 Characterization of oxygen-containing surface complexes created on a microporous carbon by air and nitric acid treatment *Carbon N. Y.* **31** 109–21

[17] Lee T, Zubir Z A, Jamil F M, Matsumoto A and Yeoh F Y 2014 Combustion and pyrolysis of activated carbon fibre from oil palm empty fruit bunch fibre assisted through chemical activation with acid treatment *J. Anal. Appl. Pyrolysis* **110** 408–18

[18] Lua A C, Yang T and Guo J 2004 Effects of pyrolysis conditions on the properties of activated carbons prepared from pistachio-nut shells *J. Anal. Appl. Pyrolysis* **72** 279–87

[19] Noor A Bin, Asri M and Mohd B 2008 Textural Characteristics of Activated Carbons Prepared from Oil Palm Shells Activated with ZnCl 2 and Pyrolysis Under Nitrogen and Carbon Dioxide
J. Phys. Sci. **19** 93–104

[20] Joseph C G, Ashri W M, Daud W, Shane Q K and Sanmugam K 2015 Parametric and Adsorption Kinetic Studies of Reactive Black 5 Removal from Textile Simulated Wastewater Using Oil Palm (Elais guineensis) Empty Fruit Bunch 1 *J. Appl. Sci.* **15** 1103–11

[21] Fiuza R A, Medeiros de Jesus Neto R, Correia L B and Carvalho Andrade H M 2015 Preparation of granular activated carbons from yellow mombin fruit stones for CO2 adsorption. *J. Environ. Manage.* **161** 198–205

[22] Sing K S W and Williams R T 2004 Physisorption hysteresis loops and the characterization of nanoporous materials *Adsorpt. Sci. Technol.* **22** 773–82

[23] Nguyen P T M, Fan C, Do D D and Nicholson D 2013 On the cavitation-like pore blocking in ink-bottle pore: Evolution of hysteresis loop with neck size *J. Phys. Chem. C* **117** 5475–84

IOP Conf. Series: Materials Science and Engineering **206** (2017) 012004 doi:10.1088/1757-899X/206/1/012004

Modeling of CO_2 Adsorption on Activated Carbon and 13X Zeolite via Vacuum Swing Adsorption

M H Zarghampoor[1], M Mozaffarian[1,*], M Soleimani[1] and M Takht Ravanchi[2]

[1] Department of Chemical Engineering, Amirkabir University of Technology, No. 424, Hafez Ave., P.O.Box: 15875-4413, Tehran, Iran.

[2] Catalyst Research Group, Petrochemical Research and Technology Company, National Petrochemical Company, Tehran, Iran

Email: mozaffarian@aut.ac.ir

Abstract. Due to carbon dioxide role in global warming, low CO_2 emission limits have been established in recent years. This has led to a variety of studies on CO_2 removal approaches. In this study, a VSA cycle consisting of two packed beds is considered for CO_2 removal from flue gas. An atmospheric stream containing 20% CO_2 and 80% N_2 is fed to the beds at 50°C. Two adsorbents, namely Zeolite 13X and activated carbon were selected to compare their performance. Due to the monolayer adsorption of CO_2 and N_2 on these adsorbents, the Toth isotherm was used for equilibrium adsorption estimation. A quasi-second order model was considered for the mass transfer rate prediction due to low CO_2 concentration. The modeling results showed that the average absolute deviation for equilibrium adsorption capacity prediction was 2%, and the CO_2 breakthrough time curve was predicted with less than 2.5% deviation. Based on the results, the VSA cycle time for zeolite 13X bed will be 3.5 times of the activated carbon bed. Another advantage of Zeolite 13X is that in each process cycle, 80% of adsorbent will be used, while only 74% of activated carbon in beds is used. The advantage of activated carbon bed is its better regeneration capability, since the activated carbon will be regenerated 5% more than zeolite 13X at a vacuum pressure of 0.02bar.

1. Introduction

The increase in greenhouse gas emission has created many environmental problems. Carbon dioxide is one of the greenhouse gases with the greatest impact on global warming [1]. In recent years, the International Energy Agency has presented a roadmap to reduce CO_2 emission, develop sensible separation technologies and limit the increase in Earth's temperature to 2°C from 2013 to 2050. However; it is expected that the Earth's temperature will rise by 4°C during this period[2]due to industrial developments, the increase in fuel consumption and low CO_2 cost. Numerous studies have been conducted in recent years on trends and developments of CO_2 separation processes [3, 4] leading to suggested processes such as absorption, adsorption, membrane and refrigeration. The selection of each process depends on various factors such as operating pressure, feed gas temperature, product purity and feed gas impurities. The comparison of these processes has indicated that amine absorption is the most widely used process in industries. However; the application of this process is not recommended due to corrosion problems and high solvent regeneration cost. Similar problems in the refrigeration process such as low life time, plastization problems, fouling in membranes and high energy requirements have attracted the researchers to the adsorption process in recent years.

In order to study the performance of adsorbers in CO_2 removal, experimental studies on different adsorbers such as activated carbon, zeolite 5A and zeolite 13X have been performed [5, 6]. Adsorbent capacity and selectivity are the most important factors in the process performance and efficiency. It should be noted that the application of very strong adsorbents will make the regeneration process more difficult. Adsorbent regeneration can be performed by heat in TSA [7], pressure in PSA [8], and electrical energy in ESA [9] processes. Vacuum can be used as the necessary driving force for regeneration when the gas feed pressure is near the atmospheric pressure. This process is known as vacuum swing adsorption (VSA). The required energy for regeneration in VSA process is 1-1.5 MJ/kg CO_2 [9]. This lower energy requirement for the VSA process compared to TSA process with 3-4 MJ/kg CO_2 [7]and PSA process with 2-3 MJ/kg CO_2[10] has drawn the attention of researchers to the VSA process in recent years.

The aim of this study is to evaluate and compare the performance of zeolite 13X and activated carbon for CO_2 removal from a CO_2/N_2 mixture in an adsorption bed with VSA cycle. For this purpose, the experimental data of equilibrium isotherms and adsorption bed parameters of Dantas and her colleagues were used[11, 12]. Finally, in order to study the process limitations, a comprehensive review of the effective parameters such as feed conditions, vacuum pressure, purge to gas ratio and cycle time, and their effect on the recovery and performance of CO_2 productivity are studied.

2. Mathematical Modeling

Each VSA process cycle consists of four steps including bed pressurization, adsorption, blow-down and purging. To improve the process performance, blow-down and purging steps are carried out in the opposite direction of adsorption step[13]. Figure 1 shows an overview of the pressure profile in a VSA cycle. As can be seen, the flow in the adsorption step is downward to eliminate the possibility of adsorbent fluidization. Also, the pressure profile is assumed to be linear in the pressurization and blow-down steps.

The adsorption process mathematical modelling predicts the unsteady state behaviour of the adsorption bed. In order to achieve this goal, a set of partial differential equations including the mass, momentum and energy balance, along with one ordinary differential equation are considered and solved simultaneously to predict the mass transfer rate [14, 15].

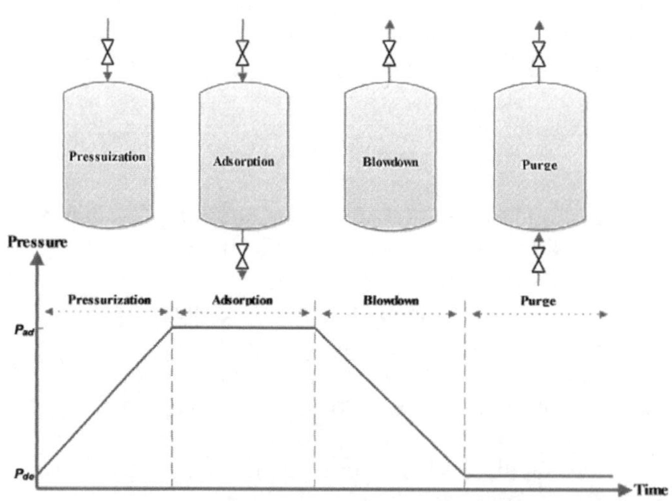

Figure 1. Pressure profile during a VSA cycle

In order to provide a mathematical model, the following assumptions are considered:

1. The adsorption bed operates under an isothermal condition.
2. The radial concentration gradients are negligible.
3. The pressure drop in the bed is calculated based on the Ergun equation.
4. The feed and purge gas conditions remain constant during the process.
5. The flow pattern is assumed to behave as an axially dispersed plug flow model.
6. The mass transfer rate is represented by a quasi-second order model.
7. The gas behaviour is considered to be ideal.

The flow model for the adsorption process is based on packed beds. The mass balance equation includes axial dispersion, convection, accumulation in gas phase, and rate of adsorption.

$$-D_{zi}\frac{\partial^2 C_i}{\partial z^2}+\frac{\partial}{\partial z}(uC_i)+\frac{\partial C_i}{\partial t}+\left(\frac{1-\varepsilon_b}{\varepsilon_b}\right)\rho_p\frac{\partial \bar{q}_i}{\partial t}=0 \quad (1)$$

where C_i (mol.m^{-3}) is the adsorbate concentration of component i in the fluid phase, u (m.s^{-1}) is the fluid velocity, z (m) represents the distance along the bed, t (s) is time, ε_b is the bed void fraction, ρ_p (kg.m^{-3}) is the particle density and $\bar{q}_i$ (mol.kg^{-1}) is the average concentration in adsorbent particle. The axial dispersion coefficient D_{zi} (m^2.s^{-1}) is defined by an empirical correlation as a function of Reynolds and Schmidt number.

$$\frac{\varepsilon_b D_{zi}}{D_{mi}}=20+0.5\,\mathrm{Sc}\,\mathrm{Re} \quad (2)$$

where D_{mi} (m^2.s^{-1}) is molecular diffusivity of component i and Sc and Re are the Schmidt and Reynolds numbers. The mass transfer rate is defined based on different adsorption kinetics. When the feed concentration is low, it is recommended to use a quasi-second order model[16]. Since CO_2 concentration in flue gas is 8 to 15 percent, it is recommended that a quasi-second order model be used to predict the mass transfer rate. The general form of this model is defined in Equation 3.

$$\frac{\partial \bar{q}_i}{\partial t}=K_{s,i}\left(q_i^*-\bar{q}_i\right)^2 \quad (3)$$

where $K_{s,i}$ (kg.mol^{-1}.s^{-1}) is the quasi-second order mass transfer coefficient and q_i^* (mol.kg^{-1}) is the adsorbed phase concentration in equilibrium with the fluid phase concentration. When the adsorption of all materials except component i is negligible, the equilibrium adsorption can be estimated with their pure isotherms at their partial pressure. Due to mono-layer adsorption of CO_2 and N_2 on zeolite 13X and activated carbon, the Toth isotherm is recommended for equilibrium adsorption prediction. This equation is defined as follows:

$$q_i^*=\frac{q_i^{\max}K_{eq}P}{\left[1+\left(K_{eq}P\right)^n\right]^{1/n}} \quad (4)$$

where $q_i^{\max}$ (mol.kg^{-1}) is the maximum amount adsorbed at equilibrium, K_{eq} (bar^{-1}) is an equilibrium constant, P (bar) is the total pressure and n is the Toth parameter. By passing the fluid through the empty spaces between adsorbent particles, a pressure drop may take place in the fluid. The Ergun equation was used to predict this pressure drop:

$$\frac{\partial P}{\partial z}=-\frac{150\mu_g\left(1-\varepsilon_b\right)^2}{\varepsilon_b^3 d_p^2}u-\frac{1.75\left(1-\varepsilon_b\right)\rho_g}{\varepsilon_b^3 d_p}u^2 \quad (5)$$

IOP Conf. Series: Materials Science and Engineering **206** (2017) 012004 doi:10.1088/1757-899X/206/1/012004

where μ_g (Pa.s^{-1}) is the gas viscosity, d_p (m) is the particle diameter and ρ_g (kg.m^{-3}) is the gas density. In order to evaluate the VSA cycle performance in CO_2 separation, it is necessary to calculate its emission to the atmosphere and the recovery. These two parameters are defined in Equations 6 and 7.

$$CO_2\ \text{Emmision} = \int_0^{t_{adsorption}} (\text{1-purge/feed ratio}).\text{Flow}_{CO_2}\,dt \quad (6)$$

$$CO_2\ \text{Recovery}=\frac{CO_2\ \text{obtained in adsorption -}CO_2\ \text{used in purging}}{CO_2\ \text{fed during pressurization \& adsorption}} \quad (7)$$

In order to predict the VSA cycle behaviour, the above equations must be solved simultaneously. The set of partial differential equations are discretized in the length direction and solved by MATLAB with the ODE15s method.

To evaluate the VSA performance in CO_2 reduction, the experimental data by Dantas et.al was used [11, 12]. Therefore, an atmospheric mixture of 20% CO_2 and 80% N_2 is introduced to the cycle at 50°C with a flow rate of 3 L.min^{-1}. The adsorbent properties and bed parameters are reported in Tables 1 and 2.

Table 1. Zeolite 13X and activated carbon properties

Zeolite 13X		Activated Carbon	
Surface area [m^2/g]	585.5	Surface area [m^2/g]	1053
Micropore volume [cm^3/g]	0.172	Micropore volume [cm^3/g]	0.097
Particle density [g/cm^3]	1.23	Particle density [g/cm^3]	1.14
Particle porosity	0.37	Particle porosity	0.46
Particle diameter [mm]	2.9	Particle diameter [mm]	3.8

Table 2. Adsorption bed parameters

Bed Length [cm]	83
Bed Diameter [cm]	2
Bed void fraction	0.52
Column wall thickness [mm]	1.5
Column wall density [kg.m^{-3}]	7280
Bed weight [kg]	0.158

3. Results and Discussion

The adsorption equilibrium determination is one of the major problems in predicting the VSA process behaviour. Figure 2 shows the results of CO_2 and N_2 equilibrium adsorption prediction with the Toth model. The results indicate that the maximum deviation error for equilibrium adsorption capacity prediction is 2.5% for CO_2 and 2% for N_2. As can be seen, the selection of Zeolite 13X is much better due to its higher capacity at low pressures.

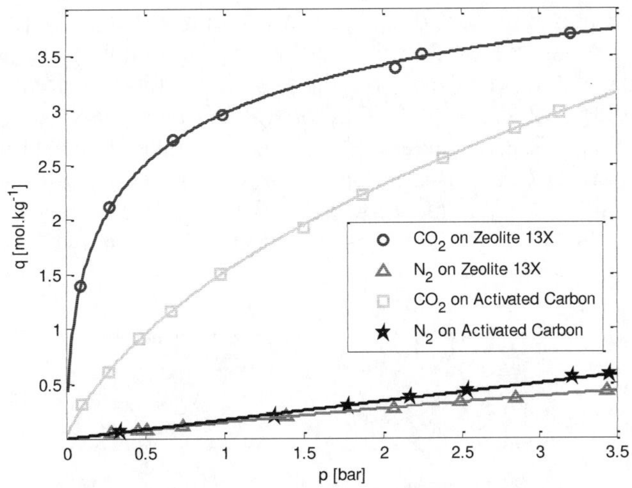

Figure 2. CO_2 and N_2 adsorption equilibrium on Zeolite 13X and activated carbon at 50°C; the solid lines represent the results of the Toth model.

Figure 3 shows breakthrough curve which is plotted to determine the proper time to carry out the adsorption cycle. Additionally, figure 3 indicates the results for modeling the VSA process in packed activated carbon and zeolite 13X beds. According to this figure; N_2 is adsorbed in the bed at first, and then the CO_2 molecules will replace N_2 due to the competitive adsorption nature of CO_2 and N_2. This phenomenon causes an overshooting peak on the N_2 breakthrough curve. The breakthrough time for zeolite 13X bed is 3.5 times greater than the one for activated carbon bed. An increase in breakthrough time reduces the necessary times for bed regeneration and improves the recovery. The results show that the average deviation error for breakthrough time prediction curve is 2.5%.

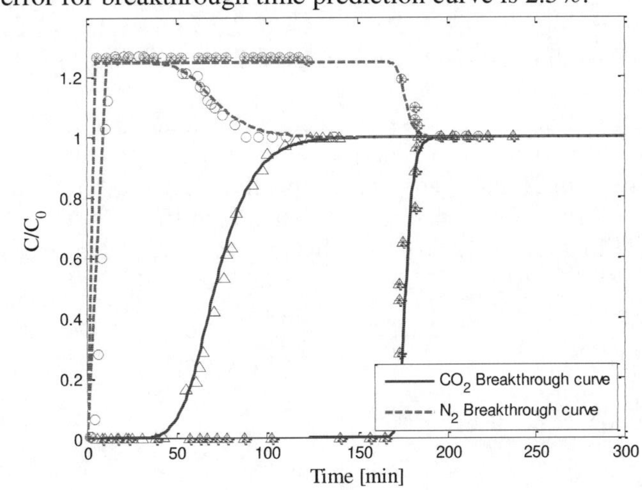

Figure 3. Breakthrough curve of CO_2/N_2 mixture on zeolite 13X(N_2:⊛,CO_2:▲) and activated carbon(N_2:○,CO_2:△); lines are obtained by a quasi-second order model.

As an additional advantage, more parts of zeolite 13X adsorbent are consumed in each cycle of the process. Figure 4 demonstrates the available area in a VSA cycle for two adsorbents. In the adsorption step, the process should be stopped at break point time due to CO_2 emission limitation. At this time, the mass transfer zone reaches the bottom of the bed. The shorter the length of mass transfer zone, the greater the amount of consumed adsorbent. The mass transfer zone in the activated carbon bed is 2.5 times of zeolite 13X bed. The results show that at the end of the adsorption step, 93% of zeolite 13X adsorbent is saturated, but only 85% of activated carbon is saturated at break point time. If the process is continued to saturate more activated carbons, the CO_2 emission to atmosphere will increase significantly. The advantage of activated carbon is the simplicity of its regeneration. At the end of blow-down to 0.1 bar and purging 15% of N_2, 89% of adsorbent will be regenerated. However, at the end of this step only 87% of zeolite 13X is regenerated. The reason is related to the strong interaction between CO_2 molecules and zeolite 13X adsorbent. This strong interaction causes a better and selective separation with more energy cost requirement for regeneration. As can be seen in figure 4, in each cycle of the process, 80% of zeolite 13X adsorbent will be consumed, as opposed to only 74% of adsorbent in the activated carbon beds.

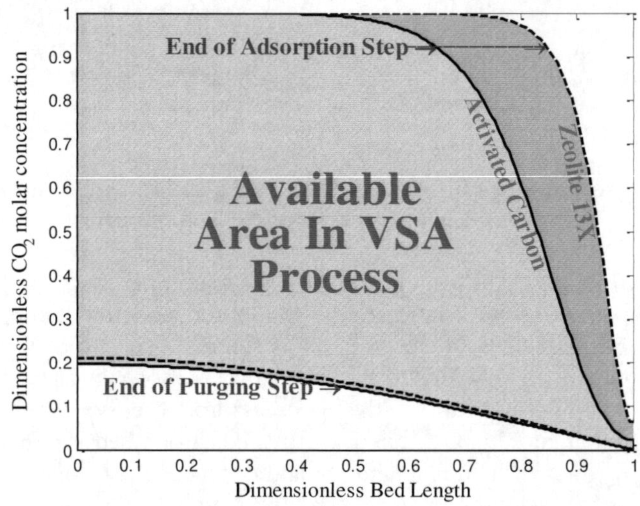

Figure 4. Adsorbent consumption in each cycle of VSA process for zeolite 13X and activated carbon; Solid line: activated carbon, Dash lines: zeolite 13X.

Since adsorption is an exothermic process, an increase in temperature will reduce the adsorbent capacity. Therefore, it is recommended to keep the temperature as low as possible in all adsorption processes to have the maximum available adsorbent capacity. Figure 5 shows the effect of temperature on the adsorption process. As can be seen, a rise of temperature from 50°C to 100°C will reduce the breakthrough time by 41% in a zeolite 13X bed and by 65% in an activated carbon bed.

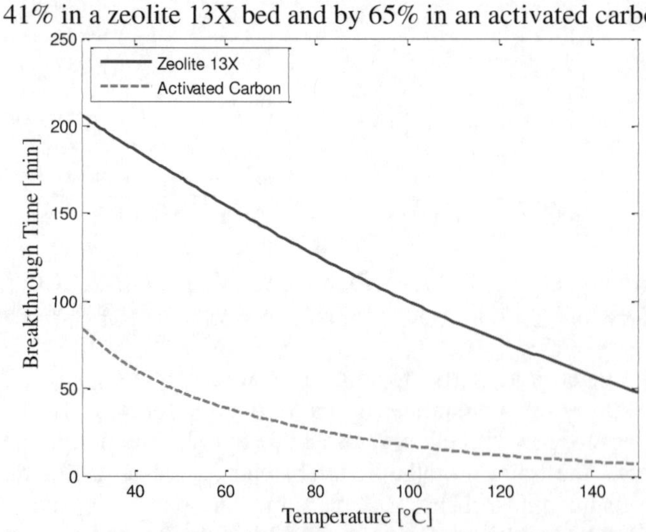

Figure 5. Temperature effects on breakthrough time of zeolite 13X and activated carbon beds.

As can be seen in Figure 6, a 33% increase in feed flow rate will decrease the breakthrough time by 25% in a zeolite 13X bed. However, this feed flow rate increase will reduce the break point time by 30% in an activated carbon bed.

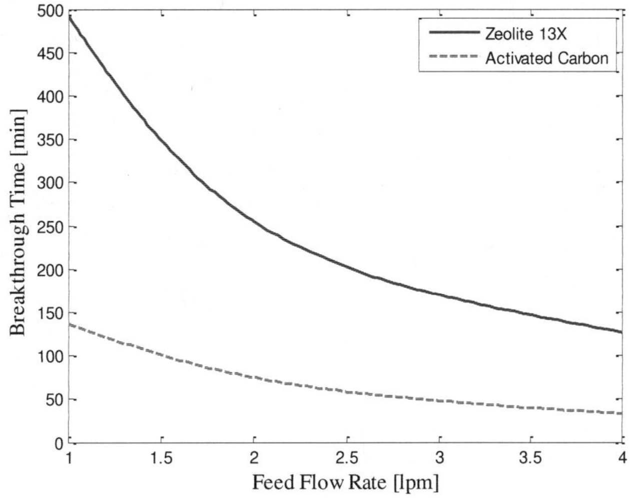

Figure 6. Feed flowrate effect on breakthrough time of zeolite 13X and activated carbon beds.

Figure 7 shows the effect of CO_2 feed concentration on the breakthrough time. An increase in CO_2 concentration will reduce the break point time and increase the number of regeneration times which will cause adsorbents' erosion and increase the pressure drop. A 15% increase in CO_2 concentration will reduce the break point time by 55% in a zeolite 13X bed, whereas this will only decrease the break point time by 26% in an activated carbon bed. Therefore, the feed concentration effect on activated carbon is less than the effect on a zeolite 13X bed.

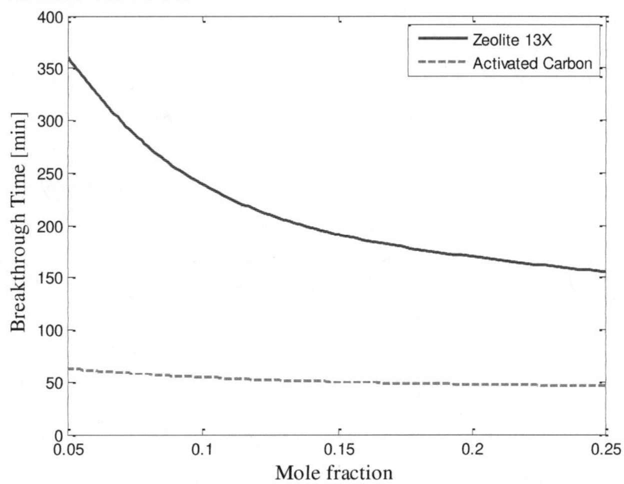

Figure 7. CO_2 feed concentration effect on breakthrough time of zeolite 13X and activated carbon beds.

One of the most important points in the design of a VSA process is its cycle time. Figure 8 shows the effect of cycle time on CO_2 emission and its recovery. In this process, 40% of cycle time is considered for the adsorption step. A reduction in cycle time increases the amount of CO_2 emission to atmosphere. The maximum cycle time is determined by the CO_2 emission limit. If the process is designed to minimize CO_2 emission, the maximum cycle time will become 280 minutes for zeolite 13X beds. However, the maximum cycle time in activated carbon beds will be only 110 minutes.

IOP Conf. Series: Materials Science and Engineering **206** (2017) 012004 doi:10.1088/1757-899X/206/1/012004

Another advantage of zeolite 13X is its higher CO_2 recovery. For a zeolite 13X bed with 280 minutes cycle time the recovery is 98%, whereas the CO_2 recovery for an activated carbon bed with a cycle time of 110 minutes is 95%.

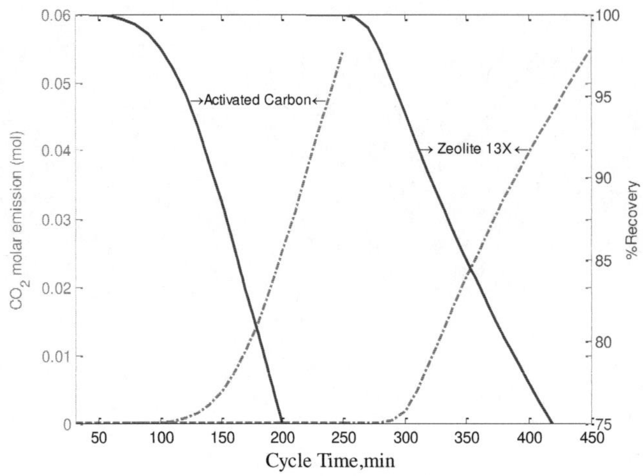

Figure 8. Cycle time effect on CO_2 emission to atmosphere and its recovery; Solid line: CO_2 recovery, Dash line: CO_2 molar emission.

A vacuum pump creates the required driving force for adsorbent regeneration in all VSA processes. Since the vacuum pump consumes energy, it is necessary to determine the required vacuum pressure. Figure 9 shows the effect of vacuum pressure in a bed regeneration process. As can be seen, any reduction of vacuum pressure increases the bed regeneration. It should be noted that both the zeolite 13X and activated carbon beds regenerated completely at 1mmHg vacuum pressure even though reaching this degree of vacuum is not cost effective. The regeneration behaviour of zeolite 13X showed that most of adsorbent regenerates at a very low pressure. The regeneration of 90% of zeolite 13X adsorbent required a vacuum pressure less than 0.01 bar. However, this amount of regeneration is possible with vacuum pressure equal to 0.06 bar in activated carbon beds.

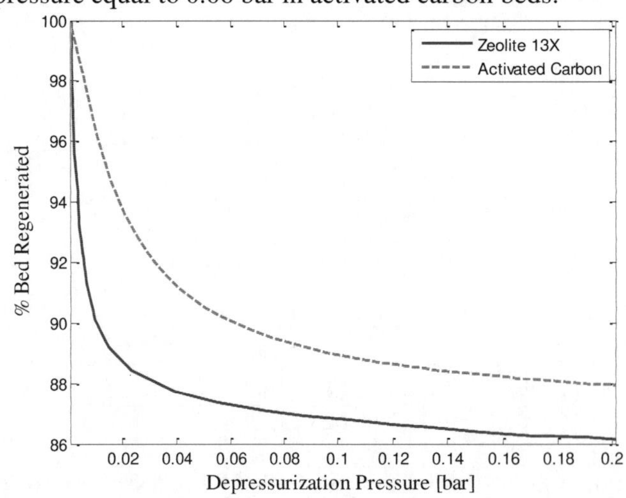

Figure 9. Vacuum pressure effect on CO_2 emission and bed regeneration.

The application of purge gas at the end of bed regeneration is another important factor in determining the process performance. As is shown in figure 10, increasing the purge to feed ratio improves the bed regeneration with little effect on CO_2 recovery. It should be noted that an increase in purge to feed ratio will reduce the quality of CO_2 product at the end of the process. In the case where the N_2 product is important, an increase in purge to feed ratio will decrease N_2 productivity and its recovery.

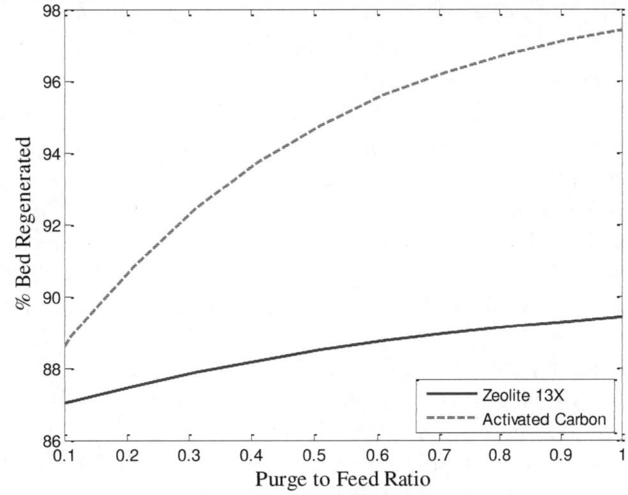

Figure 10. Purge to feed ratio effect on bed regeneration.

4. Conclusions

In order to evaluate the VSA performance in CO_2 removal, a process consisting of two adsorption beds was modelled. The equilibrium adsorptions were predicted by Toth isotherm. The comparison of the experimental results and the model showed that the maximum deviation errors for equilibrium adsorption capacity prediction are 2.5% for CO_2 and 2% for N_2. Then the beds were modelled with a quasi-second order model and the results showed that the average deviation error for breakthrough time prediction curves was 2.5%.
The results of the analysis for effective parameters can be summarized as follows:

- The breakthrough time for the zeolite 13X bed is 3.5 times greater than the time for activated carbon bed. This will reduce the necessary times for bed regeneration and improve the recovery.
- In each cycle of the process, 93% of zeolite 13X is saturated at the end of the adsorption step. Additionally, at the end of blow-down to 0.1 bar and purging 15% of product N_2, 87% of them will be regenerated. In fact 80% of adsorbent are used in each cycle. However, 85% of adsorbent are saturated at the end of adsorption step in an activated carbon bed and 89% of them will be regenerated at the end of purging. This phenomenon showed that only 74% of adsorbers are used in each cycle.
- A temperature rise of 50°C will reduce the break point time by 60% in the activated carbon bed. However, this will decrease the time by 40% in a zeolite 13X bed.
- Doubling the feed flow rate will decrease the break point time in the zeolite 13X bed by 50%, whereas this will reduce the time by 56% in an activated carbon bed.
- A 15% increase in CO_2 feed concentration reduces the breakthrough time by 55% in a zeolite 13X bed. However, this concentration change will decrease the time by only 26% in an activated carbon bed. In the case of feeding the system by several sources with different concentration, if the process steps are changed based on a time schedule rather than the usual CO_2 breakthrough, then it is better to use an activated carbon bed rather than one with zeolite 13X.
- Due to the strong interaction between CO_2 and zeolite 13X molecules and more adsorption capacity at a low pressure, most of adsorbent will be regenerated at low pressure. The regeneration of 90% of adsorbent requires a vacuum pressure of 0.01bar in a zeolite 13X bed. However, this amount of regeneration requires a vacuum pressure of 0.06bar in an activated carbon bed.

IOP Conf. Series: Materials Science and Engineering **206** (2017) 012004 doi:10.1088/1757-899X/206/1/012004

- A 10% increase in purge gas will increase the bed regeneration by 2%. It is recommended that purge gas be used as little as possible. An increase in purge gas will reduce the CO_2 product purity and N_2 recovery.

References

[1] Songolzadeh M, et al. 2014 Carbon dioxide separation from flue gases: a technological review emphasizing reduction in greenhouse gas emissions, *The Sci. World J.* **1**.
[2] IEA, 2013, Technology Roadmap: Carbon capture and storage.
[3] Mondal M.K., et al. 2012 Progress and trends in CO2 capture/separation technologies: A review, Energy, *Energy* **46**(1) 431.
[4] Yang H., et al., 2008, Progress in carbon dioxide separation and capture: a review, *J. Env. Sci (China)* **20**(1) 14.
[5] Dantas T L P, et al. 2008, Adsorção de CO2 e N2 Sobre Carvão Ativado e Zeólita 13X: Isotermas de Equilíbrio Através de Medidas Gravimétricas, *7° Encontro Brasileiro sobre Adsorção*, 25.
[6] Khazraei S et al. 2013, Experimental Study for Equilibrium Adsorption of Carbon Dioxide and Nitrogen on 5A and 13X Molecular Sieve, International Review of Chemical Engineering (IRECHE) *Int. Rev. of Che. Eng. (IRECHE)* **5** 41.
[7] Ntiamoah A et al. 2016, CO2 Capture by Temperature Swing Adsorption: Use of Hot CO2-Rich Gas for Regeneration, *I&EC Research* **55**(3) 703.
[8] Shen C, et al. 2010, Adsorption equilibria and kinetics of CO2 and N2 on activated carbon beds, *Che. Eng. J.* **160**(2) 398.
[9] Zhang J, et al. 2008, Effect of process parameters on power requirements of vacuum swing adsorption technology for CO2 capture from flue gas,Energy Conversion and Management, *Eng. Con. and Mng.* **49**(2) 346.
[10] Koga T, et al. 2011, Study on CO2 Recovery Systems by Pressure Swing Adsorption under High Pressure Condition, *J. of Novel Carbon Resources* **3** 60.
[11] Dantas T L P et al. 2011, Carbon dioxide-nitrogen separation through adsorption on activated carbon in a fixed bed, *Che. Eng. J.* **169** 11.
[12] Dantas T L P et al. 2011, Carbon dioxide–nitrogen separation through pressure swing adsorption, *Che. Eng. J.* **172**(2-3) 698.
[13] Farooq S et al. 1989, Numerical simulation of a pressure swing adsorption oxygen unit, *Che. Eng. Sci.* **44**(12) 2809
[14] Faruque Hasan M M, et al. 2012, Modeling, Simulation, and Optimization of Postcombustion CO2 Capture for Variable Feed Concentration and Flow Rate, 2. Pressure Swing Adsorption and Vacuum Swing Adsorption Processes, *I&EC Research* **51**(48) 15665.
[15] Ruthven D M, et al. 1994, Pressure Swing Adsorption, *VCH Pub. Inc.*, Chapter 3.
[16] Azizian S 2004, Kinetic models of sorption: a theoretical analysis, *J. of Colloid and Interface Sci.* **276**(1) 47.

IOP Conf. Series: Materials Science and Engineering **206** (2017) 012005 doi:10.1088/1757-899X/206/1/012005

Preparation and Characterization of Impregnated Commercial Rice Husks Activated Carbon with Piperazine for Carbon Dioxide (CO_2) Capture

S N Masoum Raman, N A Ismail and S S Jamari

Faculty of Chemical & Natural Resources Engineering, Universiti Malaysia Pahang, 26300, Gambang, Pahang, Malaysia

Email: sshima@ump.edu.my

Abstract. Development of effective materials for carbon dioxide (CO_2) capture technology is a fundamental importance to reduce CO_2 emissions. This work establishes the addition of amine functional group on the surface of activated carbon to further improve the adsorption capacity of CO_2. Rice husks activated carbon were modified using wet impregnation method by introducing piperazine onto the activated carbon surfaces at different concentrations and mixture ratios. These modified activated carbons were characterized by using X-Ray Diffraction (XRD), Brunauer, Emmett and Teller (BET), Fourier Transform Infrared Spectroscopy (FTIR) and Field Emission Scanning Electron Microscopy (FESEM). The results from XRD analysis show the presence of polyethylene butane at diffraction angles of 21.8° and 36.2° for modified activated carbon with increasing intensity corresponding to increase in piperazine concentration. BET results found the surface area and pore volume of non-impregnated activated carbon to be 126.69 m^2/g and 0.081 cm^3/g respectively, while the modified activated carbons with 4M of piperazine have lower surface area and pore volume which is 6.77 m^2/g and 0.015 cm^3/g respectively. At 10M concentration, the surface area and pore volume are the lowest which is 4.48 m^2/g and 0.0065 cm^3/g respectively. These results indicate the piperazine being filled inside the activated carbon pores thus, lowering the surface area and pore volume of the activated carbon. From the FTIR analysis, the presence of peaks at 3312 cm^{-1} and 1636 cm^{-1} proved the existence of reaction between carboxyl groups on the activated carbon surfaces with piperazine. The surface morphology of activated carbon can be clearly seen through FESEM analysis. The modified activated carbon contains fewer pores than non-modified activated carbon as the pores have been covered with piperazine.

1. Introduction

Rice husk is the outermost layer of the paddy grain that being separated throughout the milling process which consist of 20% from overall paddy production and approximately, there are 100 million tons of rice husks generated worldwide per annum [1]. Rice husk usually categorized as agricultural waste product and often disposed by dumping or burning to produce rice husk ash [2]. The high content of hydrocarbon in the rice husk enhances its potential as the feedstock for activated carbon preparation [3].

IOP Conf. Series: Materials Science and Engineering **206** (2017) 012005 doi:10.1088/1757-899X/206/1/012005

Activated carbon usually used as adsorbent for environmental control in the form of a fixed bed due to its large internal surface area and pore volume, and its ability to adsorb organic vapors for a low cost [4]. Activated carbon was also considered as one of the highly potential adsorbent for CO_2 adsorption [5]. Adsorption is a process of mass transfer occurs from bulk of gas or liquid to adsorbents and it has been thoroughly used as separation methods in chemical and other industries [6]. Activated carbon consists of acidic and basic surfaces while CO_2 gas is an acidic gas. As the CO_2 gas interacts with activated carbon, the basic groups on the activated carbon surfaces will attract CO_2 gas. To enhance the adsorption capacity of the activated carbon, amine groups have been introduced by wet impregnation method onto the activated carbon due to their high efficiency and selectivity for CO_2 capture [7]. It has been proved that the modified activated carbon with piperazine enhanced the adsorption capacity significantly [8]. Piperazine will react with CO_2 which involves the mass transfer of CO_2 from the gas phase into the aqueous phase of amine as in (1):

$$\underset{\text{carbon dioxide}}{CO_2} + \underset{\text{piperazine}}{2C_4H_{10}N_2} \rightleftharpoons \underset{\text{carbamate anion}}{C_4H_{10}NCOO^-} + \underset{\text{diethylamine cation}}{C_4H_{10}NH_2^+} \quad (1)$$

In this research work, the rice husk activated carbon was impregnated with piperazine which expected to enhance the CO_2 adsorption capacity. The research aims to study the characteristics of non-modified activated carbon and modified activated carbon with piperazine at different concentration and mixture ratio. The concentration used in this study were 2M, 4M, 6M, 8M and 10M while the mixture ratio used were 1:1, 1:2 and 2:1 of activated carbon to piperazine. The modified activated carbon then undergoes several analyses such as XRD, BET, FTIR and FESEM.

2. Materials and Methods

2.1. Materials and Chemicals

The commercial activated carbon made from rice husk ash was supplied by Sin Guan Hup Oil & Rice Mill Sdn Bhd (Penang, Malaysia). Anhydrous piperazine with a molecular weight of 86.14 g/mol was purchased from MERCK.

2.2. Activated Carbon Preparation

The activated carbon particle's size was sifted using 500nm sieve. The prepared activated carbon particles were rinsed with deionized water followed by heating in a 70°C oven for 6 hours. The sample was then being kept in a desiccator at room temperature to avoid moisture effect.

2.3. Impregnation Process

Impregnation of activated carbon was conducted with the presence of piperazine. Concentrations of piperazine being used in this work were 2M, 4M, 6M, 8M, and 10M. 50g of activated carbon was used for preparation of each sample. Ratio of the mixture was based on the weight percent of the activated carbon to piperazine (ie: 1:1, 1:2 and 2:1). Firstly, the mixture was mixed in a beaker and mechanically stirred at 60 rpm for one hour at room temperature before being heated at 70°C for 6 hours to ensure complete drying without any trace of moisture left within its particles obtained. The sample then being kept in a desiccator at room temperature.

2.4. Activated Carbon Characterization

2.4.1. X-Ray Diffraction (XRD.) Rigaku Miniflex X-Ray Diffractometer of the CuKα radiation filtered by a Ni filter was used in this research with a wavelength of 1.54 Å at 30 kV and 15 mA. The activated carbon samples were initially crushed to a fine powder (< 200 μm) to minimize inducing extra strain (surface energy) that can offset peak positions, and to randomize orientation. The specimens were placed on the glass specimen holder and pressed using a glass slide. Scanning of sample was performed starting from 3° to 80° at a speed of 1° min^{-1}. The peaks obtained from the analysis were identified using PDLX software.

2.4.2. Fourier Transform Infrared Spectroscopy (FTIR). Analysis of FTIR was conducted by FTIR (Thermo Scientific Nicolet iS10) using Attenuated Total Reflectance (ATR), Nujol and thin film. Enough samples were needed to obtain an absorption spectrum during FTIR analysis. The first step was to collect background spectra to subtract from the test spectra. Next, the sample was analysed by LTI's fully-computerized Fourier Transform Infrared Spectroscopy system which produces the absorbance spectra showing the unique chemical bonds and the molecular structure of the sample material. This profile was either in the form of an absorption/transmittance spectrum which shows peaks representing the components concentration. The wavelength used was in range of 500 – 4000 cm^{-1}. The software used to interpret the data was OMNIC.

2.4.3. Brunauer, Emmett and Teller Analysis (BET). Physical adsorption of a gas on solid surface will results in a formation of a thin layer that covers the entire solid surfaces. Analysis using BET instrument provides single sample surface area as well as pore size distribution. Firstly, around 0.3g of activated carbon was packed in the U-sample tube, and degassed at 70°C overnight and cool down to ambient temperature. The nitrogen (N_2) adsorption or desorption process was done using Thermo Scientific (Surfer) equipment. Once the TCD signal has stabilized, the sample tube was immersed in a flask fill up with liquid N_2. When the adsorption has reached the equilibrium, the sample was shifted back to ambient temperature for 15 minutes. Surface area was calculated and the effect of surface porosity and pore size was determined.

2.4.4. Field Emission Scanning Electron Microscopy (FESEM). The activated carbon was analysed by using JEOL field emission scanning electron microscope at the magnifications of 60x and 50,000x. The accelerating voltage was 5.0 kV during image capture using secondary electron and dark field detectors.

3. Results and Discussions

3.1. X-Ray Diffraction (XRD)

X-Ray diffraction analysis was used to determine the crystallographic structure and chemical composition of the non-modified activated carbon and modified activated carbon with piperazine. Figure 1 shows the difference between diffraction peaks for the non-modified and modified activated carbon. The presence of pyrazole in non-impregnated activated carbon can be seen clearly at the diffraction angle of 21.66° and 36.25° in Figure 1(A). There is pyrazole chain presence on the non-modified activated carbon surface [9]. The peaks obtained for modified activated carbon with piperazine can be observed through Figure 1(B) and Figure 1(C). Figure 1(B) shows the results for concentration of 4M of piperazine impregnated sample while Figure 1(C) shows the peaks for concentration of 10M of piperazine impregnated sample. The diffraction angle for lower concentration piperazine shows the presence of polyethylene butane at 21.86° and 36.28° while at higher concentration of piperazine which is 10M, there is presence of silicon oxide at 21.89° and polyethylene butane at 36.12°. The differences in peak height for both samples explain that there is higher content of polyethylene butane in low concentration of piperazine while the silicon oxide appears in high concentration piperazine. The intensity of the peak is higher when the amine loading on the surface of activated carbon is higher [10]. In this case, the peak of silicon oxide shows the highest intensity compared to the polyethylene butane peak, thus proved the statement made by the previous researchers.

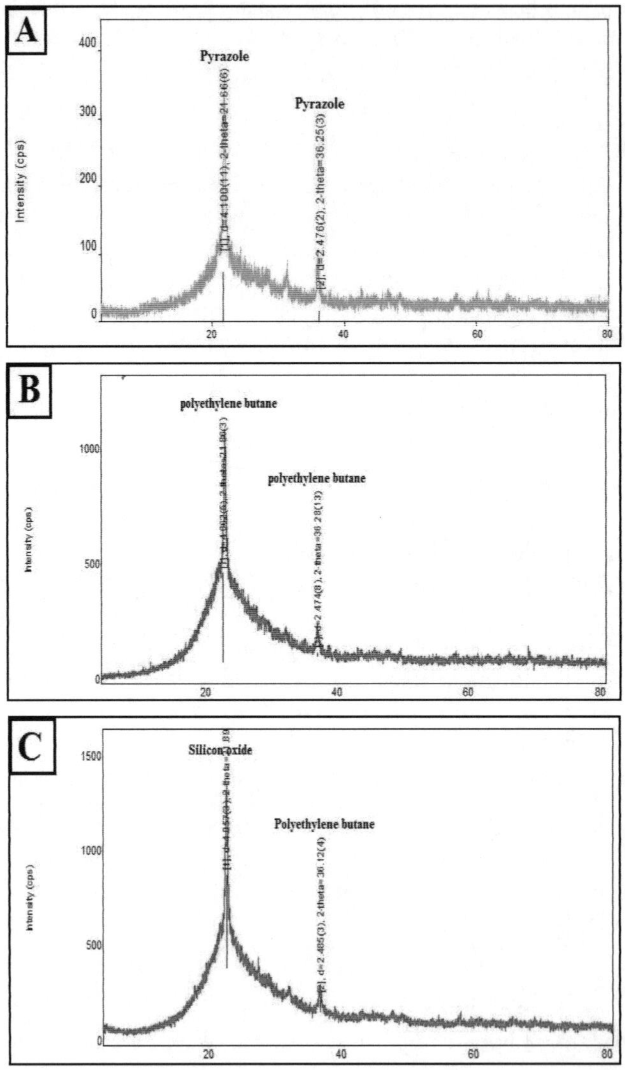

Figure 1. XRD pattern for (A) non-impregnated activated carbon, impregnated activated carbon with (B) 4M piperazine and (C) 10M piperazine.

3.2. Fourier Transform Infrared Spectroscopy (FTIR)

Based on the FTIR results, the transmission range for the samples are from 4000 – 500 cm^{-1}, however the sensitive transmission regions of the samples were detected in a few different regions. Amides regions were found across 3500 – 1500 cm^{-1}. The reaction between carboxylic acids presence on the surface of activated carbon with amine has formed amide. Figure 2 shows the difference of transmittance peaks between non-impregnated activated carbon and impregnated activated carbon. The presence of amide on the impregnated activated carbon at wavelength 3278.28 cm^{-1} with the transmittance percentage of 64% while for the non-modified activated carbon, no presence of amide's peaks observed on the sample. This results were strengthen with the research carried out which discovered the presence of NH_2 surface groups at band 3365 cm^{-1}, described as amide region [11]. The presence of amide in impregnated activated carbon is linked with presence of nitrogen functional group in the amine solution [12]. Figure 3 shows the transmittance peak for modified activated carbon with different piperazine concentration. The lowest transmittance percentage indicates the highest piperazine concentration (10M) while the lowest concentration (2M) shows a very slight curve in transmittance peak at band 1541.28 cm^{-1}. The presence of several functional groups including –OH stretch and –NH stretch on commercial activated carbon has improved CO_2 binding process onto the surface of activated carbon [13]. From Figure 4, the mixture ratio parameter shows that the modified activated carbon with high piperazine ratio which is 1:2 forms a transmittance peak at band 3278.28 cm^{-1}. The lowest quantity of piperazine used which is 2:1 has almost no peak formed at the amide region.

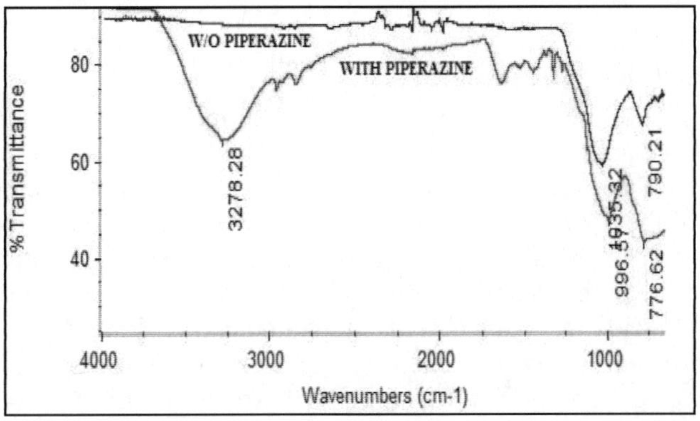

Figure 2. FTIR spectrum for non-impregnated activated carbon and impregnated activated carbon with piperazine.

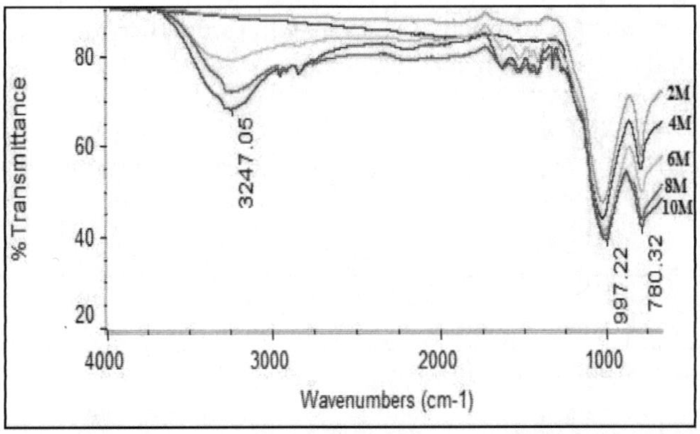

Figure 3. FTIR spectrum of impregnated activated carbon with different concentration of piperazine.

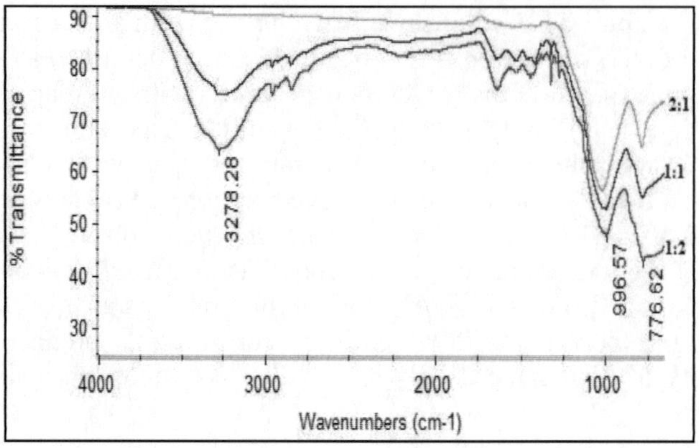

Figure 4. FTIR spectrum of piperazine impregnated activated carbon with different mixture ratio.

3.3. Brunauer, Emmett and Teller Analysis (BET)

Table 1 shows the summary of the surface area and pore volume of the non-modified activated carbon and impregnated activated carbon with 4M and 10M piperazine concentration. The BET total surface area of the non-impregnated activated carbon is 126.69 m^2/g. The result obtained is higher compared to the total surface area of the impregnated activated carbon with piperazine which is 6.77 m^2/g for 4M sample and 4.48 m^2/g for 10M sample. This suggested that the pores were blocked by the piperazine molecules and decreased the pore volume. The pore volume and surface area decreased because of the alkanolamine being filled inside the inner pores of activated carbon [14]. There was a large difference between the pore sizes of the non-impregnated activated carbon and impregnated activated carbon. The non-modified activated carbon has the highest pore size which is 0.081 cm^3/g while the impregnated activated carbon with 4M and 10M of piperazine has 0.015 cm^3/g and 0.0065 cm^3/g of pore volume respectively. High concentration of adsorbate tend to fill more pores on the adsorbent and resulting in narrow and small pore volume and surface area [15]. The surface area and pore volume become smaller with the increase of piperazine loading hence will increase the active sites for CO_2 bonding during CO_2 adsorption process. The modified adsorbent surface with piperazine has increased the capacity of CO_2 captured [16]. From the BET results, it can be concluding that the activated carbon impregnated with high concentration piperazine is a better adsorbent as the surface area and pore volume are the smallest, indicates that it has largest active sites for CO_2 adsorption.

Table 1. Surface area and pore volume analysis for non-impregnated activated carbon and impregnated activated carbon with piperazine.

Adsorbing Bed	Surface Area (m^2/g)	Pore Volume (cm^3/g)
Non-modified activated carbon	126.69	0.0810
Modified activated carbon with 4M piperazine	6.77	0.0150
Modified activated carbon with 10M piperazine	4.48	0.0065

3.4. Field Emission Scanning Electron Microscopy (FESEM)

Figure 5 shows the surface morphology of non-impregnated activated carbon and impregnated activated carbon with piperazine at different concentration. At magnification of 60 times, a major difference can be seen in the pore size and surface structure of the macropores. The non-modified activated carbon has largest pore size while the impregnated activated carbon has a stacking of piperazine on its surface. Lower concentration piperazine impregnated activated carbon shows less stacking than higher concentration piperazine impregnated activated carbon. At 50,000 times magnification, there is slight difference in the surface morphology of the mesopores. A higher concentration of piperazine causes the coating on the activated carbon to be thickened. It also causes the increment in amount of piperazine loaded in the macropores which in agreement with the previous work stated that there was a formation of octahedral stack on the surface of activated carbon impregnated with amine [17].

Figure 6 shows the morphology for different mixture ratio of impregnated activated carbon with piperazine. Figure 6(A), (B) and (C) show the surface structure for magnification of 60 times while Figure 6(D), (E) and (F) show the morphology for magnification 50,000 times. At magnification of 60 times, the mixture ratio for activated carbon to piperazine which is 1:2 has the highest piperazine loading on its surface followed by ratio of 1:1 and 2:1. At magnification of 50,000 times, the mixture ratio of 1:2 activated carbon to piperazine has slightly more covered pores compared to the other two ratios. The mesopores for 1:2 ratio activated carbon to piperazine also shows a better attachment and coating compared to ratio of 1:1 and 2:1. Higher ratio of piperazine shows the increase of amide loading on the activated carbon surface as the carboxyl groups on the activated carbon surface reacts with amine groups present in piperazine to form surface of amide [18].

IOP Conf. Series: Materials Science and Engineering **206** (2017) 012005 doi:10.1088/1757-899X/206/1/012005

Figure 5. The FESEM images for (A) non-impregnated activated carbon, (B) 60 times magnification for 4M impregnated activated carbon with piperazine, (C) 60 times magnification for 10M impregnated activated carbon with piperazine, (D) 50,000 times magnification for 4M impregnated activated carbon with piperazine and (E) 50,000 times magnification for 10M impregnated activated carbon with piperazine.

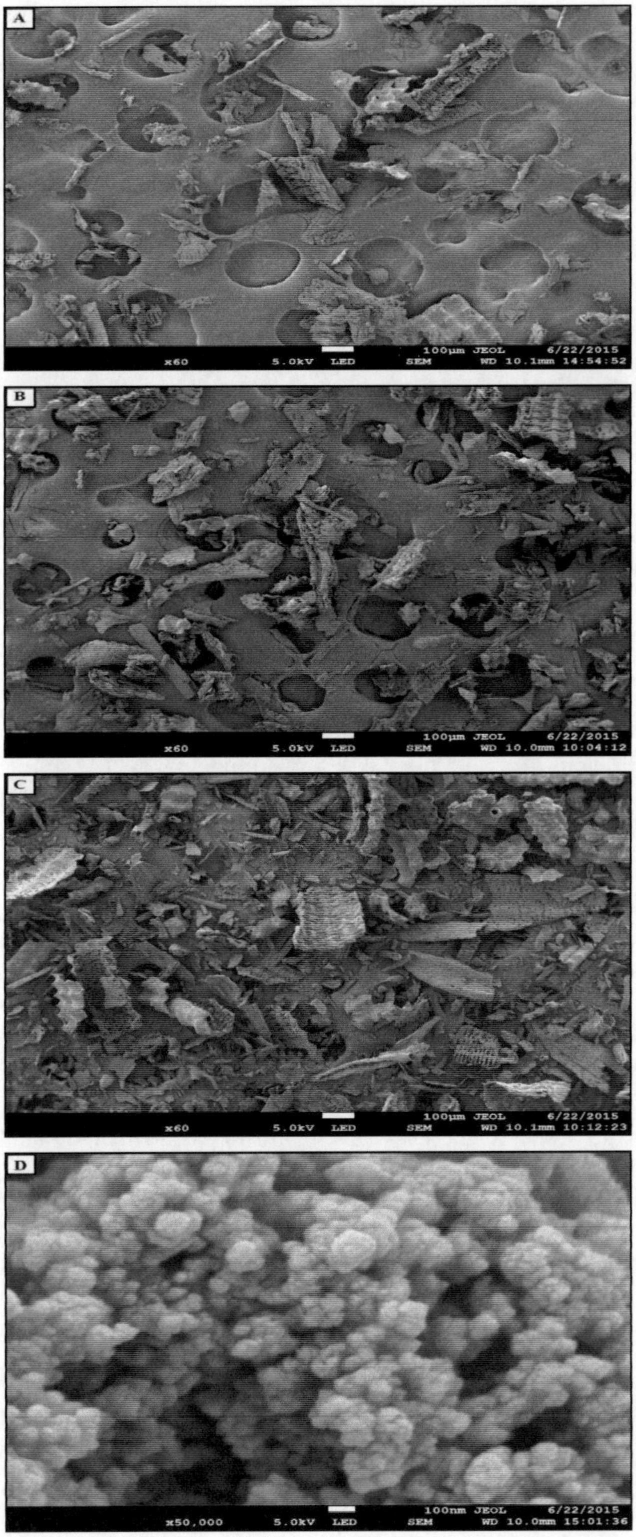
A
x60 5.0kV LED
100µm JEOL 6/22/2015
SEM WD 10.1mm 14:54:52
B
x60 5.0kV LED
100µm JEOL 6/22/2015
SEM WD 10.0mm 10:04:12
C
x60 5.0kV LED
100µm JEOL 6/22/2015
SEM WD 10.1mm 10:12:23
D
x50,000 5.0kV LED
100nm JEOL 6/22/2015
SEM WD 10.0mm 15:01:36

Figure 6. The FESEM images for impregnated activated carbon with piperazine at (A) 60 times magnification at ratio 2:1, (B) 60 times magnification at ratio 1:1, (C) 60 times magnification at ratio 1:2, (D) 50,000 times magnification at ratio 2:1, (E) 50,000 times magnification at ratio 1:1 and (F) 50,000 times magnification at ratio 1:2.

4. Conclusions

Preparation of rice husk activated carbon impregnated with piperazine was successfully performed in this research. The effects of concentration and mixture ratio are investigated by using several analytical equipments which were XRD, FTIR, BET and FESEM. In XRD analysis, diffraction angle around 21.66° and 36.25° are linked for pyrazole in non-modified activated carbon sample while in piperazine modified sample, the polyethylene butane and silicon oxide are discovered at angle 21.86° and 21.89° respectively. For FTIR analysis, amide regions are detected for all impregnated activated carbon at band 3278.28 cm^{-1} while there is an absent of amide reagent on the non-modified activated carbon sample. The higher concentration and mixture ratio have a lower transmittance percentage compared to a lower concentration and mixture ratio of impregnated activated carbon. The BET results show that the non-modified activated carbon contains large surface area and pore volume than the modified activated carbon with piperazine. The lower concentration piperazine has a higher surface area and pore volume as there is less fillings inside the pores compared to high concentration piperazine. FESEM analysis shows that the morphology of non-impregnated activated carbon contains more pores compared to modified activated carbon with piperazine. The stacking of piperazine on the activated carbon surfaces increase with the increase of concentration and mixture ratio. With all the results obtained, we can now perform the testing on the adsorbent for CO_2 capture experiment, which will be in our future work.

Acknowledgements

The authors express their sincere thanks to the Fundamental Research Grant Scheme (RDU140136), for providing the financial support for this research.

References

[1] Olawale et al 2012 Characterization of rice husk via atomic absorption spectrophotometer for optimal silica production *Int. J. of Sci. and Tech.* **2**(4) pp 210-213

[2] Nagrale S D et al 2012 Utilization of rice husk ash *Int. J. of Eng. Research and Apps.* **2**(4) pp 1-5

[3] Kumar A et al 2012 Properties and industrial applications of rice husk: A review *Int. J. of Emerging Tech. and Adv. Eng.* **2**(10) pp 86-90

[4] Jahangiri M et al 2013 Preparation of new adsorbent from activated carbon and carbon nanofiber (AC/CNF) for manufacturing organic-vapor respirator cartridge *Iranian J. of Env. Health Sci. & Eng. BioMed Central*

[5] Dali A M et al 2012 General study about activated carbon for adsorption carbon dioxide *J. of Purity, Utility Reaction and Env.* **1**(5) pp 236-251

[6] Arie A A et al 2016 Activated carbons from KOH-activation of salacca peels as low cost potential adsorbents for dye removal *Adv. Materials Letters* **7**(3) pp 226-229

[7] Wang J et al 2012 Carbon dioxide capture using polyethyleneimine-loaded mesoporous carbons *J. of Env. Sci.* **25**(1) pp 124-132

[8] Kangwanwatana W et al 2013 Study of CO2 adsorption using adsorbent modified with piperazine *Chem. Eng. Transactions* **35**(1) pp 403-408

[9] Shafeeyan M S et al 2010 A review on surface modification of activated carbon for carbon dioxide *J. of Analytical and App. Pyrolysis* **89**(1) pp 143-151

[10] Lee S C et al 2013 Adsorption by Y-type zeolite impregnated with amines in indoor air *Aerosol & Air Quality Research* **13**(1) pp 360-366

[11] Thirzha L P D et al 2012 Separation of carbon dioxide from flue gas using adsorption on porous solids *InTech Open* pp 57-80

[12] Kim Y E et al 2013 Comparison of carbon dioxide absorption in aqueous MEA, DEA, TEA, and AMP solutions *Bull. Korean Chem. Soc.* 2013 **34**(3) pp 783

[13] Anuwattana R et al 2016 Carbon dioxide adsorption using activated carbon via chemical vapour deposition process *Eng. J.* **20**(4) pp 59-68

[14] Kongnoo A et al 2014 CO2 adsorption at thermophilic temperature on alkanolamine impregnated activated carbon *Int. J. of Ad. in Chem. Eng. and Bio. Sci.* **1**(1) pp 143-148

[15] Abechi S E et al 2013 Preparation and characterization of activated carbon from palm kernel shell by chemical activation *Research J. of Chem. Sci.* **3**(7) pp 54-61

[16] Xue Q et al 2011 Improvement of amine-modification with piperazine for the adsorption of CO2 *App. Surface Sci.* **258**(8) pp 3859-3863

[17] Lin Y et al 2013 Polyethyleneimine incorporated metal-organic frameworks adsorbent for highly selective CO2 Capture *Scientific Reports* 3 pp 1-7

[18] Houshmand A et al 2011 Carbon dioxide capture with amine-grafted activated carbon *J. of Water, Air and Soil Poll.* Springer

IOP Conf. Series: Materials Science and Engineering **206** (2017) 012006 doi:10.1088/1757-899X/206/1/012006

Neoteric Media as Tools for Process Intensification

C C Beh [1], R Mammucari [2], N R Foster [1]

[1]Department of Chemical Engineering, Curtin University of Technology, Bentley 6102 WA, Australia
[2]School of Chemical Engineering, UNSW Australia, Kensington 2052 NSW, Australia

E-mail: jane.beh@curtin.edu.au

Abstract. Process intensification (PI) is a commonly used term in the chemical processing industry. When the concept of PI was first introduced in the late 1970s within the Imperial Chemical Industries (ICI) company, the main impetus was to reduce the processing cost without impairing the production rate. Neoteric media present as alternatives in chemical processing include gas-expanded liquids, ionic liquids, subcritical water, and combination of gas-expanded liquids and ionic liquids. The applications of neoteric media include particle engineering for improved bioavailability, controlled release of therapeutic implants, pharmaceutical formulations, extraction of natural products, nano-carriers for drug delivery, sterilisation of implants, and chemical reactions. This paper provides an overview of the use of these neoteric media.

1. Introduction

Neoteric can be defined as new or recent. Neoteric media in this context involves those media that are new and recently under extensive research such as gas-expanded liquids (GXLs), ionic liquids (ILs), subcritical water or superheated water, and combination of GXLs and ILs. The environmental and economic burdens from waste generation resulted from excessive usage of conventional solvents in chemical reactions is the main factor for the introduction and extensive investigation of neoteric media. Conventional solvents are commonly used in industries and laboratories for various applications which include synthetic chemistry, separations, coatings and cleaning. The annual amount of solvent waste produced is up to million tonnes and is generally being released to the environment [1]. Conventional solvents are mostly carcinogenic, and toxic to human health. Besides, they also cause disruption of ecosystems by depleting the ozone layer and resulting in tropospheric smog from chemical reactions. Although there are safety and controlled measures to regulate the use of the conventional solvents, the operations are generally lengthy and time consuming. Hence, neoteric media have emerged as a promising green alternative to conventional solvents in chemical processes [2].

The ideal alternative solvent media has to be able to achieve the following in comparison with the conventional solvents:

- retain the significant characteristics of the replaced conventional solvent such as polarity and solubility of reactants;
- increase or maintain process safety;
- be able to operate at similar or milder operating conditions;
- retain or increase product quality;

- be economically viable.

2. Neoteric Data

2.1. Gas-Expanded Liquids

A gas-expanded liquid (GXL) is a mixed solvent composed of a compressible gas and an organic solvent at pressure and temperature that are below the critical point of the mixed solvent. Among the compressible gases, carbon dioxide (CO_2) is the most commonly used in GXL studies, which is generally known as CO_2-expanded liquids (CXL). A CXL is able to combine benefits of both dense CO_2 and organic solvents, specifically good solvation abilities coupled with gas-like solubility and mass transfer properties. In addition, CXL has advantages environmentally by replacing organic solvents with CO_2 and being able to operate at milder operating conditions than supercritical CO_2 [3]. Most traditional organic solvents such as methanol, hexane, and dimethylformamide are able to dissolve large amounts of CO_2 and expand volumetrically. Hence, the mixture of organic solvents and CO_2 can experience significant changes in physical properties compared with the pre-expanded state [1]. The volumetric expansion is dependent on the mole fraction of CO_2 present in the liquid phase [3]. Gas-expanded liquids are commonly used for particle formation, chemical reactions and materials processing.

There are various methods to produce fine and homogeneous particles using GXL. Fine particles find applications in pigments, food, cosmetics, and pharmaceutical compounds. The processes involving GXL or CXL for particle formation include:

- Gas Anti-Solvent (GAS)
  In a GAS process, CO_2 is used to expand a solution of solute in organic solvent in a vessel. The solvent expansion decreases the solute solubility where its precipitation point is achieved. The associated organic solvent is then extracted by a continuous flow of CO_2. The resulting precipitate is then collected upon depressurization where the precipitate is separated from the expanded solvent [1, 4, 5]. A scanning electron microscopy (SEM) image of Insulin precipitated from dimethyl sulfoxide (DMSO) by the GAS process is illustrated in figure 1 [6].

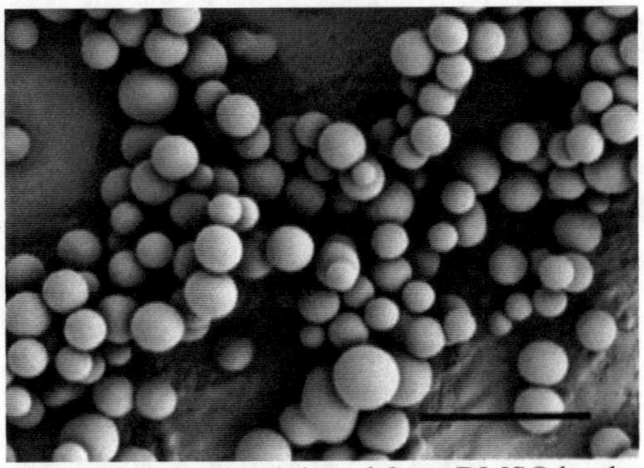

Figure 1. SEM image of Insulin precipitated from DMSO by the GAS process [6]

- Precipitation with Compressed Anti-Solvent (PCA)/Aerosol Solvent Extraction System (ASES)
  In the PCA process, a solution of solute in organic solvent is sprayed into a CO_2 pressurised vessel. Solvent expansion occurs rapidly and induces precipitation of fine solute particle. Subsequently, the remaining organic solvent is washed from the precipitated particles with CO_2 [1, 3]. An example of particle formation by the PCA process in which polystyrene was micronized at 225bar and 40°C is shown in figure 2 [7].

IOP Conf. Series: Materials Science and Engineering **206** (2017) 012006 doi:10.1088/1757-899X/206/1/012006

In an ASES process, a solution of solute in organic solvent is expanded and dissolved into supercritical CO_2. The procedure can be undertaken with CO_2 continuously flowing through the vessel concurrently or counter-currently with the organic solution to extract the organic solvent from the vessel [5]. Figure 3 is a SEM image of Budesonide micronized at 85bar and 40°C [8]. Figure 4 illustrates Cefonicid precipitated from DMSO at 150bar and 40°C [9]. Figure 5 is a SEM image of Rifampicin precipitated from DMSO at 90bar and 40°C [10].

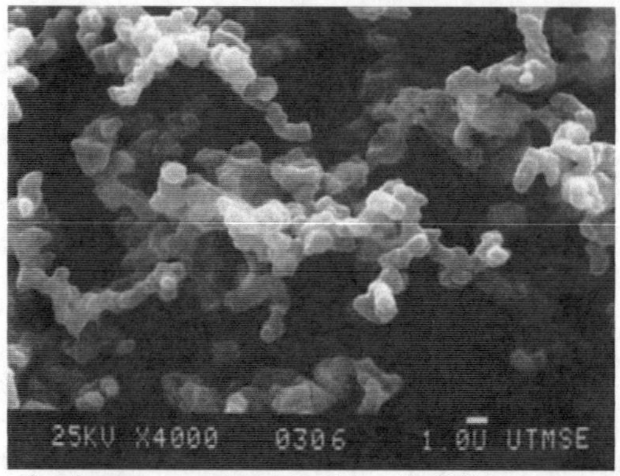

Figure 2. SEM image of polystyrene micronized by the PCA process at 225bar and 40°C [7]

Figure 3. SEM image of Budesonide micronized by the ASES process at 85bar and 40°C [8]

Figure 4. SEM images of micronized Cefonicid precipitated from DMSO at 150bar, 40°C [9]

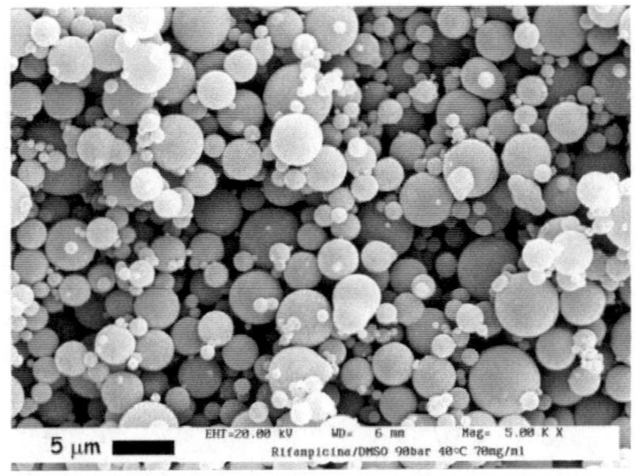

Figure 5. SEM images of Rifampicin micronized from DMSO at 90bar, 40°C [10]

- Particles from Gas-Saturated Solution (PGSS)
  Solid solute is melted by CO_2 expansion and sprayed into a vessel at atmospheric pressure via a nozzle. The rapid depressurization and cooling induce fine particle precipitation [1, 3, 11]. Figure 6 illustrates an example of cyclosporine micronized at 200bar and 25°C [11].

Apart from particle formation, GXL or specifically CO_2-expanded liquid (CXL) has advantages for hydrogenation reactions, such as safety, reaction performance, and catalysis reactions. A mixture of CO_2 and H_2 has been demonstrated to be as chemically effective as pure H_2 at the same pressure, and is a safe alternative option. In addition, accidental fires and explosions can be prevented with the presence of CO_2 in the mixture. The choice of expansion gas played a role in affecting the rate of hydrogenation [12]. The solvent expansion methods for H_2 generation are more commonly used in asymmetric hydrogenations for homogeneous catalysed reactions. It was reported that CO_2-expanded ILs are better at dissolving H_2 and H_2 is likely to diffuse rapidly into the liquid phase. With this combination, H_2 becomes more available to reactions in the IL, hence, improving rates and reaction selectivity [1].

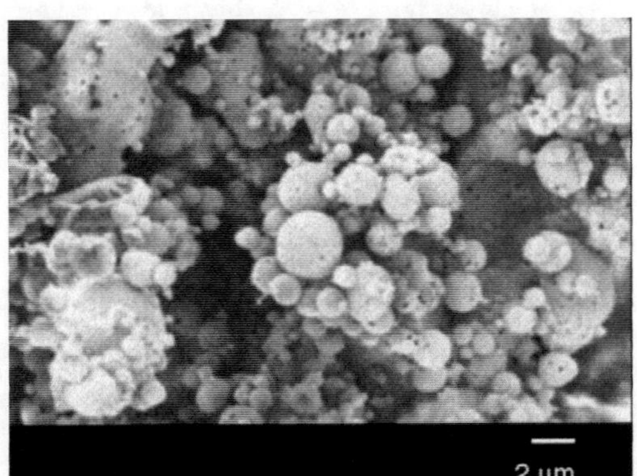

Figure 6. SEM image of cyclosporine micronized by the PGSS process at 200bar and 25°C [11]

2.2. Subcritical Water

As many chemical industries shift towards sustainability, water has become popular in many chemical and separation processes [13-15]. The strong hydrogen-bond cohesive energy between the water molecules leads to low solubility of hydrophobic compounds in water. However, the hydrogen-bond in water weakens at elevated temperature, which causes an increase of the solvating power of water for hydrophobic compounds [15]. Hence, materials such as ionic and polar compounds can be extracted using water at temperature above its ambient boiling temperature while non-polar materials can be extracted at higher temperature or at near-critical condition [16].

The term "subcritical water (SBW)" is generally used to describe water heated below its critical temperature and held at a pressure that retains its liquid state. The solvating power of subcritical water is particularly dependent on temperature and polarity. Therefore, subcritical water has increasing solvating power for various non-polar compounds with increasing temperature [15].

Subcritical water has become a popular research subject by researchers in recent years as it can be used as an alternative solvent to organic solvents for particle engineering of hydrophobic compounds, particularly Active Pharmaceutical Ingredients (APIs). The dissolution rates of APIs in the human body can be enhanced by modifying the morphology and size of APIs. Subsequently, APIs have a higher chance of reaching the targeted drug reception site. Furthermore, the efficiency of some APIs can be improved when their particle sizes are reduced, which could potentially reduce both the drug dosage required as well as the risk of adverse side effects [14]. A few examples of APIs micronized by SBW are illustrated in figure 7, figure 8 and figure 9:

Figure 7. Griseofulvin precipitated by SBW at 20bar: a) 140°C; b) 160°C; and c) 170°C [14]

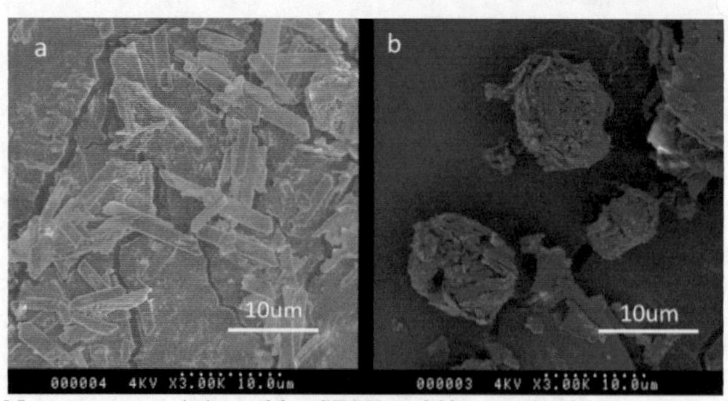

Figure 8. Naproxen precipitated by SBW at 20bar: a) 140°C; and b) 170°C [14]

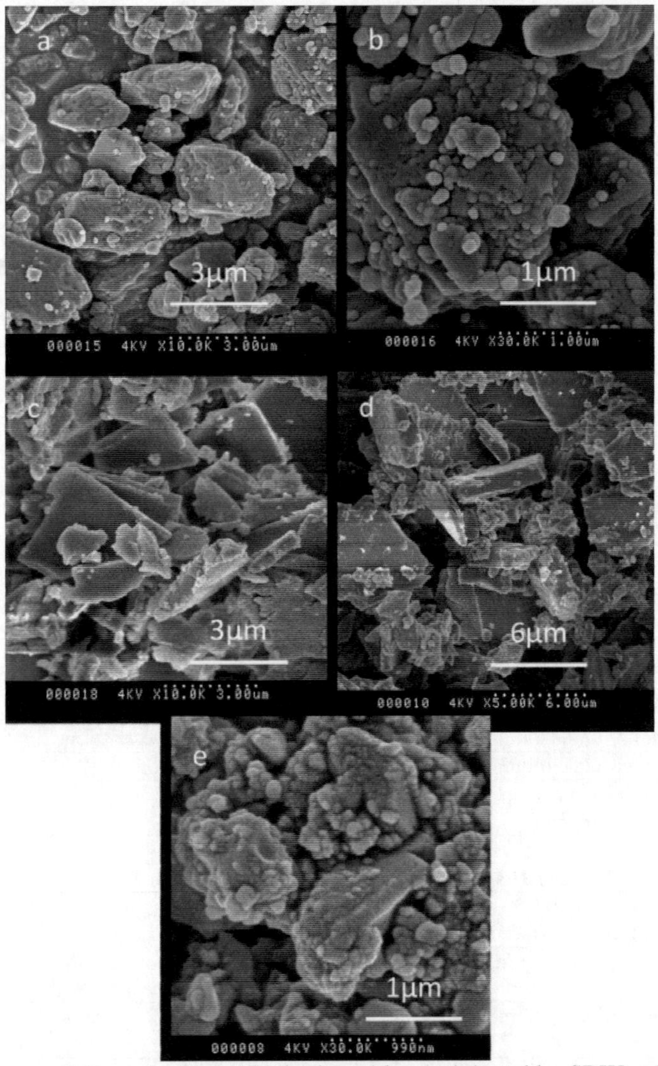

Figure 9. SEM images of a) raw budesonide; budesonide precipitated by SBW at b) 200°C; c) 130°C with 20%v/v ethanol fraction; d) 150°C with 20%v/v ethanol fraction; and e) 160°C with 10%v/v methanol fraction [14]

It has been observed that smaller particles are formed at higher temperatures of SBW. The morphology and size of APIs can be tailored accordingly by the working conditions such as temperature, co-solvents involvement, and concentration of the materials. Subcritical water has been proven to be a useful alternative and tunable solvent for particle engineering. It is a promising tool to move industry processes towards sustainability.

2.3. Ionic Liquids

Ionic liquids (ILs) are organic salts that are in liquid state below 100°C or at ambient temperature. As ILs have low, or negligible vapour pressure, they are good solvent alternatives in replacement of toxic organic solvents for a wide range of polar organic compounds and a few aromatic hydrocarbons [17-20]. The physical and chemical properties of ILs are tunable (by selecting the cation and anion constituents). Consequently, ILs are described as "designer solvents". Their potential as environmentally benign solvents to replace volatile organic solvents in chemical processes such as separation or extraction, purification, and reaction media in biochemical and chemical catalysis has been demonstrated. As ILs have low volatility, they can significantly improve the environmental impact and contamination in chemical processes. Ionic liquids are electrical conducting electrolytes, hence, they are particularly useful in electric batteries applications. Ionic liquids also find useful applications in gas handling processes as a transport medium for reactive gases, pharmaceutical production, extraction of natural and synthetic compounds, natural gas purification, carbon dioxide capture, heat transfer and storage media for solar thermal energy systems, and as dispersing agents in paints [1].

Ionic liquids are commonly used in extraction processes [21, 22]. Various types of ILs used in different extraction processes are listed in Table 1.

Table 1. Use of different types of ionic liquids in extraction of proteins, alkaloids, vitamins and antibiotics

Extraction of Protein		
Type of ILs	**Material Extracted**	**Reference**
Ammoeng 110	Albumin, lysozyme, myoglobin, trypsin	[23]
1,3-dialkylimidazolim bromide	Albumin, trypsin, cytochrome c, γ-globulin	[24]
1-butyl-3-methylimidazoli-um chloride, K2HPO4	Albumin, transferrin	[25]
Extraction of Hormones, Alkaloids and Vitamins		
Type of ILs	**Material Extracted**	**Reference**
1-methyl-3-butylimidazolium chloride, K_2HPO_4	Testosterone, epitestosterone	[26]
1,3-dialkylimidazolium chloride	Caffeine, nicotine	[27]
1-hexyl-3–methylimidazolium chloride, K_2HPO_4	Vitamin B12	[28]
Butyl-methyl-imidazolium chloride, K_3PO_4 or KH_2PO_4	Quinine	[29]
Extraction of Antibiotics		
Type of ILs	**Material Extracted**	**Reference**
1-butyl-3-methylimidazolium tetrafluoroborate, NaH_2PO_4	Penicillin G	[30]
1-butyl-3-methylimidazolium tetrafluoroborate	Asazithromycin, mydecamycin	[31]
1-butyl-3-methylimidazolium tetrafluoraborate, Na_2CO_3	Roxithromycin	[32]
1-butyl-3-methylimidazolium tetrafluoroborate, $(NH_4)_2SO_4$	Sulfadimidine	[33]
Type of ILs	**Material Extracted**	**Reference**
1-butyl-3-methylimidazolium tetrafluoraborate, NaH_2PO_4	Acetylspiramycin	[34]
1-butyl-3-methylimidazolium tetrafluoroborate, $Na_3C_6H_5O_7$	Chloramphenicol	[35]

In addition, ILs are useful for extracting metal ions. Such application usually involves addition of special ligands to increase the affinity of strongly hydrated ions to the hydrophobic ILs phase. The most commonly used ILs for this purpose are 1-alkyl-3-methylimidazolium hexafluorophosphate, tetrafluoroborate and bis (trifluoromethylsulphonyl) imide [21, 36, 37].

As for catalytic reactions, chloroaluminate (III) ionic liquids are particularly powerful solvents [17]. The Friedel-Crafts reaction is a classical synthesis reaction for commercial fragrance molecules such as Traseolide® (5-acetyl-1,1,2,6-tetramethyl-3-isopropylindane) and Tonalid® (6-acetyl-1,1,2,4,4,7-

hexamethyltetralin) [38, 39]. Ionic liquids are also being applied to the acetylation of anthracene [17]; isomerisation of polyethylene [40], stearic acid or methyl stearate [17], and oleic acid or methyl oleate [17]; dimerization and oligomerisation of olefins [41, 42]; chlorination of alkenes to produce dihaloalkanes [43]; hydrogenation of cyclohexene [44]; esterification [17]; aromatic alkylation reactions [17]; and pharmaceutical synthesis [45].

2.4. Combination of Gas-Expanded Liquids and Ionic Liquids

Since decades ago, supercritical fluids (SCF) are popular in catalysis and reaction engineering as homogeneous reaction media. Subsequently, multiphasic systems that involve a liquid phase and a compressed supercritical or subcritical gas phase have been studied [46]. Biphasic systems comprising GXL and IL at temperature and pressure below the critical conditions have recently captured the attention of researchers for the lower investment costs as simpler equipment compared to operations in SCF. Combining GXL and IL, for example, can significantly change the viscosity of the IL rich mixture [1, 47] and improve mass transfer [48].

Generally, the solubilities of ILs in CO_2 are low, even at high pressures. Hence, the CO_2 phase is free from contamination by IL in any biphasic system that involves IL and CO_2. Ionic liquids have been reported to have low volumetric expansion with CO_2 and to behave like polymers in both solubility and volumetric expansion properties [3]. As compressed CO_2 reduces the viscosity of ILs, the ion mobility of the ILs is subsequently increased, which leads to higher conductivity [1, 20].

3. Conclusion

The neoteric media are under extensive research both in academia and industry. They are considered as "green solvents" and as a potential replacement for volatile organic solvents. The applications of neoteric media have been summarised in the present article. Neoteric media are believed to present advantages in current chemical processes. In fact, some have been applied commercially [2].

References

[1] Jessop, P.G. and Subramaniam, B., 2007. Gas-expanded liquids. *Chem. rev.*, **107** (6), p.2666-2694.
[2] Welton, T., 2015, November. Solvents and sustainable chemistry. In *Proc. R. Soc. A* **471** (2183) p. 20150502. The Royal Society.
[3] Scurto, A.M., K. Hutchenson, and B. Subramaniam, Gas-expanded liquids: Fundamentals and applications, in ACS Symp. Ser. 2009, *American Chem. Soc. J.* p. 3-37.
[4] Jung, J. and M. Perrut, Particle design using supercritical fluids: Literature and patent survey. *The J. of Supercritical Fluids*, 2001. **20** (3): p. 179-219.
[5] Pasquali, I., Bettini, R. and Giordano, F., 2006. Solid-state chemistry and particle engineering with supercritical fluids in pharmaceutics. *European J. of Pharmaceutical Sci.*, **27** (4), p. 299-310.
[6] Thiering, R., Dehghani, F., Dillow, A. and Foster, N.R., 2000. Solvent effects on the controlled dense gas precipitation of model proteins. *J. of Chem. Technol. and Biotechnology*, **75** (1), p. 42-53.
[7] Dixon, D.J., Johnston, K.P. and Bodmeier, R.A., 1993. Polymeric materials formed by precipitation with a compressed fluid antisolvent. *AIChE J.*, **39** (1), p.127-139.
[8] Steckel, H., Pichert, L. and Müller, B.W., 2004. Influence of process parameters in the ASES process on particle properties of budesonide for pulmonary delivery. *European J. of Pharmaceutics and Biopharmaceutics*, **57** (3), p.507-512.

[9] Reverchon, E. and De Marco, I., 2004. Supercritical antisolvent micronization of Cefonicid: thermodynamic interpretation of results. *The J. of Supercritical Fluids*, **31** (2), p. 207-215.
[10] Reverchon, E., De Marco, I. and Della Porta, G., 2002. Rifampicin microparticles production by supercritical antisolvent precipitation. *Int. J. of Pharmaceutics*, **243** (1), p.83-91.
[11] Tandya, A., Dehghani, F. and Foster, N.R., 2006. Micronization of cyclosporine using dense gas techniques. *The J. of Supercritical Fluids*, **37** (3), pp.272-278.
[12] Thomas, C.A., Bonilla, R.J., Huang, Y. and Jessop, P.G., 2001. Hydrogenation of carbon dioxide catalyzed by ruthenium trimethylphosphine complexes Effect of gas pressure and additives on rate in the liquid phase. *Canadian J. of Chem.*, **79** (5-6), p.719-724.
[13] Canıaz, R.O. and Erkey, C., 2014. Process intensification for heavy oil upgrading using supercritical water. *Chem. Eng. Res. and Design*, **92** (10), p.1845-1863.
[14] Carr, A.G., Mammucari, R. and Foster, N.R., 2010. Solubility, solubility modeling, and precipitation of naproxen from subcritical water solutions. *Ind. & Eng. Chem. Res.*, **49** (19), p. 9385-9393.
[15] Teoh, W.H., Mammucari, R., Vieira de Melo, S.A. and Foster, N.R., 2013. Solubility and solubility modeling of polycyclic aromatic hydrocarbons in subcritical water. *Ind. & Eng. Chem. Res.*, **52** (16), p. 5806-5814.
[16] Brunner, G., 2009. Near critical and supercritical water. Part I. Hydrolytic and hydrothermal processes. *The J. of Supercritical Fluids*, **47** (3), p.373-381.
[17] Earle, M.J. and K.R. Seddon, Ionic Liquids: Green Solvents for the Future, in Clean Solvents. 2002, *American Chem. Society J.* p. 10-25.
[18] Hulsbosch, J., De Vos, D.E., Binnemans, K. and Ameloot, R., 2016. Biobased Ionic Liquids: Solvents for a Green Processing Industry. *ACS Sustainable Chem. & Eng.*, **4** (6), p. 2917-2931.
[19] Marsh, K.N., Deev, A., Wu, A.C., Tran, E. and Klamt, A., 2002. Room temperature ionic liquids as replacements for conventional solvents–A review. *Korean J. of Chem. Eng.*, **19** (3), p. 357-362.
[20] Pollet, P., Davey, E.A., Ureña-Benavides, E.E., Eckert, C.A. and Liotta, C.L., 2014. Solvents for sustainable chemical processes. *Green Chem.*, **16** (3), p.1034-1055.
[21] Flieger, J., E. Grushka, and A. Czajkowska-Zelazko, 2014. Ionic liquids as solvents in separation processes. *Austin J. Anal. Pharm. Chem.* **1** (2): p. 1-8.
[22] Salar-García, M.J., Ortiz-Martínez, V.M., Hernández-Fernández, F.J., de Los Ríos, A.P. and Quesada-Medina, J., 2017. Ionic liquid technology to recover volatile organic compounds (VOCs). *J. of Hazardous Mat.*, **321**, p. 484-499.
[23] Dreyer, S., Salim, P. and Kragl, U., 2009. Driving forces of protein partitioning in an ionic liquid-based aqueous two-phase system. *Biochemical Eng. J.*, **46** (2), p. 176-185.
[24] Pei, Y., Wang, J., Wu, K., Xuan, X. and Lu, X., 2009. Ionic liquid-based aqueous two-phase extraction of selected proteins. *Separation and Purification Technol.*, **64** (3), p. 288-295.
[25] Du, Z., Yu, Y.L. and Wang, J.H., 2007. Extraction of Proteins from Biological Fluids by Use of an Ionic Liquid/Aqueous Two-Phase System. *Chem. - A European J.*, **13** (7), p. 2130-2137.
[26] He, C., Li, S., Liu, H., Li, K. and Liu, F., 2005. Extraction of testosterone and epitestosterone in human urine using aqueous two-phase systems of ionic liquid and salt. *J. of Chromatography A*, **1082** (2), p. 143-149.
[27] Freire, M.G., Neves, C.M., Marrucho, I.M., Lopes, J.N.C., Rebelo, L.P.N. and Coutinho, J.A., 2010. High-performance extraction of alkaloids using aqueous two-phase systems with ionic liquids. *Green Chem.*, **12** (10), p. 1715-1718.
[28] Berton, P., Monasterio, R.P. and Wuilloud, R.G., 2012. Selective extraction and determination of vitamin B 12 in urine by ionic liquid-based aqueous two-phase system prior to high-performance liquid chromatography. *Talanta*, **97**, p. 521-526.
[29] Flieger, J. and Czajkowska-Żelazko, A., 2015. Aqueous two phase system based on ionic liquid for isolation of quinine from human plasma sample. *Food Chem.*, **166**, p. 150-157.

[30] Qingfen, L., Xuesheng, H., Yuhong, W., Ping, Y., Hansong, X., Jiang, Y. and Huizhou, L., 2005. Extraction of penicillin G by aqueous two-phase system of [Bmim] BF4/NaH2PO4. *Chinese Sci. Bulletin*, **50** (15), p. 1582-1585.

[31] Han, J., Wang, Y., Kang, W., Li, C., Yan, Y., Pan, J. and Xie, X., 2010. Phase equilibrium and macrolide antibiotics partitioning in real water samples using a two-phase system composed of the ionic liquid 1-butyl-3-methylimidazolium tetrafluoroborate and an aqueous solution of an inorganic salt. *Microchimica Acta*, **169** (1-2), p. 15-22.

[32] Li, C.X., Han, J., Wang, Y., Yan, Y.S., Xu, X.H. and Pan, J.M., 2009. Extraction and mechanism investigation of trace roxithromycin in real water samples by use of ionic liquid–salt aqueous two-phase system. *Analytica Chimica Acta*, **653** (2), p. 178-183.

[33] Yu, C., Han, J., Wang, Y., Yan, Y., Hu, S., Li, Y. and Ma, C., 2011. Ionic liquid/ammonium sulfate aqueous two-phase system coupled with HPLC extraction of sulfadimidine in real environmental water samples. *Chromatographia*, **74** (5-6), p. 407-413.

[34] Wang, Y., Han, J., Xie, X.Q. and Li, C.X., 2010. Extraction of trace acetylspiramycin in real aqueous environments using aqueous two-phase system of ionic liquid 1-butyl-3-methylimidazolium tetrafluoroborate and phosphate. *Central European J. of Chem.*, **8** (6), p. 1185-1191.

[35] Han, J., Wang, Y., Yu, C.L., Yan, Y.S. and Xie, X.Q., 2011. Extraction and determination of chloramphenicol in feed water, milk, and honey samples using an ionic liquid/sodium citrate aqueous two-phase system coupled with high-performance liquid chromatography. *Analytical and Bioanalytical Chem.*, **399** (3), p.1295-1304.

[36] Zhao, H., Xia, S. and Ma, P., 2005. Use of ionic liquids as 'green'solvents for extractions. *J. of Chem. Technol. and Biotechnology*, **80** (10), p.1089-1096.

[37] Zaijun, L., Junkang, L. and Xiulan, S., 2011. *Ionic Liquid As Novel Solvent For Extraction And Separation In Analytical Chem.* INTECH Open Access Publisher.

[38] Earle, M., Seddon, K. and Adams, C., 1998. Friedel–Crafts reactions in room temperature ionic liquids. *Chem. Communications*, (19), p. 2097-2098.

[39] Boon, J.A., Levisky, J.A., Pflug, J.L. and Wilkes, J.S., 1986. Friedel-Crafts reactions in ambient-temperature molten salts. *The J. of Organic Chem.*, *51*(4), p. 480-483.

[40] Adams, C.J., Earle, M.J. and Seddon, K.R., 2000. Catalytic cracking reactions of polyethylene to light alkanes in ionic liquids. *Green Chem.*, **2** (1), p. 21-24.

[41] Ellis, B., Keim, W. and Wasserscheid, P., 1999. Linear dimerisation of but-1-ene in biphasic mode using buffered chloroaluminate ionic liquid solvents. *Chem. Communications*, (4), p. 337-338.

[42] Chauvin, Y., Olivier, H., Wyrvalski, C.N., Simon, L.C. and de Souza, R.F., 1997. Oligomerization ofn-Butenes Catalyzed by Nickel Complexes Dissolved in Organochloroaluminate Ionic Liquids. *J. of Catalysis*, **165** (2), p. 275-278.

[43] Green, L., Hemeon, I. and Singer, R.D., 2000. 1-Ethyl-3-methylimidazolium halogenoaluminate ionic liquids as reaction media for the acylative cleavage of ethers. *Tetrahedron Letters*, **41** (9), p. 1343-1346.

[44] Suarez, P.A., Dullius, J.E., Einloft, S., De Souza, R.F. and Dupont, J., 1996. The use of new ionic liquids in two-phase catalytic hydrogenation reaction by rhodium complexes. *Polyhedron*, **15** (7), p. 1217-1219.

[45] Earle, M.J., McCormac, P.B. and Seddon, K.R., 2000. The first high yield green route to a pharmaceutical in a room temperature ionic liquid. *Green Chem.*, **2** (6), p. 261-262.

[46] Hintermair, U., Leitner, W. and Jessop, P., 2010. Expanded Liquid Phases in Catalysis: Gas-expanded Liquids and Liquid–Supercritical Fluid Biphasic Systems. *Handbook of green Chem.*

[47] Blanchard, L.A., Gu, Z. and Brennecke, J.F., 2001. High-pressure phase behavior of ionic liquid/CO2 systems. *The J. of Physical Chem. B*, **105** (12), p.2437-2444.

[48] Keskin, S., Kayrak-Talay, D., Akman, U. and Hortaçsu, Ö., 2007. A review of ionic liquids towards supercritical fluid applications. *The J. of Supercritical Fluids*, **43** (1), p.150-180.

Recent advancement of hybrid materials used in chemical enhanced oil recovery (CEOR): A review

M F Hamza[1,2], C M Sinnathambi[1,2*] and Z M A Merican[1,2]

[1]Center of Research in Enhanced Oil Recovery (COREOR), Universiti Teknologi Petronas, 32610 Bandar Seri Iskandar, Perak Darul Ridzuan, Malaysia.
[2]Fundamental and Applied Sciences Department, Universiti Teknologi Petronas, 32610

Bandar Seri Iskandar, Perak Darul Ridzuan, Malaysia.
E-mail: chandro@utp.edu.my

Abstract. Depletion of natural oil reserves has forced oil industries to focus on tertiary recovery methods to extract residual oil after exhausting the primary and secondary methods. Among the Enhance Oil Recovery (EOR) technologies, Chemical EOR (CEOR) is gaining popularity. Despite research efforts to increase the recovery using CEOR, increasing complexity in extraction methods are encountered. With changes in reservoir conditions (high temperature, pressure and salinity) and crude oil properties, existing chemicals used in CEOR, such as alkali, polymers and surfactants do not function desirably. These conditions have detrimental effects on the performance of EOR chemicals, like precipitation, degradation, etc. Development and utilization of effective EOR hybrids such as surfactant-polymer, polymer-nanomaterial, surfactant-nanomaterial and polymer-surfactant-nanomaterial had prevailed the effects of harsh reservoir conditions, and their applications in oil fields in recent years have increased the success of EOR. The synergistic effects between the hybrid components play major roles in improving the properties that could withstand the effect of extreme reservoir conditions and changes in crude oil properties. Therefore, this paper is aimed at reviewing recent advances in CEOR hybrid technologies, and discusses the basic concept, applications, advancement and limitations of different hybrid materials used in CEOR processes.

Keywords Reservoir conditions, hybrid, chemical enhanced oil recovery.

1. Introduction

It has been estimated that two-third of the original oil in place (OOIP) is left in the reservoir after exhausting the primary and secondary methods. Among the factors responsible for oil remaining in the rock pores are the effects of wettability, negative capillary force, and interfacial/surface tension between different phases [1-3]. Despite the discovery of new resources, recently, the United States Department of Energy reported that the world's consumption of crude oil has reached approximately 95 million oil barrels per day, and the domestic crude oil production is not sufficient enough to meet the requirement for energy [4-5]. This indicates over dependency on oil for daily activities. Thus, it has become necessary to initiate and develop other production strategies to recover the last drop of OOIP in the natural reserves. Enhanced oil recovery (EOR) technologies have been primarily designed to increase the oil production, and to bring back the lives of depleted oil fields in order to meet the world's energy demand. Among the EOR technologies, CEOR is the most effective and has the

IOP Conf. Series: Materials Science and Engineering **206** (2017) 012007 doi:10.1088/1757-899X/206/1/012007

potential to target the residual oil compared to non-chemical methods like the *in-situ* and microbial flooding. The ability of this method to minimize or overcome the wetting and interfacial/surface forces, has led to it receiving attention for applications in oil fields [6-10]. CEOR requires the injection of specific chemical fluids (usually in solutions) such as alkali, surfactants and polymers [8-9]. Each of these chemicals has a specific function when used as a flooding agent, for instance, surfactants usually promote the recovery of trapped oil in the reservoir by reducing oil/water interfacial tension [11]. The polymer is added to increase the viscosity of water, thereby reducing oil/water mobility ratio, which results in improved sweep efficiency [12]. The alkali reacts with the acid present in the oil and reduces the adsorption rate of surfactants on the rock surface [10]. With the changes in reservoir conditions like high temperature, pressure, salinity, and crude properties, flooding with either surfactant, alkali or polymer alone does not give the optimum recovery, because these conditions have detrimental effects on their individual performance (like precipitation, degradation etc.) [13-15]. However, this problem could be prevailed if hybrid materials comprising of surfactant-polymer, polymer-nanomaterial, surfactant-nanomaterial or polymer-surfactant-nanomaterial are used as CEOR agents. The synergistic effects between the hybrid components have increased the success of EOR operations by improving the rheological properties, as well as increasing the thermal and salinity resistance of CEOR hybrids [16]. The ability of CEOR hybrids to withstand the effects of changing crude properties is also associated with the combining effects of the hybrid components [16]. Therefore, in this paper, we discuss the applications, recent advancements, and limitations of different hybrid materials used in CEOR processes.

2. Basic concept

2.1 Hybrid materials

Hybrid materials contain two or more composite materials (mostly organic and inorganic) in a definite molecular ratio. The properties of hybrid materials differ from the individual properties of its corresponding components. In many cases, improved properties are usually due to the interaction of the hybrid materials, because one component provides bonding while the other imparts mechanical strength [17-18]. Depending upon the nature of hybrid components involved, hybrids can exist as liquid or solid phases and are categorized as *class I* (figure 1a and b) with weak chemical attractions (hydrogen bonding, Van der Waal force, etc.) and *class II* (figure 1c and d), which have strong interactions [18].

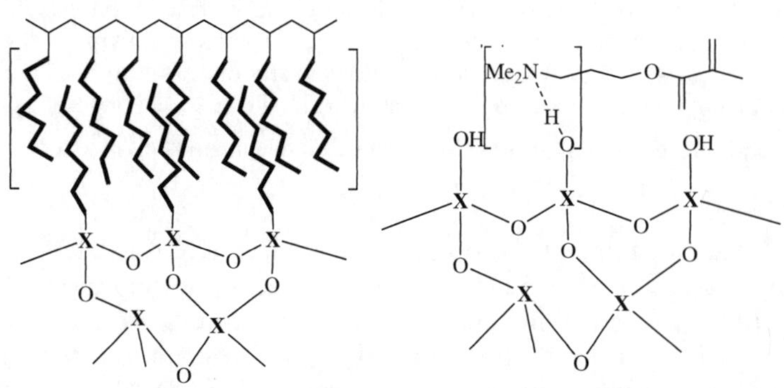

a. Vander waals b. Hydrogen bond

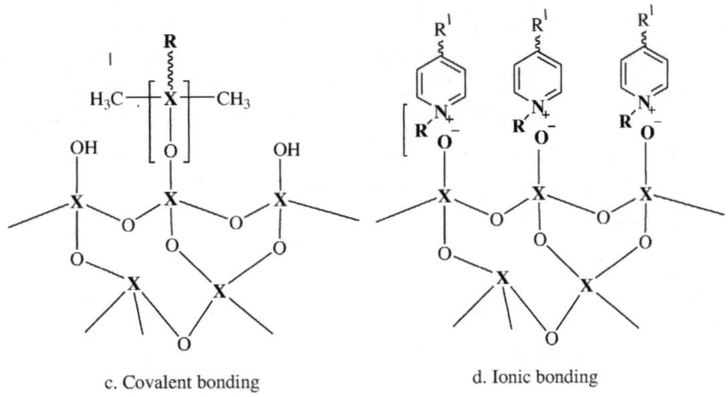

Figure 1. Classification of hybrid material: a and b are example of *class I* hybrid, c and d are example of *class II* (modified from Kickelbick (2007)).

2.1.1 Components of EOR hybrid commonly employed in CEOR

The CEOR hybrid's components mainly contain surfactants, polymers and nanomaterials in different molecular ratios depending upon the types of hybrid needed, and the nature of physicochemical properties required [18].

2.1.1.1 Surfactants

Surfactants are compounds capable of orienting themselves between the media and reduce the interfacial or surface tension between the displacing agents and oil phases [19]. Surfactants are made up of two different components; one component known as a *hydrophilic,* which has the ability to dissolve polar compounds while the other component, *hydrophobic*, dissolves non-polar compounds (figure 2). Possession of these unique properties qualifies surfactants for applications at different stages of petroleum production and processing operations [20]. Advanced technology usually combines surfactants and gases to generate foams, which is a potential chemical flooding agent for EOR applications [21]. From the literature reports, the application of surfactants in reducing the interfacial tension (IFT) between the different phases cannot be overemphasized. However, reducing IFT alone may not give the best oil recovery value, and hence, there is a need to focus on the manipulation of the overall oil sweep towards producer well.

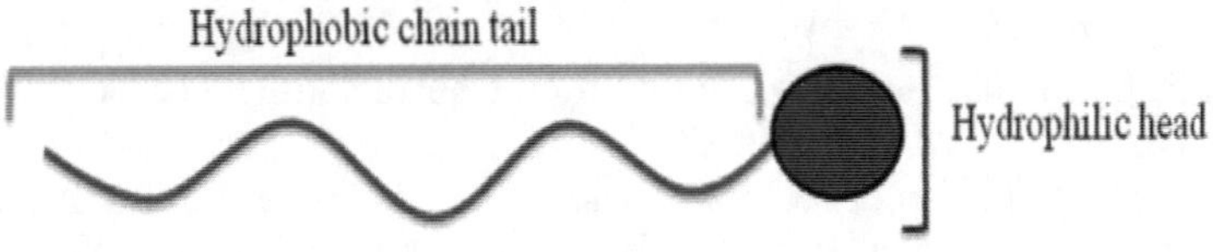

Figure 2. Schematic diagram showing the two components of surfactant.

The main challenge of injecting surfactants alone involves high economic loss, because at relatively higher temperatures their molecules are destroyed, precipitated and get deposited on the rock surface. They are also intolerant against shear stress between the rock pores, and sometimes finger into the oil bank due to less viscosity [22]. These factors affect the overall performance of surfactants which eventually lead to the recovery of the minimal amount of trapped oil.

2.1.1.2 Polymers

Polymers are used purposely to improve the rheological properties of the injected fluids (fluids containing water) by enhancing the viscosity so that the modified fluids would have the mobility less than that of oil. As a result, more oil would be pushed towards the producing wells [23]. Water soluble polymers are the most commonly employed polymers for EOR applications. Reasons being that, their physicochemical properties have been investigated and have direct relationships with the ability to increase oil production in many tested oilfields. However, despite this improvement in enhanced oil recovery, several problems associated with polymer flooding were encountered, and led to poor efficiency over time [21]. These include, but are not limited to:

a. Mechanical and thermal instability
b. Effect of high salinity
c. Change in pH
d. Cross link behaviour
e. High molecular polymers plug low permeability reservoir rocks

2.1.1.3 Nanomaterials (nanoparticles)

Researchers within the field of EOR are increasingly investigating the effects of nano-fluids for their potential applications, not only in EOR but also in pharmaceuticals, cosmetics, electronics, material and living matter [24]. The idea of employing nanotechnology in EOR fields is to economically increase oil production and delays the oil field abandonment [25-26]. For successful EOR nanotechnology, nanoparticles formulations must meet the specific requirements like penetrating into the different rock-pore sizes at relatively low concentrations to alter wettability [27]. From our literature survey, we found out that several laboratory experiments and oil fields pilot applications of nanoparticles have been documented, and their potential effects to modify rock wettability and enhanced oil recovery were reported [28-32]. The mechanisms responsible for improving oil recovery using nanoparticles have been fully understood and are no longer controversial, as nanoparticles have the ability to change rock wettability from oil-wet to water-wet conditions [33]. In addition, Dang et al [34] also described the wettability modification near water-wet conditions as the reliable mechanism to understand the chemistry of formation rock surfaces. However, the utilization of nanoparticles to develop composite hybrids is still under scientific investigations. Efforts have been made to understand the synergistic relationship between composite materials containing nanoparticles to correlate with the improved oil recovery.

3. Literature review

This section presents the recent experimental studies from the literature, and highlights some advancements, advantages, and improvements in CEOR as a result of employing hybrid materials in EOR as compared with the conventional single system flooding techniques.

3.1 Surfactant-polymer hybrid

Shedid [35] investigated the feasibility of combining the surfactant (EZEFLO F75N) with Hydrolyzed Polyacrylamide polymer (HPAM-J120) in the low permeability carbonate reservoir (>10 md) with respect to IFT, viscosity and oil recovery rate. The result indicated that apart from lowering the IFT and improving the viscosity, the surfactant-polymer slug had an additional recovery factor (10 %) compared to the injection of the single polymer slug (7 %).

Sun et al [36] studied and compared the thermal resistance and oil recovery rate of foams that were generated from surfactant (mixture of sodium dodecyl sulphate + Imidazoline) and surfactant-polymer (mixture of sodium dodecyl sulphate + Imidazoline + Xanthan polymer). After aging at 90°C, the hybrid slug of surfactant-polymer foam had four times the half-life drainage than the foam obtained from the mixture of surfactants alone. Furthermore, the oil recovery of surfactant-polymer foam was found to be twice that of surfactant foam, in core flood experiments at the same temperature.

Muhammed et al [37] had established that a surfactant-polymer hybrid increased hydrocarbon recovery compared to a single slug. Their studies revealed that the incorporating surfactant in the polymer gel solution could mitigate the effect of polymer plugging.

3.2 Polymer-nanomaterial hybrid

Zhu et al [12] developed polyacryamide (HAHPAM)/silica nanoparticles hybrid material to improve thermal stability, shear resistance, and salt tolerance behaviour of partially hydrolyzed polyacramide polymer (HPAM) for successful CEOR application. After comparing the rheological and enhanced oil recovery properties of hybrid and pure polymer at extreme reservoir conditions (temperature 85°C and 32,868 mgL-1 brine solution), they found that hybrid material of HPAM/silica nanoparticles had a better recovery factor, in addition to improved long thermal and shear stabilities as well as salt tolerance as compared to HPAM polymer alone. They concluded that the chemical interaction between silica nanoparticles and amide functional group in HPAM led to improved properties.

Similarly, Pu et al [38] synthesized a water soluble novel core-shell polymer, known as hyperbranched polymer (HBAMs) hybrid (figure 3), which is made up of polyamidoamide (subshell) and silica nanoparticles (core) via in-situ free radical polymerization reaction. Their investigations revealed that, this hybrid core-shell material could be a good CEOR agent due its unique thermal, shear degradation and salt tolerance behaviour.

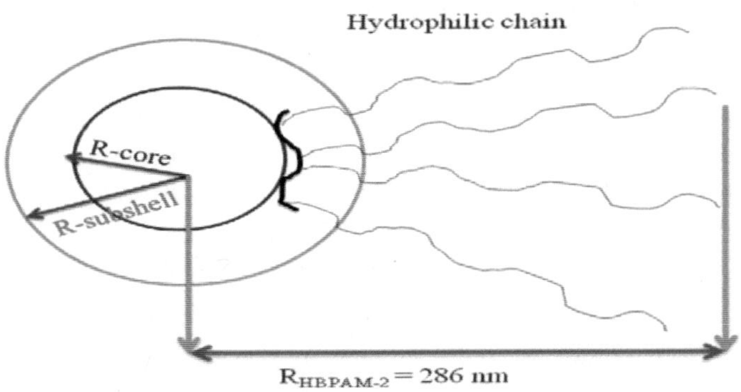

Figure 3. Schematic diagram of co-shell hybrid (modified from [38].

The review of ShamsiJazeyi et al [39] presented the applications of polymer-nanoparticles for enhanced oil recovery.

3.3 Surfactant-nanomaterial hybrid

Eftekhari et al [40] investigated the stability of nitrogen foam using nano-ash particles with alpha olefin sulfonate as a surfactant. They observed that the nano-ash particles, which are readily available as the waste product from a coal power plant, had the ability to stabilize the foam in the presence and absence of oil. Their research confirms that the nano-ash material is compatible with alpha olefin sulfonate surfactant (AOS), which could find other applications in the field of enhanced oil recovery nanotechnology.

Suleimanov et al [41] experimentally studied the effect of non-ferrous metal nanoparticles in a non-ionic surfactant solution. They discovered that the dispersion of metal nanoparticles in the surfactant solution significantly enhanced the reduction of surface tension in the oil media by nearly 90 %. And hence, recommended the use of nano fluids for application in enhanced oil recovery processes.

Worthen et al [42] revealed that silica nanoparticles could be a good composite material to obtain stable and viscous CO2 foam from caprylamidopropyl betaine (amphoteric surfactant). In their studies, they noticed that the interaction of a hydrophilic head group of the silica nanoparticles and the surfactant increased the contact angle between water, surfactant and CO2.

3.4 Surfactant-polymer-nanomaterials hybrid

El-hoshoudy et al [43] developed and successfully characterized a novel nano-composite of polyacrylamide-SiO_2 using surmers as a functional surfactant. After investigating the rheological properties of the novel nano-composite with respect to salinity, shear rate, temperature, and concentration, they observed that the novel composite showed non-Newtonian behaviour and improved properties (thermal, shear and salinity). In the core flood experiment, the novel composite recovered up to 60 % of the residual oil saturation.

Sharma et al [44] investigated the synergistic effects between polymer, surfactant and nanoparticles (SPN) in a single hybrid system to find out the justification of additional oil recovery after comparing with the surfactant-polymer system (SP). The result revealed that significant improvement of properties like viscosity, IFT and pressure drop were noticed in hybrid material (SPN). Similarly, core flood experiments were conducted by flooding with an SPN hybrid at high temperature and high pressure conditions. More than 60 % oil recovery was recorded, which could be due to capillary end effect, as compared to SP flooding.

Nguyen et al [45] synthesized surfactant/polymer/SiO_2 inorganic composite for application in EOR processes under extreme reservoir conditions. They recommended that this composite material could find suitable applications in EOR by increasing the oil displacement efficiency.

4. Limitation of CEOR hybrid materials

Despite the increasing research in finding the promising hybrid composites for successful applications in EOR, sudden changes in reservoir conditions (temperature, pressure, salinity) and crude properties are the main challenges which affect the success of CEOR hybrid flooding for improving oil recovery [46]. The compatibility between the hybrid components are strongly affected by a high temperature, as well as the concentrations of ions in brine solution. Petroleum engineers and scientists are still putting efforts to understand the mechanism of compatibility with respect to physical and chemical conditions in porous media. Apart from this, full field operational implementation of designed projects using hybrids CEOR is still facing economic and technical challenges due to the high cost of chemicals, and finally lack of good engineering technology for successful execution of the projects [47].

5. Conclusion

It is concluded that the synergistic actions between the hybrid components are the main reason that CEOR hybrids are finding wide applications in the field of EOR. This area of oil recovery requires extensive research, particularly on the compatibility studies and the overall performance of CEOR hybrids with respect to changes in physicochemical conditions of reservoirs, as well as changes in crude properties. This would help exploration Engineers to understand and implement successful EOR technologies.

Acknowledgment

The authors would like to thank Enhance Oil Recovery Mission Oriented Research Group of Universiti Teknologi PETRONAS for the encouragement & studentship support. The authors would also like to express their sincere and honest gratitude to Exploration, Production and Technical Department of PETRONAS, Malaysia for financing the project (Grant No. EPTD T1.3).

References

[1] Al-Adasani A and Bai B 2010 Recent developments and updated screening criteria of enhanced oil recovery techniques in *Int. Oil & Gas Conf. and Exhibition in China* Society of Petroleum Engineers 130726 1-24
[2] Tunio S Q, Tunio A H, Ghirano N V and El Adawy Z M 2011 Comparison of different enhanced oil recovery techniques for better oil productivity *Int. J. of Appl. Sci. and Technol.* **1**(5) 143-153
[3] Tzimas E, Georgakaki A, Garcia C and Peteres S D 2005 Enhanced oil recovery using carbon

dioxide in the European energy system Institute for energy Petten Netherlands http://ie.jrc.cec.eu.int/ 1-124

[4] United States energy report 2016 *STEO* https://www.eia.gov/forecasts/steo/index.cfm 1-57

[5] Kong X and Ohadi M 2010 Applications of micro and nano technologies in the oil and gas industry-overview of the recent progress in *Abu Dhabi Int. Petrol. Exhibition and Conf.* Society of Petroleum Engineers

[6] Salehi M M, Sahraei E and Nejad S A 2014 Comparison of oil removal in surfactant alternating gas with water alternating gas, water flooding and gas flooding in secondary oil recovery process *J. of Petrol. Sci. and Eng.* **120** 86-93

[7] Sheng J J 2010 Status of surfactant EOR technology *Petrol.* **1**(2) 97-105

[8] Alvarado V and Manrique E 2010 Enhanced oil recovery: an update review *Energies* **3**(9) 1529-1575

[9] Alagorni A H, Yaacob Z B and Nour A H 2015 An Overview of Oil Production Stages Enhanced Oil Recovery Techniques and Nitrogen Injection *Int. J. of Env. Sci. and Dev.* **6** (9) 693-701

[10] Delamaide E P, Moreau and Tabary R 2015 A New Approach for Offshore Chemical Enhanced Oil Recovery *Offshore Technol. Conf.* 1-18

[11] Barnes J R, Regalado D P, Doll M J, King T E, Petzer L E and Semple T C 2014 Essentials of Upscaling Surfactants for EOR Field Projects *Soc. of Petrol. Engineers* 179573 18

[12] Zhu, D, Wei L, Wang B and Feng Y 2014 Aqueous hybrids of silica nanoparticles and hydrophobically associating hydrolyzed polyacrylamide used for EOR in high-temperature and high-salinity reservoirs *Energies* **7**(6) 3858-3871

[13] Han M, Alsofi A, Fuseni A, Zhou X and Hassan S 2013 Development of chemical EOR formulations for a high temperature and high salinity carbonate reservoir. *Int. Petrol. Technol. Conf.* Beijin China 1-13

[14] Kapetas L, Bonnieu V, Danelis S, Rossen W R, Zadeh R, Eftekhari A A, Shafian M S R and Bahrim R Z K 2016 Effect of temperature on foam flow in porous media *J. of Ind. and Eng. Chem.* **36** 229-237

[15] Guo L, Hang M, Fuseni A and Alsofi A 2016 Laboratory Investigation of Polymeric Surfactants for EOR in High Salinity and High Temperature Reservoir *Soc. of Petrol. Engineers 7*

[16] Zhao G, Dai C, Zhang Y, Chen A, Yan Z and Zhao M 2015 Enhanced foam stability by adding comb polymer gel for in-depth profile control in high temperature reservoirs *Colloids and Surfaces A: Physicochemical and Eng. Aspects* **482** 115-124

[17] Rabu P and A 2010 Taubert, Themed issue on Advanced Hybrid Materials *J. of Mat. Chem.* 2010 **20**(42) 9279-9280

[18] Kickelbick G 2007 Introduction to hybrid materials Hybrid materials: Synthesis, characterization, and applications 1-46.

[19] Ferno M A, Gauteplass J, Pancharoen M, Haugen A, Kouscek A R and Hirasaki G 2016 Experimental Study of Foam Generation, Sweep Efficiency, and Flow in a Fracture Network Society of petroleum engineers 1-11

[20] Schramm L L 2010 Surfactants: fundamentals and applications in the petroleum industry Edinburgh (Cambridge University Press) 1-24

[21] Farzaneh S A and Sohrabi M 2013 A Review of the Status of Foam Application in Enhanced Oil Recovery, *Soc. of Petrol. Engineers*

[22] Sandersen S B, Stenby E H and Von Solms N 2012 Enhanced oil recovery with surfactant flooding Technical University of Denmark, Center for Energy Resources Engineering

[23] Hirose S, Abidin A Z, Puspasari T and Nugroho W A 2012 The International Conference on Innovation in Polymer Science and TechnologyPolymers for Enhanced Oil Recovery Technology *Procedia Chem.* **4** 11-16.

[24] Yan J, Shamim T, Chou S K, Li H, Guo K, Li H and Yu z 2015 Clean, Efficient and Affordable

Energy for a Sustainable Future The 7th International Conference on Applied Energy (ICAE2015)Metallic Nanoparticles for Enhanced Heavy Oil Recovery Promises and Challenges *Energy Procedia* **75** 2068-2073

[25] Torsater O, Li S and Hendraningrat L 2013 A coreflood investigation of nanofluid enhanced oil recovery in low-medium permeability Berea sandstone in *SPE Int. Symp. on Oilfield Chem. Soc. of Petrol. Engineers*

[26] Hendraningrat L, Li S and Torsæter 2013 A coreflood investigation of nanofluid enhanced oil recovery *J. of Petrol. Sci. and Eng.* **111** 128-138

[27] Al-Anssari S, Barifcani A, Wang S, Maxim L and Iglauer S 2016 Wettability alteration of oil-wet carbonate by silica nanofluid *J. of Colloid and Interface Sci.* **461** 435-442.

[28] Mohammed M and Babadagli T 2015 Wettability alteration: A comprehensive review of materials/methods and testing the selected ones on heavy-oil containing oil-wet systems *Advances in Colloid and Interface Sci.* **220** 54-77

[29] Esmaeilzadeh P, Sadeghi M T, Fakhroueian Z, Bahramian A and Norouzbeigi R 2015 Wettability alteration of carbonate rocks from liquid-wetting to ultra gas-wetting using TiO2, SiO2 and CNT nanofluids containing fluorochemicals, for enhanced gas recovery *J. of Natural Gas Sci. and Eng.* **26** 1294-1305

[30] Ershadi M, Rashidi A, Ramazani A and Khosravani S 2015 Carbonate and sandstone reservoirs wettability improvement without using surfactants for Chemical Enhanced Oil Recovery (C-EOR) *Fuel* **153** 408-415

[31] Wasan D, Nikolov N and Kondiparty K 2011 The wetting and spreading of nanofluids on solids: Role of the structural disjoining pressure. *Current Opinion in Colloid & Interface Sci.* **16**(4) 344-349

[32] Hashemi R, Nassar N N and Almao P P 2014 Nanoparticle technology for heavy oil in-situ upgrading and recovery enhancement: Opportunities and challenges *Appl. Energy* **133** 374-387

[33] Ehtesabi H, Ahadian M M, Taghnikhani V and Ghazanfari M H 2013 Enhanced heavy oil recovery in sandstone cores using TiO_2 nanofluids. *Energy & Fuels* **28**(1) 423-430

[34] Dang C T Q, Nghiem L X, Chen Z J and Nguyen Q P 2013 Modeling low salinity waterflooding: ion exchange, geochemistry and wettability alteration. in *SPE Annu. Technical Conf. and Exhibition Soc. of Petrol. Engineers.*

[35] Shedid S A 2015 Experimental Investigation of Alkaline/Surfactant/Polymer (ASP) Flooding in Low Permeability Heterogeneous Carbonate Reservoirs Society of Petroleum Engineers 1-16

[36] Sun L, Pu W, Wei P, Li Y and Yuan C 2015 High temperature and oil tolerance of surfactant foam/polymer–surfactant foam. *RSC Advances* **5**(30) 23410-23418

[37] Muhammed F A, Bai B and Tang T 2012 Experimental study of the interaction between surfactants and super absorbent polymer gel *J. of Petrol. Sci. and Eng.* 90–91 159-164

[38] Pu W F, Liu R, Wang K, Li K, Yang Z, Li B and Zhao L 2015 Water-Soluble Core–Shell Hyperbranched Polymers for Enhanced Oil Recovery *Ind. & Eng. Chem. Res.* **54**(3) 798-807

[39] ShamsiJazeyi H, Wang M S, Tour J M and Verduzco R 2014 Polymer- coated nanoparticles for enhanced oil recovery *J. of Appl. Polymer Sci.* **131**(15)

[40] Eftekhari A A, Krasfer, Rumen and Farajzadeh R 2015 Foam Stabilized by Fly Ash Nanoparticles for Enhancing Oil Recovery *Ind. & Eng. Chem. Res.* **54**(50) 12482-12491

[41] Suleimanov B, Ismailov F and Veliyev E 2011 Nanofluid for enhanced oil recovery *J. of Petrol. Sci. and Eng.* **78**(2) 431-437

[42] Worthen A J, Bryant S L, Huh C and Johnston K P 2013 Carbon dioxide- in- water foams stabilized with nanoparticles and surfactant acting in synergy. *AIChE J.* **59**(9) 3490-3501

[43] El-hoshoudy A N, Desouky S E M, Betiha M A and Alsabagh A M 2016 Use of 1-vinyl

imidazole based surfmers for preparation of polyacrylamide–SiO2 nanocomposite through aza-Michael addition copolymerization reaction for rock wettability alteration *Fuel* **170** 161-175

[44] Sharma T, Kumar G S, and Sangwai J S 2015 Comparative effectiveness of production performance of Pickering emulsion stabilized by nanoparticle–surfactant–polymerover surfactant–polymer (SP) flooding for enhanced oil recoveryfor Brownfield reservoir *J. of Petrol. Sci. and Eng.* **129** 221-232

[45] Nguyen P T, Do B P H, Pham D K, Nguyen Q T, Dao D Q P and Nguyen H A 2012 Evaluation on the EOR potential capacity of the synthesized composite silica-core/polymer-shell nanoparticles blended with surfactant systems for the HPHT offshore reservoir conditions *in Int. Oilfield Nanotechnology Conf. and Exhibition Soc. of Petrol. Engineers.*

[46] Gao C, Shi J and Zhao F 2014 Successful polymer flooding and surfactant-polymer flooding projects at Shengli Oilfield from 1992 to 2012. *J. of Petrol. Explor. and Prod. Technol.* **4**(1) 1-8

[47] Gharbi R, Alajmi A and Algharaib M 2012 The Potential of a Surfactant/Polymer Flood in a Middle Eastern Reservoir *Energies* 2012 **5**(1) 58-70

IOP Conf. Series: Materials Science and Engineering **206** (2017) 012008 doi:10.1088/1757-899X/206/1/012008

Enzymatic Hydrolysis of Pretreated Fibre Pressed Oil Palm Frond by using Sacchariseb C6

F S Hashim[1], H W Yussof[1], M A K M Zahari[1], R A Rahman[2], R M Illias[2]

[1]Faculty of Chemical and Natural Resources Engineering, Universiti Malaysia Pahang, Lebuhraya Tun Razak, 26300 Kuantan, Pahang, Malaysia
[2]Department of Bioprocess Engineering, Faculty of Chemical Engineering, Universiti Teknologi Malaysia, 81310 UTM Skudai, Johor, Malaysia

E-mail: hafizuddin@ump.edu.my

Abstract. Enzymatic hydrolysis becomes a prominent technology for conversion of cellulosic biomass to its glucose monomers that requires an action of cellulolytic enzymes in a sequential and synergistic manner. In this study, the effect of agitation speed, glucan loading, enzyme loading, temperature and reaction time on the production of glucose from fibre pressed oil palm frond (FPOPF) during enzymatic hydrolysis was screened by a half factorial design 2^{5-1} using Response Surface Methodology (RSM). The FPOPF sample was first delignified by alkaline pretreatment at 4.42 % (w/v) sodium hydroxide for an hour prior to enzymatic hydrolysis using commercial cellulase enzyme, Sacchariseb C6. The effect of enzymatic hydrolysis on the structural of FPOPF has been evaluated by Scanning Electron Microscopy (SEM) analysis. Characterization of raw FPOPF comprised of 4.5 % extractives, 40.7 % glucan, 26.1 % xylan, 26.2 % lignin and 1.8 % ash, whereas for pretreated FPOPF gave 0.3 % extractives, 61.4 % glucan, 20.4 % xylan, 13.3 % lignin and 1.3 % ash. From this study, it was found that the best enzymatic hydrolysis condition yielded 33.01 ± 0.73 g/L of glucose when performed at 200 rpm of agitation speed, 60 FPU/mL of enzyme loading, 4 % (w/w) of glucan loading, temperature at 55 and 72 hours of reaction time. The model obtained was significant with p-value <0.0001 as verified by the analysis of variance (ANOVA). The coefficient of determination (R^2) from ANOVA study was 0.9959. Overall, it can be concluded that addition of Sacchariseb C6 during enzymatic hydrolysis from pretreated FPOPF produce high amount of glucose that enhances it potential for industrial application. This glucose can be further used to produce high-value products.

1. Introduction

Lignocellulosic biomass (LCB) is the most abundant renewable biomass that usually can be obtained from agricultural and forest residues [1][2][3]. It comprise of three major components which are cellulose, hemicellulose and lignin [4]. Cellulose and hemicellulose can be converted into fermentable sugars such as glucose and xylose.

Bioconversion of LCB into fermentable sugars indicates high potential as a promising renewable feedstock in biorefining area. The first step in fermentable sugar production is pretreatment in order to disrupt the lignin and expose the cellulose fraction followed by enzymatic hydrolysis. The hydrolysis of LCB into fermentable sugar is a crucial stage, which mainly determines the overall process efficiency.

Enzymatic hydrolysis is carried out by cellulase enzyme which is highly specific, and the products of the hydrolysis are usually reducing sugars such as glucose and xylose. There are various methods available for the fermentable sugars production, but the chemical and enzymatic methods have been proved to be more successful. Moreover, pretreatment and enzymatic hydrolysis are the most important steps in order to convert LCB contents into fermentable sugars which are currently known to have much more chance to reduce processing cost than other processes.

In this study, FPOPF was introduced as a raw material where the hemicellulose and cellulose were converted into simple sugars concerning to maximize the utilization of OPF. In order to improve the accessibility of cellulase on FPOPF, the structure of lignocellulose must be broken down by alkaline pretreatment. FPOPF was treated with alkaline solution for delignification process before undergoes enzymatic hydrolysis [5]. Cellulase enzyme which is Sacchariseb C6 was used in enzymatic hydrolysis to promote the production of glucose by converting the cellulose.

Therefore, this study aims at assessing the effects of five important parameters which are agitation speed, glucan loading, enzyme loading, temperature and hydrolysis time on the enzymatic hydrolysis using Sacchariseb C6 and thus finding the best conditions for enzymatic hydrolysis by using statistical approach of response surface methodology (RSM).

2. Materials and Methods

2.1. Materials

Oil palm frond (OPF) was obtained from a local palm oil plantation at Kuantan, Pahang. The OPF was pressed by using sugarcane machine to remove the juices. The fibre pressed oil palm frond (FPOPF) was sun dried for 3 days before shredded into pieces and sieved into particle size less than 2 mm. Dried FPOPF was stored in sealed plastic bag at room temperature. Enzymatic hydrolysis was carried out using Sacchariseb C6, a commercial enzyme preparation kindly obtained from Advanced Enzyme Technology (India).

2.2. Compositional analysis of FPOPF

Characterization of FPOFF was carried out according to National Renewable Energy Laboratory (NREL) methods in order to determine the composition of glucan, xylan, lignin, ashes and extractives contents in untreated and alkaline pretreated FPOPF. The moisture content of the raw FPOPF was determined using a moisture analyzer (DSC, A&D MS70). The extractives content was measured using DIONEX ASE 350 (Thermo Scientific, USA) with water and ethanol as solvents for 30 minutes prior to an autoclaved-based acid hydrolysis step [6]. The recovered water extract was analyzed for soluble sugar monomers. The carbohydrates and lignin content in FPOPF were determined using two-step acid hydrolysis procedure by NREL [7]. The acid insoluble material was determined using gravimetric analysis while UV–Vis spectroscopy was used to measure acid soluble lignin in FPOPF [8].

2.3. Alkaline pretreatment

FPOPF sample was soaked in 4.42 % (w/v) sodium hydroxide (NaOH) solution. The sample was treated at 100 °C about 58.31 minutes [5]. After that, the treated FPOPF was washed thoroughly with de-ionized water until neutral. The sample was oven dried at 105 °C and stored prior to enzymatic hydrolysis.

2.4. Two-level factorial analysis experimental setup

The experimental design for factorial analysis was performed using Design Expert 7.0.0 (Stat-Ease Inc., USA) software. The effect of five independent variables as shown in Table 1 were analysed using Response Surface Methodology (RSM). The condition ranges chosen were based on the other researcher's previous work [9][10][11][12]. The factors were constructed in factorial designs of 2^{5-1} to screen their effect on the response of glucose production. All experiments consist of 16 run listed in Table 2 were carried out in triplicate and the averages were taken as responses.

Table 1. Parameters and their designated low and high value.

Factor	Units	Low value (-1)	High value (+1)
A: Agitation speed	Rpm	50	200
B: Enzyme loading	FPU/mL	20	60
C: Glucan loading	%	1	4
D: Temperature	°C	35	55
E: Reaction time	Hours	3	72

Table 2. Experimental design for screening.

Std. Order	Agitation speed (rpm)	Enzyme loading (FPU/mL)	Glucan loading (%)	Temperature (°C)	Reaction time (hours)
1	-1	-1	-1	-1	+1
2	+1	-1	-1	-1	-1
3	-1	+1	-1	-1	-1
4	+1	+1	-1	-1	+1
5	-1	-1	+1	-1	-1
6	+1	-1	+1	-1	+1
7	-1	+1	+1	-1	+1
8	+1	+1	+1	-1	-1
9	-1	-1	-1	+1	-1
10	+1	-1	-1	+1	+1
11	-1	+1	-1	+1	+1
12	+1	+1	-1	+1	-1
13	-1	-1	+1	+1	+1
14	+1	-1	+1	+1	-1
15	-1	+1	+1	+1	-1
16	+1	+1	+1	+1	+1

The validation run for factorial analysis was done to validate the experimental values synchronise with the predicted model generated by Design Expert software. The condition for the validation run was obtained from the predicted best condition developed from 2^{5-1} factorial design. The validation model was determined by comparing the predicted and experimental values. The percentage errors between these values were calculated using Equation (1).

$$\text{Error (\%)} = \frac{|\text{Calculated value} - \text{Experimental value}|}{\text{Experimental value}} \times 100\ \% \quad (1)$$

2.5. Enzymatic hydrolysis

Enzymatic hydrolysis was carried out using Sacchariseb C6 and Novozyme 188 (64 pNPGu/mL). Pretreated FPOPF (1 - 4 %) was mixed with 0.02 % (w/v) sodium azide to prevent microbial growth in the presence of 0.05 M citrate buffer at pH 4.8. The mixture was pre-incubated at certain temperature (35 - 55 °C) prior to the addition of enzymes. The enzymatic hydrolysis was then initiated by adding Sacchariseb C6 (20 - 60 FPU/ml) and Novozyme 188. The shaker started to agitate (50 - 200 rpm). At the end of the hydrolysis (3 - 72 hours) the samples were filtered and their residues were collected for further analysis. Each experiment was carried out in triplicate.

2.6. HPLC analysis

The hydrolysate was analysed by using Agilent 1200 HPLC (USA) system equipped with refractive index (RI) detector. The separation was performed using RHM Monosaccharide H^+ column. The mobile phase was prepared using Milli-Q ultrapure water (Millipore, USA). The column temperature was maintained at 60 °C. The flow rate and injection volume were at 0.4 mL/min and 5 µL, respectively. The calibration curve was generated from binary standard consists of pure glucose and pure xylose with the ranges of 1 g/L to 40 g/L.

2.7. Scanning electron microscopy (SEM) morphology

Morphological structure of FPOPF was carried out using EVO 50 SEM (Carl Zeiss, Germany). Three FPOPF samples were collected which are untreated FPOPF, pretreated FPOPF and pretreatment with enzymatic hydrolysis FPOPF. Prior to SEM analysis, the samples were oven dried (55 ± 5 °C) for overnight. The samples were fixed on the aluminium stubs and coated with carbon layer. The samples were observed at 1000x magnification power and 5 kV of working voltage.

3. Results and Discussion

3.1. Characterization of FPOPF

As a natural product, the FPOPF has various compositions in terms of glucan, xylan and lignin as reported by Zahari et al. [13] and Tan et al. [14]. This different in composition was due to the several factors such as geographic locations, plant ages, climate, and soil conditions. The FPOPF used in this study composed of 4.5 % extractives, 40.7 % glucan, 26.1 % xylan, 26.2 % lignin and 1.8 % ash for raw FPOPF. Meanwhile, pretreated FPOPF comprised of 0.3 % extractives, 61.4 % glucan, 20.4 % xylan, 13.3 % lignin and 1.3 % ash.

3.2. Scanning electron microscopy ((SEM) analysis

Untreated, pretreated and pretreated with hydrolysis FPOPF were observed using SEM under magnification at 1000x in order to evaluate the effect of alkaline pretreatment and enzymatic hydrolysis on the morphology of the FPOPF as shown in Figure 1. Surface of untreated FPOPF (Figure 1a) was smooth, flat and rigid structure with presence of undamaged surface which comparable with previous study reported in rice straw [15], empty fruit bunch (EFB) [16], and wheat straw [17]. Meanwhile, after alkaline pretreatment, the outer surface layer was destroyed and the cell wall was distorted resulting the internal structure was exposed as shown in Figure 1b. This proved that the delignification occurred during alkaline pretreatment. This structure alteration could be expected that the accessibility of enzyme increased to the cellulose in the enzymatic hydrolysis later. Meanwhile, structure of FPOPF after enzymatic hydrolysis as in Figure 1c described that structures become disrupted by the enzymatic hydrolysis.

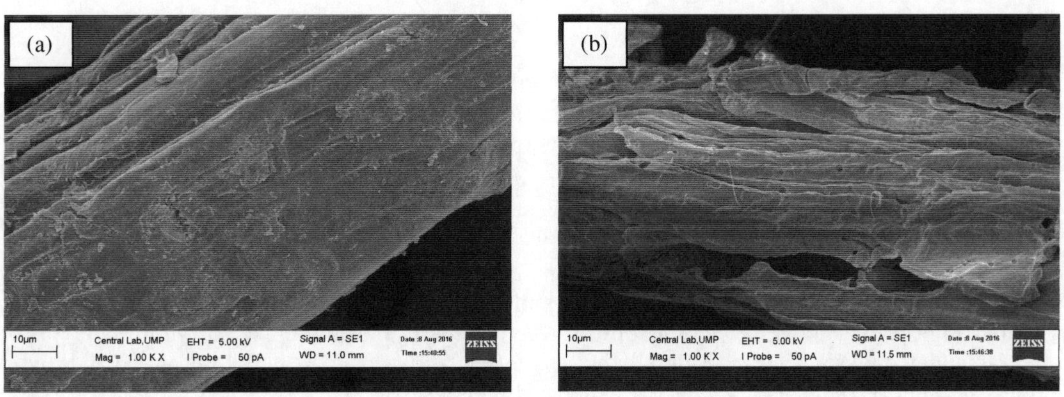

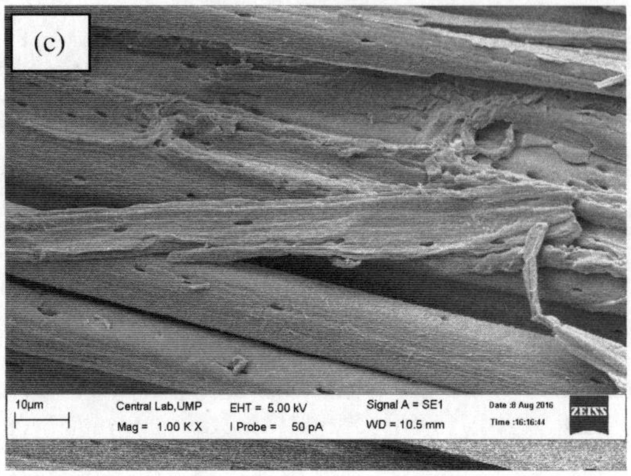

Figure 1. SEM images of FPOPF: a) untreated FPOPF, b) pretreated FPOPF, and c) FPOPF after enzymatic hydrolysis.

3.3. Factors affecting enzymatic hydrolysis

The highest production of glucose was obtained at 33.01 ± 0.73 g/L where the hydrolysis conditions at 200 rpm of agitation speed with temperature of 55 °C, 4 % of glucan loading, and 60 FPU/mL of enzyme loading for 72 hours of hydrolysis time as shown in Table 3.

Table 3. Experimental design of factor screening process using 2^{5-1} factorial design with their response.

Std.	Factors					Response
Order	Agitation speed (rpm)	Enzyme loading (FPU/mL)	Glucan loading (%)	Temp. (°C)	Reaction time (hours)	Glucose Concentration (g/L)
1	50	20	1	35	72	7.71
2	200	20	1	35	3	3.23
3	50	60	1	35	3	3.08
4	200	60	1	35	72	8.91
5	50	20	4	35	3	11.39
6	200	20	4	35	72	30.76
7	50	60	4	35	72	26.55
8	200	60	4	35	3	14.04
9	50	20	1	55	3	3.16
10	200	20	1	55	72	7.47
11	50	60	1	55	72	9.92
12	200	60	1	55	3	3.76
13	50	20	4	55	72	30.32
14	200	20	4	55	3	14.63
15	50	60	4	55	3	13.99
16	200	60	4	55	72	33.01

In factorial analysis, contribution of the main factor gives an important effect in the optimization part later. Two to three highest contributed factors will be selected from this factorial analysis. From Table 4 and Figure 2, factor C (glucan loading) gives the most contributing factor with 52.42 % to the enzymatic hydrolysis. Glucan loading indicates the availability of more cellulose can be hydrolysed to glucose. Similar results reported by Tan and Lee [12] and Nieves et al. [18] where the glucose yield was gradually increased as the substrate loading increased. Furthermore, Sun and Chen [19] reported that substrate loading is one of the main factors that contribute to the rate of enzymatic hydrolysis. Next, the second most contribution factor followed by 31.73 % of E (hydrolysis time). Glucose production increased with longer hydrolysis time because its allowed the enzyme to hydrolyse the cellulose into glucose. Similar trend was obtained by Tan and Lee [12] and Zheng et al. [9].

Table 4. Percentage of contribution of main factor in enzymatic hydrolysis.

Factors	Contribution (%)
A	0.39
B	0.087
C	52.42
D	0.47
E	31.73

From Pareto chart as shown in Figure 2, all five factors (A, B, C, D and E) gave a positive effect (refer to orange bar chart) to the enzymatic hydrolysis. It is suggested that the highest values will be used to favour the response. For example, an increase in the glucan loading (C) increases the glucose concentration. Meanwhile, the negative effect (blue bar chart) reveals that the use of the lowest range value of factor will increase conversion to glucose.

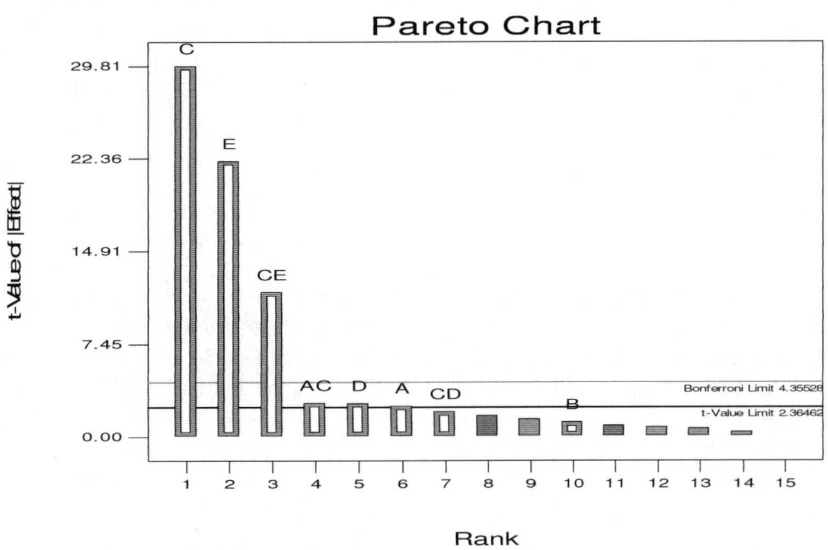

Figure 2. Pareto chart for 2^{5-1} factorial design.

From the ANOVA of this experimental design as shown in Table 5, the model obtained was significant with p-value <0.0001. The coefficient of determination (R^2) value obtained in this model was 0.9959, which is in good agreement with the adjusted R^2 value of 0.9912. The high R^2 value of 0.9959 indicates that the model was well adapted to the response.

Table 5. ANOVA table.

Source	Sum of square	Degree of freedom	Mean square	F-value	P-value	
Model	1648.46	8	206.06	212.19	<0.0001	Significant
A	5.90	1	5.90	6.07	0.0432	
B	1.32	1	1.32	1.35	0.2827	
C	862.98	1	862.98	888.65	<0.0001	
D	6.99	1	6.99	7.20	0.0314	
E	476.94	1	476.94	491.12	<0.0001	
AC	7.11	1	7.11	7.32	0.0304	
CD	3.86	1	3.86	3.97	0.0865	
CE	131.16	1	131.16	135.06	<0.0001	
Residual	6.80	1	0.97			

Equation (2) shows the response surface quadratic model for glucose production which can be presented in terms of coded factors as in the following equation:

$$Y = 14.20 + 0.61\,X_1 + 0.29\,X_2 + 8.21\,X_3 + 0.99\,X_4 + 5.46\,X_5 + 0.67\,X_1X_3 + 0.74\,X_3X_4 + 2.86\,X_3X_5 \quad (2)$$

where Y was concentration of glucose (g/L), X_1 was the agitation speed, X_2 was enzyme loading, X_3 was glucan loading, X_4 was the temperature and X_5 was the hydrolysis. The unknowns X_1, X_2, X_3, X_4 and X_5 were referred to the main effects while X_1X_3, X_3X_4 and X_3X_5 were the interaction effects contributed in the enzymatic hydrolysis process. Based on the quadratic model, coefficients of X_1 to X_5 are small compared to constant. This gives an indicator that the model equation is good with small error and can be used for further analysis.

3.4. Validation run

The validation experiments were conducted based on one suggested best condition from Design Expert 7.0 in triplicate. The experiments were performed at 160 rpm of agitation speed, 20 FPU/mL of Sacchariseb C6, glucan loading at 4 %, temperature at 56 °C and hydrolysis time at 72 hours and the result is presented in Table 6. The error from these validations runs in between 1.85 % to 4.70 %. The model was found to be reliable and reproducible as the experimental values were in good agreement with the predicted values proposed by the model with an error less than 10 %. Thus, it was proved to be an adequate model.

Table 6. Validation run for agitation speed at 160 rpm, enzyme loading at 20 FPU/ mL, 4 % of glucan loading, at 56 °C for 72 hours.

Description	Concentration of glucose (g/L)		
	Run 1	Run 2	Run 3
Predicted Value	34.024	34.024	34.024
Experimental Value	33.393	32.423	33.382
Error	1.85 %	4.70 %	1.88 %

4. Conclusion

The results obtained from this study clearly indicate that the best condition for the enzymatic hydrolysis process of FPOPF by Sacchariseb C6 at 200 rpm of agitation speed, 60 FPU/mL of enzyme loading, 4 % (w/v) of glucan loading, temperature at 55 and 72 hours of reaction time which produced 33.01 ± 0.73 g/L of glucose. In this production of glucose, there are two factors that mostly contributed in enzymatic hydrolysis which are glucan loading and temperature that will be used in the optimization part later. Overall, it can be concluded that Sacchariseb C6 is a suitable candidate for enzymatic hydrolysis of pretreated FPOPF in order to achieve higher glucose production.

References

[1] Luk G, Bekmuradov V and Luong R 2013 Pre-treatment of source separated organic waste for ethanol *J. Teknologi* **62** (2) 11–15

[2] Rambo M K D, Schmidt F L and Ferreira M C 2015 Analysis of the lignocellulosic components of biomass residues for biorefinery opportunities *Talanta* **144** 696–703

[3] Siti Sabrina M S, Roshanida A S and Norzita N 2013 Pretreatment of oil palm fronds for improving hemicelluloses content for higher recovery of xylose *J. Teknologi* **2** 39–42

[4] Harmsen P, Huijgen W, López L and Bakker R 2010 Literature review of physical and chemical pretreatment processes for lignocellulosic biomass *Food and Biobased Research* 1–49.

[5] Sukri S and Rahman R 2014 Optimization of alkaline pretreatment conditions of oil palm fronds in improving the lignocelluloses contents for reducing sugar production *Romanian Biotech. Letters* **19** (1) 9006–9018

[6] Sluiter A, Ruiz R, Scarlata C, Sluiter J and Templeton D 2008 Determination of extractives in biomass *National Renewable Energy Laboratory* 1–9

[7] Sluiter A, Hyman D, Payne C and Wolfe J 2008 Determination of insoluble solids in pretreated biomass material *National Renewable Energy Laboratory* 1–9

[8] Sluiter A, Hames B, Ruiz R, Scarlata C, Sluiter J, Templeton D and Nrel D C 2012 Determination of structural carbohydrates and lignin in biomass national *Renewable Energy Laboratory*

[9] Zheng J, Choo K, Bradt C, Lehoux R and Rehmann L. 2014 Enzymatic hydrolysis of steam exploded corncob residues after pretreatment in a twin-screw extruder *Biotech. Reports* **3** 99–107

[10] Ramos L P, da Silva L, Ballem A C, Pitarelo A P, Chiarello L M and Silveira M H L 2014 Enzymatic hydrolysis of steam-exploded sugarcane bagasse using high total solids and low enzyme loadings *Bioresource Tech.* **175** 195–202

[11] Guo H, Su R, Huang R, Qi W and He Z 2015 Co-optimization of sugar yield and input energy by the stepwise reduction of agitation rate during lignocellulose hydrolysis *Food and Bioproducts Processing* **95** 1–6

[12] Tan I S and Lee K T 2014 Enzymatic hydrolysis and fermentation of seaweed solid wastes for bioethanol production: An optimization study *J. Energy* **78** 53–62

[13] Zahari M A K M, Zakaria M R, Ariffin H, Mokhtar M N, Salihon J, Shirai Y and Hassan M A 2012 Renewable sugars from oil palm frond juice as an alternative novel fermentation feedstock for value-added products *Bioresource Tech.* **110** 566-571

[14] Tan J P, Jahim J M, Harun S, Wu T Y and Mumtaz T 2016 Utilization of oil palm fronds as a sustainable carbon source in biorefineries *International J. of Hydrogen Energy* **41** (8) 4896–4906

[15] Kim I and Han J I 2012 Optimization of alkaline pretreatment conditions for enhancing glucose yield of rice straw by response surface methodology *Biomass and Bioenergy* **46** 210–217

[16] Shamsudin S, Md Shah U K, Zainudin H, Abd-Aziz S, Mustapa Kamal S M, Shirai Y and Hassan M A 2012 Effect of steam pretreatment on oil palm empty fruit bunch for the production of sugars *J. Biomass and Bioenergy* **36** 280–288

[17] Gonzalez G, Lopez-Santin J, Caminal G and Sola C 1986 Hemicellulose at Moderate Temperature : A Simplified Kinetic Model. *Biotech. and Bioengineering* **28** 288–293

[18] Nieves D C, Ruiz H A, de Cárdenas L Z, Alvarez G M, Aguilar C N, Ilyina A and Martínez Hernández J L 2016 Enzymatic hydrolysis of chemically pretreated mango stem bark residues at high solid loading *J. Ind. Crops and Products* **83** 500–508

[19] Sun Y and Cheng J. 2002 Hydrolysis of lignocellulosic materials for ethanol production: a review *Biores. Tech.* **83** (1) 1–11

IOP Conf. Series: Materials Science and Engineering **206** (2017) 012009 doi:10.1088/1757-899X/206/1/012009

Quantitative Estimation of Risks for Production Unit Based on OSHMS and Process Resilience

D Nyambayar[1], I Koshijima[1] and H Eguchi[1]

[1]Nagoya Institute of Technology, Gokiso-Cho, Showa-Ku, Nagoya-shi 466-8555, Japan

E-mail: nyamaa0727@gmail.com

Abstract. Three principal elements in the production field of chemical/petrochemical industry are (i) Production Units, (ii) Production Plant Personnel and (iii) Production Support System (computer system introduced for improving productivity). Each principal element has production process resilience, i.e. a capability to restrain disruptive signals occurred in and out of the production field. In each principal element, risk assessment is indispensable for the production field. In a production facility, the occupational safety and health management system (Hereafter, referred to as OSHMS) has been introduced to reduce a risk of accidents and troubles that may occur during production. In OSHMS, a risk assessment is specified to reduce a potential risk in the production facility such as a factory, and PDCA activities are required for a continual improvement of safety production environments. However, there is no clear statement to adopt the OSHMS standard into the production field. This study introduces a metric to estimate the resilience of the production field by using the resilience generated by the production plant personnel and the result of the risk assessment in the production field. A method for evaluating how OSHMS functions are systematically installed in the production field is also discussed based on the resilience of the three principal elements.

1. Introduction

Three principal elements of a production process are: (i) Production Units, (ii) Production Plant Personnel and (iii) Production Support System hereafter, referred to as PSS (a computer system introduced for improving productivity). Each principal element has production process resilience [1] that is a capability to restrain disruptive signals occurred in and out of the production process. These signals affect various production activities. There are already several reports explaining the relations between the managerial subjects, the managerial objects, the managerial components and the scale of disruptive signals [2, 3, 4]. In each principal element, risk assessment is indispensable for the production field.

In a production facility of a common-law jurisdiction, employers have a common law duty to take reasonable care of their production plant persons for their general safety. Here, occupational safety and health management system hereafter, referred to as OSHMS [5], is introduced to reduce accidents and troubles in the facility [6, 7]. OSHMS can also be applied to occupational safety and health or workplace safety and health area concerned with the safety, health, and welfare of people engaged in work or employment. In this paper first half, analyzed of the OSHMS and find out for a method for evaluating how OSHMS functions are systematically installed in the production field. The latter half introduces a metric to estimate the resilience of the production field by using the resilience generated by the production plant personnel and the result of the risk assessment in the production field.

IOP Conf. Series: Materials Science and Engineering **206** (2017) 012009 doi:10.1088/1757-899X/206/1/012009

The production plant personnel is related to all production activities in the production process by quantitatively using the level of skills and knowledge [9]. The author of that paper [9] estimated the resilience generated by the production plant personnel, hereafter referred to as RPPP.

Moreover, PSS is a computer system implemented in the production process to improve the productivity, and formalized skills of the production plant personnel are introduced in a computer system as an application software. Therefore, the resilience generated by PSS, hereafter, referred to as RPSS, is estimated quantitatively by the level of skills and knowledge transferred to PSS by the technique.

As to the production facility, the risk assessment is one of the managerial objects to keep higher resilience. And all of the production activities are the objectives of a risk assessment in the production process. Therefore, the degree of potential risk is estimated in works performed by the production plant personnel. Consequently, it is possible to estimate resilience of plant units, hereafter, referred to as RPU, using RPPP and the result of the risk assessment on the production process.

2. Risk assessment on the production process

In the production facility, OSHMS (Occupational Safety and Health Management System) is introduced to reduce the accidents and troubles in the production field. OSHMS contains the risk assessment requirements to reduce the degree of potential risks in all works in the facility [7, 9, 10].

In OSHMS, PDCA (Plan, Do, Check, Act) activities are requested for continual improvement of the safety production environment. However, there is no clear statement to adopt the OSHMS standard into the production field.

2.1. Analysis of OSHMS specification

OSHMS is a standard that defines management statements of potential risks and countermeasures related to occupational health and safety concerns in a working environment, clarifies the responsibility of the production plant personnel, and requests for implementation of its program at all production field. However, while OSHMS is adopted in the actual production company, OSHMS does not specify production plant personnel who ought to perform certain actions under particular company's environment.

This paper discusses and analyses Chapter 4 of 2007[5] edition of OSHMS, which contains directives on activities relating to health and safety. In Table 1, the following items are extracted from each sentence in the chapter.

- Sentences location.
- Sentence.
- Subjects, object, and verbs within a sentence.
- Knowledge, skill, the rule for understanding subject and object, and for taking actions specified by verbs.

The results show that the entity in charge of safety is primarily the "organization" and there is no mention in the standard that 'who' should actually "act." Furthermore, when the object was extracted during the analysis, hardly any details were mentioned, and it was hence not clear what action should be taken within the organization. Table 1 shows only top a few lines as an example of the analysis of the standard clauses.

Basically, OSHMS standard was written as a rule-base. There are many rules in the standard, though knowledge and skills are necessary for applying the rules as showing Rasmussen's behavior model [11].

Skill-based behavior: This represents patterned and highly automated activities with little conscious attention.

Rule-bases behavior: This represents activities that are modularized and selected on the condition specified in advance.

Knowledge-based behavior: This represents improvisation of activities in an environment where the available rules do not exist.

Table 1. An example of analyzed clauses of the OSHMS standard (Section 4.2 of OHSAS18001:2007).

Location	Sentence	Subject	Verb	Object	Knowledge, rule, skill
29-32 row from the top, page 90	1 Hazard identification, risk assessment and determining controls The organization shall establish, implement and maintain a procedure(s) for the ongoing hazard identification, risk assessment, and determination of necessary controls.	organization	planning	risk assessment	rule
22 row from the top, page 92-16 row from the top, page 93	The procedure(s) for hazard identification and risk assessment shall take into account: a) routine and non-routine activities; b) activities of all persons having access to the workplace	organization	shall take into account	hazard risk assessment	rule
21-26 row from the top, page 95	The organization' s methodology for hazard identification and risk assessment shall: a) be defined with respect to its scope, nature and timing to ensure it is proactive rather than reactive; and b) Provide for the identification, prioritization and documentation of risks	organization	risk assessment	organization' s methodology for hazard identification	rule
24-27row from the top, page 97	For the management of change, the organization shall identify the OH&S hazards and OH&S risks associated with changes in the organization, the OH&S management system, or its activities, prior to the introduction of such changes.	organization	change	OH&S management system	knowledge, Rule
1-2 row from the top, page 98	The organization shall ensure that the results of these assessments are considered when determining controls.	organization	shall ensure	determining control	rule

2.2. Clarification of activities in OSHMS clauses

When a company implements OSHMS, it follows the relevant clauses in OSHMS standard. The standard uses PDCA cycle, which includes OH&S policy, goals, health and safety planning, its implementation and operation, daily inspections and improvements, system audits, and regular review of the system itself. This management cycle is adopted by company for continual improvement.

In Figure 4 that shows a tree diagram from factory to work, for example, certain information is needed to go PDCA cycling well. Table 1 reveals that the following three indispensable information are unclear for implementing and operating OSHMS:

- The input and output of a given activity.
- The main subject of a person who carries out the activity.
- The conditions on the limitations and resources that must be taken into account in executing the activity.

To identify these elements in each clause of OSHMS standard, IDEF0, a function modeling methodology, is used. Activities in IDEF0 can be identified as verbs in the clauses. Concomitantly, in IDEF0 drawing, feeds to execute an activity are expressed as inputs, and results of the activity can be expressed as outputs. The controls manage the activity, and mechanisms are required resources to execute the activity.

Figure 1 shows activities specified in the subsection 4.3.1 in the OSHMS standard. Existing activities in Figure 1 are the determination of the source of hazard, the risk assessment, and the establishment of steps to determine how to manage, implement, maintain, and disseminate those steps. Four steps "establishment of steps" "implementation of steps" "maintaining steps" and "dissemination of steps" form the basic structure of activities in all the OSHMS clauses.

2.3. Operational steps of resilience matrix

Figure 2 shows the Resilience Matrix (RM) used by Bracco [1]. In this matrix, Rasmussen's SRK (Skill, Rule, and Knowledge) model and organizational (Individual, Group, and Organization) levels are combined in the three-by-three chart. The steps in the RM are as follows:

Step 1: It is easy to handle signals, which are to be responded to, within normal business operations through skill-based responses.

Step 2: Although a response may occur in a reaction to unpredictable signals, a certain set of operational procedures can nonetheless be followed. Thus, it is possible to respond to certain set rules even in such cases.

Step 3: Even if the signals become harder to predict and handle, the response shifts from rule-based response to knowledge-based response.

Step 4: To respond the signals, individuals must have the superior capability to recognize difficult-to-detect signals. Therefore, it is necessary to share information with other members for the responses in such cases.

Step 5: The response in Central Step should be demonstrated as a new operational procedure in day-to-day training and work towards dropping the signal to a point where it is sufficiently easy to handle with a skill-based response.

Step 6: When the group fails to deliver an adequate knowledge-based response in Step 4, and when the signal can only be managed at a higher level, the relevant information must be reported to the organization for it to respond.

Step 7: New and difficult-to-handle signals flow in from groups in Step 5. In the case of too many incoming signals, the organization should create an operation procedure and respond by changing or updating the operational manual and the regulations.

Step 8: When operational procedures are strengthened to handle weaker signals, the new operational procedure is adopted.

Central Step: If the signals being responded to can be handled through group-level operational procedures, the response should be formed by applying business operations, regulations, and precedents

By the end of this resilience cycle, a new signal is incorporated into regulations, procedures, and precedents, thus becoming standardized. It is subsequently part of the routine business of frontline operators.

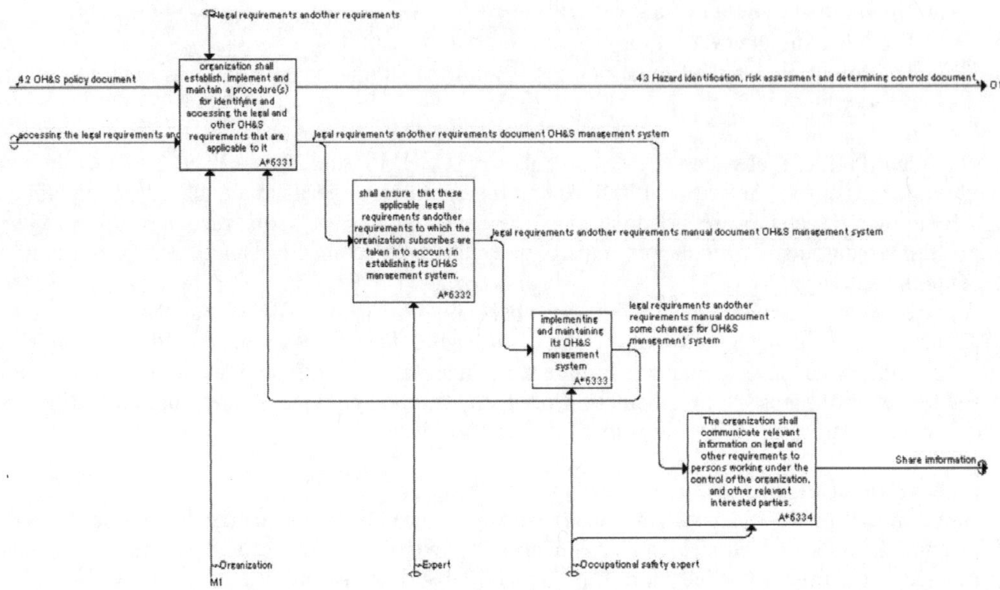

Figure 1. Sample IDEF0 diagram of OSHMS clause.

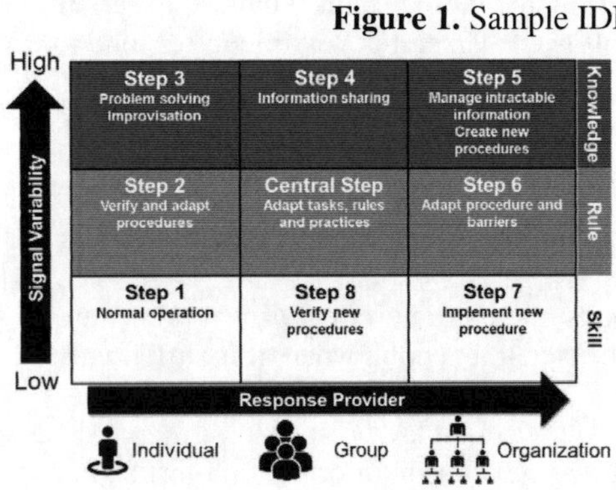

Figure 2. Resilience matrix.

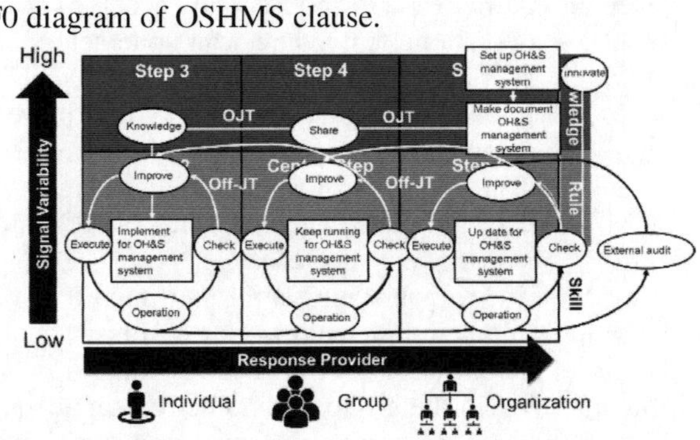

Figure 3. Resilience organization model.

2.4. Resilience matrix based OSHMS-PDCA cycle

In the factory, every unit work estimates for hazard detection. The RM can be applied to show each of unit work PDCA cycle in OSHMS standard. The risk assessment group and individual. The RM can be defined as a cycle to develop new operational procedures, for individuals to implement it, and provide feedback regarding its ease of use as a group to maintain and increase organizational resilience. Furthermore, it covers activities that occur during establishment, implementation, and maintenance. Accordingly, it seems to be appropriate to place the IDEF0 model (See Figure 1) of the OSHMS standard into the RM to specify the structure of the organizational activity cycle and identify inherent activities.

When an activity to disseminate informational output does not link up with the next activity, the production plant personnel of the organization cannot determine how to use the disseminated information based on the OSHMS clauses. Therefore, each official of the organization uses his/her discretion in using the disseminated information. Based on these issues it can be assumed that a resilient organizational model should look like that shown in Figure 3.

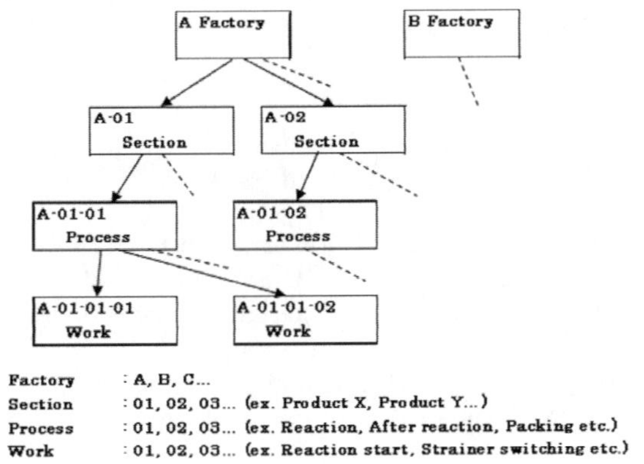

Figure 4. Tree diagram from factory to work.

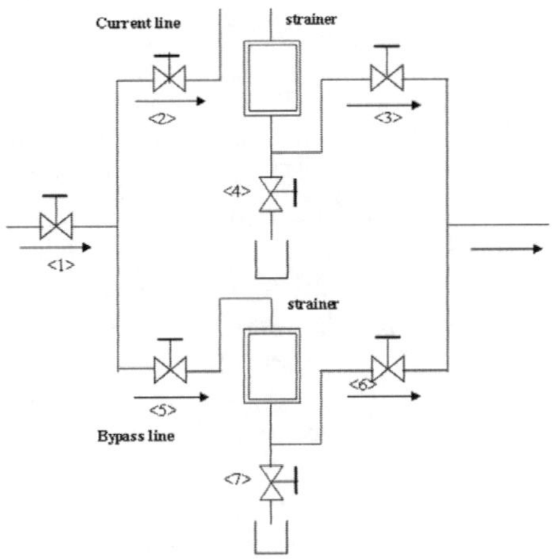

Figure 5. The structure of strainer.

PDCA cycles are shown in Figure 3 that help disseminate the OSHMS standard throughout the chemical industry. It is necessary to establish section-based PDCA cycles within the individual, group, and corporate level to run the company-wide PDCA cycles (shown in dotted line) continually without failure.

Each section goes through the implementation - maintenance - improvement stages. Subsequently, the improvement must be checked and tested. If it passes the check, one can move on to the next stage. If the improvement fails the check, one goes back to an improvement section, provides safety instructions, and returns to the cycle.

If the cycle does not work, Off-JT or OJT has to be organized as follows;

Off-JT: Education of the OHSMS standard applied to departments from the organization and to individuals from their department.

OJT: Education of own safety experience in each response provider

3. Risk assessment

Figure 6 showed a degree of potential risk. The resilience is estimated using the frequency and the degree of harm. The scale of the frequency and the degree of harm will be defined as the company's rule. The minimum potential risk is one, the maximum potential risk is 25, and higher the potential risk, the lower the resilience. We think about the education on the Rule-based, Skill-based, Knowledge bases on Table 2.

Rule-bases: Work frequency "small" and degree of harm "small."
Skill-based: Work frequency "middle" and degree of harm "middle."
Knowledge-bases: Work frequency "large" and degree of harm "large."

3.1. Identification of the work in the production process

Before the risk assessment of works in the production process, all of the works are identified to prevent the oversight. The whole structure in the factory is described as the tree diagram from the factory as the start point to the unit work in Figure 4 [12].

Figure 4 shows a typical structure of works in a factory. For developing RM, a factory in Figure 4 is an equivalent to an organization in RM. Section and unit process is reflexed as a group. Work is specified as an individual activity.

3.2. Unit work and unit operation to estimate the potential risk

All works are described according to the operation procedure of the following example [12]. In the example of the strainer to remove the foreign objects in the flow (See Figure 5), the unit work is described as the sequence of the unit operation from (1) to (7).

(1) Check and close the valve <7> on purge drain line of bypass line.
Hazard: If it opens, spout the liquid (Harm: Soaking up the liquid).
(2) Open outlet valve <6> and inlet valve <5> sequentially, and start feeding to the bypass line.
No hazard
(3) Close inlet valve <2> and outlet valve <3> sequentially, and stop feeding to the current line.
Hazard: Error of sequence causes the spout of liquid (Harm: Soaking up the liquid).
(4) Open the valve <4> slowly on purge drain line of former line, and reduce the residual pressure. Receive the residual using a bucket.
Hazard: If the bucket is too small, the liquid overflows (Harm: Soaking up the liquid).
(5) After the completion of the purge, detach the flange of the strainer, and change the strainer in the former line.
Hazard: If the size of the wrench is wrong, the slip of wrench occurs (Harm: Fall to the ground).
(6) Attach the flange to the former line.
Hazard: If the size of the wrench is wrong, the slip of wrench occurs (Harm: Fall to the ground).
(7) Close the valve <4> on purge drain line of former line.
No hazard

In this example, the potential hazard in work is identified in each unit operation (1), (3), (4), (5) and (6).

4. Estimation of resilience using the result of risk assessment

The risk assessment is executed on every unit work, and at first, the frequency of work and the degree of harm about the potential hazard are estimated. The resilience is estimated using the frequency and the degree of harm. Higher frequency of work or higher degree of harm brings to lower R_{PU}.

The degree of potential risk is shown as the component of Table 2. The scale of the frequency and the degree of harm will be defined as the company's rule. In Table 2, the minimum potential risk is 1; the maximum potential risk is 25, Component of the matrix (i, j) equals (i×j). According to the component of Figure 6, the index of resilience, hereafter referred to as IR, is defined as the inverse of the potential risk, (i.e. IR=1/potential risk).

Table 2. The matrix of potential risk.

Frequency (small → large)					
5	5	10	15	20	25
4	4	8	12	16	20
3	3	6	9	12	15
2	2	4	6	8	10
1	1	2	3	4	5
	1	2	3	4	5

small Degree of harm large

In the production process, the work of the production plant personnel reaches to all production activities concerning the production units. Therefore, the combination of IR calculated by the result of risk assessment and R_{PPP} gives R_{PU}. One production process contains several hundred or thousands of unit works, and it is possible to decide the degree of the potential risk to each unit operation. Consequently, the highest degree of the potential risk is the representative one of the whole production process. Therefore, R_{PU} is as follows.

$$R_{PU}=R_{PPP}\times \text{Minimum Value of IR} \tag{1}$$

If the degree of potential risk in the works is intolerable, the measure of risk reduction is practiced. Therefore, as the degree of risk decreases gradually, IR of the production units becomes higher by degrees. In the example (See Table 2), IR is from 0.04 to one, and R_{PU} is not greater than that of the production plant personnel. If all unit works have the minimum risk, R_{PU} equals to that of the production plant personnel (R_{PPP}). Here, in the case that R_{PPP} is 0.993 [6], suppose that the maximum value of the potential risk in work equals to 16 (Both of the frequency of work and the degree of harm equal to 4.), R_{PU} is as follows.

$$R_{PU}=0.993\times\frac{1}{16}=0.062 \tag{2}$$

4.1. Total resilience on the production process

Three components of the production process, (i) Production Units, (ii) Production Plant Personnel, (iii) PSS, have own resilience respectively. Moreover, it is possible to estimate them quantitatively. The most important resilience is RPPP. This resilience is calculated using the work hours devoted to skills (work hours engaged in the daily routine), hereafter referred to as WHDS, and the level of skills and knowledge concerning the production plant personnel [7]. The skills are classified as the three types of factors: operational skills that play an important role for the use of mainly hands and legs, memory skills that take a significant part for the use of the memory for achieving smooth operation and communication skills that have a considerable place for the attainment of effective mutual communication.

WHDS are these three factors as the duration of daily work of each production plant personnel, and the work hour is measured, and the proportion of each skill is analyzed in the daily routine. As the production plant personnel has an appropriate job according to the position in the workgroup (suitability of skill), WHDS should be weighted. The level of the skills and knowledge contains a factor of position and a factor of effort in staff duty.

$$\text{Weighted WHDS}=\text{WHDS}\times\text{suitability of skill } \textit{about} \text{ the position of staff member} \tag{3}$$

$$\text{Level of skills and knowledge}=\text{Position of the staff member}+\text{Effort in the staff duty} \tag{4}$$

Each production process has a different number of staff members, and each production plant personnel has different WHDS. Therefore the resilience is estimated by the level of skills and knowledge weighted by WHDS, i.e. R_{PPP} is the weighted-average of the level of skills and knowledge of the production plant personnel.

Each production plant personnel has own work hours in the daily routine and own WHDS; therefore, a weighted-average of the level of skills and knowledge is used to indicate R_{PPP}.

$$R_{PPP}=\frac{\sum(\text{Weighted WHDS}\times\text{Level of skills and knowledge})}{\sum(\text{Weighted WHDS})} \tag{5}$$

R_{PPP} is available to estimate the resilience regardless of the number of production plant personnel. Because a high level of skills and knowledge causes high R_{PPP}, it is indispensable to provide the education and training to improve the capability of the production plant personnel. Moreover, if the production plant personnel involves in an appropriate job for the position in the workgroup, the total work hours in the work group will be optimal. Also, the margin of work hours to devote to the education and training will be created [6, 7].

In PSS, several techniques that are the formalized skills of the production plant personnel are implemented. According to the proportion of a job replacement by PSS, PSS can take over the level of skills and knowledge [8]. R_{PPP} is the weighted-average of the level of skills and knowledge of the production plant personnel using the weighted WHDS. Therefore, R_{PSS} is deduced by the weighted-average of the replaceable level of skills and knowledge by PSS using the replaceable WHDS. R_{PSS} is as follows:

$$R_{PSS}=\frac{\sum(\text{Replaceable WHDS}\times\text{Replaceable Level of skills and knowledge})}{\sum(\text{Replaceable WHDS})} \tag{6}$$

4.2. Result total resilience on the production process

Therefore, if all skills in the production process are replaced by the function of PSS, i.e. all skills of the production plant personnel are transferred to PSS, and if the production plant personnel obtain the margin of time to master the new skills, R_{PPP} will become higher. Moreover, equation (1) shows that R_{PU} is calculated by R_{PPP} and the result of risk assessment. The higher the harm of potential risk, the lower R_{PU}. Consequently, to keep higher resilience on the production process, following measures are effective.

- The periodical and continuous education and training for the production plant personnel is effective to improve R_{PPP}.
- It is possible to reduce the total work hours in the work group by the practice of an appropriate job routine for the position. Therefore, the job reallocation should be considered periodically.
- It is necessary to implement the formalized skills as the technique in PSS to reduce the work hours of the production plant personnel.
- It is possible to use the margin of work hours to master the new skills and to practice a higher level of works.

After the estimation of potential risk in the production process, the reduction of risk to improve R_{PU} is necessary.

5. Conclusions

In the production process, the resilience that restrains the disruptive signal to prevent the accident and trouble. Analysis of the OSHMS, for a method for evaluating how OSHMS functions are systematically installed in the production field. In addition to R_{PPP} and R_{PSS}, a metric to estimate R_{PU} is proposed in this paper. The new and original contribution of this paper is that the result of the risk assessment is usable in the evaluation of the resilience of the production unit.

The measures based on the quantitative estimation that enables to improve the total resilience on the production process i.e. to increase the capacity to restrain the disruptive signal will progress the total productivity gradually of the production process.

Acknowledgments

the Ministry of Education partially supported this research, Science, Sports and Culture, Grant-in-Aid for Scientific Research (A), No.16H01837 (2016), and Council for Science, Technology and Innovation (CSTI), Cross-ministerial Strategic Innovation Promotion Program (SIP), "Cyber- Security for Critical Infrastructure" (funding agency: NEDO).

References

[1] Bracco F, Piccinno T and Dorigatti G 2013 *Proc. of 5th Symp. on Resilience Eng., Managing Trade-offs* pp 23–8

[2] Eguchi H, Aoyama T, Seki K, O'Donovan D and Koshijima I 2014 *Proc. of Joint 7th Int. Conf. on Soft Computing and Intelligent Systems and 15th Int. Symp. on Advanced Intelligent Systems* pp 861-866

[3] Eguchi H, Aoyama T, Seki K, O'Donovan D, and Koshijima I 2015 *J. JIIAE* **3** (3) pp 126–33

[4] Eguchi H, Aoyama T, Seki K, O'Donovan D and Koshijima I 2015 *J. JESTEC* pp 30–40
[5] OSHMS Project Group 2007 *OHSAS18001:2007: Occupational Health and Safety management systems* (London: BSI) pp 90–179
[6] Eguchi H, Aoyama T, Seki K, O'Donovan D and Koshijima I 2016 *J. CEJ* **49** (1) pp 1–7
[7] Eguchi H, Aoyama T, Seki K, O'Donovan D and Koshijima I 2016 *J. CEJ* **49** (7) pp 673–679
[8] Work Group of Risk Management Standard 2010 *ISO31000: 2009 Risk Management Guide* (Tokyo: Japanese Standards Association)
[9] Eguchi H 2011 *Kagakukougaku Ronbunshu* **37** (6) pp 532–8
[10] Masunaga S 2007 *Risk Gaku Nyumon* **5** (Iwanami Shoten). Tokyo
[11] Rasmussen J 1983 *IEEE Transactions on Systems, Man, and Cybernetics* pp 257–66
[12] Eguchi H and O'Donovan D 2010 *Kagakukougaku Ronbunshu* **43** (5) pp 435–42

New Acid Combination for a Successful Sandstone Acidizing

M U Shafiq[1], H K B Mahmud[1] and R Rezaee[2]

[1]Petroleum Engineering Department, Curtin University, Sarawak, Malaysia
[2]Department of Petroleum Engineering, Curtin University, Kent Street, Bentley, Perth, Australia

E-mail: umer.engr@hotmail.com

Abstract. With the development of new enhanced oil recovery techniques, sandstone acidizing has been introduced and played a pivotal role in the petroleum industry. Different acid combinations have been applied, which react with the formation, dissolve the soluble particles; thus increase the production of hydrocarbons. To solve the problems which occurred using current preflush sandstone acidizing technology (hydrochloric acid); a new acid combination has been developed. Core flooding experiments on sandstone core samples with dimensions 1.5 in. × 3 in. were conducted at a flow rate of 2 cm^3/min. A series of hydrochloric-acetic acid mixtures with different ratios were tested under 150°F temperature. The core flooding experiments performed are aimed to dissolve carbonate, sodium, potassium and calcium particles from the core samples. These experiments are followed by few important tests which include, porosity-permeability, pH value, Inductively Coupled Plasma (ICP) analysis and Nuclear Magnetic Resonance (NMR measurements). All the results are compared with the results of conventional hydrochloric acid technology. NMR and porosity analysis concluded that the new acid combination is more effective in creating fresh pore spaces and thus increasing the reservoir permeability. It can be seen from the pore distribution before and after the acidizing. Prior applying acid; the large size of pores appears most frequently in the pore distribution while with the applied acid, it was found that the small pore size is most the predominant of the pore distribution. These results are validated using ICP analysis which shows the effective removal of calcium and other positive ions from the core sample. This study concludes that the combination of acetic-hydrochloric acid can be a potential candidate for the preflush stage of sandstone acidizing at high temperature reservoirs.

1. Introduction

Acidizing is a process in which an unsteady, nonisothermal, 3D flow in porous media cause reaction of minerals and chemicals. Due to the dissolution of rock matrix and reprecipitation of formation products, acidizing can lead to continuous variation in the porosity and permeability of a formation. The main goal of acidizing a sandstone reservoir is to remove the formation damage caused by drilling, completion or production processes [1]. Acidizing can restore the original permeability of the formation or create new pores to increase the permeability.

Three steps are involved when performing acidizing operation for sandstone reservoirs; preflush, main flush and after flush. The mixture of hydrochloric acid (HCl) and hydrofluoric acid (HF), known as mud acid [2, 3] is normally used for the acidizing process to remove the damage around the wellbore. This acid has been applied by [4, 5] and results are compared with other acids. Hydrochloric acid has been applied in the preflush stage to dissolve carbonates and other positive ions [6, 7, 8, 1]. The primary reaction is presented in equation 1 [9]:

$$HCl + CaCO_3 \longrightarrow CaCl_2 + H_2O + CO_2 \quad (1)$$

However, the success rate of using hydrochloric acid varies during the preflush stage, even damages are reported by field studies during this stage [10]. The preflush problems are mainly caused by the disadvantages of hydrochloric acid. First is the presence of illite which is sensitive to HCl acid, this may cause serious issues by choking the pore throat and thus decreasing the permeability and porosity. Secondly, at high-temperature conditions, corrosion may occur and also the reaction rate will be high which may cause other problems like precipitation. Thirdly, an ineffective preflush stage may cause secondary and tertiary reaction at later stages of acidizing [9].

Because of so many problems faced using HCl acid, another approach has been studied and used as an alternative to HCl acid in the preflush stage. This new acid system is a combination of hydrochloric and acetic acid, experimented by [11]. The main aim of this study is to overcome the problems related to hydrochloric acid. While the advantages of hydrochloric acid presented by McLeod [12] should be maintained. In this study different methods such porosity and NMR measurements, pH value monitoring, and ICP analysis has been performed to monitor rock and fluid alterations at different stages of acidizing process.

2. Theory

2.1. Nuclear Magnetic Resonance

Nuclear magnetic resonance (NMR) has been widely used in chemistry, physics, and biomedicine and in clinical diagnosis for imaging the internal structure of the human body. The same principle applied in clinical imaging can also be applied for the imaging of fluid-saturated porous rock, such as reservoir rock. "NMR logging is a subcategory of electromagnetic logging [13], that can measure the induced magnet moment of hydrogen nuclei (protons) contained within the fluid-filled pore space of porous media (reservoir rocks)". Conventional logging measurements (e.g., acoustic, density, neutron, and resistivity), respond to both rock matrix and fluid properties and are strongly mineralogy dependent while NMR-logging respond to the presence of hydrogen protons. Information related to the quantities of fluids present, the properties of these fluids, and the sizes of the pores containing these fluids can be obtained using NMR analysis. From this information, it is possible to infer or estimate.

- The volume (porosity) and distribution (permeability) of the rock pore space
- Rock composition
- Type and quantity of fluid hydrocarbons
- Hydrocarbon producibility

2.1.1 BVI determination.

Determination of Bulk volume irreducible (BVI) of water in the rock sample is one of the most widely used applications of NMR logging. BVI represents the immovable or bound water in a formation. BVI is a function of pore-throat size distribution where high threshold pressure due to smaller pore throats retains the fluids in the pores. Figure 1, 2 and 3 explain the different aspects of NMR logging corresponding to T2 time decay (relaxation time).

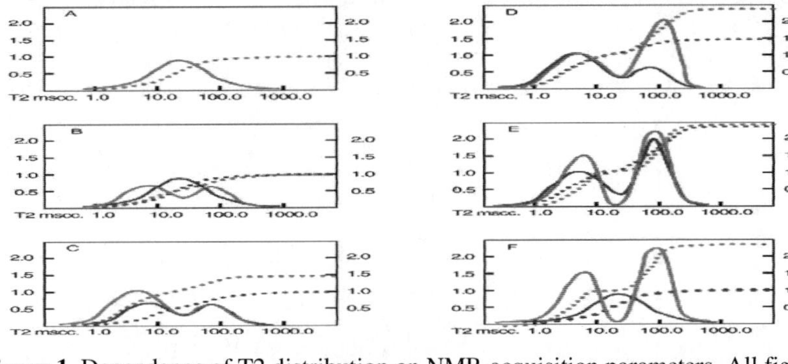

Figure 1. Dependence of T2 distribution on NMR-acquisition parameters. All figures, represents previous results and new results, solid lines represent incremental T2 (read on the left axis), and dashed lines represent cumulative T2 vs. T2 relaxation time (read on the right axis) [13]

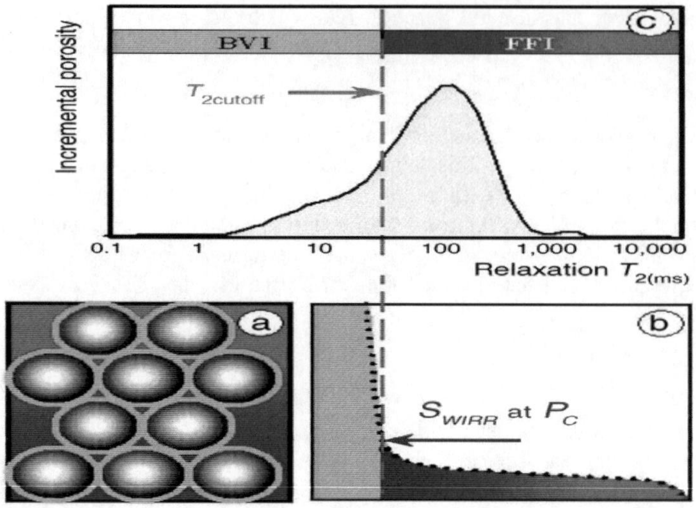

Figure 2. Correlating NMR T2 cutoff to capillary pressure: (a) represents grains, capillary-bound water and free fluids in a gradient tone blue/red; (b) capillary pressure curve (black dots) defines capillary pressure (Pc) in which water saturation becomes irreducible (Swirr); (c) a single-porosity cutoff value on the T2 distribution, corresponding to this pressure [13].

BVI can be used for indicating or estimating permeability and producibility. There are two methods currently used for BVI determination. First one is the cutoff-BVI (CBVI) model, based on a fixed-T2 value (T2cutoff) which divides the free-fluid index (FFI) and BVI. The second approach is the spectral BVI (SBVI) model, which assumes the coexistence of free and bound fluids in any pore described by the T2 distribution at water saturation (Sw) = 100% [13]. The SBVI method is used primarily for quantifying movable water and, secondarily, for estimating permeability. This research based on the CBVI model which divides BVI and FFI.

2.1.2 CBVI (cutoff) model.

The T2 signal from the rock matrix is actually very rapid, even modern logging tools cannot detect it. Accordingly, the recorded T2 distribution only represents the porosity occupied by immovable BVI, movable (FFI) water and clay-bound-water (CBW) components. Assuming that the pore size controlled the mobility of reservoir fluids (i.e., the producible fluids reside in large pores and the immobile fluids reside in small pores, a fixed-T_2 value can be related directly to a pore size at or below which fluids will not flow. This value (T2cutoff) is used to divide the T_2 distribution into movable (i.e., producible or free fluids and FFI) and immovable (i.e., bound-fluid, BVI, and CBVI) components Figure 3 [14, 15]. T_2 cutoff is a value that may vary from one formation to another and can be affected by lots of factors such as: Capillary pressure, Lithology, Grain size, Compaction, Pore characteristics.

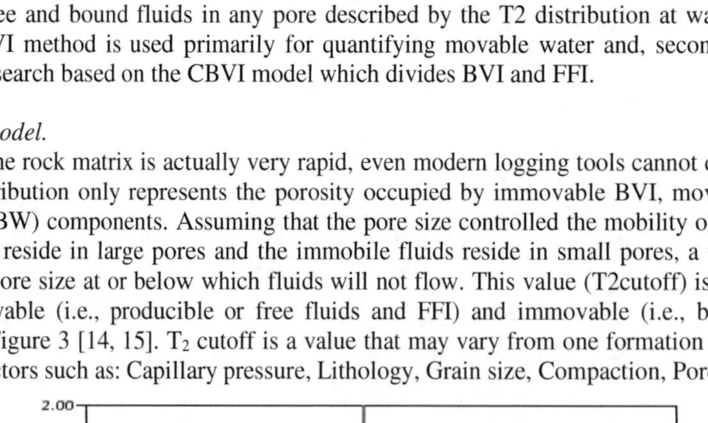

Figure 3. Illustration of the T_2cutoff used in the CBVI model to divide NMR porosity into movable and immovable components.

3. Experimental studies

3.1 Materials

Acetic acid and hydrochloric acid were purchased with concentrations of 90-100% and 37% respectively. Berea Sandstone core plugs (1.5× 3 in) were selected for this study. Core samples mineralogy consists of 93% quartz, 5% feldspar and small quantities of other minerals. All the acid solutions were prepared using distilled water prepared in the laboratory.

3.2 Core flood experiments

The core was first dried at 80°F for 24 hours and then was placed in the core holder. After the core was loaded into the core holder, it was put under a confining pressure of 1000 psi using a syringe pump. The injection rate was 2 cm^3/min in all experiments. The acid was injected at 150°F until the pressure drop was stabilized. Then the injection was stopped and distilled water was injected again to remove any acid present inside the core sample. The compositions of acids used in all core flood experiments on Berea core samples are listed in Table 1.

Table 1. Sample name and composition

Sample Name	Composition
Sample A	15% HCl
Sample B	2.5% CH_3COOH: 15% HCl
Sample C	5% CH_3COOH: 10% HCl
Sample D	3% CH_3COOH: 12%HCl

4. Result and Analysis

4.1. Porosity Analysis

The porosity is a measure of void spaces in the reservoir and gives information about the storage capacity of the reservoir rock. The porosity is one of the most important parameter for understanding the oil and gas initial in place. Determination of the porosity can be done in many ways i.e., manually and automatically. In this research Automatic Porosity Permeability Apparatus (Figure 4) is used to determine porosity and permeability values. Table 2 characterise the initial and final permeability and porosity values. The initial values are obtained before the core flooding treatment while the final values are recorded after the core flooding.

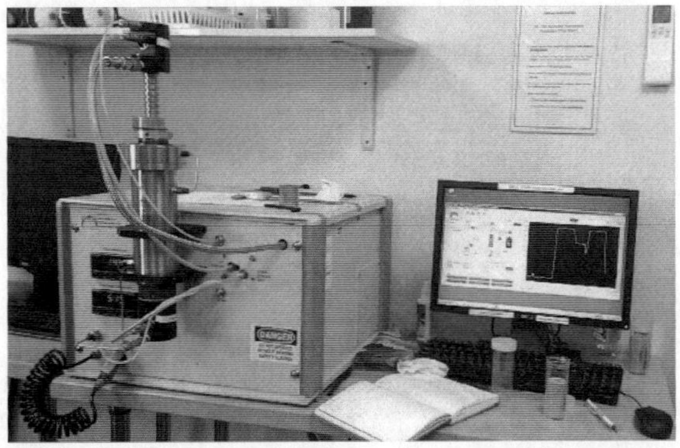

Figure 4. Automatic Porosity permeability apparatus

The main aim of the preflush stage is to remove the calcium, sodium and potassium ions that are present in the core sample. Creating new pores by dissolving these minerals increase the porosity but may have a little effect on the permeability. Therefore, the major objective of using preflush acid is to remove positive ions (Calcium, magnesium, potassium and sodium); and increase the sample porosity. Usually, sandstone samples contain very less amount of sodium, potassium and calcium ions, therefore; as a result, the change in porosity is also less. The removal of these ions is very necessary in order to get better acidizing results because these minerals can cause precipitation later during the main acid stage.

Table 2. Porosity (ø) and Permeability (K) results

Sample Name	Initial ø	Final ø	% change
Sample A	19.567	20.613	5.34
Sample B	19.659	20.652	5.05
Sample C	19.522	20.741	6.24
Sample D	19.038	19.925	4.65
Sample Name	Initial K	Final K	% change
Sample A	122.76	152.16	23.94
Sample B	166.32	169.99	02.20
Sample C	113.16	113.23	00.05
Sample D	122.76	137.49	11.99

15% HCl is the base case as this is the standard acid used in industry for the pre-flush stage of sandstone acidizing. The permeability increases when 15% HCl was applied is almost 24% as shown in Table 2, while very less variation in permeability has been observed using new acid combinations proposed. The permeability is not the major concern at this stage of acidizing. On the other hand, the porosity results are quite favourable. The change in the porosity when the new combination (5% CH_3COOH: 10% HCl) was applied is 6.2% as shown in Table 2, which is almost 1% more than the base case.

4.2 NMR measurements

NMR measurement has been applied on all four core samples before and after acidizing to investigate the change in pore size distribution using CBVI model. The results can be used to get a better understanding of the change in porosity of each sample after acidizing. Figures 5 and 6 represent the T2 spectrum and cumulative spectrum for each core sample before and after acidizing. Where series 1 (green colour) in each sample represents the initial condition while series 2 (purple colour) is the representative of the reactive sample. The spectrum analysis of each sample is presented in Figure 5 (a, b, c and d) represents that there is a shift in the curve towards the left side. It means that the new small pores have been created and now the relatively smaller pores are dominating the porosity distribution. Although there is an increase in pore space volumes for all samples studied but NMR T_2 distributions indicate that pore sizes decreased for all samples analyzed. The reduction of pore space sizes can be attributed to dissolution of minerals and re-precipitation in larger pores.

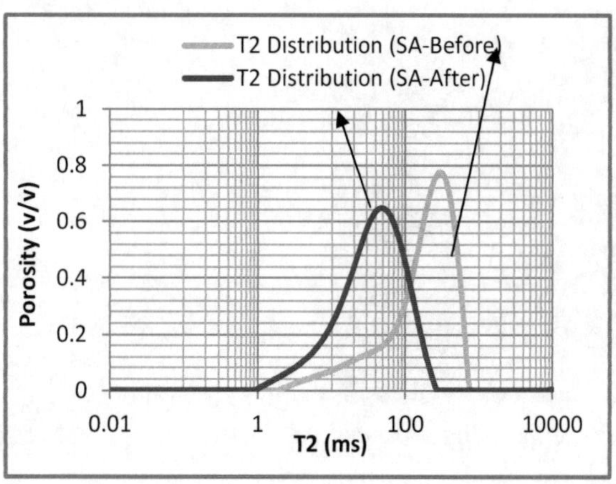

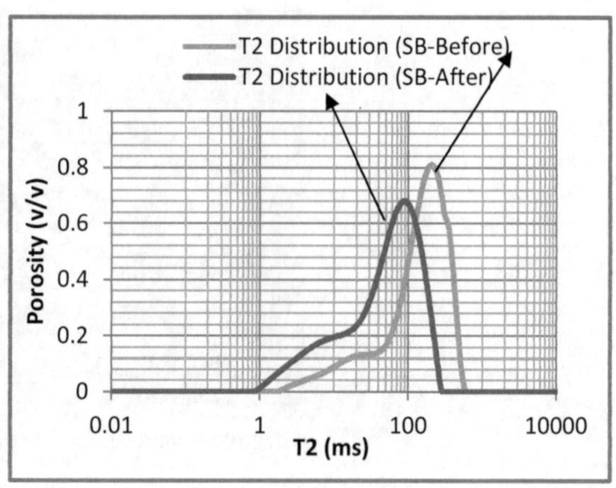

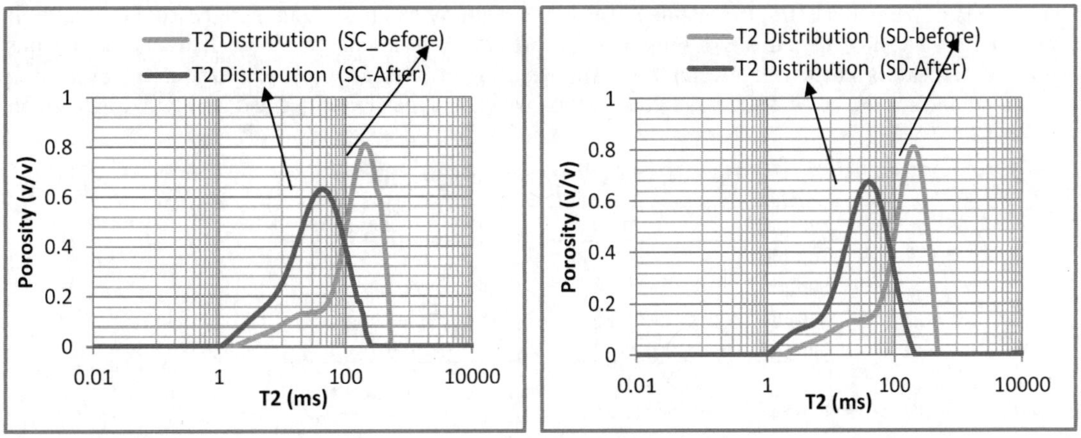

Figure 5. (a) Sample A T_2 spectrum; (b) Sample B T_2 Spectrum; (c) Sample C T_2 spectrum; (d) Sample D T_2 spectrum

The cumulative spectrum Figure 6 (a, b, c, d) clearly represents the increase in the total amount of pores in the sample before and after acidizing with the least change in the cumulative spectrum in sample D (Figure 6d); validating porosity results (Table 2).

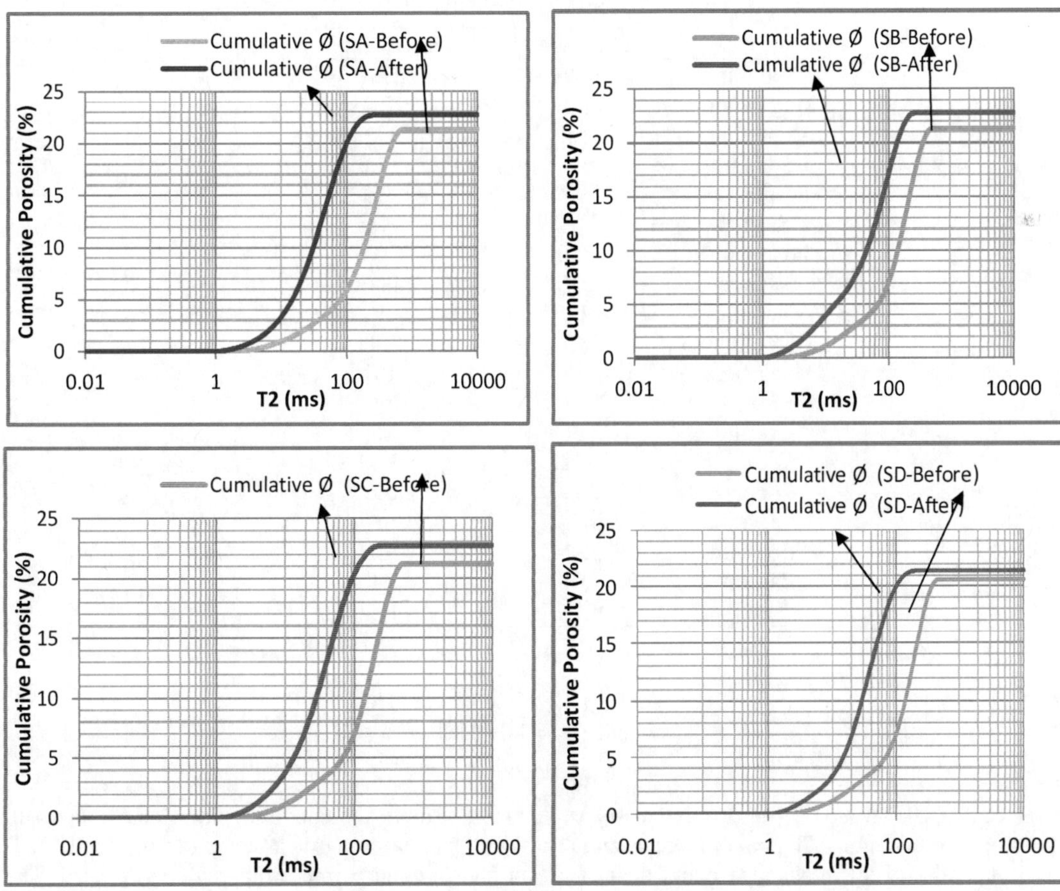

Figure 6. (a) Sample A T_2 commulative spectrum; (b) Sample B T_2 commulative Spectrum; (c) Sample C T_2 commulative spectrum; (d) Sample D T_2 commulative spectrum

IOP Conf. Series: Materials Science and Engineering **206** (2017) 012010 doi:10.1088/1757-899X/206/1/012010

Table 3 explains the porosity results obtained from NMR analysis. Figures 5 and 6 are based on these results. These results are validating the porosity results obtained previously mentioned in Table 2; where the maximum change in the porosity is noted within sample C. The advantage of NMR porosity measurement over general porosity measurement is the knowledge of pore distribution which has been explained and analysed using Figures 5 and 6.

Table 3. Porosity Results using NMR.

Sample Name	Initial ø	Final ø	% change
Sample A	21.31	22.71	6.57
Sample B	21.34	22.76	6.65
Sample C	21.17	22.70	7.22
Sample D	20.59	21.40	3.94

4.3 Inductively Coupled Plasma (ICP) Analysis

ICP analysis is used to validate the porosity results. This test investigates the concentration of different ions dissolved by the acid during the reaction. In this test core effluent samples (reacted acid) were utilized to determine the concentrations of Si, Ca, Na, K, Fe and Mg present in the acid after the reaction. The results presented are in parts per million (ppm). From the analysis presented in Table 4, it can be seen clearly that the maximum amount of each mineral is dissolved when the Concentration (A) is applied. Both samples are collected after 20 minutes of core flood experiments.

Table 4. Concentration of dissolved ions

Element	Conc (A) (ppm)	Conc (B) (ppm)
Sodium	3520	2790
Magnesium	1740	1250
Aluminium	151	97.6
Silicon	144	94.8
Phosphorous	195	152
Sulfur	43.8	48
Potassium	6550	6190
Calcium	6210	4340
Titanium	<1.2	<1.2
Chromium	18.1	8.05
Manganese	246	157
Iron	4420	2690
Cobalt	2.66	1.4
Nickel	17.2	7.12
Copper	7.19	3.2
Zinc	6.79	3.1
Strontium	5.15	5.03
Barium	3.04	2.45
Lead	6.12	4.84

where,
Conc (A) = 5% CH_3COOH: 10% HCl
Conc (B) = 15% HCl

The 5% CH_3COOH: 10% HCl can dissolve dissolves 730 ppm of more sodium, 490 ppm of more magnesium, 360 ppm of more potassium, 1870 ppm of more calcium and 1730 ppm of more iron as compared to 15% HCl. The rest of the minerals are not in abundance and their effect on porosity and permeability can be neglected. These results are validating the porosity values mentioned in Table 2, which showed that increase in porosity is more when 5% CH_3COOH: 10% HCl was applied as compared to 15% HCl. To validate ICP analysis results, next step is to analyse the change in the pH values.

4.4 pH Value Test

The pH value determination is also one of the important parameter considered during the acidizing procedure. It determines the buffer effect of acids and can give the idea of penetration power of the acids. More change in pH value describes the fast reaction of acid which subsequently lead to less penetration. Less change in pH value is favourable but it may also indicate no reaction at the late stage because most of the minerals present in the core sample may be consumed during the early stage of reaction and the rest of the acid is coming out without reacting. This may result in the low pH value of the reacted sample. The results of pH value variation for this study are presented in Table 5.

Table 5. pH value before and after reaction

	Initial pH	Final pH	% change
Sample A	0.06	0.23	283
Sample B	0.055	0.38	590
Sample C	0.09	0.41	355
Sample D	0.067	0.15	123

Sample D shows less pH value change as compared to others. This condition is favourable and an indication that the acid coming out can react further deep into the formation. The change in sample C is also almost equal to the change in sample A which is the base case.

5. Conclusion

It can be concluded that the new acid combination (5%CH_3COOH: 10% HCl) had significant effect on the porosity of the core sample and removed positive ions efficiently. It can prove to be a promising aspect in future preflush sandstone acidizing operations. The application of this acid caused more dissolution of carbonates, calcium, magnesium, and sodium ions which ultimately resulted in more increment of the porosity of core sample as compared to the conventional acid.

References

[1] Ji Q, L Zhou and H A Nasr-El-Din 2014 Acidizing Sandstone Reservoirs Using Fines Control Acid, *Soc. of Petrol. Engineers,* SPE-169395

[2] Gidley J L 1985 Acidizing Sandstone Formations: A Detailed Examination of Recent Experience, *Soc. of Petrol. Engineers,* SPE 14164.

[3] Smith C F and A R Hendrickson 1965 Hydrofluoric Acid Stimulation of Sandstone Reservoirs, *J. of petrol. Technol.* **17** (02), p 215-22

[4] Mian Umer Shafiq, Hisham Khaled Ben Mahmud and Mohamed Ali Hamid 2015 Comparison of Buffer Effect of Different Acids During Sandstone Acidizing, *IOP Conf. Series: Materials Sci. and Eng.*, **78** (2015).

[5] Mian Umer Shafiq and Hisham Khaled Ben Mahmud 2016 An Effective Acid Combination for Enhanced Properties and Corrosion Control of Acidizing Sandstone Formation, *IOP Conf. Series: Materials Sci. and Eng.*, **121** (2016).

[6] Gdanski R 1998 Kinetics of Tertiary Reactions of Hydrofluoric Acid on Aluminosilicates, *SPE Prod. and Facilities,* **13**(02) p 75- 80

[7] Gdanski R D 1999 Kinetics of the Secondary Reaction of HF on Alumino-Silicates, *SPE Prod. and Facilities,* **14** (04) p 260- 8

[8] Gdanski R D 2000 Kinetics of the Primary Reaction of HF on Alumino-Silicates, *SPE Prod. and Facilities,* **15** (04) p 279- 87

[9] A D Hill, Kamy Sepehrnoori 1994 Design of the HCl Preflush in Sandstone Acidizing, *Soc. of Petrol. Engineers,* SPE-21720

[10] Thomas R L, H A Nasr-El-Din, S Mehta, V Hilab and J D Lynn 2002 The Impact of HCl to HF Ratio on Hydrated Silica Formation During the Acidizing of a High Temperature Sandstone Gas Reservoir in Saudi Arabia, *Soc. of Petrol. Engineers,* SPE-77370

[11] Shafiq M U and M T Shuker 2013 Finding Suitable Acid for Acidizing of Low Permeable Sandstone Formation: A Research, *Soc. of Petrol. Engineers,* SPE-169641

[12] McLeod H O, Jr L B, Ledlow and M V Till 1983 The Planning, Execution, and Evaluation of Acid Treatments in Sandstone Formations, *Soc. of Petrol. Engineers,* SPE-11931

[13] Marschall D M 2000 HBVI: An NMR Method to Determine BVI as a Function of Reservoir Capillarity, *SPWLA 41st Annu. Logging Symp.*, Dallas, 4–7 June. SPWLA-2000-KK.

[14] Timur A 1969 Effective Porosity and Permeability of Sandstones Investigated Through Nuclear Magnetic Principles, *The Log Analyst,* **10** (1): 3.

[15] Coates G R., Marschall, D M Mardon, M et al 1998 A New Characterization of Bulk-Volume Irreducible Using Magnetic Resonance, *The Log Analyst,* **39** (1): 51

IOP Conf. Series: Materials Science and Engineering **206** (2017) 012011 doi:10.1088/1757-899X/206/1/012011

Surface tension profiles of nanofluid containing surfactant during microwave irradiation

K Tanaka[1], Y Asakuma[1], A Saptoro[2] and C Phan[3]

[1]Department of Chemical Engineering, University of Hyogo, 2167 Shosha, Himeji, Hyogo 671-2280, Japan
[2]Department of Chemical Engineering, Curtin University Malaysia, CDT 250, Miri 98009, Sarawak, Malaysia
[3]Department of Chemical Engineering, Curtin University, GPO Box U1987, Perth, WA 6845, Australia

E-mail: asakuma@eng.u-hyogo.ac.jp (Yusuke Asakuma)

Abstract. Manipulation of the surface tension is useful in improving heat and mass transfer performances of nanofluids in thermal systems. In our previous study, the effect of microwave irradiation on the reduction of surface tension of nanofluids (Fe_2O_3) was found even after it was turned off. In this study, a synergistic effect of microwave irradiation and surfactant addition (SDS) was investigated to obtain further surface tension reduction of nanofluid. Experimental results indicate that surfactant addition is effective for wider particle number density in reducing surface tension, and the reduction level strongly depends on the surfactant concentration. On the other hand, effect of the number density on the surface tension reduction is less significant for the same concentration of surfactant. From the obtained data, a combination of microwave irradiation and surfactant addition shows potential to be used as a promising method to manipulate surface tension of nanofluids.

1. Introduction

Nanofluids, which are the suspensions of nano-sized particles in base fluids, have been reported to possess enhanced heat transfer and mass transfer coefficients in improving heat and mass transfer systems. Calculations and analyses of these systems operated with nanofluids are regulated by key physical properties and one of these main properties is surface tension. Surface tension determines the shape of the liquid droplets and consequently, heat and mass transfer across interfaces are strongly influences by surface tension. Therefore, lower surface tension is attractive for promoting higher critical heat flux and boiling rate [1-4] and gas-liquid mass transfer especially in packed bed systems [5,6].

In our previous study [7], microwave memory effect on the surface tension of nanofluid was found for long period of time in which lower surface tension due to microwave treatment could be obtained even after the microwave was turned off. From the experimental data, it was also concluded that irradiation power and time are key factors to obtain better microwave effect. However, higher irradiation power and longer irradiation time are not desirable since they may degrade the nanofluids due to sintering of nanoparticles under superheated condition. Therefore, manipulation of surface tension using microwave irradiation should be carried out under medium operating condition to maintain nanofluid properties.

Manipulation of surface tension can also be performed by adding a chemical additive called surfactant. Surfactant addition has also been acknowledged in enhancing boiling heat transfer [8-10] and improving stability of nanoparticles in aqueous suspensions [11,12]. Nevertheless, there has no study in the literature on the synergistic effect of microwave irradiation and surfactant in reducing surface tension of nanofluid at medium operating conditions. This study, therefore, aims to investigate this synergistic effect with a hypothesis that combined benefits from surfactant addition and microwave memory effect will be able to give better reduction of surface tension.

2. Experimental

2.1. Materials

Fe_2O_3 nanoparticle-water system was selected as a nanofluid in this study. The solution containing surfactant, SDS (sodium dodecyl sulphate) was prepared by dispersing nanoparticles in distilled water at different concentrations of nanoparticles and surfactant. Fe_2O_3 with diameter of 15 nm and SDS were obtained from Kanto Chemicals Co., Japan.

2.2. Apparatus

Fig. 1 shows the microwave apparatus, which was designed by Shikoku Instrumentation Co., Inc., Japan. Pendant method was employed for the measurement of surface tension of nanofluids [7]. Teflon pipe with the dimensions of 1 mm inside diameter and 2 mm outside diameter was used to hang the droplet. Nanofluid was introduced through the pipe by injection using syringe [13,14].

Temperature sensor (model: FL-2000 Optical fiber: FS100-M), was used to measure the droplet temperature during and after microwave irradiation. This probe was inserted from the top of the reactor through the Teflon pipe after the droplet was hung. High speed camera (Sigma Koki Co., LTD Model SK-TC202USB-AT) was installed at the side of the reactor to capture the shape of the droplet. Experimental conditions of the surface tension measurement are listed in Table 1. At moderate conditions, microwave power of 300 W and irradiation time of 120 s [7], the effects of suspension density and surfactant concentration on the surface tension were investigated. Surfactant concentration was normalized by critical micelle concentration (CMC). However, surface tension measurement of nanofluids with concentration near CMC was impossible because it was difficult to hang the droplet. In addition, the measurement of droplet with higher suspension density (1 wt%) was not conducted because particles remains at the bottom of droplet.

2.3. Analytical method

Axisymmetric drop shape analysis (ADSA) was employed to measure surface tension by analyzing the edge profiles of the droplet [15]. Fig. 2 shows an example of raw image and the fitting results. ADSA software estimates the surface tension by fitting the whole droplet profile (Fig. 2(b)). Detailed procedure of shape analysis has been described previously [13].

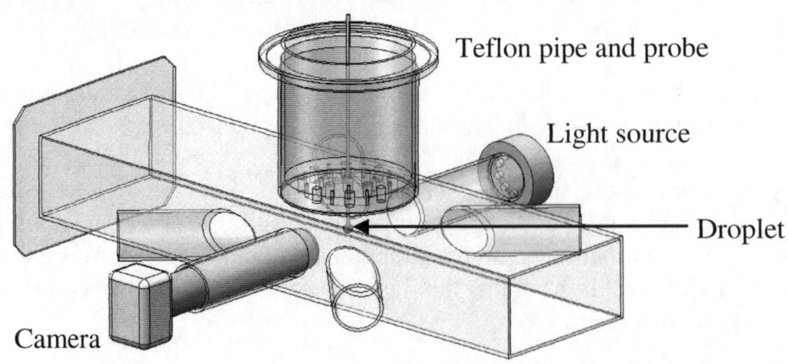

Figure 1. Microwave apparatus for surface tension measurement.

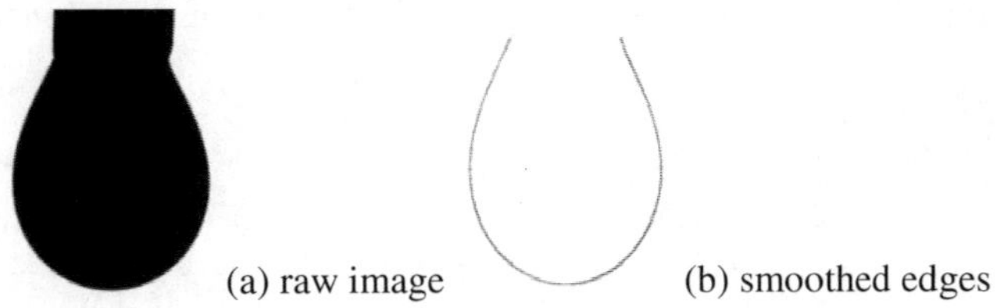

Figure 2. ADSA method for surface tension measurement.

Table 1. Experimental condition

No.	Suspension density [wt%]	Surfactant conc. [c/cmc]
1	0.01	0, 0.01, 0.1
2	0.001, 0.01, 0.1	0.1

3. Results and discussions

Fig. 3 shows surface tension profiles of nanofluids containing surfactant during and after microwave irradiation as a function of time and temperature for the experimental condition 1. Solid and unfilled symbols indicate data during and after the irradiation, respectively. Generally, surface tension during the irradiation decreased with the time because of temperature rise. Initial surface tensions were different depending on the concentrations of added surfactant, From Fig. 3(b), it is obvious that reduction levels of surface tension vary concentrations of surfactant even though maximum temperatures achieved during irradiation were approximately the same. This points out that higher surfactant concentration in the nanofluid resulted in much better surface tension reduction after microwave was turned off.

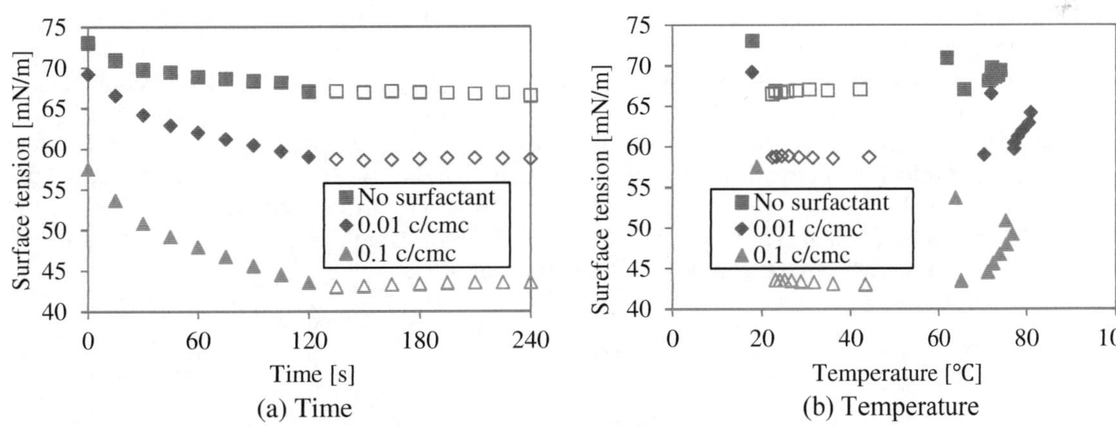

Figure 3. Surface tension profiles at different surfactant concentrations.

The mechanism of surface tension reduction is not well-understood. However, nanobubble formation is hypothesized as a driving force of surface tension reduction after microwave was switched off. Nanoparticle absorbs microwave very well, and the heat generation causes abrupt temperature rise of the solvent around nanoparticles. Then, bubbles are rapidly generated because of the evaporation [14,16]. These nanobubbles became more stable because surfactant molecules are adsorbed on the liquid-air interface. As a result, nanobubbles interrupt the tension along the interface. Fig. 4 shows surface tension of nanofluid at different suspension densities (experimental condition 2). From this figure, the effect of suspension density on the reduction of surface tension tends to be

IOP Conf. Series: Materials Science and Engineering **206** (2017) 012011 doi:10.1088/1757-899X/206/1/012011

neutral. This insignificant effect may indicate that the reduction is mostly influenced by addition of the surfactant.

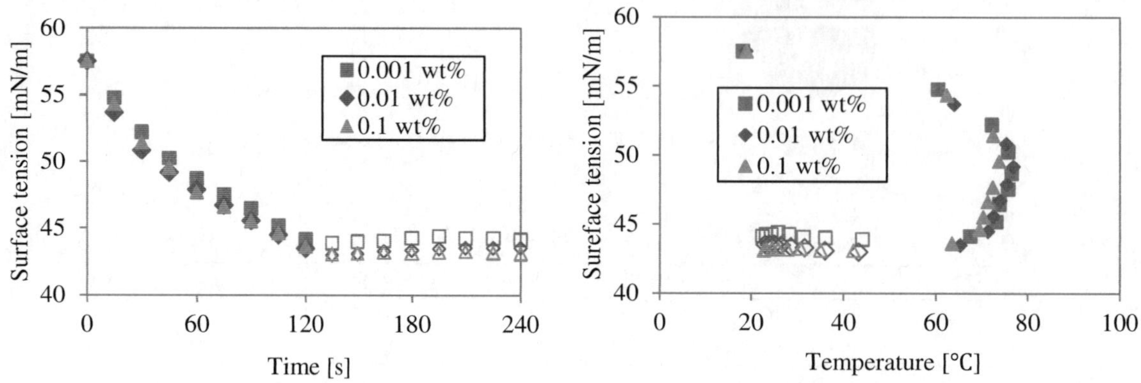

(a) Time (b) Temperature

Figure 4. Surface tension profiles at different suspension densities.

Fig. 5 summarizes surface tension reductions extracted from Figs. 3 and 4. Effect of surfactant additions under microwave irradiation was relatively saturated at around 0.01 c/cmc as shown in Fig. 5(a). Results in Fig. 5(a) also indicate that higher concentration near CMC (more than 0.01 c/cmc) is not necessarily effective. It means that 0.01 c/cmc is an optimum concentration for reducing surface tension of nanofluid. In our previous study, surface tension reduction of nanofluids was around 5 mN/m when surfactant was not added. From Figs. 5(a) and 5(b), it is evident that better surface tension reduction (up to 10 mN/m) of nanofluid with higher suspension density by microwave irradiation can be possibly achieved at moderate surfactant concentration.

The underlying mechanism on surface tension reduction by microwave irradiation is still unclear. Nevertheless, it may be explained as follows. When water is used as a base fluid, the reduction may be driven by "microwave memory effect" [13,16]. Microwave is absorbsed at the interface and consequently polar molecules heavily vibrate. Clusters of hydrogen bonds or nano-bubbles were hypothesized as one of reasons. The presence of nanoparticle in water further enhances the microwave energy absorbance. This absorbance greatly influences the surface tension [17] and causes a new conformation of molecules network [18]. Surfactant itself is adsorbed at liquid-air interface of the cluster and nano-bubble during the irradiation. These absorptions further contribute stability of the interface. As shown by Figs. 3(a) and 4(a), the surface tension reduction by the irradiation lasts for an extended period of time when nano-fluids includes surfactant.

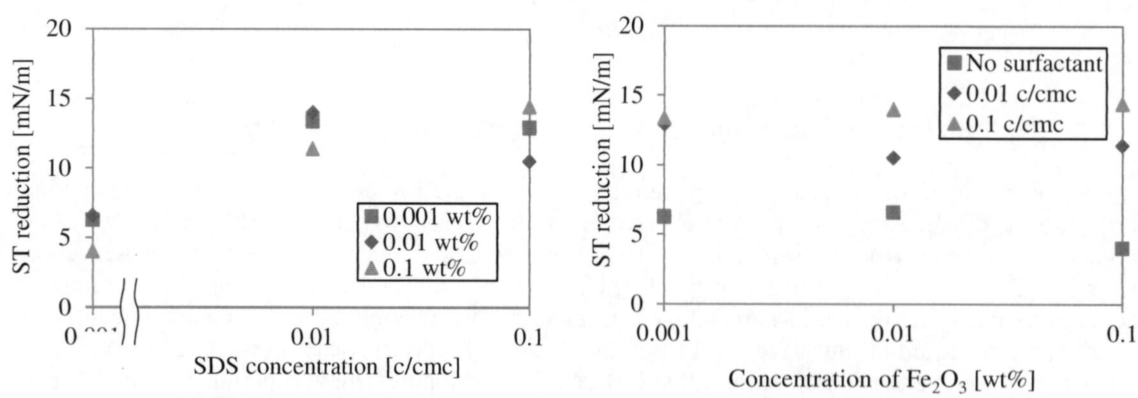

(a) Surfactant concentration (b) Nanoparticle concentration

Figure 5. Surface tension reduction after microwave was turned off.

IOP Conf. Series: Materials Science and Engineering **206** (2017) 012011 doi:10.1088/1757-899X/206/1/012011

4. Conclusions

Surface tension of Fe_2O_3 nanofluid containing surfactant during and after microwave irradiation was investigated at different surfactant concentrations and suspension densities. From the experimental data, it is apparent that a combination of microwave irradiation and surfactant addition is effective to give significant surface tension reduction. Experimental results also indicate that surfactant addition is effective for wider particle number density in reducing surface tension, and the reduction level strongly depends on the surfactant concentration. On the other hand, effect of the number density on the surface tension reduction is less significant for the same concentration of surfactant.

5. References

[1] Kakaç S, Pramuanjaroenkij A 2009 *Int. J. Heat Mass Tran.*, **52** 3187
[2] Murshed S M S, Castro C A N, Lourenço M J V, Lopes M L M, Santos F J V 2011 *Renew. Sust. Energ. Rev.* **15** 2342
[3] Wang X Q, Mujumdar A S 2007 *Int. J. Therm. Sci.,* **46** 1
[4] Wu J M, Zhao J 2013 *Prog. Nucl. Energ.* **66** 13
[5] Stemmet, C.P., Bartelds, F., Schaaf, J., Kuster, B.F.M., Schouten, J.C. (2008). Influence of liquid viscosity and surface tension on the gas-liquid mass transfer coefficient for solid foam packings in co-current two-place flow, *Chem. Eng. Res. Des.*, 86, 1094-1106.
[6] Wu H 2014 *Chem. Eng. J.* **15** 74
[7] Saptoro A, Kanazawa Y, Asada M, Asakuma Y, Phan C 2016 *Exp. Therm. Fluid Sci.* **72** 228
[8] Wu W T, Yang Y M, Maa J R 1998 *Exp. Therm. Fluid Sci.* **18** 195
[9] Hetsroni, G., Zakin, J.L., Lin, Z., Mosyak, A., Pancallo, E.A., Rozenblit, R. (2001). The effect of surfactant on bubble growth, wall thermal patterns and heat transfer in pool-boiling, *Int. J. Heat Mass Tran.*, 44, 485-497.
[10] Peng H, Ding G, Hu H 2011, *Exp. Therm. Fluid Sci.* **35** 960
[11] Ghadimi A, Saidur R, Metselaar H S C 2011 *Int. J. Heat Mass Tran.* **54** 4051
[12] Li Y, Zhou J, Tung S, Schneider E, Xi S 2009 *Powder Technol.* **196** 89
[13] Parmar H, Asada M, Kanazawa Y, Asakuma Y, Phan C M, Pareek V, Evans G M 2014 *Langmuir* **30** 9875
[14] Asakuma Y, Nakata R, Asada M, Kanazawa Y, Phan C 2016 *Int. J. Heat Mass Tran.* **103** 13
[15] Hoorfar M, Neumann A W 2004 *J. Adhesion* **80** 727
[16] Asakuma Y, Munenaga T, Nakata R 2016 *Heat and Mass transfer* **52** 1833
[17] Cai R, Yang H, He J, Zhu W 2009 *J. Mol. Struct.* **938** 15
[18] Taylor L S 1981 *Bioelectromagnetics* **2** 259

IOP Conf. Series: Materials Science and Engineering **206** (2017) 012012 doi:10.1088/1757-899X/206/1/012012

Formation, analysis and characterization of wood pyrolyzed oil

Omprakash K. Mahadwad[1], Dhanashree D. Wagh[2], Prasad L. Kokil[3]

[1,2]Department of Chemical Engineering, MGM's Jawaharlal Nehru Engineering College, N-6 CIDCO, Aurangabad 431003, Maharashtra, India

[3]Sanjay Techno Plast Pvt. Ltd., M-31 MIDC Waluj, Aurangabad 431136, Maharashtra, India

E-mail: mahadwad@gmail.com

Abstract. Pyrolysis of wood is the possible path for converting biomass to higher valuable products such as bio-oil, bio-char and bio-gas. Bio-oil or liquid biofuels have higher heating values so it can store and transport more conveniently. The by-products bio-char and bio-gas, which can be used to provide heat required in the process. This work focused on the formation, analysis and characterization of bio-oil which was obtained from the mixed wood pyrolysis. A GC-MS technique was used for the determination of families of lighter chemicals form pyrolyzed oil. Karl fisher titration and other analytical methods were used for the characterization of pyrolyzed oil. In all there were sixty-six compounds found in the GC-MS analysis of bio-oil and the major compound was acetic acid (19.06 wt%), formic acid (4.90 wt%) 1,2-benzenediol (4.43 wt%) and furfural (3.46 wt%). Along with this analysis, pyrolyzed oil was characterized by calculating its viscosity, density, calorific value, acid value, fire point, flash point, carbon, hydrogen, nitrogen, ash and water content in it. Most of the above mention properties of bio-oil matches with the properties of crude oil except it show more water content in it.

1. Introduction

Among the different thermo chemical processes, pyrolysis is a emerging and promising route to produce bio-oil or pyrolysis oil and value added chemicals from solid biomass feed stock (Bridgwater et al., 1999; Dinesh et al., 2006). Pyrolysis is essentially the thermal decomposition of organic matter under inert atmospheric conditions or in a limited supply of air, leading to the release of volatiles and formation of char. The resulting bio-oil obtained with the yield of 70 to 80 wt. % (dry feed basis) depending upon the cellulose and lignin present in the wood material (Demirbas, 2009; Lu et al., 2009).

Bio-oils are derived primarily from depolarization and fragmentation reactions of cellulose, hemicelluloses and lignin, resulting in its composition and properties of considerable difference from those of petroleum based fuel oils. Bio-oil appears in dark brown viscous state with little sulphur and nitrogen content, but has a high oxygen content nearly 45 wt. % including water. The lower heating value of bio-oil is only 40-45% of that exhibited by the hydrocarbon fuels. Moreover, organic acids in the bio-oils are highly corrosive to common construction materials. Even though hydrocarbon fuels cannot be replace by bio-oil, but its industrial production focused on the manufacture of higher value chemicals and materials other than fuels (Helena and Ralph, 2001; Mckendry, 2002).

Bio-oil is an extremely complex mixture of organic components, including various types of oxygen containing organic compounds derived from pyrolysis of biomass components. Recent detailed analyses shown that pyrolysis bio-oil contain more than 400 compounds, even though it remains a challenge to identify every compound present in it. (Serdar, 2004; Jean et al., 2001).

Because of the complexity, some simplified analytical methods were used to characterize bio-oil. Chemical characterization of pyrolysis oil has generally been based on the fractionation of oil by using solvent extraction, adsorption, molecular distillation, etc. and then the different fractions obtained were further characterized using analytical methods such as GC-MS, HPLC, gel permeation chromatography, etc. with focusing on its different chemical information (Bridgwater et al., 1999; Lu et al., 2009).

The purpose of the present work is to develop an effective quantitative analysis for light weight components of bio-oils by using GC-MS techniques. Thus yields up to 40 wt. % of the bio-oils can be quantified. Another objective of the work is to carry out the fractionation of the bio-oil using distillation technique to separate the light weight and heavy weight components to further increase the yield of bio-oil.

2. Materials and methods

2.1 Materials

Three kinds of biomass feedstock such as pine, teak, neem were selected as pyrolysis materials. All these biomass used in this study were purchased form a local wooden mills in Aurangabad, India. All this feedstock were first air dried and grounded, and then sieved to a mean partial size of 0.26-0.57 mm. Ultimate analysis of the wood samples was determined by using an analyzer. The proximate analysis, the ash content and volatile matter were determined while fixed carbon was determined by difference. The results were shown in Table 1 and Table 2.

Table 1. Ultimate analysis of three biomasses

Sample	C	H	N	O
Pine	50.8	6.4	0.3	42.5
Teak	51.3	6.1	0.2	42.4
Neem	53.9	6.3	0.1	39.7

Table 2. Proximate analysis of three biomasses

Sample	Volatile Matter	Fixed Carbon	Ash Content
Pine	85.3	12.4	2.3
Teak	80.2	18.4	1.4
Neem	77.2	22	1.6

2.2 Preparation of Bio-oils

Direct fast pyrolysis of biomass was carried out in a self designed annular packed bed reactor in Sanjay Techno Plast Pvt. Ltd. The schematic drawing is shown in Figure 1(a).

Figure 1(a). Schematic representation of experimental setup.

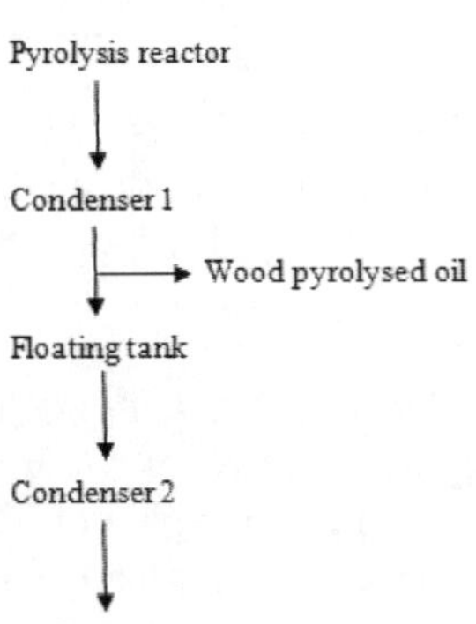

Figure 1(b). Flow diagram of formation of wood pyrolysis oil.

The annular fixed bed reactor was used to carry out the pyrolysis reaction, which can be made open from top and bottom to add and remove the content in it. The inner part of the annular reactor was used as a combustion chamber and outer part was used for the pyrolysis reaction. In combustion chamber 500 gm of wood was inserted with equal amount of all three type of wood. The heat which was generated through the burning of wood was absorbed by combustion chamber and gives it to annular space where pyrolysis reaction was carried out. The pyrolysis reaction was initiated at 200 ^{0}C in the annular reactor and continued till 650 ^{0}C. After 15 min of the reaction the gas (smoke) generated in reactor chamber was fed to the condenser-1,where tar (pyrolysis oil) get settled at the bottom of the condenser and non-condensable gases were fed to the floating tank. Floating tank was used to hold and create the high pressurized gas. Condenser-2 was filled with ¾th level of water and gases coming out from the floating tank were left 100 mm deep inside the water level. Then pressurized gas comes directly in contact with water where again the condensable gases get settled down and the non-condensable gases were feed to burner. Burners were light up continuously till the pyrolysis reaction ends. This process includes five main parts that are shown in flow diagram as follows (Figure. 1(b)).

3. Results and discussion

3.1 Experimental operating conditions and results obtained

Different experiments were conducted to set the parameter values and to obtain the best results. Following Table 3 shows the optimized parameter values used in the experiments to obtain the pyrolysis oil.

Table 3. Process parameters and its values

Parameters	Results
Wood in Annular Space (gm)	20
Initial wood for Burning (gm)	500
Feed rate used for Burning (gm/min)	28
Total Wood for Burning (gm)	4120
Pyrolysis Oil (gm)	4950
Charcoal wt (gm)	7800
Residue Wt (gm)	1450
Tank Lifting Start Time (min)	30
Weight on Tank (kg)	50
Flame Start Time after all Set up Start (min)	58
Burner -1(min)	81
Burner-2 (min)	79
Burner-3 (min)	80

3.2 Qualitative analysis

The main component of the bio-oil is a verity of small molecules with oxygen containing functional groups such as aldehydes, ketones, carboxylic acids, furans, phenols, carbohydrates, etc. (Dobele et al., 2007; Hyeon et al., 2010) Figure 2 shows the total ion chromatogram of bio-oil from mixed wood pyrolysis at 650 ^{0}C and the main compositions corresponding labels in Figure 2 were identified and listed in Table 4. Levoglucosan, a typical pyrolysis product from cellulose pyrolysis, represents incomplete cracking of carbohydrates. A large no of methylated or methoxylated phenolic compounds such as 2-methoxy-phenol, 2-methoxy-4-methyl phenol, 2,6-dimethoxy phenol, that derived from lignin pyrolysis were also detected. Ketones mainly composed of 2-hydroxy-2-cyclopenten, and large no of cyclopentadiones.

Based on an overall consideration of the qualitative results of the bio-oil from mixed wood raw material pyrolysis, following 16 compounds out of 66, were indentified and present in considerable amount i.e. acetic acid, formic acid, furfural, butanal, butyrolactone, 3-methyl-1,2-cyclopentane, 2-methyl-phenol, 2-methoxy phenol, 3-ethyl-2-hydroxy-2cyclopenten, 2-methoxy-4methyl-phenol, 1,2-

IOP Conf. Series: Materials Science and Engineering **206** (2017) 012012 doi:10.1088/1757-899X/206/1/012012

benzenediol, 3-methoxy-1,2-benzenediol, 3-methyl-1,2-benzenediol, 2,6-dimethoxy-phenol, 1,2,3-trimethosy-benzene, levoglucosan.

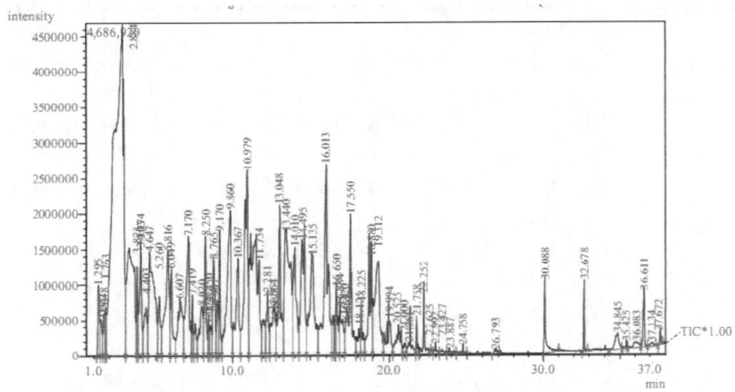

Figure 2. Total ion chromatogram of bio-oil from mixed wood pyrolysis at 650 ^{0}C.

Table 4. Identification of the main components

Peak	RT/min	Area %	Library ID
5	2.884	19.06	Acetic acid
6	3.303	4.90	Formic acid ethyl ester
10	4.647	3.46	Furfural
11	5.260	0.81	2-cyclopenten-1-one
15	7.170	2.48	Butyrolactone
25	9.860	3.44	1,2-cyclopentanedione,3-methyl
26	10.367	1.93	Phenol,2-methyl
27	10.979	3.92	Phenol,2-methoxy
28	11.734	5.69	2-cyclopenten-1-one,3-ethyl-2-hydroxy
32	13.048	2.14	Phenol,2-methoxy-4-methyl
33	13.440	4.43	1,2-benzenediol
35	14.495	2.93	1,2-benzenediol,3-methoxy
36	15.135	2.91	1,2-benzenediol,3-methyl
37	16.013	4.30	Phenol,2,6-dimethoxy
42	17.550	2.40	1,2,3-trimethoxybenzene
46	19.312	2.97	1,6-anhydro-beta-D-glucopyranose (levoglucosan)

3.3 Characterization of wood pyrolysis oil

Moisture content of the pyrolysis oil was calculated by using Karl fisher titration and it was found that mixed wood pyrolysis oil contain 66.9 % of the moisture. Along with the moisture content some of the other important properties of the bio-oil required for its characterization was also calculated and presented in Table 5.

Bio-oil obtained from pyrolysis of mixed wood is highly acidic and low dense in nature. Calorific value of the bio-oil is very less compare to petroleum based fuel oils however, bio-oil is denser than petroleum oil and contain 10-15% more energy by volume. Flash and fire value of the bio-oil is also high compared with petroleum fuels, moreover it contain high ash and sulphur contain. Bio-oil cannot be replaced with the petroleum oils but can be used to obtain the different high value chemicals (Yang et al.2007).

IOP Conf. Series: Materials Science and Engineering **206** (2017) 012012 doi:10.1088/1757-899X/206/1/012012

Table 5. Properties of wood pyrolysis oil

Property Name	Specification	Values Obtained
pH	Ph probe	2.2
Density	Specific gravity Bottle	1.02 kg/litter
Viscosity	Glass viscometer	2.1 cP
Acid Value	Titration with 0.1N KOH.	128.80 mg/KOH
Calorific Value	Calorimeter	6.113 MJ/kg
Fire value	Pensky Martin Closed Cup Apparatus	Does not catch fire up to 90 °C
Flash value	Pensky Martin Closed Cup Apparatus	Above 90 °C sample extinguishes
Ash Content	Heated in muffle furnace up to 800° C	1.63%
Carbon content	By chromatogram	10.943%
Hydrogen content	By chromatogram	10.416%
Nitrogen content	By chromatogram	0.883%

3.4 Separation of bio-oil for refinement

Bio-oil cannot be directly applied as a high-grade fuel because of its inferior properties, such as high water and oxygen contents, acidity, and low heating value. Thus, it is necessary to upgrade bio-oil to produce a high-grade liquid fuel that can be used in engines (Bridgwater et al., 1996).

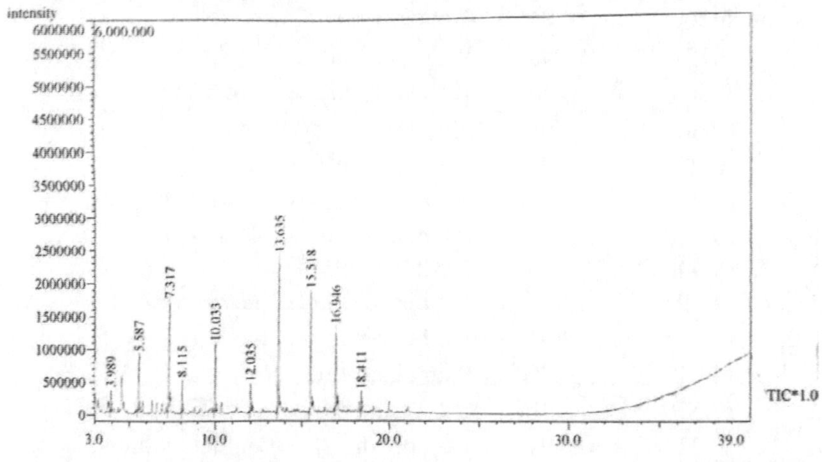

Figure 3. Total ion chromatogram of bio-oil separated by distillation process.

In view of its molecular structure and functional groups, conventional column separation technique such as laboratory distillation was used to upgrade the content of bio-oil. Due to its complex composition, the boiling of bio-oil starts below 100 ^{0}C, and then the distillation continues up to 250 ^{0}C, whereupon 35% of residue was left. The obtained residue was again analysed by using GC-MS technique and components were identified and listed in Table 6. The ion chromatogram of the residue is shown in Figure 3. Qualitative analysis of the distillate residue gives 10 major components listed in Table 6. The components presents in the residue are having the boiling points higher than 250 ^{0}C.

4. Conclusions

Overall wood pyrolysis process gives the pyrolysis oil, charcoal and gas which usually contain H_2, CH_4 and negligible amount of ash. Qualitative and chemical characterization of bio-oil is important for designing pyrolysis reaction and to find extended range of applicable fuels and chemicals. In GC-MS analysis of bio-oil in all 66 components were indentified out of which 16 components (listed in Table

4) were found in major concentration. Characterizing the properties of bio-oil like calorific value, density, viscosity, acid value, fire value and ash content one can use bio-oil as alternate source of fuel for boiler, burner, and many other uses. Distillation operation was able to separate the ten major heavy components from the complex mixture of the bio-oil.

Table 6. Identification of components present in residue

Peak	RT/ min	Area %	Library ID
1	3.989	2.77	2-cyclopenten-1-one,3-methyl-
2	5.587	9.26	1,2-cyclopentanedione,3-methyl-
3	7.317	16.54	Phenol,2-methoxy-
4	8.115	4.14	2-cyclopenten-1-one,3-ethyl-2-hydroxy-
5	10.033	9.36	Phenol,2-methoxy-4-methyl-
6	12.035	4.07	Phenol,4-ethyl-2-methoxy-
7	13.635	23.30	Phenol,2,6-dimethoxy
8	15.518	17.42	1,2,3-Trimethoxybenzene
9	16.946	10.89	Benzene,1,2,3-Trimethoxy,5-methyl-
10	18.41	2.25	2,4-Hexadienedioic acid,3,4-diethyl,dimethyl ester (z,z)

References

[1] Bridgwater A 1996 *J. Catal. Today* **29(4)** 285
[2] Bridgwater A, Meier D and Radlein D 1999 *J. Org. Geochem.* **30(12)** 1479
[3] Demirbas A 2009 *J. Energy Convers. Manag.* **50** 2782
[4] Dinesh M, Charles U, and Philip H 2006 *J. Energy Fuels* **20** 848
[5] Dobele G, Urbanovich I, Volpert A, Kampars V and Samulis E 2007 *J. Bio resources* **2** 699
[6] Helena L C, and Ralph P O 2001 *J. Fuel Process. Technol.* **71** 187
[7] Hyeon S, Hyun J, Young K, Changkook R, Dong J, Young W, et al. 2010 *J. Bioresour. Technol.* **101** S91
[8] Jean N, Hooshang P, and Christian R. 2001 J. Sep. Purif. Technol. 24 155
[9] Lu Q, Li W and Zhu, X 2009 J. Energy Convers. Manag. 50 1376
[10] Mckendry P 2002 J. Bioresour. Technol. 83 37
[11] Serdar Y 2004 J. Energy Convers. Manag. 45, 651–671.
[12] Yang H, Yan R, Chen H, Lee D and Zheng C 2007 J. Fuel 86 1781

Gasification of empty fruit bunch with carbon dioxide in an entrained flow gasifier for syngas production

N F H Rahmat and R A Rasid*

Faculty of Chemical & Natural Resources Engineering, Universiti Malaysia Pahang, Lebuhraya Tun Razak, 26300 Gambang Pahang Darul Makmur

*E-mail: ruwaida @ump.edu.my

Abstract. The main objectives of this work are to study the gasification of EFB in an atmospheric entrained flow gasifier, using carbon dioxide (CO_2) as its gasifying agent and to determine the optimum gasification operating conditions, which includes temperature and the oxidant to fuel (OTF) ratio. These were evaluated in terms of important gasification parameters such as the concentration of hydrogen (H_2) and carbon monoxide (CO) produced the syngas ratio H_2/CO and carbon conversion. The gasification reactions take place in the presence of CO_2 at very high reaction rate because of the high operating temperature (700°C - 900°C). The use of CO_2 as the oxidant for gasification process can improve the composition of syngas produced as in the Boudouard reaction. Rise of reaction temperature which is 900°C will increase the concentration of both H_2 & CO by up to 81% and 30% respectively, though their production were decreased after the OTF ratio of 0.6 for temperature 700°C & 800°C and OTF ratio 0.8 for temperature 750°C. The operating temperature must be higher than 850°C to ensure the Boudouard reaction become the more prominent reaction for the biomass gasification. The syngas ratio obtained was in the range of ≈ 0.6 – 2.4 which is sufficient for liquid fuel synthesis. For the carbon conversion, the highest fuel conversion recorded at temperature 850°C for all OTF ratios. As the OTF ratio increases, it was found that there was an increase in the formation of CO and H_2. This suggests that to achieve higher carbon conversion, high operating temperature and OTF ratio are preferable. This study provides information on the optimum operating conditions for the gasification of biomass, especially the EFB, hence may upsurge the utilization of biomass waste as an energy source.

1. Introduction

Nowadays, fossil-based fuels are the main energy source. However, renewable and clean energy sources are now under focus as the alternative and the more sustainable supplies of energy in the future due to energy crisis and environmental issues that has been plaguing the usage of the conventional fuels. There is a potential in the greenhouse gas emissions reduction by fossil fuels burning through renewable resources [1- 3].

One of the major renewable energy resources which can produce hydrogen (H_2) as a clean, sustainable, environmental–friendly energy sources is biomass. Biomass is the top four among the primary energy sources such as coal, oil and natural gas and currently caters for about 14% of the world's total energy consumption [4]. The use of biomass as renewable energy is important to alleviate global warming [5]. Biomass has high potential in produce zero net carbon dioxide (CO_2) emissions [6-7]. It can be converted into gas or liquid fuels through bio-chemical or thermochemical processes.

Oil palm is one of the most intensive producers of biomass. Oil palm mill generates three main types of waste which are solid waste (mesocarp fruit fibers (MF), palm kernel shell (PKS), EFB), liquid waste (palm oil mill effluent (POME)) and air waste (smoke from boiler and incinerators [8-10].

In Malaysia, the oil palm solid waste including EFB, PKS and fibres are cheap and abandoned materials produced during palm oil milling process. Approximately 0.07 tons PKS and 0.2 tons EFB solid waste are produced for every ton of oil palm fruit bunch during milling [11]. In palm oil mill, about 90% of the total biomass is considered as waste or 50-70 tonnes biomass residue from each hectares of production area while the other 10% constitutes to the extracted palm oil. These wastes have high fibre content and usually used as fuel in mill boilers. High moisture content of EFB cause it cannot be burn directly. Therefore, EFB usually dumped in plantation [8]. Table 1 show the quantity of major Oil Palm Wastes (OPWs) generated from Crude Palm Oil (CPO) in 2014 [9].

Table 1. Oil Palm Waste generated from Crude Palm Oil (CPO) Production

Oil Palm Waste (OPW)	% OPW	OPW generated(tons)
Empty Fruit Bunches (EFB)	22.00	15,472,600
Mesocarp Fibre (MCF)	13.50	9,494,550
Palm Kernel Shell (PKS)	11.50	8,087,950

Malaysia was the largest oil palm producer for last few decades. But since 2007, Indonesia overtook Malaysia and become the world largest producer of oil palm. This is due to high availability of oil palm plantation area and increased in cultivation of oil palm in high pace [12]. For oils and fats market in global demand, palm oil is now the most important tropical vegetable oils as measured by either production or international trade [13]. In the period 1995-2010 palm oil production is more than tripled to 46.7 million tonnes, mainly produced by two major producing countries, Indonesia (47%) and Malaysia (38%) [14]. Crude palm oil production is reaching 52.31 million tons per year globally in 2012 and Southeast Asia is the main contributor, with Indonesia accounting for 48.94%, Malaysia 36.19%, and Thailand 3.10% [15]. Indonesia and Malaysia are the two major producers of palm oil in the world, contributed 19,500 MT and 30,100 MT respectively in 2014 [16]. Figure 1 shows Certified Sustainable Palm Oil (CSPO) production capacity for 2016.

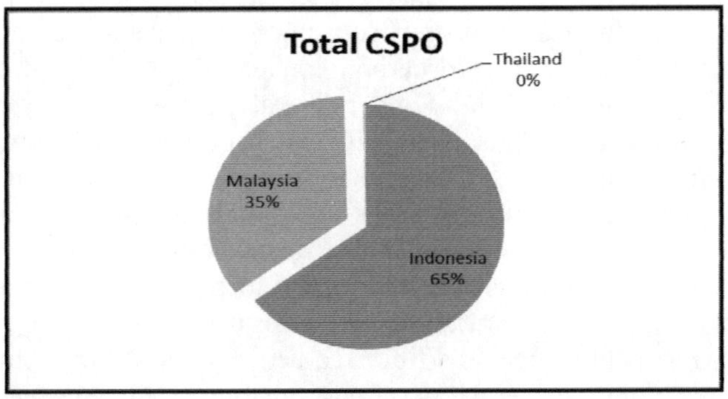

Figure 1. CSPO production capacity as for September 2016 [17]

For Southeast Asian region based on the certified sustainable palm oil as September 2016, Indonesia and Malaysia produced 6.7 and 3.6 million tonnes respectively while Thailand produced 31,811 tonnes [17].

Palm oil contributes for about 5% of the world's bio-diesel production. It has increased significantly due to increasing global demand with major importers being India, China and the European Union [18].

1.1. Biomass Gasification

Gasification is a clean technology that converts different carbonaceous feed stocks such as natural gas, coal, petroleum, coke, biomass and municipal solid wastes in a limited supply of air to gaseous products or called as synthesis gas such as H_2, carbon monoxide (CO), CO_2, water (H_2O) as well as gaseous hydrocarbons at high temperatures [19-22]. This process is one of the promising technologies that have been widely studied to exploit energy using several kind of feedstock, as coal, biomass or mixtures [23]. This process is one of the few technologies that can potentially produce carbon neutral energy with pollution-free power and also change the agricultural gas into energy [24].

Biomass and coal are considered as potential feedstock's which supply syngas though gasification for the synthesis of liquid fuels. This is due to depletion of natural gas resources which increased the necessity for reducing the consumption of natural gas [25]. However, in comparison with coal, biomass may be able to produce cleaner syngas due to its lower sulfur contents and neutral carbon footprint [26, 27]. Biomass contains more oxygen and nitrogen in its elements and has the tendency to produce tar while coal contains more ash and sulfur. Generally, the O_2 and fixed carbon content of biomass are 45 wt. % and 20-25 wt. % respectively while for coal are 2-20 wt.% and 50-85wt.%. The [H_2] / [CO] ratio for biomass is higher than coal which is in the range of between 1.7– 2.1. Therefore, biomass is said to be more biodegradable compared to coal. The gasification temperature of biomass is between (650°C - 1200°C) which is much lower than coal (1300°C - 1600°C). High energy will be consumed in high gasification operating temperature especially when steam is used as the gasifying agent [28].

Size, shape and structure of biomass affect the rate of gasification. For maximum rate of gasification and better controlling temperature, small size of biomass is preferable [29]. The composition and impurities of the syngas produced in the gasification depends on biomass feedstock, gasifier design, gasifying agent and gasification condition mainly temperature, equivalent ratio (ER), steam to biomass ratio and pressure [20, 30-31]

1.2. Entrained Flow Gasifier

Entrained flow gasifiers have the ability to gasify practically any fuels, but fuel with lower moisture and ash content are favored to reduce oxygen consumption. Due to the short residence time, high temperatures are required to ensure a good carbon conversion [32-33] and therefore entrained flow gasifiers have a high oxygen demand. High temperature also results in CO and H_2 rich gas. Soot may be formed depending on the type of fuel [33]. This reactor provides an intensive contact between the gasifying agent and biomass which result in higher conversion and higher reaction rates [20].

One of the disadvantages of entrained flow gasifiers is that the fuel particles should be very small to be completely converted in the short time that they reside inside the hot gasifier. This is especially true for solid biomass since the biomass is hard to grind to small particle sizes and this is one of the main technical barriers connected with entrained flow gasification of biomass [34].

Compared with moving bed gasifiers and fluidized bed gasifiers, entrained flow gasifiers operating at very high temperatures, can produce a clean, almost tar-free, product gas and in short residence time can achieve a high load throughput. The advantages of entrained flow gasifiers are it has a simple structure, no catalyst is required, the tar yield is low, control of the gaseous components is easy, and various kinds of biomass can be used as feedstock are available as feedstock [28].

Many researchers have studied the effect of temperature on CO_2 concentration, where both coal and biomass gasification shows similar trends. At low temperature, the CO_2 concentration is high but start decreasing when the operating temperature of the gasifiers was increased [35]. This trend is due to the increase of the endothermic reaction such as Boudouard reaction which involves the consumption of CO_2 [36]. For biomass gasification, the researcher found that the trend is the same as the coal gasification [37-38]. However, the carbon conversion is slightly higher for biomass gasification than that of the coal gasification due to the presence of high volatile matters in biomass. On the other hand, in biomass gasification, the increase of temperature will reduce the calorific values of the product gas due to the decrease in the concentration of hydrogen bond [39]. Another parameter that is highly affected by the operating temperature is the cold gas efficiency (CGE). For both biomass gasification and coal gasification, the CGE were increased with the increase of temperature [37-38]. However, its value in biomass gasification is more than that of coal gasification because of a higher yield in the gaseous product during biomass gasification as observed by [39].

The oxidant used in gasification process either in biomass gasification or coal gasification will affect the production of syngas. Traditionally, the common gasifying agent use in gasification process is air which content O_2 and N_2. From the study reported by [40], when the air become the oxidising agent, the syngas produced composition is 16.7% for H_2, 22.3% for CO, 9.61% for CO_2 and 50.8% for N_2. When the CO_2 replace the air as the oxidising agent for gasifiers, the composition for the syngas produced have change to 16.4% for H_2, 56.3% for CO, 27.3% for CO_2 and 0.06% for N_2 [40]. This shows that CO production was decreased while the CO_2 was increased.

There is huge potential in exploring the utilization of one of the greenhouse gases, CO_2 as a gasifying agent, and biomass waste, both of which are unwanted. In this study, the gasification of biomass with CO_2 by using palm oil EFB will be carried out at different temperatures and different oxidant to feed (OTF) ratio in which various concentration of CO_2 as oxidant will be varied. This is to determine the optimum temperature and OTF ratio for syngas production by means of biomass gasification.

2. Methodology

2.1. Biomass Feedstock and Characterisation

The empty fruit bunches are collected from an oil palm mill in Lepar Hilir. It has a relatively high moisture content of about 60%, hence it was cleansed and dried prior to the experiments. The ultimate analysis was done by using a CHNS/O Analyzer (model LECO TruSpec CHN, USA) following the ASTM D-5291 method. The proximate analysis of the sample to determine its moisture content (dry basis), volatile matter, fixed carbon and ash were determined using a thermo gravimetric analyser (model Mettler Toledo, TGA/SDTA851, USA). The heating value of EFB was determined by burning a weighed sample in an adiabatic oxygen-bomb calorimeter (model Parr 1341, USA) following the ASTM D240 method, while a Micromeritics helium pycnometer (AccuPyc II 1340) was used to measure the volume and apparent density of the sample. The results of the analysis are listed in Table 2.

Table 2. Properties of EFB

Proximate Analysis (wt %)		Ultimate Analysis (wt %), daf[a]	
Moisture	5.977	C	43.518
Volatile matter	67.58	H	7.255
Fixed carbon	5.893	O[b]	45.896
Ash	20.48	N	3.041
HHV (MJ/kg)	18.67	S	0.290
Density (kg/m^3)	1390.8		

[a] Dry and free ash; [b] The oxygen (O) content was determined by difference

2.2. *The Gasification Experiment Setup*

The entrained flow gasifier was designed and built in our laboratory. As shown in Figure 2, the height of the gasifier is about 50cm and the diameter is 4.5cm .The gasification test of EFB was carried out with entrained flow reactor. In a typical test, 20 g dried EFB samples without catalyst was loaded in the reactor. The carbon dioxide was used as gasification agent and nitrogen as carrier at a constant volume flow rate of 18 LPM. The reactor was heated from room temperature to a temperature range of 700°C-900°C. Once the temperature was reached, the gas flow was switched from N_2 to CO_2, the solid and water feeds were switched on. A constant gas composition was reached after 10-15s of starting the feeding. All 5 different oxidant to fuel ratio ranging from 0.2-1.0 will undergo five different operating temperatures which are 700, 750, 800, 850 and 900°C. The produced gas was collected after passing through a condenser. After the termination of the gasification process, the fraction consist of solid material recovered from the reactor bottom was defined as solid residue.

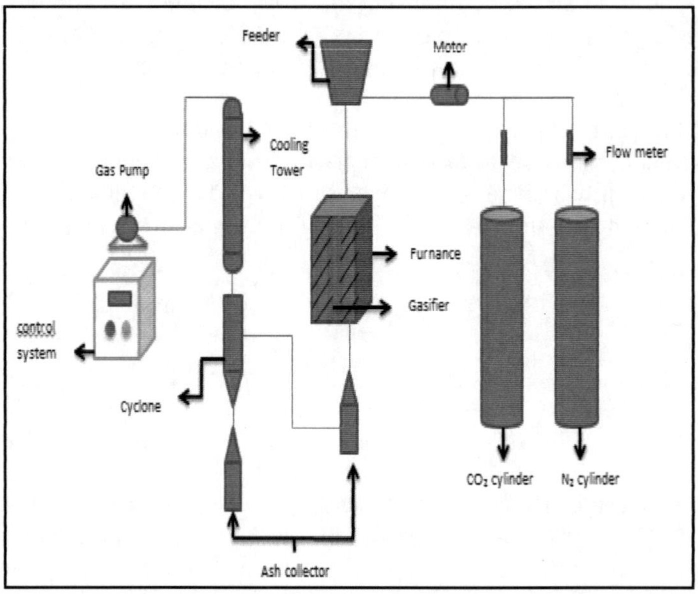

Figure 2. Schematic diagram of gasification system

2.3. *Product Analysis*

The product gas was primarily analysed using the Thermal Conductivity Detector (GC-TCD). From the values obtained, all other analysis may be done. The carbon conversion, *X* represents the percentage (%) of total carbon in the gasifier feedstock which is converted to product gases, which contain carbon (i.e. CO, CO_2, CH_4). Equation (1) is used to determine the conversion.

$$\text{Conversion}=\frac{\text{Total mol C (\%) in CO,CO2 \& CH4}}{\text{mol C (\%) obtained during ultimate analysis}} \quad \textbf{(1)}$$

The higher heating value (HHV) of the product gas is calculated based on Equation (2), based on the heating values from [42], and the obtained values were further used to calculate the CGE of the product gases as in Equation (3) [41].

H_2 = 12.76 MJ/m^3; CO = 12.63MJ/m^3; CH_4 = 39.76 MJ/m^3 [42]

Since only H_2, CO and CH_4 are combustible, the HHV of syngas is the calorific value of these three gases:

$$\Delta H = (12.76\ MJ/m^3 \times H_2\ vol\ \%) + (12.63\ MJ/m^3 \times CO\ vol\ \%) + (39.76\ MJ/m^3 \times CH_4\ vol\ \%) \quad \mathbf{(2)}$$

3. Results and Discussion

3.1 H_2 volume flowrate

In this work, the effect of OTF on the production of H_2 gas was studied from 0.2-1.0 ratio. It can be observed that H_2 is highly affected by the OTF ratio that used in gasification process. Figure 3 illustrates the trends of H_2 production with the increasing of OTF ratio for different operating temperature. It is observed that for temperature 900°C, H_2 is increased from 1.2% to 2.2% with the increase of OTF ratio from 0.2 to 0.8 and slightly decreased at higher OTF ratio. Similar trend was observed by [43] whereby the H_2 concentration declined with the rising of CO concentration in the gasification process, although the observation was for high OTF ratios, which is between 0.8 and 1.0. From this, it can be concluded that H_2 production will increased when low OTF ratio was applied and the value will be decreased when higher OTF ratio used in gasification process.

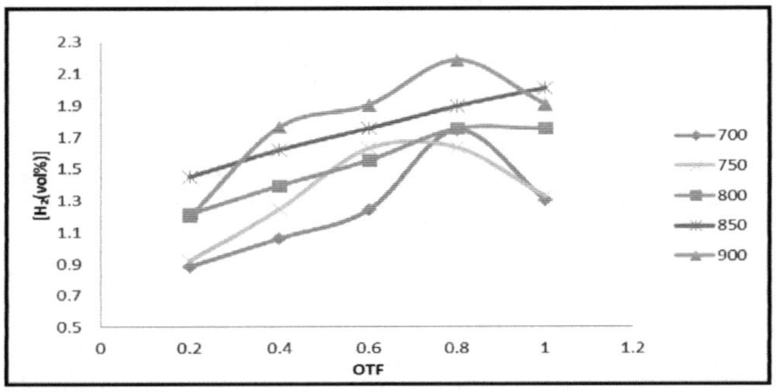

Figure 3. Effect of OTF on volume flow rate of H_2 at different operating temperature

3.2 CO volume flowrate

Figure 4 illustrates on how CO is affected by the increased of the OTF ratio from 0.2-1.0. From the figure, it can be observed that the CO production is slightly decreasing for temperature 700°C to 800°C when OTF ratio reached 0.8. On the other hand, when the operating temperature reached 900°C, CO production was increased from 0.73% to 1.0%. Similar finding was also observed by [43] studying the effect of CO_2 as the oxidizing agent in biomass gasification. The operating temperature must be higher than 800°C to ensure that the Boudouard reaction ($C + CO_2$ « $2CO$) becomes the more prominent reaction in this biomass gasification.

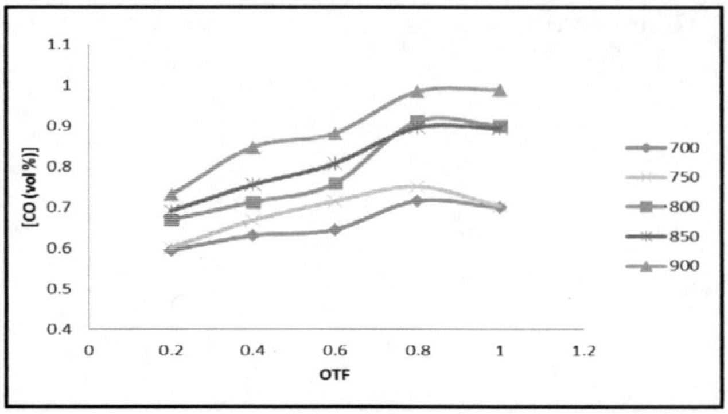

Figure 4. Effect of OTF on volume flow rate of CO at different operating temperature

3.3 CO_2 volume flowrate

The use of CO_2 as the gasifying agent will affect the vol% of the CO_2 produced in this study. The trend obtains from this study in which the effect of vol% of CO_2 with the increasing of OTF ratio is shown in Figure 5. For temperature 850°C-900°C, the concentration of CO_2 is increasing from OTF 0.2 until 0.8 and decrease when approaching 1.0. However, for temperature 700°C to 800°C, the concentration of CO_2 nearly constant after OTF 0.8.

Hanaoka et al. (2012) studied the gasification of biomass and found that the CO_2 composition will rise when the CO_2 percentage in gasifying agent increase. In this study, it was also determined that the vol% of CO_2 will increase with the increase of OTF ratio applies during gasification process. Although the CO_2 consumed in the Boudouard reaction to produced CO, the high amount of CO_2 gas feed to the reactor recover the loss of CO_2 in the reaction.

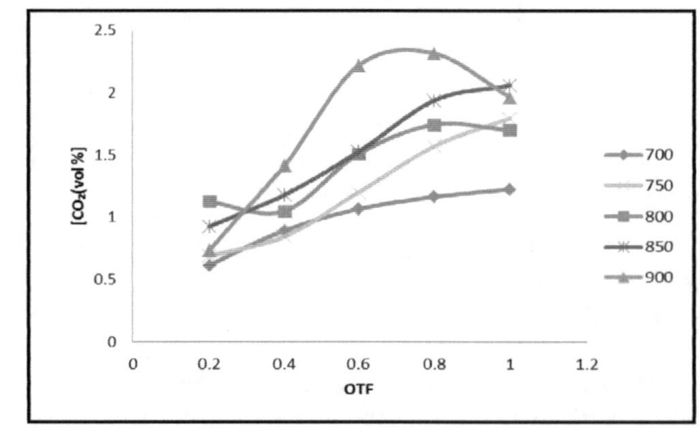

Figure 5. Effect of OTF on CO_2 volume flow rate for different temperature

3.4 HHV

HHV of produced gas from gasification process was determined using Equation (2). The graph of HHV versus temperature for different OTF ratio is shown in Figure 6. The trend shows that the increment of OTF ratio from 0.2 to 1.0 increased the HHV value of the syngas produced. This HHV value is strongly related to the amount of H_2, CO and CH_4 that obtained. As the amount of CO and CH_4 increase by the Boudouard reaction and methane production reaction at higher OTF ratio, it can be concluded that the HHV of produced syngas will increase with increase of OTF ratio.

IOP Conf. Series: Materials Science and Engineering **206** (2017) 012013 doi:10.1088/1757-899X/206/1/012013

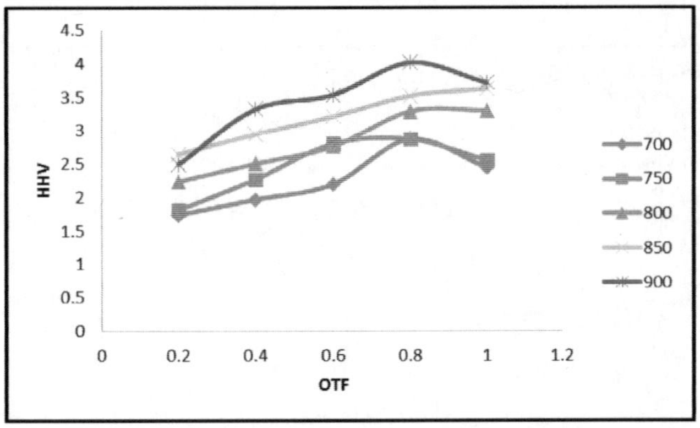

Figure 6. Effect of OTF on HHV at different operating temperature

3.5 H_2 / CO ratio

H_2/CO ratio is one of the important parameter in the analysis of the syngas produced from biomass gasification. The trends obtain from this study are shown in Figure 7. From the figure, the H_2/CO ratio was increased when the OTF was increased from 0.2 to 0.8. However, this value decreased drastically at higher OTF. This trend is almost similar to the trend obtain in the effect of OTF on H_2 volume percentage. This suggests that the H_2/CO ratio is related to the amount of H_2 produced from this gasification process. In addition, the trends obtained can be related to the moisture content of the feedstock. At higher temperature of gasifier, the moisture content decreased thus decreased the production of H_2 from the gasification process.

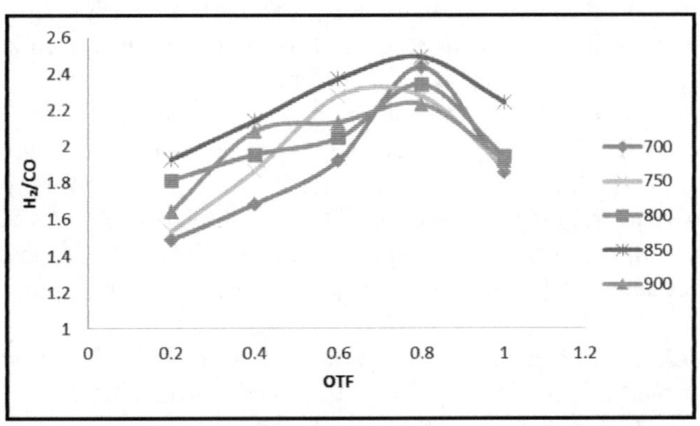

Figure 7. Effect of OTF on H_2/CO ratio at different operating temperature

3.6 Carbon conversion

Fuel conversion was determined by dividing the total amount of carbon content in CO, CO_2 and CH_4 produced from gasification process with the amount of carbon content in raw EFB from its ultimate analysis. The trend for the fuel conversion with the increase OTF ratio for different operating temperature is shown in Figure 8. It was found that the fuel conversion is positively affected by the increase of OTF ratio. This is due to the fact that higher OTF ratios will increase the production of the combustible gases: CO, CO_2 and CH_4. Higher production of these gases will cause the higher fuel conversion for this gasification. From the result obtain, volume flow rate of CO was increased due to the Boudouard reaction that is more prominent at higher operating temperature.

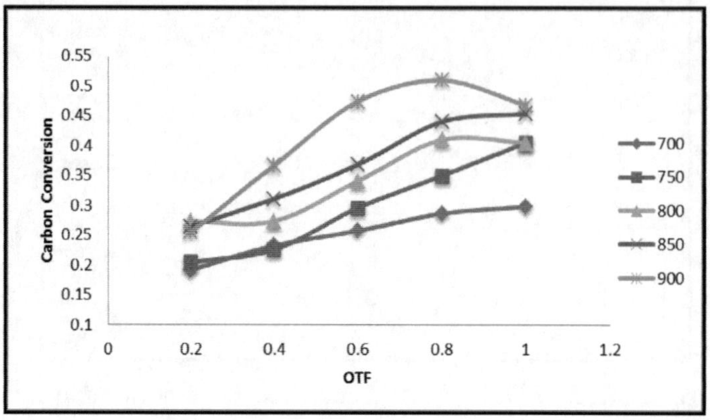

Figure 8. Effect of OTF on carbon conversion at different operating temperature

Other than that, the use of CO_2 as the oxidising agent generates the higher production of CO that effect the fuel conversion for this process. The amount of carbon present in the biomass used in the study can also affect the value of fuel conversion. The lower the carbon content in the feedstock will cause the higher value of carbon conversion for this process.

4. Conclusion

Conversion of biomass to thermochemical conversion especially by gasification process could be one of the promising methods to optimise the utilization of this abundant and valuable source of wastes.

It is important to study the best condition for gasification process in order to ensure the maximum composition of syngas produced. From this study of the influence of OTF ratio and operating temperature on biomass gasification was investigated. It can be concluded that these two parameters have significant effects to the production of syngas. The biomass gasification process using CO_2 must be conducted at high gasifier temperature which is 900°C and OTF ratio 1.0. This is to ensure all the requires reaction involve in this gasification process such as the Boudouard reaction, water gas reaction and also the methane production reaction can take place dominantly thus resulting higher syngas production.

Further research should be carried out to determine the highest syngas production among several oil palm wastes include palm kernel shell and oil palm frond. Furthermore, the continuous effort will contribute to the enrichment of new knowledge and technology whereas at the same time helps to preserve nature for the future in a more sustainable way.

Acknowledgement

This project was supported by the Ministry of Higher Education Malaysia FRGS (RDU140134). The authors also thank the Faculty of Chemical Engineering and Natural Resources, Universiti Malaysia Pahang, for providing the lab space to conduct the trials.

References

IOP Conf. Series: Materials Science and Engineering **206** (2017) 012013 doi:10.1088/1757-899X/206/1/012013

[1] Babu S P 2006 IEA bioenergy agreement task 33 thermal gasification of biomass, workshop no 1: Perspectives on Biomass Gasification **1** pp 1–46

[2] Guo F et al 2014 Effect of design and operating parameters on the gasification process of biomass in a downdraft fixed bed: An experimental study. *Int. J. Hydrogen Energy* **39**(11) pp 5625–5633

[3] Zhou J et al 2009 Biomass-oxygen gasification in a high-temperature entrained-flow gasifier. *Biotechnol. Adv.* **27**(5) pp 606–611

[4] Alauddin Z A B Z et al 2010 Gasification of lignocellulosic biomass in fluidized beds for renewable energy development: A review *Renew. and Sust. Energy Rev.* **14**(9) pp 2852–2862

[5] Mckendry P 2002 Energy production from biomass (part 1): overview of biomass *Biores. Technol.* **83** pp 37–46.

[6] Yan F et al 2010 Bioresource Technology Hydrogen-rich gas production by steam gasification of char from biomass fast pyrolysis in a fixed-bed reactor : Influence of temperature and steam on hydrogen yield and syngas composition *Biores. Technol.* **101**(14) pp 5633–5637

[7] Tijmensen M J A et al 2002 Exploration of the possibilities for production of Fischer Tropsch liquids and power via biomass gasification *Biomass and Bioenergy.* **23** pp 129–152

[8] Kurnia J C et al 2016 Advances in biofuel production from oil palm and palm oil processing wastes : A review *Biofuel Res. J.* **3**(1) pp 332–346

[9] Nyakuma B B 2015 Bioelectricity Potential of Oil Palm Waste in Malaysia *Proc. in 3rd Int. Conf. Res. & Edc. in Nat. Sci. HERTSPO 2015*

[10] Abdullah N et al 2011 Characterisation of Oil Palm Empty Fruit Bunches for Fuel Application *J. Phy. Sci.* **22**(1) pp 1–24

[11] Esfahani R M et al 2011 Hydrogen rich production from Malaysian Palm kernel shell by applying air gasification in fluidized bed reactor *Energy and Fuels* **26**(2) pp 1185-1191

[12] Awalludin M F et al 2015 An overview of the oil palm industry in Malaysia and its waste utilization through thermochemical conversion, specifically via liquefaction *Renew. and Sustain. Energy Rev.* **50** pp 1469–1484

[13] Growth W 2011 The Economic Benefit of Palm Oil to Indonesia

[14] Mahat S B A 2012 The Palm Oil Industry From The Perspective of Sustainable Development : A Case Study of Malaysian Palm Oil Industry

[15] Sime Darby 2012 A World of Infinite Possibilities

[16] Council M P O 2015 Malaysian Palm Oil : Our Role in The Global Oils & Fats Market Oils and Fats Production (1990 – 2014). Retrieved from http://www.mpoc.org.my/

[17] RSPO 2015 Global Vision Regional Action **13** pp 1-34

[18] Otieno N E et al 2016 Palm Oil Production in Malaysia: An Analytical Systems Model for Balancing Economic Prosperity, Forest Conservation and Social Welfare *Agri. Sci* pp 55–69

[19] Emami-taba L et al 2012 The effect of temperature on various parameters in coal , biomass and co-gasification : A review *Renew. and Sust. Energy Rev* **16**(8) pp 5584–5596

[20] Heidenreich S and Foscolo P U 2014 New concepts in biomass gasification *Progress in Energy and Combust. Sci.* **46** pp 72–95

[21] Rajvanshi, A K 2014 Biomass gasification *Altern. Energy in Agri.* **II**(4) pp 1–21

[22] De Filippis P et al 2015 Biomass Gasification Plant and Syngas Clean-up System *Energy Proc.* **75** pp 240–245

[23] Aznar P et al 2006 Plastic waste elimination by co-gasification with coal and biomass in fluidized bed with air in pilot plant *Fuel Process. Technol.* **87** pp 409–420

[24] Li X T et al 2004 Biomass gasification in a circulating fluidized bed *Biomass & Energy* **26** pp 171–193

[25] Droege P 2002 Renewable Energy and the City: Urban Life in an Age of Fossil Fuel Depletion and Climate Change *Bull. of Sci. Technol. & Soc.* **22**(2) pp 87–99
[26] Salleh M A M et al 2010 Gasification of Biochar from Empty Fruit Bunch in a Fluidized Bed Reactor *Energy* pp 1344–1352
[27] Leonard R et al 2001 Development Status of the Transport Gasifier at the PSDF *Gasific. Technol.* pp 1–15
[28] Ogi T et al 2013 Gasification of oil palm residues (empty fruit bunch) in an entrained-flow gasifier *Fuel* **104** pp 28–35.
[29] Emami-taba L et al 2013 Fuel blending effects on the co-gasification of coal and biomass : A review. *Biomass and Bioenergy* pp 1–15.
[30] Asadullah M 2014 Barriers of commercial power generation using biomass gasification gas : A review *Renew. and Sust. Energy Rev.* **29** pp 201–215.
[31] Berrueco C et al 2014 Pressurized gasification of torrefied woody biomass in a lab scale fluidized bed *Energy* **70** pp 68–78.
[32] Qin K et al 2012 High-temperature entrained flow gasification of biomass *Fuel* **9** pp 589–600.
[33] Van der Drift A et al 2004 Entrained flow gasification of biomass Ash behaviour, feeding issues and system analyses. Energy Research Centre of the Netherlands (ECN) report The Netherlands
[34] Risberg M 2013 Entrained Flow Gasification of Biomass – on Atomisation, Transport Processes and Gasification Reactions: Doctoral Thesis
[35] Wei L et al 2007 Steam gasification of biomass for hydrogen-rich gas in a free-fall reactor *Int. J. of Hydro. Energy* **32** pp 24–31
[36] Franco C et al 2003 The study of reactions influencing the biomass steam gasification process *Fuel* **82** (7) pp 835–842
[37] Hernández J J et al 2010 Gasification of biomass wastes in an entrained flow gasifier: Effect of the particle size and the residence time. *Fuel Process. Technol.* **91**(6) pp 681–692
[38] Kim Y J et al 2001 Coal gasification characteristics in a downer reactor *Fuel* **80**(13) pp 1915-1922
[39] Seo M W et al 2010 Gasification Characteristics of Coal / Biomass Blend in a Dual Circulating Fluidized Bed Reactor *Energy and Fuels* **24**(5) pp 3108–3118
[40] Rutberg P G et al 2011 Use of carbon dioxide in the chemical synthesis technologies , plasma gasification and carbon production. *Mater. Sci. and Eng.* **19**(1)
[41] Syed S et al 2012 Thermodynamics Equilibrium Analysis within the Entrained Flow Gasifier Environment *Int. J. Thermochem. and Environ. Eng.* **4**(1) pp 47–54
[42] Waldheim L, and Nilsson T 2001 IEA bioenergy agreement, Task 20: Thermal gasification of biomass: Heating value of gases from biomass gasification.
[43] Hanaoka T, Sakanishi K, and Okumura Y 2012 The effect of $N_2/CO_2/O_2$ content and pressure on characteristics and CO_2 gasification behavior of biomass-derived char. *Fuel Process. Technol.* **104** pp 287–294

IOP Conf. Series: Materials Science and Engineering **206** (2017) 012014 doi:10.1088/1757-899X/206/1/012014

Fixed bed slow pyrolysis of biomass solid waste for bio-char

M N Islam, M H Ali and I Ahmad
Mechanical Engineering Programme Area,Universiti Teknologi Brunei, Gadong BE 1410, Brunei Darussalam

E-mail: mohammad.gafur@utb.edu.bn

Abstract. Biomass solid waste in the form of rice husk particle is pyrolyzed in a fixed bed pyrolysis reactor. The reactor is made of stainless steel with dimensions of 76 mm in diameter and 90 cm in length. Rice husk is collected locally from Brunei–Muara district of Brunei Darussalam which is processed for pyrolysis. The particles are selected in the millimeter range. It is oven-dried at 105°C for 6 hours after being air-dried prior to pyrolysis. The reactor bed is heated by means of saw-dust in a biomass source heater. A temperature range of 390- 410°C is maintained with an apperent vapor residence time of 30 min. Nitrogen gas is passed through the reactor system to make the atmosphere inert. A water-cooled condenser is used to derive bio-oil from the condensable vapors. The system is subjected to pyrolysis for a running time of 60 min. The products obtained are solid bio-char, liquid bio-oil and pyrolytic bio-gases. The solid char yield is found to be 45% by weight of solid biomass feedstock and the liquid product yield is found to be 29% by weight of solid biomass feedstock. The rest is gas flared into the atmosphere. The bio-char is retained in the reactor and collected at the end of the experimental run. The bio-char is found to be black in color retaining its original shape. The bio-char product is subjected to energy analysis for its higher heating value (HHV) by means of an oxygen bomb calorimeter. It is found to be 20.3 MJ/kg. The density of the bio-char is found to be 238.5 kg/m^3 with an energy density of 4.85 GJ/m^3.

1. Introduction

Energy is an indispensable item in the everyday life of the present world. However, the current world is driven by fossil fuels that has been depleting at a faster rate. The demand is exceeding the supply [1]. Besides the environmental pollution from the combustion of fossil fuel is a serious concern of the world at this time. Thus, a search for alternative sources of energy appears to be an imperative option. For this purpose the renewable sources of energy is considered to be the most suitable option. Among different renewable sources of energy, biomass is considered to be a very good option. Biomass is now considered to be a contributor of bio-fuels alternative to crude petroleum oil providing clean energy and value added products [2]. Biomass has high carbon content, low ash content and it is renewable in nature. Biomass can easily be obtained from agricultural residues or green waste. Woods and other forms of biomass including energy crops and agricultural and forestry wastes are some of the main renewable energy resources available [3]. Like wind, solar and other renewable energy sources, biomass can make a positive impact on the atmosphere by reducing the dependence on climate change-inducing fossil fuels. Biomass solid wastes are available abundantly in most parts of the world. Mostly these are either unused or underutilized. The disposal of the large amount of the wastes is a problem as well. Thus, biomass is a promising eco-friendly alternative source of renewable energy in the context of current scenario [4]. From the characterization studies of biomass solid wastes it is found that these wastes have good prospects to be used as feedstock for better energy recovery in the form of solid bio-char, liquid oil and gaseous products [5, 6]. Besides value-added products can be obtained from the liquid pyrolysis oil [7]. Also, the char has its potential to be used as fuel [8]. One of the possible ways for this energy and value-added products recovery is to adopt the slow pyrolysis process for the conversion of carbonaceous biomass solid wastes into solid char [9]. Slow pyrolysis has been used to enhance char production at low temperature and low heating rates. In this process, the vapour residence time is too high in the range of 5 to 30 min and the components in the vapour phase continue to react with each other resulting the formation of solid char and other liquids [10]. This char is known as bio-char. The reaction pathway of the slow pyrolysis process is shown in Figure 1. The slow pyrolysis is considered to be one of the most promising emerging technologies for this purpose. It is the thermal degradation of the carbonaceous solid wastes into solid char at low temperature in the absence of oxygen or air so that combustion or gasification does not take place. The slow pyrolysis is meant for solid product although liquid oil and gases are produced depending upon the heating rate, reactor bed temperature and apparent vapour residence time.

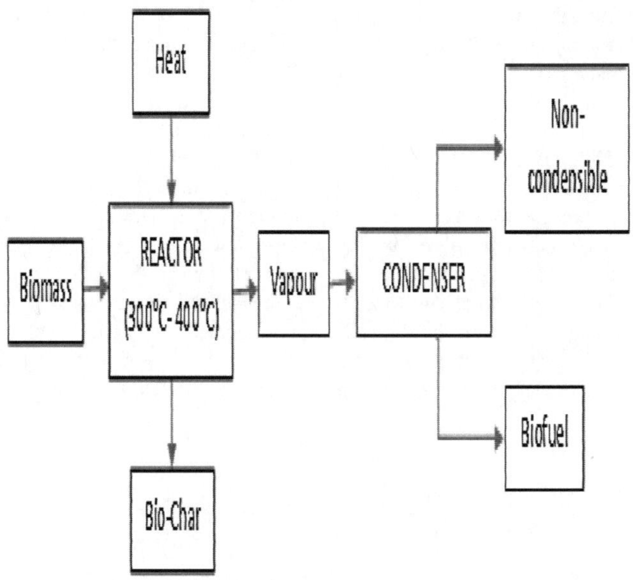

Figure 1. Reaction pathway of slow pyrolysis process.

Rice husk is an abundant agricultural by-product in countries like Brunei, Malaysia, China, Thailand, India and Bangladesh which have large paddy crops. It constitutes one fifth by weight of paddy harvested [11]. It is estimated that over 100 million tons of rice husk are generated annually with 90% accounted for by the developing countries [12]. Mostly this is under-utilized as a source of heat energy. The disposal of this waste is an environmental issue. According to Yatim [13], Malaysia generates 3.41 million cubic meter of rice husk every year. Burning the waste in an uncontrolled manner causes serious emission problems causing environmental issue. Besides rice husk waste possesses favourable physical and thermal characteristics as depicted by characterization study and TG analysis [3]. Thus, the objective of this present study is to convert biomass solid waste in the form of rice husk particles into bio-char in a fixed bed pyrolysis system recently designed and fabricated in the Manufacturing Laboratory of the Mechanical Engineering Programme Area of Universiti Teknologi Brunei (UTB), Brunei Darussalam. A few selective characterization study of the bio-char is also considered in the paper.

IOP Conf. Series: Materials Science and Engineering **206** (2017) 012014 doi:10.1088/1757-899X/206/1/012014

2. Experimental methods

2.1. Materials

The feedstock for pyrolysis considered in this study is biomass in the form of rice husk particles in the size range of millimeters as received. It is collected from the local Rice Mills at Wasan Area of Brunei - Muara district of Brunei Darussalam. The fuel for biomass heating source is saw- dust which is also renewable in nature. It is collected from the local saw-mills in Brunei Darussalam that is burnt in the biomass source heater by supplying air with the aid of a blower to provide heat for the pyrolysis reactor.

2.2. Experimental set-up

The experimental set-up is a simple fixed bed pyrolysis reactor system. It consists of a fixed bed reactor, a liquid condenser, two ice-cooled liquid collectors and a biomass source heater. The schematic diagram of fixed bed pyrolysis reactor system is shown in Figure 2. The diameter of the fixed bed reactor is 76 mm and the length is 90 cm. The pyrolysis reactor system is made of stainless steel so that it can withstand high temperature application in the corrosive atmosphere as the liquid oil is usually corrosive in nature. The condenser and liquid collectors are also made of stainless steel. Nitrogen gas is supplied from a nitrogen gas cylinder to the reactor to make inert atmosphere inside the pyrolysis system. A tube and shell type water cooled condenser is used so that the condensable vapours in the reactor are cooled rapidly to be condensed into liquid product. Two liquid collectors are placed subsequently in order to be ice-cooled for further condensation of the vapours while the non-condensable gases are flared into atmosphere.

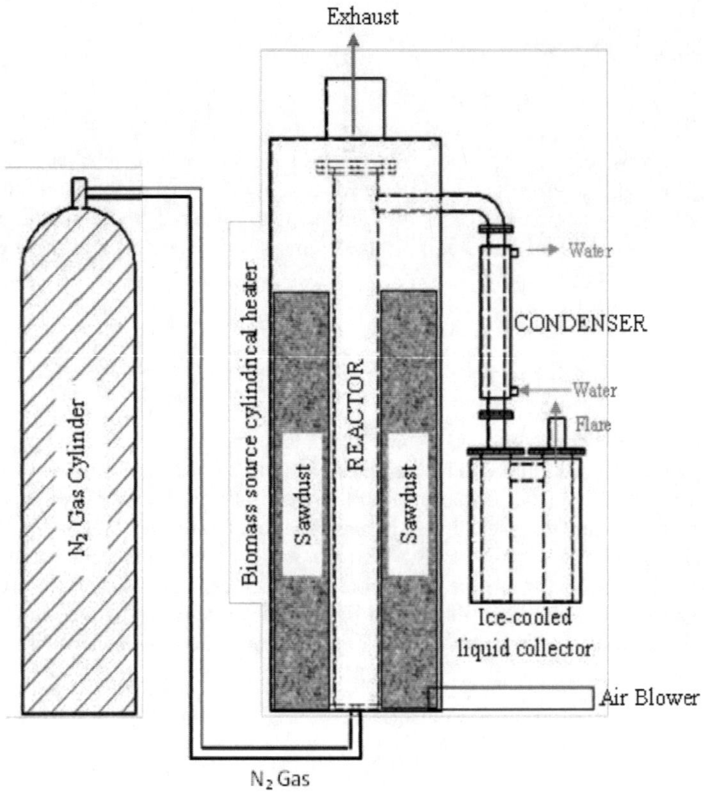

Figure 2. Schematic diagram of fixed bed pyrolysis reactor system.

2.3. Methods and procedures

In this study the slow type of pyrolysis of rice husk is considered for its simplicity and appropriateness to convert carbonaceous solid material into solid char. For this a slow pyrolysis reactor is designed and fabricated. The reactor bed temperature is maintained between 390 to 410°C with a high vapor residence time of 30 min with a slow gas flow rate. Initially the feedstock is air-dried. Then it is oven-dried at 105°C for 6 hours prior to pyrolysis. The fixed bed reactor is filled up with the rice husk to be pyrolyzed. The fixed bed pyrolysis experimental set-up is presented in Figure 3.

Figure 3. Fixed bed pyrolysis experimental set-up.

The set-up is assembled and sealed using rubber and liquid gasket. Nitrogen gas is supplied from a nitrogen gas cylinder to the reactor which is controlled by a gas flow regulator. The reactor is maintained at a low temperature range of 390 to 410°C at an apparent vapour residence time of 30 min. It is heated by means of a heater of biomass source to attain necessary heating in order to achieve the temperature in the range of 400°C. The water-cooled condenser condenses the vapours into liquid oil which is collected in the ice-cooled liquid collectors while the gas is flared into the atmosphere. After the completion of the pyrolysis run for 60 min, the system is cooled down and dismantled. The bio-char retained in the reactor is collected at the end of each experimental run. The liquid products are then collected and weighed. The amount of gaseous products is determined by difference.

3. Results and discussion

3.1. Product yields

Three types of product are obtained in this pyrolysis study. These are solid bio-char, liquid oil, and gases. The thermal degradation of lignin and hemicellulose results in considerable amount of rigid amorphous carbon matrix referred as bio-char. The solid char is found to be 45% by weight of the total solid biomass feedstock in this study. The bio-char obtained is shown in Figure 4. It is found that the char retained the original geometrical shape of the rice husk feedstock particle with a dark black colour. The product yield is found to be in good agreement with the results of other pyrolysis studies conducted by Islam [3] which was found to vary in the range of 40 to 53 wt% of solid biomass feedstock in case of rice husk and oil palm shell. However, Jahirul, Rasul, Chowdhury and Ashwath [4] reported the char yield to be a maximum of 35 wt% of feedstock. The char product yield obtained in this study is found to be higher than the solid product yields obtained by Williams and Besler [14] in their experimental studies on rice husk pyrolysis. However, their study was at higher reactor bed temperature.

Figure 4. Solid bio-char obtained from slow pyrolysis of rice-husk.

The liquid product yield is found to be 29% by weight of the total solid rice husk feedstock which is found to be in good agreement with other pyrolysis studies performed by Islam [3] and Williams and Besler [9]. The remaining non-condensable gas is determined by difference which is flared into atmosphere as shown in Figure 5.

Figure 5. Pyrolytic bio-gas flared to atmosphere.

3.2. Product characterization

The bio-char is characterized for its properties. The higher heating value (HHV) is determined by means of an oxygen bomb calorimeter following the standard test procedures. It is determined to be 20.3 MJ/kg which is found to be better than the HHV of the solid rice husk waste. The density of the bio-char is found to be 238.5 kg/m^3 and hence, the energy density of the bio-char is found to be 4.85 GJ/m^3. The properties obtained are found to be in good agreement with those obtained from similar studies for other biomass derived solid char [3]. The bio-oil is also collected as shown in Figure 6. It is found to be a single-phase, deep brown coloured liquid with an acrid smell. No phase separation is found to take place. The bio-oil is characterized for some of its properties. The density of the bio-oil is found to be 1,051 kg/m^3 and the pH value is found to be 3.0. The density of the oil is comparable with other biomass-derived pyrolysis oils [3]. The pH value indicates that the oil is acidic in nature; and hence, corrosive which is typical in case of pyrolysis oils as supported by other similar studies.

Figure 6. Liquid bio-oil obtained from slow pyrolysis of rice-husk.

4. Conclusions and recommendations

The rice husk solid waste is successfully pyrolyzed into bio-char, bio-oil and pyrolytic bio-gases. The solid bio-char product yield is found to be 45% by weight of the biomass solid feedstock at a reactor bed temperature of 400°C with an vapor residence time of 30 min for millimeter range rice husk feedstock. The liquid bio-oil yield is found to be 29% by weight of the biomass solid feedstock. Thus, the product yields in the form of bio-char and bio-oil are found to be in very good proportions. The heating value of the bio-char is found to be 20.3 MJ/kg which is quite higher than that of the rice husk solid waste. The energy density is estimated to be 4.85 GJ/m^3. Thus, it reveals the fact that the bio-char has its potential to be used as a fuel. Further studies are recommended to find the variation of product yields with reactor bed temperature and feedstock particle size. Also the liquid and solid products are recommended to be characterized for other fuel properties, their functional group compositional anaslysis by FTIR analysis and the identification of the chemical compounds by GC and GC/MS analysis.

Acknowledgments

The authors are grateful to Universiti Teknologi Brunei for providing the necessary financial assistance to procure the materials and the Laboratory facilities for the fabrication of the pyrolysis reactor system and the characterization of the liquid fuel.

References

[1] Sayigh A 1997 Renewable energy- the way forward *Proc. Int' Symp' on Advances in Alternate & Renewable Energy, ISAAE '97* (Johor Bahru: Universiti Teknologi Brunei) 1.
[2] Salema A A and Ani F N 2011 Microwave induced pyrolysis of oil palm biomass, *Bioresour. Technol.* **102** 3388 - 95.
[3] Islam M N 1999 *Pyrolysis of Biomass Solid Waste and Its Catalytic Treatment with Techno-Economic Analysis.* A PhD Thesis, Faculty of Mechanical Engg, Universiti Teknologi Malaysia, Malaysia.
[4] Jahirul M I, Rasul M G, Chowdhury A A and Ashwath N 2012 Biofuels production through biomass pyrolysis - A technological review, *Energies 2012* **5** 4952-5001.
[5] Bakar M H, Ahmad N, Islam M N and Ani F N 2014 Characterization of biomass solid wastes of Brunei Darussalam for their pyrolytic conversion into liquid oil *Proc. 5th Brunei Int' Conf' on Engineering and Technology, BICET 2014 (*Brunei Darussalam: IET).
[6] Islam M N, Ani F N, Bakar M H, and Ahmad N 2015 Characterization of saw-dust solid wastes for its pyrolytic conversion into bio-oil. *Proc. 2nd Int' Conf' on Advances in Civil, Structural and Mechanical Engineering- ACSM 2015* (Bangkok: IRED)
[7] Bridgwater A V 1996. Production of high grade fuels and chemicals from catalytic pyrolysis of biomass *Catal. Today* **29** 285-295.
[8] Islam M N, Zailani R, Ani F N1999 Pyrolytic oil from fluidised bed pyrolysis of oil palm shell and its characterization *Renew. Energ.* **17** 73- 84.
[9] Ahmad I 2016 *Design and Fabrication of a Pyrolysis System of Biomass Solid Waste for Conversion into Solid Char.* A B Eng Project Report, Mech Engg Program Area, Universiti Teknologi Brunei, Brunei Darussalam.
[10] Bridgwater A V, Czemik S, and Piskorz J 2001 An overview of fast pyrolysis, *Prog. Thermochem. Biomass Conver.s* **2** 977- 97.
[11] Faisal M, Ismail A F and Yusuf T F 1997 Rice Husk combustion using TGA. *Proc. CTCE '97: National Conference on Combustion Technologies for Cleaner Environment* (Kuala Lumpur: CTCE).
[12] Williams P T and Besler S 1993. The pyrolysis of rice husk in a thermogravimetric analyser and static batch reactor, *Fuel* **72** (2) 151- 59.
[13] Yatim B B 1996 Perception and experiences of traditional biomass energy using industries in Malaysia *FAO-RWEDP Regional Expert Consultation on Selection Criteria and Priority Rating foe Assistance to Traditional Biomass Energy Using Industries*, (Penang: FAO-RWEDP).
[14] Williams P T, and Besler S 1994 Polycyclic aromatic hydrocarbons in waste derived pyrolytic oils *J. Anal. Appl. Phys.* **30** 17-33.

IOP Conf. Series: Materials Science and Engineering **206** (2017) 012015 doi:10.1088/1757-899X/206/1/012015

Synthesis of Biodiesel in Batch and Packed-Bed Reactors Using Powdered and Granular Sugar Catalyst

J Janaun[1,2*], P M Lim[1], W S Balan[1], A Z Yaser[1], K P Chong[2]

[1] Chemical Engineering Programme, Faculty of Engineering, Universiti Malaysia Sabah, Jalan UMS, 88400 Kota Kinabalu, Sabah Malaysia.
[2] Sustainable Palm Oil Research Unit (SPOR), Faculty of Sustainable Agriculture, Universiti Malaysia Sabah, Jalan UMS, 88400 Kota Kinabalu, Sabah Malaysia

*Corresponding author: jidon@ums.edu.my

Abstract. Increasing world production of palm oil warrants effective utilization of its waste. In particular, conversion of waste cooking oil into biodiesel has obtained global interest because of renewable energy need and reduction of CO_2 emission. In this study, oleic acid used as a model compound for waste cooking oil conversion using esterification reaction catalysed by sugar catalyst (SC) in powdered (P-SC) and granular (G-SC) forms. The catalysts were synthesized via incomplete carbonization of *D*-glucose followed by functionalization with concentrated sulphuric acid. Catalysts characterizations were done for their physical and chemical properties using modern tools. Batch and packed-bed reactor systems were used to evaluate the reactivity of the catalysts. The results showed that G-SC had slightly higher total acidity and more porous than P-SC. The experimental conditions for batch reaction were temperature of 60°C, molar ratio of 1:20 (Oleic Acid:Methanol) and 2 % wt. catalyst with respect to oleic acid. The results showed the maximum oleic acid conversion using G-SC and P-SC were 52 % and 48 %, respectively. Whereas, the continuous reaction with varying feed flow rate as a function of retention time was studied by using 3 g of P-SC in 60 °C and 1:20 molar ratio in a packed-bed reactor. The results showed that a longer retention time which was 6.48 min and feed flow rate 1.38 ml/min, achieved higher average conversion of 9.9 % and decreased with further increasing flow rate. G-SC showed a better average conversion of 10.8 % at lowest feed flow rate of 1.38 ml/min in continuous reaction experiments. In a broader perspective, large scale continuous biodiesel production is feasible using granular over powdered catalyst mainly due to it lower pressure drop.

1. Introduction

In the concern of the environment issues and limited fuel resources, the discovery of biodiesel is one of the viable solutions for the problems. Biodiesel is biodegradable, non-toxic and gives a lower emission profile [1]. It is environmentally friendly as well as beneficial [2]. Biodiesel can be produced by various types of feedstocks such as vegetable oil, animal oil, and waste cooking oil or any feedstock that contains free fatty acid (FFA) or triglycerides (TG) [3]. The most favoured reaction pathways are catalytic transesterification and esterification [3]. Both of the reactions required the presence of catalyst, either acid or base catalyst. Acid catalyst can simultaneously catalyse esterification and transesterification [3], [4]. Sugar catalyst is one of heterogeneous acid catalysts which had been reported to be successful in biodiesel production [5].

IOP Conf. Series: Materials Science and Engineering **206** (2017) 012015 doi:10.1088/1757-899X/206/1/012015

Most of the reactions are carried out in a batch reactor in laboratory scale, in which the condition of the reaction is easier to be optimized in a long period of reaction, especially when using solid acid catalyst [4], [6]. For continuous production, packed bed reactor (PBR) is used. However, the pressure drop is the main concern on every continuous PBR. By then, catalyst preparation becomes a matter to reduce the resistance of the fluid flow by increasing the particle size [7].

The sugar catalyst exists fine in particle size, which is in the range of 65 μm [8]. Powdered catalyst will be likely to encounter pressure drop in the large scale production which will likely affects equilibrium compositions and the maximum conversions that can be achieved for a given feed. However, catalytic performance was poorer for larger catalyst particles, most likely due to mass transfer limitations [9].

In the present study, the effect of catalyst particle size on converting of fatty acid into fatty acid methyl ester in both batch and PBR was determined. Sugar catalyst was prepared in powdered and granular forms and the reactivity of the catalysts was investigated.

2. Methodology

2.1. Preparation of G-SC and P-SC

Dried D-glucose was pyrolysed at temperature 400 °C for 4 h with continuous supply of nitrogen gas, N_2 in a rate of 100 mL/min. The resultant material was a dark char. The char was crushed into smaller pieces by rolling method using glass rod until the largest particle size was 5-8 mm. The crushed char was sulfonated with concentrated H_2SO_4 (95 %) for 12 h. The sulfonated char was washed repeatedly using hot distilled water until no sulfate ion was detected in the wash water. Then, the washed char was dried in oven for 3 h at temperature of 100 °C. The dried sulfonated char was sieved with two meshes with openings size of 0.5 mm and 1.4 mm. The sulfonated char was divided into two different sizes, (<0.5mm) and (>1.4 mm). Sulfonated char with smaller particle size was named P-SC and larger particle size was G-SC.

2.2. Characterization of catalyst

The total acid density test was conducted by using back titration with 0.008 M NaOH and 0.02 M of HCl. Particle Size Distribution (PSD) analysis was carried out using Malvern Zetasizer series with water as dispersant for the machine refractive index of 1.330. 1.0 g of P-SC was diluted with 20 ml of distilled water and then pipetted the clear water after few hours of settling for measurement. This test was only suitable for P-SC due to its smaller particle size. Scanning Electron Microscopy (SEM) was used to analyze the surface morphology of the catalysts. It can also observe the microstructural scale which can inspect the capillary porosity and determine the particle size of G-SC as well. Energy-dispersive X-ray microanalysis (EDX) is a complementary technique to SEM and was used to determine the composition of the features. The pore size and surface area of catalyst were determined by using Automated Mercury Porosimeter, Pascal 440 series. Approximately 0.2 g of each sample was used for the analysis with the pressure of blank sample at 200 MPa and lowest increment of pressure. The functional group of catalyst was determined by using Fourier Transform Infrared (FTIR) Spectrometer, Perkin Elmer.

2.3. Catalytic Esterification Reaction

For batch system, esterification was performed in a STEM Reacto as shown in Figure 1(a) and continuous reaction was performed in PBR as shown in Figure 1(b). Figure 2 shows the apparatus setup for PBR. For laboratory scale, condenser was used as PBR. The inner column of the condenser was packed with catalyst.

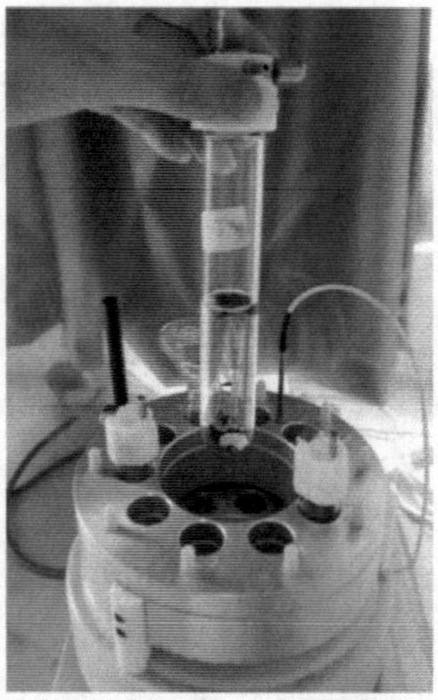

Figure 1(a). Image of STEM Reacto. Reactants and catalyst were placed in the vials.

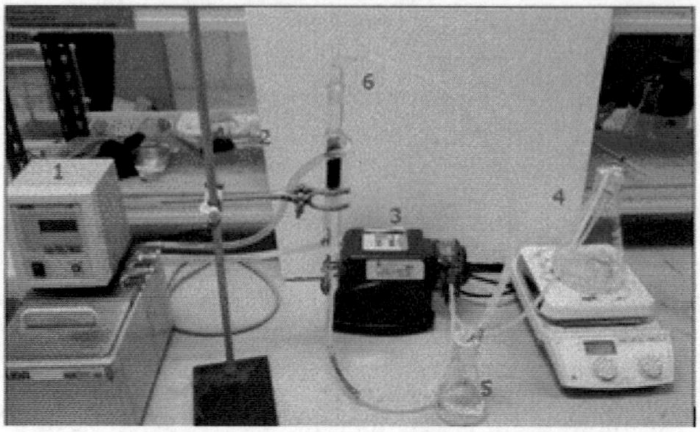

Figure 1(b). Apparatus Setup for Continuous Reaction (1. Thermal hot water bath, 2. PBR-condenser filled with catalyst, 3. Peristaltic pump, 4. Reactant mixture with hot plate, 5. Accumulation flask, 6. Product collection point).

In both systems, a molar ratio of oleic acid to methanol used was 1:20 at reaction temperature of 60 °C. For batch system, 2 wt.% of catalyst with respect to oleic acid was added and the reactants were stirred at stirring rate of 800 rpm. For every 2 hours, 100 μL of sample was pipetted from the reactor using micro pipette and placed into a clean and dry conical flask for product analysis. In continuous system, the inner column of a condenser was packed with 3 g of P-SC with and secured with glass wool. Hot thermostat water bath (60 ± 3 °C) was pumped continuously to the outer column of the condenser to heat up the reactant up to 60 °C. Reactants mixture was pumped to the reactor with a peristaltic pump with different flow rates of 1.38 ml/min, 2.64 ml/min and 3.9 ml/min. The retention times for three different flow rates were 6.48, 3.38 and 2.29 min, respectively. Reaction time for each flow rate was 1 h. The experiment was repeated with G-SC using the same conditions. The product was collected at the top of the reactor at every 15 min.

2.4. Acid value analysis

The conversion of oleic acid into methyl oleate was determined using titration method. The reduction of acid value showed the conversion of FFA into Fatty Acid Methyl Ester (FAME). A calibration curve of concentration of oleic acid against acid value was plotted. 100 μL of sample taken and diluted in 20 ml of diethyl-ether solvent and titrated with 0.1 M Potassium Hydroxide, KOH. The acid value of the sample was calculated using equation (1).

$$Acid\ value = \frac{56.1\ g}{mol} \times Normality\ of\ KOH \times \frac{Volume\ of\ titrant\ (L)}{Mass\ of\ sample\ (g)} \quad (1)$$

A blank titration is performed to determine the initial and final volume of oleic acid by using potassium hydroxide solution. The conversion of the oleic acid was calculated by using equation (2), where initial volume of oleic acid is denoted by v_i and v_f for final volume of oleic acid.

$$Conversion\ (\%) = \frac{v_i - v_f}{v_i} \times 100 \quad (2)$$

3. Results and Discussion

3.1. Characteristic of G-SC and P-SC

3.1.1. Total Acidity
G-SC was found to have slightly higher total acidity value than P-SC which is 1.778 mmol/g and 1.612 mmol/g, respectively. The results obtained are comparable to acidity of carbon catalyst reported by Zong et.al. (2007).

3.1.2. Surface Area and Porosity
Table 1 shows the properties of the granular and powdered sugar catalyst by using the mercury porosimeter. The results show that the catalysts are mesoporous with an average pore size of 1300-2300 nm (1.3 μm-2.3 μm), while the surface area is between 2.9 m^2/g to 4.0 m^2/g.

Table 1. Properties of G-SC and P-SC Obtained from Porosimeter.

	G-SC	P-SC
Total cumulative volume (mm^3/g)	46.4	29.9
Total specific surface area (m^2/g)	4.0	2.9
Average pore diameter (Nanometer)	2272.1	1326.6
Total porosity (%)	1.58	1.04
Total acidity (mmol/g)	1.78	1.61

Different particle size of catalysts does show some difference on the total porosity with a difference of 0.535 %. G-SC has a higher surface area and so does its total cumulative volume, these facts suggested that there are many layers of surface underneath the outer surface area. Therefore when mercury intrudes deep into the catalyst, there are more empty spaces and pores to be filled up, thus its total specific surface area is 1.4 times greater than P-SC. The heterogeneous catalyst activity is mainly depends on its surface area, porosity and acid group [10]. This is because the bigger the surface area, the higher the chances of attaching the acid group, so more reaction can be catalyzed, glycerol or water could block the pores which contain acid groups, thus preventing more reactions to occur [8]. G-SC has higher surface area which might also explain higher total acidity of the catalyst that P-SC.

3.1.3. Surface Morphology and Particle Size

3.1.3.1 Granular Sugar Catalyst
The size of the catalyst sample measured is approximately 3.0 x 2.7 mm. The average particle size of G-SC is 3.0 mm. In Figure 2, it can be clearly seen that glucose synthesized char is very porous, hundreds of pores can be seen on the surface of the catalyst. Pores are clearly visible under the magnification of 190. The diameter size of the pores is ranged from 4.57 μm to 119 μm.

3.1.3.2 Powdered Sugar Catalyst
Figure 3 shows the image of P-SC under SEM. The pores are not clearly seen as G-SC. Particle size of P-SC was measured by using Particle Size Distributor (PSD).The PSD result gave a range of 18.92 nm to 3219 nm in radius, thus the calculated average diameter is 843.8 nm. However, the detection scale of PSD is limited from 0.4 nm to 10000 nm in diameter. Some P-SC particle could be larger than 10000 nm (10μm). To reconfirm the analysis on the particle size, P-SC was also measured using SEM.

The result shows the particle size is in range of 100-500 μm. The smallest diameter of the visible pore determined is 2.22μm and biggest pore size is 53 μm.

Both catalysts, G-SC and P-SC, have the smallest visible pore size of 4.57 μm and 2.22 μm, respectively under SEM analysis. The determination of pore size using porosimeter shows that the average pore size of G-SC and P-SC are 2.27 μm and 1.33 μm respectively. The results was comparable with the pore size detected using SEM.

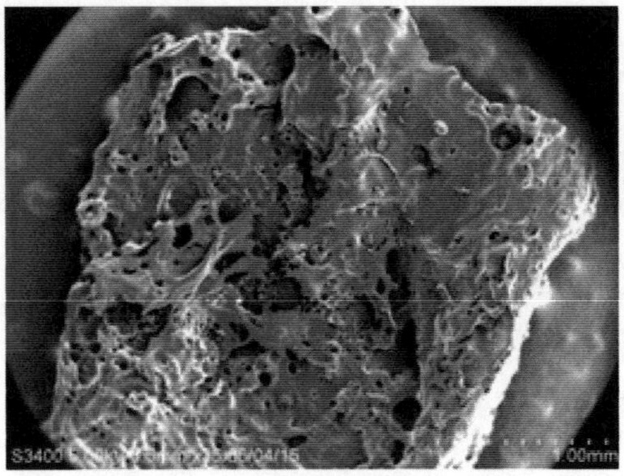

Figure 2. Granular Sugar Catalyst under SEM with magnification of 35.

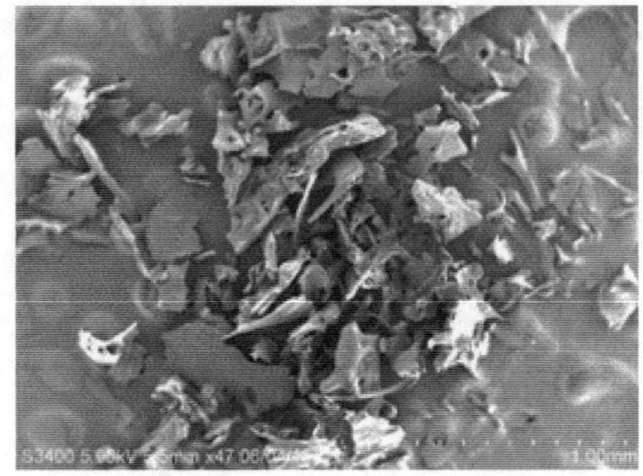

Figure 3. Powdered Sugar Catalyst under SEM with magnification of 47.

3.1.4. Surface Element Analysis

The compositions of the catalysts were determined by using Energy-Dispersive X-Ray Spectroscopy (EDX) analysis. For EDX analysis, it analyzed the composition of the elements presence on the surface of the catalyst. Table 2 shows the elemental compositions on G-SC and P-SC. Assuming all sulfurs in the catalysts are in the form of SO_3H, [4], [11] the estimated density of SO_3H is as shown in Table 2. The SO_3H density of the synthesized G-SC and P-SC are 1.380 mmol/g and 0.929 mmol/g, respectively. The SO_3H density of fresh G-SC is comparable with the result reported by Zong *et al.* (2007) P-SC shows a slightly lower SO_3H density. Reductions of sulphur content are 7.06 % and 1.82 % in mole for G-SC and P-SC, respectively.

Table 2. Composition of elements on different particle size of fresh and used sugar catalyst.

Catalyst		Mol			SO_3H
		C	O	S	(mmol/g)
Fresh	G-SC	6.395	1.174	0.138	1.380
	P-SC	6.672	1.078	0.082	0.820
Used	G-SC	5.943	1.603	0.093	0.929
	P-SC	6.551	1.176	0.078	0.780

3.1.5. Functional Group

Figure 4 shows the FT-IR spectrum of P-SC. Since the two different particle sizes show almost similar spectra, therefore the spectrum below infer the FT-IR result of both P-SC and G-SC as well.

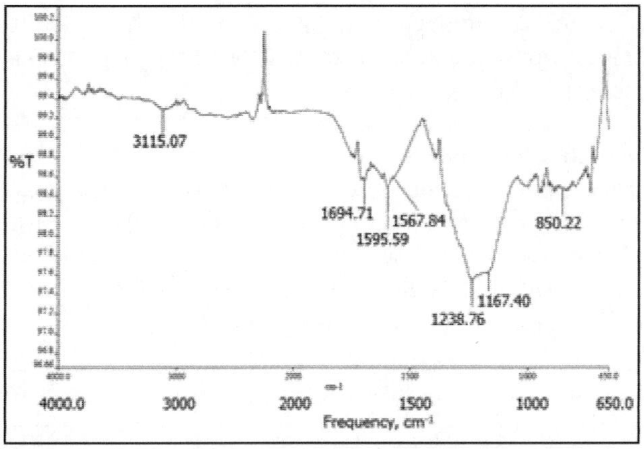

Figure 4. FT-IR spectra of powdered sugar catalyst.

Both of the catalysts were undergoing similar preparation step, therefore they have the same functional groups. The spectrum contain peak around 1000-1030 cm^{-1}, which attributes to the existence of S=O stretching [12], as well as peak at 1154 cm^{-1} attributable to the presence of SO_3H groups [13], thereby reassuring the sulfonic group existing on the carbon-based char after sulphonation. At wavenumber of 1620-1680 cm^{-1}, there is aromatic ring modes indicates C=C stretching [14]. Besides, the O-H stretching band appears at wavelength of 3200 cm^{-1}, and the band at 1702 cm^{-1} was assigned to the C=O stretching mode of the COOH group [15]. Thus, the overall FTIR analysis proven that the sulphonated char consists of polycyclic aromatic carbon sheets containing SO_3H, COOH and OH groups.

3.1.6. Thermal Stability

Based on the TGA analysis, the weight of both catalysts started to drop at 100 ˚C as this could due to the moisture content of the sample which the vaporization of water causes the decrease in the weight percent [8]. The final %weight for G-SC and P-SC are 86.21 % and 87.03 %, respectively. The different final weight% for both sugar catalysts might due to different initial moisture content as weight% of G-SC starts to decrease more than that of P-SC from 30 ˚C to 80 ˚C and the rate of decreasing %weight for both catalysts were the same after 80 ˚C until the end of analysis. Thus, the carbon-based catalysts showed almost similar thermal stability. As for DTA result, there are two significant weight loss peaks for each sample. They occurred in a range of 69-79 ˚C and 246-249 ˚C for G-SC and P-SC which can be explained by the decomposition of the functional groups. From the TGA data, it can be conclude that the G-SC and P-SC are thermally stable until 246 ˚C and 249 ˚C, respectively.

3.2. Batch Esterification Reaction

3.2.1. Effect of Particle Size of Sugar Catalyst

Figure shows the reactivity of both catalysts. From the curve fitting in Figure 5, it concluded that both catalysts have similar initial rate of reaction. However, the maximum conversion of G-SC (52 %) is higher than P-SC (48 %). The conversion values of both catalysts were supported by the findings of two journals that use sugar catalyst. The conversion values obtained was 45 % and 61 %, respectively [13]. This proves that higher total acidity of enhance the conversion rate of esterification reaction. Sugar catalyst with high SO_3H density and larger surface area is likely to have more active sites attach on it, thus G-SC had higher rate of reaction after 10 h of reaction [8]. If the heterogeneous reaction only occurs on the outer surface of the catalyst, the rate of reaction will increase substantially with decrease in particle size, since larger surface area will be obtained with decreasing particle size [16]. However, it is an interesting finding since the total specific surface area per unit volume of G-SC appears to be larger than P-SC. In this case, explanation of larger specific surface area can be incorporate with the assumption of more surface area within the granular catalyst in which there are layers of porous surface areas in G-SC catalyst.

3.2.2. Analysis of Rate Data (Kinetic Data)
Figure shows the order of reaction is second order and the k value for G-SC and P-SC are 8.0 x 10^{-7} and 7.0 x 10^{-7} $mmol^{-1}.L.min^{-1}$, respectively. The acid sites (SO_3H group) will speed up the reaction by decreasing the activation energy and increase the rate of effective collision, thus with higher total acidity value, the rate of reaction will be higher. The regressions of both lines were also in agreement with Janaun (2012) where the order of the reaction was two.

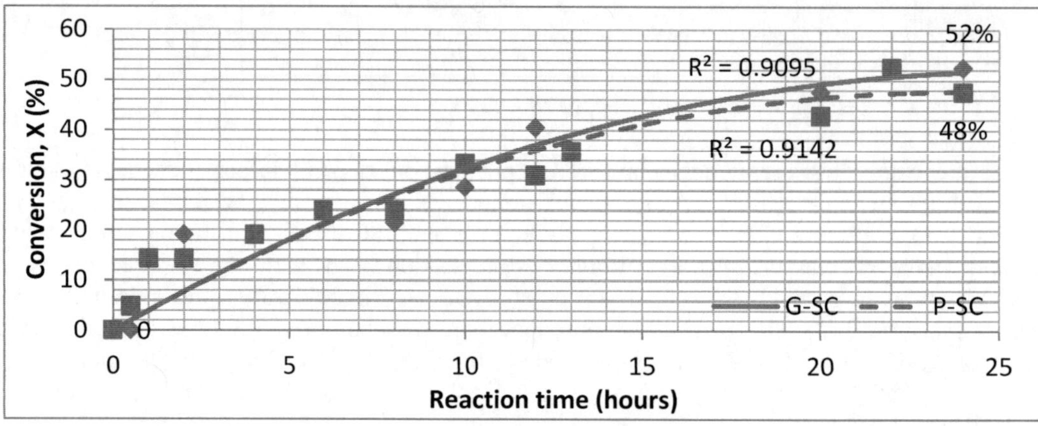

Figure 5. Conversion of oleic acid by G-SC and P-SC.

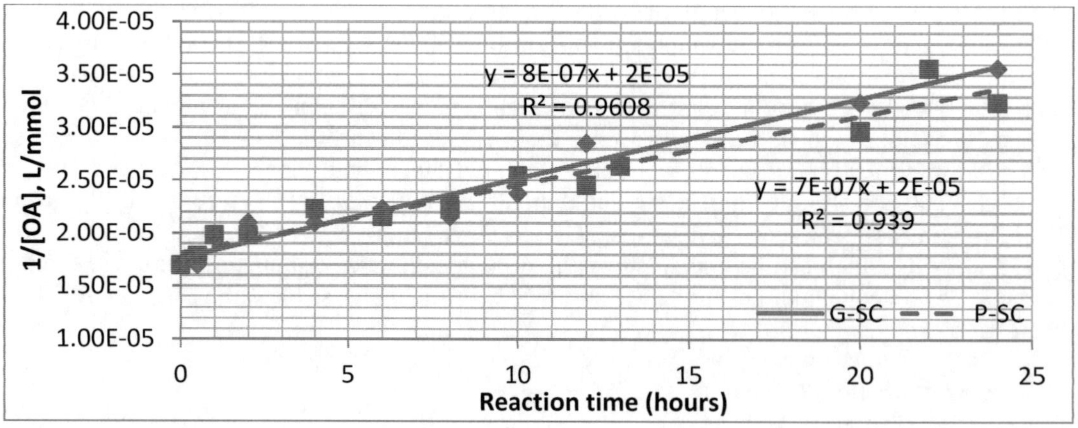

Figure 6. Concentration Differential Relationship with Concentration for k and α for G-SC and P-SC at 60˚C and 1:20 Molar Ratio.

3.3. Continuous Esterification Reaction

3.3.1. Effect of Retention Time / Feed Flow Rate
Figure 7 shows the conversion of oleic acid against time in continuous reaction at different feed flow rates (Q1=1.38 ml/min, X_{ave}=9.9 %, Q2=2.64 ml/min, X_{ave}=5.1 %, and Q3=3.90 ml/min, Xave=0.4 %,). From Figure 7, the FFA conversion of three different feed flow rates catalyzed by P-SC with a condition of 60˚C, 1:20 (OA:MeOH) molar ratio and 3g of P-SC. The maximum average FFA conversion of 9.9 % was achieved at lowest feed flow rate around 1.38 ml/min. This finding is in

agreement with the result obtained by Ren *et al.* (2012). It was reported that highest FAME conversion was obtained at lowest flow rate of 1.27 ml/min [17].

The average FFA conversion increased from 0.4 % to 9.9 % when the feed flow rate decreased from 3.90 ml/min to 1.38 ml/min (Table 3). This can be explained that the reactants have less contact time with the active sites of the catalyst when passing through the sugar catalyst column and resulting the low FFA conversion [8]. Thus, longer contact time between the reactants and the catalyst gives higher conversion.

Table 3. Summary of three different flow rates with their average conversions.

Flow rate (ml/min)	1.38	2.64	3.90
Retention time (min)	6.48	3.38	2.29
Average FFA conversion (%)	9.9	5.1	0.4

3.3.2. Effect of Particle Size of Sugar Catalyst

As expected, G-SC performed well in the continuous reaction with a slightly higher conversion of 10.8% than P-SC which convert 9.9 % of oleic acid for the same condition (feed flow rate=1.38 ml/min, temperature = 60°C, molar ratio of OA:MeOH=1:20, catalyst loading = 3 g).

In laboratory scale, pressure drop is not significant. The measurement of both inlet and outlet flow rate of the reactor, are almost similar which implies insignificant pressure drop. However, for large scale such as in industrial, P-SC is expected to encounter more pressure drop than G-SC if the length of bed catalyst is the same due to its fine particles that create more resistance to the fluid flow.

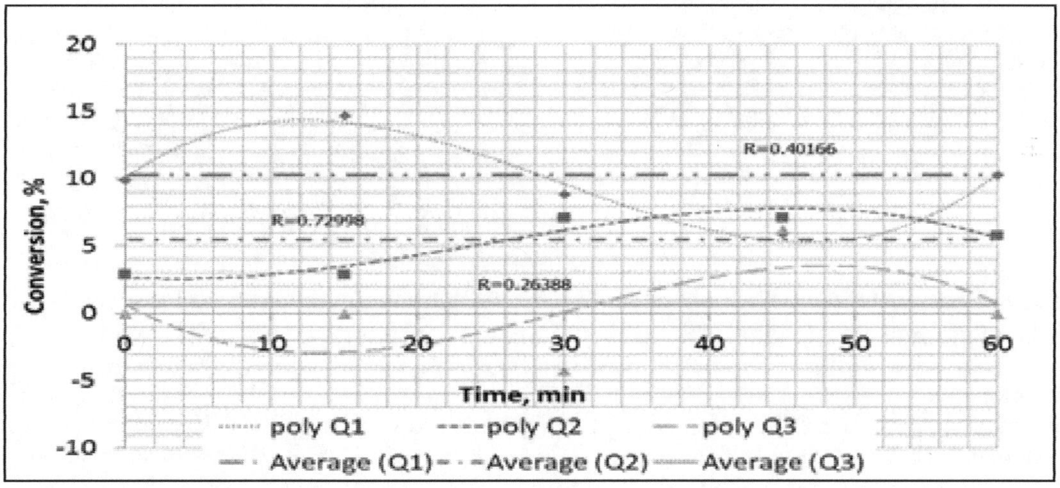

Figure 7. Conversion of oleic acid against time in continuous reaction with 3 different feed flow rates (Q1=1.38ml/min, X_{ave}=9.9%, Q2=2.64ml/min, X_{ave}=5.1%, Q3=3.90ml/min, X_{ave}=0.4%,).

3.4. Comparison Of Batch And Continuous Reactions

The esterification reaction is mostly carried out in batch condition as it is more stable and has a better control on temperature and the reactant mixtures is well-mixed throughout the entire process. Besides, the vaporization of methanol can be minimized due to the close system in batch reactor and a proper handling when pipetting the sample. Thus, the FFA conversion obtained shows a smooth curve throughout a long period of reaction. However, continuous reaction shows an undesired result data as the conversion values fluctuated during the experiment. Furthermore, difficulty to minimize the vaporization of the methanol and detection of well-mixed condition in the reactor are the factor affecting the result. Therefore, the rate of reaction is hard to determine due to the inconsistency of data.

However, continuous esterification reaction is advantageous in term of shorter reaction time. The weight percentage of catalyst used is approximately 60 times more than that in batch condition. Therefore, average FFA conversion of 9.9 % and 10.8 % were obtained in continuous reaction in a retention time of 6.48 min only, while this conversion for batch reaction shows the similar result in 2.5 h. The esterification reaction time of continuous reaction was almost 96 % faster than batch reaction in this case. Esterification reaction has a higher rate when the catalyst to reactants ratio is higher, and reaction time is shorter.

Both P-SC and G-SC catalyst works well in the continuous system as the structure of catalyst retained after 6 hours of operation as shown in Figure 8. It did not break into fine particle size due of vigorous stirring as in batch reaction shown in Figure 9. Thus, it indicates that continuous reaction is likely to create no damage to the catalyst and more preferable when it comes to time saving and more production purposes.

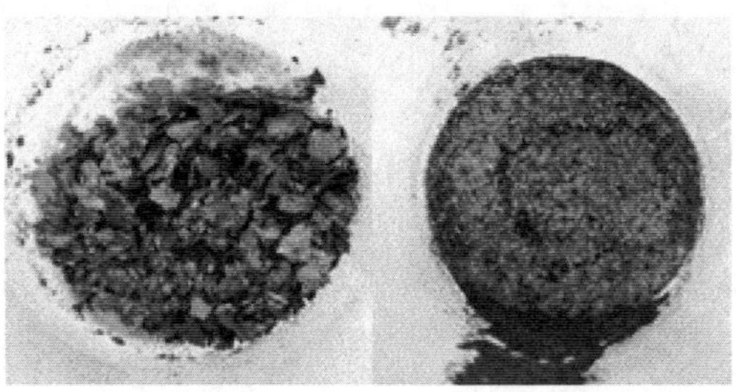

Figure 8. Condition of the G-SC (left) and P-SC (right) after continuous reaction and washing

Figure 9. Condition of the P-SC and G-SC Before and After Batch Reaction

4. Conclusions

In this research, P-SC and G-SC were successfully synthesized with the particle size ranged from 100-300 μm and 1.0-5.0 mm, respectively. The characterization of G-SC and P-SC showed that they had similarity with sugar catalyst. G-SC exhibited higher total acidity compared to P-SC. G-SC also showed better oleic acid conversion in batch and continuous reaction compared to P-SC. In short, the catalytic performance of G-SC was slightly better than P-SC. In continuous reaction, the results showed the conversion got higher as the feed flow rate got lower, this was due to the contact time of reactants with catalyst was longer.

In this research, G-SC and P-SC could be used in continuous reaction. The esterification reaction time of continuous reaction achieved 96% faster than the batch reaction. Therefore, bigger scale continuous biodiesel production by using G-SC is reasonable to perform in future for the purpose of time-saving and minimal pressure drop.

References

[1] Chen G and Fang B 2011 *Bioresour. Technol.* **102** 2635–40
[2] Huang G, Chen F, Wei D, Zhang X and Chen G 2010 *Appl. Energy* **87** 38–46
[3] Janaun J and Ellis N 2010 *Renew. Sustain. Energy Rev.* **14** 1312–20

[4] Mo X, Lotero E, Lu C, Liu Y and Goodwin J G 2008 *Catal. Lett.* **123** 1–6
[5] Cheng J K, Chao C C, Ward J D and Chien I L 2014 *J. Taiwan Inst. Chem. Eng.* **45** 76–84
[6] Liu R, Wang X, Zhao X and Feng P 2008 *Carbon* 46 1664–69
[7] Losey M W, Schmidt M A and Jensen K F 2000 *Industrial Prospects* ed P D W Ehrfeld (Berlin Heidelberg: Springer) pp 277–286
[8] Janaun J A B 2012 Development of sulfonated carbon catalysts for integrated biodiesel production
[9] Pendyala V R R, Jacobs G, Ma W, Klettlinger J L S, Yen C H and Davis B H 2014 *Chem. Eng. J.* **249** 279–284
[10] Nykulyshyn I, Voronchak T, Pikh Z and Rypka A 2012 *Cent. Eur. J. Chem.***10** 1830–41
[11] Peng L, Philippaerts A, Ke X, Van Noyen J, De Clippel F, Van Tendeloo G, Jacons P A and Sels B F 2010 *Catal. Today* **150** 140–6
[12] Xing R, Liu Y, Wang Y, Chen L, Wu H, Jiang Y, He M and Wu P 2007 *Microporous* Mesoporous Mater **105** 41–8
[13] Zong M H, Duan Z Q, Lou W Y, Smith T J and Wu H 2007 *Green Chem.* **9** 434–7
[14] Sharma R K, Wooten J B, Baliga V L, Lin X, Geoffrey Chan W and Hajaligol M R 2004 *Fuel* **83** 1469–82
[15] Chen C Y, Yeh K L, Aisyah R, Lee D J and Chang S J 2011 *Bioresour. Technol.* **102** 71–81
[16] Mekala M, Thamida S K and Goli V R 2013 *Chem. Eng. Sci.* **104** 565–573
[17] Ren Y, He B, Yan F, Wang H, Cheng Y, Lin L, Feng Y and Li J 2012 *Bioresour. Technol.* **113** 19–22

Isolation of Thermophilic Lignin Degrading Bacteria from Oil-Palm Empty Fruit Bunch (EFB) Compost

C M T Lai[1], H B Chua[1], M K Danquah[1,2] and A Saptoro[1]

[1] Department of Chemical Engineering, Curtin University Malaysia, CDT 250, Miri 98009, Sarawak, Malaysia.

[2] Curtin Sarawak Research Institute (CSRI), Curtin University Malaysia, CDT 250, Miri 98009, Sarawak, Malaysia

E-mail: chua.han.bing@curtin.edu.my

Abstract. Empty Fruit Bunch (EFB) is a potential and sustainable feedstock for bioethanol production due to its high cellulosic content and availability in Malaysia. Due to high lignin content of EFB and the lack of effective delignification process, commercial bioethanol production from EFB is presently not viable. Enzymatic delignification has been identified as one of the key steps in utilising EFB as a feedstock for bioethanol conversion. To date, limited work has been reported on the isolation of lignin degrading bacteria. Hence, there is a growing interest to search for new lignin degrading bacteria with greater tolerance to temperature and high level of ligninolytic enzymes for more effective lignin degradation. This study aimed to isolate and screen thermophilic ligninolytic microorganisms from EFB compost. Ten isolates were successfully isolated from EFB compost. Although they are not capable of decolorizing Methylene Blue (MB) dye under agar plate assay method, they are able to utilize lignin mimicked compound – guaiacol as a sole carbon on the agar plate assay. This infers that there is no correlation of ligninolytic enzymes with dye decolourization for all the isolates that have been isolated. However, they are able to produce ligninolytic enzymes (Lignin peroxidase, Manganese peroxidase, Laccase) in Minimal Salt Medium with Kraft Lignin (MSM-KL) with Lignin Peroxidase (LiP) as the predominant enzyme followed by Manganese Peroxidase (MnP) and Laccase (Lac). Among all the tested isolates, CLMT 29 has the highest LiP production up to 8.7673 U/mL following 24 h of growth.

1. Introduction

In the past two decades, extensive research has been done on second generation bioethanol production using crop residues, saw-dust, wood chips and other agriculture wastes. Most of the lignocellulosic materials are made up from agriculture wastes (rice straw, wheat straw and sugarcane bagasse) and it is cost effective and abundantly available. Malaysia is the second largest supplier of palm oil which contributed about 41% of world's supply on palm oil [1]. It has been reported that Malaysia produces 18.8 million tons of palm oil in 2012 [2]. Furthermore, production of palm oil is expected to increase up to 50 million tons per year in 2030 [1]. With the tremendous growth of palm oil production, it is predicted that 50 million tons of Empty Fruit Bunch (EFB) will be produced in 2030.

Moreover, EFB is a potential biomass material for biofuel production due to its high lignocellulosic content. Apart from that, EFB can be a promising source of renewable energy due to its abundance. In addition, bioethanol produced from EFB biomass is considered as second generation bioethanol (SGB). SGB is the ethanol derived from biomass wastes or materials of non-edible sources. Thus,

IOP Conf. Series: Materials Science and Engineering **206** (2017) 012016 doi:10.1088/1757-899X/206/1/012016

SGB is able to reduce competition of food for fuel production [3]. However, SGB production from EFB biomass is presently not viable due to high lignin content in lignocellulose structure. Lignin is a cross-linking structure (figure 1) which made up of aromatics polymers that associated with cellulose and hemicellulose as. High lignin content restricts the accessibility of enzymes involved in hydrolysis process which results in low sugar and bioethanol production. Thus, effective pretreatment process in degrading the lignin content is a vital aspect in bioethanol production from EFB as a bioethanol feedstock.

In recent years, biological method using microorganisms and enzymes has gained extensive interest in lignin pretreatment process due its cost effectiveness and environmental friendliness. Moreover, biological pretreatment process does not produce toxic compounds as much as physical and chemical process [4, 5]. Lignin peroxidase (LiP), Manganese peroxidase (MnP) and Laccase (Lac) are the major enzymes involved in lignin degradation process. To date, white rot fungi (WRF) such as *Trametes versicolor*, *Trametes hirsuta*, *P. ostreatus* and *P. radiata* are the common lignin degrading microorganisms that are capable of producing all the three major enzymes. However, ligninolytic enzymes from fungi has low stability under high working temperature, high pH condition and high substrate conditions [6, 7]. In fact, delignification process often involves high temperature up to 70 °C . Moreover, it is proven that bacterial enzymes are capable in degrading lignin and address the limitation of fungi [8].

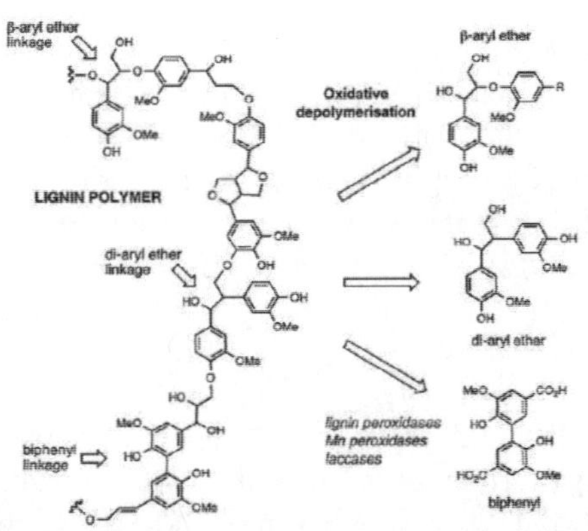

Figure 1. Lignin structure [9]

In the current prospect, *Paenibacillus* sp., *Bacillus* sp., and *Streptomyces* sp. are the common bacteria that are involved in lignin degradation process [10-12]. It is reported that Bacillus sp., *Pandoraea* sp. and *Streptomyces griseorubens* have the capability in producing LiP, MnP and Lac enzyme. Among these strains, Bacillus sp. is able to degrade wide range of lignin model and lignin derivatives such as Guaiacylglycerol- -guaiacyl ether (GGE) lignin model dimer and kraft lignin up to 70% and 35.1% (Huang et al., 2013). However, most of the isolated strains are mesophiles. Furthermore, most of the bacteria isolates are isolated from soils, wheat straw, bagasse and industrial waste. The isolation study on thermophilic bacteria from EFB for lignin degradation is relatively scarce compared to mesophiles. Only a few researchers have attempted in isolating thermophilic bacteria from EFB. *Ureibacillus terrunus*, *Nocardiopsis* sp., *S. violaceorubidus* and *Streptomyces* sp. were successfully isolated from EFB [13, 14]. However, those previous studies only demonstrated the presence of ligninolytic activities based on Remazol Brilliant Blue R (RBBR) decolouration without further examination on the LiP, MnP and Lac enzyme activities. Nevertheless, the studies above have shown the presence of thermophilic lignin degrading bacteria in the EFB compost. Thus, this study aims to isolate potential thermophilic lignin degrading bacteria from EFB compost and examine their ligninolytic enzyme activities.

2. Materials and methods

2.1. Sample collection
Young EFB (5 -10 day old) compost samples were collected from composting piles of a local oil palm mill. The compost was collected from a depth of about 20-30cm from the compost surface and stored at 4 °C . The compost samples were used in the isolation of thermophilic ligninolytic bacteria.

2.2. Isolation of bacteria
For isolation of the bacteria, compost samples were re-enriched with Minimal Salt Media which contain 0.5 g/L of kraft lignin (MSM-KL) as sole carbon source. The MSM-KL medium contained (g/L): KL (0.5); K2HPO4 (4.55); KH2PO4 (0.53); CaCl2 (0.5); MgSO4 (0.5); NH4NO3 (5) with trace elements CuSO4 (0.002), FeSO4 (0.01); MnSO4 (0.001) and ZnSO4 (0.001)[7]. 5 g of EFB compost was added to 95 mL of sterile MSM-KL media placed in 250 mL Erlenmeyer flasks at 50 °C for 7 days. The solution was stirred vigorously and then allowed to stand for the compost to settle.

One ml aliquot was aseptically, serially diluted until a dilution of was obtained. 100 μL of diluted solution was plated on MSM-KL agar plate with 30 g/L of agar content. The plates were incubated at 50 °C for 7 days until colonies were formed. The isolated bacteria were plated onto fresh MSM-KL agar plates until pure cultures were obtained. Thirty four isolates were obtained and maintained in Tryptic Soy Agar (TSA) plates. The purity of the bacteria was examined under the light microscope.

2.3. Screening of ligninolytic bacteria
The following three methods were used to screen the ligninolytic bacteria. (1) MSM-KL agar plate assays with 25 mg/L of Methylene Blue (MB) dye, 2.5 g/L Yeast extract, 2 g/L glucose; (2) TSA with 25 mg/L of MB, (3) 1 mM Guaiacol, a lignin mimicking compounds was used as a carbon source in MSM-KL agar plate assays [12]. For plate assays, agar plates were incubated at 50 °C for seven days. Growth and decolourization of the agar was monitored daily.

2.4. Enzyme profile of the selected isolates
Ten isolates which showed good growth were selected and grown aerobically in 80 mL MSM-KL medium with 10 g/L glucose, 1.0 g/L yeast extract and addition of 2mL inoculum in 250 mL Erlenmeyer flasks under submerge fermentation. Inoculum was prepared by inoculating one loopful of each selected isolates into 20 ml of Tryptic Soy Broth (TSB) and incubated at 50 °C for 16 h to obtain a final OD600 of 1.0. The MSM-KL medium with inoculum was incubated for 6 days and culture samples were removed at 24 h intervals. The un-inoculated MSM-KL medium was used as a control. The culture filtrate was centrifuged at 4,000 rpm for 20 min at 4 °C . The supernatant obtained after centrifugation was used for the enzyme assays.

2.4.1. LiP enzyme assay. The lignin peroxidase (LiP) activity were measured via the oxidation of veratryl alcohol to veratryaldehyde at 310nm. One milliliters of enzymatic assay consisted of 0.4 mL of citrate-phosphate buffer (100 mM, pH 2.7), 0.1 mL of veratryl alcohol (20 mM), and 0.5 mL of fluid sample. Forty microliters of H2O2 (20 mM), which was freshly prepared daily and added to initiate the reaction [15]. The conversion to veratraldehyde is monitored in 1 mL quartz cuvette at 310 nm wavelength using UV-vis spectrophotometer. One unit of enzyme activity corresponded to the oxidation of 1 micromole veratraldehyde converted from veratryl alcohol per minute under the assay conditions with molar extinction coefficient of $\varepsilon_{310} = 93000\ M^{-1}\ cm^{-1}$.

2.4.2. MnP enzyme assay. Manganese peroxidase (MnP) activity was measured via the oxidation of guaiacol to a coloured product using UV-Vis spectrophotometer at 465nm. The enzymatic assay of 1.0 mL consisted of 0.4 mL of sodium lactate buffer (100mM, pH 4.5), 0.1 mL of guaiacol (1 mM), 0.1 mL of MnSO4 (1 mM), and 0.4 mL of supernatant of the sample [16]. 30 µL of H2O2 (0.1 mM) which was freshly prepared daily and added to initiate the oxidation of guaiacol. The formation of the coloured product was measured in the 1 mL quartz cuvette at 465 nm wavelength. One unit of enzyme activity is defined as the amount of enzyme required to produce 1 micromole of coloured product per minute under the assay conditions with a molar extinction coefficient of $\varepsilon_{465} = 12100\ M^{-1}cm^{-1}$.

2.4.3. Lac enzyme assay. Laccase (Lac) activity was assayed via oxidation of 2,2'-azinobis-(3-ethylbenzethiazoline-6-sulphonate) (ABTS) using method developed by Wolfenden and Wilson (1982). This enzymatic mixture made up of 0.15 mL of 0.03% ABTS, 0.5 mL of 0.1 M sodium acetate buffer at pH 5.0 and 0.35 mL of supernatant of the sample [17]. The formation of oxidized ABTS was measured in 1 mL quartz cuvette (1 cm light path) and 530 nm wavelength using UV-Vis Spectrophotometer. Oxidation of ABTS was followed by absorbance increase at 420 nm with a molar extinction coefficient of $\varepsilon 420 = 32000\ M^{-1}\ cm^{-1}$.

3. Result and Discussions

3.1. Isolation and screening of bacteria

Thirty-four bacteria isolates were successfully isolated using MSM-KL agar plate. In the screening process, none of the isolates showed decolourization on MSM-KL and TSA incorporated with MB as the lignin polymeric dye. However, there were bacteria growth in all of the plate assays. This result indicated that the agar medium could sustain the growth of these isolates in the presence of glucose. Although dye-based plate assay is a quick method used to screen for lignin degrading bacteria [18], but decolourisation of MB was not observed in the present study with dye-based plate assay. Similar observation was reported for *Streptomyces psammoticus* that is capable in producing LiP, MnP and Lac but has no decolourization ability on dyes such as Poly R and RBBR [12]. The study by Machado et al. [19] showed that *Trogia buccinalis* is able to decolourize RBBR in the absence of MnP and Lac enzymes. These observations seem to suggest there could be other enzymes that are responsible in degrading the dye content.

On the guaiacol plate assays, all ten isolates showed positive growth. These isolates were maintained in TSA medium. The characteristic features, morphology and size of the ten isolates' colonies are stated in table 1. Although guaiacol is a common lignin model compound that could be degraded by LiP, MnP and Lac enzyme but no decolourization zone was observed around the colonies in any isolates. Formation of colonies on guaiacol plate assay in the absence of glucose has indicated the capability of the cultures to utilise guaiacol as the sole carbon source [12]. Table 1 summarised the growth ability of respective isolates in different agar plate assays. The bacteria colonies were visible with an average incubation time of 16 h and 24 h on TSA agar and MSM-KL agar that incorporated with MB dye respectively. The result in table 1 also revealed that isolate CLMT 20, CLMT 21 and CLMT 22 showed active growth regardless of any agar medium used. To date, several lignin degrading bacteria and actinomycetes such as *Bacillus* sp., *Ureibacillus terrenus*, *Nocardiopsis* sp., *Streptomyces violaceorubidus* and *Streptomyces* sp. were successfully isolated from oil palm plantation soils and EFB compost [7, 13, 14]. According to the result in table 1, all of the isolates are in rod shaped. Thus, it is believed that rod-shaped bacteria such as *Bacillus* sp., *Pseudomonas* sp., *Leucobacter* sp. and *Ureibacillus terrenus* could be the potential bacteria in these isolates.

3.2. Enzyme profile of the selected isolates

Ten isolates that showed positive growth on guaiacol plate assay were further screened for ligninolytic enzyme activities (LiP, MnP and Lac) using MSM-KL medium under submerged fermentation for 120 h. According to Chandra's findings [20], bacteria isolates such as *Citrobacter* sp. and *Citrobacter freundii* utilize glucose and yeast extract as carbon and nitrogen source in the initial stage and subsequently use kraft lignin as co-metabolites. Thus, glucose and yeast extract was added to promote

IOP Conf. Series: Materials Science and Engineering **206** (2017) 012016 doi:10.1088/1757-899X/206/1/012016

the bacterial growth in ligninolytic enzyme production [21]. The enzyme activities for LiP, MnP and Lac were measured at the interval of every 24 h. Interestingly, all ten isolates showed positive result in production of all three main ligninolytic enzymes (LiP, MnP and Lac) simultaneously. It was found that LiP enzyme is the predominant enzymes followed by MnP and Lac for all the isolates. As shown in figure 2 to figure 4, LiP and MnP enzymes were found to be more active than Lac activity. Apart from that, high peroxidase enzyme (LiP and MnP) activity was detected during the first 24 h of incubation but gradually decreased until termination of experiment. In fact, most of the isolates achieved maximum LiP production at 24 h except for isolate CLMT 18 and CLMT 20 on LiP enzyme production.

Table 1. Characteristic features and growth of ligninolytic bacteria isolates in the following MSM-KL,TSA and

Isolates	Morphology	size (μm) length, width	Gram staining	Colour of colonies	MSM-KL + MB[a]	TSA+ MB[b]	Guaiacol[c]
CLMT 5	Rod shape	3.00, 0.67	+ve	Yellow	+	++	+
CLMT 18	Rod shape	2.33, 0.83	+ve	Yellow	+	+++	+
CLMT 19	Rod shape	2.33, 0.67	+ve	Off-white	+	++	+
CLMT 20	Rod shape	2.00, 0.67	+ve	Yellow	++	+++	+++
CLMT 21	Rod shape	2.00, 0.50	+ve	Yellow	++	+++	++
CLMT 22	Rod shape	2.67, 0.67	-ve	Yellow	++	+++	++
CLMT 23	Rod shape	2.00, 0.83	+ve	Brown	+	++	+
CLMT 28	Rod shape	3.67, 0.50	+ve	Brown	+	++	+
CLMT 29	Rod shape	2.33, 0.83	+ve	Off-white	+	++	+
CLMT 34	Rod shape	2.67, 0.50	-ve	Off-white	+	++	+

[a] MSM-KL + 25 mg/L MB dye + 2.5 g/L Yeast extract + 2 g/L glucose
[b] TSA + 25 mg/L MB dye
[c] MSM-KL + 1 mM guaiacol
+++; very fast growth (colonies visible after 12 hours)
++; fast growth (colonies visible after 18 hours)
+; slow growth (Visible colonies after 24 hours)

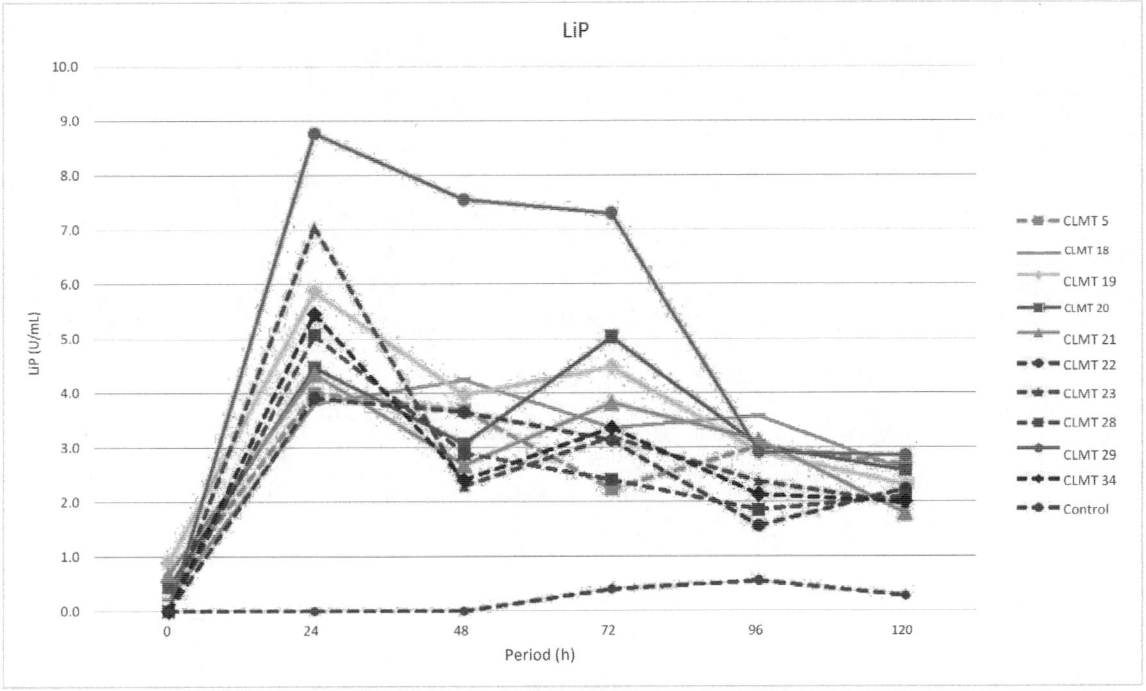

Figure 2. Enzyme profile of LiP enzymes activities for the selected 10 isolates

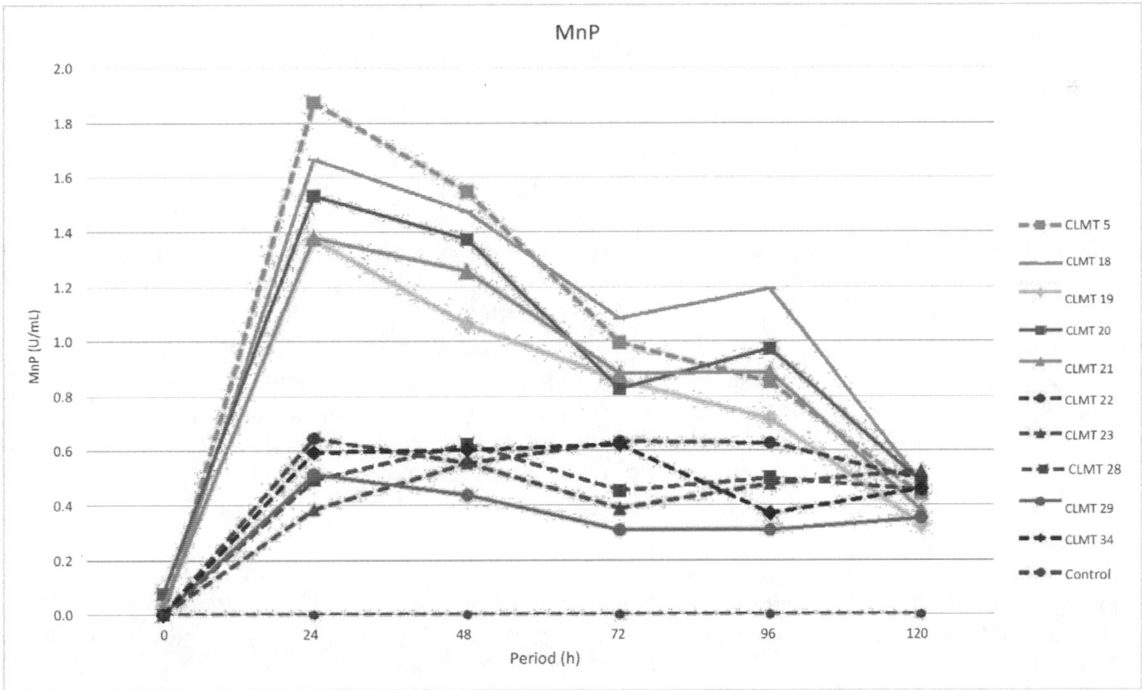

Figure 3. Enzyme profile of MnP enzymes activities for the selected 10 isolates

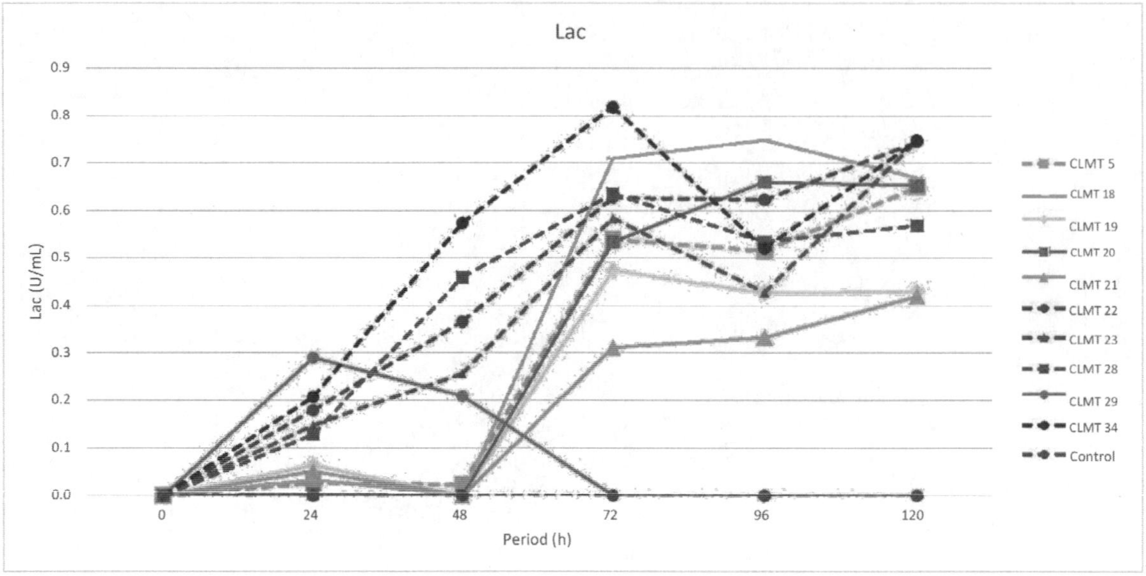

Figure 4. Enzyme profile of Lac enzymes activities for the selected 10 isolates

According to figure 4, low Lac enzyme activity was observed on the first 24 h, and it increased steadily with the maximum production at 72 h, and its activity was maintained towards the end of fermentation. The observation above is in agreement with other studies on lignin degrading bacteria such as *Comomanas* sp. B-9, *Pandoraea* sp. B-6, *B. subtilis* and *Klebsiella pneumonia* [22-24]. In fact, most of the isolates in the current study has a higher rate of enzymatic production since the maximum peroxidase production occurred after about 24 h duration of growth. Based on previous studies, peroxidase enzymes achieved maximum production in the first 96 h of incubation whereas Lac achieved its maximum production at the later stage. Both *B. subtilis* and *Klebsiella pneumonia* also achieved highest peroxidase (LiP and MnP) and Lac enzyme activities at 48 h and 96 h respectively. In comparison with the isolates obtained in this study, the rate of peroxidase enzyme production by mesophiles such as *B. subtilis* and *Klebsiella pneumonia* is relatively lower since longer period is required to achieve maximum peroxidase production. The phenomena of peroxidase enzyme being active in the early stage could be due to large lignin polymer molecules that are not able to traverse into the cell membrane in the initial stage. Thus, extracellular enzymes such as peroxidase enzymes are secreted externally to degrade lignin polymer compounds to smaller fragment during primary metabolism [22, 23]. Moreover, some strains such as *Pcynoporus cinnabarinus*, *P. sanguineus*, and *Phlebia radiata* produced maximum Lac enzymes under nitrogen-limited condition [25]. The carbon and nitrogen sources would be depleted towards the end of fermentation process. Hence, it is possible that the isolates in the present study preferred nitrogen-limited condition for Lac production, thus, Lac was secreted and reached its maximum production at about 96 h to 120 h.

In the present study, only isolate CLMT 29 showed a different trend on peroxidase and Lac enzyme production. Isolate CLMT 29 achieved maximum Lac enzyme activity at 24 h followed by declination till no Lac enzyme activity was detected from 72 h onwards. Moreover, CLMT 29 attained greatest LiP enzyme production at 8.7673 U/ml among all the tested isolates in this study. Apart from that, CLMT 29 also maintained a high level of LiP enzyme activity throughout the period of 24 h to 72 h but it has the lowest MnP and Lac enzyme activity compared to the other isolates involve.

IOP Conf. Series: Materials Science and Engineering **206** (2017) 012016 doi:10.1088/1757-899X/206/1/012016

Table 2. Highest LiP, MnP and Lac enzyme activities produced by selected isolates

Isolates	LiP		MnP		Lac	
	Enzyme activity (U/ mL)	Period (h)	Enzyme activity (U/ mL)	Period (h)	Enzyme activity (U/ mL)	Period (h)
CLMT 5	3.9923	24	**1.8759**	24	0.5392	72
CLMT 18	4.2495	48	**1.6684**	24	**0.7475**	96
CLMT 19	5.8598	24	1.3726	24	0.4728	72
CLMT 20	5.0323	72	**1.5322**	24	0.6606	96
CLMT 21	4.3613	24	1.3790	24	0.4197	120
CLMT 22	3.9140	24	0.6451	24	0.7443	120
CLMT 23	**7.0452**	24	0.5533	48	**0.7475**	120
CLMT 28	5.0658	24	0.6235	48	0.6357	72
CLMT 29	**8.7673**	24	0.5150	24	0.2907	24
CLMT 34	5.4572	24	0.6225	72	**0.8178**	72

Table 2 showed that isolate CLMT 29 and CLMT 23 produced greater LiP enzymes with maximum activities of 8.7673 U/mL and 7.0452 U/mL respectively. Other isolates produced comparative lower LiP enzyme activities which ranged from 3.9923 U/mL to 5.8598 U/mL. Result in table 2 revealed that CLMT 5, CLMT 18 and CLMT 20 have relatively high MnP enzyme production of 1.8759 U/mL, 1.6684 U/mL and 1.5322 U/mL respectively compared to other isolates. Apart from that, three isolates (CLMT 34, CLMT 18, CLMT 23) showed the highest Lac enzyme production with more than 0.7443 U/mL of enzyme activities. Based on the observation above, different isolates seemed to have different enzyme activities. The isolated bacteria isolates in the current study showed promising results since they are able to produce all major ligninolytic enzymes (LiP, MnP, Lac). Most of the lignin degrading bacteria reported only produced at most two enzymes (MnP with Lac or LiP with Lac) [22, 23, 26, 27]. To date, *Autochtonous streptomyces* is the bacteria isolate that produced highest LiP activity of about 400 U/mL. The LiP activity of CLMT 29 obtained in this study is lower compared to *Autochtonous Streptomyces*. However, *Autochtonous Streptomyces* only produces LiP and Lac enzymes, and no MnP was detected. The presence of three major ligninolytic enzymes is vital since these enzymes worked synergistically in the lignin degradation process. So far, only several bacteria such as *B. subtilis*, *Klebsiella pneumonia* and *Bacillus* sp. SHC1 were capable in producing all three enzymes [7, 24]. Due to their capability, *Bacillus* sp. is recognised as a highly potential lignin-degrading bacteria.

Comparative study in table 3 showed that LiP enzyme activity in the current study is higher than the previous studies on mesophiles such as *B. subtilis*, *Klebsiella pneumonia* and *Bacillus* sp. SHC1 but lower than *Autochtonous Streptomyces*. The LiP enzyme activities for CLMT 29 is about 50 times lower than *Autochtonous Streptomyces* (400 U/mL). In fact, the LiP enzyme activity of CLMT 29 is also less active compared to *Stereum ostrea* that achieved 270 U/g of LiP enzyme activity [28]. Interestingly, the LiP activity of all the isolates obtained in this study is significantly greater than the reported study on rod-shaped bacteria *B. subtilis* and *Bacillus* sp. SHC1. The LiP activity for CLMT 29 is about 6 times higher than LiP enzyme activity obtained from B. subtilis [24]. Moreover, it is also comparable to some of the common lignin degrader white rot fungi such as *Pleurotus ostrestus* (7.250 U/mL) [29]. To date, there is no report on isolation of thermophilic lignin degrading bacteria with high LiP enzyme activity. This is the first report on production of LiP enzyme by thermophilic bacteria that was isolated from EFB compost.

Table 3. Comparison of LiP, MnP and Lac enzyme activities with previous studies on bacteria

Isolates	Maximum enzyme activities (U/mL)			Reference
	LiP	MnP	Lac	
Comamonas sp. B-9	*ND	2.9032	1.2500	[22]
Pandoraea sp. B-6	*ND	2.2492	1.1206	[23]
Autochtonous Streptomyces	400	*ND	0.0950	[27]
B. subtilis	1.5000	0.8000	3.4000	[24]
Klebsiella pneumonia	1.1000	0.6000	0.4000	
Bacillus sp. SHC1	0.1684	2.0097	0.0021	[7]
CLMT 29	8.7673	0.5150	0.2907	Current study
CLMT 5	3.9923	1.8759	0.5392	
CLMT 34	5.4572	0.6225	0.8178	

Besides, the MnP activity for CLMT 5 (1.8759 U/mL) is also comparable to Bacillus sp. SHC1 (2.0097 U/mL) isolated from oil palm plantation soils. However, the MnP activities for the rest of the nine isolates are lower than *Bacillus* sp. SHC1. *Comamonas* sp. B-9 (2.9032 U/mL) achieved greater MnP production compared to CLMT 5 (1.8759 U/mL). This could be due to MnP being not the predominant enzyme for the isolated isolates in the current study. Apart from that, similar observation was observed on Lac enzymes. Among all the tested isolates, CLMT 34 (0.8178 U/mL) produced greatest Lac enzyme activity whereby its activity is higher than *Klebsiella pneumonia* (0.4000 U/mL) but it is about 4 times less active compared to B. subtilis (3.4000 U/mL). The variation in enzyme activities observed across different isolates implies that the production of enzyme activities could be isolate dependent and affected by the nutritional requirement in the fermentation medium.

4. Conclusions

Ten lignin thermophilic degrading bacteria isolates have been successfully isolated from the oil-palm EFB compost. This study highlights that these bacteria isolates have the capability in producing LiP, MnP and Lac simultaneously in the same fermentation medium. It was found that LiP is the predominant enzymes. In fact, most of the studies showed that MnP and Lac are the predominant enzymes in lignin degrading bacteria. Moreover, 90% of lignin is made up from non-phenolic compound that are not readily oxidised by MnP and Lac. Thus, LiP enzyme is highly useful in lignin degradation process as it has an outstanding ability in degrading non-phenolic aromatics compound. Additionally, most of the bacteria isolates obtained in this study achieved maximum enzyme production within a short period ranging from 24 h to 72 h. This preliminary study seemed to indicate that the isolated bacteria isolates could be useful in pretreatment lignin degradation process for bioethanol production from EFB. Future work on optimization of the fermentation medium and operational conditions can be performed to maximize the ligninolytic enzyme production by these isolates.

Acknowledgement

This research has been supported by Faculty of Engineering and Science, Curtin University Sarawak.

Nomenclature

ABTS	2,2'-azinobis-(3-ethylbenzethiazoline-6-sulphonate)
EFB	Empty fruit bunch
GGE	Guaiacylglycerol- β -guaiacyl ether
Lac	Laccase
LiP	Lignin peroxidase

MB	Methylene blue
MnP	Manganese peroxidase
MSM-KL	Minimal salt media with kraft lignin
RBBR	Remazol Brilliant Blue R
SGB	Second generation bioethanol
TSA	Tryptic soy agar
TSB	Tryptic soy broth
WRF	White rot fungi
ε	Extinction coefficient

References

[1] Lim J M H 2010 A case study on palm empty fruit bunch as energy feedstock *SEGi Review* **3** p 3-15

[2] Chang S H 2014 An overview of empty fruit bunch from oil palm as feedstock for bio-oil production *Biomass Bioenerg.* **62** p174-181

[3] Tye Y Y, Lee K T, Wan Abdullah W N and Leh C P 2011 Second-generation bioethanol as a sustainable energy source in Malaysia transportation sector: Status, potential and future prospects *Renew. Sustainable Energy Rev.* **15** p 4521-4536

[4] Agbor V B, Cicek N, Sparling R, Berlin A, and Levin D B 2011 Biomass pretreatment: Fundamentals toward application *Biotechnol. Adv.* **29** p 675-685

[5] Alvira P, Tomás-Pejó E, Ballesteros M, and Negro M J 2010 Pretreatment technologies for an efficient bioethanol production process based on enzymatic hydrolysis: A review *Bioresour. Technol.* **101** p 4851-4861

[6] Wang Y, Liu Q, Yan L, Gao Y, Wang Y, and Wang W 2013 A novel lignin degradation bacterial consortium for efficient pulping *Bioresour. Technol.* **139** p 113-119

[7] Abdul Rahman N H, Abdul Rahman N A, Abd Aziz S and Hassan M A 2013 Production of Ligninolytic Enzymes by Newly Isolated Bacteria from Palm Oil Plantation Soils

[8] Thambirajah J J, Zulkali M D and Hashim M A 1995 Microbiological and biochemical changes during the composting of oil palm empty-fruit-bunches. Effect of nitrogen supplementation on the substrate *Bioresour. Technol.* **52** p 133-144

[9] Bugg T D, Ahmad M, Hardiman E M and Singh R 2011 The emerging role for bacteria in lignin degradation and bio-product formation. *Curr. Opin in Biotechnol.* **22** p 394-400.

[10] Woo H L, Hazen T C, Simmons B A and DeAngelis K M 2014 Enzyme activities of aerobic lignocellulolytic bacteria isolated from wet tropical forest soils *Syst.Appl. Microbiol.* **37** p 60-67

[11] Naz S, Devtare S, Satapathy S and Gupta S 2015 Study of ligninnolytic bacteria isolation and characterization from Dhamdha Agro Field of Bhilai-Durg Region *IJERT* **4** p 258262

[12] Niladevi K N and Prema P 2005 Mangrove Actinomycetes as the source of ligninolytic enzymes *Actinomycetologica*. **19** p 40-47

[13] Ting A S Y, Hermanto A and Peh K L 2014 Indigenous actinomycetes from empty fruit bunch compost of oil palm: Evaluation on enzymatic and antagonistic properties *Biocatal. Agric. Biotechnol.* **3** p 310-315

[14] Ting A S Y, Tay H, Peh K L, Tan W S and Tee C S 2013 Novel isolation of thermophilic *Ureibacillus terrenus* from compost of empty fruit bunches (EFB) of oil palm and its enzymatic activities *Biocatal. Agric. Biotechnol.* **2** p162-164

[15] Tien M and Kirk T K 1988 Lignin peroxidase of *Phanerochaete chrysosporium Meth. Enzymol.* **161** p 238-249

[16] Li X, Jia R, Li P and Ang S 2009 Response surface analysis for enzymatic decolorization of Congo red by manganese peroxidase *J. of Mol. Catal. B Enzym.* **56** p 1-6

[17] Wolfenden B S and Willson R L 1982 Radical-cations as reference chromogens in kinetic studies of ono-electron transfer reactions: pulse radiolysis studies of 2,2'-azinobis-(3-ethylbenzthiazoline-6-sulphonate) *J. Chem. Soc. Perkin Trans. 2.* p 805-812

[18] Bandounas L, Wierckx N, de Winde J and Ruijssenaars H 2011 Isolation and characterization of novel bacterial strains exhibiting ligninolytic potential *BMC Biotechnol.* **11** p 1-11

[19] Machado K M G, Matheus D R and Bononi V L R 2005 Ligninolytic enzymes production and Remazol brilliant blue R decolorization by tropical brazilian basidiomycetes fungi *Braz. J. Microbiol.* **36** p 246-252

[20] Chandra R and Bharagava R N 2013 Bacterial degradation of synthetic and kraft lignin by axenic and mixed culture and their metabolic products *J. Environ. Biol.* **34** p 991-999

[21] Shi Y, Chai L, Tang C, Yang Z, Zhang H, Chen R, Chen Y and Zheng Y 2013 Characterization and genomic analysis of kraft lignin biodegradation by the beta-proteobacterium *Cupriavidus basilensis* B-8 *Biotechnol Biofuels* **6** p 1-14

[22] Chen Y H, Chai L Y, Zhu Y H, Yang Z H, Zheng Y and Zhang H 2012 Biodegradation of kraft lignin by a bacterial strain *Comamonas* sp. B-9 isolated from eroded bamboo slips *J. Appl. Microbiol.* **112** p 900-906

[23] Shi Y, Chai L, Tang C, Yang Z, Zheng Y, Chen Y and Jing Q 2013 Biochemical investigation of kraft lignin degradation by *Pandoraea* sp. B-6 isolated from bamboo slips *Bioprocess Biosyst. Eng.* **36** p 1957-1965

[24] Yadav S and Chandra R 2015 Syntrophic co-culture of *Bacillus subtilis* and *Klebsiella pneumonia* for degradation of kraft lignin discharged from rayon grade pulp industry *J. Environ. Sci.* **33** p 229-238

[25] Elisashvili V, Kachlishvili E and Penninckx M 2008 Effect of growth substrate, method of fermentation, and nitrogen source on lignocellulose-degrading enzymes production by white-rot basidiomycetes *J. Ind. Microbiol. Biotechnol.* **35** p 1531-1538

[26] Chen Y, Chai L, Tang C, Yang Z, Zheng Y, Shi Y and Zhang H 2012 Kraft lignin biodegradation by *Novosphingobium* sp. B-7 and analysis of the degradation process. *Bioresour. Technol.* **123** p 682

[27] Buraimoh O M, Amund O O and Ilori M O 2015 Kraft lignin degradation by *Autochtonous Strepmyces* strains isolated from a tropical lagoon ecosystem *J. Microbiol. Biotechnol. Food Sci.* **5** p 248

[28] Usha K Y, Praveen K and Reddy B R 2014 Enhanced production of ligninolytic enzymes by a mushroom *Stereum ostrea Biotechnol. Res. Int.* **2014** p 9

[29] Kheiralla Z, El-Din S, Malek S and Aziz D 2013 Optimization of cultural conditions for lignin peroxidase production by *Phanerochaete chrysosporium* and *Pleurotus ostreatus*. *Academia J. Biotechnol.* **1** p 87-95

Syngas Production from CO_2 Reforming and CO_2-steam Reforming of Methane over Ni/Ce-SBA-15 Catalyst

J S Tan[1], H T Danh[2], S Singh[1], Q D Truong[3], H D Setiabudi[1] and D -V N Vo[1, 4,*]

[1] Faculty of Chemical & Natural Resources Engineering, University Malaysia Pahang, Lebuhraya Tun Razak, Gambang 26300, Pahang, Malaysia
[2] Clean Energy and Chemical Engineering, Korea University of Science and Technology (UST), Daejeon, 305-350, Korea
[3] Institute of Multidisciplinary Research for Advanced Materials, Tohoku University, Katahira 2-1-1, Aoba-Ku, Sendai 980-8577, Japan
[4] Centre of Excellence for Advanced Research in Fluid Flow, Universiti Malaysia Pahang, 26300 Gambang, Kuantan, Pahang, Malaysia

*E-mail: vietvo@ump.edu.my (D -V N Vo)

Abstract. This study compares the catalytic performance of mesoporous 10%Ni/Ce-SBA-15 catalyst for CO_2 reforming and CO_2-steam reforming of methane reactions in syngas production. The catalytic performance of 10%Ni/Ce-SBA-15 catalyst for CO_2 reforming and CO_2-steam reforming of methane was evaluated in a temperature-controlled tubular fixed-bed reactor at stoichiometric feed composition, 1023 K and atmospheric pressure for 12 h on-stream with gas hourly space velocity (GHSV) of 36 L g_{cat}^{-1} h^{-1}. The 10%Ni/Ce-SBA-15 catalyst possessed a high specific BET surface area and average pore volume of 595.04 m^2 g^{-1}. The XRD measurement revealed the presence of NiO phase with crystallite dimension of about 13.60 nm whilst H_2-TPR result indicates that NiO phase was completely reduced to metallic Ni^0 phase at temperature beyond 800 K and the reduction temperature relied on different degrees of metal-support interaction associated with the location and size of NiO particles. The catalytic reactivity was significantly enhanced with increasing H_2O/CO_2 feed ratio. Interestingly, the H_2/CO ratio for CO_2-steam reforming of methane varied between 1 and 3 indicated the occurrence of parallel reactions, i.e., CH_4 steam reforming giving a H_2/CO of 3 whilst reverse water-gas shift (RWGS) reaction consuming H_2 to produce CO gaseous product.

1. Introduction

Approximately 80% of the present global energy requirements are supplied by the hydrocarbon-based fossil fuels which eventually lead to depletion of limited fossil energy resources and global greenhouse effect problem [1]. According to Omae et al. [2], the amount of carbon dioxide has increased by 90 ppm (480 billion tons) in about 200 years since 1800. Therefore, it is necessary to seek for an alternative solution to remedy the environmental misdeed and substituting the hydrocarbon-based energy. Syngas is referring to a mixture of H_2 and CO which has been utilised as raw material for synthetic fuels production using Fisher-Tropsch synthesis for fossil fuels replacement [3]. Generally, the catalytic steam or CO_2 reforming of hydrocarbons reaction are employed to generate syngas.

IOP Conf. Series: Materials Science and Engineering **206** (2017) 012017 doi:10.1088/1757-899X/206/1/012017

However, severe catalyst deactivation due to carbon formation in CO_2 reforming reaction is always a setback for an industrial application whilst steam reforming reaction giving a stoichiometric molar ratio of H_2/CO around 3 is unsuitable for Fischer-Tropsch synthesis [4]. Therefore, the production of syngas from combined steam and CO_2 reforming (CO_2-steam reforming) of methane seems to be a promising method reasonably due to its remarkably feasible process to control the desired H_2/CO product ratio of 2 known as a practical feed composition for the Fischer-Tropsch (FT) process by manipulating feedstock ratio of CO_2 and H_2O [5]. As seen in (1)-(3), the CO_2-steam reforming of methane reaction involves the following reactions, i.e., steam reforming of methane, CO_2 reforming of methane and water-gas shift (WGS) or reverse WGS (RWGS) reaction.

$$CH_4 + H_2O \rightarrow 3H_2 + CO \quad (\Delta H^o_{298K} = +206\ kJ\ mol^{-1}) \tag{1}$$

$$CH_4 + CO_2 \rightarrow 2H_2 + 2CO \quad (\Delta H^o_{298K} = +247\ kJ\ mol^{-1}) \tag{2}$$

$$CO + H_2O \rightleftharpoons H_2 + CO_2 \quad (\Delta H^o_{298K} = -41.2\ kJ\ mol^{-1}) \tag{3}$$

A large number of scientific publications reported that noble metals such as Ru, Rh, Pt, Pd and Ir exhibit considerable catalytic performance, stability and great coking resistance in both steam and CO_2 reforming processes [6]. However, industrial application of noble metals is unsuitable considering their high cost and restricted availability and hence, researches on non-noble metal catalyst becoming more practical and indispensable. Ni-based catalysts are one of the outstanding catalyst for reforming processes but Ni sintering of the metallic and support phases and coke formation resulting rapid catalyst deactivation is an inevitable setback at such high reaction temperature [7]. Therefore, it is necessary to develop Ni-based catalysts concerning to introduce or modify the interaction between the support and Ni metallic particles, such as using mesoporous SBA-15 as support. The mesoporous SBA-15 having an ordered structure which allows and confines Ni metallic particles in the channels and consequently, the sintering of Ni metallic particles can be mitigated and improve the stability of catalysts [8].

Mesoporous SBA-15 has attracted a significant attention in catalysis due to its hexagonal ordered structure which can be employed as an excellent host to disperse and confine nanoparticles. Apart from that, mesoporous SBA-15 also possess large specific surface areas, long-range ordering of mesoporous channels and thick pore walls with high hydrothermal stability [9]. These properties are not only beneficial to the dispersion of loading metal for creating more active sites, but also ensure the transformation of reactants and products from preventing carbon deposition and restrain the mobility of species that lead to metal sintering [10]. Furthermore, the surface silanol groups over the mesostructure provide a unique platform for immobilizing metal precursors inside the channels of SBA-15 material [11].

Ce-containing mixed oxides can be considered to be very promising supports enhancing the catalytic performance via strong metal-support interaction and improving dispersion of the active metal on catalyst support [12]. It found out that the mobility of surface oxygen species over the catalyst surface has been promoted due to the strong redox and oxygen vacancy formation between Ce^{4+} and Ce^{3+} [13]. The presence of Ce on support surface provides an additional storage capacity where oxygen can repeatedly undergoes the reduction and oxidation cycles and continuously producing mobile surface oxygen during the reaction. Subsequently, the oxygen transfer process improve the coking resistance of active metals. Thus, ceria-containing catalysts have been widely studied in CO oxidation, water-gas shift reaction and reforming processes. Shi et al. [14] revealed the addition of Ce species to Cr/SBA-15 catalysts remarkably changed the redox properties and improved the catalytic performance of Cr-based catalysts in the dehydrogenation of ethane to ethylene reaction. Wang et al. [15] have investigated the catalytic performance of Ce-SBA-15 supported Ni catalysts for methane dry reforming. They reported the incorporated Ce into SBA-15 framework promoted the nickel metallic particles formation with a smaller size compared to loaded CeO_2 on SBA-15 surface, which prevented the carbon formation and improved the catalytic activity.

It can be predicted that the combination of Ni and Ce-SBA-15 will lead to a good catalytic performance. This research intends to investigate the effect of steam addition on catalytic performance of 10%Ni/Ce-SBA-15 catalyst for CO_2 reforming of methane reaction and examine the influence of different feedstock ratios on activity of CO_2-steam reforming of methane reaction.

2. Experimental

2.1. Catalyst preparation

Ce-SBA-15 support was synthesized using triblock-poly(ethylene glycol)-block-poly(propylene glycol)-block- poly(ethylene glycol), referred as P-123 ($EO_{20}PO_{70}EO_{20}$) and tetraethyl orthosilicate (TEOS) as silica source and organic template agent, respectively. A calculated amount of $EO_{20}PO_{70}EO_{20}$ was dissolved in HCl solution with controlled pH of about 1 at 313 K and homogenized completely before the addition of TEOS and Cerium(III) nitrate hexahydrate ($Ce(NO_3)_3.6H_2O$) solutions. The mixture was further stirred at the same temperature for 24 h. Subsequently, it was transferred into a Teflon-lined autoclave and kept at 373 K for 24 h. The resulting white solid powder was filtered, washed rigorously with distilled water and dried in oven at 323 K overnight, followed by calcination in furnace at 823 K with a heating rate of 2 K min^{-1} for 5 h in order to obtain 5%Ce-95%SBA-15 support.

The 10%Ni/Ce-SBA-15 catalyst was prepared by a conventional wetness impregnation method, using as-synthesized Ce-SBA-15 as support and $Ni(NO_3)_2.6H_2O$ as metallic precursor. The measured amounts of $Ni(NO_3)_2.6H_2O$ aqueous solution and Ce-SBS-15 support were mixed and magnetically stirred at room temperature for 3 h. The mixture was then dried in oven at 373 K overnight and air-calcined again in furnace at 1073 K with a ramping rate of 2 K min^{-1} for 5 h.

2.2. Catalyst characterization

The measurement of Brunauer-Emmett-Teller (BET) surface area for support and catalyst was performed in a Micromeritics ASAP-2010 apparatus using N_2 physisorption isotherms data obtained at 77 K. The specimen was pretreated in N_2 flow at 573 K for 1 h to remove moisture and surface contamination before each measurement. X-ray diffraction (XRD) measurements of both Ce-SBA-15 support and 10%Ni/Ce-SBA-15 catalyst were also carried out in a Rigaku Miniflex II system using Cu monochromatic X-ray radiation (with wavelength, λ = 1.5418 Å) at 30 kV and 15 mA. All samples were scanned from 3° to 80° with relatively small scan speed and step size of 1° min^{-1} and 0.02^o, respectively to achieve high-resolution patterns.

The average crystallite size of NiO phase, *d(NiO)* was estimated from Scherrer's formula [16] whilst Ni^0 metallic particle size, $d(Ni^0)$ obtained after NiO reduction was approximately calculated using the relative molar volumes of Ni^0 metallic and NiO phases as shown in (4) [17].

$$d(Ni^0) = 0.84d(NiO) \tag{4}$$

By assuming spherical shape of Ni^0 particles, the metallic dispersion, *D* was estimated using (5) [18].

$$D(\%) = \frac{90nm}{d(Ni^0)} \tag{5}$$

H_2 temperature-programmed reduction (H_2-TPR) was conducted on an AutoChem II-2920 system. Prior to reduction step, calcined 10%Ni/Ce-SBA-15 catalyst was pre-treated at 373 K for 30 min in He inert gas (50 ml min^{-1}). Specimen was subsequently reduced in 50 ml min^{-1} of 10%H_2/Ar mixture from 373 to 1173 K with a heating rate of 10 K min^{-1}. Additionally, scanning electron microscopy coupled with energy-dispersive X-ray spectrometry (SEM-EDX) analyses were performed for spent catalysts in a Hitachi Tabletop Microscope TM3030Plus unit to investigate the external morphology of carbonaceous deposit on the surface of catalyst after reaction.

2.3. CO_2 or CO_2-steam reforming of methane
Both CO_2 reforming and CO_2-steam reforming of methane reactions were carried out under atmospheric pressure at 1073 K in a quartz tube fixed-bed continuous flow reactor (O.D. = 3/8 in. and length, L = 17 in.) for 12 h on-stream. Approximately, 0.1 g of catalyst was mounted by quartz wool in the middle of reactor. Additionally, the tubular reactor was placed vertically in a split temperature-controlled furnace and reaction temperature was accurately and continuously measured with time-on-stream (TOS). Gas hourly space velocity, GHSV = 36 L g_{cat}^{-1} h^{-1} was implemented for all runs to ensure the negligible internal and external transport resistances were acquired.

For CO_2 reforming of methane test, a precisely regulated amount of CH_4 and CO_2 gases was diluted in a mixer with N_2 gas employed as a tie component for material balance purpose and ensuring the total flow rate of 60 ml min^{-1} for all runs before the gaseous mixture was directed to the inlet of tubular reactor. In the case of CO_2-steam reforming reaction, purified water was also fed to the top of reactor by a KellyMed KL-602 syringe pump and vaporized before being blended with other gaseous reactants. The total flow rate of reactant mixture diluted in N_2 was also maintained at 60 ml min^{-1} for bi-reforming reaction. In addition, the composition of gaseous product stream from the bottom of fixed-bed reactor was analysed in a gas chromatograph (Agilent 6890 Series GC system) possessing TCD detector.

3. Results and discussion

3.1. Characterization results

3.1.1. Textural properties. In order to examine the mesoporous structure of 10%Ni/Ce-SBA-15 catalyst, N_2 adsorption-desorption isotherm curves of both support and catalyst are displayed in Figure 1. Based on the IUPAC classification, the isotherms of Ce-modified SBA-15 support had a type IV isotherm curve with an obvious H1-type hysteresis loop, typical for mesoporous materials possessing cylindrical channels [19]. Additionally, the sharp inflection observed in isotherm curves at p/p^0 of 0.65 to 0.85 was reasonably due to capillary condensation of N_2 within ordered mesopores (Wang et al., 2012) and suggested the existence of large mesopores [20]. Interestingly, the isotherm profiles of the 10%Ni/Ce-SBA-15 catalyst were similar to that of Ce-SBA-15 support indicating that the mesoporous structure of support was largely maintained during NiO incorporation. As seen in Table 1, despite of unchanged mesoporous structure, the unavoidable drop in both surface area and total pore volume was evident for 10%Ni/Ce-SBA-15 catalyst rationally owing to the successful penetration of NiO nanoparticles into the mesoporous channels of Ce-SBA-15 support. Additionally, some pore walls may be altered during the NiO incorporation resulting in a decrease of average pore diameter from 66.63 to 63.81 Å as given in Table 1 [21].

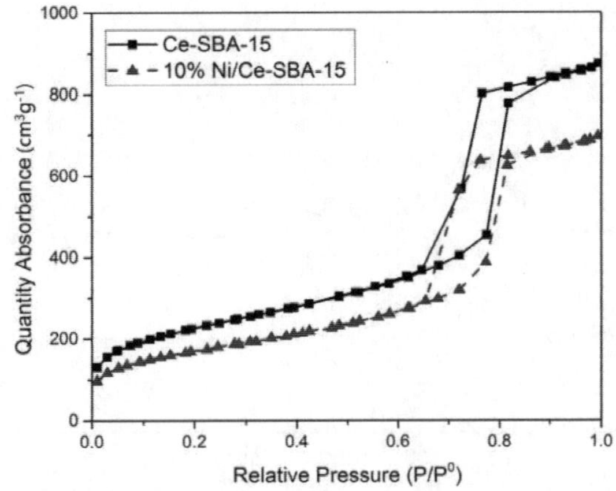

Figure 1. N_2 physisorption isotherms of Ce-SBA-15 support and 10%Ni/Ce-SBA-15 catalyst.

Table 1. Textural properties of support and catalyst.

Sample	S_{BET} (m^2 g^{-1})[a]	V_p (cm^3 g^{-1})[b]	D_p (Å)[c]
Ce-SBA-15	760.47	1.21	66.63
10%Ni/Ce-SBA-15	595.04	1.08	63.81

[a] BET specific surface area.
[b] Total pore volume obtained at $p/p^0 = 0.99$.
[c] Average pore diameter calculated by Barret-Joyner-Halenda (BJH) desorption method.

3.1.2. X-ray diffraction measurement. Figure 2 shows the XRD patterns of fresh mesoporous Ce-SBA-15 support and 10%Ni/Ce-SBA-15 catalyst. The X-ray diffractograms of both catalyst and support possessed a similar broad diffraction peak at 2θ angle of around 23° belonging to amorphous SiO_2 phase of SBA-15 framework [22]. The similarity in XRD patterns for SiO_2 phase between catalyst and support employed would further confirm the stability of mesoporous structure of Ce-SBA-15 support unaffected by embedded NiO particles. In addition, four sharp peaks with high intensity on the diffractogram of 10%Ni/Ce-SBA-15 catalyst located at 2θ = 37.46°, 43.54°, 63.03° and 75.14° were assigned to the corresponding (1 1 1), (2 0 0), (2 2 0) and (3 1 1) crystal planes of NiO phase (JCPDS card No. 47-1049). However, there was no detection of CeO_2 phase suggesting that CeO_2 particles were well dispersed into the mesoporous channels of SBA-15 support or Ce atoms isomorphously substituted Si atoms in the silica framework. Additionally, the average crystallite size of NiO particles was about 13.60 nm, estimated from the most intense NiO line at 2θ of 43.54° by Scherrer's equation. Hence, the Ni^0 metallic particle dimension and metal dispersion were computed as *ca.* 11.42 nm and 7.88% using (4) and (5), respectively.

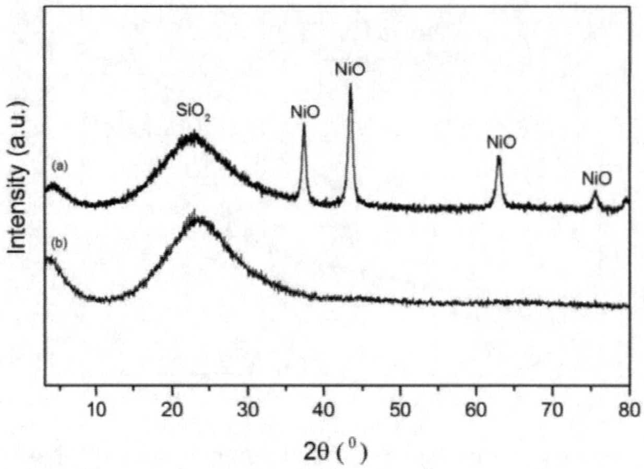

Figure 2. XRD patterns of (a) 10%Ni/Ce-SBA-15 catalyst and (b) Ce-SBA-15 support.

3.1.3. H_2 temperature-programmed reduction. Figure 3 shows the H_2-TPR profile of 10%Ni/Ce-SBA-15 catalyst with 10%H_2/Ar mixture. The reduction of NiO to Ni^0 metallic phase was widely reported as a single process [23, 24]. However, two broad peaks, P1 and P2 were observed at reduction temperature of about 650 and 728 K, correspondingly. He et al. [25] reported that nickel silicate was reduced at temperature above 973 K and nickel silicate was not detected on XRD pattern of 10%Ni/Ce-SBA-15 catalyst (see Figure 2). Thus, both P1 and P2 peaks belonged to the reduction of NiO phase to Ni^0 metallic phase. Additionally, the presence of multiple peaks in H_2-TPR was attributed to the location of NiO particles on siliceous SBA-15 support and the extent of metal-support interaction depending on crystallite size. In fact, the low temperature peak (P1) was assigned to the reduction of surface NiO phase with weak metal-support interaction [26] while the high temperature peak (P2) corresponded to the reducibility of small NiO species resided in the mesoporous framework of SBA-15 support [27].

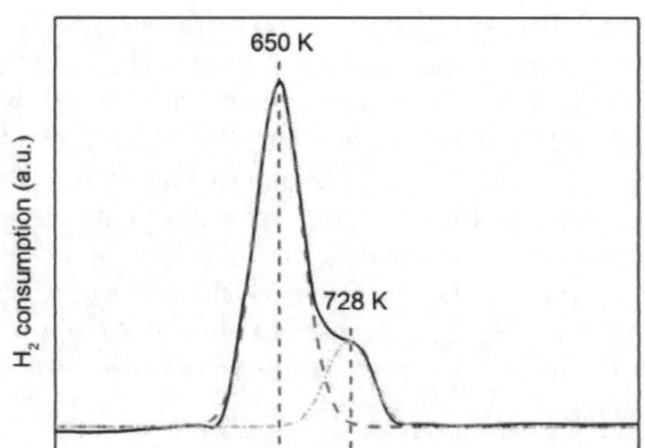

Figure 3. H_2-TPR profile of 10%Ni/Ce-SBA-15 catalyst at a heating rate of 10 K min^{-1}.

3.1.4. SEM-EDX analysis. The spent 10%Ni/Ce-SBA-15 catalysts after CO_2 reforming (at 800 K and $P_{CH4} = P_{CO2} = 20$ kPa) and CO_2-steam reforming (at 800 K, $P_{CH4} = 45$ kPa, $P_{H2O} = 30$ kPa and $P_{CO2} = 15$ kPa) of methane reactions were examined using SEM-EDX to investigate the surface morphology of carbonaceous species, elemental composition and dispersion. Figure 4 and Figure 5 display the SEM image, elemental mapping and EDX spectrum of spent 10%Ni/Ce-SBA-15 catalysts for CO_2

reforming and CO_2-steam reforming of methane, respectively. As seen in Figure 4(a) and Figure 5(a), whisker-like and graphitic carbons were present on the surface of spent catalysts from both reactions. It was evident that the dispersion of metallic Ni particles was relatively fine on Ce-SBA-15 support surface (see Figure 4(b) and Figure 5(b)). Interestingly, the carbon content of spent catalyst from CO_2-steam reforming reaction (see the inset in Figure 5(c)) was about 12.1% lower than that of used catalyst from CO_2 reforming of methane (see Figure 4(c)). The reduction in carbon content for CO_2-steam reforming reaction was reasonably due to the steam gasification of deposited carbon [28].

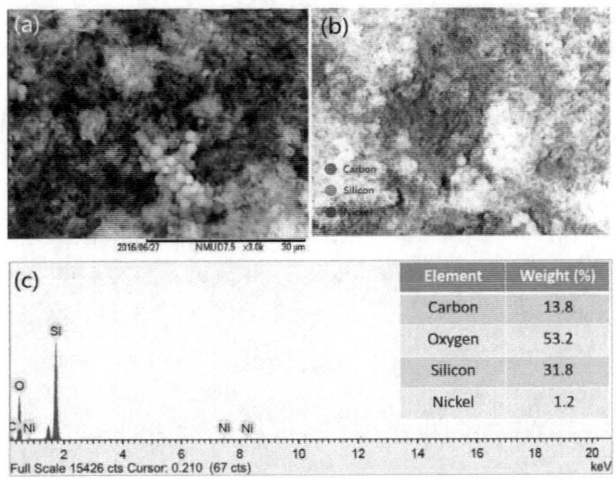

Figure 4. Spent 10%Ni/Ce-SBA-15 catalyst after CO_2 reforming of methane reaction: (a) SEM image, (b) EDX mapping image and (c) EDX spectrum.

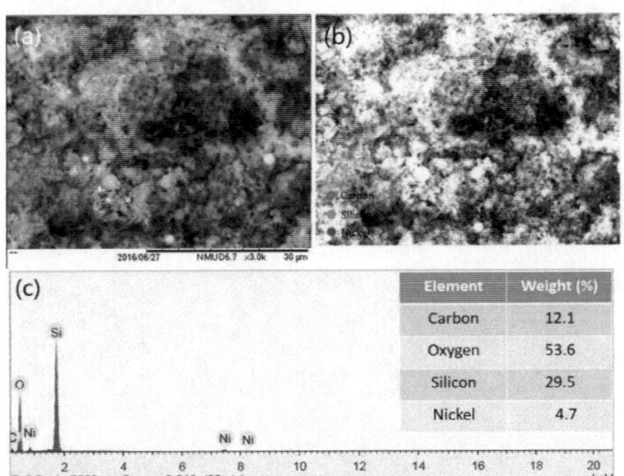

Figure 5. Spent 10%Ni/Ce-SBA-15 catalyst after CO_2-steam reforming of methane reaction: (a) SEM image, (b) EDX mapping image and (c) EDX spectrum.

3.2. Evaluation of CO_2 or CO_2-steam reforming reactions of methane

A time-on-stream test of 12 h was performed for 10%Ni/Ce-SBA-15 catalyst in order to examine the catalytic performance for both CO_2 dry reforming and CO_2-steam reforming of methane. Both reactions were carried out at stoichiometric feed composition and 1073 K. The transient profiles of CH_4 conversion and H_2/CO ratio for both reactions are shown in Figure 6 and Figure 7. It is apparent that catalyst performance was stable within 12 h on-stream for all reaction runs. Theoretically, the steam addition to CO_2 reforming of methane would increase the CH_4 conversion rationally due to the parallel occurrence of methane steam reforming reaction. In fact, as seen in Figure (6), the CO_2-steam reforming of methane exhibited a far lower CH_4 conversion compared to that of CO_2 reforming of methane. This observation was reasonably due to the H_2O, CO_2 and their decomposed products occupied the sites on the catalyst suppressing the CH_4 decomposition, as reported by Qin et al. [29]. However, the 10%Ni/Ce-SBA-15 catalyst in this study exhibited a significant CH_4 conversion of 75% for CO_2-steam reforming of methane outperforming Danilova et al. [30] synthesized porous Ni-based catalysts. In their study, porous Ni-based catalysts were only capable of achieving a satisfactory CH_4 conversion of 60% and below overall for a CO_2-steam reforming of methane with 20 h on-stream. Figure (7) shows H_2/CO ratio with time-on-stream for CO_2 reforming and CO_2-steam reforming of methane at 1073 K. The H_2/CO ratio for CO_2 reforming of methane reaction was about 0.9 at 1073 K due to the concomitant reverse water-gas shift reaction whereas a higher value of 2.4 was observed for CO_2-steam reforming of methane reaction which evident the simultaneous presence of CH_4 steam reforming during reaction runs.

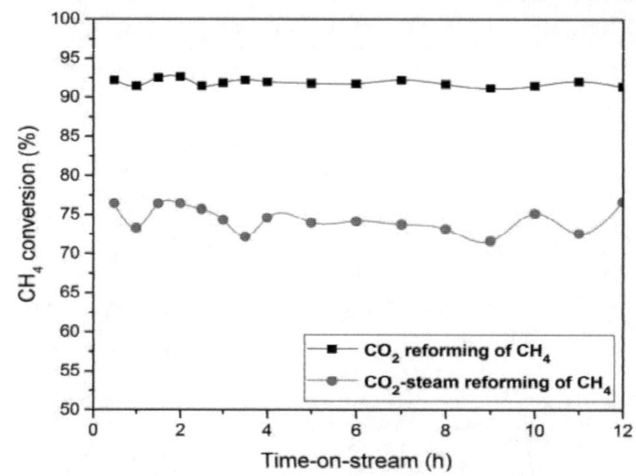

Figure 6. CH_4 conversion with time-on-stream for CO_2 reforming and CO_2-steam reforming of methane at 1073 K.

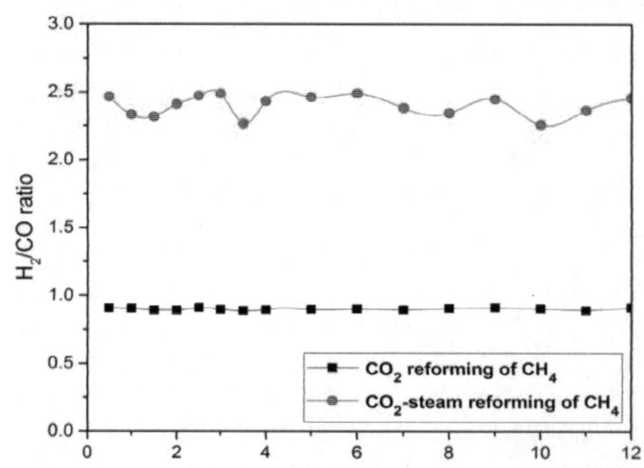

Figure 7. H_2/CO ratio with time-on-stream for CO_2 reforming and CO_2-steam reforming of methane at 1073 K.

3.3. Effect of feed ratios on CO_2-steam reforming of methane

A varying feed ratios of steam (P_{H2O}) and CO_2 (P_{CO2}) from 10 to 20 kPa were employed in CO_2-steam reforming of methane with 10%Ni/Ce-SBA-15 catalyst to investigate different feedstock ratio effect on catalytic performance and H_2/CO product ratio for synthetic fuels application. As seen in Figure 8, growing P_{H2O} was beneficial to the reaction and significantly increased the CH_4 conversion and H_2 yield by up to 11.44% and 26%, respectively. This observation was reasonably due to the presence of excess P_{H2O} enhanced the CH_4 steam reforming reaction and producing a high H_2/CO ratio.

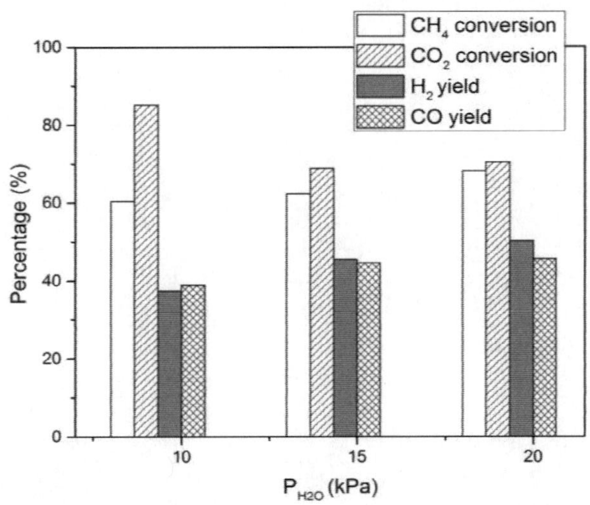

Figure 8. Effect of steam feed ratios on CH_4 conversion, CO_2 conversion, H_2 yield and CO yield for CO_2-steam reforming of methane at P_{CH4} = 45 kPa, P_{CO2} = 15 kPa and 1073 K.

In fact, in the study of thermodynamics of combined CO_2-steam reforming of methane, catalyst activity and H_2/CO ratio increased considerably with the addition of excess steam. A considerable decrease of CO_2 conversion (see Figure 8) with growing P_{H2O} was expected and supported by the works of Özkara-Aydınoğlu [31]. In the paper, the CH_4 was reportedly reacted with H_2O instead of CO_2 whenever both H_2O and CO_2 are played as co-oxidant due to the high stability of CO_2 nature. The foregoing discussion was further confirmed since the CH_4 conversion increased while CO_2 conversion decreased with rising P_{H2O} as seen in Figure 8.

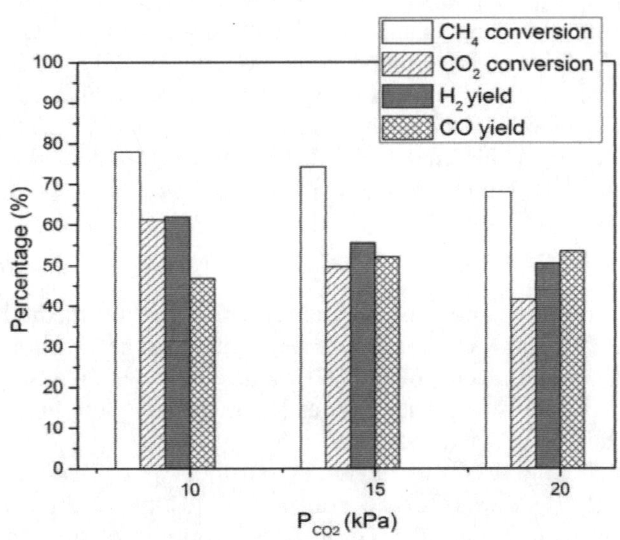

Figure 9. Effect of CO_2 feed ratios on CH_4 conversion, CO_2 conversion, H_2 yield and CO yield for CO_2-steam reforming of methane at P_{CH4} = 45 kPa, P_{H2O} = 30 kPa and 1073 K.

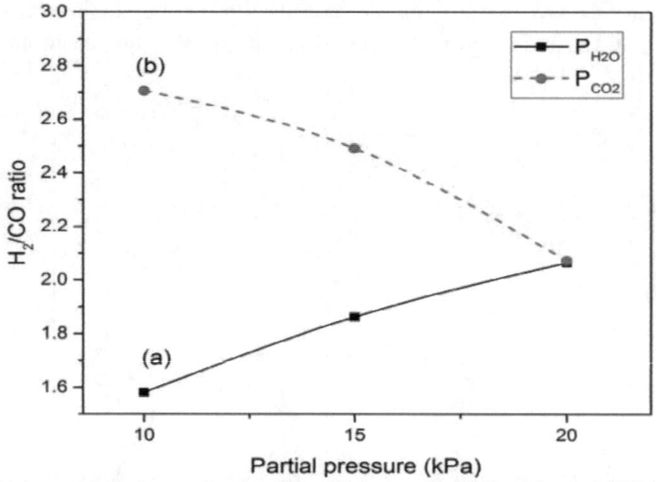

Figure 10. Effect of (a) varying steam feed ratios, P_{H2O} on H_2/CO ratio at 1073 K; P_{CH4} = 45 kPa and P_{CO2} = 15 kPa and, (b) varying CO_2 feed ratios, P_{CO2} on H_2/CO ratio at 1073 K; P_{CH4} = 45 kPa and P_{H2O} = 30 kPa.

Figure 9 shows the influence of varying CO_2 feed ratios (P_{CO2}) on the catalyst activity at 1073 K. Besides the CO yield, all the catalyst activities decreased with growing P_{CO2}, particularly H_2 yield dropped by 19.35%. As seen in Figure 9, both CH_4 and CO_2 conversion exhibited a significant decrease trend suggesting the excessive presence of CO_2 coming from CH_4 steam reforming could suppress the CH_4 adsorption and decomposition on catalyst surface and hence lessened the CO_2-steam reforming of methane reaction. Interestingly, the P_{H2O} and P_{CO2} exhibited a totally opposite trend in H_2/CO ratio as seen in Figure 10. This trend was similarly observed by Huang et al. [32] who studied on the effects of feed gas composition on the activity of 3%MgO-10%Ni/SBA-15 catalyst. The catalyst proposed in their study exhibited a high CH_4 conversion of 98.3% in CO_2-steam reforming of methane. However, in the case of production of long chain hydrocarbons a H_2/CO ratio of 2 is desirable instead. The catalyst used in this study managed to obtain a more appropriate H_2/CO ratio of 2.07 in comparison to that of Huang et al. [32]. Interestingly, the H_2/CO ratio of 2.07 could be obtained at about of 20 kPa for both P_{H2O} and P_{CO2}. In fact, the H_2/CO ratio for CO_2-steam reforming of methane varied between 1 and 3 indicated the occurrence of parallel reactions, i.e., CH_4 steam reforming giving a H_2/CO of 3 whilst reverse water-gas shift (RWGS) reaction consuming H_2 to produce CO gaseous product. In this study, the H_2/CO ratio of about 2.07 could be obtained with appropriate feed ratios of (i) CH_4:H_2O:CO_2 = 3:1.33:1 or (ii) CH_4:H_2O:CO_2 = 3:2:1.33 at 1073 K and 1 atm.

4. Conclusions

The 10%Ni/Ce-SBA-15 catalyst synthesized by wetness impregnation method had been successfully evaluated for CO_2 reforming and CO_2-steam reforming of methane reaction in a quartz fixed-bed reactor at different H_2O:CO_2 feed ratios, 1073 K and 1 atm. The 10%Ni/Ce-SBA-15 catalyst possessed high BET surface area of 595.04 m^2g^{-1}. The XRD and EDX results were in accordance that the CeO_2 were effectively embedded into ordered mesoporous SBA-15 framework. The crystallite size of NiO was reduced to metallic Ni from 13.60 to 11.42 nm and they were finely dispersed on the catalyst surface whilst H_2-TPR analysis showed that the reduction of NiO was a single process. The SEM-EDX measurement showed that the addition of steam to CO_2 reforming of methane improved the coking resistance of catalyst due to the activated oxygen from steam to the catalyst surface enhancing the gasification process of deposited carbon. Apart from that, the 10%Ni/Ce-SBA-15 catalyst experienced a gradual increase in CH_4 conversion and H_2 yield with rising P_{H2O} whilst a decrease in CH_4 conversion and H_2 yield with growing P_{CO2}. The P_{H2O} and P_{CO2} of 20 kPa seemed to be the optimal partial pressure for CO_2-steam reforming of methane in terms of conversion, H_2 yield and H_2/CO ratio of 2. Furthermore, the H_2/CO ratio of about 2.07 could be obtained with appropriate feed ratios of (i) CH_4:H_2O:CO_2 = 3:1.33:1 or (ii) CH_4:H_2O:CO_2 = 3:2:1.33 at 1073 K and 1 atm.

Acknowledgments
The authors are grateful for the financial support from Universiti Malaysia Pahang via UMP Research Grant Scheme (RDU160323) to conduct this study.

References

[1] Das D and Veziroğlu T N 2001 *India. Int. J. Hydrogen Energy* **26** 13-28
[2] Omae I 2006 *Japan. Catal. Today* **115** 33-52
[3] Vo D V N, Nguyen T H, Kennedy E M, Dlugogorski B Z and Adesina A A 2011 *Australia. Catal. Today* **175** 450-9
[4] Bae J W, Kim A R, Baek S C and Jun K W 2011 *Hungary. Reac. Kinet. Mech. Catal.* **104** 377-88.
[5] Ryi S K, Lee S W, Park J W, Oh D K, Park J S and Kim S S 2014 *South Korea. Catal. Today.* **236** 49-56
[6] Hou Z, Chen P, Fang H, Zheng X and Yashima T 2006 *China. Int. J. Hydrogen Energy* **31** 555-61
[7] Sehested J 2006 *Denmark. Catal. Today* **111** 103-10
[8] Yi N, Cao Y, Su Y, Dai W L, He H Y and Fan K N 2005 *China. J. Catal.* **230** 249-53
[9] Ungureanu A, Dragoi B, Chirieac A, Ciotonea C, Royer S, Duprez D, Mamede A S and Dumitriu E 2013 *Romania. ACS Appl. Mater. Interfaces* **5** 3010-25
[10] Bore M T, Pham H N, Ward T L and Datye A K 2004 *USA. Chem. Commun.* **22** 2620-21
[11] Liu D, Quek X Y, Wah H H A, Zeng G, Li Y and Yang Y 2009 *Simgapore. Catal. Today* **148** 243-50
[12] Zhang X and Shi P 2003 *China. J. Mol. Catal. A: Chem.* **194** 99-105
[13] Senanayake S D, Stacchiola D, Evans J, Estrella M, Barrio L, Pérez M, Hrbek J and Rodriguez J A 2010 *USA. J. Catal.* **271** 392-400
[14] Shi X, Ji S and Wang K 2008 *China. Catal. Lett.* **125** 331-39
[15] Wang N, Chu W, Zhang T and Zhao X 2012 *China. Int. J. Hydrogen Energy* **37** 19-30
[16] Patterson A L 1939 *Pennsylvania. Phys. Rev.* **56** 978
[17] Baum E J 1998 *Chemical property estimation: Theory and application* USA
[18] Xu J, Chen L, Tan K F, Borgna A and Saeys M 2009 *Singapore. J. Catal.* **261** 158-65
[19] Brunauer S, Deming L S, Deming W E and Teller E 1940 *USA. JACS* **62** 1723-32
[20] Vinu A, Sawant D, Ariga K, Hartmann M and Halligudi S 2005 *Japan. Microporous Mesoporous Mater.* **80** 195-203
[21] Wang K, Li X, Ji S, Shi X and Tang J 2009 *China. Energy Fuels* **23** 25-31
[22] Landau M, Vradman L, Wang X and Titelman L 2005 *Israel. Microporous Mesoporous Mater.* **78** 117-29
[23] Foo S Y, Cheng C K, Nguyen T H, Kennedy E M, Dlugogorski B Z and Adesina A A 2012 *Australia. Catal. Commun.* **26** 183-8
[24] Selvarajah K, Phuc N H H, Abdullah B, Alenazey F and Vo D V N, 2016 *Malaysia. Res. Chem. Intermed.* **42** 269-88
[25] He S, Jing Q, Yu W, Mo L, Lou H and Zheng X 2009 *China. Catal. Today* **148** 130-3
[26] Oemar U, Kathiraser Y, Mo L, Ho X K and Kawi S 2016 *Singapore. Catal. Sci. Tech.* **6** 1173-86

[27] Kim D, Kwak B S, Min B K and Kang M 2015 *Republic of Korea. Appl. Surf. Sci.* **332** 736-46
[28] Santos A C S F, Damyanova S, Teixeira G N R, Mattos L V, Noronha F B, Passos F B and Bueno J M C 2005 *Brazil. Appl. Catal., A* **290** 123-32
[29] Qin D and Lapszewicz J 1994 *Australia. Catal. Today* **21** 551-60
[30] Danilova M M, Fedorova Z A, Zaikovskii V I, Porsin A V, Kirillov V A and Krieger T A 2014 *Russia. Appl. Catal, B* **147** 858-63
[31] Özkara-Aydınoğlu Ş 2010 *Turkey. Int. J. Hydrogen Energy* **35** 12821-8
[32] Huang B, Li X, Ji S, Lang B, Habimana F and Li C 2008 *China. J. Nat. Gas. Chem.* **17** 225-31

IOP Conf. Series: Materials Science and Engineering **206** (2017) 012018 doi:10.1088/1757-899X/206/1/012018

Evaluation on Microalgae Biomass for Bioethanol Production

L M Chng[1,2,*], K T Lee[2], D C J Chan[2]

[1]Department of Petrochemical Engineering, Faculty of Engineering and Green Technology, Universiti Tunku Abdul Rahman, Jalan Universiti, Bandar Barat, 31900, Kampar, Perak, Malaysia
[2]School of Chemical Engineering, Universiti Sains Malaysia, Seri Ampangan, 14300, Nibong Tebal, Pulau Pinang, Malaysia

Email: chnglm@utar.edu.my

Abstract. The depletion of energy resources has triggered worldwide concern for alternative sources, especially renewable energy. Microalgae biomass offers the most promising feedstock for renewable energy because of their impressive efficient growing characteristics and valuable composition. Simple cell structure of the microalgae would simplify the pretreatment technology thus increase the cost-effectiveness of biofuel production. Scenedesmus dimorphus is a carbohydrate-rich microalgae that has potential as biomass for bioethanol. The cultivation of *Scenedesmus dimorphus* under aeration of carbon dioxide enriched air resulted 1.47 g/L of dry biomass with composition of 12% w/w total lipid, 53.7% w/w carbohydrate and 17.4% protein. Prior to ethanolic fermentation with *Saccharomyces cerevisiae*, various pre-treatment methods were investigated to release and degrade the complex carbohydrate in cell biomass thus obtaining the maximal amount of digestible sugar for ethanolic yeast. In this study, sulfuric acid was used as hydrolysis agent while amyloglucosidase as enzymatic agent. Dried biomass via hydrothermal acidic hydrolysis yielded sugar which is about 89% of total carbohydrate at reaction temperature of 125 °C and acid concentration of 4% v/v. While combination of organosolv treatment (mixture of methanol and chloroform) with enzymatic hydrolysis yielded comparable amount of sugar with 0.568 g glucose/g treated-biomass. In this study, the significant information in pre-treatment process ensures the sustainability of the biofuel produced.

1. Introduction

Bioethanol as the choices as alternatives fuels is always limited and actively debated due to fuel vs food issue. Ease of pretreatment on biomass and compatibility of treated biomass to subsequent ethanolic microbe is the core area of research, to increase the overall bioethanol production efficiency. As such, microalgae with simple cell structure become the biomass in spot for fermentation, with stipulation that treated biomass is preferred for ethanolic microbe.

Microalgae as a renewable and promising resource for biofuels production are an attention of researcher all over the world [1]. It is urgently needed as clean energy source due to heavy carbon dioxide emission that leads to severe pollution and climate change problems. Microalgae with its inherent properties, grows extremely rapidly and commonly double their biomass within 24 hours. The biomass mainly composed with lipid, carbohydrate and protein. With this, sustainable production of biofuels needs a bundle of research on the production technologies from algae strain to biofuels product (biodiesel, bioethanol, biogas). The production chain involves selection of algae strain, algae cultivation for desired requirement (higher lipid and carbohydrate productivity), harvesting, compounds extraction and biofuels production reaction [2].

Scenedesmus dimorphus is a potential microalgae species for production of bioethanol due to its capability in accumulating carbohydrate content as stored carbon in the cell. However, the cell need to be disrupted and the complex biopolymers need to be converted to simple sugars before fermentation, through pretreatment process such as hydrolysis. This is because current industrial ethanologenic yeast only consume simple sugar such as glucose and galactose. Microalgae have the beneficial for the fermentation process with its simple cell structure and some of it is unicellular species. Most of the researches examine pretreatments to degrade the carbohydrate into simple sugar that ready for fermentation by microorganism [3,4,5]. The pretreatment in study are physical, chemical and physicochemical pretreatment. Acid hydrolysis and enzymatic pretreatment are two most common and efficient steps in discuss. The quality of the hydrolysate from the treatments process have the significant effect to subsequent fermentation technology.

In this study, the potential use of *S. dimorphus* as a feedstock to produce bioethanol was evaluated. In order to provide more precise perspective of *S. dimorphus* as a promising commercial feedstock, microalgae cultivation growth and starch content were study. Few pre-treatment methods was conducted to investigate the quality of hydrolysate for yeast fermentation. Attention was placed on the fermentable sugar, i.e., glucose, sucrose and galactose, to produce bioethanol.

2. Procedure

2.1. Microalgae cultivation

Scenedesmus dimorphus(UTEX 1237, University of Texas) were cultured in a 2-L bioreactor and continuously aerated with 2 L/min of air and CO_2-enriched air. Illumination was provided continuously with 2 fluorescent lamps located at one side of the reactor. The medium consisted of 20 mL of 25 g/L $NaNO_3$, 7.5 g/L $MgSO_4.7H_2O$, 2.5 g/L NaCl, 7.5 g/L K_2HPO_4; 17.5 g/L KH_2PO_4, 11.4 g/L H_3BO_3; 2 mL of 8.82 g/L $ZnSO_4.7H_2O$, 1.18 g/L $MnCl_2.2H_2O$, 1.193 g/L $Na_2MoO_4.2H_2O$, 1 g/L $CuSO_4$, 0.401 g/L $CoCl_2.6H_2O$; 2 mL of 50 g/L EDTA-Na_2 and 4.48 g/L $FeSO_4.7H_2O$; 2 mL of 0.2 g/L Thiamine Hydrochloride B1 and 0.01 g/L Cyanocobalamin B12. The initial inoculum size was 79.2 x 106 cells. Growth was assessed daily by measuring the optical density with a UV vis spectrophotometer at a wavelength of 600 nm.

Dry weight of biomass was obtained by centrifuging and drying the collected culture solution at 60 °C oven until constant weight was achieved. The dry weight of the algae biomass was determined gravimetrically and growth was expressed in terms of dry weight per liter culture.

The pH of culture was measured daily with pH meter. Starch content was measured from late exponential phase till early stationary phase of growth by targeting higher biomass productivity at that cultivation stage. All experiments were conducted in triplicate and data reported in average value.

2.2. Biomass harvesting by natural sedimentation

Natural sedimentation was employed in order to harvest the cultured microalgae. The culture aeration was stopped and subjected to gravity sedimentation. The top liquid phase of cultures was decanted and cells that deposit at bottom phase were harvested by centrifugation at 5500 rpm for 5 min. The cells were washed twice with distilled water after centrifugation. The cell supernatant was dried at 60 °C for 24 hr and keep in desiccator. The recovery was calculated based on the OD difference at bottom point.

Recovery efficiency (%),

$$\eta = \frac{OD_i - OD_f}{OD_i} \quad (1)$$

IOP Conf. Series: Materials Science and Engineering **206** (2017) 012018 doi:10.1088/1757-899X/206/1/012018

Where OD_i is the initial optical density of medium before sedimentation at 5 cm from bottom of culture bottle and OD_f is the final optical density of medium after sedimentation at 5 cm from bottom of culture bottle.

2.3. *Biomass characterization*

2.3.1. Total carbohydrate quantification. The total carbohydrate content was determined using two-step acid hydrolysis method. The method is a modified method from National Renewable Energy Laboratory [6]. First step of analysis is sulfuric acid at 72% (w/w) was added into glass tubes that filled with 3 g of dried biomass. The solution was vortex to thoroughly mix the acid and biomass. The tubes were kept in a 30 °C ± 3 °C water bath for one hour. Subsequent step of hydrolysis was carried at concentration of 4% (w/w) for one hour at temperature of 121 °C using autoclave. The hydrolysate was neutralized to pH between 5-7 using calcium carbonate. Neutralized samples are filtered with 0.22 µm nylon filter for HPLC analysis. Monomer sugar concentrations (mg/mL) were calculated from HPLC calibration and total carbohydrate was determined by summing the monomer sugars.

2.3.2. Total starch quantification. The starch content was determined using a modified method based on the National renewable Energy Laboratory using amyloglucosidase from Aspergillus niger (aqueous solution, ≥300 U/mL, Sigma-Aldrich, Malaysia) as a hydrolysis enzyme [7]. One hundred milligrams of dried biomass was added to a boiling tube. Simultaneously, 5 mL of distilled water and 1 mL of 2N NaOH were added, and the solution was stirred for 20 min in a preheated water bath set at 90 °C. The solution was neutralized with 1 mL of 2N HCl, and the pH of the solution was maintained by adding 2 mL of acetate buffer (50 mM, pH 4.5). The reaction was started by adding 1 mL of amyloglucosidase (60 units of activity per millilitre), and the solution was stirred for 60 min at 40 °C. The resultant samples were identified with high performance liquid chromatography (HPLC).

2.3.3. Total lipid quantification. The total lipids were obtained by extracting the dry biomass with a mixture of chloroform and methanol (ratio of 1:2), which is a modification of the method proposed by Bligh and Dyer [8]. One gram of dried biomass was stirred in the solvent mixture at room temperature for 24 h. The solution was filtered with filter paper, and washed again with solvent mixture. The washing and filtration steps were repeated three times. Lastly, the solvent was evaporated by gentle heating until a constant weight was reached; this residual mass was the total lipid content.

2.3.4. Total protein quantification. Protein content was measured with the AOAC 988.05 method. Biomass was go through digestion and distillation process. The total protein was calculated based on the products of 6.25 and percentage of nitrogen content.

2.4. *Pretreatment*

2.4.1. Solvent treatment. The dried biomass of 2 g was added to a solvent mixture of methanol and chloroform (ratio 2:1). The solution was stirred (750 rpm) for 24 hours at room temperature. After the extraction process, the solution was filtered with filter paper. The extracted biomass was oven dried at 60 °C until a constant weight was achieved and stored in desiccators.

2.4.2. Hydrothermal acid hydrolysis. Acid hydrolysis was conducted using concentrated sulphuric acid (97%). One hundred milligrams of biomass was treated with 1.0 to 5.0% (v/v) of H_2SO_4 with constant working volume of 10 mL. The mixture was autoclaved for 15 minutes at 110 °C, 120 °C and 125 °C each.

2.4.3. Enzymatic hydrolysis with pre-heated biomass. Effect of heating to enzymatic hydrolysis was carried out by directly added enzyme into biomass solution without stirring solution at 90 °C.

IOP Conf. Series: Materials Science and Engineering **206** (2017) 012018 doi:10.1088/1757-899X/206/1/012018

Hydrolysis conditions were maintained at pH 4.5, enzyme concentration at 60 units/mL and temperature at 55 °C.

Enzymatic hydrolysis was carried out using the amyloglucosidase enzyme (One unit of the enzyme can convert 1 mg of starch to glucose in 3 minutes at a pH and temperature of 4.5 and 55 °C, respectively). A biomass of 100 mg was added to the boiling tube, followed by 7 mL of sodium acetate buffer solution. The solution was stirred for 15 minutes in a preheated water bath at 90 °C. One millilitre of enzyme (60 units/mL) was added and stirred for 60 minutes at various reaction temperatures (30°C, 35°C, 40°C, 55°C, and 60°C) and pH values (4.0, 4.5, 5.0, 5.5, and 6.0).

2.5. Analysis of sugar content with high performance liquid chromatography (HPLC)

The system (Agilent series 1200) consisted of an evaporative light scattering detector (ELSD) and a Hi-Plex Ca column. The nebulizer and evaporator temperature were set at 50 °C and 80 °C, respectively. The analysis was run at a flow rate of 0.6 mL/min using deionized water as the mobile phase. The sugar content was calculated based on a glucose calibration curve.

3. Results and Discussion

3.1. Scenedesmus dimorphus growth monitoring

Figure 1 demonstrates the changes of dry weight and pH during cultivation period. The curves showed typical growth pattern with exponential phase, stationary phase and death phase. The dry weight with CO_2 enrichment aeration was apparently higher than aeration with air only throughout the cultivation period. The addition of CO_2 shorten the exponential phase and extend the stationary phase with higher biomass density. The pH of the culture aerated with air continues to increase to approximately 10 at day 10th and drop gradually to 7-8, as revealed in figure 1. While the addition of CO_2 maintained the pH value fluctuated at the range of 5-6.5, which is favourable for microalgae culture [9].

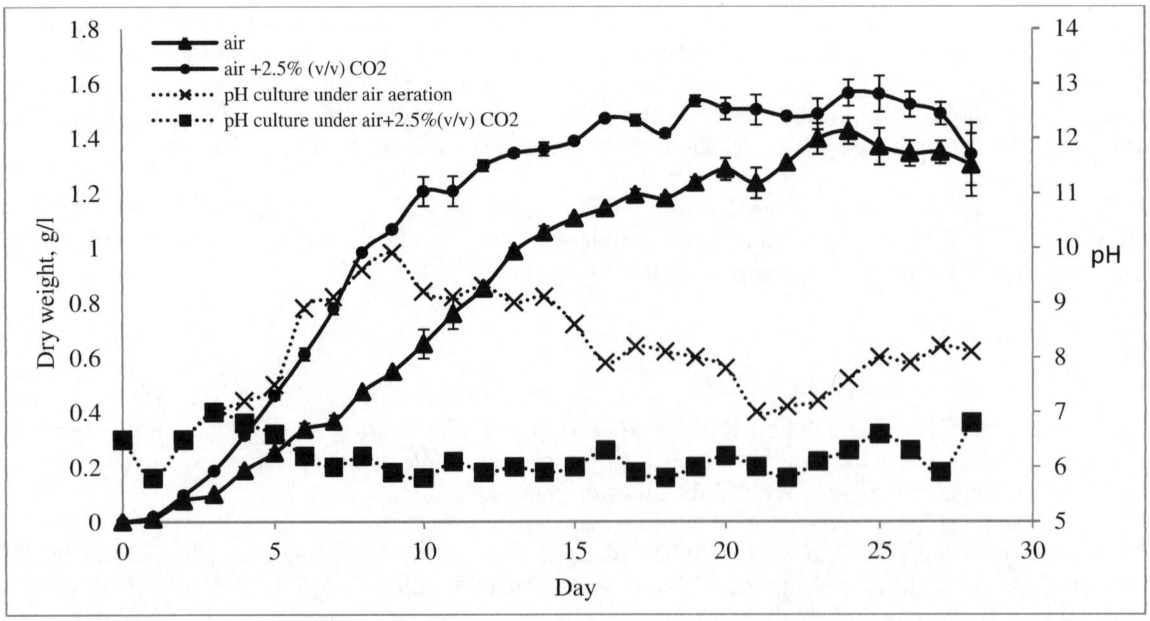

Figure 1. Dry weight and pH monitoring during cultivation time.

The starch content monitoring during cultivation period was presented in figure 2. The measurement was not calculated from day 1 to day 7 because the amount of collected biomass was too little for starch content analysis. Starch content for both aeration increase steadily from day 8 to day 16, but decrease gradually starting day 16. The starch content of 2.5% (v/v) CO_2-enriched air accumulated higher starch content compared to air aeration. The optimum value was found at day 16 with starch content of 57.36% (w/w) and 47.43% (w/w) under aeration of CO_2 enrichment air and air, respectively.

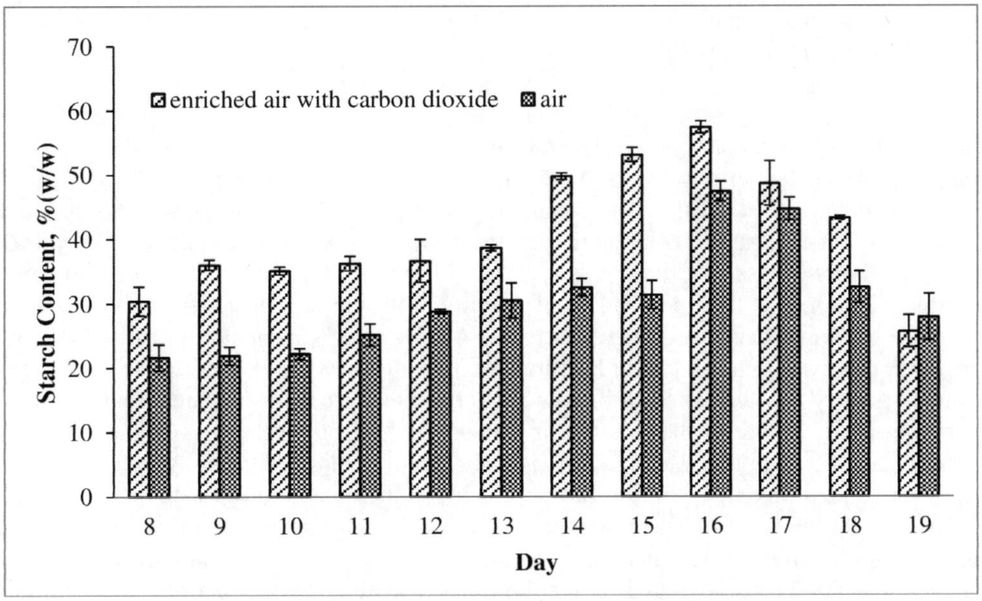

Figure 2. Starch content of biomass during cultivation time

Based on the growth monitoring presented, exponential growth occurred at initial phase when there are vast nutrient and room for cell to grow. However, growth entered stationary phase when cell density reached it saturated point where most of the nutrients had been utilized and the available spaces is limited for growth. The competition for nutrients, light and spaces for growth eventually caused the cell into death phase. The addition of carbon dioxide significantly increased the starch content and maintained a maximum growth rate. There is no additional organic or inorganic carbon sources included in the cultivation medium, thus carbon dioxide is the sole carbon source for microalgae to synthesize carbohydrates and accumulate starch in the microalgae cell [10]. As such, gaseous CO_2 first dissolves in water to form carbon dioxide(CO_2), carbonic acid(HCO_3^-) and bicarbonate ion(CO_3^{2-}) [11]. The dissociated components become the nutrient for microalgae growth. At the same time, presence of carbonic acid balanced the increasing pH that coupled with cell growth thus maintained the pH of the culture near to neutral [12,13]. This explained the minimal changes of pH for CO_2 enrichment aeration and substantial changes of pH profile for air aeration. Therefore, the CO_2 content in air is too low to support high cell growth rate. Since *S. dimorphus* have highest starch content and biomass density at day 16, the biomass was harvested for subsequent analysis and reaction.

S. dimorphus biomass was characterized and the composition was presented in table 1. The biomass contains about 53.7 % w/w of total carbohydrate, 12 % w/w of total lipid and 17.4 % w/w of protein. The total carbohydrate consisted about 44.5% (w/w) of starch, while the remainder consisted of structural polysaccharides, especially from the cell wall. Based on the HPLC chromatogram after two-step acid hydrolysis, the sugar profile indicated the presence of dextrin, sucrose, glucose and galactose. According to reported studies [14,15], green algae like *S. dimorphus* have unicellular cell wall built from cellulose and store carbon as starch. The degradation of the cellulose cell wall could yield simple sugar which is mannose, galactose and glucose [16].

Table 1. Carbohydrate content of *S. dimorphus* compared to other bioethanol feedstock.

Bioethanol feedstock	Carbohydrate content (% w/w)
Corn	59.5 [17]
Chlorella vulgaris	51.0 [18]
S.dimorphus(this study)	53.7

3.2. Biomass harvesting with natural sedimentation

S. dimorphus was harvested with natural sedimentation where aeration was ceased on day 16. The day of harvest was decided based on the starch content and dry weight of biomass as discussed earlier. Figure 3 illustrate the performance of natural sedimentation. The recovery efficiency of sedimentation was presented in figure 3 based on the optical density difference. The OD value was drop from 2.169 to 0.160 within 12 hours. At the first 4 hours of precipitation, the recovery efficiency increased from 0% to 28%. Substantial recovery is observed between 4-8 hours of sedimentation from 28% to 79.5%, but minimal increase was found afterwards from 79.5% to 90%. Microalgae cell culture potential to sediment due to gravity and formation of biological floc if the culture in stagnant situation particularly at stationery phase. In addition, the density of carbohydrates(1500 kg/m^3) is higher than those of proteins (1300 kg/m^3) and lipids (860 kg/m^3) [19]. The accumulation of carbohydrates inside the cell biomass was maximized on day 16, which further enhanced the density difference of the medium with the cell [20]. Furthermore, *S. dimorphus* cells tend to clump together in a high cell density environment. Consequently, this phenomenon increases the efficiency of sedimentation when substantial cell density is accumulated at the bottom of bottle. This explained the sedimentation pattern of *S. dimorphus* cell.

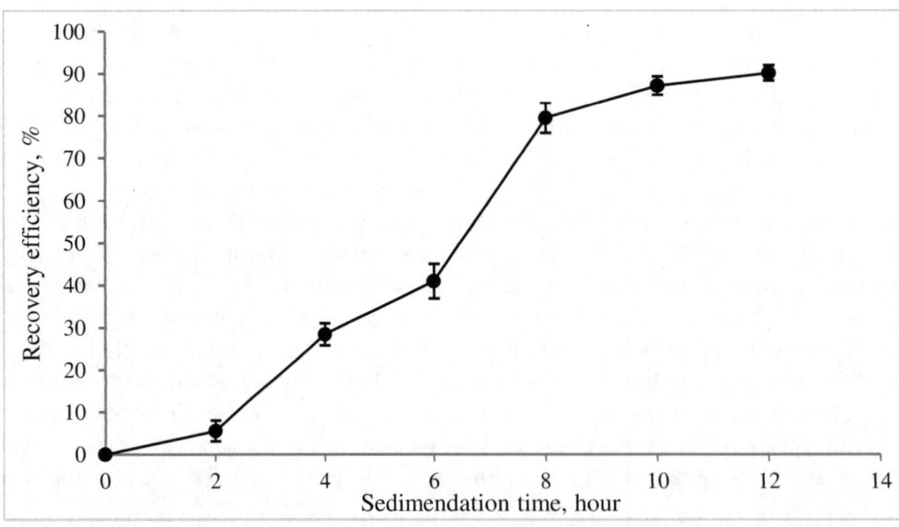

Figure 3. Recovery efficiency of microalgae with natural sedimentation.

3.3. Effects of hydrothermal acid hydrolysis at various conditions

Acid hydrolysis is a common method and sulphuric acid is the best catalyst among others after decades of study on it. Total sugar content will be presented instead of glucose since acid is capable to break most of the complex biopolymer in biomass into simple sugar. Figure 4 shows the effect of temperature and acid concentration in sulphuric acid hydrolysis. Total sugar content was increased with increasing temperature for acid concentration range of 1%-5%(v/v). Meanwhile increasing acid concentration increased the total sugar content significantly. Both parameters have interaction effects in degrading the polysaccharides to simple sugar. At higher temperature of 125 °C, considerable increase in total sugar content is observed for lower acid concentration from 1% to 2%. The yield was optimum at 4% of acid concentration and indicated declining trend at higher concentration when temperature is high enough for degradation. Therefore, the hydrolysis yield was maximized at 125 °C, which successfully produced 48% (w/w) of glucose in 15 min at an acid concentration of 4% (v/v). The data indicated that the temperature positively correlated with acid concentration in order to maximize the glucose content. Hydrolysis yield can be as high as 89.4% at optimum condition and as low as 2.4% at lowest temperature and acid concentration.

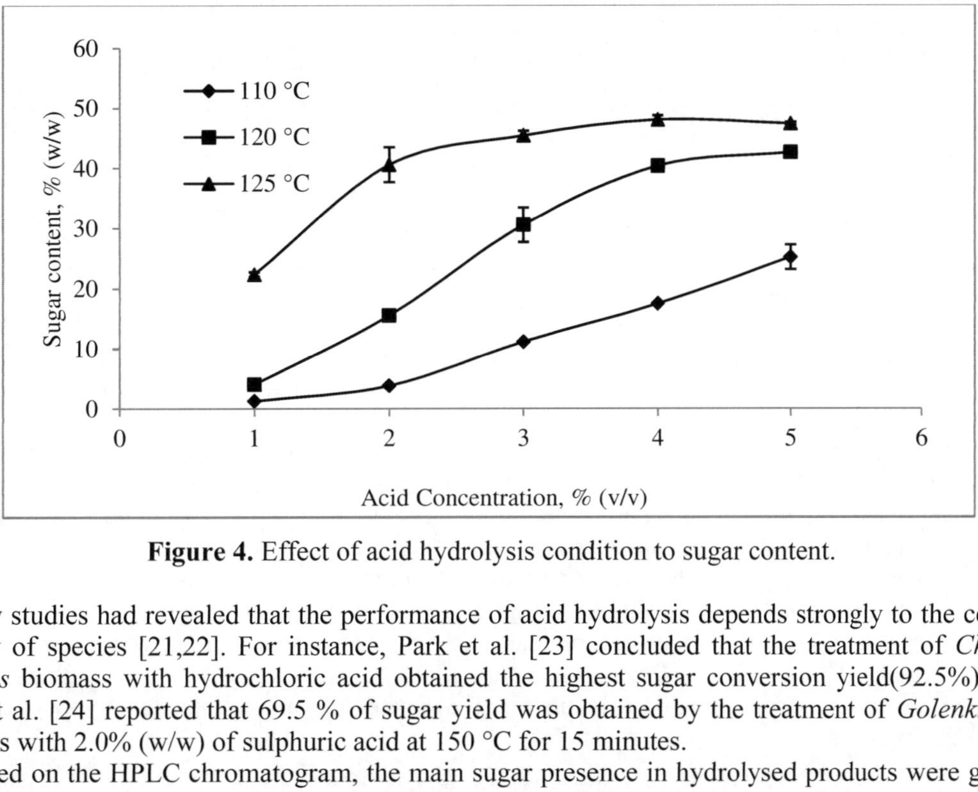

Figure 4. Effect of acid hydrolysis condition to sugar content.

Few studies had revealed that the performance of acid hydrolysis depends strongly to the cell wall rigidity of species [21,22]. For instance, Park et al. [23] concluded that the treatment of *Chlorella vulgaris* biomass with hydrochloric acid obtained the highest sugar conversion yield(92.5%); while Choi et al. [24] reported that 69.5 % of sugar yield was obtained by the treatment of *Golenkinia sp.* biomass with 2.0% (w/w) of sulphuric acid at 150 °C for 15 minutes.

Based on the HPLC chromatogram, the main sugar presence in hydrolysed products were glucose, sucrose and galactose. The hydrolysed products contains different percentage of glucose, sucrose and galactose at different hydrolysis conditions. At optimum condition, most of the sugar content attributed by glucose. The carbohydrate profile at optimum condition for sucrose, glucose and galactose are 0.053 g/g biomass, 0.363 g/g biomass, and 0.062 g/g biomass, respectively. The content of sucrose and galactose were reduced accordingly because most of them was degraded to glucose at higher temperatures and acid concentration. Further increased the temperature and acid concentration could degrade the sugar to furfural products thus reduced the yield of hydrolysis [25]. This explained the declining pattern of sugar content at higher operating temperature and acid concentration. Therefore, the acid concentration should be reduced if higher temperature was applied for optimum yield.

The usage of sulphuric acid in this treatment potential to generate side products and neutralization is required to increase the pH for yeast fermentation, mostly pH value of 4-6. The neutralization process resulting formation of waste, mostly insoluble salts that potential to create environmental issue.

3.4. Effects of enzymatic hydrolysis at various conditions

The enzymatic hydrolysis coupled with hydrothermal effect was investigated here. Table 2 presented the effect of heating to enzymatic hydrolysis. Hydrolysis without pre-heating the biomass reduced 13.9%(w/w) of glucose content. Therefore, effect of enzymatic operating conditions were evaluated by pre-heated the biomass at 90 °C in distilled water for 15 minutes. Figure 5 illustrates the effect of enzyme under various working environment. Enzymatic hydrolysis in this study was strictly targeted to liberate starch and transform it into glucose using amyloglucosidase enzyme. Thus, glucose is the only sugar found in hydrolysed products.

Table 2. Effect of pre-heating to enzymatic hydrolysis.

Treatment	Glucose content (% w/w)	Conversion yield (%)
Heating at 100 °C for 15 minutes	41.84±0.36	94.21±0.8
No heating	27.91±1.13	62.85±2.56

Figure 5(a), (b) and (c) show the effect of temperature, pH and enzyme loading to enzymatic function, respectively. In one hour time of reaction, considerable increase in glucose content from 39.7% to 42.8% is exhibited when higher temperature was applied from 30 °C to 55 °C. While temperature higher than 55 °C do not significantly improved the efficiency of enzyme. Thus, optimum temperature for enzymatic degradation was found at 55 °C with conversion yield of 96.1% based on the starch content of biomass. Therefore, effect of pH was investigated at reaction temperature of 55 °C owing to slight differences of glucose content(0.1%) between 55 °C and 60 °C.

Figure 5(b) demonstrates that the glucose content decreases with increasing pH of the medium. The glucose content gradually reduced to 34.9% when pH increased to 6. The highest glucose content of 42.6% (w/w) was obtained at pH 4. There is only 0.5% different in glucose content between pH 4 and 4.5. It is reasonable to conclude that the enzyme is active in acidic environment.

Performance of amyloglucosidase under various concentration was illustrated in figure 5(c). Enzyme activity performed well with 92.1% of glucose yield under concentration of 75 units/mL in one hour of hydrolysis period. Concentration above or below the value yielded glucose content that below 40% (w/w). This is because lower concentration did not sufficient to convert all the starch to glucose within the reaction time. Meanwhile, higher concentration of enzyme could reduce the enzyme activity.

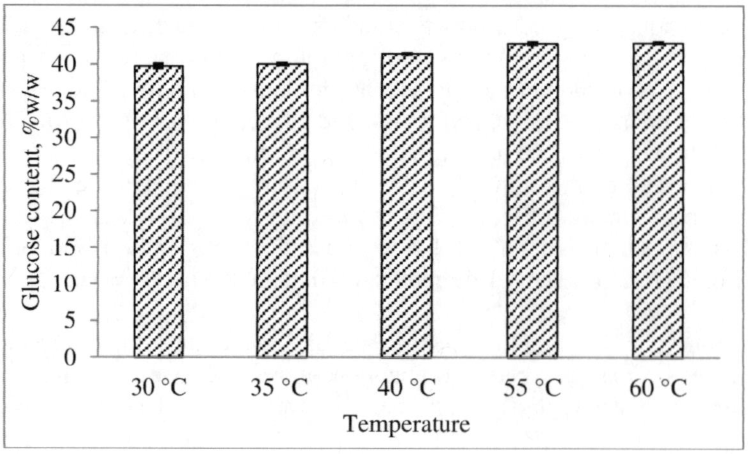

(a)

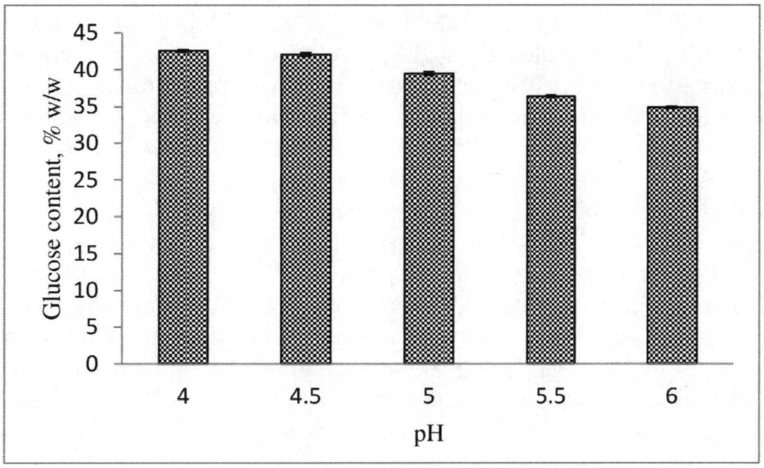

(b)

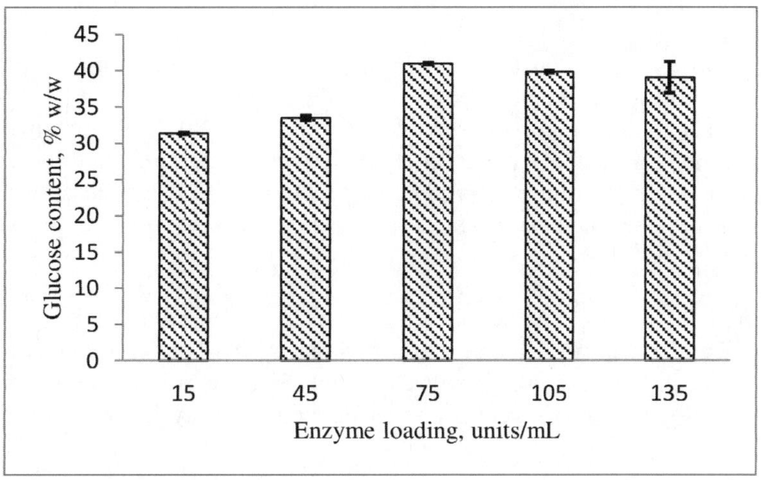

(c)

Figure 5. Enzymatic hydrolysis of *S. dimorphus* biomass at (a) various temperature, (b) various pH, and (c) enzyme loading, error bars indicate standard deviation.

As a result, it is suggested that saccharification temperature of 55 °C and a working pH of 4.5 that require less energy is reasonable for optimum performance of hydrolysis. This operating conditions agreed well with supplier suggested value and others research finding which using same type of enzyme. The performance of this enzyme over a wide range of reaction conditions is advantageous for fermentation, especially in an SSF configuration, where yeast only perform well under 30-40 °C [5]. The results clearly showed that enzymatic hydrolysis performed well under mild conditions during the glucose production step, which can be directly utilized for the subsequent fermentation process. Moreover, current production technology of enzyme have make cost of enzyme reduce significantly, thus reduce the concern on overall biofuel production cost.

3.5. Combination of solvent treatment with enzymatic hydrolysis

Single treatment with solvent only do not produce large amount of fermentable sugar when the treated biomass dissolve in distilled water. HPLC chromatogram depicted that the solution only contains 1-5% w/w of sucrose, glucose and galactose. Most of the carbohydrate remains as soluble complex biopolymer such as starch and dextrin. Therefore, enzymatic hydrolysis at optimum conditions was experimented on the solvent-treated biomass without pre-heating process. The process resulted about 0.568 g glucose/g treated-biomass. The yielded hydrolysate is a promising input for subsequent fermentation because there are no side products detected in the solution, which could harm the ethanologenic yeast.

3.6. Economic feasibility of microalgal bioethanol production

Bioethanol production has been limited by extensive economic and energy restrictions, which are considered to be linearly proportional. Economic assessment considers upstream and downstream processes to define the minimum required cost before commercialization [26]. The National Energy Laboratory (NREL) has projected the cost of bioethanol production based on a production rate of 1,339 dry U.S. ton/day of microalgae, specifically *Scenedesmus sp.* and *Chlorella sp.* [27]. Based on the analysis, the feedstock contribute the largest portion to fuel price which need significant effort to reduce it. This analysis predicted minimum bioethanol prices of $4.35/GGE [27]. Similar report have addressed the same issue that microalgae-based bioethanol production is only feasible with substantial reduction in feedstock price and improvement in conversion process [3].

4. Conclusions

Scenedesmus dimorphus is promising as an alternative source for bioethanol. The microalgae are offering outstanding feedstock for bioethanol with 1.47 g/L biomass and 53.7% of carbohydrate content. The high yield of sugar content which is more than 80 % is obtainable with simple pretreatment process, thus fermentable via counterpart fermentation. This study become the essential information for further detail study in the production technology to establish a sustainable and cost effective bioethanol. Moreover, the side products of fermentation, CO_2 could be the carbon source for microalgae cultivation.

Acknowledgements

The authors would like to acknowledge the funding given by the Universiti Tunku Abdul Rahman, Universiti Sains Malaysia(Postgraduate Research Grant Scheme no. 8046001) and Mybrain 15 MyPhD scholarship funded by the Ministry of Higher Education Malaysia for this project.

References

[1] Demirbas A 2010 *Energy Convers. Manage.* **51** (12) 2738.
[2] Brennan L and Owende P 2009 *Renew. Sustainable Energy Rev.* **14** (2) 557.
[3] Laurens L, Nagle N, Davis R, Sweeney N, Van Wychen S, Lowell A, and Pienkos P 2015 *Green Chem.* **17** (2) 1145.
[4] Kim J, Um B-H, and Kim T 2012 *Korean J. Chem. Eng.* **29** (2) 209.
[5] Choi S P, Nguyen M T, and Sim S J 2010 *Bioresour. Technol.* **101** (14) 5330.
[6] Ruiz R and Date T E, in *Laboratory Analytical Procedure No. 002,* (National Renewable Research Laboratory 1996).
[7] Ehrman T, in *Laboratory Analytical Procedure No. 016,* (National Renewable Research Laboratory 1996).
[8] Bligh E G and Dyer W J 1959 *Can. J. Biochem. Phys.* **37** (8) 911.
[9] Kumar A, Ergas S, Yuan X, Sahu A, Zhang Q, Dewulf J, Malcata F X, and van Langenhove H 2010 *Trends Biotechnol.* **28** (7) 371.
[10] Choix F J, de-Bashan L E, and Bashan Y 2012 *Enzyme Microb. Technol.* **51** (5) 300.
[11] Tang D, Han W, Li P, Miao X, and Zhong J 2011 *Bioresour. Technol.* **102** (3) 3071.
[12] Van Den Hende S, Vervaeren H, and Boon N 2012 *Biotechnol. Adv.* **30** (6) 1405.
[13] Souza S P, Gopal A R, and Seabra J E A 2015 *Energy* **81** 373.
[14] John R P, Anisha G, Nampoothiri K M, and Pandey A 2011 *Bioresour. Technol.* **102** (1) 186.
[15] Markou G, Angelidaki I, and Georgakakis D 2012 *Appl. Microbiol. Biotechnol.* **96** (3) 631.

[16] Takeda H 1996 *Phytochemistry* **42** (3) 673.
[17] Kwiatkowski J R, McAloon A J, Taylor F, and Johnston D B 2006 *Ind. Crops Prod.* **23**.
[18] Ho S-H, Huang S-W, Chen C-Y, Hasunuma T, Kondo A, and Chang J-S 2013 *Bioresour. Technol.* **135** 191.
[19] Baeyens J, Kang Q, Appels L, Dewil R, Lv Y, and Tan T 2015 *Prog. Energy Combust. Sci.* **47** 60.
[20] Günerken E, D'Hondt E, Eppink M H M, Garcia-Gonzalez L, Elst K, and Wijffels R H 2015 *Biotechnol. Adv.* **33** (2) 243.
[21] Harun R and Danquah M K 2011 *Process Biochem.* **46** (1) 304.
[22] Miranda J, Passarinho P C, and Gouveia L 2012 *Bioresour. Technol.* **104** 342.
[23] Park C, Lee J H, Yang X, Yoo H Y, Lee J H, Lee S K, and Kim S W 2016 *Bioprocess Biosyst. Eng.* **39** (6) 1015.
[24] Choi S A, Choi W I, Lee J S, Kim S W, Lee G A, Yun J, and Park J Y 2015 *Bioresour. Technol.* **190** 408.
[25] Kosugi A, Kondo A, Ueda M, Murata Y, Vaithanomsat P, Thanapase W, Arai T, and Mori Y 2009 *Renew. Energy* **34** (5) 1354.
[26] Nagarajan S, Chou S K, Cao S, Wu C, and Zhou Z 2012 *Bioresour. Technol.* **140** 550.
[27] Humbird D, Davis R, Tao L, Kinchin C, Hsu D, Aden A, Schoen P, Lukas J, Olthof B, Worley M, Sexton D, and Dudgeon D 2011 in *Technical Report NREL/TP-5100-47764. United States National Renewable Energy Laboratory, US Department of Energy*.

IOP Conf. Series: Materials Science and Engineering **206** (2017) 012019 doi:10.1088/1757-899X/206/1/012019

Bioethanol produced from *Moringa oleifera* seeds husk

E N Ali[1], and S Z Kemat[2]
[1]Senior Lecturer, Universiti Malaysia Pahang, Pahang, Malaysia.
[2]Final Year Student, Universiti Malaysia Pahang, Pahang, Malaysia.

E-mail: eman@ump.edu.my

Abstract. This paper presents the potential of bioethanol production from *Moringa oleifera* seeds husk which contains lignocellulosic through Simultaneous Saccharification and Fermentation (SSF) process by using *Saccharomyces cerevisiae*. This paper investigates the parameters which produce optimum bioethanol yield. The husk was hydrolyzed using NaOH and fermented using *Saccharomyces cerevisiae* yeast. Batch fermentation was performed with different yeast dosage of 1, 3, and 5 g/L, pH value was 4.5, 5.0 and 5.5, and fermentation time of 3, 6, 9 and 12 hours. The temperature of fermentation process in incubator shaker is kept constant at 32ºC. The samples are then filtered using a 0.20 µm nylon filter syringe. The yield of bioethanol produced was analysed using High Performance Liquid Chromatography (HPLC). The results showed that the highest yield of 29.69 g/L was obtained at 3 hours of fermentation time at pH of 4.5 and using 1g/L yeast. This research work showed that *Moringa oleifera* seeds husk can be considered to produce bioethanol.

1. Introduction

Day by day, population is increasing causing decreasing in conventional energy. Increase on world's energy demand and the progressive depletion of oil reserves are motivating the researchers to find alternative energy resources, especially for those derived from renewable materials such as biomass (Saxena et al., 2009). Agricultural raw material and biomass can be utilized to produce renewable energy in the form of liquid/gaseous fuel which can reduce CO_2 emission on combustion unlike fossil fuels which can cause greenhouse effect. The climate change and the consequent need to diminish greenhouse gases emissions is a global concern. Because of that, it has been encouraged to use bioethanol as a gasoline replacement or as an additive (Mustafa et al., 2008). Bioethanol may also be used as raw material for the production of different chemicals, thus driving a full renewable chemical industry.

Research on bioethanol production has been investigated recently as this fuel is efficient, biodegradable, environmentally friendly and cost effective alternative for the conventional fossil fuel. Bioethanol derived from lignocellulosic material is a potential renewable energy source. There are countless lignocellulosic materials that have been used in recent years to produce bioethanol such as sugarcane bagasse, paper and pulp, cassava waste, saw dust and corn (Pandey et al., 2000; Akinyele et al., 2011; Amarnath & Balakrishnan, 2007).

Moringa oleifera seeds husk is considered as new initiative to replace conventional energy resources. *Moringa oleifera* is a tropical plant belongs to the family of *Moringaceae.* It also have fourteen different species which has been identified – all of them possessing varying degrees of coagulation activity (Jahn, 1988). *Moringa oleifera* is the most widespread species, which grows quickly at low altitudes in the whole tropical belt, including arid zones (Julia, 1991; Verdcourt, 1985).

It is generally known in the developing world as a vegetable, a medicinal plant and a source of vegetable oil. However, in the Sudan, it has been traditionally used in water purification. These multiple uses of the *Moringa oleifera* plant have greatly promoted its widespread application. *Saccharomyces cerevisiae* yeast is used to produce bioethanol from *Moringa oleifera* seeds husk.
Saccharomyces cerevisiae is commonly known as baker's yeast. It has a single–celled eukaryote that is frequently used in scientific research. *Saccharomyces cerevisiae* is an excellent model microbe in biological research, and has been used for metal biosorption research (Jianlong & Can, 2006, 2009). This research work was carried out to find optimum yield of bioethanol that can be produced from *Moringa oleifera* seeds husk.

2. Material and Method

2.1 Chemicals and baker's yeast
Sulphuric Acid (H_2SO_4) and Sodium Hydroxide (NaOH) were obtained from R & M Chemicals. The fermentation was done by using baker's yeast (*Saccharomyces cerevisiae*), obtained from supermarket. Apart from that, the ultrapure water obtained from Milli–Q® Integral Water Purification System and HPLC–grade sulfuric acid obtained from Merck (Darmstadt, Germany) were mixed in order to prepare the mobile phase for HPLC. HPLC–grade ethanol was obtained from Merck (Darmstadt, Germany) for ethanol standard calibration curve.

2.2 Preparation of Moringa oleifera seeds husk
The dried *Moringa oleifera* pods were collected at Batu Berendam, Melaka, Malaysia (August 2015) (figure 1). The *Moringa oleifera* seeds (figure 2) were extracted from its pods, and the seeds were de–husked manually to get the husk for experimental work (figure 3). Around 30 g of *Moringa oleifera* seeds husk for each sample were immersed in 450 mL of pure water in 500 mL conical flasks for an overnight before experiment was started.

Figure 1. *Moringa oleifera* tree.

IOP Conf. Series: Materials Science and Engineering **206** (2017) 012019 doi:10.1088/1757-899X/206/1/012019

Figure 2. *Moringa oleifera* seeds.

Figure 3. *Moringa oleifera* seeds husk.

2.3 Pretreatment of Moringa oleifera seeds husk

Sodium hydroxide (NaOH) was added drop by drop into the samples of immersed *Moringa oleifera* seeds husk until the pH 11 was attained. The flasks were covered with cotton wool and aluminium foil (figure 4). The samples were pre–treated in an autoclave (Hirayama HV–85) at a temperature of 121°C for 2 hours.

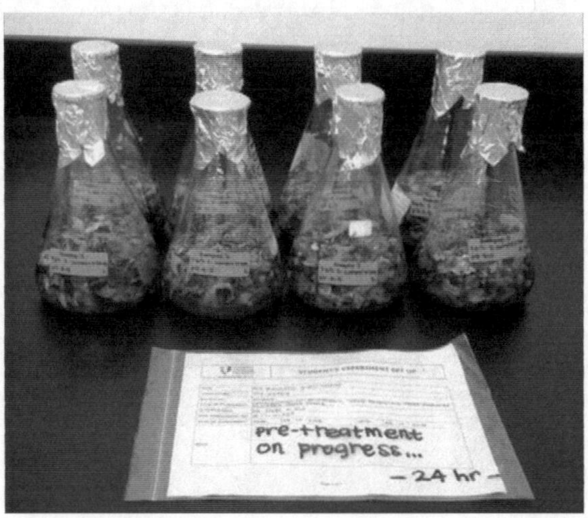

Figure 4. Pre-treatment of husk

2.4 Fermentation of Moringa oleifera seeds husk
For the first batch of experiments, 1 g/L of *Saccharomyces cerevisiae* was added into the samples. The samples were let cooled to room temperature, then sulphuric acid was added drop–by–drop in order to adjust the pH parameter which is 4.5, 5.0 and 5.5. Then, the samples were kept in a stackable incubator shaker (INFORS HT Ecotron) at 180 rpm for 12 hours at 32°C to allow a complete fermentation. The fermentation temperature was chosen based on (Marina, et al., 2009) who stated that 32°C is the ideal fermentation temperature. The fermentation temperature represents a critical step in ethanol production, since higher temperatures affect yeast behaviour, diminishing the ethanol content, which increases the consumption of energy during distillation of ethanol produced (Marina, et al., 2009).

2.5 Collection of samples
Sample of 4 mL was taken aseptically at different fermentation periods of 3, 6, 9, and 12 hours. All the collected samples were immediately stored in a freezer at - 20°C before used for chromatography analysis. The samples were filtered using 0.20 µm nylon filter syringe into 1.5 mL Agilent vials for HPLC analysis. The same procedure was followed to produce bioethanol using 3g/L and 5g/L of yeast.

2.6 Analysis of ethanol content using HPLC
The analysis of ethanol concentration from the sample was performed using Agilent 1200 HPLC equipped with Rezex ROA column (150 X 7.80 mm) and refractive index (RI) detector. The chromatography grade 0.005 N sulphuric acid was used as the mobile phase and the flow rate was set at 0.5 ml/min. The column temperature was set at 60°C and RI detector temperature at 40°C. The injection volume of samples to HPLC was 10 µL. The calibration of HPLC was done using ethanol standard. Six dilutions of ethanol standard were prepared with concentration of 5, 10, 15, 20, 25, and 30 g/L. Then, the ethanol standards were filtered using 0.20 µm nylon filter syringe before analysis. Calibration curve of the peak area (obtained from HPLC analysis) versus ethanol concentration was plotted using Microsoft ExcelTM 2013.

3. Results and Discussion

3.1 HPLC Calibration
The ethanol standards were injected to HPLC and the results are shown in figure 5 shows the calibration curve which represent the relationship between peak area and concentration of ethanol standards. The concentration of ethanol was calculated using the equation y= 70594 x + 218677 with regression correlation of R^2=0.9928.

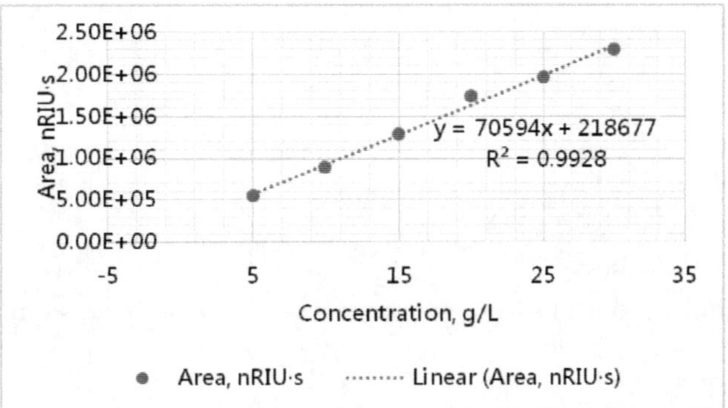

Figure 5. Ethanol standard calibration curve

Table 1 summarizes the bioethanol yield during twelve hours of fermentation period, by varying the yeast dosage and pH value. Yeast dosage is evaluated at 1, 3, and 5 g/L and pH value was varied at 4.5, 5.0 and 5.5. Fermentation temperature is kept constant at 32°C with agitation speed of 180 rpm. The fermented broth is collected at each 3 hours for twelve hours of fermentation period of 3, 6, 9 and 12 hours. The results of using 5g/L are not shown in Table 1, because there was no bioethanol produced at that yeast concentration.

Table 1. Concentration of bioethanol produced

	Concentration, g/L					
Yeast Dosage, g/L	1			3		
pH / Time (hours)	4.5	5	5.5	4.5	5	5.5
3	29.69	17.34	21.6	1.98	0.18	0
6	20.39	13.67	7.69	0	0	0
9	2.06	1.18	0	0	0	0
12	0	0	0	0	0	0

3.2 *Effect of yeast dosage on bioethanol yield*

The bioethanol yield using 1 g/L of *Saccharomyces cerevisiae* is shown in figure 6 at different pH values and fermentation time. The highest bioethanol yield was 29.69 g/L produced at pH 4.5 and 3 hours fermentation time. The bioethanol produced was decreasing to reach 0 g/L at 12 hours of fermentation time. The same trend goes by using 3g/L yeast (figure 7).
In case of adding 5g/L yeast there was no bioethanol produced.

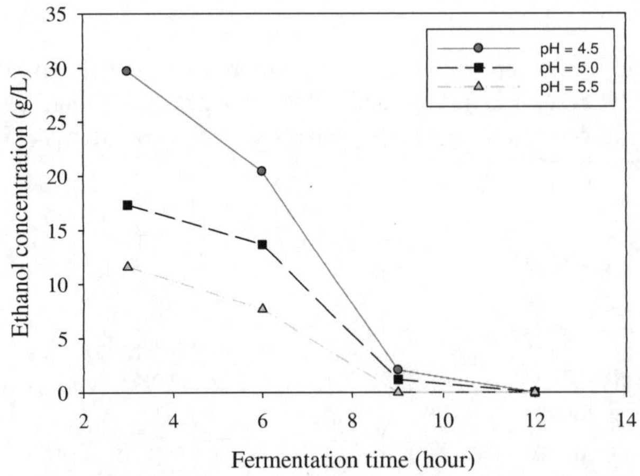

Figure 6. Bioethanol yield using 1 g/L of *Saccharomyces cerevisiae* at different pH values

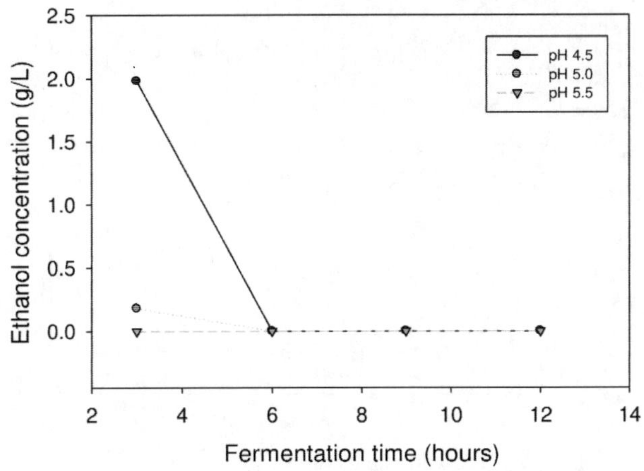

Figure 7. Bioethanol yield produced with 3 g/L of *Saccharomyces cerevisiae* at different pH.

Figure 8 shows the comparison of bioethanol yield using 1, 3 and 5 g/L of *Saccharomyces cerevisiae* at pH 4.5 and 3 hours fermentation time. The highest bioethanol yield was 29.69 g/L produced at pH 4.5 and 3 hours fermentation time for 1 g/L of yeast used. Bioethanol yield is decreased with addition of yeast to reach 1.98 g/L using yeast dosage of 3g/L, and in case of using 5g/L there was no bioethanol produced.

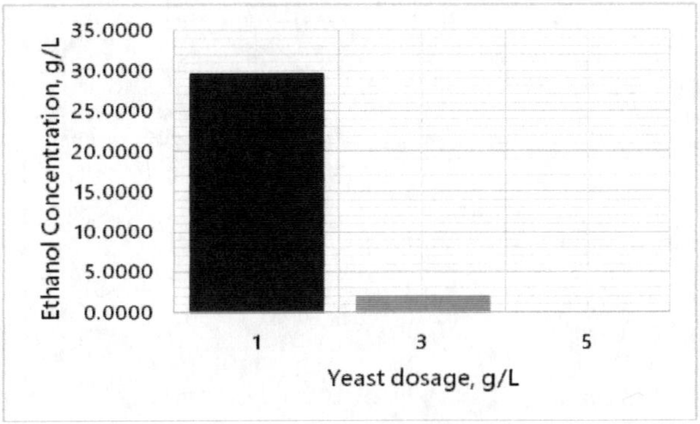

Figure 8. Comparison of bioethanol yield using different yeast dosages

3.3 *Effect of pH value on bioethanol yield*

Figure 9 shows the bioethanol yield using pH 4.5 with different yeast dosages and fermentation time. The highest bioethanol yield was 29.69 g/L produced at 3 hours fermentation time with 1 g/L of *Saccharomyces cerevisiae*. When the fermentation reached 12 hours onwards, there was no bioethanol produced.

IOP Conf. Series: Materials Science and Engineering **206** (2017) 012019 doi:10.1088/1757-899X/206/1/012019

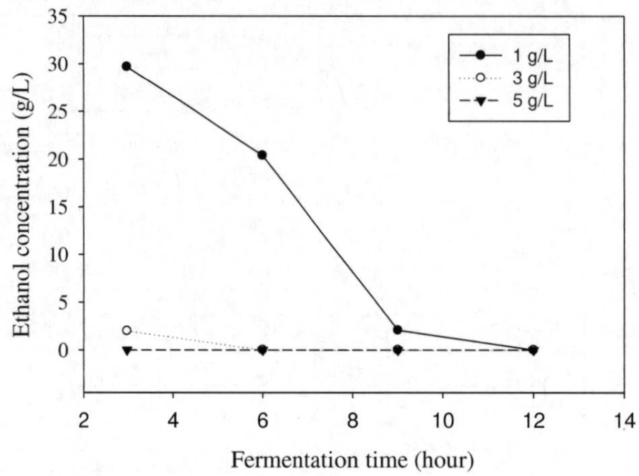

Figure 9. Bioethanol yield produced with pH 4.5

When the pH increased to 5, the bioethanol produced was decreased to reach 17.34 g/L using 1g/L yeast. At the same pH of 5 the bioethanol produced was only 0.18 g/L with 3g/L yeast, reaching to 0.00 g/L bioethanol by adding 5g/L yeast as shown in figure 10.

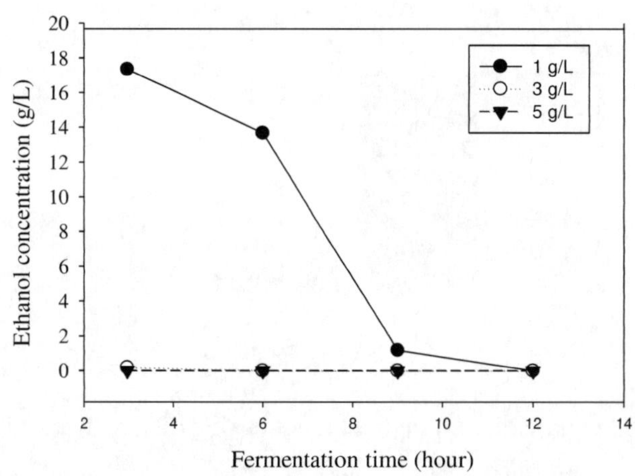

Figure 10. Bioethanol yield produced with pH 5.0

Figure 11 is showing that by doing fermentation at high pH, the bioethanol will decrease as pH increase. In this work, the bioethanol produced was 11.60 g/L at pH 5.5 and 3 hours fermentation using 1g/L yeast, compared to 29.69 g/L at pH 4.5 and 3 hours fermentation using 1g/L yeast, and 17.34 g/L at pH 5 and 3 hours fermentation using 1g/L yeast. The same trend goes when using 3g/L yeast, where the bioethanol produced was 0.00g/L at 5.5, while it was 1.98 g/L at pH 4.5, and 0.18 g/L at pH 5.

IOP Conf. Series: Materials Science and Engineering **206** (2017) 012019 doi:10.1088/1757-899X/206/1/012019

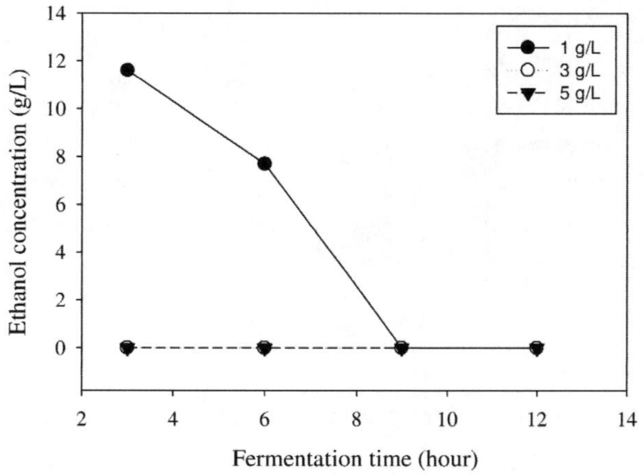

Figure 11. Bioethanol yield produced with pH 5.5

The comparison of bioethanol yield using pH 4.5, 5.0 and 5.5 on 1 g/L of *Saccharomyces cerevisiae* and 3 hours fermentation time is shown in figure 12. The highest bioethanol yield is 29.69 g/L produced at pH 4.5 and 3 hours of fermentation time for 1 g/L of yeast. Bioethanol yield is decreased with increasing pH values.

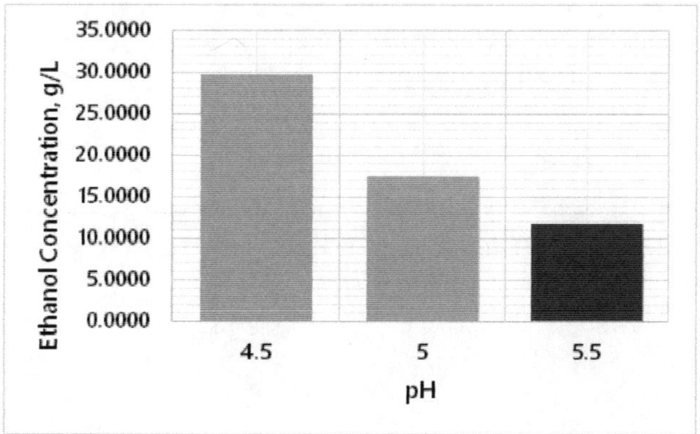

Figure 12. Comparison of bioethanol yield at different pH.

3.4 *Effect of fermentation time on bioethanol yield*

Figure 13 shows the comparison of bioethanol yield at 3, 6, 9 and 12 hours of fermentation time using pH 4.5, 5.0 and 5.5 on 1 g/L of *Saccharomyces cerevisiae*. The highest bioethanol yield is 29.69 g/L produced at pH 4.5 and 3 hours fermentation time for 1 g/L of yeast used. Bioethanol yield is decreased with increasing fermentation time.

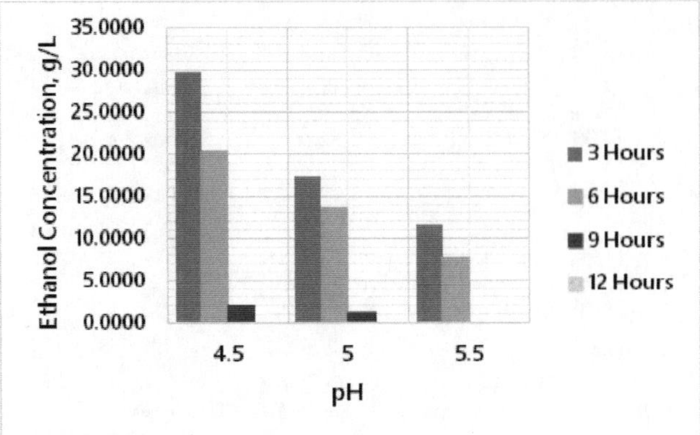

Figure 13. Comparison of fermentation time effect on bioethanol yield

4. Conclusion

Based on previous studies, it had been approved that *Moringa oleifera* seeds husk can be considered as a potential substrate for ethanol production due to their high cellulose contents (Martin et. al, 2010). The results indicated clearly that the characteristics of the husks can lead to the production of ethanol. The highest amount of the ethanol yield was 29.69 g/L at 3 hours fermentation at temperature of 32°C, agitation speed of 180 rpm, pH 4.5 and yeast concentration of 1 g/L. It can be concluded that bioethanol can be produced with a low dosage of yeast and in a short time, which both considered positive results economically. It is recommended to further this study using lower yeast concentration than 1 g/L and analyse the fermented sample each 1 hour up to 4 hours to find the optimum condition. And more studies need to be done to compare between manual de–husking and mechanical de–husking. It was proved that *Moringa oleifera* seeds husk is a good source for bioethanol production to have biodegradable fuel with no pollution problems.

Acknowledgment

The authors are grateful to the Research and Innovation Department at Universiti Malaysia Pahang, Pahang, Malaysia, for financial support under grant # **RDU 140318**.

References

[1] Akinyele B J, Olaniyi O O, and Arotupin J D 2011 Bioconversion of selected agricultural wastes and associated enzymes by *Volvariella volvacea*: An edible mushroom. *Res. J. Microbiol* **6** 63–70. DOI: 10.3923/jm.2011.63.70.

[2] Amarnath R and Balakrishnan V 2007 Evaluation of the banana (*Musa paradisiaca*) plant by–products fermentation characteristics to assess their fodder potential. *Int. J. Dairy Sci.* **2** 217–225. DOI: 10.3923/ijds.2007.217.225.

[3] Jahn S A 1988 Using Moringa Seeds as Coagulants in Developing Countries *J. Am. Water Works Assoc.* **80** (6) 43–50.

[4] Jianlong W and Can C 2006 Biosorption of heavy metals by *Saccharomyces cerevisiae*: a review *Biotechnol. Adv.* **24** (5) 427–451. DOI:10.1016/j.biotechadv.2006.03.001.

[5] Jianlong W and Can C 2009 Biosorbents for heavy metals removal and their future *Biotechnol. Adv.* **27** (2) 195–226. DOI:10.1016/j.biotechadv.2008.11.002.

[6] Julia F M 1991 The horseradish tree,Moringa pterygosperma (Moringaceae): A boon to Arid Lands *Econ. Bot.* **45** (3) 318–333.

[7] Marina O S D, Adriano V E, Silvia A N, Rubens M F, Carlos E V R, and Maria R W M 2009 Production of bioethanol and other bio–based materials from sugarcane bagasse: Integration to conventional bioethanol production process *Chem. Eng. Res. Des..* **87** 1206–1216.

[8] Martın C, Moure A, Martın G, Carrillo E, Domınguez H, and Parajo J C 2010 Fractional characterisation of jatropha, neem, moringa, trisperma, castor and candlenut seeds as potential feedstocks for biodiesel production in Cuba "https://www.researchgate.net/journal/0961-9534_Biomass_and_Bioenergy" *Biomass Bioenerg.* **34** (4) 33-538. DOI: 10.1016/j.biombioe.2009.12.019.

[9] Mustafa B, Havva B, and Cahide Ö 2008 Progress in bioethanol processing *Prog. Energ Combust.* **34** 551–573.

[10] Pandey A, Soccol C R, Nigam P, & Soccol V T 2000 Biotechnological potential of agro–industrial residues I: Sugarcane Bagasse *Biores. Technol.* **74** (1) 69–80.

[11] Saxena R C, Adhikari D K, & Goyal H B 2009 Biomass–based energy fuel through biochemical routes: A review *Renew. Sust. Energ. Rev.* **13** (1) 167–178. DOI: 10.1016/j.rser.2007.07.011.

[12] Verdcourt B 1985 A synopsis of the Moringaceae *Kew Bulletin* **40** 1–23. DOI: 10.2307/4108470.

Experimental study of bioethanol production using mixed cassava and durian seed

Q H Seer[1,2], **J Nandong**[1,2] **and T Shanon**[1]

[1] Department of Chemical Engineering, Curtin University, 98009 Miri, Sarawak, Malaysia
[2] Curtin Sarawak Research Institute, Curtin University, 98009 Miri, Sarawak, Malaysia

E-mail: seer.qiu.han@postgrad.curtin.edu.my

Abstract. The production of biofuels using conventional fermentation feedstocks, such as sugar- and starch-based agricultural crops will in the long-term lead to a serious competition with human-animal food consumption. To avoid this competition, it is important to explore various alternative feedstocks especially those from inedible waste materials. Potentially, fruit wastes such as damaged fruits, peels and seeds represent alternative cheap feedstocks for biofuel production. In this work, an experimental study was conducted on ethanol production using mixed cassava and durian seeds through fermentation by *Saccharomyces cerevisiae* yeast. The effects of pH, temperature and ratio of hydrolyzed cassava to durian seeds on the ethanol yield, substrate consumption and product formation rates were analyzed in the study. In flask-scale fermentation using the mixed cassava-durian seeds, it was found that the highest ethanol yield of 45.9% and a final ethanol concentration of 24.92 g/L were achieved at pH 5.0, temperature 35°C and 50:50 volume ratio of hydrolyzed cassava to durian seeds for a batch period of 48 hours. Additionally, the ethanol, glucose and biomass concentration profiles in a lab-scale bioreactor were examined for the fermentation using the proposed materials under the flask-scale optimum conditions. The ethanol yield of 35.7% and a final ethanol concentration of 14.61 g/L were obtained over a period of 46 hours where the glucose was almost fully consumed. It is worth noting that both pH and temperature have significant impacts on the fermentation process using the mixed cassava-durian seeds.

1. Introduction

Biofuels in general and bioethanol in particular have taken the stage in addressing energy security problem, decreasing fossil fuel dependency and most importantly in reducing greenhouse-gas (GHG) emissions. Bioethanol can be produced by fermenting sugars from biomasses and used as fuel for vehicular engine internal combustion [1]. Currently, most of the conventional ethanol fermentation feedstocks are relying on agricultural crops, such as starch-based crops and sugar-based crops. However, such a heavy reliance on agricultural crops will one day lead to a serious competition with the demand for livestock feed and human food consumption [2]. Interestingly, abundant agricultural wastes, such as lignocellulosic materials has been proposed to relieve the fuel versus food competition problems. However, the use of this kind of materials presently still requires costly pretreatment to decompose complex lignin structures predominantly encountered in the wastes [3]. For this reason, it is important to explore various other alternative feedstocks from inedible waste materials. Fruit wastes, e.g., damaged fruits, peels and seeds, represent a potential cheap alternative to be used as feedstocks for biofuel production.

IOP Conf. Series: Materials Science and Engineering **206** (2017) 012020 doi:10.1088/1757-899X/206/1/012020

In some tropical countries like Malaysia, durian (*Durio zibethinus*) seed is regarded as agricultural residue which accounts for about 20-25% of the total fruit mass [4]. Note that, durian seeds are rich source of carbohydrates in the form of starch, with the floured and peeled floured forms of the seed containing 73.9% and 76.8% of carbohydrates, respectively [5]. Meanwhile, fresh durian seeds contain high moisture (51.5%) and carbohydrate (43.6%) [6]. In this paper, an experimental study was conducted on ethanol production using mixed cassava and durian seed through fermentation by *Saccharomyces cerevisiae* yeast.

In the flask-scale fermentation, the effects of pH, temperature and ratio of hydrolyzed cassava to durian seeds on the ethanol yield, substrate consumption and product formation rates were analyzed. Enzyme-catalysed reactions are highly sensitive to changes in temperature [7]. According to [8, 9], the optimum temperature of saccharifying enzymes for fermentation and microbial growth is in the range of 25-35°C. Additionally, it has been widely recognized that the cultivation pH plays an important role in ethanol production [10]. *Saccharomyces cerevisiae* is an acidophilic organism by nature where the optimal range for this microorganism to grow lies between pH 4 and 6 [11]. A constant yeast intracellular pH has to be maintained at its optimum pH range. Maintaining the optimal intracellular pH is crucial as many pH-dependent enzymes are operating within the yeast cell during growth and metabolism [12]. In the present work, we also conducted the fermentation in a lab-scale bioreactor where the ethanol, glucose and biomass concentration profiles were studied using the proposed materials under the optimum conditions obtained from the flask-scale study. The lab-scale study was conducted in addition to the flask-scale study in order to gain a better understanding of the fermentation process using the complex mixed cassava-durian seeds.

2. Materials and methods

Hydrolyzed cassava and durian seeds were used as a carbon source for the fermentation process. The flask-scale fermentation was run for a period of 48 hours. Meanwhile, the lab-scale bioreactor experiment was executed in batch mode by using a BIOSTAT A-plus 2 L, MO-Assembly bioreactor, which was run for a period of about 48 hours until the substrate was used up.

2.1. Preparation of durian seeds

The durian seeds were washed to remove traces of durian fruit and slime and dried in dehydrator to remove water content within the seeds. Dried durian seeds were milled using a pulveriser.

2.2. Medium preparation

First, 100 g of cassava powder was mixed with 1 L of 0.5 mol/L H_2SO_4 solution. Meanwhile, 100 g of milled durian seeds was mixed with 1 L of 0.5 mol/L H_2SO_4 solution. Then, the cassava medium culture and durian seed medium culture were cooked at 121°C for 20 minutes in order to break down the starch into fermentable sugars. For the flask-scale fermentation, 200 mL of medium culture was prepared. The mixed medium cultures were prepared based on three ratios of hydrolysed cassava to durian seed where in this research the ratios: 85:15, 75:25 and 50:50. 1 g/L yeast extract, 2.5 g/L NH_4Cl, 2.91 g/L Na_2HPO_4, 3 g/L KH_2PO_4, 0.25 g/L $MgSO_4$, 0.08 g/L $CaCl_2$, 4.3 g/L citric acid and 3 g/L sodium citrate were dissolved in the mixed medium culture [13]. The medium culture was adjusted to desired pH by using solutions of 1 mol/L of NaOH and 0.5 mol/L of H_2SO_4. After adjusting the pH to a desired value, the medium culture was sterilized at 121°C for 20 minutes to avoid contamination and then cooled down to room temperature. Note that, the procedure for preparing 1.5 L of culture medium for a lab-scale bioreactor experiment was similar to that of the flask-scale experiment. The inoculum was prepared by using Baker's yeast incubated in glucose solution for about 8-10 hours under room temperature. In the experiments, 8 mL and 60 mL of inoculums were added to the flask-scale and lab-scale bioreactor fermentation respectively, prior to the fermentation start-up.

IOP Conf. Series: Materials Science and Engineering **206** (2017) 012020 doi:10.1088/1757-899X/206/1/012020

2.3. Sample analysis
The UV spectrophotometer (Lambda 25, Perkin Elmer, USA) was used to analyzed the glucose, ethanol and cell concentrations at wavelength of 340nm under room temperature. The glucose and ethanol samples were prepared by using R-Biopharm test kits. For biomass cell concentration, the samples were first diluted to appropriate concentrations so that the corresponding values of absorbance were within the range between 0.1 and 0.4.

3. Results and discussion
3.1. Flask-scale fermentation
The effects of pH, temperature and ratio of hydrolyzed cassava to durian seeds on the ethanol yield, substrate consumption and product formation rates were analyzed in the flask-scale fermentation. Tables 1 and 2 show the experimental results for the flask-scale fermentation at 30°C and 35°C for a batch period of 48 hours, respectively.

Table 1. Experimental results for flask-scale fermentation at 30°C

pH	Ratio (cassava: durian seed)	Initial glucose (g/L)	Final glucose (g/L)	Final ethanol (g/L)	Glucose consumed (%)	Ethanol Yield (%)
4.5	85:15	50.89	15.88	3.91	68.8	11.2
	75:25	75.28	12.81	7.11	83.0	11.4
	50:50	92.62	23.43	15.25	74.7	22.0
5.0	85:15	65.52	48.63	4.41	25.8	26.1
	75:25	74.10	45.73	7.46	38.3	26.3
	50:50	76.54	30.59	15.74	60.0	34.2
5.5	85:15	67.83	40.73	2.86	40.0	10.5
	75:25	69.91	47.02	3.40	32.7	14.9
	50:50	73.92	58.62	3.41	20.7	22.3

Table 2. Experimental results for flask-scale fermentation at 35°C

pH	Ratio (cassava: durian seed)	Initial glucose (g/L)	Final glucose (g/L)	Final ethanol (g/L)	Glucose consumed (%)	Ethanol Yield (%)
4.5	85:15	48.14	32.38	7.03	32.6	44.6
	75:25	58.75	32.02	9.96	45.6	37.3
	50:50	66.25	13.63	21.46	79.5	40.8
5.0	85:15	56.04	37.98	8.00	32.2	44.3
	75:25	63.72	39.43	10.76	38.1	44.3
	50:50	72.93	18.61	24.92	74.5	45.9
5.5	85:15	53.11	20.64	7.00	61.2	21.6
	75:25	62.86	32.78	7.71	47.9	25.6
	50:50	66.70	16.34	13.17	75.6	26.2

Results in Tables 1 and 2 show that the effects of ratio of cassava to durian seed on glucose and ethanol concentration profiles are significant. Notice that, the initial concentration of glucose increased when the amount of durian seeds introduced was getting larger. The culture medium with the ratio of 50:50 had more glucose content while the ratio of 85:15 had the lowest. This was due to the sugar content in durian seeds is higher than that in cassava powder. At the end of the fermentation period, the glucose was substantially consumed and converted into ethanol. Notice that, the higher the glucose content in the medium, the higher the ethanol formation during the fermentation process, which was in agreement with the literature report [8]. Therefore, the fermentation at the ratio of 50:50 was considered to be the most appropriate condition because it can lead to the highest amount of initial glucose concentration and final ethanol concentration.

As shown in Table 2, at temperature of 35°C, the percentages of glucose consumption under the ratio of 50:50 were more or less similar (about 75%) under all three different pH values. On the other hand, the percentages of glucose consumption under the ratio of 50:50 and temperature of 30°C (refer to Table 1) varied significantly across the three pH values. Under pH 5.5, the percentages of glucose consumption at temperature 30°C were lower than that at temperature 35°C. The situation was reversed under medium pH of 4.5. It can be noticed that at the ratio of 50:50, it seemed that the introduction of larger amount of durian seed into the mixed feedstock tended to increase the percentages of substrate consumption.

Apart from increasing the initial substrate concentration, the ethanol productivity can be improved by controlling a number of other parameters. One of these parameters is the medium pH in which the fermentation is taking place [14]. As investigated by [8], the ethanol production under fixed glucose concentration can strongly be affected by different pH values. A pH range of 4.0 - 5.0 is regarded as the operational limit for ethanol production. During the cell growing stage, it is crucial that yeast cells maintain its intracellular pH at an optimal value. The deviation of the extracellular pH from this optimal point requires the yeast cells to add in more effort by investing energy to either pump in or out hydrogen ions as to preserve the optimal intracellular pH [15]. Hence, the occurrence of substantial pH difference between the intracellular and extracellular environments can lead to the deactivation of enzymes, which consequently impedes the growth of yeast cells and then ultimately reduces the ethanol production efficiency [11]. In view of the importance of pH, the present studies investigated the ethanol formation at various pH values, which were 4.5, 5.0 and 5.5. From Table 1 and Table 2, it was observed that ethanol yield was at its highest when pH was fixed at 5.0 at temperature 30°C and 35°C. Interestingly, the ethanol yield was at its least at pH 5.5 among the three tested pH values. Apparently despite of diverse substrate concentrations due to different material ratios, the ethanol yields were quite comparable under the medium pH of 5.0, i.e., for the lab-scale case, the yields were within $\pm 3\%$ of the average value.

It should be noted that, yeasts are large globular proteins which act as biological catalysts. At a higher temperature, the rate of yeast activity is higher because the collisions between substrate and the enzymes active site are more frequent as molecules are moving even more rapidly at high temperature [16]. However because of several other biological limitations, the increment of yeast activity is only up to a point where the cells are at the maximum efficiency. Beyond this optimum temperature, yeast cells begin to denature due to excessive kinetic energy which breaks the bonds eventually alters the active sites shape. This results in the rapid decrease or even complete halt of the enzymatic reactions involved. On the other hand, at lower temperatures the cells specific growth rate is lower which is due to their low ethanol tolerance at lower temperatures [17]. In this study, the variation of fermentation temperatures between 30°C and 35°C revealed that the ethanol production was at its highest when the fermentation was operated at the temperature of 35°C (see Tables 1 and 2. This can be observed in all of the results obtained when comparing the experimental studies under similar pH and similar mixtures of cassava and durian seed. Moreover, it can be seen that the ethanol yield increased significantly at the temperature of 35°C compared to 30°C under all three different pH values and ratio of hydrolysed cassava to durian seed. At temperature of 35°C, the lowest ethanol yields were obtained under pH of 5.5 although with high percentages of glucose consumption. The consumption of glucose could be used for powering the tricarboxylic acid (TCA or Kreb's) cycle leading predominantly to cell respiration [18]. Hence, the optimal trade-off between respiration and fermentation seemed to occur when the highest ethanol yield was obtained under pH of 5.0.

In the flask-scale fermentation, it can be concluded that the highest ethanol yield of 45.9% and a final ethanol concentration of 24.9 g/L were achieved at pH 5.0, temperature 35°C and 50:50 volume ratio of hydrolyzed cassava to durian seed over a batch period of 48 hours. The optimum condition of the fermentation process from proposed material was used in the lab-scale bioreactor experiment.

The fermentation experiments by using solely durian seeds at pH 5.0 and temperature 35°C, over a batch period of 48 hours were conducted for a comparison purpose. Table 3 shows the experimental results for solely durian seeds fermentation.

Table 3. Experimental results for solely durian seeds fermentation

Experiments	Initial glucose (g/L)	Final glucose (g/L)	Final ethanol (g/L)	Glucose consumed (%)	Ethanol Yield (%)
A	65.39	35.79	2.72	45.3	9.2
B	63.63	31.16	3.29	51.0	10.1

From Table 3, it was observed that the ethanol yields of 9.2% and 10.1% for both experiment A and B, with initial concentration of glucose 65.39 g/L and 63.63 g/L respectively. Notice that, the initial concentrations of glucose by using solely durian seeds were slightly lower compared to the culture medium with ratio of 50:50. The initial concentration of glucose did not increase when the amount of durian seeds introduced was getting larger. This might due to the overloaded durian seeds solid suspension in the fermentation medium which caused less room for water. Thus, the hydrolysis process of durian seeds might be limited leading to low initial concentration of glucose. This in turn results in a low ethanol yield. Also, the low ethanol yield could be caused by the osmotic stress experienced by yeast cells [19]. The limitation of available water might be responsible for the obstruction of cell growth, which resulted in low ethanol production. Such stress encountered caused yeast to divert the metabolic pathway producing ethanol to other pathways such as toward the glycerol formation in order to cope with the stress.

3.2. Microaerobic lab-scale bioreactor fermentation

The ethanol, glucose and biomass concentration profiles in the lab-scale bioreactor were examined for the fermentation using the proposed materials under the optimum flask-scale conditions: pH 5.0, temperature 35°C and 50:50 volume ratio of hydrolyzed cassava to durian seeds for a batch period of about 46 hours.

The delivery of small amount of oxygen to culture broth under anaerobic fermentation is important to maintain metabolite production and growth of microorganisms [20]. Thus, a controlled microaerobic condition, i.e., 0.5 LPM, was applied in bioreactor fermentation to improve the cell viability and enhance the ethanol formation by yeast. The glucose and ethanol concentrations profiles are shown in Fig. 1, whereas the biomass concentration profile is shown in Fig. 2.

In Fig. 1, the decrement in glucose concentration and increment in ethanol concentrations are expected due to the consumption of substrate to produce ethanol throughout the fermentation process. The initial concentration of glucose was at 41.95 g/L and it dropped to about 1.00 g/L at the end of fermentation process. Meanwhile, the ethanol was produced gradually and it reached up to 14.61 g/L after batch period of 46 hours. The initial glucose concentration differed from that given by the optimum condition from the flask-scale fermentation, which could be due to the insufficient hydrolysing process to break down the long chains of starch into fermentable

IOP Conf. Series: Materials Science and Engineering **206** (2017) 012020 doi:10.1088/1757-899X/206/1/012020

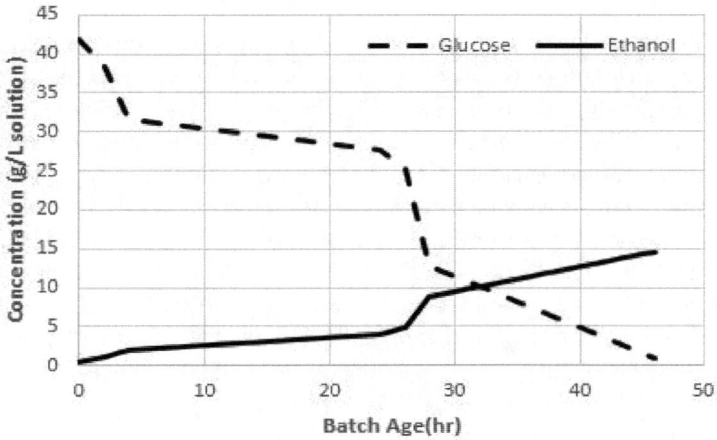

Figure 1. Glucose and ethanol concentrations profiles

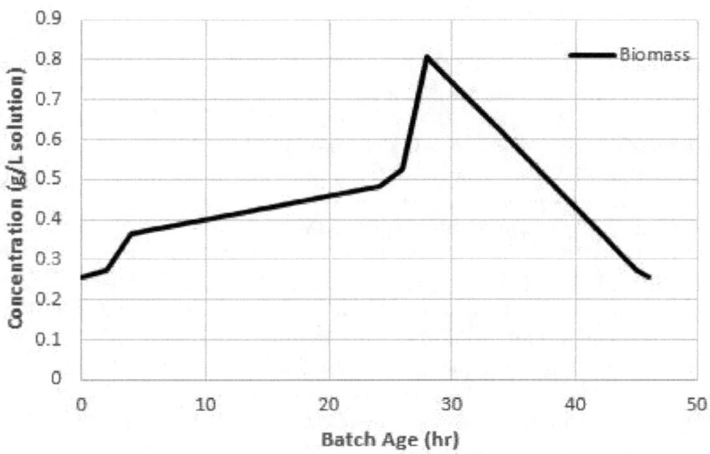

Figure 2. Biomass concentrations profile

sugar with bigger amount of raw materials (i.e., cassava and durian seeds remained). Thus, it led to lower ethanol formation as less fermentable substrate was introduced.

In Fig. 2, a slow microbial growth phase was observed for about 20 hours during which the consumption of substrate and production of ethanol were small. After the slow microbial growth phase, it was followed by a rapid growth phase in a short period, during which the cells multiplied in an exponential order after the cells eventually adapted to the medium environment. It can be noticed that, the consumption of glucose and ethanol production were increased rapidly during this period. The biomass increased throughout the experiment, but dropped towards the end of experiment. It is interesting to note that, the decrease in cell viability could be due to the exhaustion of substrate and this led to the inhibition of ATP synthesis. Another reason for the drop in cell viability could be due to the yeast cells were metabolically inactive due to the higher rate of ethanol formation and this led to the leakage of intracellular metabolites into the growth medium [21]. Hence, these circumstances caused the cell death. The ethanol yield of 35.7% and a final ethanol concentration of 14.61 g/L were obtained over a period of 46 hours where the glucose was almost fully consumed.

4. Conclusion
As a conclusion, the experimental results showed that the durian seeds can be used to supplement or even replace the conventional feedstocks (e.g. cassava) for ethanol production, which produced higher fermentable sugar content than cassava powder. In the flask-scale fermentation, the results showed that the highest ethanol yield of 45.9% and final ethanol concentration of 24.92 g/L were obtained under the conditions of pH 5.0, temperature 35°C and 50:50 volume ratio of hydrolyzed cassava to durian seed, over a period of 48 hours. The ethanol yield is about 90% of theoretical maximum of glucose conversion by yeast. Under this optimum condition, the lab-scale bioreactor fermentation resulted in 35.7% of ethanol yield of and a final ethanol concentration of 14.61 g/L over a period of 46 hours. It is interesting to note that, the ethanol fermentation using mixed cassava and durian seed is highly affected by the values of pH, temperature and ratio of hydrolysed cassava to durian seeds.

Acknowledgments
This work is supported by a grant from the Curtin Sarawak Research Institute (CSRI), no. CSRI-6009.

References
[1] Walker G M 2010 *Bioethanol: Science and Technology of Fuel Alcohol* (Dundee: Ventus Publishing Aps)
[2] Balat M and Balat H 2009 *Appl. Energ.* **86** 2273–82
[3] Cheng J (ed) 2010 *Biomass to Renewable Energy Processes* (Florida: CRC Press Inc.)
[4] Amid B T and Mirhosseini H 2012 *Food Chem.* **132** 1258–68
[5] Amin A M and Arshad R 2009 *IJPTI* **1** 367–375
[6] Brown M J 1997 *Durio - A Bibliographic Review* (New Delhi: IPGRI)
[7] Lee S H, Jung J Y and Jeon C O 2014 *Food Microbiol.* **38** 16–25
[8] Lin Y, Zhang W, Li C, Sakakibara K, Tanaka S and Kong H 2012 *Biomass Bioenerg.* **47** 395–401
[9] Zhang W, Lin Y, Zhang Q, Wang X, Wu D and Kong H 2013 *Fuel* **112** 331–337
[10] Wong C L, Yen H W, Lin C L and Chang J S 2014 *Bioresource Technol.* **152** 169–176
[11] Narendranath N V and Power R 2005 *Appl. Environ. Microb.* **71** 2239–43
[12] K C Thomas S H H and Ingledew W M 2002 *Appl. Environ. Microb.* **68** 1616–23
[13] Thatipamala R, Rohani S and Hill G A 1992 *Biotechnol. Bioeng.* **40** 289–297
[14] Fakruddin M, Quayum M A, Ahmed M M and Choudhury N 2012 *Biotechnology* **11** 248
[15] Narendranath N V, Thomas K C and Ingledew W M 2001 *J. Am. Soc. Brew. Chem.* **59** 187–194
[16] Eed J 2012 *ESSAI* **10** 19
[17] Torija M J, Rozùs N, Poblet M, Guillamón J M and Mas A 2003 *Int. J. Food Microbiol.* **80** 47–53
[18] Rao D G 2010 *Introduction to Biochemical Engineering* (New Delhi: Tata McGraw-Hill Education)
[19] Soufi B, Kelstrup C D, Stoehr G, Fröhlich F, Walther T C and Olsen J V 2009 *Mol. Biosyst.* **5** 1337–46
[20] Garcia-Ochoa F and Gomez E 2009 *Biotechnol. Adv.* **27** 153 – 176
[21] Cot M, Loret M O, Francois J and Benbadis L 2007 *FEMS Yeast Res.* **7** 22–32

IOP Conf. Series: Materials Science and Engineering **206** (2017) 012021 doi:10.1088/1757-899X/206/1/012021

Effect of flow rate, duty cycle, amplitude, and treatment Time of ultrasonic regimens towards *Escherichia coli* harbouring lipase

W S A W Omar[1,2], A Z Sulaiman[1]*, A Ajit[1], Y Chisti[3], A L T Chor[4]

[1]Faculty of Chemical and Natural Resources Engineering, Universiti Malaysia Pahang, Gambang 26300, Pahang, Malaysia
[2]Faculty of Applied Sciences, Universiti Teknologi MARA Cawangan Pahang, Bandar Jengka 26400, Pahang, Malaysia
[3]School of Engineering, Massey University, Private Bag 11 222, Palmerston North, New Zealand
[4]Faculty of Biotechnology and Biomolecular Sciences, Universiti Putra Malaysia, Serdang 43400, Selangor, Malaysia

E-mail: ziad@ump.edu.my

Abstract. A full factorial design (FFD) approach was conducted to assess the effect of four factors, namely flow rate, duty cycle, amplitude, and treatment time of ultrasonic regimens towards *Escherichia coli* harbouring lipase. The 22 experiments were performed as the following values with six replicates of centre point: flow rate (0.1, 0.2, and 0.3 L/min), duty cycle (0, 20, and 40%), amplitude (2, 6, and 10), and treatment time (10, 35, and 60 min). The FFD was employed as preliminary screening in shake flask cultivation to choose the significant factors ($P< 0.05$) for further optimisation process. In this study, zero duty cycle signified non-sonication of amplitude and no treatment time effect to the *E. coli* culture. Also, the designated flow rate and amplitude accordingly showed no effect towards the amount of dry cells weight (DCW). DCW_1 was found significantly degraded after the exposure of high duty cycle and treatment time as other factors remained constant. Whereas for the lipase activity, no significant difference was observed in any main factors or interactions. Paired samples t-test confirms the result at a *p*-value of 0.625. This experimental study suggests the direct and continuous approach of sonication caused an adverse effect on the cells culture density.

1. Introduction

Lipases, or specifically triacylglycerol acylhydrolase (EC 3.1.1.3) can be found in all living organisms. Lipases act at the oil-water edge to transform water insoluble glycerides into smaller compounds which can be readily transported across the corresponding cell wall. Due to high potential of lipases in industry [1, 2], the increasing demands of lipases are shown by many productions from wild-type [3, 4] or genetically modified organisms [5-8]. In a comparison of wild-type and recombinant lipase production, the wild-type was found capable of consuming a broader range of oil-based substrate compared to the recombinant type [9]. It was presumed to the natural behaviour of wild-type in adaptation to its resources. While for the recombinant type, it produces higher activity but only for the desired protein translated in the recombinant gene. Hence, any production of lipase should consider the aim of its process condition. Since the production of recombinant protein in bioprocess have been developed in the past 40 years [10], more strategies are needed to enhance the cultivation of the individual expression hosts. Among the current strategies are strain phenotype improvement [11], high throughput technology [12], and feeding strategy [13]. In 2003, Chisti also mentioned the feasibility of using ultrasound (US) in bioprocesses, including the recombinant protein recovery.

Despite the well-known use of sonicator for cell disruptions, degassing, cleaning, and homogenising, the use of ultrasound in enhancing cells cultivation is relatively new. Some of the recent US studies are focused on the effect of US treatments against enzymes [15-17], inactivation of enzymes and pathogens to avoid food spoilage [16], and assisting in enzymatic reaction towards insoluble substrate [18]. Owing to the potential of US in microbiology, some of the US parameters have been introduced in bioprocesses improvement from shake flask culture [19] to a laboratory bioreactor [20-22]. In this study, the use of US towards the small-scale cultivation of recombinant culture, particularly by the use of ultrasonic probes with medium flow rates, treatment time, amplitude, and the duty cycle, were discussed. To the best of our knowledge, this research was the first to study the effect of sonication towards a recombinant culture.

2. Materials and Methods

2.1 Microorganism

The recombinant *E. coli* harbouring lipase was obtained from the Laboratory of Biomolecular Medicine, Universiti Putra Malaysia. The lipase gene was gained from *Staphylococcus hyicus* and inserted into pTrcHis-TOPO vector by Hasan [23].

2.2 Growth medium and conditions

The recombinant *E. coli* was cultured in 500 mL Erlenmeyer flasks with working volume of 150 ml. Luria Bertani medium containing 10 g/L tryptone, 5 g/L yeast extract, and 5 g/L NaCl was prepared and supplemented aseptically with 10 g/L glucose. The medium was adjusted to pH 7 prior autoclaving. In addition to keeping the plasmid carries the recombinant genes, sterilised filtered ampicillin was added until the final concentration of 100 µg/mL. Ten per cent of inoculum with OD_{600nm} ~ 0.7 was added and the culture was shaken at 250 rpm under controlled temperature, 30°C. The culture was incubated for 3 h, later induced with 1 mM IPTG, and prolonged with an additional of 1 h incubation. After a total of 4 h incubation, the culture was transferred to a sterilised empty bottle with silicone bottle stopper and tubing. The tubing was connected to low volume continuous flocell (QSonica, USA) and the sonication was generated at 20 kHz. The culture was circulated using a peristaltic pump, Cole-Palmer Masterflex® L/S, USA. The schematic diagram of the experimental set is shown in figure 1. Samples were taken pre and post the treatment time (min) established by the experimental design and directly kept at 4°C for further analysis.

2.3 Full factorial design

A two-level full factorial design (FFD) was employed for the screening of important factors in enhancing the production of lipase from the recombinant *E. coli*. Flow rate (L/min), duty cycle (%), amplitude, and treatment time (min) at two different levels were selected as independent variables (table 1). The design generated 22 experiments, including six replicates of centre point, using Design-Expert software, version 7.1.6 (Stat Ease Inc., Minneapolis, MN USA). Two responses were included in the data, namely, DCW (mg/mL) and lipase activity (U/mL). The software is capable of building and analysing the experimental design by subjecting the data to the analysis of variance (ANOVA) for the determination of estimated effect and interaction and fitting the regression equation to the data as described previously in Saini *et al.* [24].

Table 1. Coded and actual values of the factors.

Factor[a]	Low (-1)	High (+1)
Flow rate (mL/min), A	0.1	0.3
Duty cycle (%), B	10	40
Amplitude, C	2	10
Treatment time (min), D	10	60

[a] centre point is the average of minimum and maximum

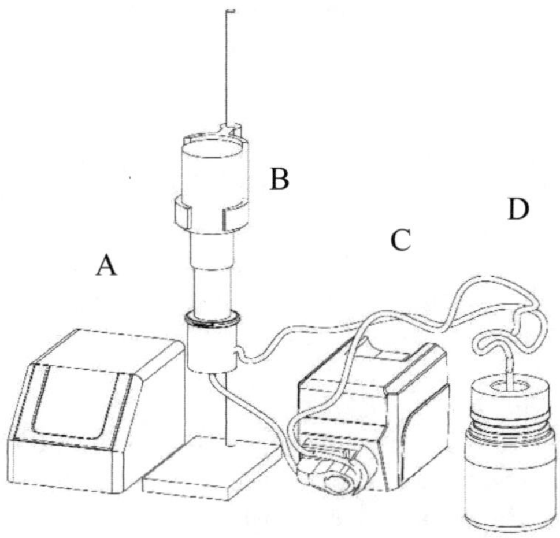

Figure 1. Experimental setup combining a transducer ultrasonicator (A), a stand with a converter and a flocell (B), a peristaltic pump with sample tubing (C), and a bottle sample (D). Throughout the experiment, the bottle sample was kept in a biosafety cabinet.

2.4 Analysis

2.4.1 Dry cell weight

The dry cell weight (DCW) of *E. coli* culture was calculated from a standard curve of optical density (OD) versus the DCW. The OD was measured by UV-Vis 1800 Spectrophotometer (Hitachi, Japan) by comparing with fresh medium as blank. The standard curve was prepared by using triplicates of *E. coli* culture. Each replicate contained 15 mL of culture at OD ranging from 0.1 to 0.7. Each sample was then filtered using cellulose acetate membrane (47 mm diameter, 0.22 µm pore size) through a vacuum filtration set. Prior to filtration, the membrane was dried in an oven at 95°C until constant weight. After the filtration, the filter membrane with cells was carefully kept in the oven until constant weight. The dry cells weight was calculated by subtracting the initial filter membrane weight from the final weighing. By having the linear regression, the DCW was measured by the following Equation (1):

$$DCW(\frac{mg}{mL}) = \frac{OD_{600nm}}{2.2341} \quad (1)$$

The standard equation in this study was found similar to the previous study by Koo *et al.* [25].

2.4.2 Lipase assay

Lipase activity was measured spectrophotometrically using p-nitrophenyl phosphate or pNPP (Sigma-Aldrich) as substrate. The assay was performed according to Kim *et al.* [26] with some modifications. The reaction mixture contained solution A (50 mM Tris-HCl buffer pH 8, 0.1% gum arabic, and 0.2% sodium deoxycholate) and solution B (10 mM pNPP in isopropanol). For the reaction, 100 µL of samples were added to 850 µL of solution A. The reaction was initiated by the addition of 50 µL of solution B and incubated at 35°C for 5 min. The reaction was stopped by the addition of 500 µL of 3 M of HCl. After centrifugation, 500 µL of clear supernatant was transferred to a cuvette and 1 mL of 2 M of NaOH was added. The mixture was read at 410 nm by UV-Vis 1800 Spectrophotometer (Hitachi, Japan) against a free enzyme mixture as blank. One lipase unit is defined as the amount of enzyme required to liberate 1 mmol of p-nitrophenol per minute.

3. Results

The recombinant cultures were cultivated in shake flasks and supplemented with ampicillin and IPTG for the expression of lipase. The factors chosen in this study were among the relevant variables discussed previously in other studies [16, 21]. ANOVA was carried out to determine the significant differences ($p<0.05$) among the factors. Throughout the experiment, the culture was initially cloudy, due to the growth of cells. After the sonication with amplitude and duty cycles, regardless the duration of engaged treatment time, the culture visually became apparent.

3.1 Effect of main factors and interactions for DCW_1

Although the responses for DCW were measured as post-treatment (DCW_1, mg/mL) and percentage of change (DCW Δ, %), both analyses have shown a similar pattern. Thus, for simplicity, only one response, specifically DCW_1, is chosen for further elaboration. Based on the ANOVA in Table 2, the model of DCW_1 is found significant with two main factors, namely duty cycle and treatment time. Whereas, in two-factor interactions, a combination of duty cycle with amplitude and treatment time are significant. The rest of factors and interactions are not significant at *p*-value >0.05. The selected interaction studies between the four factors are shown in figure 2. According to the two top graphs in figure 2, the perpendicular lines clearly projects the interaction of the high level and low level of the amplitude and duty cycle. Duty cycle and treatment time demonstrate the same pattern. Hence, both interactions are significant while other factors kept at their constant values. Moreover, in the two bottom graphs in figure 2, the plateau corresponds to no significant different at any level. Parallel curves obtained for each factor indicates a lack of interaction in the experimental study [27]. However, from equation (2), the flow rate (A) shows a slight positive response. Thus, in the validation design, this factor was suggested at the highest level to obtain maximum DCW_1.

$$\begin{aligned} DCW_1 &= +0.73 + 0.019A - 0.46B + 6.369E^{-003}C - 0.073D - 3.019E^{-003}AB + 1.719E^{-003}AC \\ &+ 7.256E^{-003}AD - 0.047BC - 0.11BD + 9.444E^{-003}CD + 2.919E^{-003}ABC - 0.023ABD \\ &+ 0.017ACD + 0.016BCD \end{aligned} \tag{2}$$

3.2 Effect of main factors and interactions for lipase

From table 2, the ANOVA shows that the effect of factors at any term or interaction for the lipase activity are not significant. A negative predicted R^2 suggests that the overall mean is a better predictor of the response. Thus, a paired sample *t*-test was conducted to measure the pre-treatment and post-treatment time of sonication towards the lipase activity. The normality test by Shapiro-Wilk has passed at P=0.777 [28]. Using IBM SPSS for Window (Version 22, Armonk, NY.), there is no significant difference in the effect for pre-treatment (M=24.14, SD=2.59) and post-treatment,

(M=23.8, SD=1.53); t (21) =0.496, p=0.625. These results agree with the finding from the ANOVA and suggest that the treatment does not affect the lipase activity regardless the results shown by DCW_1.

3.3 Profiles of sonication outputs

Four sonication outputs are summarised in figure 3. The runs with no output were left blank thus only 14 runs out of 22 were observed on the sonication regimens. The intensity is calculated by using the average power in Watt divided by the area of replaceable membrane tip, where the sonication area is engaged with the sample [29]. The elapsed time was also measured by adding the "on" sonication process from the duty cycle setup. The duty cycle was set at 2 s "on" and 8 s "off" for a 20%, while 40% was set at 4 s "on" and 6 s "off".

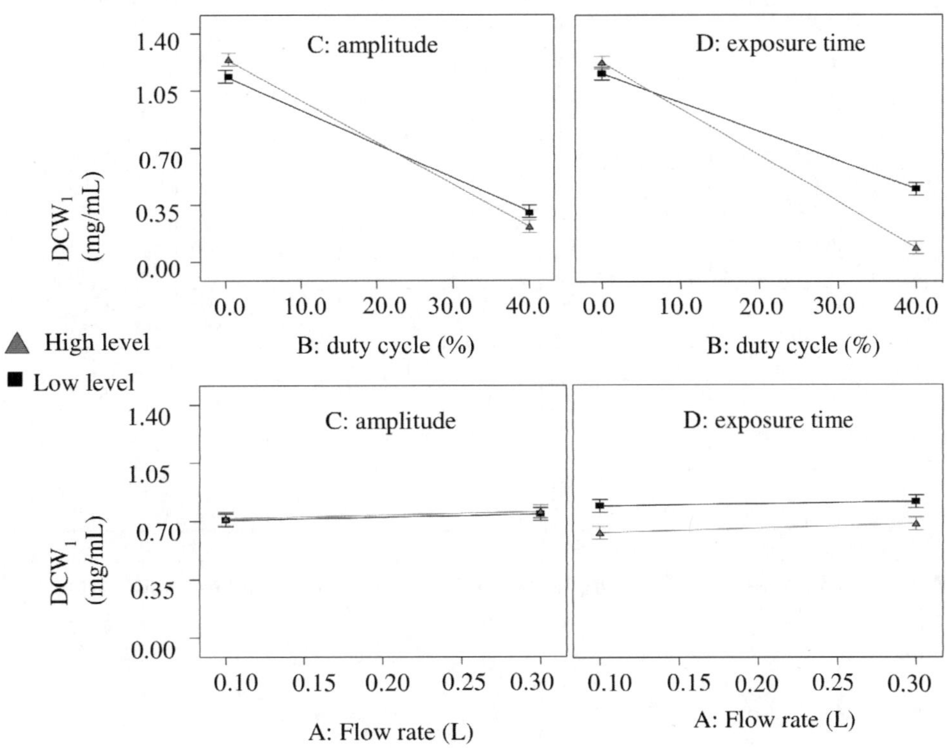

Figure 2. Interaction studies between selected variables for response DCW_1.

IOP Conf. Series: Materials Science and Engineering **206** (2017) 012021 doi:10.1088/1757-899X/206/1/012021

Table 2. Three-way interaction ANOVA for DCW_1 and lipase production in recombinant *E. coli.*

	DCW_1				Lipase			
Source	Sum of squares	Mean square	*F* value	*P*-value	Sum of squares	Mean square	*F* value	*P*-value
Model	3.73	0.27	123.69	< 0.0001[a]	37.64	2.69	1.38	0.3641
A	5.92E-003	5.92E-003	2.74	0.1487	0.088	0.088	0.045	0.8385
B	3.41	3.41	1579.33	< 0.0001[a]	4.45	4.45	2.28	0.1821
C	6.49E-004	6.49E-004	0.30	0.6031	0.080	0.080	0.041	0.8463
D	0.08	0.08	39.18	0.0008[a]	4.66	4.66	2.38	0.1735
AB	1.46E-004	1.46E-004	0.068	0.8035	1.97	1.97	1.01	0.3543
AC	4.73E-005	4.73E-005	0.022	0.8872	0.97	0.97	0.49	0.5083
AD	8.43E-004	8.43E-004	0.39	0.5550	8.50	8.50	4.35	0.0820
BC	0.04	0.04	16.37	0.0068[a]	0.069	0.069	0.035	0.8572
BD	0.18	0.18	84.37	< 0.0001[a]	1.76	1.76	0.90	0.3791
CD	1.43E-003	1.43E-003	0.66	0.4471	8.56	8.56	4.38	0.0813
ABC	1.36E-004	1.36E-004	0.063	0.8099	1.51	1.51	0.78	0.4124
ABD	8.66E-003	8.66E-003	4.02	0.0919	0.017	0.017	8.817E-003	0.9282
ACD	4.87E-003	4.87E-003	2.26	0.1837	3.28	3.28	1.68	0.2427
BCD	3.95E-003	3.95E-003	1.83	0.2245	1.74	1.74	0.89	0.3820
Curvature	0.72	0.72	333.28	< 0.0001[a]	9.109E-003	9.109E -003	4.663E-003	0.9478
Residual	0.01	2.16E-003			11.72	1.95		
Lack of Fit	4.32E-003	4.32E-003	2.51	0.1743	6.14	6.14	5.51	0.0658
R^2			0.9961		0.7626			
Adjusted R^2			0.9871		0.2086			
Predicted R^2			0.4789		-31.0285			
Adequate Precision			31.693		6.215			
Coefficient of Variance %			0.9961		5.87			

A-Flow rate, B-duty cycle, C-amplitude, D-treatment time, [a] – significant value of $p<0.05$

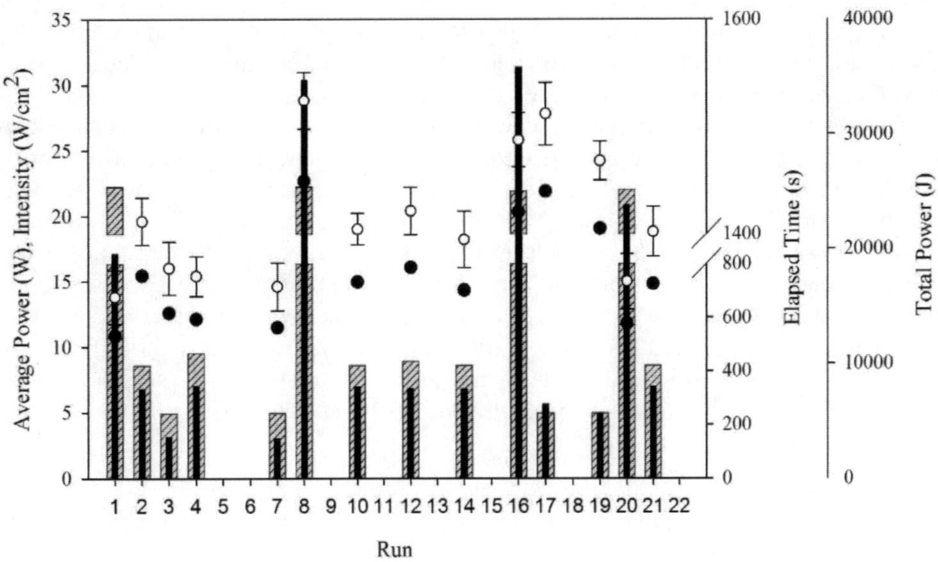

Figure 3. Sonication output profiles from experimental run, elapsed time (▧), total power (■), average power (○), and sonication intensity (●).

The rapid shift of sonication increased the intensity and with longer treatment time, the cells were damaged [16]. Table 3 shows three responses with their respective predicted and experimental data. From the table, the Cook's distance generated indicates the observations from the experimental are mostly higher than the limit. It was suggested by Altman and Krzywinski [30] that high leverage points may indicate the linear estimation must be restricted to a smaller region of the predictor region. In this study, only two main factors and two interactions are significant as described earlier in table 2. In addition to the screening factors, a Pareto chart is used to illustrate the importance of the factors complete with Bonferroni correction [31]. Therefore, figure 4 shows a Pareto chart of response coefficient for DCW_1 and lipase activity. Figure 4a describes that the factors above Bonferroni level are significant according to the positive and adverse responses, while figure 4b shows that all the factors are below the t-value limit.

3.4 Validation of FFD
The software recommended the validation experimental design according to the factors range and responses target. Triplicates of recommended runs were conducted with a desirability of 0.8. The experiments gained 98.6% of the predicted values by the combination of flow rate at 0.3 L/min, duty cycle at 0%, amplitude of 10, and treatment time of 60 min. Hence, this combination suggests that the direct sonication was excluded in aiming high cell density. A maximum flow rate was proposed due to aeration and mixing of the aerobic culture in the bottle imitating the use of incubator shaker.

4. Discussion
This study shows different outcomes as compared to other studies due to a few reasons. The first reason is the approach of sonication with direct and indirect sonication regimens by using ultrasonicator probes compared to an ultrasonic bath [20, 22]. In this study, a flocell was attached to ultrasonicator probes, thus the sample was directly exposed to the sonication effect. While using an ultrasonic bath, the samples were contained in flask and reactor. Due to this, the glass wall of flask or

reactor might have hindered the acoustic effect. Thus, the effect was reduced compared to ultrasonicator probes. Furthermore, the probe sonication has localised effect compared to bath or tank type sonication [32]. The research also suggests that the use of ultrasonicator probe is more intense and efficient in the degradation process compared to an ultrasonic bath.

Another proposed reason is the sample volume. A small volume in the Erlenmeyer flask from this study compared to a larger volume in a bioreactor [20, 21] influenced the cultivation performance. However, previous research done by Avhad *et al.* [19] also used a small volume in a glass vessel for crude broth containing fibrinolytic enzymes. The different approach was the use of ultrasonic bath instead of ultrasonicator probe. The assisted ultrasonication in their study showed that the irradiation time of 5 min was the best to obtain high fold of fibrinolytic enzymes. The enzymes activity was started to deteriorate after 5 min due to excessive irradiation. This trend supports the current study for DCW_1 which indicates that the irradiation does affect the cells and enzymes performances. In the following year, Avhad and Rathod [20] suggested that the treatment of sonication should be comparable to the volume size to minimise the decline of cells and enzymes.

Table 3. Observed and predicted responses in the experiments obtained by FFD.

Run order	DCW_1 (mg/mL)		Lipase (U/mL)	
	Exp.	Pre.	Exp.	Pre.
1	0.35	0.39	23.98	24.60
2	1.19	1.26	24.29	24.91
3	1.17	1.22	18.47	19.09
4	1.20	1.19	22.98	23.60
5	0.34	0.32	22.89	23.76
6	0.54	0.51	23.82	24.44
7	0.11	0.10	23.00	23.62
8	1.36	1.32	25.87	25.25
9	0.10	0.14	25.29	24.67
10	0.34	0.32	23.38	23.76
11	0.49	0.49	25.25	24.63
12	1.26	1.18	23.96	23.34
13	0.26	0.32	23.18	23.76
14	0.08	0.08	25.87	26.49
15	0.42	0.41	23.46	22.84
16	0.28	0.32	23.08	23.76
17	0.07	0.03	24.01	23.39
18	0.36	0.32	25.61	23.76
19	1.07	1.10	23.10	23.72
20	1.16	1.13	23.48	22.86
21	0.36	0.32	24.44	23.76
22	1.12	1.13	24.11	23.49

Exp. – Experimental, Pre. – Predicted, DCW_0 – DCW at pre-treatment time, DCW_1- dcw at the post-treatment time.

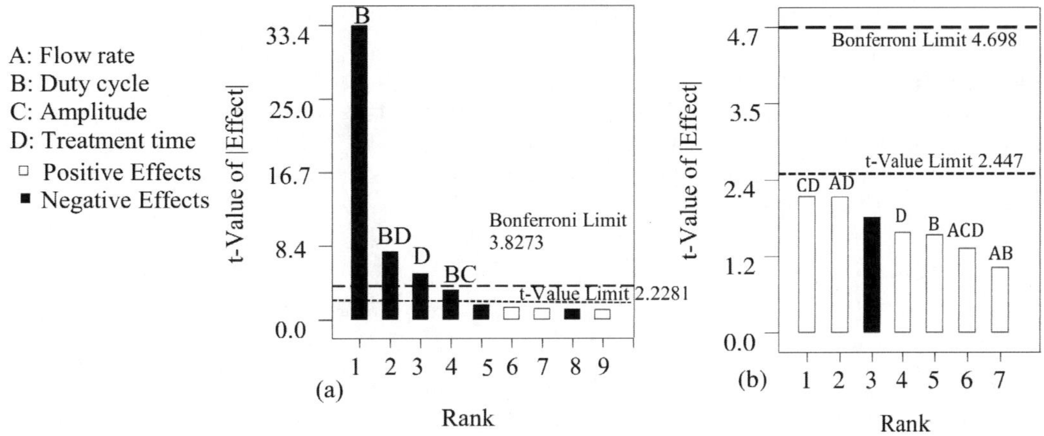

Figure 4. Response coefficient significance study on DCW_1 (a), and lipase (b).

Nevertheless, in this study, the lipase activity was found to be similar in all experimental runs. For this reason, the trend of lipase activity was not due to any of the manipulated variables. The pattern suggests that the cultivation was not fully expressing the lipase and the culture condition was not suitable for the lipase activity. It is assumed that the induction time for the lipase to express was not enough in this study. In common expression cultivation, it would take about 4 h after the induction to harvest the protein of interest. This condition might be due to the low pH throughout the fermentation duration (data not shown). As the experiments used Erlenmeyer flasks, the pH was set at 7 only before autoclaving, but was not controlled throughout the incubation period. As in conventional batch fermentation cultivation, the pH is reduced gradually from pH 7 to pH 5 because of the formation of by-products such as lactic acid and acetic acid. In other studies, by Wang *et al.* [33] and Gadge *et al.* [34], the optimum pH for lipase activity was suggested ideal between pH 6 and pH 8. Thus, the low pH cultivation in this study appears to cease the lipase activity and yield a similar pattern.

From previous discussion, the need for the controllable system includes the pH stability can be achieved using a bioreactor instead of Erlenmeyer flask. Also, the use of controlled conditions in the bioreactor aids survivability of culture against the sonication. Besides the above reasons, treatment time has also imparted an impact to the result as discussed earlier. Along with swift duty cycle regimen, the treatment time is suggested to be in short pauses as an alternative of continuous sonication. Dey and Rathod [35] and Dhanalakshmi and Nagarajan [32] recommended intermittent or pause sonication treatment rather than continuous to maintain the usability of the transducer and less power would be used.

5. Conclusions

The use of ultrasound in enhancing cell cultivations especially in harbouring recombinant protein was statistically designed and discussed in this paper. The outcome of the selected factors and level are shown to be significant to the post-treatment dry cell weight. From four factors, only duty cycles and treatment time affected the responses. While for the interactions, only two interactions, duty cycle-amplitude and duty cycle-treatment time were found to be significant. This study suggests that the use of ultrasonicator probe for the cell cultivation should be investigated in large volume with controlled parameters such as pH and dissolved oxygen for maintaining the cell growth. The ultrasonication approach also needs to be introduced carefully to avoid cell degradation and primary product denaturation. Further data collection is also required to determine exactly how the ultrasound affects the lipase activity in recombinant *E. coli* cultivation.

Acknowledgement
The research of this paper was financially supported by Ministry of Higher Education (MOHE), Malaysia under Fundamental Research Grant Scheme (FRGS). FRGS/1/2016/TK02/UMP/02/15. The first author would also like to thank Wan Atikah Najiah Wan Hasan and Jonathan Maiangwa for their technical support in this study.

References

[1] Jaeger K E and Eggert T 2002 Lipases for biotechnology *Curr. Opi. Biotech.* **13** pp 390-397
[2] Fariha H, Shah A A and Hameed A 2006 Industrial applications of microbial lipases *Enzyme Microb. Tech.* **39** 235-251
[3] Winkler U K and Stuckmann M 1979 Glycogen, Hyaluronate, and Some Other Polysaccharides Greatly Enhance the Formation of Exolipase by *Serratia marcescens J. Bacteriol.* **138** 663-670
[4] Sharma R, Chisti Y, and Banerjee U C 2001 Production, purification, characterization, and applications of lipases *Biotechnol. Adv.* **19** 627-662
[5] Akbari N *et al* 2010 High-level expression of lipase in Escherichia coli and recovery of active recombinant enzyme through in vitro refolding *Protein Expr. Purif.* **70** 75-80
[6] Li Z *et al* 2011 Expression and characterization of recombinant Rhizopus oryzae lipase for enzymatic biodiesel production *Bioresource Technol.* **102** 9810-9813
[7] Vélez A M *et al* 2013 Enhanced production of recombinant thermo-stable lipase in *Escherichia coli* at high induction temperature *Protein Expr. Purif.* **90** 96-103
[8] Huang J *et al* 2015 Biodiesel production from microalgae oil catalyzed by a recombinant lipase. *Bioresour. Technol.* **180** 47-53
[9] Saengsanga T, Siripornadulsil W and Siripornadulsil S 2016 Molecular and enzymatic characterization of alkaline lipase from *Bacillus amyloliquefaciens* E1PA isolated from lipid-rich food waste *Enzyme Microb. Tech.* **82** 23-33
[10] Ramirez O, Flores E and Galindo E 1995 Products and bioprocesses based on genetically modified organisms: Review of bioengineering issues and trends in the literature *Asia Pac. J. Mol. Bio. Biotechnol.* **3** 165-197
[11] Wang Y *et al* 2009 Industrial bioprocess control and optimization in the context of systems biotechnology *Biotechnol. Adv.* **27** 989-995
[12] Long Q *et al* 2014 The development and application of high throughput cultivation technology in bioprocess development *J. Biotechnol.* **192** 323-338
[13] Oztürk S, Çalık P and Ozdamar T H 2016 Fed-Batch Biomolecule Production by *Bacillus subtilis*: A State of the Art Review *Trends Biotechnol.* **34** 329-345
[14] Chisti Y 2003 Ultrasound—the power of a silent gong *Biotechnol. Adv.* **21** p 1
[15] O'Donnell C P *et al* 2010 Effect of ultrasonic processing on food enzymes of industrial importance *Trends Food Sci. Technol.* **21** 358-367
[16] Herceg Z *et al* 2012 The Effect of High Intensity Ultrasound Treatment on the Amount of *Staphylococcus aureus* and *Escherichia coli* in Milk *Food Technol. Biotech.* **50** 46-52
[17] Szabo O E and Csiszar E 2013 The effect of low-frequency ultrasound on the activity and efficiency of a commercial cellulase enzyme. *Carbohydr. Polym.* **98** 1483-1489
[18] Nguyen T T T and Le V V M 2013 Effects of ultrasound on cellulolytic activity of cellulase complex *Int. Food Res. J.* **20** 557-563

[19] Avhad D N, Niphadkar S S and Rathod V K 2014 Ultrasound assisted three phase partitioning of a fibrinolytic enzyme *Ultrason. Sonochem.* **21** 628-633
[20] Avhad D N and Rathod V K 2015 Ultrasound assisted production of a fibrinolytic enzyme in a bioreactor *Ultrason. Sonochem.* **22** 257-264
[21] Sulaiman A Z *et al* 2011 Ultrasound-assisted fermentation enhances bioethanol productivity *Biochem. Eng. J.* **54** 141-150
[22] Zhou Q, Zhang P and Zhang G 2014 Enhancement of cell production in photosynthetic bacteria wastewater treatment by low-strength ultrasound *Bioresource Technol.* **161** 451-454
[23] Hasan W A N W 2015 Chimeric lipase for understanding the role of key residues governing pH profile *Faculty of Biotechnology and Biomolecular Sciences* (Universiti Putra Malaysia Malaysia)
[24] Saini J K *et al* 2013 Optimization of saccharification of sweet sorghum bagasse using response surface methodology *Ind. Crop. Prod.* **44** 211-219
[25] Koo B S *et al* 2011 Enhancement of thymidine production in E. coli by eliminating repressors regulating the carbamoyl phosphate synthetase operon *Biotechnol. Lett.* **33** 71-78
[26] Kim Y O *et al* 2012 Gene cloning and catalytic characterization of cold-adapted lipase of *Photobacterium sp.* MA1-3 isolated from blood clam *J. Biosci. Bioeng.* **114** 589-595
[27] Bolton S and Bon C 2004 *Pharmaceutical statistics: practical and clinical applications* 4th Edition (New York: Marcel Dekker Inc.)
[28] Ghasemi A and Zahediasl S 2012 Normality Tests for Statistical Analysis: A Guide for Non-Statisticians *Int. J. Endocrinol. Metab.* **10** 486-489
[29] Sulaiman A Z 2011 Use of Ultrasound in Enhancing Productivity of Biotechnological Processes, *Biochemical Engineering* (Massey University Palmerston North New Zealand)
[30] Altman N and Krzywinski M 2016 Points of Significance: Analyzing outliers: influential or nuisance? *Nat. Meth.* **13** 281-282
[31] Kikwai L *et al* 2012 Effect of Various Operational Parameters on Drug Release from a 1% Hydrocortisone Semisolid Dosage Form Using the Vertical Diffusion Cell Apparatus *Dissolution Technologies* pp 6-13
[32] Dhanalakshmi N P and Nagarajan R 2011 Ultrasonic Intensification of the Chemical Degradation of Methyl Violet: An Experimental Study *Int. J. Chem. Mol. Nuclear Mat. Metal. Eng.* **5** 1019-1024
[33] Wang G *et al* 2014 Aromatic Amino Acid Mutagenesis at the Substrate Binding Pocket of Yarrowia lipolytica Lipase Lip2 Affects Its Activity and Thermostability *Scientific World J.* 1-8
[34] Gadge P P *et al* 2011 Biochemical Studies of Lipase from Germinating Oil Seeds (*Glycine max*) *Am. J. Biochem. Biotechnol.* **7** 141-145
[35] Dey S and Rathod V K 2013 Ultrasound assisted extraction of β-carotene from *Spirulina platensis Ultrason. Sonochem.* **20** pp271-276

IOP Conf. Series: Materials Science and Engineering **206** (2017) 012022 doi:10.1088/1757-899X/206/1/012022

Microalgae and Its Premises towards Sustainable Energy Development

M N Chik[1], L Yahya[1], A Zainal[1], M H Boosroh[1]

[1]Emission and Waste Management Technology Section, Renewable Energy and Green Technology Unit, TNB Research Sdn. Bhd., No. 1, Lorong Ayer Itam, Kawasan Institusi Penyelidikan Bangi, 43000 Kajang, MALAYSIA

E-mail: nazry@tnb.com.my

Abstract. This paper features the use of nature's element as a tool to combat current global issues on environment. Through research works by TNB Research Sdn. Bhd. (TNBR), marine phototrophic microalgae is used in reducing CO2 emissions from its fossil-fuel based power plants using. The research program commenced in 2011 with the aim to develop capacity, capability and facilities in biological CO2 fixation. The research program focuses on improving and enhancing the CO2 fixation through four core measures; namely species selection, nutrient optimization, flue gas admission and photobioreactor (PBR) design and engineering. The measures lead to the migration and evolution of culture facilities from laboratory conditions to outdoor, from shake flasks to 6 x 250 liter pilot PBR facility at a live coal-fired power plant, from mono species to consortium of species. Increment of CO2 fixation rates is summarized with discussion on comparisons of other achievements reported elsewhere. A considerable amount of work on analysing the bioactive compound present in the algae – protein, amino acids, carbohydrate, lipid, fatty acids - and its encouraging results, as an impetus towards sustainable development, will also be shared. Premises and observations from various microalgae research works are collated and presented in a manner sufficient to highlight the eminent roles of this tiny creature to become our mentor in providing some solutions to our worldly problems.

1. Introduction

Any utility power company nowadays has to wrestle between secure, efficient and reliable supply of electricity with safety and environmentally benign choices of power generation, transmission and distribution technologies. These variables can be conveniently summarized into the equation of sustainable energy.

Sustainable energy, as to simplify its assessment, can be reduced into seven sustainability indicators. They are: (i) price of electricity, (ii) carbon dioxide (CO2) footprint, (iii) efficiency, (iv) availability, (v) land use, (vi) water use and (vii) social impact [1]. Tenaga Nasional Berhad (TNB) constantly strives to improve sustainability ranking of its energy delivery. The recent launch of Manjung 4 power station– a new 1,000MW coal-fired power plant utilizing ultra-supercritical boiler is one example of a more sustainable energy generation [2], as the technology is 5 % more efficient than

IOP Conf. Series: Materials Science and Engineering **206** (2017) 012022 doi:10.1088/1757-899X/206/1/012022

its subcritical counterparts [3]. This marks the simultaneous attempt to reduce coal consumption thus reducing the CO2 footprint, as well as maintaining electricity affordability by using cheap fuel source.

Apart from tackling the boiler efficiency to reduce coal consumption, TNB through its research arm subsidiary company, TNB Research Sdn. Bhd. (TNBR) is also embarking on reducing the CO2 in its power generation activities head-on i.e. through application of Carbon Capture and Utilization technologies. Gasification, amine-based absorption and physical adsorption [4], [5] are among the technologies that have been developed and demonstrated at TNBR's 150 kWth Integrated Gasification Pilot Plant.

2. Microalgae research works at TNBR

Quite contrary to the above-mentioned CO2 capture technologies, another approach is through biological means using phototrophic microalgae. This approach is regarded to be more sustainable, due to its proven natural benign photosynthesis mechanism and being the fastest growing photosynthetic organisms [6]. Using this biological method, the CO2 is fixed bio-chemically into Cx-Hy-Oz compound, as a result of photosynthesis reaction that originates from CO2, water (H2O) and photon energy from the sun as the reactants. By mass ratio, approximately 2 tonnes of CO2 can be fixed by 1 ton of microalgae [7]. In positioning microalgae as the CO2 biological fixer from power plants, several research areas were identified, namely Species Selection, Culture Parameters Optimization, Nutrient Enhancement, and Photobioreactor (PBR) Design and Engineering. These areas have been identified as the concerted efforts in improving the biological fixation rates of microalgae, i.e. the amount of CO2 microalgae can fix per liter of culture per day of culture (g.CO2 per liter per day, g/L.day). It is not intended to detail out all methodologies in each of our projects from 2011 as they have been elaborated in other publications. Only pertinent highlights and results are presented.

2.1. *Species Selection*

The first attempt in identifying suitable microalgae species for the job was to sample, isolate and identify potential ones around the vicinity of a coal-fired power plant. Obviously, marine microalgae species were opted as most of thermal power plants are located by the sea shore, taking advantage of the abundance of sea water that is used for its steam condensation process. Sea water samples were taken from three different locations (North 4o 2.675'; East 100o 41.859', North 4o 8.195'; East 100o 38.103' and North 4o 9.435'; East 100o 37.079') offshore of a coastal coal-fired power plant in Seri Manjung, Perak, Malaysia. The isolation was done through streak-plating technique on agar plates continued with serial dilutions until several dominant colonies thrived. Initial media was Conway and then switched to f/2 with artificial sea water [8]. CO2 fixation rate was determined by weighing the algal biomass weight daily and applying the mass ratio between the molecular weight of CO2 and biomass, which is 1.882 [7], [9]. Through population count, it was found that the marine species *Isochrysis sp.* was the dominant species there and possessed a higher CO2 fixation rate of 0.1 g/L/day, when compared experimentally to the other species – *Tetraselmis sp.* and *Nannochloropsis Occulata* - isolated at the same vicinity.

Next initiative is to further improving the fixation rate by simply trying out culturing a consortium of microalgae species. Search for the consortium of species is based on visual observation of the TNB Janamanjung's (TNBJ's) outfall water which appears greenish. Should there be any microalgae species survives in this water, it should be robust enough to withstand the turbulence, hydrodynamic stresses and acidity prevalent at this location, as a result of rigorous aeration and desulphurization process that took place in the upstream. It is further hunched by other reports stating that growth characteristics of microalgae species could be altered when two or more microalgae species were cultured together in the same culture medium [10]. Samples taken at the outfall were further analyzed, processed and through molecular identification revealed that it contains a dominant harmonious community of *Chlorella sorokiniana*, *Chlorella pyrenedoisa* and through morphological identification a diatom *Amphora sp.* [11]. These mixed species were later combined with the existing *Isochrysis sp.* at various volumetric proportions, to form a consortium, and through fixation rate experiment using actual flue gas from the power plant, a consortium consisted of 75% mixed species and 25% *Isochrysis sp.* yielded the best fixation rate [12].

2.2. *Culture parameter optimization*

Cultivating algae in photobioreactors requires conducive environment for microalgae growth. These include pH, temperature, amount of CO2 supplied, light intensity and suitable nutrient. In order to know the optimum culture parameter to give the maximum CO2 fixation rate, these parameters need to be optimized by statistical methods using Response Surface Method (RSM). As part of experimental design, pH was varied from 4 to 8 (by regulating needle valve admitting commercialized pure CO2 gas from a cylinder), temperature from 20 ^{o}C to 40 ^{o}C (by dedicated water cooler/heater and cooling/heating coil), CO2 flowrate from 0.05 liter per minute (lpm) to 0.25 lpm, and lighting from 500 lux to 2500 lux

(from five Philips' T5 cool daylight fluorescence bulbs and Skye's PAR sensor). 21 runs were conducted altogether using a dual 10 L lab-scaled bubbling column photobioreactor. From the results of the experiments, the highest one recorded at 0.35 g/L.day. Through a 3-D surface plot, RSM then suggested that a higher rate can be achieved if temperature, pH, flowrate and lighting were set at 30 ^{o}C, pH 7.5, 0.15 lpm and 1500 lux respectively. Following the suggestion, another run was made to reveal a better fixation rate of 0.37 g/L.day, or a 5% improvement.

2.3. *Nutrient enhancement*

Using 6 x 250 liter air-lift bubbling column pilot PBR installed at the Sultan Azlan Shah 3 x 700 MW coal-fired power plant (TNBJ), *Isochrysis sp.* was cultured, using actual flue gas and sea water. Bearing in mind that actual flue gas from a coal-fired power plant can contain among others, gases like SO2, CO, NOx, unburned hydrocarbons, particulate matter, traces of fluorine and chlorine compounds [13], apart from CO2, thus, right formulation of nutrient is deemed crucial as to ensure optimum growth under these conditions. Using 240 variations of chemicals and their concentrations, formulation of new nutrient / media was carried out through statistical experiment, covering relevant constituents in macro and micronutrients. The new media formulation obtained was tested in several cases and was compared against the use of previous nutrient, the standard f/2. The test at TNBJ's power plant was done in a 250 liter air-lift bubbling column type PBR, admitting up to 50 liter per minute CO2 from actual flue gas, through KNF's diaphragm vacuum pump. The new media has demonstrated better algae growth, especially on actual flue gas. The improved specific growth rate (μ) and fixation rate were 0.2714 d^{-1} and 0.72 g/(L.day) respectively, against those obtained using the f/2 media 0.1703 d^{-1} and 0.52 g/(L.day) respectively [14]. This means with the new nutrient, algae thrived faster and better and thus more CO2 can be fixed.

2.4. *Photobioreactor design and engineering*

In this regard, attention and consideration were made to minimize the photo-inhibition effect to the microalgae, as well as peculiar design on the pilot PBR that promotes modularity, accessibility and durability.

Trending of sunlight illumination level was carried out at designated location of TNBJ power station before the pilot PBR was installed. For a week's recording, readings gave a minimum illumination of

28.5 klux (~527 $\mu mol.m^{-2}.s^{-1}$) and a maximum of 107.5 klux (~1,990 $\mu mol.m^{-2}.s^{-1}$) corresponds to morning and afternoon time respectively. Depending on geographical location, the Illuminance Saturation Level, Is, is estimated to be 140–210 $\mu mol.m^{-2}.s^{-1}$ [15] and this is the level where photoinhibition can still be coped. It is about two times lower than the minimum illumination recorded at TNBJ.

Thus detrimental photoinhibition effect was kept at its bay by adjusting the light exposure the algae cells received while it is travelling and recycling inside an air-lift bubble column. Through separate experiment, it was found that when the level of light exposure is limited to about 72% the growth rate is better [16]. This finding was incorporated in the design of the pilot PBR where the transparent outer column and opaque inner column was specified to be in pre-calculated dimensions, in order to acquire the effect obtained in the experiment as closely as possible.

The pilot PBR as installed at TNBJ is shown in Figure 1 below, was mounted on a skid of hot-dipped galvanized steel, equipped with a raised platform for added accessibility during operation and maintenance. The skid can also be used as storage compartment for the columns, underneath it, with peculiar slots, slides and latching mechanism.

Figure 1. Pilot PBR skid at TNBJ.

2.5. *CO2 fixation rate achievement*

Figure 2 above summarizes the CO2 fixation rate achievement so far, in g.CO2 per liter culture per day (g/L.day), from 2011 to 2015, with various initiatives – from indoor controlled laboratory conditions to outdoor cultivation, from mono species to consortia. The best fixation rate achieved so far is 0.965 g/L.day that was achieved through use of microalgae consortium species cultured using actual flue gas at TNBJ. Comparison, if it is to be made, with other species or initiatives, shall be done taking into account of the same CO2 fixation calculation method. Biomass analysis method is employed throughout the entire initiatives.

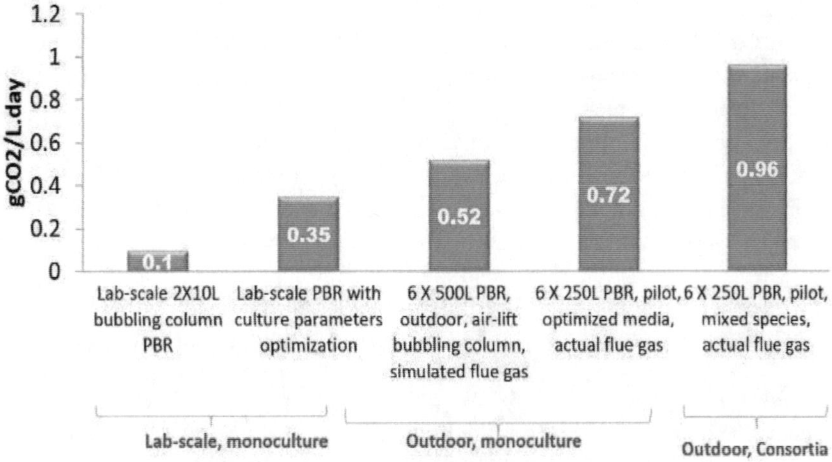

Figure 2. Summary of CO2 fixation rates.

Figure 3 below charts out the comparison of CO2 fixation rates from various species, photobioreactor types, CO2 concentration, and culture condition but with the similar fixation rate method of calculation, i.e. using biomass analysis method. [17], [18] & [13].

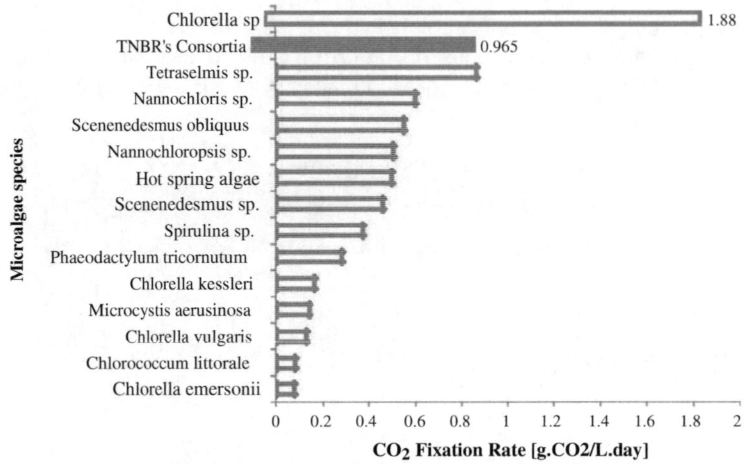

Figure 3. CO_2 fixation rates comparison.

The highest so far, by the chart is 1.88 g/L.day as achieved by *Chlorella sp.* using a mere 15 mL culture receiving 15% CO2 [18], which suggests a laboratory scale effort. An interesting point to note from that work probably is to have a higher concentration of CO2 brought to microalgae culture as a means to improve the fixation rate. This can be a daunting task as the amount of CO2 from a coal-fired power station varies from carbon contents in certain coal types in use, and often it is diluted and mixed with other gases as mentioned. To have it pure and at a uniform elevated concentration can be an expensive assignment, as this requires integration with other carbon capture method e.g. absorption or adsorption, though this is an excellent research challenge to delve into.

The level of CO2 fixation rate we achieved so far, however, is still far from being able to displace the amount of CO2 from a coal-fired power station significantly. If we were to use a baseline carbon footprint of 887 gCO2/kWh (for fuel combustion only) [19], and an annual capacity factor of 85%, a typical sub-critical 1,000-MW coal-fired power plant is capable to produce about 6.6 million tons of CO2 per year, and if biological means is the only effort made to curb this emission, it would require hundreds of hectares of land, millions of volumes of culture facility and millions of investment cost, based on current state of technology. All these are affecting the provision of electricity in accordance to sustainable development principle and components – land use, material and economy. However, as for land requirement, we see similar scenario to the palm oil plantations, rubber trees, rapeseed, and even soy beans where hectares of land are also made use to plant these trees. But why these are not questioned? It is because these plants can be turned into useful products benefitting humankind and bring commercial returns to its developers. Does not microalgae has the same quality?

2.6. *Bioactive compound analysis*

In answering the above question, we have to look into potential downstream products this consortium of species in use can offer. Bioassays of protein, amino acid, lipids and fatty acid analysis were conducted to the algal biomass as to explore its potential bioproducts. The amino acid compositions was identified using a Waters-Pico Tag Amino Acid Analyzer System. Lipid was extracted using soxhlet with hexane, in accordance to Bligh & Dyer method. Fatty acids was analyzed using gas chromatography mass spectrum. Proximate analysis revealed that the consortium content of protein (45%), lipid (21%) and carbohydrate (17%) is comparable and more balanced from other types of mono species. Its protein content of 45% is higher than other common sources of protein for cattle feed. Interestingly, essential amino acids content of 42% are encouragingly good and its composition is comparable if not exceeding some other species and soymeal [20], as displayed in Figure 4.

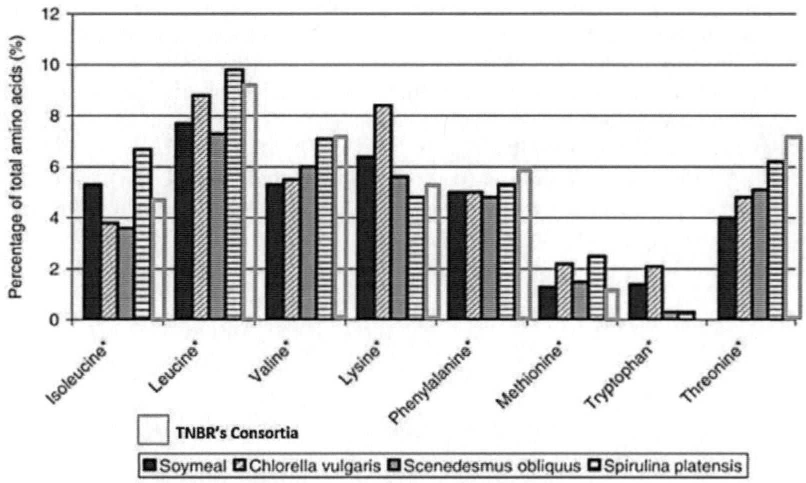

Figure 4. Essential amino content comparison.

Fatty acid analysis reveals that its content is more diverse as compared to other common sources like palm oil and soy bean. Table 1 below summarizes pertinent fatty acid content in the microalgae consortium and its potential bioproducts [21].

Table 1. Selected fatty acid analysis of consortium of species

Fatty acid content	Potential bioproducts
C18:3 Linolenic acid (ALA) -14 %	Omega 3 essential fatty acid - health supplement
C20:3 Eicosatrienoic acid – 19%	Polymer - raw material for high mechanical strength organic plastics
C16:0 Palmitic Acid - 22%	Oleo-chemicals - Suitable for detergent, glycerol, esters, fatty alcohols, and lubricants

From this initial work, this microalgae consortium is thus suitable to be further studied in terms of its viability of simultaneous application of CO2 biofixation and bioactive compound production.

3. Premises of microalgal method

Through our involvement and achievement this last four years, we have acknowledged at least the following two important traits and premises microalgae can offer towards sustainable development. They are as follows:

- Bio-fixation works on re-cycling and re-utilizing of CO2
- Living organisms relies on close integration, co-operation and complementing each other

The first premise is evidenced from familiar observations we can make to the creation and roles of plants on the earth. Photosynthesis as practiced by these plants has been providing sustenance to the earth and its aerobic forms of life for more than 3.5 billion years [18]. It is a truly sophisticated and smart process in balancing the amount of CO2 released by nature and human activities to the amount of O2 they in turn need.

The second premise requires a more careful look. Let us take an example of the photosynthesis process that occurs within microalgae's thylakoid membrane. We can see that a very close co-operation and co-ordination are orchestrated here. From photon energy received at photosystem II, it is converted into ATP (adenosine triphosphate - a reduced form of adenosine diphosphate (ADP)) and NADPH (a reduced form of nicotineamide adenine dinucleotide phosphate), assisted by electron transport chain. Later, under Calvin cycle, these ATP and NADPH convert CO2 into glucose and other sugars [18], and the process repeats itself.

It is also noticed during carbon fixation experiment using mixed species, that each of the species supports and complements each other as evidenced in Figure 5 below. All the three species survived to the end of culture period with good growth profile. Overall, they have established and maintained an ecology in the culture.

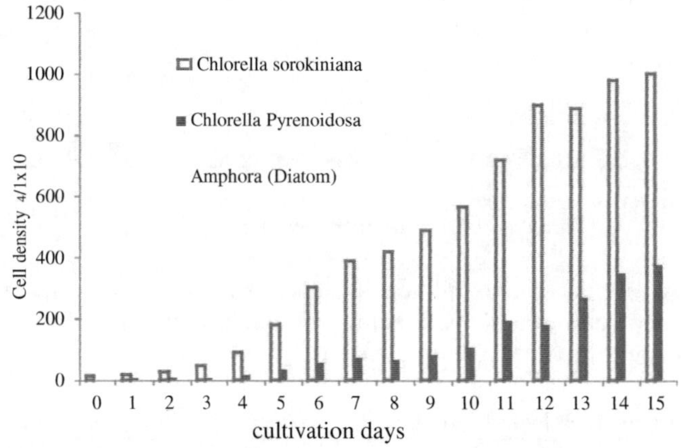

Figure 5. Population dynamics of mixed species.

Capitalizing from these premises and observations, it is thus recommended that biological CO_2 fixation – an upstream process, to be developed together with potential valuable downstream bioproducts as the end-in-mind. Similar to what microalgae's roles and behaves in its ecology, this observation should be emulated in industries towards an Industrial Ecology practise, an approach where one's waste is a commodity to other industries [22], that is by itself, is a subset of a sustainable development. Thus, within a coal-fired power plant, CO_2 as a waste from combustion process, is utilized by microalgae which in turn produces lipids, carbohydrates, proteins, pigments, which can be further processed towards productions of useful bioproducts like animal / aquaculture feed, nutraceuticals, cosmetics, biomaterials, biofuel [23], [24]. Several market values and unit price of these compounds have been reported to be high: Omega-3 – USD 25 billion in 2013 [25], Protein, a main constituent in animal feed industry is worth at USD 40 – 50 billion per year [26] and Astaxanthin Euro 7,150/kg [27], which can be justified towards commercial development thus satisfying the 'economy' component of sustainable development. After all, this recommendation is in parallel to the emerging science of 'biomimicry' where efforts are made to solve global challenges facing humanity using inspirations derived from nature, due to its efficiency, beauty, structure and functions [28].

4. Conclusion

Microalgae is seen fit as to be one of the picture in the broader framework of Carbon Capture and Utilization (CCU). CCU is becoming a new topics to pursue nowadays, with research emphasis on making CO2 utilization more efficient and economical, while at the same time widening the avenues for it. Several avenues have already been identified e.g. methanation, mineral carbonation, turning into chemical feedstock, as well as deriving downstream products from microalgae [29].

This recommendation towards CCU is and shall not be driven entirely by the allures of lucrative gains in profits, wealth or fame, but rather by a deep awareness that this is how nature solves the CO2 emission – by converting it into other substances and recycling it. We have taken microalgae, as inspired by nature, as a means to reduce CO2 emission through their photosynthesis process in our photobioreactors, in microscale, without realizing that it teaches us also a lesson or two in macroscale, on how its bioactive compounds can be turned into various useful products and be exchanged with other relevant industries. This whole framework – recycling and re-utilizing of wastes - if taken seriously from policy and regulatory level down to practitioners can ensure sustainable development agenda be fulfilled.

Acknowledgement

The authors wish to thank and appreciate the following persons and institutions for co-operation, support and expertise rendered throughout implementation and completion of various research projects above: TNBR's Technical Board Committee for approving grants in most of the research projects mentioned, MESITA committee of Ministry of Energy, Green Technology and Water, Malaysia for funding the pilot photobioreactor research facility, Managing Director and personnel of Environment and Chemical Services, TNB Janamanjung Sdn. Bhd., Professor Lokman Shamsudin and his team of Universiti Selangor (UNISEL), Dr. Rabitah Zakaria and Dr. Siti Mazlina and their team of Universiti Putra Malaysia, Dr. Khairul Adzfa and his team of Universiti Teknologi Mara (UiTM), Professor Mohd Sobri Takriff and his team of Universiti Kebangsaan Malaysia (UKM).

References

[1] Evans A, Strezov V and Evans T J 2009 Assessment of sustainability indicators for renewable energy technologies *Renew. Sustain. Energ. Rev.* **13**(5) 1082–1088.
[2] Loo S 2016 Sultan of Perak launches TNB's coal powered generating facility in Manjung.
[3] Alstom 2015 CASE STUDY: Steam Manjung Unit 4 Malaysia Two-Pass Boiler. http://www.alstom.com
[4] Muda N, Hidayu N, Boosroh M H 2015 Effect of amine blends on the pre-combustion CO2 capture process *The 7th International Conference on Clean Coal Technology*. (Krakow, Poland)
[5] Muda N and Boosroh M H 2014 Gasification of Coal-Petcoke Blends in a Pilot Scale Gasification Plant *Int. J. Auto. Mech.* **8** 1457–1466.
[6] Farrelly D J, Everard C D, Fagan C C and McDonnell K P 2013 Carbon sequestration and the role of biological carbon mitigation: A review *Renew. Sustain. Energ. Rev.* **21** 712–727
[7] Chisti Y 2007 Biodiesel from microalgae *Biotechnol. Adv.* **25**(3) 294–306
[8] Chik M N, Yahya L, Boosroh M H and Shamsudin L 2011 Scanning of microalgae species for biological CO2 fixation at a coastal coal-fired power station 2012 *OIDA Int. J. Sustain. Dev.* **04** (03) 19-28.
[9] Suryata I, Svavarsson H G and Einarsson S 2010 Geothermal CO2 bio-mitigation techniques by utilizing microalgae at the blue lagoon, Iceland. *Thirty-Fourth Workshop on Geothermal Reservoir Engineering* (March) 8

[10] Huang W, Dong B, Cai Z, and Duan S. 2011. Growth effects on mixed culture of *Dunaliella salina* and *Phaeodactylum tricornutum* under different inoculation densities and nitrogen concentrations *5th International Conference on Bioinformatics and Biomedical Engineering ICBBE 2011*, *10*(61), 13164–13173

[11] Chik M N, Yahya L, Kamarudin K F, Tao D G and Takriff M S 2015 Isolation, purification and identification of microalgae from coal-fired power plant environment. *28th Symposium of Malaysian Chemical Engineers* (Putrajaya, Malaysia)

[12] Chik M N, Yahya L, Zainal A, Kamarudin K F, Tao D G and Takriff M S 2016 Tackling Carbon Emission with Nature: Effectiveness of Indigenous Microalgae Mixed Culture *3rd International Conference on Civil and Environmental Engineering for Sustainability* (Melaka, Malaysia)

[13] Van Den Hende S, Vervaeren H and Boon, N 2012 Flue gas compounds and microalgae: (bio-) chemical interactions leading to biotechnological opportunities *Biotechnol. Adv.* **30**(6), 1405–24.

[14] Yahya L, Chik M N, Harun I, Boosroh M H, Radzun K A and Sulong A 2015 Formulation and testing of an improved nutrient for *Isochrysis sp.* culture at a live coal-fired power plant. In *International Conference on Environment 2015 (ICENV 2015)* (Penang, Malaysia)

[15] Kumar K, Dasgupta C N, Nayak B, Lindblad P, and Das D 2011 Development of suitable photobioreactors for CO2 sequestration addressing global warming using green algae and cyanobacteria. *Bioresour. Technol.* **102**(8), 4945–4953.

[16] Chik M N, Yahya L, Nadia N, Kadir A, Harun I and Pang M A M A 2014 Natural light dilution effect on microalgae growth *Inter. J. Chem. Eng. Appl.* **5** (2)

[17] Rodolfi L, Zittelli C, Bassi G, Padovani N, G Biondi, Bonini N G and Tredici, M R 2009 Microalgae for oil: Strain selection, induction of lipid synthesis and outdoor mass cultivation in a low-cost photobioreactor *Biotechnol.Bioeng.***102**(1) 100–112.

[18] Ho S H, Chen C Y, Lee D J, and Chang J S 2011 Perspectives on microalgal CO2 emission mitigation systems - A review. *Biotechnol. Adv.* **29** (2) 189–198.

[19] Varun, Bhat I K and Prakash R 2009 LCA of renewable energy for electricity generation systems - A review *Renew. Sustain. Energ. Rev.* **13** 1067–1073.

[20] Vanthoor-Koopmans M, Wijffels R H, Barbosa M J and Eppink M H M 2013 Biorefinery of microalgae for food and fuel. *Bioresour. Technol.* **135** 142–149.

[21] Chik M N, Yahya L, and Zainal A 2015 FINAL REPORT: Research on consortium of microalgae species for enhanced CO2 fixation and profiling of biochemical properties *TNB Research* (Kajang, Malaysia)

[22] Chertow M R 2008 "Uncovering" Industrial Symbiosis *J. Ind. Ecol.* **11**(1), 11– 30.

[23] Spolaore P, Joannis-Cassan C, Duran E and Isambert A 2006 Commercial applications of microalgae *J. Biosci. Bioeng.* **101**(2) 87–96.

[24] Oncel, S S 2013 Microalgae for a macroenergy world *Renew. Sustain. Energ. Rev.* **26** 241–264.

[25] Facts, P 2012 Global Market for EPA / DHA Omega-3 Products Packaged Facts, (September).

[26] Posten C and Walter C 2012 *Microalgal Biotechnology: Integration and Economy* (Ed.) (2012) De Gruyter (Berlin, Boston)

[27] Brennan L and Owende P 2010 Biofuels from microalgae - A review of technologies for production, processing, and extractions of biofuels and co-products *Renew. Sustain. Energ. Rev.* **14**(2), 557–577

[28] Gebeshuber I C and MacQueen M O 2013 What is a Physicist Doing in the Jungle? Biomimetics of the Rainforest *Appl. Mech. Mat.* **461** (July 2012), 152–162.

[29] Styring P, Jansen D, de Coninck H, Reith H and Armstrong K 2011. REPORT NO. 501: Carbon Capture and Utilisation in the green economy. Using CO2 to manufacture fuel, chemicals and materials *The Centre for Low Carbon Futures* (United Kingdom)

IOP Conf. Series: Materials Science and Engineering **206** (2017) 012023 doi:10.1088/1757-899X/206/1/012023

Biodiesel production from ethanolysis of palm oil using deep eutectic solvent (DES) as co-solvent

R Manurung[1], A Winarta[1], Taslim[1], and L Indra[1]
[1]*Department of Chemical Engineering, Engineering Faculty, University of Sumatera Utara, Medan, Indonesia*
E-mail: renitachem@yahoo.com

Abstract. Biodiesel produced from ethanolysis is more renewable and have better properties (higher oxidation stability, lower cloud and pour point) compared to methanolysis, but it has a disadvantage such as complicated purification. To improve ethanolysis process, deep eutectic solvent (DES) can be prepared from choline chloride and glycerol and used as co-solvent in ethanolysis. The deep eutectic solvent is formed from a quaternary ammonium salt (choline chloride) and a hydrogen bond donor (Glycerol), it is a non-toxic, biodegradable solvent compared to a conventional volatile organic solvent such as hexane. The deep eutectic solvent is prepared by mixing choline chloride and glycerol with molar ratio 1:2 at temperature 80 °C, stirring speed 300 rpm for 1 hour. The DES is characterized by its density and viscosity. The ethanolysis is performed at a reaction temperature of 70 °C, ethanol to oil molar ratio of 9:1, potassium hydroxide as catalyst concentration of 1.2 % wt. DES as co-solvent with concentration 0.5 to 3 % wt. stirring speed 400 rpm, and a reaction time 1 hour. The obtained biodiesel is then characterized by its density, viscosity, and ester content. The oil - ethanol phase condition is observed in the reaction tube. The oil - ethanol phase with DES tends to form meniscus compared to without DES, showed that oil and ethanol become more slightly miscible, which favors the reaction. Using DES as co-solvent in ethanolysis showed increasing in yield and easier purification. The esters properties meet the international standards ASTM D6751, with the highest yield achieved 83,67 % with 99,77 % conversion at DES concentration 2 %. Increasing DES concentration above 2 % in ethanolysis decrease the conversion and yield, because of the excessive glycerol in the systems makes the reaction equilibrium moves to the reactant side.

1. Introduction

Biodiesel is generally produced by the methanolysis of vegetable oil, waste oils, non-edible oils and animal fats. Its production from other alcohols, such as ethanol, is also possible. Methanol is generally selected because it costs less than ethanol, and the properties of the FAMEs (fatty acid methyl esters) obtained after transesterification are similar to the properties of diesel. However, biodiesel produced using methanol is not entirely renewable because methanol is predominately derived from fossil sources, including natural gas, petroleum and coal [1]. When ethanol is used in transesterification, the process can be called ethanolysis and the product is FAEEs (Fatty Acid Ethyl Esters).

Although ethanol is currently more expensive than methanol, it has the advantage of being more miscible in vegetable oils and has lower toxicity . Compared to methyl esters, ethyl esters have higher oxidation stability, lower iodine value, and improved lubricity properties . Additionally, ethyl esters have lower cloud point and pour point, which improves the engine starting at low temperatures, and the extra carbon atom provided by the ethanol molecule slightly increases the heat of combustion and the cetane number . The evaluation of exhaust emissions (NOx, CO2, soot) shows that the ethyl esters have a less negative effect on the environment, compared with that caused by the methyl esters [2].

However, the utilization of ethanol also presents inconveniences. Effectively, as it is indicated in the literature, the base-catalyzed formation of ethyl ester is difficult compared to the formation of methyl esters. Specifically, the formation of the stable emulsion during ethanolysis is a problem. Methanol and ethanol are not miscible with triglycerides at room temperature, and the reaction mixture is usually mechanically stirred to enhance mass transfer. During the course of the reaction, emulsions are usually formed. In the case of methanolysis, these emulsions break down quickly and easily to form a lower glycerol-rich layer and upper methyl ester rich layer. In ethanolysis, these emulsions are more stable and severely complicate the separation and purification of esters [3]. For an alkali-catalyzed transesterification, the glycerides and alcohol also must be substantially anhydrous because water makes the reaction partially change to saponification, which produces soap. The soap lowers the yield of esters and renders the separation of ester and glycerol and the water washing difficult. Low free fatty acids (FFAs) content in vegetable oil is required for alkali-catalyzed transesterification [4].

To overcome the disadvantages of existing biodiesel production processes, novel technologies are emerging. The use of ionic liquids (ILs) as a catalyst, a co-solvent or an extracting solvent have recently attracted attention in the biodiesel production. A new generation of ILs, called deep eutectic ILs or deep eutectic solvents (DESs), became a target of interest for many researchers studying biodiesel production due to their advantages and environmentally-friendly nature. DESs might reduce the side reactions (such as saponification), and make separation and purification processes easier. Some DESs are also efficient in removing glycerol and remaining alkaline catalyst from crude biodiesel. [5]

Recent studies had demonstrated the use of DES as co-solvent and catalyst in methanolysis can increase yield, and make the purification easier. Few studies reported the use of ChCl based DES with palm oil examples are the removal of an alkaline transesterification catalyst such as potassium hydroxide (KOH) and the separation of glycerol from biodiesel [6,7]. Recently, Hayyan et al (2011) used a novel phosphonium based deep eutectic solvent (P-DES) to produce biodiesel from industrial low-grade crude palm oil. Allyltriphenylphosphonium bromide or N,Ndiethylenethanol ammonium chloride were used as salts to prepare DES [8]. Recently, Gu, et al (2015) proposed a greener biodiesel synthesis by using ChCl : glycerol (1:2) as the co-solvent in the NaOH-catalyzed transesterification via methanolysis of rapeseed oil. Under the Optimum conditions determined by RSM based on BBD, the ester yield is higher in the presence of the DES [9]. However, The use of DES as co-solvent in ethanolysis is not yet reported.

In this work, DES was produced from choline chloride and glycerol (molar ratio 1 : 2) and used in ethanolysis of palm oil, it is expected to remove the disadvantages of ethanolysis such as low mass transfer and formation of stable emulsion which complicate the purification step by reducing its interfacial tension and side reaction (saponification).

2. Materials and Method

2.1 Materials

Choline Chloride and glycerol (98 %) were acquired from Sigma - Aldrich , Potassium hydroxide, ethanol (96 %), and phosphoric acid was acquired from Merck. crude palm oil was collected from palm oil mill PTPN IV.

2.2 Deep Eutectic Solvent Synthesis

Choline chloride and glycerol with molar ratio 1 : 2 was mixed and the mixture was stirred at 80 $^\circ$ C, stirring speed 300 rpm for 1 hour until homogeneous and transparent liquid is formed.

IOP Conf. Series: Materials Science and Engineering **206** (2017) 012023 doi:10.1088/1757-899X/206/1/012023

2.3 Ethanolysis of Palm Oil

The crude palm oil was degummed using phosphoric acid to remove the gum, before used as raw material. The ethanolysis reaction was carried out in 0.5 L flask with magnetic stirring at 400 rpm, atmospheric pressure, under reflux condition. 20 g oil was preheated until reaction temperature of 70 ^{0}C and then 9,6 g ethanol, 0,24 g potassium hydroxide, and DES were added. After 1 hour, the reaction mixture was settled to separate ethyl ester and glycerol. The ester phase was analyzed using GCMS. After separating esters from glycerol, the esters is then washed with wet washing method, and the final biodiesel product is analyzed for its density, viscosity, flash point and ester content. The experiment was done in a duplex.

2.4 Product Analysis

The free fatty acid content of oil was analyzed using AOCS Official Method Ca 5a-4. The viscosity, density, flash point, and ester content of biodiesel was analyzed using standard ASTM test method.

3. Results and Discussion

3.1 Degumming Process of Crude Palm oil

Before used in ethanolysis, the crude palm oil was degummed with phosphoric acid (acid degumming). The degummed palm oil was then analyzed for its FFA content and moisture content, Figure 1 show the result of FFA and moisture content analysis.

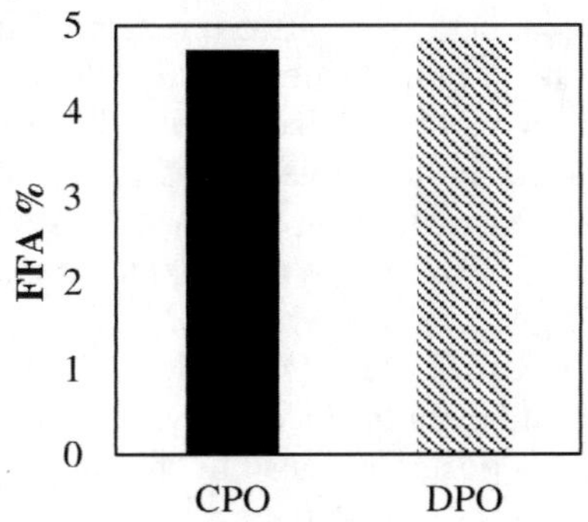

Figure 1. FFA content of CPO and DPO

Based on Figure 1, we can see that the FFA content of the oil increased from 4.71 to 4.84 %. The small rise of FFA content due to acidity of the phosphoric acid used. Phospholipid was suggested as the source of catalyst destruction, phospholipids did not carry over into methyl esters, while cause the biodiesel yield decrease, also added the difficulty separating the glycerol from the esters [10].

Although the degumming process increases the FFA and moisture content of the crude palm oil, the degumming process was needed to remove the gum, which can complicate the separation process and decrease the biodiesel yield.

3.2 Effect of DES on biodiesel yield

DES (choline-chloride and glycerol 1:2 M) was used in ethanolysis of palm oil at various dose. The reaction was carried out at 70℃, 1,2 % KOH, ethanol to oil ratio 9:1, and stirred at 400 rpm at atmospheric pressure. Figure 2 shows the effect of DES on biodiesel yield.

From Figure 2, it can be observed that the biodiesel yield was increased from 75,12 % to 83,67 %, with the increase of DES from 0,5% to 2 %, then the decreased with DES dose more than 2 % . Gu et al., (2015) proposed a reaction process for transesterification using DES as co-solvent, the reaction can maintain as two phases with the aid of DES. Since ester is not soluble in the DES/methanol mixture, the direct contact between ester and base is significantly reduced and the ester becomes one single phase; as a result, the saponification side-reaction is effectively minimized and the separation and purification process is simplified. On the contrary, when no DES is added, both FAME and base are soluble (at least slightly for base) in methanol which induces saponification, decreases the ester yield and complicates the separation and purification procedure. On the contrary, when no DES is added, both ester and base are soluble (at least slightly for base) in methanol which induces saponification, decreases the ester yield and complicates the separation and purification procedure [9].

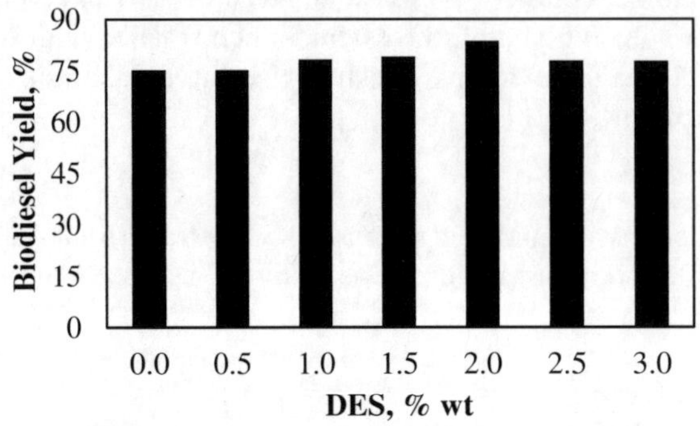

Figure 2. The effect of DES on biodiesel yield

Besides, the addition of DES increases the dissolution of base in DES/methanol mixture due to the ionic nature of DES, producing a higher concentration of methoxide needed for the transesterification. In addition, DES can capture the byproduct glycerol from reactant mixture during the reaction, which shifts the reaction equilibrium to the product side and increases the ester yield. An over-dosed concentration of DES reduced the ester yield because excess glycerol molecules in DES tend to bind with methanol molecules which results in few free methanol molecules for the reaction and thus inhibits the reaction. Moreover, the hydroxyl group in concentrated DES might also compete with methanol to react with the catalyst, which lowers the methoxide concentration and ester yield.

Figure 3 shows the effect of DES on ester content of biodiesel, DES addition from 0,5 % to 2 % didn't show many changes, however, with excess DES addition (2,5 to 3 %), biodiesel contains the trace of monoglyceride, diglyceride, and even triglyceride, indicating the ethanolysis reaction was not complete. In ethanolysis, saponification occurs because of the hydrolysis of the oil (triglyceride), free fatty acid, and ester in base condition. This reaction decreases the amount of triglyceride available for ethanolysis, decreasing ester yield and complicate the purification.

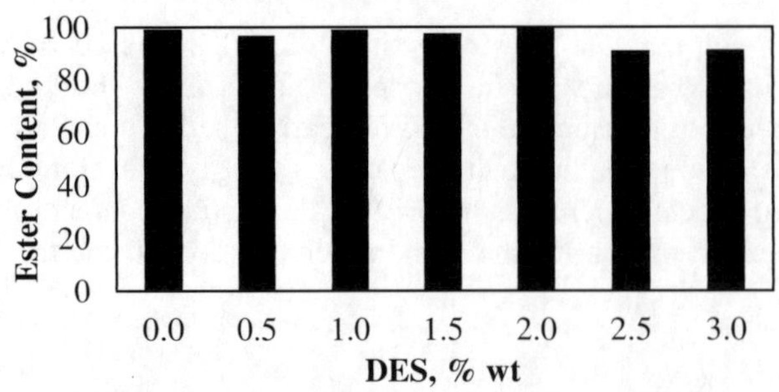

Figure 3. The effect of DES on ester content

DES mixture has a high affinity for hydroxyl group and water, which can decrease their effectiveness in hydrolysis, and prevent saponification reaction. With excess DES addition, DES not only prevent saponification, but also attract C_2H_5O, thus decreasing C_2H_5O available for ethanolysis reaction.

ethyl ester production from crude palm oil also done by Suppalakpanya, et al., (2010) in a two step transesterification using microwave system, with highest ester yield 80 % with ester content 97,4 %. The biodiesel production from crude oil usually done in two step reaction, a pretreatment process (esterification) to decrease FFA and the transesterification process. The high FFA content in oil can lead to a formation of soap and post-reaction problem. From this experiment, DES can decrease the saponification reaction, make the one-step tranesterification of crude oil or high FFA oil available without pretreatment [11]. It can be concluded that DES can increase the ester yield by decreasing saponification reaction and capturing byproduct glycerol.

3.3 Effect of DES on biodiesel purification

The effect of DES on biodiesel purification (separation and washing process) was evaluated. Figure 4. shows the effect of DES on biodiesel separation process.

(a) (b)

Figure 4. Ester and glycerol separation step (a) without and (b) with DES

It can be observed that the ester and glycerol separation step without DES form stable emulsion, which complicates the separation process. The ester and glycerol separation step with DES did not form stable emulsion. The higher mutual miscibility between the glycerol and the esters in the presence of ethanol, severely complicate the phase separation operation after the reaction. Depending upon the ethanol/oil volume fraction loaded to the reactor, the phase separation may not occur, being necessary to add glycerin or to evaporate the ethanol in order to induce phase separation. Another problem is the intensive soap formation that occurs in this system and leads to the formation of stable emulsions that also complicates the separation of phases. Therefore, the washing procedure requires large volumes of water being necessary to improve this part of the process [12].

Table 1 shows the ester and total glycerol content on ester phase after one minute of settling time, it can be observed that the reaction with DES, the ester phase did not contain bonded glycerol, this can be concluded that DES can capture the byproduct glycerol, and shortening the settling time.

Table 1. Ester content of ester phase after one minute

Reaction	After one minute settling time		
		FAEE %	Glycerol %
without DES	Ester phase	94,60 %	0,40 %
with DES		98,41%	-

Figure 5 shows the washing process of biodiesel produced without and with DES, it can be observed that biodiesel produced without DES form an emulsion in washing, which decrease the biodiesel yield. The biodiesel produced with DES form less emulsion when washed, because of the less soap formed.

Figure 5. The washing process (a) without and (b) with DES

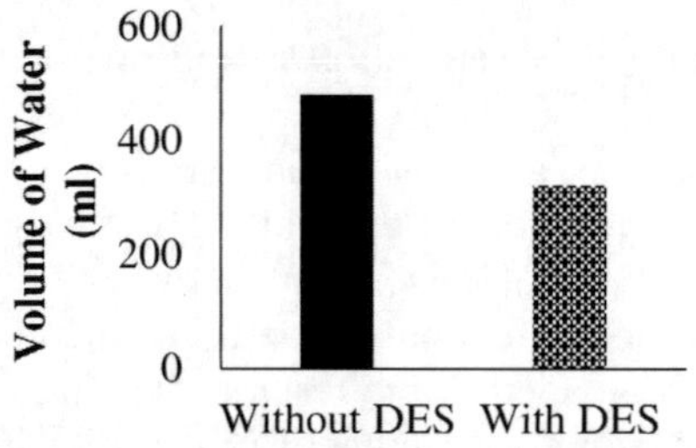

Figure 6. Volume of water needed for wet washing biodiesel

Figure 6 shows the volume of water needed for wet washing biodiesel, it can be observed that without DES, the volume of water required for wet washing is greater than with DES. Wet washing method was successful in removing residual sodium compounds and soap resulting from raw materials having free fatty acids. The saponification reaction is undesirable because it not only consumes the catalyst but also causes problems in phase separation due to the formation of a stable emulsion during water washing. The amount of soap formed during ethanolysis is three or four times higher than that formed during methanolysis under similar reaction conditions. Therefore, the washing method will require a large amount of water [13]. Wet washing biodiesel produced with DES did not form a stable emulsion, because of the less soap formed, and the amount of water needed can be reduced, then decreasing cost in biodiesel purification. It can be concluded that DES can improve the purification process (separation and washing) of biodiesel by decreasing the soap formation and capturing byproduct glycerol.

3.4 Effect of DES on Ethanol – Oil phase

The effect of DES on ethanol – oil phase was observed in a reaction tube, shown in Figure 7.

Figure 7. Effect of DES on ethanol – oil phase (a) without and (b) with DES

It can be observed that with addition of DES on ethanol – oil, the DES form a meniscus between the ethanol and oil (interfacial). Capillary forces can be considered in a general way in which one fluid phase forms a meniscus in another fluid phase capillary forces can also be caused by capillary bridges of one liquid in another immiscible liquid. The capillary force can attract the adjacent particle.

With addition of DES, the interfacial tension between ethanol and oil was decreased, increasing the mass transfer, so the reaction proceeds better and faster. DES form a new attractive force between ethanol – DES, and DES – oil, also can be a media for reaction, reducing soap formation and capturing the byproduct glycerol. However, without the addition of DES, the attractive force between ethanol – oil is very weak, thus the interfacial tension is strong and decreases the mass transfer.

3.5 Characteristic of Biodiesel produced with DES

Characteristic of biodiesel produced with DES can be seen on Table 3. The main criterion of biodiesel quality is the inclusion of its physical and chemical properties into the requirements of the adequate standard. Quality standards for biodiesel are continuously updated, due to the evolution of compression ignition engines, ever stricter emission standards, reevaluation of the eligibility of feedstocks used for the production of biodiesel, etc. The current standards for regulating the quality of biodiesel on the market are based on a variety of factors which vary from region to region, including characteristics of the existing diesel fuel standards, the predominance of the types of diesel engines most common in the region, the emissions regulations governing those engines, the development stage and the climatic properties of the region/country where it is produced and/or used, and not least, the purpose and motivation for the use of biodiesel [14].

Table 2. Characteristic of biodiesel produced

Parameter	Unit	Value	ASTM D6751 Standard	SNI Standard
Ester Content	% (m/m)	99.728	-	> 96.5
Density at 40 °C	kg/m^3	861.7	-	850-900
Kinematic viscosity at 40 °C	mm/s^2	3.7032	3.5-5	2.3 - 6
Flash Point	℃	180	> 130	>100
Free Glycerine	% (m/m)	0	< 0.02	<0.02
Total Glycerine	% (m/m)	0	<0.24	<0.24
Monoglyceride content	% (m/m)	0.0271	< 0.80	-
Diglyceride content	% (m/m)	0	< 0.20	-
Triglyceride content	% (m/m)	0	< 0.20	-

It can be concluded that the biodiesel produced with DES had met the standard both ASTM and SNI, indicating that the use of DES as co-solvent in ethanolysis to produce biodiesel is a new method to solve the disadvantage in using ethanol to produce biodiesel, as it can reduce the soap formation, and make the purification easier.

4. Conclusion

The use of DES as co-solvent in ethanolysis of palm oil to produce biodiesel indicating the interfacial tension between ethanol and oil was decreased, increasing the mass transfer, so the reaction proceeds better and faster, thus the yield increased, Also can reduce soap formation and capturing byproduct glycerol, simplifying the purification. It was found the highest yield was 83.67 % with ester content 99.7 %. This study exemplifies that the use of DES as co-solvent in ethanolysis to produce biodiesel is a new method to solve the disadvantages in using ethanol to produce biodiesel.

References

[1] Narvaez, P.C., Noriega, M.A., and Cadavid, J.G. 2015 Kinetics of Palm Oil Ethanolysis *J Energy*, **83**, 337-34.

[2] Sánchez, B.S., Benitez, B., Querini, C.A., and Mendow, G. 2015 Transesterification Of Sunflower Oil with Ethanol Using Sodium Ethoxide as Catalyst Effect of Reaction Conditions. *J Fuel Process Technol* ,**131**, 29-35.

[3] Encinar , J.M., González, J.F., and Rodríguez-Reinares. A. 2007 Ethanolysis of Used Frying Oil Biodiesel Preparation and Characterization *J Fuel Process Technol*, **88**, 513-522.

[4] Song, E.S., Lim, J.W., and Lee, H.S. 2008 Transesterification of RBD Palm Oil Using Supercritical Methanol *J Supercrit Fluids*, **44**, 35-63.

[5] Troter, D.Z., Zoran B.T., Dušica R.D., Olivera S.S., and Vlada B.V. 2010 Application of ionic liquids and deep eutectic solvents in biodiesel production : A review. *Renewable and Sustainable Energy Rev* **61**, 473 – 500.

[6] Hayyan, A., Alam, Md. Z., Mirghani, M.E.S., Kabbashi, N.A., Hakimi, N.I.N.M., Siran, Y.M., Tahiruddin, S. 2010 Sludge Palm Oil As A Renewable Raw Material for Biodiesel Production by Two-Step Processes. *J Bioresour. Technol*, **101**, 7804-7811.

[7] Shahbaz, K., Mjalli, F.S., Hashim, M.A., AlNashef, I.M. 2011 Eutectic Solvents for The Removal of Residual Palm Oil-Based Biodiesel Catalyst. *J Sep. Purif. Technol.* **81**, 216-222.

[8] Hayyan, A., Alam, Md. Z., Mirghani, M.E.S., Kabbashi, N.A., Hakimi, N.I.N.M., Siran, Y.M., Tahiruddin, S. (2011). Reduction Of High Content Of Free Fatty Acid In Sludge Palm Oil via Acid Catalyst for Biodiesel Production. *J Fuel Process. Technol,* **92**, 920-924.

[9] Gu, L., Huang, W., Tang, S.,Tian, S., and Zhang, X. 2015 A Novel Deep Eutectic Solvent For Biodiesel Preparation Using A Homogeneous Base Catalyst. *Chem Eng Journal*, **259**, 647–652.

[10] Gerpen, J.V. 2005 Biodiesel Processing and Production *J Fuel Process Technol*, **86**, 1097 – 1107.

[11] Suppalakpanya, K., Ratanawilai, S.B., and Tongurai. C. 2010 Production of Ethyl Ester From Crude Palm Oil By Two-Step Reaction with A Microwave System. *J Fuel*,**89**, 2140 – 2144.

[12] Mendow G., Veizaga, N.S., and Querini, C.A. 2011 Ethyl ester Production By Homogeneous Alkaline Transesterification: Influence of The Catalyst. *J Bioresource Technol*, **102**, 6385 -639.

[13] Stojković, I.J., Stamenković, O.S., Dragan, S.P., and Vlada, B.V. 2014 Purification technologies for crude biodiesel obtained by alkali-catalyzed transesterification. *Renewable and Sustainable Energy Rev,* **32**, 1-15.

[14] Barabas, I. and Todoruţ, I. 2011 *Biodiesel Quality, Standards and Properties, Biodiesel- Quality, Emissions and By-Products*, (Dr. Gisela Montero Ed.), ISBN: 978-953-307-784-0, InTech.

IOP Conf. Series: Materials Science and Engineering **206** (2017) 012024 doi:10.1088/1757-899X/206/1/012024

Three dimensional CFD modeling and experimental validation of a single chamber solid oxide fuel cell fed by methane

H T Nguyen[1], M V Le[1], T A Nguyen[1], T A N Nguyen[2]

[1] Faculty of Chemical Engineering, Ho Chi Minh City University of Technology, District 10, Ho Chi Minh City, Vietnam
[2] Faculty of Applied Science, Ton Duc Thang University, 19 Nguyen Huu Tho Str., District 7, Ho Chi Minh City, Vietnam

E-mail: anh.nguyen@hcmut.edu.vn, lmvien@hcmut.edu.vn

Abstract. The solid oxide fuel cell is one of the promising technologies for future energy demand. Solid oxide fuel cell operated in the single-chamber mode exhibits several advantages over conventional single oxide fuel cell due to the simplified, compact, sealing-free cell structure. There are some studies on simulating the behavior of this type of fuel cell but they mainly focus on the 2D model. In the present study, a three-dimensional numerical model of a single chamber solid oxide fuel cell (SOFC) is reported and solved using COMSOL Multiphysics software. Experiments of a planar button solid oxide fuel cell were used to verify the simulation results. The system is fed by methane and oxygen and operated at 700 °C. The cathode is LSCF6482, the anode is GDC-Ni, the electrolyte is LDM and the operating pressure is 1 atm. There was a good agreement between the cell temperature and current voltage estimated from the model and measured from the experiment. The results indicate that the model is applicable for the single chamber solid oxide fuel cell and it can provide a basic for the design, scale up of single chamber solid oxide fuel cell system.

1. Introduction

As a clean and efficient energy conversion device, the fuel cell is currently considered as one of the most promising technologies for future energy demand. Solid oxide fuel cells (SOFCs) have several advantages over other types of fuel cells, including flexibility of fuel used and relatively inexpensive materials due to high temperature operation [1]. Solid oxide fuel cell operated in the single-chamber mode, in which the fuel and oxidant are allowed to mix and anode and cathode reactions take place within the same physical chamber, have attracted recent attention. Since no separation between fuel and air is required, single chamber solid oxide fuel cells (SC-SOFCs) greatly simplify the system design and enhances the thermal and mechanical shock resistance [1]. Mathematical models of SOFCs are effective tools in understanding and investigating effects of various design and operation parameters on SOFC performances, as well as helping in SOFC developments. The modeling and simulation results can be used to optimize and control the fuel cell behaviour, increase the efficiency and performance, and design the system [2]. Therefore, a great number of researches had investigated in SOFC modelling and the internal process simulation based on physical principles. These models range from zero-dimensional (0-D) to three-dimensional (3-D) with different features and point to different research objectives. However, there are not so much report on modelling of SC-SOFCs compare to dual chamber SOFCs. The models developed so far are mainly limited to two-dimensional analysis, only few studies developed three-dimensional model. Chung and Chung [3] developed a 3D model for a Hibino B-type SC-SOFC fed by hydrogen and solved using FEMLAB. Akhtar et al. [4] model a planar type SC-SOFC fed by hydrogen using 3D analysis.

In this study, we investigate a 3D analysis of a planar type SC-SOFC fed by methane. The model is solved by COMSOL Multiphysics software and compare to the experimental data from our previous study [5].

2. The model

2.1. Experimental set up

The detail of experimental set up could be found in our study [5]. The brief summary of the cell is explained as follows. The anode-supported cell was a porous support of nickel oxide and 10 mol% Gd-doped CeO_2 (NiO+GDC). The electrolyte was $La_{1.8}Dy_{0.2}Mo_2O_9$ (LDM). There was a single interlayer of 0.5 wt% Fe_2O_3-doped $Gd_{0.1}Ce_{0.9}O_{1.95}$ (Fe-GDC), which functioned as a diffusion barrier to prevent reaction between the LAMOX and the cathode. The cathode layer was $La_{0.6}Sr_{0.4}Co_{0.8}Fe_{0.2}O_3$ (LSCF). The configuration of the fuel cell system of Ni+GDC/ LAMOX/Fe-GDC/LSCF was shown in Figure 1. The single cell was suspended on two Au wires for current collecting. The cell was positioned above a coil, which prolongs the heating time for the inflow. Both the cell and the coil were housed in a quartz tube, equipped with a neighbouring thermocouple to monitor the cell temperature. An extra Ni1GDC disk was adhered above the anode to activate the anode on-site. The geometry dimension of the system was summarized in Table 1.

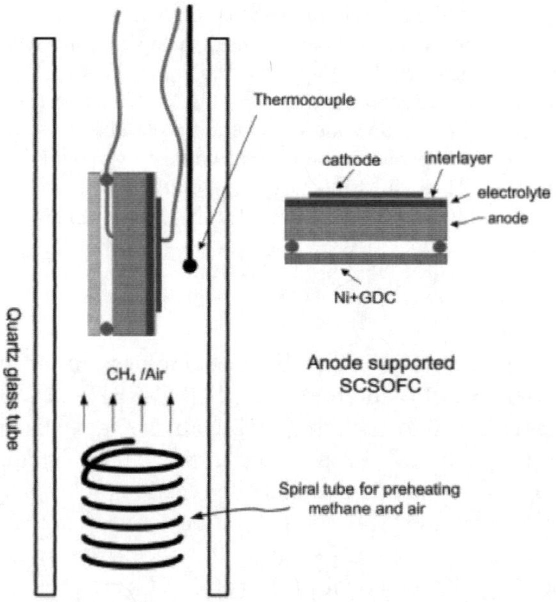

Figure 1. Configurations of single-chamber solid oxide fuel cells (SOFC) and the anode-supported cell based on LDM electrolyte [5]

Table 1. Geometry dimensions

Dimensions	Values (mm)
Tube length	45
Tube diameter	15
Extra disk, anode,	13.5
electrolyte diameter	0.55
Extra disk, anode thickness	9
Cathode diameter	0.027
Cathode thickness	0.15
Extra disk to anode distance	0.06
Electrolyte thickness	

2.2. Computational domain

2.2.1. Gas chamber

The gas chamber consists of a cylindrical tube with a planar button anode-supported cell of which the geometry was built as Figure 2.

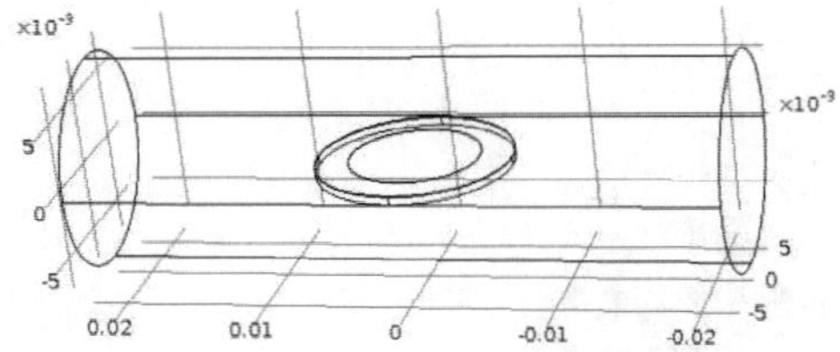

Figure 2. The geometry set up

The applicable equations in this domain are continuity equation, momentum equation, material conservation equation.

2.2.2. Gas diffusion electrodes
The gas diffusion electrodes consist of an anode and a cathode which are porous media. The equations for the domains are continuity equation, momentum equation for transport in porous media, the species conservation equation, and the charge conservation equation.

2.2.3. Electrolyte
The electrolyte is considered to be impermeable to gases. Therefore, both the mass flux and velocity normal to all surfaces of the electrolyte are zero. The other equation is the continuity of ionic current.

3. Results and discussions
The model equations were solved using COMSOL Multiphysics 5.0, a commercial finite element method (FEM) based software package.

The experimental data obtained in our earlier study [5] were used to adjust and validate the model. The cell was operated at the total gas flow rate of 350 sccm (standard cubic centimetre per minute) and the methane over oxygen ratio of 2:1. The temperature of the system is of 700 °C. Some of the parameters of the model obtained from the literature [4], some of the parameters obtained from the fitting procedure. The comparison between the experimental results and the model calculation is shown in Figure 3.

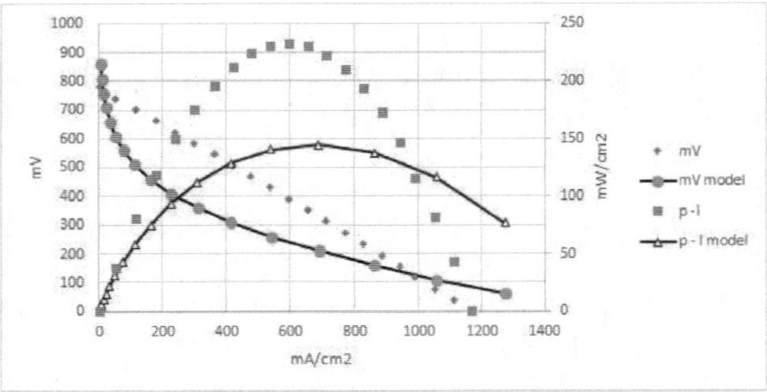

Figure 3. Comparison between experimental data and simulation result for methane gas ratio of 2:1 and temperature of 700 °C

From Figure 3, there is a difference between the experimental results and the estimation from the model. It can be explained that some factors were not fully considered in the model. In our model, the energy conservation was not yet considered. Some of the reactions in the cathode and anode domain might be missing in the model. However, the trend of the calculation results is agreed with the experimental data and it can be used for rough estimation of the performance. The agreement could be improved in the later stage when more factors is added for consideration.

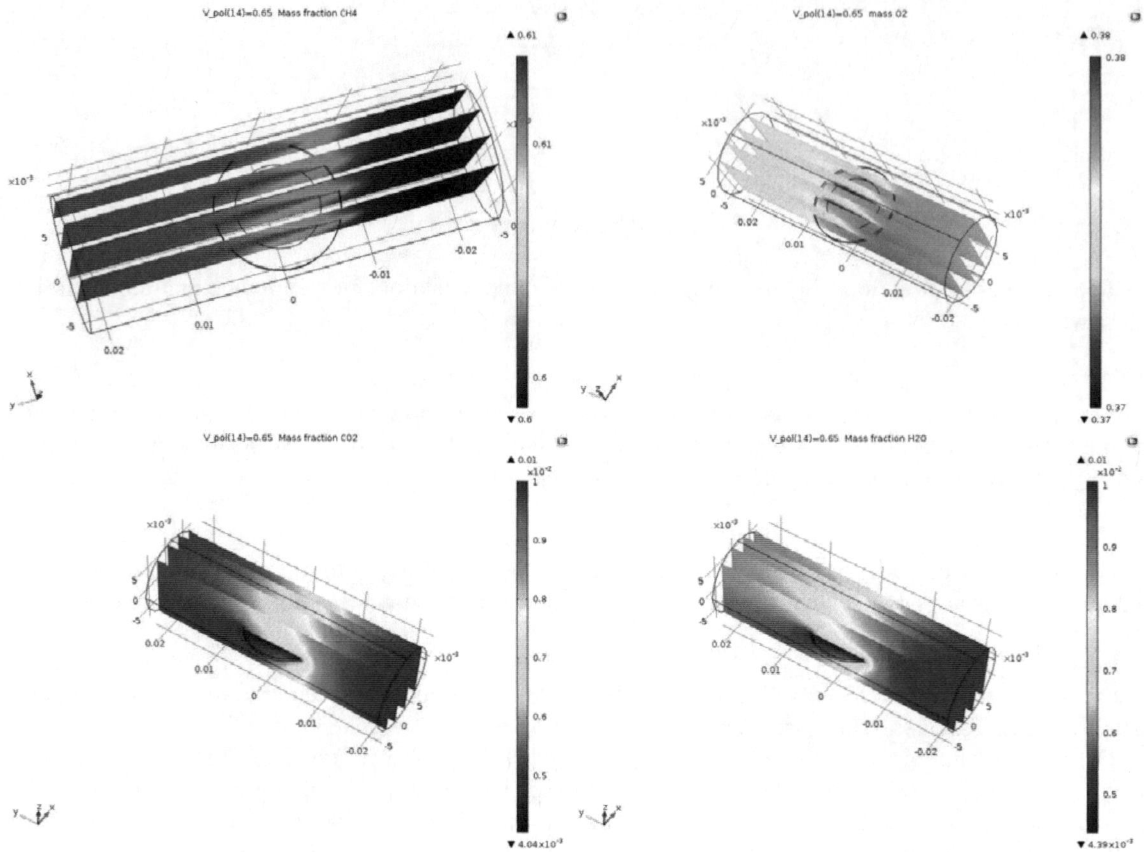

Figure 4. Mass fraction of species along the gas chamber (a) CH_4, (b) O_2, (c) CO_2, (d) H_2O

Figure 4 shows the mass fraction of different species along the gas chamber. It can be obtained that the methane is consumed along the chamber and the reactions generate the carbon dioxide and water, mainly by the full combustion.

4. Conclusions

In this study, a three-dimensional numerical model of a single chamber solid oxide fuel cell fed by methane was considered. The results show that the model might be used for roughly estimate the performance of the fuel cell. The results show the potential application of the model for the simulation of the fuel cell system. However, the poor fitting results also suggest that more factors should be taken into account in order to improve the model accuracy. The addition factors could be the heat transfer and heat conservation in the system, the other chemical reaction equations in the anode and cathode.

Acknowledgement

This research is funded by Vietnam National Foundation for Science and Technology Development (NAFOSTED) under grant number 104.03-2013.20.

References

[1] Yano M, Tomita A, Sano M and Hibino T 2007 Recent advances in single-chamber solid oxide fuel cells: A review *Solid State Ionics* **177** 3351-3359.

[2] Kakaç S, Pramuanjaroenkij A and Zhou X Y 2007 A review of numerical modeling of solid oxide fuel cells *Int. J. Hydrogen Energy* **32** 761-786.

[3] Chung C-Y and Chung Y-C 2006 Performance characteristics of micro single-chamber solid oxide fuel cell: Computational analysis *J. Power Sources* **154** 35-41.
[4] Akhtar N, Decent S P and Kendall K 2010 Numerical modelling of methane-powered micro-tubular, single-chamber solid oxide fuel cell *J. Power Sources* **195** 7796-7807.
[5] Lo J-C, Tsai D-S, Chen Y-C, Le M-V, Chung W-H and Liu F-J 2011 La2Mo2O9-Based Electrolyte: Ion Conductivity and Anode-Supported Cell under Single Chamber Conditions *J. Am. Ceram. Soc.* **94** 806-811.

IOP Conf. Series: Materials Science and Engineering **206** (2017) 012025 doi:10.1088/1757-899X/206/1/012025

Electricity generation from palm oil tree empty fruit bunch (EFB) using dual chamber microbial fuel cell (MFC)

N F Ghazali[1], N A B N Mahmood[1], K A Ibrahim[2], S A F S Muhammad[1], and N S Amalina[1]

[1]Department of Bioprocess and Polymer Engineering, Faculty of Chemical and Energy Engineering, Universiti Teknologi Malaysia, 81310 Johor Bahru, Malaysia.

[2]Department of Chemical Engineering, Faculty of Chemical and Energy Engineering, Universiti Teknologi Malaysia, 81310 Johor Bahru, Malaysia.

Email: nazlee@utm.my (N F Ghazali)

Abstract. : Microbial fuel cell (MFC) has been discovered and utilized in laboratory scale for electricity production based on microbial degradation of organic compound. However, various source of fuel has been tested and recently complex biomass such as lignocellulose biomass has been focused on. In the present research, oil palm tree empty fruit bunch (EFB) has been tested for power production using dual chamber MFC and power generation analysis has been conducted to address the performance of MFC. In addition, two microorganisms (electric harvesting microbe and cellulose degrading microbe) were used in the MFC operation. The analysis include voltage produced, calculated current and power. The first section in your paper

1. Introduction

The world today is looking for alternatives to compensate for higher demands of fossil fuel. This is due to global use of fossil fuels (coal, oil, and natural gas) increased by 1.3 percent in 2002, to 8,034 million tons of oil equivalent, according to preliminary estimates based on government and industry sources. In addition, coal, natural gas and oil accounted for 87 percent of the world's primary energy consumption based on data published in the World Watch Institute, 2013. Thus, the drive for new discoveries on renewable energy is high and is also a necessity for future generation. Renewable energy is generally defined as energy that comes from resources which are naturally refill on a human timescale such as sunlight, wind, rain, tides, waves and geothermal heat. One of most focus on in recent years is microbial fuel cell (MFC). MFC is introduced due to its sustainability and environmental friendly despite being useful to use waste to gain benefits (Franks and Nevins, 2010).

Basically, MFC converts chemical energy to electrical energy by using catalytic reaction of microorganisms in a biochemical reactor. MFC may use many types of substrates mainly any form of biodegradable organic matters such as glucose, sucrose, xylose and complex matters such as polymeric starch. A typical MFC for producing electricity consists of anodic chamber and cathodic chamber which, is separated by a proton exchange membrane (PEM). In the presence of microbes in the anaerobic anodic chamber, protons and electrons are produced by oxidation of organic substrate. The protons generated are transfer to the cathodic chamber through the PEM and subsequently combined with oxygen to form water.

In terms of substrates, MFC has been used to facilitate the conversion of simple sugar such as glucose or more complex substrates such as corn stover to electricity in laboratory scale (Logan, 2006). Lignocellulosic biomass poses a great example of natural sources for bioenergy and its abundance creates opportunity for more challenging acquisition of MFC application. Meanwhile,

Malaysia shares its richness of oil palm trees, which accumulated more than 80% (total include Indonesia) of the world oil palm tree. At the same time, the intense palm oil productions produce more than a quarter per planted land of waste or biomass from the oil palm tree. Empty fruit bunch (EFB) is one of the by-product or considered as scheduled waste.

In this study, the use of dual chamber MFC to generate electricity was studied and the MFC system ability to extract power from EFB as fuel source was observed. However, it was later demonstrated that the degradation of lignocellulose from EFB is influenced by different concentration of substrates. Complex sugar needs pre-treatment steps to be degraded into simple sugar. Hence, degradation of lignocellulose to simple sugar will generate more power compare to complex lignocellulose form.

2. Materials and Methods

2.1 Preparation of pre-treated EFB

Beforehand, dry EFB was collected from local oil palm tree plantation (Bangi, Malaysia) and stored in drying condition (approximately 50 °C) prior to use. The dried EFB was crushed and manually sieved to collect particle size between 250 to 400 mm sizes. Next, the crushed EFB was soaked in distilled water in a water bath at 80 °C. After water filtration, the EFB was then subjected to alkaline pre-treatment by addition of 2.5 M of sodium hydroxide and autoclaved for 15 min at 121 °C. The mixture was then subjected to another filtration process to separate the solid pre-treated EFB from alkaline solution and subsequently mixed with sodium hypochlorite solution (6-14 % active chlorine) for bleaching process. After the delignification process or the pH of the filtrate became approximately 3 (pre-treated EFB became almost white in color), the EFB fiber produced was washed several times with running tap water until the pH of EFB fiber became 7.0. After pre-treatment, the EFB fiber was used directly for MFC operation.

2.2 Inocula preparation

Bacillus E1 and *Clostridium Cellulolyticum* were used in this study and readily stored in the -80 ◦C as glycerol stock. Prior to use, both microorganisms were grown in Luria-Betani broth supplemented with 1.0 % (w/v) carboxymethyl cellulose (CMC) to enhance growth of both microbes. Approximately 100 mL of LB broth with optical density of 2.0 at 600 nm for each microbes was used for the MFC operation. Meanwhile, the cells were harvested through centrifugation at 5,000 rpm for 20 min and pellet was dissolved in 15 mL of potassium phosphate buffer, pH 7.0 each prior to use in MFC.

2.3 Nafion 117 pre-treatment

Nafion 117, a proton exchange membrane was used to separate the anodic and cathodic chambers. The membrane was subjected to pre-treatment procedure in order to get rid of any impurities through boiling at 80°C for one hour in distilled water, one and half hour in 3.0 % hydrogen peroxide, H_2O_2, one hour in distilled water, one and half hour in 0.5 M of sulfuric acid, H_2SO_4, and finally one hour wash with distilled water (M. Rahimnejad et al., 2011).

2.4 MFC set-up and operation

The dual chamber of MFC for the present study consists of two chambers which contain microbes, pre-treated EFB and 0.1 M potassium phosphate buffer in the anodic compartment, while in the cathodic side, potassium permanganate and the same buffer was added. Both compartments was connected with two carbon electrodes which was connected with copper wire as connector to complete the electrical circuit of the fuel cell. The electrode was placed on opposite sides in each chamber and separated by Nafion 117 as depicted in figure 1.

IOP Conf. Series: Materials Science and Engineering **206** (2017) 012025 doi:10.1088/1757-899X/206/1/012025

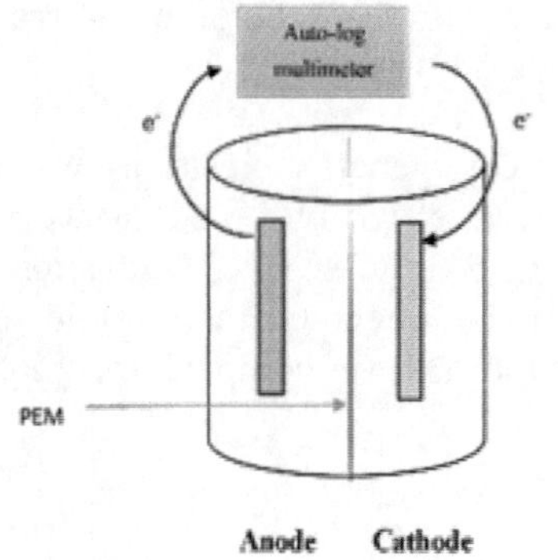

Figure 1. A schematic representation of dual chamber MFC used.

Specifically, 244 mL of 0.1 M potassium phosphate buffer solution, pH 7.0 and 6 mL of mixed microbe inoculum (15% from total volume 250 mL) were added in anodic chamber. In addition, pretreated EFB fiber was added as source of fuel. Both of the solutions were stirred in order to mix properly. While 7.902 g of potassium permanganate (electron acceptor) was diluted in 1.0 L of 0.1 M phosphate buffer solution and 250 mL of the solution was used as catholytes in cathodic chamber.

The auto-log multimeter was connected to the external circuit to obtain voltage or potential differences between the anode and cathode. The voltage data was recorded every 1.½ hours until maximum voltage was reached and the readings were stable. It took about three days to achieve maximum value of open circuit voltage (OCV). Next, the resistor was connected to obtain voltage which designated as closed-circuit voltage (CCV). The voltage data was recorded every five minutes for one hour and was repeated twice using different resistance value. All MFC trials were operated at room temperature. Meanwhile, all CCV values were then calculated to obtain both power and current using Ohm's Law formulae ($V=IR$ and $P=V^2/R$ or $P= I^2R$; V= voltage in volts, R = resistant value in Ohm, and I = current in ampheres)

3. Results and discussion

3.1 Open Circuit Voltage profile

Open Circuit Voltage (OCV) which is considered as a measurement of voltage in the MFC after some time without any current going through the system. Thermodynamically, the OCV value can be interpreted as the cell electron motive force (Emf) value which does not taking account on the losses due to potential resistance presents in the system (Logan et al., 2006) and fully working MFC cannot achieved the Emf value due to the potential losses posed within the MFC system. These include internal electrolyte which contains soluble ions, the randomized activity presents during degradation of substrates and biochemical conversion that may produces several ions that will be attracted to electrode (anode). From the results, maximum achievable OCV was averaged from maximum OCV values of 1.5, 3.5 and 5.5 g/L pre-treated EFB, and the calculated value was 0.647 ± 0.03 (V). It was reported that theoretical value of achievable Emf through glucose oxidation is 1.1 V and only half of the value was achieved from the overall OCV. It is unknown whether the complexity of the substrate or EFB is the main factor of the low value, and there is possibilities of internal hindrance such as competition of substrates between microorganisms or the removal of contaminants produced in the biomass degradation (Juang, 2012).

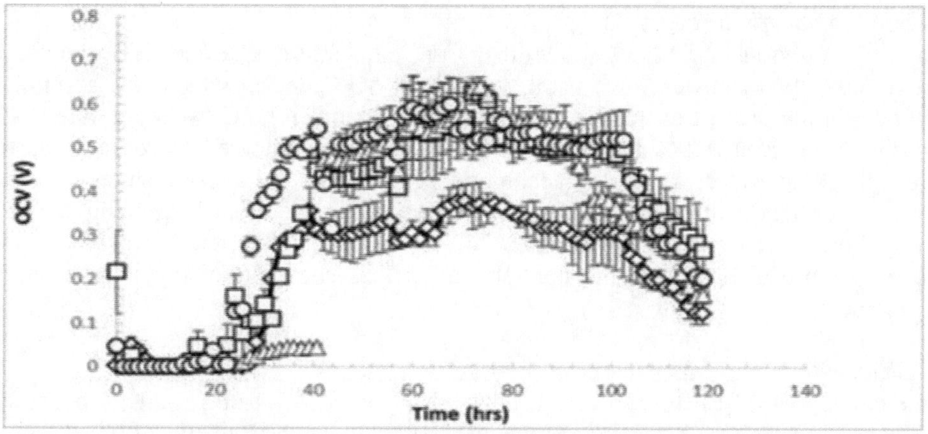

Figure 2. OCV profile for different concentration of pre-treated EFB. The symbols indicate the concentration of pre-treated EFB normalized with the volume of the anodic compartment which was 250 mL ; '◊' represents 7.5 g/mL,'□' was 5.5 g/mL, 'Δ' was 3.5 g/mL, and ' ○' was 1.5 g/mL of pre-treated EFB added.

3.2 Closed-circuit evaluation

The real evaluation of an MFC or general fuel cell, is by testing the system over several resistance value. Figure 3 shows the resistance test using varied resistant value and each voltage obtained was converted to power and current. After the maximum OCV of mixed culture was achieved as shown in Figure 3 (a), the different resistance (5 mins for each interval) was applied to the system in order to obtain power density and current density. By changing the resistance, a new voltage will be obtained and hence a new power density at that resistance as shown in Figure 3 (a). The figure shows that the 'power overshoot' phenomenon occur at the early stage of applying resistance. The power decrease quickly to 0.011W from 0.016W. However, the calculated power was stable by giving 0.012W when applied resistance of 30000 Ω to 70000 Ω. It was reported that 'power overshoot' is probably due to anode potentials changes which related to the unstable electrical activity occurs within microbial biofilm formed on the anodic electrode gradually over the time of operation (Zhu et al., 2013). However, the phenomenon is still in the process of understanding.

(a) (b)

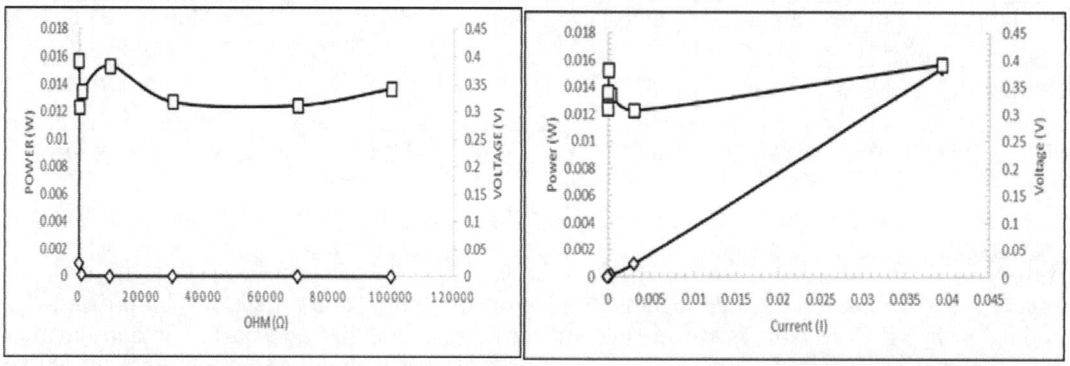

Figure 3. Power and voltage profile. (a) Power (P) and voltage (V) obtained vs resistant value applied in closed-circuit operation. (b) Power (P) and voltage (V) curve against calculated current (I) from measured voltage (V). '□' indicates power and '◊' indicates voltage respectively.

3.3 EFB vs other lignocellulosic biomass

Based on Table 1, previously we have shown that from the same pre-treated EFB, we can extracted approximately 0.002 W of power value using a different MFC design which was a single chamber with air-cathode. In the present evaluation, using the dual chamber MFC, we were able to obtain 10-fold of the single chamber MFC and in fact 100-fold higher compared to recent reported MFCs (Krisnaraj et al., 2015; Ma et al., 2015; Mahmood et al., 2015). However, since our system was conducted in a batch-mode, a total comparison with the fed batch system is quite unfair. In addition, the operating MFCs were different in terms of designs and electricigens used, which makes comparison even more difficult. On the contrary, all MFCs show promising impact on the future energy development.

3.4 EFB to Electricity

Based on the conversion of EFB to power in this study, from 1.375 g or 5.5 g/L of pre-treated EFB, maximum power of 0.016 W can be obtained which if calculated shows a value of 0.0012 W per 1 g of substrate. If the power that can be generated from 1 kilotonnes of EFB is estimated, approximately 1.16 kW of power can be extracted, which is enough to power most electrical appliances. Meanwhile, at present only few MFCs has been proven to be able to produce significant amount of power using wastewater in pilot scale which produced power in the range of 0.4 to 51 W/m3 (Tota-Maharaj and Paul, 2015; Wu et al., 2016; Jiang et al., 2010). The present MFC conducted is very promising to be scaled-up for pilot plant based system. However, the difficulties of sustaining the power produced in prolong period and considering cost efficiency of pre-treating the EFB will indeed have impacts in scaling-up or commercialization purpose.

Table 1. List of different types of lignocellulosic biomass used for MFC application.

Type of lignocellulose biomass	Type of MFC used	Mode of operation	Max power (W)	References
Rice straw	V-shaped MFC	Fed Batch	$1.1*10^{-4}$ W [a](112 mW/m^3)	Ma et al., 2015
Sugar cane bagasse	Three chamber MFC	Fed Batch	$1.3*10^{-4}$ W [a](8.78 W/m^3)	Krishnaraj et al., 2015
Corn cob	Same as above	Fed Batch	$1.0*10^{-4}$ W [a](6.73 W/m^3)	Same as above
Empty fruit bunch	Single chamber MFC	Batch	$1.9*10^{-3}$ W [a](0.678 W/m^2)	Mahmood et al., 2015
Empty fruit bunch	Dual chamber MFC	Batch	$1.6*10^{-2}$ W	This report

'a' indicates calculated power based on the reported value and volume of the anodic chambers used.

4. Conclusion

It was observed that complex substrates such as lignocellulosic biomass is possible to be utilized for power production in MFC system. However, the impact of sustainability and cost efficiency of the MFC system, in addition of the occurrence of power overshoot over high current observed in the MFC operation pose a great challenge to overcome.

Acknowledgement

We like to acknowledge Universiti Teknologi Malaysia for granting us research fund under the university grant (GUP grant no. QJ30000.2544.10H33) and under the Fundamental Research Grant Scheme (RJ130000.7844.4F616). We also like to acknowledge the technical staff of Bioprocess Engineering Department, Faculty of Chemical and Energy Engineering, UTM for their help in the process of the experiments operated.

References

[1] Franks, A. E. and Nevin, K. 2010 *Energies* **3** 899.
[2] Krishnaraj, R. N., Berchmans, S., and Pal, P. 2015 *Cellulose* **22** 655
[3] Jiang, D., Curtis, M., Troop, E., Scheible, K., McGrath, Hu, B., Suib, S., Raymond, D. and Li, B. 2011 *Int J. Hydro. Energy* **36 (1)** 876.
[4] Ma, C. Y., Wu, C. H., and Lin, C. W. 2015 A Novel Shaped Microbial Fuel Cell for Electricity Generation In Biodegrading Rice Straw Compost. *J. Adv. Agri. Technol.* **2 (1)** 57.
[5] Juang, D. F. 2012 International Conference on Environmental Science and Development (ICESD 2012) APCBEE Procedia. Vol 1 p 2.
[6] Tota-Maharaj, K. and Paul, P. 2015 *Int J Energy Environ Eng* **6** 213.
[7] Wu, S., Li, H., Zhou, X., Liang, P., Zhang, X., Huang, X. 2016 *Water Res.* **98** 396.
[8] Zhu, X., Tokash, J. C., Hong, Y. and Logan, B. E. 2013 *Bioelectrochem.* **90** 30.

IOP Conf. Series: Materials Science and Engineering **206** (2017) 012026 doi:10.1088/1757-899X/206/1/012026

Magnesium stearine production via direct reaction of palm stearine and magnesium hydroxide

M Pratiwi[1], P Ylitervo[2], A Pettersson[2], T Prakoso[1] and T H Soerawidjaja[1]

[1]Department of Chemical Engineering, Institut Teknologi Bandung, Bandung 40132, West Java, Indonesia
[2]Swedish Centre for Resource Recovery, University of Borås, 50332 Borås, Sweden University of Borås, 50332 Borås, Sweden

Email: meiti.pratiwi@itb.ac.id

Abstract. The fossil oil production could not compensate with the increase of its consumption, because of this reason the renewable alternative energy source is needed to meet this requirement of this fuel. One of the methods to produce hydrocarbon is by decarboxylation of fatty acids. Vegetable oil and fats are the greatest source of fatty acids, so these can be used as raw material for biohydrocarbon production. From other researchers on their past researchs, by heating base soap from divalent metal, those metal salts will decarboxylate and produce hydrocarbon. This study investigate the process and characterization of magnesium soaps from palm stearine by Blachford method. The metal soaps are synthesized by direct reaction of palm stearine and magnesium hydroxide to produce magnesium stearine and magnesium stearine base soaps at 140-180 °C and 6-10 bar for 3-6 hours. The operation process which succeed to gain metal soaps is 180 °C, 10 bar, for 3-6 hours. These metal soaps are then compared with commercial magnesium stearate. Based on Thermogravimetry Analysis (TGA) results, the decomposition temperature of all the metal soaps were 250 °C. Scanning Electron Microscope with Energy Dispersive X-ray (SEM-EDX) analysis have shown the traces of sodium sulphate for magnesium stearate commercial and magnesium hydroxide for both type of magnesium stearine soaps. The analysis results from Microwave Plasma-Atomic Emission Spectrometry (MP-AES) have shown that the magnesium content of magnesium stearine approximate with magnesium stearate commercial and lower compare with magnesium stearine base soaps. These experiments suggest that the presented saponification process method could produced metal soaps comparable with the commercial metal soaps.

1. Introduction

Vegetable oil or fat become one of important renewable resource (Güner et al, 2006). Those were extracted from plants, such as coconut, palm oil, sunflower, rapeseed, rubber, jatropha, etc. The main component of oil/fat is triglyceride, an ester of glycerol and three fatty acids. Fatty acids are carboxylic acid with long aliphatic chain. In addition, fatty acids can recognized as carboxylic acids contamined by carbon dioxide (R<u>COO</u>H). The basic idea of this study is to produce hydrocarbon by removing carbon dioxide from fatty acid chain.

In 1947, Chang and Wan [2] reported that by decarboxylation of metal soaps from Tung oil, it can produced of hydrocarbon. The divalent and higher valent metal soaps are water insoluble, but those metal soaps are soluble in various organics solvents [3]. Metal soaps have many application areas such

IOP Conf. Series: Materials Science and Engineering **206** (2017) 012026 doi:10.1088/1757-899X/206/1/012026

as driers in paints or inks; lubricants; stabilizers for plastics; fungicides; as catalysts for polymerization, condensation, and hydrogenation; as coatings for ceramic glazes; wire drawing; and waterproofing agents [3,4,5].

Three common processes used in the metal soap manufacture are double decomposition, fusion, and metal-acid reaction processes [6,7]. The first process involves two reactions: aqueous caustic and fatty acids reaction to form a sodium soap, and the sodium and an inorganic metal salt reaction to form the metal soap. The second process involves the fatty acid and metal oxide reaction. The third process involves the fatty acid and free metal powders to form polyvalent metal soap. The first process or double decomposition process is the most widely used in commercial scale. In the fusion process, the main concern is that the raw materials must be pure grade. The metal-acid reaction process is the least used of the three commercial processes [7].

The synthesis of metal soap was carried out with dipolar aprotic solvent, such as acetone, as liquid reaction medium, as described in [8,9]. [10] reported the production of zinc salt by stirring a mixture comprising of zinc oxide, water, molten fatty acids, and citric acids or phosporic acids as catalyst.

The raw materials in above mentioned processes are fatty acids. This experiment synthesized the metal soaps using triglyceride as the feed for saponification reaction. The abundance and low cost of triglyceride compared to the fatty acids are the main interests of this research. Most of the previous reseachs were using the metathesis method to produce metal soaps from vegetable oil/fat or triglyceride [11, 12, 13, 14, 15].

[16] reported that by heating divalent metal base soap, such as calcium and magnesium, without air at atmospheric pressure and higher temperature (200-400 °C), the decarboxylation reaction occurred to produced hydrocarbon.

This report will present the production of metal soaps via direct reaction of palm stearine and magnesium hydroxide by the Blachford [17] method. The aim of this research is to investigate the characterictics of the metal soaps produced by direct reaction of palm stearine and magnesium hydroxide by Blachford method and the produced metal soaps is compared with the commercial magnesium stearate.

2. Experimental

Palm stearine used in this research was taken from local oil refinery. The saponification and acid value of palm stearine is 190.7 mg KOH/g and 0.6 mg KOH/g. The composition of palm stearine were mainly of 68.24%-mol palmitic acid, 5%-mol stearic acid, and 20.63%-mol oleic acid. The magnesium hydroxide was prepared by co-precipitation of $MgCl_2.6H_2O$ with sodium hydroxide (modified from [18,19]). Magnesium stearine obtained from direct reaction of palm stearine and magnesium hydroxide [17]. Magnesium stearic, technical grade, was used to compare with the metal soaps products.

2.1. Preparation of magnesium hydroxide

The magnesium hydroxide was prepared by using two burrettes, the first burrette filled with 100 ml of 0,5 M MgCl.6 H_2O (A) and the second burrette with 500 ml of 0,6 M NaOH (B). The co-precipitation process controlled by the drop rate of burrette B, which was 5 times higher than the drop rate of burrette A. The co-precipitation was conducted under vigorous stirring at room temperature. The pH of the solutions was maintained at 8.5 with Triz buffer. When the co-precipitation process was finished, the suspensions were kept stirred for 30 minutes, then the suspensions was filtered with Buchner filter and washed with hot water (60 °C). Afterwards, the slurry was dried at room temperature, and then dried at 105°C for 12 hours. The dry slurry was ground, formed $Mg(OH)_2$ white powder.

2.2. The preparation of magnesium stearine soaps via direct reaction

The magnesium hydroxide was prepared by using two burrettes, the first burrette filled with 100 ml of 0,5 M MgCl.6 H_2O (A) and the second burrette with 500 ml of 0,6 M NaOH (B). The co-precipitation process controlled by the drop rate of burrette B, which was 5 times higher than the drop rate of burrette A. The co-precipitation was conducted under vigorous stirring at room temperature. The pH of the solutions was maintained at 8.5 with Triz buffer. When the co-precipitation process was finished, the suspensions were kept stirred for 30 minutes, then the suspensions was filtered with Buchner filter and washed with hot water (60 °C). Afterwards, the slurry was dried at room temperature, and then dried at 105°C for 12 hours. The dry slurry was ground, formed $Mg(OH)_2$ white powder.

The saponification reaction was carried out at 140 and 180 °C using Blachford method [17] by reacting palm stearine, magnesium hydroxide, and water. The amount of triglyceride and magnesium hydroxide were approximately same with their stoichiometric reaction (see Reaction 1 and 2). Two types of metal soaps were produced by this method, magnesium stearine ($Mg(RCOO)_2$) and magnesium stearine base soaps (Mg(RCOO)(OH)). The time of reaction were 3 and 6 hr for the magnesium base soaps, and 3 hr for magnesium stearine.

The saponification reaction produce magnesium stearine as follows :

$2\ H_2C(OOCR)CH(OOCR)CH_2(OOCR) + 3\ Mg(OH)_2 \rightarrow 2\ CH_2OH.CH(OH).\ CH_2OH + 3\ Mg(RCOO)_2$ (reaction 1)

and the saponification reaction for magnesium base soap :

$H_2C(OOCR)CH(OOCR)CH_2(OOCR) + 3\ Mg(OH)_2 \rightarrow CH_2OH.CH(OH).\ CH_2OH + 3\ Mg(RCOO)(OH)$ (reaction 2)

The reaction was performed at 100 ml mini-stainless steel autoclave. The reaction pressure was 6 bar at 140 °C and 10 bar at 180 °C. The molten layer was removed from the reactor, and then washed with hot water and filtered with Buchner filter. The obtained slurry was dried overnight at 60 °C in the oven, then ground it into magnesium soap powder.

2.3. Magnesium soaps characterization

The conversion of metal soaps was determined by weighing the wash water before and after the purification of the metal soaps. The thermal behaviour of metal soaps were obtained with TGA instrument (TGA/DSC 3^+ Star* System Mettler Toledo). The samples were heated from 25 to 700 °C at a constant heat rate of 5 K/min under 80 ml/min nitrogen gas flow. The MP-AES instrument (Agilent Technologies) was conducted to measure the magnesium contents of the metal soaps, by using standard solutions with 1-10 ppm of magnesium. The metal soap samples were digested in nitric acid solution 65% using Microwave Ethos Up (program: 12 °C/min to reach 200 °C and hold for 10 min). SEM instrumen were used to investigate the topography surface of the metal soaps and magnesium content in the metal soaps surface, with secondary electrons (LFD), backscattered electrons (SSD), and EDX.

3. Results and discussion

3.1. Saponification reaction

The saponification reaction was unsuccessfully conducted at 140 °C. The appearance of the metal soaps that produced at this operation condition were like slurry at room temperature. The preparation of metal soaps at 180 °C produced magnesium stearine and magnesium stearine base soap. However, the texture of the produced metal soaps were different to the commercial magnesium stearate. The metal soaps produced has rougher texture compared with commercial magnesium stearate, which is very powdery and fluffy. The purification was performed by washed the metal soaps with hot water. The metal soaps were insoluble in water, but the glycerol was soluble and the triglyceride slighty soluble in hot water. The purification was aimed to remove the produced glycerol and, slighty, remained triglyceride in saponification products. The reaction conversion was determined by the weight of the wash water before and after the washing step, in addition to the visual look of the wash water. There were no significant difference in weight of the washing water and no triglycerides were observed in the water after wash, indicating that the reaction conversion was nearly 100%. The following analysis (TGA, SEM-EDX, and MP-AES) were performed on magnesium stearate, magnesium stearine, magnesium base stearine (3 hr reaction), and magnesium base stearine (3 and 6 hr reaction).

3.2. TGA

The TGA result of four types of magnesium soaps, comprised of magnesium stearate, magnesium stearine, magnesium stearine base soap with either 3 (Mg(stearine)$_2$ and 3-Mg(stearine)(OH)) or 6 (6-Mg(stearine)(OH)) hr saponification reaction at 180 °C, are given in Fig.1. The thermal decomposition of magnesium soaps can be expressed as : $Mg(RCOO)_2 \rightarrow RCOR' + CO_2 + MgO$ (reaction 3)

The final residues weight for magnesium stearate and magnesium stearine were similar compared with the molecular formula of magnesium oxide. The final residue of magnesium base soaps were higher than magnesium stearate and magnesium stearine, because the amount of magnesium hydroxide as reactant in the Reaction 2 is twice from Reaction 1.

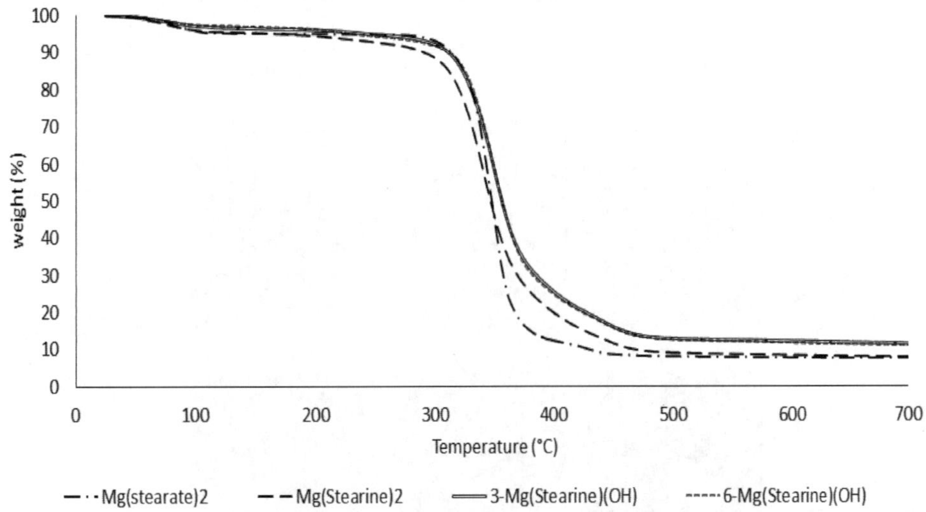

Figure 1. TGA curves of magnesium soaps

The thermogravimetric analysis (Figure 1) shows that these soaps lose their moisture up to 110 °C, then the weight was stable at 110 °C until 300°C. Between 300 and 500 °C the samples decomposed very rapidly. And finally show small weight change during further increased temperature.

Both of magnesium base stearine have similar thermal behaviour. The magnesium stearine decomposed more rapidly compared with the magnesium stearate. It may because of the difference of the fatty acids content in the metal soaps. The magnesium stearine comprise unsaturated fatty acids, oleic acids, the double bond is easier to decompose compared to the saturated bond. Typical fatty acids composition in commercial magnesium stearate is saturated fatty acids, e.g. palmitic and stearic acid.

3.3. SEM-EDX

The SEM-EDX analysis was used to determine the topography surface and magnesium content at the metal soap's surface. Figure 2 shows the elements positioning in the magnesium stearate surface, there are interactions between sodium and sulphur, marked with the red ellips, it may be the remaining of sodium sulphate, byproduct of precipitation method. The commercial magnesium stearate was manufactured by precipitation or double decomposition reaction as follow

$RCOOH_{(l)} + NaOH_{(s)} \rightarrow RCOONa_{(aq)} + H_2O_{(l)}$ (reaction 4)

$RCOONa_{(aq)} + MgSO_4.7\,H_2O_{(aq)} \rightarrow Mg(RCOO)_{2\,(s)} + Na_2SO_{4\,(aq)} + 7\,H_2O_{(l)}$ (reaction 5)

IOP Conf. Series: Materials Science and Engineering **206** (2017) 012026 doi:10.1088/1757-899X/206/1/012026

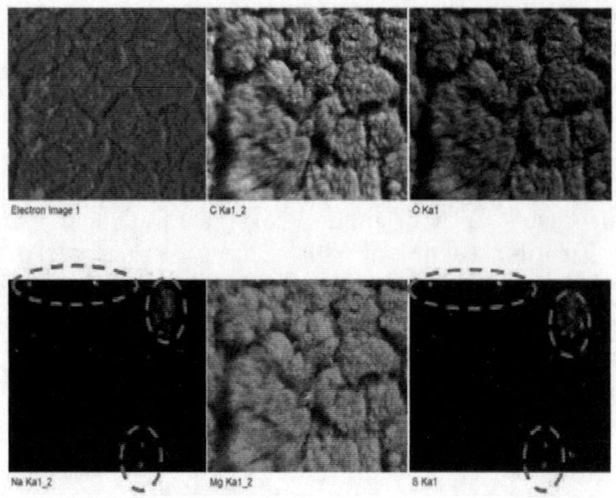

Figure 2. SEM element image of $Mg(stearate)_2$

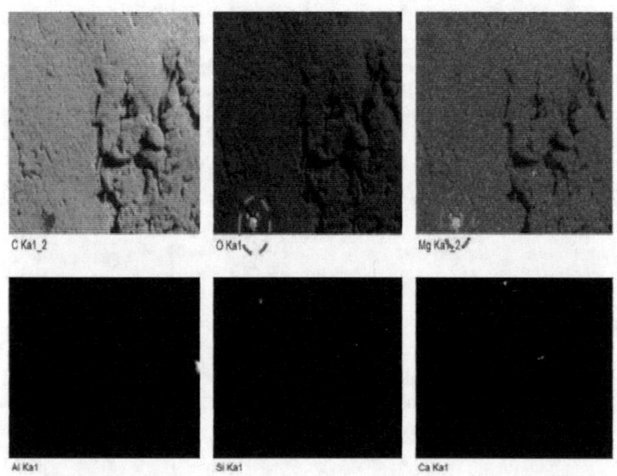

Figure 3. SEM element image $3Mg(stearine)_2$

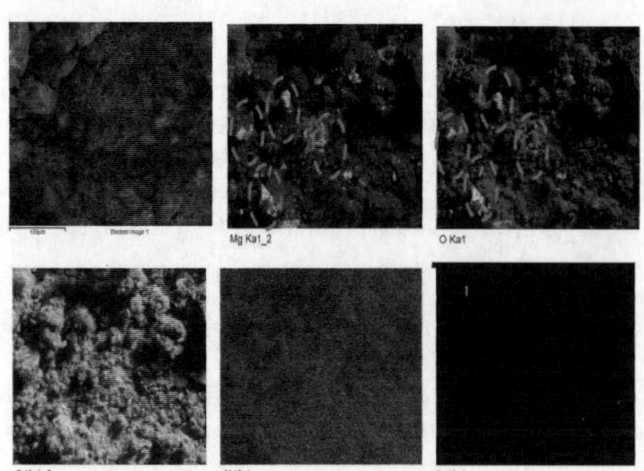

Figure 4. SEM element image of 3Mg(Stearine)(OH)

Figure 5. SEM Micrograph of (a) Magnesium stearate; (b) Magnesium stearine; and (c) Magnesium base stearine

Figures 3 and 4 show that there was interaction between magnesium and oxygen in the magnesium stearine and magnesium based stearine soaps. This indicate that there were remained magnesium hydroxide in the soaps (see Reaction 1 and 2), especially in the magnesium based stearine with excess of $Mg(OH)_2$.

Figure 5 a-c show the topography surface of the magnesium soaps. Magnesium stearate (5a) have the most homogenous topography surface compared with others soaps. The magnesium based stearine has more similar morphology to magnesium stearate, compared with magnesium stearine. From Figure 5b, magnesium stearine has morphology appearance like a molten soap. To get a topography surface similar to commercial magnesium soaps, the amount of magnesium hydroxide as reactant usually higher than the stoichiometry in Reaction 1 [12,13].

3.4. Magnesium content

The MP-AES analysis was carried out to quantify the magnesium content in the metal soaps. Before the MP-AES analysis, Mg^{2+} ions were recovered from magnesium samples by digestion with nitric acid 65%. The weight percent of Mg^{2+} ions for the metal soap samples was determined as shown in Table 1. The final residue of magnesium stearine base soaps were higher than magnesium stearine (Figure 1) due to the magnesium content in the soaps. The EDX analysis only give an indication of the ion content in the metal soaps since it measured the ion content based on the area observed. For the EDX analysis, the samples has to be homogenous in order to get reliable results.

Table 1. The magnesium content in metal soaps

Samples	%-w		
$Mg(stearate)_2$	4.45 ± 0.59[a]	5.91[b]	4.84[c]
$3\text{-}Mg(stearine)_2$	4.19 ± 0.22[a]	4.35[b]	5.00[c]
3-Mg(stearine)(OH)	6.76 ± 0.60[a]	6.71[b]	9.04[c]
6-Mg(stearine)(OH)	6.70 ± 0.61[a]	n.a.	9.04[c]

[a]MP-AES; [b]EDX; [c]theoretical; n.a not available

The MP-AES measurements were carried out three times and the average values from the analysis results are reported in Table 1. The magnesium content in $Mg(stearine)_2$ and $Mg(stearate)_2$ was similar. Despite the lower weight percent from analysis compared with theoretical, these results showed that the direct reaction of magnesium hydroxide and palm stearine could be an alternative process to produce magnesium soaps.

4. Conclusion

The direct reaction of magnesium hydroxide and palm stearine produced magnesium soap at 180 °C and reaction time from 3 hours. It was not suitable to perform the direct reaction method at 140 °C, even when the reaction time already pro-longed to 6 hours, to yield magnesium soaps. Regarding thermal behaviour, the magnesium stearine produced with this method was quite comparable with the commercial magnesium stearate. This method is still needed to be improved, such as controlling the stirring rate of reaction, grinding apparatus to get homogenous size of metal hydroxide etc.

Acknowledgement
We acknowledge the support from Linnaeus Palme Program, Sweden.

References

[1] Dandik L, Aksoy H A and Erdem-Senatalar A 1998 *Energy & Fuels* **12** pp 1148-52
[2] Chang C and Wan S 1947 *Industrial and engineering chemistri* **39** pp1543-48
[3] Akanni M S, Okoh E K, Burrows H D and Ellis H A 1992 *Thermochimica Acta* **208** pp 1-41.
[4] Elliott S B 1946 *The alkaline-earth and heavy metal soaps : American Chemical Society Monograhph Series* (New York : Reinhold Publishing Corporation) pp 111-43
[5] Rogers R H, Park P and Opem J D 1962 *U.S.Patent No.23,047,496.*
[6] Rogers R H Jr and Blew W R Jr 1959 *U.S.Patent No.2,890,232.*
[7] Gönen M, Balköse D, İnal F and Ülkü S 2005 *Ind. Eng. Chem. Res* **6** pp1627-33
[8] Akkers J B, Littler J A and Peters D C 1982 *U..S. Patent No 4,337,209*
[9] Akkers J B, Littler J A and Peters D C 1984 *U..S. Patent No 4,427,572*
[10] Chen B 2004 *U.S. Patent No 6,689,894.*
[11] Balköse D, Egbuchunam T O and Okieimen F E 2010 *J.Therm. Anal. Calorim*
[12] Folarin O and Enikanoselu O 2010 *J. Appl. Sci. Environ. Manage* **14** pp 69-73
[13] Folarin O M, Siyanbola S Y and Shittu A T 2011 *EJEAFChe* **10** pp 2623-28
[14] Isaac I O 2014 *IJARCS* **1** pp 14-22
[15] Ossai E 2014 *J. Appl. Sci. Environ. Manage* **18** pp 359-63
[16] Markley K S (editor) 1961 *Fatty acids : their chemistry, properties, production, and uses part 2* (New York : Interscience publishers Inc)
[17] Blachford J 1982 *U.S. Patent No 4,316,852.*
[18] Matter P H and Ozkan U S 2005 *J. Catal.* **234** pp 463-75.
[19] Matter P H, Braden D J and Ozkan U S 2004 *J. Catal.* **223** pp 340-51.

IOP Conf. Series: Materials Science and Engineering **206** (2017) 012027 doi:10.1088/1757-899X/206/1/012027

Effect of temperature on methanogenesis stage of two-stage anaerobic digestion of palm oil mill effluent (POME) into biogas

B Trisakti*, **Irvan, Mahdalena, Taslim, and M Turmuzi**

Chemical Engineering Department, Faculty of Engineering, University of Sumatera Utara. Jalan Almamater Komplek USU Medan, 20155, Indonesia

*E-mail: btrisakti2013@gmail.com

Abstract. This study aimed to determine the effect of temperature on methanogenesis stage of conversion of palm oil mill effluent into biogas. Methanogenesis is the second stage of methanogenic anaerobic digestion. Improved performance of the methanogenesis process was determined by measuring the growth of microorganisms, degradation of organic materials, biogas production and composition. Initially, the suitable loading up was determined by varying the HRT 100, 40, 6, and 4.0 days in the continuous stirred tank reactor (CSTR) with mixing rate 100 rpm, pH 6.7-7.5 at room temperature. Next, effect of temperature on the process was determined by varying temperature at mesophilic range (30-42°C) and thermophilic range (43-55°C). Analysis of total solids (TS), volatile solids (VS), total suspended solids (TSS), volatile suspended solids (VSS), and chemical oxygen demand (COD) were conducted in order to study the growth of microorganisms and their abilities in converting organic compound to produce biogas. Degradation of organic content i.e. VS decomposition and COD removal increased with the increasing of temperature. At mesophilic range, VS decomposition and COD removal were 51.56 ± 8.30 and 79.82 ± 6.03%, respectively. Meanwhile at thermopilic range, VS decomposition and COD removal were 67.44 ± 3.59 and 79.16 ± 1.75%, respectively. Biogas production and its methane content also increased with the increasing of temperature, but CO_2 content also increased. Biogas production at mesophilic range was 31.77 ± 3.46 L/kg-ΔVS and methane content was 75%. Meanwhile, biogas production at thermopilic range was 37.03 ± 5.16 L/kg-ΔVS and methane content was 62.25 ± 5.50%.

1. Introduction

Indonesia is the largest producer of palm oil in the world with a total production reached 33.5 million tons. Palm oil is produced from approximately 9.2 million hectares of oil palm plantations [1], which mostly located on the island of Sumatra and Borneo [2]. The process to extract oil requires significantly large amounts of water for steam sterilization of fresh palm fruit bunches (FFB) and clarification of extracted oil. As a result, palm oil mill (POM) produces large amounts of wastewater or commonly called palm oil mill effluent (POME). It is estimated in 2015 about 79.2-122.1 million tons or approximately 2.4 to 3.7 tons of POME was generated per ton of palm oil [3].

In an effort to reduce HRT, many researchers separate the anaerobic digestion process into two stages namely acidogenesis and methanogenesis stage [4]. Therefore, to improve the effectiveness of this process, the optimum operating conditions such as pH, temperature, and agitation rate of each

IOP Conf. Series: Materials Science and Engineering **206** (2017) 012027 doi:10.1088/1757-899X/206/1/012027

stage should be known [5,6]. Unfortunately, few data can be found in the literature about the effect of temperature on the growth of microorganisms in the methanogenesis stage. Therefore, this paper reported the effects of organic loading up (reduction of HRT) and temperature variations on methanogenesis stage or the second stage of two stages anaerobic digestion of conversion of POME to biogas.

Fresh POME taken from fat pit of POMs are brownish viscous mud with temperatures around 80-90°C, acidic (pH 3.8-4.5), and the concentration of organic compounds is high enough so the COD and BOD are also high [7]. POME contains large amounts of carbohydrates, protein, and fat with the composition 29.55, 12.75, and 10.21% respectively. In addition, there are also some macro and micro mineral compounds such as potassium (K), sodium (Na), calcium (Ca), iron (Fe), zinc (Zn), chromium (Cr), and others [8]. Therefore, POME can be utilized as substrate and nutrients for microorganisms in methanogenic anaerobic digestion for biogas production [9,10].

A two-stage process of anaerobic digestion was used in this study. In the first stage, the organic compounds were converted into volatile fatty acids (VFA) by hydrolysis, acidogenesis, and acetogenesis (commonly referred as acidogenesis stage) and subsequently the VFA was converted into biogas through methanogenesis stage [11–15]. Methanogenesis is the most important stage in the process of methanogenic anaerobic digestion because in this process occurs the slowest biochemical reactions.

Methanogenesis stage is greatly influenced by the operating conditions such as the composition of raw materials, feed flow rate (hidroulic retention time, HRT), temperature, agitation, and pH. Digester over loading, temperature changes, or the entry of a large amount of oxygen can result in termination of methane production [9,10].

Methanogenic microorganisms are generally more sensitive to temperature fluctuations compared to other microorganisms involved in anaerobic digestion process [16]. Methanogenic anaerobic digestion process has been reported to take place at several temperature ranges namely psycrophilic (below 25°C), mesophilic (25-45°C) and thermophilic (45-70°C) [9,17]. Temperature range commonly used in the process of metanogenic anaerobic digestion are mesophilic and thermophilic. Several researchers have reported that the mesophilic temperature optimum is in the range of 30-35°C and thermophilic temperature optimum is in the range 50-65°C [16,18].

2. Materials and Methods

2.1. Materials

The feed was effluent from acidogenesis reactor where its characteristics are presented in Table 1. The acidogenesis reactor used fresh POME collected from "a fat pit of Adolina POM, Sumatera Utara, Indonesia" as feed. Methagenic bacteria as a starter were collected from 3000 l methanogenic anaerobic digester at the University of Sumatera Utara, Medan, Indonesia. Sodium bicarbonate ($NaHCO_3$) was used for adjusting the pH.

Table 1. Characteristics of acidogenesis reactor effluent

Parameters	Units	Results
pH	-	5.0 - 5.5
Chemical Oxygen Demand (COD)	mg/L	26,900 - 43,130
Total Solid (TS)	mg/L	20,340 - 37,930
Volatile Solid (VS)	mg/L	18,980 - 36,980
Total Suspended Solid (TSS)	mg/L	11,480 - 18,440
Volatile Suspended Solid (VSS)	mg/L	10,860 - 16,920
Volatile Fatty Acids (VFA)	mg/L	5,622

2.2. Experimental Setup

The laboratory scale of continuous stirred tank reactor (CSTR) as fermenter is shown in Figure 1. The CSTR was EYELA model MBF 300ME with working volumes is 2 L. The fermenter comprised of an integrated on-line temperature and pH data recording system. No pH adjustment was applied to the fermenter. However, pH of the feed was controlled with sodium bicarbonate ($NaHCO_3$), before it was fed into the fermenter.

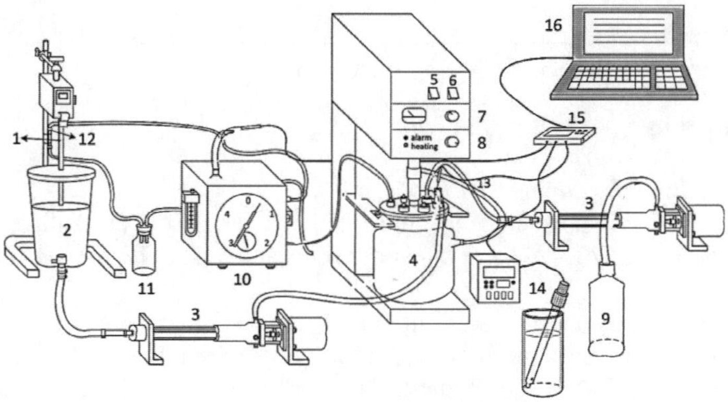

1. Mixer
2. Feed Tank
3. Screw Pump
4. Fermenter (CSTR)
5. Jacket pump-controller switch
6. Power switch
7. Mixer rate controller
8. Jacket temperature controller

09. Discharge Tank
10. Gas Meter
11. Gas holder container
12. H_2S absorber
13 Thermocouple
14. pH electrode
15. Data logger
16. Computer

Figure 1. Experimental set-up

2.3. Data Collection

The research was conducted to determine the improvement of the methanogenesis process by measuring the growth of microorganisms, the degradation of organic materials, and production and composition of biogas. Initially, the suitable loading up was determined by varying the HRT 100, 40, 6, and 4.0 days with mixing rate 100 rpm, pH 7± 0.2, and room temperature. Then the effect of temperature on the process was determined by varying temperature at mesophilic range (30-42°C) and thermophilic range (43-55°C) with HRT 4 days, pH 7 ± 0.2, and agitation rate of 100 rpm.

Analysis and methods of data collection of the liquid and gas samples are presented in Table 2. Data collected from the liquid samples were pH, alkalinity, TS, VS, TSS, VSS, and COD based on the APHA standard methods for water and wastewater [19]. Meanwhile the volume and composition of biogas were collected for the gas samples.

Table 2. Analysis methods of data collection

Parameters	Units	Methods
pH	-	
Alkalinity	mg/L	APHA 2320B
Total Solid (TS)	mg/L	APHA 2540B
Volatile Solid (VS)	mg/L	APHA 2540E
Total Suspended Solid (TSS)	mg/L	APHA 2540D

IOP Conf. Series: Materials Science and Engineering **206** (2017) 012027 doi:10.1088/1757-899X/206/1/012027

Parameters	Units	Methods
Volatile Suspended Solid (VSS)	mg/L	APHA 2540E
Chemical Oxygen Demand (COD)	mg/L	APHA 5220B
Biogas Volume	ml/day	Gas meter
Biogas composition		
• H_2S	mg/L	Gas detector tube (GASTEC, Type GV-100S)
• CO_2	%	Gas detector tube (GASTEC, 25~1600 ppm)
• Methane	%	Calculated

3. Results and Discussions

3.1. Profile of pH and Alkalinity during The Process

The pH and alkalinity profile is presented in Figure 2. The average pH at loading up, mesophilic, and thermopilic range were 6.89 ± 0.3, 7.02 ± 0.08, and 7.0 respectively. The pH was relatively constant at 7.0 as expected.

The pH was maintained by adding sodium bicarbonate into the feed. Consequently, alkalinity profile fluctuated during the process but it is still in the range where the methanogenesis process takes place i.e. in the range of 2,500 – 5,000 mg/L [17]. The average alkalinity value at loading up, mesophilic, and thermopilic range were 3,067.31 ± 569.90 mg/L, 3,213.04 ± 349.10 mg/L, and 4,440.91 ± 451.88 mg/L respectively.

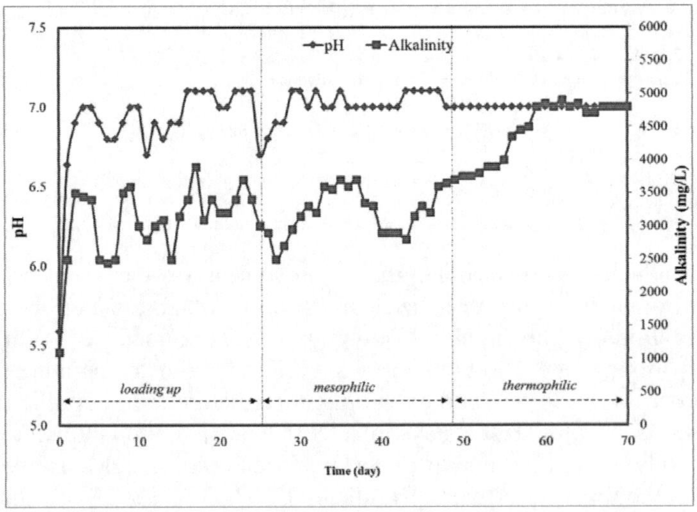

Figure 2. Profile of pH and alkalinity during the proses

3.2. Effect of Temperature on the Growth of Microorganisms

Microorganism growth is described by the change of volatile suspended solid (VSS) concentration in the fermenter. The VSS profile during the process is presented in Figure 3.

The average values of VSS at loading up, mesophilic, and thermophilic range were 21,691.89 ± 1,238.61, 19,875.73 ± 1,766.07, and 22,170.14 ± 1,281.13 mg/L, respectively which indicated that the number of microbes increased with the increase of temperature.

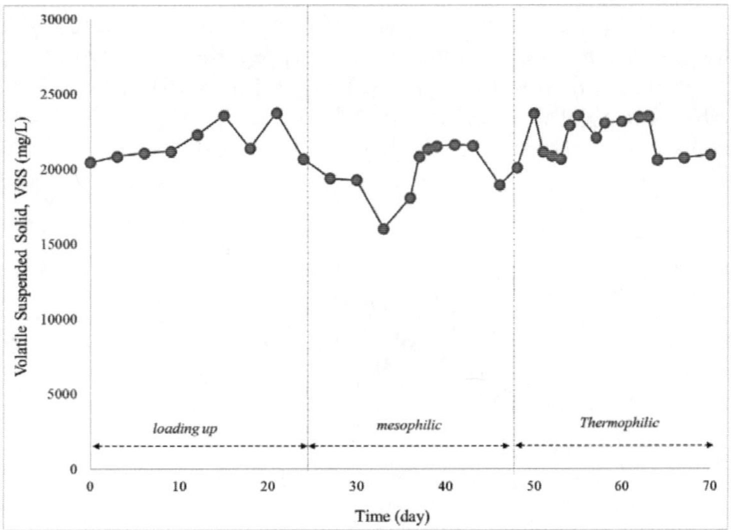

Figure 3. Profile of VSS during the proses

3.3 Effect of Temperature on the Degradation of Organic Materials

Effect of temperature on microbial activity can be observed from VS decomposition and COD removal during the process as presented on Figure 4. The average values of VS decomposition at loading up, mesophilic, and thermophilic range were 53.90 ± 3.95, 51.56 ± 8.30, and 67.44 ± 3.59%, respectively. While the COD removal at loading up, mesophilic, and thermophilic range were 87.28 ± 3.63, 79.82 ± 6.03, and 79.16 ± 1.75%, respectively, both indicating that microbial activity increased with the increase of temperature.

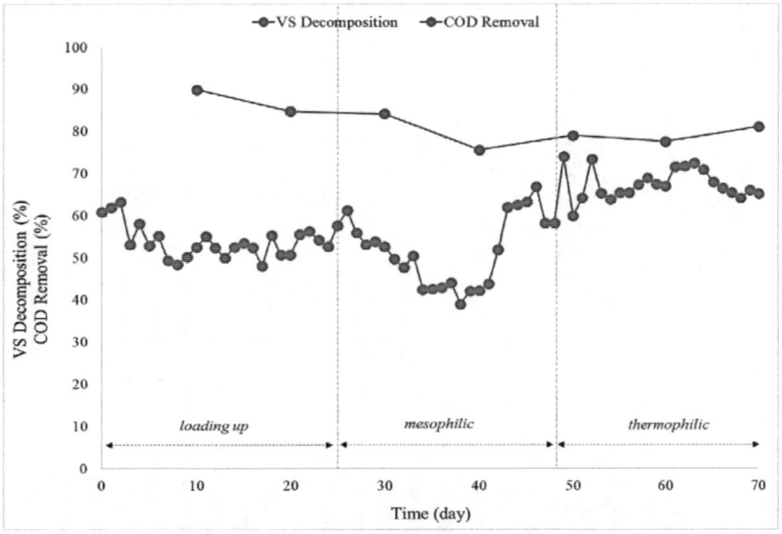

Figure 4. VS decomposition and COD removal during the process

3.4 Effect Temperature on Production and Composition of Biogas

Effect of temperature on production of biogas per volatile solid decomposed during the process is presented in Figure 5. The average biogas production at loading up, mesophilic, and thermophilic range were 18.90 ± 8.11, 31.77 ± 3.46; and 37.03 ± 5.16 L/kg-ΔVS, respectively. Biogas production is increasing with increase of temperature. Similar result was reported by previous researchers [20–24]. This is in accordance with the growth of microorganisms and degradation of organic materials which also increased when the temperature was increased.

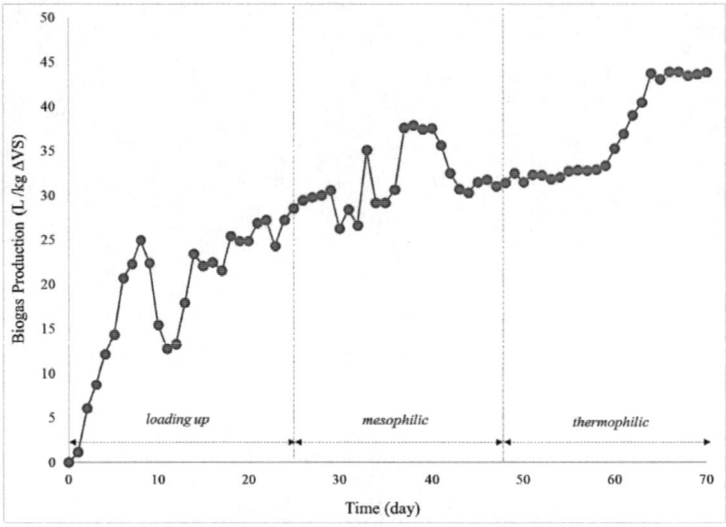

Figure 5. Biogas production during the process

The amount of H_2S, CO_2, and CH_4 contained in the biogas produced during the process is presented in Figure 6. The amount of H_2S in the biogas produced was in trace amount except at loading up it was 0.30 ± 0.30 L/kg-ΔVS. The average values of CO_2 at loading up, mesophilic, and thermophilic range were 3.75 ± 2.57, 9.88 ± 3.23, and 13.97 ± 1.98 L/kg-ΔVS, respectively. The average value of CH_4 at loading up, mesophilic, and thermophilic range were 15.08 ± 6.62, 22.30 ± 5.77, and 23.39 ± 4.82 L/kg-ΔVS, respectively.

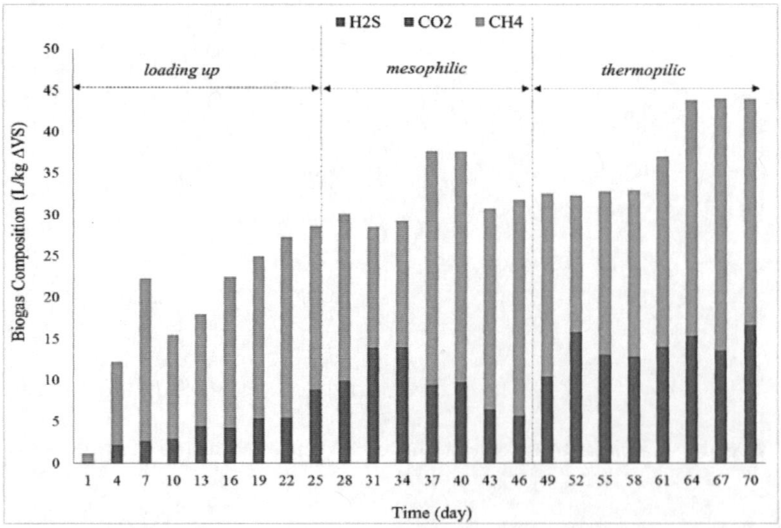

Figure 6. The amount of H_2S, CO_2, and CH_4 contained in the biogas produced during the process.

Biogas production increased with increasing temperature, but the amount of CO_2 in the biogas also increases. The CO_2 and CH_4 values at mesophilic range were 31.43 ± 12.55 and 68.57 ± 12.56% respectively, while at thermopilic range; they were 37.75 ± 5.60 and 62.25 ± 5.50% respectively.

4. Conclusions

Degradation of organic content ie VS decomposition and COD removal increased with increasing temperature. In mesophilic range, VS decomposition and COD removal were 51.56 ± 8.30 and 79.82 ± 6.03%, respectively. Meanwhile at thermopilic range, VS decomposition and COD removal were 67.44 ± 3.59 and 79.16 ± 1.75%, respectively. Biogas production and its methane content also increases with the increasing of temperature, but CO_2 content also increases. Biogas production at mesophilic range was 31.77 ± 3.46 L/kg-ΔVS and methane content was 68.57 ± 12.56%. Meanwhile, biogas production at thermopilic range was 37.03 ± 5.16 L/kg-ΔVS and methane content was 62.25 ± 5.50%.

Acknowledgment

This research was supported by Hibah Penelitian Unggulan Perguruan Tinggi No: 017/SP2H/LT/DRPM/II/2016, date February 17, 2016.

References

[1] Wright T and Rahmanulloh A 2016 *Oilseeds and Products Annual Report 2016*

[2] Jaenicke J, Rieley J O, Mott C, Kimman P and Siegert F 2008 Determination of the amount of carbon stored in Indonesian peatlands *Geoderma* **147** 151–8

[3] Trisakti B, Manalu V, Irvan, Taslim and Turmuzi M 2015 Acidogenesis of Palm Oil Mill Effluent to Produce Biogas : Effect of Hydraulic Retention Time and pH *Procedia - Soc. Behav. Sci.* **195** 2466–74

[4] Jung J-Y, Lee S-M, Shin P-K and Chung Y-C 2000 Effect of pH on phase separated anaerobic digestion *Biotechnol. Bioprocess Eng.* **5** 456–9

[5] Veeken A, Kalyuzhnyi S, Scharff H and Hamelers B 2000 Effect of pH and VFA on Hydrolysis of Organic Solid Waste **6** 1076–81

[6] Polprasert C 2007 *Organic Waste Recycling - Technology and Management* (London: Iwa Publishing)

[7] Zinatizadeh A A L, Mohamed A R, Najafpour G D, Hasnain Isa M and Nasrollahzadeh H 2006 Kinetic evaluation of palm oil mill effluent digestion in a high rate up-flow anaerobic sludge fixed film bioreactor *Process Biochem.* **41** 1038–46

[8] Salihu, A. and Alam M z. 2012 Palm oil mill effluent: a waste or a raw material? *J. Appl. Sci. Res.* **8** 466–73

[9] Seadi T A, Rutz D, Prassl H, Köttner M, Finsterwalder T, Volk S and Janssen R 2008 *Biogas Handbook* (Esbjerg: University of Southern Denmark)

[10] Irvan, Trisakti B, Sosanty F and Tomiuchi Y 2016 Effect of Discontinuing Sodium Bicarbonate on Fermentation Process of Palm Oil Mill Effluent *Asian J. Chem.* **28** 377–80

[11] Olvera R and Lopez-lopez A 2009 Biogas Production from Anaerobic Treatment of Agro-Industrial Wastewater *Biogas* ed S Kumar (Shanghai, China: InTech China) pp 91–103

[12] Li W W and Yu H Q 2011 From wastewater to bioenergy and biochemicals via two-stage bioconversion processes: A future paradigm *Biotechnol. Adv.* **29** 972–82

[13] Blonskaja V, Menert A and Vilu R 2003 Use of two-stage anaerobic treatment for distillery waste *Adv. Environ. Res.* **7** 671–8

[14] Demirer G N and Othman M 2008 Two-Phase Thermophilic Acidification and Mesophilic Methanogenesis Anaerobic Digestion of Waste-Activated Sludge *Environ. Eng. Sci.* **25** 1291–300

[15] Wong Y-S, Teng T T, Ong S-A, Norhashimah M, Rafatullah M and Lee H-C 2013 Anaerobic Acidogenesis Biodegradation of Palm Oil Mill Effluent Using Suspended Closed Anaerobic Bioreactor (SCABR) at Mesophilic Temperature *Procedia Environ. Sci.* **18** 433–41
[16] Schunurer A and Jarvis A 2009 *Microbiological Handbook for Biogas Plants* (Malmö)
[17] Zupan G D and Grilc V 2012 Anaerobic Treatment and Biogas Production from Organic Waste *Management of Organic Waste* ed S Kumar (Shanghai: InTech China) pp 3–28
[18] Verma S 2002 *Anaerobic Digestion of Biodegradable Organics in Municipal Solid Wastes* (Columbia University)
[19] APHA 1999 *Standard Methods for the Examination of Water and Wastewater* (Washington, DC)
[20] Choorit W and Wisarnwan P 2007 Effect of temperature on the anaerobic digestion of palm oil mill effluent *Electron. J. Biotechnol.* **10** 376–85
[21] Bolzonella D, Fatone F, Di Fabio S and Cecchi F 2009 Mesophilic, thermophilic and temperature phased anaerobic digestion of waste activated sludge *Chem. Eng. Trans.* **17** 385–90
[22] Bouallagui H, Rachdi B, Gannoun H and Hamdi M 2009 Mesophilic and thermophilic anaerobic co-digestion of abattoir wastewater and fruit and vegetable waste in anaerobic sequencing batch reactors *Biodegradation* **20** 401–9
[23] O-Thong S, Boe K and Angelidaki I 2012 Thermophilic anaerobic co-digestion of oil palm empty fruit bunches with palm oil mill effluent for efficient biogas production *Appl. Energy* **93** 648–54
[24] Ventura J-R S, Lee J and Jahng D 2014 A Comparative Study On The Alternating Mesophilic And Thermophilic Two-Stage Anaerobic Digestion Of Food Waste *J. Environ. Sci.* **26** 1274–83

Performance Evaluation on Otto Engine Generator Using Gasoline and Biogas from Palm Oil Mill Effluent

Irvan[1,3*], B Trisakti[1,3], T Husaini[1,3], A Sitio[2,3] and TB Sitorus[2,3]

[1]Chemical Engineering Department, University of Sumatera Utara, Medan, 20155, Indonesia
[2]Mechanical Engineering Department, University of Sumatera Utara, Medan, 20155, Indonesia
[3]Sustainable Energy and Biomaterial Center of Excellence, University of Sumatera Utara, Medan, 20155, Indonesia

E-mail: *i_v_a_n_mz@yahoo.com

Abstract. Biogas is a flammable gas produced from the fermentation of organic materials by anaerobic bacteria originating from household waste manure and organic waste including palm oil mill effluent (POME). POME is mainly discharged from the sterilization unit of palm oil processing into crude palm oil. This study utilized biogas produced from liquid waste palm oil for use as fuel in the Otto engine generator 4 – stroke, type STARKE GFH1900LX with a peak power of 1.3 kW, 1.0 kW average power, bore 55 mm, stroke 40 mm, V_d 95 × 10^{-6} m^3, V_c 10 × 10^{-6} m^3, compression ratio of 10.5 : 1 , and the number of cylinders = 1. The objective of this study is to evaluate the performance of Otto engine generator fueled with biogas that generated from POME, then comparing its performance fueled by gasoline. The performance included power, torque, specific fuel consumption, thermal efficiency, and the air-fuel ratio. Experiment was conducted by using a variation of the lamp load of 100, 200, 300, 400, and 500 W. The results revealed that the use of biogas as fuel decreased in power, torque, brake thermal efficiency, and air fuel ratio (AFR), while there is an increasing of value specific fuel consumption (SFC).

1. Introduction

Fossil fuels mentioned as the most important energy sources in the world are depleting day by day. In addition, environmental issues have become dilemma due to more fossil fuel are used. Many researchers especially in Indonesia have developed alternative fuel in order to overcome the above problems such as biogas, bioethanol, biodiesel etc. Biogas is a clean and renewable resource that can be used as fuel or other purposes. It discharges less harmfully pollutant to the atmosphere than the burning of fossil fuel. The main constituents in biogas are methane (CH_4), carbon dioxide (CO_2), and hydrogen sulfide (H_2S) [1].

Biogas can be produced using a wide range of feedstock include municipal wastewater, diary manure, food processing wastewater and agricultural waste [2]. Indonesia as tropical country has numerous biogas resources generated from agricultural production especially in palm oil industries such empty fruit bunches, fibre, shell and palm oil mill effluent (POME).

POME, which is brownish, viscous and acidic liquid waste (pH ranges between 4 and 5), cannot be discharged directly to the environment without prior treatment. POME has high chemical oxygen demand (COD), biological oxygen demand (BOD) and suspended solid by approximately 90,000,

30,000 and 25,000 mg/l, respectively and generally was discharged at high temperature (80-90°C) [3, 4, 5]. Table 1 summarizes the properties of palm oil mill effluent [6].

Table 1. Properties of fresh palm oil mill effluent

Parameters*	Values
pH	4 – 5
Total Solid (TS)	40,500 – 72,000
Volatile Solid (VS)	34,000 – 49,300
COD	44,300 – 102,000
BOD	25,000 – 65,714
Kj-N	560 – 1,200
Fat	1,700 – 9,200

** Unit for all parameter is mg/L except pH*

Typically, 1 ton of fresh palm oil fruit bunch (FFB) generates approximately 0.1 ton or 0.65 m^3 of POME [7]. Thus, there is a great potential for POME to be utilized as the fuel to produce renewable energy. Moreover, besides reducing pollutant, a by-product from the production of biogas, can be used as liquid fertilizer and pesticide.

Some researchers have performed various tests in order to study the performance of engine fueled by biogas [8, 9, 10, 11]. They focused mainly on some parameters such as: compression ratio, the torque and power. The results of their study lead to the conclusion that biogas fueled engines have a good prospect in future.

Irvan et al in 2016 conducted POME conversion into biogas at laboratory scale and pilot scale using anaerobic microbes at temperature 55°C (thermophilic) [12]. This research aimed to utilize biogas produced from pilot scale as fuel for Otto engine generator. Then the performance included power, torque, specific fuel consumption, thermal efficiency, the air-fuel ratio, exhaust emissions and combustion in spark plugs were evaluated and then comparing it using gasoline.

2. Materials and Methods

2.1. Production of Biogas

Biogas was obtained by anaerobic fermentation of POME in the digester at working volume of 3.7 m^3. Schematic drawing of pilot plant is shown in figure 1.

Fresh POME was filled into the continuous stirred feed tank, here POME was heated up to 70°C. Hot POME was then fed to the mixing tank by a centrifugal pump every two hours. The amount of daily feeding POME became 616 liters per day. Hydraulic retention time (HRT) of digester was maintained at 6 days. Hot POME and return sludge were mixed in the mixing tank for approximately five hours and still maintained at 70°C. Then this mixed waste was pumped by screw pump into an anaerobic digester every two hours. The digested slurry was overflowed from the digester to the gravity thickener where the digested slurry was concentrated and returned to the mixing tank. The effluent was discharged to the outside from the top surface of the gravity thickener. Production of biogas was measured by using a wet gas meter (SHINAGAWA, Model W-NK-0.5B). Concentrations of H_2S and CO_2 in the biogas were measured by using a suction gas injector (GASTEC, type GV-100S) and inspection tube (GASTEC, 25 ~ 1600 ppm). Characteristics of biogas produced from the pilot plant are presented in Table 2.

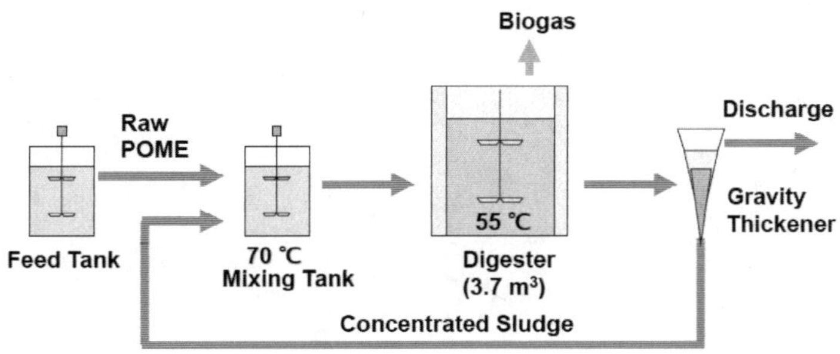

Figure 1. Schematic drawing of pilot plant

Table 2. Characteristics of biogas from the pilot plant

Parameter	Unit	Values
Gas generation	NL/gr VS*	0.68
Methane	%	60
CO_2	%	30
H_2S	ppm	40

2.2. Otto Engine Generator Fed with Biogas

The engine used in this study was 4 – stroke, type STARKE GFH1900LX with a peak power of 1.3 kW, 1.0 kW average power, bore 55 mm, stroke 40 mm, V_d 95 × 10^{-6} m^3, V_c 10 × 10^{-6} m^3, compression ratio of 10.5 : 1, and the number of cylinders = 1. Although, the objective of this work is to determine the performance of the Otto engine fed with biogas, it also used gasoline as a reference parameter.

Engine performance tests were conducted by varying the electrical power above 100 kW. When the electrical power was given then the electrical voltage and current were measured by using a multimeter. Meanwhile, a tachometer was used to determine the rotation number of the engine. By using gasoline, performance test was carried out for 5 minutes each load variation, then amount of fuel mass reduced in the tank were measured. Using biogas, performance test was performed by flowing it from the 100 L storage tank at 6 bar to the modified carburetor. The engine was turned on until the content of storage tank became empty and the time was measured by using stopwatch.

3. Results And Discussions

3.1. Engine power

Experiments in order to know performance test of the engine by using gasoline and biogas were conducted by varying the number of lamps used. Figure 2 shows the engine power curves by using gasoline and biogas.

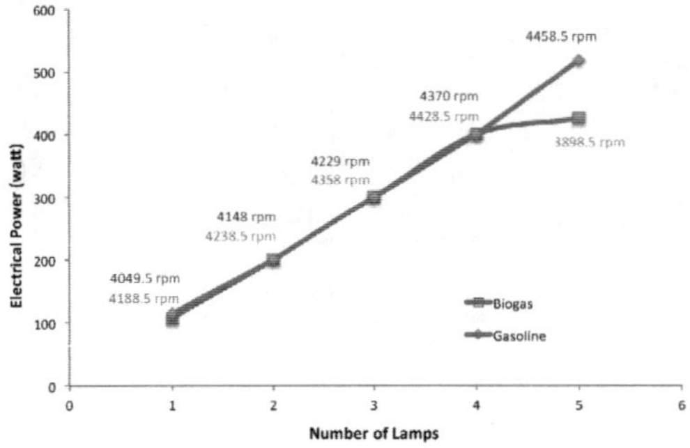

Figure 2. The engine power curves by using gasoline and biogas

By using gasoline, if number of lamps were added then rotation and power increased. A load of 1 lamp, the engine generated rotation by 4049.5 rpm with output power of 114.1 Watt, and the load of 5 lamps, rotation and power were 4458.5 rpm and 517.6 Watt respectively. Biogas used as fuel with one lamp, the engine generated a power of 107.25 Watt at 4188.5 rpm engine speed, and the load of 5 lamps were 424.7 Watt at 4398.0 rpm.

The decreasing of rotation number and output power were caused by the CO_2 content in the biogas which reaches 30 %. It affected to the combustion of biogas inside the engine which became slower and the heat value of biogas became lower. At last, biogas were accumulated inside the heat chamber, caused combustion process became inefficiency. This was indicated by the decreasing of rotation, followed with a drop in generated power generator engine.

3.2. Torque

Similar to engine power, using gasoline, if number of lamps were added then the greater the torque produced. A load of 1 lamp, the engine produced 0.269 Nm torque at 4049.5 rpm, and the load of 5 lamps, 1.019 Nm torque was obtained at 4458.5 rpm. Figure 3 shows the torque curves by using gasoline and biogas.

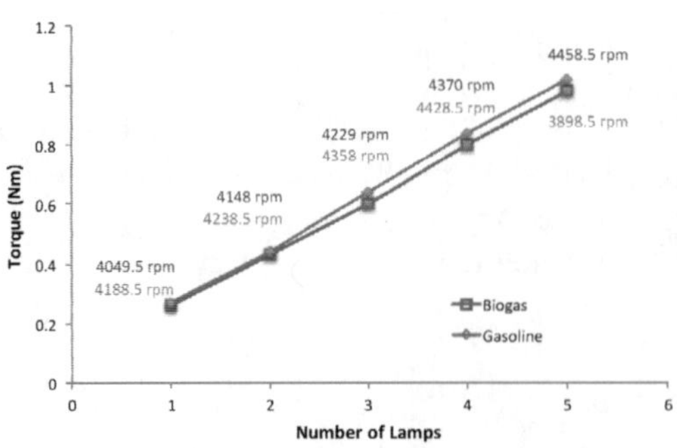

Figure 3. The torque curves by using gasoline and biogas

IOP Conf. Series: Materials Science and Engineering **206** (2017) 012028 doi:10.1088/1757-899X/206/1/012028

From Figure 3 we can see that the torque generated by engine using biogas, load of 1 lamp up to 4 lamps are not so much different, but when the generators were loaded with 5 lamps, the torque was lower than the gasoline engine. The decreasing of torques was caused by engine rotation which also decreased.

3.3. Specific Fuel Consumption

Engine performance may be described in several ways, one of the useful parameters is specific fuel consumption (SFC). It is defined as the weight of the fuel burned per unit time, per unit thrust. Generally, if the electrical load is increased then SFC value becomes smaller. Figure 4 shows the SFC curves by using gasoline and biogas. As shown on figure 4, engine with biogas has SFC value greater than the engine with gasoline. This is because the rate of fuel of biogas is greater than the gasoline.

The greatest value of SFC, indicated by engine with biogas at the load of 1 lamp generating power about 107.25 Watt at 4188.5 rpm, was around 16037.30 g/kWh. While, the lowest value of SFC was 1437.31 g/kWh, at the load of 5 lamp generating power about 517.66 Watt at 4458.5 rpm.

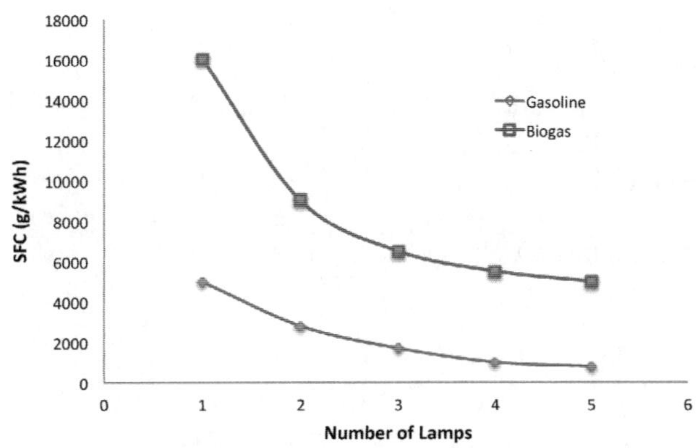

Figure 4. The SFC curves by using gasoline and biogas

3.4. Brake Thermal Efficiency

Brake thermal efficiency (BTE) can be defined as the ratio of the heat equivalent of the brake output to the heat supplied to the engine. For engine by using gasoline or biogas, if the number of lamp is increased, then the BTE value also increased. Figure 5 shows the BTE curves by using gasoline and biogas. As shown on figure 5 that gasoline engine has BTE value higher than biogas engine. This is because the fuel rate of biogas is greater than gasoline.

IOP Conf. Series: Materials Science and Engineering **206** (2017) 012028 doi:10.1088/1757-899X/206/1/012028

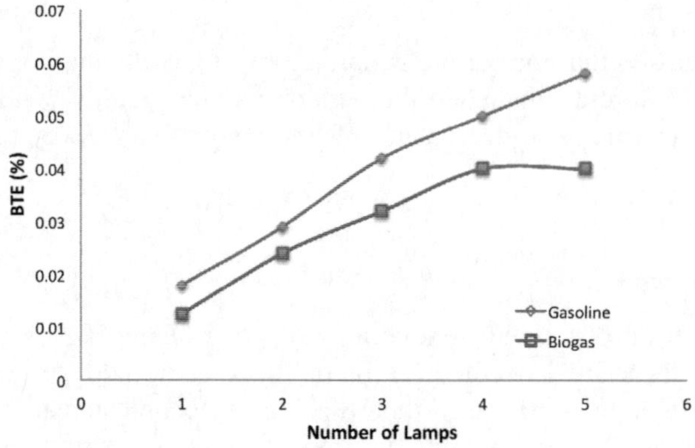

Figure 5. The BTE curves by using gasoline and biogas

Figure 5 shows the highest thermal efficiency is 5.87% by using gasoline at a load of 5 lamps with output power 517.64 at 4458.5 rpm. While, the lowest thermal efficiency was 1.31%, at a load of 1 lamp generating power about 107.25 Watt at 4188.5 rpm.

3.5. Air Fuel Ratio

Air fuel ratio (AFR) is the mass ratio of air to fuel present in a combustion process such as in an internal combustion engine. The AFR is an important measure for anti-pollution and performance-tuning reasons. In general, if the electrical load is increased then AFR value becomes smaller. Figure 6 shows the AFR curves by using gasoline and biogas. As shown on figure 6, engine with gasoline has AFR value greater than the engine with biogas.

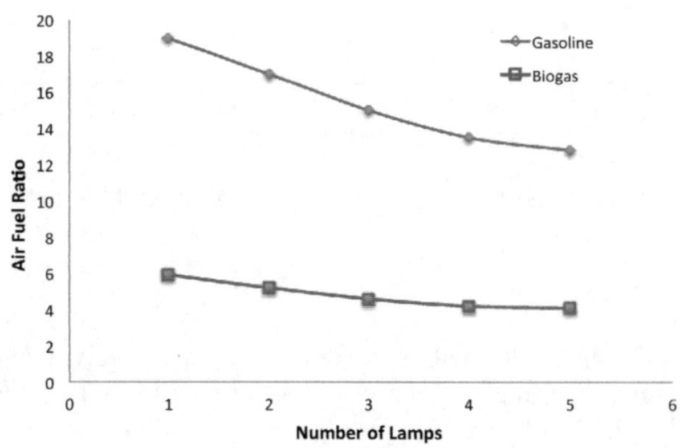

Figure 6. The AFR curves by using gasoline and biogas

Figure 6 shows the highest air fuel ratio is 18.68 by using gasoline at a load of 1 lamp with output power 114.1 at 4049.5 rpm. While, the lowest thermal efficiency was 4.5, at a load of 5 lamps generating power about 424.74 Watt at 3898.5 rpm.

4. Conclusion

Otto engine generator by using biogas as fuel could generate maximum power and torque of 424.74Watt and 1.04 Nm, respectively. These power and torque are approximately 42% and 49% of the power and torque that can be generated from the same engine by using gasoline. Because of biogas fuel rate is greater than the gasoline, then specific fuel consumption of biogas become larger and brake thermal efficiency of biogas is smaller than gasoline. In the combustion process, biogas engine required less air than gasoline engine.

Acknowledgment

This research was supported by Hibah Penelitian Unggulan Perguruan Tinggi No: 017/SP2H/LT/DRPM/II/2016, date February 17th, 2016.

References

[1] Hosseini SE and Wahid MA 2013 Feasibility study of biogas production and utilization as a source of renewable energy in Malaysia, *Renew. Sustainable Energy Rev.* **19** 454-462

[2] Dioha IJ, Ikeme CH, Nafi'u T, Soba NI and Yusuf MBS 2013 Effect of carbon to nitrogen ratio on biogas production, *Int. Res. J. of Natural Sci.* **1** 1-10

[3] Mustapha S, Ashhuby B, Rashid M and Azni I 2003 Start-up strategy of a thermophilic upflow anaerobic filter for treating palm oil mill effluent, *Process Saf. Environ. Prot.* **81** 262-266

[4] Najafpour GD, Zinatizadeh AAL, Mohamed AR, Hasnain MI and Nasrollahzadeh H 2006 High-rate anaerobic digestion of palm oil mill effluent in an upflow anaerobic sludge-fixed film bioreactor, *Process Biochem.* **41** 370-379

[5] Meesap K, Boonapatcharoen N, Techkarnjanaruk S and Chaiprasert P 2012 Microbial communities and their performances in anaerobic hybrid sludge bed-fixed film reactor for treatment of palm oil mill effluent under various organic pollutant concentrations, *J. Biomed. Biotechnol.* **2012** 1-11

[6] Ji CM, Eong PP, Ti TB, Seng CE and Ling CK 2013 Biogas from palm oil mill effluent (POME): Opportunities and challenges from Malaysia's perspective, *Renew. Sustainable Energy Rev.* **26** 717-726

[7] King LS and Yu LC 2013 A retrofitted palm oil mill effluent treatment system for tapping biogas, *Eur. Int. J. Sci. Tech.* **2** 106-114

[8] Rossetto C, de Souza, SNM, Santo RF, de Souza J, Klaus OL 2013 Performance of an Otto cycle engine using biogas as fuel, *Afr. J. Agric. Res.* **8** 5607-5610

[9] Huang J and Crookes RJ 1998 Assessment of simulated biogas as a fuel for the spark ignition engine, *Fuel* **77** 1793-1801

[10] Suzuki ABP, Fernandes DM, Faria RAP, Vidal TCM 2011 Use of biogas in internal combustion engines, *Braz. J. Appl. Tech. Agri. Sci.* **4** 221-237

IOP Conf. Series: Materials Science and Engineering **206** (2017) 012028 doi:10.1088/1757-899X/206/1/012028

[11] Munoz M, Moreno F, Morea-Roy J, Ruiz J, Arauzo J 2000 Low heating value gas on spark ignition engines, *Biomass Bioenergy* **18** 431-439

[12] Irvan, Trisakti B, Sosanty F and Tomiuchi Y 2016 Effect of discontinuing sodium bicarbonate on fermentation process of palm oil mill effluent, *Asian J. Chem.* **28** 377-380

IOP Conf. Series: Materials Science and Engineering **206** (2017) 012029 doi:10.1088/1757-899X/206/1/012029

On predicting the onset of transient convection in porous media saturated with Non-Newtonian liquid

K K Tan[1], S Y Pua[2] and A Yang[3]
[1]ELM Graduate School, HELP University, 50490, Kuala Lumpur, Malaysia
[2]Department of Chemical Engineering, University Putra, 43400 Serdang, Selangor, Malaysia
[3]Department of Mechanical Engineering, Hertfordshire University, Hatfield AL10 9EU, UK

E-mail: tan.ka.kheng@help.edu.my, kakhegtan@gmail.com

Abstract. The onset of transient convection in non-Newtonian liquid immersing porous media was simulated using a Computational Fluid Dynamics (CFD) package for the thermal boundary condition of Fixed Surface Temperature (FST). Most of the simulated values of stability criteria were found to be in good agreement with the predicted and theoretical values of transient critical Rayleigh number for non-Newtonian liquid defined by Tan and Thorpe (1992) for power-law fluids. The critical transient Rayleigh numbers for convection in porous media were found to be in good agreement with theoretical values by using apparent viscosity μapp at zero shear. The critical time and critical depth for transient heat conduction were then determined accurately that

1. Introduction

Heat transfer in porous media has been the subject of many studies due to the increasing interest in chemical catalysis, building insulation, heat exchangers, petroleum reservoirs, geothermal operations, and so on. In recent years, heat transfer in porous media saturated by non-Newtonian liquids has received considerable attention in the literature, notably for power-law fluids. An illustrative example is found in oil reservoir engineering in connection with the production of heavy crude oils which are power law fluids [20], [12] and [6]. This process involves the cyclic injection of steam into the well for the purpose of increasing the temperature of the oil reservoir. The efficiency of this hot process can be increased by studying the combined effects of convective heat transfer and convective flow in a power-law fluid filled porous media. The apparent viscosity for a power-law fluid caused by a shear rate is commonly expressed as

$$\mu_{app} = K\gamma^{n-1} \tag{1}$$

when n is less then unity for shear thinning or pseudo-plastic fluids, while for shear thickening liquids, n is greater than unity, and K is the flow consistency index [2]. The onset of transient convection caused by unsteady-state conduction in an adverse density profile under gravity has been successfully predicted by the transient instability theory of Tan and Thorpe [13], [14], [15] and [16]. Tan and Thorpe [14] derived the maximum transient Rayleigh Number for Newtonian liquid predicted the onset of convection induced by transient heat conduction in a deep fluid. The new theory has successfully predicted the maximum transient Rayleigh number, the maximum penetration depth z_{max}, the critical time and critical size of convective plumes for fixed-surface temperature (FST) boundary condition. The theory has been successfully applied to the onset of transient convection in porous media saturated by Newtonian liquids [17], whose equations can be applied to non-Newtonian liquids by assuming an apparent viscosity for zero-shear [13]. The theoretical value of onset of convection between two plates of 39.5 was first calculated by Horton and Rogers [4]. The relevant equations for the onset of transient convection in porous media saturated with non-Newtonian liquids under FST boundary condition summarized as follows.

$$Ra_{max} = 39.5 = \frac{0.83 K_e g \alpha \Delta T_c}{\nu_{app}} \sqrt{\frac{t_c}{\kappa_m}} \tag{2}$$

$$z_{max} = 2\sqrt{\kappa_m t_c} \tag{3}$$

$$\lambda_c = 4\sqrt{\kappa_m t_c} \tag{4}$$

$$t_c = 2262 \left(\frac{\nu_{app}\sqrt{\kappa_m}}{K_e g \alpha \Delta T_c} \right)^2 \tag{5}$$

The thermal diffusivity of porous media is defined by convention for heat conduction or energy diffusion, which is before the onset of convection [5]. Tan et al. [17] pointed out the lack of a rational scientific basis for the introduction of the modified thermal diffusivity, which arbitrarily altered the experimental values of Rayleigh number substantially. Nield [9] erroneously believed that the thermal energy equation for flowing fluid should be the only basis of the definition of thermal diffusivity for porous media, which is opposite to the conventional definition of the thermal diffusivity based on a stationary fluid. The thermal diffusivity of any material is defined by convention for heat conduction or energy diffusion, which is used in the linear stability analysis for determining a critical Rayleigh number for predicting the onset of convection from the stationary state. Therefore, the thermal energy equation for a flowing fluid is not the basis for defining thermal diffusivity and Rayleigh number for porous media [20].

The simulations of the onset of transient convection in deep fluid and porous media have been successfully conducted by Tan et al. [19] and Tan et al. [17] respectively, with simple Boussinesq's approximation. In this study, we employed a commercial Computational Fluid Dynamics (CFD) package FLUENT to simulate accurately the onset of convection caused by unsteady-state heat conduction and the transport phenomena in porous media saturated with non-Newtonian liquids. The results of simulations will be compared with theoretical predictions.

2. Computational methods

2.1. The simulation model

In this study, the onset of convection in porous medium saturated by non-Newtonian liquid (CMC-5) was investigated. The deep layer was bounded above and below by two rigid impermeable conducting boundaries. 2-D unsteady-state simulations were carried out for the Fixed Surface Temperature (FST) boundary condition. The top of the deep layer was cooled instantaneously under unsteady-state, while its bottom surface was maintained at FST. The non-Newtonian liquid filling the porous medium initially with higher temperature T_0. In the simulation the top surface was suddenly cooled down by a temperature difference of $\Delta T_s = (T_0 - T_s)$.

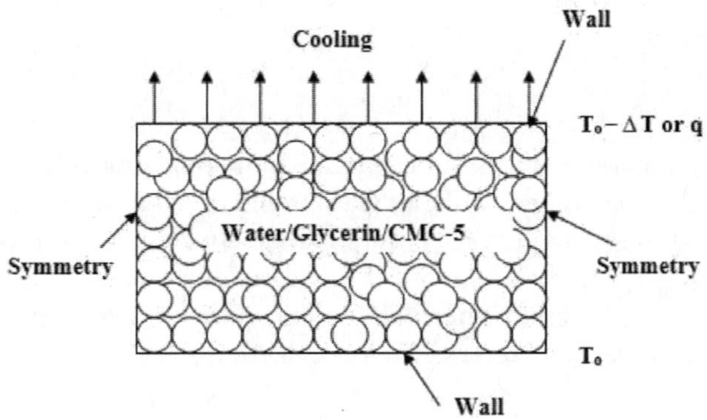

Figure 1. Schematic of the boundary layer setting of the domains used in the simulation.

The physical properties of the porous medium applied in this simulation were based on the glass beads and CMC-5. The density, thermal conductivity and specific heat capacity of the glass beads was assumed to be a constant. However, density, thermal conductivity and specific heat capacity of the CMC-5 were based on the mean temperature of the top and bottom surface temperature. Since CMC-5 is non-Newtonian fluid thus its viscosity depends strongly on shear rate. FLUENT will calculate the shear rate during the flow of fluid and compute the apparent viscosity from the power law. The physical properties of the porous media were presented in Table 1. The permeability of different sizes of glass beads were obtained from the Kozeny-Carmen relation [3].

Table 1. Properties of porous media consisted of glass beads in case1 in CHF boundary condition.

Properties	Unit	Glass bead (27°C)	CMC-5 (45°C)	Porous media
Density, ρ	kg/m^3	2500	988.7	1978.6
Diameter, d_s	m	n/a	n/a	0.05
Porosity, ϕ		n/a	n/a	0.345
viscosity, μ	Pa.s	n/a	3.5	3.5
kinematic viscosity, ν	m^2/s	n/a	3.54×10^{-3}	3.54×10^{-3}
Specific heat, c_p	J/kg.K	750	4217	n/a
Expansion coeff., α	K^{-1}	n/a	5.51×10^{-4}	5.51×10^{-4}
Permeability, K_e	m^2	n/a	n/a	1.385×10^{-6}
Thermal conductivity, k_m	W/m.K	1.4	6.15×10^{-1}	1.1293
Thermal diffusivity, κ_m	m^2/s	7.467×10^{-7}	1.476×10^{-7}	4.235×10^{-7}
Darcy number, Da		n/a	n/a	4.222×10^{-6}

$$K_e = \frac{d_S{}^2 \phi^3}{172.8(1-\phi)^2} \quad (6)$$

The Darcy number *Da [11]* should be infinitesimally small with less than 10^{-5} in porous media condition. This mean critical depth layer has to exceed the glass bead size before convection occurring.

$$Da = \frac{Ke}{Z_{max}^{2}} \tag{7}$$

2.2. Setup for Simulation

The design of the computation cells is vital to the numerical convergence of the solution. The boundary layer occurring at the cooled walls temperature and velocities have their largest variation in this region. Therefore, the mesh should be optimized to adequately solve this region to provide more accurate results. For unsteady-state simulation, the critical time and size of plumes predicted by the theory of Tan et al. [17] can be used to determine the computation cell size, their detailed set-up were adopted in this study. The readers can refer to their paper for details.

3. Results and discussions

The CFD simulations were first conducted for water, glycerin and non-Newtonian liquids to ascertain the validity of zero-shear model for predicting the onset of transient convection before we proceed to the case of porous media.

3.1. Unsteady-state convection in porous media saturated by Newtonian liquids

Simulations were first carried out to model the onset of convection in porous media saturated by Newtonian liquids. The top cooling of water and glycerin under FST condition bounded by two rigid surfaces are simulated with a surface temperature difference, ΔT of 10ºC and 27ºC. The permeability of water in FST condition is 1.99 x 10^{-8} m^2, while glycerin is saturated for permeability 2.22 x 10^{-7} m^2 in both with porosity of 0.345.

The simulated critical times and maximum Rayleigh numbers at the onset of convection are found to be accurate when compared to those values calculated from equations (5) and (2) of Tan et al.[17], as shown in table 2. The close agreement of theoretical and simulated critical times validated the setup of the simulation. Consequently, more simulations were conducted for porous media saturated with non-Newtonian liquids.

3.2. The Formation and Development of the Thermal Plume

The formation and development of the fingering thermal plumes in porous media saturated by non-Newtonian were clearly seen from the temperature contour and velocity vector profile at various time steps, as shown in figure 2. They were the mirror image of rising fingering plumes generated by bottom heating observed by Tan, Sam, and Jamaludin [17]. The various stages of formation and development of the transient falling fingering plume were described table 3. At the onset of convection, the non-Newtonian liquid displayed the same development with Newtonian liquid in porous media as table 3. The main difference between the plumes in porous media and continuous fluid is the shape of the plumes, fingering plumes were formed in the porous media (figure 3) and mushroom shaped plumes for normal liquids. The high viscosity and porous media contributed to the long critical times of convection. The small bead size also extended the long critical times owing to the low permeability.

Table 2. Comparison between the critical time and critical Rayleigh number for Newtonian and non-Newtonian liquid at FST boundary condition saturate porous media.

liquid	Bead size, d(cm)	ΔT (ºC)	t_c (s) Predicted	t_c (s) Simulation	Ra_c Theory	Ra_c Simulation	Deviation (%)
Water	0.6	10	2899	2500	39.5	36.7	7.2
Glycerin	2	27	151000	110000	39.5	33.7	14.6
CMC-5	5	27	246000	220000	39.5	37.4	5.4

Table 3. The stages of plumes formation of non-fingering plumes in non-Newtonian liquids in porous media induced by transient top cooling.

Stage	Description
I	(a)Surface contraction due to cooling and draining down of cooled water from the top layers. Thermal boundary layer starts to form and thickening but no fluid motion is observed, figure 2a. (b)The mechanism of heat transfer is purely heat conduction.
II	(a) The local temperature gradient exceeds the point of stability, thus begins the onset of convection, velocity of fluid increases rapidly, figure 2b. The fluid starts to accelerate. (b) Formation of the tail side of the plumes, figure 2c.
III	(a) Filamentous plumes extend deep in bulk fluid like 'finger', figure 2d. (b)The scale of plumes size and the amplitude of plumes are dependent on the rate of the heat conduction. (c) Thermal plumes continuous moving downwards, and begin to coalesce with adjacent plumes. The bulk fluid. Occasionally growing plumes (d)Thermal plumes dissipate heat to the surrounding and disappear.

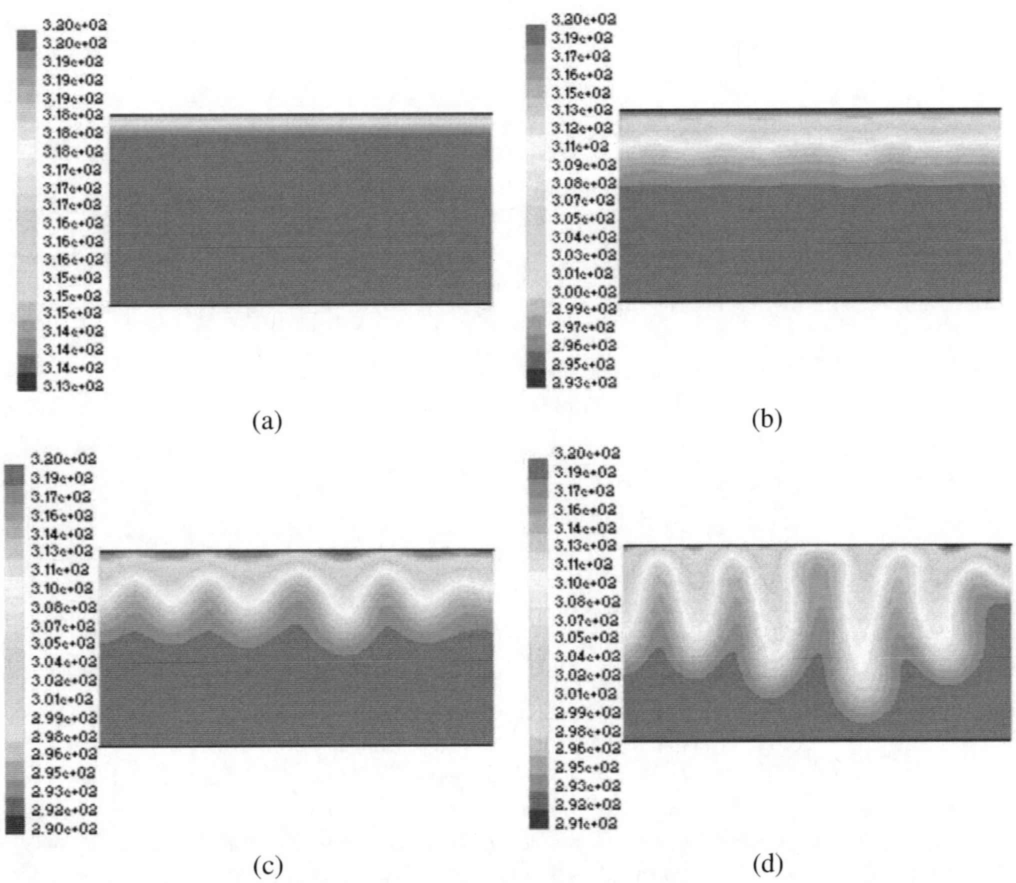

Figure 2. The development of thermal plumes for 5-cm glass beads saturated with CMC-5 for FST top cooling ($\Delta T = 27^{o}C$) with $t_c = 220,000$ s.

3.3. Critical Times and Critical Rayleigh Number

The critical times of onset of transient convection in the simulations were determined from the temperature contour and velocity magnitude (figure 3). They are compared to the values predicted from equation (5) for various temperature difference T, as shown in table 4 and figure 5, which show very good agreement. The critical times of onset of convection is inversely proportional to the temperature difference and permeability.

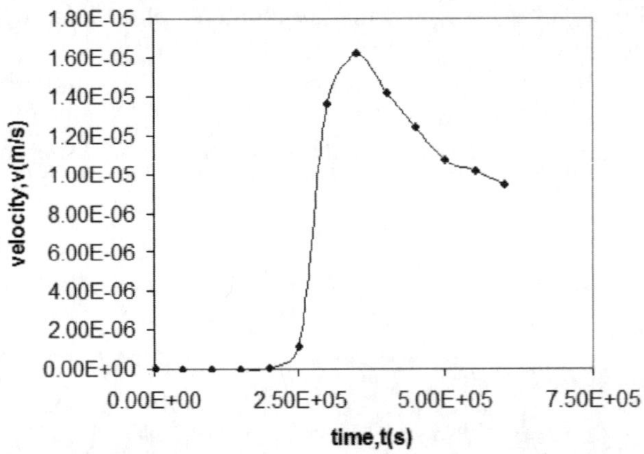

Figure 3. Maximum magnitude of velocity at various times for top cooling of non-Newtonian in FST boundary condition. (case1, 5cm glass beads, $t_c = 220,000$ s).

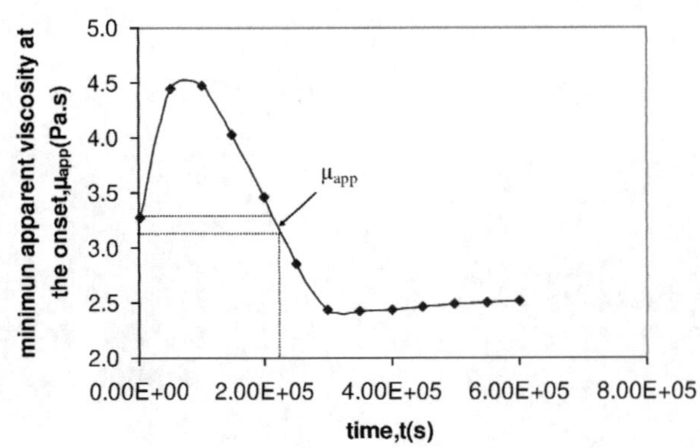

Figure 4. Minimum viscosity versus time for 5-cm glass beads saturated with CMC-5 FST top cooling at, $\Delta T = 27°C$, $t_c = 220,000$ s. The average viscosity at initial and critical time is the apparent viscosity.

The maximum transient Rayleigh number is calculated from the critical time and the corresponding apparent viscosity (figure 4) in simulation for top cooling of CMC-5. Its average value is 38.4, which is close to the theoretical value 39.5 for FST (solid-solid) boundary condition.

Table 4. Comparison of the critical times and Rayleigh number from the simulations and redictions for non-Newtonian liquids in porous media for FST boundary condition.

bead size, d (cm)	ΔT, (^{o}C)	Permeability Ke ($\times 10^{-6}$ m^2)	t_c (s) Predicted	t_c (s) Simulation	Ra Simulation	z_{max} (m) Predicted	z_{max} (m) Simulation
5	27	1.38	2.5E+05	2.2E+05	37.4	0.7	0.6
5	24	1.38	3.4E+05	2.9E+05	36.4	0.8	0.7
5	21	1.38	4.6E+05	4.0E+05	36.9	0.9	0.8
5	18	1.38	6.4E+05	6.2E+05	38.9	1.0	1.0
5.5	27	1.68	1.6E+05	1.4E+05	37.2	0.5	0.5
5.5	24	1.68	2.1E+05	2.0E+05	38.3	0.6	0.6
5.5	21	1.68	2.9E+05	2.7E+05	37.8	0.7	0.7
5.5	18	1.68	4.0E+05	4.2E+05	40.5	0.8	0.8
6	27	1.99	9.1E+04	1.0E+05	41.4	0.4	0.4
6	24	1.99	1.3E+05	1.2E+05	37.7	0.5	0.5
6	21	1.99	1.8E+05	2.0E+05	41.2	0.6	0.6
6	18	1.99	2.7E+05	2.4E+05	37.6	0.7	0.6

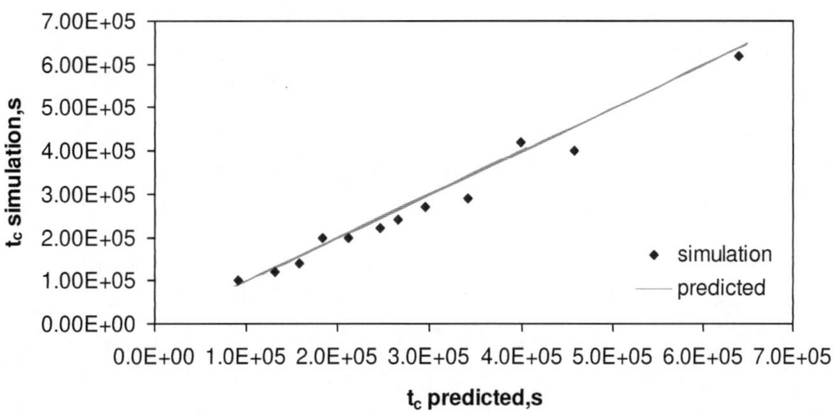

Figure 5. The critical times from the simulations and predictions for non-Newtonian liquids in porous media.

IOP Conf. Series: Materials Science and Engineering **206** (2017) 012029 doi:10.1088/1757-899X/206/1/012029

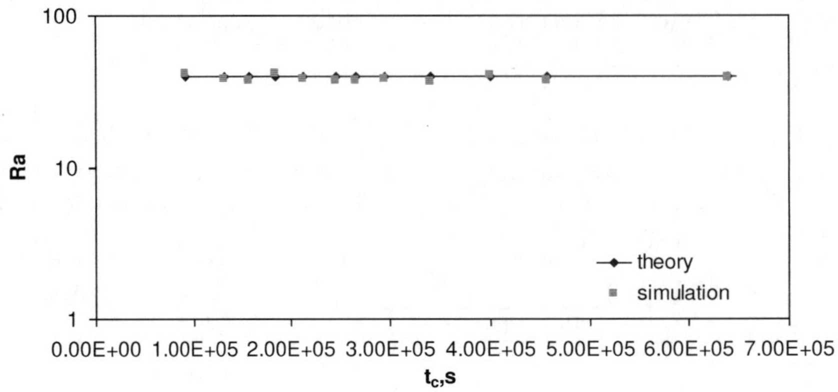

Figure 6. Maximum transient Rayleigh number for various simulations run for FST boundary condition.

3.4. Critical Thermal Depth

The critical depth of penetration just before the onset of convection in simulations were determined by measuring the distance from the top surface. They compare well with predictions by equation (3) as shown in table 4 and figure 7. The results of simulations showed that the domain size for the formation of fingering plumes are sufficient for them to form and develop freely without hindrance.

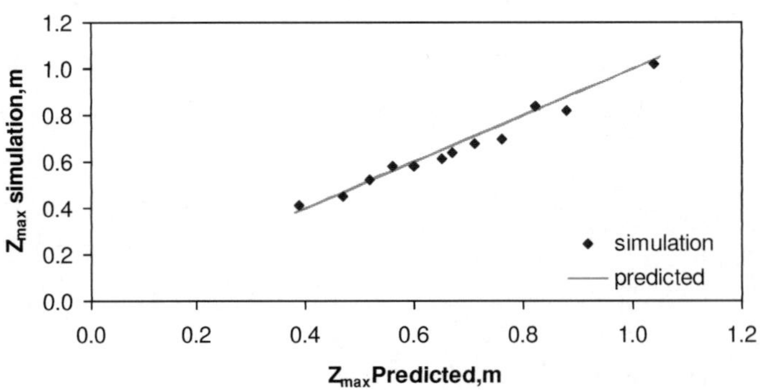

Figure 7. Comparison of the critical thermal depth, Z_{max} from the simulations and predicted for top cooling of FST boundary condition

4. Conclusions

The CFD simulations of the onset of transient convection caused by top cooling of non-Newtonian liquids in porous media have been successfully conducted. They provide good verification of the key stability parameters predicted by transient instability theory advanced by Tan and Thorpe [15] and [16].

The critical times of simulations were accurately predicted with equation (5), which allow the computation of the maximum transient Rayleigh number. The average value of 38.4 was close to that of theory 39.5 [4].

The maximum critical thermal depths were predicted accurately for various bead diameters and permeability.

References

[1] Carslaw H S and Jaeger J C 1973 *Conduction of Heats in Solid* Chap. II 62-75 (Oxford University Press)

[2] Chi T and Tsuei H S 1969 Thermal instability of a horizontal layer of non-Newtonian fluid heated from below *Int. J. Heat Mass Transfer* **12**(9) 1173-1178

[3] Combarnous M A 1975 Hydrothermal convection in saturated porous media *Adv. Hydrosci.* **10** 231-307

[4] Horton C W and Rogers Jr F T 1945 Convection currents in a porous medium *J. App. Phys.* **16**(6) 367-370

[5] Kaviany M 1995 *Principles of Heat Transfer in Porous Media* (2nd ed.), Springer, New York.

[6] Lamsaadi M, Naimi M and Hasnaoui M 2005 Natural convection of non-Newtonian power law fluids in a shallow horizontal rectangular cavity uniformly heated from below *Heat Mass Transfer* **41**(3) 239-249

[7] Lapwood E R 1948 Convection of a fluid in a porous medium *J. Fluid Mech.* **44** 508-521

[8] Nield D A and Bejan A 1999 Convection in Porous Media, 2nd edn., Springger-Verlag, New York

[9] Nield D A 2004 Letter to the Editor of *Int. J Heat Mass Transfer* **47** 641-643

[10] Rayleigh L 1916 LIX. On convection currents in a horizontal layer of fluid, when the higher temperature is on the under-side *The London, Edinburgh, and Dublin Philosophical Mag. J. Sci.* **32**(192) 529-546

[11] Shattuck M D, Behringer, R P, Johnson G A and Georgiadis J G 1997 Convection and flow in porous media. Part 1. Visualization by magnetic resonance imaging *J. Fluid Mech.* ***332*** 215-245

[12] Shenoy A V 1994 Non-Newtonian fluid heat transfer in porous media *Adv. Heat Transfer* **24** 101-190

[13] Tan . K and Thorpe R B 1992 Gas diffusion into viscous and non-Newtonian liquids *Chem. Eng. Sci.* **47**(13-14) 3565-3572

[14] Tan K K and Thorpe R B 1996 The onset of convection caused by buoyancy during transient heat conduction in deep fluids *Chem. Eng. Sci.* **51**(17) 4127-4136

[15] Tan K K and Thorpe R B 1999a The onset of convection driven by buoyancy effects caused by various modes of transient heat conduction: Part I. Transient Rayleigh numbers *Chem. Eng. Sci.* **54**(2) 225-238

[16] Tan K K and Thorpe R B 1999b The onset of convection driven by buoyancy caused by various modes of transient heat conduction: Part II. The sizes of plumes *Chem. Eng. Sci.* **54**(2) 239-244.

[17] Tan K K, Sam T and Jamaludin H 2003 The onset of transient convection in bottom heated porous media *Int. J. Heat Mass Transfer* **46**(15) 2857-2873.

[18] Tan K K, Tan K Y and Tan Y W 2010 3D Time-Dependent Simulations of Onset of Transient

Convection in Evaporative Cooling of Liquids *Eng. Appl. Comput. Fluid Mech.* **4**(1) 106-115
[19] Tan K K 2016 Thermal diffusivity and Rayleigh number for Porous Media. https://www.researchgate.net/publication/307378081_Thermal_diffusivity_and_Rayleigh_number_for_Porous_Media
[20] Wang C and Tu C 1989 Boundary-layer flow and heat transfer of non-Newtonian fluids in porous media *Int. J. Heat Fluid Flow* **10**(2) 160-165

Analysis of solar water heater with parabolic dish concentrator and conical absorber

G Rajamohan[1*], P Kumar[2], M Anwar[3] and T Mohanraj[4]

[1&3]Department of Mechanical Engineering, Curtin University Malaysia, CDT 250, Miri 98009, Sarawak, Malaysia
[2]Department of Chemical Engineering, Curtin University Malaysia, CDT 250, Miri 98009, Sarawak, Malaysia
[4]Department of Mechanical Engineering, Sastra University, Thanjavur 613403, Tamilnadu, India

Email: rajamohan.g@curtin.edu.my/rajamohan71@hotmail.com

Abstract. This research focuses on developing novel technique for a solar water heating system. The novel solar system comprises a parabolic dish concentrator, conical absorber and water heater. In this system, the conical absorber tube directly absorbs solar radiation from the sun and the parabolic dish concentrator reflects the solar radiations towards the conical absorber tube from all directions, therefore both radiations would significantly improve the thermal collector efficiency. The working fluid water is stored at the bottom of the absorber tubes. The absorber tubes get heated and increases the temperature of the working fluid inside of the absorber tube and causes the working fluid to partially evaporate. The partially vaporized working fluid moves in the upward direction due to buoyancy effect and enters the heat exchanger. When fresh water passes through the heat exchanger, temperature of the vapour decreases through heat exchange. This leads to condensation of the vapour and forms liquid phase. The working fluid returns to the bottom of the collector absorber tube by gravity. Hence, this will continue as a cyclic process inside the system. The proposed investigation shows an improvement of collector efficiency, enhanced heat transfer and a quality water heating system.

1. Introduction

Limitation of fossil fuel and environmental issues has increased the demand on renewable energy. Through renewable energy, a country could optimize per capita energy consumption maintaining the environmental impact. Solar energy is one of the potential renewable energy and this could be used for domestic water heating [10-11]. Though solar powered water heating system comprising flat plate solar collector and parabolic dish solar collector have been well established, the heat transfer rate is low due to single faced heat absorber [13]. In order to solve such issue, a novel conical heat absorber solar collector water heating system will be developed. Conical heat absorber will be able to absorb heat from multi directions, and hence, enhance the heat transfer rate. It is expected that rate of heat exchange would be 20% higher than the flat plate solar water heater.

IOP Conf. Series: Materials Science and Engineering **206** (2017) 012030 doi:10.1088/1757-899X/206/1/012030

Normally this type shows better efficiency than flat plate type. The solar flat plate collectors and evacuated tube collectors [1-2] are used as water collector in the different climatic regions. In conventional flat plate collector solar water heaters, copper, or mild steel plate embedded with galvanized iron, mild steel pipes are used as absorber, and glass wool used as insulation material [3-5]. However, to supply hot water to domestic and industrial applications, only little efforts have been made to develop building integrated solar water heating technology [6-7]. Normally, in flat plate and vacuum tube water heaters, collector and storage tank are two separate parts. An integrated collector storage technique is adopted to make the system simple and economical [8]. In integrated collector storage solar water heater, solar collector and water storage tank are integrated as a single unit. It is simpler in construction, lower in cost compared to flat plate or vacuum tube thermosiphon type water heaters and capable of fulfilling hot water demand of 100-200 liters per day [9]. Zekai Sen [10] presented a comprehensive review of solar energy in progress and future research trends. Ong and Tong [11] experimentally studied the system performance of U-Tube solar water heaters and the paper reports the results of outdoor tests conducted on several evacuated U tube solar collectors under natural and forced convection.

Experimental study and analysis of parabolic trough collector with various reflectors has been investigated [12]. In this experimental setup, the reflected solar radiations were focused on absorber tube which was placed at focal length of the parabolic trough. The design and development of a parabolic dish solar water heater for domestic hot water application was investigated [13]. Their research the heater is to provide 40 liters of hot water a day for a family of four, assuming that each member of the family requires 10 liters of hot water per day. For effective performance, the design requires that the solar water heater track the sun continuously, and an automatic electronic control circuit was designed and developed for this purpose. The design, construction and study of frusto-conical solar collector were carried out using locally available materials [14]. The conical shaped concentrator was made from galvanized iron sheet with a thickness of 0.644 mm. The inner side of the cone shape with thickness of 10 mm was gummed with insulating foam material. A critical review on literature shows that considerable research has been carried out by many researchers on parabolic dish collector, flat plate, point focus, line focus and evacuated tube solar collector and mostly for urban usage. However, efficiency of those collectors is low due to unidirectional collection which leads to reduction in heat transfer coefficient. In this work, a novel solar water heating systems with conical heat absorber is proposed in order to accommodate multi directional solar collector. This would significantly contribute towards enhancing the collector efficiency through enhanced heat transfer rate providing the pave way to the usage in rural, domestic and industrial application.

2. System Description

The solar water heating system consists of a parabolic dish concentrator, conical heat absorber, water heater and water as a working fluid. A schematic of solar water heater with parabolic dish and conical heat absorber is shown in Fig 1. The experimental model is suitable for open ground to have a better view of the sun's radiation. The parabolic dish collector was constructed with aluminium sheet because of low weight, ease of fabrication and energy efficiency. To reduce the weight of the dish, a light glass mirror of 2 mm thickness, of high surface quality and good specular reflectance was used.
A glass mirror was selected over polished aluminium surface because its reflectivity of 95% is better than that of aluminium (85%). Also, glass surface is easier to clean than aluminium surface [15]. The spiral copper tube is used for circulation of vaporized working fluid in the heat exchanger (water heater). The reflecting solar radiations from the parabolic dish collector are focused on the conical absorber tube absorb the radiation from the solar, both joint together will give significant heat improvement on absorber tubes. In this setup, water is used as working fluid which collects heat from absorber tube to heat exchanger. Initially, the working fluid is stored at the bottom of the absorber tubes. Due to solar radiations the absorber tubes get heated, when the heat transfer inside the tube, increases the temperature of a working fluid inside of the absorber tube and causes the working fluid to evaporate.

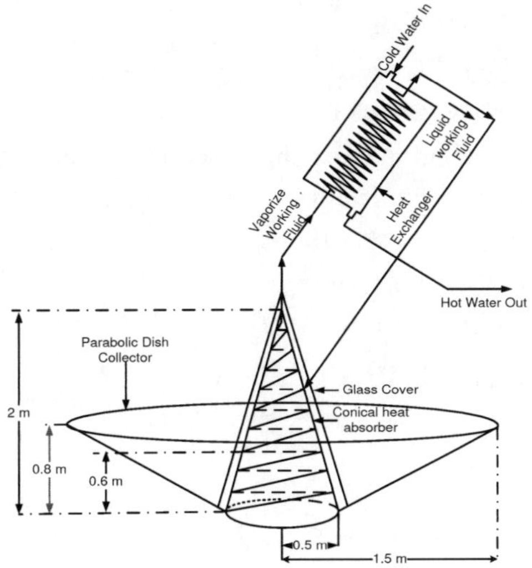

Figure 1. Schematic of solar water heater system.

Due to volume expansion and density gradient of air resulted from the buoyancy effect, the vaporize working fluid move upward direction and spontaneously enter to the heat exchanger. For extracting heat energy, the fresh cold water is circulated inside the heat exchanger which would decrease the temperature of the vapor and finally condensate the vapor to form liquid phase. Gravity driven working fluid returns to the bottom of the collector absorber tube. Hence, this will continue as a cyclic process inside the system. In this research, combined effect of conical absorber tube direct absorption and parabolic dish concentrator reflection from multi direction would provide better heat gain for solar water system compare to existing systems.

3. Analysis of Useful Heat Gain

The performance of collector could be determined by an energy balance that indicates the distribution of incident solar radiation into the useful energy gain and thermal losses. In order that the performance of the solar collector be as high economically practical design and operating factors that increases the solar heat flux absorbed and reduce the thermal losses are determined. Under steady condition, the useful heat collected by solar collector can be stated below [10].

Heat collected = Energy absorbed – Heat losses

$$Q_c = A_c[HR(\tau\alpha) - U(T_p - T_a)] \quad (1)$$

where

R	Conversion factor
A_c	Area of collector, m^2
H	Radiation incident rate, W/m^2
U	Overall heat loss coefficient, W/m^2K
Q_C	Heat collected from the solar collector
T_p	Tube temperature, $^\circ C$
T_a	Ambient temperature, $^\circ C$
α	Absorptivity
τ	Transmissivity

Alternatively the energy balance equation on the whole collector can be written as:

$$A_c[HR(\tau\alpha)_b + HR(\tau\alpha)_d] = Q_c + Q_L + Q_s \quad (2)$$

IOP Conf. Series: Materials Science and Engineering **206** (2017) 012030 doi:10.1088/1757-899X/206/1/012030

Collector efficiency is defined as the ratio of useful heat absorbed over solar radiation available over the collector surface.

$$\eta = \frac{Q_c}{A_c HR} \tag{3}$$

An attempt is made to drive the equations, which is used to describe the thermal performance of parabolic concentrator and conical absorber. The useful heat gain is the heat transfer rate to the fluid after considering all resistances and heat losses [15].
Useful heat gain to the fluid is obtained is,

$$Q_u = WLF_p\left(S - U\left(T_f - T_a\right)\right) \tag{4}$$

Useful heat gain collector efficiency is,

$$\eta = \frac{Qu}{A\ HR} \tag{5}$$

4. Analysis of Fluid

In this analysis, the energy balance is performed with three control volumes to study the temperature distribution and behaviour of fluid at subcooled, saturated and superheated regions. Consequently, it yields three differential equations as follows.
Subcooled condition:

$$\frac{dT_f}{dy} = \frac{WF_p}{\dot{m}_L C_{PL}}\left[S - U\left(T_f - T_a\right)\right] \tag{6}$$

Saturated condition:

$$\frac{d\dot{m}_v}{dy} = \frac{WF_p}{h_g - C_{PL}T_{f2}}\left[S - U\left(T_f - T_a\right)\right] \tag{7}$$

Superheated condition:

$$\frac{dT_{f3}}{dy} = \frac{WF_p}{\dot{m}_v C_{PV}}\left[S - U\left(T_f - T_a\right)\right] \tag{8}$$

The three differential equations are solved using the fourth Order Runge-Kutta method written in MATLAB program. To run the analysis and calculated the numerical values of some parameters. Using the predetermined variables, the validated differential equations are solved to numerically investigate the working fluid temperature distribution in subcooled, saturated and superheated regions. The data generated are plotted in graphs to show the correlation between the tube length and the refrigerant temperature along the tube. Figure 2 illustrates the temperature distribution of water along the heated tube from 20°C to 80°C. It covers the three regions as mentioned above. The working fluid temperature at the inlet is 20°C. The heat energy from the absorber increases the temperature of the subcooled liquid from 20 to 58°C in 1.58 m. At the saturation point, the refrigerant starts to evaporate at constant temperature at corresponding saturation pressure. The distance needed for the evaporating refrigerant (working fluid) to reach superheated region is 6.17 m. From there (at about 7.75 m from the tube inlet), the superheated refrigerant experiences a rise in temperature to a pre-set limit of 80°C. Besides, as expected the length needed for sensible heating in subcooled and superheated regions are relatively short as compared to the length needed for liquid evaporation.

The analysis has found three governing parameters which determine the dimensions and performance of the solar collector and these factors could be ranked based on their dominant effect on the solar collector. The governing parameters are, mass flow rate, overall heat loss coefficient and solar heat flux. An investigation on equation (4) gave a clear explanation. Increased the mass flow increases the tube length needed to superheat the liquid refrigerant. Figure 2 shows the result when the flow rate is increased from 0.010 kg/s to 0.018 kg/s while retaining other parameters. The result shown the length needed is about 80m. Although the working fluid will reach the required temperature at the end, the collector area is far too high.

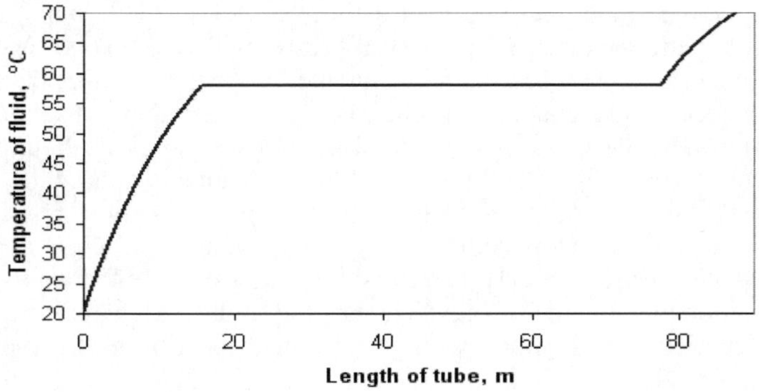

Figure 2. Temperature distribution of working fluid, when the mass flow rate is increased.

Similarly while retaining other variables reduced heat loss coefficient allows better absorption of heat energy in the collector. Equation (4) shows higher overall heat loss coefficient, U increases the heat loss product which results in greater reduction of heat energy collected. From Figure 3, the heat loss coefficient which is reduced from 8 to 4 W/m^2K by incorporating double glazing (16) into the collector increases the heat energy inside the collector and thus the tube length needed for the whole process is relatively short in comparison to the single glazing flat-plate collector.

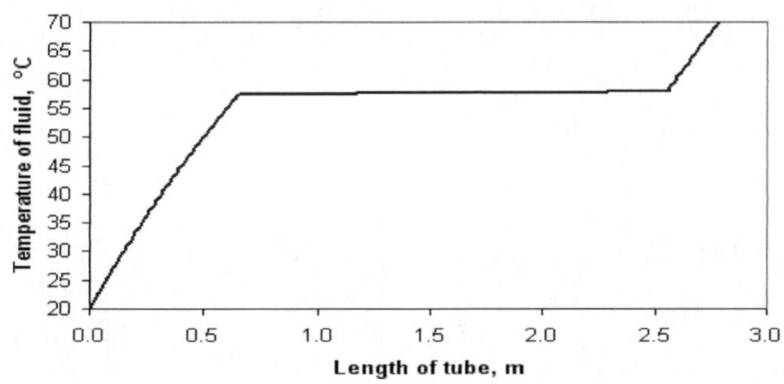

Figure 3. Temperature distribution of working fluid, when the overall heat loss coefficient is reduced.

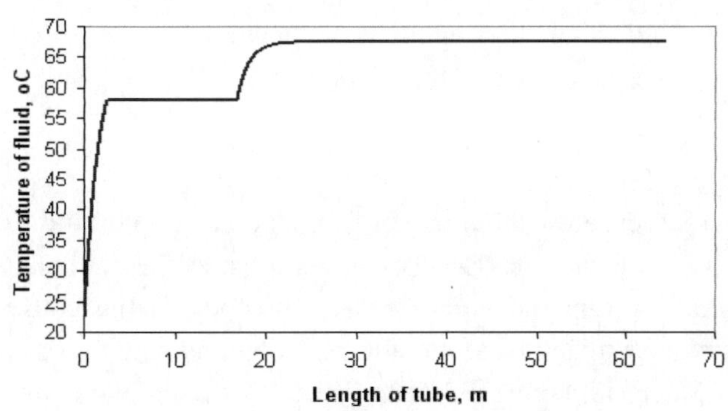

Figure 4. Temperature distribution of working fluid when solar heat flux is reduced.

In Figure 4 shows the similar effect is anticipated with the decrease of solar heat flux quantity. This three factors the solar heat flux is the most dominant factor that governs the potential and practicality of the solar collector. Though the performance of the collector is very much affected by the changes in mass flow rate and overall heat loss coefficient, insufficient solar heat flux close the system down totally. As shown, the reduction of solar heat flux from 400 to 300 W/m^2 altered the refrigerant temperature distribution pattern and the effect is even greater with lower solar heat flux.

Apart from the governing parameters, reducing the thermal resistances in collector efficiency increased the performance of the collector. From calculations the thermal resistance for a thin tube wall of good thermal conductivity and for the bond is almost negligible. Fouling resistances due to scale formation is about 0.01061 K/W and the fluid film resistance is about 0.0796 K/W. As assumed that the collector is well bonded and made from high thermal conductivity material, the tube thermal resistance is the most dominant factor in the collector performance which is about 0.8651 *K/W*. For this reason improving the tube design and material could extensively enhance the collector efficiency regardless of other minor thermal resistances. Though the ambient temperature is an uncontrollable factor, the fact that it varies with time and with geographical location means that collector efficiency depends on this factor as well.

From the specific heat transferred to the working fluid in the conical absorber and the specific heat transferred from the working fluid in the water heater are found using the following formulas [17].

Mass flow rate of cold water (kg/s) entering inside the water heater

$$m_{cold} = \frac{Q_{cond}}{[Cp_{water}(T_{c2} - T_{C1})]} \tag{9}$$

heat transfer rate (kJ/s)

$$Q_{cond} = m_C(q_A) \tag{10}$$

Surface area of the Condenser (m^2)

$$A_{cond} = \frac{Q_{cond}}{\{U_{heater}[(T_1 + T_4)/2 - (T_{c1} + T_{c2})/2\}} \tag{11}$$

Where

T_1 Temperature of vaporized working fluid entering the water heater
T_2 Temperature of partially condensed working fluid exiting from the water heater
T_c Temperature of the cold water entering the water heater
T_h Temperature of the hot water exiting the condenser

5. Result and Discussions

The first experimental study focused without parabolic dish concentrator and second experiment study focused with parabolic dish concentrator. Finally, the both experimental results are used to calculate the efficiency of the overall system and compare the efficiency of the both model. Figure 5 and 6 shows the variation of temperature and solar radiation for continuous time of two days experiment the month of August in Miri, Malaysia. Figure 5 results for without parabolic dish concentrator and Figure 6 results for with parabolic dish concentrator respectively. The water temperature, atmospheric temperature and solar radiation results are similar for both cases, so when comparing the efficiency of both cases will give reasonable outcome in this study.

Figure 7 shows the outlet temperature of water for with parabolic and without parabolic solar concentrator. The difference between the two cases differed by around 20 %. It is observed that parabolic dish concentrator will help to improve the temperature of the absorber tube. Combined effect of conical absorber tube direct absorption and parabolic dish concentration from multi direction would provide better heat gain for solar water system compare to without parabolic dish concentrator.

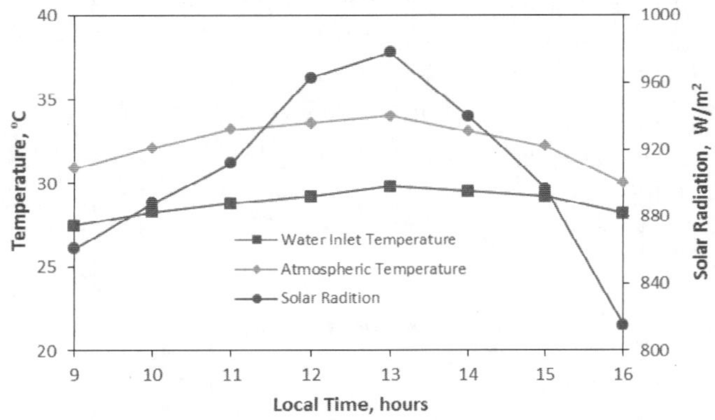

Figure 5. Variation of Temperature and solar Radiation for time of day (hours).

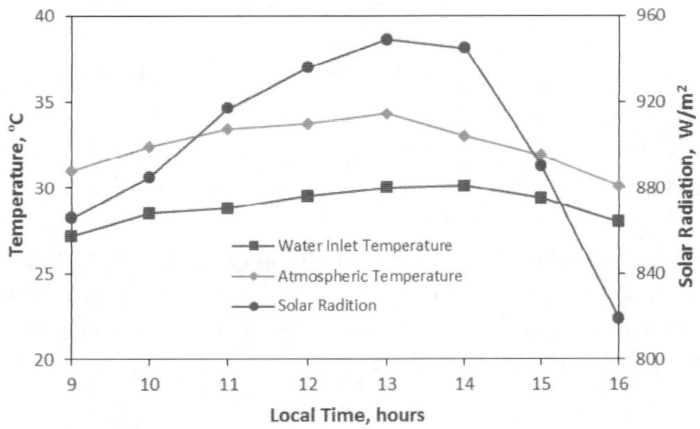

Figure 6. Variation of absorber tempertaure without conical collector for time of day (hours).

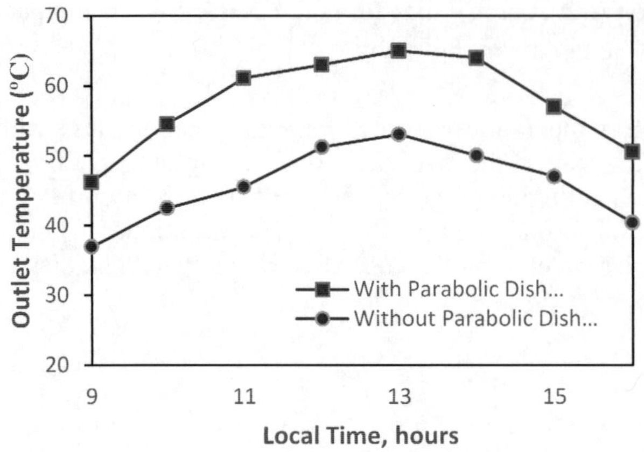

Figure 7. Variation of water hetar outlet temperature for time of day (hours).

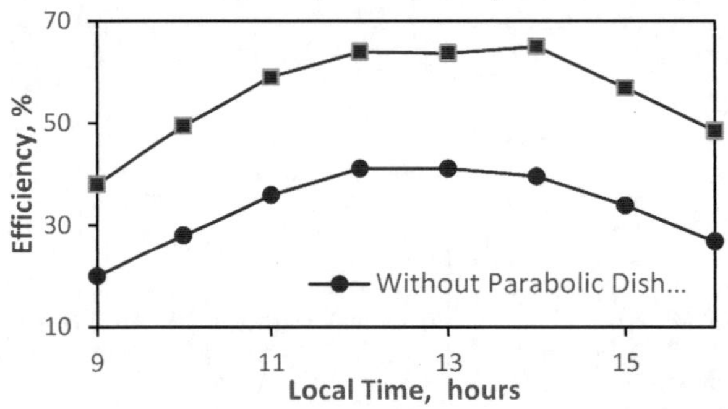

Figure 8. Efficiency of solar collector with and without parabolic for times of day (hours).

Comparison of the both experiment analysis with parabolic and without parabolic dish system efficiency is shown in Figure 8. The first day (9 to 16 hr) experiment without parabolic dish concentrator the average efficiency is 33.25% and the second day (9 to 16 hr) with parabolic dish concentrator average system efficiency is 55.40%. The difference between the two system differed by around 22.00%. The two continuous day experiment, in a particular time the maximum efficiency was achieved 65.00% for with parabolic dish concentrator and 41.00% achieved for without parabolic dish concentrator. The above system efficiency difference shows the unidirectional collection leads better collector efficiency than normal solar collector.

6. Conclusion

The analysis used in this paper is an alternative method of water heating from renewable energy resource with the use of parabolic dish concentrator and conical absorber. The heat transfer analysis performed in this paper is an effort to obtain the useful heat gain which governs the fluid temperature distribution along the tube. It has been concluded that the potential and practicality of the solar collector is dominantly influenced by many factors such as; solar collector area, absorber tube arrangements, the solar heat flux, mass flow rate and overall heat loss coefficient. The generation of maximum 65.00 % efficiency achieved for with parabolic dish collector and 41.04 % efficiency achieved for without parabolic dish collector. The difference between the two system differed by around 23.96 %. Combined effect of parabolic dish and conical absorber would provide better heat gain for solar water heating system. To work out large amount of hot water from heat energy source, a parabolic dish collector and conical absorber is an extremely attractive option.

IOP Conf. Series: Materials Science and Engineering **206** (2017) 012030 doi:10.1088/1757-899X/206/1/012030

References

[1] Chow T T, Dong Z, Chan L S, Fong K F and Bai Y *2011 Energ. Buildings* **43** 3467–3474.
[2] Hang Y, Qu M and Zhao F 2012 *Energ. Buildings* **45** 181–188.
[3] Zondag H A 2008 *Renew. Sust. Energ. Rev.* **12** 891–959.
[4] Sopian K and Zulkifli R 2002 *J. Mater. Proc.Technol.* **123** 179–184.
[5] Bilgen E and Richard M A 2002 *Solar Energ.* **72** 405–413.
[6] Nahar N M 2003, Year round performance and potential of a natural circulation type of solar water heater in India, *Energ. Buildings* **35** 239–247.
[7] Sowmy D S and Prado R T A 2008 Energy and Buildings, 40, 2128–2132.
[8] Souliotis M and Tripanagnostopoulos Y 2008 *Renew. Energ.* **33** 846–58.
[9] Tripanagnostopoulos Y, Souliotis M and Nousia T H 2002 *Solar Energ.* **72** 327–50.
[10] Zekai S 2004 *Prog. Energ.Combust. Sci.* **30** 367–416.
[11] Ong K S and Tong W L 2012 *J.Appl.Sci.Eng.* **15** 105-110.
[12] Avadhesh Y, Manoj k and Balram 2013 *Int. J. Physical Nat. Sci. Eng.* **7** 1-5.
[13] Ibrahim L M 2012 *Int. J. Eng. Res.App.* **2** 822-830.
[14] Vijayan G, Sebastian M V, Umarani K and Karunakaran R 2013 *Int. J. Eng. Res.Technol.* **2** 1220-1226.
[15] Ballaney P L 2003 *Thermal Engineering* Khanna Publishers Delhi.
[16] Goswami D Y, Kreith F and Kreider J F 2000 *Principles of Solar Engineering* Taylor and Francis USA.
[17] Nguyen T, Johnson P, Akbarzadeh A , Gibson K and Mochiziki 1994 *J. Heat Recover. Syst. CHP* **15** 333-346.

IOP Conf. Series: Materials Science and Engineering **206** (2017) 012031 doi:10.1088/1757-899X/206/1/012031

Effects of heavy metals (Fe^{3+}/Cr^{6+}) on low-level energy generation in a microbial fuel cell

A R Caparanga[1], **A S Balatbat**[1] **and L Tayo**[1]

[1]School of Chemical Engineering and Chemistry, Mapúa Institute of Technology, Manila, 1002 Philippines

E-mail: arcaparanga@mapua.edu.ph

Abstract. A dual-chamber microbial fuel cell (MFC) was constructed with *Pseudomonas aeruginosa* as biocatalyst to facilitate substrate conversion and, consequently, low-level energy generation. To simulate a wastewater situation with BOD and heavy metals contamination, glucose and Fe^{3+} and Cr^{6+} were used as substrate and heavy-metal spikes, respectively. The effects of varying substrate concentrations (150 ppm, 300 ppm, 600 ppm) and heavy metal loads (10 ppm, 50 ppm, 100 ppm) on overall power generation were evaluated. The presence of Cr^{6+} in the anode compartment decreased the potential from 565 to 201 mV (*i.e.*, lowest value achieved at highest Cr^{6+} concentration of 300 ppm). On the other hand, replacing Cr^{6+} with Fe^{3+} as electron acceptor resulted in substantial increase in measured potential (*i.e.,* from 565 to 703 mV). Increasing glucose concentrations resulted in longer time to reach constant open circuit voltage. A maximum potential of 606 mV was achieved at 1200 ppm glucose. Incorporating *Pseudomonas aeruginosa* increased the potential from 256 to 592 mV. On the basis of these results, a microbial fuel cell feeding on wastewater can be an important potential technology for generating low-level energy

1. Introduction

There is a need to develop alternative energy source to partially contribute to addressing the impending energy crisis predicted to happen in a few decades in the future. One such technology is microbial fuel cell (MFC), an electrochemical device that can generate low-level electricity from the energy effects associated with the bioconversion of a substrate. MFC has cathode and anode separated by an electrolyte. The microbes oxidize the organic substrate anaerobically in the anode. The electrons lost via oxidation (at the anode) is transported via a resistor to the cathode, where reduction occurs. This transfer is facilitated by membrane-associated components or soluble electron shuttle [1].

MFC technology is not novel. The use of microorganisms a fuel cell started some decades ago [2, 3]. It was used in wastewater treatment [4]. However, interest on the electricity produced during the oxidation/reduction process has been of interest recently [5, 6, 7]. Current inorganic fuel cells such as polymer–electrolyte, direct-methanol, and solid-oxide systems [8] have already reached innovation phases. Such systems, however have drawbacks in cost and waste management. Among technologies for generating energy from organic waste, MFC has the following advantages: (a) direct utilization of substrate via biologically-mediated, non-temperature-sensitive chemical reactions that consequently generate energy [1, 9]; low or no energy requirement for a sufficiently ventilated cathode chamber [5, 10]; (c) a possible widespread application in regions where energy resources are wanting. However, the study concerning this technology is still on its early stage. MFCs at present show great potential in harvesting energy from wastes. But, little is known about how presence of other factors affects waste

IOP Conf. Series: Materials Science and Engineering **206** (2017) 012031 doi:10.1088/1757-899X/206/1/012031

treatment and the subsequent generation of low-level electricity. In this study, the effect of heavy metals, which may be present in the waste stream, on the low-level energy generation is in focus. It is construed that the electrons generated from the redox reactions between the heavy metals and the microorganism might affect or contribute to the electricity generation. Fe^{3+} and Cr^{6+} were used as models.

Pseudomonas aeruginosa, a biocatalyst with wide-range biosoption capability was used. *Pseudomonas aeruginosa* has attractive features: being inexpensive, simple mass cultivation, and its being capable of producing pyocyanin, a mediator that serves as electron shuttle in the anode chamber of MFC [11]. *Pseudomonas aeruginosa* catalyzes the transformation of biochemical energy to generate electrical energy. The anaerobic fermentation pathway of *Pseudomonas aeruginosa* enables the conversion of pyruvate to ethanol. When NAD^+ is reduced to NADH it produces two ATP molecules, two H^+ ions and two electrons [12]. The electrons generated at the anode can then traverse via external circuit to produce power. The electrons then move into the cathode chamber to react with the protons transported via ion exchanger. *Pseudomonas aeruginosa* can simultaneously act as a good biosorbent. It possesses metal-sequestering properties and can reduce the heavy metal ions concentration in a given solution. At high efficiency, it can remarkably sequester dissolved metal ions out of dilute complex solutions [13, 14].

The main objective of this study is to harvest energy from wastewater using MFC while simultaneously treating the heavy metal content of the wastewater. To accomplish this, determination of the following were taken into consideration: energy generated by the MFC at different time intervals; the effect of adding *Pseudomonas aeruginosa*, whether it enhances the production of energy and reduction of heavy metal by quantifying concentrations of the specific heavy metals using spectrophotometric assay. This study specifically determines the effect of heavy metals present in the substrate in the generation of energy upon reduction (by the microorganism).

This study, however, does not include attempt at storing the energy generated.

2. Materials and Method

2.1. Microbial fuel cell assembly

The microbial fuel cell used in this study was the dual chamber type, adapted from the design of Rabaey [15]. The compartments were constructed using plexi glass (600 ml volume) for anode chamber and 300 ml for cathode chamber. The anode and cathode chambers are separated using an ion exchange membrane made of 10% agar and 5% potassium nitrate solution [16]. The electrodes were constructed using carbon brass (3 in) and copper wire was used to hold each carbon rod. Minimal oxygen content was maintained in the anode using an air-tight cover. The cathode comprised of 50 mM $K_3Fe(CN)_6$. Electro-analysis was conducted using a Proskit multimeter.

2.2 Preparation and cultivation of the cultures

The microorganism (obtained from the Department of Science and Technology, Manila, Philippines) was grown in test tubes. The medium for the pure culture comprised of peptone (20 g/L), $MgCl_2$ (1.4 g/L), NaCl (10 g/l) and agar (10 g/L) (Sigma-Aldrich™). The medium was autoclaved at 121°C for 20 min. The *Pseudomonas aeruginosa* cells were inoculated into sterilized media at ambient temperature and incubated at 30°C. The pure culture was grown for 2 days until pyocyanin pigment appeared. A drop of concentrated HCL would confirm the presence of pyocyanin.

2.3 Microbial fuel cell (MFC) operation

In the operation of MFC, 100 ppm of glucose was used in the anode chamber. To test the effect of heavy metal reduction in the low-level energy generation, simulated sample was used. The heavy metal tested was chromium (as potassium dichromate). The concentration was varied (10 ppm, 50 ppm and 100 ppm) and a blank solution using deionized water was considered. The effects of incorporating chromium (VI) in the cathode and substituting potassium hexacyanoferrate by potassium dichromate as electron acceptor to potassium dichromate were determined. To test the effect of the substrate concentration, glucose concentration was varied (150 ppm, 300 ppm, 600 ppm). The concentration after a 1-day study period was evaluated using atomic absorption spectroscopy. For an actual system study, mixed culture inoculum, waste water from a brook near the meat section a public market was used.

2.4 Measurement and calculations

The cell voltage across the external load was continuously measured using Proskit multimeter. Power and current will be calculated according to $P = IV$ and $I = (V/R_{ext})$, where P (W) is the generated power, V (V) is the measured cell voltage, R_{ext} (Ω) is the external load, and I (A) is the current. The current and power produced will be normalized to the surface area of the electrode. The data gathered at hourly time interval (1 day) were plotted accordingly. Also the efficiency of the microbial fuel cell was evaluated via the energetic efficiency (μ).

$$\mu = \frac{V}{\Delta H(nF)^{-1}} \tag{1}$$

where V is the voltage across the load, ΔH is the calorific value of glucose, n is the number of electron moles transferred across the load, and F the Faraday constant.

3. Results and Discussion

3.1 P. aeruginosa cultivation and pyocyanin synthesis

Mediator is critical towards the power generation in MFC as it shuttles the electrons from the bioconversion of glucose to the anode. Usually this mediator is externally supplied, but *P. aeruginosa* is capable of synthesizing its own pyocyanin. During the cultivation period, *P. aeruginosa* would form colonies surrounded by a blue-green zone due to pyocyanin production. This could be due to the composition of the growth media. The peptone provided the essential nitrogenous nutrients, carbon, sulfur and trace elements; its low phosphorous content minimized the inhibitory action on pyocyanin production. Potassium sulfate and magnesium chloride are key components to induced pyocyanin synthesis. Although the components of the media enhances the production of pyocyanin, some pigmentation like red (pyorubin), yellow-green (pyoverdine, fluorescein) or brown (pyomelanine) pigments were observed and masked the pyocyanin; hence, to validate that majority of the pigment synthesized were pyocyanin a drop of concentrated HCl was incorporated turning the media red. This, as suggested by King et al. [17, 18], is a reliable method indicating that *P. aeruginosa* was able to synthesize pyocyanin; this was aside from the preliminary observations (blue-green colonies).

3.2 Microbial fuel cell performance

In a given MFC, the oxidation of substrate like glucose is the main source of energy. The specific potentials of the anode and the cathode are determined through the redox potential of the reactions at the electrode. Theoretically, the maximum open circuit voltage that can be acquired due oxidation of glucose and reduction of oxygen is 1.216 V but, since the oxidation occurs within the microbial cell, which is external to the electrode, the theoretical value is not fully acquired.

Under an oxygen limited environment, *P. aeruginosa* proceeds a fermentation process where NADH is being oxidized by a specific enzyme. Certain amount of NADH will permeate through the extracellular environment. This will facilitate a redox couple NADH / NAD+ on the boundary of the anode where electrons and protons are being dissipated [19], but NADH is significantly hindered because of the presence of the cellular membrane. This is the primary reason a mediator such as the pyocyanin is crucial in the electron transfer (with respect to the electrode). Pyocyanin undergoes reduction by NADH and then oxidized once it is contacted with the electrode. The reactions on the anode are summarized below:

IOP Conf. Series: Materials Science and Engineering **206** (2017) 012031 doi:10.1088/1757-899X/206/1/012031

Cell: NADH → $NAD^+ + H^+ + e^-$ (2)
Electrode contact: $Pyocyanin_{(reduced)}$ → $Pyocyanin_{(oxidized)} + 2e^- + H^+$ (3)

On the other hand, on the cathode compartment; the electrons from the anode reduces $K_3Fe(CN)_6$. Then the protons (H^+), which passes through the ion exchange membrane, reduces the oxygen molecules present in the cathode to form water molecules.

Electrode contact: $4\ K_3Fe(CN)_{6\ (oxidized)} + 4e^- \rightarrow 4\ K_3Fe(CN)_{6\ (reduced)}$ (4)
$4K_3Fe(CN)_{6\ (reduced)} + O_2 + 4H^+ \rightarrow 2H_2O + 4\ K_3Fe(CN)_{6\ (oxidized)}$ (5)

Figure 1 shows that open circuit voltage increased with time. A previous growth phase study reported that this could be due to the number of microorganisms undergoing metabolism increasing as well [20]. The voltage across the membrane increased almost linearly with time until the 20th hour. This period could coincide with the logarithmic phase of the bacterial growth, where the number of bacterial cells growing in the anode chamber would have rapidly increased. In the succeeding hours, the voltage across the MFC slowly decreased, which could be a period coinciding with stationary phase of the bacterial growth, where in the number of cells dying would be equal to the cells growing in the anode compartment. Possible source of discrepancy in the overall potential could be due to bio-fouling of the microorganism in the membrane; it could develop an internal resistance resulting to a decrease in the overall potential.

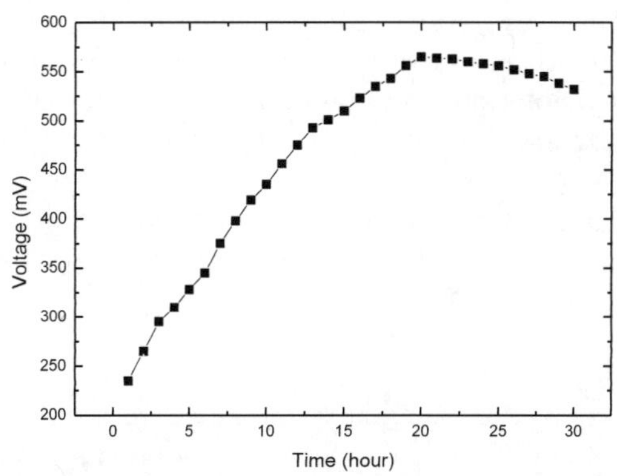

Figure 1. Voltage output upon substrate utilization with time

3.3 Effect of heavy metal concentration in the anode compartment

Figure 2 shows that as the concentration of the metal (Cr^{6+}) increases from 50 ppm to 300 ppm the voltage across the MFC decreases. It could be that presence of such metal could disrupt the cell membrane of the microorganism once adsorbed; hence, the number of cells growing in the anode chamber would decrease and so was the voltage across the MFC [7]. In addition, the protons and electrons generated from the utilization of glucose could have reduced metal ions present in the anode chamber resulting in disruption of the ion flow gradient (from anode to cathode); hence, decreasing in the overall potential in the MFC. On this assumption, Cr^{6+} was incorporated in the cathode compartment to serve as an alternative electron acceptor.

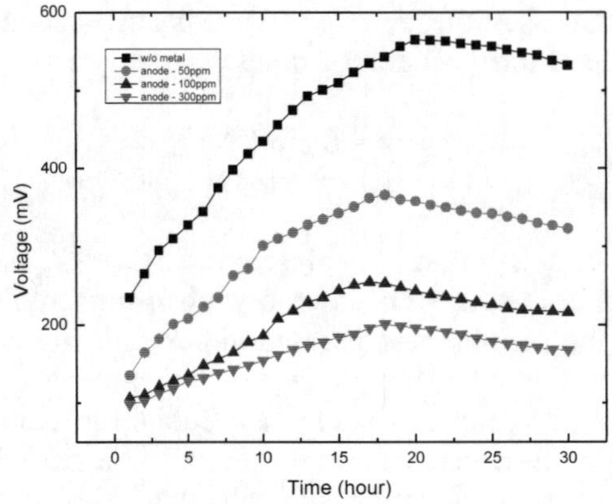

Figure 2. Effect of adding heavy metal (Cr^{6+}) in the anode on voltage generation

3.4 Effect of heavy metal concentration in the anode compartment

Here, instead of using potassium hexacyanoferrate as an electron acceptor, Cr^{+6} from potassium dichromate was used instead. This was on the basis that, theoretically the oxidation potential of Cr^{6+} (1.33V) is higher compared to potassium hexacyanoferrate (0.36 V) [6]; hence, as seen in Figure 3, substitution of potassium hexacyanoferrate to potassium dichromate led to a substantial increase in voltage generated (558 mV to 702 mV). As mentioned, this could be attributed to the higher redox of potential $Cr_2O_7^{2-}$.

The chromium ions could have electro-deposited on the surface of the cathode, decreasing the chromium concentration in the solution (pH = 2) [6]. For this experiment, the maximum percent removal of chromium was approximately 90.6%.

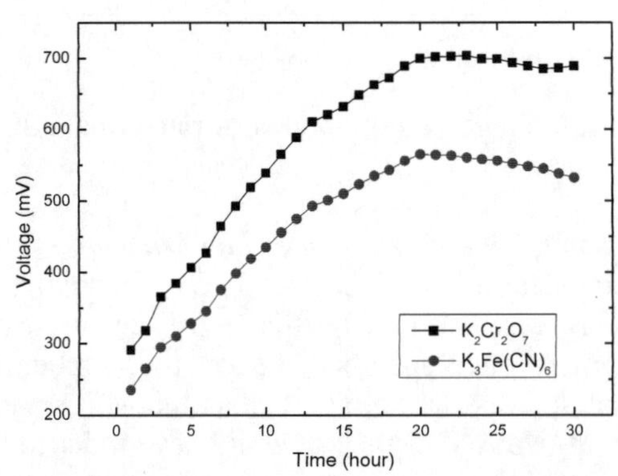

Figure 3. Effect of adding heavy metal (Cr^{6+}/Fe^{3+}) in the cathode on voltage generation

3.5 Effect of substrate concentration

The results indicated that an increase in the substrate concentration increased the time needed to reach constant OCV (open circuit voltage). Liu et al. [21] suggested that once the sufficient amount of biomass is formed, the use of an excessive substrate becomes unnecessary; since, there is a specific threshold value for the metabolic reaction rate of microorganisms on electrode surface to be optimal.

As seen in Figure 4, as the concentration of the substrate increases the maximum open circuit voltage increases as well.

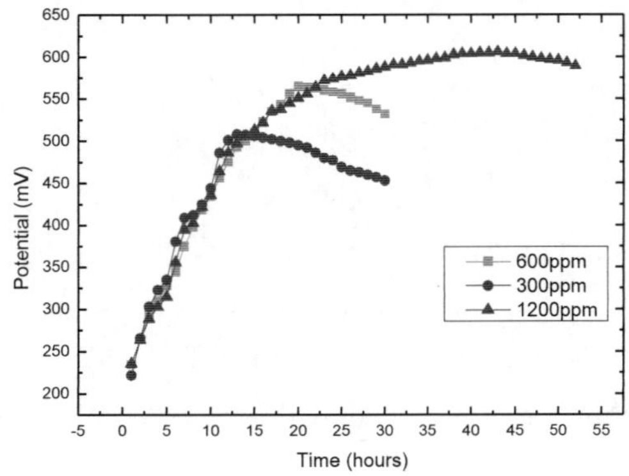

Figure 4. MFC performance at varying substrate concentrations

3.6 MFC performance using actual wastewater

To elucidate the performance of the MFC in an actual waste water. A sample from a creek near a food chain and a public market was considered. This water sample is believed to have lipids, carbohydrates and etc., which can serve as nutrient source for the organisms already present in the sample and to *P. aeruginosa* which is supplied.

Figure 5 shows the sample without *P. aeruginosa* was still able to produce a steady open circuit voltage; with a maximum value of 263 mV, primarily because the flora of microorganisms innately present in the sample were still metabolizing the substrates present in the sample; but due to the lack of electron mediator (like pyocyanin), the voltage output was relatively lower as compared to sample with *P. aeruginosa* (capable of producing its own mediator), which had maximum voltage output of 622 mV.

The higher voltage output can attributed to the possibility that the pyocyanin produced by *P. aeruginosa* served as an electron shuttle molecule to other microorganisms present in the sample (see Figure 6).

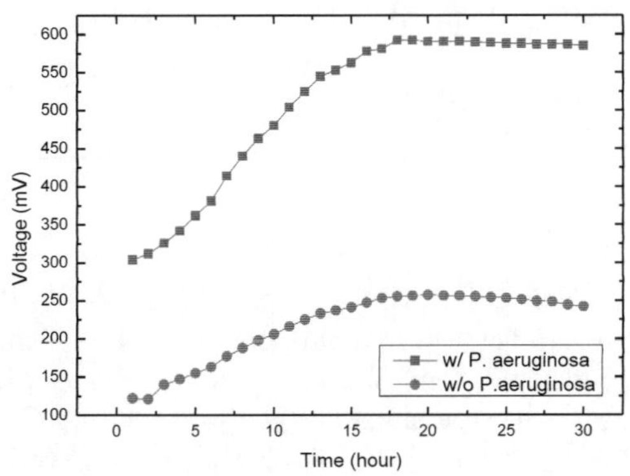

Figure 5. MFC performance on actual wastewater

IOP Conf. Series: Materials Science and Engineering **206** (2017) 012031 doi:10.1088/1757-899X/206/1/012031

It should be noted that not all microorganisms present in the sample are compatible with pyocyanin; it is strain specific, meaning, pyocyanin can be toxic to other microorganisms. A study by Rabaey et al. [1] showed that *E. coli* ATCC4157 had illustrated an adverse effects when pyocyanin was incorporated, while no effects were noted on *A. faecalis* KRA.

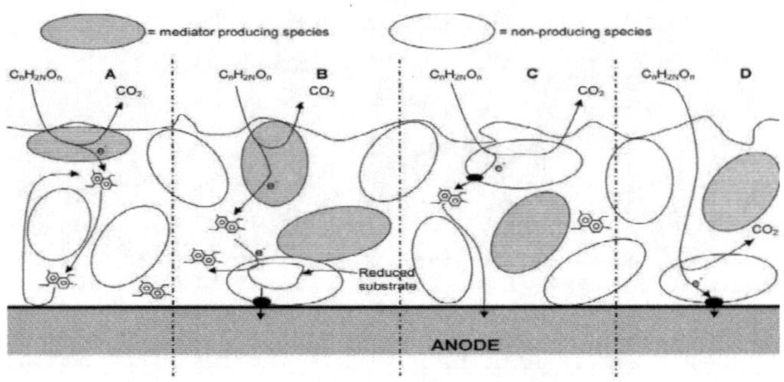

Figure 6. Possible mechanism of self-mediated MFC in a mixed microbial flora [1]

3.7 MFC efficiency

The efficiency of an MFC is dependent on the nature of reaction with the substrate. Thermochemically, the utilization efficiency of a substrate can be deduced from Gibbs free energy and enthalpy of formation (efficiency = $\Delta G/\Delta H$).

For glucose, $\Delta G_{formation}$ = -911 kJ/mole, and $\Delta H_{formation}$ = -1267 kJ/mole; hence, 71.8% of energy is obtainable for utilization. Based on the experimental data, the maximum energetic efficiency per experiment was calculated and the results are shown in Table 1. Incorporation of metal in the anode compartment decreased the efficiency from 65.4 % (normal operation) to 23% (with 300ppm metal). Substitution of electron acceptor (from potassium hexacyanoferrate to potassium dichromate) enhanced the MFC performance up to a maximum energetic efficiency of 80.6 % as mentioned, this can be attributed to the higher redox potential of $Cr_2O_7^{2-}$. Finally the presence of *P. aeruginosa*, as biocatalyst, enhanced the efficiency from 29.4% to 67.4 %.

Table 1. MFC efficiencies of different substrate / electrode systems systems

MFC System	Efficiency
Normal operation	0.654
Heavy metal at anode compartment (50 ppm)	0.419
Heavy metal at anode compartment (100 ppm)	0.292
Heavy metal at anode compartment (200 ppm)	0.230
Electron acceptor substitution	0.806
Actual wastewater, without biocatalyst added	0.294
Actual wastewater, with biocatalyst added	0.674

4. Conclusion

The dual-chamber microbial fuel cell (MFC) constructed, with *Pseudomonas aeruginosa* as biocatalyst, facilitated substrate conversion and produced low-level energy. The effects of varying substrate concentrations (150 ppm, 300 ppm, 600 ppm) and heavy metal loads (10 ppm, 50 ppm, 100 ppm) on overall power generation were evaluated. The presence of Cr^{6+} in the anode compartment decreased the potential from 565 to 201 mV (*i.e.*, lowest value achieved at highest Cr^{6+} concentration of 300 ppm). On the other hand, replacing Cr^{6+} with Fe^{3+} as electron acceptor resulted in substantial increase in measure potential (*i.e.,* from 565 to 703 mV). Increasing glucose concentrations resulted in longer time to reach constant open circuit voltage. A maximum potential of 606 mV was achieved at

IOP Conf. Series: Materials Science and Engineering **206** (2017) 012031 doi:10.1088/1757-899X/206/1/012031

1200 ppm glucose. Incorporating *Pseudomonas aeruginosa* increased the potential from 256 to 592 mV. The MFC also worked well with actual wastewater. On the basis of these results, a microbial fuel cell feeding on wastewater can be an important potential technology for generating low-level energy.

References

[1] Rabaey K and Verstraete 2005 *Trends Biotechnol.* **23** 292
[2] Suzuki S 1976 *Hospital Hygiene, Gesundheitswesen und desinfektion* **159**
[3] Roller S 1984 *J. Chem. Technol. Biotechnol. B Biotechnol.* **34** 3
[4] Habermann W and Pommer E 1991 *Appl. Microbiol. Biotechnol.* **25** 128
[5] Liu H and Logan B 2004 *Environ. Sci. Technol.* **38** 4040
[6] Li Z, Xingwang Z and Lei L 2008 *Process Biochemistry* **43** 1352
[7] Abourached C, Catal T and Liu H 2014 *Water Res.* **51** 228
[8] Larminie J and Dicks A 2003 *Fuel Cell Systems Explained* (Chichester, West Sussex: J. Wiley)
[9] Pant D, van Bogaert G, Diels L and Vanbroekhoven K 2010 *Bioresour. Technol.* **101** 1533
[10] Liu H 2004 *Environ. Sci. Technol.* **38** 2281
[11] Sayed E, Tsujiguchi T and Nakagawa N 2012 *Bioelectrochemistry* **86** 97
[12] Nelson D and Cox M 2005 *Principles of Biochemistry* (New York: W H Freeman)
[13] Veglio F and Beolchini F 1997 *Hydrometallurgy* **44** 301
[14] Volesky B (1990) *Biosorption of Heavy Metals* (Florida: CRC Press)
[15] Rabaey K 2003 Biotechnol. Lett. **25** 1531
[16] Sevda S and Sreekrishnan T 2012 *J. Environ. Sci. Health A Tox. Hazard Subst Environ. Eng.* **47** 878
[17] King E, Ward M and Raney D 1954 *J. Lab. Clin. Med.* **44** 301
[18] El-Fouly M, Sharaf A, Shahin A, El-Bialy H and Omara A 2015 *J. Radiat. Res. Appl. Sci.* **8** 36
[19] Shukla A, Suresh P, Berchmans S and Rejendran A 2004 *Curr. Sci.* **87** 455
[20] Merina P 2013 *Int. J. Chem Tech Res.* **5** 2193
[21] Liu J, Qiao Y, Guo C, Lim S, Song H and Li C 2012 *Bioresour. Technol.* **114** 275

IOP Conf. Series: Materials Science and Engineering **206** (2017) 012032 doi:10.1088/1757-899X/206/1/012032

Stability studies of immobilized lipase on rice husk and eggshell membrane

R Abdulla[1,2]*, S A Sanny[1] , E Derman[1]
[1]Faculty of Science and Natural Resources, Universiti Malaysia Sabah, 88400 Kota Kinabalu, Malaysia

[2]Energy Research Unit, Universiti Malaysia Sabah, 88400 Kota Kinabalu, Malaysia

*E-mail: rahmahabdulla@gmail.com

Abstract. Lipase immobilization for biodiesel production is gaining importance day by day. In this study, lipase from *Burkholderia cepacia* was immobilized on activated support materials namely rice husk and egg shell membrane. Both rice husk and eggshell membrane are natural wastes that holds a lot of potential as immobilization matrix. Rice husk and eggshell membrane were activated with glutaraldehyde. Lipase was immobilized on the glutaraldehyde-activated support material through adsorption. Immobilization efficiency together with enzyme activity was observed to choose the highest enzyme loading for further stability studies. Immobilization efficiency of lipase on rice husk was 81% as compared to an immobilization efficiency of 87% on eggshell membrane. Immobilized lipase on eggshell membrane exhibited higher enzyme activity as compared to immobilized lipase on rice husk. Eggshell membrane also reported higher stability than rice husk as immobilization matrix. Both types of immobilized lipase retatined its activity after ten cycles of reuse. In short, eggshell membrane showed to be a better immobilization platform for lipase as compared to rice husk. However, with further improvement in technique of immobilization, the stability of both types of immobilized lipase can be improved to a greater extent.

1. Introduction

The arising of global dilemmas such as depletion of fossil fuels, increasing fuel prices and global warming has caused a great concern and led the search for alternative, sustainable, renewable, efficient and cost-effective resources [1]. Biofuels are renewable energy resources produced from biomass material which can be used to replace fossil fuels [2]. Biodiesel is an attractive liquid biofuel which can replace existing fossil derived diesel to a greater extent. Lipase catalyzed biodiesel production is an attractive, environmental friendly technique. However, the high cost of lipase serves as a main hurdle for its commercialization. This can be overcome to a greater extent by immobilizing the lipase in a suitable matrix thus increasing its reusability [3].

Immobilization can be done through a number of ways. Out of these the main methods are adsorption, covalent binding, entrapment and cross-linking [4] The methods mentioned can be coupled together to further increase the stability of the enzyme. Immobilizing an enzyme is proven to be more efficient than normal enzymes because of their properties. It is more stable as it is bound to a solid support which may increase its stability. Moreover, immobilized enzymes can be reused [5].

The carrier matrix used must be cheap or affordable. However, it must have certain characteristics for it to be an ideal support [6]. A carrier must be coupled with proper immobilization technique for better design of an immobilized enzyme [7-8].

Rice husk is able to serve as a good matrix for immobilization due to its properties of resistance towards fungi and microbial growth. Rice husk is one of the major agricultural wastes that have been produced around the globe especially from countries such as China, India, Indonesia, Malaysia, Bangladesh and Vietnam. The proper disposal of rice husk is a main challenge in waste management. Furthermore handling rice husk and rice husk ash takes a lot of work because of its density [9].

On the other hand, eggshell membrane (ESM) is a thin white film located between the eggshell and albumin of eggs. The eggshell membrane consists of crosslinked collagen, glycosaminoglycans, egg white protein and eggshell matrix protein. The membrane is not permeable to water as it is made up of disulfide bonds which gives it an advantage in certain field of application. It has a large surface area, high porosity and contain amino acids. It is reported that the content of eggshell membrane is acidic [10]. Eggshell membrane is highly adsorbent and it is used as a material in removing heavy metals from aqueous solution [11-12].

The objective of this research is to utilize two important waste materials namely rice husk and eggshell membrane as a matrix for immobilization of lipase for biodiesel production. However before it can be applied for production, it is very important to study the stability of the immobilized lipase. Thus, the main stability parameters were studied and characterized for design of an immobilized lipase for its application to biodiesel production.

2. Materials and methods

2.1. Materials. Rice husk was collected from a local rice mill in Tuaran, Sabah, Malaysia. The egg shells were collected from our faculty's cafeteria. *Burkholderia cepacia* lipase was kindly gifted by Amano Enzymes, Japan. All other reagents and chemicals used were of analytical grade and was obtained commercially.

2.2. Immobilization of lipase on rice husk. Immobilization of lipase on rice husk was done by modifying the method of [13]. At first, 500µl of 25% glutaraldehyde was added into a beaker containing rice husk and incubated for 5 minutes. This was followed by addition of different lipase concentrations separately (100µg/ml, 200µg/ml, 300µg/ml, 400µg/ml, 500µg/ml) into 1g of rice husk and incubated for 20 minutes.

2.3. Immobilization of lipase on eggshell membrane. Immobilization of lipase onto eggshell membrane (1g) was performed by modifying the method of [13]. 500µl of 25% glutaraldehyde was added into a beaker containing eggshell membrane and incubated for 5 minutes. After the incubation, different lipase concentration (100µg/ml, 200µg/ml, 300µg/ml, 400µg/ml, 500µg/ml) was added into different beakers containing 1g of rice husk and incubated for 20 minutes.

2.4. Determination of lipase activity. The activity of the immobilized lipase on rice husk and egg shell membrane were determined by using titrimetric assay with olive oil emulsion as the substrate[14]. The reaction mixtures contained 10ml of olive oil emulsion (1ml of olive oil and 9ml of arabic gum at 10%), 20 ml of distilled water and immobilized lipase (1g). The reaction was carried out at 37°C for 30 minutes using orbital shaker. The amount of free fatty acid released during hydrolysis were estimated by titration with 0.01N NaOH solution. One unit of activity is defined as the amount of enzyme, which release 1µmole fatty acid per min per ml under specified assay conditions.

2.5. Determination of pH and thermal stability. This was performed by modifying the method of [15]. The pH stability of immobilized lipase on rice husk and eggshell membrane were studied by incubating 1g of immobilized lipase in Tris HCl buffer at different pH (6-10) for 1 hour at $4^{o}C$. The hydrolytic activity were determined using olive oil hydrolysis. The thermal stability of immobilized lipase were done by incubating them at various temperatures in the range of $20\text{-}60^{o}C$ for 1h. The activities were measured using olive oil hydrolysis.

2.6. Determination of organic solvent stability. Stability of immobilized lipase were tested in different solvents like ethanol, methanol, hexane, n-butanol and 1-propanol. The immobilized lipase was incubated for 1h in the above mentioned solvents at $4^{o}C$. The immobilized lipase were filtered, washed and the catalytic activities were determined using olive oil hydrolysis assay [16].

2.7. Determination of storage and reusability studies. The storage stability of immobilized lipase on rice husk and eggshell membrane were done by storing in Tris HCl buffer (pH 7) at $4^{o}C$ for one week. The activity of immobilized lipase was determined on a daily basis.

The immobilized lipase on rice husk and eggshell membrane were repeatedly used ten times for hydrolysis of olive oil emulsion. Followed by hydrolysis, the immobilized lipase was filtered, washed and suspended again in fresh olive oil emulsion in order to measure its lipase activity.

3. Results and discussion

3.1. Immobilization efficiency. The immobilization efficiency of lipase on rice husk and eggshell membrane are shown in Figure 1. An immobilization efficiency of 81% was achieved with a lipase concentration of 500 µg/ml on rice husk. On the other hand, a lower lipase concentration of 400 µg/ml gave highest immobilization efficiency of 87% on eggshell membrane.

Rice husk is known to be a highly adsorbent natural material because of its porosity. Wan Ngah and Hanafiah [17] has reported rice husk to be a good adsorbent material for metals and dyes. In addition to this, rice husk has greater surface area which makes it as a suitable carrier for immobilization [18]. The low lipase activity on rice husk as compared to eggshell membrane might be due to the fact that the enzyme's catalytic site being blocked or lipase enzyme is deactivated [19].

On the other hand, immobilized lipase on eggshell membrane exhibited an efficiency of 87%. This can be attributed to the properties of eggshell membrane. The eggshell membrane surface has high permeability towards substrates and products. The amino acids present on the membrane increases its chances of getting crosslinked with glutaraldehyde which inturn leads to increased immobilization efficiency [13].

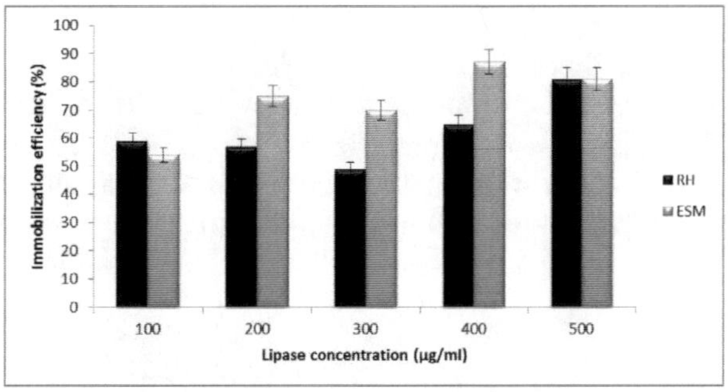

Figure 1. Immobilization efficiency on rice husk and eggshell membrane

3.2. Enzyme Loading.

Figure 2 shows the results of enzyme loading at different concentrations on rice husk and eggshell membrane. It is clear that 300 µg/ml of lipase concentration gives the highest enzyme activity on rice husk while 400 µg/ml gave the highest enzyme activity for eggshell membrane. Therefore, immobilized lipase on rice husk was chosen at 300µg/ml enzyme loading and 400µg/ml enzyme loading for eggshell membrane for further investigation in stability studies.

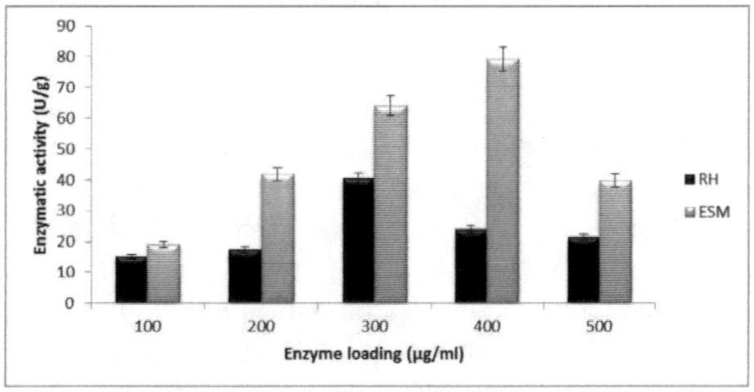

Figure 2. Enzyme loading on rice husk and eggshell membrane

3.3. pH stability.

As shown in the Figure 3, enzyme activity for rice husk is shown to be the highest at pH 8 indicating the optimum pH for immobilized lipase. As for immobilized lipase on eggshell membrane the optimum pH was found to be 7. With further increase in pH, both immobilized lipase showed a decrease in enzymatic activity. Similar results for pH studies were also reported [20] [21]. The decreased enzymatic activity with increase in pH can be attributed to the loss of H^+ ions from side chain groups of the matrix [18]. The pH of immobilized lipase will depend to a greater extent on the properties of the carrier and also the method of immobilization employed [22].

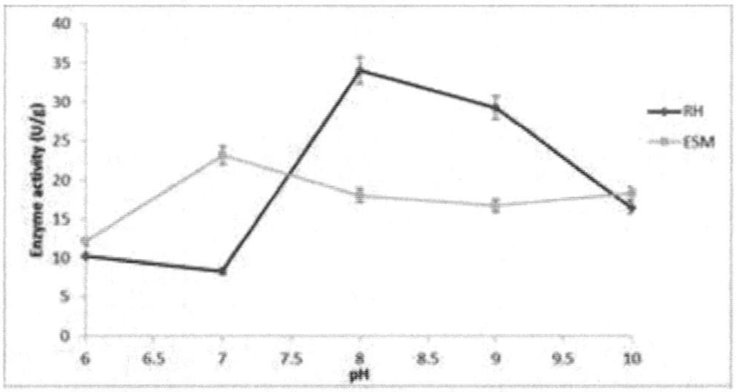

Figure 3. pH stability of immobilized lipase on rice husk and eggshell membrane

3.4. Temperature Stability.
The enzymatic activity of immobilized lipase on rice husk and eggshell membrane at different temperatures are shown in Figure 4. The optimum temperature for both types of immobilized lipase were found to be 40°C. The activity of immobilized lipase decreased with further increase in temperature. This is due to the deactivation of the enzyme as well as disintegration of the carrier material [18]. The rice husk showed better stability at high temperature as compared to the eggshell membrane.

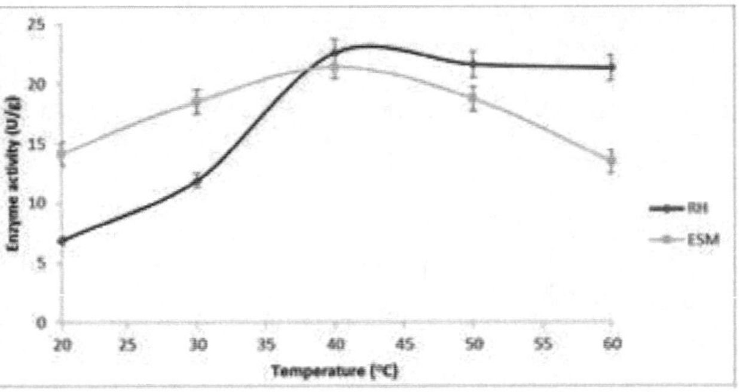

Figure 4. Temperature stability of immobilized lipase on rice husk and eggshell membrane

3.5. Organic solvent stability.
The stability of immobilized lipase on rice husk and eggshell membrane in different organic solvents are shown in Figure 5. It can be seen that immobilized lipase on eggshell membrane exhibited higher activity as compared to immobilized lipase on rice husk. Immobilized lipase on eggshell membrane gave higher enzymatic activity in ethanol as a solvent. On the other hand, immobilized lipase on rice husk showed maximum enzymatic activity in methanol. The stability of immobilized lipase in organic solvents is very important towards its application in industry. The stability inturn depends on the matrix used, mode of attachment of lipase to the matrix and also polarity of the solvent [23]. The higher stability of lipase immobilized on eggshell membrane as compared to the lipase on rice husk can be attributed to the better properties of eggshell membrane.

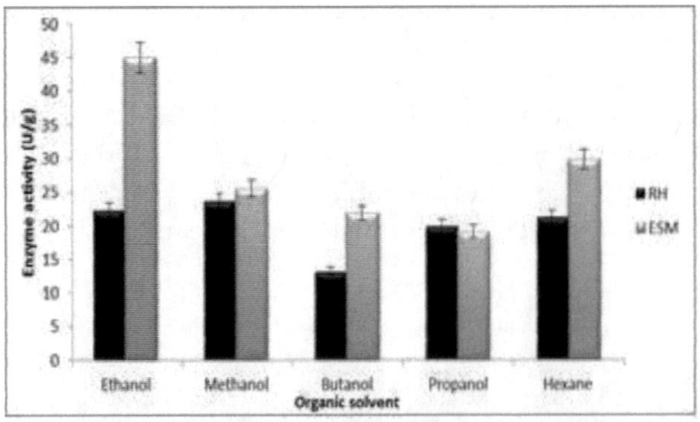

Figure 5. Organic solvent stability of immobilized lipase on rice husk and eggshell membrane.

3.6. Storage stability. The storage stability of immobilized lipase on eggshell membrane and rice husk for a period of seven days is shown in Figure 6. Both types of immobilized lipase were stored at 4°C in tris-HCl buffer (50mM, pH 7). In both cases, there was loss of enzyme activity with increase in number of days. The reasons can be attributed to leakage of enzyme from the carrier during storage

[21]. With respect to storage stability both types of immobilized lipase showed more or less similar characteristics.

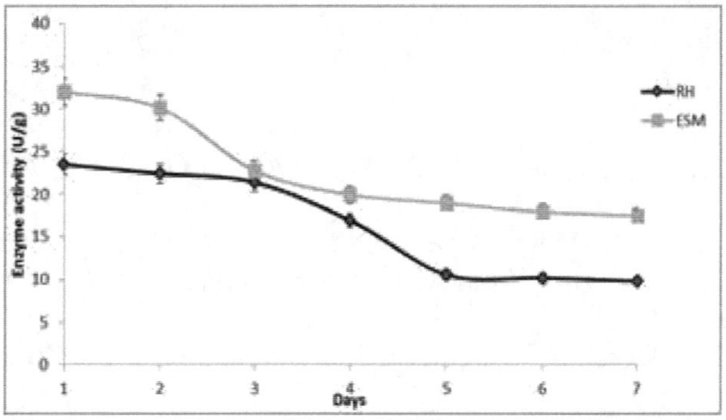

Figure 6. Storage stability of immobilized lipase on rice husk and eggshell membrane at 4^oC.

3.7. Reusability.
Reusability is one of the most significant parameters in design of an immobilized enzyme for industrial application. The enzyme activity of immobilized lipase on eggshell membrane and rice husk measured under the same storage conditions for over a period of ten reuses is shown in Figure 7. The enzymatic activity was shown to decrease with each cycle of reuse. After 10 number of cycles, immobilized lipase on eggshell membrane retained an activity of 13.5% and immobilized lipase on rice husk retained an activity of 11.3%. The loss of activity with each cycle of reuse can be explained in terms of enzyme leakage. Moreover, some enzymes might desorb from the matrix during the process of washing and reuse [24].

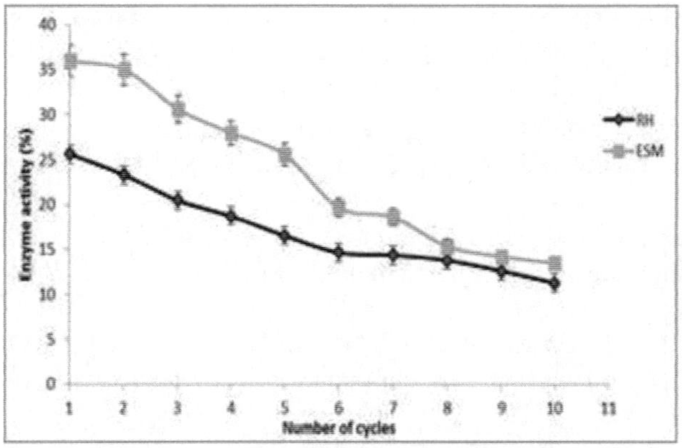

Figure 7. Reusability of immobilized lipase on rice husk and eggshell membrane

4. Conclusion

Agricultural and domestic wastes such as rice husk and eggshell membrane were tested as lipase immobilization matrices. Lipase from *B.cepacia* was successfully immobilized on eggshell membrane and rice husk and used for stability studies mainly by hydrolysing olive oil emulsion. Both types of immobilized lipase were proven to be quite faster and cheaper. In short, this work demonstrates a simple, easy and economical method to immobilize lipase for biodiesel production by utilizing domestic wastes such as egg shell and rice husk.

Acknowledgement
This work was partially supported by the Fundamental Research Grant Scheme (FRG0366-SG-1/2014).

References

[1] Jambo S A, Abdulla R, Azhar S H M, Marbawi H, Gansau J A and Ravindra P 2016 *Renew. Sustain. Energy Rev.* **65** 756.
[2] Galik C S, Abt R C, Latta G, Méley A and Henderson J D 2016 *Appl. Energy.* **172** 264.
[3] Fessner W D 2000 *Biocatalysis: From Discovery to Application* (Germany: Springer Science & Business Media)
[4] Mateo C, Palomo J M, Lorente G F, Guisan J M, Lafuente R F 2007 *Enzyme Microb. Technol.* **40** 1451.
[5] Barbosa O, Ortiz C, Murcia A B, Torres R, Rodrigues R C, Lafuente R F 2015 *Biotechnol. Adv.* **33** 435.
[6] Sirisha V L, Jain A, and Jain A 2016 *Adv. Food Nutr. Res.* **79** 179.
[7] Rueda N, Santos J, Ortiz C, Torres R, Barbosa O, Rodrigues R C and Fernandez-Lafuente R 2016 *Chem. Rec.* **16** 1436.
[8] Nastasovic A 2012 *Bioresour. Technol.* **115** 126.
[9] Kumar A, Mohanta K, Kumar D and Parkash O 2012 *Int. J. Emerging. Technol. Adv. Eng.* **2** 86.
[10] Aini B N, Siddiquee S, Ampon K, Rodrigues K F and Suryani S 2015 *Sens. Biosensing Res.* **4** 46.
[11] Vijayaraghavan K, Jegan J, Palanivelu K and Velan M 2005. *Min. Eng.* **18** 545.
[12] Tsai W T, Hsien K J, Hsu H C, Lin C M, Lin K Y and Chiu C H 2008 *Bioresour. Technol.* **99** 1623.
[13] Tembe S, Kubal B S, Karve M and Souza S F D 2008 *Anal. Chim. Acta.* **2** 212.
[14] Dhake K P, Deshmukh K M, Patil Y P, Singhal R S, Bhanage B M 2011 *J. Biotechnol.* **156** 46.
[15] Chattopadhyay S, Sen R 2012 *Process Biochem.* **47** 749.
[16] Xie C, Wu B, Qin S and He B 2016 *Bioprocess Biosyst. Eng.* **39** 59.
[17] Wan Ngah W S and Hanafiah M A K M 2008 *Bioresour. Technol.* **99** 3935.
[18] Seshabala P, Rao A V S P and Mukkanti K 2012 *Int. J. Eng. Resour. Technol.* **1** 1.
[19] Sheldon R A and Van Pelt S 2013 *Chem. Soc. Rev.* **42** 6223.
[19] Dalal S, Singh P K, Raghava S, Rawat S and Gupta M N 2008 *Biotechnol. Appl. Biochem.* **51** 23
[20] Abdulla R and Ravindra P 2013 *J. Taiwan. Inst. Chem. Eng.* **44** 545.
[21] Zhang S, Shang W, Yang X, Zhang S, Zhang X and Chen J 2013 *Bull. Korean Chem. Soc.* **34** 2741.
[22] Atadashi I M, Aroua M K, Aziz A A and Sulaiman N M N 2013 *J. Ind. Eng. Chem.* **19** 14.
[23] Jobanputra A H, Karode B A and Chincholkar S B 2011 *Biotechnol. Bioinformatics. Bioeng.* **1** 529.

IOP Conf. Series: Materials Science and Engineering **206** (2017) 012033 doi:10.1088/1757-899X/206/1/012033

The flow dynamics behind a flexible finite cylinder as a flexible agitator

T H Yong[1], H B Chan[2], S S Dol[2], S K Wee[2] and P Kumar[3]

[1]Department of Mechanical Engineering, Curtin University Malaysia
[2]Department of Petroleum Engineering, Curtin University Malaysia
[3]Department of Chemical Engineering, Curtin University Malaysia

Email: wee.siaw.khur@curtin.edu.my

Abstract. This paper investigates the flow dynamics behind a flexible finite cylinder in a single-phase flow using a water tunnel. The cylinder was individually submerged in water at Re_D = 4000, 6000 and 8000. The cylinder investigated has a *AR* = 10 and 16 and is made of EVA in order to achieve the lower stiffness for flexibility. A same *AR* of its aluminium rigid cylinder was investigated to serve as a benchmark to the flow dynamics behind a flexible cylinder. The results the downwash that hinders the transportation of vortices to the downstream was diminished. As a direct consequence of this phenomenon, the turbulence production has seen significant improvement for flexible finite cylinder.

1. Introduction

As a common passive turbulence enhancement method, a rigid cantilever is a very popular candidate to generate turbulence in the wake as it is well-known that flow separates when in contact with bluff body object. The study of flow dynamics behind a rigid cantilever shows highly complicated and three-dimensional flow around the free end. Regular Kármán vortices are supressed near the free end by tip effects but with a good cause as these tip vortices generated have more turbulent energy than the regular Kármán vortices [1]. Rostamy et al. [2] found a higher turbulence intensity and Reynolds stress at the free end. Apart from that, the downwash phenomenon associated with the rigid cantilever draws the fluid downwards to the ground instead of downstream, making most of the high kinetic energy content vortices impinge in the near wake, unable to travel downstream to a greater distance [3]. There have been consistent findings on the flow dynamics behind a rigid cantilever [4-7]. However, there are very little studies reporting on flexible cantilever. The only notable investigation was by Govardhan and Williamson [8] who reported that the Reynolds stress of an elastically-mounted rigid cylinder do increase significantly than that of a rigid cylinder.

In this study, flexibility is introduced into a finite cylinder (hereby refers as 'cylinder') by utilising low stiffness material (*i.e.* ethylene vinyl acetate). The cylinder's flexibility allows the cylinder to oscillate due to the alternating lift force resulted from vortex shedding. It is expected that the oscillating motion of the flexible cylinder alters the fluid dynamics hence generating stronger turbulence level than the rigid (static) cylinder. This study investigates the turbulence enhancement achieved by flexible cylinder by comparing the turbulence production between the rigid and flexible cylinders under the exact same flow condition.

IOP Conf. Series: Materials Science and Engineering **206** (2017) 012033 doi:10.1088/1757-899X/206/1/012033

2. Experimental Apparatus and Procedure

2.1. Water Tunnel and Test Models

A circulating open channel water tunnel with a test section of 200 mm x 200 mm in cross section and 750 mm in length was used. Figure 1(a) shows the rigid aluminium test cylinder of *AR* = 10 with a diameter of d = 13 mm and figure 1(b) shows identical cylinder but with *AR* = 16 and a diameter of d = 11 mm. The flexible cylinder in Figure 1(c) is made of ethylene vinyl acetate, EVA and has a diameter of d = 11 mm for both *AR* = 10 and 16. All cylinder has a flat and sharp edge at its free end.

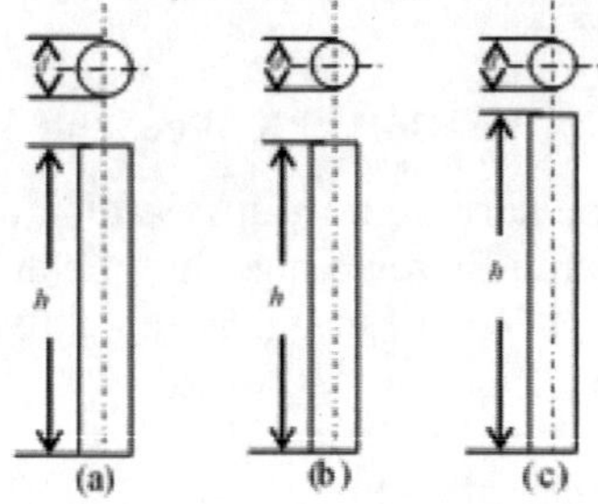

Figure 1. The cylinder models used in this study.

2.2. Experimental set-up

The experimental set-up within the test section is illustrated in Figure 2. A polished smooth surface end plate with 30° chamfer on the leading edge was employed to enhance the development of zero pressure gradient boundary layer. This design was also employed by Park and Lee [4] in their experimental work to investigate the free end effects on the near wake flow structure behind a finite circular cylinder. The origin is set at the centre of the cylinder placed perpendicularly to the flow on the end plate at 350mm distance from the leading edge to ensure the boundary layer was fully developed.

Measurements in the near wake (x-z plane) of the cylinder was measured over a 5 mm uniform grid at the top 5D span of the cylinder at streamwise location of x/D = 1 to x/D = 5. At the span below 5D, the measurement point in the wall-normal direction (z plane) increased to a 10 mm uniform grid until 10 mm from the ground. A more refined measurement points was given from the free end to 5D span of the cylinder because tip effects are generally extended up to 3D to 4D from the free end [4, 6, 9]. The cylinder was partially immersed in the boundary layer on the ground plane rating at most of $\delta/H = 0.43$. The blockage ratio was at most 4.8% and no correction was made. The Reynolds number based on the cylinder diameter is $Re_D = 4x10^3, 6x10^3, 8x10^3$.

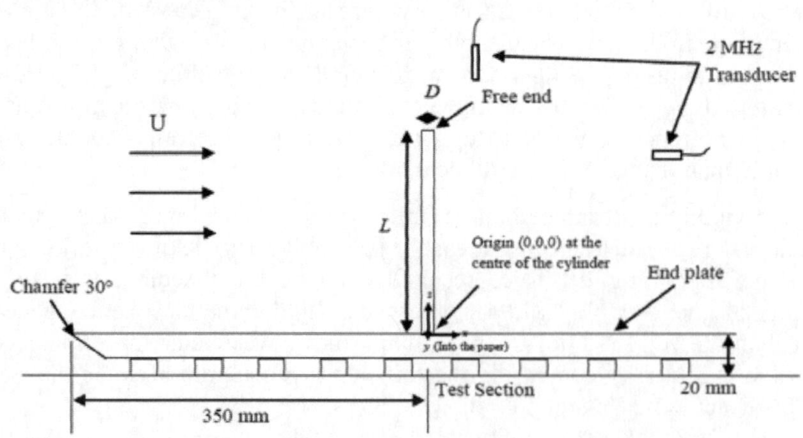

Figure 2. Schematic diagram of experimental set-up in the water tunnel

2.3. Ultrasonic Velocity Profiler

Ultrasonic Velocity Profiler, UVP manufactured by MET-FLOW is used for the measurements of time-averaged flow fields in the wakes. A UVP is a non-intrusive equipment that measures velocity as a function of both space and time in a one-dimensional Eulerian frame. When combined with two or more probes, it can measure a two or three-dimensional flow [10]. The UVP was mounted perpendicular to each other as shown in Figure 2 in the wake behind the cylinder. The horizontal probe was always positioned in a distance 8D downstream the cylinder with reference to the vertical probe to avoid interference in the near wake. Time-averaged streamwise (x-plane) and wall-normal (z-plane) velocity flow fields were measured. The measurement was done at a sampling rate of 100Hz. A total of 4096 profiles were taken for each velocity component.

3. Experimental Results and Discussion

The u- and w- components velocity profiles of AR=10 and AR=16 cases are shown in Figure 3 and 4 respectively.

3.1. Velocity profiles of rigid cylinder in the wake of x-z plane

The velocity components (u/U, w/U) of rigid cylinder at 5D downstream of the cylinder are presented. With reference to the u component, there is a very steep velocity gradient at the region near the free end for both AR. These changes are caused by the tip effects where the tip vortex plays a role. In general, the strong influence of tip effects can be seen taking effective to approximately 3D from the free end. The same phenomenon is consistently reported by Rostamy et al. [2]; Park & Lee [4]; Okamoto & Yagita [6]; Roh & Park [9].

Below the steep velocity gradient (indicated by double-dot dashed arrow) is a strong velocity defect that has a rather consistent profile where regular Kármán vortex is likely to exist. The strong velocity defect can serve as a 'body' of the wake region. The velocity defect happens due to the presence of vortical motion in the flow field which extract the energy from the mean flow, resulting in the lower fluid velocity [11]. Therefore, the wake region is approximately 3D shorter than the free end. When examined together with the w component, the higher value of localized negative w-component velocity at approximately $x/D < 3$ shows that the flow is brought downwards instead of downstream. It is a common phenomenon called downwash. The vortices generated are brought toward the floor immediately since they follow the path of the mean flow. Since the motion of vortex is hindered at near wake, it causes the region of augmented turbulence region effective in the very near wake. The variation in Re contributes minor influence to the velocity profiles behind a rigid cylinder.

3.2. Velocity profiles of flexible cylinder in the wake of x-z plane

To the best of author's knowledge, the velocity profiles of passively-oscillating flexible cylinder are uncommon, if not none. It is worth noting that since the flexible cylinder was deflecting during measurements, measurements in $x/D \leq 3$ were hindered by the deflected cylinder and could not be measured (depending on the degree of deflection). However, the earliest available measurements were put into comparison with rigid cylinder at $x/D = 1$ as it is the nearest position to the free end.

In general, as can be seen with the cylinders that deflect (AR = 16 at Re 4000, 6000 and 8000), the most noticeable observation is that the localized w-component velocity is weaker in the case of flexible cylinder. The lower value of localized w component as opposed to u component in the very near wake signifies that the influence of downwash is diminished. As a result, the vortices generated can be transported downstream instead of downwards. This observation suggests that the turbulence energy carried by the vortices can be delivered to downstream more effectively, resulting an increasing turbulence production at downstream. This shall be seen from the turbulence production in the under section.

IOP Conf. Series: Materials Science and Engineering **206** (2017) 012033 doi:10.1088/1757-899X/206/1/012033

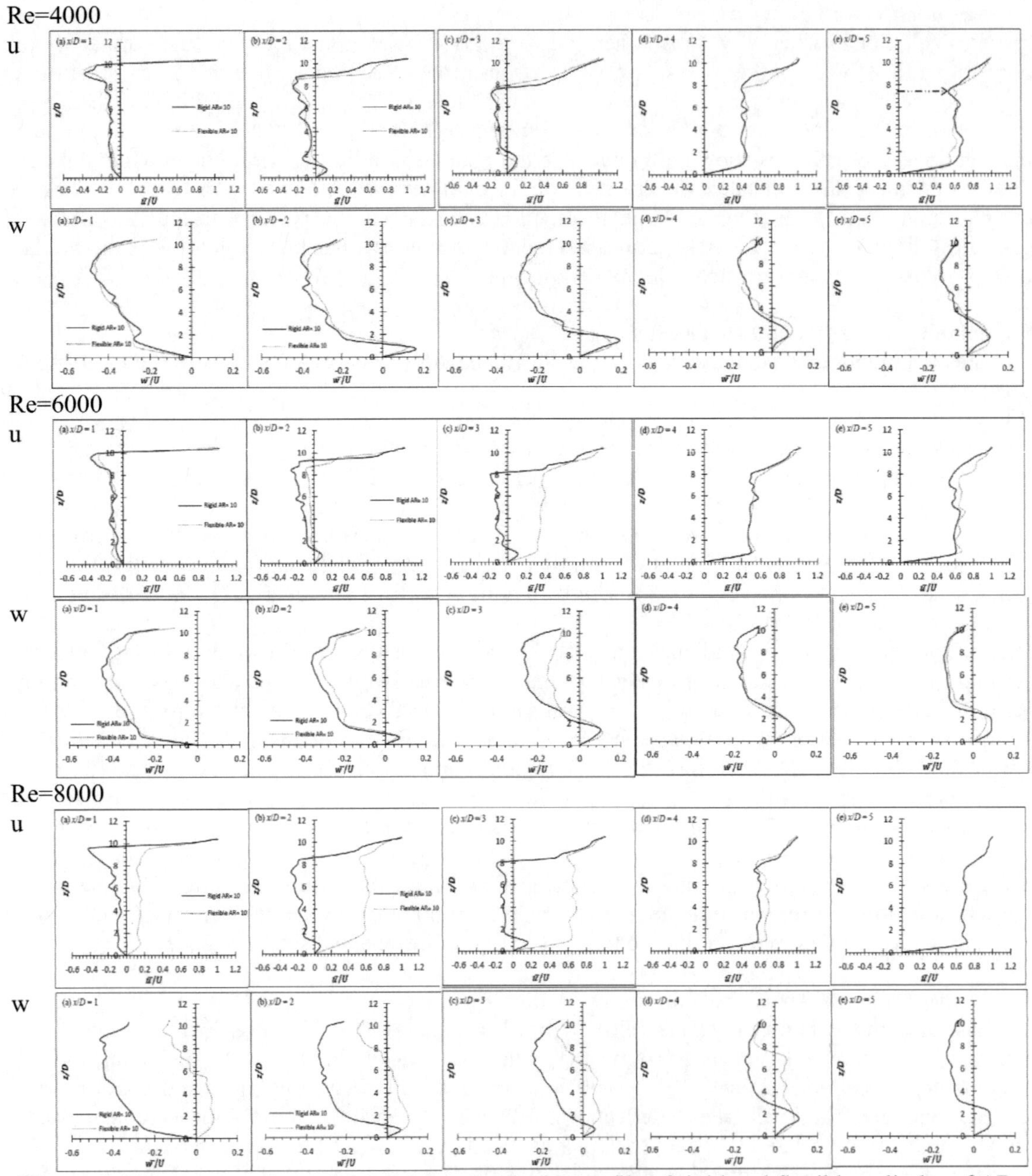

Figure 3. Time-averaged u and w component velocity profile of rigid and flexible cylinder of AR=10 at Re=4000, 6000 and 8000.

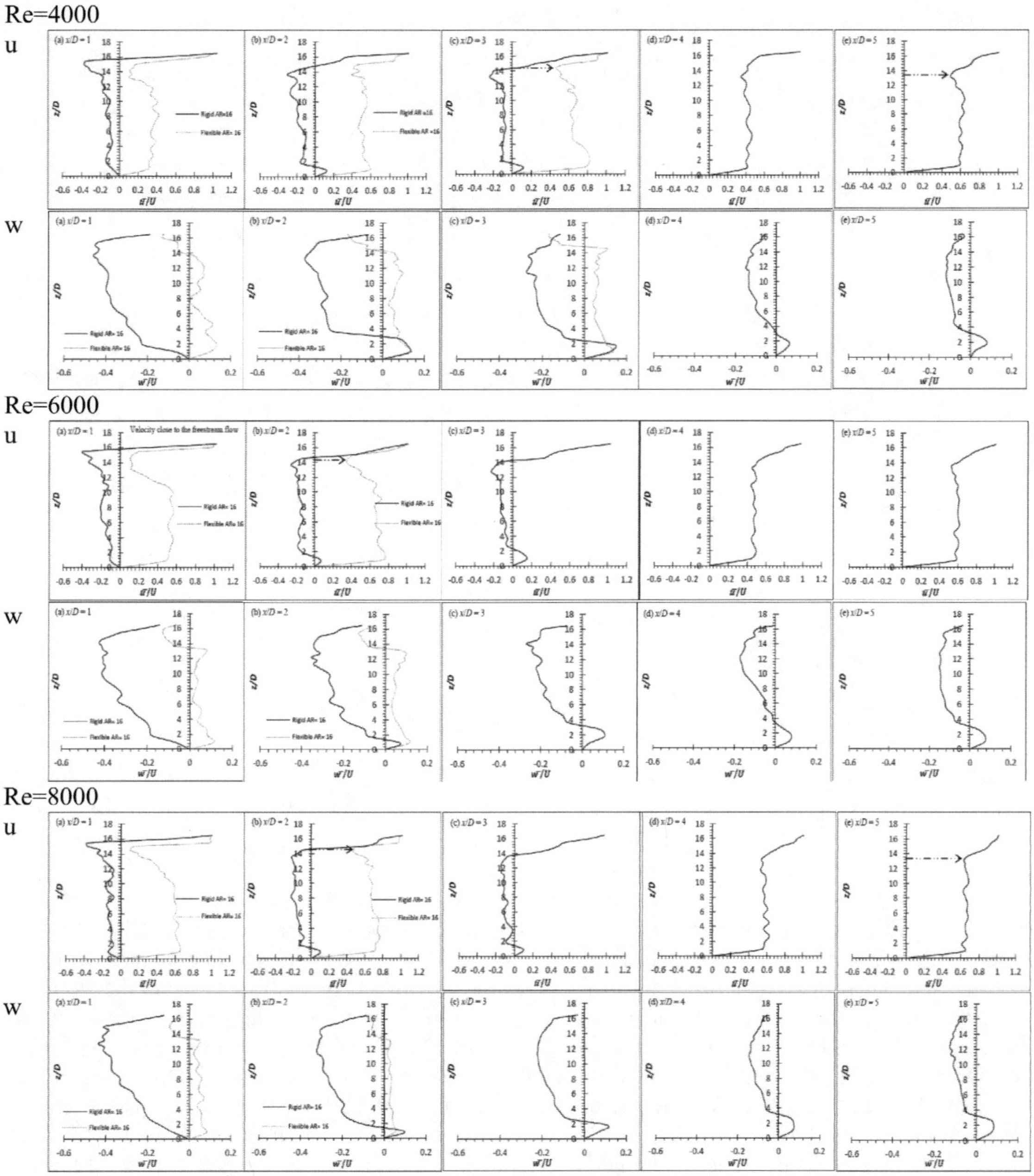

Figure 4. Time-averaged u and w component velocity profile of rigid and flexible cylinder of AR=16 at Re=4000, 6000 and 8000.

3.3. Turbulence production of rigid and flexible cylinders in the x-z plane

As mentioned earlier, the turbulence production between rigid and flexible case are compared to identify the greater turbulence enhancement achieved by flexible cylinder. With reference to this, the evolution of turbulent kinetic energy production is investigated in this section. The production term, (labelled as ***P*** for convenience) which is part of the kinetic energy budget equation, analyses the amount of energy that is contributed to the turbulence formation from the mean flow. The ***P*** term that is being evaluated here is:

$$\boldsymbol{P} = -[(u'\,u')\cdot\partial u/\partial x+(u'\,w')\cdot\partial U/\partial z+(u'\,w')\cdot\partial w/\partial x+(w'\,w')\cdot\partial w/\partial z] \quad (1)$$

The ***P*** of rigid case is as shown in Figure 5; flexible case is as shown in Figure 6.

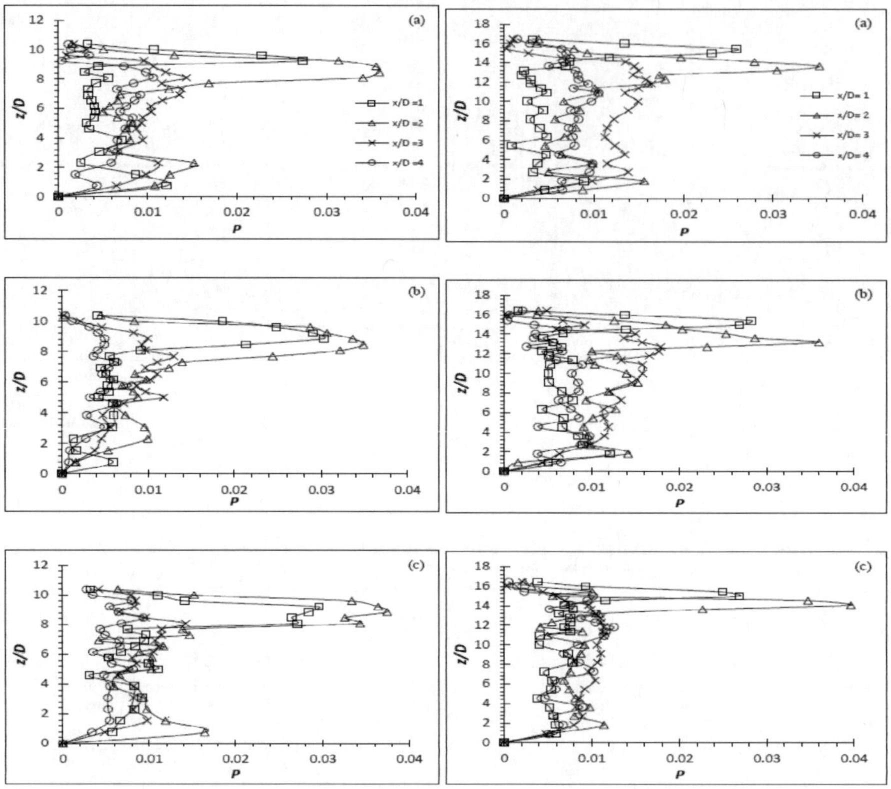

Figure 5. Production term for rigid cylinder of (left) AR=10 and (right) AR=16 at Re= (a) 4000, (b) 6000 and (c) 8000.

Based on the ***P*** behind the rigid cylinder as shown in Figure 5, it is observed that the ***P*** term varies at different *z*-location. The greatest ***P***, *i.e.* ~0.036m^2/s^3, is located approximately 2D below the free end of the cylinder and is impinged in near the wake, x/D ≤ 2. This is in good agreement with the mean flow velocity profiles (Figure 3 and 4) where the downwash transports most of the vortices downwards (*i.e.* right behind the cylinder). In other words, the vortices are temporally trapped in the near wake (x/D<2) due to downwash resulting the maximum ***P*** to occur in near wake region.

The ***P*** term for the flexible cylinder on the other hand (Figure 6), do see significant increase in value due to the effectiveness of VIV in increasing the strength of vortices shed [12]. The maximum increment achieved by flexible cylinder is 127% (2.27 times greater), which can be seen at x/D=4 of the AR=16 flexible cylinder. These elevated ***P*** values clearly show the stronger turbulence level behind a flexible cylinder than that of a rigid cylinder. Hence, proving that a flexible cylinder is indeed a better turbulence generator.

Besides, this is also an expected result as the previous finding (*i.e.* vortices are transported downstream instead of downwards due to the diminishing of downwash) has suggested that the energy carried by the vortices generated from the flexible cylinder can be transported to the downstream directly and more effectively. The fact that the ***P*** term is seen expanding to downstream further than that of rigid cylinder (*i.e.* maximum happened at x/D=4 instead of 2 as shown in rigid case) signifying the previous explanation is correct and the augmented turbulence level is indeed covering a greater region as discussed previously.

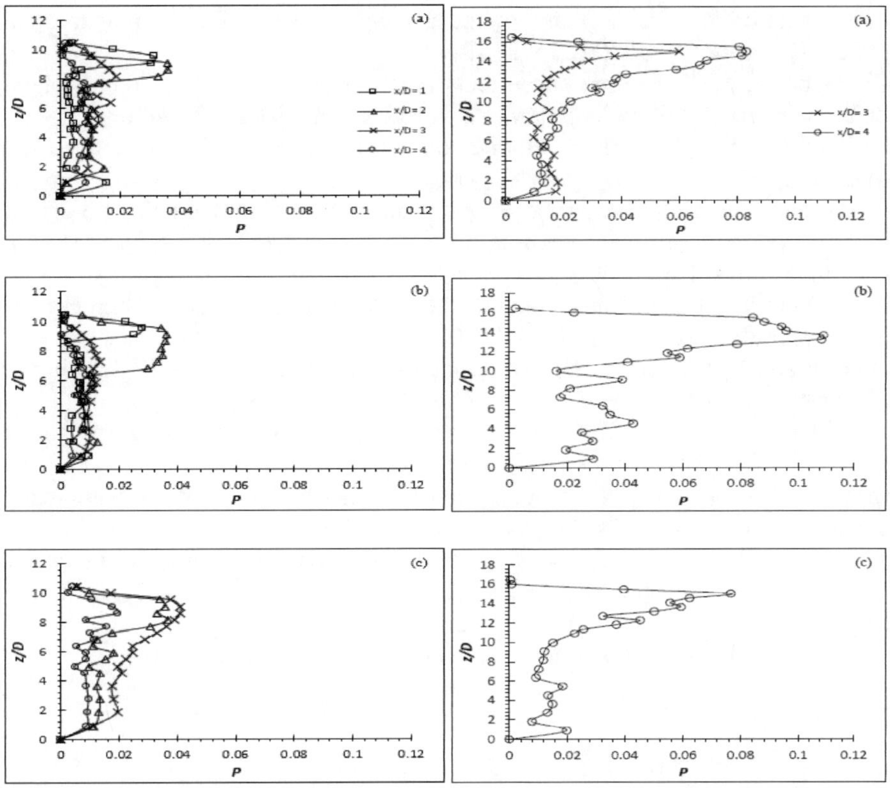

Figure 6. Production term for flexible cylinder of (left) AR=10 and (right) AR=16 at Re= (a) 4000, (b) 6000 and (c) 8000.

4. Conclusion

The flow dynamics behind flexible and rigid cylinder is experimentally studied and compared in this research. The flexible cylinder has shown to be a promising alternative to conventional rigid cylinder as agitator. Based on the current turbulence production (***P*** term) analysis, the flexible cylinder can generate turbulence that is, at maximum 2.27 times, of the amount generated by the rigid cylinder under the same flow condition. This clearly shows that the turbulence production of the flexible cylinder is significantly greater than that of the rigid cylinder. Based on the current analysis, it is found that one of the contributors of such elevation is due to the weakening of downwash behind the flexible cylinder as shown from the u- and w-component velocity profiles. As a result, the flow can be transported downstream more effectively, so are the vortices generated as they generally follow the path of the mean flow. Subsequently, the turbulence production, ***P*** shows a greater value at the downstream of the flexible cylinder, signifying stronger turbulence.

Acknowledgement

This project is supported by Fundamental Research Grant Scheme (FRGS) and Curtin University Malaysia.

References

[1] Sumner D 2013 Flow above the free end of a surface-mounted finite-height circular cylinder: a review *J.l of Fluids and Struct.* **43** 41-63

[2] Rostamy N, Sumner D, Nergstrom DJ, and Bugg JD 2012 Local flow field of a surface-mounted finite circular cylinder *J. of Fluids and Struct.* **34** 105-22

[3] Kawamura T, Hiwada M, Hibino T, Mabuchi I and Kumada M 1984 Flow around a finite

circular cylinder on a flat plate: cylinder height greater than turbulent boundary layer thickness *Bulletin of JSME* **27**(232) 2142-51

[4] Park CW and Lee SJ 2000 Free end effects on the near wake flow structure behind a finite circular cylinder *J. of Wind Eng. and Ind. Aerodynamics* **88**(2) 231-46

[5] Park CW and Lee SJ 2004 Effects of free-end corner shape on flow structure around a finite cylinder *J. of Fluids and Struct.* **19**(2) 141-58

[6] Okamoto T and Yagita M 1973 The experimental investigation on the flow past a circular cylinder of finite length placed normal to the plane surface in a uniform stream *Bulletin of JSME* 19(**95**) 805-14

[7] Sumner D, Heseltine JL, and Dansereau OJP 2004 Wake structure of a finite circular cylinder of small aspect ratio. *Exp. in Fluids* **17**(5) 720-30

[8] Govardhan R and Williamson CHK 2001 Mean and fluctuating velocity field in the wake of a freely-vibrating cylinder *J. of Fluids and Struct.* **15**(3) 489-501

[9] Roh S and Park S 2003 Vortical flow over the free end surface of a finite circular cylinder mounted on a flat plate *Exp. in Fluids* **34**(1) 63-67

[10] Yokoyama K, Kashiwaguma N, Okubo T and Takeda Y 2004 Flow measurement in an open channel by UVP *proceedings ISUD* **4** 201-10

[11] Chan HB, Yong TH, Kumar P, Wee SK and Dol SS 2016 The numerical investigation on the effects of aspect ratio and cross-sectional shape on the wake structure behind a cantilever *ARPN J. of Eng. and App. Sci.* **11**(16) 9926-32

[12] Thulukkanam K 2013 *Heat exchanger design handbook* CPC Press

IOP Conf. Series: Materials Science and Engineering **206** (2017) 012034 doi:10.1088/1757-899X/206/1/012034

Determination of sugars composition in abscission zone of oil palm fruit

Y M Thang[1,5], A A Ariffin[1,3,4], D R Appleton[6], A J Asis[4], M N Mokhtar[2] and R Yunus[1,2]

[1] Institute of Advanced Technology, Universiti Putra Malaysia, 43400 UPM Serdang, Selangor, Malaysia
[2] Faculty of Engineering, Universiti Putra Malaysia, 43400 Serdang, Selangor, Malaysia
[3] Faculty of Food Science and Technology, Universiti Putra Malaysia, 43400 UPM Serdang, Selangor, Malaysia
[4] Dolphin International Berhad, 17 & 19, Jalan Puteri 5/20, Bandar Puteri, 47100 Puchong, Selangor, Malaysia
[5] Sime Darby Research Sdn Bhd, R & D Centre Carey Island, 42960 Kuala Langat, Selangor, Malaysia
[6] Sime Darby Technology Centre Sdn Bhd, 1st floor, Block B, UPM-MTDC Technology Centre III, Universiti Putra Malaysia, 43400 Serdang, Selangor, Malaysia
E-mail: robiah@upm.edu.my

Abstract. Fresh oil palm fruit bunches (FFB) arriving at a palm oil mill are graded manually and randomly for ripeness classification by counting the number of empty fruit sockets (EFS) found in each bunch before processing. FFBs with at least ten EFS are classified as ripe bunch, FFBs with less than ten EFS are classified as under-ripe, while bunches without any EFS are classified as unripe. The aim of the present study is to determine the composition of sugars in the abscission of these three groups of FFBs by monitoring their sugars composition. The bunches were grouped according to the number of empty fruit sockets: (i) nil; (ii) 1-9; (iii) ≥10 as unripe, under-ripe and ripe bunches, respectively. Non-structural, structural and water-soluble sugars extracted from the abscission zone were analyzed. The principal component analysis (PCA) based on various sugars compositions revealed some natural clustering among the samples. Bunches with more than one empty fruit sockets were distinguished from the others using glucose, sucrose and oligomers. In conclusion, analysis of sugars composition of the abscission zone could potentially be used as a chemical marker to differentiate those bunches at different stages of ripeness.

1. Introduction

In oil palm plantations, the detachment and fall of the ripe fruit from its oil palm fruit bunch is taken as an indication of readiness for harvesting [1, 2, 3]. Additionally, the change of mesocarp color is also used as another indicator [2, 3, 4]. Orange bunches are classified as ripe while purplish or yellowish bunches are classified as unripe or under-ripe. Precise determination of ripeness is important during harvesting because milling of unripe and under-ripe bunches results in a lower oil extraction rate while over-ripe bunches reduce the quality of palm oil produced [4, 5]. In palm oil mills, fresh fruit bunches (FFBs) are checked randomly via visual inspection and counting the numbers of empty fruit sockets (EFS) per bunch by the experienced grader on duty. According to Malaysian Palm Oil Board (MPOB), FFBs with at least ten EFS are classified as ripe bunch, FFBs with less than ten EFS are classified as under-ripe, while bunches without any EFS are classified as unripe [6]. Therefore, the present study is to find out the difference amongst these three groups of FFBs by monitoring their sugars composition.

Colorimetry approaches such as ultraviolet-visible photometric [2, 5, 7, 8, 9, 10], near infrared spectroscopic [11] and multi-parameter fluorescence sensor [12] were investigated to classify the ripeness of FFBs. Other researchers, taking the metabolomics approach, revealed the changes in amino acids, organic acids, sugars and nucleosides [13, 14], lipid content [13, 15, 16] and polyamines [17] during the oil palm fruit development. [17, 18, 19, 20] reported that the transcription factors expressed might lead to the identification of ripening. These studies monitored the controlled samples collected from 12 to 24 weeks after anthesis. No study has sought a profiling pattern between these characteristics and the manual ripeness determinations of FFBs delivered to the palm oil mill. The characteristics of FFBs delivered to the mill might be different from those of controlled bunches from trial plot because of a certain degree of damage due to the handling during transportation [21] .

Fruit ripening is reported to associate with biochemical changes in cell wall materials such as cellulose, hemicellulose and pectin via hydrolytic processes [22, 23]. The chemical changes, especially at the abscission zone have been found to result in separation of cells, as well as in separation of the fruit from the stalk [24]. The abscission zone is the place from which the oil palm fruit is shed (Figure 1). Hydrolase were suggested to be involved in cell wall degradation might be responsible for co-ordinating the shedding of the oil palm fruit [20, 25]. Younger palm fruits found to exhibit slightly thicker cell walls and middle lamellae at the abscission zone than those mature fruits [24]. Fruit separation is determined by the age or ripeness of the fruits [26].

In this study, the abscission zones of unripe, under-ripe and ripe FFBs were evaluated based on the non-structural, structural and water-soluble sugar contents in the present study. The sugars data obtained were analyzed using chemometric techniques to classify and differentiate according to the ripeness of the FFBs. This information could potentially be used as a reaffirming biochemical transformation to show a similar classification pattern using the empty fruit socket monitoring grading system in palm oil mills.

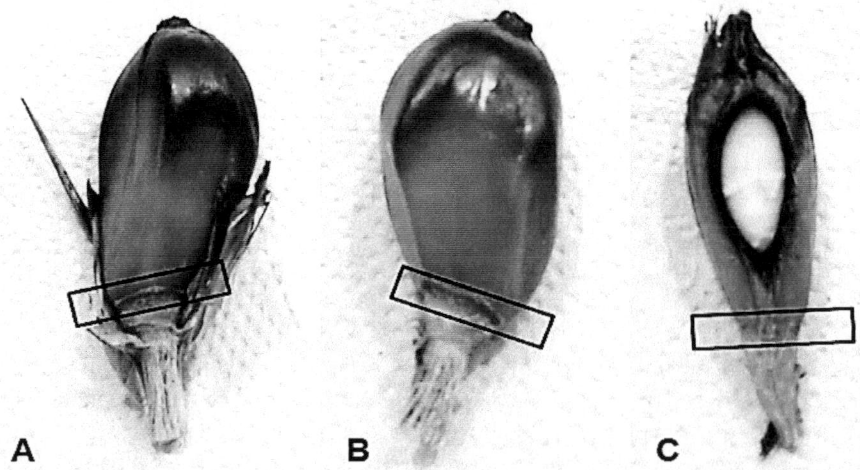

Figure 1. Abscission zone (rectangle area) of an oil palm fruit. A, with tepal attached; B, without tepal; C, halve view.

2. Materials and methods

2.1. Fresh fruit bunches

Oil palm fresh fruit bunches (FFBs) were collected from Labu Palm Oil Mill, Labu, Negeri Sembilan. In total, 27 FFBs were collected: 10 ripe ($\geq$ 10 empty fruit socket); 8 under-ripe (1 to 9 empty fruit socket); 9 unripe (no empty fruit socket). The weight of FFBs collected were ranging from 15kg to 45kg. FFBs were cut to obtain the spikelets of aprical (top), equatorial (middle) and basal (bottom)

IOP Conf. Series: Materials Science and Engineering **206** (2017) 012034 doi:10.1088/1757-899X/206/1/012034

regions according to [3]. All samples were stored at -80°C until analysis. Samples were thawed and cut to remove the abscission zone [27]. The abscission zone was ground in liquid nitrogen using an IKA Batch Mill A11 Basic (Wilmington, North Carolina, USA) and then freeze-dried with a ScanVac CoolSafe freeze dryer (Lunge, Denmark). The freeze-dried samples were then kept at -80°C prior to analysis.

2.2. Standards and reagents

Analytical grade ethanol was purchased from Merck (Darmstadt, Germany), sulfuric acid and phenol were purchased from Fisher Scientific (Geel, Belgium). Calcium carbonate and sugar standards, namely xylan, sucrose, glucose, xylose, arabinose, galactose, mannose and fructose were purchased from Sigma Aldrich (St Louis, MO, USA).

2.3. Determination of non-structural sugar (NS)

The non-structural sugar (NS) was extracted and analyzed with slight method modification [28]. About 300mg of dried samples were extracted with 5mL of 80% ethanol. An Eppendorf 5355 thermomixer (Hamburg, Germany) was used to maintain the temperature at 40°C with a mixing speed of 750rpm for 10 min. After each extraction, the tubes were centrifuged at 4000rpm for 5 min with an Eppendorf 5810R centrifuge (Hamburg, Germany), the supernatants of three extractions were combined for sugar analysis. Approximately 50μL of solution containing sugar was added with 150μL of distilled water. Approximately 400μL of 2% phenol was added into the diluted sugar solution followed by additional of 1mL concentrated sulfuric acid. The solution was stood in dark for 10 mins and then left in ice water for another 10 mins. About 200μL of brown solution containing sugar derivatives was placed in a 96-well plate and analyzed at 490nm using a BioTek ELx800 microplate reader (Winooski, USA). The sugar content in each sample was quantified against glucose standard solution with concentration ranging from 0.05% w/v to 1.5% w/v.

2.4. Determination of structural sugar (SS)

Structural sugar (SS) extraction was conducted according to the NREL method [29] with slight modification. Approximately 3mL of 72% sulfuric acid was added into the tube containing the residue from the ethanol extraction and vortex for 1 min. The tube was shook at 400rpm at 30°C in a Thermo Scientific MaxQ400 incubator shaker (Waltham, Massachusetts, USA) for 1 hour After that, about 84mL of purified water was added into the tube. Then, the sample was autoclaved for 1 hour at 121°C. After completion of the autoclave, the hydrolyzed solution was allowed to cool to room temperature. About 12mL of the hydrolyzed solution was transferred to another vial; about 0.6g of calcium carbonate was added slowly to obtain pH 5-6. The total sugar content of supernatant was analyzed at 490nm similar to aforementioned NS procedures.

2.5. Determination of water-soluble sugar (WS)

Approximately 300mg of freeze-dried samples were autoclaved for 1 hour at 121°C. After completion of the autoclave, the samples were extracted with 5mL of water for three times. The supernatants, namely water-soluble sugar (WS) of three extractions were pooled and analyzed at 490nm as per aforementioned NS procedures.

2.6. High performance liquid chromatography (HPLC) analysis

The water-soluble sugar (WS) were filtered at 0.2μm and analyzed the sugar composition using an Agilent 1260 Infinity LC system coupled with 1260 Infinity Evaporative Light Scattering Detector (HPLC-ELSD) (Santa Clara, USA). Separation was carried out using an Aminex HPX-87P column (9μm, 300mm x 7.8mm i.d) purchased from Bio-Rad (Philadelphia, USA). The column was set at 80°C. The temperature of the drift tube and sensitivity of detector were set at 75°C and 5, respectively with nitrogen gas flow of 1.6L/min. Filtered water used as the mobile phase. Total run time was 25 mins. A series of sugar standard solution, consisting of xylan, sucrose, glucose, xylose, arabinose, galactose, mannose and fructose were prepared in concentration ranging from 0.01 % w/v to 0.3 % w/v.

2.7. Principal component analysis and statistical analysis
The data is expressed as a mean ± standard deviation (SD) of triplicate analyses. All statistical analyses were performed using Microsoft office Excel 2010. Analysis of variance (ANOVA) and significance, defined at $P<0.05$ were applied. Principal component analysis (PCA) was performed using Simca-P, version 13 (Umetrics).

3. Results and discussion

3.1. Statistical analysis
All values are presented as mean±standard deviation (SD). The data were analyzed using ANOVA. No significant difference was observed between the samples collected from the regions of top, middle and bottom from a bunch. The fruits on a bunch do not ripen simultaneously, with ripening usually begins from the top to the bottom of the bunch [3]. Samples collected in the present study are abscission zone which were also different from those fruit samples from the previous study. Abscission zone containing cell wall hydrolytic enzymes that enhance the degradation of cell wall components for fruit separation, which is different from the adjacent tissues [20]. Abscission zone were found to have no oil body in their cells, that also indicating that abscission zone is different from other cells of a ripe fruit mesocarp [24, 25].

3.2. Non-structural, structural, water-soluble sugars composition
The grading standard implemented in a palm oil mill is based on total numbers of empty fruit socket (EFS), i.e unripe, nil; under-ripe, 1-9 EFS; ripe, ≥ 10 EFS. The bunches collected regardless of the aforementioned ripeness stage, they are ready-to-harvest with maximum oil content. Therefore, in the present study, the sugar content at the abscission zone was used as a chemical marker to distinguish those bunches containing different numbers of empty fruit socket.

The sugars composition, namely non-structural sugar (NS), structural sugar (SS) and water-soluble sugar (WS) of each sample were analyzed and and summarized in Table 1. Sugar-based carbohydrates can be classified as non-structural and structural [29]. The non-structural carbohydrates are the extractive which can be removed using extraction or washing step. On the other hand, structural carbohydrates are those bound in the matrix of the biomass, such as cellulose and hemicellulose. The NS, SS, WS were significantly different due to the differences in the treatment method which resulted in certain degree of hydrolysis of the samples.

From the ANOVA analysis, the data for unripe, under-ripe and ripe bunches were significantly different since the p value <0.05. On the other hand, no difference was observed in between under-ripe and ripe bunches. Therefore, sugars composition, such as NS, SS, or WS could be used to distinguish unripe bunches from those under-ripe and ripe bunches. In addition, NS, SS, or WS could be used to distinguish those FFBs with and without empty fruit socket, because all fruits are intact in unripe bunches while under-ripe and ripe bunches contain at least one detached fruit.

3.2.1. Non-structural sugar (NS) content
The NS content found in abscission zone of unripe bunches were ranging from 1.78% to 3.61% with a mean of 2.47±0.39%. For under-ripe bunches, the NS content of 1.80-3.29% were detected. This also recorded a mean of 2.68±0.37% of NS in under-ripe bunches. On the other hand, about 2.08-3.84% of NS were detected in ripe bunches with a mean of 2.86±0.47%. From the mean value, the highest amount of NS was found in the ripe bunches, followed by under-ripe and then unripe bunches. Water extracted sugar in oil palm was detected as 1.3% w/w at week 23 after pollination [18]. Sugars content detected in the samples of present study was slightly higher as compared to that reported [18], this might due to the FFBs collected in the present study are mostly with slight degree of damage

IOP Conf. Series: Materials Science and Engineering **206** (2017) 012034 doi:10.1088/1757-899X/206/1/012034

during transportation from the plantation to the mill. Furthermore, the samples used in the present study were obtained from the abscission zone, which contain mostly the cell wall materials.

Table 1. Sugars composition (% w/v) in FFB samples, namely non-structural (NS), water-soluble (WS) and structural (SS) sugars.

	NS	WS	SS
Unripe (n= 9)			
Top	$2.452{\pm}0.315^{ab}$	$3.197{\pm}0.561^{ab}$	$5.277{\pm}0.644^{ab}$
Middle	$2.524{\pm}0.337^{ac}$	$3.141{\pm}0.478^{ad}$	$5.229{\pm}1.252^{ac}$
Bottom	$2.424{\pm}0.510^{bc}$	$3.342{\pm}0.409^{ce}$	$4.443{\pm}1.009^{bc}$
Under-ripe (n=8)			
Top	$2.487{\pm}0.457^{ab}$	$3.525{\pm}0.531^{ab}$	$5.929{\pm}0.714^{ab}$
Middle	$2.883{\pm}0.314^{ac}$	$3.469{\pm}0.378^{ac}$	$5.857{\pm}0.950^{ad}$
Bottom	$2.677{\pm}0.192^{bc}$	$3.702{\pm}0.611^{bc}$	$6.291{\pm}0.422^{ce}$
Ripe (n=10)			
Top	$2.852{\pm}0.460^{ab}$	$3.650{\pm}0.578^{ab}$	$6.048{\pm}0.936^{ab}$
Middle	$2.880{\pm}0.534^{ac}$	$3.723{\pm}0.187^{ac}$	$6.305{\pm}0.886^{ac}$
Bottom	$2.841{\pm}0.454^{bc}$	$3.457{\pm}0.417^{bc}$	$5.767{\pm}0.665^{bc}$

Means within a column segment followed by different superscript letters (a-e) are significantly different ($p<0.05$).

3.2.2. Water-soluble sugar (WS)

The water-soluble sugar (WS) obtained after autoclave were found slightly higher than those NS content. The WS content found in abscission zone of unripe bunches were ranging from 2.52% to 4.53% with a mean of 3.23±0.48%. In under-ripe bunches, the WS content of 2.75-4.82% (mean: 3.57±0.51%) were detected in bunches. Besides, about 2.68%-4.68% (mean: 3.61±0.43%) of WS were found in ripe bunches. From the mean value, similar trend of WS was observed, highest WS amount was found in ripe bunches, followed by under-ripe and then unripe bunches. The amounts of WS were found slightly higher than those NS contents, due to the hydrolysis or degradation of cell wall materials that took place during high temperature autoclave before extraction. Elevated temperatures with sufficient water enhanced the hydrolysis of cell wall materials and resulted in the formation of five-carbon and six-carbon simple sugars [24]. In addition, high temperature hydrolysis happened in the palm fruits during sterilization would degrade the semi-stable hemicellulose [24].

In order to confirm the degradation of of cell wall materials, the composition of WS of each sample was analyzed using HPLC-ELSD and quantified based on the calibration curves of respective standards, namely glucose, xylose, arabinose, galactose, mannose and fructose. One of the chromatograms is compared to the standards as shown in Figure 2. Sucrose and glucose were found clearly in the chromatogram, both ranging from 0.34% to 0.54% for all unripe, under-ripe and ripe samples. Higher sucrose (1.3% w/w) than glucose (0.058% w/w) and fructose (0.15% w/w) were found in the ripe oil palm samples [18]. The values found were different from those reported [18] , it could be due to (i) different samples (abscission zone vs palm fruits), (ii) different extraction method and (iii) different analytical methods used in both study. Formation of glucose, sucrose and fructose could be due to sucrose catabolism inside the palm fruit before the harvesting [30]. In the present study, no fructose was detected maybe because of the detection limitation of the analytical instrument.

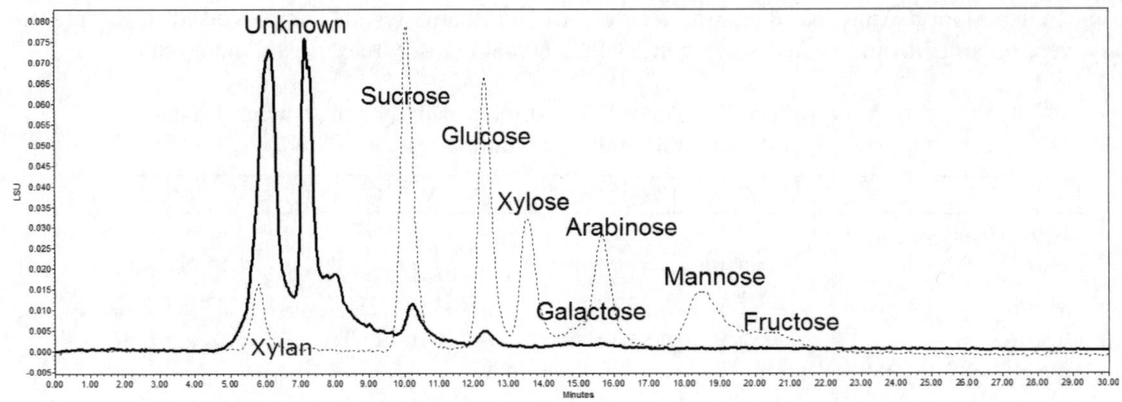

Figure 2. HPLC-ELSD chromatogram of sample (black line) as compared to that of standards (dotted line).

Table 2. Sugar composition of WS fraction (% w/v) in FFB samples.

	Glucose	Sucrose	Unknown
Unripe (n= 9)			
Top	0.432±0.065[ab]	0.416±0.041[ab]	5.521±1.253[ab]
Middle	0.437±0.071[ac]	0.420±0.059[ac]	5.776±1.260[ac]
Bottom	0.398±0.055[bc]	0.392±0.044[bc]	7.917±2.354[bc]
Under-ripe (n= 8)			
Top	0.407±0.052[ac]	0.391±0.036[ac]	6.657±1.954[ab]
Middle	0.480±0.089[bd]	0.444±0.073[bd]	6.491±2.625[ac]
Bottom	0.411±0.079[cd]	0.391±0.069[cd]	6.968±1.519[bc]
Ripe (n= 10)			
Top	0.429±0.062[ab]	0.418±0.047[ab]	5.578±1.426[ab]
Middle	0.436±0.066[ac]	0.423±0.039[ac]	6.098±1.529[ac]
Bottom	0.412±0.041[bc]	0.399±0.033[bc]	6.954±1.736[bc]

Means within a column segment followed by different superscript letters (a-e) are significantly different ($p<0.05$).

A group of unknown peaks was observed at retention time between 4.5 min to 8.5 min. These peaks were unable to be identified due to lack of standards. The unknown peaks were suspected as the oligo-intermediates by comparing the chromatogram and retention time with the study of Schwald and team [31], they found that the oligomers eluted before glucose and sucrose using similar column and HPLC setting.

Therefore, by assuming the unknown peak is oligo-intermediates, it can be assumed that the hydrolysis did take place during the autoclave to degrade some of the hemicellulose into the smaller oligo-intermediates. However, autoclaving at 121°C for 1 hour was not sufficient to convert them further to simple monomers may be due to insufficient quantity of water and missing acid catalyst. Thus, none of the hemicellulose monomer such as xylose, arabinose and galactose was detected by HPLC-ELSD. Hemicellulose in palm fruits is labile to heat if the heating condition is supplied with sufficient water [24] . In addition, [32], [33], [34] found that the degradation of cell wall materials formed oligo-intermediates before formation of the monomers such as glucose and xylose. Due to the amount of unknown peaks are significant in the chromatogram; therefore, they were quantified as xylan equivalent, as the first unknown peak having the retention time similar as xylan (5 min to 6 min). The unknown compounds found in unripe, under-ripe and ripe bunches were 6.40±1.97%, 6.71±2.02% and 6.21±1.62%, respectively. Oligomers are hydrolyzed by concentrated sulfuric acid during the phenol-sulfuric assay to form monomers [28]. Therefore, although none of xylose, arabinose, nor galactose was detected by HPLC, but higher amount of water soluble sugar were detected as WS due to the hydrolysis of unknown compounds into reducing sugar.

3.2.3. Structural sugar (SS)
With the addition of sulfuric acid, more sugars in the form of structural sugar (SS) were extracted as a result of more cell wall materials were hydrolyzed. The amount of SS obtained were 2.54% to 7.83%, with a mean of 4.98±1.04% in unripe. About 4.05% to 7.53%, with a mean of 6.03±0.73% of SS were detected in under-ripe bunches. For the ripe bunches, about 4.18-7.42%, with a mean of 6.04±0.84% of SS were detected. From the mean value, higher WS amount was found in ripe and under-ripe bunches as compared to those of unripe bunches. Cell wall materials were hydrolyzed more easily with the aid of acid than with water alone. Hydrothermal treatment was not as good as diluted acid hydrolysis of wood material to produce their respective monomers [34], [35], This also in agreement with the results of structural sugars obtained in the present study where the structural sugars were found higher than water-soluble sugars.

3.3. Principal component analysis (PCA)
Principal component analysis method or PCA was performed on the autoscaled data matrix, to provide a partial visualization of data structure in order to identify the distribution among the samples with different ripeness. PCA was conducted on the sugars composition data of the 27 FFBs with different ripeness and bunch region of top, middle and bottom. The data matrix consisted of 90 rows and 6 variables (NS, SS, WS, glucose, sucrose, unknown). The 2D scatter plot shown in Figure 3 did not show any cluster present. However, after rearranging it into a 3D plot, the PCA on the data revealed the presence of some natural clustering as shown in Figure 4. The first and second PCA factors together reveals 63% of the variance in the data set. From the PCA scatter plot, samples formed a narrower cluster, ripe samples (green) on top of unripe samples (blue), while under-ripe samples (red) appeared at the bottom along the Num-axis. Although some misclassifications were observed, the output of the PCA plot was very satisfactory. This could be due to certain degree of damage of the samples when transporting from the plantation to the mill. By comparing the PCA scores plots with the loading plots, the component mostly contributed to the separation of the ripeness classification is revealed. The unknown compounds are the chemical markers which are responsible for differentiating the ripe samples from the others. WS components, namely glucose, sucrose present in the unripe samples are markedly different from the WS in the under-ripe and ripe samples. While both non-structural and structural sugars are the factors that differentiate under-ripe samples from the rest as shown in Figure 4.

4. Conclusions
The classification of FFB ripeness by measuring the sugars composition seems feasible to be used as the reaffirming chemical markers to support the ripeness grading system according to the total number of the empty fruit socket that is currently practised in palm oil mills. Sugars composition such as non-structural, structural, or water extract could be used to differentiate in between bunches with loose fruits and those without any loose fruit. Compositions such as glucose, sucrose and unknown compounds could be used to confirm the ripe bunches with at least 10 empty fruit sockets. In combination with chemometrics method namely principal component analysis (PCA), the ripeness classification could be visualised. Our chemometrics approach was based on an unbiased autoscaled classification approach. The present method is interesting as it is able to distinguish the ripeness stage of the oil palm fresh fruit bunches based on the clustering of certain chemicals compounds, thus provides an in-dept understanding of the components of cell at different ripeness stages.

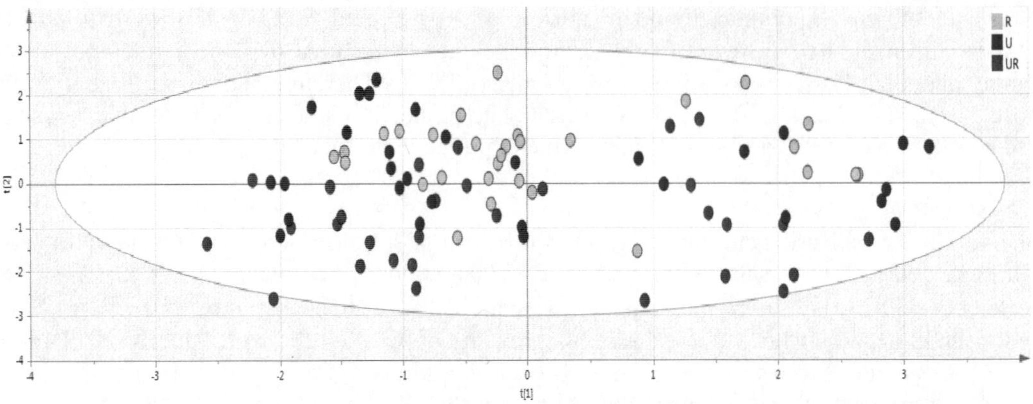

Figure 3. First two factors of the PCA scores plot on the different sugars composition (auto-scaled data) of FFBs with different ripeness, size and bunch region: green, ripe; blue, unripe; red, under-ripe.

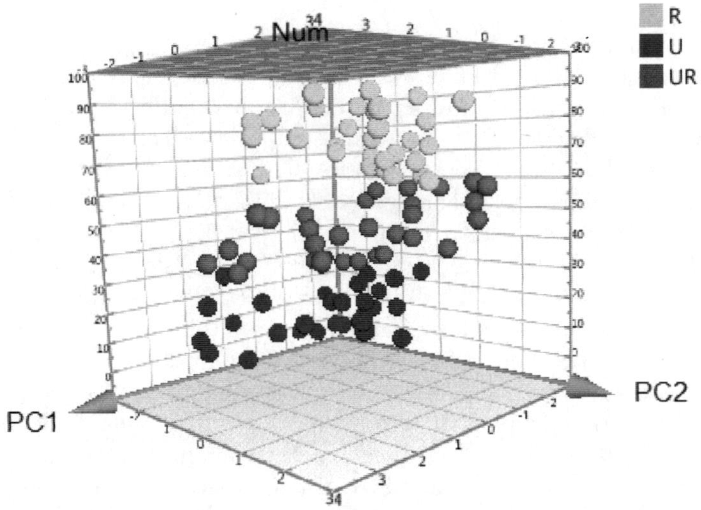

Figure 4. 3D PCA scores plot of first two factors on the different sugars composition (auto-scaled data) of FFBs with different ripeness, size and bunch region: green, ripe; blue, unripe; red, under-ripe.

Acknowledgements
We are grateful to Sime Darby Plantation Sdn Bhd and University Putra Malaysia LRGS for granting the project and also thank to Labu Palm Oil Mill for the samples provided and Sime Darby Technology Centre Sdn Bhd for providing the experimental station. We would also like to thank Dr Teh Huey Fang, Dr Umer Rashid, Mohd Zairey and Alena Sanusi for the guidance and editing.

Reference
[1] Ariffin A A, Rosnah M S, Mohamadiah B, Wan Zailan W O and Chen K W 1991 Ripeness standard: any sign of loose fruit and with one loose fruit per bunch on the minimum standard *Proceedings of the 1991 PIPOC, Agric Conf,* (Bangi:PORIM) pp 120-9
[2] Abdullah M Z, Guan L C and Azemi M 2001 Stepwise discriminant analysis for colour grading of oil palm using machine vision system *IChemE* **79** 223-31

[3] Razali M H, Halim A S M A and Roslan S 2012 A review on crop plant production and ripeness forecasting *IJACS* **4**(2) 54-63
[4] Abdullah M Z, Guan L C, Lim K C and Karim A A 2004 The applications of computer vision system and tomographic radar imaging for assessing physical properties of food *J. Food Eng.* **61** 125-35
[5] Abdullah M Z, Lim C G, Mohamed A M D and Noor M A M 2002 Color vision system for ripeness inspection of oil palm elaeis guineensis *J. Food Process Pres.* **26** 213-35
[6] Malaysian Palm Oil Board (MPOB) 2003 *Oil Palm Fruit Grading Manual* (2nd Ed.). Kajang: Malaysian Palm Oil Board (MPOB) pp1-47
[7] Shariff R, Adnan N A, Mispan R, Mansor S, Halim R and Goyal R 2004 Correlation between oil content and DN values *GISdevelopment.net*
[8] Alfatni M S M, Shariff A R M, Shafri H Z M, Saaed O M B and Eshanta O M 2008 Oil palm fruit bunch grading system using red, green and blue digital number *J. Appl. Sci.* **8**(8) 1444-52
[9] Sunilkumar K and Babu D S S 2013 Surface color based prediction of oil content in oil palm *African .J Agr. Res.* **8**(6) 564-9
[10] Makky M and Soni P 2013 Development of an automatic grading machine for oil palm fresh fruits bunches (FFBs) based on machine vision *Comput. Electron. Agr.* **93** 129-39
[11] Saeed O M B, Sankaran S, Shariff A R M, Shafri H Z M, Ehsani R, Alfatni M S and Hazir M H M 2012 Classification of oil palm fresh fruit bunches based on their maturity using portable four-band sensor system *Comput. Electron. Agr.* **82** 55-60
[12] Hazir M H M, Shariff A R M and Amiruddin M D 2012 Determination of oil palm fresh fruit bunch ripeness- based on flavonoids and anthocynin content *Ind. Crops Prod.* **36** 466-75
[13] Neoh B K, Teh H F, Ng T L M, Tiong S H, Thang Y M, Ersad M A, Mohamed M, Chew F T, Kulaveerasingam H and Appleton D R 2013 Profiling of metabolites in oil palm mesocarp at different stages of oil biosynthesis *J. Agric. Food Chem.* **61**(8) 1920-7
[14] Teh H F, Neoh B K, Hong M P L, Low J Y S, Ng T L M, Ithnin N, Thang Y M, Mohamed M, Chew F T, Yusof H M, Kulaveerasingam H and Appleton D R 2013 Differential metabolite profiles during fruit development in high-yielding oil palm mesocarp *PLoS ONE* **8**(4) e61344
[15] Prada F, Ayala-Diaz I M, Delgado W, Ruiz-Romero R and Romero H M 2011 Effect of fruit ripening on content and chemical composition of oil from three oil palm cultivars (Elaeis guineensis Jacq.) grown in Colombia *J. Agric. Food Chem.* **59**(18) 10136-42
[16] Sambanthamurthi R, Sundram K and Tan Y A 2000 Chemistry and biochemistry of palm oil *Prog. Lipid Res.* 39 507-58
[17] Teh H F, Neoh B K, Wong Y C, Kwong Q B, Ooi T E K, Ng T L M, Tiong S H, Low J Y S, Danial A D, Ersad M A, Kulaveerasingam H and Appleton D R 2014 Hormones, polyamines, and cell wall metabolism during oil palm fruit mesocarp development and ripening *J. Agric. Food Chem.* **62** 8143-52
[18] Bourgis F, Kilaru A, Cao X, Ngando-Ebongue G, Drira N, Ohlrogge J B and Arondel V 2011 Comparative transcriptome and metabolite analysis of oil palm and date palm mesocarp that differ dramatically in carbon partitioning *PNAS* **108**(30) 12527-32
[19] Tranbarger T J, Dussert S, Joët T, Argout X, Summo M, Champion A, Cros D, Omore A, Nouy B and Morcillo F 2011 Regulatory mechanisms underlying oil palm fruit mesocarp maturation, ripening, and functional specialization in lipid and carotenoid metabolism *Plant Physiol.* **156** 564-84
[20] Roongsattham P, Morcillo F, Jantasuriyarat C, Pizot M, Moussu S, Jayaweera D, Collin M, Gonzalez-Carranza Z H, Amblard P, Tregear J W, Tragoonrung S, Verdeil J and Tranbarger T J M 2012 Temporal and spatial expression of polygalacturonase gene family members reveals divergent regulation during fleshy fruit ripening and abscission in the monocot species oil palm *BMC Plant Biol.* **12**(150) 1-10
[21] Chong C L and Sambanthamurthi R 1993 Effects of mesocarp bruising on the rate of free fatty acid release in oil palm fruits *Int Biodeterio Biodegradation* **31** 65-70

[22] Payasi A, Mishra N N, Chaves A L S and Singh R 2009 Biochemistry of fruit softening: an overview *Physiol. Mol. Biol. Plants* **15**(2) 105-13
[23] Prasanna V, Prabha T N and Tharanathan R N 2007 Fruit Ripening Phenomena-An Overview *Crit. Rev. Food Sci. Nutr.* **47** 1-19
[24] Ariffin A A 1991 Chemical changes during sterilization process affecting strippability and oil quality *Seminar on Developments in Palm Oil Milling Technology and Environmental Management* (Genting Highlands: PORIM) pp 2-12
[25] Henderson J and Osborne D J 1994 Inter-tissue signalling during the two-phase abscission in oil palm fruit *J. Exp. Bot.* **45**(276) 943-51
[26] Henderson J and Osborne D J 1990 Cell separation and anatomy of abscission in the oil palm, Elaeis guineensis Jacq. *J. Exp. Bot.* **41**(223) 203-10
[27] Henderson J, Davies H A, Heyes S J and Osborne D J 2001 The study of a monocotyledon abscission zone using microscopic, chemical, enzymatic and solid state 13C CP/MAS NMR analyses *Phytochemistry* **56** 131-9
[28] Chow P S and Landhäusser S M 2004 A method for routine measurement of total sugar and starch content in woody plant tissues *Tree Physiology* **24** 1129-36
[29] Sluiter A, Hames B, Ruiz R, Scarlata C, Sluiter J, Templeton D and Croker D 2008 Determination of structural carbohydrates and lignin in biomass *Laboratory Analytical procedure* (Colorado: National Renewable Energy Laboratory)
[30] Dennis D T and Blakeley S D 2000 Carbohydrate Metabolism *Biochemistry and Molecular Biology of Plant*, eds B B Buchanan, W Gruissem and R L Jones (Maryland: American Society of Plant Physiologists) chapter 13 pp 630-75
[31] Schwald W, Concin R and Bonn G 1985 Analysis of oligomeric and monomeric carbohydrates from hydrothermal degradation of cotton-waste materials using HPLC and GPC *Chromatographia* **20**(1) 35-40
[32] Garrote G, Domínguez H and Parajó J C 1999. Mild autohydrolysis: an environmentally friendly technology for xylooligosaccharide production from wood *J. Chem. Technol. Biotechnol.* **74** 1101-9
[33] Mittal A, Chatterjee S G, Scott G M and Amidon T E 2009 Modeling xylan solubilization during autohydrolysis of sugar maple and aspen wood chips: reaction kinetics and mass transfer *Chem. Eng. Sci.* **64** 3031-41
[34] Jacobsen S E and Wyman C E 2000 Cellulose and hemicellulose hydrolysis models for application to current and novel pretreatment processes *Appl. Biochem. Biotechnol.* **84-86** 81-96
[35] Lloyd T and Wyman C E 2003 Application of a depolymerization model for predicting thermochemical hydrolysis of hemicellulose *Appl. Biochem. Biotechnol.* **105-108** 53-67

IOP Conf. Series: Materials Science and Engineering **206** (2017) 012035 doi:10.1088/1757-899X/206/1/012035

Extraction of antioxidants from *Chlorella sp.* using subcritical water treatment

S M Zakaria,[1] S M Mustapa Kamal,[1] M R Harun,[2] R Omar,[2] and S I Siajam[2]

[1]Department of Process and Food Engineering, Faculty of Engineering, Universiti Putra Malaysia, 43400 Serdang, Selangor. Malaysia
[2]Department of Chemical and Environmental Engineering, Faculty of Engineering, Universiti Putra Malaysia, 43400 Serdang, Selangor. Malaysia

E-mail: smazlina@upm.edu.my

Abstract. *Chlorella sp.* microalgae is one of the main source of natural bioactive compounds used in the food and pharmaceutical industries. Subcritical water extraction is the technique that offers an efficient, non-toxic, and environmental-friendly method to obtain natural ingredients. In this work, the extracts of *Chlorella sp.* microalgae was evaluated in terms of: chemical composition, extraction (polysaccharides) yield and antioxidant activity, using subcritical water extraction. Extractions were performed at temperatures ranging from 100^{0}C to 300^{0}C. The results show that by using subcritical water, the highest yield of polysaccharides is 23.6% that obtained at 150^{0}C. Analysis on the polysaccharides yield show that the contents were highly influenced by the extraction temperature. The individual antioxidant activity were evaluated by in vitro assay using a free radical method. In general, the antioxidant activity of the extracts obtained at different water temperatures was high, with values of 31.08-54.29%. The results indicated that extraction by subcritical water was effective and *Chlorella sp.* can be a useful source of natural antioxidants.

1. Introduction

Microalgae are particularly attractive as natural sources for bioactive compounds since they have the potential to produce the compounds in culture which enables the production of structurally complex molecules [1-4]. The capability of microalgae to synthesize variety of bioactive chemicals is due to its multitude of physiological, biochemical and molecular strategies to cope with stress. As photosynthetic organisms, microalgae contain chlorophyll that can be used for food and cosmetic purposes [5]. They can also be used in pharmaceutical industries, as some species of microalgae produce bioactive compounds such as antioxidants, anticancer, anti-inflammatory, antiviral, antibiotics and antitoxins [6]. Besides, microalgae are used as nutrient supplements for human consumption as they are high in protein, vitamins and polysaccharides contents [7]. They are also have been described to secrete a wide range of compounds that could be potentially employed as functional ingredients including polyphenols, carotenoids and other antioxidant pigments [8].

Chlorella sp. is a type of microalgae which can normally be found in freshwater. It is unicellular photosynthetic algae which contains green photosynthetic pigments chlorophyll in its chloroplast and also contains lutein and other primary carotenoids such as carotene [9]. Recently, *Chlorella sp.* is known and commercialized because of its nutrients and benefits. This kind of microalgae has the capabilities

IOP Conf. Series: Materials Science and Engineering **206** (2017) 012035 doi:10.1088/1757-899X/206/1/012035

to produce secondary metabolites including polysaccharides. Polysaccharides extracted from algae can be considered as source of dietetic fiber which has identical functionality as prebiotics since they are indigestible in the human digestive system [10].

Extraction processes for the separation of bioactive compounds have been developed to obtain highly purified products and rendering them useful in a wide range of applications. These technologies could provide an innovative approach to increase the production of the desired compounds including extraction of bioactive compounds from microalgae. Many extraction techniques have been developed in extracting the bioactive components. The current techniques mostly involve solvents and heating [11] such as soxhlet, microwave, ultrasonic and supercritical extraction technique.

Subcritical water extraction (SWE) is one of the potential technique that can be applied to enhance the selectivity and the yield of desirable bioactive compounds. This extraction using hot water under pressure, has recently emerged as a useful tool to replace the traditional extraction methods. SWE is an environmental-friendly technique that can provide higher extraction yields from solid samples. SWE is carried out using hot water (from 100 to 374°C) under high pressure (usually from 10 to 60 bar) to maintain water in the liquid state. In general, the use of SWE provides a number of advantages over traditional extraction techniques. These are, mainly low extraction times, higher quality of the extracts, lower costs of the extracting agent, and an environmentally compatible technique [12].

In the last decade, subcritical water has gained much interest from researchers as a green method for extracting compounds from various materials. The goal of the present investigation was to evaluate the extracts from the *Chlorella sp.* microalgae using subcritical water as method of extraction at different temperature conditions. For this purpose, the extracts were evaluated based on extraction yield and polysaccharides contents. A study on the effect of the temperature on the extraction efficiency of the antioxidant compounds was also performed.

2. Materials and Method

2.1. Samples and chemicals

Chlorella sp. bluegreen algae (derived from *Chlorella vulgaris*) was purchased from PureBulk, USA. Ethanol, methanol, Folin–ciocalteu reagent, sodium carbonate, gallic acid, diphenyl-1-Picryl-Hydrazyl-Hydrate (DPPH) was purchased from Sigma Aldrich, Malaysia. All chemicals were of analytical grade and used as received, without further purification.

2.2. Proximate analysis

The proximate compositions of *Chlorella sp.* was determined according to the Official Analytical Chemists Association (AOAC) method [13]. Moisture was determined by drying a sample in an air oven for 5 h at 105°C. The Kjeldahl method was used to determine the crude protein content using Tecator Kjeltec protein analyzer (FOSS, Hillerod, Denmark). The fat content was determined using the soxhlet method. Ash was determined by heating the samples in a furnace at 550°C for 8–12 h.

2.3. Subcritical water extraction

Extractions were performed with a batch subcritical fluid extraction system (Figure 1) at extraction temperature between 100 to 300 °C. Previous to each extraction, the molten salt bath heat-up was carried out for a few minutes. Likewise, all extractions were performed in stainless steel batch reactor (SUS316, 7.5mm x 150.4mm) cells, containing 0.5 g of sample.

The extraction procedure was as follows: (i) sample was loaded into reactor cell; (ii) cell was filled with 5 mL of distilled water; (iii) the air was drawn out by purging argon gas into the reactor cell; (iv) the reactor was then placed into the preheated molten salt bath to initiate the reaction; (v) after reaching 10 minutes, the reactor cell was taken out and quickly quenched in flowing tap water at room temperature to terminate the reaction. After that, the reactor content were centrifuged at 5000 rpm, for 10 minutes. Then, filtration process was took place using Whatman No.1 filter paper and filter funnel into a conical flask. The supernatant and residue were collected for further analysis.

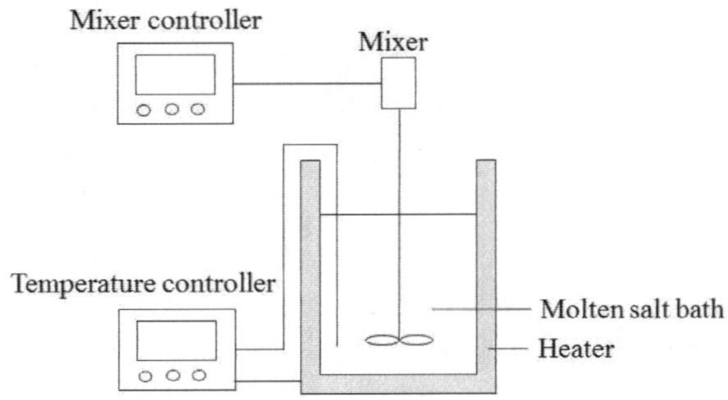

Figure 1. Subcritical fluid extraction system

The residue were dried at 60°C in drying oven for overnight and the weight was determined. The crude extracts yield was calculated by following equation:

$$Crude\ extract\ yield\ (\%) = \frac{W_{microalgae} - W_{residue}}{W_{microalgae}} x\ 100\% \quad (1)$$

Where $W_{residue}$ is the mass of residue (g) and $W_{microalgae}$ is the total mass of microalgae (g)

2.4. Polysaccharides yield determination
The supernatant was treated by using three volumes of absolute ethanol (v/v) and was then precipitated for incubation overnight at 4°C. The precipitate was centrifuged at 4000 rpm for 5 min and was washed three times using acetone. Afterward, the precipitate was freeze-dried using a vacuum freeze dryer (SCANVAC Freeze Dryer) at −108 °C for 72 hrs and the weight was determined. The yield of polysaccharides was calculated using the following equation:

$$Polysaccharides\ yield\ (\%) = \frac{W_{polysaccharides}}{W_{microalgae}} x\ 100\% \quad (2)$$

Where $W_{polysaccharides}$ is the mass of polysaccharides (g) and $W_{microalgae}$ is the total mass of microalgae (g)

2.5. Total phenolic compound
The total phenolic content of the extract was determined by modified Folin–Ciocalteu method. Briefly, 200 µL of extract were made up to 3 mL with distilled water, mixed thoroughly with 0.5 mL of Folin–Ciocalteu reagent for 3 min, followed by the addition of 2 mL of 20% (w/v) sodium carbonate. The mixture was allowed to stand for a further 60 min in the dark, and absorbance was measured at 765 nm by UV- visible spectrophotometer. The total phenolic content was calculated from the gallic acid calibration curve.

2.6. Antioxidant analysis (DPPH)
DPPH (2,2-diphenyl-1-picryl-hydrazyl-hydrate) free radical scavenging assay [14] was performed to determine the antioxidant activity. 0.1 mM solution of DPPH in methanol was prepared and 1 ml of this solution was added to 3mL extracts and incubated in the dark place for 30 minutes. After that, the absorbance was measured using UV-visible spectrophotometer at 517 nm. 4mL of distilled water was used as blank. While, 3mL of methanol solution and 1mL of DPPH solutions were mixed together which act as control. The changes in color of the mixture were observed. Then, the absorbance was measured. Lower absorbance of the reaction mixture indicated higher free radical scavenging activity. The scavenging activity of DPPH radical in percentage calculated using following equation:

$$Antioxidant\ activity\ (\%) = \frac{A_O - A_1}{A_O} x\ 100\% \quad (3)$$

Where A_O is the absorbance of the control sample and A_1 is the absorbance of the extracts.

3. Result and Discussion

3.1. Chemical composition of Chlorella sp.
The chemical composition of natural ingredients is very important not only from the perspective of human nutritional health, but also for assessment of the potential development and application of materials in food and pharmaceutical system. Proximate analysis were conducted on *Chlorella sp.* (Table 1) and the results showed that the microalgae contained 37.05 ± 1.38% carbohydrate, 52.25 ± 0.19% crude protein, 3.73 ± 1.21% moisture, 2.51 ± 0.14% crude oil, and 4.45 ± 0.06% ash.

Table 1. Composition percentage of *Chlorella sp.*

Component	Content (%)
Moisture content	3.73 ± 1.21
Carbohydrate	37.05 ± 1.38
Crude protein	52.25 ± 0.19
Crude oil	2.51 ± 0.14
Ash	4.45 ± 0.06

The analysis shows that *Chlorella sp.* was high in carbohydrate content (37.05%). Previous studies have shown that the carbohydrate content of few microalgae varies from 18.5 to 45.8% [15]. *Chlorella sp.* was within this range, indicating that it possessed considerable polysaccharides content.

3.2. Extraction yield
SWE of the *Chlorella sp.* was carried out at 100-300 °C for 10 min. Figure 2 shows the crude extraction yield of microalgae with different temperatures. It shows that extraction temperature of 200°C gave a maximum amount of crude yield. The crude yield at 100°C was 25.59% and it reached 50.10% yield at 200°C. The crude yield of the extracts start to decrease at 250°C until it reach 300°C with 16.07% of yield extracted.

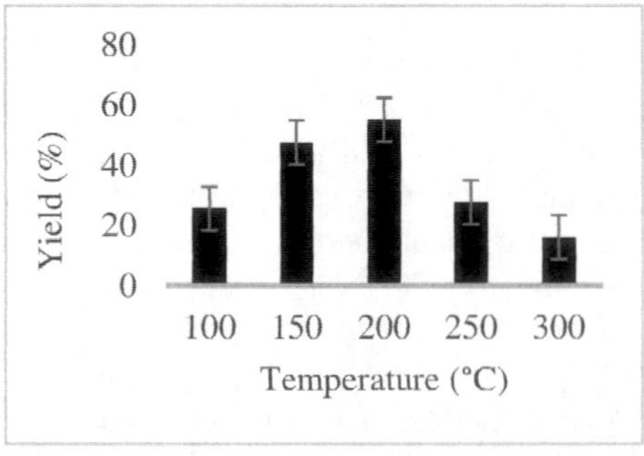

Figure 2. Yield of *Chlorella sp.* crude extracts at different temperature

Figure 3 shows the effects of extraction temperature on the polysaccharides yield. As described in Figure 3, the effect of temperature on polysaccharides yield was significant. The polysaccharides yield increased with the rise of extraction temperature from 100°C to 150°C and decreased from 150°C to

IOP Conf. Series: Materials Science and Engineering **206** (2017) 012035 doi:10.1088/1757-899X/206/1/012035

300°C. The minimum and maximum polysaccharides yield were 5.1% at 300°C and 23.6% at 150°C, respectively.

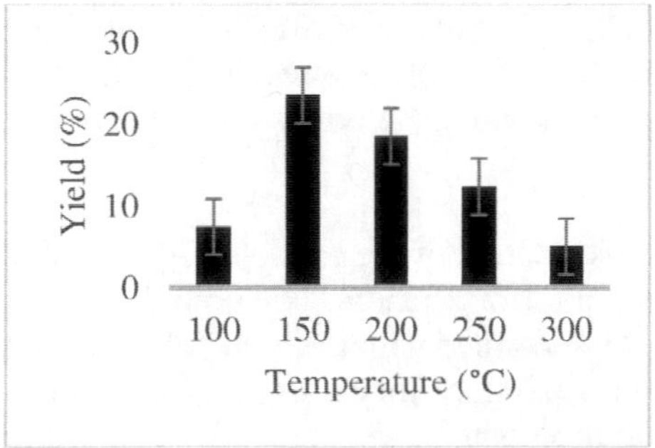

Figure 3. Yield of *Chlorella sp.* polysaccharides extracted at different temperature

When temperature increased from 100 to 150°C, the dielectric constant of subcritical water significantly decreases. At such high temperature, polysaccharides could dissolve in subcritical water as much as they dissolve in the organic solvents. However, polysaccharides yield decreased at 200°C and above possibly due to further degradation of polysaccharides and some compounds and decreasing of dielectric constant provided less effective contribution to the polysaccharides recovery. In general, it has been observed that amorphous polysaccharides to water soluble products increases rapidly as the temperature increase from 100 to 150°C and it decrease when the temperature reaches 200°C. In addition, other studies reported that polysaccharides like celluloses remain stable below 200°C [16]. Polysaccharides in microalgal cell wall were complex and diversified. Moreover, the chemical structures of the polysaccharides were vary for different species. It was difficult to utilize an ambient temperature for extraction of polysaccharides due to complexity and inherent tensile strength of the cell wall. Therefore, additional detailed study is required to explain the characteristics of the extracted polysaccharides and the effects on every point of temperature within subcritical phase.

3.3. Antioxidant analysis

Once the different extracts were obtained, the next step was their functional characterization in terms of antioxidant activity. As mentioned, free radical method (DPPH) was used to measure the antioxidant activity of the extracts. Figure 4 shows the antioxidant activities of the extracts obtained, at different subcritical water temperatures.

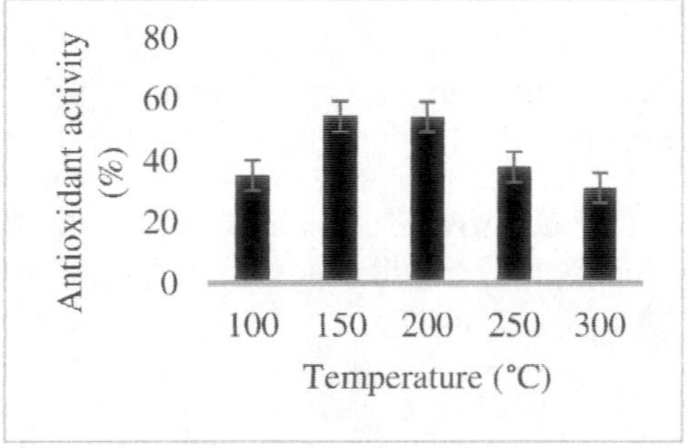

Figure 4. Antioxidant activity of the extracts at different temperature

As can be seen the highest antioxidant activities for *Chlorella sp.* extracts was obtained at 150°C with 54.29% inhibition. When the temperature increased from 100 to 150°C, the antioxidant activity was also increased from 35.06% to 54.29%. The antioxidant activity of the extracts were decreased when the extraction temperature reach 250°C and above with 31.08% inhibition at 300°C. This may due to the decomposition of most of the component in the extracts at very high temperature. Comparatively, for the extracts obtained at 200°C, the antioxidant activities (54.11%) found were almost the same as those observed at 150°C.

Figure 5 shows the total phenolic content of the extracts at different temperature. As the temperature increases from 100 to 150°C, the total of phenolic compound also increases. At 150°C, the highest recovery of phenolic compound was obtained with the value of 57.1mg/g. Overall, the results show that extraction yield increases as the temperature increases as a result of increased solubility of phenolic compounds in water. However, it drop when it reached 200°C and it might due to the degradation of phenolic compound. In addition, previous studies have reported that, degradation of phenolic compounds was observed above 180°C [17].

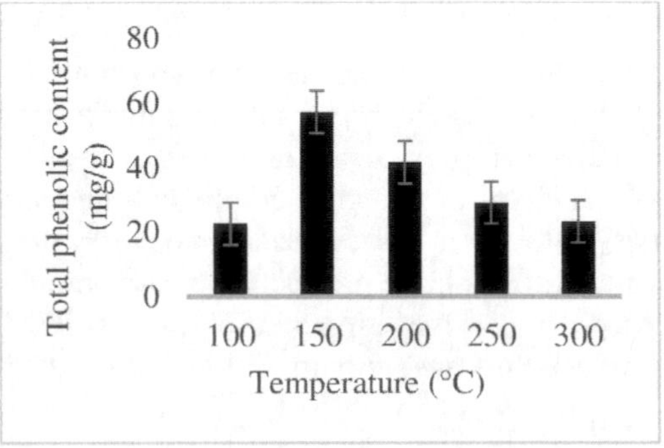

Figure 5. Total phenolic content of the extracts at different temperature

This suggested that other compounds should have a positive influence on the total antioxidant activity of the extracts at 200°C. The presence of phenolic compounds alone, cannot explain the antioxidant capacity observed. Some studies have reported that Maillard reaction occurred during the extraction at high temperature which resulted in the formation of new antioxidant compounds [18]. Chemical composition of this microalgae with the presence of carbohydrates together with amino acids could generate some compounds during the extraction at high temperature. These compounds could have an effective influence on the total antioxidant activity of the extract produced.

4. Conclusion

This study showed that extraction by subcritical water, produces polysaccharides-rich extracts from *Chlorella sp*. The recommended operating condition for subcritical water extraction to obtain the highest concentration of polysaccharides is 150°C with yield 23.6%. Antioxidant was efficiently extracted from *Chlorella sp.* with environmentally, non-toxic subcritical water. The extracts, which are free from organic solvent residue and exhibit high free radical scavenging activities should be ready for used as antioxidants for food and pharmaceutical products.

Acknowledgement

The authors gratefully acknowledge Universiti Putra Malaysia and financial support from the Fundamental Research Grant Scheme (FRGS) provided by the Ministry of Higher Education Malaysia.

References

[1] Cifuentes A, Herrero M, Ibáñez E, Jaime L and Martín Álvarez PJ 2006 Optimization of the extraction of antioxidants from *Dunaliella salina* microalga by pressurized liquids *J. Agric. Food Chem.* **54** 5597

[2] Jaime L, Mendiola JA, Herrero M, Soler-Rivas C, Santoyo S, Senorans FJ, Cifuentes A and Ibanez E 2005 Separation and characterization of antioxidants from *Spirulina platensis* microalga combining pressurized liquid extraction, TLC, and HPLC-DAD *J. Sep. Sci.* **28** 2111

[3] Mendiola JA, Marín FR, Hernández SF, Arredondo BO, Señoráns FJ, Ibañez E and Reglero G 2005 Characterization via liquid chromatography coupled to diode array detector and tandem mass spectrometry of supercritical fluid antioxidant extracts of *Spirulina platensis* microalga *J. Sep. Sci.* **28** 1031

[4] Rodriguez-Meizoso I, Jaime L, Santoyo S, Cifuentes A, Garcia-Blairsy Reina G, Senorans FJ and Ibanez E 2008 Pressurized fluid extraction of bioactive compounds from *Phormidium* species *J. Agric. Food Chem* **56** 3517

[5] Spolaore P, Joannis-Cassan C, Duran E and Isambert A 2006 Commercial applications of microalgae *J. Biosci. Bioeng.* **101** 87

[6] Garcia-Casal MN, Ramírez J, Leets I, Pereira AC and Quiroga MF 2009 Antioxidant capacity, polyphenol content and iron bioavailability from algae (*Ulva sp., Sargassum sp.* and *Porphyra sp.*) in human subjects *Br. J. Nutr.* **101** 79

[7] Carballo-Cárdenas EC, Tuan PM, Janssen M and Wijffels RH 2003 Vitamin E (α-tocopherol) production by the marine microalgae *Dunaliella tertiolecta* and *Tetraselmis suecica* in batch cultivation *Biomol. Eng.* **20** 139

[8] Plaza, M, Herrero M, Cifuentes A and Ibáñez E 2009 Innovative natural functional ingredients from microalgae *J. Agric. Food Chem* **57** 7159

[9] Young AJ 1993 *Carotenoids in Photosynthesis* ed Young AJ and Britton G (Netherlands: Springer) pp 72-95

[10] Ibañez E, Herrero M, Mendiola J and Castro-Puyana M 2012 Extraction and Characterization of bioactive compounds with health benefits from marine resources: macro and micro algae, cyanobacteria, and invertebrates *Marine Bioactive Compounds* ed Hayes M (United States: Springer) pp 55-98

[11] Joana Gil-Chávez G, Villa JA, Fernando Ayala-Zavala J, Basilio Heredia J, Sepulveda D, Yahia EM and González-Aguilar GA 2013 Technologies for extraction and production of bioactive compounds to be used as nutraceuticals and food ingredients: An overview *Compr. Rev. Food. Sci. Food Saf.* **12** 5

[12] Herrero M, Cifuentes A and Ibanez E 2006 Sub-and supercritical fluid extraction of functional ingredients from different natural sources: Plants, food-by-products, algae and microalgae: A review *Food Chem.* **98** 136

[13] AOAC Association of Official Analitical Chemist 2005 *Official Methods of Analysis of AOAC International* (Gaithersburg Md: AOAC International)

[14] Villano D, Fernández-Pachón M, Moyá M, Troncoso A and García-Parrilla M 2007 Radical scavenging ability of polyphenolic compounds towards DPPH free radical *Talanta* **71** 230

[15] Tibbetts SM, Milley JE and Lall SP 2015 Chemical composition and nutritional properties of freshwater and marine microalgal biomass cultured in photobioreactors *J. Appl. Phycol.* **27** 1109

[16] Bobleter O 1994 *Prog. Polym. Sci.* **19** 797

[17] Singh PP and Saldaña MDA 2011 Subcritical water extraction of phenolic compounds from potato peel *Food Res. Int.* **44** 2452

[18] Plaza M, Amigo-Benavent M, Del Castillo MD, Ibáñez E and Herrero M 2010 Facts about the formation of new antioxidants in natural samples after subcritical water extraction *Food Res. Int.* **43** 2341

IOP Conf. Series: Materials Science and Engineering **206** (2017) 012036 doi:10.1088/1757-899X/206/1/012036

Application of Artificial Neural Network to Predict Colour Change, Shrinkage and Texture of Osmotically Dehydrated Pumpkin

S Y Tang[1], J S Lee[2], S P Loh[3], H J Tham[1*]

1 Chemical Engineering Program, Faculty of Engineering, Universiti Malaysia Sabah, Jalan UMS 88400 Kota Kinabalu, Sabah, Malaysia

2 Faculty of Food Science and Nutrition, Universiti Malaysia Sabah, Jalan UMS 88400 Kota Kinabalu, Sabah, Malaysia

3 Department of Nutrition and Dietetics, Faculty of Medicine and Health Sciences, Universiti Putra Malaysia, 43400, Serdang, Selangor, Malaysia

*E-mail: hjtham@ums.edu.my

Abstract. The objectives of this study were to use Artificial Neural Network (ANN) to predict colour change, shrinkage and texture of osmotically dehydrated pumpkin slices. The effects of process variables such as concentration of osmotic solution, immersion temperature and immersion time on the above mentioned physical properties were studied. The colour of the samples was measured using a colorimeter and the net colour difference changes, ΔE were determined. The texture was measured in terms of hardness by using a Texture Analyzer. As for the shrinkage, displacement of volume method was applied and percentage of shrinkage was obtained in terms of volume changes. A feed-forward backpropagation network with sigmoidal function was developed and best network configuration was chosen based on the highest correlation coefficients between the experimental values versus predicted values. As a comparison, Response Surface Methodology (RSM) statistical analysis was also employed. The performances of both RSM and ANN modelling were evaluated based on absolute average deviation (AAD), correlation of determination (R^2) and root mean square error (RMSE). The results showed that ANN has higher prediction capability as compared to RSM. The relative importance of the variables on the physical properties were also determined by using connection weight approach in ANN. It was found that solution concentration showed the highest influence on all three physical properties.

1.Introduction

Pumpkin (Cucurbita pepo L.) has been used for human consumption and animal feed [1]. It is a fruit that can be taken fresh or cooked, or made into various kind of food such as soups, pies and bread. However, pumpkin is still considered as underrated fruits in the food industry. Nowadays, the increase of public awareness for a healthy lifestyle leads to the consumption of high nutritional and health related food material including processed pumpkin. Pumpkin is a good source of carotenoids, contains mainly α- carotene and β-carotene as well as minerals e.g potassium, phosphorus, magnesium, iron and selenium. Besides, it is also a low caloric food [2] and contains phenolics, flavonoids, polysaccharides, vitamins, and other substances that are good for health [3]. However, fresh pumpkin is very sensitive to microbial spoilage, even at refrigerated conditions, hence, it must be processed to dried or frozen products [4].

Drying is the most widely applied method to preserve fruits and vegetables. However, high heat can cause excessive damage to the foods. Hence, osmotic dehydration (OD) is often used as a pre-treatment before drying process. Osmotic dehydration is a method to partially reduce the moisture content of fruits or vegetables, with the purpose to extend the shelf life of food materials [5]. Due to the mild condition used, it can reduce the damage of heat to the flavour and colours of fruits or vegetables, prevent spoilage and reduce energy consumption. It had been applied to fruits such as apple, banana, kiwi fruits, mango and etc [6–10].

Dehydration process leads to alteration of chemical properties, nutritional values and physical properties of food products such as colour, food shape (shrinkage) and texture [9]. Colour of a food product is the most important quality parameter considered by the consumers, and it has high influence in the acceptance of a product [11]. This quality may be altered during processing, depending on the water content in foods, particularly dried foods. Studies conducted on the colour changes of fruits during OD were reported. For example, Falade et al. [12] investigated the effects of OD on the colour change of watermelon. It was observed that the colour intensity increase with the osmotic solution concentration. Krokida et al. [13] studied the colour changes of apple and banana during OD. Osmotically treated samples show less browning than untreated sample and lightness L decreased slightly, while redness a, yellowness b increased slightly. Another physical change of food during OD is shrinkage. The volume and food shape of products will be altered during the dehydration process [14]. Among studies reported on the volumetric shrinkage during OD, include strawberries [15], mango [16] and tomatoes [17]. Texture is a multi-parameter attributes and also one of the sensory property [18]. Several studies investigated the effect of OD on texture of some fruits, such as pineapple [19], apple [20], cucumber [21], strawberry [22], guava, melon and papaya [23].

Artificial neural network (ANN) modelling is commonly utilized due to its capability of relating the input and output parameter by learning from examples through iteration, without requiring a prior knowledge on the relationships between the process variables [24]. ANN has been applied successfully in modelling of physical properties of foods. Zenoozian et al. [25] used ANN and image analysis to predict the colour intensity (DE), shrinkage percentage as well as Heywood shape factor of osmotically dehydrated and air-dried pumpkin pieces. Zenoozian and Devahastin [26] also applied wavelet transform coupled with ANN to predict physicochemical properties of osmotically dehydrated pumpkin. Chen et al. [27] used multi-layer ANN models with three inputs (concentration of osmotic solution, temperature, and contact time) to predict five outputs (drying time, colour, texture, rehydration ratio, and hardness) during osmo-convective drying of blueberries. Youssefi et al.[28] employed ANN with three inputs (carrier type, carrier concentration, and concentration of crystalline cellulose) to predict the quality parameter of spray-dried pomegranate juice with five outputs (drying yield, solubility, colour change, total anthocyanin content, and antioxidant activity).

Response surface methodology (RSM) is a statistical technique used for experimental design, data analysis and modelling. It is widely used to investigate the relationship between process parameters and output, optimization of the operating conditions and responses in order to fulfil customer demands and target specifications [29]. There are numerous studies on the optimized conditions for osmotic dehydration process using RSM, such as papaya [30], diced green pepper [31], cantaloupe [32], pumpkin [33] and so on. However, only several studies on the prediction of physical properties of osmotic dehydration process have been published, such as potato [34] and Carrot [35].

2.Materials and Methods

2.1. Fresh Material

IOP Conf. Series: Materials Science and Engineering **206** (2017) 012036 doi:10.1088/1757-899X/206/1/012036

Pumpkins were purchased from a local market. The initial moisture content of the fresh pumpkin varied within 92 ± 0.2% (w.b.). Commercial sucrose and salt were also purchased from the local supermarket.

2.2. Experimental Procedure

At the beginning of each experiment, pumpkin was cleaned, peeled and cut into slab with dimensions of 60 mm x 15 mm x 5 mm. Osmotic solution with the desired concentration using commercial sucrose, sodium chloride solution and distilled water was prepared. The independent variables were sucrose concentration (30, 45 and 60°Brix), solution temperature (35, 50 and 65°C) and immersion time (90, 150 and 210 min). The range of process variables chosen were the typical range used in OD processes [9,36,37]. All osmotic solutions consist of 5%w/w sodium chloride. The concentration of osmotic solution was obtained using a refractometer. The experiment was conducted in a water bath (Memmert, WB 14) to maintain a constant temperature. The pumpkin to solution ratio was fixed at 1:10. The osmotic solution was agitated manually at every 30 minute interval to avoid localize dilution of sucrose solution. After the osmotic dehydration treatment, pumpkin slabs were removed from the osmotic solution and gently blotted with kitchen towel to remove excess sucrose solution, and kept in a sealed bag until experimental determinations. All the experiments were performed in triplicate, and the average value was used for the determination of colour changes, shrinkage and texture.

2.3. Determination of colour

Colour of pumpkin slabs were measured using Hunterlab Colorflex (Hunter Associates Laboratory Inc, Mumbai). The meter was calibrated using the manufacturer's standard white and black plate. The colour of L*a*b* values were obtained directly from the meter. The net colour difference (ΔE) was calculated with the equation 1 [38]:

$$\Delta E = \sqrt{(L_2^* - L_1^*)^2 + (a_2^* - a_1^*)^2 + (b_2^* - b_1^*)^2} \quad (1)$$

where L* represents lightness, a* represents redness and b* represents yellowness.

2.4. Shrinkage

To determine the shrinkage of osmotically dehydrated pumpkin slices, volume displacement method was used. Shrinkage is expressed in terms of the percentage change of the sample's volume as compared with its original volume [39].

$$\%S = \left(\frac{V_i - V_f}{V_i}\right) \times 100 \quad (2)$$

where V_i and V_f are the volumes of the samples at the starting and at the end of osmotic dehydration experiment.

2.5. Texture Profile Analysis (TPA)

For determination of sample's firmness, Texture Analyzer (TA.XT Plus, Stable Micro System Ltd, US) was used to measure the hardness of samples. All measurements were conducted at room temperature of 25°C. The maximum force measurement was carried out using a 5 kg loading cell and a knife blade probe (width 7 cm, thickness 3 mm). The test speed is 5.0 mm/s. The knife blade probe cut the samples placed on a mounted fixed table. The blade was lowered until almost contacted the surface of the plate. Each sample was cut one time and repeated 3 times with 3 samples for each condition.

2.6. Artificial Neural Network

In this study, Neural Network Toolbox 7.12.0 of MATLAB (Mathwork, 2011) software was used. A feed forward multi-layered ANN trained by back propagation (BP) algorithm was selected and trained with three input variables which were sucrose concentration, solution temperature and immersion time and colour changes, shrinkage and texture as the output of the model. The general scheme of the ANN network is shown in Fig1.

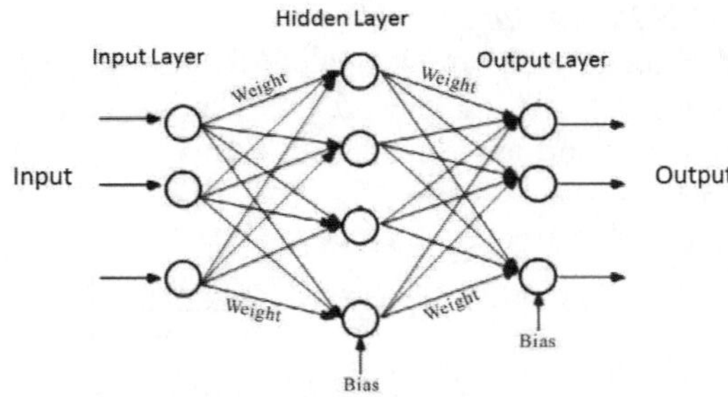

Figure 1. Structure of feed forward multi-layer ANN.

In this study, the tangent sigmoid transfer function (equation 3 and equation 4), and pure linear transfer function were applied at the hidden layer and output layer by trial and error method. The tangent sigmoid transfer function was selected as the activation function for both the hidden layer and the output layer, due to the lower calculated mean square error (MSE) when compared with linear function.

$$\mathrm{sig(x)} = \frac{1}{1+e^{-x}} \tag{3}$$

$$\tanh(x) = \frac{e^{x}-e^{-x}}{e^{x}+e^{-x}} \tag{4}$$

In order to find the best networks, 27 sets of data were used for three different concentration, three temperatures and three immersion time. The data were randomly divided into three partitions, 70% in the training set, 15% in the validation set and 15% in the test set. The number of neurons varied from 3 to 12 were used in the hidden layer. The adopted learning function was "trainlm". "Trainlm" is an iterative technique that updates the connection weight and bias values according to Levenberg-Marquardt (LM) algorithm. The training process was conducted for 1000 epochs. Both input and output variables were normalized between 0 to 1 for reduction of network error. The norrmalized equation applied is as follows:

$$Y_n = \frac{Y_a - Y_{min}}{Y_{max} - Y_{min}} \tag{5}$$

where Y_n, Y_a, Y_{min} and Y_{max} are normalized value, actual value, minimum value and maximum value, respectively. The evaluation of performance of the ANN was based on MSE and the highest correlation coefficients between the experimental values versus predicted values (R^2). This is to ensure the accuracy of the neural network to produce the output that are closer or equal to the experimental values.

2.7. Response Surface Methodology

A full factorial experimental design was generated with the Design-Expert 8.0.7.1 software. The same independent process variables were selected, i.e. sucrose concentration (X_1), solution temperature (X_2), and immersion time (X_3). A second order polynomial equation was used to fit experimental data and the quadratic model which includes the linear model can be described as:

$$Y = \beta_0 + \textstyle\sum_{j=1}^{k} \beta_j x_j + \sum_{j=1}^{k} \beta_{jj} x_j^2 + \sum_{j=1}^{k} \beta_{ij} x_i x_j + e_i \tag{6}$$

where Y is the response; $\mathbf{x_i}$ and $\mathbf{x_j}$ are independent variables, $\boldsymbol{\beta_0}$, $\boldsymbol{\beta_i}$, $\boldsymbol{\beta_{ii}}$, $\boldsymbol{\beta_{ij}}$ are the regression coefficients for intercept, linear, quadratic and second-order terms, respectively; k is the number of parameters, and $\mathbf{e_i}$ is the error [40].

2.8. Comparison of RSM and ANN Performance

In order to evaluate the performance of RSM models and ANN models, error analysis was performed in terms of Root Mean Square Error (RMSE), Model Predictive Error (MPE), and Correlation of Determination (R^2) between predicted and experimental values.

$$\mathrm{RMSE} = \sqrt{\frac{\sum_{i=1}^{n}(Y_{i,exp}-Y_{i,pred})^2}{n}} \tag{7}$$

$$\mathrm{MPE}(\%) = \frac{100}{n}\sum_{i=1}^{n}\left|\frac{(Y_{i,exp}-Y_{i,pred}}{Y_{i,pred}}\right| \tag{8}$$

$$R^2 = 1 - \frac{\sum_{i=1}^{n}(Y_{i,pred}-Y_{i,exp})^2}{\sum_{i=1}^{n}(Y_{i,pred}-Y_{i,m})^2} \tag{9}$$

where $Y_{i,exp}$ was the experimental value of the *i*th experiment; $Y_{i,pred}$ was the predicted value of the *i*th experiment, n was the number of experiment and $Y_{i,m}$ was the average of the experimental value.

3.Result and Discussion

3.1. ANN modelling

The optimal number of neurons in the hidden layer of the neural network are examined by varying the number of neurons in the hidden layer using trial and error method. ANN optimization process need training to reduce the error function (MSE) by searching for a set of connection weights and bias that can enable the ANN to predict the more accurate outputs that are identical or closer to the experimental values. The error measures and correlation of determination (R^2) associated with different ANN configurations for estimations are presented in Table 1.

Table 1. Errors and correlation of determination (R^2) in prediction of net colour difference, shrinkage and texture using ANN with different numbers of neurons for osmotically dehydrated pumpkin slab.

No.	**No. of neurons**	**Color Changes**		**Shrinkage**		**Texture**	
		MSE	R^2	MSE	R^2	MSE	R^2
1	3	0.1977	0.9199	6.7322	0.9502	0.0066	0.9581
2	4	0.2210	0.9199	4.9030	0.9647	**0.0014**	**0.9926**
3	5	0.2033	0.9189	3.8146	0.9730	0.0025	0.9845
4	6	0.2916	0.8970	5.7380	0.9682	0.0142	0.9183
5	7	0.3276	0.8661	7.1610	0.9454	0.0020	0.9888
6	8	0.1406	0.9432	2.9180	0.9778	0.0041	0.9788
7	9	0.2371	0.9235	6.9280	0.9474	0.0023	0.9859
8	10	0.5140	0.8477	**2.6089**	**0.9804**	0.0074	0.9626
9	11	**0.1322**	**0.9457**	2.9397	0.9780	0.0052	0.9679
10	12	0.3448	0.8583	6.4621	0.9522	0.0120	0.9302

After several repetitions, it is found that a network with 11 hidden layers (colour changes), 10 hidden layers (shrinkage) and 4 hidden layers (texture) produce the best performance. The correlation of determination (R^2) for estimation of colour changes, shrinkage and texture (0.9457, 0.9804, and 0.9926, respectively) were revealed good agreement between predicted and experimental values.

The neural net weight matrix can be used to evaluate the relative importance of the input variables on the output variables. Garson et al. [41] proposed an equation based on the partitioning of connection weights:

$$I_j = \frac{\sum_{m=1}^{m=N_h}\left(\left(|W_{jm}^{ih}|/\sum_{k=1}^{N_i}|W_{km}^{ih}|\right)\times|W_{mn}^{ho}|\right)}{\sum_{k=1}^{k=N_i}\left\{\sum_{m=1}^{m=N_h}(|W_{km}^{ih}|/\sum_{k=1}^{N_i}|W_{km}^{ih}|)\times|W_{mn}^{ho}|\right\}} \tag{10}$$

where I_j is the relative importance of the jth input variable on the output variable, N_i and N_h are the numbers of input and hidden neurons, respectively, W is the connection weight, the superscripts "k", "m" and "n" refer to input, hidden and output neurons, respectively. The relative importance of the three input variables is shown in Table 2.

Table 2. Relative importance of input variables

Output Variable	Colour Changes	Shrinkage	Texture
Input Variable	Importance (%)	Importance (%)	Importance (%)
Concentration	41.47	36.06	48.88
Temperature	33.46	28.04	47.08
Time	25.07	35.90	4.04
Total	100	100	100

It can be seen that concentration has significantly affected the colour change of the samples compared to immersion temperature and time. For shrinkage, concentration and immersion time showed almost equal relative importance. As for the case of texture, concentration and temperature dominantly influence the property with immersion time showing only minimal effect.

3.2. RSM modelling

A quadratic polynomial equation was fitted with the experimental results obtained and the resulting RSM model equation is following:

$$\text{Colour} = 9.48974 - 0.1747 \times X_1 + 0.060052 \times X_2 - 9.88215 \times 10^{-3} \times X_3 + 2.01177 \times 10^{-3} \times X_1X_2 + 7.95101 \times 10^{-4} \times X_1X_3 + 2.86333 \times 10^{-4} \times X_2X_3 - 1.44471 \times 10^{-4} \times X_1^2 - 1.16503 \times 10^{-4} \times X_2^2 - 9.29399 \times 10^{-5} \times X_3^2 \quad (11)$$

$$\text{Shrinkage} = -46.9656 + 0.83909 \times X_1 + 1.90649 \times X_2 + 0.18509 \times X_3 + 5.5811 \times 10^{-3} \times X_1X_2 + 1.50907 \times 10^{-3} \times X_1X_3 - 4.74356 \times 10^{-4} \times X_2X_3 - 6.16985 \times 10^{-3} \times X_1^2 - 0.019982\, X_2^2 - 4.31191 \times 10^{-4} \times X_3^2 \quad (12)$$

$$\text{Texture} = 2.46958 - 0.0178 \times X_1 - 0.026628 \times X_2 + 3.88043 \times 10^{-3} \times X_3 + 4.76255 \times 10^{-4} \times X_1X_2 + 1.26218 \times 10^{-4} \times X_1X_3 - 8.85556 \times 10^{-5} \times X_2X_3 - 7.85213 \times 10^{-5} \times X_1^2 - 1.74842 \times 10^{-5} \times X_2^2 - 5.38906 \times 10^{-6} \times X_3^2 \quad (13)$$

Table 3. ANOVA for the experimental results of the RSM modeling

Source	**DF**	**Colour Changes**			**Shrinkage (%)**			**Texture (kg)**		
		CE	SS	P value	CE	SS	P value	CE	SS	P value
Model	9	9.89	48.56	0.0012	60.92	3391.77	<0.0001	1.85	4.13	<0.0001
X_1	1	0.48	4.19	0.0526	11.84	2522.59	<0.0001	0.27	1.29	<0.0001
X_2	1	1.16	24.03	0.0001	1.32	31.59	0.0634	-0.30	1.66	<0.0001
X_3	1	0.74	9.85	0.0053	6	647.05	<0.0001	0.21	0.80	<0.0001
X_1X_2	1	0.45	2.46	0.1288	1.26	18.92	0.1426	0.11	0.14	0.0013
X_1X_3	1	0.72	6.14	0.0218	1.36	22.14	0.1147	0.11	0.15	0.0008
X_2X_3	1	0.26	0.8	0.3761	-0.43	2.19	0.6080	-0.080	0.076	0.0111
X_1^2	1	-0.033	6.34E-03	0.9363	-1.39	11.56	0.2459	-0.018	1.873E-03	0.6607
X_2^2	1	-0.26	0.41	0.5220	-4.5	121.28	0.0012	-3.934E-03	9.286E-05	0.9219
X_3^2	1	-0.33	0.67	0.4156	-1.55	14.46	0.1967	-0.019	2.258E-03	0.6300
Std. Dev		0.98			2.83			0.097		
Mean		9.47			55.96			1.83		
CV%		10.37			5.06			5.30		
PRESS		41.34			392.95			0.40		
Adeq.Prec		10.252			23.612			26.550		
R^2		0.7476			0.9614			0.9628		
Adj R^2		0.6140			0.9410			0.9431		
Pred R^2		0.3636			0.8886			0.9059		

DF – degree of freedom; CE – coefficient; SS – sum of square

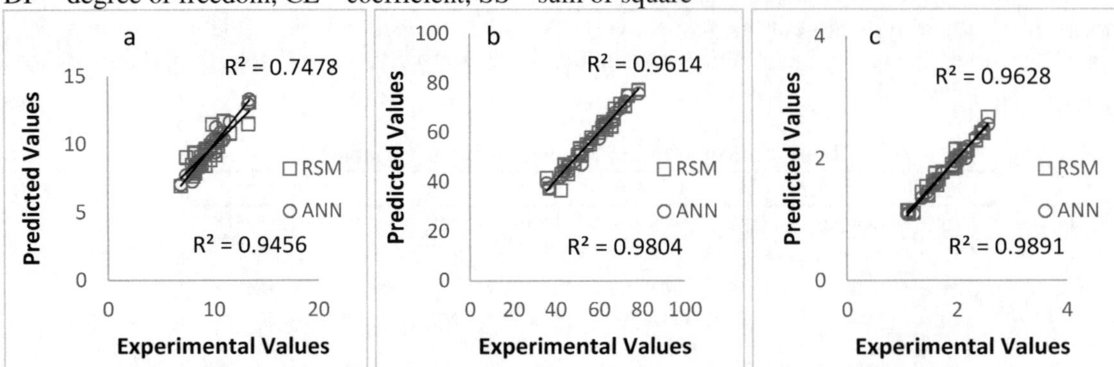

Figure 2. Experimental vs. predicted values for (a) colour change, (b) shrinkage percentage, (c) texture, by optimum RSM and ANN configuration respectively.

From equation 11, it can be seen that sucrose concentration and immersion time have negative effect on the colour changes while solution temperature has positive effect on colour changes of osmotically pumpkin slabs. From equation 12, both the three process variables have positive effect on the shrinkage percentage. However, from equation 13, sucrose concentration and solution temperature show negative effect, while immersion time shows positive effect on texture.

Table 3 shows the results of the second order polynomial model in the form of Analysis of Variance (ANOVA) and the equation indicated adequately relationship between the independent variables and the output variables. The ANOVA result for the color changes, shrinkage and texture indicates F-value of 5.59, 47.07 and 48.86 which proves that the model is significant. Coefficient of determination (R^2) was calculated to check the significance of the model and the values of R^2 were obtained to be 0.7476, 0.9614 and 0.9628 for color changes, shrinkage and texture respectively which imply the model is highly significant. In addition, p-value was obtained to be 0.0012 for colour changes, <0.0001 for shrinkage and texture, which indicates the proper fitting for the model. The coefficient of variation (CV%) measures the dispersion of the experiments values from the predictions values of the quadratic polynomial models [42].The values of CV are 10.37, 5.06 and 5.3 and these indicate the low deviation between the experimental and predicted values. Adequate precision (AP) refers to measures of the experimental signal to-noise ratio and AP exceeds 4 is desirable [43]. In this study, AP is found to be greater than 10, which implies good prediction of the model. Linear regression analysis was performed between the output variables (colour changes, shrinkage percentage and texture) values estimated by both RSM and ANN with their experimental values as shown in Figure 2.

3.3. Comparison of ANN and RSM models

The comparative values of RMSE, MPE (%) and R^2 of ANN and RSM has been calculated and listed in Table 4. The root mean square error (RMSE) for colour change by RSM and ANN is 0.779 and 0.364, the Model Predictive Error (MPE) is 6.319% and 2.874%, and the coefficient of determination (R^2) is 0.745 and 0.946. The RMSE for shrinkage by RSM and ANN is 2.245 and 1.615, the MPE is 3.360% and 2.092%, and the coefficient of determination (R^2) is 0.961 and 0.980. The RMSE for texture by RSM and ANN is 0.078 and 0.038, the Model Predictive Error (MPE) is 3.796% and 1.569%, the coefficient of determination (R^2) is 0.963 and 0.992. The results when compared with RSM showed that ANN has much better performance in correlating non-linear relationships. The same observation was also reported by Youssefi et al. [28].

Table 4. Comparison between RSM and ANN

Statistical parameters	Color Changes		Shrinkage		Texture	
	RSM	ANN	RSM	ANN	RSM	ANN
RMSE	0.779	0.364	2.245	1.615	0.078	0.038
MPE (%)	6.319	2.874	3.360	2.092	3.796	1.569
R^2	0.745	0.946	0.961	0.980	0.963	0.992

The results of error prediction by RSM are larger than ANN, indicating that the ANN model has higher modelling and predictive ability than RSM. The higher predictive accuracy of ANN can be attributed to its universal ability to model non-linear system, while RSM only restricted to quadratic polynomial. RSM also requires a standard experimental design in order to construct the model. However, ANN model may require a greater number of experimental data and higher computation time than RSM [29,44].

4.Conclusion

This study compares the performance of RSM and ANN for their prediction capabilities using the experimental data of osmotically dehydrated pumpkin. The ANN models was found to have better predictive capabilities in color changes, shrinkage and texture after comparing RMSE, MPE and R^2 of both models. The structured nature of RSM provides the predicted quadratic equation to exhibit the factors contributions from the coefficient regression of the models. This ability is robust in identifying the significant and insignificant terms in the model and hence can reduce the complexity of the models. However, the ANN presents a better alternative in modelling and prediction.

Acknowledgement

The authors would like to acknowledge the financial support from University Malaysia Sabah and Ministry of Higher Education Malaysia (grant: RACE009-TK-2013).

References

[1] Kowalska H, Lenart A and Leszczyk D 2008 The effect of blanching and freezing on osmotic dehydration of pumpkin *J. Food Eng.* **86** 30–8
[2] Ciurzyńska A, Lenart A and Greda K J 2014 Effect of pre-treatment conditions on content and activity of water and colour of freeze-dried pumpkin *LWT - Food Sci. Technol.* **59** 1075–81
[3] Yang X, Zhao Y and Lv Y 2007 Chemical composition and antioxidant activity of an acidic polysaccharide extracted from Cucurbita moschata Duchesne ex Poiret. *J. Agric. Food Chem.* **55** 4684–90
[4] Doymaz İ 2007 The kinetics of forced convective air-drying of pumpkin slices *J. Food Eng.* **79** 243–8
[5] Assis F R, Morais R M S C and Morais A M M B 2015 Mass Transfer in Osmotic Dehydration of Food Products: Comparison Between Mathematical Models *Food Eng. Rev.* **8** 116–33
[6] Chiralt A, Martínez-Navarrete N, Martínez-Monzó J, Talens P, Moraga G, Ayala A and Fito P 2001 Changes in mechanical properties throughout osmotic processes *J. Food Eng.* **49** 129–35
[7] Garcia C C, Mauro M A and Kimura M 2007 Kinetics of osmotic dehydration and air-drying of pumpkins (Cucurbita moschata) *J. Food Eng.* **82** 284–91
[8] Lee J.S, Lim L.S 2011 Osmo-dehydration pretreatment for drying of pumpkin slice *Int. Food Res. J.* **18** 1223–30
[9] Mayor L, Cunha R L and Sereno A M 2007 Relation between mechanical properties and structural changes during osmotic dehydration of pumpkin *Food Res. Int.* **40** 448–60
[10] Krokida M K, Karathanos V T and Maioulis Z B 2000 Effect of Osmotic Dehydration on Viscoelastic Properties of Apple and Banana *Dry. Technol.* **18** 951–66

[11] Ciurzyńska A, Andrzej L and Gręda K J 2014 Effect of pre-treatment conditions on content and activity of water and colour of freeze-dried pumpkin *LWT - Food Sci. Technol.* **59** 1075–81
[12] Falade K O, Igbeka J C and Ayanwuyi F A 2007 Kinetics of mass transfer, and colour changes during osmotic dehydration of watermelon *J. Food Eng.* **80** 979–85
[13] Krokida M K, Karathanos V T and Maroulis Z B 2000 Effect of Osmotic Dehydration on Color and Sorption Characteristics of Apple and Banana *Dry. Technol.* **18** 937–50
[14] Fernández L, Castillero C and Aguilera J M 2005 An application of image analysis to dehydration of apple discs *J. Food Eng.* **67** 185–93
[15] Viberg U, Freuler S, Gekas V and Sjiiholm I 1997 Pretreatment of Strawberries Effects and Shrinkage *J. Food Eng.* **35** 135–45
[16] Giraldo G, Talens P, Fito P and Chiralt A 2003 Influence of sucrose solution concentration on kinetics and yield during osmotic dehydration of mango *J. Food Eng.* **58** 33–43
[17] de Souza Silva K, Caetano L C, Garcia C C, Romero J T, Santos A B and Mauro M A 2011 Osmotic dehydration process for low temperature blanched pumpkin *J. Food Eng.* **105** 56–64
[18] Szczesniak A S 2002 Texture is a sensory property *Food Qual. Prefer.* **13** 215–25
[19] Silva K S, Fernandes M A and Mauro M A 2014 Effect of calcium on the osmotic dehydration kinetics and quality of pineapple *J. Food Eng.* **134** 37–44
[20] Garcia Loredo A B, Guerrero S N, Gomez P L and Alzamora S M 2013 Relationships between rheological properties, texture and structure of apple (Granny Smith var.) affected by blanching and/or osmotic dehydration *Food Bioprocess Technol.* **6** 475–88
[21] Dermesonlouoglou E K, Pourgouri S and Taoukis P S 2008 Kinetic study of the effect of the osmotic dehydration pre-treatment to the shelf life of frozen cucumber *Innov. Food Sci. Emerg. Technol.* **9** 542–9
[22] Dermesonlouoglou E K, Giannakourou M and Taoukis P S 2016 Kinetic study of the effect of the osmotic dehydration pre-treatment with alternative osmotic solutes to the shelf life of frozen strawberry *Food Bioprod. Process.* **99** 212–21
[23] Pereira L M, Ferrari C C, Mastrantonio S D S, Rodrigues A C C and Hubinger M D 2006 Kinetic Aspects, Texture, and Color Evaluation of Some Tropical Fruits during Osmotic Dehydration *Dry. Technol.* **24** 475–84
[24] Cevoli C, Cerretani L, Gori A, Caboni M F, Gallina Toschi T and Fabbri A 2011 Classification of Pecorino cheeses using electronic nose combined with artificial neural network and comparison with GC-MS analysis of volatile compounds. *Food Chem.* **129** 1315–9
[25] Zenoozian M S, Devahastin S, Razavi M A, Shahidi F and Poreza H R 2007 Use of Artificial Neural Network and Image Analysis to Predict Physical Properties of Osmotically Dehydrated Pumpkin *Dry. Technol.* **26** 132–44
[26] Shafafi Zenoozian M and Devahastin S 2009 Application of wavelet transform coupled with artificial neural network for predicting physicochemical properties of osmotically dehydrated pumpkin *J. Food Eng.* **90** 219–27
[27] Chen C R, Ramaswamy H S and Alli I 2007 Prediction of Quality Changes During Osmo-Convective Drying of Blueberries Using Neural Network Models for Process Optimization *Dry. Technol.* **19** 507–23
[28] Youssefi S, Emam-Djomeh Z and Mousavi S M 2009 Comparison of Artificial Neural Network (ANN) and Response Surface Methodology (RSM) in the Prediction of Quality Parameters of Spray-Dried Pomegranate Juice *Dry. Technol.* **27** 910–7
[29] Prakash Maran J, Sivakumar V, Thirugnanasambandham K and Sridhar R 2013 Artificial neural network and response surface methodology modeling in mass transfer parameters predictions during osmotic dehydration of Carica papaya L. *Alexandria Eng. J.* **52** 507–16
[30] El-Aouar Â A, Azoubel P M, Barbosa J L and Murr F E X 2006 Influence of the osmotic agent on the osmotic dehydration of papaya (Carica papaya L.) *J. Food Eng.* **75** 267–74
[31] Ozdemir M, Ozen B F, Dock L L and Floros J D 2008 Optimization of osmotic dehydration of diced green peppers by response surface methodology *LWT - Food Sci. Technol.* **41** 2044–50

[32] Corzo O and Gomez E R 2004 Optimization of osmotic dehydration of cantaloupe using desired function methodology *J. Food Eng.* **64** 213–9
[33] Mokhtarian M, Heydari Majd M, Koushki F, Bakhshabadi H, Daraei Garmakhany A and Rashidzadeh S 2014 Optimisation of pumpkin mass transfer kinetic during osmotic dehydration using artificial neural network and response surface methodology modelling *Qual. Assur. Saf. Crop. Foods* **6** 201–14
[34] Oztop M H, Sahin S and Sumnu G 2007 Optimization of microwave frying of osmotically dehydrated potato slices by using response surface methodology *Eur. Food Res. Technol.* **224** 707–13
[35] Kargozari M, Moini S and Emam-Djomeh Z 2010 Prediction of Some Physical Properties of Osmodehydrated Carrot Cubes Using Response Surface Methodology: Osmodehydrated Carrot Cubes *J. Food Process. Preserv.* **34** 1041–63
[36] Mayor L, Moreira R, Chenlo F and Sereno A M 2007 Osmotic Dehydration Kinetics of Pumpkin Fruits Using Ternary Solutions of Sodium Chloride and Sucrose *Dry. Technol.* **25** 1749–58
[37] Yadav A K and Singh S V 2014 Osmotic dehydration of fruits and vegetables: a review. *J. Food Sci. Technol.* **51** 1654–73
[38] Amiryousefi M R, Mohebbi M and Khodaiyan F 2014 Applying an intelligent model and sensitivity analysis to inspect mass transfer kinetics, shrinkage and crust color changes of deep-fat fried ostrich meat cubes. *Meat Sci.* **96** 172–8
[39] Kerdpiboon S, Devahastin S and Kerr W L 2007 Comparative fractal characterization of physical changes of different food products during drying *J. Food Eng.* **83** 570–80
[40] Prakash Maran J, Mekala V and Manikandan S 2013 Modeling and optimization of ultrasound-assisted extraction of polysaccharide from Cucurbita moschata *Carbohydr. Polym.* **92** 2018–26
[41] Garson G D 1991 Interpreting neural-network connection weights *AI Expert* **6** 46–51
[42] Jeganathan P M, Venkatachalam S, Karichappan T and Ramasamy S 2014 Model Development and Process Optimization for Solvent Extraction of Polyphenols from Red Grapes Using Box–Behnken Design *Prep. Biochem. Biotechnol.* **44** 56–67
[43] Prakash Maran J and Manikandan S 2012 Response surface modeling and optimization of process parameters for aqueous extraction of pigments from prickly pear (Opuntia ficus-indica) fruit *Dye. Pigment.* **95** 465–72
[44] Geyikçi F, Kiliç E, Çoruh S and Elevli S 2012 Modelling of lead adsorption from industrial sludge leachate on red mud by using RSM and ANN *Chem. Eng. J.* **183** 53–9

IOP Conf. Series: Materials Science and Engineering **206** (2017) 012037 doi:10.1088/1757-899X/206/1/012037

Application of Empirical Peleg Model to Study the Water Adsorption of Full Cream Milk in Drying Process

S Abd Hashib[1]*, H Rosli[1], M U H Suzihaque, [2], N A Md Zaki[2], U K Ibrahim[2]

[1]Department of Chemical Engineering, Faculty of Chemical Engineering, Universiti Teknologi MARA, 40450 Shah Alam, Selangor, Malaysia

[2]Department of Bioprocess Engineering, Faculty of Chemical Engineering, Universiti Teknologi MARA, 40450 Shah Alam, Selangor, Malaysia

*Email: syafiza0358@salam.uitm.edu.my

Abstract. The ability of spray dryer in producing full cream milk at different inlet temperatures and the effectiveness of empirical model used in order to interpret the drying process data is evaluated in this study. In this study, a lab-scale spray dryer was used to dry full cream milk into powder with inlet temperature from 100 to 160°C with a constant pump speed 4rpm. Peleg empirical model was chosen in order to manipulate the drying data into the mathematical equation. This research was carry out specifically to determine the equilibrium moisture content of full cream milk powder at various inlet temperature and to evaluate the effectiveness of Peleg empirical model equation in order to describe the moisture sorption curves for full cream milk. There were two conditions set for this experiments ; in the first condition (C1), further drying process of milk powder in the oven at 98°C to 100°C while the second condition (C2) is mixing the milk powder with different salt solutions like Magnesium Chloride (MgCl), Potassium Nitrite (KNO_2), Sodium Nitrite ($NaNO_2$) and Ammonium Sulfate ($(NH_4)_2SO_4$). For C1, the optimum temperature were 160°C with equilibrium moisture content at 3.16 % weight dry basis and slowest sorption rates (dM/dt) at 0.0743% weight dry basis/hr. For C2, the best temperature for the mixture of dry samples with MgCl is at 115°C with equilibrium moisture content and sorption rates is -78.079 % weight dry basis and 0.01 % weight dry basis/hr. The best temperature for the mixture of milk powder with KNO_2 is also at 115°C with equilibrium moisture content and sorption rates at -83.9645% weight dry basis and 0.0008% weight dry basis/hr respectively. For mixture of dry samples with $NaNO_2$, the best temperature is 160°C with equilibrium moisture content and sorption rates at 84.1306% weight dry basis and 0.0013% weight dry basis/hr respectively. Lastly, the mixture of dry samples with $((NH4)_2SO_4$ where the best temperature is at 115°C with equilibrium moisture content -83.8778% weight dry basis and sorption rates at 0.0021% weight dry basis/hr. The best temperature selected best on the lowest moisture content formed and also the slowest sorption rates.

1. Introduction

According to [1] the model implied that the moisture equilibrium was somewhat higher than that determination on the assumption that the sample reached constant weight, but there was no conclusive evidence that this was really the case. This is because, the weight of the full cream milk powder under two conditions which are soaked in different salt solutions and being dried inside the oven at 98°C to 100°C never reached a constant weight. The reading was taken every 30 minutes within 270 minutes. The weight of the full cream milk powder under this two conditions were decreasing but never achieved the constant weight that can be indicated the moisture equilibrium was achieved.

Not only the weight of the full cream milk samples was investigated, but after plotting a graph of moisture content Wet Basis (%) versus moisture content Dry Basis (%), the graph obtained was in linear regression form. This prove that the statement from M.Peleg, 1988 where published sorption curves in the form of moisture versus time relationship of milk powder and rice, exposed to moist atmosphere or soaked in water, were fitted by a two parameter, non-exponential empirical model was true. Unlike other agriculture products like bananas [13], green beans and olive cake [4] the equilibrium of moisture content can be investigated by using Mathematical Model like Lewis, Henderson and Pabis, Wang and Singh, Since the graph of moisture content (wet basis) (%) versus moisture content (dry basis) (%) obtained was in the exponential form, therefore, Peleg equation was chosen in order to determine the equilibrium moisture content for full cream milk powder under two different conditions since this model enabled prediction of moisture contents after long exposure from experimental data obtained in relatively short time exactly well before the moisture level appeared to reach a plateau.

The challenges that rise up regarding this research are the ability of the spray dryer to dry up the full cream milk at different temperatures and the effectiveness of empirical model used in order to interpret the drying process data for the description of moisture sorption curves. In order to overcome these problems, the lab-scale spray dryer was used to dried up the full cream milk into powder with the selection of temperatures of 100°C, 115°C, 130°C, 145°C and 160°C with a constant pump speed which is 4 rpm while the Peleg empirical model was chosen in order to manipulate the drying data into the mathematic equation.

The objectives of this study is, to determine the equilibrium moisture content of full cream milk powder at various inlet temperature and evaluate the effectiveness of Peleg empirical model equation in order to describe the moisture sorption curves for full cream milk powder

1.1. Drying process of full cream milk

As we know, the pathogens like bacteria, gems and bugs growth rapidly in a moisture condition. Therefore, drying will solve the problem. Drying is the process designated in order to remove the water or any types of solvents by evaporation from a solid, semi-solid or liquid surface. While drying process takes place, water molecule was migrate from the interior surface of the product to the outer surface of the product by a molecular diffusion. Following the Second Law of Thermodynamics, the water will remove from the higher moisture content to the area of lower moisture content. For the drying process, moisture content plays the important role for the drying process to take place. Milk can be present as powder to solve the following above problems. The milk powder can be produced by using spray drying and roller drying. Spray Drying Process

The process of spray dryer is starting from the introduction of concentrated milk into a chamber (usually a fine mist) through which hot air is circulating. The milk droplets soon will lose their water and fall to the floor as fine powder after finish the process. Spray-dryer can be used to dry many form of milk such as whole milk, skim milk, whey and fat-enriched milk. Increased demand for milk powder especially in food industry has led the spray-dryer to widening its application. A milk spray-dryer is usually used in a large capacity with a million liters of milk per day therefore make it more economic. Usually, the inlet temperature chose around 200°C (400°F) to maintain the powder quality. The important properties for the final products of milk spray-dryer are flow properties, wettability,

IOP Conf. Series: Materials Science and Engineering **206** (2017) 012037 doi:10.1088/1757-899X/206/1/012037

sinkability, dispersability and solubility. For whole milk powder, it demands for fast reconstitution ability by the addition of additives called lecithin.

1.2. Mathematical models for drying process

The results from the drying process of full cream milk was determined for further steps in order to determine the best fit mathematical model for this product. There are many mathematical models used in agricultural field to identify the best fit mathematical models that can be used to ensure the effectiveness of the drying process. Many researches for agriculture applied several types of mathematical models to interpreted and explained each of them was the best fit for the drying process besides claimed the calculation for the convection drying kinetics.

The examples of mathematical models that are usually used for drying process of agriculture products are Lewis, Henderson and Pabis, Wang and Singh, Peleg, Page and Silva et al. Table 1 below shows varies empirical models to describe drying kinetics.

Table 1. Empirical models to describe drying kinetics

	Mathematical Model	Empirical Expression
1	Lewis	$X^* = e^{-at}$
2	Henderson and Pabis	$X^* = ae^{-bt}$
3	Wang and Singh	$X^* = 1 + at + bt^2$
4	Peleg	$X^* = 1 - t(a+bt)$
5	Page	$X^* = e^{-at\wedge b}$
6	Silva et al	$X^* = e^{-at-b\sqrt{1}}$

From Table 1, it can be seen that all types of mathematical model have always been used together with the result from the drying process of the agricultural products. The examples of drying process that cluster with these variations of mathematical models are Mathematical models to describe thin-layer drying and to determine drying rate of whole bananas [11]. Modelling of olive cake thin-layer drying [15], thin-layer modeling of black tea drying process and drying behavior of green beans [15] are the proof of all types of mathematical models such as Lewis, Page, Modified Page, Henderson and Pabis and also Logarithmic are reasonable and logical to be used.

Among several journals stated before, a thin-layer drying model for agricultural products was used where it described well that the drying process of agricultural materials mainly fall into three categories. The three categories involved namely as theoretical, semi-theoretical and the last one is empirical. Theoretical takes into account only the internal resistance to moisture transfer while for both semi-theoretical and empirical consider only external resistance to moisture transfer between product and air. All the types of mathematical models stated in Table 1 were declared as semi-theoretical thin layer drying models. As they were applied this models to the agricultural products, it might also be suitable for the milk drying process. The steps of drying process and also the calculation involved the moisture ratio can be applied also to the drying process of milk.

IOP Conf. Series: Materials Science and Engineering **206** (2017) 012037 doi:10.1088/1757-899X/206/1/012037

2. Methodology

2.1 Materials

The equipment used in this research is laboratory scale spray dryer. Besides that, the materials involved in this experiment are full cream milk (liquid form), Magnesium Chloride (MgCl), Potassium Nitrite (KNO_2), Sodium Nitrite ($NaNO_2$) and Ammonium Sulfate ($(NH_4)_2SO_4$). The apparatus used in this experiment are 50mL beaker, 1000mL beaker, 1000mL conical flask, Cryovac® plastic bag, aluminum foil and electronic weighing balance.

2.2 Moisture content measurement

Full cream milk was stored in a refrigerator prior to spray drying. The inlet temperature of the spray dryer was set up with ranges from 100°C to 160°C. The amount of the powder collected in cyclone collector (cyclone recovery) and manual sweeping the spray dryer's wall was packed and stored in the desiccator. Empty beaker was weighed on electronic weighing balance and recorded as W1. Then 2g of full cream milk powder were added and recorded as W2.

2.3 Moisture Content Calculation

The initial weight of the full cream milk powder, W_o was calculated by using Equation 2.1:

$$W_o = W2 - W1 \tag{1}$$

For condition one, the samples were left inside the oven at 98°C to 100°C until final weight is achieved. The final weight, W_D of these samples was recorded for every 30 minutes. The moisture content in wet basis, M_{wb} and moisture content in dry basis, M_{db} was calculated by using this Equation 2.2 and 2.3.

$$M_{wb} = \frac{(Wo\text{-}WD)}{Wo} \tag{2}$$

$$M_{db} = \frac{(Wo\text{-}WD)}{WD} \tag{3}$$

For condition two, full cream milk powder was mixed with four types of salts ; Magnesium Chloride (MgCl), Potassium Nitrite (KNO_2), Sodium Nitrite ($NaNO_2$) and Ammonium Sulfate ($(NH_4)_2SO_4$). All samples were stored at the evacuated desiccators over P_2O_5.

2.4 Empirical Peleg Model equation

The graph t/(Mt-Mo) (hour/M) against time (hour) was constructed from tabulated data and constant k1 and k2 was determined. With the value of k1 and k2 constant, initial moisture content (Mo) (% weight dry basis), moisture content at time t (M(t)) (% weight dry basis), equilibrium moisture content (ME) (% weight dry basis), the sorption rate (dM/dt) and ratio (R) was calculated and extracted from the graph . The time taken (t_R) required to achieved ratio, R was calculated using Equation 2.4- 2.8 :

$$M(t) = Mo + t/(k_1+k_2t) \tag{4}$$

$$ME = Mo + 1/k_2 \tag{5}$$

$$dM/dt = k_1/(k_1+k_2t)_2 \tag{6}$$

$$R = (Mt\text{-}Mo)/(ME\text{-}Mo) \tag{7}$$

$$t_R/t_{1/2} = R/(1\text{-}R) \tag{8}$$

IOP Conf. Series: Materials Science and Engineering **206** (2017) 012037 doi:10.1088/1757-899X/206/1/012037

3. Results and Discussion

For most of the results, at each different temperature, the final weights, WD was in a range between 1.8532g to 1.9776g. The highest final weight, WD value was found to be at 1.9776g for sample number 1 at temperature 160°C while the lowest final weight, WD was found to be at 1.8532g for sample number 10 at 100°C. In addition, the average dry basis moisture content, MDB % also shows the lowest value with 1.1314% weight dry basis. Overall, the trend whereas the temperature increases, the value for final weight, WD are also increases. At high final weight, WD, the value for average dry basis moisture content, MDB % will decreases at best temperature 115 °C and 160 °C .

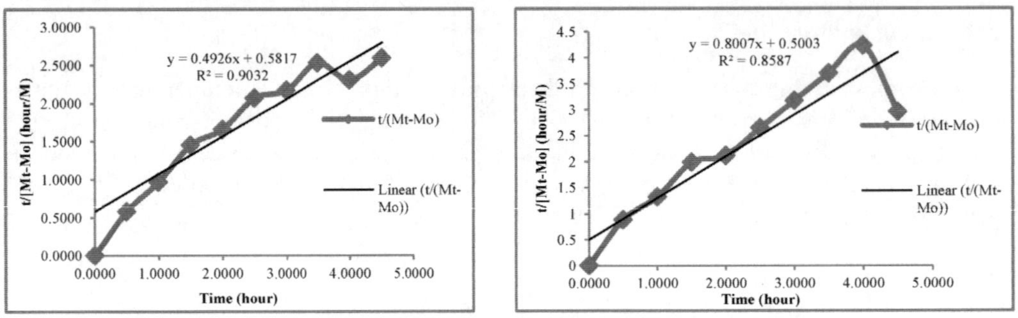

(a) (b)

Figure 1. Graph of t/[M(t)-Mo] (hour/M) versus Time (hour) for (a) temperature 160°C and (b) temperature 115°C.

The graph of t/[M(t)-Mo] (hour/M) versus Time (hour) for dry sample of full cream milk powder mixed with MgCl showed in Figure 2 (a) and (b) at temperature 160 °C and 115 °C respectively resulted in linear regression line. In (a), the value for t/[M(t)-Mo] (hour/M) was suddenly drop due to sudden increased of dry samples weight at time 4.5 hours. This is because by the exposure of the sample number 10 towards the environment thus make it adsorb the moisture content or humidity from the environment. Therefore, it is very important to tighten up the samples beaker. From the equation, the value of constants k1 (hr/%weight) and also k2 (1/%dry weight basis) are 16.488 and 16.55 respectively. In (b) the weight dry samples of full cream milk powder mixed with MgCl decreases, with the increases of t/[M(t)-Mo] (hour/M) within the time 0.0 hour to 4.5 hours. The value of constants k1 (hr/%weight) and also k2 (1/%dry weight basis) directly extracted from the linear equation were 116.49 and 161.37 respectively. Both k1 and k2 constant values can be further used in Peleg equation in order to determine the equilibrium moisture content (ME) and also the sorption rate (dM/dt) for every sample from 0.0 hour to 4.5 hours.

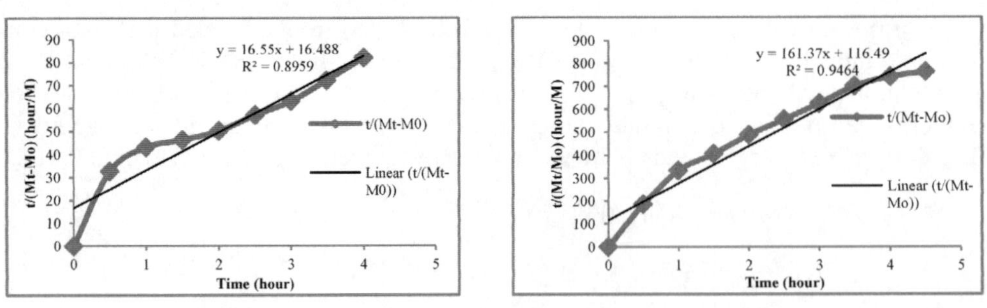

(a) (b)

Figure 2. Graph of t/[Mt-Mo] (hour/M) versus Time (hour) at temperature (a) 160 °C and (b) 115°C in MgCl solution.

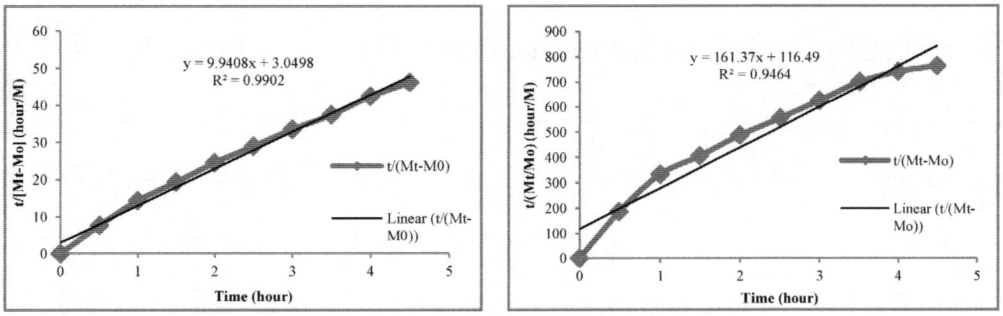

(a) (b)

Figure 3. Graph of t/[Mt-Mo] (hour/M) versus Time (hour) at temperature (a) 160 °C and (b) 115°C in NaNO2 solution.

Figure 3 shows the graph of t/[M(t)-Mo] (hour/M) versus Time (hour) for dry samples of full cream milk powder mixed with NaNO2 at temperature 160°C and 115°C. The graph plotted was in a best fit linear regression line. In graph (a) the equation of the graph is y = 9.9408x + 3.0498. Therefore, where the constants k1 (hr/%weight) and also k2 (1/%dry weight basis) are 3.0498 and 9.9408 respectively. In graph (b) y = 40.664x + 24.198. Thus, the constant values required in Peleg equation can be extracted directly from the equation where the constants k1 (hr/%weight) and also k2 (1/%dry weight basis) are 24.198 and 40.664 respectively.

4. Conclusion

In this research, empirical Peleg model was selected in order to describe the moisture sorption curves of full cream milk powder under certain condition. The graph of t/(M(t)-Mo) versus time (hour) formed a linear regression thus make it suitable to use the empirical Peleg model. The equilibrium moisture content (ME) was easily calculated by using this model as compared with the Page and Lewis model. This is because, the data sets of full cream milk powder after drying process was differ with other agriculture products like whole banana, olive cake, ginger and peach slice as mentioned in many journals.

Following steps in determination of equilibrium moisture content (ME) by using Page and Lewis model, a graph of moisture content wet basis % versus moisture content dry basis % was illustrated. Unfortunately, the graph obtained was not in exponential form due the weight of dry samples of full cream milk at certain conditions is changed with time. There is no constant weight was recorded in during 270 minutes experiment running. The graph of moisture content wet basis % versus moisture content dry basis % was in a linear regression form. This make it difficult in order to determine the equilibrium moisture content (ME) by using Page and Lewis model since the constant required in the equation is absence.

Acknowledgement

Support for this research was provided by Research Acculturation Grant Scheme (RAGS/1/2014/TK05/UITM//4) from the Ministry of Higher Education of Malaysia and Research Management Institute (RMI), Universiti Teknologi Mara (UiTM) Shah Alam. Thanks to all who may contribute directly or indirectly towards making this research successful.

References

[1] Peleg M, An Empirical Model for the Description of Moisture Sorption Curves, 1216 *J. of Food Sci.*-Volume **53** No. 4, 1988

[2] Kockel T K, Allen S, Hennigs C, Langrish T A G, An experimental study of the equilibrium for skim milk powder at elevated temperatures, **51** 291-297.

[3] Wang S, Langrish T, A review of process simulations and the use of additives in spray drying, **42** 13-25.

[4] Chen X D and Lin S X Q, Air Drying of Milk Droplet under Constant and Time-Dependent Conditions, **51** 1790-1799.

[5] Gharsallaoui A et al 2007 Applications of spray-drying in microencapsulation of food ingredients: An overview, **40** 1107-1121.

[6] Midilli A , Kucuk H & Yapar Z, A new model for single-layer drying, **20** 1503-1513.

[7] Zafer E, Filiz I, A Review of Thin Layer Drying of Foods: Theory, Modeling, and Experimental Results, **50** 441-464.

[8] Langrish T A G, Chan W C, Kota K, Comparison of maltodextrin and skim milk wall deposition rates in a pilot-scale spray dryer, **179** 84-89.

[9] Chiou D, Langrish T A G, 2007 Development and characterisation of novel nutraceuticals with spray drying technology, **82** 84-91.

[10 James P, George A.K, Datta, Development and validation of heat and mass transfer models for freeze-drying of vegetable slices, **52** 89-93.

[11] Silva W P *et al* 2013 Mathematical models to describe thin-layer drying and to determine drying rate of whole bananas **13** 67-74.

[12] Mezhericher M, A. Levy a & I. Borde, Model of Single Droplets Containing Insoluble or Dissolved Solids, **25** 1025-1032.

[13] Zhu A S, Shen X Q, The model and mass transfer characteristics of convection drying of peach slices, **72** 345-351.

[14] Athanasia M G, Thodoris D K, Dimitris S A, Konstantinos G A 2008 Water sorption isotherms and glass transition temperature of spray dried tomato pulp, **85** 73-83.

[15] Nalan A A, Doymaz I, Modelling of olive cake thin-layer drying process, **68** 455-461.

IOP Conf. Series: Materials Science and Engineering **206** (2017) 012038 doi:10.1088/1757-899X/206/1/012038

A study on friability, hardness and fiber content analysis of fiber enriched milk tablet

M U H Suzihaque, M H Irfan and U K Ibrahim

Faculty of Chemical Engineering, Universiti Teknologi MARA, 40450 Shah Alam, Selangor, Malaysia

Email: suzihaque@salam.uitm.edu.my

Abstract. This study was performed to analyze the friability, hardness and fiber content of fiber enriched milk tablet derived from five different local fiber sources such as carrot, spinach, dragon fruit, mango and watermelon. Cow milk was mixed to complement with the tablet as a protein source. The powder were spray dried at 100°C, 120°C and 140°C and freeze dried at -60°C. The mixture of fruits and milk were made into equal ratio with the addition of 15% maltodextrin as a carrier. Tablets formed were used for friability and hardness test while dried powder were used for fiber content analysis. Dragon fruit tablet dried at 140°C have the highest friability with 11. 42% of weight loss. The second highest friability was spinach tablet dried at 100°C and 120°C drying temp erature with 9.30% and 9.28% respectively. The lowest friability was exhibited by carrot, mango and watermelon tablet at 100°C and dragon fruit at 120°C while carrot and spinach at 140°C. In contras t, none of the freeze dried tablets showed any weight loss hence they are not friable. For hardness test, all of the freeze dried showed to have higher tensile strength than spray dried, where carrot showed to be the highest at 2.27 Newton and the lowest were spray dried mango at 0.16 Newton. In fiber content analysis, freeze dried mango have the highest fiber content followed by freeze dried carrot and 140°C s pray dried carrot. It can be concluded that the higher the spray dry temperature, the more friable is the tablet. While, high friability leads to lower hardness of tablets. In terms of fiber content, the higher the spray dry temperature, the lower the fiber content found.

1. Introduction

Recently, supplement is extensively developed to solve malnutrition which includes kids. Kids are prone to lack of fiber and protein which may retard their growth physically and mentally. Nowadays, kids have consumed unhealthy food like candy and junk food more than the food should be required like vegetables or fruits. Deficient of fiber has caused many gastrointestinal problems like constipation, diarrhea, inflammatory bowel disease and hemorrhoids. While the insufficient of milk has led to stunting, rickets, Kwashiorkor and even having lower IQs. Therefore, a more attractive way of approaching kids to have balanced food consumption is by introducing fiber enriched milk tablet. Previously, fiber enriched milk supplement has been invented to enhance the dairy products positive effects toward obese people. People are consuming dairy product like milk to control their body weight[1]. The enrichment of fiber in the product has added more beneficial effects as it enhance the insulin sensitivity in obese people. Furthermore, fiber can also enhances the satiety of consumer due to its viscosity-producing and bulking effect which eventually helps people in controlling the body weight. A research has been proved that increasing dietary calcium of obese patients for one year resulted in a 4.9 kg loss of body fat[2].

On top of that, concerning on the advantages of fiber and milk itself, they promote a combination of useful effects toward kids in term of health and growth. Dietary fiber which comprised of soluble and insoluble form has individual roles in preventing Irritable Bowel Syndrome, inflammatory bowel disease, diabetes, heart disease, constipation, colon cancer, hemorrhoids and diverticulosis. Same goes with milk as it can helps prevent eye problems, promotes a healthy immune system, and essential for the growth and development of cells, mucous membranes, skin, bones and tooth health [3]. Fiber source for fiber enriched milk tablet could be derived from variety of plants or fruits. Different plants or fruits have particular content of fiber and it would not be the same. The problems arise in determining how the fiber content would vary between a source with another. Moreover, supplement made in tablet form has more benefits as it has longer shelf life, easy for transportation and storing. However, it is crucial to know how the different fiber sources would affect the structure of produced tablet like its hardness and friability. Therefore in this experiment, five different fiber sources are selected from vegetables and fruits to determine those parameters. Spray drying and freeze drying methods are used to transform the fiber and milk mixture liquid into powder form before tableting. Different temperatures of spray drying would also affect the tablet friability, tablet hardness and its fiber content. Therefore, analysis is done by varying the spray drying temperature to know how it would affect the resulting parameters and how would it be if compared with freeze drying technique.

2. Methodology

2.1. Materials

Fiber source like carrot, spinach, dragon fruit, mango and watermelon were purchased at Giant, Seksyen 7, Shah Alam, Selangor. Local fruits and vegetables selections are based on different colours as follow:

Table1. Colours represented by five different fiber sources

Fruits/Vegetables	Representative for colour
Spinach	Green
Watermelon	Peach
Mango	Yellow
Carrot	Orange
Dragon fruit	Purple

2.2. Spray Drying

500 g carrots are cut into smaller pieces and blended with 500 g milk (equal ratio) and mixed with 15% of maltodextrin using blender Pensonic PB-325. The mixture is transferred into beaker and filtered by filtering cloth to eliminate any solid material that might cause clogging through the spray dryer nozzle. Then, it is dried with spray drier (SD-Basic Lab Plant) with inlet air temperature of 100°C, 120°C and 140°C with compressed air pressure of 0.10 Mpa [4]. The powder produced is collected and stored in freezer before tableting. Repeat all the steps for different fiber source.

2.3. Freeze Drying

500 g carrots are cut into smaller pieces and blended with 500 g milk (equal ratio) and mixed with 15% of maltodextrin using blender Pensonic PB-325. The mixture is transferred into a freeze dryer round bottomed flask and covered with parafilm. Then, it is stored in freezer overnight before dried with freeze drier (SASTEC) with -60 C and vacuum pressure. After 2-3 days when it is sufficiently dried, the powder produced is collected and punched by mortar to reduce the particle size. Then, it is stored in freezer before sent to tableting. Repeat all the steps by different fiber source A subsubsection. The paragraph text follows on from the subsubsection heading but should not be in italic.

2.4. Tableting

The powder made from spray dry and freeze dry are sent to Faculty of Pharmacy of University Malaya for tableting. It used direct compression machine called GlobePharma MTCM-1 model which produced averagely 0.7 gram tablet with 6.3 mm thickness and diameter of 12.7 mm. The pressure used is 3000 atm.

2.5. Friability

Friability test was run at Faculty of Pharmacy at UniversitiTeknologi Mara (UiTM). The equipment used is ElectrolabFriabilator which set with 60 rotations. Every tablet was cleaned with brush and weighted by weighing balance before inserted into the friabilator. Then the equipment will rotate 60 times and once it finished, the tablet is collected. The tablet is cleaned again by brush and weighted by the same weighing balance. The percentage of weight loss is calculated as formulae follow:

$$\text{Percentage of weight loss} = \frac{\text{initial weight of tablet} - \text{final weight of tablet}}{\text{initial weight of tablet}} \times 100 \qquad (1)$$

2.6. Hardness

Hardness test is run at Faculty of Pharmacy in UniversitiTeknologi Mara (UiTM). The equipment used is Pharmatest. Once the tablet is held vertically on the machine platform, it will be compressed against the wall to measure the weight, diameter and thickness. Then the tablet will slide and brought horizontally before being compressed once again until it break to measure the force needed to break the tablet. All the reading is recorded for every tablet. The hardness of the tablet is measured by knowing its tensile strength which is calculated as following formulae.

$$\text{Tensile strength} = \frac{2F}{\pi dv} \qquad (2)$$

F = force needed to break the tablet (Newton)
d = tablet diameter
v = tablet thickness

2.7. Fiber Content

Association of Official Agricultural Chemists (AOAC) method [5] 985.29 is a standard method invented by United States Department of Agriculture is used to study the content of total dietary fiber. The sample of fiber enriched milk powder is tested with three different enzymatic digestions which are alpha-amylase to gelatinize the solution, protease to remove protein, and amyloglucosidase to remove starch. Next is precipitation using ethanol to precipitate the residue. Then the precipitated residue is filtered by vacuum filter and washed by three different chemical which are 78% ethanol, 95% ethanol and acetone. Next, one set of duplicate of residue was tested protein analysis and another set of duplicate was tested for ash analysis. The total fiber content will be calculated by subtracting the initial weight residue with weight of (protein + ash)[6].

i. Alpha-Amylase digestion

Duplicate of 1 gram of each fiber source powder is weighted and recorded as initial weight before put into 250 mL conical flask. 10 mL phosphate buffer (pH 6.0) is added into the beaker and check pH with pH meter. Adjust if pH does not equal 6.0±0.1. 0.1 mL heat-stable alpha-amylase solution is added. The sample is then transferred into boiling tube. The tube is covered with aluminium foil and placed in boiling water bath. The tube must be left at 95-100°C for 1 5 minutes and is shaked gently at every 5 minutes intervals. Then, the solution is cooled to room temperature.

ii. Protease digestion

The pH is first adjusted to 7.5±0.1 by adding drops of 0.275 M NaOH solution. pH is checked with pH meter. 0.1 mL of protease solution is added. The conical flask is covered with aluminium foil and incubated at 60°C with continuous agitation for 30 minutes. Then, the solution is cooled to room temperature.

iii. Amyloglucosidase digestion

Drops of 0.325 M HCl solution are added to adjust pH to 4.5±0.2. pH is checked with pH meter. 0.3 mL amyloglucosidase is added, then covered with aluminium foil, and incubated for 30 minutes at 60°C with con tinuous agitation. Then let it cool.

iv. Precipitation of residue by Ethanol

60 mL of 95% EtOH is first pre-heated to 60°C (measure volume before heating) then added into the solution from previous step. Let precipitate form at room temperature for 60 minutes.

v. Filtration of Residue

Weigh 1 gram of celite to nearest 0.1 mg, then wet and distribute bed of celite in fitted crucible by using stream of 78% ethanol. Celite efficiently separates the fiber from the fritted glass of the crucible, to allow easy removal of the crucible contents. By using vacuum pump, suction is applied to draw celite onto fritted glass as even mat. The suction is maintained and quantitatively transferred the precipitate from enzyme digest to crucible. The residue is washed successively with three 20 mL portions of 78% Ethanol, two 10 mL portions of 95% Ethanol, and two 10 mL portions of acetone. In some cases, gums might form during filtration, trapping liquid in residue. If so, break surface film with spatula to enhance filtration.

vi. Drying of Filtered Residue

Residue is transferred into ceramic crucible and dried overnight in 105°C air oven. Cooled in desiccator a nd weigh to nearest 0.1 mg. Subtracted (crucible + celite + residue) with initial (crucible + celite) weights to determine weight of residue. The residue is analysed from one sample of set of duplicates for protein content and the other set of duplicate for ash content. Protein analysis is done by Kjeldahl Method, where the sample is get by cautiously scraping the celite and the fiber mat onto a proper piece of filter paper and kept it for Kjeldahl method.

IOP Conf. Series: Materials Science and Engineering **206** (2017) 012038 doi:10.1088/1757-899X/206/1/012038

vi. Fiber Content Calculation

Incinerate residue from second sample of duplicate for 5 hours at 500°C. The samples in the ceramic crucible s are arranged in the furnace and label first before started the incineration. After 5 hours, the furnace is left for another 5 hour for the temperature to cool before take out the samples. When ready, the samples are weighed to 0.1 mg. and recorded the reading. Subtract the reading with crucible and celite weights to determine ash.

vi. Ash Analysis

The fiber content was calculated by the equation below[5]:

$$\text{Total dietary fiber} = \left(\frac{\frac{R1+R2}{2}}{\frac{W1+W2}{2}} - P - A\right) \times 100 \qquad (3)$$

R1= residue weight of first duplicate R2 =
residue weight of second duplicate P =
Protein weight
A = Ash weight
W1 = weight of initial sample of first duplicate, W2 =
weight of initial sample of second duplicate

vi. Kjeldahl Method

a)Digestion. The sample from previous procedure is placed into a digestion flask, along with 12 ml of concentrated sulfuric acid (H2SO4). 7 grams of potassium sulfate is added and a catalyst, usually copper to increase the boiling point. The digestion flask is heated to 390 C until white fumes can be seen, and then continue the heating for about 60 minutes. The flask is cooled and cautiously adding 250 ml of water.

b)Distillation. The pH of the mixture is raised using sodium hydroxide (45% NaOH solution). This has effect of changing the ammonium (NH4+) ions which are dissolved in the liquid to ammonia (NH3), which is a gas. The nitrogen is separated away from the digestion mixture by distilling the ammonia (converting it to a volatile gas, by raising the temperature to boiling point) and then the distilled vapors is trapped in a special trapping solution of about 15 ml hydrochloric acid in 70 ml of water. The trapping flask is removed and the condenser is rinsed with water so as to make sure that all the ammonia has been dissolved.

*c)Titration.*5 drops of phenolphthalein is added to the trapping solution. The dye turned into colorless indicating significant amount of the original acid still exists. A standard solution of NaOH is put into the burette, and slowly is added to the acid solution. The point at which the dye turns pink, indicate the "endpoint" has reached which all the acid has been neutralized by the base. The volume of the neutralizing base is recorded.

d)Calculation of Protein Content. One mole of ammonia coming from the digestion mixture will neutralize exactly one mole of acid in trapping flask. Therefore, the number of ammonia moles which equivalent to moles of nitrogen is found by steps as follow:

1. The number of moles of acid in the trapping flask originally (before any ammonia was trapped) is calculated as: moles of acid = molarity of acid x volume used trapping flask (molesA = M x V)
2. The number of moles of base (NaOH) that were added from the burette to neutralize the remaining acid (that NOT neutralized by the ammonia) is calculated as:
   Moles of base = molarity of base x volume added from burette (molesB = M x V)

3. The number of "moles of ammonia" is calculated by subtracting mole base by mole of acid.
4. The number of "moles of ammonia" are coming from the protein is similar to the "moles of nitrogen"
5. The number of nitrogen (gram) in the original sample of protein is calculated as:
   Grams of nitrogen = moles nitrogen x atomic mass
6. The percentage of nitrogen found in the original sample can now be calculated as:
   %nitrogen =(gms nitrogen / gms sample) x 100%
7. Crude protein , CP = %nitrogen x 6.25 (6.25 as conversion factor)

3. Results and Discussion

3.1. Friablity Test

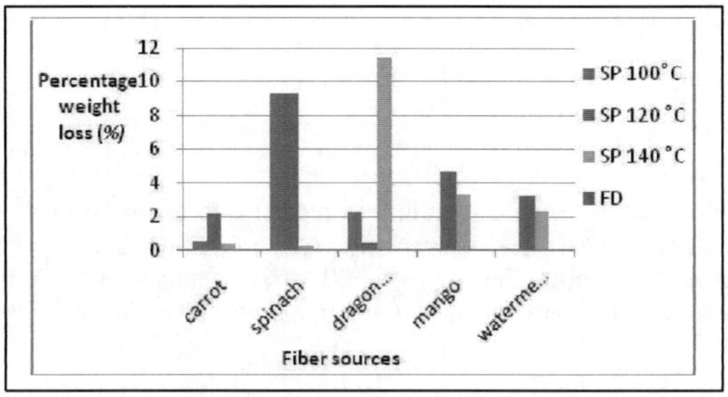

Figure 1. Percentage weight loss (*%)* vs Fiber sources

During the rotations, the surface of the tablet will gradually rupture and the tablet will lose some of its weight due to hitting force. Thus, this test will measure the friability of the tablet by determining the percentage weight loss. The higher the percentage of weight loss, the higher the tablet friability. Based on figure 1, at spray drying inlet air temperature of 100 C and 120 C, the spinach tablet shows the highest percentage of weight loss which around 9.3%. It shows quite big gap when compared with other fiber source tablet. This is because spinach has the lowest carbohydrate content, but high fiber content. Initially during the preparation for spray drying, the mixture of spinach and milk is filtered to eliminate any solid substance like pulp. A fully liquid suspension is obtained to prevent any clogging in the spray drier and also to get high quality of powder. The filtration somehow has eliminated most of fiber contained in the spinach [7]. Same goes to some of carbohydrate content which had left behind, along with the fiber during filtration. Basically, carbohydrate is comprised of sugar and starch [8]. Both substance acts as natural binding agent which increase the hygroscopicity and interparticulate bonding strength within the tablet [9]. Therefore, the higher the carbohydrate content, the higher the hardness of tablet. When more fiber is removed, carbohydrate content will also be removed. Therefore, based on figure 2, the spinach shows quite low value of tensile strength which is 0.52% at spray drying temperature of 100 C and 120 C. This is due to the fact of the higher the friability, the lower the tensile strength and thus the hardness.

At inlet air temperature of 140 C, dragon fruit tablet shows the highest percentage of weight loss which around 11.5%. It shows great gap when compared with inlet air temperature of

100 C and 120 C and even with other tablet at the same temperature. This is because at high inlet a temperature, the residual moisture content within powder particle is reduced. When the moisture content reduced, there will be less water to bind powder particle together by its surface tension or known as water holding capacity [10]. This give rise to reduced internal cohesion and shear strength between particles in the tablet. Shear strength is defined as the ability of solid particle to resist friction [11]. This is because the existing solid particle of powder tends to flow against each other when external force is applied and thus cause internal friction. When the shear strength is reduce, the powder particle become less cohesive. As a result, the flowability of the particle will increase(A-sun 2015). Flowability is defined as the ability of powder to flow[11]. If moisture content is decreases, the flowability of the particle will increase[12]. As the powder particles are free to flow, the tablet will exhibit a friable characteristic. Therefore, it can be concluded that the tablet becomes more friable at higher air inlet temperature.

While at inlet air temperature of 120 C and 140 C; mango, watermelon and carrot tablet are ranked the second, third, and fourth of decreasing friability respectively. At 120 C, the percentage of weight loss of mango is 4.66%, watermelon is 3.21% and carrot is 2.18%. At 140 C, the percentage of weight loss of mango is 3.35%, watermelon is 2.35% and carrot is 0.36%. Mango has the second highest friability because it has high sugar and fiber content. Since, a lot of fiber has separated out during filtration, it has caused the high amount sugar to left behind as well as the amount fiber. Therefore, there is lack of remaining sugar content which make the mango tablet more friable than watermelon and carrot. While, watermelon has moderate friability because it has moderate sugar content and least fiber content. Therefore, even the sugar content is moderate, the fiber content is too low to drag along much of sugar to be separated during filtration. Thus, the watermelon is less friable than mango. Next, carrot tablet is less friable than mango and watermelon because it contains more fiber that has been separated through filtration, but the remaining sugar content is being concentrated instead of drag along with fiber. This is due to the structure of carrot sugar which in form of dilution instead of mango and watermelon sugar which is present in pulp or semi liquid. Hence, the carrot sugar will pass through the filter medium more efficiently while mango and watermelon sugar are partially passed through it due to its semi liquid structure. It is even trapped by the filter aid. Thus, the carrot tablet is less friable than mango and watermelon.

The figure also shows that the friability of most of tablet is lower at inlet air temperature of 100 C, than at 120 C and 140 C. This may due to larger moisture content that is still present in the powder. Moisture content is aid in cohesiveness of the powder particle especially when it contains sugar. The water will bind more tightly with carbohydrate rather than lipid, protein or fiber. Therefore, it can be seen that the spinach in contrast has the most friable because it contain least sugar. It is followed by dragon fruit, carrot, mango and watermelon. The ranking is however does not justifying the rank of sugar content for each of fiber source. This may due to size of fruit particle itself which has different effect towards the friability of the tablet. Particle size is majorly affects the cohesiveness of the tablet. Lower particle size leads to lower degree of size distribution of particle. So, the particle will be arranged between them compactly where they are in close form and there are fewer spaces between particles [13]. As a result, the bulk density will be increased. This indirectly has reduce the flowability because there will be more internal friction that cause the particle hard to flow. Basically, particle size would not be affected by the inlet air temperature, otherwise after added by high concentration of carrier such as maltodextrin like 30% and above.

Based on the figure 1, all of freeze dried tablet is found to have no percentage of weight loss. This is because the powder formed exhibit flaky or irregular shape particle [14].This can be observed during the powder collection where the powder structure is obviously different from spray dried powder. The flaky structured is observed when the powder is being punched by mortar to reduce the particle size into finer. The freeze drying mechanism has caused sublimation of solid into gas without entering liquid phase. Eventually, in effect of the collapse and shrinkage of the product is prevented thereby resulting in a porous dried material which make the particle formed to be flaky [15]. Flaky particles have a high surface to volume ratio and as a result, the particles have high inter-particle friction. This has led to a decrease in flowability and become less or not friable at all.

3.2. Hardness Test

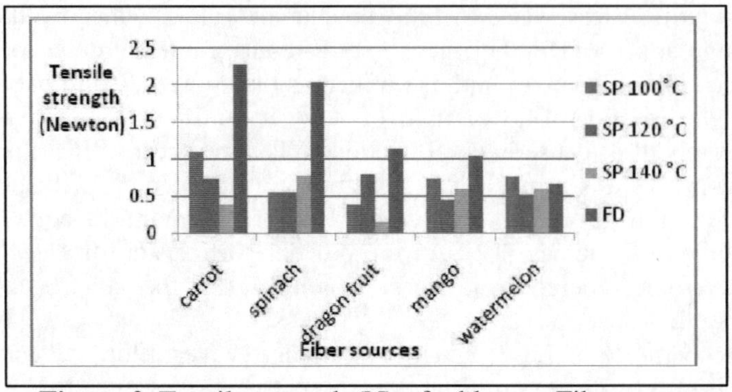

Figure 2. Tensile strength (N) of tablets vs Fiber sources

Hardness is largely influenced by the friability of tablet. The higher the friability, the lower the hardness of the tablet. As stated earlier, flowability is the ability of powder to flow under applied pressure. When the powder particle is more cohesive toward internal friction, the tablet will be less friable and eventually exhibit higher hardness. When comparing figure 1 and 2, the hypothesis earlier is justified. At inlet air temperature of 140 C, the highest percentage of weight loss exhibited by dragon fruit in figure 1 shows the lowest tensile strength in figure 2 which is 0.16%. Same goes with the lowest weight loss percentage exhibited by spinach shows the highest tensile strength in subsequent figure which is 0.78%. While, the very low weight loss percentage of carrot, mango and watermelon at inlet air temperature of 100 C shows a higher tensile strength in the subsequent figure respectively.

Meanwhile, there is no percentage of weight loss exhibited by most of freeze dried tablet except for spinach which account only tiny amount. Therefore, the hardness or tensile strength of all of freeze dried tablet is seen drastically high at figure 2 when compared with spray dried tablet. This is because the irregular shape of the powder formed where the more irregular the shapes of the particles, the poorer its flowability [14]. This may due to the compression during the tablet formation, where the irregular shape of the particles forming a high compaction and inter-particle friction. Thus, flowability is reduced and high in cohesiveness. Eventually, the tablet gain higher tensile strength and become harder than spray dried tablet.

3.3.Fiber Content Analysis

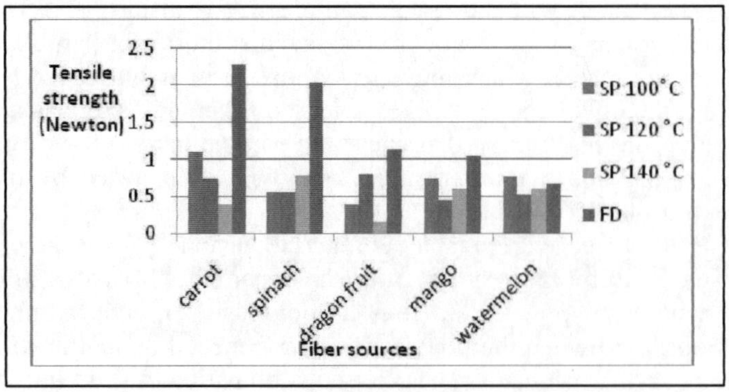

Figure 3. Fiber Content (%) vs Fiber sources

Theoretically, the highest fiber content of those five fiber source tablet is carrot, followed by mango, spinach, dragon fruit and watermelon . However, the results produced are not justifying the theory. This may due to the procedure used in spray drying which the mixture of fiber and milk is filtered first before sucked by the equipment for powder formation. Based on figure 3, there is inconsistent trend of fiber content through the increasing inlet air temperature of spray dryer for each of the fiber source powder. Only mango powder is justifying the theory of decreased fiber in increased inlet air temperature. Basically, total dietary fiber contains insoluble and soluble fiber. Insoluble fiber is insignificantly affected by the changes of inlet air temperature. Soluble fiber instead, is significantly changes as its amount is decreases in increased temperature. This is because soluble fiber is comprised majorly of mucilage around 93% [16]. Mucilage will be degraded as early as 45 C as it is entering the glass transition temperature. Glass transition temperature is where an amorphous material transforms from hard and relatively brittle state into molten and rubber-like state in increasing temperature [17]. In other word, the mucilage started to degrade and disintegrate at its own temperature. Therefore, the fiber content of mango powder is decreasing at increasing inlet air temperature. Higher temperature can also reduce total dietary fiber as it disrupts the cellular matrix of the fiber source or it is called as thermal degradation [18].

For the other fiber source powder, like carrot it shows the lower fiber content at inlet air temperature of 120 C than 100 C and 140 C. While, spinach and dragon fruit shows increasing fiber content in increased inlet air temperature. Watermelon on the other hand shows quiet insignificant changes among those temperatures. These trends are not justifying the previous theory maybe due to the age of the fiber source used. The content of soluble fiber decreases with age but insoluble fiber increase instead[16]. This shows how the alteration of fiber content which has affected by the maturity of the fiber source used. For every inlet air temperature of spray drying procedure, the fiber source used are slightly not in the same maturity, therefore the final total dietary fiber would be varied between them. The fiber content get is somehow different for freeze dried method. Since freeze drying is not introducing high temperature toward the sample, hence no thermal degradation is occurred. As a result, the fiber content is found higher than spray dry method. This is shown obviously by mango and carrot powder in figure 3 which have 9.86% and 9.57% respectively. Dragon fruit and spinach have shown moderate and the lowest fiber content with 8.75% and 4.75% respectively. While watermelon shows insignificant changes between spray dry and freeze dry method with average 7.34%. In conclusion, there are many other factors that would not have known that has significant impacts towards the findings of friability, hardness and fiber content analysis. Hence further studies need to be performed.The flowability of powder could be further analyzed with equipment like angle of repose, compressibility index, flow rate through an orifice and shear cell. Thus, these will prove and strengthen the support towards the findings of friability and hardness test. Same goes with the details fiber content of vegetables and fruits used. It is importance to know how the different composition of elements under dietary fiber would affect the final fiber content.

4. Conclusion

In a nut shell, fiber enriched milk powder and tablet has varying characteristics in term of friability, hardness and fiber content depending on the properties of the fiber itself. There is a huge range of knowledge about vegetables and fruits need to be studied. Different fiber source own different fiber structure, sugar content, carbohydrate content and moisture content. These relate with final fiber content, hygroscopicity, flowability and particles size of powder or tablet formed. Eventually it link with cohesiveness and interpaticle strength within the powder. Based on findings, for freeze dried tablet, carrot is the hardest while for spray dried tablet, dragon fruit is the most friable at inlet air temperature of 140 C. On top of that, dragon fruit shows the highest fiber content at 140 C inlet air temperatures while mango is the highest fiber content for the freeze dried tablet.

Acknowledgement
Support for this research was provided by a grant(RAGS/1/2014/TK05/UITM//4) from the Ministry of Higher Education of Malaysia. Thank you to Universiti Teknologi MARA and to all who may contribute directly or indirectly towards making this research successful.

IOP Conf. Series: Materials Science and Engineering **206** (2017) 012038 doi:10.1088/1757-899X/206/1/012038

References

[1] N Lummela R A K, Tiina J, Taru K P, Tuula T, Salme J, Johan G E and Riitta K 2009. Effects of a fibre-enriched milk drink on insulin and glucose levels in healthy subjects. *Nut. J.*: 1-5.

[2] Michael B Zemel H S, Betty, Greer, Douglas, Dirienzo P C Z 2000. Regulation of adiposity by dietary calcium. *The Faseb J.* 1132-1138.

[3] Early R. 1998. Characteristics of Milk. New York, *Thompson Sci.*

[4] Ramachandra H G, RAO P M G 2002. Development of spray dried orange juice blended skim milk powder. *EDP Sci.* 1-8.

[5] AOAC A O M 1985. Total Dietary Fiber in Foods. *AOAC Official Method* 985.29.

[6] Chemists A o O A 1985. United States of America.

[7] Reinagel M 2013. QuickandDirtyTips.com. Juicing: Healthy Habit or Blood Sugar Bomb? [8] Decker F 2016. SFGate. *Differences Between Starches & carbs*

[9] Patel Shailendra A S, Lodhi B S 2012. Natural Binding Agents in Tablet Formulation. *Int. J. of Pharma. & Bio.l Arc.* 467.

[10] GeoCities 2009. GeoCities.com. Factors Affecting Angle of Repose

[11] Prescott J K 2000. On Powder Flowability. *Pharma. Tech.* 64.

[12] A-sun K. Thumthanaruk Lekhavat S, Jumnongpon R 2015. Effect of spray drying conditions on physical characteristics of coconut sugar powder, *I n t . Food Res. J.l* **23**(3): 1317.

[13] Ng L T, Chan P H 2012. Physicochemical and Nutritional Properties of Spray-dried Pitaya Fruit Powder as Natural Colorant. *Food Sci. Biotechnol*: 679.

[14] Ekenlebie E P 2014. Pharmaceutical process of optimisation of bulk lyophilisates: implications of powder handling, ASTON UNIVERSITY.

[15] Caparino J T,.Nindo C I, Sablani S S, Powers J R, Fellman J K 2012. Effect of drying methods on the physical properties and microstructures of mango (Philippine 'Carabao' var.) powder.

[16] Contreras-Padilla M 2004. Effects of Drying Process on the Physicochemical Properties of Nopal Cladodes at Different Maturity Stages. *Plant Foods for Human Nut.*.

[17] Overney R M, Luginbuhl C B R, Dinelli F 2000. Glass and structural transitions measured at polymer surface on the nanoscale. **59**.

[18] Onifade T B, Ige M T, Ajay A O, 2013. Some physical properties and thin layer drying characteristics of local varieties of tomatoes (Lycopersicon lycopersicum). *Agri. and Bio. J. of North America* 278

IOP Conf. Series: Materials Science and Engineering **206** (2017) 012039 doi:10.1088/1757-899X/206/1/012039

Effect of temperature to the properties of sago starch

M Mustafa Kamal[1], R Baini[1*], S Mohamaddan[2], O S Selaman[3], N Ahmad Zauzi[1], M R Rahman[1], N Abdul Rahman[1], Chong, K H[5], M F Atan[1], N A S Abdul Samat[1], S N L Taib[3], A K Othman[4]

[1]Chemical and Energy Sustainability Department, Faculty of Engineering, Universiti Malaysia Sarawak (UNIMAS), Kota Samarahan 94300, Sarawak, Malaysia
[2]Mechanical and Manufacturing Department, Faculty of Engineering, Universiti Malaysia Sarawak (UNIMAS), Kota Samarahan 94300, Sarawak, Malaysia
[3]Civil Department, Faculty of Engineering, Universiti Malaysia Sarawak (UNIMAS), Kota Samarahan 94300, Sarawak, Malaysia
[4]Electrical and Electronic Department, Faculty of Engineering, Universiti Malaysia Sarawak (UNIMAS), Kota Samarahan 94300, Sarawak, Malaysia
[5]Faculty of Engineering, Computing and Science, Swinburne University of Technology, Sarawak Campus, Jalan Simpang Tiga, 93350 Kuching, Sarawak, Malaysia

Email: ruby@feng.unimas.my

Abstract. Recently, the importance of sago starch has increased, as it has become one of the main economically important agricultural crops to the most Southeast Asia countries. In the present work, an analysis on drying process of sago starch (Metroxylon sagu Rottb.) underwent various temperature has been made by using four empirical equations. The main goal of this analysis is to suggest the most accurate equation, in order to model and simulate the mechanical drying of sago starch. The experimental investigations were carried out in a gravity convection lab oven; and ±50g of sago starch (sample heights of 1 cm) was dried through four different temperatures, which were 50, 60, 70 and 80°C. The effect of drying temperature on the drying kinetics, as well as various qualities attributes of sago starch, such as microstructure, colour and functional properties were investigated. The results suggested that drying temperature has significant effect on sago starch drying kinetic; therefore, drying temperature would be the basis to select drying condition. Meanwhile, it was found that the various drying temperature ranging from 50 to 80°C affected the product quality especially in term of colour.

1. Introduction

Sago starch is the product of sago palm (Metroxylon sagu) or commonly called as "rumbia". It is a tropical plant with huge trunk that may reach height up to 25 m and a diameter of 40cm; with pinnate-leaves up to 9 m long. It can tolerate wet-swampy lands such as peat soil. Sago palm reaches its maturity after 9 to 12 years of planting [5]. In term of productivity, sago palm produces the highest among the starchy crops in the world which is 250kg per palm; therefore it is also known as the 'starch crop of the 21st century' by many scientists [4]. Sago starch main content is carbohydrate which is higher than rice and wheat flour. For this reason, it has become the staple food since ancient times in

Asian region. In the food industry, sago starch has become an important raw material and it is predicted that the demand of sago in the future will increase. Malaysia produces among the highest starch in the world and Sarawak is the largest sago-growing areas in Malaysia [9].

Sago processing in modern factory starts by feeding a segment of sago log into a slicer to separate between pith and bark. Then, the debarked section is fed into a mechanical rasper to produce finer pieces, which are then fed in hammer mill via conveyor belt. The resulting starch slurry, which still containing coarse fibre, is passed through a series of centrifugal sieves. Further purification is done by the separation process in a nozzle separator through sieve bends, as well as a series of cyclone separators, to obtain very pure starch. Most importantly, dewatering of starch is carried out using a rotary vacuum drum dryer followed by a hot air drying process [5].

Meanwhile, traditional processing of sago is made manually from 'skinning' the bark to chopping the pith and grating it into material resembling sawdust. The pith is then kneaded with water and filtered through sieves to extract the relatively large (20–60 mm) starch granules. After being subjected to several washing and straining, the starch milk is allowed to run into troughs. The starch settles in the troughs, which is subsequently sun or kiln-dried.

Our previous study [13] revealed that the resulted starch from modern and traditional factory differ from each other in terms of moisture content, colour, microstructure, particle size and infrared absorbance. It is suspected that the differences are due to several factors such as the extraction and the drying process. Researches on correlation between extraction and starch yield have been done by some researchers[6, 11]. However, the drying mechanism of sago starch has not been studied, widely.

Drying is a crucial process in the food industry. In modern sago processing factory, the starch is dried to increase their shelf life, to reduce the packaging cost and to improve the appearance. However, some of the processing stages in the modern sago processing had changed the properties of the sago starch [7]. In the traditional factory, the quality of starch is far from demand but very close to its natural and beneficial properties. Therefore, this study aimed to understand the effect of temperature on the properties of sago starch.

2. Materials and Methods

2.1 Sample Preparation

Sago log were harvested from a nearby farm in Kota Samarahan, Sarawak. The log was kept in a refrigerator at a temperature under -80ºC to avoid changes in its properties until the day it is ready for processing. The processing method involved chopping, blending, sieving and settling the starch.

First, the sago log was chopped with knife into small pieces. Then, the chunks were blended to produce sawdust. The sawdust was then mixed with water and sieved several times with kitchen siever, fabric siever (napkin) and lastly the 100um siever, respectively. After each sieving, the solution was left at room temperature for 3 hours, 1 hour and 30 min, respectively, before the precipitated starch can be collected and used in the drying experiment.

The sample preparation can be concluded as shown in figure 1.

IOP Conf. Series: Materials Science and Engineering **206** (2017) 012039 doi:10.1088/1757-899X/206/1/012039

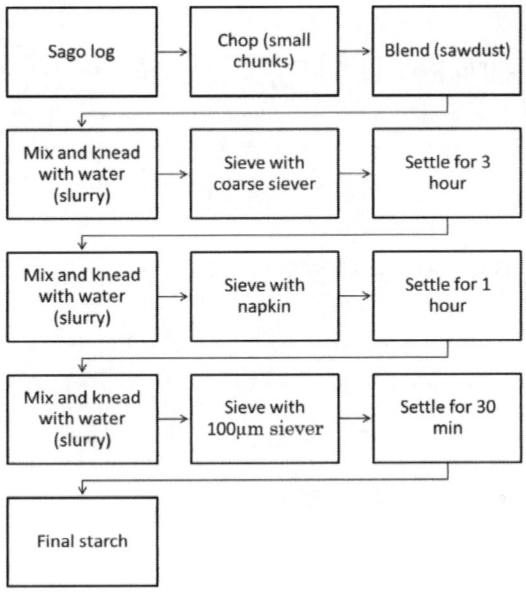

Figure 1. Step for extracting sago starch

2.2 Drying Experiment

A basic gravity convection oven was used for the drying experiment. For each experiment, the oven was preheated to the desired temperature. ±50g of sago starch was placed in an aluminium foil-container and then it was placed in the oven. The treatment duration was dependent upon sample mass and oven temperature. The drying experiments were conducted at 50, 60, 70 and 80°C. The dried sago starch was weighed every hour until a constant mass was reached. During the sample mass measurement, the container was removed from the oven and it was placed on a weighing balance with least count of 0.001 mg, and then it was put back in the oven, quickly. This weighing process took place in less than one minute. Dried sample are ready for subjecting into psychochemical and Fourier transform infrared (FT-IR) analysis.

2.3 Mathematical Modelling and Data Analysis

Microsoft Excel 2010 version was used to analyse all drying data. All experiments were performed in three replications. The initial moisture content (M_i) was determined according to the AACC2000 method. The equilibrium moisture content (M_e) of the sago starch at various drying temperature was determined by following a dynamic method. The moisture content at which the sample weight was unchanged with drying time was the equilibrium moisture content. These values, together with moisture content at particular drying time (M) were used to calculate the moisture ratio (MR), as given in equation 1. This empirical model commonly used to evaluate the drying kinetics of food (T. Dikbasan, 2007).

$$\mathrm{MR}=\frac{\mathrm{M}-\mathrm{M_e}}{M_i - M_e} \tag{1}$$

The moisture ratio (MR) was then subsequently regressed with Lewis model equation 2 to become equation 4.

$$\text{Lewis model, MR= exp(-kxt)} \tag{2}$$

$$\text{Drying rate; } -\frac{dX}{dt} = k_d(X - X_e) \tag{3}$$

Here;

MR= $(M-M_e)/(M_o-M_e)$; dimensionless moisture ratio

k_d = drying constant

t=drying time

The equation was integrated to give the following ones,

$$MR = \frac{X - X_e}{X_o - X_e} = \exp(-kdt) \tag{4}$$

$$X = (Xo-Xe)\exp(-kdt) + Xe \tag{5}$$

$$\text{Standard error, } (SE) = \left[\frac{\sum_{n-1}^{i}[(X_{Exp} - X_{Pred})^2]}{n-1}\right]^{\frac{1}{2}} \tag{6}$$

Where;

X_{Exp} = the Exp^{th} observed moisture ratio

X_{Pred} = the $Pred^{th}$ moisture ratio

n = number of observations

These equations express first-order kinetics in terms of the moisture ratio equation (3) to give a predicted content or fitted moisture content, X, equation (4) that decreases exponentially with time. X_e is the equilibrium moisture content expressed on a dry basis. Standard error (SE) which means the deviation between predicted and experimental values was then calculated to confirm the model.

2.4 Physicochemical analysis

Phsychochemical analysis involved the analysis of moisture content and colour analysis. The determination of moisture content in sago flour and lemantak were analysed by drying 3g of sample in gravity oven at 105°C until constant weight according to the AOAC, 2000 standard method. The moisture content was calculated in drying basis formula;

$$MC = \frac{M_i - M_t}{M_i} \times 100\% \tag{7}$$

Where M_i is the initial mass of sample (g) and M_t is the mass of sample (g) at a time (t), in this case, the final mass.

Meanwhile, an analysis of colour was carried out by taking the images of the samples manually, and analysed the images by using Matlab software that will give the pixel data of samples image in RGB value. The RGB values were then converted into CIE-L*ab value by using a RGB-to-Lab converter program available from the internet at http://colormine.org/convert/rgb-to-lab. The parameter 'L' measures the luminosity of the sample from (L = 0) black to (L = 100) white. The parameters 'a' and 'b' are measured from -60 to +60. The degree of measurement for 'a' is from green to red, while 'b' is from blue to yellow. In another word, ''L'' represents lightness, ''a'' represents redness (+) or greenness (-), and ''b'' represents yellowness (+) or blueness (-). Whiteness index were calculated according to (Yadav et al., 2006).

$$WI = 100 - \sqrt{(100 - L)^2 + a^2 + b^2} \tag{8}$$

The initial colour of sago starch before drying was also measured and the value was used to calculate the total amount of colour change by using the following formula (Baini & Langrish, 2009).

$$E = \sqrt{(L_o - L)^2 + (a_o - a)^2 + (b_o - b)^2} \tag{9}$$

2.5 FT-IR

Fourier transform infrared (FT-IR) spectroscopy was used to evaluate the structural differences in the samples. The infrared spectra of the commercial and traditional sago starch were recorded on a Shimadzu Fourier Transform Infrared Spectroscopy (FTIR) 81001 Spectrophotometer. An appropriate amount of starch powder was pressed into a crystal window and the samples were analysed with a resolution of 4cm^{-1} and average scanning time of 1 min. Furthermore, the spectral resolution range of scan was 4000 to 400 cm^{-1} (mid-infrared region) with ATR mode being utilized. The obtained spectra were transferred into a data analysis package. All optical measurements were performed at room temperature under ambient conditions.

3. Results and Discussions

3.1 Equilibrium moisture content (EMC)

EMC is termed as the moisture content of a wet solid in equilibrium with air of given humidity and temperature. The EMC of sago starch drying in different temperature are shown in table 1. Drying sago starch at 50°C resulted in sago starch with highest EMC (10.36%), while drying at 80% resulted in lowest EMC which is 1.31%. As for the whole result, the EMC was found to be decreased with the increase in drying temperature. Furthermore, the EMC value of starch drying at 50°C was significantly higher than that of 60, 70 and 80°C. This result can be explained due to decrease in the relative air humidity at higher temperature; in this case 60, 70 and 80°C [10].

IOP Conf. Series: Materials Science and Engineering **206** (2017) 012039 doi:10.1088/1757-899X/206/1/012039

Table 1. Equilibrium moisture content (EMC) in dry basis (db) at different temperatures

Drying temperature, °C	EMC, % db [a]
50	10.36±0.10
60	3.45±0.14
70	2.47±0.13
80	1.31±0.18

[a] All values are means of three measurements. Values are means ± standard deviation.

3.2 Drying Behaviour of sago starch

The changes in moisture content with time for different drying temperatures are shown in figure 2.

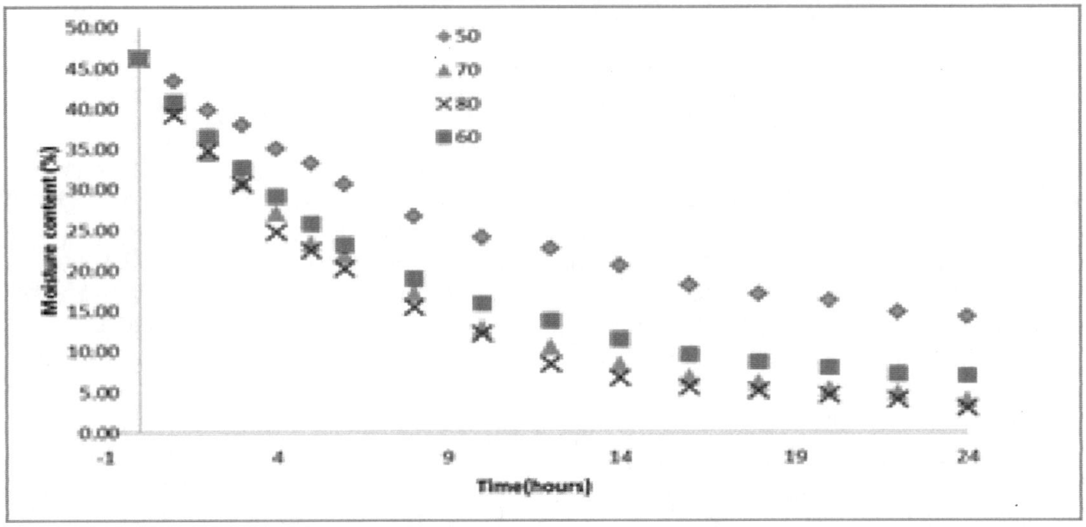

Figure 2. Moisture content of sago starch drying at different temperature (50, 60, 70 and 80°C)

The final moisture content of samples dried under different temperatures (50, 60, 70, 80°C) ranges from 14.30%, 6.99%, 4.01% and 3.07%, respectively. The result indicated that drying at higher temperature had produced sago starch with lower moisture content which suggested that drying temperature has an important effect on the final moisture content of sago starch.

In previous work, moisture content of commercial sago starch (rotary vacuum drum dryer and hot air drier at 50°C or above) had been compared with traditional sago starch (sun dry) which revealed that the former meet Malaysian standard (MS470 1992) for sago starch which is 13% maximum. On the other side, the latter did not meet the standard as the moisture content was 38.8% [7]. It should be noted that higher moisture content lead to high risk of microbial growth and spoilage while 0% drying until 0% moisture content will change the properties of the starch. In order to relate the previous result with this result, it is suggested that drying at temperature between 50 and 60°C will be the best to meet the standard for sago starch moisture content.

Figure 3 shows drying rate plotted over moisture content in different temperature (50, 60, 70 and 80°C). Sago starch display only falling rate period as many other foods and agricultural products which do not display constant rate period at all. This also indicates that the mechanism of mass transfer in sago starch is a moisture diffusion process. Therefore, during sago starch drying process, the internal heat and mass transfer rates determine the rate at which water becomes available at the exposed evaporating surface. In addition, the drying surface becomes first partially saturated and then fully saturated until it reaches equilibrium moisture content [9].

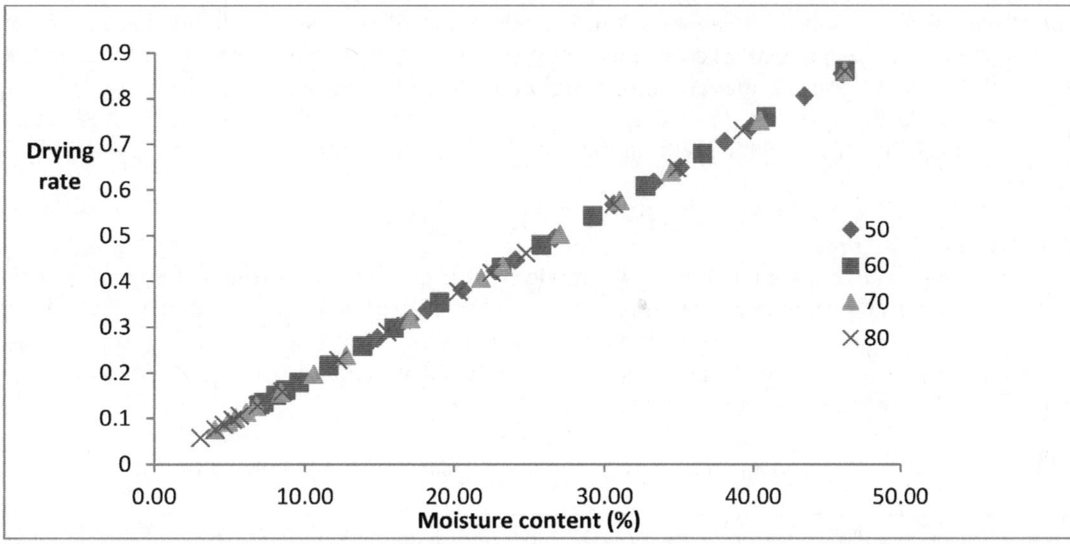

Figure 3. Sago starch drying rate over moisture content at different temperatures (50, 60, 70 and 80°C)

The drying rate of sago starch over time is shown in figure 4 below. It can be observed that at any of the time, drying curve at higher temperature is steeper than that at lower temperature, which means drying rate at higher temperature is higher than that of lower temperature, resulting in lower moisture content. The increase in drying temperature increases the heating rate and water vapour pressure inside the starch, thus accelerating the water migration inside it, consequently leading to higher drying rates and lower moisture content [3].
Besides, the result also shows that drying rate at 50°C decrease more gradually than at 60, 70 and 80°C are significantly steeper than 50°C. This shows that sago starch drying at 50°C exhibit consistent evaporation process for surface evaporation and inner moisture transportation.

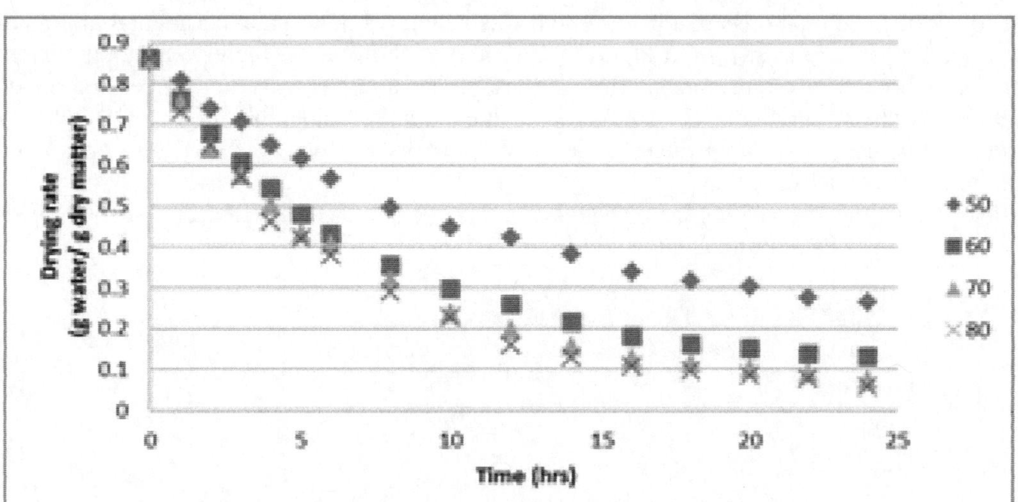

Figure 4. Sago starch drying rate at different temperatures (50, 60, 70 and 80ºC)

Furthermore, the drying rate at 70 and 80°C are quite similar with each other indicated that sago starch reacts in the same manner when drying at both temperatures. Briefly, both temperatures allow high amount of near-surface moisture evaporation in the beginning of process. As the drying progresses, both drying rate decreases gradually because moisture is transported from the inner layer to the surface before it evaporates. As a result, the falling rate region expressed an increase in resistance to heat and mass transfer inside the material. In addition, rapid evaporation of surface moisture could lead to formation of hard layer surface which will inhibit the moisture removal from starch; thus, reducing the drying rate.

3.3 Mathematical Modelling

The first order approach, equation (2) by Lewis model was fitted to the experimental data of moisture ratio of sago starch at different temperatures. This model has been found to fit the experimental moisture content for materials that were dried continuously very well [12]. Standard error (SE) was used to determine the average difference that the experimental values fall from the regression line of Lewis model and the results are showed in table 2.

Table 2. Standard error (SE), correlation coefficient and drying constant results

Drying temperatures (ºC)	Standard error (SE)	Correlation coefficient (R^2)	Drying constant (k)
50	0.77	0.9979	0.105
60	0.76	0.9993	0.144
70	0.63	0.9988	0.152
80	0.64	0.9981	0.162

The result from table 2 shows that low values of standard errors (SE<2.5) were found for Lewis model. The average value also indicated that the experimental data were 0.7% closer to the regression line. In another word, 99.3% of the experimental data fall within plus/minus 2* standard error of the regression from the regression line, which is also mean quick approximation of a 99.3% prediction interval.

In order to confirm the relationship, a linear graph was generated between experimental and predicted data for each temperature as shown in Figure 3. The results showed that there were good agreements between the predicted moisture content from Lewis model and experimental values for all drying temperatures as shown in figure 5. All the R^2 values were considerably high (R^2 >0.98) which indicated that this model was applicable for describing the drying behaviour of sago starch at drying temperature of 50, 60, 70 and 80°C.

In addition to this, the exponential curve between experimental and predicted data also shows good relationship in term of data fitting as shown in figure 6.

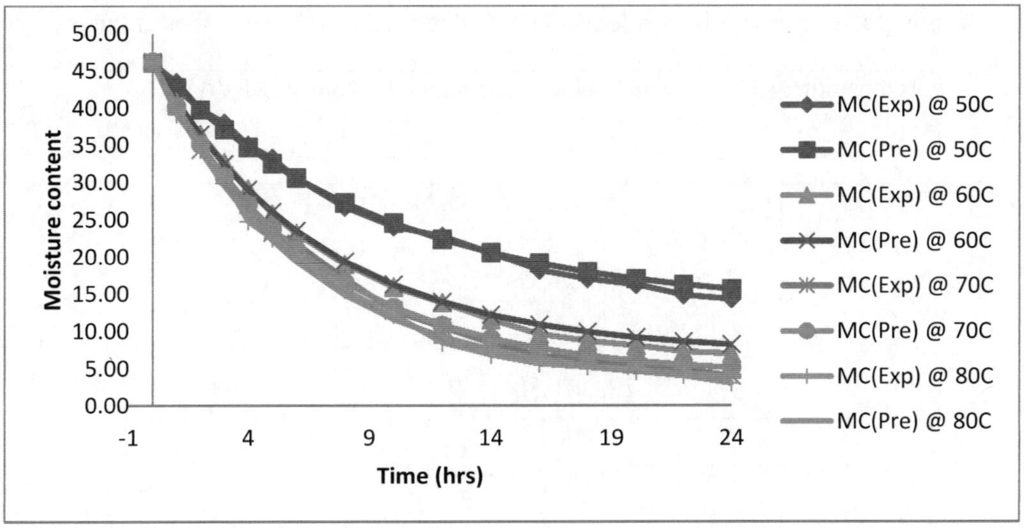

Figure 5. Comparison of experimental and predicted moisture content (mc) by Lewis model

The results obtained from figure 5 confirmed the applicability of Lewis model to be used for estimating drying behaviour of sago starch. In this preliminary work of drying sago starch, Lewis model as an empirical equation was selected to give better fit to the experimental data even though without any understanding of the transport process involved. Interestingly, this model has been used widely by many other researchers to predict the drying kinetic of fruits and vegetables [2].

3.4 Physicochemical Analysis

The effect of drying temperature on the properties of sago starch was observed through the colour changes analysis and the results are shown in table 3. As mentioned in Section 2.4, the parameter 'L' measures the luminosity of the sample from (L = 0) black to (L = 100) white. The parameters 'a' and 'b' are measured from -60 to +60. The degree of measurement for 'a' is from green to red, while 'b' is from blue to yellow. In another word, ''L'' represents lightness, ''a'' represents redness (+) or greenness (-), and ''b'' represents yellowness (+) or blueness (-).

The results show that increase in temperature will increase the darkness (browning) of sago starch. Sago starches obtained from 50°C drying temperature showed more lightness (L) and whiteness (WI) as well as less browning than those dried at higher temperature. The changes in colour were relatively smaller, which suggested that drying sago starch at 50°C is suitable to preserve the natural colour properties of sago starch. In contrast, sago starches obtained from 80°C drying temperature showed more darkness (L) and browning than those dried at lower temperature. It is also can be observed that during the sago starch drying, the whiteness level decrease while the browning values increase in higher temperature.

Table 3. The measured final colours of sago starch dried at different temperatures

Temperature (°C)	Final colour parameters for sago starch drying[a]				
	WI	E	L	a	b
Predried	54.92	0	55.24	-0.18	5.38
50	52.02	3.95	52.77	-0.27	8.47
60	48.24	8.22	49.47	-0.17	11.23
70	44.13	12.24	44.59	5.76	4.22
80	38.78	16.80	39.82	4.11	10.48

[a] WI=whiteness; E=browning; L=lightness; a=redness; b=yellowness

In previous study [7], the whiteness value of commercial sago flour was significantly higher than traditional sago flour even though it has been processed with high temperature (rotary vacuum drum dryer and hot air drying). This can be explained with the bleaching treatment applied in commercial sago flour factory [8]. However, in this study, it is confirmed that drying sago starch in high temperature result in unwanted color changes or the degradation of starch color. In fact, the higher the temperature, the darker the color.

Browning of sago starch is linked to its enzyme activity of isolated-DL-epicathechin and D-catechin compounds. These polyphenol compounds are hard to remove during extraction. The consequent activity of polyphenol oxidase will lead to color changes. The amount of (+)–catechin and (–)–epicatechinoxidized will increase when mix with water containing iron and copper, especially un-filtered water. Furthermore, the amount of these polyphenol compounds will also increase with an increase in temperature above 30°C [5].

3.5Fourier Transform Infrared Spectroscopy (FT-IR)

FT-IR spectra of sago starch in different drying temperature, including pre-dried, are shown in figure 7. The x-axis represents wavelength with range of 4000-600 cm^{-1} region and the y-axis represent percentage transmittance which show absorbance value or intensity.

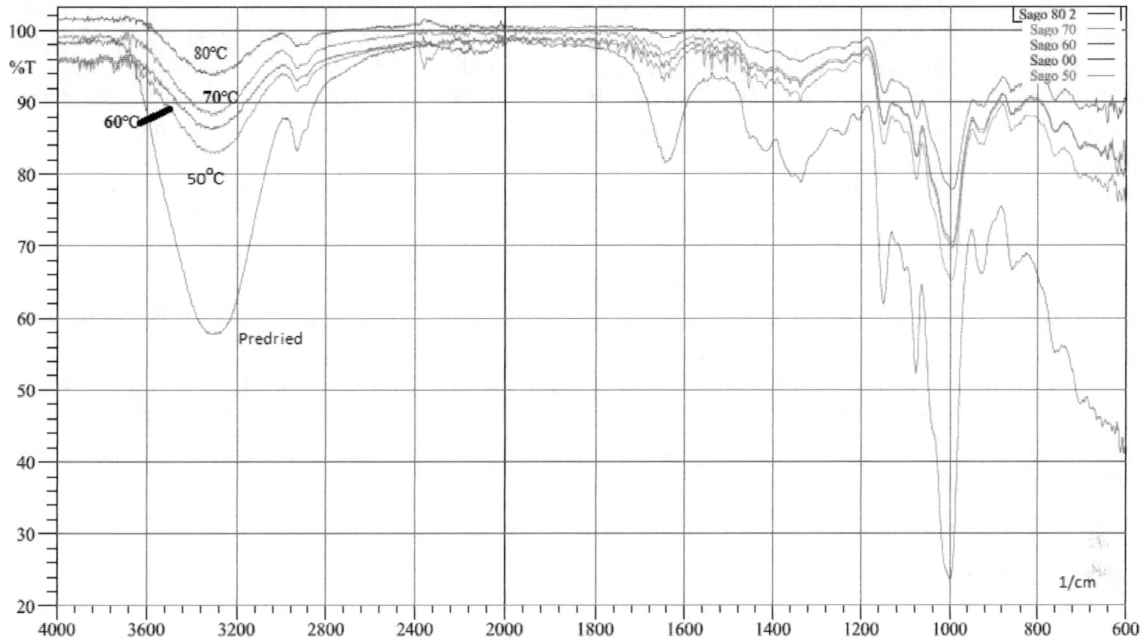

Figure 6. FT-IR spectra of sago starch drying at different temperature 50, 60, 70 and 80°C

The result shows that the pre-dried, as well as all dried starches exhibited a similar pattern and showed 11 major peaks. The absorbance peaks are near 760, 880, 960, 1000, 1080, 1160, 1360, 1440, 1640, 2880 and 3240 wavelength. Peak absorbance in starch have been assigned and matched with the vibrational modes of the chemical bonds and the structures of starch molecules.

Table 4. Peaks absorbance and the assigned chemical bonds/structures [14]

Peak modes (result) cm^{-1}	Peak attribution
3240, 2880	O-H and C-H bond
1640	H_2O bending vibration
1440	H-C-H, C-H and O-H bending modes
1160, 1360	C-O-H stretching
1080	C-O bond stretching
1000	The ordered and amorphous structures of starch
960	D-glucopyranosyl ring vibrational modes
880	C-H absorbances of the D-glucopyranosyl rings
760	D-glucopyranosyl ring stretching

Peaks absorbance in figure 6 and their attribution are listed in table 4. These 11 peaks absorbance play crucial role for the main structure of starch which is O-H and C-H bond, as well as the ordered and amorphous structures of starch. Therefore, the change in size of FTIR peaks may indicated either the change amount of bonds exist in the sample or the difference in polarity. Besides, it may also indicated the presence of asymmetric stretching bond that has the most energy, yet unstable.

Result from figure 5 shows that there were significant differences of peak intensities and width between pre-dried and dried sago starch which also can be observed at all the major peaks. Dried sago starch exhibited significantly less absorbance width and intensity compared to pre-dried starch, possibly due to oxidation during drying process. It also can be observed that the peaks width and intensities decrease with increase in the drying temperature. Increase in temperature may increase the oxidation, thus explained the decreasing in peak width and intensities.

4. Conclusions

In this study, the drying kinetics and its effects of drying temperatures (50, 60, 70 and 80ºC) on sago starch were investigated. The results obtained showed that drying at higher temperature produced sago starch with lower moisture content. It is suggested that drying temperatures has an essential role in the characterization of sago starch drying behaviour in terms of equilibrium moisture content and drying kinetics. Lewis model was found to fit the drying of sago starches very well. Drying process also had changed the colour and chemical bonding in sago starch. Increase in temperature resulted in darkness increment. However, since drying process is one of a crucial process in sago starch industry, it is suggested that drying at 50ºC will be the best temperature in terms of starch quality of sago starch.

Acknowledgements

This work was supported by FRGS/TK02(04)/1304/2015(210) from Ministry of Higher Education. The authors thank the Muhammad Rifqi Harraz bin Abdullah from CRAUN Research Sarawak for supplying the sample.

References

[1] Baini R, and Langrish T A G 2009 Assessment of colour development in dried bananas - measurements and implications for modelling *J. Food Eng.*, 93(**2**), 177–182

[2] Dikbasan T 2007 *Determination of effective parameters for drying of apples* (Graduate School of Engineering and Sciences of Izmir Institute of Technology, Turkey)

[3] Dixit O, Treppe K, Khaidurova A, and Mollekopf N 2012 Experimental investigation and theoretical modelling of absolute convective drying of starch *J Food Eng*, 111(**1**), 14–20

[4] Flach M, 1997 Sago palm *Metroxylon sagu Rottb.Promoting the conservation and use of underutilized and neglected crops* (**13**) Institute of Plant Genetics and Crop Plant Research, Gatersleben/International Plant genetic resources Institute, Rome, Italy

[5] Karim A A, Tie P L, Manan D M A and Zaidul I S M, 2008 Starch from the Sago (Metroxylon sagu) Palm Tree — Properties , Prospects , and Challenges as a New Industrial Source for Food and Other Uses *Compre. Reviews in Food Sci. Food Safety* (**7**), 215-226

[6] Mazlina S, Kamal M, Mahmud S N, Hussain S A, & Ahmadun F R 2007 Improvement of sago flour processing *Int. J. Eng & Tech*, 4(**1**), 8–14

[7] Arun S, Mujumdar and Devahastin S, *Fund. principles of drying* p 1-22

[8] Ranjith D and Balakrishnan P A 2015 Study of Bleaching Property of Sago Starch Using Ozonation *National Conf. on Research Advances in Comm., Comp., Elec. Sci. Struct. (NCRACCESS-2015)*

[9] Rekha S S *et al* 2008 Review Industrial production, processing, and utilization of sago palm-derived products *Carbohyd. Polym.* **72**, 1–20

[10] Thao H M and Noomhorm A 2011 Modeling and Effects of Various Drying Methods on Sweet Potato Starch Properties *Walailak J.* 8(**2**) 139–158

[11] Wang X and Kito K 2014 Development of Sago Starch Extractor with Stirrer Rotary Blade for Improving Extraction Performance *Int. J. Sci. Tech* 6(**5**) 2472–2481

[12] Wilton P D S *et al* 2012 Empirical models to describe thin layer drying of lima bean (Phaseolus lunatus L.). *African J. Agr. Res.,* 7(**48**), 6376–6382

[13] Ramesh A Y, Manisha G, Tharanathan R N and Ramteke R S, 2006 Influence of drying conditions on functional properties of potato flour *Eur. Food Res. Technol.* **223** 553–60
[14] Zeng J, Li G L, Gao H Y and Ru Z G, 2011 Comparison of A and B starch granules from three wheat varieties *Molecules* 16 (**12**)

IOP Conf. Series: Materials Science and Engineering **206** (2017) 012040 doi:10.1088/1757-899X/206/1/012040

Local Fruit Wastes as a Potential Source of Natural Antioxidant: An Overview

U K Ibrahim[1*], N Kamarrudin[1], M U H Suzihaque [1] and S Abd Hashib[1]
[1]Faculty of Chemical Engineering, Universiti Teknologi Mara, 40450 Shah Alam, Selangor, Malaysia

[*]Email: ummi985@salam.uitm.edu.my

Abstract. Food industry in Malaysia which used fruits as one of the raw material such as the production of fruit juices, concentrates, jams and dried fruits, the main wastes of the production are the peel and the seed of the fruit. Nowadays, people have shown the interests to study the antioxidant content in the fruit wastes. All kind of fruits are believed to contain high amount of natural antioxidant properties such as vitamins, phenol, flavonoid and carotenoid. Thus, this paper presented the work done by researcher on antioxidant activity in the peel especially on local fruit such as mango peel, watermelon rind, banana peel and mangosteen pericarp. The review shows that the peel of the fruit is a good source of antioxidant and other bioactive compounds which have many benefits especially towards human health.

1. Introduction

The use of fruit and vegetable waste recently has becoming more popular to be studied by taking into account that these residues are one of the important sources of polyphenols. Researchers become interested to study about the residue because agricultural and industrial residues are attractive source of antioxidants and dietary fiber besides reducing environmental pollution [1,2]. For example, fruit wastes are so many in Egypt but the constraint is lacking information regarding their activity and content of antioxidant compounds [3].

Research has been done in some vegetable and fruit wastes to investigate the extraction and identification of antioxidant compounds existence [4]. Researchers are now inventing new by-products applications and interested to turn them into useful products and to have a positive environmental impact. The functional properties of some peel components such as pectin, flavonoids, carotenoids, limonene and polymethoxy flavones should be considered accordingly [5].

Based on the epidemiological studies, it has been shown that consumption of fruits frequently is related with a lower risk of chronic diseases [6]. Vitamins and polyphenols contained in fruits and vegetables are natural source of antioxidants and considered to be responsible for human health benefits [7,8]. One of the most important categories of natural antioxidants of interest is phenolic compounds [9]. There is much evidence is derived on the antioxidant potency together with their prevention of diseases capability [10]. Based on the recent studies that have been done, the peel and seed of some fruits contain high amount of phenolic compounds and antioxidant potency [11]. This studies indicate that fruit residue has the potential to be utilized as one of the source of bioactive compounds such as natural antioxidants.

Malaysia is one of the countries that having a fast-economic growth and high population of people. Along with the rise of the economy, Malaysia produced different varieties of vegetables and fruit. This locally grown agriculture is not only consumed by Malaysian but also being export to other countries. The vegetables and fruits that are being grown in Malaysia also undergo some industrial processing for example the production of canned fruit, fruit juice and also as the flavoring. Due to the high consumption and industrial processing of the edible parts of the fruits, fruits wastes such as watermelon rind, mango peel, rambutan skin and other fruit residues principally the peels and the seeds are generated in large quantities especially at the food industrial area and in big cities for example Kuala Lumpur. In fact, one of the main sources of municipal solid waste (MSW) is fruit waste which has been an increasingly tough environmental issue.

Furthermore, some synthetic antioxidants have the potential risks towards human health and due to that the attention to identify natural source of antioxidants that is possibly more economical and effective is increasing. Synthetic antioxidants that are commonly used are butylatedhydroxytoluene (BHT) and butylatedhydroxyanisole (BHA). These synthetic antioxidants are used in order to reduce free radical damage to the human body. Their usage as a food preservatives are significant and thus consumed in appreciable quantities by humans. The problem with the usage of synthetic antioxidants is that, such compound has been related to health risks. Besides that, synthetic antioxidants have been suspected as one of the cause of liver damage and carcinogenesis. That is why strict regulations over their use are applied especially in foods.

For all the stated reason, the municipal solid waste need to be reduced and one of the way is reduce the fruit waste especially the peel. Besides that, the use of synthetic antioxidant should also be replaced with natural antioxidants to lower the risk towards human health. One of the beneficial approaches is to recover the bioactive constituents in the fruit waste especially the phenolic compounds and antioxidant properties, making full use of them in some sectors and industry such as food, pharmaceutical as well as cosmetics industry. Thus, sources of bioactive from fruit wastes utilization might be considerable in terms of human health, environment and also economics.

2. Antioxidant in Fruit Wastes

In food industry that used fruits as one of the raw material such as the production of fruit juices, concentrates, jams and dried fruits, the main wastes of the production are the peel of the fruit. For examples, in the production of mango juice or mango concentrate, the used part of the fruit is only the flesh of the mango whereas the peel will become the waste. Researchers have becoming more interested to study the antioxidant content in the fruit waste such as the peel. This is because, as people knows, any kinds of fruits contained high amount of natural antioxidant such as vitamins, phenol, and carotenoid, thus, researchers believed that, the peel of the fruits also have the antioxidant activity. Researchers have done some study on the peel of various fruit to prove that antioxidant activity exist in the peel of the fruit. As a result, it shows that the peel of the fruit is a good source of polyphenols, carotenoids and other bioactive compounds which have many benefits especially towards human health [12].

As early in the year of 1994, researches have started the study of antioxidant activity in the peel of plant. Potato peel has been studied on the antioxidant activity in 1994 by Rodriguez de Sotillo et al. [13]. The result is positive as the aqueous extract of potato peel is a good source of phenolic acid and exhibit a good antioxidant activity. This study has set a benchmark in later studies to see whether other fruit peel also contained a good antioxidant activity. Not long after that, mango peel has become the subject for antioxidant activity study. As the result, it was found that mango peel also contains

IOP Conf. Series: Materials Science and Engineering **206** (2017) 012040 doi:10.1088/1757-899X/206/1/012040

polyphenols which is a type of antioxidant and a good source of dietary fiber [14]. Banana and tomato peels also reported to be a good source of antioxidant that is carotenoids [15].

Based on all the research that has been done, it proves that the peel of the fruit contained antioxidant activity in a significant amount. More research should be done to investigate any other peel that might also contain antioxidant activity and also maybe can be utilize the antioxidant contained in the peel of the fruit.

3. Local Malaysian Fruit Wastes

3.1. Water Melon Rind

Watermelon (*Citrulluslanatus Thunb.*) is from the family of cucumber (*Cucurbitacea*). Watermelon is a large fruit having a shape of oval, round or oblong and a type of tropical fruit [16]. Watermelon can be found in many tropical countries. The skin of the watermelon is smooth having dark green rind or sometimes pale green stripes and when the fruit is fully ripe, the green stripes turn into yellowish green. Commonly found watermelon has either red or yellow flesh. Watermelon fruit is a very refreshing fruit to be eaten in hot weather because it contained vast amount of water and the sweet taste is irresistible.

In food industry, such as the production of watermelon juice, the rind (Figure 1) is usually discarded as wastes. The application of watermelon rind is up to fertilizer application or applied to animal feeds only. Less people know that watermelon rind is also edible and can be used as a vegetable. In China for instant, the watermelon rind is de-skinned and de-fruited before it is stir-fried together with olive oil, garlic, chili peppers, scallions, sugar and rum. In China, the watermelon rind is also available in powdered forms and Chinese used the watermelon rind as a traditional medicine extensively to clear away heat in the body and eliminate toxic substances. Watermelon rind is also can be found in pickled form and is commonly consumed in the Southern US, Russia, Ukraine, Romania and Bulgaria. Watermelon rinds are also fermented, blended and consumed as juice in Nigeria. Based on what other countries applied watermelon rind, it shows that watermelon rind also has its own value and not just a waste like people always did.

Figure 1. Watermelon rind

Watermelon has been reported to have therapeutic effect which has been ascribed to antioxidant compounds [17]. The major antioxidant activity in watermelon rinds is citrulline. Because of the antioxidant activity of citrulline in watermelon rind, it has the ability to protect human from free-radical damage besides an efficient hydroxyl radical scavenger. Moreover, citrulline can be converted to arginine which is a type of amino acid that is vital to human heart, circulatory system as well as immune system [12]. Researches also believed that watermelon rind has the potential in relaxing blood vessels as well as cancer and cardiovascular diseases besides improving sexual stamina and erectile function.

Researchers are becoming more interested to study the watermelon rinds properties and moving towards utilizing it in food products. Recently, watermelon rinds had been utilized by using it in the making of bakery products by turning the watermelon rinds in the form of powder and mixed with the flour to make a cake batter [3]. The result was very promising as the cake that forms from the mixture of watermelon rinds powder and flour had a longer shelf-life. This shows that the antioxidant activity in watermelon rinds can slow down the oxidizing process in the cake and as a result prolonged the shelf-life of the cake.

Citrulline is a non-essential amino and was first identified from the juice of the watermelon (*Citrullus vulgaris*) [18]. The richest known source of citrulline is watermelon [19]. Watermelon as reported has the therapeutic effect and this effect has been ascribed to antioxidant compounds [20]. The antioxidant property that comes from the citrulline in the watermelon rinds helps to protect human being from free-radical damage. Besides that, an amino acid that is being produced form the conversion of citrulline to arginine plays an important role to the heart as well as circulatory system and immune system. Furthermore, some researchers also speculate that watermelon rind has the potential in relaxing blood vessels including cancer and cardiovascular diseases.

3.2. Mango Peel

Besides watermelon, *Mangiferaindica* L. or commonly known as mango is also one of the most important tropical fruits consumed in fresh or processed form globally. Typically, mango has a bright yellow-flesh with either green or yellow skin, together with having a sweet-sour taste. Mango can be consumed freshly even though it is not fully ripe.

Most people consumed the mango flesh only but other parts of mango such as peel and kernel are being thrown as waste. Apparently, mango waste contains a very significant number of phytochemicals, which can be utilized for value-added applications in functional foods and nutraceuticals. Mango peel is a rich source of pectin, lipids, proteins, carotenoids, cellulose, hemicellulose, vitamins, and polyphenols with excellent health-promoting properties, mainly antioxidant activity [21].

In the manufacturing of processed mango such as mango juice, concentrate or puree, massive amount of peel is being generated and become the major waste in industry. Although the peel waste is biodegradable, huge amount of it takes time to be decomposed and can also cause environmental pollution. Studies have shown that mango peel contains various numbers of valuable compounds such as polyphenols, carotenoids, enzymes and dietary fibres together with excellent antioxidant activity and functional properties [22].

Based on the antioxidant activity contained in mango peel (Figure 2), it has been utilized and produced in a powdered form that is mango peel flour. However, the use of the mango peel flour has not yet been widely applied. Mango peel flour has many potential as it can be used as a functional ingredient especially in developing healthy food products for examples bread, biscuits, cakes, and as well as noodles or other bakery products. Mango peel flour can also be the essential ingredients in making baby foods [23].

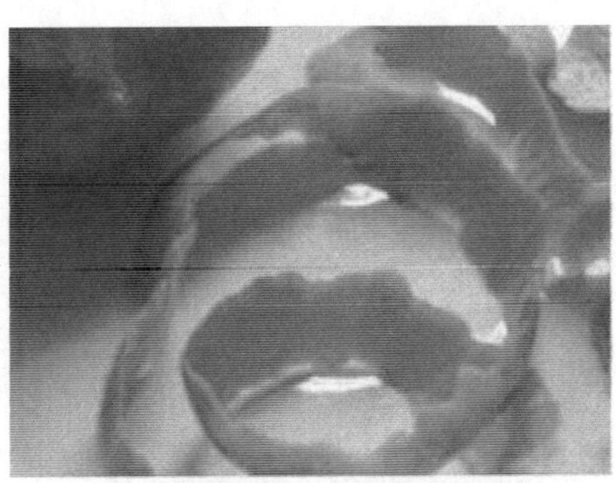

Figure 2. Mango peel

3.3. Banana Peel

Banana is an edible fruit that has grown worldwide nowadays. Banana plants are the largest herbaceous plant in the world which has grown in large quantities in many developed countries [24]. Banana also is the one of the most inspired fruits in tropical and subtropical regions [25]. Banana is obtainable in many types according to their size, colour and inflexibility. In 1753, Linnaeus who has published his book which is Species Plantarum, the origin of botanical nomenclature has given the first scientific name for banana which is *Musa Paradisiaca Linn*. The straightforward explanation has made based on plantain cultivar manners and lean fruits that maintain its own starchy even when fully ripe. The dried relics on the male bud rachides are firm and remain as their male flowers and bracts of plantains.

Fruits from banana plant are the main purposed in the banana cultivation. The flesh of banana fruits has variety of taste from starchy to sweet depending to their own cultivar and ripeness. Many people have used banana as a food as banana is easy to obtain and banana also is free from toxics affect. A few countries have labelled the banana fruit as the golden fruit of the nature. It is contributing to their behaviour that can promote the natural beauty by providing the body with essential nutrients. In fact, banana fruit also prevent constipation, diarrhoea and other digestive problems because a nutrient in banana improves digestive system. Banana also has been reported that its fruit can prevent many diseases. Banana should be well thought-out to be a good basis of natural antioxidant for foods and functional food source in opposition to cancer and heart disease [26].

The fruit is protected by its peels (Figure 3) which are leftover as a waste after the inner flesh has been eaten [27]. Peels are often being a waste of all fruits. Instead of recycled, usually all fruit waste especially peels always been discharged. This is because that many people do not know that there are many benefits of commercial use can be extracted for those peels. 30% from banana fruit is peels which act as main by-product. Environmental problems can be in serious condition from peels because it contains large quantities of nitrogen and phosphorus and its high-water content makes it vulnerable to adaptation by microorganisms [28].

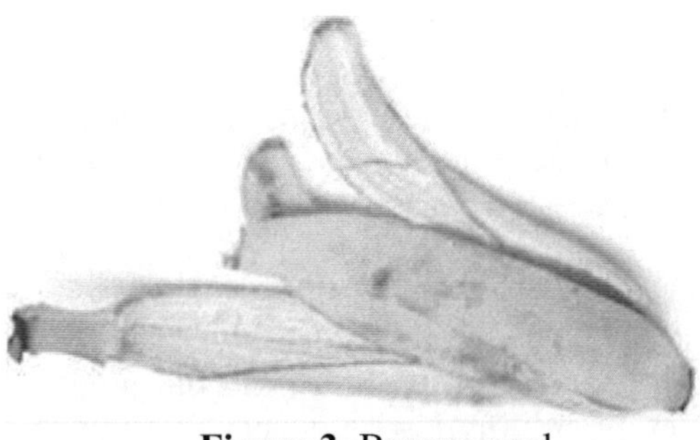

Figure 3. Banana peel

Surprisingly, many researchers have found that there are higher antioxidants activities and total phenolic contents from some of various fruits such as banana are from their peel and seed fractions, not in their pulp. The banana peels chemical composition influence the potential application of the banana peel. It has been found that total phytochemical compounds especially antioxidants are wealthy in banana peel [28]. The banana peel can be the major source of obtaining natural antioxidant. Nowadays, banana peel also is often use as a home therapy for quite a few skin problems such as allergies, bruises and skin irritation. Acne, poison ivy rashes irritation from mosquito bites, wrinkles and many more can be treated and managed by banana peels. Furthermore, banana peel helps skin to look much healthier, fresh and radiant as it can manage wrinkles although it will not have instantly effect.

Kanazawa and Sakakibara [29] have classified banana as one of the antioxidative foods. Banana fruits also can protect themselves from the oxidative stress that caused by intense sunshine and high temperature because they have strong ability to increase their own antioxidant levels. Banana is acknowledged by people as weak primary antioxidant source but in terms of secondary antioxidant they are powerful [30, 31, 32]. Ascorbic acid, tocopherol, beta carotene, phenolic groups, dopamine and gallocathechin are antioxidant compounds identified in banana peel [33].

There is gallocathechin contain at a concentration 160 mg/100g dry weight in the banana peel [26]. Anthocyanins, delphinidin, cyaniding and catecholamines also contain in ripe banana peel. Other than that, carotenoids such as α-carotene, β-carotene and xanthophylls have been discovered in banana peel as well as sterols and triterpenes.

The synthesis of silver nanoparticles also used of the banana efficiently. Currently, silver nanoparticles are utilized in the production of antibacterial and anti-fungal agent's biotechnology and bioengineering, textile engineering, water treatment, and silver based consumed product. There are methods that can be used to synthesize the nanoparticles which are chemical, physical and biological methods. There are facts that mentioned about nanoparticles were one of the useful medium in opposition to the fungal and bacteria.

Currently, the used of banana peel in the silver nanoparticles production is according to their composition as they are naturally rich in polymers such as lignin, hemicelluloses and pectin that contribute to the synthesis of silver nanoparticles as mentioned well by A. Bankar et al. [34]. Banana is the one of the new method in order to obtain silver nanoparticles that produced from its peels for antimicrobials production nowadays.

3.4. Mangosteen Pericarp

Garcinia mangostana or also usually known as mangosteen, is a symbol of "Queen of Fruits" in Asia because of its delicious tasting tropical fruits. The mangosteen is a tropical evergreen tree which usually can be found in tropical countries such as Southeast Asia, Northern Australia, Brazil, Central America, Hawaii, Southern India, Indonesia, Malaysia and Thailand. The tree can grow until 6-25 m and it has leathery, glabrous leaves, and takes a long time to grow. When the mangosteen fruit ripe, it will become deep reddish purple in colour. The fruit diameter is around 5-7 cm. However, the pericarp is around 6-10 mm thickness [35]. At Southeast Asia, they mainly used the fruit hull of mangosteen as a traditional medicine. They treated diarrhea, inflammation, dysentery, skin affection, wounds, and ulcers with the mangosteen's hull [36, 37].

Nowadays, mangosteen become one of the popular ingredient that been used in nutritional supplements. There a lot of mangosteen products available and very famous because of their benefits in enhancing the human health. For example, mangosteen fruit juice was a top-selling "botanicals" product on the market [36]. Mangosteen consists of 17% of outer pericarp, 48% of inner pericarp, 31% of flesh and 4% of cap. Xanthones can be isolate from all parts of mangosteen which are from pericarp, whole fruit, bark, and leaves. Several studies have shown that each part of mangosteen contain xanthones [38].

Figure 4. Mangosteen pericarp

Xanthones are major secondary metabolites that are found in the pericarp of mangosteen (Figure 4). α-mangostin and β-mangostin are compounds of the xanthones. These two compounds that are found in the pericarp of mangosteen shown cancer chemopreventive agent. α-mangostin was found to have efficacy in inhibiting preneoplastic lesion in rat colon. α-mangostin also reported to have antimicrobacterial properties and act as selective inhibitor against borine brain-derived acidic sphingomyelinase [39]. α-mangostin are believed to decrease human LDL oxidation induced by copper or peroxyl radical which will increase the consumption of α-tocopherol. Extraction of mangosteen also significantly reduces the ROS production of polymorphonuclear leucocytes (PML). Besides that, xanthones from the pericarp significantly have anti-inflammatory properties [40].

4. Conclusion

The growing demand for natural antioxidants observed in food and cosmetic industries forces the search for new sources of these compounds. Numerous scientific investigations point at consecutive rich sources of antioxidants, especially among fruits, but only few of them involve waste parts of plants such as fruit rinds and peels. Plant wastes which are formed in great amounts during industrial processing represent a serious problem, as they exert an influence on environment and need to be managed and/or utilized. Furthermore, they are very rich in bioactive components, which are considered to have a beneficial effect on health.

For the last decade, efforts have been made to improve methods and ways of re-using fruit rinds wastes. The important purpose is the valorisation of the antioxidants and other biocomponents in by-products from the industries. It has been attempted to discuss which plants and their by-products can be considered as a rich source of natural antioxidants and what methods should be used for their efficient extraction. The competition between natural and synthetic antioxidants, in terms of consumer acceptance, legal needs for market access, toxicity and thermal stability, is a big problem.

Therefore, research study must be carried out to identify the potential of bioactive/antioxidant compounds sources in selected local fruit wastes in Malaysia such as watermelon rind, mango peel, banana peel and mangosteen pericarp. Besides the antioxidant activity, other studies such as total phenolic content, total flavonoids content and antimicrobial activities also can be carried out. The data obtain from the study hope will give a beneficial input especially to food, pharmaceutical and nutraceutical industries and generally to public and researchers.

Acknowledgement

Thanks to Research Acculturation Grant Scheme (RAGS) [File No: 600-RMI/RAGS 5/3 (213/2014)], Ministry of Education (MOE) and Institute of Research Management & Innovation (IRMI), UiTM Shah Alam, Malaysia for research funding and all of the contributions in completing the research study.

References

[31] Hemaida M 1994 Isolation of natural antioxidants from vegetables waste by-products *Agri. Sci.* Mansura University **19** pp 2953-2960

[32] Larrosa M, Llorach R, Espın J C and Tomás-Barberán F A 2002 Increase of antioxidant activity of tomato juice upon functionalisation with vegetable byproduct extracts LWT - *Food Sci. and Tech.* **35**(6) pp 532–542

[33] Al-Sayed H M A and Ahmed A R 2013 Utilization of watermelon rinds and sharlyn melon peels as a natural source of dietary fiber and antioxidants in cake *Annals of Agri. Sci.* **58**(1) pp 83-95

[34] Nawal N, Zeitoun M and Barbary O 2008 Utilization of some vegetables and fruits waste as natural antioxidants Alex *J. of Food Sci. & Tech.* **5**(1) pp 1-11

[35] Lia S, Lambrosa T, Wanga Z, Goodnowa R and Ho C T 2007 Efficient and scalable method in isolation of polymethoxy flavones from orange peel extract by supercritical fluid chromatography *J. of Chromatography* B **846**(1-2) pp 291-297

[36] He F J, Nowson C A, Lucas M and Macgregor G A 2007 Increased consumption of fruit and vegetables is related to a reduced risk of coronary heart disease *J. Hum Hypertens.* **21**(9) pp 717-728

[37] Ruxton C H, Gardner E J and Walker D 2006 Can pure fruit and vegetables juices protect against cancer and cardiovascular disease too? A review of the evidence *Int. J. of Food Sci. Nut.* **57**(3-4) pp 249-272

[38] Saura-Calixto F and Goni I 2006 Antioxidant capacity of the Spanish Mediterranean diet *Food Che.* pp **94**(3) pp 442-447

[39] Scalbert A, Johnson I T and Saltmarsh M 2005 Polyphenols: Antioxidants and beyond *The American J. of Clinical Nut.* **81** pp 215S-217S

[40] Fresco P, Borges F, Diniz C and Marques M P 2006 New insights on the anticancer properties of dietary polyphenols *Med Res Rev.* **26**(6) pp 747-766

[41] Ajila C M, Bhat S G and Prasada Rao U J S 2007 Valuable components of raw and ripe peels from two Indian mango varieties *Food Che.* **102**(4) pp 1006-1011

[42] Wolfe K, Wu X and Liu R H 2003 Antioxidant activity of apple peels Journal of Agricultural and *Food Che.* **51**(3) pp 609-614

[43] Rodriguez de Sotillo D, Hadleyn M and Holm E T 1994 Potato peel waste stability and antioxidant activity of freeze dried extract *J. of Food Sci.* **59**(5) pp 1031-1033

[44] Larrauri J, Ruperez P, Borroto B and Saura-Calixto F 1996 Mango peel as a new tropical fibre: preparation and characterization LWT - *Food Sci. and Tech.* **29**(8) pp 729-733

[45] Baysal T, Ersus S and Starmans D A J 2000 Super critical CO2 extraction of b-carotene and lycopene from tomato paste waste *J. of Agri. and Food Che.* **48**(11) pp 5507-5511

[46] Koocheki A, Razavi S M A, Milani E, Moghadam T M, Abedini M, Alamatiyan S and Izadkhah S 2007 Physical properties of watermelon seed as a function of moisture content and variety *Int. Agrophysics* **21** pp 349-359

[47] Leong L P and Shui G 2002 An investigation of antioxidant capacity of fruits in Singapore markets *Food Che.* **76**(1) pp 69-75

[48] Mandel H, Levy N, Izkovitch S and Korman S H 2005 Elevated plasma citrulline and arginine due to consumption of Citrullus vulgaris (watermelon*) J Inherit Metab* **28**(4) pp 467-472

[49] Rimando A M and Perkins-veazie P M 2005 Determination of citrullin in watermelon rind *J. of Chromatography* A **1078**(1-2) pp 196-200

[50] Lewinsohna E, Sitritb Y, Bara E, Azulaya Y, Ibdaha M, Meira A, Yosef E, Zamird D and Tadmora Y 2005 Not just colors-carotenoid degradation as a link between pigmentation and aroma in tomato and watermelon fruit *T. in Food Sci. and Tech.* **16**(9) pp 407-415

[51] Manthey J A and Perkins-Veazie P 2009 Influences of harvest date and location on the levels of b-carotene, ascorbic acid, total phenols, the in vitro antioxidant capacity, and phenolic profiles of five commercial varieties of mango (Mangifera indica L.) *J. of Agri. and Food Che.* **57**(22) pp 10825-10830

[52] Subaigo A, Morita N and Sawada S 1996 Carotenoids and their fatty acid esters in banana peel *J. of Nutrisci. and Vitaminology* **42**(6) pp 553-566

[53] Noor Aziah Abdul Aziz, Lee M W, Bhat R and Lai H C 2012 Evaluation of processed green and ripe mango peel and pulp flours (Magnifera indica var Chokanan) in term of chemical composition, antioxidant compounds and functional properties *J. of the Sci. of Food and Agri.* **92**(3) pp 557-563

[54] Aurore G, Parfait B, Fahrasmane L 2009 Bananas, raw materials for making processed food products *T. in Food Sci. & Tech.* **20**(2) pp 78-91

[55] Alkarkhi A F M, Saifullah Ramli Yeoh S Y, and Azhar Mat Easa 2010 Physicochemical properties of banana flour as influenced by variety and stage of ripeness: multivariate statistical analysis *Asian J. of Food and Agro-Ind.* **3**(3) pp 349-362

[56] Someya S, Yoshiki Y, and Okubo K 2002 Antioxidant compounds from bananas (Musa Cavendish) *Food Che.* **79**(3) pp 351-354

[57] Anhwange B A, Ugye T J and Nyiatagher T D 2009 Chemical composition of Musa Sepientum (banana) peels *Elec. J. of Environ. Agri. and Food Che.* **8** pp 437-442

[58] González-Montelongo R, Lobo M G, and González M 2010 Antioxidant activity in banana peel extracts: Testing extraction conditions and related bioactive compounds *Food Che.* **119**(3) pp 1030-1039

[59] Kanazawa K and Sakakibara H 2000 High content of dopamine, a strong antioxidant, in Cavendish banana Journal of Agricultural and *Food Che.* **48**(3) pp 844-848

[60] Haripyaree A, Guneshwor K and Damayanti M 2010 Evaluation of antioxidant properties of some wild edible fruit extracts by cell free assays *Elec. J. of Environ., Agri. and Food Che.* **9**(2) pp 345-350

[61] Lim Y Y, Lim T T and Tee J J 2007 Antioxidant properties of several tropical fruits: A comparative study *Food Che.* **103**(3) pp 1003-1008

[62] Lim Y Y, Lim T T and Tee J J 2006 Antioxidant properties of guava fruit: Comparison with some local fruits *Sunway Aca. J.* **3** pp 9-20

[63] Qusti S Y, Abo-Khatwa A N, Ladwa M A 2010 Free radical scavengers enzyme of fruit plant species cited in Holy Quran World *App. Sci. J.* **9**(3) pp 338-344

[64] Bankar A, Joshi B, Kumar A R, Zinjarde S 2010 Banana peel extract mediated novel route for the synthesis of silver nanoparticles Colloids and Surfaces A: *Physicoche. and Eng. Aspects* **368**(1-3) pp 58-63

[65] Zadernowski R, Czaplicki S, and Naczk M 2009 Phenolic acid profiles of mangosteen fruits (Garcinia mangostana) *Food Che.* **112**(3) pp 685-689

[66] Chin Y W, Jung H A, Chai H, Keller W J and Kinghorn A D 2008 Xanthones with quinone reductase-inducing activity from the fruits of Garcinia mangostana (Mangosteen) *Phytoche.* **69**(3) pp 754–758

[67] Ji X, Avula B and Khan I A 2007 Quantitative and qualitative determination of six xanthones in Garcinia mangostana L. by LC-PDA and LC-ESI-MS *J. of Pharma. and Biomed. Analysis* **43** pp 1270-1276

[68] Pedraza-Chaverri J, Cárdenas-Rodríguez N, Orozco-Ibarra M and Pérez-Rojas J M 2008 Medicinal properties of mangosteen (Garcinia mangostana) *Food and Chem. Toxicology* **46**(10) pp 3227–3239

[69] Jung H A, Su B N, Keller W J, Mehta R G and Kinghorn A D 2006 Antioxidant Xanthones from the Pericarp of Garcinia mangostana (Mangosteen) *J. Agric. Food Chem.* **54** pp 2077−2082

[70] Itoh T, Ohguchi K, Iinuma M, Nozawa Y and Akao Y 2008 Inhibitory effect of xanthones isolated from the pericarp of Garcinia mangostana L. on rat basophilic leukemia RBL-2H3 cell degranulation *Bioorg Med Chem.* **16**(8) pp 4500-4508

IOP Conf. Series: Materials Science and Engineering **206** (2017) 012041 doi:10.1088/1757-899X/206/1/012041

The Effect of Fermentation and Drying on Clicanthus Nutans Lindai Extract Chemical Constituents and Bioactivity

Y Y Lee,[1] P S Teh[1], S Kumaresan[1*] and C H Tiong[1]

[1]Faculty of Engineering, Universiti Malaysia Sabah, Jalan UMS, Kota Kinabalu 88400, Sabah, Malaysia

[*]Email: shiva@ums.edu.my

Abstract. In this study, two processes (fermentation and drying) were conducted on Clicanthus Nutans Lindau, which is well-known for its antioxidant activities. The aim of this research is to investigate the effects of these processes towards the Total Phenolic Content (TPC), antioxidant activity and the crude yield of the leaves. For the fermentation process, the fermentation time is varied for 6, 12, 24, 48 and 72 hours. For drying, oven drying (40, 45 and 50°C), freeze-drying and shade drying are investigated. Blanching was also conducted for oven dried samples (90°C for 1 minute). It was found that 12 hours fermentation produced the highest yield corresponding to Total Phenolic Content at 0.33% while 6 hours fermentation produced the highest Antioxidant Activity and yield at 2.218 x 10-10 mg/mL and 0.64%. On the other hand, 48 hours fermentation produced the highest overall extract yield at 0.64%. Analysis for antioxidant activity was carried out using 2,2-diphenyl-1-picrylhydrazyl (DPPH) and Folin-Ciocalteau method for Total Phenolic Content. Thermal inactivation kinetics (blanching) has been investigated for oven drying samples. Results showed that 40°C oven drying and 50°C oven drying with blanching has better retention of bioactive constituents. Blanching has no significant effect on phenolic content but has increased the crude extract for higher temperature oven drying. It has no positive effect on antioxidant activity. Drying is also a better processing method compared to fermentation as drying has slightly favorable results in all three aspects. Through comparison with extraction time of 60 minutes for 40°C oven drying and 6 hour fermented samples, yield of total phenolic content and overall yield of oven dried sample is higher than fermented sample by 4.14% and 0.39% respectively as well as showing better antioxidant activity.

1. Introduction

Medicinal herbs are fast gaining popularity in many countries [1]. Such products have become more widely available commercially, particularly in developed nations. Although modern medicine may exist side by side with such traditional practice, herbal medicine has often kept their popularity for historical and ethnic causes. According to the World Health Organization (WHO) figures, 80% of the world's population depend on herbal medicine for primary health care [2]. Tea is considered as one of the most popular beverage consumed around the globe [3]. A growing body of research suggests that drinking tea actually helps to prevent cardiovascular disease, burn calories and ward off some types of cancer [4].

Today, tea is grown on a commercial scale in some three dozen countries and, in each country the product makes a substantial contribution to the economy of the agriculture sector. The thousands of different varieties of teas available across the globe world only vary by the region it is grown, the time of year it is picked, and how it is manufactured. There are many types of teas available in the market, however the well known ones are commonly green tea, black tea, white tea, oolong tea, and pu-erh tea. They are all derived from the Camellia sinensis plant, a shrub native to China and India and they are produced based on the degree of fermentation [5].

All teas contain unique antioxidants called flavonoids which are antioxidants that help neutralize free radicals that can destroy cells and contribute to chronic diseases [6]. Caffeine and theanine are also found in teas whereby these two compounds affect the brain and heighten mental alertness [6]. Polyphenols are the secondary metabolites of plants and have a great potential as an alternative source of treatment of chronic diseases. The polyphenolic content of tea is mainly attributed to flavonoids including flavan-3-ols and flavanols, which account for 30% of the dry weight of fresh tea leaves. The major metabolites in tea leaves are flavonoids such as epicatechin (EC), epigallocatechin (EGC), epicatechin gallate (ECG), and epigallocatechin gallate (EGCG) [7]. EGCG is an effective flavan-3-ol component that is known for its antioxidant and anticancer activities among all the flavonoids [7]. The simple substrates for example catechins are acted upon by the oxidative enzyme, polyphenol oxidase and peroxidase, to form theaflavins and thearubigins during fermentation [7].

Tea is a promising natural source for these bioactive ingredients which play a significant role in different nutritional and therapeutic effects of tea to delay the onset of risk factors associated with the development of diseases like cancer and diabetes [8]. It has been proposed to have antimutagenic, antiviral, anti-oxidant and anti-inflammatory properties in various biological systems [9].

Sabah Snake Grass is a herb that belongs to the acanthaceae family. Its scientific name is Clinacanthus nutans Belalai Gajah or Tarum Siam in Malay or Dandang Gendis in Indonesia. It can be commonly found in countries with tropical weather such as in Malaysia and Thailand. The leaves of Clinacanthus nutans Lindau have long been traditionally used in Thailand as an anti-inflammatory drug for the treatment of insect bites, herpes infection and allergic responses [10]. Recently, it was discovered that it possess anti-oxidant and anti-proliferative properties that are clinically proven effective against cultured cancer cell lines [8]. A research study carried out by Yong et al [8] has shown that it possess antioxidant and anti-proliferative properties against cultured cancer cell lines, suggesting an alternative adjunctive regimen for cancer prevention or treatment.

2. Experimental Setup

Two sets of procedures were carried out separately, namely fermentation and drying. The processed C. Nutans leaves will be extracted using the same parameters and analyzed.

2.1. Raw Materials Preparation

Sabah Snake Grass is obtained from the plot planted in Universiti Malaysia Sabah. The leaves are plucked and ensured to be free from damage. The leaves are rinsed with distilled water and pat-dried.

2.2. Fermentation procedure

3 grams of C. Nutans leaves are withered first and fermented after that. After fermentation, the leaves will be dried to halt the fermentation process. The leaves will then be extracted and analyzed accordingly.

2.2.1. Withering. The fresh leaves were withered in the incubator at 28°C in an incubator to bring about adequate physical and chemical withering. The leaves were withered till total moisture content is between 60%-70% of the initial moisture content of the leaves.

2.2.2. Fermentation. The sample was then crushed and wrapped in aluminium foil and then placed in ceramic beaker at the desired duration. The fermentation was conducted for different duration of fermentation (3, 6, 12, 24, 48, 72 hours) in an incubator at 30°C.

2.2.3. Drying. The leaves were then oven-dried at 40°C until the moisture content of tea leaves is between 3-5% to stop the fermentation activity [8]. This is the optimum range of temperature drying for C. nutans.

2.2.4. Extraction. There was only one extraction method used in this study; hot water extraction using distilled water. The fermented Sabah Snake Grass leaves were soaked in distilled water for 1 hour with magnetic stirrer for which the solvent to solid ratio (w/w) was 100:1. 300 g of distilled water at 80°C were used in the extraction. The experiment tested the effect of extraction time on Sabah Snake Grass for DPPH, Total Phenolic Content and Crude. Each experiments set were replicated for accuracy of results. Sampling was conducted at 5, 10, 15, 25, 40, 60 and 90 minutes.

2.3. Drying Procedure

A separate batch of C.Nutans leaves were conducted for drying. In this research, freeze drying, shade drying and oven drying was carried out. Blanching is also observed for the drying process.

2.3.1. Total Moisture Content of C. Nutans leaves. 5 grams of C. Nutans leaves are dried at 105°C for 24 hours according to the A.O.A.C. method. The weight before and after drying were measured and the total dry mass of the leaves are determined using the formula below. The total moisture content is the total mass subtracted with the total dry mass.

$$Moisture\ Content\ (\%) = \left[\frac{M_i - M_f}{M_i}\right] \times 100\% \quad (1)$$

Where M_i is the initial mass and M_f is the final mass.

2.3.2. Drying Kinetics. The drying kinetics for oven dried samples and shade dried samples are developed by drying the samples and weighing the samples at constant time intervals.

2.3.3. Blanching. Blanching is carried out for oven dried samples for 1 minute at 90°C. The samples are immediately immersed in ambient temperature distilled water to stop the blanching process. The samples are then oven dried accordingly.

2.3.4. Drying. The samples are oven dried at 40, 45 and 50°C, freeze dried and shade dried. The drying is carried out in duplicates. 3 grams of C. Nutans leaves were used for each sample.

2.3.4.1. Oven Drying. The samples are dried for 4 hours for 50°C, 7 hours for 45°C and 17 hours for 40°C, based on the drying kinetics.

2.3.4.2. Freeze Drying. The fresh leaves are frozen in a -20°C freezer overnight. The leaves are then placed into the freeze dryer for 48 hours at -50°C and a pressure of 500 mbar.

2.3.4.3. Shade Drying. The samples are placed in paper boxes with punctured holes to maximize air flow. The boxes are placed in well ventilated and airy areas but free from sunlight and rain.

2.4. Extraction.

The dried samples are placed in a teabag and extracted with 250 ml of distilled water at 80°C and extracted for 90 minutes. 3ml is sampled at the time intervals of 5, 10, 15, 25, 40, 60 and 90 minutes for analysis, similar to the extraction method of the fermentation process.

2.5. Analysis

The extracted samples were analysed using the methods shown below and the results are tabulated and compared between fermentation and drying.

2.5.1. Crude Concentration. 0.3ml of extract is taken from the 3ml sample that is extracted at every time interval. It is diluted into 3ml using distilled water. The samples for each time interval is analysed using the UV-Vis Spectrophotometer.

2.5.2. Total Phenolic Content. 0.2ml of extract is used for the total phenolic content analysis. Folin-Ciocalteau reagent is diluted 10 times. A sodium carbonate solution of 60g/L is also prepared. With 0.2ml of extract, 1.5ml of diluted Folin-Ciocalteau reagent is added and left for 10 minutes. After that, 1.5ml of the prepared sodium carbonate solution is added and left for 120 minutes. The samples are analysed using the UV-Vis Spectrophotometer.

2.5.3. Total Antioxidant Activity. 0.02% of 2,2-diphenyl-1-picrylhydrazyl (DPPH) solution is prepared using 97-99% ethanol. 1ml of DPPH is added to 2.5ml of extract and left in the dark for 1 hour. The solution is analysed using the UV-Vis Spectrophotometer and the Radical Scavenging Activity is calculated based on the formula shown below.

$$DPPH\ Radical\ Scavenging\ Activity\ (\%) = \left[\frac{A_o - A_s}{A_o}\right] \times 100 \qquad (2)$$

2.5.4. Overall Yield. A selected sample extract was freeze dried and a calibration curve was made from freeze dried extract on UV-Vis Spectrophotometer to measure concentration. In each experiment, the same calibration curve was used to measure extract concentration. Overall yield is calculated based on the formula shown below:

$$Yield = \frac{volume\ of\ extract \times extract\ concentration}{weight}\ of\ sample \qquad (3)$$

3. Results and Discussion

3.1. Fermentation Process

The results of the analysis carried out from the fermentation process were discussed in this section.

3.1.1. Crude Extract. Figure 1 shows the average yield produced from Sabah Snake Grass. The yield for Sabah Snake Grass is calculated by multiplying the volume of solution left with the concentration of crude, both for every time interval to obtain the crude extract weight within the solution. The calculated extract weight is then divided by the amount of tea leaves used, which is 3g to obtain the yield (wt/wt %) produced. The calculations were done in a Microsoft Excel spread sheet.

Based on Figure 1, 40 minutes has been set as the cut-off point since the yield nearly reaches the equilibrium where most of the value compound could have been extracted. It can be observed that 48 hours fermentation produced the highest yield at a value of 0.62% while 72 hours fermentation produced the lowest yield. This may due to further enzymatic activities in sample which degrade desired compounds such as phenols and theaflavins in leaves after the optimum fermentation period. The low yield for 12 hours fermentation during 10 minutes extraction may be due to solid diffusion factor that hinder the compound from diffusing out from the leaves.

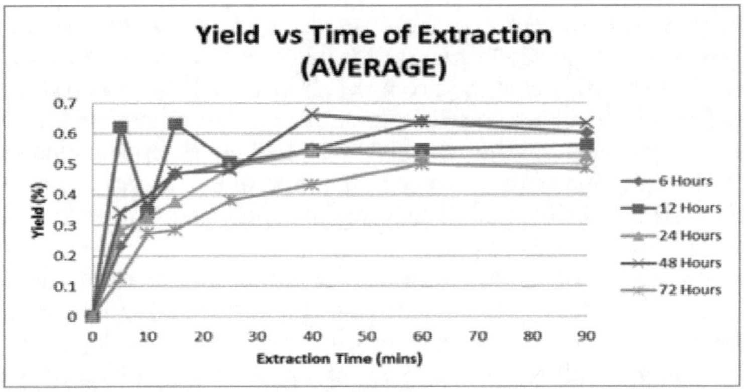

Figure 1. Average Yield produced vs Time of Extraction

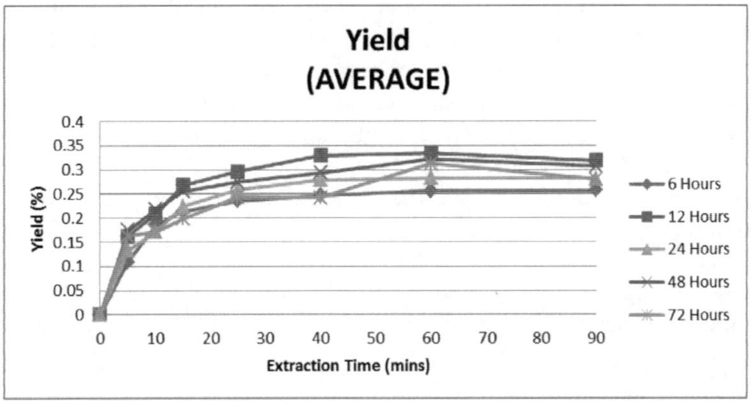

Figure 2. Average Yield of Various Time of Fermented Sabah Snake Grass corresponding to Total Phenolic Content

3.1.2. Effect of Fermentation Duration on Sabah Snake Grass for Total Phenolic Content. For higher accuracy, the average values of yield corresponding to total phenolic content of dried Sabah Snake Grass extract at various duration of fermentation and extraction for both replicates was calculated and plotted as shown in Figure 1. 60 minutes has been set as the cut-off point as the yield started decreasing after that point assuming that it has nearly reach equilibrium where most of the value compound could have been extracted. 60th minute has been set as the cut-off point since the yield started decreasing. It can be observed that the highest yield was found to be 12 hours of fermentation peaking at 60 minutes with a value of 0.33%.

It can be said that most of the phenolic content have already been extracted for the first 60 minutes and started decreasing slightly onwards. In other words, high amount of phenolic compound is being extracted by hot distilled water during this period and started degrading after 60 minutes.

A previous study by Ariffin et al. [11] explained that tea fermentation induces polyphenols oxidation reactions by the enzyme of polyphenols oxidase which resulted in the breakdown of phenolic compounds as main contributor to the antioxidant activities. This may explain the inconsistencies of the results for the first replicate and second replicate. The duration of fermentation affects the level of phenolic compounds [12]. The result obtained indicated that time for tea fermentation is indeed quite crucial which may affect the level of phenolic compounds and potential antioxidant activities of the tea plants [13].

It can be concurred that 12 hours fermentation is the optimum time for fermentation of C. nutans since it is able to produce the high total phenolic content as well as yield.

3.1.3. Effect of Fermentation Duration on Sabah Snake Grass for Antioxidant Content (DPPH). For higher accuracy, the average of the IC50 values of the two replicates were calculated. The IC50 values (concentration of the extract that is able to scavenge half of the DPPH radical) are presented in Table 1. Based on Table 1, it can be observed that 6 hours fermentation recorded the lowest IC50 value which is an indicative of good antioxidant potential. While 48 hours fermentation recorded the highest IC50 value. Low IC50 value is indicative of good inhibition of the enzyme. The antioxidant potential of a compound can be attributed to its radical scavenging ability, and in order to evaluate the ability of the plant extracts to serve as antioxidants [14].

Based on Table 1, it can be observed that 6 hours fermentation produced the highest antioxidant activity since it has the lowest IC50 value compared to the rest. The IC50 values also increased with increasing fermentation time. This is in agreement with that study by Heong et al.[12], also observed that the longer fermentation in full-fermented of tea had lower level of total antioxidant activities than in partial-fermented tea.

The extracts for various fermented time showed a propensity to extinguish the free radicals, as indicated by the dose dependent increase in percentage inhibition. This corresponded to a rapid decrease in absorbance in the presence of a plant extract, indicating high antioxidant potency of the extracts in terms of electron or hydrogen atom-donating capacity [15].

Most of the antioxidant potential of medicinal plants is due to the redox properties of phenolic compounds, which enable them to act as reducing agents, hydrogen donors and singlet oxygen scavengers [16]. In addition, the hydrogen-donating substituents (hydroxyl groups) attached to the aromatic ring structures of flavonoids enable them to undergo a redox reaction, which in turn, help them scavenge free radicals [17].

Table 1. Antioxidant activity represented by IC50 . Various Duration of Fermented Sabah Snake Grass Extract, measured by DPPH Radical Scavenging Test by taking the average.

Duration of Fermentation	**DPPH Test IC_{50} (mg/mL)**
6 Hours	2.218×10^{-10}
12 Hours	4.942×10^{-0}
24 Hours	12.273×10^{-0}
48 Hours	2.403×10^{11}
72 Hours	22.414×10^{-0}

Figure 3 illustrates the overall yield produced corresponding to the total antioxidants produced against the time of extraction. 60 minutes has been set as the cut-off point since the yield for most fermentation time started decreasing thereafter, indicating that all constituents containing in the fermented Sabah Snake Grass tea has been extracted. It can be observed that both 6 hours fermentation and 48 hours fermentation produced the highest yield with a value of 0.64% and 0.64% respectively, with 48 hours fermentation registered a slightly higher value. On the other hand, 72 hours fermentation registered the overall lowest yield.

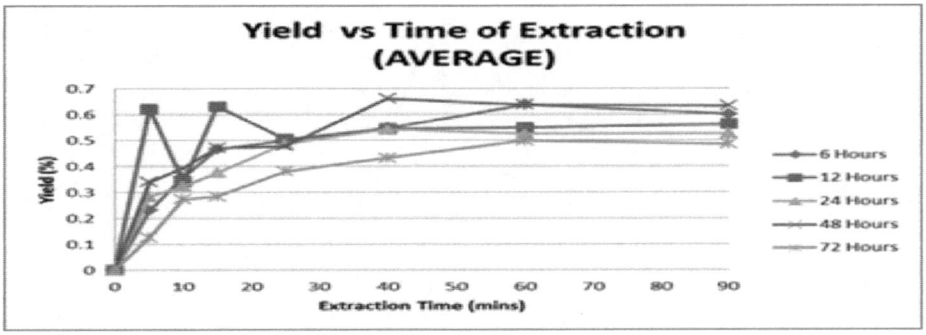

Figure 3. Overall Yield corresponding to Total Antioxidant vs Time of Extraction

3.1.4. Analysis of TPC per Crude in Sabah Snake Grass. Analysis of TPC per Crude is to see the relation between TPC and crude in fermentation of C. nutans. Based on the plot, it can be determined at which point does the crude and total phenolic content is released. Figure 4 shows the overall TPC per Crude activity of various fermented Sabah Snake Grass. The range of yield (wt/wt %) of TPC/Crude activity can be observed to be between 0.39% to 1.91%.

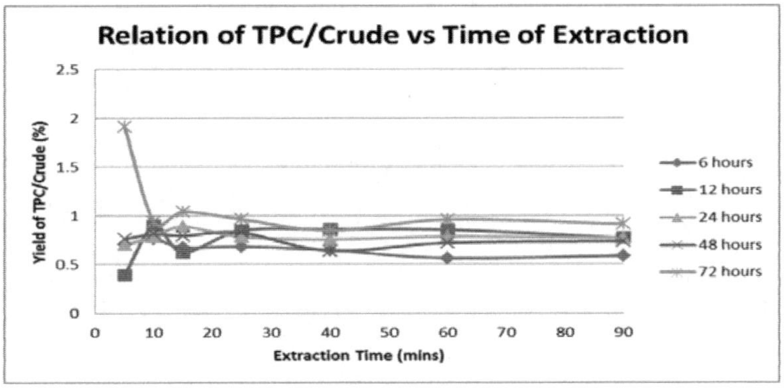

Figure 4. Relation of TPC/Crude against Time of Extraction

3.2. Drying Procedure

3.2.1. Total Moisture Content of C. Nutans leaves. The total moisture content is found to be 79.61±0.0424% based on duplicate samples.

3.2.2. Drying Kinetics. The drying kinetics are shown in Figure 5. It is shown that for 40°C sample, the drying time is 17 hours, 7 hours for 45°C and 4 hours for 50°C. The shade dried sample takes 3 days to dry completely, which is shown in Figure 6.

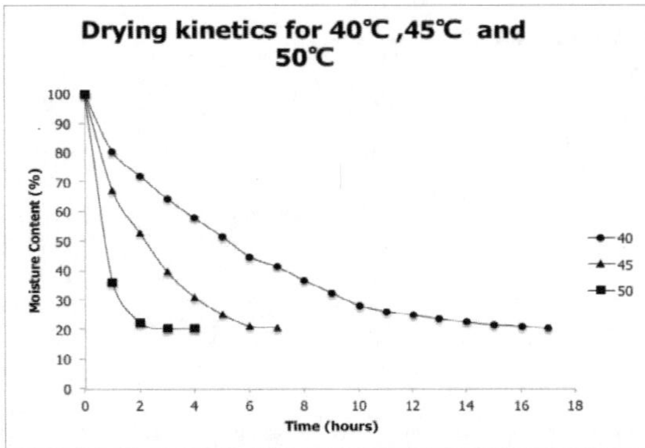

Figure 5. The drying kinetics for 40, 45 and 50 °C oven drying.

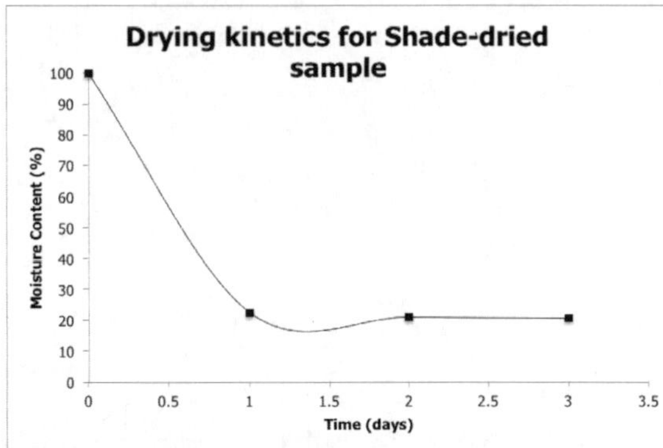

Figure 6. The drying kinetics for shade-dried sample.

3.2.3. Crude Extract. Based on Figure 7, it can be observed that the graphs all follow the first order curves, increasing rapidly at the first 20 minutes of extraction, then starts to slow down until 40 minutes. After 60 minutes, the curve starts to flatten out, which means that it has started to stabilize, and also that all of the extracts have been extracted. Hence, the 60 minute point is the equilibrium point for the crude concentration. From the graphs above, it can be seen that the total yield ranges from around 0.8% to 1.1% (at the 60 minute point).

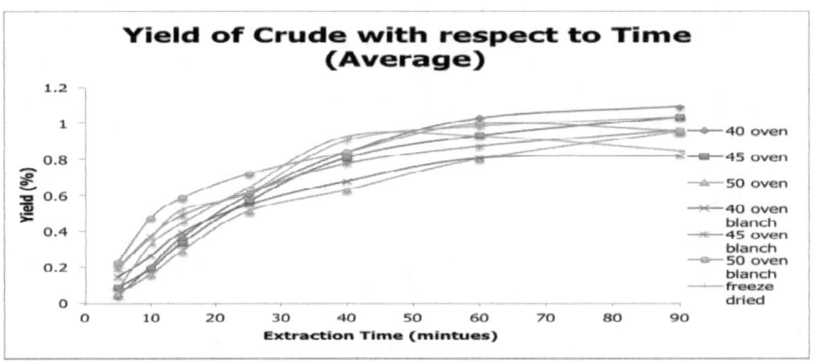

Figure 7. The yield of crude with respect to time

As shown in Figure 7, the trend is observed to be as follows:
40°C oven > 50°C oven blanched > Freeze dried > 50°C oven > 45°Coven > 45°C oven blanched > 40°C oven blanched > 50°C oven

Comparing the trends shown above, there may be a slight difference in the arrangements, but referring to the plots of the yields shown, it can be generalized that for non-blanched oven samples, the lower the temperature the higher the yield. However for blanched oven samples, the higher the temperature, the higher the yield, which is more desired. Whereas the freeze dried samples have a higher yield compared to the shade dried samples. Both the shade dried and freeze dried samples have relatively high yields as compared to the oven dried samples. Based on the trend above, it can be noticed that blanching had an effect on the extractability of crude. Blanching is used for enzyme inactivation and retention of the quality of fresh plants. Although there may be possibilities that the volatiles and some constituents may leech into the water from the blanching process, the structure of the leaves for each plant and their constituents vary from one another [18]. Hence, blanching may have a different effect on different samples. For the case of C.Nutans, the change was visible in crude

IOP Conf. Series: Materials Science and Engineering **206** (2017) 012041 doi:10.1088/1757-899X/206/1/012041

samples. Hence this may mean that blanching was unfavorable for low temperature drying but increasingly beneficial for constituents retention for increasing temperature.

3.2.4. Total Phenolic Content. Based on Figure 8, it has a same first order curve, with rapid increase at the first 30 minutes, slowing down at 40 minutes and eventually stabilizes after 60 minutes. This has the same pattern as the curves for yield of crude extracted. It can be seen that the yield of phenolic content ranges from 0.25% to 0.35%. The 45°C oven blanched curve has a decrease in phenolic yield, different from the other drying methods. This may be due to the deterioration of phenolic content after extraction of 80°C for an extended 30 minutes. At 60 minutes, the descending order of phenolic content is as shown below.

Shade dried > 50°C oven blanched > 40°C oven > Freeze dried > 40°C oven blanched > 50°C oven > 45°C oven > 45°C oven blanched

Although there may be a difference in the arrangement in the trends for both replicates, generally for the shade dried, freeze dried, 40°C oven-dried and 50°C oven-dried are quite close in yield. It can also be seen that there is a similar trend in descending order for both blanched and not blanched oven samples.

Oven-dried samples

40°C oven > 50°C oven > 45°C oven

Blanched oven-dried samples

50°C oven blanched > 40°C oven blanched > 45°C oven blanched

Based on Figure 9, the 45°C oven drying has the lowest phenolic yield for both. Hence this should be looked into. For both cases, the 45°C drying has the lowest yields. Based on a research, convective drying resulted in a lower content of rosmarinic acid due to the long time exposure to oxygen during the drying process [19]. They have also proposed in another research that rosmarinic acid degradation is associated to the temperature of the air for drying [20]. It can be observed from the bar charts above, for both blanched and unblanched, the phenolic contents drop from 40°C to 45°C then increases at 50°C which can be explained by the findings stated above. Although 40°C drying has the longest drying time and it is most exposed to oxidation, less volatile constituents were evaporated due to its lower temperature. Hence it could be concluded that the phenolics in the extract are heat-sensitive. However, 45°C drying has higher temperature and longer drying time than 40°C. Hence, 45°C drying has shown to have the lowest phenolic yield among all drying methods. 50°C however has an increased phenolic content. This indicates that more phenolic contents are released by higher heat. It has also shown that blanching has no significant effect on the phenolic contents as there are no differences in the trend between the blanched and unblanched samples. The freeze dried and shade dried samples have a relatively high phenolic content compared to the oven dried samples. Freeze-dried is a favorable drying method for samples possessing thermolabile.

From the trends above, the shade dried samples have very high phenolic contents. This corresponds to the research done on different herbs where the air-dried samples have the highest total phenolic content [21]. As the leaves were dried using the longest period of time amongst all drying methods, metabolically active plants lose moisture slowly and sensed moisture loss as a stress. As a defence mechanism, plants in general produce phenolic compounds as a stress response [22].

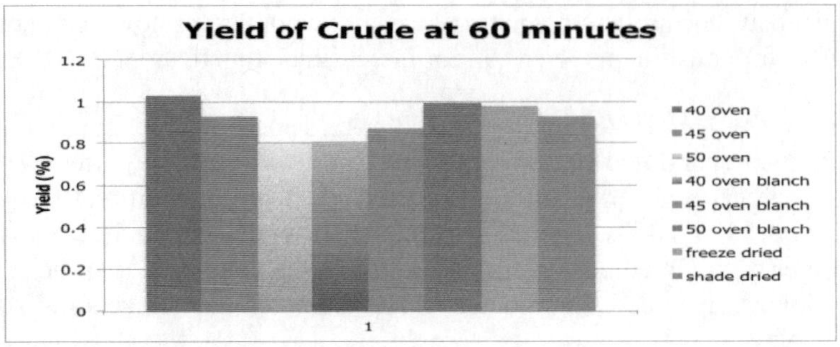

Figure 8. The yield of crude at 60 minutes.

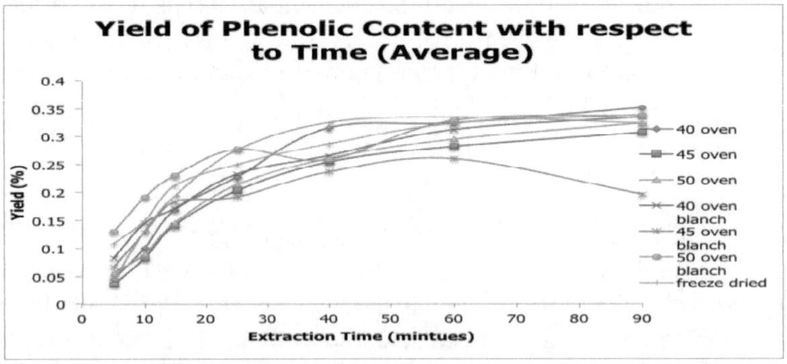

Figure 9. The yield of Phenolic Content with respect to time.

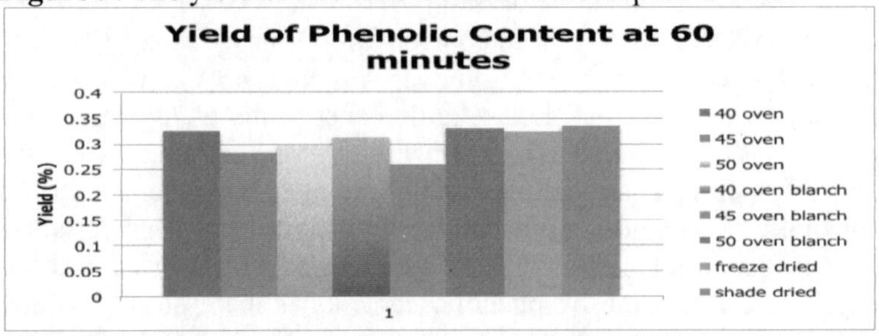

Figure 10. The yield of Phenolic Content at 60 minutes.

3.2.5. Total Antioxidant Activity. From Figure 11, the radical scavenging activity ranges from 73%-97%. At 10 minutes, the radical scavenging activity ranges from 85%-97%. It reaches an equilibrium point at around 60 minutes. It can be observed that the freeze drying samples have lower than expected antioxidant contents, as many researches stated that freeze drying has been one of the best ways of preservation [23-25]. It has also been said to show a less pronounced damaging effect on tissue structure than other drying methods [26]. From the graphs of radical scavenging activity with respect to extraction time below it can be observed that the freeze-dried method does not yield the highest radical scavenging activity which corresponds to the antioxidant contents in C. Nutans leaves. This finding corresponds to what have been found on the research on lemon balm, where the essential oil contents of dried samples were far lower than dried samples particularly samples of freeze drying and convective drying. They suggested that by processing the leaves at a pressure lower than atmospheric pressure is one of the causes of essential oil deterioration because of the volatility, polarity and chemical structure of the constituents in the leaves [27]. The decreasing of radical scavenging activity may due to the degradation of compounds that scavenge free radicals in hot water extraction condition as the extraction period increases.

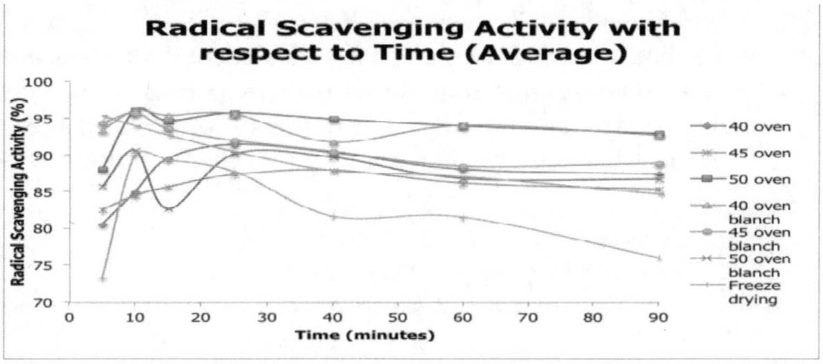

Figure 11. The Radical Scavenging Activity with respect to Time.

3.2.6. Total Phenolic Content per Crude. Based on Figure 8, it can be seen that generally the phenolic contents are extracted faster than the crude. For both replicates, the 40°C oven dried samples and to 50°C oven dried samples, the phenolics are extracted faster compared to other samples. For all the samples, after 40 minutes, the graphs starts to flatten out meaning that the phenolics have mostly been extracted by that time. So the equilibrium ratio of total phenolic content to crude is in the range of 0.4 to 0.7 for all samples. The phenolics exit faster due to the structure of the leaves that contains these compounds may be more vulnerable to heat.

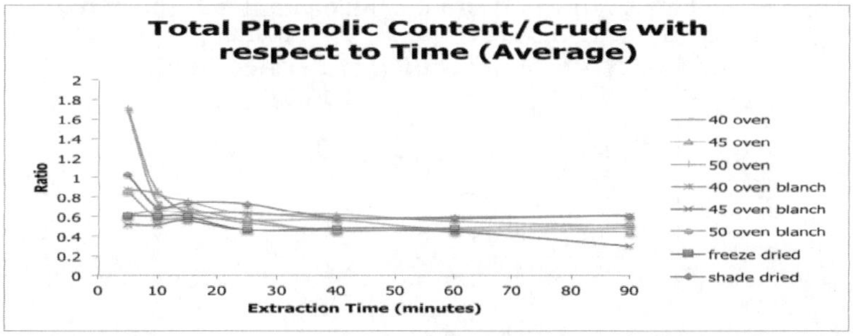

Figure 12. The Total Phenolic Content per Crude with respect to Time.

3.3. Comparison between Drying and Fermentation for Optimum Condition
The drying of C. nutans leaves was carried out by the other researcher who is studying the effect of drying on C. nutans leaves. The drying and fermentation data were compared at optimum condition.

3.3.1. Total Phenolic Content. Figure 13 shows the total phenolic content for fermentation and drying of C. nutans leaves at optimum condition whereby 12 hours fermentation data was compared against drying datas of C. nutans at 40°C. It can be observed that the total phenolic content for fermented C. nutans is higher compared to drying only for the first 40 minutes. During fermentation, biotic and abiotic stress factors such as wounding, low temperature and pathogen attacks may trigger defences mechanisms by the synthesis of phenylpropanoid compounds such as flavonoids, isoflavonoids, psoralens, coumarins, phenolic acids, lignin and suberin [28-29].

By comparing at the cut-off point (t=60 mins) for various fermented times of C. nutans, the yield corresponding to total phenolic content for drying at 40°C is significantly higher compared to fermentation as shown in Table 2. The findings for this is supported by Abu Bakar et al. [30] where the unfermented leaves of Strobilanthes crispus showed better antioxidant and antiradical activities as compared to fermented leaves due to deterioration of phenolics content during fermentation.

Other previous studies also reported that some phenolics compound present in leaves might be converted or degraded during fermentation [31-34]. The formation of colour and flavour compounds during fermentation of C. nutans leaves also might reduce the polyphenol concentration [35].

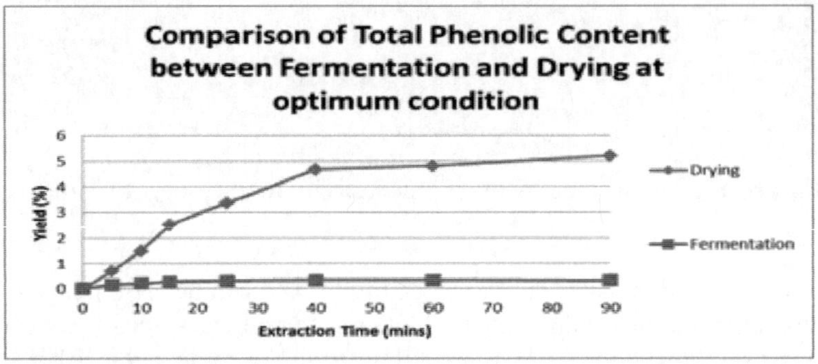

Figure 13. Comparison of Total Phenolic Content between Drying and Fermentation.

Table 2. Comparison of TPC between drying and fermentation for Sabah Snake Grass extraction at optimum condition. (t=60 mins)

Yield (%)	Drying	Duration of Fermentation (Hrs)				
	Temp (40°C)	6	12	24	48	72
	4.80	0.64	0.55	0.52	0.64	0.50

3.3.2. Antioxidant Activity (DPPH). The drying and fermentation data are compared at optimum condition. Table 4 shows the comparison of antioxidant activity, represented by IC50 of various duration of Fermented Sabah Snake Grass extract, measured by DPPH Radical Scavenging Test between Drying at 40°C and 6 hours fermentation. Based on Table 4, the IC50 values for drying of C. nutans leaves is significantly lower compared to the fermentation of C. nutans leaves. A low IC50 value is indicative of good inhibition of the enzyme.

The antioxidant potential of a compound can be attributed to its radical scavenging ability, and in order to evaluate the ability of the plant extracts to serve as antioxidants [14]. The findings for this experiment suggested that there is a reduction in antioxidant capacity in tea fermentation as supported by Kim et al. [5]. The decline of other phytochemicals including flavonol. glycosides, caffeine, saponin and ascorbic acid by oxidation or heat exposure during the fermentation process might lead to the reduction of antioxidant capacity [15]. This implies that drying method on C. nutans can preserve more antioxidant on this C. nutans leaves.

IOP Conf. Series: Materials Science and Engineering **206** (2017) 012041 doi:10.1088/1757-899X/206/1/012041

Table 3. Comparison of Antioxidant activity, represented by IC50 of Various Duration of Fermented Sabah Snake Grass Extract, measured by DPPH Radical Scavenging Test between Drying and Fermentation.

Method	DPPH Test IC_{50} (mg/mL)
Drying	1.670×10^{-11}
Fermentation	2.218×10^{-10}

Table 4. Comparison of Antioxidant activity, represented by IC50 of Various Duration of Fermented Sabah Snake Grass Extract, measured by DPPH Radical Scavenging Test between Drying and Fermentation at optimum condition. (t=60 mins)

IC_{50} (mg/mL)	**Drying Temp (40°C)**	**Duration of Fermentation (Hrs)**				
		6	**12**	**24**	**48**	**72**
	1.670×10^{-11}	2.218×10^{-10}	4.942×10^{-0}	12.273×10^{-0}	2.403×10^{11}	22.419×10^{-0}

3.3.3. Crude Yield. Figure 14 shows the comparison of extract yield between fermentation and drying of C. nutans at optimum condition whereby the average yield of 48 hours fermentation data was compared against drying conducted at 40°C. It can be observed that the yield produced by drying is higher than fermentation from 20 minutes onwards. It is also significantly higher when compared with the yield produced from fermentation at various times as shown in Table 5. During drying, enzymatic activities in samples were halted which prevents the deterioration of phenolic compounds which results in higher phenolic yield compared to those undergoes fermentation [21]. The yield range of these experiments is within the range of most herbal extracts which are between 1.0 to 5% yield.

Fermentation of tea leaves involves the chemical breakdown of substance by bacteria, yeast or other microorganism. There is a possibility that some compound present in C. nutans tea leaves might have been converted or degraded during fermentation, hence resulting in a lower yield produced compared to drying. Further study is required to be carried out in order to examine the reasons behind this phenomenon.

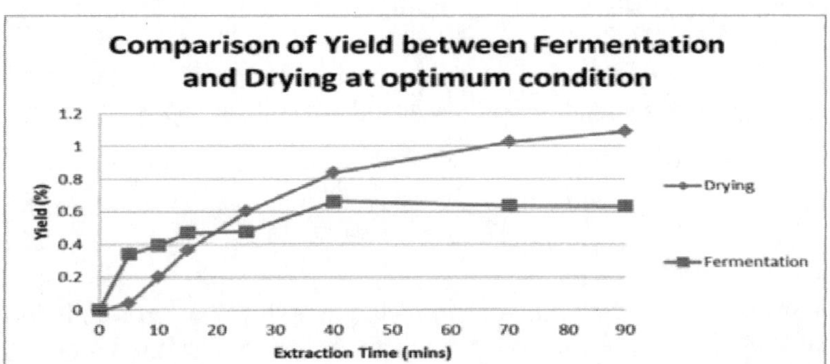

Figure 14. Comparison of Extract Yield between Drying and Fermentation.

Table 5. Comparison of Extract Yield between Drying and Fermentation for Sabah Snake Grass extraction experiments at optimum condition (t=60 mins)

Yield (%)	Drying Temp (40°C)	Duration of Fermentation (Hrs) 6	12	24	48	72
	1.0300	0.6368	0.5484	0.5241	0.6371	0.4987

4. Conclusion

In this study, two processes which are drying and fermentation are conducted on C. Nutans leaves to investigate how these processes affect the total phenolic content, antioxidant activity and the yield of crude. The fermentation time is varied for the fermentation process while different drying methods were carried out for the drying process. Blanching was also conducted on the drying process. For the fermentation process, to produce a tea with high phenolic content as well as highest yield, C. nutans leaves will need to be fermented for 12 hours. On the other hand, to produce a tea with high antioxidant content as well as the highest yield. C. nutans leaves will need to be fermented for 6 hours. To produce with the overall highest yield without taking the antioxidant and total phenolic content into consideration, 48 hours fermentation will be sufficient. It can also be concurred that C. nutans tea leaves should be brewed for 40 to 60 minutes in order to reap the maximum antioxidant and phenolic compounds. For the drying process, the leaves were found to dry completely when oven dried for 4 hours for 50°C, 7 hours for 45°C and 17 hours for 40°C.The extraction with time for crude and phenolic contents similar to a first-order kinetics graph and reaches equilibrium at around 60 minutes. The yield of crude was found to be in the range 0.8%-1.1% for all drying methods and has a concentration of 0.0942mg/ml – 0.1302mg/ml. 40°C oven drying and 50°C oven drying with blanching is more favorable for crude. All the drying methods yields around 0.25%-0.35% of phenolic content equivalent to 0.0016mg/ml GAE – 0.0020mg/ml GAE. The best methods for favorable phenolic retention are shade drying, 40°C oven drying and 50°C oven drying with blanching. The IC50 values for antioxidant activity shows that with higher processing temperature, the better the antioxidant content, but more research has to be carried out on this for a better justification as the results were not consistent. Blanching has an effect of increasing crude extract with increasing temperature, whilst it is vice versa without blanching. There is no significant effect of blanching towards the phenolic content of C. Nutans. When compared to fermentation based on the results from another researcher, drying showed a more favorable edge in all three aspects. In a nutshell, 40°C oven drying and 50°C oven drying with blanching showed more favorable results for both crude and phenolic content. There were no significant differences in the radical scavenging activities of both. Hence, 40°C oven drying and 50°C oven drying with blanching are the best methods of processing C. Nutans with high bioactive constituent retention. Further research should be carried out at different blanching temperatures and higher oven drying temperatures.

Acknowledgement

The authors would like to express their sincere appreciation and deepest gratitude to their supervisors, Assoc. Prof Dr. Sivakumar Kumaresan and Associate Professor Dr. Ivy Wong Nyet Kui for their valuable guidance and advice throughout the year. The author would like to acknowledge University Malaysia Sabah (UMS) and Faculty of Engineering for providing a good environment and research facilities to complete this project and the Ministry of Higher Education funding through the

Fundamental Research Grant Scheme (FRG0402-TK-2/2014. "Clicanthus Nutans (Sabah Snake Grass) Solid State Fermentation Study").

References

[1] Biber-Klemm S, Cottier T and Berglas D S 2006 *Rights to plant genetic resources and traditional knowledge: basic issues and perspectives* (Wallingford: CABI)

[2] Kayne S B 2008 *Complementary and Alternative Medicine* (Pharmaceutical Press)

[3] Hicks A 2001 Review of Global Tea Production and the Impact on Industry of the Asian Economic Situation

[4] Ungar L 2013 Tea's health benefits boost its popularity *Washington Post*

[5] Kim Y, Goodner K L, Park J-D, Choi J and Talcott S T 2011 Changes in antioxidant phytochemicals and volatile composition of Camellia sinensis by oxidation during tea fermentation *Food Chem.* **129** 1331–42

[6] Edgar J 2009 Types of Teas and Their Health Benefits *WebMD* 1–2

[7] Muthumani T and Senthil Kumar R S 2007 Influence of fermentation time on the development of compounds responsible for quality in black tea *Food Chem.* **101** 98–102

[8] Yong Y K, Tan J J, Teh S S, Mah S H, Ee G C L, Chiong H S and Ahmad Z 2013 Clinacanthus nutans Extracts Are Antioxidant with Antiproliferative Effect on Cultured Human Cancer Cell Lines. *Evid. Based. Complement. Alternat. Med.* **2013** 1–8

[9] Jain A, Manghani C, Kohli S, Nigam D and Rani V 2013 Tea and human health: The dark shadows *Toxicol. Lett.* **220** 82–7

[10] Sakdarat S, Shuyprom A, Dechatiwongse T, Ayudhya N, Waterman P G and Karagianis G 2008 Chemical composition investigation of the Clinacanthus nutans Lindau leaves บทคั ดย□ อ *Thai J. Phytopharm.* **13** 13–24

[11] Ariffin F, Heong Chew S, Bhupinder K, Karim A A and Huda N 2011 Antioxidant capacity and phenolic composition of fermented Centella asiatica herbal teas *J. Sci. Food Agric.* **91** 2731–9

[12] Shio Heong C, Huda N, Karim A A and Fazilah A 2011 Effect of Fermentation on the Composition of Centella asiatica Teas *Am. J. Food Technol.* **6** 581–93

[13] Hunaefi D and Smetanska I 2013 The effect of tea fermentation on rosmarinic acid and antioxidant properties using selected in vitro sprout culture of Orthosiphon aristatus as a model study **2**

[14] Adewusi E A, Moodley N and Steenkamp V 2011 Antioxidant and acetylcholinesterase inhibitory activity of selected southern African medicinal plants *South African J. Bot.* **77** 638–44

[15] Barek L and Fadzelly M 2015 Effect of different drying methods on phytochemicals and antioxidant properties of unfermented and fermented teas from Sabah Snake Grass (Clinacanthus nutans Lind.) leaves *Int. Food Res. J.* **22** 661–70

[16] Lukmanul Hakkim F, Gowri Shankar C and Girija S 2007 Chemical composition and antioxidant property of holy basil (Ocimum sanctum L.) leaves, stems, and inflorescence and their in vitro callus cultures *J. Agric. Food Chem.* **55** 9109–17

[17] Brand-Williams W, Cuvelier M E and Berset C 1995 Use of a Free Radical Method to Evaluate Antioxidant Activity *LWT- Food Sci. Technol.* **28** 25–30

[18] Lin L, Lei F, Sun D-W, Dong Y, Yang B and Zhao M 2012 Thermal inactivation kinetics of Rabdosia serra (Maxim.) Hara leaf peroxidase and polyphenol oxidase and comparative evaluation of drying methods on leaf phenolic profile and bioactivities *Food Chem.* **134** 2021–9

[19] Argyropoulos D and Müller J 2014 Effect of convective-, vacuum- and freeze drying on sorption behaviour and bioactive compounds of lemon balm (Melissa officinalis L.) *J. Dermatol. Sci.* **1** 59–69

[20] Argyropoulos D and Müller J 2014 Changes of essential oil content and composition during

IOP Conf. Series: Materials Science and Engineering **206** (2017) 012041 doi:10.1088/1757-899X/206/1/012041

convective drying of lemon balm (Melissa officinalis L.) *Ind. Crops Prod.* **52** 118–24

[21] Hossain M.B. B-R C, Martin-Diana A.B. and Brunton N.P. 2010 Effect of drying method on the antioxidant capacity of six Lamiaceae herbs *Food Chem.* **123** 85–91

[22] McClure J W 1979 *Biochemistry of Plant Phenolics* ed C F V S Tony Swain, Jeffrey B. Harbone (New York: Plenum Press)

[23] Bušić A, Vojvodić A, Komes D, Akkermans C, Belščak-Cvitanović A, Stolk M and Hofland G 2014 Comparative evaluation of CO 2 drying as an alternative drying technique of basil (Ocimum basilicum L.) — The effect on bioactive and sensory properties *Food Res. Int.* **64** 34–42

[24] Katsube T, Tsurunaga Y, Sugiyama M, Furuno T and Yamasaki Y 2009 Effect of air-drying temperature on antioxidant capacity and stability of polyphenolic compounds in mulberry (Morus alba L.) leaves *Food Chem.* **113** 964–9

[25] Martínez-Las Heras R, Heredia A, Castelló M and Andrés A 2014 Influence of drying method and extraction variables on the antioxidant properties of persimmon leaves *Food Biosci.* **6** 1–8

[26] Yousif A N, Scaman C H, Durance T D and Girard B 1999 Flavor Volatiles and Physical Properties of Vacuum-Microwave-and Air-Dried Sweet Basil (Ocimum basilicum L.) *J. Agric. Food Chem.* **47** 4777–81

[27] Calín-Sánchez Á, Szumny A, Figiel A, Jałoszy Ski K, Adamski M and Carbonell-Barrachina Á A 2011 Effects of vacuum level and microwave power on rosemary volatile composition during vacuumâ€"microwave drying *J. Food Eng.* **103** 219–27

[28] Dixon R A and Paiva N L 1995 Stress-Induced Phenylpropanoid Metabolism *Plant Cell Online* **7** 1085–97

[29] Rabeta M S and Lai S Y 2013 Effects of drying, fermented and unfermented tea of Ocimum tenuiflorum Linn. on the antioxidant capacity *Int. Food Res. J.* **20** 1601–8

[30] Abu Bakar, Mohd Fadzelly, Arnida Hani Teh A R and N H 2005 Antioxidant tea from leaves of Strobilanthes crispus *J. Trop. Med. Plants* **5** 199–204

[31] Joubert E 1996 HPLC quantification of the dihydrochalcones, aspalathin and nothofagin in rooibos tea (Aspalathus linearis) as affected by processing *Food Chem.* **55** 403–11

[32] Standley L, Winterton P, Marnewick J L, Gelderblom W C A, Joubert E and Britz T J 2001 Influence of Processing Stages on Antimutagenic and Antioxidant Potentials of Rooibos Tea *J. Agric. Food Chem.* **49** 114–7

[33] Bramati L, Aquilano F and Pietta P 2003 Unfermented Rooibos Tea: Quantitative Characterization of Flavonoids by HPLC−UV and Determination of the Total Antioxidant Activity *J. Agric. Food Chem.* **51** 7472–4

[34] Schmandke H 2005 Antioxidants in rooibos tea: Dihydrochalcones and flavonoids *Ernährungs-Umschau* 18–20

[35] Vanderhaegen B, Neven H, Coghe S, Verstrepen K J, Verachtert H and Derdelinckx G 2003 Evolution of Chemical and Sensory Properties during Aging of Top-Fermented Beer *J. Agric. Food Chem.* **51** 6782–90

Comparison of rheological properties of graphene / carbon nanotube hydrogenated oil based biodegradable drilling fluid

Y H Chai[1], S Yusup[1], V S Chok[2], S Irawan[3], J D B S Singh[4], B L F Chin[5]

[1] Biomass Processing Laboratory, Center of Biomass and Biofuels Research, MOR Green Technology, Chemical Engineering Department, Universiti Teknologi PETRONAS, Seri Iskandar, 32160 Perak, Malaysia

[2] Chemical Engineering Department, Heriot-Watt University Malaysia, No. 1, Jalan Venna P5/2, Precinct 5, 62200 Putrajaya, Malaysia

[3] Petroleum Engineering Department, Universiti Teknologi PETRONAS, Seri Iskandar, 32160 Perak, Malaysia

[4] Graphene Nanochem Plc, Suite 0.2, Level 9, WORK@Clearwater Changkat Semantan, Off Jalan Semantan, Damansara Heights, 50490, Kuala Lumpur, Malaysia

[5] Department of Chemical Engineering, Curtin University Sarawak Campus, CDT 250, Miri 98009, Sarawak Malaysia

E-mail: drsuzana_yusuf@utp.edu.my

Abstract. An experimental investigation has been carried out to investigate the rheological properties of graphene / carbon nanotube hydrogenated oil based biodegradable drilling fluid at different nanoparticle loadings. The rheological behaviours of interest in this investigation are the viscosity and shear stresses of two different nanofluids respectively. The limiting parameters in this study are 25 ppm, 50 ppm and 100 ppm weight concentration at operating temperature ranging from 30°C to 50°C. Both nanofluids are subjected to shear rate ranging from 0 – 140 s^{-1} for comparison of rheological behaviours. Both samples' viscosity reduces to base fluid's viscosity value at higher shear rate with carbon nanotube-hydrogenated oil yielding higher viscosity compared to graphene-hydrogenated oil for all nanoparticle loadings at lower shear rate. Shear stress analysis also shows similar results with carbon nanotube based samples showing higher stress between the two at all particle loadings. Both samples show Newtonian behaviour that is similar to base fluid even at higher particle loadings. Analysis revealed both nanofluids yields close to zero yield stress even with the presence of graphene or carbon nanotube particles. The significance of this study shows that addition of low nanomaterials for enhancement of drilling fluids can improve its thermophysical properties without compromising the quality of drilling fluids such as viscosity and shear stress properties.

1. Introduction

Drilling fluids are categorized into three general types, namely water based mud (WBM), oil based mud (OBM) and synthetic based mud (SBM). While such drilling fluids are widely available in the market, the need for an environmental friendly drilling fluids which possess low toxicity and biodegradable attributes are highly desired [1].

In the drilling sector, the formulation of a potentially suitable drilling fluid is greatly sought after. Drilling fluids readily deteriorate when subjected to high temperature and high pressure environment

IOP Conf. Series: Materials Science and Engineering **206** (2017) 012042 doi:10.1088/1757-899X/206/1/012042

hinders drilling operations [2]. The variations in conditions of reservoir types and drilling processes requires suitable drilling fluids to be customized accordingly. The incorporation of nanotechnology to develop and improve drilling fluid formulations is regarded as one of the solutions to such problems [3].

Nanofluids are made up of binary systems where solid and liquid phases are present which contains metallic or non-metallic nanoparticles suspensions that have the average size of 100 nm or less [4]. Numerous investigations exhibited nanofluids with enhanced thermophysical properties such as thermal conductivity, electrical conductivity and viscosity. Kole and Dey [5] obtained up to 15% thermal conductivity enhancement with 0.395 vol% functionalized graphene in ethylene glycol. Hadadian et al. [6] observed an anomalous enhancement in electrical conductivity properties in distilled water at 25,678% enhancement at 0.0006 mass fraction of graphene oxide at room temperature. Furthermore, Ding et al. [7] dispersed 0.5 wt% carbon nanotubes into distilled water to achieve over 350% enhancement in heat convection at laminar flow.

Addition of nanoparticles have been proven useful in heat transfer applications especially at higher nanoparticle loadings but the trade-off point resulted in the increase of nanofluid viscosity. Mishra et al. [8] reported that viscosity of nanofluids increases with respect to the volume concentration of nanoparticles. While effect of viscosity could be unfavourable to heat transfer systems [8], benefit of high viscosity properties in drilling fluids enable solid cuttings to be kept at suspension at low shear rate to prevent sagging when channelling drilling muds out from the wellbore [9]. However, Vajjha and Das [10] showed that particle loadings greater than 3 vol% exhibits penalty drop and increment in pumping costs.

Despite numerous publications regarding the thermo-physical properties of nanofluids, there has been little publications on the rheological behaviour of nanofluids [11-13] especially on the rheological behaviours of vegetable oil containing nanoparticles.

The focus of this paper investigates the rheological properties of hydrogenated oil-based drilling fluid which is dispersed with graphene nanosheet and carbon nanotube nanoparticles respectively by using hydrodynamic and acoustic cavitation (HAC) combination as two-steps dispersion method.

2. Materials and Methods

2.1. Materials

All materials in this study, namely hydrogenated oil-based drilling fluid, graphene nanosheet and carbon nanotube powders, were procured from Platinum Green Chemicals Sdn. Bhd., Malaysia. Fourier Transform Infrared (FTIR) spectroscopy analysis of graphene nanosheet and carbon nanotube is as shown in Fig. 1 and Fig. 2 respectively. Both FTIR spectra shows the presence of O-H functional groups which falls within the range of 2800 to 3500 cm^{-1} which shows good agreement with the presence of stretching vibrations of O-H bonds [14].

There are trace presence of halogen atoms exists in both nanoparticles. C-N stretching at 1100 cm^{-1} is found on graphene nanosheet surfaces while CH_2X groups is found on the surface of carbon nanotube to form alkyl halides. The presence of nitrogen as halogen maybe derived during the preparation and exfoliation of graphene nanosheets and carbon nanotubes using nitric acid [15].

Hydrogenated oil is comprised of a mixture of straight chain and branched paraffin of carbon chain length from C15 to C18. The thermophysical properties of hydrogenated oil is highlighted in Tab. 1.

2.2. Nanofluids preparation

The targeted concentration of nanoparticles in this study are fixed at very low nanoparticle loadings, namely 25 ppm,50 ppm and 100 ppm for both graphene nanosheet and carbon nanotube respectively to ensure no alteration on the rheological behaviour of hydrogenated oil-based nanofluids.

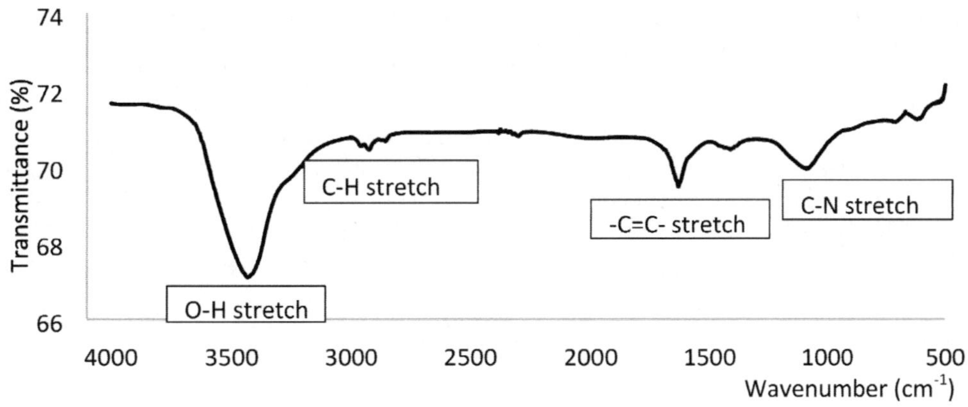

Figure 1. FTIR spectra of graphene nanosheet.

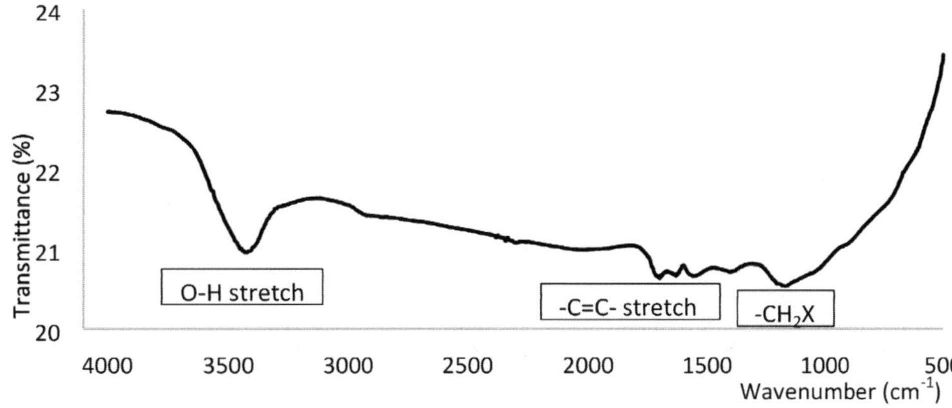

Figure 2. FTIR spectra of carbon nanotube.

Table 1. Thermophysical properties of hydrogenated oil.

Density (kg/m^3)	Viscosity (cP)	Boiling Point (°C)	Oxidizing Properties
780 (at 15°C)	1.5 – 2.0 (at 40°C)	330	None

Two-steps dispersion method was used to physically disperse nanoparticles into hydrogenated oilbased drilling fluids via hydrodynamic and acoustic cavitation (HAC) combination. Before each HAC dispersion at all concentrations, nanoparticles are pre-homogenized via high-speed shearing at 800 rpm constantly for 15 minutes with the schematic set-up as shown in Figure 5.

Pre-homogenized nanofluids were transferred to a single-stage hydrodynamic cavitation (HC) unit for further dispersion. A variable frequency drive (VFD) pump with the range of 0 to 50 Hz frequency was used to control the inlet flow of nanofluid to one-hole orifice with outer diameter of 1 mm. HC dispersion was carried out for 3 hours at 10 bar with flow rate of 1.5 L min^{-1} averagely.

Bath Ultrasonic Branson 8510E-DTH with 320 W power and 40 kHz frequency was used to homogenize nanofluid samples. Samples carried out in HC dispersion were collected and transferred to ultrasonic bath for 3 hours of ultrasonication process to further homogenize the suspended nanoparticles in hydrogenated oil-based nanofluid.

2.3. Rheological analysis
Malvern Bohlin Gemini II Rheometer was used to measure the viscosity and shear stress values of hydrogenated oil-based nanofluids. The rheometer is connected to a PC controlled Peltier plate temperature controller to vary the temperature. The pre-set gap for the spindle to the plate is set at 30 μm. The selected spindle in this study is a stainless steel cone-and-plate spindle with 2° angle and 40 mm diameter.

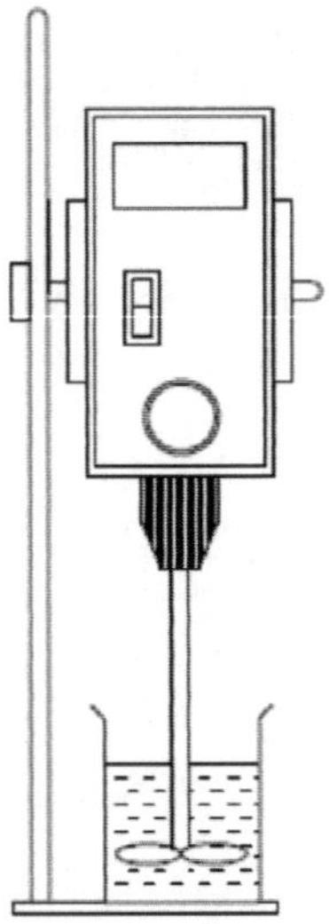

Figure 3. Schematic set-up of high speed stirrer.

All measurements have a delay time and integration time of 10 seconds to record data at steady state conditions. In this study, the data collection of the rheometer is viscosity and shear stress parameters with respect to temperature, shear rate, nanoparticle concentration and types. The shear rate in this study is within the range of 0 to 140 s^{-1} at the temperature range of 30°C to 50°C with 10°C step increase. Each measurement was repeated 3 times with a new sample to ensure the hydrogenated oil-based nanofluids do not deform when subjected to stress. Each samples were subjected to 20 minutes of ultrasonication to ensure homogeneity in each samples before performing rheological analysis.

3. Models
The application of rheological models to determine the shear stress – shear rate relationships are used to predict rheological behaviours due to an array of applications with various rheological characteristics. Non-Newtonian models considered in this study are Bingham Plastic model and Power Law model that are widely used for prediction of rheological behaviours.

3.1. Bingham Plastic model
Bingham Plastic fluids exhibit an "infinite" viscosity until a sufficiently high stress is applied to initiate flow. Eq. 1 shows the Bingham Plastic model.

$$\sigma = \sigma_o + \mu \quad (1)$$

where σ is the shear stress, σ_0 is the limiting shear stress, μ is the viscosity and γ is the shear rate.

The limiting shear stress is often referred to as Bingham yield stress of the material [16]. The Bingham Plastic model is calculated by a simple regression analysis due to its linear equation model.The Bingham Plastic model is suitable for colloidal systems that shows Bingham behaviours.

3.2. Power Law model
The Power Law model is also referred to as Ostwald model. Numerous non-Newtonian materials behaves differently as the shear rates increase. The changes made to viscosity at increasing shear rate can be categorised into two groups, namely shear thinning and shear thickening.

Shear thinning or pseudoplastic effect refers to the decrease in viscosity at higher shear rate while shear thickening has the opposite effect. The Power Law model is shown in Eq. 2.

$$\sigma = \mu . \gamma^n \tag{2}$$

where σ is the shear stress, , μ is the fluid viscosity, γ is the shear rate and n is the power law index of the material. When $n < 1$, the material possess shear thinning behaviour while $n > 1$ shows a shear thickening fluid

The Power Law model is suitable for material that flows under small shear rate range between 10^1 to 10^4 s^{-1} [16]. At higher shear rate range, the materials will deviate from this relationship.

4. Results and Discussions

4.1. TEM analysis of nanoparticles
Transmission electron microscope (TEM) images of nanoparticles types are taken with Zeiss Libra 200FE Microscope. Fig. 4 shows an image of graphene nanosheet which was taken at 20,000x magnifications. Several dark layers are superimposed on one another which denotes stacking of several graphene nanosheet layers. At Fig. 5(a), image of carbon nanotubes are taken at 80,000x magnification with average outer diameter within the range of 10 to 12 nm.

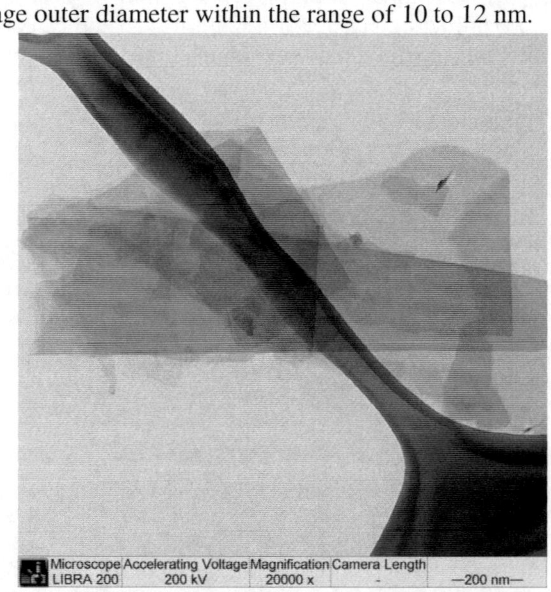

Figure 4. TEM image of graphene nanosheets at 20,000x magnification.

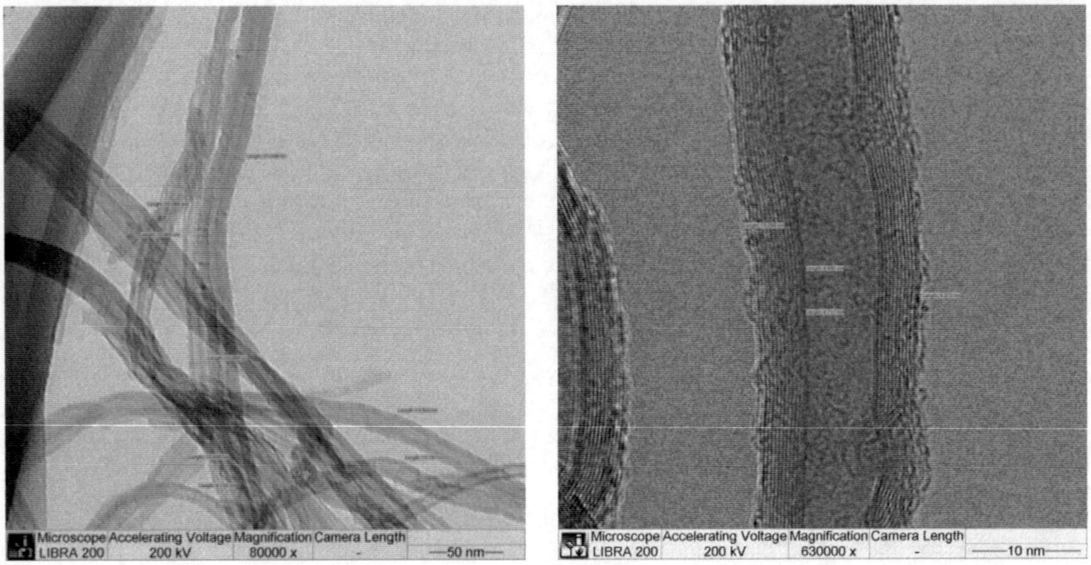

(a) (b)

Figure 5. TEM image of carbon nanotubes at (a) 80,000x and (b) 200,00x magnification.

Furthermore, it can be seen from Fig. 5(a) carbon nanotubes are easily entangled with one another due to its high length-to-diameter (L/D) ratio. The mean diameter outer layer thickness is 4.7nm with inner diameter within the range of 7 nm as shown in Fig. 5(b).

4.2. Rheological properties of hydrogenated oil-based drilling fluid

The rheological behaviour of hydrogenated oil-based drilling fluid can be deduced from the relationship between viscosity, shear stress and shear rate parameters. Fig. 6(a) and Fig. 6(b) shows the viscosity and shear stress with respect to shear rate at 30°C respectively.

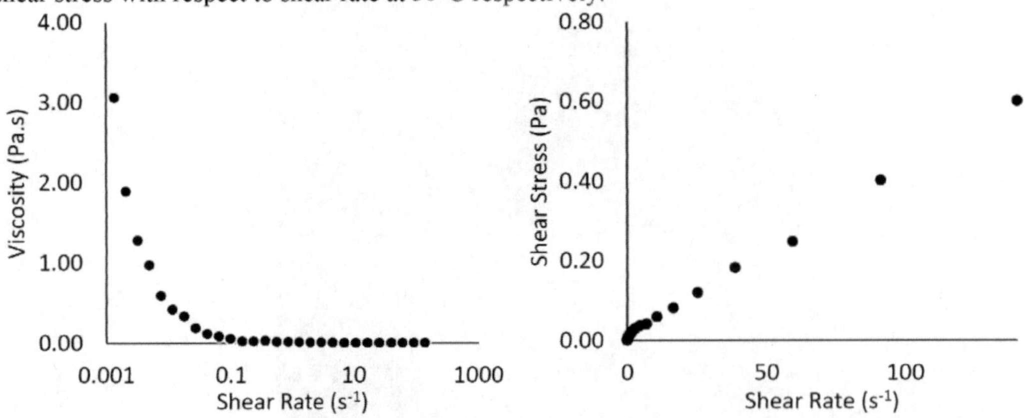

(a) (b)

Figure 6. Graphical illustrations of (a) viscosity and (b) shear stress of hydrogenated oil-based fluid at 30°C.

Bingham fluids behave rigidly at very low stresses but flows as viscous fluids when sufficient stress is applied. Although Fig. 6(b) depicts similar rheological profile to a Newtonian fluid, the viscosity shown in Fig. 6(a) decreases exponentially with increasing shear rates. Viscosity of Newtonian fluid remains constant regardless of shear rate applied, while viscosity of Bingham fluid varies with respect to shear rate.

Therefore, hydrogenated oil-based drilling fluid used in this study exhibits Bingham fluid properties with shear thickening behaviour and near zero shear stress.

4.3. Effect of nanoparticle concentration on viscosity of hydrogenated oil-based nanofluids

As observed from Fig. 7(a) and Fig.7(b), the addition of nanoparticles does not influence the viscosity profile of hydrogenated oil-based nanofluid. At very low shear rates, the viscosity of nanofluids are closely similar and overlapped at all concentrations at the higher shear rate range.

The addition of nanoparticles increases the viscosity properties of hydrogenated oil with respect to concentration. Similar findings by other researchers also show viscosity of nanofluids increases with respect to concentrations but decreases at higher shear rates [5; 17-18]. The increase in solid particle concentrations within liquid suspension increases the inter-particle frictions which increases the resistance of fluid to flow and subsequently increases the viscosity of nanofluids [17].

The effect of nanoparticle "jamming" at the bottom of the cone-and-plate spindle should be taken into account contributing in minor fluctuations of viscosity as seen in Fig. 8.

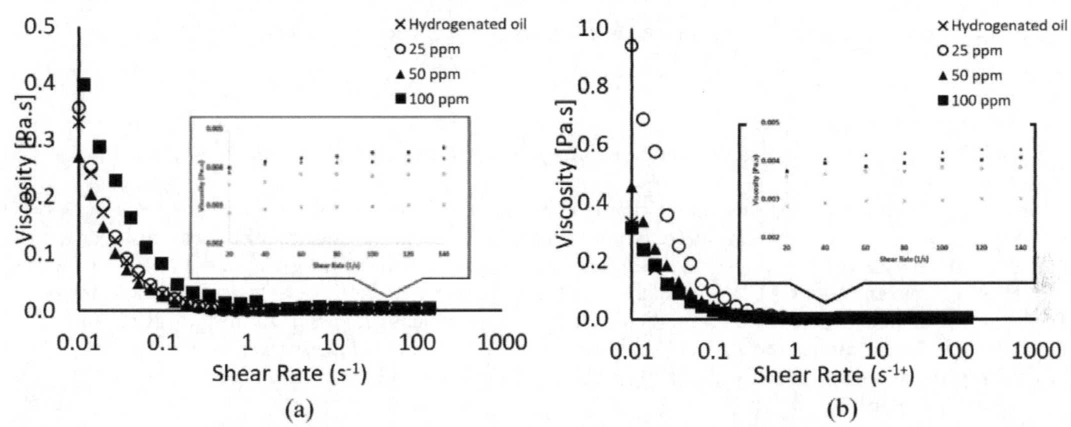

Figure 7. Effect of (a) graphene nanosheet and (b) carbon nanotube concentration at 50°C on viscosity of hydrogenated oil-based nanofluid.

Furthermore, it can be observed from Fig. 7 (a) and Fig. 7(b) hydrogenated oil-based nanofluids exhibit slight shear thickening behaviour at higher shear rate. Ijam et al. [19] and Kinloch et al. [20] both attributed this profile to the percolation structure of nanoparticles suspended in the base fluid. The percolation structure formed is broken down when high level of shearing is achieved to form primary particles and subsequently the stress and viscosity increases with increasing rate.

4.4. Effect of temperature on viscosity of hydrogenated oil-based nanofluids

Viscosity properties is also influenced by temperature. At higher temperature, the viscosity of all fluids yielded lower viscosity values as compared to lower temperature conditions at similar concentrations. Fig. 8 shows a linear declining relationship between viscosity and temperature regardless of the types of nanoparticles dispersed into hydrogenated oil.

The intermolecular attractions between nanoparticles and the particles of base fluids are highly influenced by temperature. The increase in temperature provides higher energy into the system, which subsequently decreases the interparticle and intermolecular adhesive forces of the particles [21] leading to the decrease of fluid's viscosity. Similar results by other researchers [19; 22] exhibit identical behaviour as well.

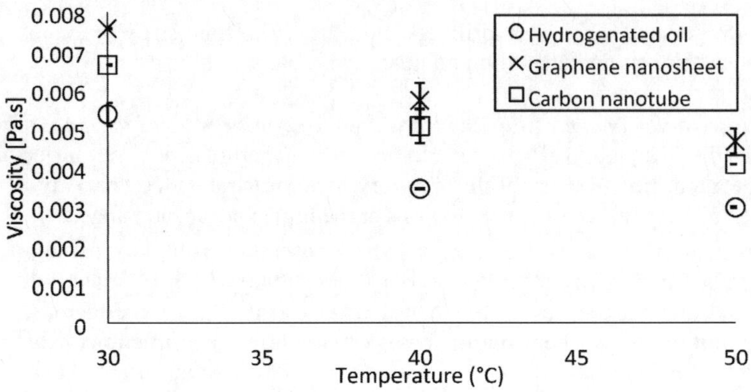

Figure 8. Effect of temperature on viscosity of hydrogenated oil-based nanofluid

4.5. Effect of nanoparticle concentration on rheological behaviour of hydrogenated oil-based nanofluids

Figure 9(a) and Figure 9(b) shows the comparison between experimental data and rheological models of graphene nanosheets and carbon nanotubes nanofluids with respect to different nanoparticle concentration respectively. As observed, Bingham model gave a better fitting compared to Power Law model at all concentrations.At lower shear rate, the experimental data are able to fit closely to the Power Law model. However, the predicted shear stress values deviated further away from the experimental data at higher shear rate. This is attributed to the flow behaviour index, $n > 1$, as there is slight shear thickening at higher shear rates as observed in Figure 7(a) and Figure 7(b).

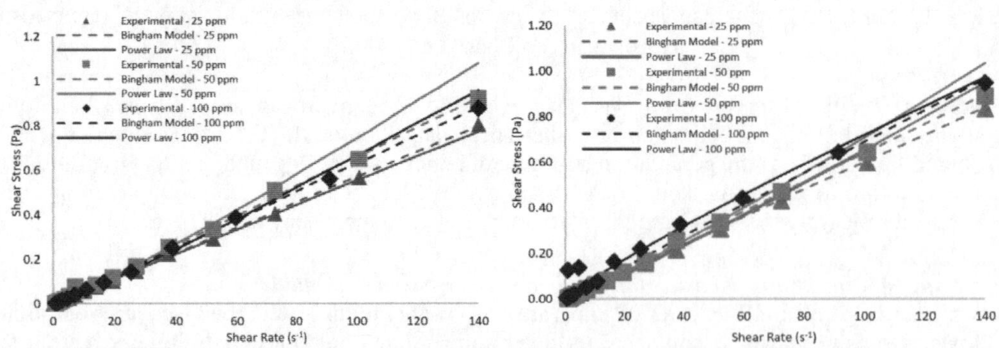

(a) (b)

Figure 9. Comparison of experimental data and rheological models of (a) graphene nanosheet and (b) carbon nanotube – hydrogenated oil-based nanofluid at 30°C

IOP Conf. Series: Materials Science and Engineering **206** (2017) 012042 doi:10.1088/1757-899X/206/1/012042

5. Conclusion

From this study, hydrogenated oil-based fluid exhibit a non-Newtonian behaviour with shear thinning properties. The presence of very low nanoparticles loadings increases the viscosity of hydrogenated oilbased nanofluids and does not change its rheological behaviour. The slight increase in viscosity and shear stress properties maybe useful in the oil and gas industry in suspending solid cuttings during drilling operations while simultaneously improving the thermophysical properties of drilling fluids.

Acknowledgement

The authors would like to gratefully acknowledge the materials and equipment provided by Platinum Green Chemicals Sdn. Bhd. to carry out this investigation. This research is in collaboration with Platinum Green Chemicals Sdn. Bhd., Malaysia. The authors would also like to thank Centralized Analytical Laboratory (CAL), Universiti Teknologi PETRONAS for their kind assistance in carrying out FTIR analysis for this study.

References

[1] Xie S, Jiang G, Chen M, Deng H, Liu G, Xu Y, Wang J and Qiu K 2011 *Petrol. Explor. Dev.* **38** 379-78
[2] Ismail A R, Rashid N M, Jaafar M Z, Sulaiman W R W and Buang N A 2014 *J. Appl. Sci.* **14** 1192-97
[3] Choi S U S 2009 Nanofluids: From Vision to Reality Through Research *J. Heat Transfer* **131** 033106
[4] Mukherjee S and Paria S 2013 *IOSR – JMCE* **9** 63-9
[5] Kole M and Dey T K 2013 Investigation of thermal conductivity, viscosity and electrical conductivity of graphene based nanofluids *J. App. Phys.* **113**
[6] Hadadian M, Goharshadi E K and Youssefi A 2014 *J. Nanopart. Res.* **16**
[7] Ding Y, Alias H, Wen D and Williams R A 2006 *Int. J. Heat Mass Tran.* **49** 240-50
[8] Mishra P C, Mukherjee S, Nayak S K and Panda A 2014 *Int. Nano Lett.* **4** 109-120
[9] Agarwal S, Phuoc T X, Soong Y, Martello D and Gupta R K 2013 *Can. J. Chem. Eng.* **91** 1641-49
[10] Vajjha R S and Das D K 2012 *Int. J. Heat Mass Tran.* **55** 4063-4078
[11] Chen H, Ding Y and Lapkin A 2009 *Powder Technol.* **194** 132-41
[12] William J K M, Ponmani S, Samuel R, Nagarajan R and Sangwai J. S. 2014 *J. Pet. Sci. Technol.* **117** 15-27
[13] Mehrali M, Sadeghinezhad E, Latibari S T, Kazi S N, Mehrali M, Zubir M N M and Metselaar H S C Investigation of thermal conductivity and rheological properties of nanofluids containing graphene nanoplatelets 2014 *Nanoscale Res.* Lett. **9**
[14] Farbod M, Ahangarpour A and Etemad S G Stability and thermal conductivity of water-based carbon nanotube nanofluids 2014 *Particuology* **22**
[15] Haddad Z, Abid C, Oztop H F and Mataoui A 2013 *Int. J. Therm. Sci.* **76** 168-89
[16] Rao M A 2014 Rheology of Fluid, Semisolid, and Solid Foods, *Food Engineering Series*
[17] Ahammed N, Asirvatham L G and Wongwises S 2016 *J. Therm. Anal. Calorim.* **123** 1399-1409
[18] Ho C Y, Yusup S, Chok V S and Arpin M T 2016 *Procedia Eng.* **148** 49-56
[19] Ijam A, Saidur R, Ganesan P and Golsheikh A M 2015 *Int. J. Heat Mass Tran.* 87 92-103
[20] Kinloch I A, Roberst S A, Windle A H 2002 *Polymer* **43** 7483-91
[21] Thomas S and Balakrishna P S C 2011 *Nanoscale Res. Lett.* **6** 377
[22] Ma W, Yang F, Shi J, Wang F, Zhang Z and Wang S 2013 *Colloids Surf.*, A **431** 120-26

IOP Conf. Series: Materials Science and Engineering **206** (2017) 012043 doi:10.1088/1757-899X/206/1/012043

Comparison on tribological properties of vegetable oil upon addition of carbon based nanoparticles

S S K Kiu[1], S Yusup[1], V S Chok[2], A Taufiq[3], R N M Kamil[4], S Syahrullail[5] and B L F Chin[6]

[1] Biomass Processing Laboratory, Center of Biomass and Biofuels Research, MOR Green Technology, Chemical Engineering Department, Universiti Teknologi PETRONAS, Seri Iskandar, 32160 Perak, Malaysia
[2] Chemical Engineering Department, Heriot-Watt University Malaysia, No. 1, Jalan Venna P5/2, Precinct 5, 62200 Putrajaya, Malaysia
[3] Scomi Platinum Sdn. Bhd., LOT15-19 & PT1409, Senawang Industrial Estate, Batu 4, Jalan Tampin, 70450 Seremban, Negeri Sembilan Darul Khusus, Malaysia, Malaysia
[4] Fundamental & Applied Science, Universiti Teknologi PETRONAS, Perak, Malaysia
[5] Faculty of Mechanical Engineering, Universiti Teknologi Malaysia, 81310 UTM Skudai, Johor, Malaysia
[6] Department of Chemical Engineering, Curtin University Sarawak Malaysia, CDT 250, 98009 Miri, Sarawak, Malaysia

E-mail: drsuzana_yusuf@utp.edu.my

Abstract. Carbon-based nanoparticles have gained much interest as lubricant additive due to their remarkable properties in mechanical, chemical and electrical field. In this research, graphene nanosheets (GN), carbon nanotubes (CNT), and graphene oxide (GO) were used as lubricant additives to investigate their effect on tribological properties. Friction coefficient and wear scar diameter were studied as parameters to determine the effectiveness of lubricant. In this study, vegetable oil (VO) was used as base fluid lubricant. GN, CNT and GO were added at 50ppm and 100ppm respectively to VO to study their optimum concentration when compared to pure VO. All nanoparticles were well dispersed by using a homogenizer. Results showed that addition of 50ppm GN has the most positive effect in improving the tribological properties of vegetable oil.

1. Introduction

In recent years, nanofluids have attracted more attention due to its wide range of applications. The main driving force for nanofluids research is cost effective. It is because by adding a relatively small

IOP Conf. Series: Materials Science and Engineering **206** (2017) 012043 doi:10.1088/1757-899X/206/1/012043

quantity of suitable nanoparticles, targeted properties of the base fluid can be improved remarkably. Specifically for lubricant with nano-additives, numerous researchers reported a reduction of coefficient of friction [1], increment in load carrying capacity [2] and reduction of wear [3]. Generally, there are three main differences which distinguish each research on nano-lubricants from one another. For instance, researchers often vary the type of nanoparticles used as additive, types of base fluid used as lubricant, and the homogenization mechanisms between nanoparticles with base fluid. In this study, vegetable oil was selected as lubricant to replace conventional lubricant which generally contains non environmental friendly compounds such as sulphur, phosphor, and chlorine [4]. Biolubricant production using vegetable oil is renewable, cheap, biodegradated and no adverse effects on nature [5].

In this study, graphene nanosheets (GN), carbon nanotubes (CNT), and graphene oxide (GO) were selected as additive in lubricant due to their outstanding performance in improving nanofluid properties [6-8]. A high speed homogenizer was chosen as homogenizing mechanism in this study. It utilized high shear force which promotes mixing between solid and liquid. 50ppm and 100ppm of GN, CNT and GO were added to vegetable oil (VO) to study the optimum concentration of the additives for lowest friction coefficient and wear scar diameter. The tribological properties of nanolubricants were tested using four-ball machine, following ASTM 4172.

2. Materials and Methods

2.1. Materials

GN, CNT, GO nanoparticles and vegetable oil (VO) were purchased from Scomi Sdn. Bhd, a local company which manufactures high performance chemical product. The properties of VO were shown in Table 1.

Table 1. Physical properties of VO

Flash Point	205°C
Density	1100 kg/m^3
Iodine Value (Wijs)	56.7
Saponification Value	128.0 mg KOH/g
Kinematic viscosity	455.8 mm^2/s (25°C)

2.2. Nanofluids preparation

Six nanofluid samples were prepared in this study. GN, CNT and GO were added to VO at 50ppm and 100ppm respectively. Each sample was homogenized using IKA homogenizer for an hour, at 8000 rpm for one hour. IKA T10 basic homogenizer are used as homogenizing apparatus in this study. It consists of a rotor-stator design with an outerstationary tube and an innerturning rotor which is connected to a motor. The high shearing force was used to provide an intensive mixing between solid and liquid phases. The schematic diagram of IKA T10 homogenizer is as shown in Fig. 1.

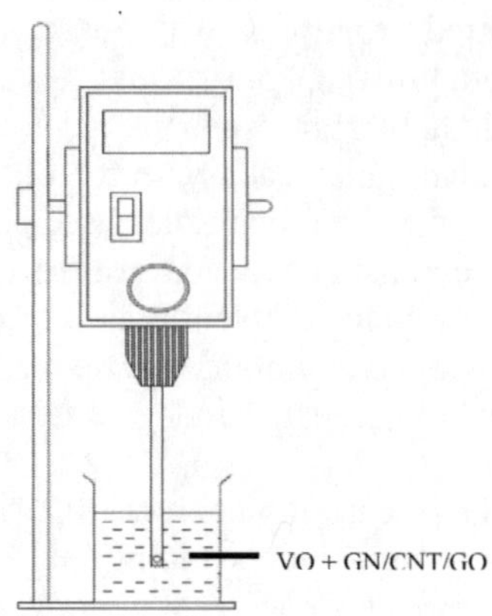

Figure 1. Schematic set-up of high speed homogenizer.

2.3. *Four ball test*

Pure VO and six prepared nanofluids were tested under ASTM D4172, Wear Preventive Characteristics of Lubricating Fluid (Four-Ball Method) using DUCOM fourball tribotester. Three 12.7 mm diameter steel balls were clamped together and covered with the lubricant to be evaluated as shown in Fig. 2 by following the setup procedure which was discussed by Chiong and his team [9]. By rotating the top ball against three bottom ball at operating condition of 1200 rpm, 75°C, 392N for one hour, wear scar is produced on the contact surfaces of the balls, and the average friction coefficient between the contact surfaces was recorded using Winducom 2010.

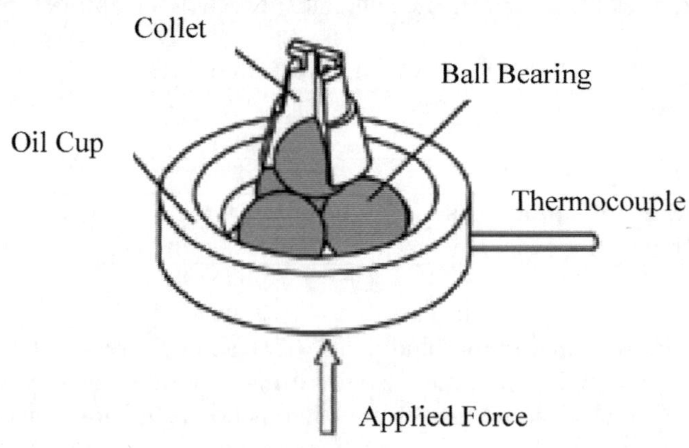

Figure. 2. Schematic diagram of four ball setup in experiment [10]

2.4. *Wear Scar Diameter*

The wear scar diameter was observed by optical microscope attached to a computer. The measurement of wear scar diameter was carried out with the aid of dimension software. The image and diameter of the wear scar was captured and measured as shown in Fig. 3. The average reading of wear scar diameters for three balls was recorded.

Figure. 3. Wear scar diameter measured on one of the sample ball

3. Results and Discussion

3.1. Nanofluid Dispersion

Fig. 4 showed the dispersion of GN, CNT and GO nanofluid by using high speed homogenizer. There were no observation of sediments of nanoparticles at the bottom of the measuring cylinder after 45 days of observations.

Figure. 4. Samples containing GN (left), CNT (middle) and GO (right) respectively

3.2. Friction Coefficient and Wear Scar Diameter

From Fig. 5 and Fig 6, the addition of GN, CNT and GO recorded unique trend of coefficient of friction and wear scar diameter against increasing concentration of nanoparticles. For GN, friction coefficient and wear scar diameter initially decreased at 50ppm. However, 100ppm concentration showed increment in both friction coefficient and wear scar diameter. Similarly, Eswariah et.al found that friction coefficient initially decreases with increase GN concentration but further increases with higher GN concentration [11]. They concluded that initial decrement in frictional coefficient can be attributed to the layered structure of GN and the optimum concentration is 0.025 mg/mL. Since GN are made up of planar structure, it is possible that 100 ppm of GN caused stacking and agglomerations which indirectly caused more wear and friction.

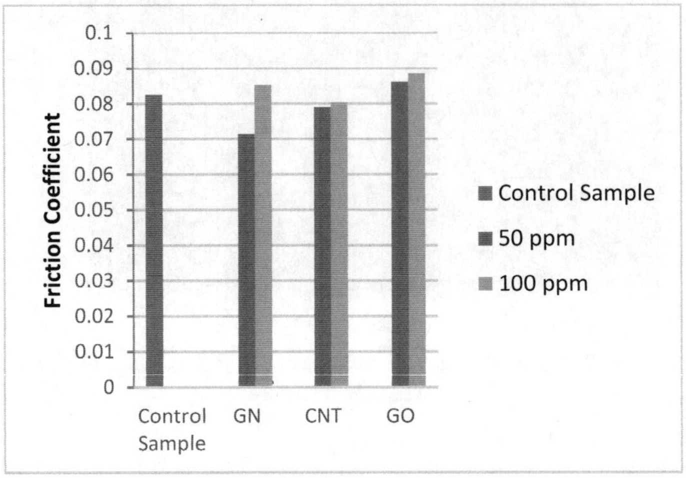

Figure. 5. Friction coefficient against 50ppm and 100ppm GN, CNT, and GO

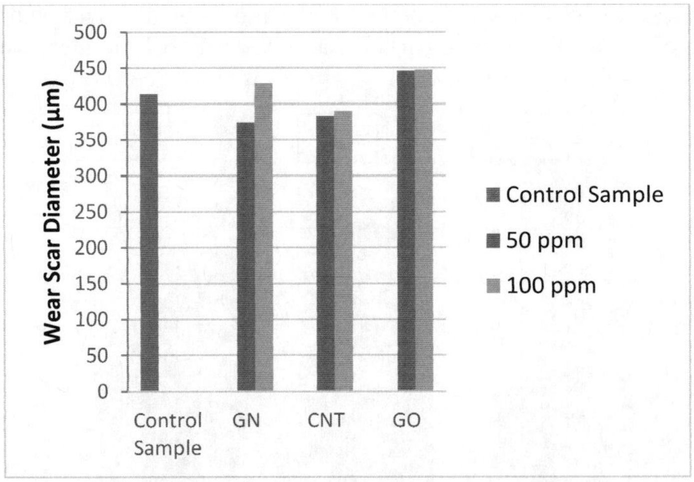

Figure. 6. Wear scar diameter against 50ppm and 100ppm GN, CNT and GO

For CNT, both 50ppm and 100ppm CNT samples showed greater reduced friction coefficient and wear scar diameter as compared to pure VO, which signifies the positive impact of addition of CNT into VO. However, both 50ppm and 100ppm CNT samples showed similar intensity in reducing the friction coefficient and wear value, which signifies the ineffectiveness of increasing concentration from 50ppm to 100ppm. Similarly, Antonio and his colleagues found that by increasing CNT concentration from 0.01 to 0.05 %, the COF value varied from 0.063 to 0.066, which is insignificant as well [9].

Meanwhile, both 50ppm and 100ppm of GO addition recorded higher friction and wear scar diameter than pure VO. It was probably due to the immiscibility of GO with the base fluid. GO have higher tendency to disperse in polar organic solvents due to their functional group on surface such as hydroxyl and carboxyl group [12]. However, VO which made up of long carbon group is non-polar in nature. The difference in polarity causes immiscibility and GO tends to form agglomerates. Hence, more friction and wear were produced. Generally, GO was used as water based lubricant additive [13]. Besides that, other researchers carried out surface modification [14] or mix GO within composite to improve the miscibility of nanoparticles into their base fluid [15] before carrying out tribological studies.

In comparison among six prepared nanofluids and pure VO, 50ppm of GN addition showed the lowest COF and WSD. Therefore, the optimum concentration and most effective nanoparticle type in this study is 50ppm GN.

4. Conclusion

In conclusion, 50ppm of GN addition showed the lowest coefficient of friction and lowest wear scar diameter among all seven samples tested including pure VO. Furthermore, 50ppm of GN showed reduced friction coefficient and wear but further addition up to 100ppm caused higher friction and wear due to stacking of GN. Addition of CNT at both 50ppm and 100ppm demonstrated reduction in friction coefficient and wear, but there are small difference between the two concentrations. On contrary, addition of GO at both 50ppm and 100ppm caused more wear and friction as compared to VO, which is detrimental towards tribological properties of VO. In future, more experiments involving nanoparticles concentration at smaller stepwise increment are needed to understand better the trend for each graphene, graphene oxide and carbon nanotube.

Acknowledgement

The authors wish to thank Universiti Teknologi Petronas, Scomi Sdn. Bhd and Universiti Teknologi Malaysia Mechanical Department for their support during this study. The authors would also like to thank BIOMASS Grant (0153AB-F02) for their financial support to carry out this study.

References

[1] Antonio J, Cornelio C, Andrea P, Hoyos-palacio L M, Lara-romero J & Toro A 2015. Tribological properties of carbon nanotubes as lubricant additive in oil and water for a wheel – rail system ☆. *Integrative Med. Res.* **5** 68–76

[2] Arbain N H & Salimon J 2010. Synthesis And Characterization Of Ester Trimethylolpropane Based Jatropha Curcas Oil As Biolubrificant Base Stocks. *J. of Sci. and Techno.* 47–58.

[3] Chiong I T, Rafiq A K M, Azli Y & Syahrullail S 2012. Tribological behaviour of refined bleached and deodorized palm olein in different loads using a four-ball tribotester. *Scientia Iranica*, **19** 1487–1492

[4] Elomaa O, Singh V K, Iyer A, Hakala T J & Koskinen J 2015. Graphene oxide in water lubrication on diamond-like carbon vs. stainless steel high-load contacts. *Diamond and Related Mat.*, **52** 43–48

[5] Eswaraiah V, Sankaranarayanan V & Ramaprabhu S 2011. Graphene-based engine oil nanofluids for tribological applications. *ACS App. Mat. and Interfaces*, **3** 4221–4227

[6] Ilie F & Covaliu C 2016. Tribological Properties of the Lubricant Containing Titanium Dioxide Nanoparticles as an Additive. *Lubricants*, **4** 1–13

[7] Ing T C, Rafiq M & Kadir A 2011. Experimental Evaluation on Lubricity of RBD Palm Olein Using Fourball Tribotester. *Tribology - Lubricants and Lubrication*, *2011*, 175–185

[8] Jia Z, Pang X, Li H, Ni J & Shao X 2015. Synthesis and wear behavior of oleic acid capped calcium borate/graphene oxide composites. *Tribology Int.* **90** 240–247

[9] Kinoshita H, Nishina Y, Alias A A & Fujii M 2014. Tribological properties of monolayer graphene oxide sheets as water-based lubricant additives. *Carbon*, **66** 720–723

[10] Meng Y, Su F & Chen Y 2015. Synthesis of nano-Cu/graphene oxide composites by supercritical CO2-assisted deposition as a novel material for reducing friction and wear. *Chem. Eng. J.*, **281** 11–19

[11] Peng D X, Kang Y, Hwang R M, Shyr S S & Chang Y P 2009. Tribological properties of diamond and SiO2 nanoparticles added in paraffin. *Tribology Int.*, **42** 911–917

[12] Taha-Tijerina J, Venkataramani D, Aichele C P, Tiwary C S, Smay J E, Mathkar A & Ajayan P M 2015. Quantification of the particle size and stability of graphene oxide in a variety of solvents. *Particle and Particle Systems Characterization*, **32** 334–339

[13] Yang G B, Chai S T, Xiong X J, Zhang S M, Yu L G, & Zhang P Y 2012. Preparation and tribological properties of surface modified Cu nanoparticles. *Trans. of Nonferrous Metals Society of China (English Ed.)*, **22** 366–372

[14] Zhang C, Zhang S, Yu L, Zhang P, Zhang Z & Wu Z 2012. Tribological behavior of 1-methyl-3-hexadecylimidazolium tetrafluoroborate ionic liquid crystal as a neat lubricant and as an additive of liquid paraffin. *Tribology Let.*, **46** 49–54

[15] Zhang W, Zhou M, Zhu H, Tian Y, Wang K, Wei J & Wu D 2011. Tribological properties of oleic acid-modified graphene as lubricant oil additives. *J of Phy. D: App. Phy.* **44** 1–4

Recycling of surfactant template in mesoporous MCM-41 synthesis

J Y Lai[1], F Twaiq[1], L H Ngu[1]

[1]Faculty of Engineering, Computing and Science, Swinburne University of Technology Sarawak Campus, Jalan Simpang Tiga, Kuching 93350, Sarawak, Malaysia

Email: ftwaiq@swinburne.edu.my

Abstract. The recycling of surfactant template is investigated through the reuse of the surfactant template in the mesoporous MCM-41 synthesis process. In the synthesis of MCM-41, tetraethylorthosilicate (TEOS) solution in water was utilized as the silica source while hexadecyltrimethylammonium bromide (CTAB) solution in ethyl alcohol was used as a surfactant template. The synthesized gel is formed thoroughly by mixing the two solutions under acid conditions with a pH value of 0.5 for 1 hour and kept for crystallization for 48 hours. The as-synthesized MCM-41 powder is recovered by filtration while the filtrate (mother liquor) was then reused for the second synthesis cycle. The synthesis procedure was repeated till no further solid product was formed. The synthesized gel was not produced in the unifying solution in the fifth cycle of MCM-41 synthesis. The quality of the calcined MCM-41 powder produced in each synthesis cycle was evaluated by calculating the amount of MCM-41 produced and the surface area of the powder product. The result showed that 1.28, 0.37, 1.64, 1.90 and 0.037 g were obtained in the 1st, 2nd, 3rd, 4th and 5th synthesis cycle, respectively. The surface area of the powder produced was found to be 1170, 916, 728, and 508 m^2/g for 1st, 2nd, 3rd and 4th respectively. The concentration of the surfactant template has reached value lower than the critical micelle concentration (CMC) and remained constant after the 4th cycle. There was no further formation of gel due to low availability in the interaction between silicate anions and surfactant cations when the amount of TEOS was fixed for every synthesis cycle.

1. Introduction

Mesoporous silica materials were synthesized and found with great potential as catalysts and adsorbents in many chemical industrial applications because of their thermal stability, high specific surface area and high pore volume [1]. MCM-41 is a hexagonal mesophase material that is synthesized using silica source and a surfactant template which is then removed to form the structured mesoporous materials.

Two methods were reported for the synthesis of MCM-41 that are hydrothermal synthesis and liquid templating methods. The first method used normal heating or microwave heating in an autoclave to synthesize the mesoporous material at high temperature under alkaline condition in the presence of surfactants for up to 72 hours. The synthesis consists of two methods of treatment that are direct hydrothermal synthesis or post-hydrothermal synthesis [2]. The second method synthesized the MCM-41 using cationic surfactants as template at different surfactant/silica molar ratios [1]. The synthesis is conducted under a basic or an acidic condition by initiating the aggregation of surfactant micelles at below the critical micelle concentration (CMC) such as 0.0009 mol·L^{-1} in water system, 0.022 mol·L^{-1} in ethanol-water system (95 vol% ethanol) and 0.24 mol·L^{-1} in ethanol system (100 vol% ethanol) [3]. The silica source was condensed around the surfactant template to form silica walls while the surfactant template was then removed from the mesopores [4]. The use of silica source, includes sodium silicate (Na_2SiO_3) and fumed silica were commonly used in hydrothermal synthesis, whereas TEOS was used in the liquid templating method. Three types of surfactant such as the anionic, cationic and neutral surfactant are required in the synthesis of MCM-41. Anionic surfactants consist of high potent detergency and lower cost of carboxylate, sulfate, sulfonate and phosphate. The typical anionic surfactant used as the surfactant template for mesoporous silica material synthesis is sodium dodecylbenzenesulfonate (SDBS), sodium dodecylsulfate (SDS) and lauric acid sodium salt (LAS) [5]. The neutral surfactant is the surfactant contains neutral primary amine with structural formula of $C_nH_{2n+1}NH_2$ such as dodecylamine ($C_{12}H_{27}N$) is generally used as the surfactant template in the synthesis process [6]. Other than anionic and neutral surfactant, the cationic surfactant includes n-alkyltrimethyl ammonium bromides, or chloride with different lengths of alkyl chain (C_{12} to C_{20}). For instance, CTAB is typically used for mesoporous material synthesis as the surfactant template [7].

Conventionally, the removal of surfactant template was carried out by calcination at high temperature that causes the transformation of mesophase, though the surfactant template was effectively burnt off under high temperature [7]. Alternative surfactant template removal methods, including solvent extraction, supercritical fluid extraction (SFE) enhanced with sonication, oxidation by hydrogen peroxide (H_2O_2) or ammonium perchlorate (AP) and ultraviolet light (UV)-ozone treatments were investigated in order to achieve efficient removal of the surfactant template and have less negative effect on the structure of the material [4]. These methods have enhanced the removal process of the surfactant template from the as-synthesized mesoporous material by obtaining a more order structure of material and effectively recovered the surfactant template [8].

Some attempts have been reported in previous studies that solvent extraction is efficient in removing the surfactant template and causes less destructive effect on the mesostructure [8]. The resulting surfactant template extracted is being recovered by refluxing the as-synthesized MCM-41 powder in acidified ethanol solution at 70°C for 30 hours and the filtrate is then undergone distillation at 80 °C, neutralization and recrystallization using acetone/ethanol mixture in order to separate out the ethanol and acid as well as obtaining the surfactant powder for reuse in the next synthesis cycle [9]. Other than that, surfactant template recovered through sublimation under vacuum was investigated. The result showed that 67% of the surfactant template was recovered by sublimating the sample using 60wt% non-surfactant, menthol as a surfactant template in ethanol under vacuum at 50 °C [10]. The menthol solid was collected and reused for the next mesoporous material synthesis. Besides solvent extraction and sublimation, the application of ultrasound irradiation with methanol has greatly improved the efficiency of the removal by disrupting the aggregation of the surfactant micelles in the MCM-41 structure to break into monomers [11]. The monomers, that are the surfactant molecules, are released into the methanol and can be recovered to reuse in further MCM-41 synthesis. Furthermore, supercritical fluid extraction (SFE) was evaluated to be an effective surfactant template removal method with a very high extraction efficiency, enhanced by methanol and water [12]. High surfactant recovery (over 90%) was achieved from mesoporous material such as MCM-48 through the water-enhanced supercritical CO_2 extraction over pressure ranging from 80 to 140 bar at temperature 85-105 °C for 3 h [12]. The extracted aqueous surfactant was recovered from water by stripping, distillation or adsorption [13] and could be reused as a surfactant template for further mesoporous material synthesis. Although these removal methods have increased the efficiency of removal and enhanced the recovery of the surfactant template, however the usage of mother liquor from filtration during the synthesis of MCM-41 that contained the surfactant molecules was not reported. Therefore, this issue urges us to investigate a way to recycle the surfactant in the mother liquor and the quality of the MCM-41 synthesized by utilizing the recycled surfactant template.

In the present paper, the recycling of surfactant template is performed by utilizing the surfactant template in the filtrate to synthesize MCM-41 mesoporous material for consecutive MCM-41 synthesis cycle. The ethanol produced by the TEOS hydrolysis will be utilized as an extraction solvent. The number of reuse will be examined by the amount of calcined MCM-41 produced. The quality of the produced product will be evaluated by surface area analysis and nitrogen adsorption and desorption isotherms.

2. Experimental procedure

2.1. Materials

TEOS ($C_8H_{20}O_4Si$, 98%, Sigma-Aldrich) and CTAB powder ($C_{19}H_{42}BrN$, R&M) were used as silica source and surfactant template respectively. The commercially available nitric acid (HNO_3, 65%, Fisher) was used as catalyst in hydrolysis process and pH adjustment. Ethanol (C_2H_5OH, 95%, DULAB) was used as solvent.

2.2. Synthesis of MCM-41

TEOS with total volume of 5 mL was initially dissolved in 15 mL of distilled water refered to as Solution 1 (S1). While, 3.17 g of CTAB was added to ethanol and the solution was named Solution 2 (S2). S1 and S2 were stirred vigorously for 30 min. After stirring, the mixture was tested to evaluate the concentration of surfactant template, ethanol and TEOS using PerkinElmer LAMBDA 35 UV/Vis spectrophotometer. The mixture was kept for homoginization for 1.5 h. 2 mL of nitric acid (HNO_3) was added and the mixing was continued for an hour. The pH was monitored and the mixture was kept for 48 h without mixing for gel formation. The resulting gel was then filtered to separate the resulting powder (as-synthesized MCM-41) and the mother liquor. The as-synthesized powder was further calcined at 550 °C for 8 h. The mother liquor was reused for the 2nd synthesis cycle by adding 5 mL TEOS. The procedure for successive MCM-41 synthesis batches was continued until no significant powder is produced. Figure 1 illustrates the complete synthesis and surfactant template reuse cycles.

2.3. Characterization

The synthesized MCM-41 powder from first (1st) MCM-41 synthesis cycle to fourth (4th) MCM-41 synthesis cycle were characterized using nitrogen (N_2) adsorption-desorption isotherms by performing ASTM D3663-03(2008) test method using Autosorb Quanta Chrome 6B gas sorption analyzer under vacuum at 77.35 K. The Brunauer-Emmett-Teller (BET) analysis was used to determine the specific surface area. The pore size and the pore volume of the MCM-41 powder produced in the respective synthesis cycles were determined using Barrett-Joyner-Halenda (BJH) method. Besides, scanning electron microscopy (SEM) was carried out using scanning electron microscope model FEI Quanta 200 FESEM with accelerating voltage ranging from 2 kV to 30 kV to characterize the morphology of the synthesized MCM-41 powder.

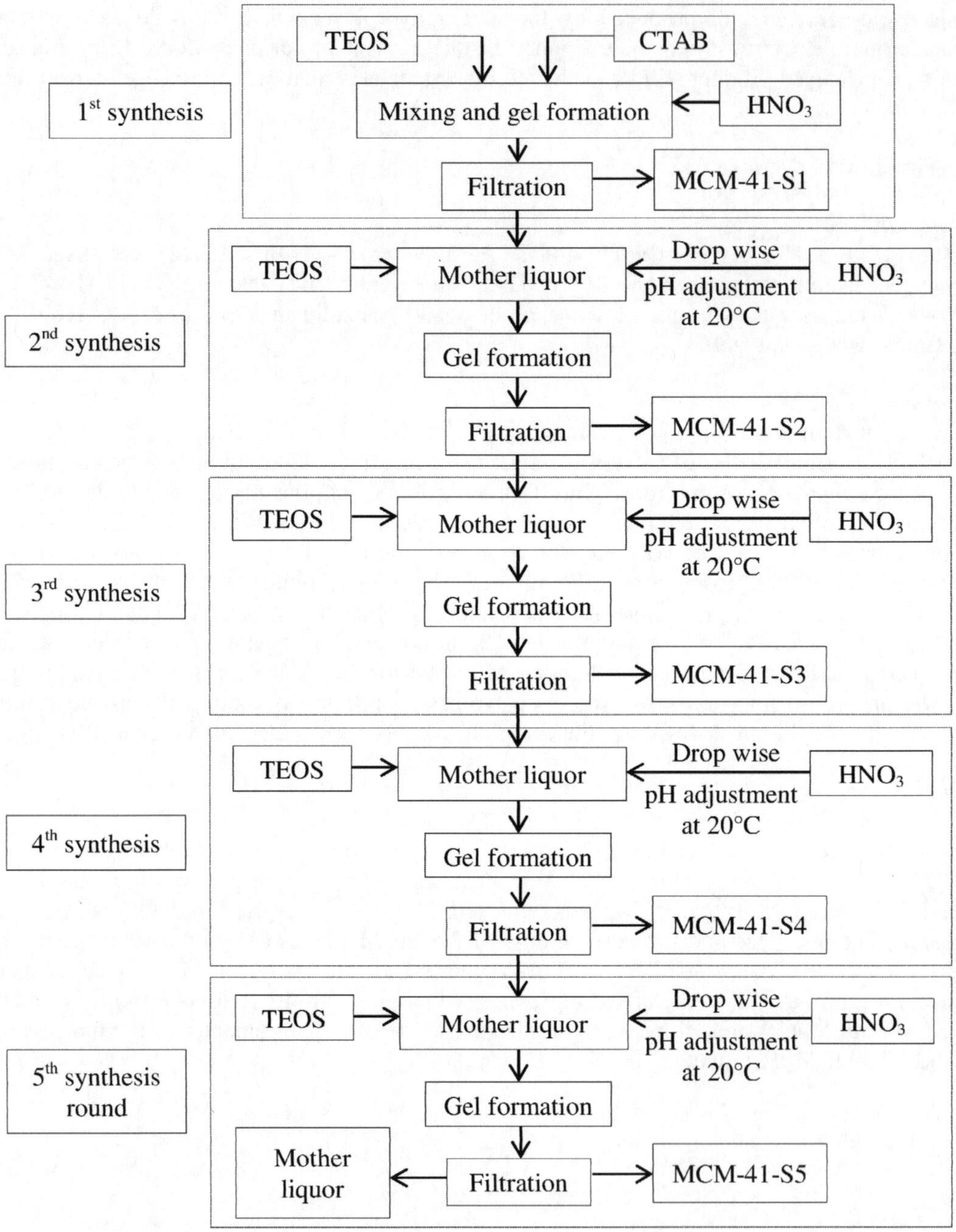

Figure 1. Illustration of MCM-41 synthesis and mother liquor reuse cycles.

3. Result and discussion

3.1. Relative yield produced

In the previous studies as stated in section 1, the synthesis of MCM-41 was performed up to three subsequent batches through the recycling of the extracted CTAB powder [9]. In comparison to the present work, the surfactant template in the filtrate was found to be utilized up to four synthesis cycles of MCM-41 synthesis. The relative amount of the calcined white powder produced in each synthesis cycle is shown in Table 1. As in observed the relative yield of calcined white powder tabulated in Table 1, a large quantity of calcined white powder was being produced in 1st, 3rd and 4th MCM-41 synthesis cycles than in 2nd and 5th synthesis cycle. This can be explained that the effect of CTAB concentration influenced the formation of synthesis gel. In the MCM-41 synthesis experiment, more gel formation was observed in 1st, 3rd and 4th synthesis cycle when the concentration of CTAB increases that becomes the molar ratio of CTAB/ethanol for the respective three cycles were 0.044, 0.027 and 0.026 as shown in Table 1. As compared to 2nd cycle, the resulting gel formation observed in the 2nd synthesis cycle was lesser than 1st, 3rd and 4th cycle when the molar ratio of CTAB/ethanol was 0.025. When the CTAB concentration steadily decreased until it reaches constant, neither gel or powder was formed in the 5th synthesis cycle. This is because an increase in CTAB concentration resulted more interaction between surfactant cations and silicate anions which increases the rate of silicate polymerization and causes more gel formation to occur when the amount of TEOS was fixed for every synthesis batch [14]. Therefore, a total mass of 0.37 g and 0.037 g calcined white powder were obtained in the respective MCM-41-S2 and MCM-41-S5 which was lesser than the resulting 1.28 g MCM-41-S1, 1.64 g MCM-41-S3 and 1.90 g MCM-41-S4.

Besides, the formation of MCM-41 was found to occur at low concentration of surfactant [14]. As shown in Table 1, low molar ratio of CTAB/ethanol was resulted for all samples. Therefore, the calcined white powder in all samples exhibited the characteristic type of MCM-41 mesoporous material with hexagonal shape in nature.

Table 1. Experimental results of MCM-41 synthesis.

Sample Name	Cycle No.	Mass of Calcined MCM-41 (g)	TEOS/ Ethanol Ratio	CTAB/ Ethanol Ratio
MCM-41-S1	1st	1.280	1.2	0.044
MCM-41-S2	2nd	0.370	0.6	0.025
MCM-41-S3	3rd	1.640	0.6	0.027
MCM-41-S4	4th	1.900	0.6	0.026
MCM-41-S5	5th	0.037	0.6	0.026

3.2. Structural and physical characteristics

The SEM micrograph of MCM-41-S1 is shown in Figure 2. The calcined white powder in MCM-41-S1 showed the smooth surface morphology composed of an array of uniformly regular spherical shaped particles. Thus, the sample of MCM-41-S1 was identified as the white hexagonal shape of mesoporous material. Since the synthesis procedure repeated the same throughout the five cycles except different amount of HNO_3 was added when the pH was adjusted to 0.5, hence the calcined white powder of MCM-41-S2 and MCM-41-S4 was assumed to have the same characteristics as MCM-41-S1 except MCM-41-S3, which exhibited irregular shape particles.

Figure 2. SEM images of MCM-41-S1.

The specific surface area, pore volume and pore size calculated using the BET method for the respective MCM-41-S1, MCM-41-S2, MCM-41-S3 and MCM-41-S4 were tabulated in Table 2. The concentration of CTAB also affected the surface morphology of calcined white powder for four samples. An increase in the CTAB concentration causes more spherical particles to form [14]. The highest molar ratio of CTAB/ethanol in sample MCM-41-S1 as observed in Table 1 has resulted the formation of spherical shape calcined powder with high surface area of 1170 m^2/g. As compared to the other three samples such as MCM-41-S2, MCM-41-S3 and MCM-41-S4, less spherical calcined powder with their respective specific surface area of 916 m^2/g, 728 m^2/g and 508 m^2/g were formed respectively when the molar ratio of CTAB/ethanol was found steadily decreased and remained nearly constant.

The isotherms of MCM-41-S1, MCM-41-S2, MCM-41-S3 and MCM-41-S4 are shown in Figure 3. MCM-41-S1 showed the type I (b) isotherms whereas MCM-41-S3 has the typical type II isotherms according to IUPAC classification. For MCM-41-S2 and MCM-41-S4, type IV isotherms based on the IUPAC classification [15] are illustrated in Figure 3. The isotherms of MCM-41-S1 stated that the calcined white powder produced in 1st MCM-41 synthesis cycle has a narrow mesopore size with 1.96 nm while the calcined white powder in the consecutive 2nd, 3rd and 4th synthesis cycles indicated mesoporous materials with larger pore size with value of 2.89 nm, 5.02 nm and 3.41 nm respectively. The calcined white powder produced in all four consecutive synthesis cycles has the pore diameter within the mesoporous size ranging from 2 to 50 nm [16]. Therefore, the calcined white powder in the sample MCM-41-S1, MCM-41-S2, MCM-41-S3 and MCM-41-S4 were identified as mesoporous material.

At a high relative pressure, a hysteresis loop with steep parallel adsorption and desorption isotherms was illustrated in the isotherms of MCM-41-S1 which was indicating the calcined white powder produced in the 1st synthesis cycle has a uniform pore structure. Whereas, the isotherms of MCM-41-S2 and MCM-41-S4 showed a smooth adsorption step with sharp desorption step in the hysteresis loop at higher relative pressure. This can be explained that the calcined white powder obtained from 2nd and 4th cycle of MCM-41 synthesis were mesoporous materials with non-uniform pore sizes. As compared to the first three isotherms, type H3 of hysteresis loop was shown in the isotherms of MCM-41-S3 at high P/P_0 according to IUPAC classification. The calcined white powder produced in the 3rd MCM-41 synthesis cycle consists of an agglomeration of plate-like particles with irregular slit-shaped mesopore structure [15]. Agglomeration that resulted due to the altering pH value in the mixture causes the enlargement of particle size and the shape alteration to occur in sample MCM-41-S3 without changing the surface area of particles [17]. Therefore, the resulting calcined white powder obtained in sample MCM-41-S3 has larger size of pores and larger pore volume with irregular shape particles than the other three samples as shown in Table 2.

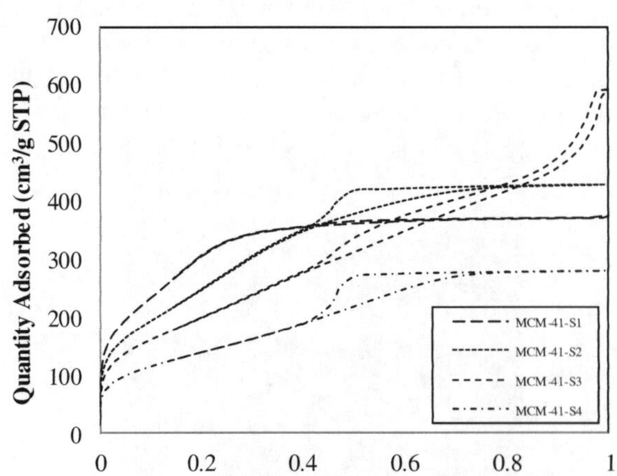

Figure 3. N_2 adsorption-desorption isotherms.

Table 2. Physical characteristics data.

Sample Name	Specific Surface Area (m^2/g)	Pore Volume (cm^3/g)	Pore Size (Å)
MCM-41-S1	1170	0.57	19.6
MCM-41-S2	916	0.66	28.9
MCM-41-S3	728	0.91	50.2
MCM-41-S4	508	0.43	34.1

4. Conclusions

In this current work, the recycling of surfactant was achieved up to five MCM-41 synthesis cycles. The effect of CTAB concentration on the gel formation and the calcined white powder surface morphology was studied. The resulting calcined white powder with a total mass of 1.28 g, 1.64 g and 1.90 g were obtained in the respective 1st, 3rd and the 4th synthesis cycle when the concentration of CTAB increases to reach the molar ratio of CTAB/ethanol with a value of 0.044, 0.027 and 0.026. Whereas, 0.37 g and 0.037 g of calcined white powder were formed in the 2nd and 5th cycle due to the decrease in molar ratio of CTAB/ethanol that was 0.025 in 2nd cycle as compared to 1st, 3rd and 4th cycle while remained constant after the 4th cycle that causes no further formation of gel to occur in the 5th cycle. The formation of spherical shape mesoporous material was resulted in 1st, 2nd and 4th synthesis cycle with their respective specific surface area value of 1170 m^2/g, 916 m^2/g and 508 m^2/g except MCM-41-S3. Agglomeration occurred in the 3rd cycle causes MCM-41-S3 exhibited irregular slit-shaped mesopore structure with a surface area of 728 m^2/g, a larger pore size and a larger pore volume. Both SEM result and the measured molar ratio of CTAB/ethanol confirmed that the calcined powder for all five samples possessed the MCM-41 characteristics and hexagonal shape in nature.

Acknowledgement

The Research Fund by Swinburne University of Technology Sarawak Campus (SSRG 2-5513) is acknowledged. The authors thank NANOCAT, University Malaya for sample characterization.

References

[1] Barrera D, Villarroel-Rocha J, Marenco L, Oliva M I and Sapag K 2011 Non-hydrothermal synthesis of cylindrical mesoporous materials: influence of the surfactant/silica molar ratio *Adsorption Science & Technology* **29** ed Müller E A (UK: Multi-Science Publishing Co Ltd) pp 975-988

[2] Xu R, Pang W, Yu J H, Huo Q S and Chen J S 2009 Synthesis and structure *Chem. Zeolites and Relat. Porous Mater.* (New Jersey: Wiley) pp 533-535

[3] Li W, Han Y C, Zhang J L and Wang B G 2005 Effect of ethanol on the aggregation properties of cetyltrimethylammonium bromide surfactant *Colloid J.* **67** 159-63.

[4] Twaiq F, Nasser M S and Onaizi S 2014 Effect of the degree of template removal from mesoporous silicate materials on their adsorption of heavy oil from aqueous solution *Front. Chem. Sci. Eng.* **8** 488-97.

[5] Steen E V, Callanan L H and Claeys C 2004 Recent advances in the science and technology of zeolites and related materials *Proc. 14th Int. Zeolite Conf.* Cape Town South Africa 25-30th April 2004 (Amsterdam: Elsevier).

[6] Twaiq F, Nasser M, Al-Ryiami S and Al-Ryiami H 2012 Performance of mesoporous organosilicates on the adsorption of heavy oil from produced water *Int. Conf. on Fundamental and Applied Sciences 2012:(ICFAS2012)* **1482** 579-584.

[7] Du P D, Khieu D Q and Hoa T T 2013 Removal of organic template from mesoporous MCM-41 *Tạp chí Khoa học Đại học Huế* **69.**

[8] Jabariyan S and Zanjanchi M A 2012 A simple and fast sonication procedure to remove surfactant templates from mesoporous MCM-41 *Ultrason. Sonochem.* **19** 1087-93.

[9] Ng E P, Goh J Y, Ling T and Mukti R 2013 Eco-friendly synthesis for MCM-41 nanoporous

IOP Conf. Series: Materials Science and Engineering **206** (2017) 012044 doi:10.1088/1757-899X/206/1/012044

materials using the non-reacted reagents in mother liquor *Nanoscale Res. Lett.* **8** 1-8.

[10] Mukherjee I, Chen J, Yin H and Wei Y 2011 Use of sublimation for template removal from menthol templated mesoporous organosilica prepared via sol–gel process *Powder Technol.* **206** 214-17.

[11] Zanjanchi M A and Jabariyan S 2014 Application of ultrasound and methanol for the rapid removal of surfactant from MCM-41 molecular sieve *J. Serb. Chem. Soc.* **79** 25-38.

[12] Huang Z, Luan D Y, Shen S C, Hidajat K and Kawi S 2005 Supercritical fluid extraction of the organic template from synthesized porous materials: effect of pore size *J. Supercrit. Fluids* **35** 40-8.

[13] Scovazzo P, Chen W Y, Wang L and Shammas N 2006 Solvent extraction, leaching and supercritical extraction *Advanced Physicochemical Treatment Processes* ed Wang L, Hung Y T and Shammas N (New York: Humana Press).

[14] Ahmed S and Ramli A 2011 Effect of surfactant concentration on the physico-chemical characteristics of mesoporous molecular sieve *J. Appl. Sci.* **11** 1178-84.

[15] Thommes M, Kaneko K, Neimark A V, Olivier J P, Rodriguez-Reinoso F, Rouquerol J and Sing K 2015 Physisorption of gases, with special reference to the evaluation of surface area and pore size distribution (IUPAC Technical Report) *Pure Appl. Chem.* 1-19

[16] Othman A L and Zeid A 2012 A review: fundamental aspects of silicate mesoporous materials *Mater.* **5** 2874-902.

[17] Forsch D 2013 *Nanomaterials* (Germany: Wiley-VCH).

IOP Conf. Series: Materials Science and Engineering **206** (2017) 012045 doi:10.1088/1757-899X/206/1/012045

Mechanical properties of unidirectional oil palm empty fruit bunch (OPEFB) fiber reinforced epoxy composite

C S Hassan[1,*], C W Yeo[1], B Sahari[2], M S Salit[2] and N Abdul Aziz[2]

[1]Mechanical Engineering Department, UCSI University, Cheras 56000, Kuala Lumpur, Malaysia
[2]Department of Mechanical and Manufacturing Engineering, Universiti Putra Malaysia,
Serdang 43400, Selangor, Malaysia

*E-mail: suhana@ucsiuniversity.edu.my

Abstract. Natural fibers have proven to be an excellent reinforcement material for various polymers. In this study, OPEFB fiber with unidirectional alignment was incorporated in epoxy and an investigation on tensile and flexural characteristics of the composite has been carried out. A fiber surface modification utilizing alkaline treatment with 1% sodium hydroxide solution was used in order to increase the fiber matrix bond in the composite. The investigation was carried out for 0°, 45° and 90° fiber orientation. Result showed that the higher the angle of the fiber orientation, the higher the tensile strength and flexural strength the composite will yield.

1. Introduction

Natural-fiber-reinforced polymer composites, or termed as bio-composites, have been gaining interest over the years due to the increases in environmental awareness and sustainability concept. Biocomposites materials consist of natural fibre embedded in a matrix. In bio-composites, fibres are the principal load carrying members, while the surrounding matrix keeps them in the desired location and orientation. Matrix also acts as a load transfer medium between the fibres, and protects them from environmental damages due to elevated temperatures, humidity and corrosion.

Malaysia has about 3.5 million hectares of oil palm producing annually over 10 million tons of crude palm oil. However, crude palm oil and its economic products such as palm kernel oil and palm kernel cake constitute only 10% of the crop, leaving the rest of the biomass to waste. This biomass also includes OPEFB fiber, being the residual bunch after removal of the fruit constituting 20% to 22% of the weight of the fresh fruit bunches (Ridzuan et al. 2002). Hence, utilization of OPEFB fiber not only create value added product but will also solve disposal problem.

Various researches have been conducted to produce OPEFB fiber reinforced polymer. (Ewulonu & Igwe, 2012) studied the properties of OPEFB fibre filled high density polyethylene by varying the filler content between 0 to 1.5 weight percent. Three particle size of OPEFB fibre were utilized, that is 0.15 mm, 0.212 mm and 0.3 mm. In addition, the effect of compatibilizer, Maleic anhydride-gpolyethylene (MAPE), on the composite properties also investigated. It is reported that for all three particle size, the tensile strength of the composite were found to be decreases with the increases of filler content. This is attributed to the poor bonding between the OPEFB fibre and the matrix, which might be due to the physical structure of the OPEFB fibre. The incorporation of MAPE was found to be able to improve the tensile strength of the composite. It is believed that there is reaction between maleic anhydride molecules from MAPE with the hydroxyl groups in the fibre which resulted in better bonding between the fibre and the matrix.

(Ngo W.L. et. al 2014) utilized OPEFB fibre as reinforcement filler for polylactic acid (PLA) and polystyrene (PS). The length and diameter of the OPEFB fibre used is in the range of 6-10 mm and

IOP Conf. Series: Materials Science and Engineering **206** (2017) 012045 doi:10.1088/1757-899X/206/1/012045

60100μm, respectively. The composites were fabricated using conventional vacuum press moulding with fibre loading of 20%, 40% and 60%. The composites were subjected to tensile and flexural test. It is reported that the tensile and flexural strength of the OPEFB fibre reinforced PLA composite increases with the increases of fibre loading up to 40%, however decreases with further increases of fibre loading. The high content of fibre would tend to aggregate in the composite which weaken the bonding between the fibre and matrix. Meanwhile, the utilization of OPEFB fibre to reinforce PS has resulted in even lower tensile and flexural strength as compared to the pure PS.

However, present researches are limited to the usage of discontinuous OPEFB fiber as the reinforcement material. Utilization of continuous OPEFB fiber to reinforced polymer has yet to be studied. Discontinuous-fiber composite are normally random in alignment, while continuous-fiber generally have preferred orientation. The random alignment of the discontinuous-fiber dramatically reduces their strength and modulus (Tanwer 2014).

One method of reinforcing as continuous-fiber is by unidirectional fiber alignment. Unidirectional fibers could be distinguished between its reinforcement angles. The reinforcement angles vary in the direction of load applied. The composites are said to be reinforced with 0° fiber alignment if the composite is loaded in the direction parallel to its fiber, or to be reinforced with 90° fiber alignment if the composite is loaded in the direction perpendicular to its fiber, as shown in figure 1. It is believed that the unidirectional fiber alignment would be able to provide the greatest strength in the direction of the fibers.

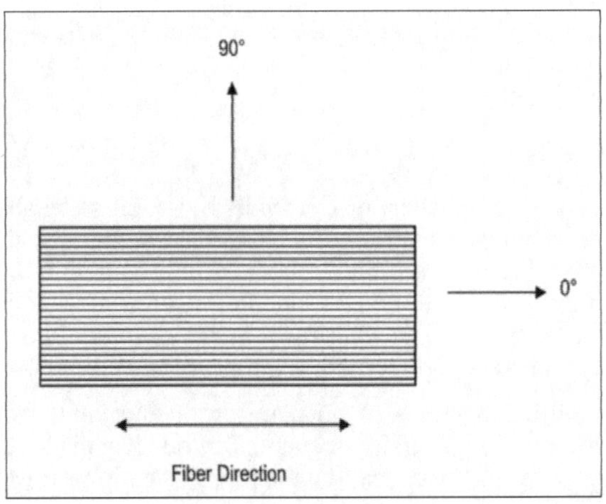

Figure 1. Unidirectional continuous-fiber angle definition (Campbell 2010).

In the applications where stiffness and strength is required, the unidirectional reinforcement of natural fibre is preferable. Many researches have been carried out utilizing unidirectional natural fibre to reinforce polymer in order to generate a stiffer yet lightweight material. As example, (Aldousiri, 2013) has fabricated composite composed of date palm fibre and recycled high density polyethylene (HDPE). Tensile behaviour of such composites were tested and analysed. The composite samples, pure HDPE and HDPE + 6% date palm fibre, were prepared using hot press technique. The date palm fibres were aligned in 0° alignment. It has been found that the incorporation of date palm fibre into the recycled HDPE with 0° fibre alignment has resulted in increases of tensile strength from 28 MPa to 38 MPa. Minimal fibre pull-out was observed through scanning electron microscope (SEM) which indicates that the fibre, when oriented parallel with the load applied, has effectively preserved the ductility of the HDPE.

The objective of this paper is to investigate the effect of fiber orientation on the tensile strength and flexural strength of the OPEFB fiber reinforced epoxy composite. It is expected that the unidirectional composite with 0° fiber alignment to be stronger and stiffer as compared to composite with other fiber alignment angles. As fiber orientation changes from 0° to 90°, the properties of the fibers decline and the properties of the matrix dominate (Hedge 2015), hence, the load will be carried by the much weaker polymeric matrix (Campbell 2010).

2. Methodology

2.1. Fiber preparation

Fibers were washed in distilled water to remove all sorts of impurities such as residue oil as well as other large particles and were allowed to dry for 20 hours in room temperature. The fibers were then soaked in a 1% sodium hydroxide solution for 1 hour and were allowed to dry for another 24 hours in room temperature. The fibre treatment was performed in order to roughen the fiber surface which subsequently will contribute to a better interfacial bonding with the matrix.

The dried fibers were then straightened using a comb and by applying pressure at both ends of the fiber while placing them in straight position for approximately 8 hours duration.

2.2. Composite preparation

Epoxy and hardener were mixed in a ratio of 10:1, respectively, to form matrix. The fibers were aligned manually on the mold prepared prior to this process. The mixture of epoxy and hardener were then poured into the mould with OPEFB fibers. The mould was closed with designed lid and small pressure was applied to prevent air bubbles during the curing process. The composite were allowed to cure for 24 hours duration.

The composites were then cut based on the ASTM D638 Type I standard and ASTM D790 for tensile testing and flexural testing, respectively. Five composite samples were prepared for each fiber orientation; 0°, 45° and 90°.

2.3. Tensile test

The obtained tensile specimens were tested using INSTRON 3366 10kN universal testing machine at a crosshead of 5 mm/sec and the data is recorded down.

2.4. Flexural test

The obtained flexural specimens were tested using INSTRON 3366 5kN universal testing machine at a crosshead of 2 mm/sec and the data is recorded down.

3. Results and Discussion

Figure 2 and figure 3 shows the stress strain curve obtained through tensile testing of the OPEFB fibre reinforced epoxy composite for 0° and 90° fibre orientation angle, respectively. Stress strain curve measure the relationship between the magnitude of the applied stress on a material and the resulting strain or elongation. Both stress strain curve shows nonlinear response, indicating there are reductions in modulus of elasticity before fracture. These reductions of modulus of elasticity reflect drops in stiffness of the composite. 0° fibre orientation angle exhibits brittle failure mode while 90° fibre orientation angle exhibits nonlinear behaviour before complete fracture. After the peak strength, both fibre and matrix are damaged and cracks are thoroughly connected. As can be seen from figure 4, the fractured surface of composite with 0° fibre orientation angle is rather smooth as compared to the composite with 90° fibre orientation angle which is brushier and dislodging.

IOP Conf. Series: Materials Science and Engineering **206** (2017) 012045 doi:10.1088/1757-899X/206/1/012045

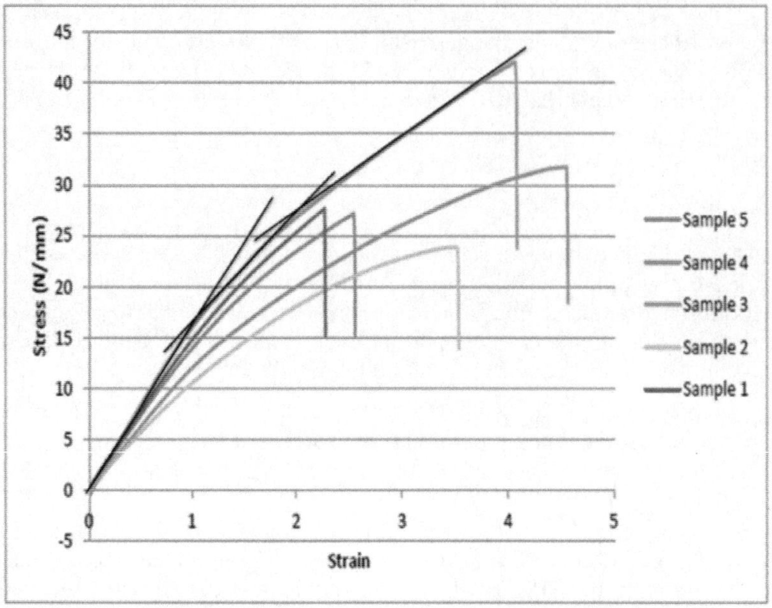

Figure 2. Tensile Stress Strain Curve of the Unidirectional OPEFB Fiber Reinforced Epoxy Composites with 0˚ Fibre Orientation Angle

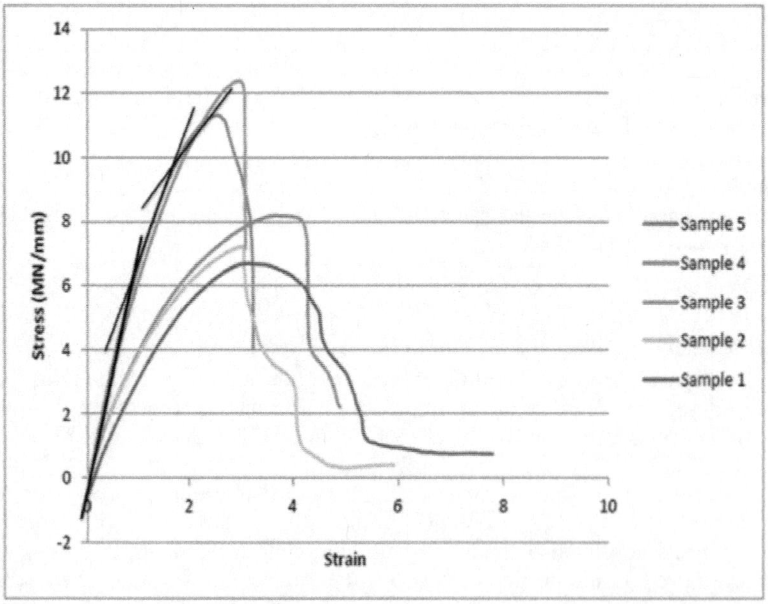

Fig ure 3. Tensile Stress Strain Curve of the Unidirectional OPEFB Fiber Reinforced Epoxy Composites with 90˚ Fibre Orientation Angle

IOP Conf. Series: Materials Science and Engineering **206** (2017) 012045 doi:10.1088/1757-899X/206/1/012045

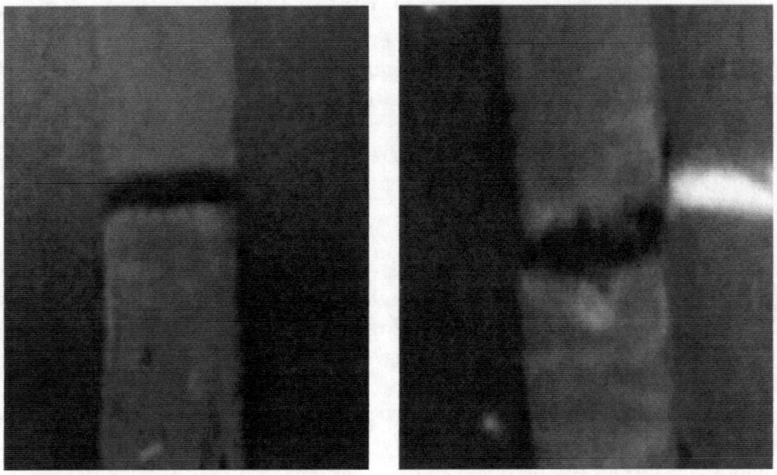

(a) (b)

Figure 4. Fractured Surface of the Unidirectional OPEFB Fiber Reinforced Epoxy Composites with (a) 0° Fibre Orientation Angle, and (b) 90° Fibre Orientation Angle.

Figure 5 shows tensile strength of the unidirectional OPEFB fiber reinforced epoxy composite. It can be seen that the tensile strength increases when the angle of the fiber orientation increases. The trend is similar to the finding by (Aldousiri, 2013) as elaborated earlier. It is observed that for all fibre orientations under tensile load, the composites have failed in shear which involving interfacial debonding and fibre pull out.

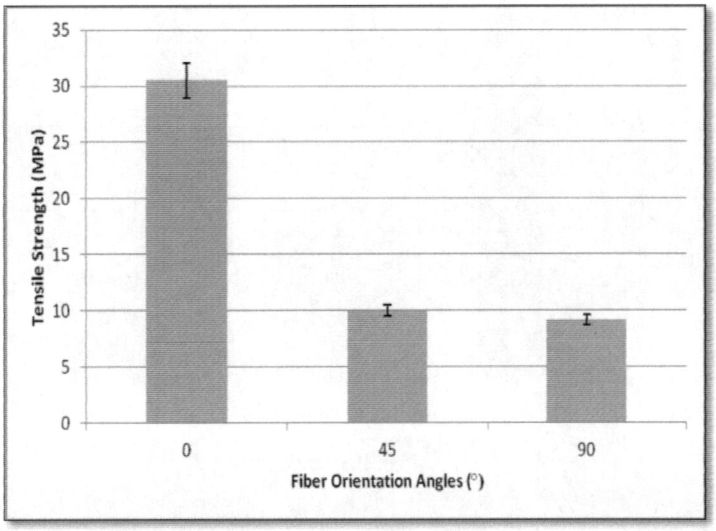

Figure 5. Tensile Strength of the Unidirectional OPEFB Fiber Reinforced Epoxy Composites

Fiber orientation at angle of 0° has the highest tensile strength value of 30.5 MPa followed by fiber orientation angle at 45° with value of 10 MPa. Fiber orientation at angle of 90° indicated the lowest tensile strength of 9.2 MPa. This shows that the composite with fiber oriented at 90° are weakest when the tension of force pull perpendicular to the fiber direction and it takes shorter time to deform and eventually break apart. The transversely oriented fibers acts as a barrier and prevents the distribution of stresses thorughout the matrix and this in turn causes higher concentration of localized stresses (Manikandan et al. 1996). This explains the reduction in tensile properties of the composite at 45° and 90° oriented fiber composite. Contrary, when fibers are aligned longitudinally and in the direction of the force applied, the properties of the fibers will be reflected significantly (Murray 1997). As such, the strength of the OPEFB fiber oriented in the longitudinal direction has produced a higher tensile strength for the composite.

The decreases in tensile strength of the OPEFB fiber reinforced epoxy composite is about 67% when the orientation of fiber in the composite changed from 0 to 45 . eanwhile, the composite with 45 fiber orientation angle exhibited tensile strength higher by around 20 as compared to the composite with 90 fiber orientation. This finding is also in agreement with study done by other researcher. (de Olieveira et. al 2012) studied behaviour of isophthalic polyester matrix composites reinforced with unidirectional curaua fibers. The author reported that the strength of the composites experiences gradual depreciation when the reinforcement angle increased. There was a decrease of nearly 60 in strength when fiber oriented at 15 orientation as compared to the 0 orientation angle and another decreases of around 30 when the fiber changed from 15 to 30 .

Figure 6 shows the flexural strength of the unidirectional OPEFB fiber reinforced epoxy composite. It can be seen that the difference in the fiber orientation angles affect the flexural strength of the composites. Fiber orientation at angle of 0° has the highest flexural strength value of 213.04 MPa followed by fiber orientation angle at 45° with value of 89.38 MPa. Fiber orientation at angle of 90° indicated the lowest flexural strength of 63.18 MPa.

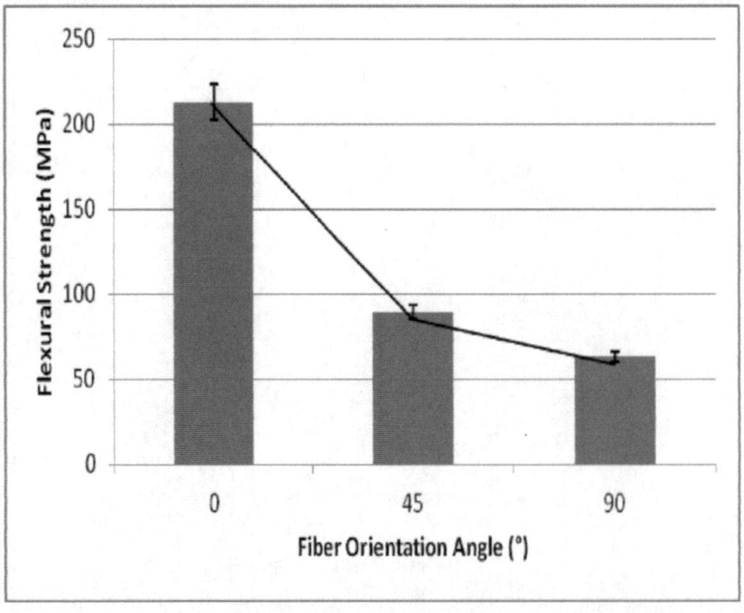

Figure 6. Flexural Strength of the Unidirectional OPEFB Fiber Reinforced Epoxy Composites

Fibers that are diverging from longitudinal axis resulted in stress transmission to the matrix. Hence composites with fibers parallel to the load are capable to withstand loads better than those with obliquely oriented fibers (Callister 1997).

4. Conclusions

Tensile and flexural test has been conducted on the OPEFB fiber reinforced epoxy composites with 0°, 45° and 90° fiber orientation. It has been found that varying fiber orientation angle affect the properties of the composite. Fiber aligned in the direction of force applied, 0° orientation, showed higher tensile and flexural properties as compared to the fiber aligned in 45° and 90° orientation.

Acknowledgments

The authors would like to thank the UCSI University engineering faculty lab tutors and Mr. Muhammad Wildan Ilyas for their assistance and support.

References

[1] Aldousiri B, Alajmi M and Shalwan A 2013 Mechanical properties of palm fibre reinforced recycle HDPE *Adv. in Mat.Sci. and Eng.* 508179
[2] Callister Jr WD 1997 *Materials science and engineering: an introduction. Composites* 3rd ed (New York: Wiley) chapter 17 p 513
[3] Campbell FC 2010 *Struct. Composite Mat.* **9** ASM International USA
[4] de Oliveira FH, Helfer AL and Amico SC 2012 Mechanical Behavior of Unidirectional Curaua Fiber and Glass Fiber Composites 91-92
[5] Ewulonu CM and Igwe IO 2012 Properties of Oil Palm Empty Fruit Bunch Fibre Filled High Density Polyethylene. *Eng. and Techno.* **3**(6) 458–471
[6] Hedge A, Darshan RS, Mulla F, Shoeb M and Rajanish M 2015 Tensile properties of unidirectional glass/epoxy composites at different orientations of fibers, *Int. J. of Eng. Res. and App.* **5**(3) 150-153
[7] Manikandan Nair KC, Diwan SM and Thomas S 1996 Tensile properties of short sisal fibre reinforced polystyrene composites *J. Appl. Polym. Sci.* **60** 1483
[8] Murray GT 1997 *Handbook of materials selection for engineering applications*, Marcel (Dekker Inc. New York)
[9] Ngo WL, Pang MM and Yong LC KYT 2014 Mechanical Properties of Natural Fibre (Kenaf, Oil Palm Empty Fruit Bunch) *Adv. in Environ. Bio.* **8(1)** 2742–2747
[10] Ridzuan R, Shaler S and Jamaludin MA 2002 Properties of medium density fibreboard from oil palm empty fruit bunch fiber *J. Oil Palm Res.* **14**(2) 34-40
[11] Tanwer AK 2014 Mechanical properties of uni-directional and bi-directional glass fibre reinforced epoxy based composite *Int. J. Res. in Advent Technol.* **2**(11) 34-39

IOP Conf. Series: Materials Science and Engineering **206** (2017) 012046 doi:10.1088/1757-899X/206/1/012046

Development of epoxide compound from *kapok* oil for enhanced oil recovery

M K Anam, Supranto, B Murachman and S Purwono*

Chemical Engineering Department, Universitas Gadjah Mada
Jalan Grafika No. 2, Kampus UGM, D.I. Yogyakarta, Indonesia.

*E-mail: spurwono@chemeng.ugm.ac.id

Abstract. Epoxide compound is made by reacting *Kapok* Oil with acetic acid and hydrogen peroxide with *in situ* method. The epoxidation reaction was varied at temperatures of 60 °C, 70 °C and 80 °C, while the time of reaction time was varied at 15 minutes, 30 minutes, 60 minutes and 90 minutes. The reaction rate coefficient for the epoxide was obtained as $\boldsymbol{k} = \mathbf{124{,}82\ exp}\left(\frac{\mathbf{-24{,}14}}{\boldsymbol{RT}}\right)$. The addition of the epoxide compound 0.5% w/w in the formulation of SLS was able to reduce the IFT value up to 9.95 x $10^{-2}$$m$N/ m. The addition of co-surfactant (1-octanol) was varied between 0.1 and 0.4 % of the total mass of the main formulation (SLS + epoxide + water formation). The smallest interfacial tension value is obtained on the addition of co-surfactants as much as 0.2% w/w, with the IFT value is 2.43 x $10^{-3}$$m$N/m. The effectiveness of the chemicals was tested through micro displacement using artificial porous medium. The experimental results show that some chemicals developed in the laboratory can be used as EOR chemicals. The oil displacement experiments show that as much as 20 % to 80 % of remaining oil can be recovered by flooding it with the chemicals. The results also show that the oil recovery depends on type of chemicals and chemical concentration.

1. Introduction

The increasing demand for petroleum-based fuels and petrochemicals, along with the diminishing supply of new sources for oil, has escalated interest in improving the extraction of oil from existing reservoirs. Among the most promising processes is enhanced oil recovery (EOR). Currently, various researches in EOR technology are underway with intensity. One of the related researches that are given much attention is finding of chemicals for optimizing the enhancement of oil recovery. Even though many experiments have been done in this area, the applications in the field rarely succeeded. Expensive chemicals is certainly not the main reason for the failure, but more related to the process as application in the field is not properly guided by processes in the laboratory. There are many problems that need to be solved in the area of EOR basic research to allow results of research to be applied in the field.

From the earliest days, it was recognized that capillary forces caused large quantities of oil to be left behind in well-swept zones of water flooded oil reservoirs. Capillary forces are the consequence of the interfacial tension (IFT) between the oil and water phase that resists externally applied viscous forces and causes the injected and banked-up connate waters to locally bypass oil. Similarly, early efforts of enhanced oil recovery strove to displace this oil by decreasing the oil-water IFT. Though many techniques have been proposed and field tested, the predominant EOR technique for achieving low IFT is surfactant flooding.

IOP Conf. Series: Materials Science and Engineering **206** (2017) 012046 doi:10.1088/1757-899X/206/1/012046

Surfactants can be manufactured from various raw materials that are ultimately derived from petroleum or agricultural feedstock. Currently, most EOR surfactants are based on petrochemical feedstock. In the future a vicious cycle would be created, as the demand for more petroleum-based surfactants would increase proportionally to the demand for more oil for fuel and petrochemicals. To address this challenge, research is currently under way to develop surfactants, suitable for EOR, that are partially or completely derived from renewable agricultural crops. These offer the advantage of providing green, biodegradable, nontoxic, environmentally friendly alternatives that are not dependent on the availability of petrochemicals. There are several possible sources of chemicals for EOR such as lignin extracted from palm oil industrial waste or modification of the chemicals in order to reduce the IFT. One example is Sodium Ligno Sulfonate (SLS) which is useful in decreasing IFT for EOR industry. SLS can be made from Lignin of empty fruit bunches (EFB). However, the value of IFT of this surfactant is not low enough to be used in the reservoir. Surfactants which can be used as Surfactant Flooding should have a value of IFT$\leq 10^{-3}$*m*N/ m, while SLS only have IFT value around 1 - 4 *m*N/m [1].

1.1. Literature review

One of the ways to improve the performance of SLS is modify with some chemicals or adding with some co-surfactants. Improving the ability of SLS in reducing the interfacial tension can be done by adding epoxide compound and 1-octanol. Kapok oil has unsaturated fatty acids (double bond) up to 57%, thereby it has potential as epoxide compound. The bonding of SLS and epoxide compound is formed as a result of the opening of the oxirane ring [2]. The addition of 1-octanol can fill the empty space between the surfactant and reducing repelling effect surfactant [1]. Therefore the combination of epoxide compound and 1-octanol in the formulation of the SLS are expected to improve the quality of surfactants for EOR.

1.1.1. Epoxidation. Epoxidation is defined as a reaction that aims to transform the double bond into epoxies. Epoxidation is an exothermic reaction, therefore it should be avoided the use of a high concentration of peroxide acid. The composition of fatty acid will influence kinetics of epoxidation reaction [2].

1.1.2. Co-Surfactant. Co-surfactant is chemical added to a process to enhance the effectiveness of a surfactant for reducing the surface tension of the liquid mixture. In the oil industry, co-surfactants are often used to increase the oil-solubilising capacity of micro emulsion surfactant systems. So that the layer formation stage between micelles and micro-emulsion can occur continuously. The addition of co-surfactant can produce low IFT because co-surfactant is able to enter into monolayer, thereby changes the curvature or stiffness of surfactant [3].

1.1.3. Kinetics of Epoxidation. Epoxidation is formation reaction oxirane ring of the double bond. There are two main reactions in the epoxidation reaction. First, is the formation of peroacetic acid. Second, is the epoxidation reaction itself.

Step 1 : $AA + H_2O_2 \leftrightarrow PAA$ (1)

Step 2 : $DB + PAA \rightarrow EO + AA$ (2)

Where:

AA= acetic acid EO = epoxide oil
PAA = peroacetic acid DB = double bond

Acetic acid mass balance can be written as:

IOP Conf. Series: Materials Science and Engineering **206** (2017) 012046 doi:10.1088/1757-899X/206/1/012046

$$\frac{d[AA]}{dt} = -k_1[AA][H_2O_2] + k_2[PAA] + k_3[DB][PAA] \tag{3}$$

2. Research Methodology

2.1. Raw Material

Sodium Lignosulfonate (SLS) was obtained from Weifang Longstar Chemical Inc. Shandong, China. *Kapok* oil was obtained from PT. Subali Makmur, Semarang. Co-surfactant: 1-octanol 99% was obtained from Sigma-Aldrich Co. LLC. Saponification materials were NaOH 5 M and Ethanol 96%. Materials for epoxidation are hydrogen peroxide 50% and acetic acid glacial. The catalyst is sulfuric acid 98%. Water injection to the formulation is taken from the Rantau Field PT. Pertamina (Persero).

2.2. Research Procedure

2.2.1. Saponification process. Saponification process is carried out at temperatures 85°C (Campbell, et. al., 2009). The first step, NaOH solution was dissolved in ethanol, and then further reacted with Kapok oil which has been heated before. Its products are soap and glycerol. Separation of the two products is done with the addition of saturated NaCl solution [4].

2.2.2. Epoxidation Stage. Epoxidation process is using in-situ methods, where the reactants are reacted simultaneously. The catalyst used is sulphuric acid (H_2SO_4). Raw material are used is soap, hydrogen peroxide and acetic acid. The ratio of hydrogen peroxide and acetic acid is 1 : 1.4. The Weight of soap is 5 grams. The addition of the catalyst is as much as 2% of the total reactants [5].

2.2.3. Formulation Stage. Raw materials for developing surfactant formula are SLS, epoxies, co-surfactant, and injection water as a solvent. Formulation process was carried out using a hot plate at the temperature of 70 °C with the rotational speed on the scale 2.

2.3. Analysis

2.3.1. Concentration of double bonds. The measurement of iodine value is carried out by Wijs method. The concentration of double bond can be approached with iodine value [6].

$$[DB] = \frac{(B-S).N.12.69}{w} \tag{4}$$

Where:

[DB] = double bond concentration (mol/ml)
N = normality of $Na_2S_2O_3$
B = blank titration (ml)
S= volume titration of sample (ml)
w = weight of sample

The conversion of fatty acids is calculated by equation from [7]:

$$\bar{x} = \frac{[DB]_0 - [DB]_t}{[DB]_0} \tag{5}$$

Where:

x = conversion
$[DB]_0$ = initial concentration and $[DB]_t$ = concentration at time t

2.3.2. Epoxide concentration. The concentration of the epoxide formed is calculated from oxirane oxygen experiment, through the method of [8].

$$[OO]_e = \frac{1.6 . V . N}{w} \tag{6}$$

Where:

$[OO]_e$	= oxirane oxygen experiment;	N	= normality of HBR
V	= HBr solution for titration;	w	= mass of sample (gram)

2.4. Experimental set-up for flooding

The effectiveness of the chemicals was tested through micro displacement using artificial porous medium. The flow rates of oil, water and surfactant were controlled using pressurized gas. The porous media used in the oil displacement process was prepared from stainless steel cylinder and contain fine crushed core from Rantau formation, Aceh. It is proposed to be an artificial reservoir with dimension about 3.75 cm ID x 10.8 cm L. The porosity of the porous media was 0.4461, the density was 3.24 g/cm^3, and the permeability was 107 mD; and it was water wet. The cores were crushed to 30 / 60 mesh size which resulted in a mean particle size of 2.41 x 10^{-4} m in diameter. Porosity of a rock depends on the pore or grain size distribution, in which the small grains can fill the space between larger grains. Porosity is a strong function of grain size distribution (sorting).

The procedure is as follows: initially the reservoir model was filled with brine (1 to 3 % salt concentration) until it was 100 % saturated. Then, to represent oil migration, oil was injected into the medium until minimum water saturation (S_{wc}) of about 30 % is reached. After this, the medium was flooded by the same brine until minimum oil saturation, S_{or}, which was about 10 %. The oil remaining in the reservoir after this water flood was then subjected to various injections of different chemicals for further oil recovery.

3. Result and discussion

3.1. Effect of Temperature and Reaction Time on Epoxidation

Effect of temperature and reaction time were observed by comparing the results of the initial conversion double bond and conversion double bond at the end of the process. Comparison of temperature and reaction time on the conversion values can be found in Figure 1.

The increase in conversion indicates that the epoxidation reaction goes toward the product. Increasing reaction time will increase the conversion rates of the double bond into an epoxide compound. However increase in reaction time would also lead to the degradation of the oxirane ring [9]. At temperature 70 °C, the increase occurred at the beginning of the process of conversion. The conversion reaches a maximum value at the time of the reaction run for 30 minutes, and after that, the conversion decreases slowly. This is caused by the formation of by products from the reaction of epoxidation. According to [9], at temperatures above 65 °C, it will decrease the epoxide compounds due to opening of the oxirane ring. At temperatures of 80 °C, it is shown the decrease of conversion with increasing reaction time. This situation is due to a side reaction during the process of epoxidation.

3.2. Kinetics of Epoxidation

Based on Equation 5, the relationship $\ln\{[H_2O_2]_0 - [EO]\}$ = ln X and reaction time for any temperature can be plotted. Figure 2 shows the experiment at temperature 60, 70 and 80 °C.

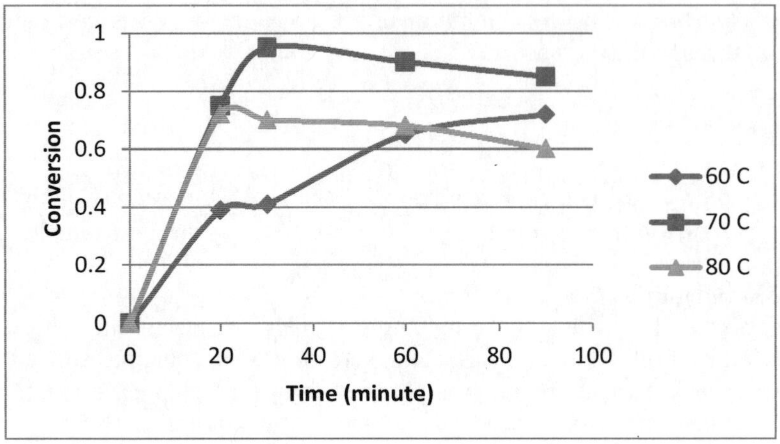

Figure 1. Conversion at various operating conditions

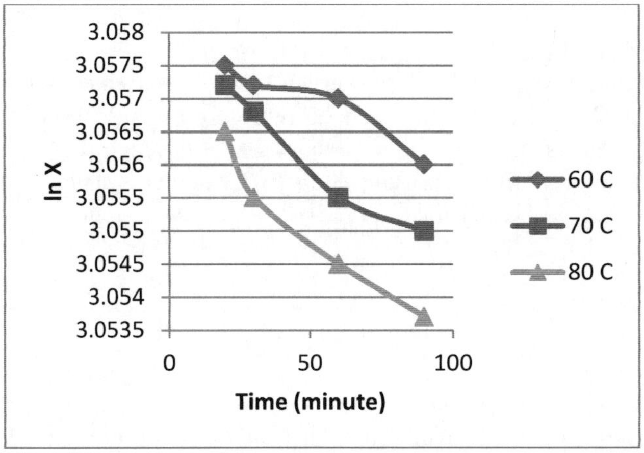

Figure 2. Correlation of time and ln X

Based on Figure 2, it shows that the formation of the epoxide compound [EO] increases with time. The formation rate of epoxide compound is different in each reaction temperature, where the reaction temperature is higher (70 °C and 80 °C) gives a faster rate of formation than lower reaction temperature (60 °C).

The experiment at various temperatures was linearized to determine the reaction rate constant and comparison between the data and the model equations. The suitability was measured by the value of R Square (R^2). Linear equations that can be used to obtain the value of the constant of reaction rate (k). The values of reaction rate constants are shown in Table 1.

Table 1. Reaction rate constant (k)

T (°C)	Linear equation	R^2	Value of $k.10^{-6}$ (L $mol^{-1}menit^{-1}$)
60	$y = -2.1553.10^{-5}t + 3.0579$	0.9888	1.24344
70	$y = -3.3140.10^{-5}t + 3.0576$	0.9968	1.91192
80	$y = -3.5181.10^{-5}t + 3.0568$	0.9724	2.02967

IOP Conf. Series: Materials Science and Engineering **206** (2017) 012046 doi:10.1088/1757-899X/206/1/012046

The value of the reaction rate constant at each temperature of the experiment can be used to determine the activation energy (*Ea*) and the frequency factor (*Ar*). Equation 7 can be used to determine the value of Ea and Ar.

$$\ln k = \ln A - \frac{Ea}{RT} \tag{7}$$

Table 2. Value of $\frac{1}{T}\left(\frac{1}{Kelvin}\right)$ and $\ln k$

$\frac{1}{T}$	$\ln k$
0.003001651	-13.59762697
0.002914177	-13.16740098
0.002831658	-13.10763582

Plot ln k versus 1/T will give the value of reaction rate constants.

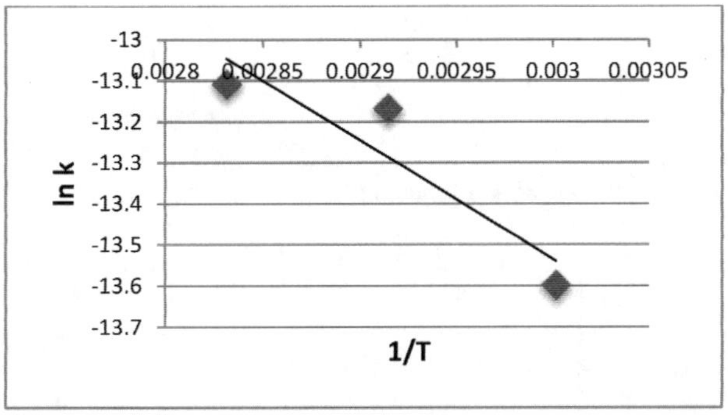

Figure 3. Temperature and reaction rate constant

The activation energy in the epoxidation was of 24.14 kJ / mol and the frequency factor of 124.82. Based on the Arrhenius equation, then the equation of the reaction rate constants can be formulated as Equation 8.

$$k = 124.82 \exp\left(-\frac{24.14 kJ/mol}{RT}\right) \tag{8}$$

3.3. Effect of Addition of Epoxide and Co-surfactant in Decreasing IFT Value

The addition of the epoxide compound and co-surfactant on the SLS are expected to reduce the value of IFT from SLS formulation. The results of measurement IFT value are presented in Table 3.

Table 3. IFT Value of Formulation

Type of Formula	Composition				IFT Value (mN/m)
	SLS (g)	Epoxide (g)	Co-surfactant: 1-Octanol (g)	Formation water (g)	
Formula A	0.15	-	-	29.7	2.34
Formula B	0.15	0.15	-	29.7	9.95×10^{-2}
Formula C	0.15	0.15	0.03	29.7	6.08×10^{-3}
Formula D	0.15	0.15	0.06	29.7	2.43×10^{-3}
Formula E	0.15	0.15	0.09	29.7	2.53×10^{-3}
Formula F	0.15	0.15	0.12	29.7	2.53×10^{-3}

Table 3 shows the IFT value of Formulation of SLS. The difference of Formula A and Formula B is in the addition of the epoxide. Formula C is modification of the Formula B with the addition of co-surfactant 0.1%. In order to learn about effect of co-surfactant to IFT value, then the addition of co-surfactant on the Formula D was increased to 0.2%.The addition of co-surfactant in Formula E and F as much as 0.3% and 0.4% did not reduce in value of IFT. This is because of excess of the need of co-surfactant in formulations. This accords well with research done by [10] which states that the addition of co-surfactant generally ranges from 0.1 to 0.2% of the formulation.

3.4. Oil recovery
In the EOR process, after injection of water (water flooding), some oil is still left in the reservoir due to high viscosity or high interfacial tension (IFT) between oil and the reservoir rocks. Therefore, after injection of chemicals which has higher mobility ratio, more oil can be recovered.

Figure 4 shows the process of water flooding followed by surfactant flooding. From the figure, it can be seen that after injection of water (2 % brine) some oil is still left in the reservoir. After injection of surfactant, more oil can be recovered. The oil displacement experiments show that as much as 20 % to 80 % of remaining oil can be recovered by flooding it with the chemicals.

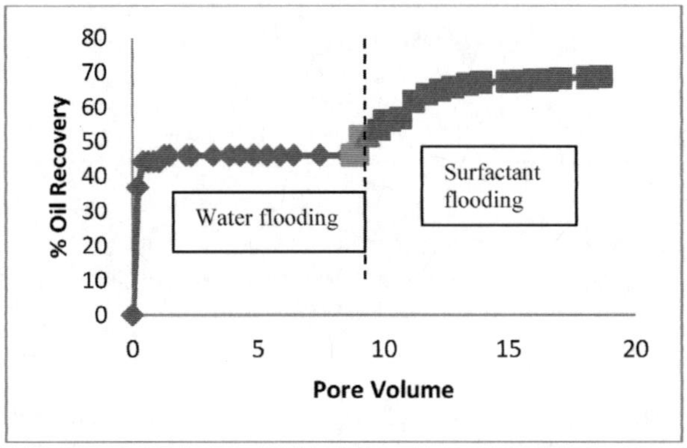

Figure 4. Example of water and surfactant flooding

Another studies of micro-emulsion demonstrate that if the mixture among oil, brine, surfactant and co-surfactant are in the form of micro-emulsion, the interfacial tensions (IFT) between oil/micro-emulsion and brine/micro-emulsion are very low (close to zero). Therefore, the oil which can be recovered after water flooding is high. The formation of a single-phase micro-emulsion with water-to-oil ratio close to unity requires a very high concentration of surfactant. In many cases, the addition of a fourth component, mainly alcohol, can facilitate the formation of a micro-emulsion at a lower surfactant concentration. Therefore, in this experiment, the results also show that for the same kind of surfactant, the amount of micro-emulsion will be different when the co-surfactant added is different.

Further study shows that the oil recovery increases when the chemical concentration is raised. The increasing oil recovery is caused by the increasing amount of micro-emulsion due to the lowering IFT.

4. Conclusions
It can be concluded that rising temperature up to 70 °C with reaction time of 30 minutes can increase the conversion of *Kapok* oil to epoxide compound. Temperatures above 70 °C would lead to reduce in the conversion value, which caused by side reaction.

Epoxide compounds and co-surfactants added to the formulation can reduce IFT value. The optimum results can be obtained on additional co-surfactant 0.2%. The optimum results addition of co-surfactant can be obtained on 0.2%.The addition above 0.2% did not give a reduction in the IFT value. The results show that these chemicals have potential uses as EOR chemicals.

Acknowledgment

The authors would like to express sincere gratitude to Upstream Technology Centre Pertamina for providing fund for this research.

References

[1] Teke J 2014 The Effect of Castor Oil Risinoleat Acid Epoxidation in Decreasing IFT of Sodium Ligno Sulfonate for EOR Processes Thesis Chemical Engineering Department Universitas Gadjah Mada Yogyakarta.

[2] Ariska C R 2011 Development of Epoxide compound from Oleic Acid Thesis Chemical Engineering Department Universitas Gadjah Mada Yogyakarta.

[3] Dantas C S T, Neto D A A, Moura E L, Neto B, Forte K R and Leite R H L 2003 Heavy Metals Extraction by Microemulsions Universitas Federal Do Rio Grande Do Norte, Centro De Tecnology PPGEQ Campus Universitario Brazil.

[4] Gebelein C G 2005 *Chemistry and our World* Brown Publishers USA.

[5] Goud V V, Narayan C P and Anand VP 2006 *JAOCS* **83(7)** AOCS Press.

[6] Siggia S and Hanna J G 1979 Quantitative Organic Analysis via Functional Groups Wiley Interscience New York.

[7] Sinadinovi´c-Fiˇser S, Milovan J and Zoran S P 2001 *JAOCS* **78(7)** AOCS Press.

[8] Official Methods and Recommended Practices of the American Oil Chemists' Society (AOCS) edited by David Firestone 4th edition *American Oil Chemists*' Society Champaign 1989 Method Cd9-57.

[9] Mungroo R, Narayan C P, Vaibhav V G and Ajay K D 2008 Epoxidation of Canola Oil with Hydrogen Peroxide Catalyzed by Acidic Ion Exchange Resin *J Am Oil ChemSoc* **85** 887–896.

[10] Green D W and Paul W 1998 Enhanced Oil Recovery Society of Petroleum Engineers Inc Texas USA.

Study on CO_2/ N_2 separation: the effect of rubbery polymer coating on PVDF membrane

M Z Zuwairi And S A Rahman

Faculty of Chemical & Natural Resources Engineering, Universiti Malaysia Pahang, 26300 Gambang, Pahang,Malaysia.

Email: sunarti@ump.edu.my

Abstract. The emission of harmful gases such as carbon dioxide (CO_2) via gas processing plant and daily human activities gave negative impacts to the environment and global inhabitant. Flat sheet asymmetric membranes were produced from homogenous solution of Poly(vinylideneflouride) (PVDF) via phase inversion method using N-methyl-2-pyrrolidone (NMP) as the solvent. While the poly ether b-amide (PEBAX) was dissolve by using of (70% ethanol and 30% water) as a solvent and and lithium chloride as a additives. The morphology and cross section of the produced membranes were observed by Scanning Electron Microscope (SEM). Then, the membranes were tested for chemical analysis to define the presence of PEBAX in the membrane by using Fourier Transform Infrared (FTIR) spectroscopy. The permeation performances of the membranes were evaluated in terms of permeability and selectivity of the membranes by using gas permeation test. Increasing the PEBAX content significantly increased the selectivity of the PVDF membrane to separate the CO_2/N_2 gases but decreased the amount of the gases that passed through the membrane.

1. Introduction

One of the top problems the world is facing nowadays is how to reduce greenhouse gas. Carbon dioxide (CO_2) is one of the main greenhouse gases. More than six and a half billion people burn fuel to keep warm, to provide electricity to light their homes and to run industry, and to move about using cars, buses, boats, trains, and airplanes. The burning of fuel produces carbon dioxide, which is released to the atmosphere. The burning of fuels adds about 6 gigatons of carbon to the atmosphere each year [1].

There are three primary methods for reducing the amount of CO_2 in the atmosphere: employing energy efficiency and conservation practices; using carbon-free or reduced-carbon energy resources; and capturing and storing carbon either from fossil fuels or from the atmosphere. CO_2 capture and sequestration (CCS) could play an important role in reducing greenhouse gas emissions, while enabling low-carbon electricity generation from power plants. CCS technologies are currently available and can dramatically reduce (by 80-90%) CO_2 emissions from power plants that burn fossil fuels. The amount of GHG emissions avoided (with a 90% reduction efficiency) would be equivalent to [2]:

- Planting more than 62 million trees, and waiting at least 10 years for them to grow

IOP Conf. Series: Materials Science and Engineering **206** (2017) 012047 doi:10.1088/1757-899X/206/1/012047

- Avoiding annual electricity-related emissions from more than 300,000 homes

Being one of the main greenhouse gases, CO_2 still can be treat by remove it from the atmosphere and store it deep in the oceans and the soil surround by the roots of many plants. In industries, the emission of CO_2 is through the combustion of fossil fuel such as coal, oil and natural gas. CO_2 is capture from the flue gas or syngas and will be passed through a series of membrane in order to separate it from nitrogen gas (N2), sulfur oxide (SO_2) and nitrogen oxide. The separation occurs based on the different molecular size of the gas compositions that are passed through the membrane pores. Therefore, in order to remove the CO_2 in the flue gas or syngas, the thin film composite membranes were developed consists of PVDF and PEBAX in this research. From literature, the PEBAX membranes are very competitive in performance and economic. It also suitable membranes formers in terms of its chain rigidity, chain interactions, stereoregulity and polarity of its functional group and offer the ability to separate CO_2 from syngas and in particular hydrogen with reasonable and selectivity [3]. Meanwhile, PVDF polymer has excellent, almost universal chemical resistance. It can be used both at high and very low temperatures (-260 to +260°C). It also possesses outstanding resistance to weathering (UV-stability).

The hydrophobicity of the PVDF membranes can be improved through various methods such as physical blending, chemical grafting, and surface modification. However, physical blending with inorganic materials like lithium chloride, glycerol, polyvinyl pyrrolidone or polyethylene glycol provide a great interest due to the ease of handling the materials and good performances. The producer of composite membranes of PEBAX and PVDF membranes were observed with SEM for the morphology and tested with FTIR for the chemical analysis while the performances of the membranes were investigated using gas permeation test.

2. Methodologies

2.1 Chemicals

Chemicals use in this study is Poly (vinylideneflouride) (PVDF) polymer pellets, poly ether b-amide(PEBAX), N-methyl-2-pyrrolidone (NMP) with 99.5% purity, ethanol and Lithium Chloride (LiCl). The pure CO_2 and N_2 gases were ready in Chemical Engineering Laboratory, Universiti Malaysia Pahang.

2.2 Membranes Preparation

To prepare the microporous support membrane, a casting solution of 10 wt % PVDF in 80 wt % NMP and 10 wt % lithium chloride was prepared. This membrane was cast at a casting speed of 3.5 m/min and casting thickness of 178 μm. The membranes were precipitated in water, washed for 24 hours, and dried in an oven at 80°-100° C. After that, the casting membranes of PVDF membranes was coating with the PEBAX solution which is 90 wt % of (70% ethanol and 30% water), 4 wt % of PEBAX and 6 wt % of lithium chloride which is (90:4:6) for first ratio and followed by (90:6:4), and (90:2:8) for the second and third ratio as shown in the Table 1.

Table 1. The PEBAX and LiCl compositions in the membranes

Composite Membrane	PEBAX content (wt%)	LiCl content (wt%)
A(90:2:8)	2	8
B(90:4:6)	4	6
C(90:6:4)	6	4

2.3 Chemical and Physical Characterization

The thin film composite (TFC) membrane will be analyzed physical and chemically by using Scanning Electron Microscopy (SEM) and Fourier Transform Infrared Spectroscopy (FTIR) devices. The SEM device was used to examine the membrane morphology. The FTIR is use for chemical characterization of the membrane.

2.4 CO_2/N_2 Separation Test

The TFC membrane was then tested with CO_2/N_2 separation test. The pressure-normalized fluxes were measured for the desired gases at certain temperature and pressure. According to researchers, (1) was used to determine the permeance through membranes [4,5]. The calculation for the gas permeability, P is shown in (2). Meanwhile, selectivity was calculated using (3).

$$\left(\frac{P}{l}\right) = \frac{V}{(At\Delta P)} \qquad (1)$$

where P is permeability, l is the thickness of membrane, A is the effective membrane area, V is the volume displaced in time(s) and ΔP is the pressure differences across the membrane.

$$P = \frac{Vl}{At\Delta P} \qquad (2)$$

The membrane selectivity is the membrane ability to separate 2 gases which is A and B. It is the ratio of the permeabilities of A and B as shown in (3).

$$\alpha_{AB} = \frac{P_A}{P_B} \qquad (3)$$

To study the effect of feed pressure to the performance of the membrane, the feed pressure was varied from 0.5 to 1 bar. Similar type of method was applied to get the permeability and selectivity of the produced membrane.

3. Results and Discussion

3.1 Effect of different PEBAX ratio on the PVDF Membrane Morphology

The top layer of a membrane usually consist of a finger-like macro void structure that can be observed in most types of membrane that is fabricated through phase inversion method as the exchange rates of solvent and non-solvent occur rapidly. Figure 1 (a-c) shows the cross-sectional morphology of the prepared PVDF membrane with 90:2:8 ratio of PEBAX. The structure is more like a macro porous structure in the membrane cross-section with addition of additive on PEBAX ratio to the membrane casting solution.

According to [6], only small amount of LiCl added to the casting solution can significantly increase the viscosity of the solution. The viscosity of the polymer casting solution is the main parameter which affects the kinetics of the phase inversion that affects the structure and performance of the produced membrane. The cross-section morphology of the PVDF membranes indicated that with more additives on PEBAX ratio to the casting solution have improved the structure of the membrane to be more sponge-like structure as illustrated in Figure 1 (b-c). This phenomenon could be caused by the strong interaction between LiCl and NMP. As reported by [7], a great increase in viscosity was observed when NMP was used as solvent instead of DMAc although same amount of LiCl added to the solutions and this indicates that LiCl exhibit a stronger interaction with NMP than DMAc.

Increasing the additive on PEBAX ratio concentration demonstrated a better morphology and the finger-like macro voids are suppressed. The above observations were believed to be associated with the change of the thermodynamic and kinetic properties of the casting solution after the addition of additive on PEBAX ratio. The strong interactions between LiCl, polymer and solvent have increased the viscosity of casting solution that caused the delayed in liquid-liquid demixing process and lower the phase separation rate. This will result in the suppression of the macro voids and provided less porous sponge-like structure [8,9].

3.2 Chemical Analysis using FTIR

FTIR is an important analysis as it provides the support data for the results obtain in this research. FTIR will demonstrate the bands that show the functional group of the membranes and provide information about the molecular interaction between the components contain inside the polymer solution. The significant peaks formed indicated the molecular orientation of the molecules. The distribution of the functional groups present in the membrane and the allocation of the peaks of interest are tabulated in Table 2.

The same bands of PVDF functional group will appear at all the membrane as the same type of membrane was used. The C-H stretches vibration in vinylidene groups of the PVDF chain have wavelength of 3000 – 3020 cm^{-1} were detected in all the PVDF membranes together with C-F stretch alkyl halides that appeared at 1400 cm^{-1}. A sharp and strong peak detected around 837 cm^{-1} which belong to the C-Cl group indicated the presence of LiCl in the membranes while pure PVDF membrane without additive shows a rough and weak peak at the same wavelength [10,11,12].

3.3 The Performance of Membrane using Gas Permeation Test

Figure 2 illustrate the CO_2/N_2 gas permeance and CO_2/N_2 selectivity. The test was carried out at room temperature with the feed pressure of 1.0 bar. As can be seen, higher amount of additive

IOP Conf. Series: Materials Science and Engineering **206** (2017) 012047 doi:10.1088/1757-899X/206/1/012047

results in increase of CO_2/N_2 selectivity. This could be caused by the compactness of the membrane at high addition of LiCl that leads to a decrease in the fractional free volume of the membrane and this will consequently higher CO_2/N_2 selectivity [13].

Polymeric membrane exhibit the performance of quite high CO_2/N_2 selectivity but quite low CO_2 permeability and most of the polymeric membrane for gas separation in post-combustion will worked at below 200°C due to low permeability.

Table 2. Characteristic IR absorption for the functional groups present in the membrane

Sample	Allocation	Functional Group	Wavenumber (cm^{-1})
a	C-H stretch	Alkanes (vinylidene group) (ether b-amide)	3000
	C-F stretch	Alkyl Halide	1400
	C-Cl stretch	Alkyl Halide (presence of LiCl)	838
b	C-H stretch	Alkanes (vinylidene group) (ether b-amide)	3020
	C-F stretch	Alkyl Halide	1400
	C-Cl stretch	Alkyl Halide (presence of LiCl)	837
c	C-H stretch	Alkanes (vinylidene group) (ether b-amide)	3019
	C-F stretch	Alkyl Halide	1400
	C-Cl stretch	Alkyl Halide (presence of LiCl)	837

IOP Conf. Series: Materials Science and Engineering **206** (2017) 012047 doi:10.1088/1757-899X/206/1/012047

(a)
(b)
(c)

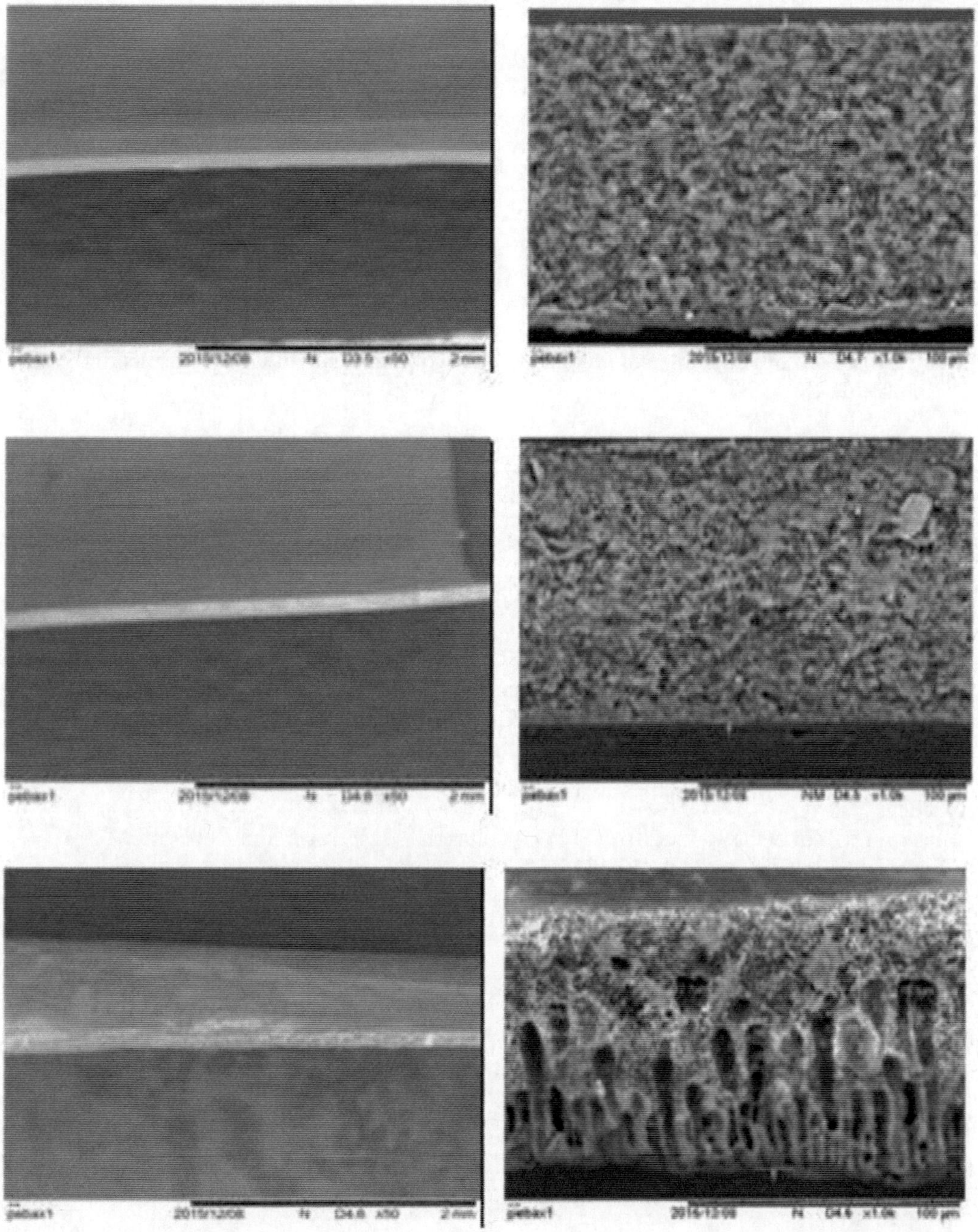

Figure1. Cross-section of flat sheet membranes in different PEBAX ratios (a) 90:2:8, (b) 90:4:6, (c) 90:6:4.

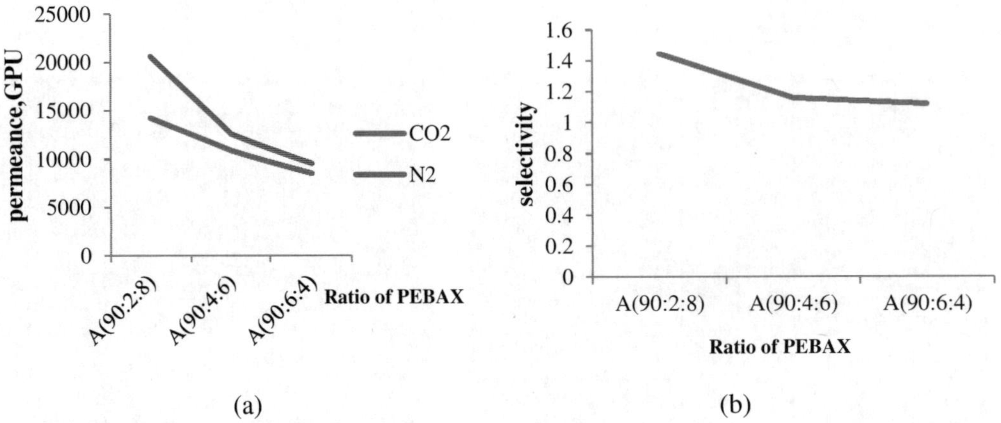

Figure 2. (a) Graph of permeance of CO_2/N_2 vs Ratio of PEBAX, (b) Graph of selectivity CO_2/N_2 vs Ratio of PEBAX.

4. Conclusion
This study successful developed the TFC PVDF/ PEBAX membrane. More sponge-like asymmetric structures were formed across the membrane wall and the macro void formation is suppressed. This is due to the higher viscosity of the casting solution that delayed the phase inversion rate (kinetic effect) which results in a sponge-like structure.

Acknowledgments
This research is fully supported by LRGS grant (RDU150901) and 304/PJKIMIA/6050296/U124) and UMP Grant (RDU150311). The authors fully acknowledged Ministry of Higher Education (MOHE) and Universiti Malaysia Pahang for the approved fund which makes this important research viable and effective.

References
[1] Burling I R, Yokelson R J, Griffith D W T, Johnson T J, Veres P and Roberts J M 2010 *Atmospheric Che. and Phy.* **10** 1115
[2] Bastianoni S, Pulselli F M and Tiezzi E 2004 *Ecological Eco.* **49** 253
[3] Zhao L, Riensche E, Menzer R, Blum L and Stolten D 2008 *J. Membrane Sci..* **325** 284
[4] Du N, Park H B, Dal-Cin M M. and Guiver M D 2012 *Ener. Environ. Sci..* **5** 7306
[5] Freeman B D 1999 *Macromolecules* **32** 375
[6] Shi L, Wang R, Cao Y, Feng C, Liang D T and Tay J H 2007 *J. Membrane Sci.* **305** 215
[7] Striegel A M 2003 *J. Chil. Chem. Soc.* **48** 30
[8] Liu F, Hashim N A, Liu Y, Abed M R M and Li K *J. Membrane Sci.* **375**(1)
[9] Powell C E and Qiao G G 2006 *J. Membrane Sci.*: **279** (1)
[10] Elashmawi I S 2011 *Materials Che. and Phy.*: **107** 96.
[11] Seo J B, Jeon S, Choi W J, Kim J W, Lee G H, and Oh K J 2011 *Korean J. Chem. Eng.*: **28** 170
[12] Shen Y and Lua A C 2012 *Chem. Eng. J.* **192** 201
[13] Speight JG, 2012 *Coal-Fired Power Generation Handbook*: **760** 2123

IOP Conf. Series: Materials Science and Engineering **206** (2017) 012048 doi:10.1088/1757-899X/206/1/012048

Nano silver diffusion behaviour on conductive polymer during doping process for high voltage application

A Mohammad[1*], A Mahmood[1], K T Chin[1], M K Danquah[2], S van Stratan[1]

[1]Department of Mechanical Engineering, Curtin University Sarawak Campus, Faculty of Engineering and Science, Miri 98009, Sarawak, Malaysia

[2]Department of Chemical Engineering, Curtin University Sarawak Campus, Faculty of Engineering and Science, Miri 98009, Sarawak, Malaysia

*Email: mohammad.asyraf@postgrad.curtin.edu.my

Abstract: Conductive polymer had opened a new era of engineering for microelectronics and semiconductor applications. However, it is still a challenge for high voltage applications due to lower electrical conductivity compare to metals. This results tremendous energy losses during transmission and restricts its usage. In order to address such problem a novel method was investigated using nano silver particle doped iodothiophene since silver is the highest electrical conductive material. The experiments were carried out to study the organometallic diffusion behaviour of nanosilver doped iodothiophene with different concentration of iodothiophene. Five different mixing ratio between nanosilver and the solution of iodothiophene dissolved in diethyl ether were used which are 1:1.25, 1:1.5, 1:2.5, 1:3 and l:5. It was revealed that there is an effective threshold concentration of which the nano silver evenly distributed and there was no coagulation observed. These parameters laid the foundation of better doping process between the nano silver and the polymer significantly which would contribute developing conductive polymer towards high voltage application for industries that are vulnerable to corrosive environment.

1. Introduction

Conductive polymers are recently discovered polymers that were found to conduct electricity. Some conductive polymers had found the purpose in the semiconductor and microelectronics (Kar, 2013a). Also, the conductive polymer resistance to corrosion are deemed to be good substitutes for metal conductors.

However, the application in the high voltage industries is still a challenge due to limited charge transfer capabilities that only provide lower conductivities (Taherian, 2016). Polythiophene (PT) is one of such conductive polymer. PT has a good solubility in organic solvents like diethyl-ether and good environmental stability properties (Ansari, 2006). Figure 1 below shows the structure of PT.

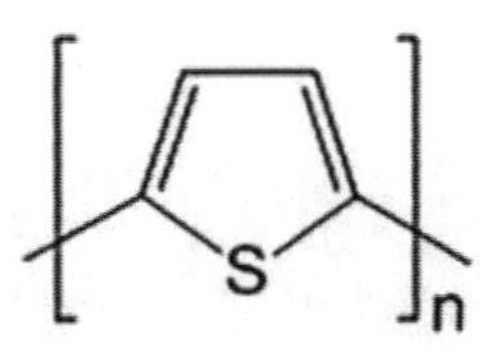

Figure 1. Polythiophene structure

Intrinsically conductive polymers are highly resistive for electricity flow (Kar, 2013b). In order to make it conductive, they had to undergo doping process (Kar, 2013c). Doping process defined as introducing impurities into a pure substance to enhance its electrical conductivity (Kar, 2013b). PT had been doped with both metallic as well as non-metallic known dopants such as copper (Shanmugapriya and Velraj, 2016) and HClO4 (S Masubuchi et al., 1993), etc. There were also attempts to dope PT with other techniques such as using microwave plasma deposition (Paosawatyanyong et al., 2010) and binary organic solvent such as dichloromethane (Jeon et al., 2010) to study the electrical conductivities of that end product.

Among metallic dopant, silver is known as highest electrical conductivity (Fackler, 2011). Multiple attempts had been done before in doping of PT with silver. But these studies had been focused on other area such as photocurrent enhancement (Szeremeta et al., 2014, Bagheri et al., 2016), solar properties enhancement (Liu et al., 2014), and electrode fabrication (Ghanbari, 2014), etc. It is hypothesized that small sizes particle of silver will enhance the charge transfer in the resulted organometallic compound of the PT during doping process in monomer stage (Kong and Jang, 2006). i.e. High electrical conductivity could be achieved through doping nano silver particle with iodothiophene, a monomer of PT.

However, the coagulation and uniform distribution of the nano particle in the iodothiophene monomers are the issues that hamper the effective diffusion during doping process. In order to address such issue, the effect of iodothiophene concentrations to the nano particle is one of the important parameters.

In this paper, the effect of iodothiophene concentrations to the nano particle was investigated. The homogenous nano particles distribution in different iodothiophene concentrations would qualitatively indicate the effective threshold concentration of iodothiophene monomers. It is expected this finding would laid the foundation of better doping process between the nano silver and iodothiophene monomers significantly in order to developed conductive polymer for high voltage application .

2. Experimental

2.1 Materials and preparation

All materials used in this study were supplied by Sigma- Aldrich. The iodothiophene assay is 98% in liquid form. It is stored in temperature around 2-8°C. The nano silver powders are 99% trace metal. 1 mg of nano silver was taken inside the test tube with diethyl ether as a solvent. Then, 1.25mL iodothiophene were added into the test tube. The processes were repeated with 1.50mL, 2.5mL, 3.00mL and 5.00mL of iodothiophene. The test tubes were then placed in the water bath of ultrasonic cleaner for 15 minutes and left for air drying for 60 minutes. Then, the samples were oven dried for 24 hours.

2.2 Design of experiment

Five different samples concentrations were prepared. The amount of nano silver, diethyl ether and iodothiophene used are displayed in the Table 1 Experiment design for effect of concentration of iodothiophene below.

IOP Conf. Series: Materials Science and Engineering **206** (2017) 012048 doi:10.1088/1757-899X/206/1/012048

Table 1. Experiment design for effect of concentration of iodothiophene

Sample	Nano Silver (mg)	Diethyl Ether (mL)	Iodothiophene (mL)
KRX1B	1	0.02	1.25
KRX2B	1	0.02	1.50
KRX3B	1	0.02	2.50
KRX4B	1	0.02	3.00
KRX5B	1	0.02	5.00

3. Result and Discussion

The diffusion behaviour with distribution and coagulation of nano silver particles are shown in the micrographs Figure 2. It is interesting to note there are distinct distributions of nano silver particles between concentrations of iodothiophene solutions. Several black spots were observed on sample KRX1B (Figure 2) which indicates significant coagulations of the nano silver particles.

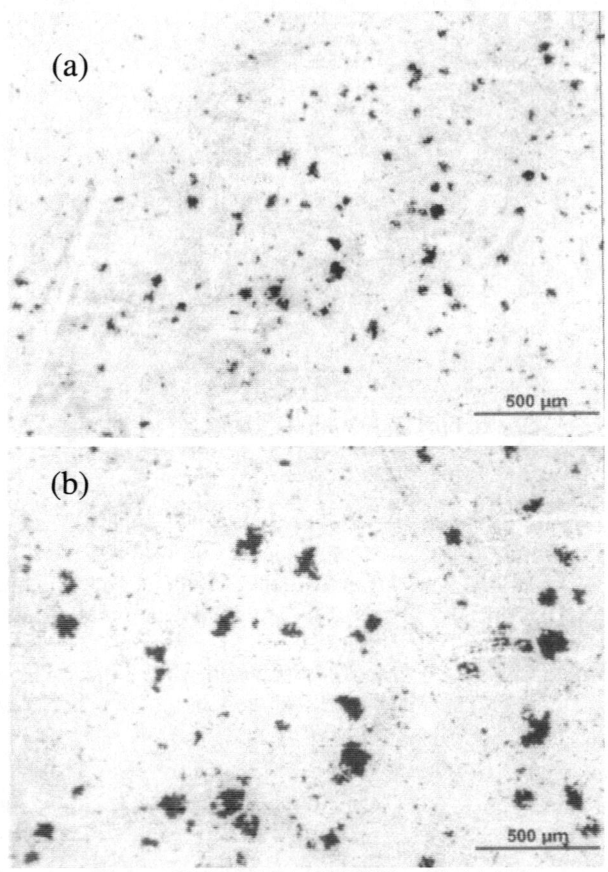

Figure 2(a) and **(b).** KRX2B with (a) 5X magnification and (b) 10X magnification

Similar behaviour was observed for sample KRX2B (Figure 3). It is shown black spots indicating coagulations of nano silver particles (Figure 3). This indicates that the nano silver had not been fully diffused due to the low concentration of iodothiophene.

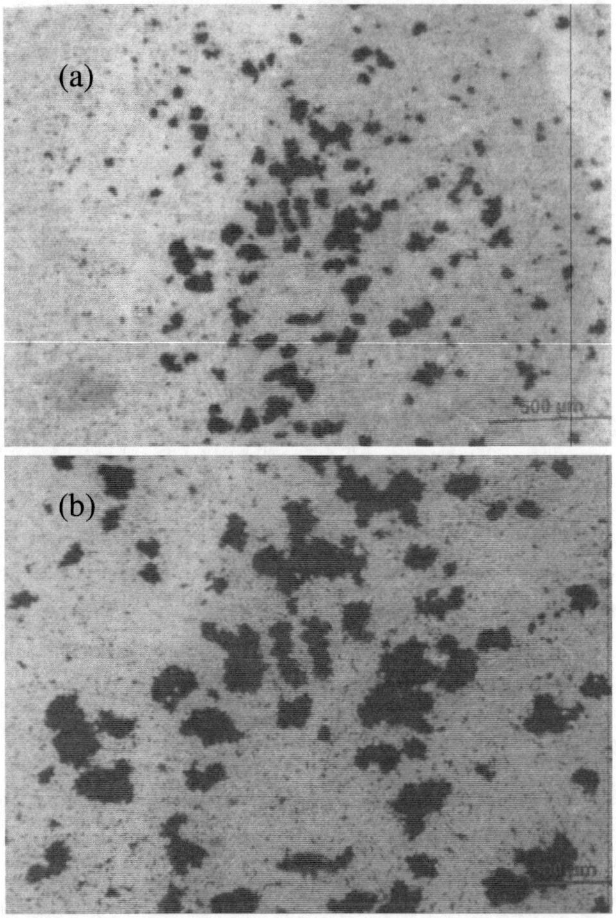

Figure 3(a) and **(b).** KRX2B with (a) 5X magnification and (b) 10X magnification

As per stoichiometry, it can be deduced that the nano silver diffusion had been influenced by iodothiophene concentration. Reaction had stopped when all of the iodothiophene had been saturated with nano silver. The remaining nano silver particles were coagulated and form the black spots. This would indicate that the 1.25mL and 1.5mL of iodothiophene are not sufficient to react completely with 1mg of nano silver. The nano silver is considered to be in excess. However, the sample KRX3B as shown in Figure 4 below shows no black spots visible from the microscopy images. This would indicate that the nano silver had been evenly distributed and there is no coagulation occurred.

IOP Conf. Series: Materials Science and Engineering **206** (2017) 012048 doi:10.1088/1757-899X/206/1/012048

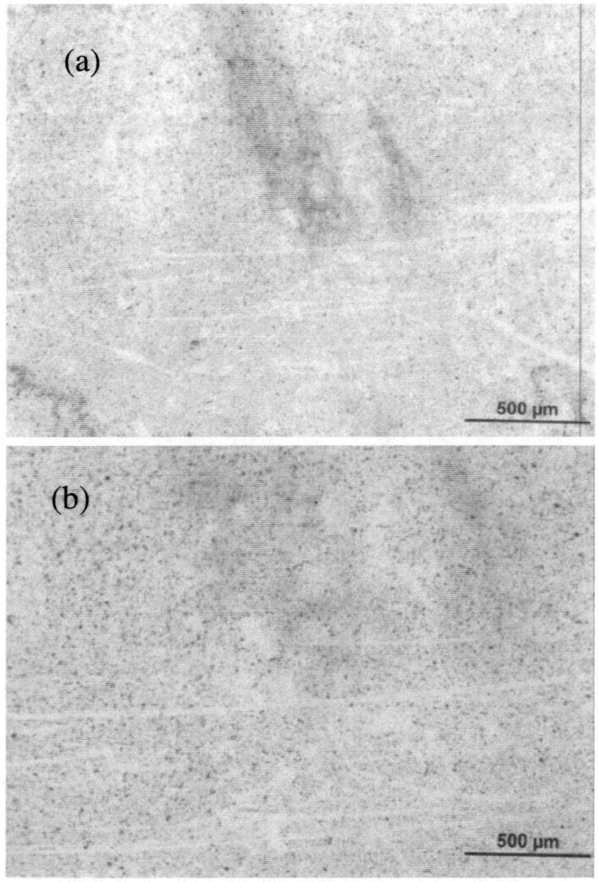

Figure 4(a) and (b): KRX3B with (a) 5X magnification and (b) 10X magnification

It is revealed that this is the concentration which the right concentration of iodothiophene is needed for 1mg of nano silver. Moreover, KRX4B (Figure 5) and KRX5B (Figure 6) shows the similar behaviour of even distribution of the particles as in the KRX3B (Figure 4). However, it can be deduced that all of the nano silver particles had reacted to the iodothiophene. Hence, it is considered as excess reactants, where all of the nano silver had been consumed but the iodothiophene are in excess. However, the two large void spots on both KRX4B (Figure 5) and KRX5B (Figure 6) are not coagulations. They are marks of labelling sticker at the back of the slides and can be ignored.

Figure 5: KRX4B with 10X magnification

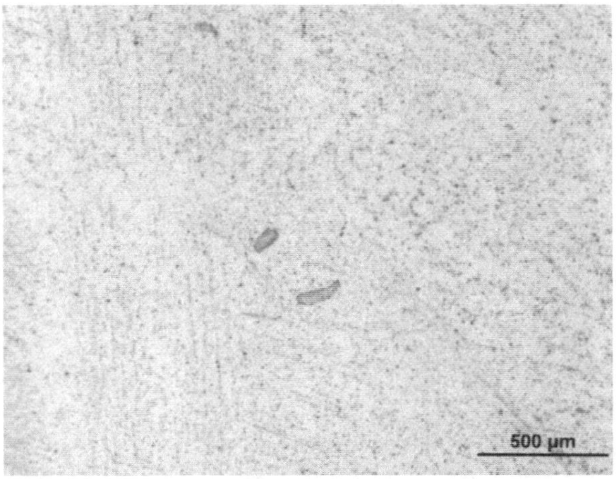

Figure 6: KRX5B with 10X magnification

The following Table 2 were constructed to simplify the results.

Table 2. Result of various concentration of Iodothiophene in nano silver

Sample	Ratio of nano silver to iodothiophene	Observation
KRX1B	1.25	Coagulation
KRX2B	1.50	Coagulation
KRX3B	2.50	No coagulation
KRX4B	3.00	No coagulation, excess reactants
KRX5B	5.00	No coagulation, excess reactants

4. Conclusions

In this study, nano silver particles were added into iodothiophene solution. The effects of iodothiophene concentration on the coagulation and distribution of nano silver were studied. It is revealed that the threshold concentration for 1mg of nano silver is 2.5mL of iodothiophene in 0.02mL. Coagulation will occur in lower concentration of iodothiophene as in 1:1.25 and 1:1.5 ratio of nano silver to iodothiophene. Homogenous distribution was observed at higher concentrations of iodothiophene in 1: 2.5 onwards such as in 1:3.00 and 1:5.00. It is also found that there is an effect of concentration of iodothiophene towards homogenous distribution of nano silver during doping process.

References

[1] Kar P, Influence of Dopant on the Applications of Conjugated Polymer 2013 Doping in Conjugated Polymers pp. 113-129: John Wiley & Sons, Inc

[2] Taherian R, Experimental and analytical model for the electrical conductivity of polymer-based nanocomposites 2016 *Composites Sci. and Techno.* **123** 17-31

[3] Ansari R, Polypyrrole Conducting Electroactive Polymers 2006 *E-J. of Che.* **3** 186-201

[4] Kar P, Introduction to Doping in Conjugated Polymer 2013 Doping in Conjugated Polymers 1-18: John Wiley & Sons, Inc

[5] Kar P, Classification of Dopants for the Conjugated Polymer 2013 Doping in Conjugated Polymers, pp. 19-46: John Wiley & Sons, Inc

[6] Shanmugapriya C and Velraj G, Investigation on structural and electrical properties of FeCl3 doped polythiophene (PT) blended with micro and nano copper particles by mechanical mixing 2016 *Optik – Int. J. for Light and Electron Optics,* **127** 8940-8950

[7] S Masubuchi S K, K Mizoguchi M H, K Kume, and Matsuyama R M a T 1993 Metallic transport properties in electrochemically as-grown and heavily doped polythiophene and poly(3-methylthiophene).pdf

[8] Paosawatyanyong B, Kamphiranon P, Bannarakkul W, Srithana-anant Y and Bhanthumnavin W, Doping of polythiophene by microwave plasma deposition 2010 *Surface and Coatings Techno.,* **204** 3053-3058

[9] Jeon S S, Yang S J, Lee K-J and Im S S, A facile and rapid synthesis of unsubstituted polythiophene with high electrical conductivity using binary organic solvents 2010 *Poly.,* **51** 4069-4076

[10] Fackler J P, Silver: Organometallic Chemistry 2011 *Encyclo. of Inor. and Bioinor.Che.* John Wiley & Sons, Ltd

[11] Szeremeta J, Nyk M and Samoc M, Photocurrent enhancement in polythiophene doped with silver nanoparticles 2014 *Optical Mat.,* **37** 688-694

[12] Bagheri H, Banihashemi S and Jelvani S, A polythiophene–silver nanocomposite for headspace needle trap extraction 2016 *J of Chromatography A* **1460** 1-8

[13] Liu X-H, Hou L-X, Wang J-F, Liu B, Yu Z-S, Ma L-Q, Yang S-P and Fu G-S, Plasmonic-enhanced polymer solar cells with high efficiency by addition of silver nanoparticles of different sizes in different layers 2014 *Solar Ener.* **110** 627-635

[14] Ghanbari K, Fabrication of silver nanoparticles–polypyrrole composite modified electrode for electrocatalytic oxidation of hydrazine 2014 *Synthetic Metals* **195** 234-240

[15] Kong H and Jang J, One-step fabrication of silver nanoparticle embedded polymer nanofibers by radical-mediated dispersion polymerization 2006 *Chem. Comm.* **28** 3010-3012

IOP Conf. Series: Materials Science and Engineering **206** (2017) 012049 doi:10.1088/1757-899X/206/1/012049

Correlating the synthesis protocol of aromatic polyimide film with the properties of polyamic acid precursor

P C Tan, B S Ooi, A L Ahmad and S C Low*

School of Chemical Engineering, Engineering Campus, Universiti Sains Malaysia, Seri Ampangan, 14300 Nibong Tebal, Pulau Pinang, Malaysia.

*E-mail: chsclow@usm.my; siewchun@gmail.com

Abstract. Thousands of different copolyimide combinations render it technically impossible to have a single universal synthesis method to produce aromatic polyimide film. This study aimed to outline the selection of synthesis protocol, either through the casting of chemically imidized polyimide solution or thermal imidization of polyamic acid (PAA), to produce the polyimide film. The rheological behaviour, molecular weight, and solubility of five structurally different PAA were analysed and correlated to both imidization methods. In this work, a tough polyimide film was successfully synthesized by casting the chemically imidized polyimide derived from high viscosity (> 81 cP) and high molecular weight ($\geq 1.35 \times 10^6$ g/mol) PAA. On the contrary, both low viscosity (< 13 cP) and high viscosity (> 81 cP) PAA demonstrated the possibility to produce polyimide film via thermal imidization route. The longer molecular chain of ODPA-6FpDA:DABA (3:2) polyimide produced from thermal imidization had restricted the passage of CO_2 across the polyimide film when it was applied in the gas separation application. The outcome from this work serves as a guideline for the selection of suitable polyimide film synthesis protocol, which will minimize the time and chemical consumption in future exploration of new polyimide structure.

1. Introduction

In the advent of technology advancement, global demands in searching high performance materials for industrial applications are increasing. In view of that, aromatic polyimide with attractive features such as high dimensional stability, excellent thermal and thermo-oxidative stability, good dielectric and optical properties is found to be a great potential material to apply in many fields [1-3]. At present, polyimides are extensively used as the microelectronic parts, high temperature insulator for aircraft, substrate for flexible printed circuit board, structural adhesive and gas separation membrane [4-7].

In general, polyimide can be processed into film, mold, fiber and foam. Among the polyimide forms, polyimide film constitutes the largest end user market [8]. It is evidenced by the wide application of Kapton polyimide film produced by Du Pont. Besides, Apical introduced by Kaneka as well as Upilex fabricated by Ube are among the reputable commercially available polyimide films [9]. The properties of polyimide film can be easily tailored through molecular design, in which the dianhydrides and/or diamines monomers used in polyimide synthesis are systematically selected to fulfill the specific application requirement. However, the diversity of dianhydride-diamine combinations makes it impractical to have a single universal synthesis protocol of polyimide film attributed to different processability.

Pioneering works had proposed different routes to produce polyimide film. One of the common synthesis routes is the casting of polyimide dope solution [10-13]. In this approach, polyimide powder is first obtained by performing the conventional two-step reaction, in which the polyamic acid (PAA) precursor is formed through the polycondensation of dianhydride and diamine; followed by chemical imidization of PAA to produce polyimide. Typically, chemical imidization is carried out by adding a basic catalyst (triethylamine or pyridine) and dehydrating agent (acetic anhydride) into the PAA

IOP Conf. Series: Materials Science and Engineering **206** (2017) 012049 doi:10.1088/1757-899X/206/1/012049

solution to promote the ring closure process (conversion of amide to imide functional group with the loss of water molecule). From the aspect of energy efficiency, the chemical imidization route is preferable as the ring closure process is readily achieved under ambient temperature [2].

However, owing to the insolubility of certain polyimides, the previous route is not recommended for polyimide film synthesis [14]. In this context, PAA solution is directly casted onto a substrate. Thereafter, thermal imidization is carried out by heating the thin PAA layer gradually up to 250-350 °C to convert the PAA precursor to polyimide, hence producing polyimide film [8, 15]. This method is especially useful when the final product is desirable in the form of film. However, a controlled heating protocol is needed as too rapid heating will cause the formation of bubbles in the casted PAA layer [2]. This will subsequently cause the failure in the film production. Basically, this approach is relatively simple and time saving since it omits the extra step of producing polyimide powder as compared to the previous chemical imidization route. Besides, no dangerous reagents are involved in this method.

The present study aimed to outline the selection of synthesis protocol, either through the casting of chemically imidized polyimide solution or thermal imidization of the casted PAA thin layer, to produce the polyimide film. Since PAA acts as the precursor of polyimide, its rheological behavior and molecular weight were correlated to the synthesis protocols. By taking membrane gas separation as the final application, permeation tests were performed to assess the properties of the polyimide films that were produced through the two different imidization approaches.

2. Materials and methods

2.1. Materials

Monomers comprise of 4,4′-(hexafluoroisopropylidene) diphthalic anhydride (6FDA, 99 % purity), benzophenone-3,3′,4,4′-tetracarboxylic dianhydride (BTDA, 96 % purity), 4,4′-oxydiphthalic anhydride (ODPA, 97 % purity), pyromellitic dianhydride (PMDA, 97 % purity), 2,4,6-trimethyl-mphenylenediamine (DAM, 96 % purity), 4,4′-(hexafluoroisopropylidene) dianiline (6FpDA, 98 % purity) and 3,5-diaminobenzoic acid (DABA, 98 % purity) were purchased from Sigma-Aldrich (USA). 1-methyl-2-pyrrolidinone (Sigma-Aldrich, USA), acetic anhydride (Acros Organics, Belgium), triethylamine (Sigma-Aldrich, USA) and methanol (Merck, Germany) were used as the solvents in polyimide synthesis. Besides, tetrahydrofuran (THF) was obtained from Fisher Scientific (UK). The gases used in permeation test were attained from Wellgas (Malaysia) with purity of > 99 %. All the chemicals were used as received without further purification.

2.2. Synthesis of PAA precursor

PAA precursor was synthesized via polycondensation of dianhydride and diamine. Owing to the water sensitive nature of reaction, all monomers were dried overnight at 40 °C. Throughout the reaction, the set-up was purged continuously with purified nitrogen to remove the trapped moisture. Two diamines ($diamine_1$=6FpDA or DAM and $diamine_2$=DABA) with molar ratio of 3:2 were first dissolved in 1methyl-2-pyrrolidinone (NMP). Stoichiometric amount of dianhydride (6FDA, BTDA, ODPA or PMDA) was then added slowly into the diamines solution. The solution was stirred for 24 h under ambient temperature (25 °C) to yield 20 wt. % PAA solution. The PAA precursor was then converted to polyimide through either chemical or thermal imidization. A few combinations of dianhydride and diamine were studied, where the polyimide structures are displayed in figure 1. In the discussion, the polyimide was named as A-B:C where A is the dianhydride, B is $diamine_1$, C is $diamine_2$ and the ratio between B:C was 3:2.

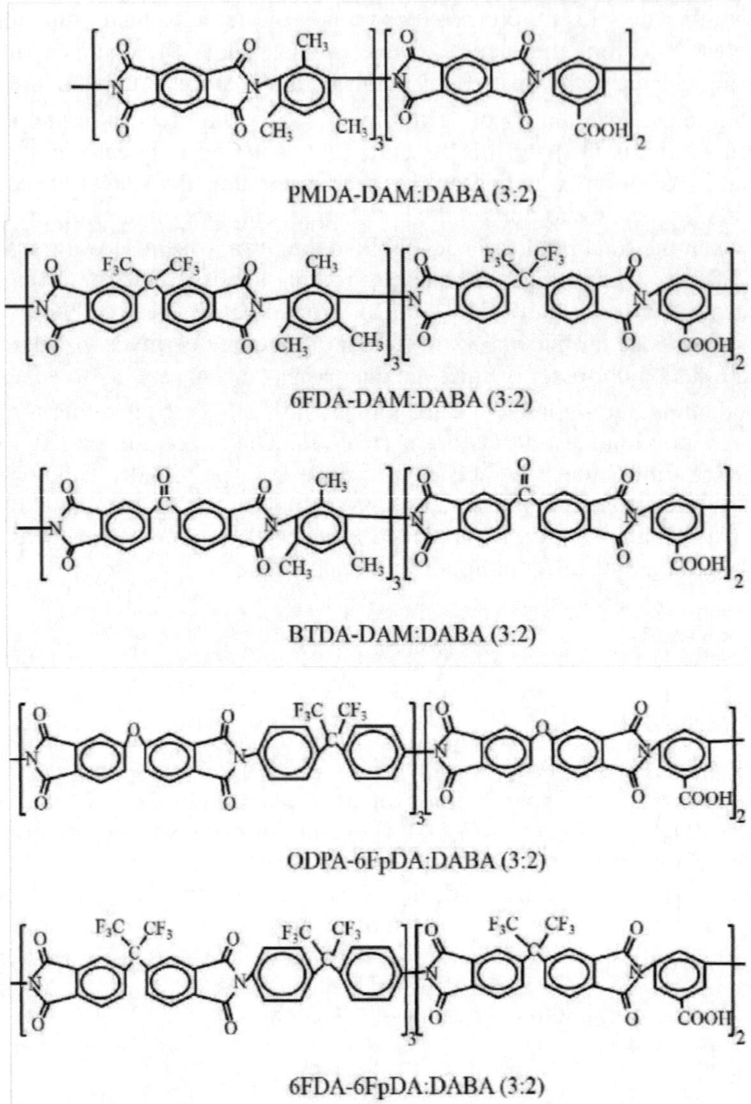

Figure 1. Polyimide structures studied in this work.

2.3. Film preparation using chemically imidized polyimide

The PAA was chemically imidized through the addition of equimolar triethylamine (catalyst) and acetic anhydride (dehydrating agent) into the PAA solution. The solution was stirred for 24 h under nitrogen purging and ambient temperature (25 °C) to produce polyimide solution. Thereafter, the polyimide was precipitated and washed for several times using methanol. The precipitated polyimide was then air-dried in fume hood for 16 h followed by drying at 200 °C for 24 h under vacuum condition.

10 wt. % dope solution was prepared by dissolving the polyimide in THF (6FDA and ODPA based polyimide) or NMP (PMDA and BTDA based polyimide). The dope solution with volume corresponding to initial film thickness of 1300 μm was poured onto a glass petri dish placed inside a glove box with controlled environment. After two days of phase inversion, the film was dried in vacuum oven at 180 °C for 24 h to eliminate the residual solvent. Afterward, the film was thermally treated in a dual zone split tube furnace (MTI- OTF-1200X-80-II, USA) under nitrogen purging to

IOP Conf. Series: Materials Science and Engineering **206** (2017) 012049 doi:10.1088/1757-899X/206/1/012049

promote decarboxylation crosslinking. The thermal treatment was performed in three steps: heating from room temperature to 330 °C at fast ramping rate of 5 °C/min, slow ramping rate of 1 °C/min to 370 °C and dwelling at 370 °C for 1 h.

2.4. Film preparation via thermal imidization of PAA precursor

20 wt. % PAA solution was diluted to concentration of 10 wt. % using NMP solvent. Specific volume of 10 wt. % PAA solution corresponding to initial film thickness of 1300 μm was poured onto a glass petri dish placed inside vacuum oven. The casted PAA thin layer was then thermally imidized to polyimide film by successive heating at 60 °C, 150 °C and 200 °C for 1 h each. The produced polyimide film was further thermally treated in a dual zone split tube furnace with similar thermal treatment protocol described in section 2.3. However, an additional 30 min heating step at 300 °C was included to ensure the complete imidization of the PAA precursor.

2.5. Characterization

The viscosity of 10 wt. % PAA solution was measured by Brookfield viscometer (model: DV2TLV, USA). The measurement was carried out under room temperature (25 °C). Multiple point averaging method was selected to obtain the average viscosity at rotation speed of 10 rpm.

Gel permeation chromatography (GPC, Agilent Technology 1260 Infinity, USA) equipped with PLgel 5 μm MIXED-C column (300 mm x 7.5 mm, 5 μm particle size) was employed to determine the average molecular weight (M_w) of PAA precursors and polyimides. Samples (PAA precursors and polyimides) were prepared in NMP or THF at 0.05 wt. % concentration. Throughout the analysis, the flow rate of mobile phase was controlled at 1 mL/min. The elution time of samples was detected using the variable wavelength detector (VWD), which was then correlated to the M_w of PAA and polyimide.

Gas separation performances of the synthesized polyimide films were analyzed using a membrane cell with effective area of 7.07 cm^2 at 25 °C. The whole system was thoroughly purged with purified nitrogen prior to the permeation test. After achieving steady condition, the permeate flow rate was measured by a bubble flow meter. The gas permeability was calculated based on the steady permeate flow rate according to equation (1).

$$P = \frac{\dot{V}pl}{TA\,\Delta p} \times 2695.93 \tag{1}$$

where P is gas permeability in Barrer (1 Barrer = 1 x 10^{-10} cm^3 (STP) $cm/cm^2 \cdot s \cdot cmHg$), V indicates permeate flow rate (cm^3/s), p refers to downstream pressure (Pa), l signifies membrane thickness (μm), T denotes operating temperature (K), A represents effective membrane area (cm^2) and Δp stands for differential pressure (cmHg).

3. Results and discussion

The five PAA precursors studied in this work demonstrated two extremely different rheological behaviors which can be categorized into two groups: high viscosity (> 81 cP) and low viscosity (< 13 cP), as shown in figure 2. PAA combinations of 6FDA-DAM:DABA, 6FDA-6FpDA:DABA and ODPA-6FpDA:DABA showed significantly higher viscosity in the range of 81.65- 91.33 cP as compared to PAA combinations of BTDA-DAM:DABA and PMDA-DAM:DABA, which showed viscosity at only 5.78 cP and 12.38 cP respectively. Indeed, both categories exhibited viscosity difference of at least 560%.

IOP Conf. Series: Materials Science and Engineering **206** (2017) 012049 doi:10.1088/1757-899X/206/1/012049

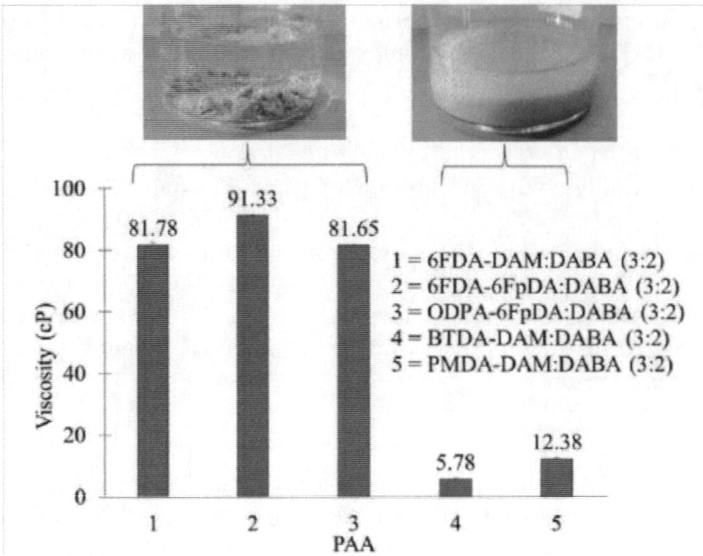

Figure 2. Viscosities of PAA precursors formed from different combinations of dianhydride and diamine.

Principally, the viscosity of PAA solution is closely related to the PAA molecular weight, which can be qualitatively interpreted from its precipitated form. Figure 2 displays the big clusters of PAA precipitant when the high viscosity PAA precursors (>81 cP) were precipitated in deionized (DI) water. On the contrary, the precipitation of low viscosity PAA precursors (< 13 cP) yielded a fine, powdery deposit. By comparing both precipitated forms, it can be deduced that the bigger lump of PAA (left image in figure 2) visually indicated higher molecular weight than the powdery deposit (right image in figure 2). It is suggested that PAA of higher M_w should have more restricted molecular mobility attributed to the higher internal friction between molecules. The greater flow resistance hence increased the viscosity of the solution. The result was in well agreement to the findings by other researchers, where a more viscous polymer solution always reflected in higher molecular weight of polymer [16, 17].

Quantitatively, the molecular weight of PAA was analyzed by GPC, as summarized in table 1. The high viscosity PAA (>81 cP) had consistently revealed higher molecular weight than that of the low viscosity PAA (< 13 cP). As for the less viscous BTDA-DAM:DABA and PMDA-DAM:DABA, the molecular weight of PAA were an order of magnitude smaller than those of high viscosity PAA. Indeed, the GPC result supported well the previous discussion on figure 2, where higher molecular weight PAA was present in the more viscous solution.

Table 1. Molecular weight of PAA precursors.

PAA	M_w (g/mol)
6FDA-6FpDA:DABA	5.28×10^6
ODPA-6FpDA:DABA	2.49×10^6
6FDA-DAM:DABA	1.35×10^6
BTDA-DAM:DABA	0.54×10^6
PMDA-DAM:DABA	0.30×10^6

IOP Conf. Series: Materials Science and Engineering **206** (2017) 012049 doi:10.1088/1757-899X/206/1/012049

In this work, the solubility of PAA precipitant was tested using THF, the preferable solvent used to synthesize polyimide film due to its high volatility. It is hypothesized that if PAA is undissolved in THF, the THF will also be unable to dissolve the latter chemically imidized polyimide since their chemical properties are literally identical (same monomer combinations). In such case, the chemically imidized polyimide approach would not be feasible to produce the polyimide film. Instead, thermal imidization route is preferable. After 24 h of stirring, the PAA precipitant of 6FDA-DAM:DABA, 6FDA-6FpDA:DABA and ODPA-6FpDA:DABA were shown to be fully dissolved in THF, where clear and homogeneous solutions were observed (data not shown). Meanwhile, suspensions were observed when dissolving the PAA of BTDA-DAM:DABA and PMDA-DAM:DABA, which indicated the poor solubility of both PAA precursors in THF. Hence, it was deduced that further chemical imidization of BTDA-DAM:DABA and PMDA-DAM:DABA PAA precursors was not recommended. Both PAA precursors should proceed for membrane casting followed by thermal imidization.

For fair comparison between the different synthesis routes, all the five PAA combinations were subjected to both chemical and thermal imidization in producing polyimide film. NMP was used instead as the casting solvent for those polyimide undissolved in THF. As illustrated in figure 3, defect free polyimide thin films were obtained for the chemically imidized 6FDA-DAM:DABA, 6FDA6FpDA:DABA and ODPA-DA:DABA polyimide (figures 3a, 3c and 3e). The successful formation of defect free polyimide films had a direct relationship to the M_w of polyimide. It is reasonable to assume that the trend of polyimide M_w behaves similarly to that of PAA M_w, as polyimide is transformed from PAA precursor with only the loss of water molecule [18]. Therefore, the high M_w of 6FDADAM:DABA, 6FDA-6FpDA:DABA and ODPA-6FpDA:DABA PAA precursors in table 1 reflected to the high M_w of the corresponding polyimides. During film formation, mass exchange occurred between the solvent (THF/NMP in casting dope solution) and non-solvent (water vapor in atmosphere), where polyimide molecules in the casting dope tended to attract to each other. The higher M_w of polyimide had promoted better interaction between the molecules, hence forming a tough and defect free polyimide thin film [19].

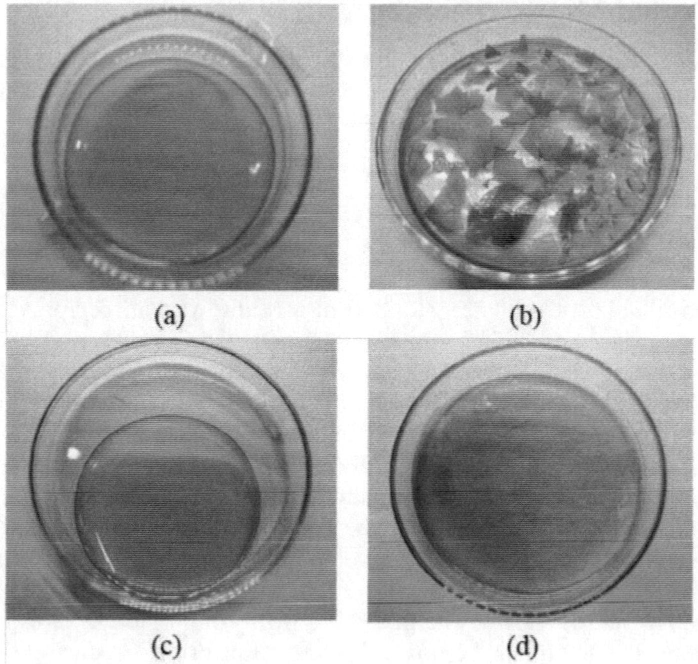

(a) (b)

(c) (d)

IOP Conf. Series: Materials Science and Engineering **206** (2017) 012049 doi:10.1088/1757-899X/206/1/012049

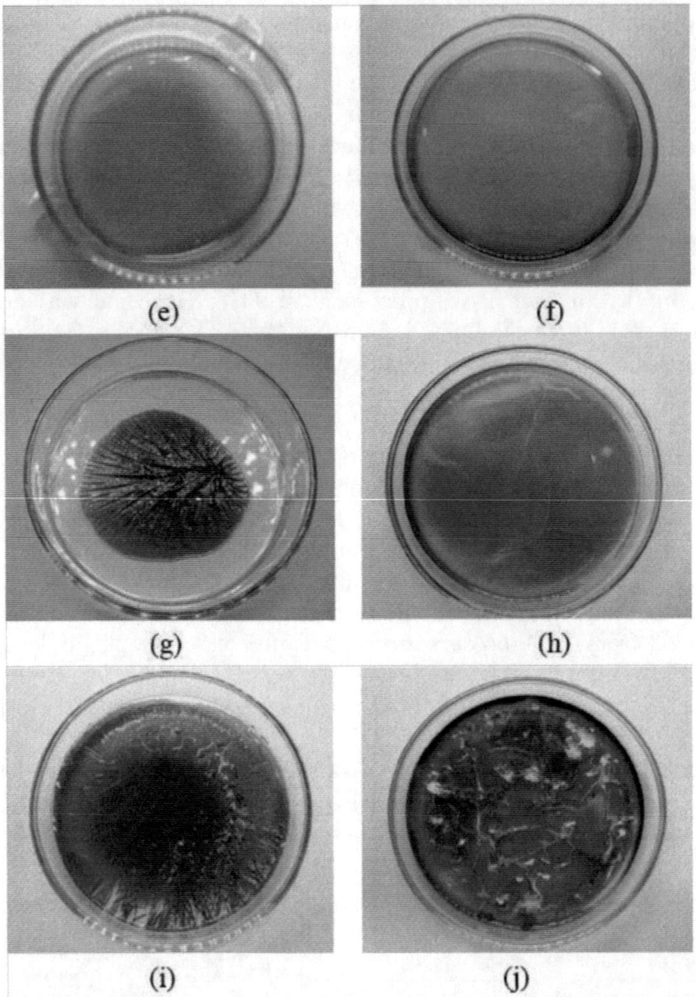

Figure 3. Polyimide film of (a-b) 6FDA-DAM:DABA, (c-d) 6FDA-6FpDA:DABA, (e-f) ODPA6FpDA:DABA, (g-h) BTDA-DAM:DABA and (i,j) PMDA-DAM:DABA synthesized by casting of chemically imidized polyimide solution (left) and thermal imidization of casted PAA precursor (right). Experiments repeated at least thrice to confirm the repeatability of film formation.

The significant contribution of polymer M_w in film-forming was further proven as the polyimide with lower M_w failed to produce a tough thin film (figures 3g and 3i). BTDA-DAM:DABA polyimide film showed serious shrinkage and cracked (figure 3g). Meanwhile, a fragile cracked film was also observed for the PMDA-DAM:DABA polyimide film (figure 3i). It was believed that the interaction between the low molecular weight polyimide was insufficient to ensure a proper film formation. Up to this point, the experimental observation stressed the necessity of high viscosity and high molecular weight of PAA precursor for successful film fabrication via chemically imidized polyimide approach.

Unfortunately, the polyimide film-forming via thermal imidization of casted PAA thin layer did not correlate well with the viscosity and/or molecular weight of PAA precursors. Either high viscosityhigh molecular weight PAA (ODPA-6FpDA:DABA, figure 3f) or low viscosity-low molecular weight PAA (BTDA-DAM:DABA, figure 3h) showed the possibility to form a nice piece of polyimide film. The results hinted that another factor might be governing the film formation during thermal imidization. Thence, the stiffness of thermally imidized polyimides were tested by dissolving the thermally imidized polyimide films (figures 3b, 3d, 3f, 3h and 3j) in THF. As found, the nicely formed polyimide films in figures 3f and 3h (ODPA-6FpDA:DABA and BTDA-DAM:DABA) were insoluble in THF while other cracked films (figures 3b, 3d and 3j) were soluble in THF. This indicated that stiffer polyimide chains were formed by the ODPA-6FpDA:DABA and BTDA-DAM:DABA PAA throughout the thermal imidization process [20]. It is well accepted that the stiffness of polyimide chain increased with the degree of imidization [2], hence it can be inferred that both ODPA6FpDA:DABA and BTDA-DAM:DABA polyimide films had achieved the sufficient imidization degree. Owing to the stronger mechanical strength, the stiffer ODPA-6FpDA:DABA and BTDADAM:DABA polyimide chains were thus able to produce the nice pieces of polyimide films (figures 3f and 3h).

It is of interest to analyze the properties of polyimides obtained from the two different imidization approaches. Figure 4 presents the molecular weight of polyimides synthesized from chemical imidization and thermal imidization. Generally, thermally imidized polyimides consistently demonstrated higher molecular weight than that of chemically imidized polyimides. This phenomenon can be explained from the aspect of PAA hydrolytic stability. PAA precursor is well-known to be hydrolytically unstable, in which PAA chain cleaves under humid condition hence lowers the molecular weight of the resulting polyimide [21]. In chemical imidization, the water liberated during ring closure process (imidization) was removed by acetic anhydride. However, the removal of water molecules occurred by chance based on the effective contact between acetic anhydride and water molecules during stirring. This meant the possibility of chain cleavage by water molecules during chemical imidization still existed. Thus, relatively shorter polyimide chains were observed (figure 4). Since the thermally imidized polyimides possessed the higher molecular weight, it could be postulated that thermal heating was more effective in removing the water by-product from system, hence minimized the chain scissoring that reduced the molecular weight of polyimide.

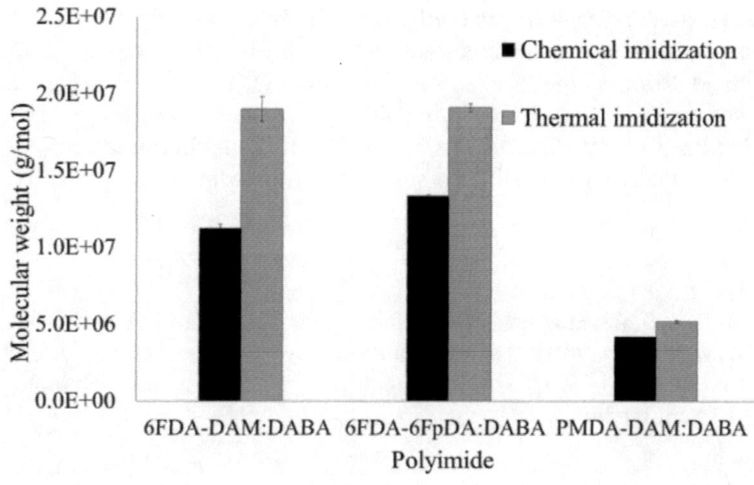

Figure 4. Molecular weight of polyimides synthesized from chemical and thermal imidization.

The molecular weight of polyimide will not only affect the film formation but also its effectiveness in gas separation. In view of the increasing interest on carbon dioxide removal, by taking ODPA6FpDA:DABA as example, the gas transport behavior of the polyimide films prepared from the two approaches was evaluated. As denoted in figure 5, the chemically imidized ODPA-6FpDA:DABA film had demonstrated higher CO_2 permeability over the tested pressure range (up to 6 bar). As discussed earlier, the chemically imidized polyimide exhibited lower molecular weight than the thermally imidized polyimide (figure 4). In regards, more polymer chain ends existed within the chemically imidized polyimide film matrix [22]. In fact, the higher number of polymer chain ends contributed towards higher free volume of the polyimide film [23]. Hence, more free spaces were available for the transport of CO_2 across the chemically imidized ODPA-6FpDA:DABA film, which subsequently increased the CO_2 permeability.

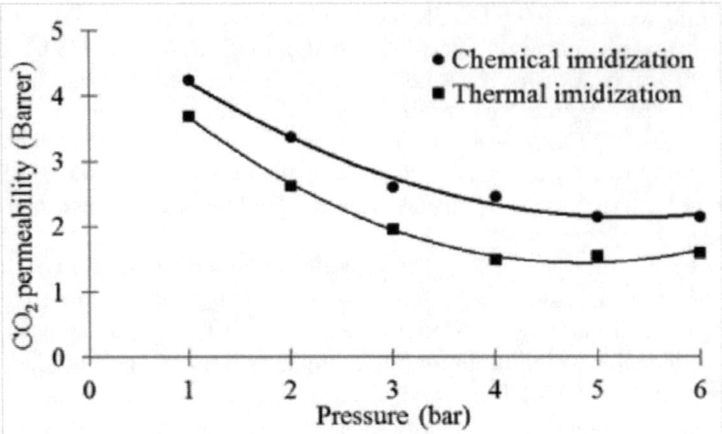

Figure 5. CO_2 permeability of ODPA-6FpDA:DABA polyimide film prepared by casting of chemically imidized polyimide solution and thermal imidization of casted PAA precursor.

4. Conclusions

The present work provided a guidance on the selection of synthesis protocol to produce polyimide film, by correlating the synthesis protocol to the molecular weight and rheological behavior of PAA precursor. A high viscosity (> 81 cP) and high molecular weight ($\geq 1.35 \times 10^6$ g/mol) of PAA precursor was necessary to ensure the formation of defect free polyimide film through the casting of chemically imidized polyimide solution. On the contrary, the production of polyimide film via thermal imidization of casted PAA layer was feasible by using both low viscosity (< 13 cP) and high viscosity (> 81 cP) PAA precursor. It was speculated that high imidization degree was the prerequisite to obtain a defect free polyimide film via the thermal imidization route. In comparison, the chemically imidized polyimide had consistently showed lower molecular weight than that of thermally imidized polyimide. The chemically imidized ODPA-6FpDA:DABA (3:2) film with shorter chain length had reflected in higher free spaces for the transport of CO_2 across the film, which subsequently increased the CO_2 permeability.

Acknowledgments

The authors gratefully acknowledge the financial support from Long Term Research Grant Scheme (LRGS) (304/PJKIMIA/6050296/U124). All authors are affiliated to the Membrane Science and Technology Cluster of USM. P C Tan is financially assisted by the Human Life Advancement Foundation (HLAF) Scholarships.

References

[1] Liaw D J, Wang K L, Huang Y C, Lee K R, Lai J Y and Ha C-S 2012 *Prog. Polym. Sci.* **37** 907-74

[2] Ghosh M K and Mittal K L 1996 *Polyimides: Fundamentals and Applications* (New York: Marcel Dekker)

[3] Chang C C and Chen W C 2002 *Chem. Mater.* **14** 4242-8

[4] Peng X, Wu Q, Jiang S, Hanif M, Chen S and Hou H 2014 *Mater. Lett.* **133** 240-2

[5] Lv M, Zheng F, Wang Q, Wang T and Liang Y 2015 *Tribol. Int.* **92** 246-54

[6] Naito K, Onta M and Kogo Y 2012 *Int. J. Adhes. Adhes.* **36** 77-85

[7] Kraftschik B, Koros W J, Johnson J R and Karvan O 2013 *J. Membr. Sci.* **428** 608-19

[8] Hsiao S H and Chen Y J 2002 *Eur. Polym. J.* **38** 815-28

[9] Pardos F 2004 *Plastic Films: Situation and Outlook: a Rapra Market Report* (UK: Rapra Technology Limited)
[10] Kratochvil A M and Koros W J 2008 *Macromolecules* **41** 7920-7
[11] Tong H, Hu C, Yang S, Ma Y, Guo H and Fan L 2015 *Poly.* **69** 138-47
[12] Qiu W, Xu L, Chen C-C, Paul D R and Koros W J 2013 *Poly.* **54** 6226-35
[13] Chua M L, Xiao Y and Chung T-S 2014 *Sep. Purif. Technol.* **133** 120-8
[14] Ohya H, Kudryavsev V V and Semenova S I 1996 *Polyimide Membranes: Applications, Fabrications and Properties* (Tokyo: Kodansha Ltd.)
[15] Wang Y C, Huang S H, Hu C C, Li C L, Lee K R, Liaw D J and Lai J Y 2005 *J. Membr. Sci.* **248** 15-25
[16] Fox T G and Flory P J 1954 *J. Polym. Sci.* **14** 315-9
[17] Fox Jr T G and Flory P J 1948 *J. Am. Chem. Soc.* **70** 2384-95
[18] Han H, Gryte C C and Ree M 1995 *Polym.* **36** 1663-72
[19] Liaw D J, Hsu P N, Chen W H and Lin S L 2002 *Macromolecules* **35** 4669-76
[20] Kausar A, Zulfiqar S, Yavuz C T and Sarwar M I 2011 *Polym. Degrad. Stabil.* **96** 1333-41
[21] Krishnan P S G, Vora R H, Chung T-S, Uchimura S-I and Sasaki N 2005 *J. Polym. Res.* **11** 299-308
[22] Santangelo P G and Roland C M 1998 *Macromolecules* **31** 4581-5
[23] Jang J H, Ozisik R and Mattice W L 2000 *Macromolecules* **33** 7663-71

IOP Conf. Series: Materials Science and Engineering **206** (2017) 012050 doi:10.1088/1757-899X/206/1/012050

Antimicrobial activity of coconut shell liquid smoke

SI Kailaku[1*], M Syakir[2], I Mulyawanti [1], ANA Syah[3]

[1]Indonesian Center for Agricultural Postharvest Research and Development, Jl. Tentara Pelajar 12 Bogor, Indonesia
[2]Indonesian Agency for Agricultural Research and Development, Jl. Ragunan 29, Jakarta, Indonesia
[3]Center for Estate Crops Research and Development, Jl. Tentara Pelajar 1 Bogor, Indonesia

*E-mail: sari.kaylaku@gmail.com

Abstract. Coconut shell liquid smoke is produced from the pyrolysis and condensation of smoke from the burning process of coconut shell. It is known to have considerably high content of polyphenol. Beside acting as antioxidant, polyphenol is also a good antimicrobial. This research was conducted in order to study the antimicrobial activity of coconut shell liquid smoke. Coconut shell liquid smoke used in this study was produced from three different processing stages, which obtained three different grades of liquid smoke (grade 1, 2 and 3). Each sample of coconut shell liquid smoke was extracted using ethyl alcohol and petroleum ether. The extract was then analyzed for its antimicrobial activity against *S. aereus, E. coli* and *C. albicans* using well diffusion method. Total phenol and microbial microscopic structure of the liquid smoke were also examined. The results showed that there was influence of coconut shell liquid smoke on the inhibition of *S. aureus, E. coli* and *C. albican* growth. This fact was marked by the forming of clear area surrounding the well on the dish agar media. The highest percentage of inhibition showed by the extract of grade 3 coconut shell liquid smoke. This may be explained by the highest total phenol content in grade 3 liquid smoke. Microscopic examination showed that there was a breakage of microbial cell walls caused by the antimicrobial property of the liquid smoke. It was concluded that coconut shell liquid smoke was beneficial as antimicrobial agent, and while all grades of liquid smoke contains polyphenol, the content was influenced by the processing stage and thus influenced its level of microbial growth inhibition.

1. Introduction

Coconut shell is a part of coconut and not commonly utilized into high value products. Traditionally coconut shell is burned and used for cooking. This practice may result in dangerous waste and increase air pollution [1]. Pyrolisis, followed by condensation, can produce high quality liquid smoke from coconut shell. Coconut shell chips is smoldered and then condensation is carried under limited oxygen to obtain liquid smoke [2].

Coconut shell liquid smoke is known to be a good food preservation agent, substituting formalin, due to its antimicrobial and antibacterial activities, thus has inhibitory effects on pathogens [3,4,5]. Antimicrobial and antioxidant compounds presenting in coconut shell liquid smoke are aldehyde, carboxylic acids and phenols [6,7].

Despite these potential in coconut shell liquid smoke, there is reluctance in using it, especially for certain products where odour, flavor and color are easily affected. For raw food, these qualities may be

IOP Conf. Series: Materials Science and Engineering **206** (2017) 012050 doi:10.1088/1757-899X/206/1/012050

recovered by cooking and other handling process. Other products may not have this advantages. Therefore, refining process is often done to liquid smoke in order to reduce the undesirable odour, flavor and color [2]. Refined liquid smoke therefore is more flexible in applications to various food or non food products, compared to non refined liquid smoke [8].

There are three grading of coconut shell liquid smoke, i.e. grade 1, 2 and 3. The difference between each grade lies on the extent of filtration stages. The more refined was called Grade 1 and the less refined was Grade 3. Surely, Grade 1 has the best characteristics for consumers convenience and liking, while Grade 3 may posses some undesirable characteristics, although these characteristics were not necessarily less beneficial. Grade 1 is very slightly yellowish and has the highest clarity compared to the other grades (Figure 1). It is more commonly used as food preservatives for meatball, noodle, sausage, tofu, etc. Grade 2 is yellowish and usually used as antioxidant and antimicrobial agent, while Grade 3 is very dark with sharp odor, often used as wood preservatives, rubber coagulant and odor absorber.

Figure 1. Coconut shell liquid smoke from different grades.

The growing interest in research and industrial world encouraged a great attention on evaluation of antimicrobial activity. Various methods have been developed for this purpose. One of the recommended methods is the agar well diffusion method. This method is widely used to evaluate the antimicrobial activity of plants or microbial extracts [9,10].

The objective of this research was to evaluate the antiumicrobial activity of coconut shell liquid smoke produced from three different processing stages.

2. Material and Method

2.1. Coconut shell liquid smoke

Coconut shell liquid smoke (CS-LS) with three different grades were obtained from Indonesia Liquid Smoke, Ciracas, East Jakarta. It was produced by processing coconut shell in furnace with 400-500°C for 1-2 hours and followed by condensation of the smoke (Figure 2).

The produced CS-LS from the process above was labeled as Grade 3 CS-LS. Grade 2 was produced from distillation and filtration of the condensed smoke using zeolite. Grade 1 was the result of further filtration using activated carbon.

Zeolite was prepared by calsination at 400°C for 4 hours. Afterwards, it was activated using KOH solution (1N) for 24 hours and centrifuged (1500 rpm, 5 minutes) and oven-dried at 40-60°C for 24 hours. Meanwhile, activated carbon was prepared using $CaCl_2$ (25%) through the same process.

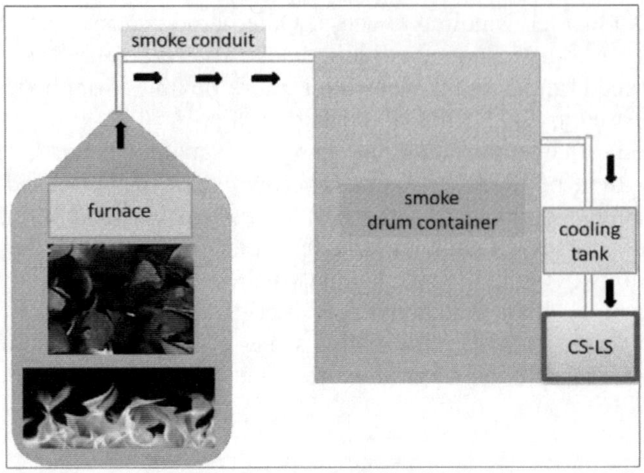

Figure 2. Scheme of coconut shell liquid smoke production.

2.2. Agar well diffusion

Agar well diffusion method was conducted by a modified method of Balouiri et al. [11]. Agar plate surface was inoculated by spreading a volume of microbial inoculum over the its entire surface. A hole with a diameter of 6-8 mm was then punched aseptically with a tip, followed by a volume of 20-100 μL of the antimicrobial agent with the concentration of 25, 50, or 75% being dripped into the well. Then, the plates of agar were incubated in 37°C for 24 hours. The growth of the microbial strain was evaluated by observing the formation zones and the diameter was measured in mm. Microbial strains tested were *S. aureus, E. coli* and *Candida sp.*

2.3. Total phenolic content

The total phenolic content of CS-LS was determined by Folin-Ciocalteu method as was done by Kaur and Kapoor [12]. 200 μL of sample (1 mg/mL) were made up to 3 mL with distilled water, followed by the addition of 0.5 mL of Folin–Ciocalteu reagent and the mixture were mixed thoroughly for 3 min. 2 mL of 20% (w/v) sodium carbonate was then added prior to 60 min storage in the dark. Using spectrophotometer, absorbance was measured at 650 nm. The total phenolic content was calculated from the calibration curve, and the results were expressed as mg of gallic acid equivalent per g dry weight.

2.4. Physicochemical characteristics

Color of CS-LS was evaluated using Chromameter. Chromameter was calibrated using white standard material. The color was valued by Hunter color system (L, a, and b value) with white as standard color (L1, a1, b1).

Clarity was analyzed using Spectrophotometer. Distilled water in cuvet was used as 100% clarity control. CS-LS sample in other cuvet was evaluated with 420 nm wavelength.

pH was measured using pHmeter. The electrode of pHmeter was simply dipped into the CS-LS sample until the instrument showed the measurement result.

3. Results and Discussion

The antimicrobial effects of CS-LS as shown by the agar well diffusion analysis results were only significant for the growth of *Staphylococcus aureus* and *E. coli*. There was no antimicrobial effect was shown for the growth of *Candida sp.* This proved that antimicrobial activity in CS-LS was more effective on bacteria then on yeast. Moreover, it was more obvious on the test for *S. aureus* where all samples of CS-LS showed inhibitory effects, leading to conclusion that CS-LS was more effective to inhibit the growth of gram-positive bacteria. Some studies reported that gram positive bacteria were more sensitive than gram negative ones to some plants extract [13]. On gram-positive bacteria, all grades of CS-LS with each concentration provided good inhibition effect (shown as the forming of clear area around the wells) (Figure 3).

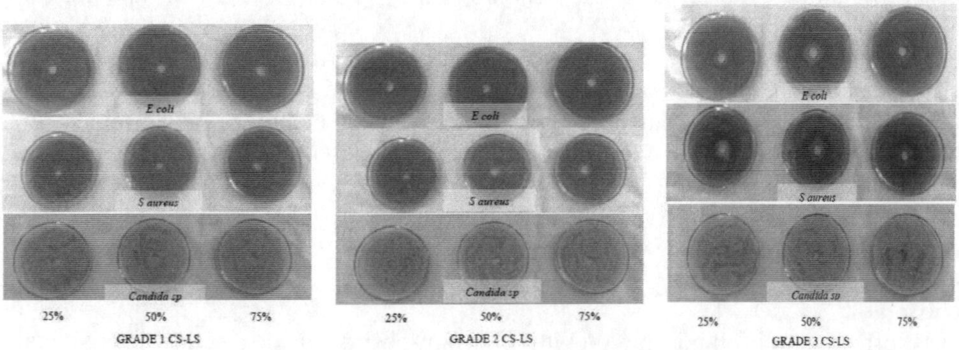

Figure 3. Inhibitory effects of coconut shell liquid smoke (CS-LS) from three different grades and three different concentration on *E coli, S aureus* and *Candida sp.*

Results also showed that higher concentration of liquid smoke gave better inhibitory effect to the growth of bacteria. Antimicrobial effect of liquid smoke was contributed by its phenolic content. The higher total phenolic content in CS-LS, the better its antimicrobial activity. This fact was demonstrated in this research. Table 1 showed that Grade 3 CS-LS had the highest total phenolic content (Table 1) and it was correlated with the best inhibitory effect on microbes. Filtration using zeolite and activated carbon can decrease phenol content [14]. Grade 3 CS-LS did not undergo filtration process, thus contains more phenol.

Polyphenols are bioactive molecules. Biological activities are related to the molecules structures; by their hydroxyl groups or by phenolic ring, phenolic compounds have the capacity to link with proteins and bacterial membranes to form complexes [15]. Phenolic compounds, such as phenol, 2-methoxylphenol (guaiacol), 3,4dimethoxylphenols, and 2-methoxy-4-methyl phenol are prominent in CS-SL and play a major contribution to antibacterial activity [16]. The antimicrobial mechanism from phenolic compounds on gram positive bacteria might be described as disrupting cell wall, increasing cell membrane permeability, and then leading to the leakage of cell constituents [13].

Table 1. Physicochemical properties of coconut shell liquid smoke

Parameters	Coconut shell liquid smoke		
	Grade 1	Grade 2	Grade 3
Total phenolic content (%)	0.18±0.004	0.30±0.002	1.5±0.1
Lightness (L)	95.76	93.01	29.87
a	1.97	-1.26	39.97
b	5.50	20.82	47.92
Clarity (%)	93.70	64.23	0.00
pH	2.79	3.12	2.76

Zeolite has adsorber ability due to its porous structure, allowing tar and benzopyrene compounds in liquid smoke to be caught in its pores. Zeolit can absorb a considerable amount of molecules with the same or smaller size than zeolite's pores [17].

Activated carbon has a vast carbon surface and porous structure due to its granules form. Aromatic compounds with the same or smaller molecules size than its pores can be adsorbed during filtration. This mechanism allows liquid smoke to lost its smoke aroma and flavor and making it more acceptable as food preservatives.

Table 1 also presented the different physicochemical properties of CS-LS based on their grades. Grade 1 had the highest lightness (L value) and clarity compared with the other grades. Grade 3 had a considerably dark and cloudy appearance. These properties alone could be a determinant values in the utilization of each grade of CS-LS. As mentioned above, Grade 1 is more common as food preservatives, probably because of its better aroma and appearance, Grade 2 is used as antimicrobial agent and Grade 3 for wood preservatives and rubber coagulant, where aroma and appearance are not ranked very high in the selection of material.

4. Conclusions

CS-LS was effective to inhibit the growth of gram-positive bacteria due to its total phenolic compound. Higher total phenol gave better inhibition effect. The filtration process can reduce the phenol content of CS-LS.

References

[1] Lombok JZ, Setiaji B, Trisunaryanti W and Wijaya K 2014 Effect of pyrolisis temperature and distillation on character of coconut shell liquid smoke *Asian J. Sci Tech* **5(6)** 320-5

[2] Varlet V, Serot T and Prost C 2010 Smoke flavoring technology in seafood *Handbook of seafood and seafood products analysis* ed LML Nollet and F Toldra (Boca Raton, FL: CRCPress) pp 233–54

[3] Milly PJ, Toledo RT and Chen J 2008 Evaluation of liquid smoke treated ready-to eat (RTE) meat products for control of listeria innocua *J Food Sci* **73** 179-183

[4] Soldera S, Sebatianuso N and Bortolomeazzi R 2008 Composition of phenolic compound and antioxidant activity of commercial aqueous smoke flavoring. *J Agri Food Chem* **56** 2727-34

[5] Toledo RT 2007 Wood Smoke Components and Functional Properties *Proc. Int. Conf. on Smoked Seafood (Alaska: Sea Grant College Program)*

[6] Leroi F and Joffraud JJ 2000 Salt and smoke simultaneously affect chemical and sensory quality of cold-smoked salmon during 5°C storage predicted using factorial design *J Food Protection* **63** 1222-27

[7] Rorvik LM 2000 *Listeria monocytogenes* in the smoked salmon industry *Int J Food Microbiol* **62** 183-190.

[8] Martinez O, Salmero J, Guillen MD and Casas C 2007 Textural and physicochemical changes in salmon (*Salmo salar*) treated with commercial liquid smoke flavourings *Food Chem* **100** 498-503

[9] Magaldi S, Mata-Essayag S and Hartung de Capriles C 2004 Well diffusion for antifungal susceptibility testing *Int J Infectious Dis* **8** 39–45

[10] Valgas C, DeSouza SM and Smânia EFA 2007 Screening methods to determine antibacterial activity of natural products *Braz J Microbiol* **38** 369–80

[11] Balouiri M, Sadiki M and Ibnsouda SK 2016 Methods for in vitro evaluating antimicrobial activity: A review. *J Phar Anal* **6** 71-9

[12] Kaur C and KapoorHC 2002 Antioxidant activity and total phenolic content of some Asian vegetables. *Int J Food Sci Tech* **37** 153–61

[13] Cheng L and Hu Q 2015 Antibacterial activities and mechanism of action of acetone extracts from Rabdosia rubescens *Aceh Int J Sci Tech* **4** 1-6

[14] Mubarokah I 2010 Gabungan metode aerasi dan adsorbsi dalam menurunkan kadar fenol dan COD pada limbah cair UKM Batik Purnama Desa Kliwonan, Kecamatan Masaran, Kabupaten Sragen Tahun 2010 *Public Health Department Dissertation* (*Semarang Indonesia)* (Fakultas Ilmu Keolahragaan/Universitas Negeri Semarang)

[15] Nitiema LW, Aly S, Jacques S, Dayeri D and Alfred ST 2012 In vitro antimicrobial activity of some phenolic compounds (coumarin and quercetin) against gastroenteritis bacterial strains *Int J Microbiol Res* **3**(3) 183-7

[16] Zuraida I, Sukarno and Budijanto S 2011 Antibacterial activity of coconut shell liquid smoke (CS-LS) and its application on fish ball preservation *Int Food Res J* **18** 405-10

[17] Setiaji B 2000 Pemanfaatan Zeolit untuk Adsorpsi Benzopiren sebagai Senyawa Racun dalam Asap Cair *Majalah Iptek* **11**

IOP Conf. Series: Materials Science and Engineering **206** (2017) 012051 doi:10.1088/1757-899X/206/1/012051

Effect of ZnO addition on the purity and densification of forsterite ceramic

C Y Tan [1*], Y M Tan [1], S Ramesh [1] and B K Yap [2]

[1]Centre of Advance Manufacturing and Materials Processing, Department of Mechanical Engineering, Faculty of Engineering, University of Malaya, 50603, Kuala Lumpur, Malaysia.

[2]Department of Electronics and Communication, College of Engineering, Universiti Tenaga Nasional, Km-7, Jalan Ikram-UNITEN, 43009, Kajang, Selangor, Malaysia

*E-mail: chouyong@um.edu.my

Abstract: Forsterite was found to the next potential candidate for bone implant application due to its superior mechanical properties as compared to the commonly used hydroxyapatite (HA). Various methods including two-step sintering and improvise synthesis method were introduced in hope to further improve the mechanical properties of forsterite. In this work, sintering additives, particularly zinc oxide (ZnO), was introduced into forsterite to provide enhancement on the densification of forsterite at lower sintering temperature. Forsterite powder was synthesized via solid-state reaction method with heat treatment at 1000 °C for 2 hours with 10 °C/min ramping rate. Addition of ZnO was conducted using milling process with composition of 0.5 wt% and 1.0 wt% ZnO. Green bulk samples were prepared prior to sintering process at 1200 °C to 1500 °C for 2 hours with 10 °C/min ramping rate. Characterization was conducted in terms of phase stability and densification of forsterite with morphology examination to observe the grain surface of all samples. It was revealed that the addition of ZnO showed a more superior densification as compared to the undoped samples at all sintering regime studied with a maximum relative density obtained at 97.7% by 0.5 wt% ZnO doped sample sintered at 1500 °C.

1. Introduction

In the context of biomedical applications, hydroxyapatite (HA) was generally popular among other bioceramics owing to its excellent biocompatibility to the human body [1,13]. Nevertheless, for many years HA was not used in load-bearing application due to its low mechanical properties, particularly fracture toughness (0.7 – 1.2 MPa m1/2) when compared to the human bone (2 – 12 MPa m1/2) [2,10]. This has motivated researchers to move towards other potential bioceramics and thus forsterite was introduced. In the early years, forsterite (Mg2SiO4) was not thoroughly investigated by researcher until recently due to the relatively high fracture toughness (2.4 MPa m1/2) and good biocompatibility [2].

Various methods were introduced in producing forsterite with enhanced mechanical properties such as solid-state reaction method, sol gel method, molten salt approach and aqueous route [3-7]. Even different sintering methods were introduced throughout the research to further enhance the mechanical properties such as spark plasma sintering [8] but most of these methods introduced were either costly or complex. Hence, sintering additives was introduced and was widely used by many researchers especially on HA, alumina and zirconia [9]. Doping with another chemical substance was known to improve the chemical and mechanical properties of bioceramics. It was reported by Mukhopadhyay and Basu (2007) that dopant was found to restrict possible grain growth consequently producing nano-scaled grains. However, it is possible that adding dopant into forsterite could deteriorate the thermal stability of forsterite causing the formation of secondary phases known as enstatite and periclase which is undesirable to the mechanical properties (Sarve and Hummel, 1962; Meilcarek et al., 2004).

Zinc oxide (ZnO) was known to provide positive stimulatory on the bone construction by subduing osteoclastic bone resorption (Yamaguchi, 2010; Miao et al. 2005). It was conducted by Bandyopadhyay et al. (2007) that doping ZnO on HA produced significantly higher density as compared to the undoped HA. Owing to this appealing result obtained, sintering additives in particular ZnO, was introduced into forsterite. In this work, forsterite powder was first synthesized via mechanochemical method (solid state method) with heat treatment prior to the doping of ZnO. Investigation was conducted on the thermal stability and densification rate of both undoped and doped samples.

2. Materials and methods

2.1. Synthesis of forsterite powder

Forsterite was produced via solid-state method. Two precursors were prepared (Magnesium carbonate basic; Merck, 98% and Mg3Si4O10(OH)2; Sigma-Aldrich, 99%) and mixed with a weight ratio of 1.302:1, respectively. The solvent used for the mixing is ethanol. Each precursor was subjected to ultrasonic for 2 minutes to reduce agglomeration of powders followed by another 30 minutes of ultrasonic after mixing both of the precursors together. Upon ultrasonication, milling was carried for 5 hours using attrition milling machine with 550 rpm. Zirconia balls with a diameter of 3 mm were used as a milling media with ball to powder ratio of 30:1. Prior to the completion of milling (5 mins before), 1 mol of ammonium chloride (Merck) was added into the mixture. Upon drying at 60 ^{o}C for 24 hr, the mixture was crushed and sieved using a metal mesh (212 um). As-obtained powder was then heat treated at 1000 ^{o}C for 2 hours with a ramp rate of 10 ^{o}C/min.

2.2 Doping Stage

Forsterite powder was mixed with zinc oxide (ZnO; Systerm) with weight ratio of 0.5 wt% and 1.0 wt% into a beaker filled with 200 ml ethanol, respectively. Ultrasonic vibration was conducted on the mixtures for 30 min prior to attrition milling for 1 hour at 550 rpm. Similar steps during the synthesis of forsterite powder were taken i.e. milling, drying and sieving. Upon obtaining ZnO doped forsterite powders for both compositions, uniaxial press at 2.5 MPa was done on the powder to form disc-shaped samples. Sintering at 1200 oC to 1500 oC for 2 hours with ramp rate of 10 oC/min was done on the samples. Subsequently, grinding and polishing using silicon carbide (SiC) and diamond paste was carried out prior to body characterization, respectively.

2.3 Characterization Stage

Thermal stability of forsterite was analysed via X-ray diffractometer (XRD) (Empyrean, PANalytical, Netherlands) operating at 45 kV and 40 mA via Cu-Kα radiation source. Step scan and scan speed of 0.02o min-1 and 0.5o min-1 with 2Θ scanning range from 20o to 50o. Joint Committee on Powder Diffraction Standards (JCPDS) reference cards were used to observe the stability of the samples. The lists of the cards used are as follows: forsterite (00-034-0189), periclase (00-043-1022), and enstatite (00-011-0273).

The bulk density of forsterite was measured using water immersion technique, based on Archimedes principle, with distilled water as the medium for immersion. The relative density was calculated by comparing it with the theoretical density of forsterite (ρ = 3.271 g cm−3) [2].

3. Results and discussions

The XRD analysis in Figure 1 showed that upon heat treatment on the milled mixture at 1000 °C for 2 hours with ramp rate of 10 °C/min, only forsterite peaks were observed. No sign of secondary phases such as enstatite and periclase was seen.

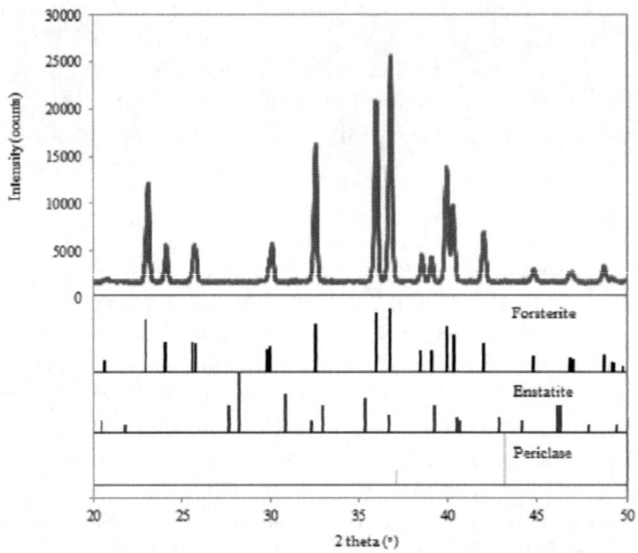

Figure 1. Thermal stability of forsterite powder upon heat treated at 1000 °C/2 hours at 10 °C/min.

As a controlling factor, Figure 2 showed the XRD result of undoped samples upon sintered at its lowest and highest sintering temperatures. Comparative study was done with both 0.5 wt% and 1.0 wt% ZnO doped forsterite samples shown in Figure 3 and 4. It was found that addition of ZnO into forsterite did not lead to any decompositions or chemical changes that could harm the overall densification of forsterite. It was reported by Tavangarian and Emadi (2009) that the presence of either periclase and/or enstatite would harm the overall mechanical performance of forsterite ceramics [12].

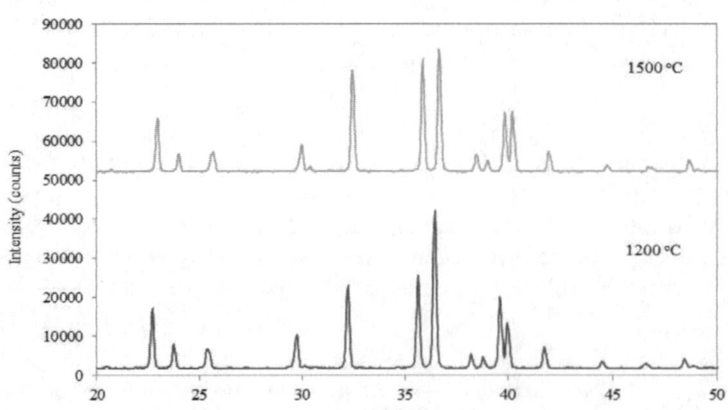

Figure 2. Thermal stability of undoped forsterite bulk upon sintering at the 1200 °C and 1500 °C.

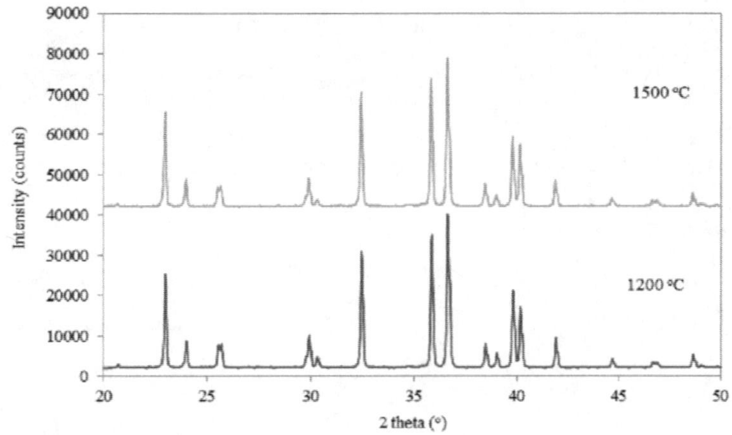

Figure 3. Thermal stability of 0.5 wt% ZnO doped forsterite bulk upon sintering at the 1200 °C and 1500 °C.

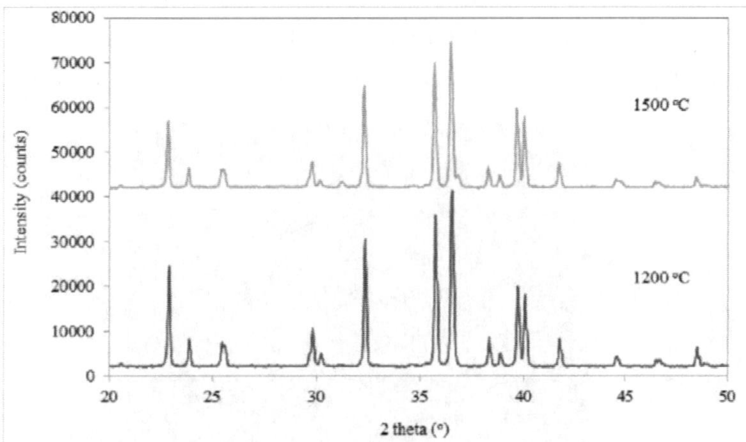

Figure 4. Thermal stability of 1.0 wt% ZnO doped forsterite bulk upon sintering at the 1200 °C and 1500 °C.

In terms of densification of forsterite bulk, all samples showed an increase in density with sintering temperature. Nevertheless, ZnO doped samples showed higher relative density as compared to undoped sample with a maximum densification of 97.7% followed by 95.5% obtained by 0.5 wt% and 1.0 wt% ZnO doped sample, respectively, at 1500 °C. At sintering temperature of 1200 °C to 1300 °C, all samples showed higher rate of densification owing to the formation of grains stage. At this point, the porosities were slowly removed and grains were formed. The relative density obtained in this study is higher than the relative density reported by Ramesh et al. [11] by 7%. The effect of doping ZnO had significantly enhance the densification of forsterite and the microstructure of of the dense 0.5 wt% ZnO doped forsterite was shown in Figure 6. At 1350 °C, all the doped samples have reached its plateau as the relative density began to settle down within the range of 97-98% relative density. The relative density of undoped samples was still increasing and achieved its highest densification at 1500 °C with value of 93.5%.

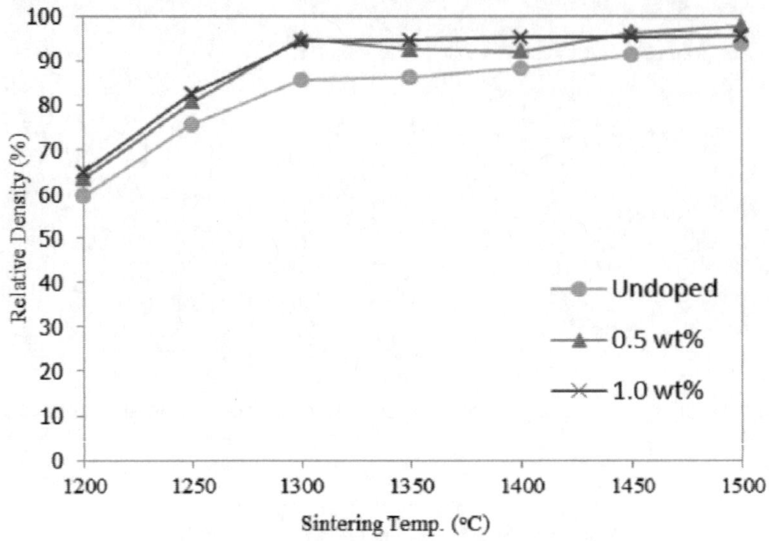

Figure 5. Relative density of undoped and doped forsterite bulk with respect to sintering temperatures.

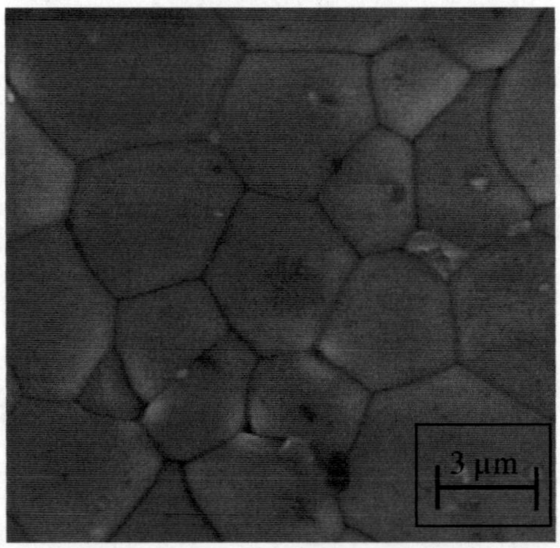

Figure 6. Morphology of 0.5 wt% ZnO doped forsterite sample sintered at 1500°C.

The addition of ZnO into forsterite had proven to be beneficial to the densification of forsterite ceramics. With this appealing result obtained, further study can be conducted to observe on the hardness and fracture toughness of forsterite.

4. Conclusion

In conclusion, we managed to doped ZnO into forsterite without affecting the thermal stability of forsterite and producing phase-pure forsterite upon sintering at 1200 oC to 1500 oC. The highest relative density was obtained by 0.5 wt% ZnO doped sample when sintered at 1500 oC with a value of 97.7% whereas undoped sample only obtained 93.5%. The densification of ZnO doped forsterite showed promising result that could lead to enhancement not only in densification but also other mechanical properties.

IOP Conf. Series: Materials Science and Engineering **206** (2017) 012051 doi:10.1088/1757-899X/206/1/012051

References

[1] Guha A.K., Singh S, Kumaresan, R Nayar, S. and Sinha A 2009 Mesenchymal cell response to nanosized biphasic calcium phosphate composites. *Colloids Surf. B* **73** 146-151

[2] Ghomi H, Jaberzadeh M and Fathi M H 2011 Novel fabrication of forsterite scaffold with improved mechanical properties. *J. Alloys Compd.* **509** 63-68

[3] Tavangarian F, Emadi R, & Shafyei A 2010 Influence of mechanical activation and thermal treatment time on nanoparticle forsterite formation mechanism. *Powder Technol.* **198**(3) 412-416

[4] Cheng L, Liu P, Chen X, Niu W, Yao G, Liu C and Zhang H 2012 Fabrication of nanopowders by high energy ball milling and low temperature sintering of Mg_2SiO_4 microwave dielectrics. *J. Alloys Compd.***513** 373-377

[5] Naghiu M A, Gorea M, Mutch E, Kristaly F, & Tomoaia-Cotisel M 2013 Forsterite Nanopowder: Structural Characterization and Biocompatibility Evaluation. *J. Mater. Sci. Technol.* **7** 628-632

[6] Sun H T, Fujii M, Nitta N, Mizuhata M, Yasuda H, Deki S, & Hayashi S 2009 Molten-Salt Synthesis and Characterization of Nickel-Doped Forsterite Nanocrystals. *J. Am. Ceram. Soc.* **92**(4) 962-966

[7] Douy A 2002 Aqueous Syntheses of Forsterite (Mg_2SiO_4) and Enstatite ($MgSiO_3$). *J. Sol-Gel Sci. Technol.* **24**(3) 221-228

[8] Zhong H, Wang L, Fan Y, He L, Lin K, Jiang W, Jiang C and Chen L 2011 Mechanical properties and bioactivity of β-Ca2SiO4 ceramics synthesized by spark plasma sintering. *Ceram. Int.* **37**(7) 2459-2465

[9] Evis Z and Webster T J 2011 Nanosize hydroxyapatite: doping with various ions. *Adv. Appl. Ceram.* **110**(5) 311-321

[10] Ni S, Chou L, and Chang J 2007 Preparation and characterization of forsterite (Mg2SiO4) bioceramics. *Ceram. Int.* **33**(1) 83-88

[11] Ramesh S, Yaghoubi A, Lee K Y, Chin K M, Purbolaksono J, Hamdi M, & Hassan M A. 2013 Nanocrystalline forsterite for biomedical applications: synthesis, microstructure and mechanical properties. *J. Mech. Behav. Biomed. Mater.***25** 63-69

[12] Sanosh K P, Balakrishnan A, Francis L & Kim T N 2010 Sol–gel synthesis of forsterite nanopowders with narrow particle size distribution. *J. Alloys Compd.* **495**(1) 113-115

[13] Suresh Kumar G, Thamizhavel A, Yokogawa Y, Narayana Kalkura S and Girija E K 2012 Synthesis, characterization and in vitro studies of zinc and carbonate co-substituted nanohydroxyapatite for biomedical applications. *Mater. Chem. Phys.* **134** 1127-1135

IOP Conf. Series: Materials Science and Engineering **206** (2017) 012052 doi:10.1088/1757-899X/206/1/012052

Response surface method (RSM) for optimization of ionic conductivity of membranes polymer electrolyte poly (vinylidene fluoride) (PVDF) with polyvinyl pyrrolidone (PVP) as pore forming agent

E R Dyartanti.[1,2,*], H Susanto[2,*], I N Widiasa[2] and A Purwanto[1]

[1]Department of Chemical Engineering, Universitas Sebelas Maret, Indonesia.

[2]Department of Chemical Engineering, Diponegoro University, Indonesia.

*E-mail: endah_rd@uns.ac.id

Abstract. The Membranes Polymer Gel Electrolyte (MPGEs) based poly (vinylidene fluoride) (PVDF) was prepared by a phase inversion method using polyvinyl pyrrolidone (PVP) as a pore-forming agent and N, N-dimethyl acetamide (DMAc) as a solvent and water as non solvet. The membranes were then soaked in 1 M lithium hexafluorophosphate (LiPF6) in ethylene carbonate (EC) / dimethyl carbonate (DMC) / Diethyl carbonate (DEC) (4:2:4 %vol) solution in order to prepare polymer electrolyte membranes. The MPEGs PVDF/PVP/Nanoclay was applied using central composite design (CCD) experimental design to obtain a quantitative relationship between selected membranes prepared parameters namely (PVDF, PVP as pore forming agent and nanoclay filler concentration) and Ionic conductivity MPEGs. The model was used to find the optimum ionic conductivity from polymer electrolyte membranes. The polymer electrolyte membranes show good ionic conductivity on the order of 6.3 - 8.7 x 10^{-3} S cm^{-1} at the ambient temperatures. The ionic conductivity tended to increase with PVP and nanoclay concentration and decrease with PVDF composition. The model predicted the maximum ionic conductivity of 8.47 x 10^{-3} S cm^{-1} when the PVDF, PVP and nanoclay concentration were set at 8.01 %, 8.04 % and 10.12%, respectively. The first section in your paper.

1. Introduction

Lithium ion battery has many excellent properties such as high working voltage, high energy density, long cycle life, low self-discharge rate, low pollution, no memory effect [1]. However, the use of liquid electrolytes which has highly reactive properties in conventional lithium ion batteries will increase their internal pressure in the cell and lead to unsafe conditions. It conditions will also lead to a leakage problem. The use of polymer gel electrolyte instead of liquid electrolyte is an effective way to overcome the weakness of the liquid electrolyte because GPE consisting of a polymer matrix, plasticizer and additives.

The gel polymer electrolyte membrane with a pore size of nanometers up to micrometers have good electrochemical properties when soaked in an organic solution lithium salts. The solution containing sufficient charge carrier as a medium for ion mobility. Ionic conductivity of polymer electrolyte membranes depends on not only the polymer type, but also pore size, porosity, tortuosity, and pore distribution. Its characteristic is determined by processing method.

Phase inversion is a technology that is widely used in the manufacturing of porous polymer electrolyte. The porous polymer electrolyte with pore size within micro-meter range prepared by phase inversion can entrap large amounts of fluid electrolyte which increase the ratio of uptake and improves ionic conductivity [2]. However, the abundant micro-pores and absorption of liquid electrolyte will weaken the mechanical strength of the polymer matrix. Adding some inorganic particles such as ZrO_2, TiO_2, MgO into the polymer matrix was confirmed some effects in solving this problem [3–7].

Poly (vinylidene fluoride) or Poly (vinylidene difluoride) (PVDF) is one of the most promising host matrix for the polymer electrolyte because it has many excellent properties such as good chemical resistance, high dielectric constant, and thermal stability [8–10]. The use of pore-forming agent can enhance the ionic conductivity of membrane at room temperature by improving the pore structure within the membranes. Some literature have reported that the polyvinyl pyrrolidone (PVP) can be used as a pore-forming agent to improve polymer electrolyte ionic conductivity [11–13]. Xiao and his coworkers explored the effect of PVP, on the properties of PVDF based gel polymer electrolyte. The result showed that PVP can be used as pore forming agents for PVDF polymer. Further, the results showed that the physicochemical and electrochemical properties of electrolyte with PVP, indicate a promising pore forming agent for polymer electrolyte in rechargeable lithium-ion polymer battery[14]. The polymer electrolyte has a weak mechanical strength, because of the plasticization of the entrapped liquid electrolyte in the polymer matrix. Several studies have been performed to balance the ionic conductivity at room temperature with its mechanical strength, such as by adding inorganic particles (ZrO_2, Al_2O_3, TiO_2, and MgO) into a polymer matrix [15,16].

The addition of Al_2O_3 nanoparticles can also increase the ionic conductivity, because Lewis acidbase effects of the - OH groups can weaken the interaction between the Li^+ ion and F atoms of the polymer chain as has been shown by Zhaohui[17]. Li et al. successfully prepared a P(VDF-HFP) polymer electrolyte with urea as pore-forming agent and nano-SiO_2 as additives. Their research found that urea and nano-SiO_2 can improve the performance of polymer electrolyte, such as ionic conductivity at room temperature, mechanical strength, electrochemical stability, lithium-ion transference number, and interfacial properties. The performance of Battery lithium ion cell assembled with GPEs also demonstrated excellent properties[18].

Design of experiment (DOE) method is a statistically based approach that creates the numerical model, determines the importance of parameters, and optimizes a process . In this case, response surface methods (RSM) suitable to be applied because of its simplicity and its ability to consider the interaction between the parameters [19,20]. The RSM is a straightforward strategy that might consider collaborations around the parameters. RSM consists of five steps: preference of requisite parameters and their levels, backbone of necessary tests by suitable design, performing experiments and

regression analysis, computation of optimum operating condition, and validation of the model [21–24]. The analyses for optimizing the PVDF/PVP ionic conductivities were decided by the Central composite design (CCD) method. The PVDF, PVP and Nanoclay concentration were selected as influential parameters. The ionic conductivity was chosen as response.

The purpose of this study is to investigate the quantitative relationship among three parameters, i.e., PVDF, PVP and nanoclay concentration and their influence on ionic conductivity of MPGEs using the method of RSM. The Central Composite Design (CCD) used in this testing performed to get the optimal parameters values, which resulted the maximum ionic conductivity of MPEG, in the quadratic model with three levels for each parameter.

2. Experimental

2.1Materials

PVDF (Sigma Aldrich , MW 534,000) was used as polymer matrix in this work. Polyvinyl pyrrolidone (PVP) with 25,000 g/mol molecular weight from Merck were used as pore forming

IOP Conf. Series: Materials Science and Engineering **206** (2017) 012052 doi:10.1088/1757-899X/206/1/012052

agents in the preparation processes of GPEs. N,N-DimethylAcetamide (A.R., DMAc) was used as solvent directly without further purification and octadecylamine-modified montmorillonite nanoclay (MMT, SigmaAldrich) was used as filler. The electrolyte solution of $LiPF_6$ in solvent mixture of ethylene carbonate (EC) and dimethyl carbonate (DMC) 1 M lithium hexafluorophosphate ($LiPF_6$) in ethylene carbonate (EC)/ dimethyl carbonate (DMC)/ Diethyl carbonate (DEC) (4:2:4 vol.%) solution was supplied by MTI Company.

2.2Membrane preparationo

The nano-clay particles (10% wt of PVDF) were mixed with DMAc solvent (2000 rpm, 30 °C) for 2 h. Amount of PVDF was dissolved in dimethylacetamide (DMAC) under continuous stirring for 4 h at 45°C with the ratio with DMAc of 8-12. PVP powder (4 - 10% wt of PVDF) were then added to the above solution, and kept for 4 h. The solution mixture was cast using an automatic film coater (Adjustable Doctor Blade - MSK-AFA-II, MTI corp). In this experiment, the casting thickness was set to be 250 μm. The membranes were then immediately soaked in deionized water at room temperature for 48 h. The membranes were dried in the air at room temperature for 24 h. The resultant membrane was vacuum dried at 60°C for 12 h and transferred to a dry box at 30°C (H_2O level < 10 ppm).

2.3 Ionic conductivity measurement

PVDF/PVP/Nanoclay membrane gel electrolytes were prepared by soaking the membrane in the liquid electrolyte, 1 M $LiPF_6$ in (EC) / (DMC) / (DEC) (4:2:4 % vol), MTI Corp. The ionic conductivities were measured by electrochemical impedance spectroscopy (EIS). The samples were measured in coin cells CR-2032, where the gel membranes activated with liquid electrolyte were sandwiched between two stainless steel electrodes. Impedance data were obtained with Hioki 3532-50 LCR in the frequency range 50 Hz to 1 MHz at 25°C. The ionic conductivity was calculated from (1):

$$\sigma = \frac{1}{(\mathrm{Rb\ A})} \quad [\mathrm{in\ mS\ cm^{-1}}] \tag{1}$$

where σ is the ionic conductivity, Rb is the bulk resistance (Ω), l is the thickness (cm) of the membrane separator, and A (cm^2) is the area of the stainless steel electrode. Rb was obtained using the high frequency intercept on the real axis. The effective area was determined by the circular stainless steel electrodes, and the thickness was around 300 μm.

2.4 Response Surface Method

To examine the relationship between response (Y) and independent variables, and optimizing the value variables to predict the maximum value the response in this research used RSM method. Central Composite Design (CCD) is an advantageous design for sequential experiments. Therefore, CCD was applied to determine the effect of operational variables on the ionic conductivity in this work.

3. Results And Discussions

3.1 Ionic conductivity study

Ionic conductivity of the electrolyte membrane depends on the number of ion carriers and ion mobility. The concentration of dissolved ions will affect the effective number of ion carriers. While the mobility of the polymer chain segments will facilitate the mobility of ions in the polymer electrolyte. A membrane that has a high absorption liquid electrolyte having a structure are fully interconnected porous. The process of absorption takes place rapidly, and maximum electrolyte absorption will be reached in just a few minutes. The membrane porosity, which absorbs the liquid electrolyte and the gelation of the membrane, is important factor that influences electrolyte absorption [23].

Table 1. Experimental design parameters and properties of membrane polymer electrolyte

No.	Factors						Ionic conductivity $x10^{-3}$ S cm^{-1}	
	PVDF (ratio with DMAc)		PVP (%wt PVDF)		Nano clay (%)			
	Actual values	Coded values	Actual values	Coded values	Actual values	Coded values	Exp	Calculated from the model
1	8	-1	4	-1	4	-1	6.6130	6.652
2	8	-1	4	-1	12	1	8.0101	7.792
3	8	-1	10	1	4	-1	7.9609	7.761
4	8	-1	10	1	12	1	8.2235	8.136
5	12	1	4	-1	4	-1	6.3475	6.365
6	12	1	4	-1	12	1	7.5737	7.505
7	12	1	10	1	4	-1	7.1173	7.474
8	12	1	10	1	12	1	7.9547	7.761
9	6.63641	-1.68179	7	0	8	0	8.4756	8.712
10	13.36359	1.68179	7	0	8	0	8.3801	8.229
11	10	0	1.95462	-1.68179	8	0	6.2372	6.344
12	10	0	12.04538	1.68179	8	0	7.5806	7.566
13	10	0	7	0	1.27283	-1.68179	6.5162	6.363
14	10	0	7	0	14.72717	1.68179	7.3890	7.637
15	10	0	7	0	8	0	8.0285	8.026
16	10	0	7	0	8	0	8.0408	8.026
17	10	0	7	0	8	0	8.0285	8.026
18	10	0	7	0	8	0	8.0469	8.026
19	10	0	7	0	8	0	7.9978	8.026
20	10	0	7	0	8	0	8.0347	8.026

The EIS results show a sloping straight line, representation of the electrolyte / electrode double layer capacitance behavior observed in the entire frequency evaluation range. It can be concluded that there is only a small contribution to the resistance of the resistance boundary total. Intercept of a straight line is a bulk resistance of the electrolyte (Rb). The resultant polymer electrolyte composed of three parts: (1) Solid polymer, (2) partially amorphous swelling membranes, and (3) a liquid electrolyte enclosed in the membrane pores.

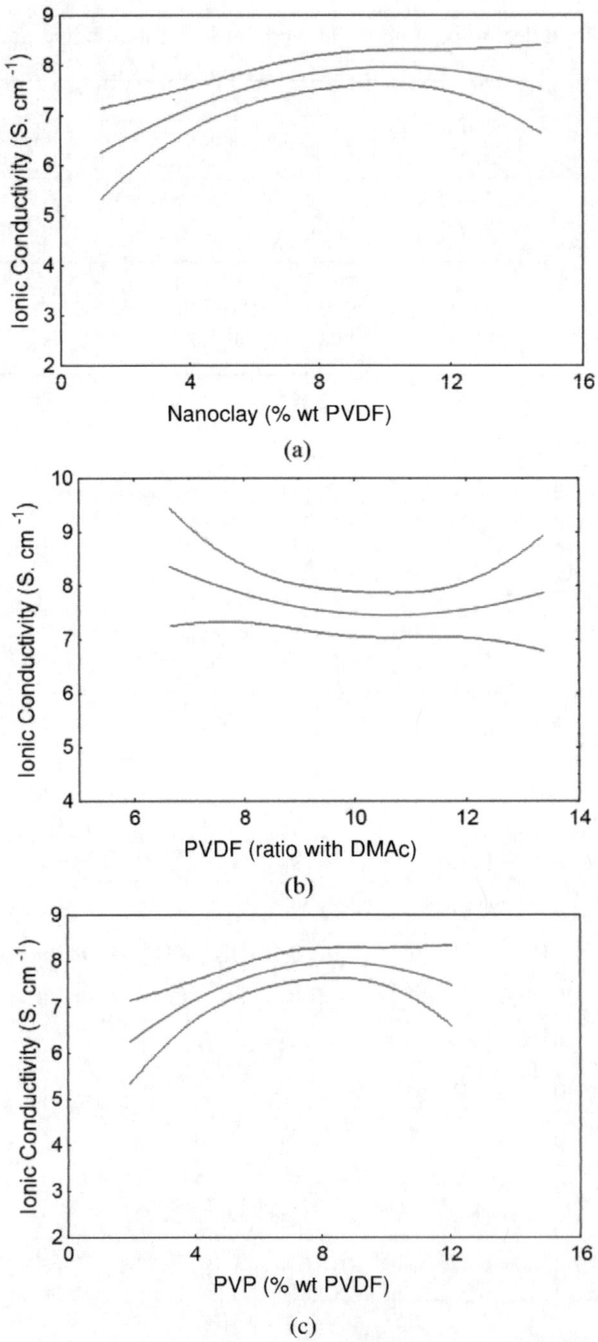

Figure 1. Effect of process parameters on ionic conductivity of PVDF/PVP/Nanoclay polymer electrolyte. a). PVDF concentration. b). PVP concentration. c). Nanoclay concentration. The black lines indicate the 95% confidence level

Two ionic conduction mechanisms occur in PGE: one is the fast conduction along the liquid electrolyte and the other is slower conduction in swelling polymer. The overall ionic conductivity of the polymer electrolyte is strongly influenced by the electrolyte liquid phase. The amount of liquid electrolyte that is adsorbed in the porous of PGE influences more ionic conductivity. Meanwhile its higher conductivity than the solution bounded in the swollen phase of porous polymer is observed.

The effects each parameter on ionic are shown in Figure.1. As shown in Figure. 1(a) ionic conductivity versus the content of nanoclay when PVP and PVDF concentration was maintained at 8 (wt% PVDF) and 1: 9 (ratio PVDFwith DMAc), the increased in ionic conductivity is observed as nanoclay filler was added until reached a maximum conductivity (8.476 mS cm^{-1}) at 8 wt% PVP. However, beyond this amount it tended to decrease. Several researches reported that lewis acid base interactive between particle site on the surface and the electrolyte could be the reason. Association between filler and polymer chain have been same reason. The higher ionic conductivity (8.476 mS $cm^{1}$from Figure. 1(a)) of PVDF/PVP/Nanoclay membrane suggested that membranes contained more individual nanoclay filler than its intercalated assemblies. The primary reasons for the enhancement of ionic conductivity have been reported as the Lewis acid–base interactions between the functional group located on the surface of clay platelets and the electrolyte polar groups, and association between the filler particle and the polymer chains [25–27]. There were two distinct stages of liquid electrolyte adsorption in membranes with porous structure. The first, it filled the porous of membrane, next the electrolyte penetrated polymer chain resulted swollen phase amorphous. In Figure. 1(b), it is observed that, at constant total polymer concentration, the ionic conductivity of the membranes decreased with increasing PVDF composition; this is due to the decrease in solution viscosity with increasing PVDF composition.

In order to investigate the effect of PVP on the properties of the polymer electrolyte, the ionic conductivity of polymer electrolyte based on the PVP and PVDF was measured as a function of PVP content. Figure 1(c), shows the ionic conductivity versus the content PVP. The ionic conductivity increased with increasing PVP content until reached its maximum value at PVP = 7 wt%. Beyond this value it decreased. This is related to the pore volume of the polymer film. With PVP content increase, the pore volume and the electrolyte uptake also increase, so the conductivity of the polymer electrolyte also increases [12]. The results showed the membranes pore structure makes ion pass easily between the electrodes. As compared to the value of ionic conductivity in the electrolyte solution $LiPF_6$ is 10.5 mS cm^{-1} (MTI SDS).

3.2 Response surface development

To evaluate the full quadratic response surface model from experimental design, an analysis of variance (ANOVA) was used. The result is presented in Table 2. From the test of significant model in this table, P values of the model <0.0001 indicates that the model is significant. Exactly between the data and model indicated by $R^2 = 0.9587$. The fitted second-order model for ionic conductivity was :

$$Y = +6.55256\text{E-}003 - 8.46438\text{E-}004 * A + 9.22844\text{E-}004 * B + 5.07913\text{E-}004 * C - 8.54167\text{E-}006 * AB + 6.09375\text{E-}006 * AC - 1.59375\text{E-}005 * BC + 3.92836\text{E-}005 * A^2 - 4.20555\text{E-}005 * B^2 - 2.26618\text{E-}005 * C^2 \quad (2)$$

where Y is variable response, A, B, and C are the values of parameter (PVDF concentration, PVP concentration, and Nanoclay concentration), respectively. Significance of the factors was analyzed in order to understand the level of influence of each parameter. The parameter A (PVDF), B (PVP) and C (Nanoclay) are significant as they have P values of less than 0.0001, respectively. The interaction terms AB and AC have a large P value of 0.4719 and 0.4932 indicating that this term is insignificant. The other interaction factor BC and the second-order terms A^2, B^2 and C^2 have small P values (<0.0001).

Eliminating the insignificant terms will enhance the efficiency of the model. After removing insignificant parameters in the model (AB and AC), the modified equation in uncoded variables, it is given by Eq. 3

$$Y = +6.66297 E\text{-}003 - 8.57480 E\text{-}004 * A + 8.37427 E\text{-}004 * B + 5.68851 E\text{-}004 * C \ (3) - 1.59375 E\text{-}005 * BC + 3.92836 E\text{-}005 * A^2 - 4.20555 E\text{-}005 * B^2 - 2.26618 E\text{-}005 * C^2 \quad (3)$$

Table 2. Estimated regression coefficients for the quadratic polynomial model and the analysis of variance for the experimental results.

Ionic conductivity ANOVA for response surface quadratic model					
Source	Sum of Square	dF	Mean Square	F value	p-value Prob>F
Model	8.717E-006	9	9.685E-007	25.76	<0.0001
A-PVDF	2.817E-007	1	2.817E-007	7.49	0.0209
B-PVP	1.804E-006	1	1.804E-006	47.99	<0.0001
C-clay	1.960E-006	1	1.960E-006	52.13	<0.0001
AB	2.101E-008	1	2.101E-008	0.56	0.4719
AC	1.901E-008	1	1.901E-008	0.51	0.4932
BC	2.926E-007	1	2.926E-007	7.78	0.0191
A^2	3.558E-007	1	3.558E-007	9.47	0.0117
B^2	2.065E-006	1	2.065E-006	54.92	<0.0001
C^2	1.895E-006	1	1.895E-006	50.40	<0.0001
Residual	3.759E-007	10			
Lack of Fit	179.4407	5			
Pure Error	2.083E-009	5			
Cor Total	9.093E-006	19			
Std. Dev.	1.939E-004		R^2	0.9587	
PRESS	2.952E-006		Adj R^2	0.9214	
			Pred R^2	0.6753	

Based on the ANOVA results for this modified model (Table 3), the significant terms in model can be shown from the low of P values in the new model. The terms number in the model will always increase as R^2. Hence, smaller R^2 values is thought for the new model, for the sake of smaller the terms.

In Table 1, the ionic conductivity counted by the modified model (Eq. 3) is presented. Figure. 2 shows the plotted of the experimental versus the predicted of ionic conductivity from the modified model. The experimental and models values around the range of parameters under evaluation suggests that the results were well verified.

Making use of the model, the maximum ionic conductivity is equal to 8.47 $mS.cm^{-1}$ corresponding to the test in Table 1. The model predicted the maximum ionic conductivity when the PVDF concentration set at 8.01 %, the PVP concentration set at 8.04 wt.%, and the nanoclay composition set at 10.12 wt.%. Wang produced the polymer electrolyte based on poly(vinylidene fluoride), PVDF, poly(ethylene glycol), PEG, and Li-exchanged vermiculite (Li-VMT) [28]. The maximum ionic conductivity achieved in their work for the membranes was 4.2×10^{-3} $S.cm^{-1}$. Xiao et al. prepared the poly(vinylidene fluoride – co - hexa fluoropropylene) (PVDF-HFP) - based gel polymer electrolytes (GPEs) with polyvinylpyrrolidone (PVP) and urea as a pore forming agent by phase inversion method [14].

Table 3 Statistical parameters obtained from ANOVA for ionic conductivity

Ionic conductivity ANOVA for response surface quadratic model					
Source	**Sum of Square**	**dF**	**Mean Square**	**F value**	**p-value Prob>F**
Model	8.677E-006	7	9.093E-006	35.76	< 0.0001
A-PVDF	2.817E-007	1	2.817E-007	8.13	0.0146
B-PVP	1.804E-006	1	1.804E-006	52.05	< 0.0001
C-clay	1.960E-006	1	1.960E-006	56.53	< 0.0001
BC	2.926E-007	1	2.926E-007	8.44	0.0132
A^2	3.558E-007	1	3.558E-007	10.27	0.0076
B^2	2.065E-006	1	2.065E-006	59.56	< 0.0001
C^2	1.895E-006	1	1.895E-006	54.66	< 0.0001
Residual	4.159E-007	12			
Lack of Fit	141.89	7			
Pure Error	2.083E-009	5			
Cor Total	9.093E-006	19			
Std. Dev.	1.862E-004		R^2	0.9543	
PRESS	1.858E-006		Adj R^2	0.9276	
			Pred R^2	0.7957	

IOP Conf. Series: Materials Science and Engineering **206** (2017) 012052 doi:10.1088/1757-899X/206/1/012052

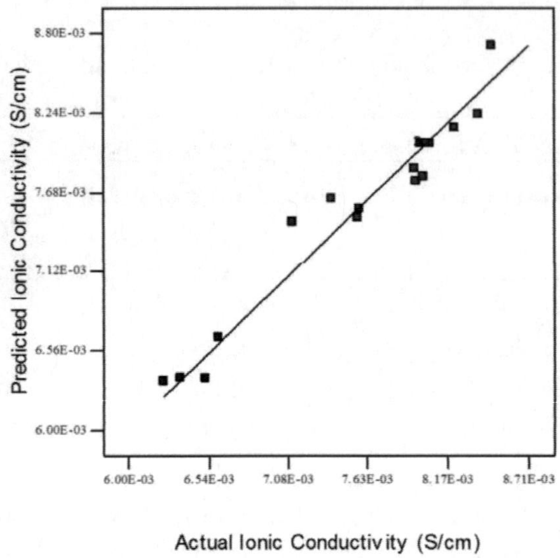

Figure 2. Plot of model predicted ionic conductivity against experimental ionic conductivity

The measured the ionic conductivity at room temperature was 2.212 x 10^{-3} $S.cm^{-1}$ for PVP pore forming and 2.823 x 10^{-3} Scm^{-1} for urea pore forming . Figures 3-5 show the interaction of variables and the optimum level of each parameter for maximum ionic conductivity. The ionic conductivity with different nanoclay and PVP contents at PVDF (ratio with DMAc) 1:10 is presented in Figure 3. It is clear that the nanoclay filler has positive effect on the ionic conductivity. As nanoclay increases from 4 to 12 (%wt PVDF), the ionic conductivity increases from 6.23 to 8.47 x 10^{-3} S. cm^{-1}.

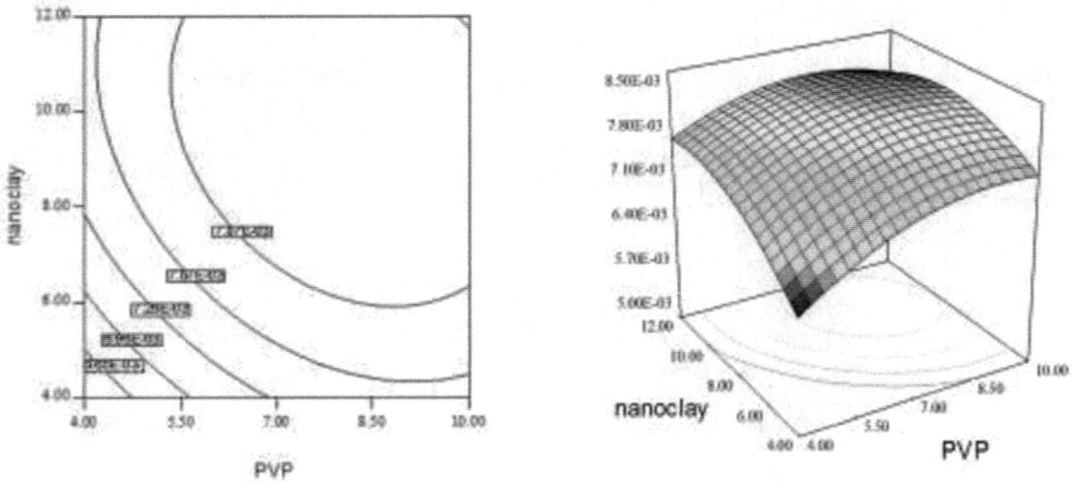

Figure 3. Response Surface and contour plot for different PVP and Nanoclay content at PVDF ratio

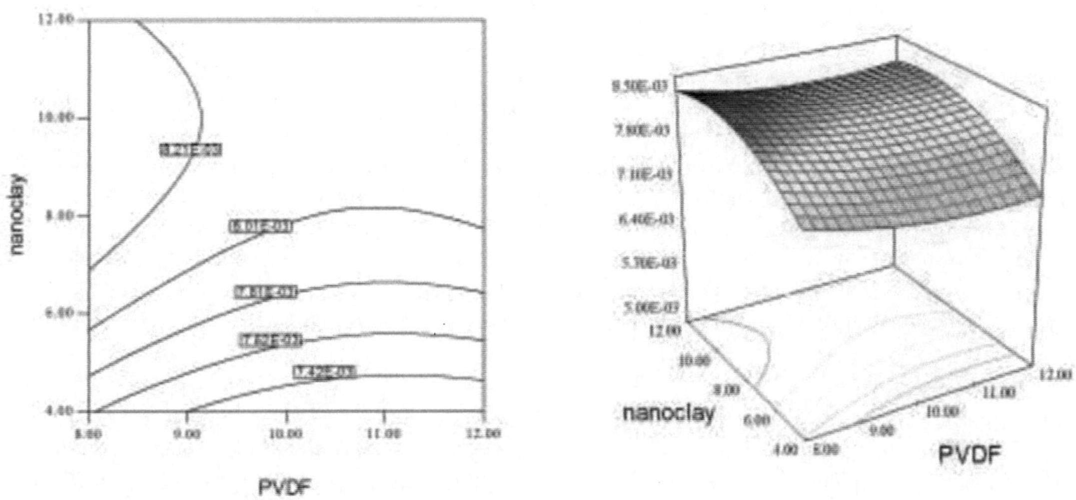

Figure 4. Response Surface and contour plot for different Nanoclay and PVDF (ratio with DMAc) at PVP 8 %wt PVDF

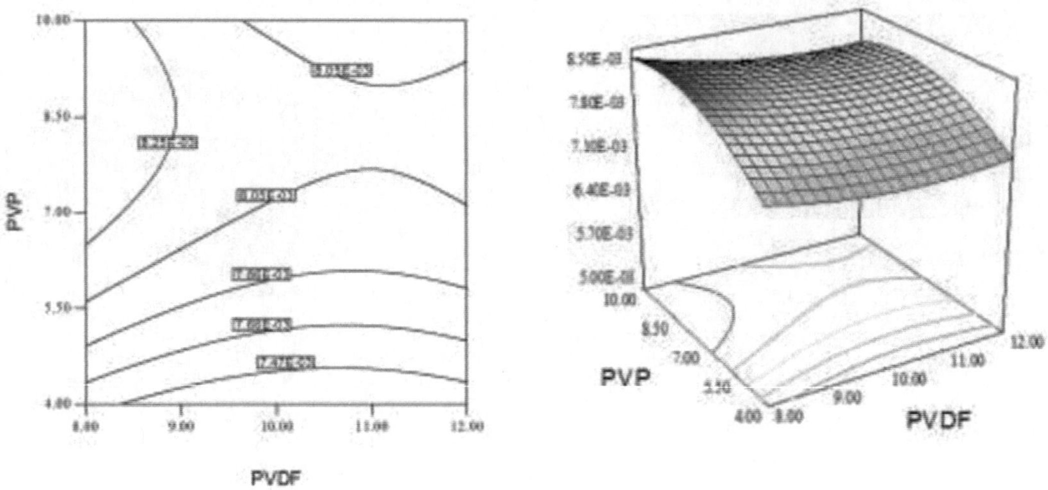

Figure 5. Response surface and contour plot for different PVP and ratio PVDF with DMAc at Nanoclay 10% wt PVDF

Figure. 4 shows the 3D response surfaces and contour plots as a function of nanoclay and PVDF concentration, while the PVP content was kept constant (8 %wt PVDF). The ionic conductivity increases with nanoclay content. The ionic conductivity of PVDF membranes decreases as the PVDF concentration is away from 4 (ratio with DMAc). An interaction between the two mentioned parameters can be noticed in these figures. Figure. 5 shows the effect of PVDF and PVP content on ionic conductivity at constant nanoclay on content of 10 (%wt PVDF). The decrease in the PVDF concentration and the increase in PVP content lead to higher ionic conductivity. Based on the response surface plot shown in Figure. 5, applying high PVP content and low PVDF concentration results in the maximum ionic conductivity.

4. Conclusions

The Polymer electrolyte membranes were prepared by phase inversion method. Here PVDF concentration, PVP and Nanoclay concentration were chosen as parameter with ionic conductivity as a response variable. To obtain the maximum ionic conductivity Central Composite Design (CCD) as a method of RSM was used. For validating the model was investigated by analysis of variance (ANOVA) results. The ionic conductivity of the MPEGs polymer electrolyte increased as PVP and nanoclay concentration were increased. On the contrary the PVDF decreased concentration resulted in the decrease in ionic conductivity, as resulted from the higher uptake of the electrolyte solution in the membranes modified with pore forming agents and filler.

Acknowledgements

This work has been financially supported by the Ministry of Research, Technology and Higher Education (Kemenristekdikti) for Hibah PUPT 2016, Grant number 353 /UN27.21/PN/2016 . Endah R Dyartanti is thankful to the Sebelas Maret University for the scholarship doctoral program. The authors also thankfully acknowledge to The Center for Science and Technology Advanced Materials, National Nuclear Energy Agency Indonesia for helpful and assistance to this experiment.

References

[1] Liao Y H, Zhou D Y, Rao M M, Li W S, Cai Z P, Liang Y, et al 2009 Self-supported poly (methyl methacrylate–acrylonitrile–vinyl acetate)-based gel electrolyte for lithium ion battery *J. Power Sources* **189** 139–44

[2] Pu W H, He X M, Wang L, Jiang C Y and Wan C R 2006 Preparation of PVDF-HFP microporous membrane for Li-ion batteries by phase inversion 2006 *J. Membr. Sci.* **272** 11–4

[3] Aravindan V and Vickraman P 2007 Effects of TiO2 and ZrO2 nanofillers in LiBOB based PVdF/PVC composite polymer electrolytes (CPE) *J Phys D Appl Phys.* **40** 6754

[4] Cao J, Wang L, Shang Y, Fang M, Deng L, Gao J, et al 2013 Dispersibility of nano-TiO 2 on performance of composite polymer electrolytes for Li-ion batteries *Electrochim Acta* **111**674-9

[5] Huang X. A lithium-ion battery separator prepared using a phase inversion process 2012 *J. Power Sources* **216** 216–21

[6] Zhao Y-H, Qian Y-L, Zhu B-K and Xu Y-Y 2008 Modification of porous poly(vinylidene fluoride) membrane using amphiphilic polymers with different structures in phase inversion process *J Memb Sci* **310** 567–76

[7] Kim K M, Park N-G, Ryu K S and Chang S H 2006 Characteristics of PVdF-HFP/TiO 2 composite membrane electrolytes prepared by phase inversion and conventional casting methods *Electrochim Acta* **51** 5636–44

[8] Costa C M, Silva M M and Lanceros-Méndez S 2013 Battery separators based on vinylidene fluoride (VDF) polymers and copolymers for lithium ion battery applications *RSC Adv.* **3**11404

[9] Yang M and Hou J 2012 Membranes in lithium ion batteries *Membranes (Basel)*;**2**:367–83.

[10] Huang X and Hitt J 2013 Lithium ion battery separators: Development and performance characterization of a composite membrane*J Memb Sci* **425–426** 163–8

[11] Chen N and Hong L 2002 Surface phase morphology and composition of the casting Films of PVDF - PVP Blend *Polymer* **43** 1429–36

[12] Wang Z and Tang Z 2003 Characterization of the polymer electrolyte based on the blend of poly (vinylidene fluoride-co-hexafluoropropylene) and poly (vinyl pyrrolidone) for lithium ion battery *Mater.Chem. Phys.* **82** 16–20

[13] Susanto H and Ulbricht 2009. Characteristics, performance and stability of polyethersulfone ultrafiltration membranes prepared by phase separation method using different macromolecular additives 2009 *J. Memb. Sci.* **327** 125–35

[14] Xiao W, Li X, Wang Z, Guo H, Li Y and Yang B 2012 Performance of PVDF-HFP-based gel polymer electrolytes with different pore forming agents *Iran Polym J* **21** 755–61

[15] Li Y, Wang J, Tang J, Liu Y and He Y 2009 Conductive performances of solid polymer electrolyte films based on PVB/LiClO 4 plasticized by PEG 200, PEG 400 and PEG 600 *J Power Sources* **187** 305–11

[16] Padmaraj O, Venkateswarlu M and Satyanarayana N 2013 Effect of ZnO filler concentration on the conductivity, structure and morphology of PVdF-HFP nanocomposite solid polymer electrolyte for lithium battery application *Ionics* **19** 1835–42

[17] Li Z, Su G, Wang X and Gao D 2005 Micro-porous P(VDF-HFP)-based polymer electrolyte filled with AlO nanoparticles. *Solid State Ionics* **176** 1903–8

[18] Li Y, Xiao W, Li X, Miao C, Guo H and Wang Z 2014 Study on performance of a novel P(VDFHFP)/SiO2 composite polymer electrolyte using urea as pore-forming agent. 2014 Ionics **20** 1217–24

[19] Hunter W G and Hunter JS 1978, *Statistics for experimenters: an introduction to design, data analysis, and model building.* Wiley New York.

[20] Myers R H, Montgomery D C and Anderson-Cook C M 2016 *Response surface methodology: process and product optimization using designed experiments*. John Wiley & Sons

[21] Hakkak F and Rafizadeh M 2013 Optimization of electrospun polyacrylonitrile/poly(Vinylidene fluoride) nanofiber diameter using the response surface method. *J Macromol Sci Part B*; **52** 1250–64

[22] Reza M, Mehdi M and Farhad R 2016 Effect of electrospinning on the ionic conductivity of polyacrylonitrile / polymethyl methacrylate nanofibrous membranes : optimization based on the response surface method. *Iran Polym J*, **25(6)** 525-37

[23] Hakkak F, Rafizadeh M, Sarabi A A and Yousefi M 2015 Optimization of ionic conductivity of electrospun polyacrylonitrile/poly (vinylidene fluoride) (PAN/PVdF) electrolyte using the response surface method (RSM) *Ionics* **21** 1945–57

[24] Nunes-Pereira J, Kundu M, Gören A, Silva M M, Costa C M, Liu L, et al 2016 Optimization of filler type within poly(vinylidene fluoride-co-trifluoroethylene) composite separator membranes for improved lithium-ion battery performance *Compos Part B Eng*. **96** 94–102.

[25] Prasanth R, Shubha N, Hng H H and Srinivasan 2013 Effect of nano-clay on ionic conductivity and electrochemical properties of poly(vinylidene fluoride) based nanocomposite porous polymer membranes and their application as polymer electrolyte in lithium ion batteries *Eur Polym. J*. **49** 307–18.

[26] Raghavan P 2010 Electrochemical performance of electrospun poly(vinylidene fluoride-cohexafluoropropylene)-based nanocomposite polymer electrolytes incorporating ceramic fillers and room temperature ionic liquid *Electrochim Acta* **55** 1347–54.

[27] Priya L and Jog J P 2003 Intercalated poly (vinylidene fluoride)/clay nanocomposites: structure and properties. *J. Polym. Sci. Part B Polym. Phys.* **41** 31–8.

[28] Wang Y-P, Gao X-H, Li H-K, Li H-J, Liu H-G and Guo H-X 2009 Effect of Active Filler Addition on the Ionic Conductivity of PVDF-PEG Polymer Electrolyte *J. Macromol. Sci. Part A* **46** 461–7

IOP Conf. Series: Materials Science and Engineering **206** (2017) 012053 doi:10.1088/1757-899X/206/1/012053

Synthesis and characterization of hybrid composite aerogels from alginic acid and graphene oxide

C J U Co[1,2], A T Quitain[2], J Q Borja[1], N P Dugos[1], M Takafuji[2], T Kida[2]

[1]Department of Chemical Engineering, De La Salle University, Malate, Manila, Philippines 1004

[2]Department of Applied Chemistry and Biochemistry, Kumamoto University, Kumamoto, Japan 860-0862

*E-mail: clarence_co@dlsu.edu.ph

Abstract. Aerogels are one class of solid adsorbents that are gaining considerable attention because of their very high porosity, high specific surface area, and extremely low density. However, most aerogels being studied and used recently are synthetic in nature, which are usually mesoporous silica and metal-organic frameworks (MOFs). As research focus is geared towards sustainable engineering, it is desired to utilize biomass to synthesize aerogels. This study thus aims to produce alginic acid-graphene oxide hybrid composite aerogels and compare them with its existing synthetic counterparts. Alginic acid (AA) is an abundant marine biopolymer that easily forms gels, while graphene oxide (GO) is a nanomaterial consisting of many functional groups. Aerogels made up of AA and GO were successfully synthesized using a sol-gel method. The hydrogel was converted into an aerogel by drying with supercritical carbon dioxide. The percentage of graphene oxide was varied from 0 to 20%. The aerogels were characterized by scanning electron microscopy (SEM), Fourier Transform Infrared Spectroscopy (FTIR), X-ray diffraction (XRD), thermogravimetric analysis (TGA) and nitrogen adsorption–desorption measurements. The addition of GO increased the specific surface area of the aerogel up to a certain point, after which it decreased. The 10% GO-AA aerogel showed the most favourable porosity characteristics with a specific surface area of 177.26 m^2/g and average pore diameter of 53.2 nm. There had been no observable difference in the thermal behaviour of the aerogels with a change in the concentration of graphene oxide.

1. Introduction

Solid adsorbents are being used nowadays in numerous applications: adsorption of metal ions from solution [1], adsorption of dyes [2][3], selective adsorption of gases from bulk gas [4][5], and even as drug carrier systems [6][7], among several others. They are also being eyed as a potential instrument in wastewater treatment [8].

There are several kinds of solid adsorbents being studied in literature, the prominent ones being mesoporous silica, metal-organic frameworks (MOFs), and activated carbon [9]. Yet another class of adsorbents, aerogels, are gaining attention because of their high specific surface area, porosities near unity and very well connected pore space. Aerogels are hydrogels which the liquid content (usually water) contained in the solid framework of the gel is replaced with a gas, leaving behind a porous structure. This cannot be accomplished through air or ambient drying, as passing through the liquid-gas interface would cause capillary pressure to collapse the structure. In order to avoid it, the liquid-gas interface is circumvented by using either freeze drying or supercritical drying [10].

Hybrid composites are materials which have two or more components mixed in the nanoscale level. Although at that level, the mixture seems heterogeneous, at the macroscopic level, they have uniform properties. In addition, formations of hybrid composites have sometimes found to produce a

IOP Conf. Series: Materials Science and Engineering **206** (2017) 012053 doi:10.1088/1757-899X/206/1/012053

synergistic effect, that is, the characteristics of the composite turns out to be better than either starting species [11]. Thus, it is of great interest that this principle be applied in the current study.

In the selection of the materials to be used for the hybrid composite aerogel, it was deemed by the authors that the current trend today is towards biomass utilization, as greener and environmentally-safe industrial products are pursued, and nanoscale technology, which has brought about new materials with unique properties and applications.

As such, in this study, hybrid composite aerogels were formed from the combination of a bio-based polymer, alginic acid, and graphene oxide. Alginic acid is cheap, abundant and can be easily gelated. On the other hand, graphene oxide is a flat molecule with numerous oxygen-containing functionalities, which has a high surface area and acts as a nanofiller [12]. The aerogels formed from this study is promising as it can be used as an adsorbent for different applications.

2. Experimental Procedure

2.1. Reagents and Chemicals

For the synthesis of graphene oxide, powdered graphite, sodium nitrate, potassium permanganate, concentrated sulfuric acid, hydrogen peroxide (30%) and concentrated hydrochloric acid were purchased from Wako Pure Chemical Industries, Ltd. All were used as supplied except for the concentrated hydrochloric acid, which was diluted to 5%.

For the synthesis of the hybrid composite aerogel, alginic acid and sodium hexametaphosphate were purchased from Wako Pure Chemical Industries, Ltd. Calcium hydrogenphosphate dihydrate was purchased from Nacalai Tesque, Inc. and D-(+)-glucono-1,5-lactone (GDL) from Tokyo Chemical Industry Co.

2.2. Methods

2.2.1. Synthesis of Graphene Oxide. The first step was the synthesis of graphene oxide (GO) which was done using a modified Hummer's Method as reported by Koinuma et al. [13]. To produce one batch of GO, four grams of graphite powder and four grams of sodium nitrate were dissolved in 184 mL of concentrated sulfuric acid in a 2-L beaker under an ice bath with vigorous stirring using a magnetic stirrer for 30 minutes. Twenty grams of potassium permanganate was slowly added into the solution under continued stirring. The reaction mixture was transferred into an oil bath at 35°C and was allowed to stand for 40 minutes. The beaker was returned to the ice bath, and 184 mL of water was added very slowly dropwise through the use of a separatory funnel. The beaker was again transferred to the oil bath, whose temperature was now set to 95°C. After 15 minutes, the beaker was returned to the ice bath, after which 400 mL of water was added dropwise. Thereafter, hydrogen peroxide (30%) was added slowly, which caused the changing of the color of the solution from dark brown to yellow as well as the formation of bubbles, until such time that no more bubbles were formed. The solution was centrifugated at 4000 rpm for 10 minutes, after which the liquid portion was drawn off. In order to remove the metal ions, the residue, containing the graphite oxide, was washed with hydrochloric acid for three times by addition of 5% HCl and centrifugation for 4000 rpm for 10 minutes. Washing with water was performed afterwards, with the same procedure as that of HCl, except that water was added and centrifugation was done for 30 minutes. After each stage, the liquid upper portion was discarded. This was also done three times. After washing with water, the solution was sonicated for 5 hours to ensure the complete exfoliation of graphite oxide into graphene oxide. The solution was again centrifugated for 30 minutes at 10000 rpm. This time, the upper liquid portion from each tube was collected and is called the "supernatant" while the residue was discarded. Each batch produced approximately one liter of supernatant, which was a GO suspension in water.

IOP Conf. Series: Materials Science and Engineering **206** (2017) 012053 doi:10.1088/1757-899X/206/1/012053

2.2.2. Synthesis of Hybrid Composite Aerogel. This suspension was then mixed with the bio-based polymer, which was chosen to be alginic acid (AA). The hydrogel was synthesized by a sol-gel method similar to that of Rassis et al. [14] but modified to incorporate the addition of GO as in the method of Alhwaige et al. [12].

An amount of supernatant was obtained and diluted with enough water to make a 25-mL solution of desired concentration of GO. Alginic acid weighing 0.625 g was added and left under constant stirring for at least 7 hours. Sodium hexametaphosphate (0.375 g) was added and stirred for half an hour, followed by the addition of calcium hydrogenphosphate dihydrate (0.474 g) and was mixed for another hour at 40°C. Lastly, GDL weighing 0.375 g was added and stirred for five minutes at room temperature. The solution was transferred into moulds and stored at 4°C for at least 15 hours for gelation.

The hydrogels were subjected to a four-step solvent exchange from water to ethanol, using a series of increasingly concentrated mixtures of ethanol and water: 50% (v/v) for 2 hours, 75% for 2 hours, 100% for 12 hours, and another 100% for 24 hours to ensure the complete exchange of solvent from water to ethanol.

The hydrogels were then converted into aerogels using supercritical drying using $scCO_2$. The solvent exchange was necessary because water is not miscible in liquid CO_2. The gels, now containing ethanol, are submerged in liquid CO_2 for 12 hours to replace the ethanol with CO_2. Afterwards, the pressure was increased to 10 MPa and the temperature to 40°C to turn the CO_2 into its supercritical state. The pressure was then released very slowly until atmospheric, and the aerogels were retrieved.

2.2.3. Characterization Techniques. The HCAs were then subjected to several characterization techniques, such as scanning electron microscopy (SEM), X-ray diffraction (XRD), thermogravimetric analysis (TGA), nitrogen adsorption-desorption measurements (also known as BET method), and Fourier Transform Infrared Spectroscopy (FTIR). SEM images were taken using JEOL JSM-7600F Field Emission Scanning Electron Microscope with an acceleration voltage of 3.0 kV. XRD was conducted using Rigaku MiniFlex with a voltage of 40 kV and a current of 15 mA using Cu-Kα as source. The thermogravimetric analyzer used was Seiko TG/DTA 6300 operating from 35°C to 550°C with a ramp rate of 10 C°/min and aluminum pan as holder of the sample. For the BET measurements, BELSORP-mini II was used. The samples were first dried by heating to 90°C for 6 hours under high vacuum to drive off all the excess moisture and other adsorbed material. The measurements were then conducted at -196°C using nitrogen gas. The FTIR spectra of the samples were obtained using JASCO FT/IR-4100. The sample was crushed with anhydrous KBr in a ratio of roughly 1:9. The mixture was made into a pellet before insertion into the machine.

3. Results and Discussion

3.1. Characterization of the Graphene Oxide

The XRD pattern of graphene oxide is shown in Figure 1. A characteristic peak was observed at around

10°, which is consistent with that reported in numerous studies found in literature. The weak and broad peak at 21° is due to the glass holder that was used in the characterization and is present in all diffraction patterns in this study.

The FTIR spectrum of the graphene oxide is shown in Figure 2. The wide band found in the 37003000 cm^{-1} range indicates the presence of hydroxyl groups, while the several small peaks on the right of 2000 cm^{-1} are indicative of other oxygen-containing groups, such as carbonyl and epoxides. The weaker intensity possessed by the latter groups as compared to the hydroxyl band suggests the relative abundance of the hydroxyl groups, which is indeed the case in graphene oxide.

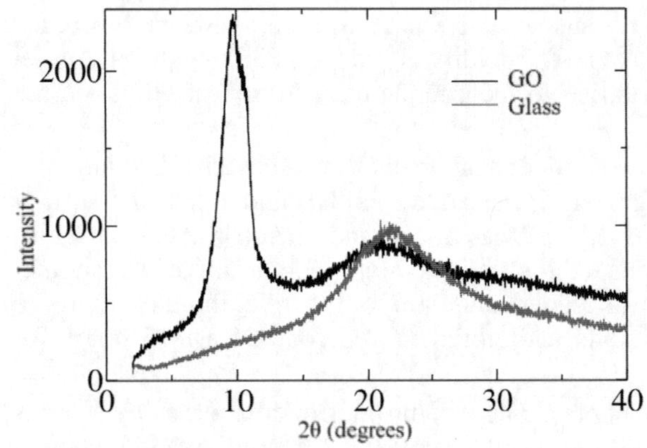

Figure 1. XRD Pattern of Graphene Oxide

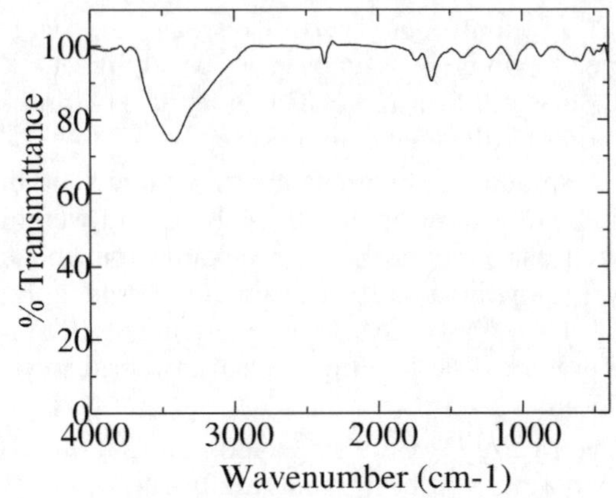

Figure 2. FTIR Spectrum of Graphene Oxide

3.2. Characterization of the Hybrid Composite Aerogel

The supernatant of graphene oxide whose properties have been discussed in the previous section was mixed with alginic acid (AA) in different proportions and made into aerogels. The aerogels were very light, as expected of an aerogel, has a visibly porous structure, and was relatively brittle. Figure 3 shows the aerogels produced with different concentrations of graphene oxide. On average, each monolith is 15.2 mm in diameter and 14.7 mm in height. They have masses ranging from 0.25-0.30 grams, and rough calculations provide a density of 0.10-0.15 g/cm^3. Qualitatively, it can be observed that the color of the monoliths became darker with increasing GO concentration.

Figure 3.. Hybrid Composite Aerogels with varying GO concentration

In order to determine the functional groups present in the hybrid composite aerogel, Fourier Transform Infrared Spectroscopy (FTIR) was performed. The FTIR spectra of aerogels with different GO concentrations are as shown in Figure 4. The graphs show a progression, with the peaks generally becoming more intense with increasing GO concentration.

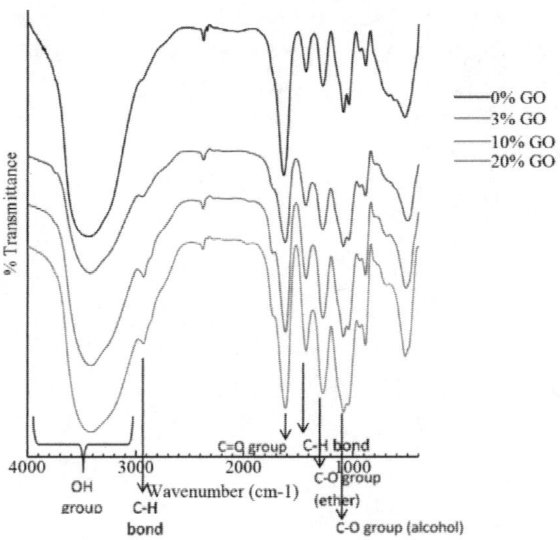

Figure 4. FTIR Spectra of HCAs with Various GO concentrations

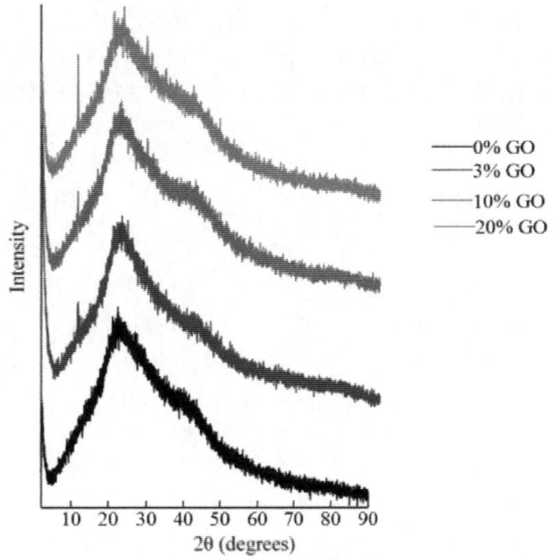

Figure 5. XRD Patterns of HCAs with Various GO concentrations

The FTIR spectra show consistent peaks across all concentrations of graphene oxide. Many of the peaks are attributable to the functionalities present in alginic acid and graphene oxide, such as the hydroxyl group, carbonyl group, and ethers.

The presence of graphene oxide in the aerogel was verified using X-ray diffraction (XRD). The diffraction patterns of the aerogels containing different GO concentrations are shown in Figure 5. The patterns are consistent, with peaks at 2° and 24° and a soft peak at around 42°. A very sharp peak at 11° is not present in pure alginic acid aerogel, yet is present in samples containing graphene oxide. This peak is characteristic to that of graphene oxide, as is shown in Figure 1 and is an indication that it is indeed contained in the aerogel. As the concentration of graphene oxide is increased from 0% to 20%, it can be observed that the peak at 11° also increases in intensity, which is supportive to the claim that the GO is incorporated in the aerogel.

The thermal stability of the aerogels is important because it determines the range of applications it can be intended for. In order to determine the thermal stability, thermogravimetric analysis (TGA) was performed on the samples. The temperature range tested was from 35°C to 550°C. The TGA curves show the percentage of the original mass remaining upon reaching a certain temperature, while the differential thermogravimetric curve (DTG) plots the intensity of the change in the mass of the sample with respect to the temperature. The TGA curves of the different samples are shown in Figure 6. In addition, the TGA and DTG curves of pure GO is added for comparison. The curves across all aerogel samples are similar, ending at roughly 50% of the original weight after reaching 550°C. Furthermore, there is no observable trend in the remaining weight of the sample at a particular temperature with respect to the concentration of GO.

The differential thermogravimetric curve shows the fastest decrease in mass occurring at 50-60°C and from 200-300°C. The same curve for GO show peaks at 150-200°C and 350-400°C. This suggests that GO did not increase the thermal stability of the aerogel, nor does it significantly affect the thermal performance of the aerogel at any specified temperature within the range tested. From this information, it can be inferred that the hybrid composite aerogel cannot be used at temperatures higher than 150°C.

The curves obtained are consistent with the findings of Soares et al. [15]. The first peak corresponds to the dehydration of alginic acid, while the second peak corresponds to its decomposition. Dehydration reaction occurs between carboxylic acid groups, forming an acid anhydride. The residue obtained after the decomposition process was a carbonaceous material, which is also consistent with the results obtained, which was a black powdery substance. In theory, the black carbonaceous residue is largely composed of the carbon backbone of the aerogel. Due to the heating, the functional groups present in the calcium alginate structure have broken off, and a residue containing mostly carbon remains, thereby accounting for its black color.

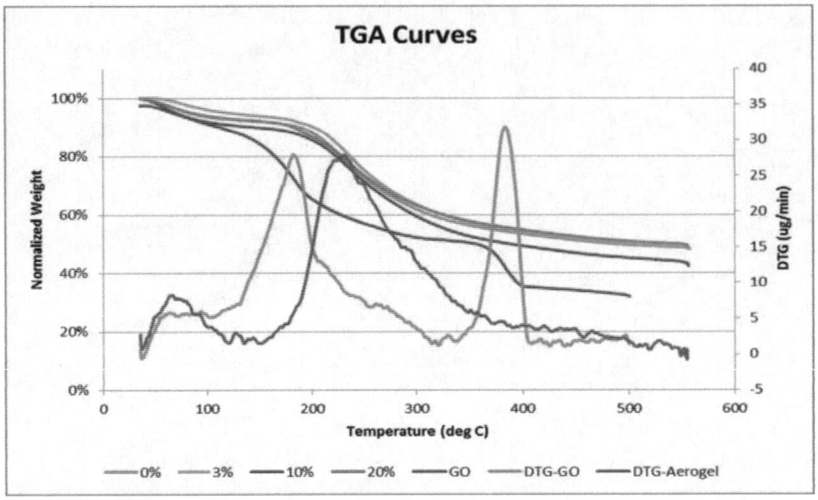

Figure 6. TGA Curves of pure GO and HCAs with various GO concentrations

From the SEM images as shown in Figure 7, all of the samples were observed to exhibit mesoporous and macroporous cavities, which strongly suggest the successful creation of an aerogel. The increase in GO concentration appears to create a more spongy morphology (Fig. 7.c and 7.d), as opposed to the rather dense appearance of the GO-free sample (Fig. 7.a). GO is known to act as a nanofiller, which probably assisted in the uniformity and mesoporosity of the material as it was synthesized during the sol-gel process. This can also explain the increase in the BET surface area as GO is introduced in the alginic acid aerogel, which is discussed further below.

The results from the nitrogen adsorption-desorption measurements are summarized in Table 1. It shows that addition of graphene oxide, in general, increases V_m, the monolayer nitrogen volume adsorbed on the surface of the material, the surface area and the pore volume. The highest values of these characteristics were observed at 10% concentration of graphene oxide. The observed lowering of these characteristics when the concentration was further increased to 20% may be explained by graphene oxide being too concentrated inside the aerogel, thereby being "crowded" and reducing the effective surface area and pore volume of the hybrid composite aerogel. Too small pore diameters pose difficulty for adsorbates to enter into pores and be adsorbed, whereas too large diameters cause the lowering of selectivity and general activity of the material, thus, mesopores, characterized by pore diameters ranging from 2-50 nm, are ideal for this application. The synthesized aerogels are in close range with this and are satisfactory.

(a) (b) (c) (d)

Figure7SEM images of aerogels with various GO concentrations: (a) 0%, (b) 3%, (c) 10%, and (d) 20%.

Table 1. Data from Nitrogen Adsorption-Desorption Measurements

Sample	V_m (cm^3 STP/g)	S Area (m^2/g)	Pore Volume (cm^3/g)	Average Pore Diameter (nm)	Apparent Porosity
0% GO	23.250	101.19	1.3695	54.133	0.9250
3%	29.657	129.08	1.2814	39.707	0.9292
10%	40.726	177.26	2.9584	53.219	0.9268
20%	32.871	149.07	1.5561	43.505	0.9081

4. Conclusions

Hybrid composite aerogels composed of alginic acid and graphene oxide of various concentrations were successfully synthesized. Findings from analyses using FTIR, XRD, BET and SEM support the successful incorporation of graphene oxide in the alginic acid aerogel. This aerogel can be used as an adsorbent because of its porosity, and is a promising material for various applications.

References

[1] Sitko R, Turek E, Zawisza B, Malicka E, Talik E, Heimann J Wrzalik R 2013 Adsorption of divalent metal ions from aqueous solutions using graphene oxide. *Dalton Trans.*42 5682-5689

[2] Sun L, Yu H, & Fugetsu B 2012 Graphene oxide adsorption enhanced by in situ reduction with sodium hydrosulfite to remove acridine orange from aqueous solution. *J. Hazard. Mat.* **203** 101-110

[3] Li Y, Du Q, Liu T, Sun J, Wang Y, Wu S Xia L 2013 Methylene blue adsorption on graphene oxide/calcium alginate composites. *Carbohydrate Poly.* **95** 501-507

[4] Kizzie A, Wong-Foy A, & Matzger A 2011. Effect of humidity on the performance of microporous coordination polymers as adsorbents for CO2 capture. *Langmuir* **27** 6368-6373.

[5] Yu K, Kiesling K, & Schmidt J 2012 Trace Flue Gas Contaminants Poison Coordinatively Unsaturated Metal–Organic Frameworks: Implications for CO2 Adsorption and Separation. *J. Phy. Che. C* **116**(38) 20480-20488.

[6] Garcia-Gonzalez C, Alnaief M, & Smirnova I 2011 Polysaccharide-based aerogels – Promising biodegradable carriers for drug delivery systems. *Carbohydrate Poly.* **86** 1425-1438.

[7] Mehling T, Smirnova I, Guenther U, & Neubert R 2009 Polysaccharide-based aerogels as drug carriers. *J. of Non-Crystalline Solids* **355** 2472-2479.

[8] Gupta V K, Carrott P, Ribeiro Carrott M M L, and Suhas 2009 Low-Cost Adsorbents: Growing Approach to Wastewater Treatment—a Review. *Critical Rev. in Environ. Sci.***35**(5) 783-842

[9] Yu C H, Huang C H, & Tan C S 2012 A Review of CO2 Capture by Absorption and Adsorption. *Aerosol and Air Quality Res.* **12** 745-769. doi:10.4209/aaqr.2012.05.0132

[10] Biedermann A (n.d.). *Aerogel Drying*. Retrieved April 4, 2015, from Introduction to Aerogels and Sol-Gel Science: http://people.virginia.edu/~amb4ht/aeroDrying .html

[11] Husing N, & Hartmann S 2009. Inorganic-Organic Hybrid Porous Materials. In L. Merhari, *Hybrid Nanocomposites for Nanotechno.: Elect., Opt., Magnetic and Biomed. App.* (pp. 131-172). New York: Springer Science+Business Media.

[12] Alhwaige A A, Agag T, Ishida H, & Qutubuddin S 2013. Biobased chitosan hybrid aerogels with superior adsorption: Role of graphene oxide in CO2 capture. *RSC Adv.* **31** 6011-16020.

[13] Koinuma M, Ogata C, Kamei Y, Hatakeyama K, Tateishi H, Watanabe Y, Matsumoto Y. 2012. Photochemical engineering of graphene oxide nanosheets. *The J. Phy. Che. C* **116** 19822-19827.

[14] Rassis D, Nussinovitch A, Saguy I S 1997 Tailor-made porous solid foods. *Int. J. Food Sci. Techno.* **32**(4) 271–278

[15] Soares J P, Santos J E, Chierice G O, & Cavalheiro E T 2004 Thermal behavior of alginic acid and its sodium salt. *Ecletica Quimica* **29**(2) 53-56

IOP Conf. Series: Materials Science and Engineering **206** (2017) 012054 doi:10.1088/1757-899X/206/1/012054

Filter aids influence on pressure drop across a filtration system

S Hajar [1], M Rashid [1,*], A Nurnadia [1], M R Ammar [2], C M Hasfalina [3]

[1]Air Resources Research Laboratory, Malaysia-Japan International Institute of Technology, 54100 UTM Kuala Lumpur, Malaysia

[2]AMR Environmental Sdn. Bhd., Taman Sri Pulai Perdana, 81110 Johor Bahru, Malaysia

[3]Department of Biological and Agricultural Engineering, Universiti Putra Malaysia, 43400 Serdang, Selangor, Malaysia

*E-mail: rashidyusof.kl@utm.my

Abstract. Filter aids is commonly used to reduce pressure drop across air filtration system as it helps to increase the efficiency of filtration of accumulated filter cake. Filtration velocity is one of the main parameters that affect the performance of filter aids material. In this study, a formulated filter aids consisting of PreKot™ and activated carbon mixture (designated as PrekotAC) was tested on PTFE filter media under various filtration velocities of 5, 6, and 8 m/min at a constant material loading of 0.2 mg/mm2. Results showed that pressure drop is highly influenced by filtration velocity where higher filtration velocity leads to a higher pressure drop across the filter cake. It was found that PrekotAC performed better in terms of reducing the pressure drop across the filter cake even at the highest filtration velocity. The diversity in different particle size distribution of non-uniform particle size in the formulated PrekotAC mixture presents a higher permeability causes a lower pressure drop across the accumulated filter cake. The finding suggests that PrekotAC is a promising filter aids material that helps reducing the pressure drop across fabric filtration system.

1. Introduction

Fabric filter is defined as any permeable material upon which or within which particles are deposited by the filtration process [1]. It is widely known as one of the excellent air pollution control system with 99% collection efficiency that possesses the ability to capture fine particles and collect a variety of dusts [2]. However, the main issue associated with this application is its short life span due to wear and tear problem in association with the pressure drop which is closely related to cleaning cycle, energy consumptions and operational costs. Thus, the application of filter aids material during filtration process is one of the simplest techniques that can be used to overcome this problem. Filter aids material works as a pre-coating material that coats on the surface of the filter media with a fresh layer of material during filtration process [1, 3].

It was reported that a good filter aids material should has the ability to form a porous filter cake with low surface area and non-uniform particle size distribution [4]. PrekotAC is a newly developed filter aids material that consists of activated carbon and PreKot™. PrekotAC is introduced as a two in one filter aids material, a pre-coating material and an adsorbent. PrekotAC has the capability to form a porous cake with low pressure drop due to its diverse particle size distribution [5, 6].

However, there are a few factors that need to be considered when studying the pressure drop across the filtration system which is the filtration velocity. Studies had been reported on the effects of filtration velocities on pressure drop where it was found that higher filtration velocity leads to higher pressure drop across a filter cake [7-9].

Hence, this study investigates on the influences of a formulated filter aids on pressure drop across a PTFE filter media under various filtration velocities with a constant material loading.

2. Methodology

2.1. Filter Aids Material

The filter aids material or PrekotAC is a combination of 90% activated carbon and 10% PreKot™ were dried in an oven at 110°C for 24 hours before mixing (ASTM D2867-09). Activated carbon is a commonly used filter aids material works as an adsorbent in the industry while PreKot™ is an amorphous hydrated glassy volcanic rock fused aluminium silicate potential used as a pre-coating material. Table 1 presents the specifications of both materials used in the formulation of the filter aids.

Table 1. Properties of activated carbon and PreKot™

Activated Carbon	PreKot™
Form and color: Powder and black Origin: Coal based pH: 9-11 Ash content: 8%-max Surface area: 850 m²/g	Form and color: Powder and snowy white Fusion point: 1300-1400°C Softening point: 900-1100°C Thermal conductivity: Less than 0.0500kcal/mh°C at 0°C

** PreKot™ is proprietary of AMR Environmental Sdn. Bhd.*

2.2. Experimental Procedures

Figure 1 shows the fabricated laboratory scale fabric filtration system that had been designed in order to study and evaluate the performance of filter aids across a PTFE filter media. The unit consists of dust feeder, filter media holder, pressure manometer, particle counter, rotameter as well as vacuum [10]. Pressure manometer (Extech Instrument Model HD755) and Grimm Portable Laser Aerosol Spectrometer (Model 1.109) were installed in the system in order to measure the pressure drop and particle penetration across a filter cake, respectively.

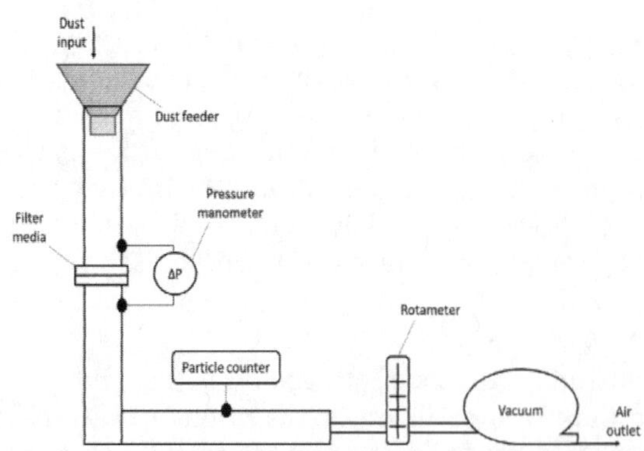

Figure 1. Laboratory scale fabric filtration system

A PTFE filter media with its basic weight of 800, thickness of 1.3 mm and working temperature of less than 250C was used in the study. Three different filtration velocities of 5, 6, and 8 m/min were applied under a constant filter aids loading of 0.2 mg/mm² where the pressure drop and particle penetration across the filter media was observed in each run. Table 2 presents the summary of the operating parameters involved in this study.

IOP Conf. Series: Materials Science and Engineering **206** (2017) 012054 doi:10.1088/1757-899X/206/1/012054

Table 2. Operating parameters used in the study

Filter media	PTFE
Total filtration area	755 mm²
Material loading	0.2 mg/mm²
Filtration Velocity	5, 6, and 8 m/min

3. Results and Discussions

3.1. Effects of filtration velocity on pressure drop

Figure 2 presents the effects of filter aids materials, activated carbon, PreKot™ and PrekotAC on PTFE filter media on pressure drop under a constant material loading of 0.2 mg/mm^2 and three different filtration velocities which showed that the activated carbon and PreKot™ registered the highest and the lowest pressure drop, respectively. On the contrary, PrekotAC showed a consistent pattern of having lower pressure drop compared to activated carbon alone even at the highest filtration velocity of 8 m/min.

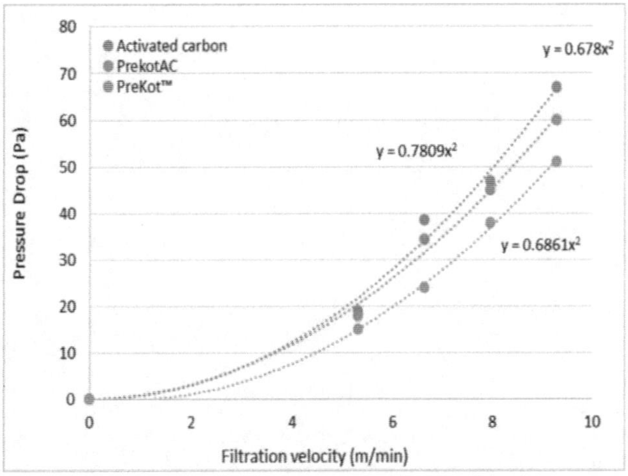

Figure 2. Effects of three different filter aids on pressure drop under different filtration velocities.

Previous studies reported that the difference in particle size distribution of the filter aids material influences the pressure drop performance of the material [6, 10]. As in this case, the activated carbon which consists of fine particles (80% of its particles <75 μm) compared to PreKot™ (20% of its particles <75 μm), resulting in a higher pressure drop across its more compact filter cake characteristic [11, 12]. Thus, a combination of activated carbon and PreKot™ presents a diverse particle size distribution material which has the ability to form a more porous filter cake that helps lowering the pressure drop across it.

As in Figure 2, it was observed that pressure drop increases with high filtration velocity due to the compactness of the filter cake that reduces the permeability (airflow) of the filtration [7, 13, 14]. The compacted filter cake creates a higher resistance which corresponds to the pressure drop performance. [9, 15, 16].

The influences of activated carbon, PreKot™ and PrekotAC filter aids on pressure drop with mass loading of 0.2 mg/mm² can be explained by Eq. 1, 2, and 3, respectively;

$$\Delta P_{AC} = 0.61V_f^2 \quad (1)$$
$$\Delta P_P = 0.58V_{f2} \quad (2)$$
$$\Delta P_{PC} = 0.54V_{f2} \quad (3)$$

Where ΔP_{AC} , ΔP_P , ΔP_{PC} are the pressure drop (in Pa) across the activated carbon, PreKot™, and PrekotAC filter aids, respectively; and V_f (in m/min) is the filtering velocity. As observed, the slope of Eq. 1 (activated carbon) is higher than the other two filter aids illustrating that the change in the pressure drop using activated carbon as filter aids alone is much more greater compared to the other two materials. Nevertheless, the above equations can be used estimate the pressure drop across the PTFE media with filter aids loading of 0.2 mg/mm^2. It anticipated that the slope of the curve will be much greater as filter aids loading increases.

Figure 3 presents the effect of filtration velocity on the permeability of accumulated filter aids cake which showed that permeability decreases as filtration velocity increases. It is observed that the permeability of the filter cake is inversely proportional to its pressure drop due to compactness of filter aids as stated in the previous statement. Permeability indicates the relative ease for the fluid to travel through the pore space in the filter media which is closely related to porosity of the accumulated filter cake [17-18]. Hence, a more compacted filter cake is formed under high filtration velocity that gives high pressure drop due to low permeability across the filter cake.

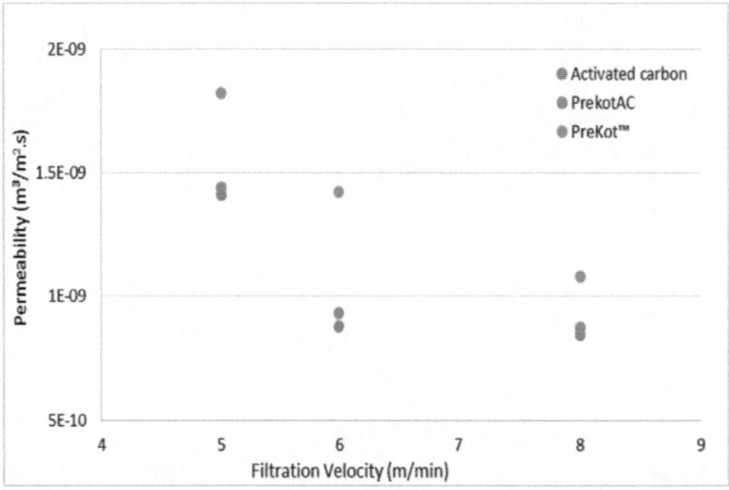

Figure 3. Effects of filtration velocity on filter aids permeability.

The experimental results indicate that filtration velocity play an important role affecting the pressure drop across filter aids material in a filtration system. The study suggests that PrekotAC is a promising filter aids material as it is not merely two on one filter aids but at the same time it presents better performance in terms of pressure drop across its filter cake compared to activated carbon.

4. Conclusions

A study on the influence of filter aids material on pressure drop under different filtration velocities has been reported. Filtration velocity in a filtration process is one of the parameters that could influence the pressure drop performance where a higher filtration velocity leads to higher pressure drop across a filter cake. However and as observed in this study, the characteristics of filter aids also influence the pressure drop across the media in filtration process. The finding suggests that PrekotAC is a promising two in one filter aids material to be applied in the actual filtration process.

Acknowledgement

The authors, S. Hajar and A. Nurnadia, both are PhD students, wish to express their gratitude to MJIIT, Universiti Teknologi Malaysia for financial fellowship.

References

[1] Purchas D 1996 *Handbook of Filter Media: 1st Edition*. Elsevier Advanced Technology

[2] Cooper C D and Alley F C 2002 *Air Pollution Control; A Design Approach. 3rd Edition* (Waveland Press, Inc.) p 177-208

[3] Purchas D B and Sutherland K 2002 *Handbook of Filter Media (Second Edition)*. (Elsevier Science). p 411-452

[4] Cheremisinoff N P 1998 *Liquid Filtration (Second Edition)*. (Butterworth-Heinemann). p 19-58.

[5] Hajar S, Rashid M, Nurnadia A, Norelyza H and Ammar M 2014 PrekotAC as Filter Aids for Efficient Dust Separation in a Fabric Filter. *J. Techno.* **67**(4) 29–31

[6] Nurnadia A, Rashid M, Hajar S and Ammar M R 2015 Effect of Filter Aids on Pressure Drop and Permeability of Two Different Filter Media. *Adv. Mat. Res.* **1125** 245-249

[7] Thomas D, Contal P, Renaudin V, Penicot P, Leclerc D and Vendel J 1999 Modelling Pressure Drop in HEPA Filters during Dynamic Filtration. *J. Aerosol Sci.* **30**(2) 235-246

[8] Fabbro L D, Laborde J C, Merlin P and Ricciardi L 2002 Air flows and pressure drop modelling for different pleated industrial filters. *Filtr. Sep.* **39**(1) 34-40

[9] Saleem M, Krammer G, Khan R U and Tahir M S 2012a Influence of Operating Parameters On Cake Formation in Pilot Scale Pulse-Jet Bag Filter. *Powder Technol.* **224** 28-35

[10] Hajar S, Rashid M, Nurnadia A and Ammar M R 2015 The Characteristics of a Formulated Filter Aids for Fabric Filters. *Powder Technol.* **283** 315-320

[11] Al-Otoom A Y 2005 Prediction of the Collection Efficiency, the Porosity, and the Pressure Drop across Filter Cakes in Particulate Air Filtration. *Atmospheric Environ.* **39** 51-57

[12] Song C B, Park H S and Lee K W 2006 Experimental Study of Filter Clogging with Monodisperse PSL Particles. *Powder Techno.* **163** 152-159

[13] Wang L K, Pereira N C and Hung Y T 2004 *Air pollution control engineering*. (Humana Press). p 59-95.

[14] Saleem M and Krammer G 2007 Effect of Filtration Velocity and Dust Concentration on Cake Formation and Filter Operation in a Pilot Scale Jet Pulsed Bag Filter. *J. Hazard. Mat.* **144** 677-681

[15] Chen Y S and Hsiau S S 2009 Influence of filtration superficial velocity on cake compression and cake formation. *Chem. Eng. and Proc.* **48** 988-996

[16] Saleem M, Krammer G and Tahir M S 2012b The Effect of Operating Conditions On Resistance Parameters of Filter Media and Limestone Dust Cake for Uniformly Loaded Needle Felts in A Pilot Scale Test Facility at Ambient Conditions. *Powder Technol.* **228** 100–107

[17] Schmidt E and Pliz T 1996 Raw gas condition and other additional techniques for improving surface filter performance. *Filtr. Sep.* **33** 409-415

[18] Ma Y and Shishoo R 1999 Permeability Characterization of Different Architectural Fabrics. *J. Composite Mat.*. **33**(8) 729-750

Promotion of nucleation for nano-particle formation by two-stage microwave irradiation

S Nishijima [1], S Matsumura [1], Y Asakuma [1*] and A Saptoro [2]

[1]Department of Chemical Engineering, University of Hyogo, 2167 Shosha, Himeji, Hyogo 671-2280, Japan
[2]Department of Chemical Engineering, Curtin University Sarawak Campus, CDT 250, Miri 98009, Sarawak, Malaysia

*E-mail: asakuma@eng.u-hyogo.ac.jp

Abstract. Microwave has been widely used for nano-particle synthesis because rapid growth and mono-dispersed particle size can be obtained. In our previous work, it was found that bubble formation during the irradiation is greatly affected by particle size and suspension density. Nevertheless, the underlying mechanism has not been clearly understood, especially pertinent to superheat behavior caused by the higher power when colloidal particle of ferric hydroxide was produced by heating ferric chloride solution under the irradiation. In this study, to prevent superheat behavior, two-stage irradiation was proposed in nano-particle formation process. Based on in-situ measurement data, such as the profiles of bubble size, final particle size and brightness of scattering light of suspension, it is evident that nucleation of nano-particle is promoted by higher power of the first irradiation. As a result, particle number density became higher, and then microwave absorbance energy was evenly distributed to each particle. Due to suppression of heat generation in a particle, bubble size became smaller. Two-stage irradiation became more advantageous in obtaining smaller particle than continuous irradiation because lower power of the second irradiation prevents superheat behavior.

1. Introduction

Microwave has been widely used for nano-particle production in the process of hydrothermal synthesis method due to its ability to result in fine particle of mono-dispersed distribution [1], nano-crystalline [2-5] and novel physical property [6]. Benefits of using microwave such as rapid heating, selective heating and local heating are known as the thermal effect. Recently, the mechanism of particle formation has been studied through in-situ measurement data [7] and the importance of bubble growth around nuclei during the irradiation was observed. Swift bubble growth was promoted by the hot spot [8] due to superheat behavior of longer duration and higher power microwave irradiation at non-equilibrium condition of local heating [9]. Thus, appearance of hot spot is strongly related to the bubble size. However, the occurrence of hot spot could be undesirable because local temperature distribution by evaporation heat may affect the particle growth. To prevent superheat behavior, apart from microwave power and duration, it was found that the particle number density, the type of particle and particle size are very influential [10]. For example, higher suspension density and non-magnetic property are advantageous because of lower heat generation within particles. Moreover, eddy current, which is the origin of superheat, can be easily induced in bigger particle. Therefore, higher suspension density and fine particle must be obtained in real scale nanoparticle synthesis process. Based on data of nano-bubble size profiles under microwave irradiation from our previous study [11,12], bubble size during microwave irradiation can be controlled by the particle number density. Because the microwave absorbance energy is distributed to each particle, the energy of each particle, which is used for the evaporation of solvent, becomes smaller. Although bubble size during the irradiation is strongly related to superheat behavior, it is difficult to detect the appearance and to estimate the frequency. In our previous study [7], rapid bubble growth due to the hot spot was measured as flash brightness of the scattering light and then profiles of the bubble size and the occurrence were predicted. In this study, to prevent superheat behavior, two-stage irradiation is proposed in the nano-particle formation process. In-situ profiles such as maximum bubble size, final particle size and the brightness of nano-particle formation were observed and measured. Firstly, higher power, shorter duration of the first irradiation was expected to promote the particle nucleation. Then, the particles continuously grew due to longer duration, higher power of the second irradiation to prevent superheat because of the higher suspension density. Optimum operation conditions of two-stage irradiation were also investigated by varying the microwave power of the first irradiation.

2. Experimental

In this experiment, ferric chloride was chosen as a solute because colloid particle can be formed through the heating process. Figure 1 shows an overview of the microwave reactor (Micro Denshi Co., Ltd., 2.45GHz) [11]. Ferric chloride was dissolved to distilled water at 0.01 mol/L, and the solution was poured into an optical quartz cell (10 mm × 10 mm × 40 mm; 4 mL). The cell was placed on a sample holder made of polyether ether ketone resin in the center of a microwave reactor. When the laser irradiates a sample containing nano-particles or bubbles, the light was scattered. Finally, the number of scattered photons was counted by a photomultiplier (Hamamatsu Photonics Co., Ltd., model) and the size was calculated though the time-averaged autocorrelation function of photon count (DLS; Dynamic Light Scattering). The dynamic changes in size during or after microwave irradiation were measured every 10–15 s. The irradiation modes were varied to separate nucleation and particle growth process. Two-stage irradiation started with higher irradiation power (400-1000 W) for a short period (2 s). Then, lower power (50 W) was used to irradiate for longer duration. A fluorescent temperature probe (Anritsu Meter Co., Ltd., FL-2000; Optical fiber: FS100-M) was used to measure the temperature near the surface throughout the experiment. Duration of second irradiation was decided until solution temperature reaches to 90 °C. Table 1 summarizes the operating conditions of higher power of first irradiation.

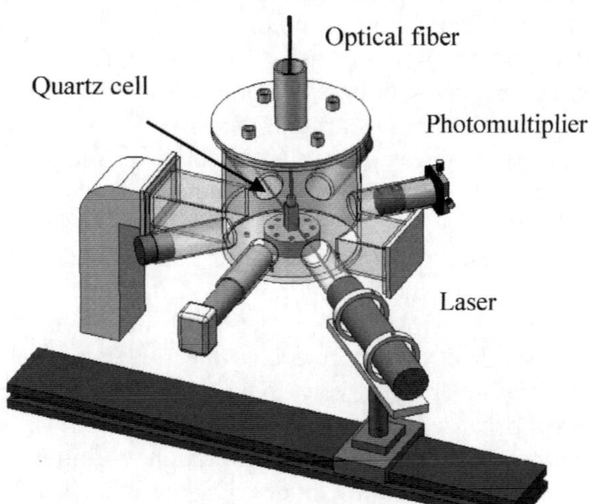

Figure 1. Microwave reactor with DLS

Table 1. Operating conditions

Concentration [mol/L]	Power of the first irradiation [W]	Duration of the first irradiation [s]
0.01	400, 600, 800, 1000	2

Movies of scattering light were captured and the intensity of red line, which is related to the suspension density and number of particle, was extracted to investigate bubble formation behavior around produced particles. For example, Figure 2 shows appearances of the solution inside the cell before and during irradiation. Red line of scattering light was unclear before the irradiation because of no particle formation. Intensity of red line at the center became brighter as the particle nucleation and growth promoted by the irradiation. Bright red spots in Figure 2 are refractions of the incident light on the cell wall. Red color was extracted from the digital movie data among their spots [13], and thus appearance of the particle formation can be detected. Superheat behavior was estimated from the fluctuation of the brightness due to the bumping behavior caused by the reflection of bubble growth during rapid evaporation.

(a) Before irradiation

(b) During irradiation

Figure 2. Photos of solution in cell

3. Results and discussion

3.1. Effect of microwave power and concentration on sizes

Figure 3 shows the profiles of size and temperature in the solution during and after microwave irradiation. Solid and unfilled symbols indicate sizes during and after microwave irradiation, respectively, and the lines indicate the temperature profiles. From this figure, it is apparent that the size increased during microwave irradiation. Temperature also increased with irradiation time, and then gradually dropped after irradiation. In the case of two-stage irradiation, heating rates are different for lower and higher power conditions, and there is a point, where the slope for the temperature profile changed. The maximum size was obtained at around the time when the microwave source was turned off. The gradually decreasing sizes indicated that the microwave irradiation effect indirectly continued for some time due to condensation of vapor. Finally, the size tends to be constant for an extended period of time. Accordingly, the maximum size should be the bubble growth around the nucleus during the irradiation even though the particle grows at the same time. The maximum bubble size and final particle size were plotted against the irradiation power in Figure 4. Solid symbols indicate sizes obtained by continuous mode in previous study [7] and unfilled symbols shows data of two-stage irradiation in the current study. The x-axis represents the power of continuous mode and the first irradiation of two-stage irradiation, respectively. Nuclei or fine particle produced by microwave heating absorbs more energy because of its magnetic property [10]. The maximum bubble size is proportional with the power because more heat generation within the particle is consumed for evaporating liquid (for the case of continuous mode). On the other hand, due to rapid generation of nuclei caused by higher power of the first irradiation, energy from the second irradiation is distributed to the particle. Consequently, energy used for evaporation around each particle became lower. The higher number density resulted in smaller bubble size. Therefore, two-stage irradiation is useful in promoting nucleation and shortening the irradiation time. These advantageous will be promising in addressing the issues of continuous mode operation at higher power (above 300 W) which is greatly affected by bumping behavior [7].

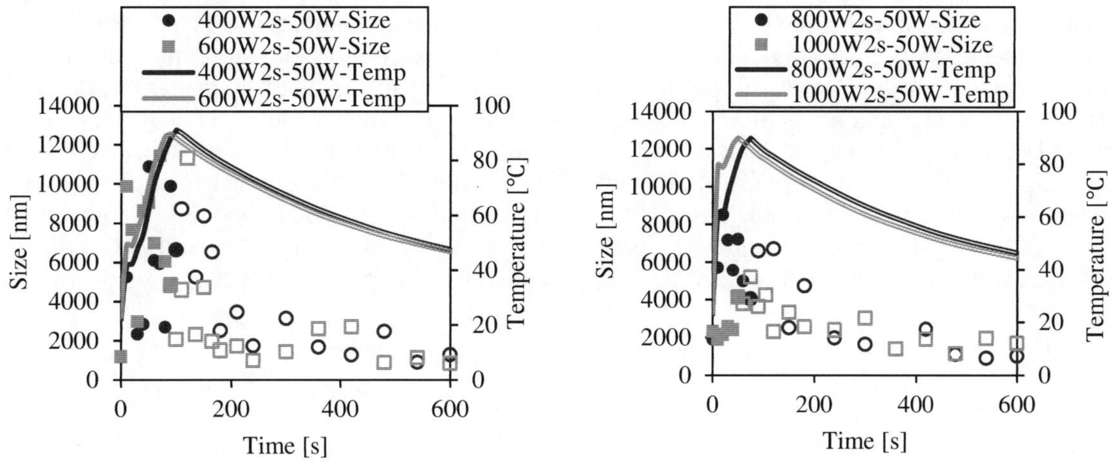

(a) Lower power (b) Higher power

Figure 3. Profiles of size and temperature in water during and after microwave irradiation

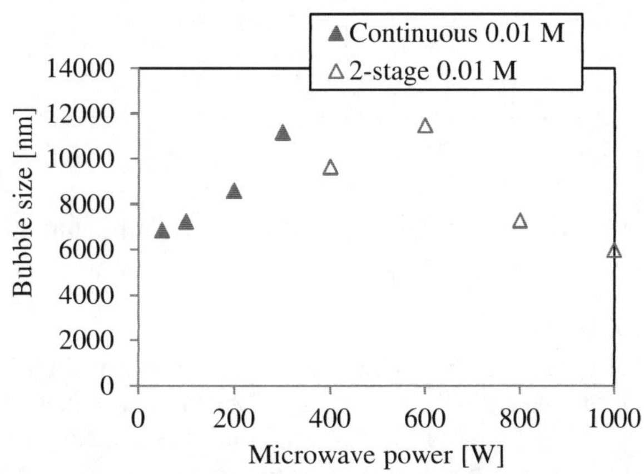

Figure 4. The maximum bubble size as a function of irradiation power

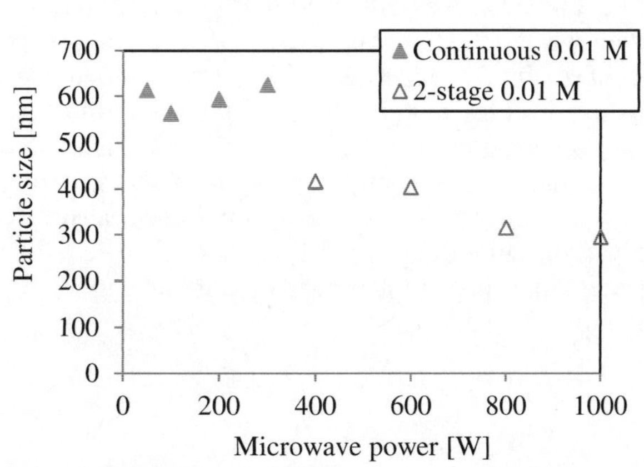

Figure 5. The particle size as a function of irradiation power

IOP Conf. Series: Materials Science and Engineering **206** (2017) 012055 doi:10.1088/1757-899X/206/1/012055

Figure 5 shows the final particle sizes measured at room temperature and they are plotted against irradiation power. The x-axis is defined as the first irradiation power for the case of two-stage irradiation. From this figure, it is evident that smaller bubble sizes were produced by two-stage irradiation. The size became lower for higher power of the first irradiation. The suspension density became higher because nucleation quickly occurred in the first stage of irradiation. This phenomenon may correspond to the previous finding that higher suspension density resulted in smaller bubble [11]. Thus, in this study, smaller particle size was also produced at higher suspension density. It is hypothesized that the maximum bubble size during the first irradiation may be related to the final particle size. Figure 6 indicates the relation between the maximum bubble size and final particle size. Suppression of the maximum bubble size during the irradiation is postulated to play roles in obtaining smaller particle size. Local temperature distribution caused by evaporation around the particle, therefore, should be prevented because heat is required for the stable particle growth.

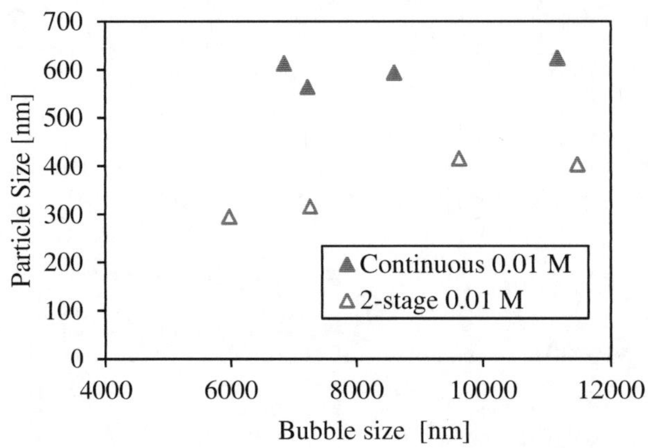

Figure 6. The relation between final particle size and maximum bubble size

3.2. Behavior of particle growth during and after irradiation

The intensity of red line during and after irradiation for different irradiation modes can be seen in Figure 7. For the case of continuous irradiation (50 and 300 W), over the time, the intensity became higher. Particularly, the intensity of higher power for continuous irradiation rapidly increased and dramatically fluctuated during the temperature rise. The higher intensity means that particles were quickly produced. Flash of the scattering light, which corresponds to the rapid evaporation and bubble growth due to the hot spot of superheat, was observed during the irradiation. Although superheat was suppressed when lower power was applied, it took longer time for nano-particle formation process to reach the designated temperature. On the other hand, the intensity of two-stage irradiation was lower than of continuous mode. Although final temperatures are the same, the suspension density remained lower for the case of two-stage irradiation (1000 W – 50 W). Rapid and more nucleation were observed at the first stage of irradiation when higher power and shorter irradiation time were used. In addition, superheat did not occur during the second irradiation because of lower power and higher suspension. Finally, stable particle growth by lower power in the second irradiation was confirmed from size profiles and brightness intensity of the scattering light during the irradiation.

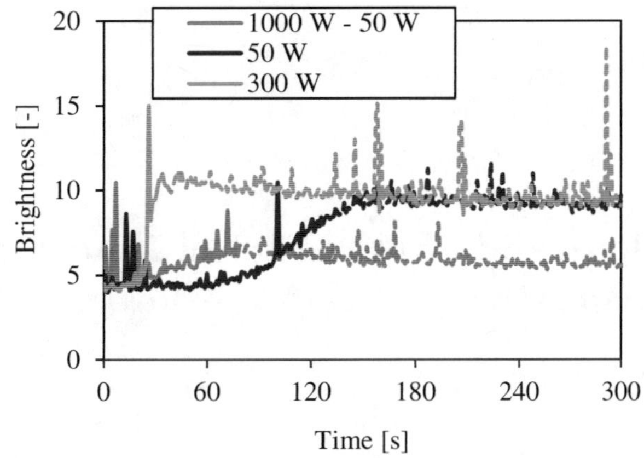

Figure 7. Brightness of the scattering light as a function of time

4. Conclusions

Sizes of bubble and particle during and after two-stage microwave irradiation were investigated and measured using a reactor equipped with a DLS apparatus. From the experimental data, bubble size generated in the two-stage irradiation is smaller than that of continues mode. Although the first irradiation power was higher, the suspension density also became higher due to the quick particle nucleation and shorter irradiation time. Due to distribution of microwave absorbance energy to each particle at higher suspension density in the first irradiation, bubble size became smaller. Therefore, two-stage irradiation is useful to separate nucleation and particle growth. This is very promising since longer period and higher power of the continuous irradiation was impractical. Through this study, effectiveness of two-stage irradiation was confirmed by the data of higher suspension density and smaller particle obtained from in-situ measurement of size profiles and brightness of the scattering light. Two-stage irradiation is advantageous for preventing superheat behavior such as quick bubble growth, which greatly affects local temperature distribution around the particle. Finally, if perfect control of suspension density and particle size can be achieved by two-stage irradiation, industrial applications of microwave technology will come into reality for the stable operation of nano-particle formation process.

References

[1] Hu S, Lu C, Wang W, Ding M, Ni Y and Xu Z 2013 *J. Rare Earths*, **31** 490
[2] Borja-Urby R, Diaz-Torres L A, Garcia-Martinez I, Bahena-Uribe D, Casillas G, Ponce A, Jose-Yacaman M 2015 *Catalysis Today* **250** 95
[3] Tee T S, Hui T C, Yi C W, Chin Y C, Umar A A, Titian G R, Beng L H, Sing L K, Yahaya M, Salleh MM 2016 *Sensors and Actuators B: Chem.* **227** 304
[4] Rai P, Song H-M, Kim Y-S, Song M-K, Oh P-R, Yoon J-M, Yu Y-T 2012 *Mat. Lett.* **68** 90
[5] Li H, Shao, F-Q Huang H, Feng J-J, Wang A-J 2016 *Sensors and Actuators B: Chem.l* **226** 506
[6] Zhu X H, Hang Q M 2013 *Micron* **44** 21
[7] Matsumura S, Nakata R, Asakuma Y 2015 *The 6th Asian Particle Techno. Symp.*, P01-11
[8] Horikoshi S, Osawa A, Sakamoto S, Serpone N 2013 *Chem. Eng. and Processing: Process Intensification* **69** 52
[9] Tsukahara Y, Higashi A, Yamauchi T, Nakamura T, Yasuda M Baba A, Wada Y 2010 *J. Phys. Chem. C* **114** 8965
[10] Asakuma Y, Nakata R, Matsumura S, Saptoro A 2016 *Pro. Eng. J.* **157** 411
[11] Asakuma Y, Munenaga T, Nakata R 2016 *Heat and Mass Transfer* **52** 1833
[12] Asakuma Y, Nakata R, Asada M, Kanazawa Y, Phan C 2016 *Int. J. Heat and Mass Transfer* **103** 13
[13] Asakuma Y, Maeda Y, Araki N 2015 *Int. J. Chem. Kinetics* **47** 811

IOP Conf. Series: Materials Science and Engineering **206** (2017) 012056 doi:10.1088/1757-899X/206/1/012056

Surface morphology and physicochemical properties of ordered mesoporous silica SBA-15 synthesized at low temperature

M H Koh, S A Haji Azaman, B H Hameed, A T Mohd Din*

School of Chemical Engineering, Engineering Campus, Universiti Sains Malaysia, 14300 Nibong Tebal, Pulau Pinang, Malaysia.

*E-mail: chazam@usm.my

Abstract. The effects of process parameters on the surface morphology and physicochemical characteristics of ordered mesoporous silica SBA-15 synthesized at low temperature have been investigated in this study. SBA-15 particles were synthesized through sol-gel method using non-ionic surfactant Pluronic P123 and TEOS as a silica source with aqueous hydrochloric acid (HCl) as a catalyst under the following conditions: HCl concentration (1.0-2.5 M), ageing temperature (40-70°C) and ageing time (12-48 hours). A series of physicochemical characterizations and material analyses were performed on SBA-15 particles including Scanning Electron Microscopy (SEM), Energy Dispersive X-ray Spectroscopy (EDX), Transmission Electron Microscopy (TEM), BET surface area analysis, Fourier transform infrared (FTIR) analysis and X-ray Diffraction (XRD) analysis. From the experimental observation, the conditions of HCl concentration, ageing temperature and ageing time were able to influence the surface morphology of SBA-15 particles. The presence of the ordered structures in SBA-15 particles was observed through the formation of 1-D cylindrical channels and 2-D hexagonal pores, inspected by using TEM. The detected XRD peak at (100) reflection signified the presence of ordered meso structures within the SBA-15 particles. Therefore, synthesis of SBA-15 particles through sol-gel method at low temperature is feasible and more sustainable if compared to the energy intensive hydrothermal method.

1. Introduction

Porous materials are defined as a continuous and solid network material filled with voids [1]. In the past few decades, demand on the usage of advanced structural materials has led to abundance of research carried out on porous solids such as porous carbon, synthetic silicate zeolites, mesoporous silicates and ordered porous metal oxides [2]. According to the International Union of Pure and Applied Chemistry (IUPAC), porous materials can be classified into three major classes based on their pore diameter sizes (d), microporous $d < 2$ nm, mesoporous $2 \leq d \leq 50$ nm and macroporous $d > 50$ nm. Microporous materials such as zeolites and metal-inorganic frameworks possess good stability, selectivity and activity due to their crystallinity and the presence of incorporated hetero elements in the structure. However, size limitation is a problem when it comes to large molecular application using microporous materials [3].

In 1972, Mobil Corporation explored the conversion of methanol into gasoline using microporous Zeolite ZSM-5 (Zeolite Socony Mobil) to obtain cheaper gasoline from acid-base reactions taking place within the micropores of zeolites [4]. However, this approach failed due to the fact that zeolite pore size was too small to enable the entry of larger organic molecules and for reactions to take place. Due to its size limitation, porous solids industry moved on to explore the possibility for mesoporous materials as substitutes to the microporous zeolites.

Generally, mesoporous materials possess high surface area of 400 – 1000 m^2 /g, large pore volume and excellent thermal stability at 500 – 600°C. Mesoporous solids can be prepared either by soft template or hard template method, in which organic molecules act as surfactant in soft template technique, while porous solids such as porous carbon is used in place of surfactant in hard template technique [1]. Mesoporous materials are highly favourable in large-molecular applications, such as cracking of heavy oil, polymer separation, enzyme immobilization and controlled-release of drugs [5].

The call for synthesis of ordered mesoporous materials stems from the fact that these materials consist of well-defined structural features such as pore sizes, pore shapes, pore arrangement and connectivity, which can be precisely-controlled by its synthetic conditions [5]. In fact, ordered mesoporous materials possess better hydrothermal stability, mechanical stability and catalytic activity in comparison to the disordered mesoporous materials [6]. Ordered mesoporous materials exhibit distinct, ordered-arrays of mesoporous channels in TEM images. Examples of ordered mesoporous materials are ordered metal oxides, ordered mesoporous silica, ordered mesoporous carbon, carbon nanotubes and mesoporous anodic alumina [7-8].

Ordered mesoporous silica SBA-15 has tunable pore size through modification on its synthesis parameters. A wide variety of SBA-15 morphologies has been reported in literature; including rods, fibers, spheres, gyroids, discoid-like and doughnut-like shape. Spherical hollow silica with porous shell finds great usage in controlled delivery of biomedical materials. They allow higher loading capacity to encapsulate drugs, genes or biological molecules in the shell [9]. Sakamoto and co-researchers reported a successful synthesis of rod-shaped SBA-15 in acidic media using Pluronic P123 as surfactant and tetramethyl orthosilicate as silica source; to give better adsorption of enzymes than the spherical SBA-15 particles [10].

There are various methods to synthesize SBA-15 including direct synthesis, synthetic grafting and impregnation, sol gel and immobilization. For direct synthesis method, it involves the reaction between co-condensation of silica source and silane groups or with different groups of metal such as transition, alkaline earth, rare earth and poor metals. Post grafting technique is the reaction between organosilane with silanol groups to produce covalent attachment of functional groups on surface of material using solvent under reflux condition [11]. Sol-gel is one of the methods that have been widely investigated in materials science. Orderly pore structure, size and shape of the SBA-15 can be achieved through this simple method [12]. Another advantage is the availability of high purity raw materials to support sol gel method [13]. This approach provides an important pathway to synthesize porous silica without requiring extremely high temperature as compared to the conventional hydrothermal treatment that requires high temperature [12]. For that reason, synthesis of SBA-15 particles through sol-gel method at low temperature is more sustainable.

In this work, SBA-15 was prepared by varying process parameters including ageing temperature, ageing time and HCl concentration, using sol-gel method. The surface morphology and physicochemical properties of SBA-15 prepared under low ageing temperature will be assessed in detail.

2. Materials and methods

2.1. Synthesis of SBA-15

The chemicals used in this work include Pluronic P123 ($PEO_{20}PPO_{70}PEO_{20}$) as surfactant, a silica source tetraethyl orthosilicate (TEOS)(98 %), and aqueous hydrochloric acid (HCl) were purchased from Sigma-Aldrich (Malaysia), Acros Organics (Malaysia), and R & M Chemicals (Malaysia), respectively. 150 mL of 2.5 M HCl was prepared and mixed with 5 g of Pluronic P123 in a beaker.

The mixture was then stirred until a colourless solution formed. At this point, 9.67 g of tetraethyl orthosilicate (TEOS) was added dropwise into the solution. The mixture was then stirred again for another 2 h until a white solution was formed. This mixture was left for ageing in a water bath shaker for 48 h at ageing temperature of 40 °C. White precipitate is filtered out and washed continuously with deionized water. The filtered white solid was then left for natural drying before calcination was carried out at 550°C for 4 h with heating rate of 5 °C/min. Subsequent trials were carried out at different synthesis conditions through one-variable-at-time (OVAT) approach summarized in Table 1. Samples 3, 5 and 9 are the replication points representing the maximum MB uptake capacity, respectively. Variation on other parameters such as surfactant and precursor concentration ratio, types of surfactant or precursor, calcination temperature etc, are not carried out, and these parameters remain constant whenever applicable.

Table 1. SBA-15 synthesis conditions.

Sample	[HCl] (M)	Ageing Temp (°C)	Ageing Time (h)
1	1.0	40	48
2	1.5	40	48
3*	2.0	40	48
4	2.5	40	48
5*	2.0	40	48
6	2.0	50	48
7	2.0	60	48
8	2.0	70	48
9	2.0	40	12
10	2.0	40	24
11	2.0	40	36
12*	2.0	40	48

*Synthesis under similar conditions.

2.2. *Adsorption Test*

A simple methylene blue (MB) adsorption test was carried out on the SBA-15 samples to determine the maximum MB uptake capacity (mg MB/g adsorbent). A mass of 0.1 g SBA-15 was added to 100 ml MB solution (100 ppm) and left for 48 h to achieve equilibrium. The MB concentration was analyzed using UV-Vis spectrophotometer (Shimadzu UV1800, Japan).

2.3. *Materials Characterizations*

Surface area properties of SBA-15 was determined using surface area analyser (Micrometrics ASAP 2010, USA) at 77 K. SEM with EDX analysis (SEM Quanta FEG 450, USA) was used to study surface topology of SBA-15 and textural images were captured using TEM (Philips TEM CM12, The Netherlands) Surface chemistry of SBA-15 and the crystal structure of SBA-15 are determined using FTIR spectrophotometer (Shimadzu IRPrestige-21, Japan) and Materials Research Diffractometer, respectively.

3. Results and discussion

The surface morphology of SBA-15 synthesized at different HCl concentrations are shown in Figure 1. The formation of SBA-15 rods and spherical-shaped particles can be observed at 1.0 M HCl, evolving from shorter rods (2.0 M, 1.5 M) to longer rectangular rods (2.5 M), with increasing HCl concentrations. This is evidenced that the particle shape of SBA-15 can be greatly influenced by the addition of different HCl concentrations. This can be explained by colloidal phase separation mechanism (CPSM) proposed by Zhao and co-workers [14]. The theory suggests that when phase separation occurs, precipitates will be observed in the solution and the rate of this phase separation affects the final morphology of ordered mesoporous structures. Besides, it is better to synthesize SBA-

15 at highly acidic condition when using TEOS as a silica source. Xiao Ying and co-researchers studied the effect of HCl concentration using three different silica sources: 1,2-bis(trimethoxysilyl)ethane (BTMSE), 1,2-bis(triethoxysilyl)ethane (BTESE) and tetraethyl orthosilicate (TEOS), all accompanied by the same surfactant, Pluronic P123. The results shown that organosilica precursor (BTMSE/ BTESE) interacts poorly with surfactant and condenses faster than the silica precursor such as TEOS. Since hydrochloric acid acts as a catalyst to improve hydrolysis and condensation rate of the precursors, high acid concentration leads to rapid condensation of organosilica framework, which results in poorly defined pores. Lowering the acid concentration has proven to reduce the condensation rate of BTMSE and BTESE and only then a well-defined hexagonal pore arrangement in SBA-15 could be obtained. In contrast to the synthesis using TEOS, condensation happens very slowly in aqueous solution and high acid concentration is generally required to increase the condensation rate of silicate framework [15]. In another study, a formation of rod-shaped SBA-15 particles prepared using TEOS at different HCl concentrations without the use of any additives has been previously observed [16]. It was reported that there was no significant differences in the rod-shaped and length of SBA-15 synthesized at 2.0 M HCl, which is consistent to the current observation in Figure 1(c).

Figure 2 shows the surface morphology of SBA-15 synthesized at different ageing temperatures of 40, 50, 60 and 70°C, using 2.0 M HCl and ageing time of 48 h. It is observed that as the ageing temperature increases, the shape of SBA-15 particles changes from short rods (40 and 50°C) to thinner fibre-like structure (60 and 70°C). This phenomenon can be explained using the CPSM theory. Free energy of mesostructure self-assembly (ΔG), dominantly affects the final morphology of SBA-15. As mesostructure is continuously being formed, to minimize ΔG, more crystal-like structures of SBA-15 can be obtained. In this work, SBA-15 particles tend to take up fibre-like structure at 60 and 70°C. The average length and width were measured at 700 nm and 500 nm for the SBA-15 sample prepared at ageing temperature of 40°C. SBA-15 is commonly formed by grains of relatively uniform size; typically in the range of 300–500 nm (width) and 700–1000 nm (length) [17]. A previous study has reported the average particle length and width of 800 nm and 500 nm, respectively for the SBA-15 synthesized at 35°C, 2.0 M HCl and ageing time of 10 h [16].

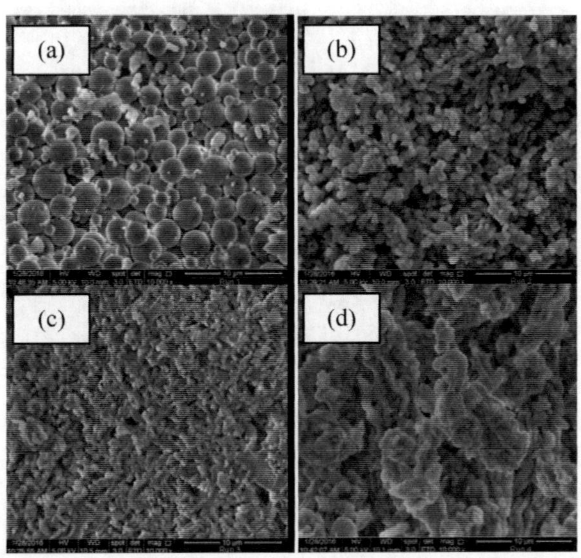

Figure 1. Surface morphology of SBA-15 synthesized at different HCl concentrations; (a) 1.0 M (b) 1.5 M (c) 2.0 M and (d) 2.5 M (×10k magnification).

IOP Conf. Series: Materials Science and Engineering **206** (2017) 012056 doi:10.1088/1757-899X/206/1/012056

Figure 2. Surface morphology of SBA-15 synthesized at different ageing temperature; (a) 40°C (b) 50°C (c) 60°C and (d) 70°C (×10k magnification).

Figure 3 shows surface morphology of SBA-15 prepared at different ageing times of 12, 24, 36 and 48 h (HCl concentration: 2.0 M, ageing temperature: 40°C). From the images, it is evidence that the surface morphology of SBA-15 can be influenced by the ageing time parameter. The SBA-15 particles prepared at 12 and 24 h ageing times were random in size and not uniform in shape. More well-defined short rod-shape SBA-15 particles are noticed when ageing time is increased to 36 and 48 h. This could be explained since longer ageing time has allowed more polymerization and condensation of silica species on the walls to form ordered mesostructure of SBA-15 [18].

Figure 3. Surface morphology of SBA-15 synthesized at different ageing time; (a) 12 h (b) 24 h (c) 36 h (d) 48 h (×10k magnification).

A similar observation has also been recorded in our previous work [19]. However, a shorter ageing time is not unusual [20]. An investigation on the effect of ageing time on the mesoscopical order and pore structure of functionalized mesoporous silica SBA-15 has shown that longer ageing time has contributed to a better ordered pore structure formation [18].

A simple MB adsorption test was performed to select the SBA-15 sample with the highest MB uptake capacity. MB adsorption test is a common method to determine the mesoporosity attribute of a porous solid. Maximum MB uptake capacities for all the SBA-15 samples, synthesized in this study are summarized in Table 2. It was found out that sample 3, prepared under the respective conditions: 2.0 M HCL, 40°C and 48 h ageing time, has exhibited the highest MB uptake of 196.3 mg MB/g adsorbent. Hence, sample 3 SBA-15 has been selected for further physicochemical analyses.

Figure 4 shows the textural imaging of sample 3 SBA-15. From the TEM images, 1-D cylindrical hollow channels and 2-D hexagonal pore arrangement can be clearly observed, a common attribute for SBA-15. The hollow channels were measured at 2.5 nm in width. The value is smaller to a measurement of 5 nm in width for hollow channels of SBA-15 prepared through hydrothermal treatment at 80°C [17]. Hydrothermal method normally requires the process to be sustained at temperature ranging from 80 - 120 °C, which means more energy utilization [12].

Table 2. Maximum MB uptake capacity.

Sample	MB uptake capacity (mg/g)
1	113.6
2	99.6
3*	196.3
4	132.7
5*	196.3
6	58.9
7	36.1
8	2.60
9	99.0
10	95.1
11	35.7
12*	196.3

*Synthesis under similar conditions.

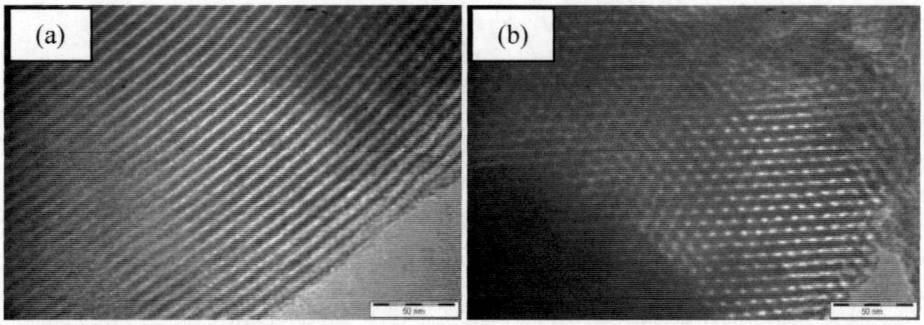

Figure 4. TEM images of Sample 3 SBA-15 (a) cross-sectional view and (b) perpendicular to pore channels.

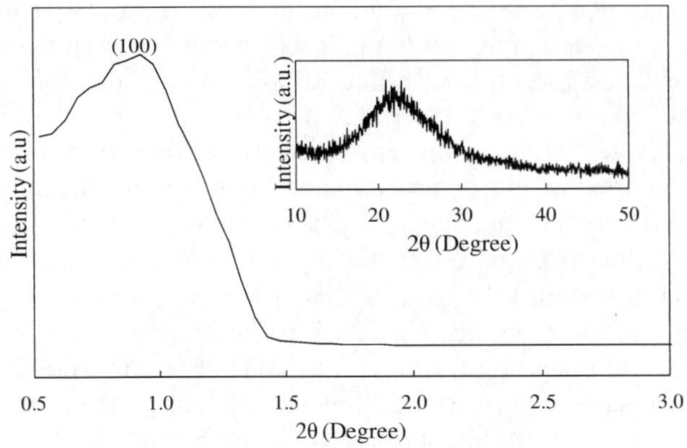

Figure 5. XRD pattern for Sample 3 SBA-15.

Figure 5 shows the XRD profile for Sample 3 SBA-15. At low angle, a well-resolved diffraction peak is detected corresponding to the Miller index (100) reflection around 1° point. This peak is a characteristic of 2-D hexagonal pore arrangement, a typical signature for SBA-15 material.

2-D hexagonal symmetry structure (p6mm) corresponding to (110) and (200) reflections are not detected in the XRD profile. Unit cell parameter, a_o, is calculated at 10.88 nm based on the first Bragg peak position, d_{100} = 9.42 nm. The wall thickness, w_t, of SBA-15 is measured at 6.98 nm. The formation of distintive peak between 20-25° represents the amorphous nature of the synthesized siliceous material (Figure 5 - inset).

The EDX spectra (Figure 6) shows the respective weight percentage contributed by each element in SBA-15 sample. The weight percent for carbon, silica and oxygen are given at 8.33 % (C), 58 % (Si) and 33.67 % (O), respectively. The intensity of the detected peaks for Si and O elements is comparable to the literature findings [21].

From Figure 7, a strong broad peak observed at 1094 cm^{-1} can be assigned to siloxane bond (-Si-O-Si-) stretch. -OH stretching vibrations mode of the silanol groups involved in hydrogen interactions with the adsorbed water molecules is observed at 3454 cm^{-1} peak. Bending H_2O band at 1637 cm^{-1} is also ascribed to the adsorbed water molecules on the material. The presence of these two functional groups confirmed the silica formation. Another obvious band at around 802 cm^{-1} can be ascribed to symmetric stretching from -Si-O bonds, while another peak at around 466 cm^{-1} can be attributed to bending of -Si-O-Si- bonds [22]. The absence of peaks at 3000-2850 cm^{-1} range (-CH_2- stretching) and 1470-1450 cm^{-1} (-CH_2- bending) confirmed the complete removal of Pluronic P123 in SBA-15 sample [20].

Figure 8 shows the nitrogen adsorption-desorption isotherm performed on Sample 3 SBA-15 at 77K. The respective sample exhibits Type IV adsorption isotherm with H1-type hysteresis loop. H1 is commonly associated with porous materials exhibiting cylindrical-like pores [23]. It has been observed from literature that the well-formed SBA-15 always shows Type IV adsorption isotherm with H1-type hysteresis loop [17]. Due to the capillary condensation process in hysteresis loop region, larger pressure drop is required during desorption in order to overcome the van der Waals interactions among the adsorbed liquid molecules in the confined pores or capillaries. This is a typical phenomenon observed for adsorption-desorption isotherm of mesoporous materials [24].

The steep increase of N_2 adsorption at P/Po = 0.52 indicates the mesopore uniformity. This is highly consistent with the Barret-Joyner-Haleda (BJH) pore size distribution plot (figure is not shown) having a sharp peak at pore diameter of 4.06 nm [15]. The average pore size was measured at 3.90 nm based on the BET adsorption branch. The total pore volume, V_{Total}, and micropore volume, V_{μ}, for this sample are 0.36 and 0.021 cm^3/g, respectively. Total BET surface area is calculated at 364 .71 m^2/g.

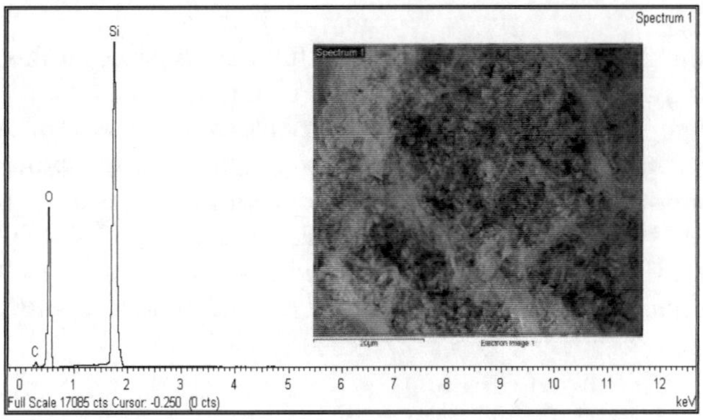

Figure 6. EDX spectra of of Sample 3 SBA-15.

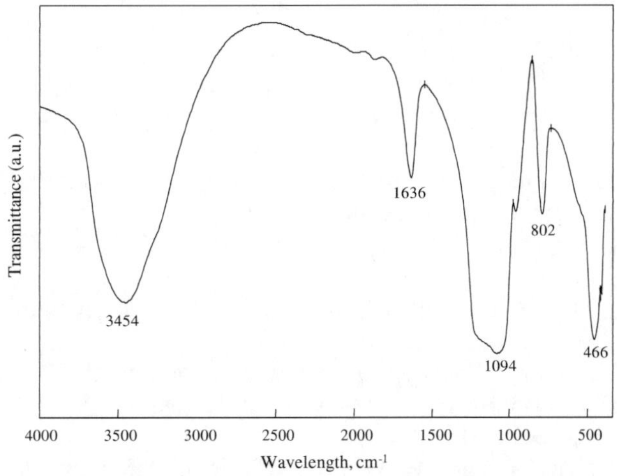

Figure 7. FTIR spectra of Sample 3 SBA-15.

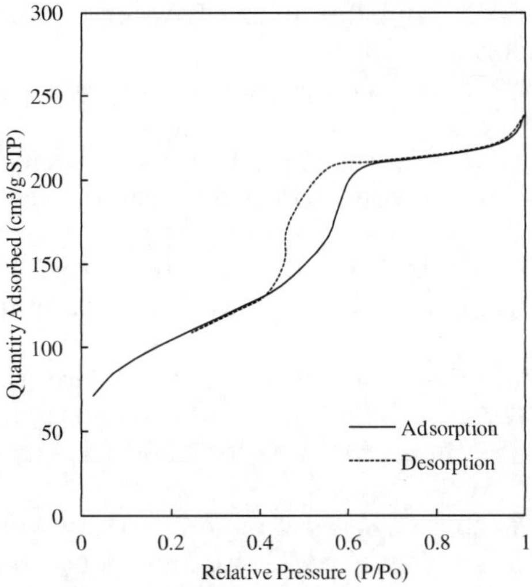

Figure 8. Nitrogen adsorption-desorption isotherm of Sample 3 SBA-15 at 77K.

4. Conclusions

Ordered mesoporous silica SBA-15 has been successfully synthesised in this present work. At lower HCl concentrations, the synthesized SBA-15 particles tend to form a spherical shaped; while at higher HCl concentration range, a rod-like shape nanoparticles were observed. Ageing temperature and ageing time also have been found to influence the shape of SBA-15 particles. Upon increasing the ageing temperature from 40 to 70°C, SBA-15 changed from short rods to thin-fibre like structure. The presence of the ordered structures in SBA-15 sample was observed through the formation of 1-D cylindrical channel and 2-D hexagonal pores, inspected using TEM. The results were further validated with the detected XRD peak at (100) reflection; a sign of ordered meso structures formation in SBA-15 particles.

Acknowledgements

The authors are thankful to Universiti Sains Malaysia and Ministry of Higher Education, Malaysia (MOHE) for providing research facilities and financial support for this research. This research is funded by MOHE under the Fundamental Research Grant Scheme (FRGS - Project number: 6071330).

References

[1] Pal N and Bhaumik A 2013 Soft templating strategies for the synthesis of mesoporous materials: inorganic, organic-inorganic hybrid and purely organic solids *Adv. Colloid. Interface. Sci.* **189-190** 21-41

[2] Mejia E E 2013 *Characterization of Some Natural and Synthetic Materials with Silicate Structures* (Sweden:Universitetstryckeriet, Luleå)

[3] Buckley R W and Benito P 2007 *Solid State Chemistry Research Trends* (Nova Science Publishers) pp 173-225

[4] Vallet-Regí M 2012 Mesoporous Silica Nanoparticles: Their Projection in Nanomedicine *ISRN. Mat. Sci.* **2012** 1-20

[5] Zhang Q and Wei F 2014 *Adv. Hierar. Nanostruct. Mat.* (Wiley) p 512

[6] Bonneviot L, Béland F, Danumah C, Giasson S and Kaliaguine S 1998 *Mesoporous Molecular Sieves* (Elsevier Science) p 614

[7] Nhut J M, Pesant L, Tessonnier J P, Winé G, Guille J, Pham-Huu C and Ledoux M J 2003 Mesoporous carbon nanotubes for use as support in catalysis and as nanosized reactors for one-dimensional inorganic material synthesis *App. Catalysis A: General.* **254** 345-363

[8] Bruschi L, Mistura G, Park S J and Lee W 2014 Adsorption of argon onmesoporous anodic alumina *Adsorption.* **20** 889-897

[9] Prokopovich P 2015 *Biological and Pharmaceutical Applications of Nanomaterials* (CRC Press) p 444

[10] Sakamoto Y, Kaneda M, Terasaki O, Zhao D Y, Kim J M, STucky G, Shin H J and Ryoo R 2000 Direct imaging of the pores and cages of three dimensional mesoporous materials *Nature*. 449-453

[11] Rahmat N, Abdullah A Z and Mohamed A R 2010 A review: Mesoporous Santa Barbara amorphous-15, types, synthesis and its applications towards biorefinery production *Am. J. Appl. Sci.* **7** 86-1579

[12] Rivera-Muñoz E M and Huirache-Acuña R 2010 Sol Gel-derived SBA-16 mesoporous material *Int. J. Mol. Sci.* **11** 86-3096

[13] AlOthman Z A 2012 A review: Fundamental aspects of silicate mesoporous materials. *Materials. (Basel).* **5** 2874–2874

[14] Zhao D, Wan Y and Zhou W 2012 *Ordered Mesoporous Mat.* (Weiheim: Wiley)

[15] Xiao Ying B, X S Z, Xu L, Pai Ann C and Jun L 2004 A Novel Route toward the Synthesis of High-Quality Large-Pore Periodic Mesoporous Organosilicas *J. Phys. Chem. B.* **108** 4684-4689

[16] Wang Y, Zhang F, Wang Y, Ren J, Li C, Liu X, Guo, Y, Guo Y and Lu G 2009 Synthesis of

length controllable mesoporous SBA-15 rods *Mat. Che. and Phy.*. **115** 649-655

[17] Klimova T, Esquivel A, Reyes J, Rubio M, Bokhimi X and Aracil J 2006 Factorial design for the evaluation of the influence of synthesis parameters upon the textural and structural properties of SBA-15 ordered materials *Microporous and Mesoporous Mat.* **93** 331-343

[18] Wei Q, Hao Z N Y, Liu L and Zou Z C J 2006 Effect of synthesis conditions on the mesoscopical order of mesoporous silica SBA-15 functionalized by amino groups *J. Sol-Gel. Sci. Techn.* **39** 9-103

[19] Mohd Din A T, Ahmad M A and Hameed B H 2015 Ordered mesoporous carbons originated from non-edible polyethylene glycol 400 (PEG-400) for chloramphenicol antibiotic recovery from liquid phase *Chem. Eng. J.* **260** 730-739

[20] Gandhi S, Sethuraman S and Krishnan U M 2013 Heterogeneous mesoporous SBA-15 silica as catalyst towards the synthesis of various biodegradable aliphatic polyesters *Macromolecular. Res.* **21** 833-842

[21] Tomer V K, Duhan S, Adhyapak P V, Mulla I S and Gouma P 2015 Mn-Loaded Mesoporous Silica Nanocomposite: A Highly Efficient Humidity Sensor *J. American Ceramic Soc.*. **98** 741-747

[22] Yang L, Jiang Z, Lai S, Jiang C and Zhong H 2014 Synthesis of Titanium Containing SBA-15 and Its Application for Photocatalytic Degradation of Phenol *Int. J. Chem. Eng.* **2014** 1-7

[23] Thommes M 2010 Physical adsorption characterization of nanoporous materials *Chemie-Ingenieur-Technik.* **82** 73-1059

[24] Naumov S 2009 *Hysteresis Phenomena in Mesoporous Mat* (Leipzig, Leipzig University)

IOP Conf. Series: Materials Science and Engineering **206** (2017) 012057 doi:10.1088/1757-899X/206/1/012057

Finite element simulation of mechanical properties of graphene sheets

N Khandoker*, S Islam, Y S Hiung

Department of Mechanical Engineering, Curtin Malaysia, CDT 250, Miri 98009, Sarawak, Malaysia.

*E-mail: noman.khandoker@curtin.edu.my

Abstract. Graphene is the material for the twenty first century applications. In this paper, the elastic properties of monolayer and double layer Graphene sheets, typically less than 10nm in size are investigated through linear finite element simulations. The effect of aspect ratio, sizes and chirality of the Graphene sheet on the Young's modulus, Shear modulus and Poisson's ratio are studied. By using structural mechanics approach combining atomistic and equivalent continuum techniques, the Young's modulus, shear modulus and the Poisson ratio were found and they slightly increase with the aspect ratio but decrease with the size of the Graphene sheet. These simulated properties compliment the mechanical properties of Graphene found in literature.

1. Introduction

Graphene has been discovered for about ten years since 2004 and its area of research grew extremely fast that hundreds of laboratories all over the world deal with different aspects of graphene research nowadays, which is driven by its uniqueness of possessing remarkable properties such as high electrical and thermal conductivity [1]. In addition, graphene is also considered to be a promising material for spintronics and spin qubits which will allow for massive increase in future computational power [2].

Besides, graphene also gave birth to a new class of crystal such as fluro-graphene, boron nitride, [3] and molybdenum disulphide [4] that are also just one atom thick. Recently, graphene related two-dimensional crystals and its hybrid systems were found to be available in energy conversion and storage for future applications [5]. Furthermore, composite materials with graphene as reinforcements are also feasible and beneficial for many applications. For example, in case of corrosion prevention, graphene grown on carbon steel with Nickel catalysts, provide almost 7 times more corrosion resistance than that of the original steel [6]. Hence, it is important to understand the mechanical behavior of graphene sheets in facilitating a better characterization of the resulting nano-composites in its future applications of different industries.

Therefore, the objectives of this current study are to investigate the effect of sizes, chirality and aspect ratios of graphene on its mechanical properties such as Young's modulus, shear modulus and Poisson's ratio. Furthermore, the effects of Van der Waal's interactions in double layered graphene on its mechanical properties are also investigated.

2. Literature Review

Calculations for the elastic modulus of single layer graphene sheet can range from quantum mechanics calculations to molecular dynamic, atomistic simulations and continuum models. Molecular dynamics and ab initio methods are quite popular methods in determining the elastic properties of nano materials such as carbon nanotubes and graphene.

Molecular dynamics method has been used by Lu [7] to determine the Poisson's ratio and Young's modulus of carbon nanotube. In this model, the force constants were empirically determined by fitting to measured elastic constants and the phonon frequencies. This phonon spectrum of graphite and the Poisson's ratio are calculated by minimizing the strain energy with respect to both the radial compression and the axial extension with the wall thickness of 0.34 nm. The results showed an averaged Poisson ratio of 0.28 and Young's modulus of 0.974 TPa for the nanotubes of various sizes.

Hernandez et al. [8] did a similar construction for the single wall carbon nanotubes and some other composite nanotubes using a non-orthogonal tight binding model. The average Poisson's ratio and Young's modulus of different sizes of nanotubes were 0.262 and 1.24 TPa respectively. Meanwhile, the ab initio study of Gregory et. al used an implementation of the theory of Hartree–Fock 6-31G* level, and an ab-initio multiplicative integral approach for the calculations [9]. The results revealed that Young's Modulus of graphene is 1.11 TPa. Ricardo et al. also used ab initio method to investigate the structural, electronic and mechanical properties of zigzag graphene nano-ribbons in the presence of stress by applying density functional theory within the GGA-PBE (Generalized Gradient Approximation-Perdew–Burke–Ernzerhof) approximation [10]. The average Young's modulus generated was 0.96 TPa with the thickness of 0.335 nm and the Poisson's ratio found was 0.17.

A more recently used method is the structural mechanics method combining atomistic and continuum modelling techniques. In this modelling technique, the graphene structure are modelled at nano-scale with every detail concerning the position of atoms and the force field connecting them. However, the connections for each atom of the lattice molecular structure are replaced with a continuum medium of an specific element type representing the force field.

A. Sakhaee-Pour [11] adopted an atomistic simulation model comprised of structural beam which simulates the covalent forces of the carbon atoms using potential energy of bond bending, bond stretching, out of plane and dihedral torsional of bond. The model was considered for the shape of the graphene sheet and investigations were performed for various chirality.

3. Methodology

For this study, a structural mechanics method is adopted with atomistic and continuum techniques approach to investigate the elastic characteristics of defect-free single-layered and double layered graphene sheet. The atomistic model used finite beam elements to simulate the covalent bonds between the carbon atoms. The elasticity of the beam elements are characterized into the molecular and structural potential energies in term of stretching, bending, and torsional potential energies based on Odegard's methodology. [20]

From the work of Raji Heyrovska, it is described that a graphene hexagon which consists of 7 carbon atoms, has a carbon-carbon bond length of approximately 1.42 Å and the bond angle between the covalent bond is 1200 to each other [12]. Hence using this information, the graphene sheet structure can then be modelled with respect to the Cartesian coordinate system (x, y, z) as shown in figure 1 and figure 2. The modelling of the graphene is done by Visual Molecular Dynamic (VMD) software and the scale used in the model is in Angstrom or 0.1nm. The interatomic distance between

IOP Conf. Series: Materials Science and Engineering **206** (2017) 012057 doi:10.1088/1757-899X/206/1/012057

the adjacent carbon atoms which belong to the same plane is modelled as 1.418 Å and the interlayer spacing or the thickness of the graphene sheet is approximately equal to 0.34 nm.

Neglecting the van der Waal forces and the interatomic electrostatic interaction, the total molecular potential energy, UT can be considered as the following equation:

$$U_T = \sum U_r + \sum U_\theta + \sum U_\tau \quad (1)$$

where U_r is the bond stretching energy, U_θ is the angle bending energy and U_τ is the torsional potential energy which includes the out of plane and dihedral angle torsions. Each of these molecular potential energies is related to its respective degree of freedom by a force field constant known as stiffnesses and are independent to each other. The relationships with the stiffnesses are as shown in equations 2, 3, and 4 below:

$$U_r = \frac{1}{2} k_r (r - r_o)^2 = \frac{1}{2} k_r (\Delta r)^2 \quad (2)$$

$$U_\theta = \frac{1}{2} k_\theta (\theta - \theta_o)^2 = \frac{1}{2} k_\theta (\Delta\theta)^2 \quad (3)$$

$$U_\tau = \frac{1}{2} k_\emptyset (\emptyset - \emptyset_o)^2 = \frac{1}{2} k_\tau (\Delta\emptyset)^2 \quad (4)$$

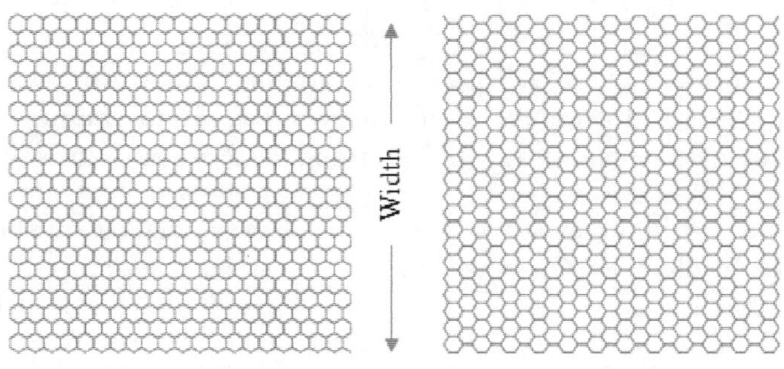

Figure 1. Armchair (left) and Zigzag (right) shapes of single layered graphene sheets

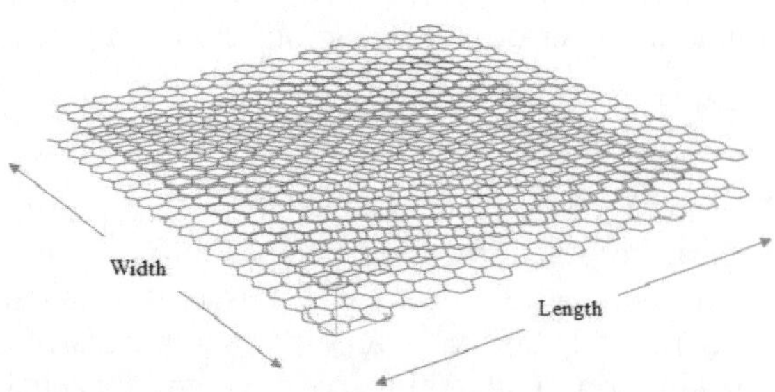

Figure 2. Armchair shape of double layered graphene sheet

where k_r, k_θ, and $k_\emptyset$ are the bond stretching, angle bending and torsional stiffnesses respectively and Δr, Δθ, and Δ∅ are the corresponding displacement in its respective degree of freedom. Based on structural mechanics, the potential energy for strain energy for a uniform beam is expressed as equation 5 below. The bending potential energy under a pure bending load is expressed as the equation 6 while the torsional potential energy for pure torsional load is expressed as equation 7:

$$U_A = \frac{1}{2} \int_0^L \frac{N^2}{EA} dl = \frac{1}{2} \frac{N^2 L}{EA} = \frac{1}{2} \frac{EA}{L} (\Delta L)^2 \quad (5)$$

$$U_M = \frac{1}{2}\int_0^L \frac{M}{EI}dl = \frac{2EI}{L}\alpha^2 = \frac{1}{2}\frac{EI}{L}(2\alpha)^2 \tag{6}$$

$$U_T = \frac{1}{2}\int_0^L \frac{T^2}{GJ}dl = \frac{1}{2}\frac{T^2L}{GJ} = \frac{1}{2}\frac{GJ}{L}(\Delta\beta)^2 \tag{7}$$

where A is the cross section of the beam, L is the beam length, ΔL is the displacement in length and E the Young modulus of the considered beam for equation 5. In equation 6, I and α represent the moment of inertia and rotational angle of the beam ends while in equation 7, $\Delta\beta$ is the relative bending angle, G represents the shear modulus, and J is the polar moment of inertia.

In order to simulate the molecular potential by structural means, the molecular potential energy are imposed into the structural potential energies by equating equations 2, 3 and 4 with 5, 6, and 7 respectively. This can be done since the potential energy terms in each system are independent to each other. By doing this, the stiffness of the equivalent structure beam can then be related in term of the force field constants as shown in equations 8 below:

$$\frac{EA}{L} = k_r \quad \frac{EI}{L} = k_\theta \quad \frac{GJ}{L} = k_\emptyset \tag{8}$$

The force field constant of graphene is taken from J. Medina based on the linear density approximation results as shown in table 1 below. [13] It is noted that with the stiffness related to the force field constant, the stiffness matrix can then be applied to the structural beam. However, in this study, that approach is not going to be followed. Instead, the approach will be the finite element analysis method where the characteristic of the beam is fulfilled by the beam elastic moduli and diameter.

$$A = \frac{\pi d^2}{4} \quad I = \frac{\pi d^4}{64} \quad J = \frac{\pi d^4}{32} \tag{9}$$

Hence, the Young's modulus, shear modulus and the beam diameter is calculated by substituting the relationships as shown in equations 9 to the beam characteristic relationship in terms of the force field. The derived equations for the beam elastic moduli and the corresponding diameter are as shown in equations 10 and the values are tabulated in Table 2.

$$d = 4\sqrt{\frac{k_\theta}{k_r}} \qquad E = \frac{k_r^2 L}{4\pi k_\theta} \qquad G = \frac{k_r^2 k_\emptyset L}{8\pi k_\theta^2} \tag{10}$$

Table 1. Values of the force field constants

Force field constant	Values
Bond stretching stiffness, $\boldsymbol{k_r}$	740 N/m
Bond bending stiffness $\boldsymbol{k_\theta}$	7.69 x10-19 Nm/ rad^2
Bond torsional stiffness $\boldsymbol{k_\emptyset}$	2.78 x10-19 Nm/ rad^2

Table 2. Values of beam characteristics

Beam Characteristics	Values
Beam's diameter, d	1.29x10-10 m
Beam's elastic modulus, E	8.05 x1012 N/m^2
Beam's shear modulus, G	1.45x1012 N/m^2

For double layered graphene sheet, the Van der Waal forces of interaction are modelled as truss beam structure with approximate non-linear characteristics. The non-linear relationship of the Van der Waal forces between the graphene sheets is usually modelled by using the Lennard-Jones 6-12 potential energy equation as shown below:

$$U_{vdw}(r) = 4\varepsilon_{vdw}\left[\left(\frac{\sigma_{vdw}}{r}\right)^{12} - \left(\frac{\sigma_{vdw}}{r}\right)^{6}\right] \quad (11)$$

where σ_{vdw} is the finite distance at which the inter-particle potential is zero (in case of graphene, σ is 0.34 nm), r is the distance between the particles, and ε_{vdw} is the depth of the potential well. For graphene and carbon nanotube, it is given by 3.8655×10^{-13} N nm from the work of Odegard et.al. [20]. In this study, the non-linear characteristic of the Van der Waal forces is compared to the structural strain energy of the truss element used in the model and the expression of the Young's modulus is obtained as has done by A. Parashar et.al. [14] The expression for the Young's modulus is as shown below:

$$E(r) = \frac{8R_{eq}\varepsilon_{vdw}}{A_t(r-R_{eq})^2}\left[\left(\frac{\sigma_{vdw}}{r}\right)^{12} - \left(\frac{\sigma_{vdw}}{r}\right)^{6}\right] \quad (12)$$

where R_{eq} is equilibrium distance for Van der Waals forces (0.3816nm), A_t is cross sectional area for a truss element (0.0907 nm2) and $E(r)$ is the estimated Young's modulus for the truss element [14]. For this simulation, the average interatomic distance of the double layered graphene is taken to be the value of r which is found to be 0.36378 nm. Hence, the value of the average Young's modulus for the truss element is evaluated as 9.10×10^{-9} N nm^{-2}. For simplicity of the simulation, this material property is defined for the all truss elements in the model.

In the model, each of the atoms in the upper layer is connected to either four of the atoms (type 1 interaction) or six of the atoms (type 2 interaction) in the lower layer via the truss elements. The Van der Waals interaction for type 1 and type 2 are as shown in Figure 3. The double layered graphene model with the truss elements as the Van der Waal forces is shown in Figure 4.

4. FEA Simulation

By having all the information required for the beam, the model can then be simulated. However, the structure of single layered graphene are 2-dimenisonal and hexagonal in shape which have many empty spaces in between the structure. This implies many difficulties in applying forces or boundary conditions on the graphene sheet to obtain the elastic moduli of the graphene.

Coupling constraints were used for the boundary conditions of this model. The boundary conditions applied to both ends of the graphene sheet were done by using kinematic coupling constraints with a control point at a reference point located at the middle of both ends of the graphene. This constraint was applied to make sure the boundary conditions are applied to every point of the edge as equivalence to a continuum structure.

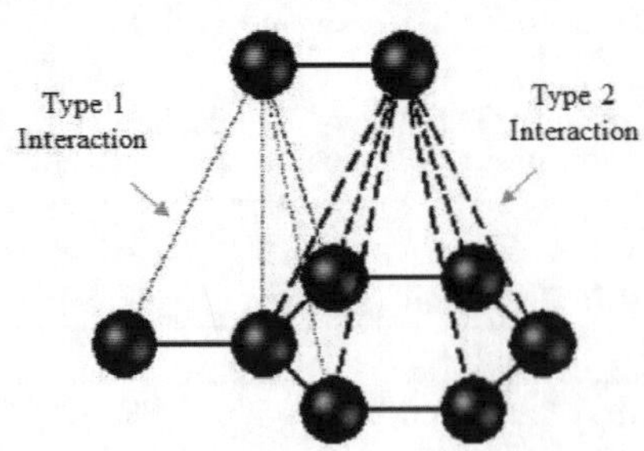

Figure 3. Truss Representing Van der Waals Forces

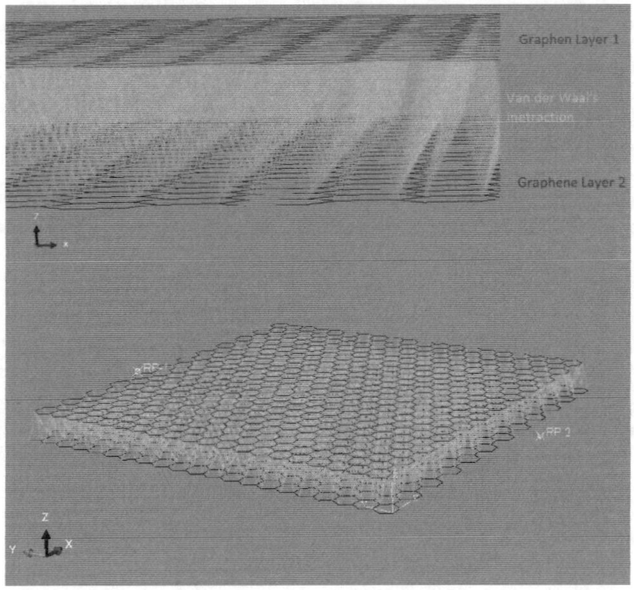

Figure 4. Double Layered Graphene Model

In order to understand the elastic behaviour of the graphene layer under tension, perpendicular forces needed to be applied to the free edge of the graphene sheet. However, instead of applying forces, displacement boundary conditions are set in this study. This can be done by applying Hooke's law which stated that the linear elastic displacement was directly proportional to the forces acted in the displacement direction as long as the elastic limit is not exceeded.

The boundary condition was set to be fixed in all degree of freedom at one end while the other end was subjected to a displacement of 10% of its width in the y-direction for tensile load or length in the x-direction for shear load.

In this study, the elastic properties of the graphene sheet were predicted to be dependent on the chirality. Therefore, zigzag and armchair shaped graphene sheet were modelled and investigated. The variables considered in this study were basically the size and shape. However, in this study, the length to width ratio of graphene sheet is also varied. The size considered includes square shaped graphene with length of 3nm, 5nm, 7 nm and 10nm. For length to width ratio analysis, five lengths to width ratio of approximate 0.33, 0.5, 2 and 3 are done in this simulation. A double layer was also done to analyse the effect of number of layer to the Young's modulus.

The model can be equalled to a continuum model, it can then be said that the normal stress, σ of the continuum sheet can be found using the classic stress definition given by the ratio of force to cross sectional area as Equation 13 below:

$$\sigma = \frac{f_n}{bt} \tag{13}$$

where f_n is the magnitude of the tensile forces computed from simulation, b is the graphene length and t is the thickness of the graphene. Hence, the Young's modulus of the graphene sheet can be computed by using the relationship of normal stress and strain as the following equation 14:

$$E = \frac{\sigma}{\varepsilon} = \frac{f_n w}{bt\Delta w} \tag{14}$$

where ε represents the tensile strain that is defined as the ratio of elongation, Δw to the graphene sheet initial width, w. On the other hand, the shear stress of the continuum sheet can also be found using the similar manner as the normal stress shown equation 11. However, the forces required are the

IOP Conf. Series: Materials Science and Engineering **206** (2017) 012057 doi:10.1088/1757-899X/206/1/012057

tangential forces to the free edge of the graphene sheet. The expression for shear stress is as shown in equation 15 below:

$$\tau = \frac{f_s}{bt} \tag{15}$$

where, f_s is now the magnitude of the shear forces computed from simulation. Therefore, the shear modulus, G of the graphene sheet is also able to be computed by using the following equation 16 where γ is the shear strain of the graphene sheet under shear stress:

$$G = \frac{\tau}{\gamma} = \frac{f_s}{\gamma bt} \tag{16}$$

The shear strain of the graphene sheet can be computed using the boundary conditions set for the shear stress condition as shown in equation 17 below, where Δb is the displacement in the length.

$$\gamma = \frac{\Delta b}{w} \tag{17}$$

By obtaining the Young's modulus, shear modulus and the results from the simulation, the Poisson ratio can finally be calculated for the zigzag and armchair graphene sheet. The Poisson ratio, based on Hooke's law on elastic properties can be computer via the Equation 18 as shown below:

$$v = \frac{E}{2G} - 1 \tag{18}$$

5. Results and Discussion

5.1. Size and Shape Analysis

In the first parametric study, a single layer armchair graphene was studied by increasing the size (length and width) of the graphene. Then, a single layer zigzag graphene was studied by using the same size as the studied armchair graphene.

From the simulations, the reaction forces obtained are plotted against the length of the graphene as in Figure 5. It shows that, the reaction force has a linear relationship to the length of the graphene. These results are actually well predicted since the loading displacements are fixed 10% of its original height. Therefore, since the force is directly proportional to the number of nodes, the bigger the size (length) of the graphene, the higher the number of nodes, hence, the higher the reaction forces.

However, the more obvious result from the reaction force obtained is that the armchair graphene has a higher reaction force when compared to zigzag graphene with the same size. This may because the direction of the load applied to all the armchair graphene's covalent bonds (beams) are tilted at an angle of 600 while in the zigzag graphene, the direction of the load applied is parallel to some of the covalent bonds. With the tilted covalent bonds, more forces are required to increase the distance between them compared to a straight covalent bond as bending of the bonds are involved too in the tilted condition. Therefore, higher force was observed in the armchair graphene.

The reaction forces obtained from the simulation were utilized to calculate the Young's Modulus and the Shear Modulus and these results are plotted with respect to their corresponding sizes in Figure 6 and 7. From Figure 6, it is clear that the Young modulus are in the range of 0.8TPa to 0.95TPa. [16], [17], [18] and [21]. These results are slightly lower than those obtained from the literature which reported to be around be around 1.0 TPa. However, since different force constants are used, these results are deemed acceptable. For the shear modulus, the calculated results are in the same order as those reported in literature, approximate 0.25TPa to 0.45TPa. Although slightly lower, but low shear modulus also has been reported as in [11] and [15].

IOP Conf. Series: Materials Science and Engineering **206** (2017) 012057 doi:10.1088/1757-899X/206/1/012057

Besides that, it is also obvious that the Young's modulus of armchair graphene is greater than that of zigzag graphene when comparing the same size. This is as discussed in the reaction force before, since the armchair graphene required a higher force to have a 10% strain, the Young's modulus will certainly be higher. In other word, the armchair graphene will require more stress or force per unit area for every 1% strain increment. Moreover, these findings are consistent with the published literature [19].

Based on the graph in Figure 6, it can also be noted that the Young's modulus for both armchair and zigzag decreases with size due to hexagonal structure of graphene sheet. As the size of graphene sheet increases, the effect of side shrinkage becomes more obvious. The shrinkage of side of the graphene sheet may reduce the force required to increase the defection. Hence, the Young's modulus decreases as the size of graphene increases. However, it is reported in [19] that these phenomena seem to disappear as the length and width of graphene increases. Hence, further study and research on this matter is still necessary.

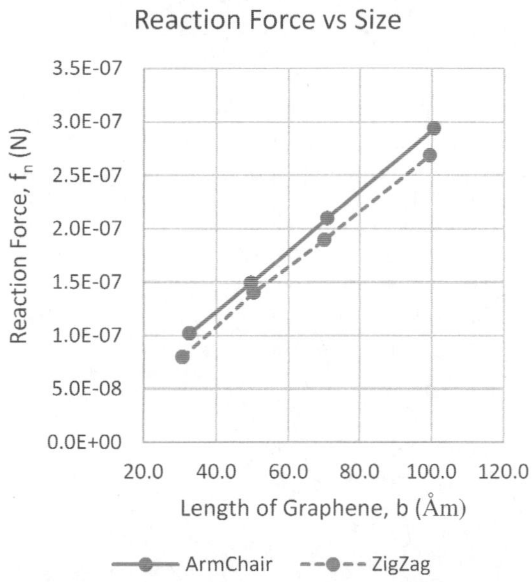

Figure 5. Graph of Reaction Force vs Size of Armchair and Zigzag Graphene

From the Young's modulus and shear modulus, the Poisson's ratio is then calculated as mentioned earlier. The calculated Poisson's ratio is then plotted against the size as shown in Figure 8. Based on the graph, the Poisson's ratio seems to be constant at around 0.8 for armchair graphene and 0.7 for zigzag graphene. The deviation of the constant Poisson ratio for the 3x3 nm graphene sheet may due to the aspect ratio slightly deviated from 1 to 1 ratio.

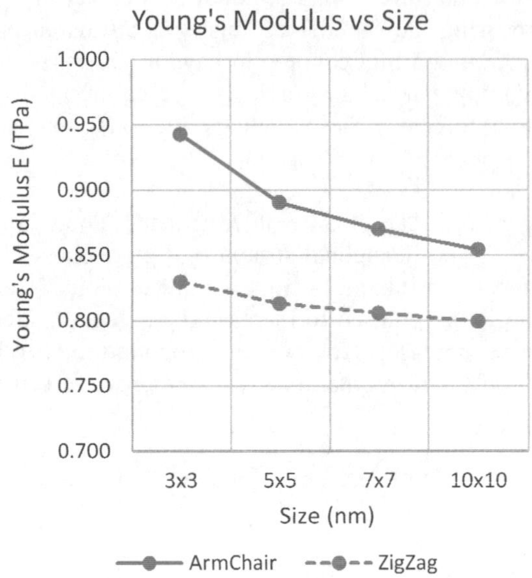

Figure 6. Graph of Young's Modulus vs Size of Armchair and Zigzag Graphene

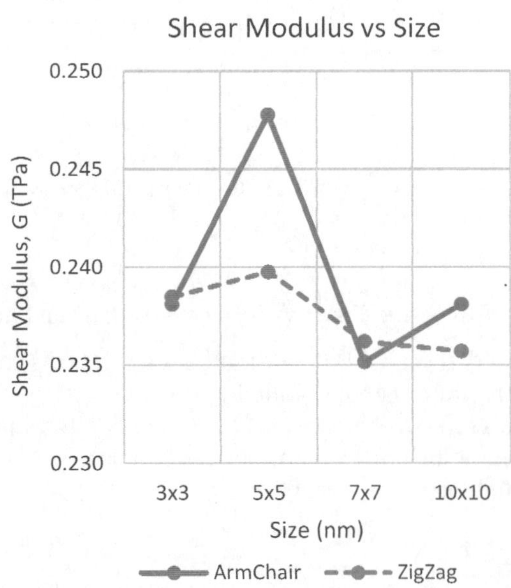

Figure 7. Graph of Shear Modulus vs Size of Armchair and Zigzag Graphene

5.2. Aspect Ratio and Shape Analysis:

In the second parametric study, a single layered armchair and zigzag graphene were studied by increasing the value of aspect ratio (length to width ratio) of the graphene. Similar to the study for effect of size, the reactions forces are obtained and the graph of reaction forces against aspect ratio for both armchair and zigzag graphene are plotted as shown in Figure 9 below. The graph plotted does not show any significant relation between the force and the aspect ratio of the graphene. Instead, the reaction forces vary according to the length of the graphene as mentioned in previous section. Since the graphene with the aspect ratio 2 has the largest length value, thus it has a highest reaction forces. Meanwhile, the graphene with the aspect ratio of 0.33 has the lowest length value, hence it shows the lowest reaction force.

Based on the reaction force obtained, the Young's Modulus and the Shear Modulus are calculated for the corresponding graphene aspect ratio. The results are then plotted against the aspect ratio as shown in Figure 10 and 11 below. From Figure 10, it can be noted the Young's modulus seems to increase with the aspect ratio of the graphene and this is observed for both armchair and zigzag graphene. The gradient of the graph for armchair and zigzag are found to be almost similar to each other at approximate 0.06TPa for every 1 increment of the aspect ratio. This indicated that the Young's modulus of a single layer graphene sheet is also dependent on the aspect ratio. Besides, it is also worth to mention that the Young's modulus for armchair is higher when compared to zigzag graphene. This indicated that the results obtained for this analysis matches well with the previous section.

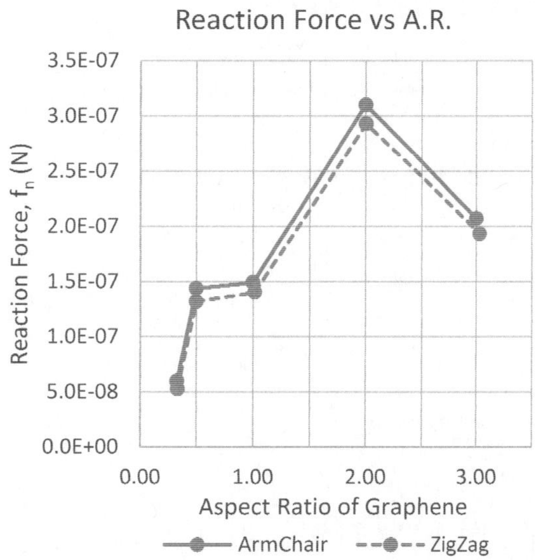

Figure 8. Graph of Reaction Force vs Aspect Ratio of Armchair and Zigzag Graphene

On the other hand, the effect of the aspect ratio to the Poisson ratio is remarkable. Similarly, to the previous section, using the Young's modulus and the shear modulus obtained, the Poisson's ratio are calculated for both armchair and zigzag and the results are plotted against the aspect ratio as shown in Figure 12. From Figure 12, the Poisson's ratio also increased with the aspect ratio, however, the gradient of the graph for the armchair graphene sheet is lower than that of the zigzag graphene. This shows that the Poisson's ratio of a single layer graphene sheet is also dependant on the aspect ratio with zigzag graphene having a higher effect. From the graph, it is also noticed that the line intersection for the armchair and zigzag graphene is at the aspect ratio value of 3. Hence, it is believed that as the aspect ratio greater than 3, the zigzag graphene would have a higher Poisson's ratio than armchair graphene.

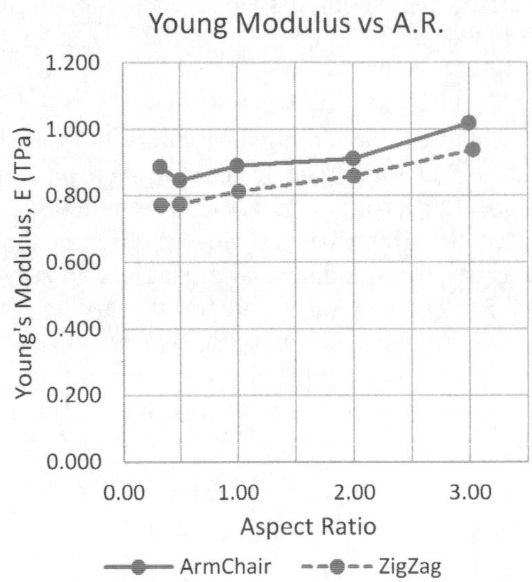

Figure 9. Graph of Young's Modulus vs Aspect Ratio of Armchair and Zigzag Graphene

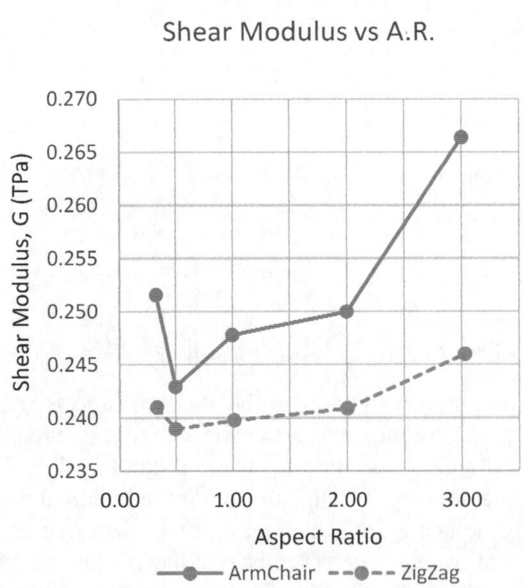

Figure 10. Graph of Shear Modulus vs Aspect Ratio of Armchair and Zigzag Graphene

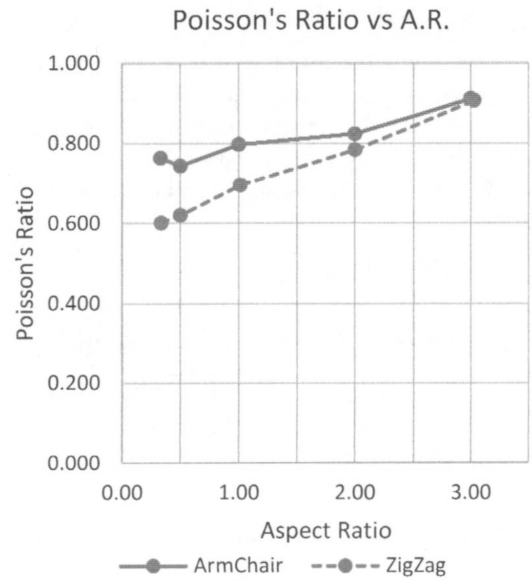

Figure 11. Graph of Poisson's Ratio vs Aspect Ratio of Armchair and Zigzag Graphene

5.3. Analysis of Double Layer Graphene Sheet:

A double layered graphene is also analysed in this study to determine any differences in the Young's modulus and Poisson ratio between a double layered and a single layered graphene sheet(s). For this parametric study, a 5x5 armchair double layered graphene sheets are modelled and simulated using the same procedure as the single layered graphene. The results obtained are compared with a single layered graphene with the same size, chirality and aspect ratio. The exact dimension of the double layered graphene and the results from simulation is as shown in Table 3.

Table 3. Comparison of single layered and double layered graphene properties

Parameter	Single Layered	Double Layered
Length (nm)	4.9630	5.0339
Width (nm)	5.0349	5.0349
Thickness (nm)	0.34	0.68
Reaction Forces (N)	1.492e-7	3.03e-7
Young's Modulus (TPa)	0.890	0.891
Shear Modulus (TPa)	0.248	0.238
Poisson's Ratio	0.794	0.872

From the table above, it can be seen that the Young's modulus for both single and double layered graphene are very similar to each other. This indicates that the Young's moduli of the graphene are independent to the number of layer. However, the Poisson's ratio of the double layered graphene is slightly higher than the single layered graphene. Hence, the Poisson's ratio might also dependant on the number of layer of the graphene sheet.

IOP Conf. Series: Materials Science and Engineering **206** (2017) 012057 doi:10.1088/1757-899X/206/1/012057

5.4. Comparison with Experimental Result:

Lastly it is of utter importance that the simulation results are compared with experimental results. Hence, the simulation results obtained in this study are compared with the available experimental results in the literature. Elastic properties of Graphene obtained from direct measurement experiments are preferred for this comparison. These comparisons are presented in Table – 4 below. It can be seen that the simulation results obtained in this study are in good agreement with the experimental results.

Table 4. Comparison of experimental and simulation results

Source	Approach	Sample	Young's Modulus (TPa)
Lee et. al [22]	Experimental (AFM)	Mono-layer graphene	1.0
Politano et. al [23]	Experimental (Phonon dispersion)	Macroscopic graphene sample	1.0
Current study	Simulation	Single layer graphene	0.890
Current study	Simulation	Double layer graphene	0.891

6. Conclusions

In conclusion, finite element simulations using atomistic modelling technique and continuum technique has been conducted on single layered and double layered graphene with a range of sizes and aspect ratios to determine the effect of size and aspect ratio to the Young's modulus, shear modulus and Poisson's ratio. In order to model the nanostructure, the mechanical properties of the carbon-carbon bonds were characterized as a circular beam element in which its elastic constants were properly defined. These elastic properties are determined by equating the potential energies of the molecular structure based on linear Hooke's law to the structural potential energy of the beam element.

In order to prove the validity of the model, the Young's modulus, shear modulus and Poisson's ratio found in this simulation were compared to the result found in literature and it matches well. Based on the simulation, the effect of size and aspect ratio on the Young's modulus and the Poisson ratio were identified. It is found that the Young's modulus increases slightly with the aspect ratio but decreases with the size while the Poisson's ratio also increases slightly with the aspect ratio but constant with size. The numbers of graphene layers affect the Poisson's ratio value but Young's modulus and shear modulus of this material remain unchanged within investigated parameters. These results are consistent with the current literature.

References

[1] Bolotin K I, Sikes K J, Jianga Z, Klimac M, Fudenberg G, Hone J, Kim P, and Stormer H L 2008 *Solid State Comm.* **146**(9-10) 351-355

[2] Novoselov K S, Blake P, and Katsnelson M I 2008, *Encyclo. of Mat.: Sci. and Techno. (Second Ed.)*, pages: 1-6

[3] Liu L, Park J W, David A Siegel, McCarty K F, Clark K W, Deng W, Basile L, Idrobo J C, Li A P, Gu G2014, *Sci.* **343**(6167) 163-167

[4] Najmaei S, Liu Z, Zhou W, Zou X L, Shi G, Lei S D, Yakobson B I, Idrobo J C, Ajayan P M, and Lou J. 2013, *Nature mat.* **12** (8) 754-759

[5] Bonaccorso F, Colombo L, Yu G H, Stoller M, Tozzini V, Ferrari A C, Ruoff R S, Pellegrini V. 2015, *Sci.* **347**(6217)

[6] Ye X H, Lin Z, Zhang H J, Zhu H W, Liu Z, and Zhong M L 2015, *Carbon* **94** 326-334

IOP Conf. Series: Materials Science and Engineering **206** (2017) 012057 doi:10.1088/1757-899X/206/1/012057

[7] Lu J P 1997, *Phy Rev. Lett.* **79(7)** 1297-1300
[8] Hernández E, Goze C, Bernier P, and Rubio A. 1998, *Phy Rev. Lett.* **80**(20) 4502-4505
[9] Van Lier G, Van Alsenoy C, Van Doren V, and Geerlings P 2000, *Chem. Phy. Lett.* **326**(1-2) 181-185
[10] Faccio R, Denis P A, Pardo H, Goyenola C, and Mombru A W. 2009 *J. of Phy.: Condensed Matter* **21**(28) 1-7
[11] Sakhaee-Pour A 2009, *Solid State Comm.* **149**(1-2) 91-95
[12] Heyrovska R, 2008, *arXiv:0804.4086v1.*
[13] Medina J, Avilés F & Tapia A. 2015, *Molecular Phy.* **113**(11) 1297–1305
[14] Parashar A, Mertiny P 2013, *Solid State Comm.* **173** (2013) 56–60
[15] Scarpa F, Adhikari S, and Srikantha Phani A 2009, *Nanotechno.* **20**(6) 1-11
[16] Georgantzinos S K, Giannopoulos G I, Katsareas D E, Kakavas P A, and Anifantis N K 2011 *Comp.Mat. Sci.***50** (7) 2057–2062
[17] Giannopoulos G I, Liosatos I A, and Moukanidis A K 2011, *Phy. E* **44** (1) 124–134
[18] Shokrieh M M, and Rafiee R 2010, *Mat. and Design* **31**(2) 790-795
[19] Georgantzinos S K, Giannopoulos G I, and Anifantis N K 2010, *Mat. and Design* **31**(10) 4646–4654
[20] Odegarda G M, Gatesb T S, Nicholsonc L M, and Wised K E 2002, *Composites Sci. and Techno.* **62**(14) 1869-1880
[21] Alzebdeh K. 2012, *Int. J. of Mech. and Mat. in Design* **8**(3) 269-278
[22] Lee C G, Wei X D, Kysar J W and Hone J, 2008, *Sci.* **321**(5887) 385-388
[23] Politano A, Marino A R, Campi D, Farı´as D, Miranda R, Chiarello G 2012 *Carbon* **50** 4903–4910

IOP Conf. Series: Materials Science and Engineering **206** (2017) 012058 doi:10.1088/1757-899X/206/1/012058

Simulation of process identification and controller tuning for flow control system

I M Chew [1,2], F Wong[2], A Bono[2] and K I Wong[1]

[1] Faculty of Engineering, Curtin University Malaysia, CDT 250, Miri 98009, Sarawak, Malaysia

[2] Faculty of Engineering, Universiti Malaysia Sabah, Jalan UMS, 88400 Kota Kinabalu, Sabah, Malaysia.

Email: chewim@curtin.edu.my (I M Chew)

Abstract. PID controller is undeniably the most popular method used in controlling various industrial processes. The feature to tune the three elements in PID has allowed the controller to deal with specific needs of the industrial processes. This paper discusses the three elements of control actions and improving robustness of controllers through combination of these control actions in various forms. A plant model is simulated using the Process Control Simulator in order to evaluate the controller performance. At first, the open loop response of the plant is studied by applying a step input to the plant and collecting the output data from the plant. Then, FOPDT of physical model is formed by using both Matlab-Simulink and PRC method. Then, calculation of controller's setting is performed to find the values of Kc and τi that will give satisfactory control in closed loop system. Then, the performance analysis of closed loop system is obtained by set point tracking analysis and disturbance rejection performance. To optimize the overall physical system performance, a refined tuning of PID or detuning is further conducted to ensure a consistent resultant output of closed loop system reaction to the set point changes and disturbances to the physical model. As a result, the PB = 100 (%) and τi = 2.0 (s) is preferably chosen for setpoint tracking while PB = 100 (%) and τi = 2.5 (s) is selected for rejecting the imposed disturbance to the model. In a nutshell, selecting correlation tuning values is likewise depended on the required control's objective for the stability performance of overall physical model.

1. Introduction

In the new era of technology, most of the controller in manufacturing plants are automated control system. The mentioned control system is extended to various industries such as food processing, refinery, chemical, and power generation plants. The control systems consist of a single instrument or a group of instruments which are designed, developed, installed and operated to control a process.

Control system is suffered with the issues of instability and limited capability to resist external disturbances [1]. It typically requires continuous monitoring and controlling by operators so as to maintain output response of system. Even so, the full-time availability of the operators in the plant is unachievable due to other assigned duties and tasks that need completion as well. Therefore, the idea of utilizing instruments to automate the control system is becoming a popular discussion in many industries.

Bi et al (1999) explains that control loop is exactly a feedback control loop that compares the process variable, *PV* of plant to its set point, *SP*. This is followed by generating a manipulated value, *MV* to drive *PV* for the new *SP*. At first, control loop must utilize a sensor to measure *PV*. Secondly, control loop must have a controller block with an actuator that control parameters of the process. A typical automated control system is depicted in block diagram as shown in Figure 1.

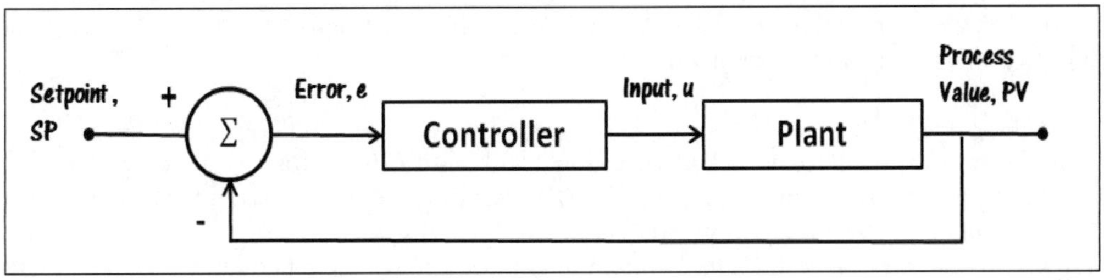

Figure 1. A typical automated control loop.

Plant is the model to be controlled. A plant incorporates a group of physical instruments to produce the products. The plant is inherently fixed and does not react to any changes of process parameters as well as interferences from external environment. *PV* is varies relatively to any changes of parameters in the plant. The imposed changes of *PV* are feedback and compared at summing point. This is a symbol in the diagram that conceptually adding two or more input signals, and produces a single sum output signal as error *e*. The e is transmitted to the controller for generating control signals or *MV* that regulates the initial changes of *PV*. In a nutshell, the main objective of control loop is to regulate *PV* close to SP at all times.

A reliable control loop needs a sophisticated controller, which is able to regulate consistently any factor that destabilized *PV* in the plant. There is various applied control methodologies whereas PID controller is the most widely used [2] because of its flexibility and simplicity to change the settings cope with various requirements in the system. There are elements of control actions for a PID controller [9],[10] comprises proportional action, *P*, integral action, *I* and derivative action, *D*.

1.1. Proportional Action P

Proportional action *P* is the most common form of control action. It is depended on magnitude of *e* that is generated after *PV* is compared with changes of *SP*. In common, the *P* is represented as proportional gain *Kc* which implies the ratio of changes of *PV* to the changes of *SP*.

Equation (1) shows that the output of controller, *u* is equal to multiplication of *Kc* with *e*.

$$u(t) = K_c e(t) \tag{1}$$

The greater value of *Kc* means to more magnificent control actions given for similar amount of *e* existed in the process. The controller will take greater stage action for just a small amount of *PV* deviated from *SP*. It is depicting the purpose of *P* that improving capability of controller so to reduce steady state error when the *PV* attains a new steady state condition. Nevertheless, *Kc* is unable to eliminate steady state error.

There is an applied term that is closely related to *Kc*, which is known as proportional band, *PB*. It expresses the gain of the controller as a percentage of the span of the instrument. In mathematical form, *PB* is reciprocal to *Kc* as described in equation (2).

$$PB = \frac{100\%}{K_p} \tag{2}$$

The lower value of *PB* means a greater *Kc* setting for the controller. In essence, the response is increasingly robust when *PB* is reduced until the limit and eventually the controller will become over-responsive. At this moment, the controller is reacting as an ON/OFF controller and consequently the action of actuator is limited to fully open and fully closed.

In contract, a greater *PB* means a smaller *Kc* and subsequent control actions become 'inaccurate' due to a lack of responsiveness to the *e*. In this extent, the actuator is not responding to minor signal *e* which is just a tiny parameter change in the process. Therefore, appropriate setting range of *PB* is important to allow robust and efficient controller.

1.2. Integral Action I

The second term of control action is known as integral action *I*. It is prominently to govern *PV* to stay tuned with *SP*. *I* reacts by gradually shifting the *PV* once *e* existed. The action is tended to eliminate *e* so that *PV* will attempt to converge towards *SP*.

The gain of integration is known as integral gain *Ki*. It reviews capability of the controller to overcome e versus to the SP setting. The greater the *Ki*, the controller becomes more responsive to process and typically to produce more oscillatory output response due to changes of *PV*.

Mathematically, *I* is the sum of the instantaneous *e* over time and gives the accumulated offset that should have been corrected previously. *I* responses to changes of e by multiplying *Ki* with net area of *e* from time equals to 0. The relationship between the output of the controller, *u* and the *e* is described in equation (3).

$$u(t) = \int_0^t K_i e(t).d(t) \tag{3}$$

where

$$I = Ki = \frac{K_c}{\tau_i}$$

The drawback of *I* is associated with a sort of storehouse of remembered past error that can continuously act towards *e* even after the *PV* reaches the desired value. As the *PV* converges back to the *SP*, the proportional term begins to decrease. But the *I* is continuing to act because of the stored *e* under the error curve, even though *SP* had been reached. *I* will drives the *PV* past through *SP* and to shed its accumulated *e* and stay continuously away from *SP* until accumulated *e* of the opposite direction accumulated and canceling the previous accumulated *e*. Ultimately, resultant *PV* is increased oscillatory and gradually contributes to "Integral Wind up". In dealing with this issue, the controller is recommended to be temporary turned OFF once control element is saturated.

1.3. Derivative Action D

Derivative action *D* is the third element in PID control system. *D* reacts to the rate of change of the *e* at the current time. Mathematically, the derivative of the error e is determined by computing the slope of the *e* over time and then multiplying the rate of change with the gain, which is also known as derivative gain *Kd*.

The mathematical representation of *D* action is described in equation (4).

$$u(t) = K_d \frac{de(t)}{d(t)} \tag{4}$$

where

$$D = Kd = Kc \cdot \tau d$$

IOP Conf. Series: Materials Science and Engineering **206** (2017) 012058 doi:10.1088/1757-899X/206/1/012058

D provides a sudden shift in output power level as the result of rate changes (the slope of error curve) in measured value. If *PV* changes, *D* immediately produces control signal in an attempt to correct the perturbation before the *e* signal goes too far.

This immediate action reflects high sensitivity of *D* relatively to the noise, which is considered as drawback of *D*. Any appeared noises in the process give the slope of error a flip-flop up and down with severe magnitudes, and consequently slight change of *D*. The flicking control signals of controller drives the final control element vigorously to one direction, and then move to another direction. As a result, *PV* becomes oscillatory and unstable. In dealing with this issue, *e* has to be filtered with hardware filter or a software filter prior to be compared with *SP*.

2. The application of proportional, integral and derivative action

Combination of two or more than two control actions in parallel produces more comprehensive controller such as proportional-plus-integral controller (PI) and proportional-integral-derivative controller (PID). Each respective control action reacts to e. The greater the *Kc*, the more final controller element moves so as to drive *PV* to a new steady state condition. However, *Kc* is unable to get rid of steady state error. This is the reason of *I* being added to controller so the controller is constantly acting to regulate *PV* until it achieves *SP* at the end. Besides, *D* enables the controller to predict the potential *e* based on the current trends of process and respond in advance. It sees the *e* approaching and starts controlling to stop the growth of e before it becomes larger.

2.1. PI Controller

PI controller produces more energized control besides regulating process without steady state error when the controlled system is obtaining new *SV*. The combined strength of each control element subsequently makes controller attached to the process that requires high speed response without compromising any steady state error. Figure 2 illustrates the close loop process, which constantly compares SP with feedback of PV so to produce e to controller.

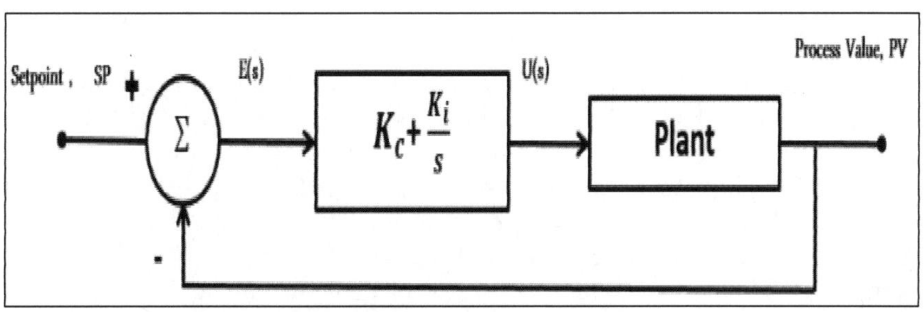

Figure 2. A PI controller in closed loop system.

To increase seems to improve robustness of response by reducing rise time and besides to reduce steady state error as well. Nevertheless, over setting of this element leads to the increase of oscillatory response from controller that destabilizes the process outcome. In essence, the setting of this element must lead to minimum steady state error where oscillatory response will start. Meanwhile, K_i is initiated purposely to eliminate steady state error. It leads to higher controller capability to eliminate the steady state error. Nonetheless, over setting of this element leads to oscillatory response as well.

Mathematically, the PI controller is represented in equation (5).

$$u(t) = K_c e(t) + K_i \int_0^t e(t)d(t) \tag{5}$$

PI controller is not able to predict the future *e* thus PI controller is unable to reduce the settling time of process response. Nonetheless, PI controller is still preferably used in industries compared to other controller modes [3]. PI controller is widely applied to system with large capacity where the speed is

IOP Conf. Series: Materials Science and Engineering **206** (2017) 012058 doi:10.1088/1757-899X/206/1/012058

not an important tuning criterion but precise control is required such as level control of water in tank. Besides, PI controller is also applied to the process where overall process response is fast comparatively to sensor and controller such as flow control system.

3. Process identification

A proper controller tuning is imperatively needed for satisfactory performance of close loop control system. Nonetheless, the presumed performance is less likely to be achieved without a good approach to identify the dynamic of process behavior. Identifying process behaviour typically describes the relationship in between input and output of process, which is represented in mathematical model. This is also known as process identification through modeling.

Modelling process is not an easy task because of complicated processes in industries and high probability of non-linearity, causes difficulties in tuning the controller. There are quite many literature studies had been undertaken for process identification [4],[6]. Among all the studies, one the most common process identification approach is known as Open loop method [9].

This method identifies process behavior by using Process Reaction Curve PRC method. Data such as process gain, Kp, time constant τp, and process time delay, θp are to be replaced into first order plus dead time, *FOPDT* model and then to calculate correlation tuning values of controller that gives optimum control to the process.

The *FODPT* algorithm of process is represented in mathematical model shown in equation (6).

$$Y(s) = \frac{K_p e^{-\theta_p s}}{1+\tau_p s} \tag{6}$$

where

Kp = process gain,
τp = time constant and
θp = time delay.

There are several stages in applying this method [5 - 7] as described in the following steps:

1. To design an experiment and accumulating input-output process data. Prior to experiment test, *PV* is ensured to be in physically stable and controller is in Manual mode.
2. To impose a small percentage of step change as *MV* and observes the change of *PV*. Transient response of *PV* is recorded and plotted in Process Reaction Curve accordingly.
3. To calculate the Kp, τp and θp. Replace those parameters into the *FOPDT* expression.
4. To compute the correlation tuning parameters of controller and after that to apply these values to controller. Set the controller to Automatic mode.
5. Evaluate the new process response with another applied step input and performing re-fined tuning accordingly.

Open loop method is prominently used in modeling a process. Most of the higher order and complicated process can be approximated into *FOPDT* method [3],[8]. This is recommended to repeat the similar experiment for several times to get average of process response for analysis. The similar stage will improve reliability of developed *FOPDT* model. Figure 3 illustrates a unit step *MV* applied and *PV* is observed.

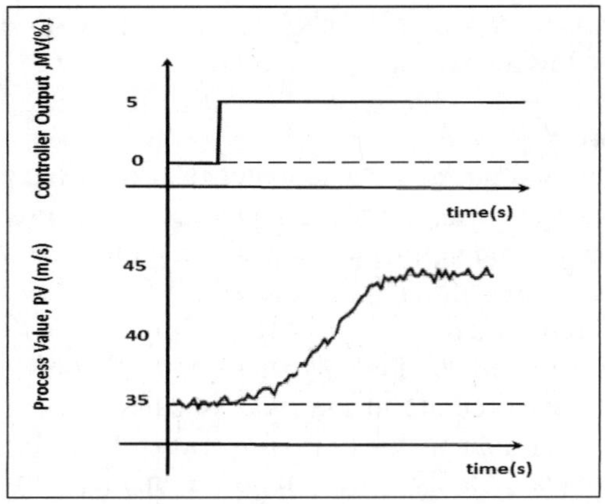

Figure 3. Response of open loop process towards a step change.

3.1. *Process reaction curve (PRC)*

PRC proposes a set of *PV* is accumulated from empirical data and plotted in graph by using software program (for instance Microsoft Excel). Graph of respective *MV* and *PV* versus time are drawn and compared as illustrated in figure 4.

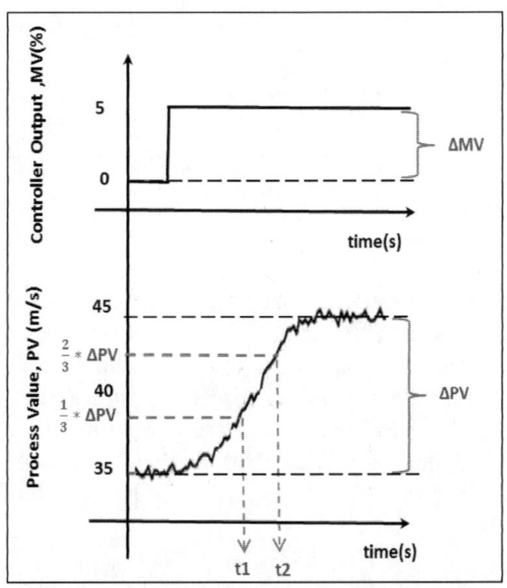

Figure 4. Process reaction curve method.

Determination of all parameters for developing FOPDT model is depicted in equation (7), (8) and (9).

$$\text{Process gain, } K_p = \frac{\Delta PV}{\Delta MV} \tag{7}$$

$$\text{Time constant, } \tau_p = 1.5\,(t_2 - t_1) \tag{8}$$

$$\text{Time delay, } \theta_p = 0.5(t_2 - \tau_p) \tag{9}$$

In calculating K_p , the ratio of *PV* change over *MV* change is determined. Two points of time scales (t_1 and t_2) are recorded where the *PV* increases to approximately 1/3 times and 2/3 times of its final steady-state value. Afterwards, the relative *FOPDT* model is developed by computing K_p, τ_p and θ_p with formulas of Cohen-Coon tuning method.

3.2. Cohen-Coon tuning method

Cohen-Coon tuning method is introduced a decade later than the Ziegler-Nichols method. It uses the mathematical equations to determine K_c, τ_i and τ_d. The stages used to determine those parameters are exactly similar to Ziegler-Nichols tuning method. Selecting either one of tuning method is fine as far as the limitation of each tuning method is in considered. In application, Ziegler-Nichols method is more commonly used compared to Cohen-Coon method because this method is superior for higher order system, whereas Cohen-Coon method is only used for single order system. Nonetheless, Ziegler-Nichols method will become less accurate in the process where ratio of θ_p to τ_p is less than 1/2. In contrast, Cohen-Coon method is able to tolerate with ratio of θ_p to τ_p until 3/4 and even can be extended further. That is the reason for Cohen-Coon tuning method to become as dominant method in most control textbooks (Tavakoli and Fleming, 2003).

Calculating K_c and τ_i parameters for a PI controller by using Cohen-Coon's formula (Bi et al, 1999) is illustrated in table 1.

Table 1. Proportional gain and integral time of PI controller (Cohen Coon).

Controller Type	K_c	τ_i	τ_d
PI	$\frac{1}{K_p}\frac{\tau_p}{\theta_p}\left(\frac{9}{10}+\frac{\theta_p}{12\tau_p}\right)$	$\theta_p\left(\frac{30+3\frac{\theta_p}{\tau_p}}{9+20\frac{\theta_p}{\tau_p}}\right)$	-

4. Methodology

4.1. Equipment

In this paper, the Process Control Simulator (Model: 201) is used as physical plant to produce experiment data for analysis as shown in figure 5. The differential pressure value of airflow between both side of orifice plate was measured and converted to electrical signal (mA) by pressure transducer. Furthermore, controller of the simulator is able to control airflow either Manual or Auto mode. The values of all parameters of *SP*, *MV* and *PV* can be recorded and converted to Microsoft excel datasheet for further analysis. This is ideal equipment used to analyze the system response of various PID parameters setting.

Figure 5. Physical plant for experiment.

IOP Conf. Series: Materials Science and Engineering **206** (2017) 012058 doi:10.1088/1757-899X/206/1/012058

4.2. Process identification and least square curve fitting

Experiment data was accumulated by using open loop test as depicted in figure 6. A step function (very small % changes for *MV*) is imposed to the system (showed in the blue line). The resultant output response is likely to be a S-curve. From the graph, the resultant output response is lifted slowly and then gradually speeded up more until it attained the maximum value (approximately 10 seconds), which is also called as new steady state condition.

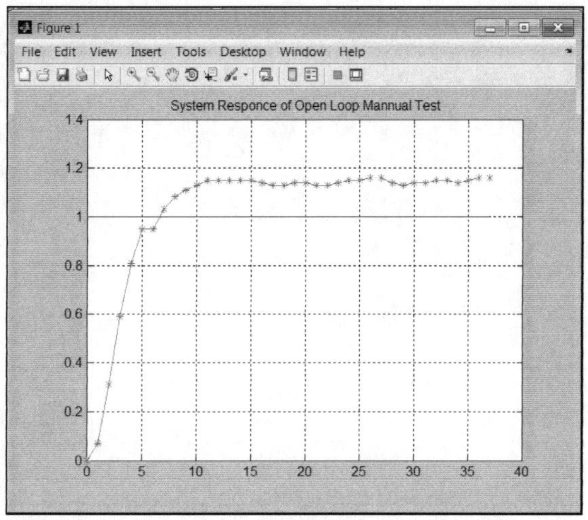

Figure 6. Output response vs unit step input.

The values for K_p is 1.146, τ_p is 2.306 (s) and θ_p of 1.3 (s) are obtained by using *PRC* method. This yields the developed *FOPDT* model as represented in equation (10).

$$G(s) = \frac{1.146e^{-1.3s}}{1+1.2306s} \tag{10}$$

Least Square Curve Fitting is applied to validate the reliability of the developed *FOPDT* model. Figure 7 shows graph of *FOPDT* model, which is compared with the experimental data. In essence, the developed *FOPDT* model is fitted to experimental data to prove that *FOPDT* model expresses the approximated dynamic of process.

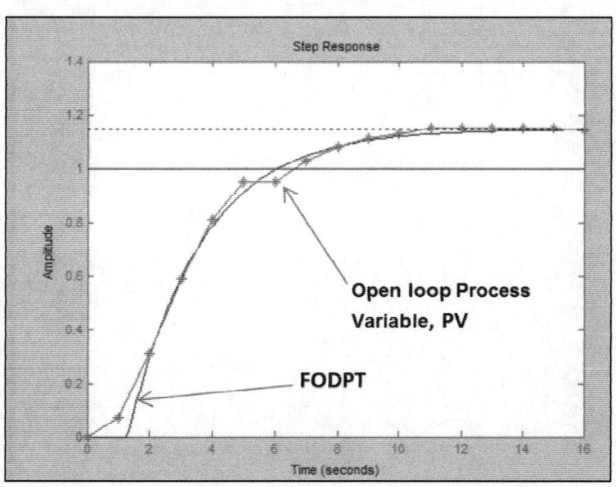

Figure 7. FOPDT and least square curve fitting.

4.3. Flow control diagram with PI controller

A closed loop control diagram is designed and simulated by using Matlab-Simulink as depicted in figure 8. Recall back parameters in FOPDT model, where $K_p = 1.146$ and $\tau_i = 2.306$ (s) and $\theta_p = 1.3$ (s). Replacing those parameters into the formulas of Cohen-Coon tuning formulas yields Kc is 1.5 and τi is 2.0 (s). Both K_c and τ_i are set to the PI controller. The result of simulation is generated after RUN button is pressed.

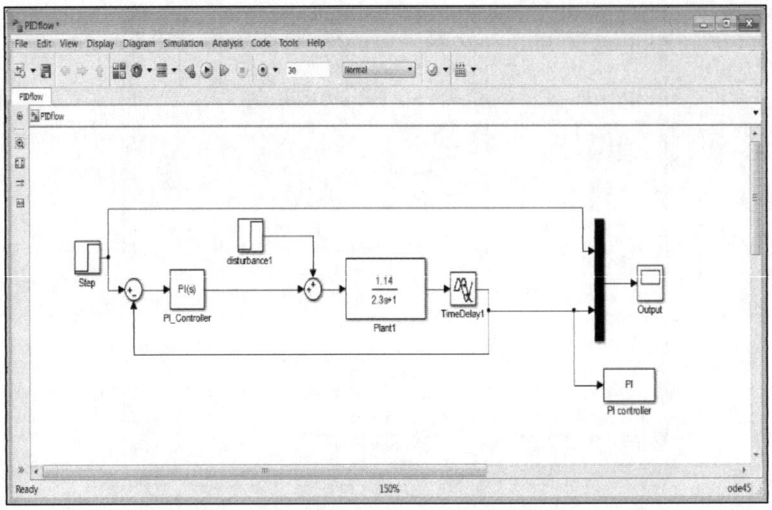

Figure 8. Close loop diagram for flow rate control.

4.4. Comparison of simulation and experimental data

The experimental data from PI controller ($K_c = 1.5$ and $\tau_i = 2.0$ (s)) setting is compared with simulated data as illustrated in figure 9. This is purposely to validate the performed physical system is in line with simulation data. The minor deviation in between simulation data with process data is possibly due to the non-linearity of system or friction of moving part such as motor fan.

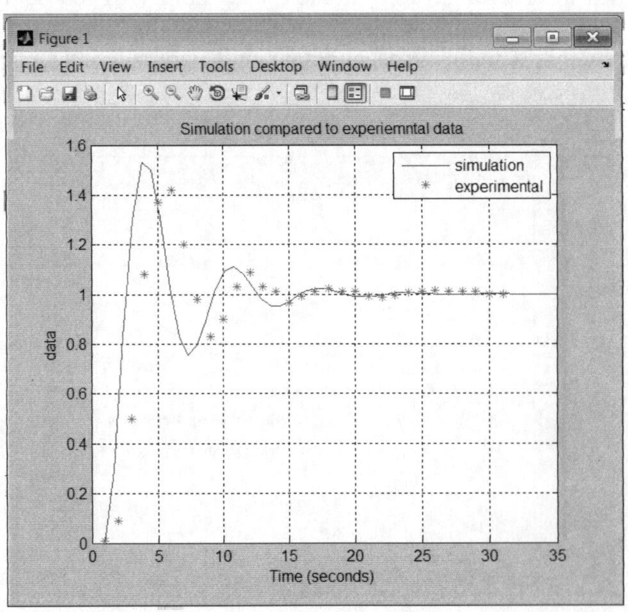

Figure 9. Simulated value vs experimental value.

An optimized control system requires the correlation tuning values for *Kc* and τi to the best fitted for the consistent control response. Ideally, the controller regulates *PV* aggressively without producing any overshoot [7]. The well-known approach is known as refined-PID tuning. It suggests a minor tuning for *Kc* and τi parameters so as to attain a satisfactory close loop response with minimum overshoot and rise time besides settling time less than 2%. Refined-PID tuning optimizes all features of control actions without deteriorating other features. However, refined-PID tuning might takes minutes or hours for a point setting to produce a stable effect at the end.

5. Analysis for setpoint tracking and disturbance rejection performance

5.1. Set point tracking of flow control system

For a proper controlled system, the output response should be magnificent, and should not oscillate in any new condition of set point or applied disturbance. Robustness of controller response can be enhanced by increasing *Kc* value which subsequence more oscillatory response as well. In contrast, reducing *Kc* value reduces oscillatory response. In obtaining empirical result, each correlation tuning values is applied and then followed by a step change for recording output response simultaneously.

After trials on various setting for *Kc* and τi parameters, the selected value for PB = 100 (Kc = 1 instead of 1.5) whereas the τi varies in re-fined PID tuning. The *SP* change and related output response is illustrated in figure 10, which shows that process response with τi = 2.0 (s) is somehow more robust with smaller overshoot incurred. Furthermore, these is no overshoot occurring after τi is increased to 2.5 (s). This implies that higher value of τi will slow down the system response to the *SP* change. Therefore, PB = 100 (%) and τi = 2.0 (s) is preferably chosen.

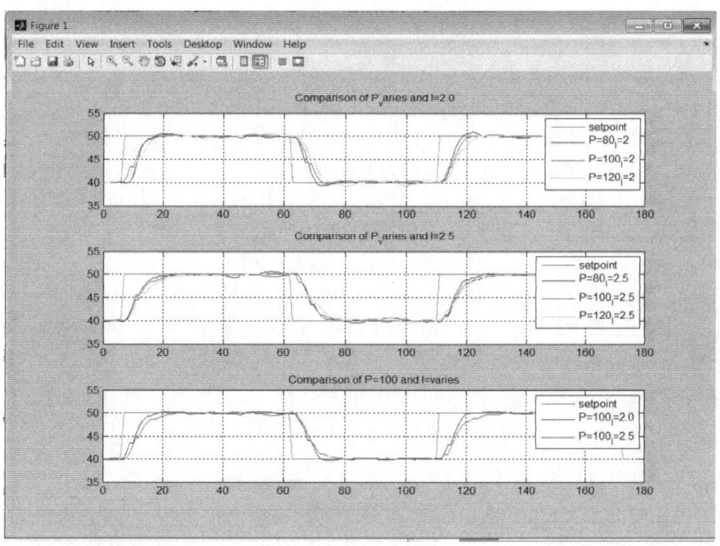

Figure 10. Output response of setpoint tracking analysis.

5.2. Disturbance rejection performance of flow control system

Some systems require strong capability to resist instability of process response due to high frequency of external disturbances rather than the *SP* change. This is specifically referred to systems, which suffered from noises or external interferences. A proper tuned control system provides reasonably damped response with minimum overshoot and fast recovery to the set point after imposed with disturbances. This is achievable by tuning PI controller so as to produce consistent response with minimum oscillation.

Figure 11 illustrates that PI controller with PB = 100 (%) and τi = 2.0 (s) caused more oscillatory response compared to τi = 2.5 (s). The larger the τi, the better the system performs to resist disturbances imposed to the process. Therefore, PI controller with *PB* = 100 (%) and τi = 2.5 (s) is preferably chosen.

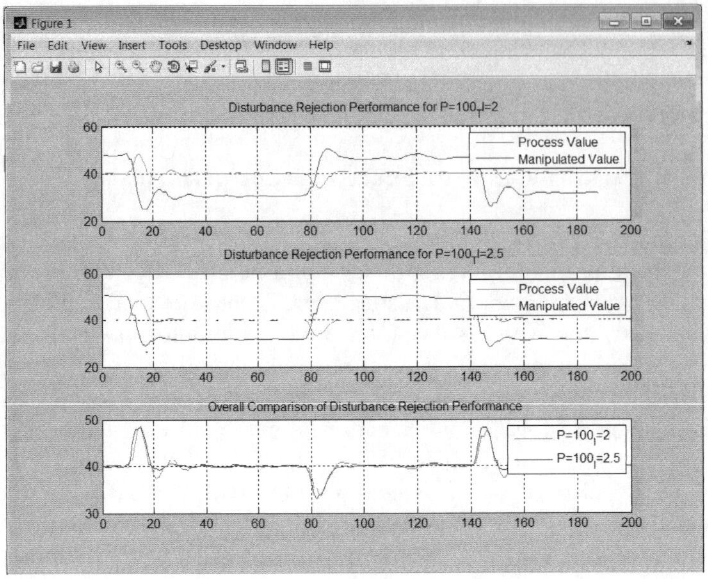

Figure 11. Output response from disturbance rejection performance.

6. Conclusion

This paper discusses the three elements of control actions and improving effectiveness of controllers through combination of these control actions in various forms. A Process Control Simulator had been used as model to analyze process response of various applied control actions and even combination of them. Due to the reaction of physical process is relatively fast to sensor and controller for flow control, PI controller is chosen as control model. With identified *FOPDT* model, correlation tuning parameters of PI controller were determined by using Cohen-Coon tuning method.

The analysis of process response is further extended to setpoint tracking analysis and disturbance rejection performance. The analysis is purposely to justify the best value of Kc and τi for optimized system performance with PI controller. In conclusion, *PB* = 100 (%) and τi = 2.0 (s) is preferably chosen for changes of *SP* value. Meanwhile, *PB* = 100 (%) and τi = 2.5 (s) is preferably selected for rejection to the noises and external inferences imposed to the system.

References

[1] Astrom K J and Hagglund T 2001 *Control Engineering Practice.* **9** 1163
[2] Ho M T, Datta A and Bhattacharyya S P 1997 *American Control Conference* **1** 571
[3] Tavakoli S and Fleming P 2003. *European Control Conference* 2196
[4] Ahmed S, Huang B and Shah S L 2006 *Journal of Process Control* **16** 323
[5] Bi Q, Cai W J, Lee E L, Wang Q G, Hang C C and Zhang Y 1999 *Control Eng. Practice* **7** 71
[6] Rohit R, Lakshminarayanan S, and Rangaiah G P 2005 *Journal of The Institution of Engineers*, Singapore, **45**
[7] Liu G P and Daley S 2001 *Control Eng. Practice* **9** 1185
[8] Ou L, Zhang W, and Yu L 2009 *IEEE Transactions in Automation Control* **54** 774-84
[9] Riggs J B and Nazmul Karim M 2006 *Chemical and Bio-Process Control-Third Edition.*(US: Pearson Educational International) **3** 205-18; 235-49; 262-302
[10] King M 2011 *Process control - A practical approach* (UK: John Wiley and Sons Ltd) 29-39

IOP Conf. Series: Materials Science and Engineering **206** (2017) 012059 doi:10.1088/1757-899X/206/1/012059

Optimization of an auto-thermal ammonia synthesis reactor using cyclic coordinate method

T A-N Nguyen[1], T-A Nguyen[2], T-D Vu[2], K-T Nguyen[2], T K-T Dao[2] and K P-H Huynh[2]

[1] Faculty of Applied Science, Ton Duc Thang University, 19 Nguyen Huu Tho Str., District 7, Ho Chi Minh City, Vietnam

[2] Faculty of Chemical Engineering, Ho Chi Minh City University of Technology, District 10, Ho Chi Minh City, Vietnam

E-mail: anh.nguyen@hcmut.edu.vn

Abstract. The ammonia synthesis system is an important chemical process used in the manufacture of fertilizers, chemicals, explosives, fibers, plastics, refrigeration. In the literature, many works approaching the modeling, simulation and optimization of an auto-thermal ammonia synthesis reactor can be found. However, they just focus on the optimization of the reactor length while keeping the others parameters constant. In this study, the other parameters are also considered in the optimization problem such as the temperature of feed gas enters the catalyst zone, the initial nitrogen proportion. The optimal problem requires the maximization of an objective function which is multivariable function and subject to a number of equality constraints involving the solution of coupled differential equations and also inequality constraint. The cyclic coordinate search was applied to solve the multivariable-optimization problem. In each coordinate, the golden section method was applied to find the maximum value. The inequality constraints were treated using penalty method. The coupled differential equations system was solved using Runge-Kutta 4th order method. The results obtained from this study are also compared to the results from the literature.

1. Introduction

Ammonia is one of the most important synthetic chemical product in the world. Most of the commercially produced ammonia is used in fertilizers with the remainder used in a variety of applications such as plastics, synthetic fibers and resins, pharmaceuticals, explosives, papers and refrigeration [1]. Therefore, modelling and optimization of ammonia synthesis have received a considerable attention by industries and researcher. Most of the world consumption is manufactured from the elements nitrogen and hydrogen in a catalytic process originally developed by Haber and Bosch using a promoted iron catalyst as follows:

$$N_2 + 3H_2 \rightleftarrows 2NH_3 \tag{1}$$

The reaction is reversible and exothermic with considerable release of heat. Therefore, the reaction is usually carried out in an auto-thermal synthesis reactor, which utilizes the heat of reaction to pre-heat the feed gas and maintain the appropriate temperature inside. The production of ammonia depends on several factors such as the reactor length, temperature of feed gas, temperature of the reacted gas mixture at the exit of the reactor, and the mass flow rate of nitrogen. The optimization problem consists of maximizing the economic return from the process. There are many study discussing the modelling, simulation, and optimization of an auto-thermal ammonia synthesis reactor. Some of them can be mentioned here as in Babu and Angira [2], Babu et al. [3], Carvalho et al. [4], Edgar et al. [5], Ksasy et al. [6], Murase et al. [7], Upreti and Deb [8], Yusup et al. [9]. However, in the studies of Edgar et al. [5], Murase et al. [7], the model used has some uncorrected points and have been modified in Upreti and Deb [8]. Furthermore, the studies mainly focus on the optimization of reactor length corresponding to a particular reactor top temperature, usually of 694 K [3, 4], or in only several temperatures [6].

In this study, not only the reactor length but also the reactor top temperature is included in the optimization variables for finding the maximum profit of the process. The coupled differential equations which simulate the synthesis reactor are solved using Runge-Kutta 4th order method. Cyclic coordinate search method was applied for the multivariate optimization problem. In each coordinate axis search, the problem becomes single variable optimization and the golden section search was employed to find the maximum of the profit. The box-constraints of the parameters were guaranteed by adding the penalty to the objective function when any of the limits is violated. The results from the study are also compared to the other reports found in literature.

2. Problem formulation

The problem formulation in this work is similar to the model in Murase et al. [7] and includes the modifications mentioned in Babu and Angira [2], Upreti and Deb [8]. The heat released by the reaction is used to heat the incoming gas mixture in counter current flow. The production of ammonia depends on the temperature of feed gas at the entrance of the reaction zone (top temperature), the partial pressures of the reactants (nitrogen and hydrogen), and the reactor length. The optimal design problem aims to obtain the optimal parameters yielding maximum economic return from the reactor operation.

2.1. Objective function

The objective function for the reactor optimization is the profit of the process based on the difference between the value of the product gas (heating value and the ammonia value) and the value of feed gas (as a source of heat only) less the amortization of reactor capital costs. Other operating cost are neglected.

$$F = 1.3356\times10^{7} - 1.708\times10^{4} N_{N_2} + 704.09\left(T_g - T_0\right) - 699.27\left(T_f - T_0\right) - \left(3.4566\times10^{7} + 1.9837\times10^{9} x\right)^{\frac{1}{2}} \quad (2)$$

2.2. Equality constraints

The model of ammonia reactor is obtained considering the energy balance for feed gas and reacting gas, and the mass balance for nitrogen, respectively.

$$\frac{dT_f}{dx} = \frac{US_1}{WC_{pf}}\left(T_g - T_f\right) \quad (3)$$

$$\frac{dT_f}{dx} = -\frac{US_1}{WC_{pg}}\left(T_g - T_f\right) + \frac{\left(-\Delta H\right)S_2}{WC_{pg}}\left(-\frac{dN_2}{dx}\right) \quad (4)$$

$$\frac{dN_2}{dx} = -f\left(k_1 \frac{p_{N_2} p_{H_2}^{1.5}}{p_{NH_3}} - k_2 \frac{p_{NH_3}}{p_{H_2}^{1.5}}\right) \quad (5)$$

in which

$$k_1 = 1.78954\times10^{4} \exp\left(\frac{-20800}{RT_g}\right) \quad (6)$$

$$k_2 = 2.5714\times10^{16} \exp\left(\frac{-47400}{RT_g}\right) \quad (7)$$

$$p_{N_2} = \frac{286 N_{N_2}}{2.598 N_{N_2}^{0} + 2 N_N} \quad (8)$$

$$p_{H_2} = 3p_{N_2} \tag{9}$$

$$p_{NH_3} = \frac{286\left(2.23N_N^0 - 2N_{N_2}\right)}{2.598N_N^0 + 2N_N} \tag{10}$$

The notations T_f^0, T_g^0, and $N_{N_2}^0$ denote the initial value at x=0 for T_f, T_g, and N_{N_2}, respectively. The initial values are given by

$$T_f^0 = T_g^0 = T_0,\ \ N_{N_2}^0 = 701.2\,\text{kmol/hm}^2 \tag{11}$$

2.3. Box constraints

As usual in industry, the following physical restrictions are imposed over the variables:

$$0 < x \le 10,\ \ 400 \le T_f \le 800,\ \ 0 \le N_{N_2} \le 3220 \tag{12}$$

From the system model, it can be obtained that the length of the reactor and the top temperature can be chosen as the independent variables. The remaining variables (T_f, T_g, and N_{N_2}) can be calculated from the three differential equations. From that, the objective function is estimated and is considered as a function of two variables T_0 and x. From the constraints, the variable T0 is also set to be within 400 and 800.

The optimal design problem is summarized as follows

$$\max F\left(x, T_0\right)$$

$$s.t. \begin{cases} \dfrac{dT_f}{dx} = \dfrac{US_1}{WC_{pf}}\left(T_g - T_f\right) \\ \dfrac{dT_f}{dx} = -\dfrac{US_1}{WC_{pg}}\left(T_g - T_f\right) + \dfrac{(-\Delta H)S_2}{WC_{pg}}\left(-\dfrac{dN_2}{dx}\right) \\ \dfrac{dN_2}{dx} = -f\left(k_1\dfrac{p_{N_2}p_{H_2}^{1.5}}{p_{NH_3}} - k_2\dfrac{p_{NH_3}}{p_{H_2}^{1.5}}\right) \\ T_f^0 = T_g^0 = T_0,\ \ N_{N_2}^0 = 701.2 \\ 400 \le T_f \le 800,\ \ 0 \le N_{N_2} \le 3220 \\ 0 < x \le 10,\ \ 400 \le T_0 \le 800 \end{cases} \tag{13}$$

3. Optimization strategy

The procedure to solve problem (13) consists of obtaining approximate solutions of differential equation system (3), (4) and (5), with initial conditions (11) through the fourth-order Runge–Kutta method. The initial condition is clearly defined when the top temperature (T_0), which is also a variable, is assigned. The interval of integration is defined when the reactor length is assigned. From these computed values, the objective function is evaluated and the procedure above is repeated until an optimal solution is found. Therefore, the problem can be considered as two-variable optimization and the simple cyclic coordinate search can be employed. In order to guaranty the constraints, the objective function is forced to assume an undesired value whenever any of variable limits is violated during the optimization process. Such approach consists of defining a penalty or barrier to the objective function when any of the limits is violated. In each direction search of cyclic coordinate method, a single-variable optimization has to be performed and a derivative-free algorithm, the golden search method, is chosen. The proposed strategy is detailed as follows.

3.1. Cyclic coordinate search

Often in the solution of multivariable optimization problems it is desired to be done with a gradient-free algorithm. This may be the case when gradient evaluations are difficult, or in fact gradients of the

IOP Conf. Series: Materials Science and Engineering **206** (2017) 012059 doi:10.1088/1757-899X/206/1/012059

underlying optimization method do not exist. Such a method that offers this feature is the method of the cyclic coordinate search, the simplest method for nonlinear optimization [10].

The minimization problem considered is: $\underset{\mathbf{x}\in\mathbb{R}^n}{\text{optimize}}\, f(\mathbf{x})$

The method in its basic form uses the coordinate axes as the search directions. Therefore, along each search direction $i = 1:n$, the corresponding variable x_i is changed only, with all remaining variables being kept constant to their previous values.

The optimization is carried out in order over all variables with indices 1,…,*n* at each iteration of the algorithm. The task is repeated until the convergence condition satisfied, such as no significant improvement after one cycle.

Step 1. Initialization

Select a tolerance $\varepsilon > 0$, to be used in the termination criterion. Select an initial point $\mathbf{x}_{(0)}$ and initialize by setting $\mathbf{z}_{(1)} \leftarrow \mathbf{x}_{(0)}$. Set k = 0 and i = 1, where k indicates the number of iterations, i is the pointer variable for the direction.

Step 2. Main iteration

Search the optimal solution $f(z^*)=\text{opt}\, f(x)$, which is the single-variable optimization of variable x_i. Update $\mathbf{z}_{(i+1)}=z^*$.

If j < n, then increase i to i +1 and repeat step 2. Otherwise, if j = n, then go to step 3.

Step 3. Termination check

Set $\mathbf{x}_{(k+1)} = \mathbf{z}_{(n)}$. If the termination criterion is satisfied, for example $\text{abs}(x_{(k+1)} - x_{(k)}) < \varepsilon$, then stop. Else, set $\mathbf{z}_{(1)} = \mathbf{x}_{(k+1)}$. Increase k to k + 1, set i = 1 and repeat step 2.

3.2. Barrier method for constrained optimization

In barrier or penalty method, a constrained optimization problem is replaced by an unconstrained problem. The solution will favour satisfaction of the constraints by adding to the objective function a term that produces a high cost for violation of the limits. Therefore, the objective function can be re-defined as

$$F = \begin{cases} F \text{ if } 400 \le T_f \le 800,\ \ 0 \le N_{N_2} \le 3220 \\ -10^5 \text{ otherwise} \end{cases} \tag{14}$$

3.3. Golden section search

The golden section search is a technique for finding the optimum point of a function by iteratively narrowing the interval containing the extremum similar to bisection brackets the zero of a function. The algorithm determining the maximum of a function *F* is briefly described below.

Step 1. Let $[x_l, x_u]$ be an interval containing the maximum, $\varepsilon > 0$ be a tolerance. Set k=0.

$$\begin{cases} x_1 = x_u - \phi(x_u - x_l) \\ x_2 = x_l + \phi(x_u - x_l) \end{cases} \quad \text{where } \phi = \left(\sqrt{5}-1\right)/2 \tag{15}$$

Step 2. Evaluate $F(x_1)$ and $F(x_2)$

If $F(x_1) \ge F(x_2)$, set $x_u=x_2$, $x_2=x_1$, $x_1=x_u-\phi(x_u-x_l)$.

Else, set $x_l=x_1$, $x_1=x_2$, $x_2=x_l+\phi(x_u-x_l)$.

Step 3. If $x_u - x_l < \varepsilon$, then the maximum occurs at $(x_l + x_u)/2$ and stop iterating, else go to Step 2.

3.4. Parameters

In this work, the parameters were obtained from the literature and summarized in Table 1.

Table 1. Used parameters

Parameter	Value	Unit
C_{pf}	0.707	kcal/kg K
C_{pg}	0.719	kcal/kg K
f	1.0	
ΔH	−26000	kcal/kmol
R	1.987	kcal/kmol K
S_1	10	m
S_2	0.78	m^2
U	500	kcal/h m^2 K
W	26400	kg/h

4. RESULTS AND DISCUSSIONS

4.1. Temperature and flow rate profiles

In order to simulate and optimize the reactor, several Matlab programing files was developed based on the discussed algorithm. Figure 1 shows the profiles obtained at different top temperature (T_0). It is evident from the graph that the profiles are smooth and well agree with the others found in literature. It is also obtained that the proposed numerical method is stable even at the top temperature as high as 800 K, which is different from the conclusions from [8].

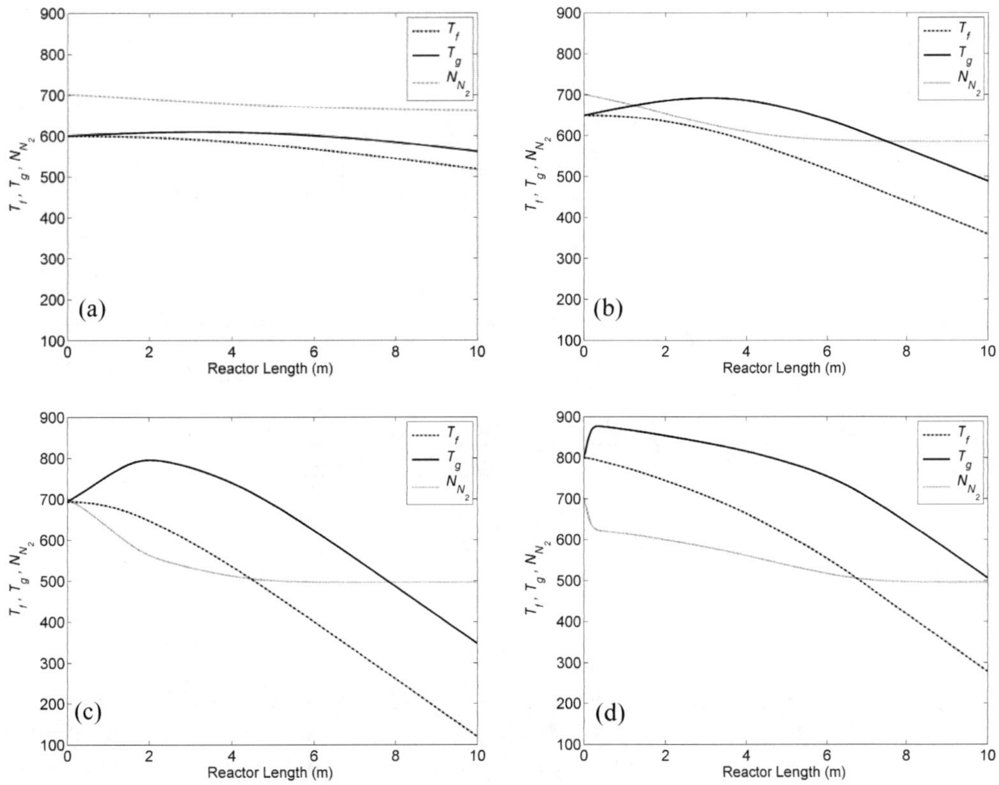

Figure 1. The profiles at different top temperatures
(a) T_0=600 K (b) T_0=650 K (c) T_0=694 K (d) T_0=800 K

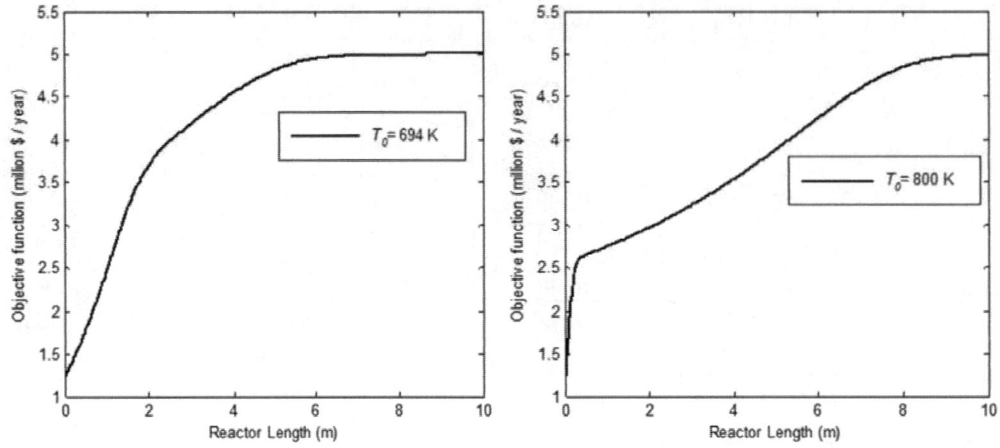

Figure 2. Objective function at different top temperatures

4.2. Optimization of the profit function

Figure 2 shows the variation of objective function with reactor length at different top temperatures. It can be obtained that the objective function has single maximum point at each top temperature. Therefore, the golden section search can be applied for optimization of the profit.

The optimum objective function at the top temperature of 694 K is usually found in literature [2-4, 8]. Therefore, the optimum profit at the top temperature of 694 K was also conducted and compared to the other reports. The results are showed in Table 2.

Table 2. Optimum length at top temperature of 694 K

Parameters	Babu and Angira [2]	Yusup et al. [11]	Ksasy et al. [6]	Carvalho et al. [4]	Current study
x (m)	6.586	6.695	6.425	6.694	6.695
T_f (K)	406.55	399.85	308.36	400.00	400.02
F (10^6\$/year)	5.014	5.017	5.664	5.016	5.016

Table 3. Optimum results

Variables	Interval	Optimal solution
x (m)	[0,10]	6.724
T_0 (K)	[600,800]	700.27
T_f (K)	[400,800]	400.00
T_g (K)		629.94
N_{N2}	[0,3220]	490.68
F (10^6\$/year)		5.018

The profit return from the process depends not only the length of the reactor but also the top temperature. Therefore, the cyclic coordinate search was used for optimization the objective function. It was found that the optimum reactor length is 6.724 m and the top temperature should be operated at 700.27 K. The profit obtained from the process is 5.018×10^6 \$ per year. The other parameters of the process are shown in Table 3. The profit value is higher than the others found in literature. The result suggests that the temperature of the feed entering the catalyst zone should be slightly higher and the reactor length is also slightly longer the other found in literature.

5. Conclusions

In this study, we presented a Runge-Kutta 4th order method for solving system of differential equation modeling the ammonia synthesis reactor. The maximum profit return was considered as multivariable optimization problem. The optimization strategy included cyclic coordinate search method for the top temperature and reactor length, golden section search for each coordinate search and barrier function for the constraints. Numerical results suggested the accuracy and efficiency of the method. The optimum solution satisfied all the constraints and higher than other reports. The synthesis system should be operated at top temperature of 700.27 K with the reactor length of 6.724 m. The corresponding economic return is 5.018×10^6 \$ / year.

Acknowledgement

The authors would like to thank Associate Professor Yoshikawa Shiro and Tokyo Institute of Technology, Tokyo, Japan for the kind support during the preparation of the manuscript.

References

[1] Appl M 1999 *Ammonia: principles and industrial practice* (Weinheim: Wiley-VCH)

[2] Babu B V and Angira R 2005 Optimal design of an auto-thermal ammonia synthesis reactor *Computers & Chemical Engineering* **29** 1041-1045.

[3] Babu B V, Angira R and Nilekar A 2004 Optimal design of an auto-thermal ammonia synthesis reactor using differential evolution. in *The Eighth World Multi-Conf. on Systemics, Cybernetics and Informatics* (Orlando, Florida, USA).

[4] Carvalho E P, Borges C, Andrade D, Yuan J Y and Ravagnani M A S S 2014 Modeling and optimization of an ammonia reactor using a penalty-like method *Applied Mathematics and Computation* **237** 330-339.

[5] Edgar T F, Himmelblau D M and Lasdon L S 2001 *Optimization of Chemical Processes* (McGraw-Hill)

[6] Ksasy M S M, Areed F, Saraya S and Khalik M A 2010 Optimal Reactor Length of an Auto-Thermal Ammonia Synthesis Reactor *International Journal of Electrical & Computer Sciences* **10** 6-15.

[7] Murase A, Roberts H L and Converse A O 1970 Optimal Thermal Design of an Autothermal Ammonia Synthesis Reactor *Industrial & Engineering Chemistry Process Design and Development* **9** 503-513.

[8] Upreti S R and Deb K 1997 Optimal design of an ammonia synthesis reactor using genetic algorithms *Computers & Chemical Engineering* **21** 87-92.

[9] Yusup S, Zabiri H, Yusoff N and Yew Y C 2006 *Modeling and optimization of Ammonia reactor using shooting methods*, in *Proc. of the 5th WSEAS Int. Conf. on Data networks, Communications and Computers*. 2006, World Scientific and Engineering Academy and Society (WSEAS): Bucharest, Romania. p. 258-268.

[10] Floudas C A and Pardalos P M 2008 *Encyclopedia of Optimization* (Springer US)

[11] Yusup S, Zabiri H, Yusoff N and Yew Y C 2006 *Modeling and optimization of Ammonia reactor using shooting methods*, in *Proceedings of the 5th WSEAS international conference on Data networks, communications and computers*. 2006: Bucharest, Romania. p. 258-268.

IOP Conf. Series: Materials Science and Engineering **206** (2017) 012060 doi:10.1088/1757-899X/206/1/012060

Stabilization and analytical tuning rule of double-loop control scheme for unstable dead-time process

B Ugon[1], J Nandong[2] and Z Zang[3]

[1] Department of Foundation in Engineering and Science, Curtin University Malaysia, CDT 250, Miri 98009, Sarawak, Malaysia
[2] Deparment of Chemical Engineering, Curtin Univerisity Malaysia, CDT 250, Miri 98009, Sarawak, Malaysia
[3] Department of Electrical and Computer Engineering, Curtin University Malaysia, CDT 250, Miri 98009, Sarawak, Malaysia

Email: bejay@curtin.edu.my (B Ugon)

Abstract. The presence of unstable dead-time systems in process plants often leads to a daunting challenge in the design of standard PID controllers, which are not only intended to provide close-loop stability but also to give good performance-robustness overall. In this paper, we conduct stability analysis on a double-loop control scheme based on the Routh-Hurwitz stability criteria. We propose to use this unstable double-loop control scheme which employs two P/PID controllers to control first-order or second-order unstable dead-time processes typically found in process industries. Based on the Routh-Hurwitz stability necessary and sufficient criteria, we establish several stability regions which enclose within them the P/PID parameter values that guarantee close-loop stability of the double-loop control scheme. A systematic tuning rule is developed for the purpose of obtaining the optimal P/PID parameter values within the established regions. The effectiveness of the proposed tuning rule is demonstrated using several numerical examples and the result are compared with some well-established tuning methods reported in the literature.

1. Introduction

There exist many different types of design and classical control methods such as Nyquist/Bode plot, Routh stability criteria and roots locus analysis or pole placement. In this research, we use Routh stability criteria as a measurement of stability for PID controller. The stability of a system depends on the characteristic polynomial equation of the denominator of the closed-loop transfer function. The system is stable if the roots of it characteristics equation are all on the left-half plane (LHP) on the complex plane. Based on the Routh Stability criteria, instead of solving the characteristics equation, we just to examine the coefficient of the polynomial, where all the coefficients must be positive numbers – a necessary criterion for stability.

In this research, we assume that a process is represented by the first-order delay unstable process (FODUP) model – this model is widely used to represent real processes. Since it is an unstable process, the controlled process tends to give large overshoot in response to a set-point change, or having inverse response [1]. The stability for an unstable process can be determined by finding the mean of it optimum values for controller gain [2]. Over the past decades, researchers have proposed various PID tuning rules for stable systems by using different control theories [3].

In the present work, we propose to first pre-stabilize the unstable system via an inner feedback loop forming an augmented plant which is stable. Then, the overall system stability is rigorously addressed based on the Routh stability criteria. We establish several ranges (regions) of the loop gain, integral time and derivative time, within which the parameters for stabilizing PID controller can be found. For the dead-time approximation, we use both first-order Taylor series and Padé formulas. Both Taylor and Padé formulas serve as a reasonable approximation of $e^{-\theta s}$ for small values of $-\theta s$. It is interesting to mention that several others researchers used different approximation formulas for the dead-time, such as Laurent series by Panda [4] and Maclaurin series by Lee et al. [5].

The paper is organized as follows. Section 2 gives details of PID stability analysis and the tuning algorithm for the double-loop structure, Section 3 presents the simulation results on different methods and Section 4 presents some conclusions of the current study.

2. PID Stability

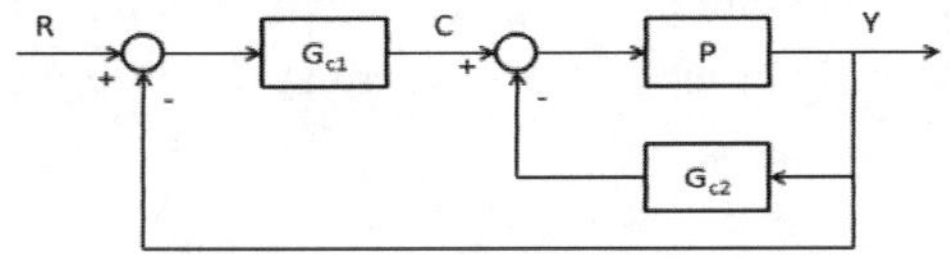

Figure 1. The doube-loop structure of feedback control scheme.

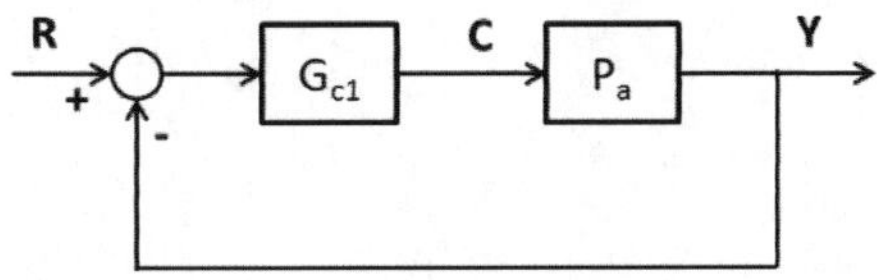

Figure 2. The equivalent single-loop structure of the double-loop control scheme.

Let us assume that a process is given by the first-order delay unstable process (FODUP) model

$$P(s) = \frac{K_p e^{-\theta s}}{(\tau s - 1)} \tag{1}$$

Here, K_p, τ and θ represent the process gain, time constant and deadtime for the unstable process in (1).

Consider a double-loop feedback control structure shown in Figure. 1 is used to control the process (1). Note that, the double-loop control structure consists of two controllers: a primary controller G_{c1} and a secondary controller G_{c2}. The purpose of G_{c2} is to pre-stabilize the process (1). The G_{c1} is to provide desired robustness and performance of the overall closed-loop system. For the G_{c1}, it is assumed that an ideal PID controller is used, i.e., of the form given by

$$G_{c1}(s) = \frac{K_c}{\tau_I}\left(\tau_I \tau_D s^2 + \tau_I s + 1\right) \tag{2}$$

where K_c, τ_I and τ_D denote the controller gain, reset or integral time and derivative time respectively.

Note that, for the purpose of conducting stability analysis on the secondary loop, the dead-time term is approximated using the first-order Taylor series

$$e^{-\theta s} \cong 1 - 2\alpha s, \qquad \alpha = \frac{\theta}{2} \tag{3}$$

while for the overall system stability analysis, the first order Padé approximation is used instead

$$e^{-\theta s} \cong \frac{1-\alpha s}{1+\alpha s}, \quad \alpha = \frac{\theta}{2} \tag{4}$$

2.1. *Double-loop secondary controller design*

Referring to figure 1, notice that the closed-loop (augmented plant) transfer function from C to Y is given by

$$P_a(s) = \frac{G_{c2}}{1+G_{c2}P(s)} \tag{5}$$

Interestingly, Figure 1 can simply be reduced to the standard single-loop feedback control structure shown in Figure 2. Hence based on Figure 2, one can use several established methods to design the primary controller G_{c1}. Now, let us assume that the secondary controller is selected as a P controller, i.e., $G_{c2} = K_{c2}$. From (5), the closed-loop characteristic equation for the secondary loop is identified as follows

$$1+G_{c2}P(s) = 0 \tag{6}$$

It follow that using (3) to approximate the deadtime, the close-loop characteristics in (6) can be expressed as

$$(\tau - K_{c2}K_p\,\theta)s - 1 + K_{c2}K_p = 0 \tag{7}$$

Based on the necessary condition of Routh stability, for the inner-loop stability the secondary loop gain must be bounded as follows

$$1 < K_{c2}K_p < \frac{\tau}{2\alpha} \tag{8}$$

where the secondary loop gain is $K_{c2}K_p$.

It should be noted that, the augmented plant (5) can be expressed as a first-order stable process given by

$$P_a(s) = \frac{K_p^0 e^{-\theta s}}{\tau_c s+1} \tag{9}$$

Let use assume, assume that $K_{c2}K_p = r\tau/(2\alpha)$, with is $r \in (r_{\min}, 1)$ a positive scalar parameter which can be used to tune the value of $K_{c2}\ K_{c2}$. For the close-loop stability it is important that $r_{\min}$ must be large enough that the upper limit on the secondary loop gain is always larger than its coressponding lower limit. To ensure that it is so, the minimum value of the scaling parameter must be

$$r_{\min} = \frac{2\alpha}{\tau} \tag{10}$$

Notice that, K_p^0 and τ_c are given in the forms of

$$K_p^o = \frac{2K_p\alpha}{r\tau - 2\alpha} \tag{11}$$

$$\tau_c = \frac{2\alpha\tau(1-r)}{r\tau - 2\alpha} \tag{12}$$

IOP Conf. Series: Materials Science and Engineering **206** (2017) 012060 doi:10.1088/1757-899X/206/1/012060

2.2. *Overall controller design*

The close-loop set-point tracking transfer function in Figure 2, can be written as

$$P_b(s) = \frac{G_{c1}P_a(s)}{1+G_{c1}P_a(s)} \tag{13}$$

Let use assume the primary controller is selected as an ideal PID controller (2). From (13), the overall close-loop characteristic equation is

$$1+G_{c1}P_a(s) = 0 \tag{14}$$

After simplication, the closed-loop characteristic polynomial equation corresponding to (14) becomes

$$\left[\tau_I\alpha(\tau_c - \overline{K}\tau_D)\right]s^3 + \left[\tau_I(\tau_c+\alpha)+\overline{K}\tau_I(\tau_D-\alpha)\right]s^2 + \left[\tau_I+\overline{K}(\tau_I-\alpha)\right]s + \overline{K} = 0 \tag{15}$$

Based on the (15), the limits on the overall loop gain ($\overline{K} = K_cK_p^o$) for the entire system can be determined by first applying the necessary criterion of Routh stability. The Routh array is

$$\begin{vmatrix} \tau_I\alpha(\tau_c - \overline{K}\tau_D) & \tau_I + \overline{K}(\tau_I-\alpha) \\ \tau_I(\tau_c+\alpha)+\overline{K}(\tau_I\tau_D-\tau_I\alpha) & \overline{K} \\ b_1 & \\ c_1 & \end{vmatrix} \tag{16}$$

From the Routh array, the sufficient condition for the stability is given by

$$b_1 = \frac{\left[[\tau_I(\tau_c+\alpha)+\overline{K}(\tau_I\tau_D-\tau_I\alpha)]\right]\left[\tau_I+\overline{K}(\tau_I-\alpha)\right] - \overline{K}[\tau_I\alpha(\tau_c-\overline{K}\tau_D)]}{\tau_I(\tau_c+\alpha)+\overline{K}(\tau_I\tau_D-\tau_I\alpha)} > 0 \tag{17}$$

$$c_1 := \overline{K} > \overline{K}_{Llmt,1} = 0 \tag{18}$$

Meanwhile based on the necessary condition, from the coeffiecient of s^3 an upper limit on the primary loop gain is obtained

$$\overline{K} < \overline{K}_{Ulmt,1} = \frac{\tau_c}{\tau_D} \tag{19}$$

The notations: $\overline{K}_{Limt,1}$ and $\overline{K}_{Ulmt,1}$ denote the lower and upper limit on the overall loop gain beyond which the closed-loop stability is not possible for the PID controller.

Thus, based on the necessary condition the closed-loop is stable if and only if the loop gain

$$0 < \overline{K} < \frac{\tau_c}{\tau_D} \tag{20}$$

The element of b_1 in the Routh matrix gives rise to four possible condition (C.1,C.2,C.3 and C.4) detailed as follows:

- Condition C.1: $\tau_I - \alpha > 0$ and $\tau_D - \alpha > 0$
- Condition C.2: $\tau_I - \alpha < 0$ and $\tau_D - \alpha < 0$
- Condition C.3: $\tau_I - \alpha > 0$ and $\tau_D - \alpha < 0$
- Condition C.4: $\tau_I - \alpha < 0$ and $\tau_D - \alpha > 0$

Corresponding to the condition C.1, there are two extra lower limits on $\overline{K}$,

$$\overline{K} > \overline{K}_{Llmt,1.1} = \frac{-\tau_I}{\tau_I - \alpha} \tag{21}$$

$$\overline{K} > \overline{K}_{Llmt,1.2} = \frac{-(\tau_c + \alpha)}{\tau_D - \alpha} \tag{22}$$

Based on the condition C.2, we can have two extra upper limits on $\overline{K}$,

$$\overline{K} < \overline{K}_{Ulmt,2.1} = \frac{\tau_I}{\alpha - \tau_I} \tag{23}$$

$$\overline{K} < \overline{K}_{Ulmt,2.2} = \frac{\tau_c + \alpha}{\alpha - \tau_D} \tag{24}$$

With repects to the condition C.3, the extra lower limit on the loop gain is similar to that in (21) while another extra upper limit is as that of (24). The condition C.4 yeilds an extra upper limit as in (24) and a lower limit as that of (22). It is important to note that, when there are multiple lower limits and upper limits exist, the loop gain must be bounded from below by the maximum lower limit and bounded from above by the minimum upper limit.

2.3. *Cases of closed-loop stability*

The closed-loop meets a necessary condition of stability when the value of primary loop gain is bounded between its maximum lower limit and minimum upper limit, i.e.

$\overline{K}_{min} < \overline{K} < \overline{K}_{max}$. Alternatively, we can express the actual loop gain within the range above as follows

$\overline{K} = r_p\left(\overline{K}_{max} - \overline{K}_{min}\right) + \overline{K}_{min}$, $0 < r_p < 1$ with is $r_p \in (r_{min}, 1)$ a positive scalar parameter which can be used to tune the value of $\overline{K}$. Please note that, in the present work we only consider the stability analysis corresponding to the conditions C.1 and C.2 only. The remaining two conditions are described in another separate paper.

Condition C.1

There are three lower limits and one upper limit on the primary loop gain. The minimum lower limit is $\overline{K}_{min} = \max\left(\overline{K}_{Llmt,1}, \overline{K}_{Llmt,1.1}, \overline{K}_{Llmt,1.2}\right) = 0$.The upper limit is only one, so the maximum loop gain is $\overline{K}_{max} = \overline{K}_{Ulmt,1} = \frac{\tau_c}{\tau_D}$.Hence, the range of $\overline{K}$ is denoted by

$$0 < \overline{K} < \frac{\tau_c}{\tau_D} \tag{25}$$

Condition C.2

For this condition, there exists only one lower limit ($\overline{K}_{Llmt,1}$) and three upper limits ($\overline{K}_{Ulmt,2.1}$ $\overline{K}_{Ulmt,2.2}$, and $\overline{K}_{Ulmt,1}$). Hence, the minimum and maximum loop gains are $\overline{K}_{min} = \overline{K}_{Llmt,1} = 0$ and $\overline{K}_{max} = \min\left(\overline{K}_{Ulmt,1}, \overline{K}_{Ulmt,2.1}, \overline{K}_{Ulmt,2.2}\right)$. From the condition C.2, we can identify 4 different cases described as follows.

Case 2.1

Note that, $\overline{K}_{Ulmt,2.1} < \overline{K}_{Ulmt,2.2}$, if and only if the integral time is bounded from above

$\tau_I < \tau_{I,Ulmt,2.1} = \dfrac{\alpha(\tau_c + \alpha)}{2\alpha + \tau_c - \tau_D}$. Hence, two sub-cases are considered.

Case 2.1.1

Assuming Case 2.1, we can ensure that $\overline{K}_{Ulmt,1} < \overline{K}_{Ulmt,2.1}$ if and only if $\tau_I > \tau_{I,Llmt,2.1.1} = \dfrac{\alpha\tau_c}{\tau_c + \tau_D}$.

For this sub-case, the primary loop gain is bounded as in (25), Otherwise, we encounter another sub-case as follows.

Case 2.1.2

This case leads to $\overline{K}_{Ulmt,2.1} > \overline{K}_{Ulmt,2.2}$ when the derivative time is bounded from above

Then, the loop gain is bounded as $0 < \overline{K} < \dfrac{\tau_c + \alpha}{\alpha - \tau_D}$.

Case 2.2

This case leads to $\overline{K}_{Ulmt,2.1} > \overline{K}_{Ulmt,2.2}$ when the derivative time is bounded from above as

$\tau_I > \tau_{I,Llmt,2.2} = \dfrac{\alpha^2 + \tau_c\alpha}{2\alpha - \tau_D + \tau_c}$. Furthermore, suppose that $\overline{K}_{Ulmt,1} < \overline{K}_{Ulmt,2.2}$, then the integral time must take $\tau_I > \tau_{I,Llmt,2.2.1} = \dfrac{\alpha\tau_c}{\tau_D + \tau_c}$.

So, the range of primary loop gain becomes $0 < \overline{K} < \tau_c / \tau_D$, i.e., similar to (25).

Case 2.2.2

Suppose that $\overline{K}_{Ulmt,1} > \overline{K}_{Ulmt,2.2}$, and then the integral time must be bounded from above given by

$\tau_I < \tau_{I,Ulmt,2.2.2} = \dfrac{\alpha\tau_c}{\tau_D + \tau_c}$. Hence the loop gain become $0 < \overline{K} < \dfrac{\tau_I}{\alpha - \tau_I}$.

2.4. *Sufficient condition for closed-loop stability*

Provided that $\overline{K}$ and τ_D are set as to meet the necessary criterion of Routh stability, based on the sufficient condition of stability we find a lower limit on integral time as

$$\tau_{I,Llmt,1} > \left(\frac{\overline{K}}{1+\overline{K}}\right)\left[\alpha + \frac{\alpha(\tau_c - \overline{K}\tau_D)}{\alpha + \tau_c + \overline{K}(\tau_D - \alpha)}\right]$$

2.5. *Stability regions*

In this research, the FODUP stability regions are based on the Routh stability criteria. Based on the previous chapter analysis, one can conclude that the stability regions as follows. Based on the Condition C.2, there are 5 possible stability regions (*SRs*) depending on conditions imposed (choice of parameter values). The regions can be summarised as follows:

Case 1

$$SR_1 : \begin{cases} \tau_D > \alpha \\ 0 < \overline{K} < \tau_c / \tau_D \\ \tau_I > \tau_{\min} \\ \quad \tau_{I,\min} = \max(\alpha, \tau_{I,Llmt,1}) \end{cases}$$

IOP Conf. Series: Materials Science and Engineering **206** (2017) 012060 doi:10.1088/1757-899X/206/1/012060

Case 2

$$SR_{2.1.1}:\begin{cases}\tau_{D,Llmt,2.1.1} < \tau_D < \alpha \\ 0 < \overline{K} < \tau_c / \tau_D \\ \tau_{I,\min} < \tau_D < \tau_{I,\max} \\ \quad \tau_{I,\min} = \max\left(\tau_{I,Llmt,1}, \tau_{I,Llmt,2.1.1}\right) \\ \quad \tau_{I,\max} = \alpha \end{cases}$$

$$SR_{2.1.2}:\begin{cases}\tau_D < \min\left(\alpha, \tau_{D,Ulmt,2.1.2}\right) \\ 0 < \overline{K} < \dfrac{\tau_c + \alpha}{\alpha - \tau_D} \\ \tau_{I,\min} < \tau_I < \tau_{I,\max} \\ \quad \tau_{I,\min} = \tau_{I,Llmt,1} \\ \quad \tau_{I,\max} = \min\left(\alpha, \tau_{I,Ulmt,2.1}\right) \end{cases}$$

$$SR_{2.2.1}:\begin{cases}\tau_D < \alpha \\ 0 < \overline{K} < \tau_c / \tau_D \\ \tau_{I,\min} < \tau_I < \tau_{I,\max} \\ \quad \tau_{I,\min} = \max\left(\tau_{I,Llmt,1}, \tau_{I,Llmt,2.2}, \tau_{I,Llmt,2.2.1}\right) \\ \quad \tau_{I,\max} = \alpha \end{cases}$$

$$SR_{2.2.2}:\begin{cases}\tau_D < \alpha \\ 0 < \overline{K} < \dfrac{\tau_I}{\alpha - \tau_I} \\ \tau_{I,\min} < \tau_I < \tau_{I,\max} \\ \quad \tau_{I,\min} = \max\left(\tau_{I,Llmt,1}, \tau_{I,Llmt,2.2}\right) \\ \quad \tau_{I,\max} = \min\left(\alpha, \tau_{I,Ulmt,2.2.2}\right) \end{cases}$$

2.6. Stability regions

We propose a tuning algorithm for the double-loop control system based on the stabilizing region 1 (case 1).

Step 1. Tune the P-only controller in the inner loop as

$$K_{c2} = \left(\frac{1 + r_{\min}}{2}\right)\frac{\tau}{2\alpha}$$

Step 2. Set the derivative time as

$$\tau_D = r_D\alpha \, , \; r_D > 1$$

Use initial value $r_D = 1.2$

IOP Conf. Series: Materials Science and Engineering **206** (2017) 012060 doi:10.1088/1757-899X/206/1/012060

Step 3. Set the controller gain as

$$K_c = r_p\left(\frac{\tau_c}{\tau_D}\right), \quad 0 < r_p < 1$$

Use initial value $r_p = 0.5$

Step 4. Calculate the lower limit from the sufficient condition of Routh stability.

Step 5. Set the integral time as follows

$$\tau_I = r_i\, \tau_{I,min}\,, \; r_i > 1$$

Use initial value $r_i = 2$

Step 6. Determine gain margin (GM) and phase margin (PM) from the Nyquist plot. If the target GM and PM ranges $5.0db < GM < 6dB$ and $35^o < PM < 50^o$ are met, then stop. Otherwise, repeat Steps 2 – 6 by further adjusting for r_D, r_p, r_i until those targets GM and PM are met.

Step 7. Design the a setpoint pre-filter as

$$F_r = \frac{\eta\tau_I s + 1}{\tau_I s + 1}, \quad 1.5 < \eta < 5$$

Gradually increase η from small value until the desired overshoot is met. This completes the entire control system design.

3. Result and Discussion

In this part we will illustrate the present tuning rule and compare it with others existing works.

Example 1. Consider a (FODUP) system

$$P(s) = \frac{e^{-\theta s}}{s-1} \tag{26}$$

We compare the existing work with that from Nandong [6] and Lee et al. [7]. Table 1 described all parameters for the PID controllers.

Table 1. Parameters for the PID controllers.

Method	K_c	τ_I	τ_D
Proposed	1.7	0.5	0.12
Nandong [6]	4.468	1.1	0.091
Cho, Lee and Edgar [7]	2.5957	3.0041	0

Figure 3 show the nominal response with set-point at $t = 1$ and $\theta = 0.2$. The proposed method in term of IAE is better compare to both existing works. Figure 4 show the response under the perturbed condition based on [6].

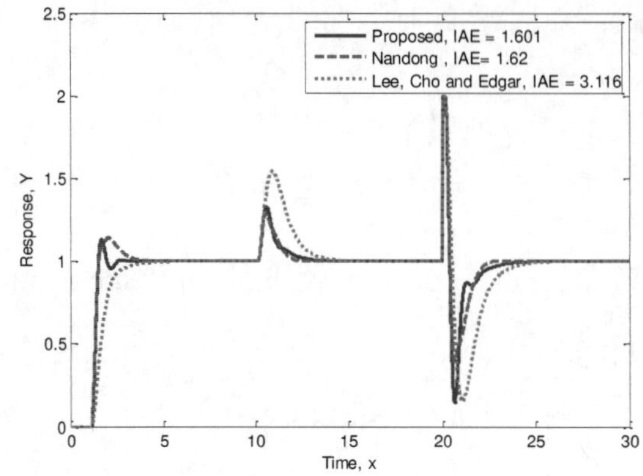

Figure 3. Nominal condition for close-loop system.

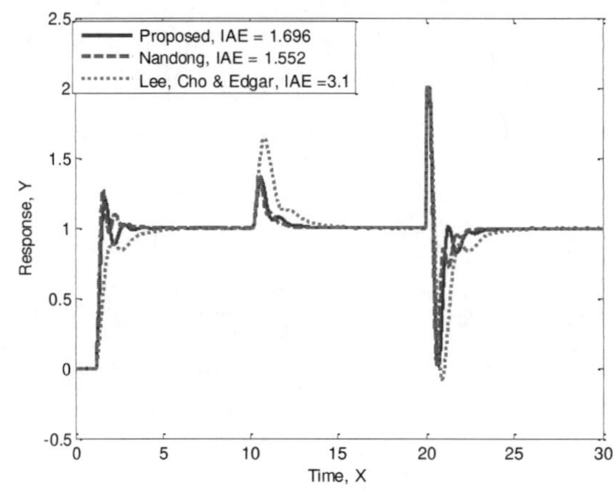

Figure 4. Perturbed condition for the close-loop system.

Example 2.
Consider another (FODUP) system similar with (26) and the $\theta = 0.4$. Figure 5 and 6 show the nominal condition and perturbed condition respectively. The proposed methods are compare with [7] and [8]. Perturbed condition response is based on [6]. Since the dead-time is increased, it has more oscillation compared to example 1. For the perturbed condition case, it does not show degradation in the closed-loop performance. The proposed double-loop control scheme obviously outperforms the other two single-loop PID/PI controllers.

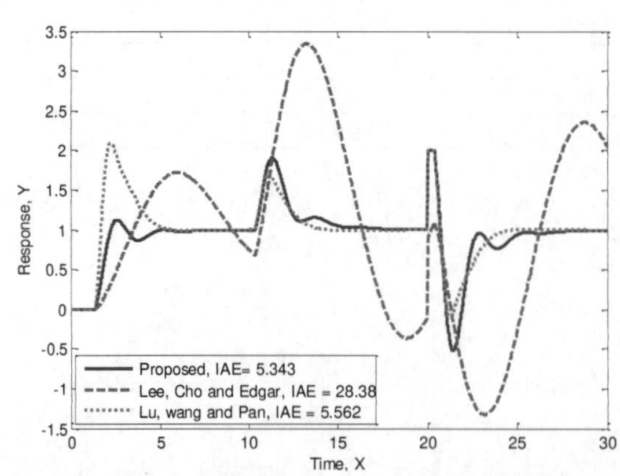

Figure 5. Nominal condition for close-loop system.

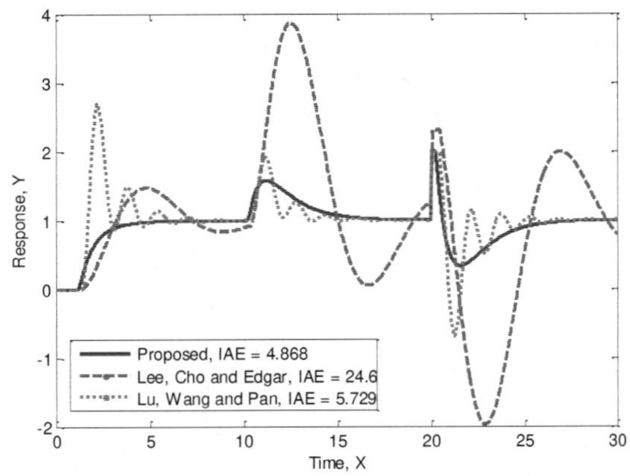

Figure 6. Perturbed condition for the close-loop system.

4. Conclusions

The rigours stability analysis has been conducted on the double-loop control scheme employing P/PID controllers for unstable process given as FOPDT model. The stability analysis has established five stabilizing regions for the PID controller parameters. Based on one of the regions, a simple P/PID tuning algorithm has been proposed. The effectiveness of the new P/PID tuning algorithm for the double-loop control scheme has been demonstrated using two numerical examples. It is worth highlighting that, the double-loop control scheme designed using the proposed algorithm can provide superior performance-robustness over several well-established PID tuning methods for the class of unstable processes of interest.

References

[1] Padma Sree R and Chidambaram M 2006 *Control of unstable systems* Alpha Science Int'l Ltd.

[2] Latha K, Rajinikanth V and Surekha P M 2013 PSO-based PID controller design for a class of stable and unstable systems *ISRN Artificial Intelligence*

[3] O'Dwyer A 2009 *Handbook of PI and PID controller tuning rules* World Scientific

[4] Panda R C 2009 Synthesis of PID controller for unstable and integrating processes *Chem. Eng. Sci.* **64**(12) 2807-16

[5] Lee Y, Lee J and Park S 2000 PID controller tuning for integrating and unstable processes with time delay *Chem. Eng. Sci.* **55**(17) 3481-93

[6] Nandong J 2015 Guaranteed stable PID controller tuning rules for First-Order Dead-time Unstable Processes. In *Industrial Electronics and Applications (ICIEA)*, IEEE 10th Conference on 2015 Jun 15 (pp. 1443-1448). IEEE

[7] Cho W, Lee J and Edgar T F 2014 Simple analytic proportional-integral-derivative (PID) controller tuning rules for unstable processes *Ind. Eng. Chem. Res.* **53**(13) 5048-54

[8] Wang Q, Lu C and Pan W 2016 IMC PID controller tuning for stable and unstable processes with time delay *Chem. Eng. Res. Des.* **105** 120-9

Selection of optimum ionic liquid solvents for flavonoid and phenolic acids extraction

N R A Rahman[1,2], N A Yunus[1,2] and A A Mustaffa[1,2,3]

[1] Faculty of Chemical and Energy Engineering, Universiti Teknologi Malaysia, 81310 UTM Johor Bahru, Johor, Malaysia.
[2] Process Systems Engineering Centre (PROSPECT), Research Institute for Sustainable Environment (RISE), Universiti Teknologi Malaysia, 81310 UTM Johor Bahru, Johor, Malaysia.

Email: azizulazri@utm.my (A A Mustaffa)

Abstract. Phytochemicals are important in improving human health with their functions as antioxidants, antimicrobials and anticancer agents. However, the quality of phytochemicals extract relies on the efficiency of extraction process. Ionic liquids (ILs) have become a research phenomenal as extraction solvent due to their unique properties such as unlimited range of ILs, non-volatile, strongly solvating and may become either polarity. In phytochemical extraction, the determination of the best solvent that can extract highest yield of solute (phytochemical) is very important. Therefore, this study is conducted to determine the best IL solvent to extract flavonoids and phenolic acids through a property prediction modeling approach. ILs were selected from the imidazolium-based anion for alkyl chains ranging from ethyl > octyl and cations consisting of Br, Cl, [PF_6], BF_4], [H_2PO_4], [SO_4], [CF_3SO_3], [TF_2N] and [HSO_4]. This work are divided into several stages. In Stage 1, a Microsoft Excel-based database containing available solubility parameter values of phytochemicals and ILs including its prediction models and their parameters has been established. The database also includes available solubility data of phytochemicals in IL, and activity coefficient models, for solid-liquid phase equilibrium (SLE) calculations. In Stage 2, the solubility parameter values of the flavonoids (e.g. kaempferol, quercetin and myricetin) and phenolic acids (e.g. gallic acid and caffeic acid) are determined either directly from database or predicted using Stefanis and Marrero-Gani group contribution model for the phytochemicals. A cation-anion contribution model is used for IL. In Stage 3, the amount of phytochemicals extracted can be determined by using SLE relationship involving UNIFAC-IL model. For missing parameters (UNIFAC-IL), they are regressed using available solubility data. Finally, in Stage 4, the solvent candidates are ranked and five ILs, ([OMIM] [TF_2N], [HeMIM] [TF_2N], [HMIM] [TF_2N], [HeMIM] [CF_3SO_3] and [HMIM] [CF_3SO_3]) were identified and selected.

1. Introduction

Herbal plants have played a major role in the treatment of human ailments since ages ago. It also known as main resources for food, flavouring, medicines and even function as fragrances. Herbs can synthesize wide range of bioactive compounds known as phytochemicals that can act as important biological function such as antibacterial, antiviral, anticancer and antifungal. Due to their high efficacy in treatment, natural herbs gain acceptances all over the world where 88 percent of human population now depends on traditional herbal medicine as their main daily health care [1]. The high demand of natural

IOP Conf. Series: Materials Science and Engineering **206** (2017) 012061 doi:10.1088/1757-899X/206/1/012061

herbs worldwide have encouraged alternatives and improvements to produce high quality herbal products.

Labisia pumila (*L. pumila*) or locally known as *Kacip Fatimah* is the most common herb used in Malaysian community for a variety of illnesses and also used as health supplements *L. pumila* is well-acknowledged for their antioxidant and anticancer activity. Phenolic acids and flavonoids are the main phytochemicals of *L. pumila* that are responsible for variety of pharmacological activities of the herb [2].Typically, phytochemicals can be identified and characterized from various plant parts such as leaves, stem, flower and fruits showed that the leaves of *L. pumila* exhibit higher antioxidant activities and total saponins amounts compared to the plant roots and stems [3][4].

The qualitative and quantitative of phytochemicals relies on the proper selection of the extraction method [5]. Commonly, solvents are used to extract phytochemical from herbs. Therefore, solvent selection is highly important to ensure the effectiveness of the solvent to extract phytochemicals from herbs [6]. Different solvents have different capacities of extraction depending on the solubility and polarity of the phytochemicals [7].

Current herbal extraction processes used organic solvents such as ethanol, methanol, chloroform and acetone. However, these conventional solvents are volatile, toxic and the selectivity of extraction is low. Therefore, ionic liquid (IL) has been considered as an alternative solvents in order to efficiently achieved high quality herbal products. IL has a wide range of applications and has the ability to separate a wide range of organic and inorganic compounds. Other than their unique properties, it can be designed according to the specific needs by matching the anion and cation while considering their bonds and other properties exhibited. In addition, IL has been experimentally proven to extract higher amount of phytochemicals compared to conventional solvents [8].

This study is to select the highest potential IL solvent to extract main flavonoids and phenolic acids from *L. pumila* through a property prediction modeling approach. Main phytochemicals considered are kaempferol, myricetin and quercetin for the range of flavonoids and gallic acid and caffeic acid of phenolic acids range. These main phytochemicals were selected based on previous studies of [9] and [10].

As an extraction solvent, ILs are responsible to stimulate the interactions between the analyte and the solvent [11]. The types of anion and cations in the ILs determined the solubility of the targeted phytochemicals. Phytochemical extraction solvents involved in this study are various IL with imidazolium-based cation for alkyl chains ranging from ethyl to octyl and anions consisting of bis(trifluoromethylsulfonyl)imide [Tf_2N], tetrafluoroborate [BF_4], hexafluorophosphate [PF_6], bromide [Br], chloride [Cl], acetate [Ac], ethylsulfate [$EtSO_4$], trifluoroacetate [CF_3COO] and trifluoromethanesulfonate [CF_3SO_3].

Due to thousands possibilities of IL anions and cations interactions, property prediction methods such as activity coefficient models, group contribution methods, COSMO-RS and molecular simulations to reduce the cost of experimentation [12]. However, activity coefficient models, group contribution methods and COSMO-RS still requires some experimental results to optimize their parameters and validate the findings.

The solubility parameter is one of the effective screening tools for solvent selection due to the basic principle 'like dissolves like' [13]. According to the principle, when the solubility parameter of the solvent and solutes approach with each other, the solubility of the solute is higher in that solvent. The solubility parameter can be estimated experimentally or by using theoretical methods [14][15]. The experimental methods include measuring the heat of vaporization by calorimetry, inverse gas chromatography (IGC), measuring melting temperatures, intrinsic viscosity measurements, or by using the activation energy of viscosity [16][17]. While, theoretical approach in determining solubility parameters include using intrinsic viscosity, enthalpy of vaporization, PC-SAFT such as COSMO [18]. However, some of these theoretical methods have some limitations and still requires parameter from experimental data.

IOP Conf. Series: Materials Science and Engineering **206** (2017) 012061 doi:10.1088/1757-899X/206/1/012061

2. Methodology

2.1. *Stage 1: Database and Property Models Library Development*

An Excel-based database has been developed, containing physicochemical data of IL and phytochemicals obtained and collected from literatures, ILThermo NIST database [19] and an unpublished IL database from Technical University of Denmark (DTU). The collected data include molecular weight, melting point, heat of fusion, viscosity, density and Hansen solubility parameters (HSPs).

Besides physicochemical data, property models with its parameters are also included in the database. An extensive review of applicable property models were done through literature search. Available models were collected and analysed further to determine the accuracy of the equations for IL/phytochemical systems. In the current state of the database, HSPs models for IL involving an anion-cation correlation model [15] and Marrero-Gani (MG) group contribution model [20] for phytochemicals have been included. For solid liquid equilibrium (SLE) calculations, parameters of the MG model for melting point and heat of fusion are also included with the UNIFAC-IL activity coefficient model [21]. For these models, missing parameters are predicted based on available data.

2.2. *Stage 2: Solubility Parameter Determination*

In this section, HSPs of the phytochemicals (kaempferol, myricetin, quercetin, gallic acid and caffeic acid) are determined using the MG model. Meanwhile, the HSPs of imidazolium-based ILs were estimated using the anion-cation correlation model. The MG model can be used to estimate the properties such as melting point, heat of fusion, HSP and boiling point of pure organic compounds such as phytochemicals. The mathematical expression is shown in equation (1) where f(X) is a function of property X (HSP in this study) and contain the additional adjustable model parameters and constants depending on the property involved.

$$f(X) = \sum_i N_i C_i + w \sum_j M_j D_j + z \sum_k E_k O_k \tag{1}$$

C_i is the contribution of first order group that occurs N_i times, M_j is the contribution of second order group that occurs D_i times and E_k is the contribution of third order group that occurs O_k times. Equation equation (2), (3) and (4) describe the partial solubility parameters based on MG group contribution model. The value of partial solubility of dispersion, (δd), polar (δp), and hydrogen bond (δh) were obtained using the Integrated Computer Aided System (ICAS) Version 18.0 software.

$$\delta_d = \left(\sum_i N_i \delta_{d1i} + \sum_j M_j \delta_{d2j} + \sum_k O_k \delta_{d3k}\right) MPa^{1/2} \tag{2}$$

$$\delta_p = \left(\sum_i N_i \delta_{p1i} + \sum_j M_j \delta_{p2j} + \sum_k O_k \delta_{p3k}\right) Mpa^{1/2} \tag{3}$$

$$\delta_h = \left(\sum_i N_i \delta_{h1i} + \sum_j M_j \delta_{h2j} + \sum_k O_k \delta_{h3k}\right) MPa^{1/2} \tag{4}$$

where δ_{d1i}, δ_{d2j}, δ_{d3k}, δ_{p1i}, δ_{p2j}, δ_{p3k} and δ_{h1i}, δ_{h2j}, δ_{h3k} denotes the group contribution parameter for first-order group, second-order group and third-order group respectively.

In this study, HSPs of ILs were determined by using a correlation of the weighted sum of attractions and repulsions between the anions and cations [15]. By considering IL as a mixture of cation and anion, using the concept of group contribution, the solubility parameters of various anions and cations have been computed through molecular simulations. The correlation model is shown in equation (5), where δ_{IL} represents the HSPs of ILs, δ_{ca} represents solubility parameter contribution for cations and δ_{an} represents solubility parameter of anions.

$$\delta_{IL}^2 = 0.6084\delta_{ca}\delta_{an} - 0.1958\delta_{ca}^2 - 0.1958\delta_{an}^2 \tag{5}$$

According to this correlation, HSPs of IL can be determined by having solubility parameter contributions of both cations and anions; or by the experimental solubility parameter of ionic liquids. Thus, this correlation can be used to determine wide range of ionic liquid with available data.

However, certain parameter such as for cations ([DMIM], [PMIM], [PeMIM] and [HeMIM]) were not available. Therefore, in this study, we determined the missing parameters using the systematic correlation/relationship between the molecular weight of cation and solubility parameters. The missing anions (acetate [Ac], ethylsulfate [$EtSO_4$], and trifluoromethanesulfonate [CF_3SO_3]) solubility parameter contribution were determined using anion-cation relationship using the experimental solubility parameter data.

Using both HSP values of the ILs and phytochemicals, the total solubility of phytochemicals in the ILs can be predicted. ILs candidates screened by the total phytochemicals solubility value. The model for total solubility of phytochemical in a solvent is shown as in equation (6).

$$\Delta\delta = \left[\left(\delta_{d,P} - \delta_{d,S}\right)^2 + \left(\delta_{p,P} - \delta_{p,S}\right)^2 + \left(\delta_{h,P} - \delta_{h,S}\right)^2 \right] \tag{6}$$

where $\delta_{(d,P)}$ ($MPa^{1/2}$) is partial dispersion solubility of phytochemical, $\delta_{(d,S)}$ ($MPa^{1/2}$) is partial dispersion solubility of solvent, $\delta_{(p,P)}$ ($MPa1^{/2}$) is partial polar solubility of phytochemical, $\delta_{(p,S)}$ ($MPa^{1/2}$) is partial polar solubility of solvent, $\delta_{(h,P)}$ ($MPa^{1/2}$) is partial hydrogen bond solubility of phytochemical and $\delta_{(h,S)}$ ($MPa^{1/2}$) is partial hydrogen bond solubility of solvent. According to the equation above, it can be concluded that small value of Δδ indicates that the solvent is more efficient for the extraction of that compound [22].

Only ILs with total phytochemicals value below 10 $Mpa^{1/2}$ which indicates 90% solubility of solute in a solvent were selected for further analysis. This value was selected according to published experimental data where phytochemicals recoveries were around 97 percent in ionic liquid solvent [23].

2.3 Stage 3: UNIFAC-IL Binary Interaction Parameter Regression and SLE Calculations

The SLE thermodynamic model was used together with MG model (to predict melting point and heat of fusion when the experimental data is missing) and UNIFAC-IL (to predict activity coefficient). The SLE of a system can be determined by involving pure component properties of the system such as melting point and enthalpy of fusion of the solid (solute) into the system's real behaviour in the liquid phase [24] such as in equation (7).

$$\ln(x_1\gamma_1) = \frac{\Delta h_{fus}(Tm)}{R}\left(\frac{1}{T_m} - \frac{1}{T}\right) \tag{7}$$

where, χ_1 is mole fraction of dissolved solute, γ_1 is activity coefficient of solute in solute-solvent system, Δh_{fus} is an enthalpy of fusion, R is a gas constant and T_m is the melting point.

The UNIFAC-IL model equation (8) consists of the combinatorial contribution (γ_i^c), due to differences in molecular size and shape; and the residual contribution (γ_i^R) arising due to the differences in intermolecular forces of attraction. The activity coefficient is expressed as functions of composition and temperature.

$$\text{In}\ \gamma_i = \text{In}\ \gamma_i^c + \text{In}\ \gamma_i^R \tag{8}$$

UNIFAC required ILs to be decomposed into different groups. However, in the study, the skeletons of cation (without the alkyl chain) and anion are treated as a whole because ionic pair has a strong electrostatic interaction [25] such as [BMIM][Br] were decomposed into one CH_3 group, three CH_2 groups and one [MIM][Br] group. Based on the screened ILs (after Stage 2), missing binary interaction parameters (BIPs) from the original UNIFAC-IL parameter table were determined. Using available experimental data, parameter regression work was conducted in order to extend the UNIFAC-IL parameter matrix. The parameter regression were done using the Thermodynamic Model Parameter Equation (the TML) module in the ICAS 18.0 software using available published experimental data. Regressed parameters were then used for the estimation of solubility of the selected phytochemicals in the screened ILs.

*2.4 Selection of the Best Ionic Liquid Solvent Candidate*s

UNIFAC-IL was used together with the SLE equation to predict the solubility of flavonoids and phenolic acids solubility in the selected imidazolium based ILs. In Stage 2, the HSPs calculation had assisted in the screening and reducing the large number of ILs. Based on the screened ILs SLE was used to determine that amount of the flavonoids and phenolic acids that can be extracted by the imidazolium-based ILs. Based on the predicted solubility values of these phytochemicals, the ILs were ranked to determine the best solvent.

3. Results and Discussion

This study focused on determining the best imidazolium-based IL solvent for the extraction of phytochemicals (flavonoid and phenolic acid) from *L. pumila*. First it has been observed that the selections of 163 imidazolium-based IL are able to extract the main flavonoid and phenolic acid compound from the selected herbs [23][26]. In this case, the main flavonoid and phenolic acid identified from *L. pumila* are kaempferol, quercetin and myricetin for flavonoid; and gallic acid and caffeic acid of the phenolic acids group.

3.1. Prediction of Phytochemicals/ILs HSPs and Determination of Missing Cation and Anion Solubility Parameters.

Solubility parameter and other physicochemical properties for *L. pumila* phytochemicals are predicted by MG model and presented in Table 1. Phytochemicals properties such as solubility parameter, heat of fusion and melting point were determined using ProPred of the Integrated Computer Aided System (ICAS) Version 18.0 software.

According to previous study, increase of alkyl chain length in ILs enhanced the solubility efficiency of both flavonoid and phenolic acid in the extraction solvent [23]. This is due to the increasing molecular weight of the IL as extraction solvent [27]. On the other hand, solubility parameter of IL observed to be decreasing with the increasing molecular weight of the IL's cation (Figure 1) and viscosity of the solvent candidates. The solubility parameter value decreased when the alkyl chain length and viscosity increased. The relationship can be included as HSPs equation as presented by equation (9).

$$\delta = \left[\frac{\Delta H_v - RT}{V_m} \right]^{0.5} \tag{9}$$

where δ is the solubility parameter, ΔH_v is the heat of vaporization, R is a gas constant, T for temperature and V_m is the molar volume.

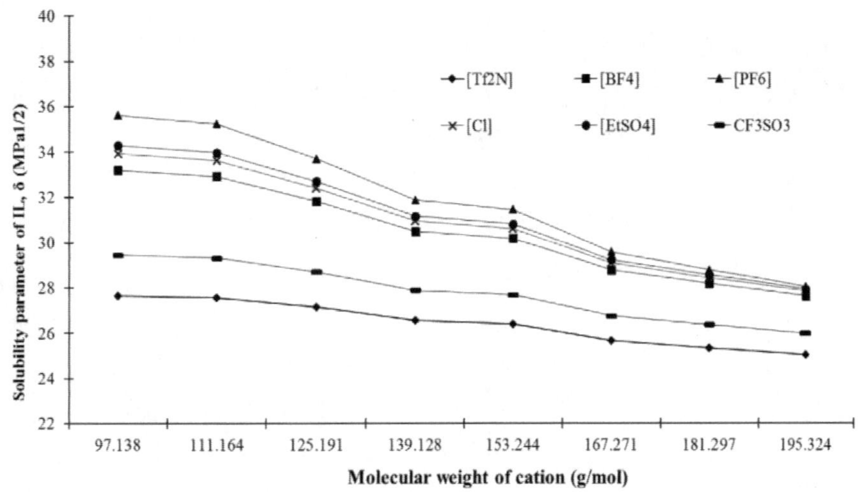

Figure 1. Relationship of IL solubility parameter with cation molecular weight

Table 1. Phytochemicals properties predicted by Marrero – Gani group contribution method.

Phytochemicals	FLAVONOID			PHENOLIC ACID	
	Kaempferol	**Quercetin**	**Myricetin**	**Gallic Acid**	**Caffeic acid**
Structure	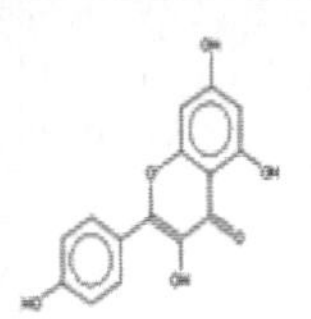	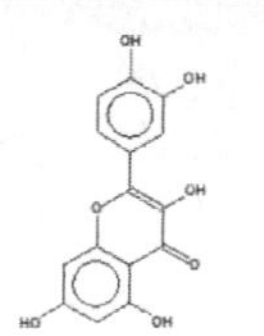	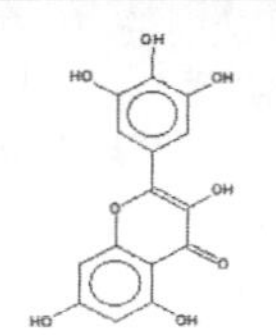	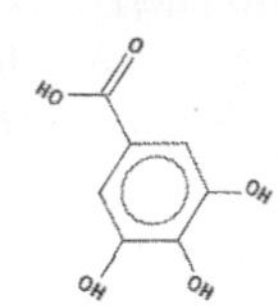	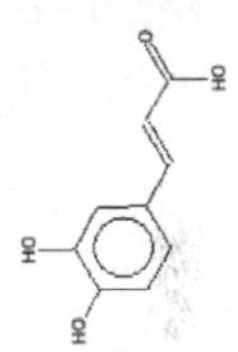
Molecular weight (g/mol)	286.24	302.24	318.23	170.12	180.16
Melting point, T_m (K)	503.59	520.29	537.53	492.89	443.75
Boiling point, T_b (K)	729.78	751.01	771.02	641.52	634.29
Enthalpy of fusion (KJ/mol)	120.69	133.7	146.71	89.66	81.05
Hansen solubility parameter, δ ($Mpa^{1/2}$)	32.07	34.29	36.51	28.82	28.34

However, no systematic relationship was found between anion molecular weight towards solubility parameter of IL. Figure 2 was constructed to show the anion's influence on the solubility parameter of ILs based on 1, 3-dimethylimidazolium [DMIM] cations. The solubility parameter increases in the following order: [Tf_2N] < [CF_3SO_3] < [BF_4] < Br] < [Cl] < [$EtSO_4$] < [Ac] < [CF_3COO] < [PF_6]. The highest values of δ^2 are for [$EtSO_4$], [Ac], [CF_3COO] and [PF_6] anions, whilst the lowest value is for the [Tf_2N] anion.

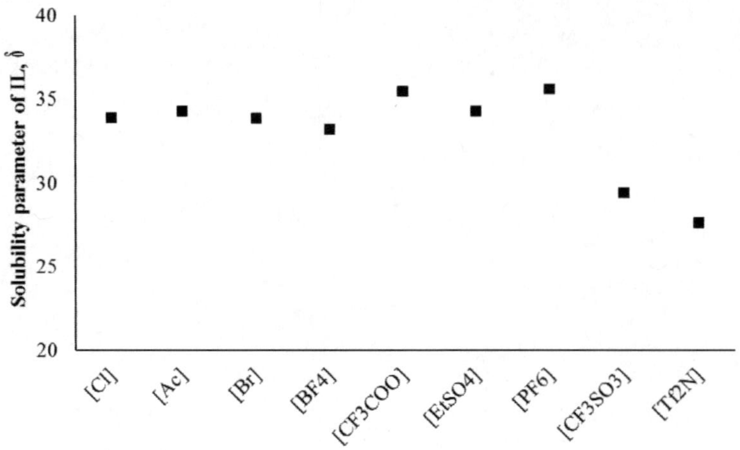

Figure 2. Influence of anions of 1, 3-dimethylimidazolium [DMIM] to solubility parameter of IL

In the list of 163 imidazolium-based ILs, from the developed Excel database, it was found that 4 different cations of 1,3-dimethylimidazolium [DMIM], 1-methyl-3-propylimidazo-lium [PMIM], 1-methyl-3-pentylimidazolium [PeMIM] and 1-heptyl-3-methylimidazolium [HeMIM] with different alkyl chain length had an unknown solubility parameters, when checked with the available cation solubility parameters in the Sistla anion-cation correlation model. As observed, IL solubility parameters decreased linearly with the increasing value of the molecular weight of cation due to the increasing of alkyl chain as well. Based on the systematic relationship of the cation's solubility parameter with its molecular weight as shown in figure 3, a fitted linear equation is obtained. Missing cation solubility parameters were determined from the equation based on its known molecular weight.

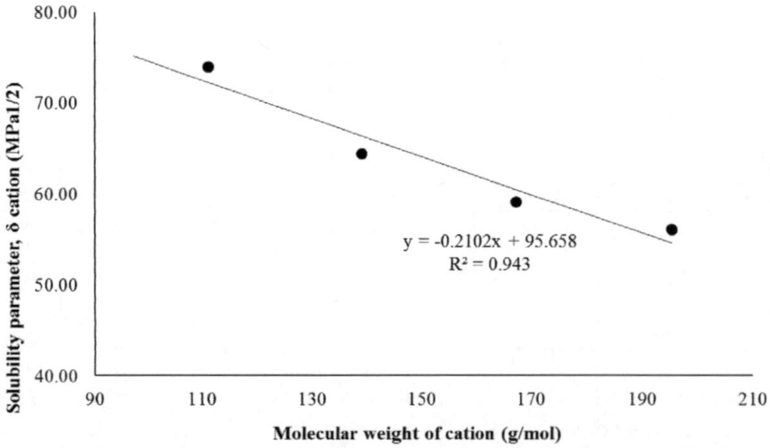

Figure 3. Linear regression of imidazolium based cations.

For the missing anion parameters, since no direct relationship can be found between its molecular weight and the IL solubility parameter, it is determined based on the relationship between cation solubility parameter and the solubility parameter of ILs containing the same cation and the missing anions (obtained from database). The relationship is based on equation (5). However, if the solubility parameter of the ILs containing related anion is not available, then the missing anion parameter cannot be determined.

Three new anions ([EtSO$_4$], [Ac] and [CF$_3$SO$_3$] solubility parameter were determined. Table 2 recorded the solubility parameters of cations and anions (reported in [15] and determined from this study) used for phytochemical extraction of *L. pumila* in this study.

Table 2. Solubility parameters of cations and anions

Cation		δ cation	Anion	δ anion
***1,3-dimethylimidazolium**	DMIM	75.24	**[Tf$_2$N]**	55.80
1-ethyl-3-methylimidazolium	EMIM	74.00	**[BF$_4$]**	73.25
***1-methyl-3-propylimidazolium**	PMIM	69.34	**[Br]**	76.09
1-butyl-3-methylimidazolium	BMIM	64.50	**[Cl]**	76.44
***1-methyl-3-pentylimidazolium**	PeMIM	63.45	***[EtSO$_4$]**	78.15
1-hexyl-3-methylimidazolium	HMIM	59.20	***[Ac]**	78.32
***1-heptyl-3-methylimidazolium**	HeMIM	57.55	**[CF$_3$COO]**	84.70
1-methyl-3-octylimidazolium	OMIM	56.10	**[PF$_6$]**	85.80
			***[CF$_3$SO$_3$]**	60.51

*cation with new predicted parameter value from this study

3.2. Total Solubility of Phytochemicals

Total solubility parameter is determined to distinguish the interactions of solute and solvent, thus the efficiency of the solvent used in extracting phytochemicals can be also identified [28]. Total solubility of *L. pumila* flavonoid and phenolic acid was essential for solvent screening. According to equation (6), it can also be deduced that total solubility of phytochemical in a solvent can be determine by the difference of ionic liquid and phytochemical solubility parameter as shown in Table 3.

The total solvent candidates of 163 imidazolium-based ILs initially available in the database were screened down to 72 candidates due to the unavailable data of solubility parameter for ILs and also their anions. At this stage, by considering the total solubility of phytochemical in the solvent, solvent candidates were screened down to 51 ILs for flavonoid extraction and 32 ILs for phenolic acid extraction. This study consider total solubility less than value of 10 Mpa$^{1/2}$ due to previous published experimental studies of phytochemical extraction using ILs which have higher extraction value compared to organic solvent [26]. In this case, we only considered 32 screened ILs as we are targeting for IL solvent that efficiently extract both flavonoid and phenolic acids. Therefore, the screened candidates were further analysed using SLE relationship to determine the total mole fraction of solute that is soluble in the selected solvents.

3.3. UNIFAC-IL Binary Interaction Parameter Regression and Solubility Determination from SLE

As mentioned in Section 2.3, UNIFAC-IL was used to predict the activity coefficient term in the SLE equation to predict solubility. In order to use UNIFAC, the BIPs need to be available. Therefore the 32 screened ILs from Stage 2, were decomposed into smaller groups and the groups are compared with the published UNIFAC-IL parameter table. Based on analysis, it is found that there were still few missing parameters compared to UNIFAC parameter matrix of [25] such as [MIM][Tf$_2$N], [MIM][Ac], [MIM][Br] and [MIM][CF$_3$COO]. Therefore, these missing groups need to be added in the initial parameter table and their parameters need to be regressed from available experimental data.

$$F = K_{set}^{\gamma} K_j^{\gamma} \sum_i K_i^{\gamma} \left(\frac{\gamma_i - \gamma_i^{Exp}}{\gamma_i^{Exp}} \right)^2 \quad (10)$$

The group BIPs were obtained by correlating the activity coefficients of binary systems involving organic compounds in imidazolium based ILs. ICAS 18.0 was used for the BIPs regression with the objective function shown in figure 10, where γ_i is activity coefficient of solute, γi^{Exp} is the activity coefficient of solute obtained from the experimental data and K_i^{γ} is the equilibrium ratio (K-value).

Table 3. Total solubility parameter of phytochemicals in ionic liquids

Ionic liquid	δ predicted ($Mpa^{1/2}$)	Total solubility, $\Delta\delta$ Flavonoid ($Mpa^{1/2}$)				Total solubility $\Delta\delta_{Phenolic\ acid}$ ($Mpa^{1/2}$)		
		Kaempferol	Quercetin	Myricetin	Mean	Gallic acid	Caffeic acid	Mean
[OMIM][Tf_2N]	25.02	3.05	2.92	3.33	3.10	6.65	4.85	5.75
[HeMIM][Tf_2N]	25.33	3.36	3.23	3.64	3.41	6.96	5.16	6.06
[HMIM][Tf_2N]	25.66	3.69	3.56	3.97	3.74	7.29	5.49	6.39
[OMIM][CF_3SO_3]	25.98	4.01	3.88	4.29	4.06	7.61	5.81	6.71
[HeMIM][CF_3SO_3]	26.36	4.39	4.26	4.67	4.44	7.99	6.19	7.09
[PeMIM][Tf_2N]	26.39	4.42	4.29	4.70	4.47	8.02	6.22	7.12
[BMIM][Tf_2N]	26.55	4.58	4.45	4.86	4.63	8.18	6.38	7.28
[HMIM][CF_3SO_3]	26.76	4.79	4.66	5.07	4.84	8.39	6.59	7.49
[PMIM][Tf_2N]	27.15	5.18	5.05	5.46	5.23	8.78	6.98	7.88
[EMIM][Tf_2N]	27.56	5.59	5.46	5.87	5.64	9.19	7.39	8.29
[OMIM][BF_4]	27.62	5.65	5.52	5.93	5.70	9.25	7.45	8.35
[DMIM][Tf_2N]	27.64	5.67	5.54	5.95	5.72	9.27	7.47	8.37
[PeMIM][CF_3SO_3]	27.68	5.71	5.58	5.99	5.76	9.31	7.51	8.41
[OMIM][Br]	27.82	5.85	5.72	6.13	5.90	9.45	7.65	8.55
[OMIM][Cl]	27.84	5.87	5.74	6.15	5.92	9.47	7.67	8.57
[BMIM][$CF3SO_3$]	27.88	5.91	5.78	6.19	5.96	9.51	7.71	8.61
[OMIM][$EtSO_4$]	27.92	5.95	5.82	6.23	6.00	9.55	7.75	8.65
[OMIM][Ac]	27.93	5.96	5.83	6.24	6.01	9.56	7.76	8.66
[OMIM][PF_6]	28.04	6.07	5.94	6.35	6.12	9.67	7.87	8.77
[OMIM][CF_3COO]	28.05	6.08	5.95	6.36	6.13	9.68	7.88	8.78
[HeMIM][BF_4]	28.17	6.20	6.07	6.48	6.25	9.80	8.00	8.90
[HeMIM][Br]	28.41	6.44	6.31	6.72	6.49	10.04	8.24	9.14
[HeMIM][Cl]	28.43	6.46	6.33	6.74	6.51	10.06	8.26	9.16
[HeMIM][$EtSO_4$]	28.55	6.58	6.45	6.86	6.63	10.18	8.38	9.28
[PMIM][CF_3SO_3]	28.70	6.73	6.60	7.01	6.78	10.33	8.53	9.43
[HeMIM][CF_3COO]	28.77	6.80	6.67	7.08	6.85	10.40	8.60	9.50
[HMIM][BF_4]	28.77	6.80	6.67	7.08	6.85	10.40	8.60	9.50
[HeMIM][PF_6]	28.77	6.80	6.67	7.08	6.85	10.40	8.60	9.50
[HMIM][Br]	29.05	7.08	6.95	7.36	7.13	10.68	8.88	9.78
[HMIM][Cl]	29.08	7.11	6.98	7.39	7.16	10.71	8.91	9.81
[HMIM][$EtSO_4$]	29.22	7.25	7.12	7.53	7.30	10.85	9.05	9.95
[HMIM][Ac]	29.22	7.25	7.12	7.53	7.30	10.85	9.05	9.95

Table 4. Volume, R_k and Surface Area, Q_k for parameter regression

New subgroups	R_k	Q_k
[MIM][CF_3COO]	7.22	1.67
[MIM][$EtSO_4$]	9.37	2.01

Table 4 described the value for volume, R_k and surface area, Q_k for the new sub-groups and table 5 showed the tabulated regressed a_{mn} and a_{nm} BIPs with tolerance of 1 x 10^{-6} and maximum iteration steps of 1000. The table listed the value of missing parameters as an additional to the UNIFAC parameter matrix of [25]. However, parameter regression were not done for [MIM] [$EtSO_4$] due to the lack of experimental data point. As according to the BIPs, there were only interaction between organic group with ILs group and none interaction between the ILs group in the system. All values were in the range to fit the parameters.

The activity coefficient estimated using UNIFAC-IL using the available and newly regressed parameters were obtained and applied to the SLE equation (7). The solubility of flavonoid and phenolic acid in 5 different type of ILs (the most optimum from the list of 32 ILs) are presented in Figure 4. As observed, ILs with [HMIM] [Tf_2N], [OMIM][Tf_2N] and [HeMIM][Tf_2N] anions and longer alkyl chain were the best solvents for extracting flavonoid and [HMIM][CF_3SO_3] and [OMIM][CF_3SO_3] for phenolic acid. ILs with organic fluorinated anions tend to have higher electrostatic contribution (equal to the collective contribution of coulombic and hydrogen bonding) and higher thermal stability [15].

In addition, the increase of alkyl chain length increased the hydrophobicity of the ionic liquid [29], therefore suitable in flavonoid and phenolic acid extraction. Furthermore, according to [30], the solubility of the solutes increases as the length of alkyl substitute on the ionic liquid increases as this study showed that the longer alkyl chain had the highest value of solubility data.

Table 5. UNIFAC-IL group interaction parameters (amn ≠ amn)

	CH2	C=C	ACH	ACOH	OH	COOH	[MIM] [Tf_2N]	[MIM] [Ac]	[MIM] [Br]	[MIM] [CF_3COO]
CH2							523.31	49.07	20.80	0.00
C=C							-0.17	-13.78	23.17	-1.87
ACH							-686.62	-31.72	-38.15	0.26
OH							6.96	-25.93	-50.98	0.00
AACOH							-756.67	-52.55	-50.83	-10.08
COOH							-2850.23	-2.03	10.62	0.00
[MIM] [Tf_2N]	-220.02	0.27	-2774.24	-0.09	-3243.87	-2410.81				
[MIM] [Ac]	61.51	-6.01	-32.23	-21.62	-94.66	-6.21				
[MIM] [Br]	71.58	-11.21	-0.78	-25.94	-101.91	-7.43				
[MIM] [CF_3COO]	0.02	0.04	21.45	-2.20	-12.86	-7.91				

No interactions between group parameters
According to previously published parameters from [25]

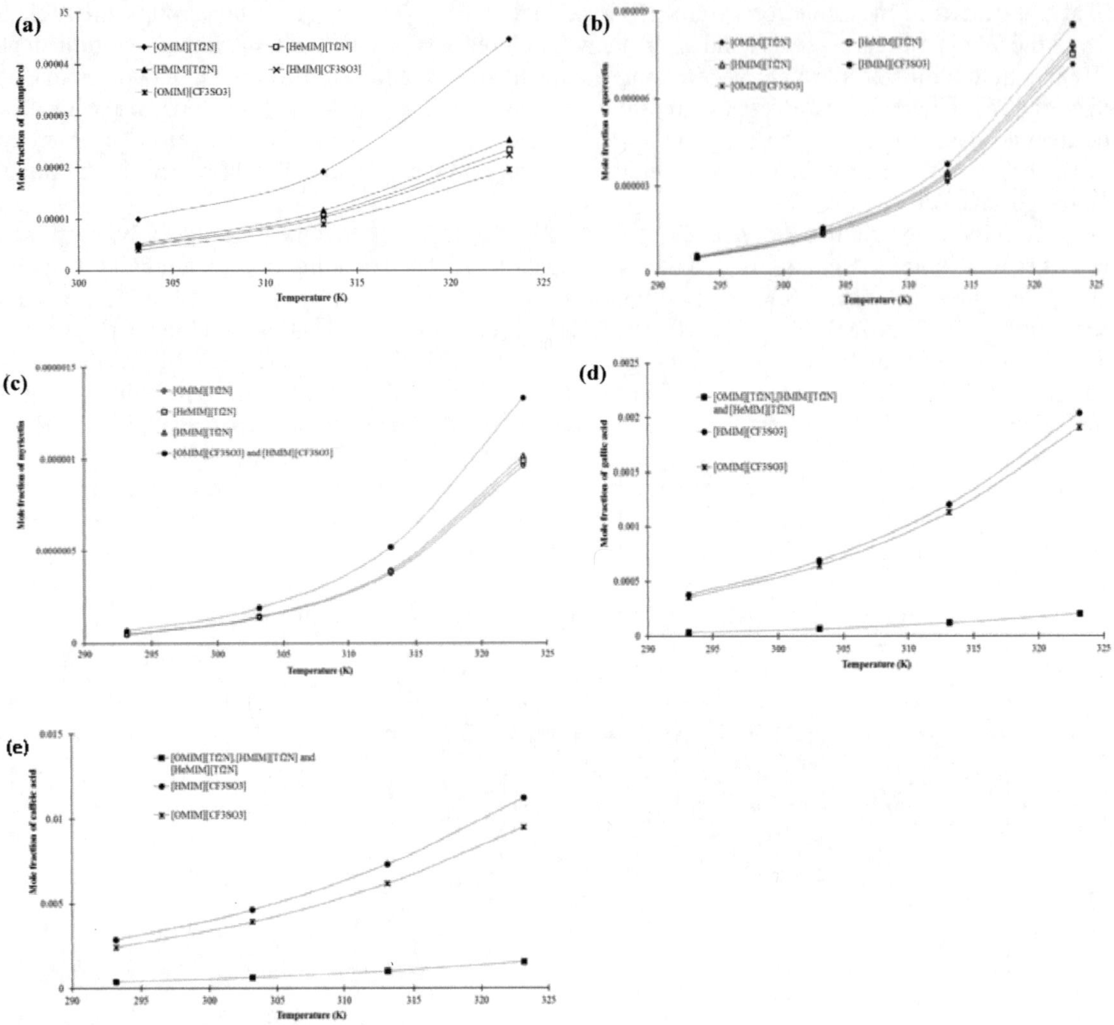

Figure 4. Solubility of a) Kaempferol b) Quercetin c) myricetin d) gallic acid and e) caffeic acid in six different ionic liquids

4. Conclusions

Anion-cation relationship is a convenient method to determine an unknown solubility parameter of IL, with available data of either anion and/or cation and/or solubility parameter of ionic liquid. In this study, four solubility parameters of cation ([DMIM], [PMIM], [HMIM] and [HeMIM]) and three solubility parameter of anion ([$EtSO_4$], [Ac] and [CF_3SO_3]) were added into the database using data regression. Solubility parameter is important for screening and reducing solvent candidates. Screened candidates were further assessed using UNIFAC-IL to determine the best IL candidates to extract flavonoid and phenolic acid from *L. pumila*.

5. References

[1] Doughari J H 2012 *IntechOpen* 1–33
[2] Zhang Q and Ye M 2008 J Chromatogr A **1216** 1954–69
[3] Azmir J, Zaidul I S M, Rahman M M, Sharif K M, Mohamed A, Sahena F and Omar K M 2013 *Journal of Food Engineering* **117** 426–36

[4] Karimi E, Jaafar H Z E and Ahmad S 2011 *Molecules* **16(6)** 4438–50
[5] Sasidharan S, Chen Y, Saravanan D, Sundram K M and Yoga Latha L 2011 *African Journal of Traditional, Complementary and Alternative Medicines* **8(1)** 1–10
[6] Aziz R A, Sarmidi M R, Kumaresan S and Taher Z M 2006 *Jurnal Kejuruteraan Kimia Malaysia*, **3** 45–60
[7] Cowan M M 1999 *Clinical Microbiology Reviews* **12(4)** 564–82.
[8] Xu W, Chu K, Li H, Zhang Y, Zheng H, Chen R and Chen L 2012 *Molecules* **17(12):143** 23–35
[9] Chua L S, Latiff N A, Lee S Y, Lee C T, Sarmidi M R and Aziz R A 2011 *Food Chemistry* **127(3)** 1186–92
[10] Jaafar H Z, Ibrahim M H and Karimi E 2012 *Molecules* **17(6)** 6331–47
[11] Tang B, Bi W, Tian M and Row K H 2012 *J. Chromatogr. B. Analyt. Technol. Biomed. Life Sci.* **904** 1–21
[12] Eckert F and Klamt A 2002 *AIChE J.* **48** 369–85
[13] Marciniak A 2011 *Int. J. Mol. Sci.* **12** 3553
[14] Gupta J, Nunes C, Vyas S and Jonnalagadda S 2011 *The Journal of Physical Chemistry B* **115(9)** 2014–23
[15] Sistla S Y, Jain L and Khanna A 2012 *Separation and Purification Technology* **97** 51–64
[16] Wang Q, Chen Y, Deng L, Tang J and Zhang Z 2013 *Journal of Molecular Liquids* **180** 135–8
[17] Marciniak A 2010 *International Journal of Molecular Sciences* **11** 1973–90
[18] Weerachanchai P, Wong Y, Lim K H, Tan T T Y and Lee J M 2014 *Chem Phys Chem* **15(16)** 3580–91
[19] Dong Q, Muzny C D, Kazakov A, Diky V, Magee J W, Widegren J A, Chirico R D, Marsh K N and Frenkel M 2007 *J. Chem. Eng. Data* **52(4)** 1151–9
[20] Marrero J and Gani R 2002 *Industrial & Engineering Chemistry Research* **41(25)** 6623–33
[21] Lei Z, Zhang J, Li Q and Chen B 2009 *Industrial & Engineering Chemistry Research* **48(5)** 2697–704
[22] Van Krevelen D W and Klaas T N 2009 Properties of polymers: Their correlation with chemical structure; their numerical estimation and prediction from additive group contributions (Netherlands: Elsevier) 1021
[23] Zhao C, Lu Z, Li C, He X, Li Z, Shi K, Yang L, Fu Y and Zu Y 2014 *The Scientific World Journal* Volume 2014
[24] You S -H, Jeong I -Y, Hwang I –C and Park S -J 2014 *Fluid Phase Equilibria* **383** 21–6
[25] Lei Z, Dai C, Liu X, Xiao L and Chen B 2012 *Industrial & Engineering Chemistry Research* **51(37)** 12135–44
[26] Xu W, Chu K, Li H, Zhang Y, Zheng H, Chen R and Chen L 2012 *Molecules* **17(12)** 14323–35
[27] Qilong R E N, Huabin X I N G, Zongbi B A O, Baogen S U, Qiwei Y A N G, Yiwen Y A N G and Zhang Z 2013 *Chinese Journal of Chemical Engineering* **21(9)** 937–52
[28] Mohammad A S N H, Mustaffa A A, Wan Alwi S R, Manan Z A and Chua L S 2014 *Proceedings of the 24th European Symposium on Computer Aided Process Engineering – ESCAPE 24*, ed J J Klemeš, P S Varbanov and P Y Liew (Hungary: Elsevier) p 1225–30
[29] Bogel-Lukasik R. Ed. 2015 *Royal Society of Chemistry* **36** 186
[30] Panteli E K and Voutsas E K 2009 *J. Chem. Eng.* **54** 812–8

Acknowledgement
The financial supports from Fundamental Research Grant Scheme (FRGS), Malaysia Ministry of Higher Education (Vote number: R.J130000.7844.4F670) and MyBrain15 scholarship by Ministry of Higher Education Malaysia is greatly appreciated. Data on IL supplied by Prof. Rafiqul Gani from Department of Chemical and Biochemical Engineering, Technical University of Denmark is highly appreciated.

A study of palm biomass processing strategy in Sarawak

S J Y Lee[1], W P Q Ng[1] and K H Law[2]

[1] Department of Chemical Engineering, Curtin University Malaysia, CDT 250, Miri 98009, Sarawak, Malaysia.
[2] Department of Electrical and Computer Engineering, Curtin University Malaysia, CDT 250, Miri 98009, Sarawak, Malaysia.

Email: wendyngpq@curtin.edu.my (W P Q Ng)

Abstract. In the past decades, palm industry is booming due to its profitable nature. An environmental concern regarding on the palm industry is the enormous amount of waste produced from palm industry. The waste produced or palm biomass is one significant renewable energy source and raw material for value-added products like fiber mats, activated carbon, dried fiber, bio-fertilizer and et cetera in Malaysia. There is a need to establish the palm biomass industry for the recovery of palm biomass for efficient utilization and waste reduction. The development of the industry is strongly depending on the two reasons, the availability and supply consistency of palm biomass as well as the availability of palm biomass processing facilities. In Malaysia, the development of palm biomass industry is lagging due to the lack of mature commercial technology and difficult logistic planning as a result of scattered locality of palm oil mill, where palm biomass is generated. Two main studies have been carried out in this research work: i) industrial study of the feasibility of decentralized and centralized palm biomass processing in Sarawak and ii) development of a systematic and optimized palm biomass processing planning for the development of palm biomass industry in Sarawak, Malaysia. Mathematical optimization technique is used in this work to model the above case scenario for biomass processing to achieve maximum economic potential and resource feasibility. An industrial study of palm biomass processing strategy in Sarawak has been carried out to evaluate the optimality of centralized processing and decentralize processing of the local biomass industry. An optimal biomass processing strategy is achieved.

1. Introduction

In a few decades time the area of oil palm plantation have increased drastically in Malaysia. The main product of oil palm is crude palm oil (CPO) which is extracted from palm fresh fruit bunch (FFB). FFB is typically processed in palm oil mill (POM). Along with FFB processing, waste is produced. The organic wastes generated are collectively termed palm biomass. The quantity of palm biomass increases in proportion with the increasing number of POM, POM processing capacity or amount of FFB being processed.

With the increasing volume of palm biomass, the disposal of palm biomass has become one source of pollution which has raised the concern of environmentalists and citizens. Due to the increased concern on environmental impacts and sustainability performance, researchers are contributing efforts to investigate the utilization of palm biomass for recovery and conversion through waste-to-wealth strategy to convert biomass into bio-energy and value-added products. Despite numerous researches that have been carried out by individuals, research institutions and universities, there is only limited number of

IOP Conf. Series: Materials Science and Engineering **206** (2017) 012062 doi:10.1088/1757-899X/206/1/012062

palm biomass technologies being commercialized due to the lack of confidence towards palm biomass value-added products as well as the lack of investment from investors. Besides, the scattered location of palm biomass sources has also needs to be tackled if biomass processing plant were to be set-up and operated efficiently. Biomass availability has to be secured and these processing facilities have to be logistically connected. The low density of biomass and its high moisture content have contributed to the infeasibility of logistic activity to collect biomass for centralize processing or product distribution in most countries. These factors have resulted in the slow development of biomass industry in Malaysia.

The key issue for the production system is the transportation of biomass, storage as well as supply chain system design. Centralize and decentralize processing of biomass is a major concern for effective supply network design and biomass industry development. [1] asserted that the general decomposition scheme was proposed for the generation of smaller sub-systems for the solution of worldwide optimality. The study aims to solve for each decentralized problem to the optimality as well as the original problem by integration of optimal schedule of each sub-system [1].

[2] had carried out a study on the economic evaluation of decentralized and centralized gasification plant. It was concluded that the decentralized strategies results in the high capital and operating costs of fast pyrolysis process and was hindering any benefits from the cost-effective transportation of the bio-oil from biomass plant to the gasification plant [2]. Then, [3] made comparison between the centralized and decentralized strategy on the processing system of the cellulosic biomass biorefining systems. The cost analysis showed better performance for centralized small-scale biorefineries and decentralized large-scale biorefineries [3]. Whereas, the environmental impact of the processing plant increased proportionally with the size of factory [3].

Research works have been done on the economic evaluation on the centralized and decentralized on the processing plant. However, the performances of cases vary with the properties of biomasses and localized scenario. None or less local study on the performance centralized and decentralized strategies towards palm biomass industry in Sarawak, Malaysia. In this work, the local performance of centralized and decentralized strategies and feasibility of the palm biomass processing strategy is simulated and evaluated.

2. Problem statement

The problem to be stated in this work is addressed as follows: FFB are processed in a set of palm oil mill source $a \in A$ and the biomass is send to a set of centralized or decentralized processing hubs $b \in B$ and eventually to sink $c \in C$. The problem is divided into two parts: (i) industrial study of the feasibility of decentralized and centralized palm biomass processing in Sarawak, and (ii) development of an optimized palm biomass processing strategy for the development of palm biomass industry in Sarawak, Malaysia.

3. Model formulation

Figure 1 denotes the summary of the methodology which the processed FFB is processed in POM a and the biomass is sent to processing hub b for conversion of value-added product and eventually to sink c for exportation purpose. The details formulation is shown in the following section.

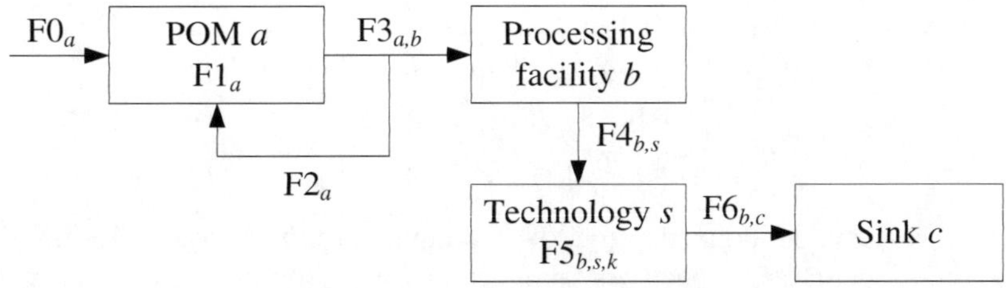

Figure 1. Summary of Methodology.

3.1. Flowrate of biomass in POM a

Palm biomass consists of multicomponent, such as empty fruit bunch (EFB), palm oil mill effluent (POME), palm kernel shell (PKS), oil palm frond (OPF) and so on. The availability of the biomass in POM is represented as below.

$$F1_a = F0_a \times COM_i \qquad a \in A,\ i \in I \qquad (1a)$$

where a is POM, $a \in A$, $A=\{1, 2, 3, \ldots, 39\}$; $F1_a$ is the flowrate of EFB in the POM a (t/h); $F0_a$ is the flowrate of FFB in POM a (t/h); and COM_i is the fraction of the fraction of multicomponent of palm biomass i generated from FFB processing.

However, EFB is the only raw material used in the case study. Therefore, equation (1a) can be simplified to the following:

$$F1_a = F0_a \times COM \qquad a \in A \qquad (1b)$$

where COM is the fraction of FFB to generate EFB.

Typically, palm biomass is either dumped for mulching or recovered for further utilization. Biomass can be incinerated directly as fuel source to generate heat and/or electricity or recovered for value-added product generation. The amount of palm biomass recovered for further processing is represented below:

$$F2_a = F1_a \times REC \qquad a \in A \qquad (2)$$

where $F2_a$ is the flowrate of EFB retained for consumption in POM a (t/h); and REC is the fraction of palm biomass (assumed to be 0.8) retained for POM's self-consumption.

3.2. Flowrate from POM a to processing hub b

The remaining palm biomass from POM a is sent to the processing facility b for further processing. The centralized palm biomass processing facilities, processing hub b are planned to be located in the center of a cluster of POM a.

$$\sum_{b \in B} F3_{a,b} = F1_a - F2_a \qquad a \in A,\ b \in B \qquad (3)$$

where $F3_{a,b}$ is the palm biomass flowrate from POM a to processing hub b (t/h).

3.3. Conversion of EFB into value-added product

In the palm biomass processing facilities, palm biomass is sent to selected biomass processing technology with flowrate $F4_{b,s}$ (t/h) for additional processing to convert the biomass into value-added products.

$$F4_{b,s} \times PR_{s,k} = F5_{b,s,k} \qquad b \in B,\ s \in S,\ k \in K \qquad (4)$$

where s is the technology for biomass processing, $s \in S$, $S=\{$Drying, Densification$\}$; k is the pam biomass products, $k \in K$, $K=\{$Dry Long Fibre (DLF), Pellet (PLT)$\}$; $PR_{s,k}$ is the rate of conversion of technology s to produce product k; and $F5_{b,s,k}$ is the flowrate of product in palm biomass processing hub b produced from technology s (t/h).

The amount of palm biomass is sent to technology s and the amount is limited with an upper boundary for a maximum operating capacity as well as a lower boundary to constraint for the minimum operating capacity to achievement an optimal operating efficiency of the equipment:

$$F4_{b,s} \leq UB_s \qquad b \in B,\ s \in S \qquad (5a)$$

$$F4_{b,s} \leq LB_s \qquad b \in B,\ s \in S \qquad (5b)$$

where UB_s and LB_s are the upper boundary and lower boundary of the operating capacity for technology s. 25% of the maximum operating capacity is taken as the lower boundary as the minimum operational flowrate of the equipment; while the upper boundary of the maximum operating capacity is taken to be 2.5 t/h.

3.4. *Flowrate of value-added product to port*

The final product will be sent to sink *c* for exportation which is represented below:

$$\sum_{b \in B, s \in S, k \in K} F5_{b,s,k} = \sum_{c \in C} F6_{b,c} \qquad b \in B,\ c \in C,\ s \in S,\ k \in K \tag{6}$$

where *c* is the set index of sink, $c \in C$, $C=\{1, 2, 3, 4\}$; and $F6_{b,c}$ is the flowrate of product from palm biomass processing hub *b* to sink *c*.

3.5. *Flowrate from POM a to processing hub b*

The distance travelled to deliver EFB from source *a* to hub *b* is considered in this work. The great-circle distance of EFB delivered from source point to the processing facilities to the sink *c* is estimated using Spherical Law of Cosines. Consideration of the location of processing hub *b* is based on the two aspects, i) location of POM *a*, which its location is within 150 km from the processing hub, and ii) capacity of processing hub *b*, which its capacity is not higher than the upper bound set.

$$\text{Distance}_{a,b} = \text{acos}[\sin(\text{lat}_a) \times \sin(\text{lat}_b) + \cos(\text{lat}_a) \times \cos(\text{lat}_b) \times \cos(\text{lon}_b - \text{lon}_a)] \times \text{r} \qquad a \in A,\ b \in B \tag{7a}$$

$$\text{Distance}_{b,c} = \text{acos}[\sin(\text{lat}_b) \times \sin(\text{lat}_c) + \cos(\text{lat}_b) \times \cos(\text{lat}_c) \times \cos(\text{lon}_c - \text{lon}_b)] \times \text{r} \qquad b \in B,\ c \in C \tag{7b}$$

where lat_a, lat_b, lon_a and lon_b are the latitude and longitude coordinates of the corresponding POM *a* and the centralized processing hub *b*;

The upper boundary is estimated to be 150 km due to large coverage area of Sarawak, Malaysia; lat_c and lon_c are the latitude and longitude coordinates of sink *c*; and r is the radius of earth (6371 km).

3.6. *Economic evaluation*

The transportation cost of EFB, Cd (MYR/d), is estimated and represented by the equation below:

$$\text{Cd} = \left(\sum_{a \in A, b \in B} \text{Distance}_{a,b} \times F3_{a,b} + \sum_{b \in B, c \in C} \text{Distance}_{b,c} \times F6_{b,c}\right) \times (\text{TCP}/30) \times t^h \tag{8}$$

where Cd is the total transportation cost to deliver biomass from POM *a* to processing facility *b* to sink *c* per day (MYR/d); t^h is the operating hour per day of the processing facility (8 h/d); and TCP is the transportation cost parameter (172 MYR/t for a distance of 30 km [4]).

The total cost of the palm biomass processing facilities with accountability of capital and logistic cost can be represented with the equation below:

$$\text{CT} = \text{Cc} + 10\ \text{y} \times \sum_{a \in A, b \in B} (F3_{a,b} \times \text{Cr} \times t^h \times t^d + F3_{a,b} \times \text{Cop} \times t^h \times t^d + \text{Cd} \times t^d + \text{Cm} \tag{9}$$

where CT is the total cost incurred to set up and operate the palm biomass processing facility which considers the capital cost, raw material cost, operating cost, logistic cost and maintenance cost in 10 years duration (MYR); Cc is the capital cost (MYR); Cr is the raw material cost (MYR/t); t^d is the number of operating days of the processing facility per year (d/y); Cop is the operating cost (MYR/t); and Cm is maintenance cost which is assumed to be 10% of the capital cost and is imposed once every three years as (MYR/3y) [5].

The calculation for payback period is represented as below:

$$\text{Pb} = \text{CT}/\text{AR} \tag{10}$$

where Pb is the payback period (y); and AR is the annual revenue (MYR/y).

4. Model illustration

Three scenarios are investigated in this work: (i) centralized palm biomass processing facility with single product, (ii) centralized palm biomass processing facility with multiple products, and (iii) decentralized palm biomass processing facility with single product. The generation rate of palm biomass from FFB in Sarawak is listed in Table **Error! Reference source not found.**1. Four ports are identified as product sinks and the locations of the sinks are listed in Table 2. Thirty-nine out of seventy-six POMs are identified in Sarawak, Malaysia. The locations and the capacities of the generalised sources of POMs are listed Table 3.

Table 1. Palm Biomass General Composition in Sarawak [6].

Palm Biomass Component	Biomass Generation Rate (% from FFB)
EFB	15.07
OPF	8.90
PKS	3.77
POME	44.52
Others	27.74

Table 2. Location and Capacity of Sink *c*.

No.	Latitude (°)	Longitude (°)
*c*1	1.615826	110.452764
*c*2	2.289373	111.823615
*c*3	3.267837	113.078303
*c*4	4.563727	114.040497

Table 3. Location and capacity of POMs *a*.

No.	Latitude (°)	Longitude (°)	Capacity (t/h)	No.	Latitude (°)	Longitude (°)	Capacity (t/h)
*a*1	2.855475	112.444793	45	*a*21	3.316018	113.450317	30
*a*2	2.414777	111.826769	60	*a*22	2.140779	112.103255	60
*a*3	3.810816	113.845825	40	*a*23	4.137053	113.977446	90
*a*4	3.390203	113.345417	60	*a*24	3.809446	113.859558	80
*a*5	3.822167	114.025194	45	*a*25	1.334718	112.000122	60
*a*6	3.507938	113.609619	90	*a*26	2.906984	112.313262	60
*a*7	1.153186	110.676056	60	*a*27	3.400940	113.347730	60
*a*8	3.163611	113.599750	30	*a*28	2.463753	111.928695	40
*a*9	1.444549	110.121460	40	*a*29	1.499153	109.905671	45
*a*10	3.191754	113.039500	40	*a*30	2.593907	112.338923	40
*a*11	3.585771	113.431968	60	*a*31	3.182660	113.505036	60
*a*12	2.487052	112.225823	60	*a*32	3.768750	113.696250	30
*a*13	2.850447	112.335189	60	*a*33	1.257834	111.445313	60
*a*14	3.872687	113.862575	120	*a*34	3.448807	113.769950	40
*a*15	3.610556	113.661667	60	*a*35	3.524878	113.747263	60
*a*16	1.669686	109.819336	40	*a*36	3.031650	113.882551	60
*a*17	1.062866	110.895996	60	*a*37	3.638889	113.504444	40
*a*18	1.397469	110.446100	45	*a*38	4.034047	114.303772	90
*a*19	3.260240	113.671980	30	*a*39	4.839639	115.257353	40
*a*20	3.727067	114.281073	45				

Dried long fibre (DLF) is taken as the prior product for the conversion of biomass into value-added product while DLF and biomass pellet (PLT) are taken into consideration for multiple products biomass processing facility. DLF is highly demanded for mattress and fibre mat production; while the increasing demand of EFB pellet for utilization as fuel in biomass power station. They are opted for the product choice due to their demand.

IOP Conf. Series: Materials Science and Engineering **206** (2017) 012062 doi:10.1088/1757-899X/206/1/012062

In the first scenario, single product production facility is studied and DLF is the sole product being produced in the biomass processing facilities. The feasibility of the centralized processing hub *b* with single product generation in Sarawak is investigated.

In the second scenario, multiple products production in centralized processing facilities is investigated. DLF and PLT are the products generated in the centralized processing facilities. Therefore, the consistent demand of PLT is secured and it is then suitable to be taken as one value-added products to be produced practically.

In the third scenario, decentralized processing facility of single product production is investigated. DLF is the lone product in this case. The feasibility of the decentralized processing hub *b* with single product generation in Sarawak is investigated.

The lifespan of each processing facility is taken to be 20 years. However, the payback period is calculated using the gross profit of the palm biomass processing facility. The upper boundary of the palm biomass processing facility taken to be 2.5 t/h due to the less production of EFB in Sarawak. Furthermore, the absence of highway also becomes one of the factor of capacity limitation. In addition, the real time market prices of DLF and PLT are the major factor that influences the economic potential of the plant. The fluctuation of the market price is mainly affected by the FFB yield which controls the raw material availability and the consistency supply of products which affects the stabilisation of products' market prices. As such, palm biomass processing facilities with 2.5 t/h is taken to be the optimum capacity. The parameters used to simulate the case studies are tabulated in Table 4 and Table 5.

Table 4. Parameter for Case Studies.

Parameters	Value
EFB (%)	15.07
Recovery Fraction (%)	80
DLF Conversion Rate (%) *	37.52
PLT Conversion Rate (%) **	26.64
Earth's Radius (km)	6371
Transportation Cost Parameter (MYR/km)	5.73
USD to MYR Exchange Rate	4.03
Operating Day per year (d)	330
Operating Hours per day (h)	8
Plant Life Span	10

* The conversion rate of raw EFB to DLF is 1 ton of raw EFB is required to produce 0.67 ton of wet long fiber (WLF) and 0.24 ton wet short fiber (WSF), while 1 ton of WLF is required to produce 0.56 ton of DLF [7] with the assumption that all WSF is screened out to be disposed.

** The conversion rate of raw EFB to PLT is 1 ton of raw biomass consists of EFB and WSF, is needed to produce 0.33 ton of PLT, while 1 ton of WLF is needed to produce 0.11 ton of WSF [7] and it is assumed that all WSF is recovered for PLT production.

Table 5. Raw materials and products selling prices, production costs, operating costs and capital costs [7].

Material	Cost (MYR/t)	Product	Price (MYR/t)	Operating Cost (MYR/t)	Capital Cost (MYR)
EFB	16.10	PLT	402.59	88.57	905,823
		DLF	805.18	74.48	2,173,975

5. Result and discussion

The model results are summarized in Table 6. Figure 2 to Figure 4 detailed the capacity and performance of the three models.

Table 6. Average processing capacity of biomass facility for each model.

		Model 1	Model 2	Model 3
Processing capacity of biomass facilities	DLF flowrate (t/h)	1.36 - 2.48	1.13 - 1.78	0.34 - 1.36
	PLT flowrate (t/h)	-	0.80 - 1.26	-
Number of processing facility hub *b*		12	8	39
Estimated payback period (y)		1.07 - 9.10	0.68 - 1.63	1.01 - 4.68

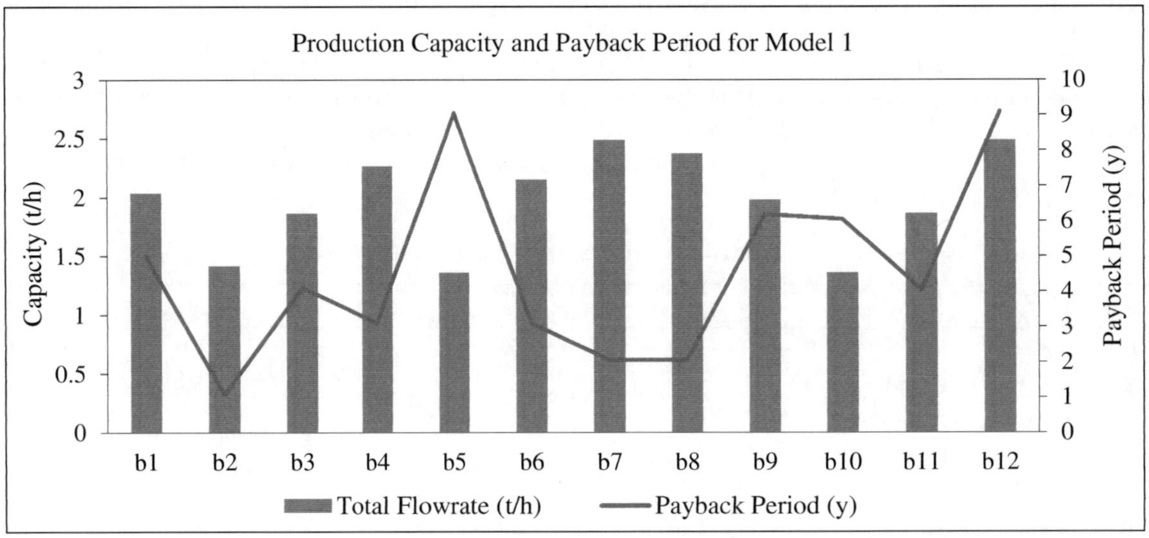

Figure 2. Result of case study 1.

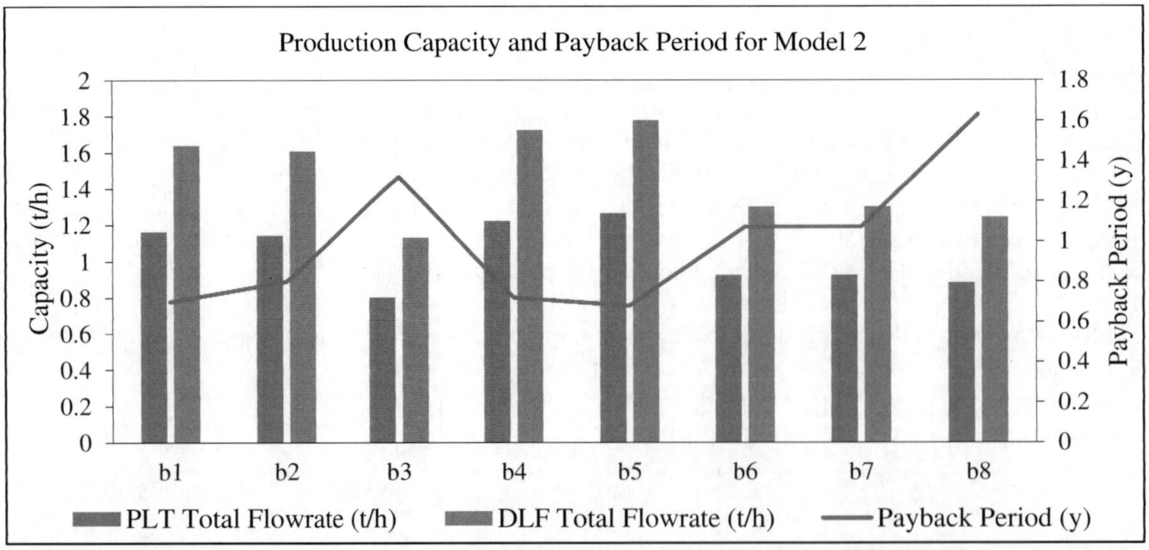

Figure 3. Result of case study 2.

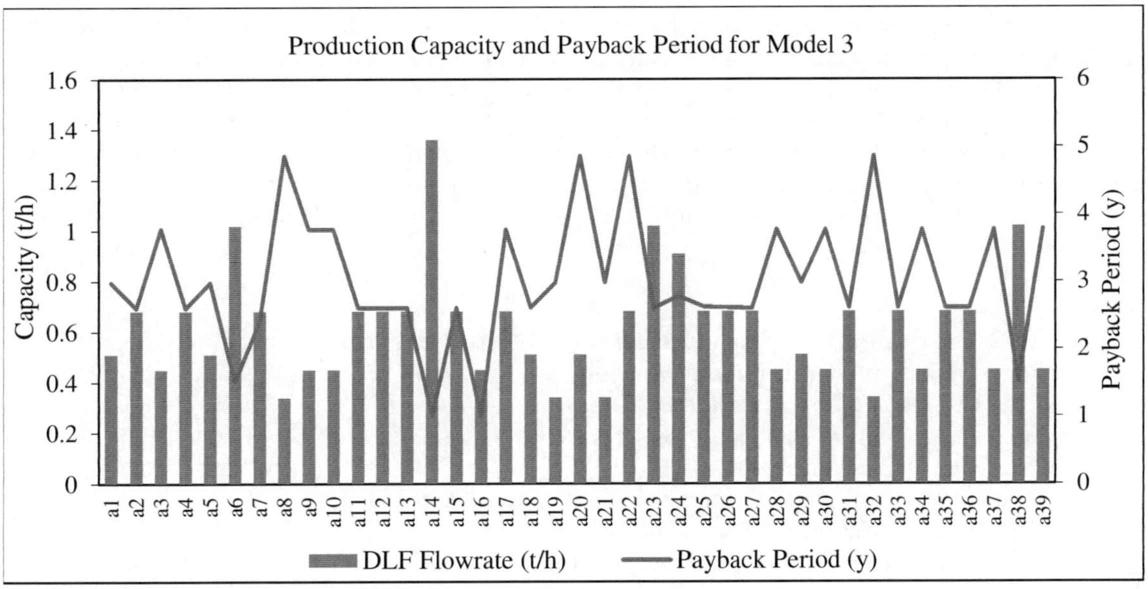

Figure 4. Result from case study 3.

6. Model limitation

In this work, assumptions are made in developing the models: (i) The locations of the facilities are represented in two-dimensional form and the straight-line distance between the facilities are calculated. This may result in inaccuracy in estimating the logistic cost which brings great impact to the performance of Model 1 and Model 2. The limitation can be improved by considering three-dimensional coordinates in estimating distances or taking the real distance between the facilities based on existing roadway availability; (ii) The capital costs of the processing equipment are taking the average value of equipment cost of all capacity. It shall be noted that lower capacity equipment generally incurs higher cost per unit capacity and higher capacity equipment generally incurs lower cost per unit capacity; (iii) Only 39 out of 76 of the POMs in Sarawak, Malaysia are considered in this work due to the unavailability of POM information. If all POMs are taken into consideration, a different scenario to develop the biomass industry may be obtained; (iv) The prices products are assumed to be the same from previous publication. It shall be noted that product prices for exportation generally fluctuates depending on stock supply and customer demand. Besides, the drastic currency fluctuation of the country affects the potential of the models. Sensitivity analysis can be carried out to account for this uncertainty; (v) The models assumed a constant supply of EFB from POMs. This is less realistic as FFB supply, and therefore EFB supply, fluctuates over the year due to local weathers. Proper production scheduling and stocking may be taken into consideration to tackle this scenario.

Nevertheless, the main objective of this work is to provide a framework which introduces a potential strategy of centralising and decentralising the palm biomass processing facility for the development of regional palm biomass industry.

7. Conclusions and future works

This research investigates the feasibility of the centralized or decentralized palm biomass processing facilities in Sarawak, Malaysia. Three scenarios are developed in this work, which are centralized palm biomass processing facilities with single product, centralized palm biomass processing facilities with multiple products and decentralized palm biomass processing facilities. Cost analysis is carried out to evaluate the economic performance of these three scenarios. It is observed that centralized palm biomass processing facilities with multiple products scenario has the highest potential to be developed.

This work can be extended to resolve the model limitations. The models can be further expanded to include all the biomass available in a POM such as PKS, OPF, OPT and et cetera. Business tendency can be introduced for this development of palm biomass processing facility supply network. An industrial symbiotic strategy can be introduced and promoted to optimize the usage and recovery of resources among the processing facilities. Besides, further research can be done study the effect of integrating carbon footprint towards the performance of centralized processing facilities. Besides, the potential of each biomass processing facilities can be evaluated to provide a rating for a systematic development of the industry.

References

[1] Shah N, Saharidis G K D, Jia Z and Ierapetritou M G 2009 Centralized–decentralized optimization for refinery scheduling *Comput Chem. Eng.* **33** 2091-105 doi:10.1016/j.compchemeng.2009.06.010.

[2] Braimakis K, Atsonios K, Panopoulos K D, Karellas S and Kakaras E 2014 Economic Evaluation of Decentralized Pyrolysis for the Production of Bio-Oil as an Energy Carrier for Improved Logistics towards a Large Centralized Gasification Plant *Renew. Sust. Energ. Rev.* **35** 57-72 doi:10.1016/j.rser.2014.03.052.

[3] Kim S and Dale B E 2015 Comparing Alternative Cellulosic Biomass Biorefining Systems: Centralized Versus Distributed Processing Systems *Biomass Bioenerg.* **74** 135-47 doi:10.1016/j.biombioe.2015.01.018.

[4] Yunus R, Omar R, Abidin Z Z and Biak D R A 2012 Oil Palm as Bioenergy Feedstock *Palm Oil as Bioenergy Feedstock* Elsevier Inc p 653-92 doi:10.1016/b978-0-9818936-9-3.50025-3.

[5] Mani S, Sokhansanj S, Bi X and Turhollow A 2006 Economics of Producing Fuel Pellets from Biomass *Appl. Eng. Agric.* **22**(3) 421-6 doi:10.13031/2013.20447.

[6] Sarawak . Energy 2013 *Palm Oil Biomass* Accessed April 20, 2016 <www.sarawakenergy.com.my/index.php/r-d/biomass-energy/palm-oil-biomass>

[7] Ng P Q W and Lam H L 2014 A Supply Network Optimisation with Functional Clustering of Industrial Resources *J. Cleaner Prod.* **71** 87-97 doi:10.1016/j.jclepro.2013.11.052.

Nomenclature

a	Set index of palm oil mill source points
AR	Annual revenue (MYR/y)
b	Set index of palm biomass processing facilities
c	Set index of sink c
Cc	Total capital cost (MYR)
Cd	Total transportation cost (MYR)
Cm	Maintenance cost in every three years (MYR/ 3y)
Cop	Total operational cost (MYR)
Cr	Total raw material cost (MYR/t)
COM_i	Composition of component i
CPO	Crude palm oil
CT	Total cost (MYR)
DLF	Dried long fiber
EFB	Empty fruit bunch
$F0_a$	Flowrate of palm biomass in palm oil mill a (t/h)
$F1_a$	Flowrate of component i in palm oil mill a (t/h)
$F2_a$	Recyclables flowrate in palm oil mill a (t/h)
$F3_{a,b}$	Leftover palm biomass's flowrate in palm oil mill a to processing hub b (t/h)
$F4_{b,s}$	Flowrate of palm biomass in processing hub b sent to technologies s for conversion (t/h)
$F5_{b,s,k}$	Product flowrate in processing hub b produced from technology s into palm biomass product k (t/h)

IOP Conf. Series: Materials Science and Engineering **206** (2017) 012062 doi:10.1088/1757-899X/206/1/012062

$F6_{b,c}$	Flowrate of products from processing hub *b* to sink *c* (t/h)
FFB	Palm fresh fruit bunch
i	Set index of type of palm biomass
k	Set index of palm biomass product
lat	Latitude (°)
LB_s	Lower boundary of the operating capacity of technology *s*, 25% of the maximum operating capacity (t/h)
lon	Longitude (°)
OPF	Oil palm fronds
Pb	Payback period of processing hub *b* (y)
PKS	Palm kernel shell
PLT	Pellet
POM	Palm oil mill
POME	Palm oil mill effluent
PR_s	Conversion rate of technology *s*
r	Radius of earth (km)
REC	fraction of palm biomass retained for POM's self-consumption
s	Set index for palm biomass processing technology
t_d	Total operational days for the processing hub
t_h	Total operational hours for the processing hub
UB_s	Upper boundary of the operating capacity of the technology *s* (t/h)
WLF	Wet long fiber
WSF	Wet short fiber

IOP Conf. Series: Materials Science and Engineering **206** (2017) 012063 doi:10.1088/1757-899X/206/1/012063

Kinetic Modelling of the Pyrolysis of Biomass for the Development of Charcoal Briquette

Y R Idris, H T Bayu, J Wintoko, B Murachman, A T Yuliansyah and S Purwono

Chemical Engineering Department, Universitas Gadjah Mada
Jalan Grafika No. 2, Kampus UGM, D.I. Yogyakarta, Indonesia.

E-mail: spurwono@chemeng.ugm.ac.id

Abstract. Waste of biomass can be utilized as an energy alternative such as a charcoal briquette. In the waste of biomass, there is carbon element bonded in the cellulose which can be utilized as an energy source of solid fuel. Charcoal briquette from waste of biomass can be developed via pyrolysis process. *Terminalia Catappa L.* and *Myristica fragrans* (nutmeg seeds shells) shells were used as raw material for the manufacture of charcoal briquettes. Pyrolysis process took place under isothermal conditions at a temperature of 350°C, 400°C, 450°C, 500°C, and 550°C with variation of times were 30 minutes, 60 minutes and 90 minutes. During the pyrolysis process, there were three main components observed, namely liquid (bio oil), gases and solids (char). Data obtained for measuring the kinetics of liquids and gases were taken in interval of 5 minutes. The results showed that the rise in temperature will increase the rate of pyrolysis process and increase the yield of gases and liquids as well as lowering the yield for solid. The best fitted kinetic model is the representation of biomass pyrolysis process involving secondary decomposition of the liquid. The results of briquette development showed that these two biomasses can be used as raw material of energy alternative.

1. Introduction

1.1. Background

Biomass can be converted into a material that can be used as other forms of energy such as carbon for briquettes production. Charcoal briquette is charcoal (one type of fuel) of various kinds of biological material or biomass, such as wood, twigs, grass, straw and other agricultural waste. As raw material of briquettes production, it should be chosen a material that has carbon element.

The benefit of this research is to increase the economic value of biomass waste by converting it into a more useful product. The purpose of this study is to obtain data the effect of temperature and time toward pirolysis products yield and to obtain the value of the pyrolysis kinetic parameters of biomass and to determine the effect of temperature and pirolysis time on the quality of the briquettes. *Terminalia Catappa L.* and *Myristica fragrans* shells (nutmeg seeds shells) were used as raw material for the manufacture of charcoal briquettes.

1.2. Theoretical basis

Pyrolysis is the chemical decomposition of organic materials through a process of heating without oxygen or under conditions of little air, causing the release of volatiles and char formation [1]. Thurner and Mann [2] proposed a parallel reaction mechanism, in which each of these reactions occurs separately. The reaction mechanism is as follows:

IOP Conf. Series: Materials Science and Engineering **206** (2017) 012063 doi:10.1088/1757-899X/206/1/012063

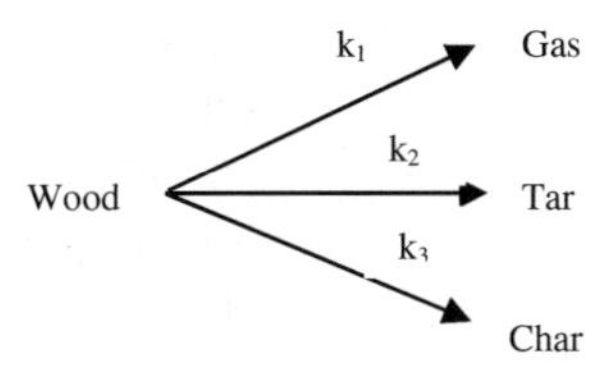

Figure 1. The mechanism of pyrolysis reaction fromTurner and Mann [2].

Mass balance equation for the decomposition of biomass and formation of product are as follows:

$$\frac{dm_w(t)}{dt} = -(k_1 + k_2 + k_3)m_w(t) \tag{1}$$

$$\frac{dm_G(t)}{dt} = k_1 m_w(t) \tag{2}$$

$$\frac{dm_T(t)}{dt} = k_2 m_w(t) \tag{3}$$

$$\frac{dm_C(t)}{dt} = k_3 m_w(t) \tag{4}$$

In this experiment, three reaction models were proposed to obtain the kinetic parameters.

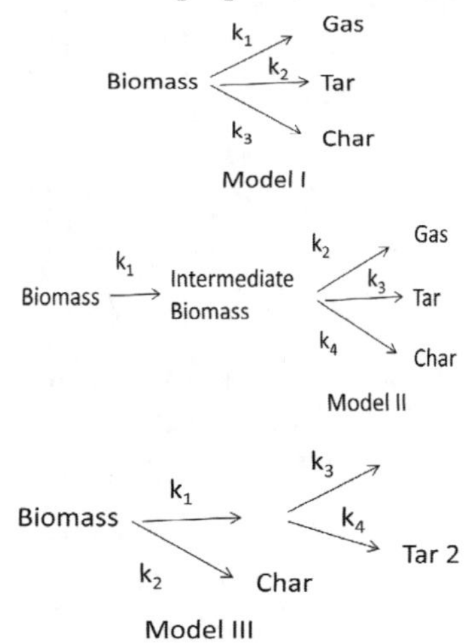

Figure 2. Models of Pyrolysis Reaction Mechanism

Arrhenius equation is used to calculate reaction rate constants in those equations above.

$$(k) = A_0\left(\exp\left[-\frac{E_a}{RT}\right]\right) \tag{5}$$

2. Research Methodology

2.1. Materials

Terminalia Catappa L. biomasses used was taken from several places in the city of Yogyakarta, which is at the Gadjah Mada University and Tempel region. Nutmeg seeds shells were from Gurabati village in Tidore islands. As the adhesive it was used 'Rose Brand' Tapioca flour and water.

2.2. Experimental set-up

Pyrolysis apparatus consists of a reactor made of metal pipe, the upper part of the reactor was covered by insulation. The outside of the reactor is insulated with a furnace insulator. The temperature inside the reactor was measured with thermocouples. The liquid product from the reactor is cooled through a condenser pipe and collected into different flask.

2.3. Procedure of operation

Biomass was crushed in crusher to form granules, and then it was weighed and loaded into the reactor and sealed. Then, it was heated up to the desired temperature of 350-550°C. During pyrolysis process, the volumes of liquids and gases were measured every 15 minutes. After 60 minutes of the process, gas samples were taken and stored in a vacuum tube to be analyzed its content of CO, CO_2 and CH_4. The gas pressure was measured using a manometer. After reaction time was completed, weigh the solid as charcoal. Then the charcoal was mixed with adhesive (2.5-5% by weight), stirred evenly and was pressed using hydraulic press equipment.

2.4. Research analysis

The analysis includes proximate analysis, lignocelluloses, gases and solids. The purpose of proximate analysis is to determine the amount of fixed carbon (FC), volatile matters (VM), moisture content (MC) and ash in the sample of briquettes in units of weight percent (wt). Lignocelluloses analysis was intended to determine levels of hemicelluloses, cellulose and lignin. Gas content data was used to calculate the mass of the gas using the following equation 6.

$$Gas\ Mass\ (G) = \frac{PV_G}{RT}\ x\ Average\ MW \tag{6}$$

3. Result and discussion

3.1. Material characteristic

From proximate analysis test results it was found that the content of volatiles in the biomass was very high. With 62.66 % of the total composition of materials, volatile content is higher than the fixed carbon which was only 18.81 %. The calorific value of the raw material itself was quite high, up to 4471.26 cal / g.

The analysis results showed that the biomass was dominated by lignocelluloses. The content of lignin in the material used was 43.4 5%, which is higher compared with cellulose and hemicelluloses. The amounts of those components were 16.60% and 24.69 % respectively.

3.2. The effect of temperature and time toward pyrolisis product

The example of the effect of temperature to yield product of gases, liquids and solids can be seen in Figures 3 to 5.

In the pyrolysis process, there were three products, gas, liquid and solid. The solid product was then presurized to be charcoal briquette. From the preliminary research, it was found that the increasing temperature and pyrolysis time will decrease the solid product. The thermal decomposition of hydrocarbons involves a series of primary and secondary reactions leading to a complex mixture of products. Studies show that the distribution of pyrolysis products varies considerably with the pyrolysis conditions and the type of reactor used. A study by Tsai and Albright [3] reported that at least seven surface reactions occur during most industrial pyrolysis processes.

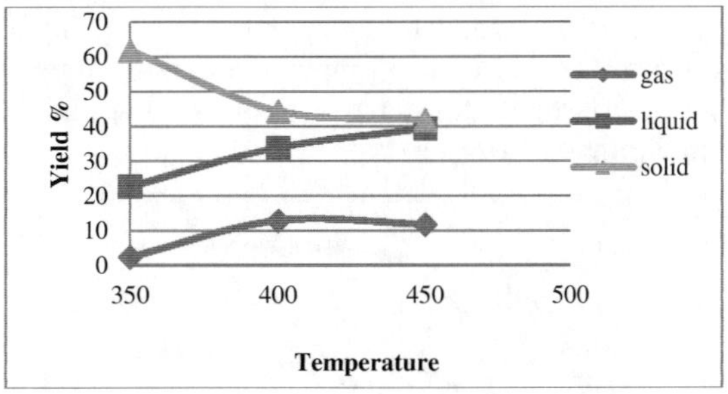

Figure 3. Yield products

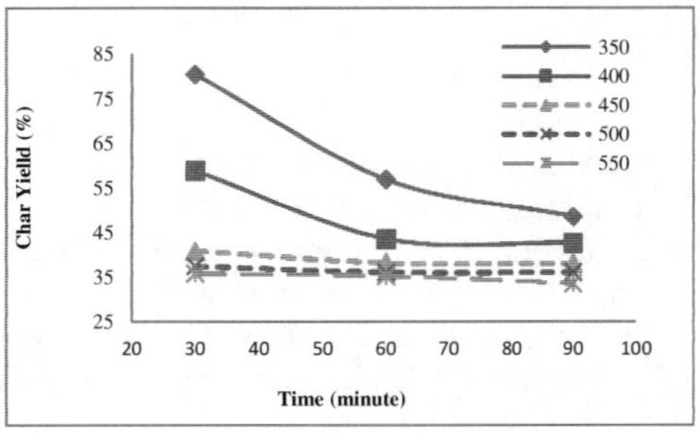

Figure 4. Char yield

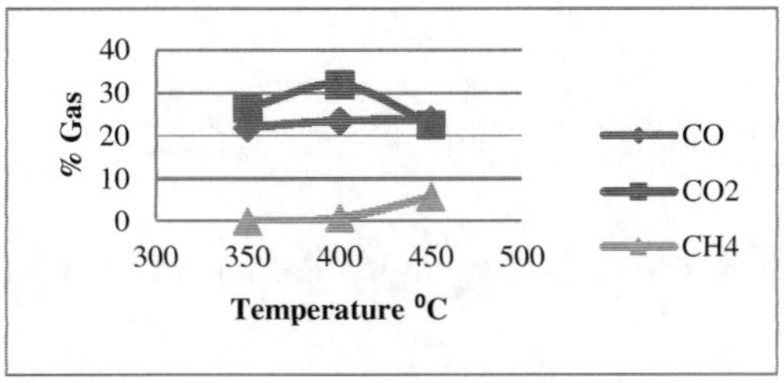

Figure 5. Percentage of gas product

In the pyrolysis experiments, the condensate formed is in the range of 25 to 45 percent of the original weight of biomass while the gas formed is in the range of 10 to 20 percent, and the rest is the solid product. This study demonstrated that more than 36 components; mainly alcohol like methanol, ethanol and propanol; were produced during the pyrolysis process. The experimental results also show that the condensate and gas formed are affected by the temperature in the reactor and the heating rate. Further studies show that the reaction rate of the decomposition process depends on temperature. Therefore, the gas and liquid pyrolysis products increase when the pyrolysis temperature increases.

3.3. Determining reaction rate constant

The primary data were obtained from the laboratory in the form of liquid and gas mass for 5 minutes interval. This data was then processed using computer program to get some of other data needed to determine the reaction rate constant of biomass pyrolysis. Example data from experimental and calculation can be seen in Figure 6 below.

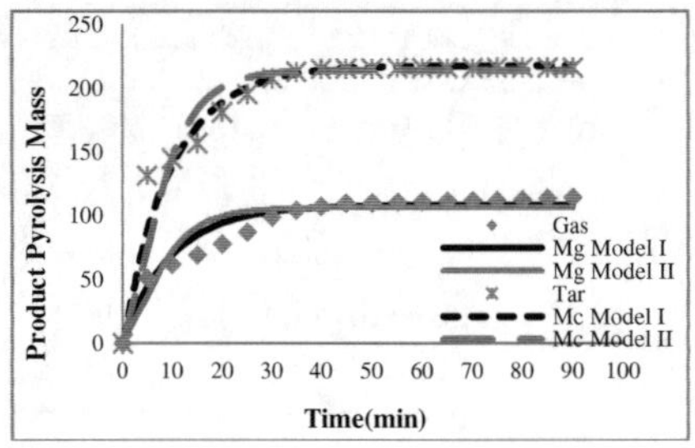

Figure 6. Example of pyrolysis product

Compared to the second model, the first model is closer to the experimental data. To assure that first model is the appropriate model, it was then developed a third model as a comparison against the first model. The third model is a compilation model, by modifying the first and second models. In this third model, biomass is not experiencing biomass intermediate stage. There is a secondary reaction in the next stage, where the secondary reaction will produce both gas and liquid. Comparison of the experiments results with the model I and III can be seen in Figure 7 below.

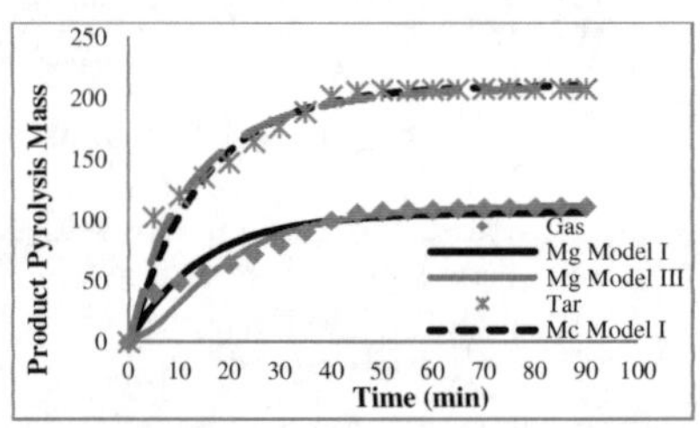

Figure 7. Correlation Chart of Time and Pyrolysis Product mass on Simulation of Model I and III for 500 °C

As discussed previously, model I is much closer to the experimental data than model II. Therefore, model I is then compared with a modified model. From R^2 value obtained, the average R^2 for model III is better than model I.

The effect of temperature on reaction rate constant follows the Arrhenius equation below

$$\ln k = \ln A_0 - \frac{E_a}{RT} \tag{7}$$

By plotting ln k and 1/T the value of Ea and A_0 can be obtained. The correlation of ln k and 1/T can be seen in Figure 8 below.

Reaction rate constant k_1, k_2 and k_3 are determined in five different reaction temperatures within the range of 350-550°C. In Figure 8, ln k is reported as a function of 1/T, and graphically they form a straight line correlation. It proves that the reaction occurred was still in line with Arrhenius equation in experimental temperature range. Ea values for various models are presented in Table 1 below.

The activation energies obtained are within the range of Ea values given in the literature. The activation energy of Ketapang seed shell is lower than the activation energy of Casew, Pistachio and Walnut which has a value of E above 100 kJ/mol. But still between the activation energy of almonds (42.4 to 99.7 kJ/mol), Brazil nut (47.2 to 82.0 kJ/mol), coconut (58.9 to 114.8 kJ/mol), and Peanut (44.3 to 71.5 kJ/mol) [4, 5 and 6].

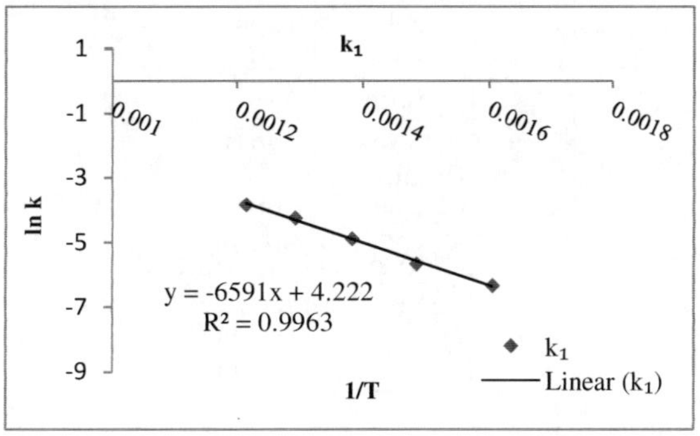

Figure 8. Correlation of ln k and 1/T

Table 1. Value of Ea for Various Model

Reaction Rate Constant (min^{-1})	Model I	Model II	Model III
Ea(kJ/mol)	71.329 2	26.5562	65.0602

3.4. Quality of briquette

The quality of briquettes is better when moisture content, ash content and volatile matter is lower. Standard SNI No. 1/6235/2000 stated that the moisture content and ash content should be less than 8%, while the volatile matter should be less than 15%.

Figures 9 show the calorific value for the briquette developed. This result is good enough if compare to the calorific value of wood (4491.2 cal/g) or young age coal (1887.3 cal/g). However this value is still lower than the value of charcoal from wood which has calorific value of 7047.3 cal/g). Table 2 shows the calorific value of some solid fuels [7].

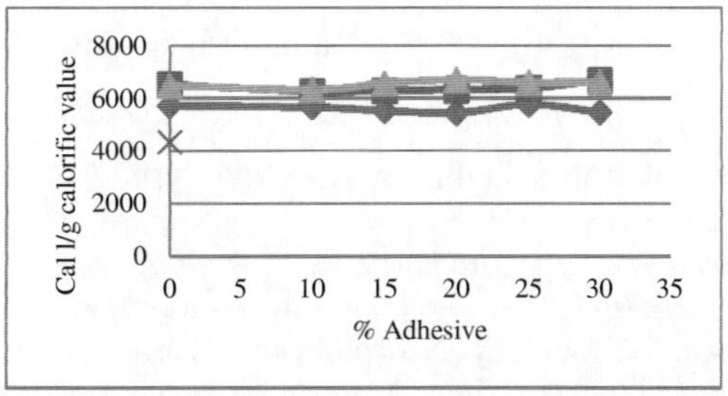

Figure 9. Calorific value for different pyrolysis temperature

3.5. Gas emission testing

The testing of gas emission resulting from the burning of charcoal briquette can be seen in Table 3. From the table, it can be seen that the pyrolysis process will affect the production of emission gases.

Table 2. Calorific value of several types of fuel

Type of fuel	Calorific value (cal/g)
Dry wood	4491.2
Young coal (lignite)	1887.3
Coal	6999.5
Charcoal	7047.3
Crude oil	10008.,2
Fuel oil	10224.6
Natural gas	9722.9

Table 3. Results of gas emission test from burning the charcoal briquette, % volume.

Time (min)	Gas CO Temperature of pyrolysis		Gas CO_2 Temperature of pyrolysis	
	250°C	300°C	250°C	300°C
0	0.09	0.09	2.0	2.0
30	0.08	0.08	2.1	2.2
60	0.06	0.04	2.2	2.5
90	0.04	0.03	2.4	2.7

4. Conclusions

It can be concluded that the *Terminalia Catappa L.* and *Myristica fragrans* (nutmeg seeds shells) shells can be used as raw material for the manufacture of charcoal briquettes. The most suitable model is a model III with the reaction rate constant equation and the activation energy for solids as adalah $k_2 = 487{,}8461 \exp\frac{-7825}{T}$ and Ea = 65.0602 (kJ/mol). The calorific value obtained was better than some of solid fuels.

References

[1] Sinha S, Jhalani A, Ravi MR, Ray A. Modelling of pyrolysis in wood: a review. *SESI J.* 2000; **10** (1) pp 41-62.
[2] Thurner F and Mann U 1981 *Ind .Eng. Chem. Process Des. Dev.* **20** pp 482–488.
[3] Tsai CH, Albright FL. Industrial and Laboratory Pyrolysis, ACS Symp. Series, Washington DC. 1976 pp 275-311.
[4] White J E, Catallo W J and Legendre B L 2011 *J. Anal. Appl. Pyrolysis* **91** pp 1-33.
[5] Thomson L A J and Evans B 2006 *Permanent Agriculture Resources (PAR)* Holualoa Hawai
[6] Inbaraj BS, Sulochana N. Mercury adsorption on a carbon sorbent derived from fruit shell of Terminalia catappa. *J. of Hazardous Materials.* 2006 May 20; **133** (1) pp 283-90.
[7] Rousset P, Caldeira-Pires A, Sablowski A, Rodrigues T. LCA of eucalyptus wood charcoal briquettes. *J. of Cleaner Prod.* 2011 Sep 30; **19** (14) pp 1647-53.

IOP Conf. Series: Materials Science and Engineering **206** (2017) 012064 doi:10.1088/1757-899X/206/1/012064

Screening of Catalyst and Important Variable for The Esterification of Acrylic Acid with 2 Ethylhexanol

M A A Ahmad [1], S Y Chin[1,2*]

[1]Faculty of Chemical & Natural Resources Engineering, Universiti Malaysia Pahang, Lebuhraya Tun Razak, 26300, Gambang, Kuantan, Pahang, Malaysia

[2]Center of Excellence for Advanced Research in Fluid Flow, Universiti Malaysia Pahang, Lebuhraya Tun Razak, 26300, Gambang, Kuantan, Pahang, Malaysia

E-mail: chin@ump.edu.my

Abstract. The global demand of 2-ethylhexyl acrylate (2EHA) market has witnessed a significant growth in the past few years and this growth is anticipated to increase in the coming years. 2EHA is one of the basic organic building blocks that mainly used in the production of coatings, adhesives, superabsorbents, thickeners and plastic additives. Homogenous acid-catalysed esterification of acrylic acid (AA) with 2-ethylhexanol (2EH) is commonly used for the production of 2EHA. The homogeneous catalysts such as sulfuric and para-toluene sulfonic acid have resulted the costly and complicated downstream process that generates acidic, corrosive and non-environmental friendly waste. Therefore, it is importance to develop a cheaper process that employing heterogeneous catalysts and alternative raw material from wastewater containing acrylic acid. In this research, the study for the esterification of AA with 2EH catalysed by ion-exchange resin was conducted. The best sulfonic acid functional cation-exchange resin among SK104, SK1B, PK208, PK216, PK228, RCP145, and RCP160 was screened. PK208 outperformed the other resins and it was used subsequently in the parametric studies. The effect of important parameters (initial concentration of acrylic acid (AA), temperature, molar ratio of reactant (AA and 2EH), catalyst loading, and polymerisation inhibitor loading) was studied using 2 factorial design to determine the significant parameters to the esterification. It was found that the initial concentration of AA and temperature were most significantly affecting the esterification of AA with 2EH.

1. Introduction

2-Ethylhexyl Acrylate (2EHA) is an acrylate ester produced in a large scale owing to the wide application of its polymer in the manufacturing of coating materials, adhesives, and impregnating agents. 2EHA appears in clear and volatile liquid naturally. It is slightly soluble in water and completely soluble in alcohols, ethers, and solvents.

2EHA is manufactured through the direct esterification of acrylic acid (AA) with 2-ethyl hexanol (2EH). Traditionally, sulphuric or p-toluenesulfonic acid have been used as the catalysts to accelerate the reaction rate in production of 2EHA [1]. However, the use of these strong acidic compounds as catalysts has resulted the elevated downstream processing and operation costs. Furthermore, the unrecycled homogeneous acidic catalysts also cause environmental problem. In recent years, the research interest has been shifted to the heterogeneously catalysed AA esterification reaction study to overcome these problems.

One of the important characteristics of the heterogeneous catalysts is water resistance since the by-product during the esterification of AA with alcohol is water. A few of heterogeneous catalysts have been found to be water resisted which including organic functional polymers , zeolites, salts of heteropolyacids, some of the sulfated oxides , and other solid oxides [2][3]. Some of it such as salts of heteropolyacids were not used in the commercial process due to the leaching problems [4].

Ion exchange resins (IER) have attracted considerable attention in the researches about the esterification of organic acid with alcohol. Amberlyst, Dowex, Nafion, and Purolite are among the commonly used IER catalysts in the esterification of organic acid with alcohol. Altiokka and Odes [5], Essayem et al. [6] and Okuhara et al. [3] employed several type of heterogeneous catalysts in the esterification of AA with propylene glycol, butane and butanol respectively. They found that IER type of catalyst outperformed the other type of catalyst for the esterification reaction.

The research on heterogeneously catalysed 2EHA synthesis is still rare. Komon et al. [4] performed the best catalyst screening among the Amberlyst group (Amberlyst 39, Amberlyst 46, Amberlyst 70, and Amberlyst 13) before carrying out the kinetics study of the esterification of AA with 2EHA catalysed by Amberlyst 70. Meanwhile, Chin et al. [7] has done the kinetic and one factorial at time (OFAT) parametric study for the identical reaction catalysed by Amberlyst 15.

The objective of this study was to screen the best IER catalyst among SK104, SK1B, PK208, PK216, PK228, RCP145 and RCP160. Afterward, the effect of important parameters such as AA initial concentration, temperature, molar ratio, catalyst loading, and polymerisation inhibitor loading on the production of 2EHA with AA were evaluated. The 2 factorial design were used to determine the interaction factors accurately in a timely and cost saving mode.

2. Methodology

2.1. Materials

Both reactants, 2-ethyl hexanol (99.99%) and acrylic acid (99.99%) as well as polymerisation inhibitor, phenothiazine were supplied by Sigma Aldrich. Another polymerisation inhibitor, 4-Methoxyphenol (MEHQ) was supplied by ReagentPlus. Diaion SK104, SK1B, PK208, PK216, PK228, RCP145, and RCP160 were sponsored by Mitsubishi Chemical Corporation. All these chemicals and catalyst were used without further purification. Ultra-pure water was obtained from ultra-pure water system MILLIPORE model SAS 67120.

2.2. Apparatus

All the experiments were conducted in a 500 mL 3-necked round bottom flask fitted with reflux condenser and thermocouple. Rotamantle was the source of heating and stirring. The temperature was maintained within desired temperature (±1 K) by PID temperature controller. In a typical procedure, a pre-determined amount of alcohol and acrylic acid with the corresponding amount of polymerisation inhibitor were preheated separately until achieving the desired temperature before it was mixed. Subsequently, catalyst with a pre-fixed loading was transferred to the reactor and the reaction was started. A product sample of circa 0.5 mL was withdrawn every 1 hour interval. To avoid any further reaction, the collected samples were rapidly cooled down in 277 K chiller prior to the GC analysis in spite of the very slow reaction rate for the un-catalysed reaction.

2.3. Catalyst characterisation

2.3.1. Ion Exchange Capacity (IEC).

0.1 g of catalyst was immersed in 100 ml of 0.1 M NaCl solution for 24 hours. The NaCl solution was then titrated using 0.01 M KOH with phenolphthalein as the indicator. The ion exchange capacity in milliequivalents/gram (meq/g) is calculated using Equation (1):

$$\text{IEC} \vee \text{TLC} = \frac{V_{KOH} \times M_{KOH}}{\text{mass}_{\text{catalyst}}} \quad (1)$$

where V is volume and M is molarity.

2.3.2. Leaching test.

0.5 g of catalyst was immersed in 50 ml of water and 2 ethyl hexanol (2EH) at 95 ^{0} C for 4 hours (imitate the reaction duration) respectively for sulphonic acid leaching analysis. The solution was then titrated using 0.01 M KOH with phenolphthalein as indicator. The total leaching capacity (TLC) in meq/g is calculated based on Eq. (2.1) while the percentage of leaching (L%) is calculated using Equation (2) respectively:

$$L = \frac{\text{TLC}}{\text{IEC}} \times 100 \quad (2)$$

2.3.3. Performance test.

For the performance test, 7 types of DIAION resins (SK104, SK1B, PK208, PK216, PK228, RCP145, and RCP160) were tested for their activity at the temperature of 368 K, catalyst loading of 10% w/w (catalyst/acrylic acid) and molar ratio of AA:2EH is 1:3. The samples were withdrawn at every 1 h for 4 h. To avoid any further reaction, the collected samples were cooled rapidly at 277 K prior for gas chromatography (GC) analysis. The best catalyst was selected in subsequent experimental work.

2.3.4. Screening of Important operating parameters.

Design Expert DX7 that is statistical software is used to generate the design of experimental (DOE) for screening and optimization studies to identify the vital factors that affect the esterification process. It is advisable to proceed with two level factorial design first as the factors is more than 3. Two level factorial fractional designs (half factorials) which is available to explore many factors is selected. This design setting each factor to only two levels for the screening studies. It is sufficient and efficient designs used to screen many factors to find the few that are critically significant and to estimate the effects of all interactions. After the screening studies has identified the critical variable that affect the esterification performance, response surface method (RSM) central composite design (CCD) is proceeded. In this design, the consideration of three groups of design points that is two-level factorial or fractional factorial design points, axial points and center points is proceed to quantify the relationships between that chosen critical factors for optimization purpose. Table 1 and 2 shows the high and low level of each variable and the list of design experiment from the software tools respectively.

Table 1. Value for high and low level

Parameters	**High level**	**Low level**
Innitial AA concentration	100% w/w	10% w/w
Temperature	100°C	80°C
Molar ratio	1:1	1:5
Catalyst loading	3% w/w	10% w/w
Inhibitor loading	0% w/v	0.5% w/v

Table 2. Experimental design for 2 factorial analysis (half factorial)

Experiment	AA Conc. (% w/w)	Temperature (°C)	Molar Ratio (AA:2EH)	Cat. Loading (% w/w)	Inhibitor Loading (% w/v)
1	10	80	1:1	3	0.5
2	100	80	1:1	3	0
3	10	100	1:1	3	0
4	100	100	1:1	3	0.5
5	10	80	1:5	3	0
6	100	80	1:5	3	0.5
7	10	100	1:5	3	0.5
8	100	100	1:5	3	0
9	10	80	1:1	10	0
10	100	80	1:1	10	0.5
11	10	100	1:1	10	0.5
12	100	100	1:1	10	0
13	10	80	1:5	10	0.5
14	100	80	1:5	10	0
15	10	100	1:5	10	0
16	100	100	1:5	10	0.5
17	55	90	1:3	6.5	0.25
18	55	90	1:3	6.5	0.25
19	55	90	1:3	6.5	0.25

3. Results and Discussion

Figure 1 displays the divinylbenzene crosslinkage (DVBCL) percent, ion exchange capacity (IEC), and sulfonic acid (SO3-) leaching properties of each IER used in this studies. Three types of ion exchange resins were tested. It included highly macroporous resins (RCP145 and RCP160), macroporous resins (PK208, PK216, and PK228) and gelular resins (SK104 and SK1B).

All IER (SK104, SK1B, PK208, PK216, PK228, RCP145, and RCP160) has been tested for their catalytic performance in the synthesis of 2EHA via esterification of AA with 2EH. The yield of 2EHA obtained through the esterification reaction catalysed by different types of resin is shown in Figure 2. PK208 has outperformed all the other catalysts over 4 hours of reaction by showing the highest yield of 2EHA (41.2% yield). The yield of 2EHA obtained for the other catalysts is arranged in the descending order as follow: SK1B (35.5%), RCP145 (33.6%), SK104 (31.3%), PK216 (26.1%), RCP 160 (15.8%) and PK228 (15.1%).

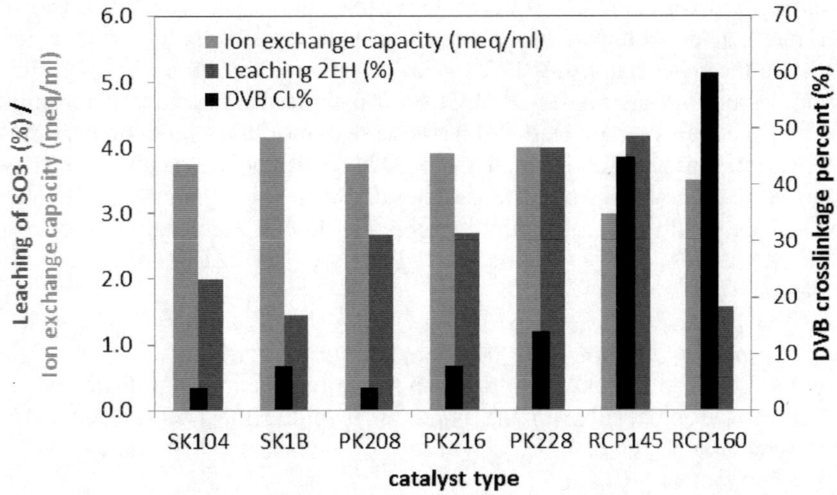

Figure 1. Percentage of DVB crosslinkage for the IER catalysts. Percentage of IEC for the IER catalyst. Leaching of sulfonic acid for the IER catalysts

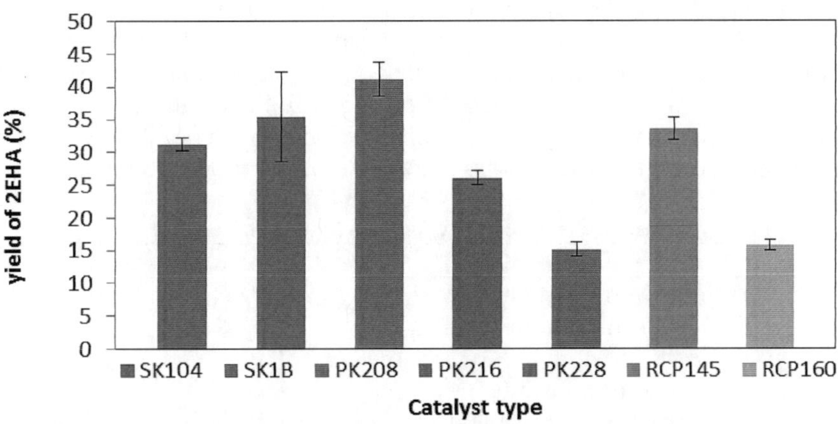

Figure 2. Yield of 2EHA for the reaction catalysed by different types resin catalysts for 4 h. Operating condition: Initial conc. AA: 100%; Temperature: 95°C; Catalyst loading of 10% w/w (catalyst/AA); molar ratio of AA:2EH is 1:3

Figure 1 shows that RCP145 and PK228 possessed the poorest sulfonic acid leaching properties as compared to the others. Despite the poor sulfonic acid leaching property, RCP145 has shown the substantial yield as shown in Figure 2 due to the reaction catalysed by leached sulfonic acid in homogeneous phase. So, these heterogeneous catalysts are not considered as the good catalysts due to its severe leaching. The leaching properties for the other catalysts were within the acceptable levels.

It is generally known that for IER catalysts, the catalytic performance was typically attributed by the IEC of the resins. Nevertheless, Figure 1 shows that the IEC of IER catalysts did not differ much (ranged from 3-4.15), implying that the catalytic performance of IER catalysts in the present study was not solely affected by IEC but the DVB crosslinkage percentage (CL%) and sulfonic acid leaching percentage (L%).

DVB crosslinkage percent (CL%) generally have impact on the structure (porosity and surface area) and swelling properties of the catalyst. The higher DVB CL%, the more porous the catalyst would be thus increasing the surface area of the catalyst. The degree of crosslinkage of resins is also closely related to its pore sizes. Resins with the lower degree of crosslinkage have a larger pore size.

IOP Conf. Series: Materials Science and Engineering **206** (2017) 012064 doi:10.1088/1757-899X/206/1/012064

Reactants can easily diffuse and travel to the active sites due to its lower mass transfer resistances. Thus, a higher rate can be obtained for the reaction catalyzed by resins with a lower degree of crosslinkage [8]. On the other hand, DVB CL% has negative effect on swelling properties, in which the catalysts swell lesser with the higher DVB CL%. It is noticeable that the increasing of DVB CL% for each category of the resins would lead to a negative response in its catalytic performance (Figure 1 and 2). The increase of surface area in this case does not enhance the catalytic activity as the active site distribution over surface area would be decreased. Moreover, the size of 2EH which is bigger than the other alcohols would have difficulty to penetrate though the porous site of resin with higher DVB CL%. PK208, the best catalyst giving the highest yield of 2EHA, was selected to be used in subsequent studies.

3.1. Screening of important operating parameter via 2 factorial design.

The effect of some of the factors to the reaction can sometimes be less significant. These insignificant factors would lead to wrong results on the importance of the most significant variables [9]. The purpose of using screening design is to be economical, by saving time and money on chemicals, and to identify the significant factors [10].

Table 3 shows that the initial AA concentration and temperature significantly affected the 2EHA yield in the esterification of AA with 2EH (p-value < 0.0001). However, other factors such as 2EH to AA molar ratio, catalyst loading and inhibitor loading were insignificant (p-value > 0.0001). The initial AA concentration was the major contributing factor (66.52%), followed by the temperature (11.9%) as shown in Table 4. The interaction of initial AA concentration with temperature was also prominent (9.13%). The yield of 2EHA increased from 1.3 to 14.7 % with the increase of initial AA concentration from 10 to 100 % w/w (Figure 3(A)). In addition, the yield of 2EHA also increased from 5.2 to 10.8 % when the reaction temperature was increased from 80 to 100 ^{0} C (Figure 3(B)).

Table 3. P-values of responses against factors.

Factor	p-value for response
Initial AA concentration	<0.0001
Temperature	<0.0001
Molar ratio	0.0344
Catalyst loading	0.0003
Polymerisation inhibitor loading	0.0381

Table 4. The percentage contribution of factors on response variables.

Factor	Percentage contribution
A-AA concentration	66.52
B-Temperature	11.90
C-Molar ratio	0.50
D-Catalyst loading	2.58
E-Inhibitor loading	0.47
AB	9.13
AC	4.02×10^{-3}
AD	1.31
AE	0.21
BC	0.36
BD	1.04×10^{-3}
BE	0.016
CD	-
CE	0.02
DE	0.08

The significance of the initial AA concentration and reaction temperature was implied by the steeper slope of the line plots for these factors as shown in Figure 3. A higher initial AA concentration is preferred in esterification reaction of AA with 2EH to produce 2EHA. The dilute AA with large amount of water would swollen the resin catalyst and subsequently blocking the access of the reactant to internal pore of the catalyst. Thus would limited the access to effective active site which also lead to decrease of effective active site over surface area. Besides, the water in the reactant would also shift the reaction to reactant side since esterification is a reversible reaction. Moreover, it was believed that the poor accessibility of reactants to acid sites might be occured in the diluted system. The presence of huge amount of water would interrupt the adsorption of the reactant to the active site of catalyst [11][12][13]. The affinity of IER to the polar solvent, water, is stronger than the affinity to the organic solvent, 2EH.

The increase of temperature also gave a better yield. This is attributed to the increasing kinetic energy of the reactant molecules and hence increasing effective collision between AA and 2EH molecules. Thus, the minimum amount of energy required to form products [7].

The other factors like molar ratio of AA:2EHA, catalyst loading and polymerisation inhibitor loading are the insignificant factors ($p>0.0001$). The effect of molar ratio of AA:2EHA to the yield of 2EHA was negligible in the present study due to the presence of water in the reactant. The water in the reactant has traded off the positive effect of using excess amount of 2EH in driving the reaction equilibrium to 2EHA side.

A)

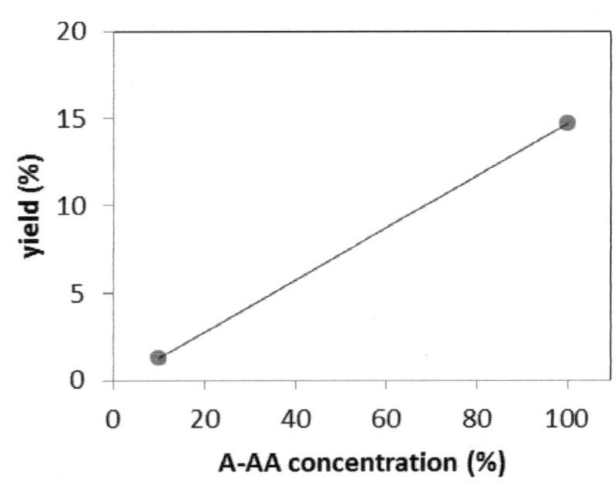

B)

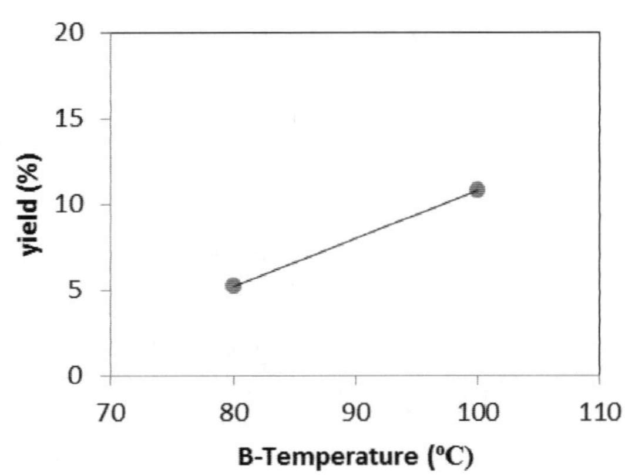

IOP Conf. Series: Materials Science and Engineering **206** (2017) 012064 doi:10.1088/1757-899X/206/1/012064

C)

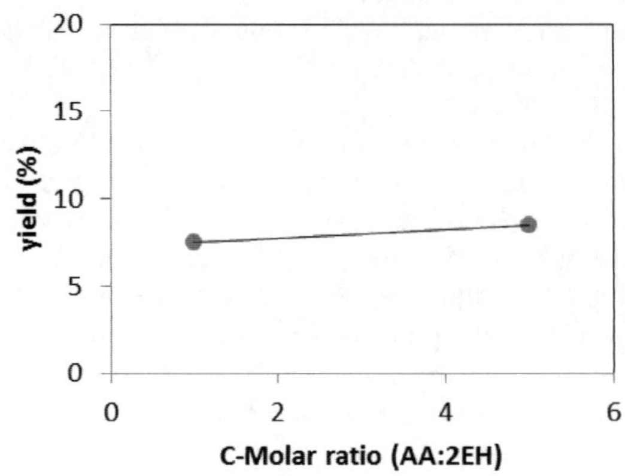

D)

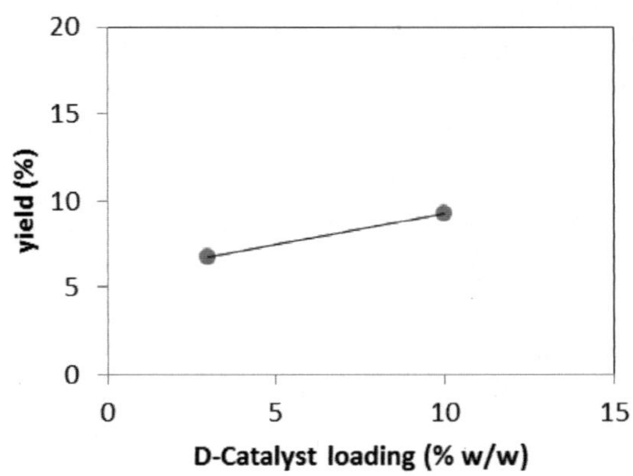

E)

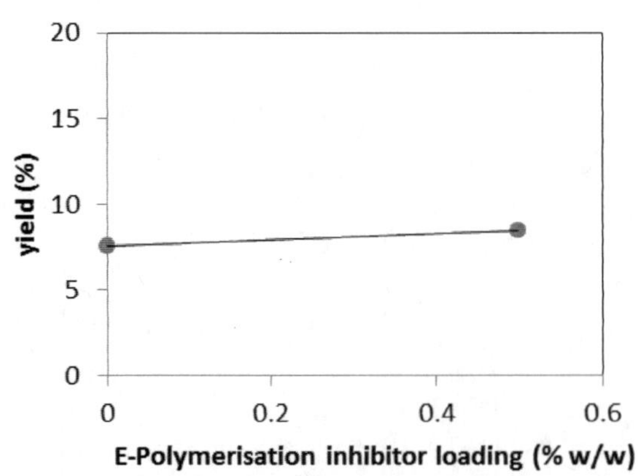

Figure 3. Effect of the initial AA concentration (A), temperature (B), molar ratio of AA:2EH (C), catalyst loading (D) and polymerisation inhibitor loading (E) on the performance in term of yield of esterification AA with 2EH.

The catalyst amount used has provided a sufficient number of active site. Similarly, the amount of the polymerisation inhibitor used was adequate to diminish the AA polymerisation reaction since this side reaction is minimal at the reaction temperature below 100^0 C.

Statistically, the model F-value of 133.31 in Table 5 implies the model is significant. The model equation is shown in Equation (3). There is only a 0.01% chance that a "Model F-Value" this large could occur due to noise. The "Curvature F-value" of 122.71 infers there is significant curvature (as measured by difference between the average of the center points and the average of the factorial points) in the design space. The "Lack of Fit F-value" of 5.82 indicates the Lack of Fit is not significant relative to the pure error. This model equation can be used to predict the yield of 2EHA accurately at different reaction operating conditions. There is a 15.44% chance that a "Lack of Fit F-value" this large could occur due to noise. Non-significant lack of fit is good.

$$\text{Yield}=-0.17480 - 0.39279*A- 0.020243*B + 0.29406*C + 0.049829* D + 2.29400*E + 5.60806E\text{-}003*A*B + 6.07063E\text{-}003*A*D \quad (3)$$

Table 5. Analysis of variance (ANOVA) table for 2 factorial studies.

Source	Sum of Squares	df	Mean Square	F-Value	
Model	740.35	7	105.764	133.310	significant
A-AA Concentration	660.13	1	660.128	832.056	
B-Temperature	118.13	1	118.129	148.895	
C-Molar Ratio	4.92	1	4.919	6.201	
D-Catalyst Loading	25.65	1	25.652	32.333	
E-Inhibitor Loading	4.68	1	4.678	5.896	
AB	90.58	1	90.577	114.167	
AD	13.00	1	13.002	16.388	
Curvature	97.36	1	97.356	122.712	significant
Residual	7.14	9	0.793		
Lack of Fit	6.81	7	0.972	5.822	not significant
Pure Error	0.33	2	0.167		
Cor Total	844.84	17			

4. Conclusions

The best catalyst was PK208 ascribing to its complementary IEC, DVB CL%, and leaching properties. Two level half factorial design screening was successfully employed to determine the significant factors. The initial AA concentration and temperature were found to significantly ($p<0.0001$) affect the yield of 2EHA in the esterification of AA with 2EH. These significant factors underlie the process optimisation study in the future.

Acknowledgement

The authors would like to thank Universiti Malaysia Pahang, for the short-term grant scheme (RDU130313) and the Postgraduate Graduate Research Scheme (GRS150397). Gratitude is given especially to Mitsubishi Chemical Holding Group for sponsoring the catalyst.

References

[1] Otera J and Nishikido J 2010 Esterification Methods, Reactions and Applications, seconded., *Wiley-vch verlag GmbH & Co.*, KGaA, Weinheim,.

[2] Izumi Y. 1997 Hydration/hydrolysis by solid acids. *Catalysts Today* **33** 371-409.

[3] Okuhara T, Kimura M, Kawai T, Xu Z, and Nakato T 1998 Organic reactions in excess water catalyzed by solid acids. *Catalysis Today.* **45** 73–77.

[4] Komon T, Niewiadomski P, Oracz P, and Jamróz M E 2013 Esterification of acrylic acid with 2-ethylhexan-1-ol: Thermodynamic and kinetic study. *J. Appl. Catalysis A: General.* **451** 127– 136.

[5] Altıokka M R, and Ödeş E 2009 Reaction kinetics of the catalytic esterification of acrylic acid with propylene glycol. *Appl. Catalysis A: General.* **362(1-2)** 115–120.

[6] Essayem N, Martin V, Riondel A, and Védrine J C 2007 Esterification of acrylic acid with but-1-ene over sulfated Fe- and Mn-promoted zirconia. *Appl. Catalysis A: General.* **326(1)** 74–81.

[7] Chin S Y, Ahmad M A A, Kamaruzaman M R , and Cheng C K 2015 Kinetic studies of the esterification of pure and dilute acrylic acid with 2-ethyl hexanol catalysed by Amberlyst 15. *Chem. Eng. Sci.* **129** 116–125

[8] Bhandari V M, Sorokhaibam L G, and Ranade V V 2016 Ion Exchange Resin Catalyzed Reactions—An Overview Ed. Joshi, S.S., Ranade, V.V. *Ind. Catalytic Processes for Fine and Specialty Chem.* 393-426.

[9] Jalbani N, Kazi T G, Arain B M, Jamali M K, Afridi I, and Irfraz R A 2006 Application of factorial design in optimization of ultrasonic-assisted extraction of aluminum in juices and soft drinks. *Talanta* **70 (2)** 307–314.

[10] Prasad K N, Kong K W, Ramanan R N, Azlan A, and Ismail A 2012 Selection of experimental domain using two-level factorial design to determine extract yield, antioxidant capacity phenolics, and flavonoids from Mangifera pajang Kosterm. *Sep. Sci. Technol.* **47 (16)** 2417–2423.

[11] Haas M J 2005 Improving the economics of biodiesel production through the use of low value lipids as feedstock: vegetable oil, soapstock. *J. Fuel Process Technol.* **86** 1087–1096.

[12] Rat M, Zahedi-Niaki M H, and Kaliaguine S T O D 2008 Sulfonic acid functionalized periodic mesoporous organosilicas as acetalization catalysts. *Microporos Mesoporous Mat.* **112**, 26–31.

[13] Ahmad M A A, Kamaruzzaman M R, Chin S Y 2014 New method for acrylic acid recovery from industrial waste water via esterification with 2-ethyl hexanol. *J. Process Safety and Env. Protection.* **92** 522–531

IOP Conf. Series: Materials Science and Engineering **206** (2017) 012065 doi:10.1088/1757-899X/206/1/012065

Optimization and Performance parameters for adsorption of Cr^{6+} by microwave assisted carbon from *Sterculia foetida* shells

N Gnanasundaram[1], M Loganathan[1], A Singh[1*]

[1]Department of Chemical Engineering, , School of Civil & Chemical Engineering, Vellore Institute of Technology University, Vellore, 632014, TamilNadu India

[*]Mass Transfer Laboratory, School of Civil & Chemical Engineering, Vellore Institute of Technology University, Vellore, 632014, TamilNadu India

E-mail: arunasingh@vit.ac.in

Abstract. Modeling of adsorption of Cr^{6+} on to activated carbon prepared from *Sterculia foetida* dried seed shells under different drying techniques namely sun, oven, and microwave drying (450W, 600W, 900W power). Optimization of process parameters such as pH, adsorbent dosage (g/ml), temperature (°C), contact time (min) were evaluated using Central Composite Rotatable Design (CCRD) of Response Surface Methodology (RSM). For batch adsorption studies at pH 3, adsorbent dosage of 1.5 g/ml, temperature 35°C and contact time 90 min were found to be optimum for the system under consideration and Microwave Activated Carbonized *Sterculia foetida* (MACSF) at 450W was found to be best suited for the adsorption of Cr^{+6} ions. The system was found to follow Langmuir type monolayer adsorption for the given operational parameters. SEM analysis was used to study the surface morphology of the carbon samples and the effect of pretreatment on carbonization.

1. Introduction

Metals and their derivatives are very harmful when released into the atmosphere and can be particularly problematic due to their stability and mobility. The hexavalent chromium (Cr^{+6}) in effluents generated from various industries, such as electroplating, steel making, leather tanning, mining etc., is a major concern due to its adverse effects to human health and other living organisms [1]. Hexavalent [Cr^{+6}] ions are highly toxic and can lead to mutagenic and carcinogenic health hazards[2]. Numerous techniques have been practiced to eliminate Cr^{+6} ions from wastewater, the principal processes being chemical processes [3,4], ion exchange [5], membrane separation [6] and adsorption [7]. Adsorption is one of the most promising and technically feasible methods as it involves low cost adsorbents to remove heavy metal ions because of its high affinity for trace ions.

In order to make the process of adsorbent activated carbon preparation more attractive in terms of production costs, various bio-waste materials have been used as low-cost precursors. *Sterculia foetida*, a soft wooded tree that can grow up to 115 feet tall, is such a material found in abundance in most parts of Southern India and hence easily available. Compared to the conventional heating methods, microwave heating has been used as energy source in the preparation of activated carbon from several biomasses [8-11]. Among several advantages, microwave-assisted activation method leads to a more efficient carbonization, resulting in a material with higher surface area and well-developed porosity [8-11].

In the present study, carbon made from seed shells of *Sterculia foetida* was used as adsorbent to remove chromium ions from waste water sources. Further, the effects of the process parameters such as pH, adsorbent dosage, temperature, contact time and initial concentration on removal of Cr^{6+} were

IOP Conf. Series: Materials Science and Engineering **206** (2017) 012065 doi:10.1088/1757-899X/206/1/012065

investigated and optimized by the application of the Response Surface Methodology (RSM), which is a combination of mathematical and statistical tool. Since conventional method of optimizing the process variables requires a very large number of experimental runs, which is highly expensive and time consuming, RSM is used for developing, improving and optimizing the process variables and to evaluate the relative significance of several process parameters in the presence of complex interactions. In order to have better understanding of the adsorption characteristics Langmuir and Freundlich isotherm models at various operating parameters were studied.

2. Materials and methods

Fresh sterculia flowers were collected from VIT University, Vellore Tamilnadu India. The shells were further taken for drying and used for preparing activated carbon discarding the seeds. *Sterculia foetida* shells were subjected to drying to remove all the moisture content under five different conditions which include direct sun drying, oven drying and microwave drying of three different powers (450W, 600W, 900W) before carbonization.

Dried sterculia samples were broken into uniform size (1inch pieces) and sealed in earthen pots. The pots were kept in a muffle furnace and heated from room temperature to 550°C at an increment of 10°C/min and was held at 550°C for 2 h. The carbon samples obtained were cooled to room temperature. The ash formed while carbonization was removed by washing the samples thoroughly with double distilled water. The wet carbon samples were dried in an oven drier for 24 h at 110°C for further use as adsorbent.

A stock solution of Cr^{+6} (1000 mg/L) was prepared from potassium dichromate salt by adding 2.829 g of $K_2Cr_2O_7$in 1000 mL distilled water. Other working solutions of different concentrations were also prepared by appropriate dilution of the stock solutions.

2.1. Batch adsorption studies

The adsorption experiments were carried out in batch mode by mixing a specific amount of adsorbent in 50 mL of Cr^{6+} solution under constant shaking (120 rpm) in a thermostat shaker. The effects of contact time, amount of adsorbent, initial Cr^{6+} concentration, pH and temperature of the solutions were investigated. The samples were analyzed using UV-visible spectrophotometer to estimate the Cr^{6+} ion concentration in the solution at 540nm against a standard curve already plotted.

2.2. Experimental optimization

The experimental optimization was done for batch adsorption studies on different activated carbon samples in order to find out the best pre-treatment method for carbonizing that has the maximum adsorption capacity. Simultaneously, the parameters which include pH, adsorbent dosage, temperature, contact time and initial concentration were also optimized. The amount of Cr^{6+} adsorbed at equilibrium q_e (mg/g) was calculated using(1).

$$q_e = \left(\frac{(C_0 - C_e)V}{W} \right) \tag{1}$$

$$\%\ removal = \left(\frac{(C_0 - C_e)}{C_0} \times 100 \right) \tag{2}$$

where C_0 and C_e are the initial and equilibrium liquid phase concentrations of Chromium. V is the volume of the solution and W is the mass of the adsorbent.

2.3. Central Composite Rotary Design (CCRD)

Response surface methods (RSM) provide superb statistical design tools and analysis of experimental runs aimed at process optimization. CCRD was used because of its ability to analyze the complex

IOP Conf. Series: Materials Science and Engineering **206** (2017) 012065 doi:10.1088/1757-899X/206/1/012065

interaction between process parameters. The CCRD was employed to fit a quadratic equation by multiple regression method which needs a minimum number of experimental test runs for modeling.

The CCRD model consists of a 2^n factorial runs ($\pm$ notations) with 2n axial runs ($\pm\alpha$,0…0), (0, $\pm\alpha$,0…0),…., (0,0,…. $\pm\alpha$) and n_c center runs (six replicates,0,0,….0). As the number of factors n increases the number of experimental runs for a complete replicate design also increases according to the following relation

$$N = 2^n + 2n + n_c \tag{3}$$

An empirical model was developed so as to correlate the response to the adsorption of Cr^{6+} on to MACSF based on second order polynomial that follows the equation below;

$$Y = \beta_0 + \sum_{i=1}^{n} \beta_i X_i + \sum_{i=1}^{n}\sum_{j=1}^{n} \beta_{ij} X_i X_j \tag{4}$$

where Y is the predicted response, β0 is the constant coefficient, βi is the linear coefficient, βii is the interaction coefficient and βij is the quadratic coefficient and Xi, Xj are the coded.

3. Results and discussions

3.1. Batch adsorption studies

3.1.1. Optimization of drying pre-treatments. Among the drying processes considered maximum adsorption capacity of 71% removal was obtained for microwave activated carbonized *Sterculia foetida* (MACSF) at 450W (figure 1). In general, microwave drying had a better efficiency in Cr^{6+} removal compared to other drying processes. The chemical bonds can be vigorously vibrated and ruptured by microwave heating, enhancing the pore structure by molecular diffusion thus leading to higher adsorption in the thermo-chemical process.

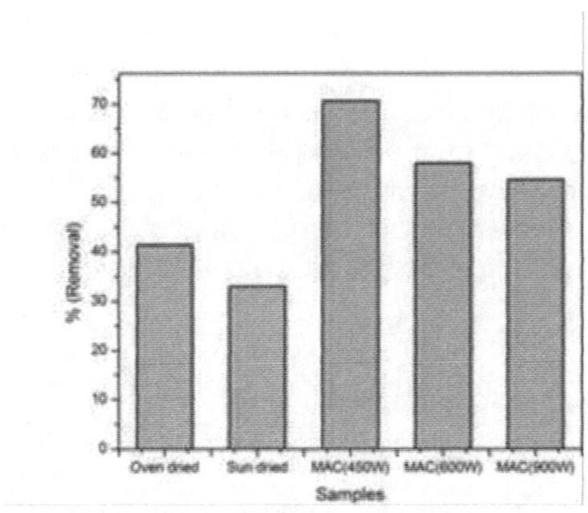

Figure 1. % Removal of Cr+6 for different activated carbon samples

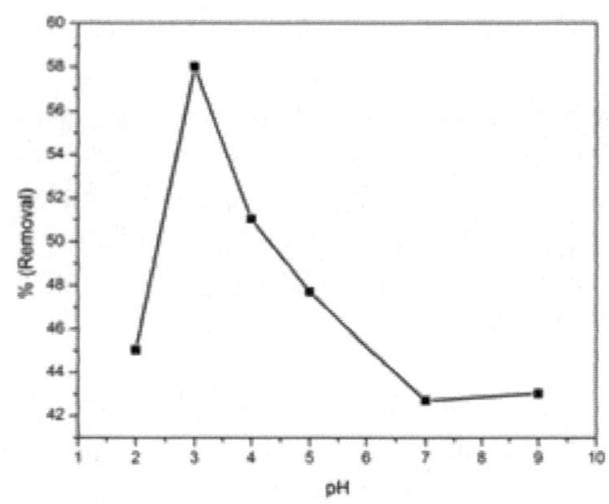

Figure 2. Effect of pH on % removal.

3.1.2. Effect of pH. The effect of pH on chromium removal is shown in figure 2. The adsorption of chromium increased from 45.4% to 58.4% for increase in pH from 2 to 3. For pH from 3 to 9 adsorption capacity tremendously decreased indicating a strong dependency on pH. The maximum adsorption capacity for the chromium ions occurred at pH 3. As reported earlier, acidic pH is more

IOP Conf. Series: Materials Science and Engineering **206** (2017) 012065 doi:10.1088/1757-899X/206/1/012065

effective for Cr^{6+} adsorption [12].

3.1.3. Effect of adsorbent dosage. Different dosages (0.5 g, 1 g, 1.5 g, 2 g) of activated carbon were prepared in 50 mL of chromium solution at pH 3 at room temperature. Figure 3 demonstrated that adsorption increased with increase in adsorbent dosage due to greater availability of pores and surface area of the adsorbent. The significant increase was observed when dosage was increased from 0.5 g to 1.5 g. The maximum adsorption capacity of 61.3% removal of Cr^{6+} was obtained at adsorbent dosage 1.5g/50mL. Equivalent trends in Cr^{6+} adsorption were reported by other researchers [13].

3.1.4. Effect of temperature. The chromium solution at pH 3 and adsorbent dosage of 1.5g/50mL was fixed and prepared. The samples were kept in rotary shaker at different temperatures (30°C, 35°C, 40°C, 45°C) for 2 h at 150 rpm. From figure 4, it is observed that the maximum adsorption capacity of 65% removal was found to be at temperature 35°C. The significant increase in percent removal from 45% to 66% was observed when temperature was increased from 30°C to 35°C.

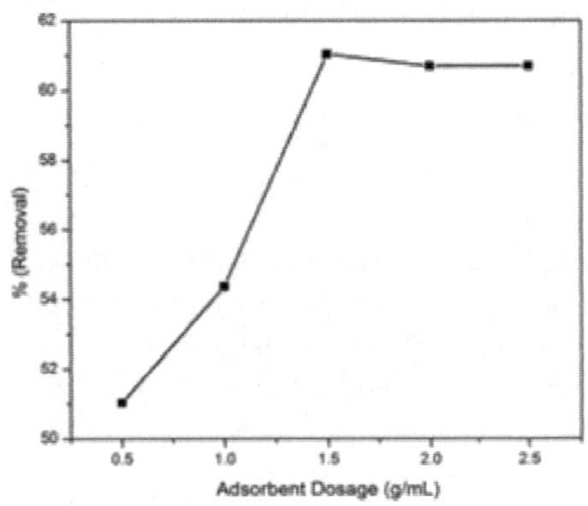

Figure 3. Effect of adsorbent dosage on % removal.

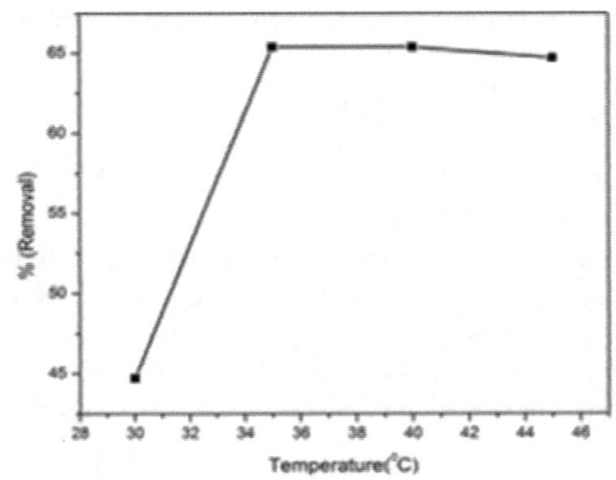

Figure 4. Effect of temperature on % removal.

3.1.5. Effect of contact time. The time dependency was planned at pH 3, adsorbent dosage of 1.5g/50mL and temperature at 35ºC process condition as maximum adsorption was attained at these levels. The effect of contact time was correlated for the adsorption of Cr^{6+} ions at regular time intervals ranging from 30min. to 180 min. Figure 5 revealed that the % removal increased and became constant after 90 min. The maximum adsorption capacity of 64.6% removal was observed at 90 min.

3.1.6. Effect of initial concentration. The effect of initial concentration in the range of 300 mg/L to 600mg/L on adsorption was investigated under specified conditions (pH 3, dosage 1.5g/50mL, temperature 350C, Contact time 90 min.) as shown in figure 6. It was observed that the amount of chromium removed is dependent on initial concentration; at low initial concentration of adsorbate increased percent removal of Cr^{6+} ions was observed. The maximum removal was obtained at 300 mg/L which was 70.11%.

IOP Conf. Series: Materials Science and Engineering **206** (2017) 012065 doi:10.1088/1757-899X/206/1/012065

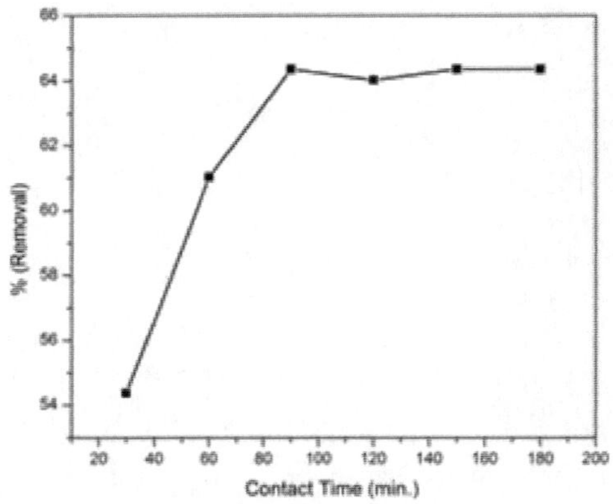

Figure 5. Effect of contact time on % removal

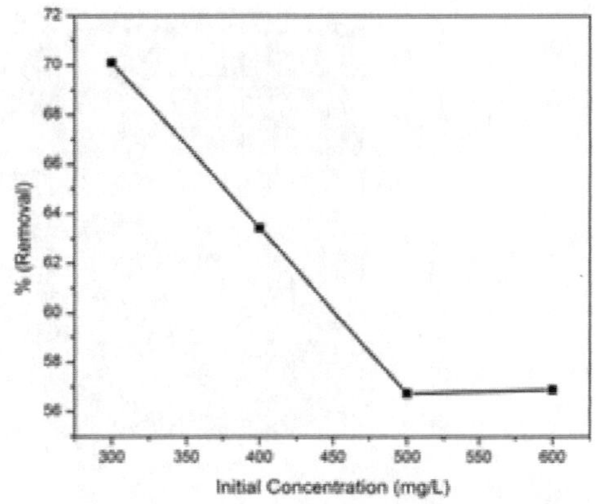

Figure 6. Effect of initial concentrations on % removal

3.1.7. Characterization. The surface morphology and developed pores for the carbonized samples were analyzed using Scanning electron microscope (SEM) with 20kV electron source, make Zeiss EV018 High resolution SEM, Germany. The SEM images figure 7a for Sun dried sample of *Sterculia foetida* showed few active sites on the periphery whereas the image of MACSF figure 7b had irregular surfaces. This is considered helpful for the diffusion of Cr^{+6} ions to activated carbon, as greater the surface area of pores more will be adsorption. From figure 7b and figure 7c, it is observed that the pores present in figure 7b were not visible in figure 7c suggesting a great deal of crystal being adhered to the surface. Thus, these images prove that Cr^{+6} ion was adsorbed by activated carbon into its pores and developed a layer of Cr^{+6} ions on the surface.

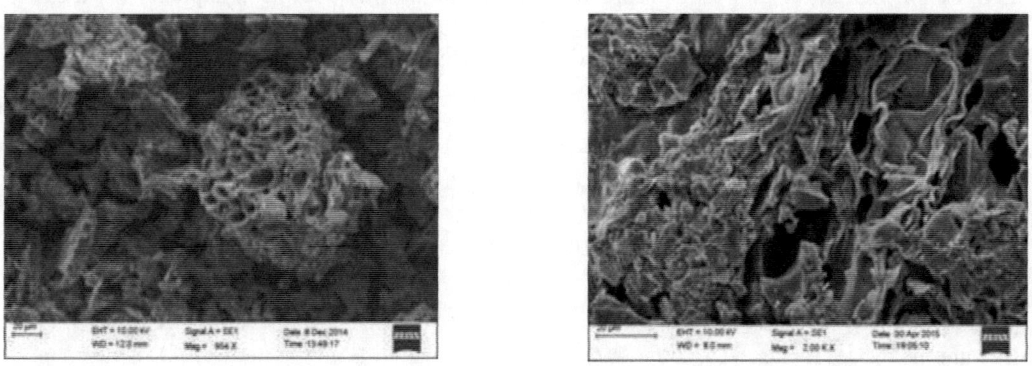

(a) (b)

IOP Conf. Series: Materials Science and Engineering **206** (2017) 012065 doi:10.1088/1757-899X/206/1/012065

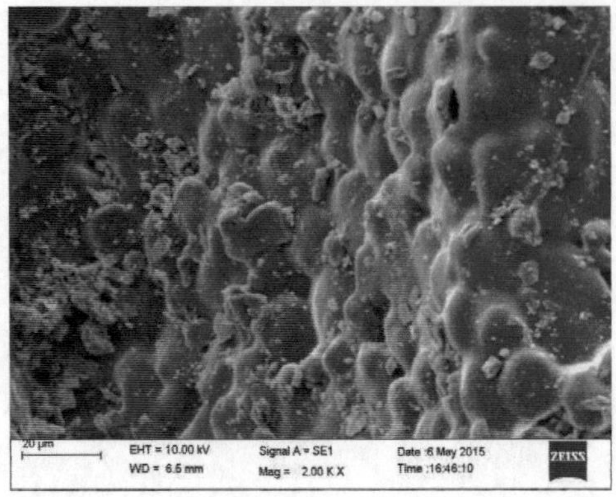

(c)

Figure 7. SEM images of (a) Sun Sterculia carbon dried, (b) MACSF (450MW) before adsorption, and (c) MACSF (450MW) after adsorption.

3.1.8. Isotherm models. Adsorption isotherm studies are based on the assumptions that relate homogeneity and heterogeneity of adsorbents, type of coverage and possibility of interaction between adsorbate species. The equilibrium adsorption data were tested for suitability for Langmuir and Freundlich isotherm plot (table 1 & table 2).

Langmuir model equation is given as follows

$$\frac{1}{q_e} = \frac{1}{q_m} + \frac{1}{q_m K_L C_e} \tag{5}$$

where q_m and K_L are the Langmuir constants, representing the maximum adsorption capacity for the solid phase loading and the energy constant related to the heat of adsorption respectively.
The Freundlich isotherm is an empirical model based on adsorption on heterogeneous surface and is given by the following equation

$$\log q_e = \log K_L + \frac{1}{n} \log C_e, \tag{6}$$

where Kf and n are the Freundlich constants, the characteristics of the system. Kf and n are the indicators of the adsorption capacity and adsorption intensity, respectively.

The Tempkin isotherm model is based on the assumption that the decline of the heat of adsorption as a function of linear rather than logarithmic, as given by the Freundlich isotherm. A linear form of the Tempkin model is expressed as

$$q_e = \frac{RT}{b} \ln A + \frac{RT}{b} \ln C_e \tag{7}$$

$$q_e = B \ln A + B \ln C_e \tag{8}$$

Where $B=\frac{RT}{b}$, R is gas constant (8.314 J/molK-1) and T is temperature (in K)

IOP Conf. Series: Materials Science and Engineering **206** (2017) 012065 doi:10.1088/1757-899X/206/1/012065

The linearized plot for Langmuir, Freundlich and Tempkin equations are plotted (fig. 8-10). Table 1 with R2 value of 0.93 suggests a Langmuir monolayer type of adsorption with adsorption capacity of 32.7963 mg/g, at 35ºC compared to R2 value of 0.88 for Freundlich. This is in accordance with other reported literature [14].

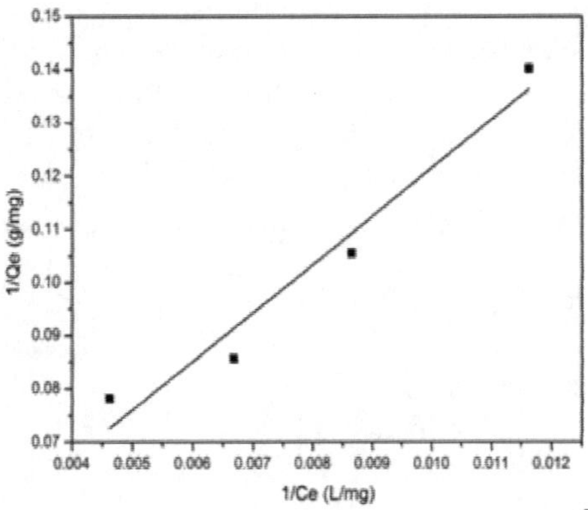

Figure 8. Langmuir isotherm plot at 35^{0}C.

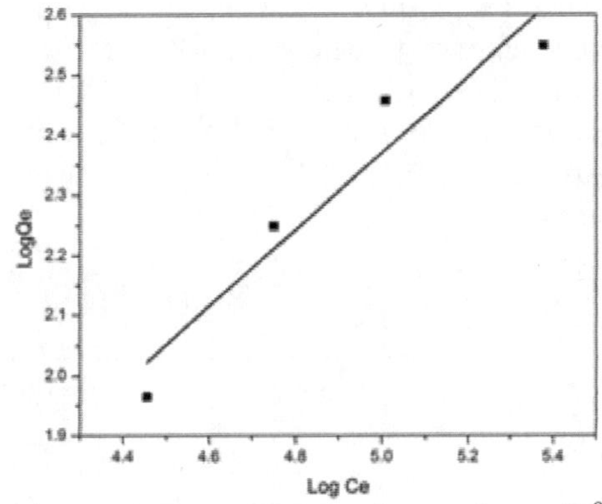

Figure 9. Freundlich Isotherm plot at 35^{0}C.

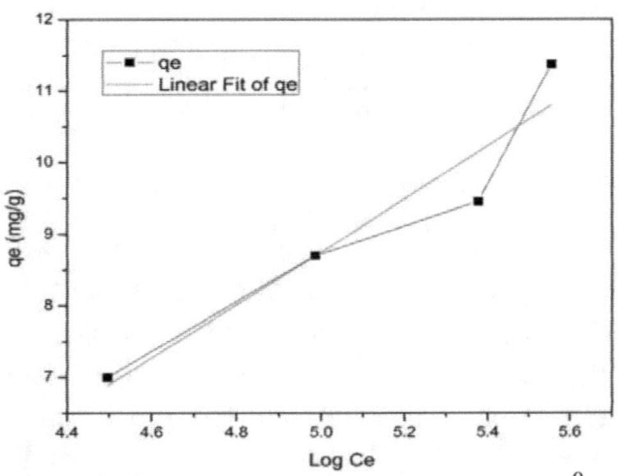

Figure 10. Tempkin isotherm plot at 35^{0}C

Table 1 Isotherm Model Factors

Isotherm	Parameters	Parameter values
Langmuir	Coefficient of determination(R^2)	0.9399
$\frac{1}{q_e} = \frac{1}{q_m} + \frac{1}{q_m K_L C_e}$	Maximum adsorption capacity(q_m)	32.7963 (mg g^{-1})
	Energy Constant(Langmuir) K_L	0.003381 (L mg^{-1})
Freundlich	Separation Factor(R_L)	0.22
$\log q_e = \log K_f + \frac{1}{n}\log C_e$	Coefficient of determination(R^2)	0.8861
	Adsorption Intensity(n)	1.571
	Freundlich Isotherm constant(K_f)	0.442 (L mg^{-1})
Tempkin	Coefficient of determination(R^2)	0.907
$q_e = B\ln A + B\ln C_e$	Constant related to heat of sorption (B)	6.07831 (J/mol)
	Equilibrium Binding constant (A)	0.04128

From the above studies, it was concluded that monolayer adsorption took place as Langmuir isotherm best fits the experimental data.

3.1.9. Optimization studies. The standard RSM design known as CCRD (Central Composite Rotatable Design) was employed to determine the optimum process variables for adsorption of Cr^{6+} on MACSF using software (Design Expert 9.0). The CCRD of 3 variables 5 levels; plus and minus alpha (axial points), plus and minus 1 (factorial points) and the center point are taken as independent variables as shown in table 2.

Table 2. Range of optimization variables

Variables	Temperature °C X_1	Adsorbent Dosage g/mL X_2	Contact Time Min X_3
Low	30	1	60
High	40	2	120
+α	43.41	2.34	140.45
-α	39.55	0.66	26.59

From table 3, R^2 value for the model is calculated to be 0.98 which signifies that the generated model is in agreement with the experimental results, as revealed by fig 10 -13. "Adeq Precision" measures the signal to noise ratio with a ratio greater than 4 as desirable. This ratio of 21.680 indicates an adequate signal and the fitting of data to the model. This model can be used to navigate the design space with the parametric ranges chosen. Further, from ANOVA, a high significance of the regression model was suggested because of the low probability P value (<0.0001) with the F value for the model at 71.36 implying that this model is significant for adsorption capacity design. There is only a 0.01% chance that an F-value this large could occur due to noise.

The optimum is found to be maximum adsorption capacity of 31.432 mg/g at temperature 34.3°C, adsorbent dosage 1.576g/50mL and contact time 101.266 min with 99.15 desirability. This is comparable to results reported by Sun et al (12).

Table 3 ANOVA

PRESS	34.53
R-Squared	0.9847
Adj R-Squared	0.9709
Pred R-Squared	0,8586
Adeq Precision	21.68

IOP Conf. Series: Materials Science and Engineering **206** (2017) 012065 doi:10.1088/1757-899X/206/1/012065

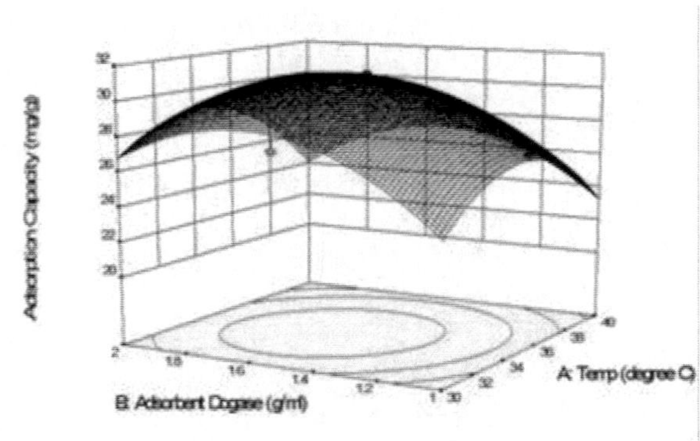

Figure 10. RSM for adsorbent dosage vs. temperature

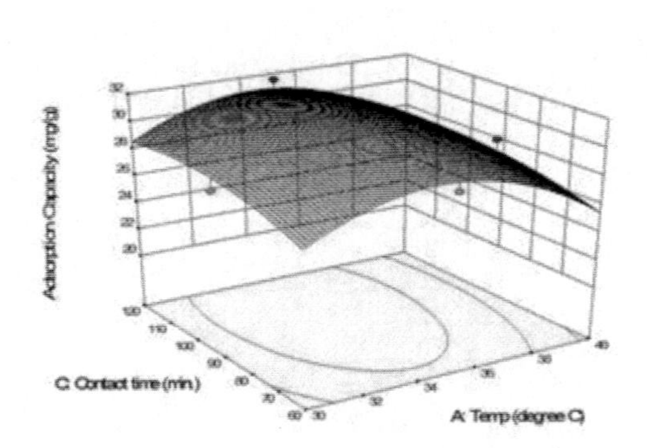

Figure 11. RSM for contact time vs. temperature

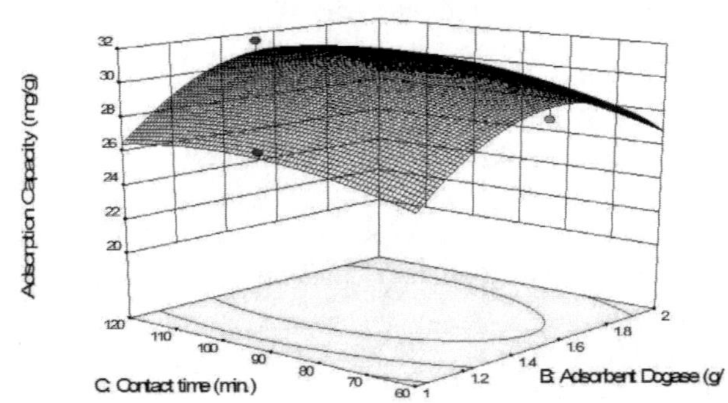

Figure 12. RSM for contact time vs. adsorbent dosage.

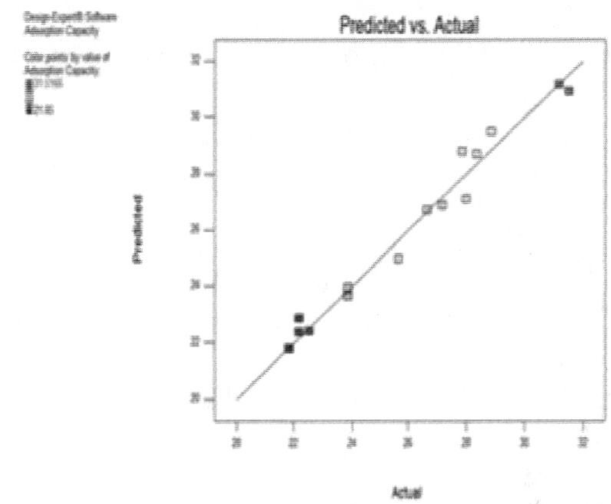

Figure 13. Predicted vs. actual adsorption capacity

.

The regression equation for the system under consideration was evaluated as

Adsorption capacity $Y = -175.25 + 9.19471X_1 + 48.612X_2 + 0.214X_3 - 0.198X_1X_2 - 2.68E{-}04X_2X_3 + 8.264E{-}03X_3X_1 + 0.129X_{1\,2} - 13.532X_{2\,2} - 1.074E{-}03X_{3\,2}$

The analysis of the regression model suggested a quadratic relation with a clear optimum. The interaction between the parameters was negligible as observed with a very low coefficient with an insignificant contribution to the outcome of the relation but the interaction between temperature and dosage showed a negative effect. Temperature was assessed to be one prominent parameter.

4. Conclusions

The effectiveness of using microwave assisted *Sterculia foetida* activated carbon for the removal of Cr^{6+} ions from aqueous solutions was studied. The experimental optimization of batch adsorption parameters revealed maximum adsorption at pH 3, adsorbent dosage 1.5 g/50mL, temperature 35°C for a contact time of 90 minutes and initial concentration at 300 ppm. Langmuir and Freundlich equations were used to describe the adsorption of Cr^{6+} onto prepared activated carbon. Langmuir has better correlation coefficient than Freundlich model in the concentration studied at different temperatures. The optimized process conditions for the removal of Cr (VI) ions under adsorption onto MACSF adsorbent is predicted using RSM as an effective tool for optimizing process variables.

Acknowledgments
Authors acknowledge the support and infrastructure facilities extended by VIT University to carry out the research.

References

[1] Van der Geer J, Hanraads J A J and Lupton R A 2000 *J Scientific Comm.* **163** 51.
[2] Strunk Jr W and White E B 1979 *The Elements of Style*, Third ed., Macmillan, New York.
[3] Mettam G R and Adams L B 1999 In: B.S. Jones, R.Z. Smith (Eds.), *Introduction to the Electronic Age*, E-Publishing Inc., New York. 281.
[4] Malik R, Ramteke D S and Wate S R 2007 *Waste Manage.* **27** 1129.
[5] Couto Jr. O M, Matos I, Fonseca I M, Arroyo P A, Silva E A and Barros M A S D 2015 *Canadian J. Chem. Eng.* **93** 68.
[6] Thamilarasu P and Karunakaran K 2013 *Canadian J Chem. Eng.* **91** 9.
[7] Lamine S M, Ridha C, Mahfoud H M, Mouad C, Lotfi B and Al-Dujaili A H 2014 *Energ. Proc.* **50** 393.
[8] Saucier C, Adebayo M A, Lima E C, Cataluña R, Thue P S, Prola L D T, Puchana-Rosero M J, Machado F M, Pavan F A and Dotto G L 2015 *J. Hazard. Mat.* **28** 18.
[9] Liao P, Ismael Z M, Zhang W, Yuan S, Tong M, Wang K and Bao J 2012 *Chem. Eng. J.* **195**, 339.
[10] Foo K Y and Hameed B H 2012 *Chem. Engg. J.* **180** 66.
[11] Yao S, Zhang J, Shen D, Xiao R, Gu S, Zhao M and Liang J 2016 *J. Colloid Interf. Sci.* **463** 118.
[12] Sun Y, Yue Q, Mao Y, Gao B, Gao Y and Huang L 2014 *J. Hazard. Mat.* **265** 191.
[13] Donmez G C, Aksu Z, Ozturk A and Kutsal T 1999 *Process Biochem.* **34** 885.
[14] Demirai H, Demirai I, Tümsek F and Karabacakoğlu B 2008 *Chem. Eng. J.* **144**, 188.

IOP Conf. Series: Materials Science and Engineering **206** (2017) 012066 doi:10.1088/1757-899X/206/1/012066

Adsorption of diastase over natural halloysite nanotubes (HNTs)

F Twaiq*, K X Chang, J Y W Ling

Faculty of Engineering, Computer and Science, Swinburne University of Technology Sarawak campus, 93350, Kuching, Sarawak, Malaysia Sarawak

E-mail: ftwaiq@swinburne.edu.my

Abstract. Adsorption of diastase over natural halloysite nanotubes is studied in order to evaluate the adsorption capacity of diastase. The halloysite surface characteristics were assessed using nitrogen adsorption, x-ray diffraction (XRD), thermal gravimetric analysis (TGA) and Fourier transformed infrared (FTIR). The surface area of the natural halloysite is found to be 51 $m^2 \cdot g^{-1}$, with total pore volume of 0.106 $cm^3 \cdot g^{-1}$. The natural halloysite has a basal spacing (*d*001) of 10 Å confirming the structure of the natural halloysite material. TGA results indicated that halloysite loses its interlayer water in the range of 30 to 105 °C and the dehydration in the structural layer above 150 °C. The dehydroxylation of halloysite has occurred at approximately 460 °C. The FTIR result of the thermally treated halloysite sample indicated that the bands observed are assigned to Si-O and Al-O bonds. The effects of solution pH and temperature were studied on the adsorption capacity and percent removal of diastase from the solution. The adsorption kinetic found to fit well with both the Pseudo first-order and Pseudo second-order models, and the values of the kinetic constant were found to be 0.173 min^{-1} and 0.00018 $g \cdot mg^{-1} \cdot min^{-1}$ respectively. The Langmuir isotherm model is found to fit well to the adsorption data and a kinetic value is found to be 0.00059 $m^3 \cdot g^{-1}$. The maximum adsorption capacity was found to be 370 $mg \cdot g^{-1}$, indicating the potential for applications of the natural nanostructured halloysite material as an effective adsorbent for diastase.

1. Introduction

1.1. Enzyme adsorption

Enzymes are able to work as a particularly specialized molecular machine to break down, ensemble or convert molecules [1]. Starch-converting enzymes α- and β-amylase produced by plants, animals and microorganisms belong to the group designation of diastase. Diastase is greatly used as a biocatalyst in the production of bioethanol from various sources such as malt and fungal. It is used to promote the hydrolysis of starch using yeast to a fermentable dextrose as shown in the chemical equation (1) [2],

$$(C_6H_{10}O_5)_n \text{ (Starch)} + nH_2O \text{ (Water)} \rightarrow nC_6H_{12}O_6 \text{ (Dextrose)} \qquad (1)$$

Normally, liquefaction and saccharification are the two separate steps involved in the hydrolysis of starch. Soluble oligosaccharides are produced from the hydrolysing of starch in the liquefaction step. Then, glucose is produced from the hydrolysis of soluble oligosaccharides in the following saccharification step [3]. However, it is difficult to recover the α-amylase from the reaction media once it is used. Thus, it is highly recommended to use immobilized α-amylase instead of a free α-amylase [4]. There are numerous techniques for enzyme immobilization which can be separated into two types: physical techniques and chemical techniques. Physical techniques include adsorption and entrapment and encapsulation whereas chemical technique includes cross-linking [5]. Advancements in technology have allowed the enzyme immobilization onto a support material by physical methods to increase the stability of the enzyme and allow long-term operation in the industries. As the "binding mechanism" of enzyme is generally reversible, its recovery by using adsorption technique is simple and non-invasive. Therefore, physical adsorption is the recommended immobilization technique for numerous industrial applications because no chemical additives are necessary and the carrier can be generated for repetitive uses [6].

Inorganic supports materials exhibit benefits of high stability, resistibility against organic solvents and microbial attacks, and ease of disposal or reusability in contrast with organic supports materials [7]. Hydrotalcites and zeolite are the classic inorganic materials that have been studied in enzyme immobilization. However, these materials possess drawback of restriction of pore diameter and surface area, thus only the loading of certain specific enzymes is allowable. For instance, zeolite have been recommended as prospective supports for enzyme immobilization. Nonetheless, zeolite with pore diameter ranging from 0.5 and 1.1 nm, it proves that the access of large molecules to the microporous surface can be restricted by diffusion [8]. Due to the large pore diameters of HNTs, it appeared that HNTs are more attractive for enzyme immobilizations as compared to zeolites [7]. The amount of urease adsorbed onto the HNTs was determined to be 11.7 $mg{\cdot}g^{-1}$ whereas for α-amylase was 15.4 $mg{\cdot}g^{-1}$. The amount of enzymes adsorbed onto the HNTs is reported to be higher as compared to molecular sieve MCM-41 by applying the similar physical adsorption technique for enzyme immobilization. HNTs can offer sufficient space for large amounts of enzymes as compared to MCM-41 as it processes a larger pore size of than the diameter of α-amylase and urease.

Halloysite is a double-layered aluminosilicate clay mineral composing of one alumina octahedron sheet and one silica tetrahedron sheet in 1:1 stoichiometric ratio. It has similar chemical properties as kaolin, diverging primarily in the morphology of crystals [7]. Moreover, it has a structural formula of $Al_2(OH)_4Si_2O_5{\bullet}nH_2O$, where a monolayer of water molecules splits up the unit layers in HNT, which owns a one-dimensional tubular porous structure of the mesoporous range (2-50 nm). Such property allows the multipurpose potential uses, including as a nano-sized support for the loading of functional guests [9]. In terms of length, HNTs differ from the submicron scale to various microns and some even greater than 30 μm. Its external diameter varies from about 30 to 190 nm, whereas its internal diameter varies from about 10 to 100 nm [10]. HNT is a natural occurring low cost material and available in abundance. Most significantly, when the pH is less than 8.5, positive charged is retained in the inner surface of HNTs which foster the loading of HNTs with a negatively charged macromolecules [11]. On the other hand, it is reported that a maximum adsorption capacity of 84.32 $mg{\cdot}g^{-1}$ of methylene blue (MB) was accomplished in the adsorption of methylene blue onto HNTs which is much higher as compared to other clay minerals [12]. It has been proven in this study that HNTs is an effective nano-support in the aid in removing cationic dye through physical adsorption from aqueous solution. There was no leaching out of the enzyme α-amylase and urease occurred during the reaction and no free enzymes desorbed from the HNTs. Therefore, HNTs can be considered as nanoreactors as it offers as a stable support matrix for the enzyme immobilization.

A research conducted on the adsorption of diastase onto acid-treated bentonite between the concentrations of 0.05 to 0.5 $g{\cdot}mL^{-1}$. It was found out that the increased in the initial diastase concentration increased the diastase adsorption which eventually saturated the initial α-amylase concentration beyond 0.5 $g{\cdot}mL^{-1}$. It was claimed that a huge amount of diastase molecule moved towards the interface with increasing diastase concentration [13,14]. At the present time, the immobilization of diastase onto the natural halloysite nanotubes (HNTs) through physical absorption is not well reported. The immobilized diastase hinges on the features and efficiency of the HNTs and the diastase stability in immobilized state.

During a study on the immobilization of enzyme biocatalyst on HNTs conducted by Zhai et al. (2010) [7], it was found out that a tetrahedral layer of Si and O is condensed with an octahedral layer of A1 and OH. As the result of the surface potential of SiO_2 with a slight influence from the positive Al_2O_3 internal surface as the chemical properties of the internal cylinder core might be related with Al_2O_3, the zeta potential behaviour of HNTs particles is generally negative at pH 6 to 7 [15]. Due to the isoelectric point of urease is at pH 6 and α-amylase is at pH 4.6, α-amylase and urease retained negatively charged during the process of immobilization at pH 6. Hence, negatively charged α-amylase and urease were immobilized on the positively charged inner lumen of HNTs at pH 6 via ionic adsorption.

The effect of temperature on the adsorption of diastase onto acid-treated bentonite clay surfaces was reported [13]. The adsorption test was carried out by varying the temperature ranging from 5 to 45 °C. The results showed that as the temperature increased, the diastase adsorption increased. It was claimed that the motion of the diastase molecules increased with increasing temperature as adsorption is usually a process controlled by diffusion. It was pointed out that there are several other forces other than electrostatic forces granting towards the adsorption. It was suggested that hydrophobic forces might be the force taking affecting the process.

1.2. Adsorption kinetics

Kinetics and mechanism of adsorption onto the surface of the adsorbent are usually determined by numerous models. The rate of adsorption (dq_t/dt) is assumed to be proportional to the change amid the amount of the adsorption capacity of adsorbent and adsorption at time (t) which is ($q_e - q_t$) in an adsorption process and k_n is the proportionality constant. Whereby q_e and q_t are the adsorption capacity at equilibrium and at time t, respectively. Rate law is of arbitrary order n and q follows in (2),

$$(dq_t)/d_t = k_n(q_e - q_t)^n \tag{2}$$

The rate of occupation of sorption sites is proportional to the number of unoccupied sites in pseudo first-order power model. The Lagergren pseudo first-order is normally expressed as in (3),

$$(dq_t)/dt = k_1(q_e - q_t) \tag{3}$$

k_1 is the rate constant of pseudo first-order kinetic model in (min^{-1}). The integrated form of (3), after applying boundary conditions t = 0 to t = t and $q_t = 0$ to $q_t = q_t$, is presented in (4),

$$ln(q_e - q_t) = ln(q_e) - k_1t \tag{4}$$

A plot of $ln(q_e - q_t)$ against t contributes a linear relationship whereby k_1 and q_e can be identified from the slope and y-intercept of the plot respectively [16]. The non-linear form of (4) is shown in (5),

$$q_t = q_e[1 - e^{(-k1t)}] \tag{5}$$

The appropriateness of the pseudo first-order power equation to experimental values usually varies in two ways; the number of available sites is not represented by the parameter $k_1(q_e - q_t)$. The parameter $ln(q_e)$ is a variable parameter and frequently determined to be unequal to the y-intercept of the plot $ln(q_e - q_t)$ against t, while in the exact first order, $ln(q_e)$ must be equal to the y-intercept. In pseudo second-order power model that the rate-limiting stage can be chemical sorption relating to valence forces via sharing or exchange electrons amid adsorbent and adsorbate. The pseudo second-order adsorption kinetic rate equation (6) is expressed as follow,

$$dq_t/dt = k_2(q_e - q_t)^2 \tag{6}$$

whereby k_2 is the pseudo second-order kinetic rate constant (g-adsorbent·mg-adsorbate^{-1}·min^{-1}). The integrated form of equation (6), after applying the boundary conditions t = 0 to t = t and $q_t = 0$ to $q_t = q_t$, is presented in (7) or in linear form equation (8),

$$1/(q_e - q_t) = 1/q_e - k_2t \tag{7}$$

$$(t/q_t) = 1/h + 1/q_e t \quad (8)$$

where $h = k_2 q_e^2$ is the initial adsorbate rate (mg-adsorbate/g-adsorbent·min). A plot of (t/q_t) against t of (8), gives a linear relationship whereby q_e and k_2 can be identifed from the slope and y-intercept of the plot respectively [16].

Adsorption isotherm is commonly used to investigate the mechanisms of adsorption and the surface properties as well as the affinities of the absorbent [17]. Adsorption isotherm is known as the correlation between the quantities of a substance being adsorbed at constant temperature and its concentration in the equilibrium solution [18]. The reversible adsorption of an enzyme onto a solid surface is described by Langmuir theory, which can be applied to the adsorption from solution of diastase, D onto a surface site, S of HNTs [19]. To describe the adsorption mechanism by Langmuir model, the following assumptions are made; the adsorption is occurring only in physisorption mechanism; the adsorbate acts as an ideal solute in liquid phase; the adsorbate molecules stay at specific adsorption sites; the adsorbent surface is homogeneously dispersed with adsorption sites; each adsorption site has the same energy of interaction and, there are no interactions between the adsorbed molecules [20]. The adsorption and desorption kinetics are shown in (9) and (10).

$$\text{Rate of adsorption, } r_{ad} = k_{ad}\, C_D C_v \quad (9)$$

$$\text{Rate of desorption, } r_d = k_d\, C_{(D\bullet S)} \quad (10)$$

where C_D is the molar concentration of diastase, C_v is the molar concentration of free vacant sites (moles of S per g-HNTs), $C_{(D\bullet S)}$ is the concentration of diastase-halloysite complex (moles/g-HNTs). k_{ad} is the forward adsorption rate constant (L·mole^{-1}·min^{-1}), k_d is the desorption rate constant (min^{-1}). At equilibrium, whereas the rate of adsorption is equivalent to the rate of desorption (r_{ad}=r_d) and rearranging gives (11),

$$C_{(D\bullet S)}/(C_{(Deq)} C_v) = k_{ad}/k_d = K_{eq} \quad (11)$$

where K_{eq} is the adsorption equilibrium constant (volume of solution per mole of S). The total concentration of all sites, C_t=C_v+$C_{(D\bullet S)}$ is combined with (11) result in (12),

$$C_t = C_{(D\bullet S)}/(C_{(Deq)} K_{eq}) + C_{(D\bullet S)} = (1+K_{eq}C_{(Deq)})/(K_{eq}C_{Deq})(C_{(D\bullet S)}) \quad (12)$$

Relating fraction of surface sites, θ=$C_{(D\bullet S)}/C_t$ gives Langmuir adsorption isotherm as shown in (13)

$$\theta = (K_{eq}\, C_{(Deq)})/(1+K_{eq}\, C_{(Deq)}) \quad (13)$$

When θ=q_e/q_{max}, the alternative form of (13) is given in (14),

$$q_e = (q_{max} \cdot K_L[D])/(1+K_L[D]) \quad (14)$$

where q_{max} and K_L are the maximum adsorption (mg·g^{-1}) and the Langmuir isotherm constant (mg^{-1}) respectively. q_e is the number of binding sites that are occupied by the adsorbate at the concentration [D] [19].

To identify the best fitting isotherm, linear regression is widely used. The correlation coefficients are used to judge and compare the applicability of isotherm equations [21]. A swift increase in the amount of adsorbed adsorbate as the adsorbate concentrations increase at low concentrations can be predicted by Langmuir isotherm model. The amount of adsorbed adsorbate attains a maximum level which is equivalent to the number adsorption sites, q_{max} once the adsorption sites are fully filled. It is pointed out that the maximum adsorption capacity ought to concur with saturation of the monolayer coverage. As a result, it does not depend on temperature [22]. Langmuir model delivers details on uptake qualifications and is qualified of reflecting the common process behaviour of equilibrium adsorption even though it sheds no light on the mechanistic features of adsorption [23]. It is assumed by Langmuir model that chemical unsaturated atoms (sum of binding sites) exert forces that do not prolong more than the diameter of one adsorbed molecule. Consequently, adsorption is limited to a

IOP Conf. Series: Materials Science and Engineering **206** (2017) 012066 doi:10.1088/1757-899X/206/1/012066

monolayer. Commonly, equilibrium behaviour is not described precisely by Langmuir isotherm, particularly with heterogeneous adsorption systems where adsorption sustained further than a monolayer [23].

2. Experimental

2.1. Characterization

The natural halloysite nanotubes (HNTs, sigma-aldrich) were characterized using nitrogen adsorption-desorption using Quantachrome Autosorb Automated Gas Sorption System, thermal gravimetric analysis (TGA) using Mettler Toledo 851, Stare SW 10, Fourier transformed infrared (FTIR) and X-ray diffraction (XRD) using Bruker AXS D8 advance X-Ray Diffreactometer. The purpose of the characterization is to determine the surface area, pore sizes, crystal structures and thermal stability.

The nitrogen adsorption-desorption was employed for pore structure analysis and pore size distribution using BJH method. This technique determines over 0 – 1 range of relative pressure at a temperature of –196 °C. BET method is used for the measurement of the specific surface area of material. TGA analysis was performed by heating about 0.1 g of the halloysite sample in an alumina crucible with a heating rate of 10°C·min^{-1}, under high purity N_2 flow (50 mL·min^{-1}). FTIR spectra was recorded in the wavelength range of 4000 – 400 cm^{-1}. The XRD patterns of the halloysite sample were recorded at 2θ in the range of 2 – 90°.

2.2 Adsorption test

The dynamic batch process is applied in the current study for immobilizing diastase onto HNTs. Three parameters were tested in adsorption test which include the effect of contact time, solution temperature and pH. Standard solutions of 0.5, 0.25, 0.125 and 0.0625 g·L^{-1} were prepared. The standard solutions were scanned at a wavelength of 280 nm to determine the UV-Vis spectrophotometer absorbance. The absorbance values for each standard solutions were read at the determined wavelength and the data were recorded. The concentration of diastase in solution was identified by measuring the absorbance at a single wavelength of 280 nm. The concentration of an absorbing diastase solution is linearly interrelated to the absorbance.

Two grams of HNTs were placed into an oven at 105 °C for drying overnight. A 500 mL of diastase solution was adsorbed using 0.5 g of HNTs. Samples were taken at pre-set time intervals of 5 min (for the first 20 min) and 10 min time intervals (for 30 to 120 min). The samples were filtered instantly after collection and the absorbance of the samples were measured in triplicate using UV-Vis spectrophotometer at 280 nm at once. Each adsorption experiment was repeated for 3 times and the average result was obtained.

For effect of temperature study, 50 mL of 1 g·L^{-1} of the diastase solution in distillate water (pH 7) was used with 0.05 g of HNTs. The mixtures were stirred slowly at temperatures of 25, 35 and 45 °C respectively, for 2 h until equilibrium is achieved. The solutions were filtered and the absorbance of the supernatant for remaining diastase was measured in triplicate by a UV-Vis spectrophotometer. The experiment was repeated for 3 times in order to obtain the average result. Similar experiments were conducted at 25 °C by varying the pH of a distillate water in the range of (3 – 8) and diastase concentration of 1 g·L^{-1}. The pH was adjusted by adding a few drops of 0.1 M NaOH or 0.1 M HCl before each run.

The amount of diastase adsorbed onto the HNTs at any time, q_t (mg·g^{-1}) was calculated by using mass balance relationship as shown in (15),

$$q_t = (C_0 - C_t)(V/W) \tag{15}$$

where C_0 and C_t are the initial and concentration of diastase solution at time t respectively (mg·L^{-1}), V is the volume of the solution (L) and W is the dry weight (g) of the HNT. $\ln(q_e - q_t)$ against time (t) was plotted for pseudo first-order power model. Likewise, (t/q_t) against time (t) was plotted for pseudo second-order power model. The kinetic parameters, k_1 and k_2 and regression coefficients were determined from the plots.

3. Results and discussions

3.1. Characterization

The textural characterization of untreated halloysite was carried out using nitrogen adsorption-desorption at –196 °C. According to the International Union of Pure and Applied Chemistry (IUPAC), type IV classification with H3 hysteresis loop is identified. The hysteresis involves filling and emptying of the mesopores by capillary condensation. Such behaviour does not exhibit any limiting adsorption at high P/P_0 because it results from the existence of non-rigid aggregates of plate-like particles or assemblages of slit-shaped pores [24]. This explains that the natural halloysite exhibits the characteristic of conventional micro- and mesoporous materials such as MCM- 41 as shown in figure 1. Using the BET method, the surface area of the natural halloysite was calculated to be 51 $m^2 \cdot g^{-1}$. Meanwhile, the total pore volume of the natural halloysite was found to be 0.11 $cm^3 \cdot g^{-1}$ using BJH method. This reviewed that the untreated halloysite possesses a relatively high specific surface area and total pore volume.

TGA was used to determine the mass loss of the natural halloysite as a function of temperature at a constant heating rate. To evaluate the thermal stability of the natural halloysite compound, the TGA analysis is presented in figure 2 in the temperature range of 30 to 700 °C. The first peak is detected in the temperature ranging from 30 to 105 °C accompanied with mass loss of 2.1%. This is due losing adsorbed water on the external surfaces of the halloysite. The second and third peaks are observed at approximately 150°C and 260 °C respectively. Such weight losses are attributed to the thermal dehydration of halloysite in the structural layer. Small deviation theoretical values were observed at the temperatures ranging from 300 to 400 °C because of the presence of intercalated water, eliminated slowly from the interlayer spaces of halloysite. Dehydroxylation is where the hydroxyl group in the structural layer of halloysite is being removed when continuous heat above 480 °C. This contributes to 12.4 % of the total mass of the sample.

Fourier transformed infrared (FTIR) spectra provides an evidence of the changes of the structure due to thermal effects. The FTIR was characterized for halloysite at 400 °C. Figure 3 reports the FTIR spectrum of the halloysite which has shown three bands in the region of 3700 – 3600 cm^{-1}. Through the FTIR spectrum, the bands are located at 3695 and 3626 cm^{-1}. This region is assigned to the stretching vibrations of hydroxyl groups. It can be observed that there are no significant changes in the functional group thermal treatment of halloysite at 400 °C. This is because the band at 3695 and 3626 cm^{-1} are attributed to the inner surface hydroxyl groups. The deformation of H–O–H is observed in the region of 1700 – 1500 cm^{-1}. The interlayer water is indicated by the vibration band at 1653 cm^{-1} in the natural halloysite. The band observed 1036 cm^{-1} is resulted by the stretching vibrations of Si–O–Si. The intensity of the vibration of O–H deformation of the inner Al-OH groups are found at 913 cm^{-1} of the thermally treated halloysite. The bands at 753 cm^{-1} exhibits the typical OH translational vibrations.

XRD was used to characterize the natural halloysite as shown in figure 4. The XRD patterns of the HNTs show the diffraction peaks which are indexed to the characteristics of the HNTs. The natural halloysite has a basal spacing (*d*001) of 10 Å confirming the structure of the natural halloysite material.

3.2. Adsorption test

The results and discussions of the kinetic by pseudo first-order and pseudo second-order models are presented in this section. Diastase uptake rate controls with its residence time uptake at solid and solid interface is described by this sorption kinetics study. The contact time between adsorbate and adsorbent is extremely significant as it hinges on the nature of the system used. Figure 5 shows the variation of adsorption intake of diastase by HNTs with contact time. It is observed that the diastase adsorption intake was rapid in the initial 20 min, thereafter proceeded at a negative adsorption rate and finally achieved equilibrium. The abundant number of vacant adsorption sites available on the surface of the HNTs at the initial stage is the main cause of higher rates at the beginning. The high driving force for the relocation of diastase molecules from solution to the HNTs' surface which is resulted from the high concentration gradient in the beginning of adsorption process. Nonetheless, as time proceeds, it is noticed that the concentration gradient decreases because of the increases in the occupation of vacant sites by the diastase molecules in the solution. At 20 min, the adsorption process reached saturation where the vacant site on the HNTs were fully occupied with maximum adsorption capacity of 169 mg-diastase per g-HNT. Meanwhile, the forward adsorption rate is zero at this stage.

From 20 to 40 min, the diastase adsorption capacity decreases gradually to 136 mg-diastase per g-HNT and with further small decreases from 40 to 60 min. This implies that desorption commences in the solution where the diastase molecules release from the surface of HNTs in order to achieve the state of sorption equilibrium. Concurrently, the forward adsorption rate is negative at this stage as the adsorption became less efficient due to the active binding sites were being used up with time, thus limited free sites for diastase molecules to attach to. At about 60 min onwards, the adsorption process attained equilibrium where the adsorption capacity remains approximately constant.

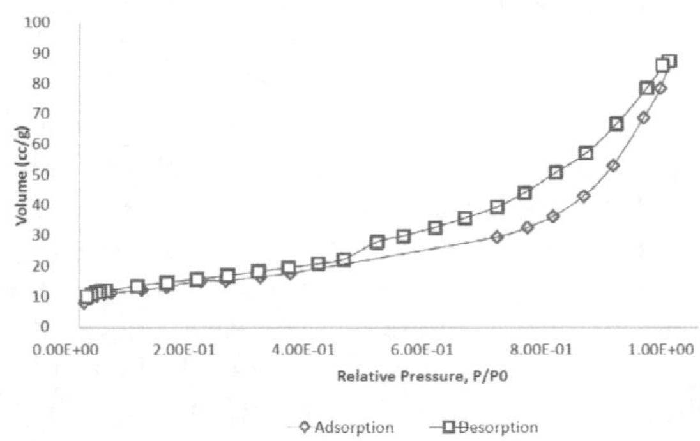

Figure 1. Nitrogen adsorption-desorption isotherms of NHNs.

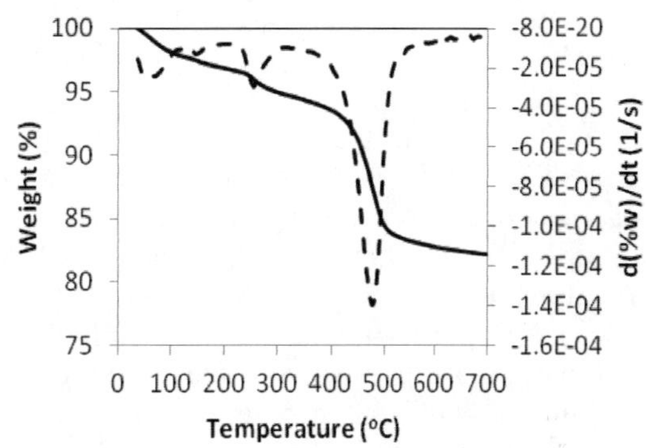

Figure 2. TGA analysis of NHNs.

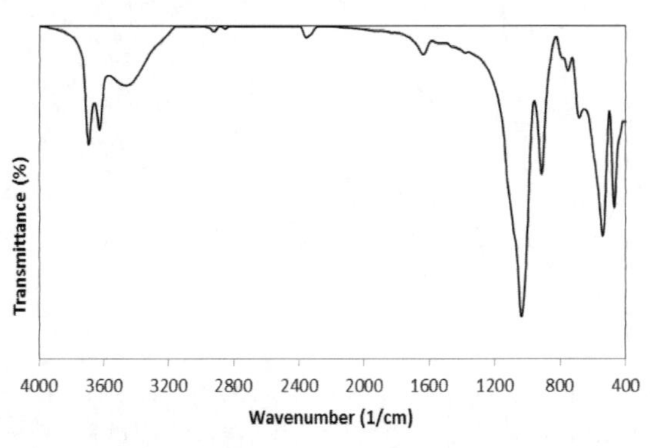

Figure 3. FTIR spectra of thermal-treated halloysite at 400°C.

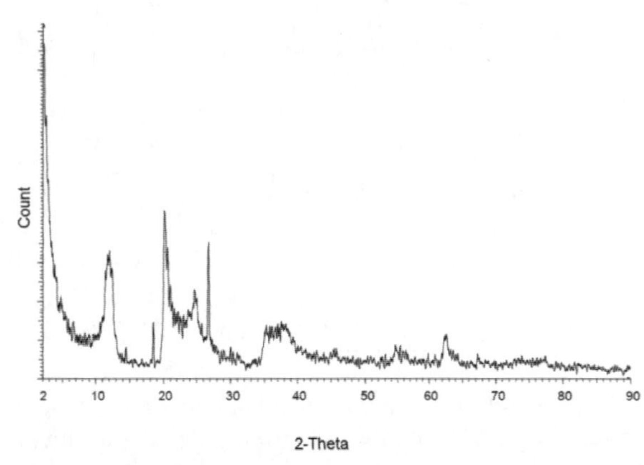

Figure 4. XRD scans of NHNs.

The rate of adsorption of diastase onto the HNTs is said to be equal to the rate of desorption off the HNTs surface. In other words, the amount of diastase being adsorbed onto the HNTs was equivalent to the amount of diastase being desorbed from the HNTs. Consequently, it is observed from the contact time curve that the state of sorption equilibrium is reached at 60 min. From this outcome, in order to make sure the state of sorption equilibrium is accomplished, the agitation time was fixed at 120 min of time for the rest of the batch experiments. Potential monolayer coverage of the diastase on the HNTs surface is denoted as the contact time curve attained is smooth and continuous, leading to saturation. The rapid diastase intake is particularly significant in process design and operation in practical usages.

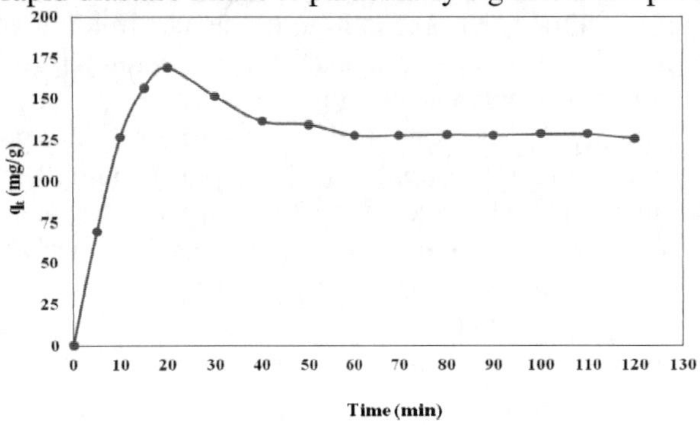

Figure 5. Effect of contact time on the adsorption of diastase by HNTs (diastase concentration: 1 g·L^{-1}; T: 25 °C; pH 7).

3.3. Kinetic study

In order to examine the controlling mechanism of the adsorption process, pseudo first-order and pseudo second-order power models were used to test the kinetic data. The best fit values of $(q_e)_{cal}$ and k_1 of pseudo first-order equation (4) and h, $(q_e)_{cal}$ and k_2 of the pseudo second-order equation (8) determined from figures 6 and 7 and are presented in table 1 along with their corresponding correlation coefficients respectively. For the pseudo second-order power model, it is perceived that the $(q_e)_{exp}$ varies significantly from $(q_e)_{cal}$, while the values are much closer to that of the pseudo first-order power model. By comparison, the correlation coefficient is closer to unity for pseudo first-order kinetic than pseudo second-order kinetics. Nevertheless, the kinetic data were said to fit well with both the pseudo first-order and pseudo second-order power models with a correlation coefficient of 0.97 and 0.95 respectively.

Table 1. Adsorption kinetic parameters of diastase onto HNTs.

Kinetics Model	Parameters	Values
Pseudo first-order	$(q_e)_{cal}$ (mg·g^{-1})	200
	k_1 (min^{-1})	0.1732
	$(q_e)_{exp}$ (mg·g^{-1})	169
	R^2	0.97
Pseudo second-order	$(q_e)_{cal}$ (mg·g^{-1})	323
	k_2 (g·mg^{-1}·min^{-1})	0.00018
	h (mg·g^{-1}·min^{-1})	19
	$(q_e)_{exp}$ (mg·g^{-1})	169
	R^2	0.95

The mechanism of the diastase adsorption process in which a single site on the surface of HNT is liable for acquiring a single molecule of the diastase on its surface is supported by pseudo first order power model [16]. As a result, the HNTs' surface area processes can be expounded as mono-site-occupancy adsorption mechanism. In addition, the physisorption process by which weaker interactions such as electrostatic forces and van der Waals forces necessitate between the adsorbate and the surface of the adsorbent is supported by pseudo first-order power model. Furthermore, the chemical properties of both adsorbate and the adsorbent do not change and the binding is not strong in physisorption. Hence, the physisorption mechanism is rapid and absolutely reversible through desorption (Bulut et al. 2008). On the other hand, pseudo second-order power model supports the mechanism that chemisorption dominates the comprehensive rate of diastase adsorption process by which two-site occupancy adsorption mechanism is followed. Stronger particular forces encompassing in the development of chemical linkage such as a single ionic and covalent bond are involved in chemisorption. Chemically adsorbed molecules are unable to transfer to the surface within an interface, thus chemisorption mechanism is completely irreversible. Ultimately, the good agreement of both models points out that the adsorption process can be controlled by both physiosorption and chemisorption.

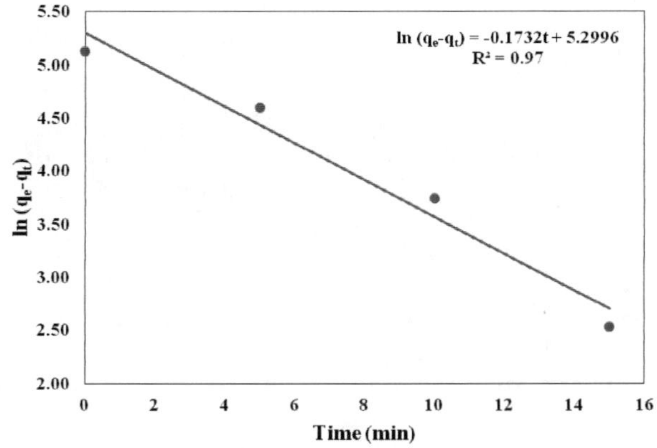

Figure 6. Pseudo first-order plot for the adsorption of diastase by HNTs (diastase concentration: $1 g \cdot L^{-1}$; T: 25°C; pH 7).

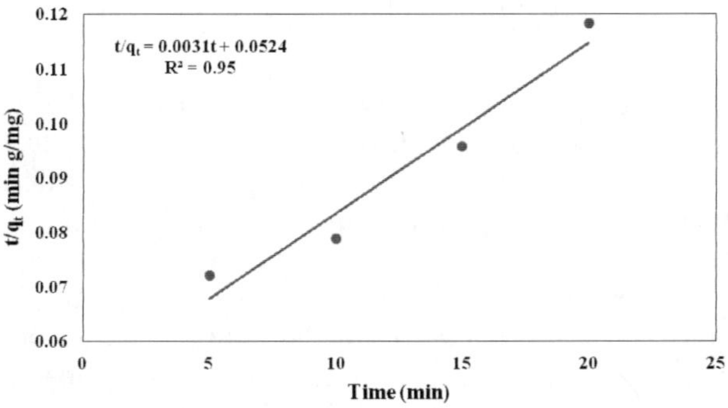

Figure 7. Pseudo second-order plot for the adsorption of diastase by HNTs (diastase concentration: $1 g \cdot L^{-1}$; T: 25 °C; pH 7).

3.4. Langmuir adsorption isotherm

The initial diastase concentration has an effect on the adsorptive intake of diastase by HNTs as shown in figure 8. The adsorption capacity at equilibrium, q_e increased from 126 to 128 mg-diastase per g-HNT with an increase in the initial diastase concentration from 900 to 1200 $mg \cdot L^{-1}$, signifying that the higher the adsorbate to adsorbent ratio, the superior will be the quantity of diastase adsorbed. The high diastase concentration gradient generating a high driving force to transport the diastase molecules from solution to the surface of HNTs. Henceforth, the adsorption process was conducted much more sufficient, resulting in the rise in the diastase adsorption capacity. Langmuir adsorption isotherm was proposed for a quantitative understanding of the interaction between diastase and specific surface of HNTs. The best fit values of q_{max}, K_L of Langmuir adsorption model equation (14) determined by plotting the linear correlation $1/C_e$ against $1/q_e$ and are presented in table 2. The Langmuir adsorption capacity, q_{max} was obtained as 370 $mg \cdot g^{-1}$ from the experimental data. The correlation coefficient shows a strong positive proof that the adsorption of diastase onto HNTs follows the Langmuir adsorption isotherm whereby a monolayer coverage occurs on the surface of HNTs and only physisorption occurs. This proposes that the adsorption happens to the active sites of the adsorbent, confirming the suitability of the Langmuir adsorption model. Hence, it can be presumed that the molecules are absorbed at a fixed number of distinct sites with each site accept only one molecule of

IOP Conf. Series: Materials Science and Engineering **206** (2017) 012066 doi:10.1088/1757-899X/206/1/012066

adsorbate. Additionally, the vacant sites are also expected to be enthusiastically equivalent and distant to one another. Therefore, there are certainly no interactions between adsorbed molecules occur in the adjacent regions [4].

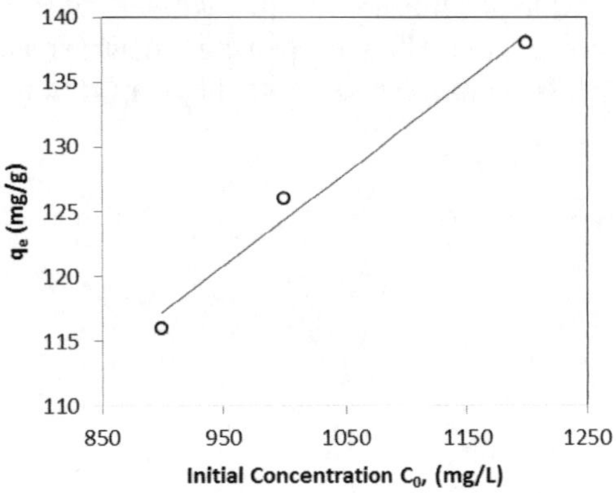

Figure 8. Effect of initial diastase concentration on the adsorption over HNTs (T: 25°C; pH 7).

Table 2. Langmuir adsorption isotherm constants of diastase onto HNTs.

Parameters	Values
q_{max} ($mg \cdot g^{-1}$)	370
K_L ($L \cdot mg^{-1}$)	0.0006
R^2	0.98

3.5. Effect of temperature

The adsorption studies were performed over a temperature range of 25 to 45 °C in order to determine the effect of temperature on the diastase adsorption. As illustrated in figure 9, it is observed that the equilibrium adsorption capacity was clearly affected by temperature, with the amount of diastase adsorbed increasing from 127 to 160 mg-diastase per g-HNT when the temperature was raised from 25 to 35 °C. This can be described by the rise in the rate of diffusion of the adsorbate molecules crosswise the exterior boundary layer as well in the interior pores of the adsorbent molecules which results from the decline in the viscosity of the solution as well as the alteration in the equilibrium adsorption capacity of the adsorbent for a specific adsorbate [25]. It is worth proclaiming that a lower adsorption capacity is expected as the electrostatic attractions are generally weakened at higher temperature and this would be only occurred at temperatures above 35 °C. However, a higher adsorption capacity at 35 °C absolutely designate that other than electrostatic forces, there are numerous other forces such as hydrophobic forces ought to be contributing towards the adsorption [13]. This can be evidenced by the reported data that protein retention in hydrophobic interaction is improved with increasing in temperature [14]. Apart from that, the increase is due to the increasing in van der Waals interactions with increasing temperature [4]. Furthermore, the amount of active sites increases on the HNTs surface with increasing temperature and consequently, the amount of adsorbed diastase increased. It is postulated that the motion of the diastase molecules increased with increasing temperature since adsorption is generally a diffusion controlled process, thus causing in higher adsorption whereby simplifying the development of surface monolayers [13]. The decline in adsorption capacity was observed at temperatures over 35 °C is due to the conformational alterations to the surface of diastase or structural distortion of HNTs [14]. Hydrogen bonds identifies the shape of proteins and they are effortlessly disrupted via alterations of temperature. Below 35 °C, the bonds which define the shape of protein are inflexible enough to allow the shape alterations essential for fitting into a site. Above 35 °C, the bonds are excessively weak to clutch the protein in proper position as well as sustain its shape. Eventually, denaturation occurs as the proper shape of protein is lost and enzyme is demolished.

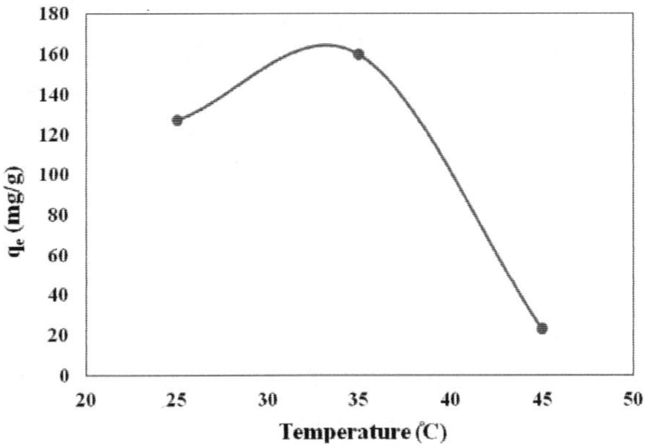

Figure 9. Effect of temperature on the adsorption of diastase by HNTs (diastase concentration: 1g·L^{-1}; pH 7).

3.6. Effect of pH

The two main factors influencing the solution pH dependence of adsorption are the dispersal of adsorbate in the solution phase and the overall charge of the adsorbent. On the other hand, the factor that is responsible for the bonding or enzyme immobilization includes the carrier's surface activity in league with a functional moiety or characteristic group on the enzyme's protein surface. The existence of the bonds such as ionic, hydrogen and covalent hydrophobic amid the enzyme protein and the support hinges on the nature of the surface of enzyme. Enzyme can be immobilized either externally or internally on the support's surface via bonding. The hydrophilic residues in the diastase are usually discovered on the surface of the molecule. The type of interaction suggested are the hydrophobic interactions amid hydrophobic siloxane layer (Si-O-Si) of the surface of clay and enzyme molecule and the electrostatic interactions amid positive charged aluminols (Al-OH) of the internal surface of clay and negatively charged amino acid residues of the enzyme [13]. The effect of solution pH on the adsorption of diastase by HNTs is shown in figure 10. The adsorption capacity of the diastase onto HNTs increased with decrease solution pH. This is due to the electrostatic attraction of negatively charged amino acid residues of diastase with positive charged surface of HNTs at low pH. The amount of hydroxyl groups (OH) which are the key reactive groups that are able to interact with charges is contained in HNTs. The zeta potential of HNTs was identified from 2.2 to 11 and the pH_{zpc} (pH at the zero point of charge) was discovered to be 2.7. Due to Al-OH groups experience protonation, the surface of HNTs is positively charged under pH 2.7 whereas a negatively charged is acquired beyond this pH. The generation of the positive or negative charge surface is due to the attribution. Plentiful of Si-OH and Al-OH are contained on the surface of HNTs that undergo ionization in (16)

$$\mathrm{Si/Al} \quad \mathrm{OH_2}^+ \xleftarrow{\mathrm{H}^+} \mathrm{Si/Al} \quad \mathrm{OH} \xrightarrow{\mathrm{OH}^-} \mathrm{Si/Al} \quad \mathrm{O}^- \qquad (16)$$

The reaction on the right hand side of equation (16) is caused by silica, an acidic oxide to be leading over a great range of pH values whereas an amphoteric behavior is displayed by alumina. The surface is negatively charged over a wide range of pH values as the surface of the HNTs is primarily occupied by silica. Consequently, negatively charged diastase is attached readily onto HNTs in aqueous solution. The positive charged surface of HNTs is saturated by electrons with the increasing of pH > pH_{zpc}. Competition of negatively charged diastase with hydroxide ions for the adsorption sites occurs and thereby decreases the adsorption capacity of diastase. An increase in the adsorption capacity of diastase is caused by the increasing of electrostatic attraction amid adsorption sites and negatively charged diastase with decreasing solution pH.

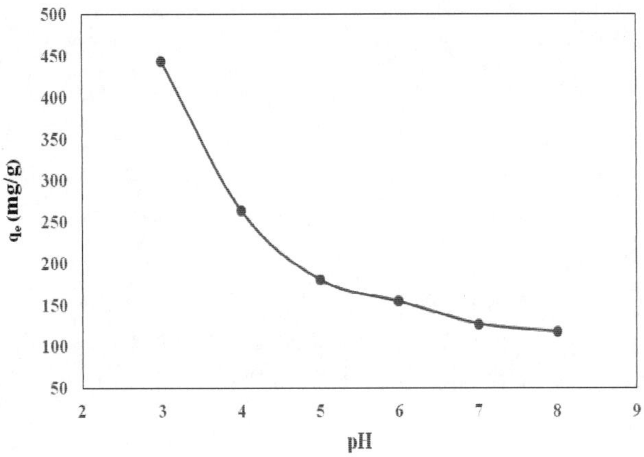

Figure 10. Effect of pH on the adsorption of diastase by HNTs (diastase concentration: 1 g·L^{-1}; T: 25 °C).

4. Conclusions

The adsorption kinetic and mechanism of diastase over natural halloysite nanotubes (HNTs) was studied and found to be reliant on the pH and temperature. The kinetic data were found to fit well with both the pseudo-first order and pseudo-second order model with kinetic constants of 0.1730 min^{-1} and 0.00018 g·mg^{-1}·min^{-1} respectively. An agreement of both model points out that the adsorption process can be controlled by both physiosorption and chemisorption. The fit of the Langmuir adsorption model is achieved, indicating that a monolayer coverage occurs on the surface of HNTs and only physisorption occurs. The maximum adsorption capacity is 0.370g diastase per gram of halloysite, indicating the potential for a possible use of the natural nanostructured material as an effective immobilizing support for diastase. The reversible adsorption of diastase onto HNTs is well described by Langmuir theory.

Acknowledgement

The Research Fund by Swinburne University of Technology Sarawak Campus (SSRG 2-5513) is acknowledged. The authors thank NANOCAT, University Malaya for sample characterization

References

[1]. Talens P, Marín-Navarro J and Polaina J 2016 *Encyclopedia of Food and Health* (Elsevier) pp 532–538
[2]. Gupta R B and Demirbas A 2010 *Gasoline, diesel and ethanol biofuels from grasses and plants* (New York: Cambridge University Press)
[3]. Park D H, Jang S, Ahn K and Kim W S 2005 *Process Biochem* **40** pp 53–61
[4]. Baysal Z, Bulut Y, Yavuz M and Aytekin C 2014 *Starch-Stärke* **66** pp 484–490
[5]. Zhang B, Weng Y, Xu H and Mao Z 2012 *Appl. Microbiol. Biotechnol.* **93** pp 61–70
[6]. Reshmi R, Sanjay G and Sugunan S 2006 *Catal. Commun.* **7** pp 460–465
[7]. Zhai R, Zhang B, Liu LX, Zhang H and Liu 2010 *Catal. Commun.* **12** pp 259–263
[8]. Corma A, Fornes V and Rey F 2002 *Adv. Mater.* **14** pp 71–74
[9]. Yuan P, Tan D and Annabi-Bergaya F 2015 *Appl. Clay Sci.* **112 pp** 75–93
[10]. Joussein E *et al* 2005 *Clay Miner.* **40** pp 383–426
[11]. Yuan P *et al* 2013 *Appl. Clay Sci.* **83** pp 68–76
[12]. Zhao M and Liu P 2010 *Microporous Mesoporous Mater.* **112** pp 3040–3046
[13]. Bajpai K and Sachdeva R 2002 *Colloid. Polym. Sci.* **280** pp 892–899
[14]. Tüzmen N, Kalburcu T and Denizli A 2012 *J. Mol. Catal. B: Enzym.* **78** pp 16–23
[15]. Shchukin D G *et al* 2008 *Weinheim an der Bergstrasse* **1** pp 510–513
[16]. Twaiq F A *et al* 2012 *AIP Conference Series* **1482** pp 579–584
[17]. Bulut E, Özacar M and Şengil I A 2008 *Microporous Mesoporous Mater.* **115** pp 234–246

[18]. Kiani G 2014 *Appl. Clay Sci.* **90** pp 159–164
[19]. Ghabbour E A and Davies G 2011 *Colloids Surf A* **381** pp 37–40
[20]. Epstein J *et al* 2015 *J. chromatogr. A* **1380** pp 81–87
[21]. Vasanth K K and Sivanesan S 2007 *Dyes and Pigments* **72** pp 130–133
[22]. Kresic N 2006 *Hydrogeology and Groundwater Modeling.* (USA: 2nd edn CRC Press)
[23]. Wang L K *et al* 2010 *Environmental bioengineering: handbook of environmental engineering* (New York: Springer Science & Business Media)
[24]. Yuan P *et al* 2008 *J. Phy. Chem. C* **112** pp 15742–15751
[25]. Foo K Y and Hameed B H 2010 *Chem. Eng. J.* **156** pp 2–10

IOP Conf. Series: Materials Science and Engineering **206** (2017) 012067 doi:10.1088/1757-899X/206/1/012067

Humic acid batteries derived from vermicomposts at different C/N ratios

R M Shamsuddin*, **A Borhan, W K Lim**

Chemical Engineering Department, Faculty of Engineering, Universiti Teknologi PETRONAS, 32610 Bandar Seri Iskandar, Perak, Malaysia

E-mail: mrashid.shamsuddin@utp.edu.my

Abstract. Humic acid is a known fertilizer derived from decomposed organic matters. Organic wastes are normally landfilled for disposal which had contributed negatively to the environment. From waste-to-wealth perspective, such wastes are potential precursors for compost fertilizers. When worms are added into a composting process, the process is termed as vermicomposting. In this work, humic acid from vermicompost derived from campus green wastes was developed into a battery. This adds value proposition to compost instead of being traditionally used solely as soil improver. This research work aimed to study the correlation between electrical potential generated by humic acid at different Carbon to Nitrogen (C/N) ratios of vermicompost at 20, 25, 30 and 35. The temperature and pH profiles of composting revealed that the compost was ready after 55 days. The humic acid was extracted from compost via alkaline extraction followed by precipitation in a strong acid. The extracted humic acid together with other additives were packed into a compartment and termed as vermibattery. Another set of battery running only on the additives was also prepared as a control. The net voltage produced by a single vermibattery cell with Zn and PbO electrodes was in the range of 0.31 to 0.44 V with compost at C/N ratio of 30 gave the highest voltage. The battery can be connected in series to increase the voltage generation. Quality assessment on the compost revealed that the final carbon content is between 16 to 23 wt%, nitrogen content of 0.4 to 0.5 wt%, humic acid yield of 0.7 to 1.5 wt% and final compost mass reduction of 10 to 35 wt%. Composting campus green wastes carries multi-fold benefits of reducing labour requirement, generating fertilizer for campus greenery and green battery construction.

1. Introduction

Batteries are a potable form of energy and they are in demand parallel to modern lifestyle. Batteries generate electricity via electron movement from negative terminal (anode) to external positive terminal (cathode) through an electrolyte [1]. A primary battery is a single use battery whereas a secondary battery can be recharged for multiple usages. Spent batteries need a proper disposal as they contain harmful chemicals such as cadmium, lithium, mercury and nickel [2, 3]. Therefore when possible, the use of secondary battery is more preferable to reduce the disposal hazards. Nonetheless, even the simplest form of secondary battery that is acid-lead battery is environmentally hazardous. Thus, there is a need for a greener battery such as the ones derived from bio-based compounds.

Current bio-batteries generate energy from a reaction between simple starches such as glucose with enzymes instead of using harmful chemicals [4]. These batteries have many environmental advantages but lacking reliable power generation due to natural degradation of biological compounds. Bio-batteries can potentially generate about 50 miliwatts of electricity, proving them useful for small devices application [4]. To make bio-batteries more appealing, variety of electrolytes should be developed using natural compounds like humic acid.

Humic acid (HA) is typically used as a fertilizer in agriculture and horticulture due to its rich nutrients content. It is not a single acid but rather a complex combination of many weak acids. Humic acid typically behaves as negatively charged colloid due to dissociation of carboxyl and phenolic groups of these weak acids [5]. Therefore, humic acid could potentially generate a small amount of electrical potential. Major components of natural organic matters in soil and water which includes geological organic deposits such as lake sediments, brown coals and peats are humic substances [6]. Hence, humic acid can be obtained from composting process of organic wastes.

Green wastes such as leaves, branches, grass clipping and vegetable scraps can be effectively and efficiently reduced via composting. It is a biological decomposition process of organic matters into humus through microbial activities [7]. The simplest form of composting is to pile organic wastes and leave them to break down naturally over a period of time [8]. The time required for the green wastes to degrade depends on several factors such as the Carbon (C) and Nitrogen (N) content of the raw materials, temperature, moisture content, particle size, oxygen level and piling techniques [9]. Earthworms are sometimes added to increase the efficiency of composting as well as to enhance the compost quality. Typical earthworms used include red wigglers (*Eisenia fetida*) and European nightcrawlers (*Eisenia hortensis*). Thus, the process is sometimes referred to as vermicomposting.

From vermicompost, humic substances can be separated into three different fractions which are humic acid, fulvic acid and humin [10]. Humic acid can be interpreted as liquid humus which is beneficial to plant growth [11]. It is a mixture of dibasic acids with carboxylic and phenolic groups contributing the most to the surface charge with pK values at pH 4 and pH 9 respectively [5]. At typical compost pH of 7, humic acid exists as an ampholyte with carboxylic groups becoming proton donors and phenolic groups are proton acceptors.

Karim et al. (2011) work revealed that vermicompost can produce electricity of up to 1.5 volts in just a single cell. Furthermore, it was shown that vermicompost contains high nitrate content and other essential growth elements for plants. The scope of the research however was limited to spent tea waste only [12].

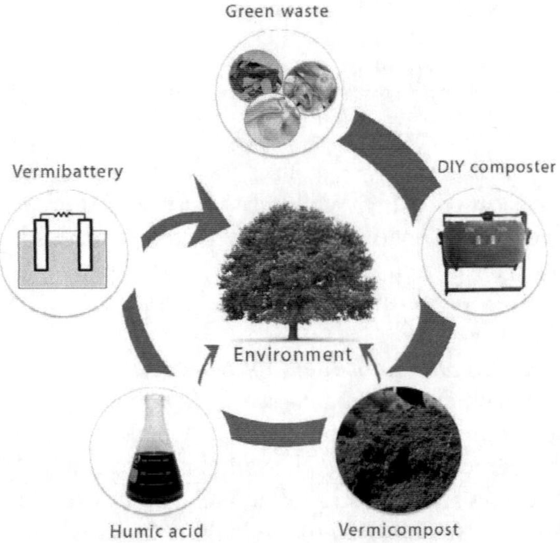

Figure 1. Overview of the project.

IOP Conf. Series: Materials Science and Engineering **206** (2017) 012067 doi:10.1088/1757-899X/206/1/012067

In this work, abundant campus green wastes were utilized as compost precursors. They were added at specific Carbon to Nitrogen (C/N) ratios of 20, 25, 30 and 35 as the compost starters. Red wiggler worms were added to reduce the composting duration as well as to enhance the compost quality. Humic acid was extracted from compost at each formulation for quality and voltage analysis. It is intended for the findings to propose for an alternative to typical landfilling waste management for organics and to develop a mini bio-battery to power model cars for green-themed competition such as the annual Chem-e-car competition [13]. Figure 1 depicts the bigger picture of the project where strong emphasis is given on being green thoroughly.

2. Experimental

2.1. Materials and equipment

Dry leaves and vegetable scraps were used as the carbon and nitrogen source organics. They were collected within the Universiti Teknologi PETRONAS (UTP) campus. The raw materials were mixed at specific weight percentage (wt%) based on their suggested C and N content [14, 15]. The formulations are given in table 1. They were loaded in custom-made rotary composters accordingly. The composters were made from recycle HDPE chemical drums of 250L and 12mm thickness. They were mounted on 4-legged steel stands. Handles were installed for manual turning. Holes were made on the drum wall and on the internal spinal steel for passive aeration. Red wiggler worms were added at about 20 numbers, which is excluded from the total weight of compost because they were removed during compost harvesting. The composters were loaded at approximately 15% of the total capacity.

Table 1. Initial compost formulations.

Batch	C source		N source		Total weight (kg)	C/N ratio
	Type	Weight (kg)	Type	Weight (kg)		
A	Dry leaves	0.170	Vege scraps	0.830	1	20
B		0.270		0.730		25
C		0.375		0.625		30
D		0.480		0.520		35

2.2. Temperature and pH monitoring

Temperature and pH were recorded using WalkLab Microcomputer pH Meter TI9000 from Trans Instruments. The readings were taken once a week at six different places in the composting bins and the readings were averaged. The data were used to plot profile curves over time in order to determine the maturation period.

2.3. Total Organic Carbon (TOC) and Total Kjeldahl Nitrogen (TKN)

Samples were collected from each compost batch at six different locations in the bins and mixed in a beaker to get a good representative of the compost. For TOC, about 100 mg of the samples were treated with 10% HCl to remove inorganic substances. The samples were dried at 105°C for 3 hours in an oven prior to analysis using an Analytik Jena HT 1300. For TKN, 2g of each sample was weighed and added into a mixture of 25 ml of 97% H_2SO_4 and 10g of Kjeldahl catalyst tablets. The mixture was stirred and heated at 370°C for 1 hour using a BUCHI Digest Automat K-438. Once cooled to room temperature, each sample was distillated with 35% NaOH for 4 minutes and titrated with 0.25M H_2SO_4 to pH 4.65 using a BUCHI Auto Kjeldahl K-370 unit.

2.4. Compost harvesting and mass yield calculations

IOP Conf. Series: Materials Science and Engineering **206** (2017) 012067 doi:10.1088/1757-899X/206/1/012067

Matured composts were removed from the composters and sieved to a maximum particle size of 1cm x 1cm to remove odd objects and coarse particles and dried under direct 12 hours sunlight. The final mass of composts was recorded after the drying process and they are kept in tight plastic bags for further analysis. Mass yield of vermicompost and extracted humic acid were calculated in terms of weight percentage (wt%).

2.5. *Humic acid extraction*

2.5.1. Sample preparation. About 100 g of the composts were ground using a mortar and pestle, sieved to a maximum size of 250 μm (equivalent to US standard sieve mesh No. 60) and dried in an oven at 95°C for overnight.

2.5.2. Alkaline extraction. About 50g of the compost was weighed and transferred into a 1L Erlenmeyer flask. NaOH at 0.1M was added into the flask slowly while stirring until the final volume of 1L was reached. The mixture was further stirred at 250 rpm for 6 hours. Next, the mixture was centrifuged at 5000 rpm for 15 minutes, the supernatant was collected and pallet was discarded.

2.5.3. Acid precipitation. The leached solution or the supernatant was titrated with concentrated HCl until pH 1 in an Erlenmeyer flask while stirring. The solution was left mixing to equilibrate for further 1 hour and the pH was monitored. The stirring was stopped and the HA was let to precipitate to the bottom of the flask for overnight. The solid HA particles were recovered via filtration, rinsed using distilled water and dried in an oven at 95oC for overnight. The dried mass of HA was recorded for yield calculation.

2.6. *Vermibattery construction and electricity generation*

Three types of battery were constructed that are a standard battery, vermibattery using vermicompost and vermibattery using humic acid. The standard battery contained only additives while vermibatteries consisted of additives together with either vermicompost or humic acid from each batch of compost. The standard battery was constructed as a control. The full formulations of the batteries are given in table 2. Simple battery cells as depicted in figure 1 (inset) were constructed with zinc and lead oxide plates as anode and cathode respectively. Voltage readings were recorded at time intervals for two consecutive days using a handheld digital multimeter.

Table 2. Vermibattery compositions.

	Additives	Weight (g)
	Standard battery	
1	Activated carbon, s	10
2	1M H_2SO_4, aq	60
	Vermibattery	
	Additives 1 and 2	70
	Vermicompost of each batch, s	50
	Vermibattery	
	Additives 1 and 2	70
	Humic acid of each batch, s (extracted from 50g compost)	0.35

3. Results and discussion

3.1. Physical observation

Moving from batch A to D, the following observations were made throughout the composting period- organics decomposed slower; rotting smell reduced; worms survived longer; composts were less watery; the final mass reduction decreased and mould presence decreased. These are typical trends of natural composting, attributed to the C and N content of the raw materials. In general, N-based organics are the elemental units for protein that promote growth while C-based organics provide energy and act as the building block for organism. Too much of N will promote ammonia gas generation, responsible for bad odour while excess of C will slow down the degradation due to lack of growth element [16]. The worms multiplied faster in batch A until a certain period which they later reduced in numbers due to lack of food. This trend diminishes in the order of batch B, C and D due to higher carbon content that can sustain worm population longer. C-based materials can be viewed as a slow-release energy provider.

3.2. Temperature and pH profiles

From figure 2, the temperature ranges from 27 to 32°C with no clear distinction of three characteristic composting phases- mesophile, thermophile and maturation. The reason is because the composters were located outdoor under a shade, which exposed them to weathering effect. The temperature range agrees to typical local temperature of 25 to 35°C. Even though the readings were taken approximately at the same time of the day, the change in weather pattern have affected the temperature profile greatly. The maturation stage cannot be reliably determined by the temperature profile due to fluctuations. However, it can be approximated at day 55 as reflected by the constant pH readings, in line with findings from many natural composting work [17-19].

Unlike the temperature profile, pH (figure 3) is not affected too much by the weather as it is mainly the result of bacterial and worms' activity in the bins. As part of the natural process, living organisms tend to rectify the pH to a favourable condition of pH 7-8. The low initial pH reflects the natural pH of raw materials. Towards day 10, the pH increased as the result of bacterial growth in aerobic process. The increase in population and consumption of organics had resulted in higher release of organic acids and therefore lowering the pH from day 10 to 30. This trend is reflected by high presence of mould at lower pH and reduced towards the end.

After day 30, there are periods of fluctuation in the pHs until they stabilized to pH 7.5 for all the batches. This occurrence is due to "bacterial growth vs. organic acid accumulation" cycle where high population resulted in more acidic environment; hindering the bacterial growth; lowering the pH and as the population decreased; the pH was ideal again; therefore the bacteria regenerated [20]. The cycles continued until the food sources depleted towards the maturation period at day 55. Batch A can be seen to reach maturation earlier at about day 48 followed by the rest, attributed to its higher N content.
These findings indicated a healthy composting pH where it is suggested that with sufficient air the pH range can naturally be maintained at 5.5 to 8.5. This condition is ideal to promote high bacterial activity and hence faster composting process [9].

IOP Conf. Series: Materials Science and Engineering **206** (2017) 012067 doi:10.1088/1757-899X/206/1/012067

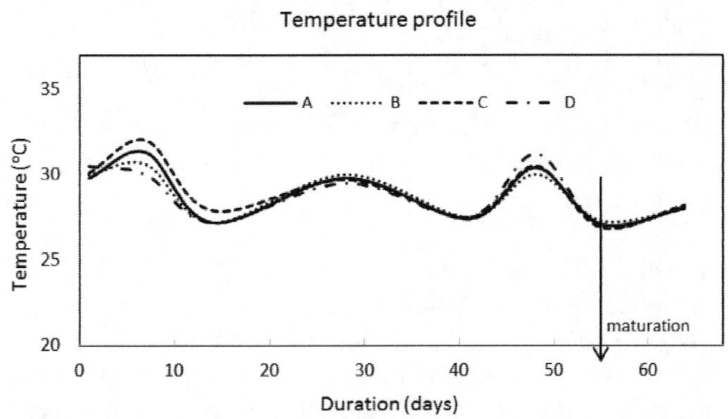

Figure 2. Temperature profiles for batch A to D over time.

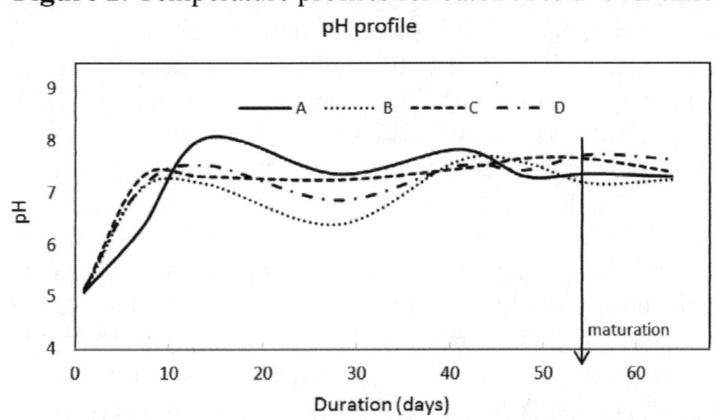

Figure 3. pH profiles for batch A to D over time.

3.3. TOC, TKN, HA mass yield and compost mass reduction

Compost quality was analysed via TOC and TKN as well as mass yield of humic acid. Comparison was made to the readily available vermicompost in the market. The results are tabulated in table 3.

Table 3. TOC, TKN and HA yields of Vermicomposts.

Analysis	Batch				Commercial compost
	A	B	C	D	
Initial C/N	20	25	30	35	-
TOC (wt%)	16.2	13.3	14.3	23.6	25.7
TKN (wt%)	0.48	0.45	0.50	0.56	2.5
Final C/N	33.8	29.6	28.6	42.1	10.3
Weight of HA (g/50g compost)	0.35	0.70	0.74	0.65	-
Yield of HA (wt%)	0.70	1.40	1.48	1.30	-
Initial mass (kg)	1	1	1	1	-

Final mass, dried and sieved (kg)	0.65	0.75	0.75	0.90	-
Mass reduction (wt%)	35	25	25	10	-

The final C/N ratio is the highest for batch D (42.1), followed by A (33.8), B (29.6) and C (28.6). It was expected for D to display such character due to its high initial C content (dry leaves). However, batch A with the least initial C source gave the second highest final C/N ratio, suggesting an uneven decomposition process. Many literature suggest for C/N ratio of 25 to 35 for a healthy composting activity of which outside the limits could be environmentally disadvantageous for bacteria to break down the waste [21]. Composting at extreme C/N ratios will result in uneven decomposition of C and N sources as N is always favoured naturally by the bacteria and worms.

The commercial compost has C/N ratio of only 10.3 with significantly high N content at 2.5 wt%. This indicates that the compost was prepared with a very high amount of initial N and little C sources. From business point of view, this practice is advantageous due to the ability to produce compost in a short time. However, the compost quality could be lower as the result of uneven decomposition.
Batch B and C resulted in ideal compost C/N ratios as suggested by their comparable initial and final values. The finding reflects the C and N sources of the raw materials were decomposed evenly.
Humic acid content in compost has a direct relation to the quality as it serves as growth promoter in plants [22]. Batch C has the highest yield of humic acid at 1.48 wt% followed by B (1.4 wt%), D (1.3 wt%) and A (0.7 wt%). These yields are considerably high compared to humic acid extracted from vermicompost of paddy straw (0.081 wt%), water hyacinth (0.053 wt%) and sawdust (0.034 wt%) [23]. Based on humic acid content, batch C with initial C/N ratio of 30 resulted in vermicompost with higher quality.

In terms of compost mass reduction, batch A scored the highest at 35 wt% while batch D is the lowest at 10 wt%. This is attributed to their initial C and N contents where N sources are consumed faster than C. From the results, it is clear that composting process is beneficial to reduce the mass of organics into nutrient-rich compost.

3.4. Voltage generation from vermibatteries

Figure 4 summarizes the voltage generation of vermibatteries using electrolytes from vermicompost (figure 4A) and humic acid (figure 4B) together with other additives. The readings after 24 hours were discarded due to the equilibrium was reached at 12 hours for vermicompost and the additives alone while 6 hours for humic acid. The voltage readings between 24 to 48 hours are within 2% differences. Comparing to the control battery running only with the additives (denoted by X), humic acid battery gave almost identical readings compared to vermicompost battery. For vermicompost battery, the presence of additives helped to serve as an electrolytic medium for ions from humic acid to flow. The solid vermicompost has provided resistance to the overall flow of the ions and therefore the voltage generation can be sustained for much longer with equilibrium voltage was reached at 12 hours (figure 4A), similar to the battery running on the additives alone.

In the case of humic acid battery, the internal resistance was removed and the active compound was utilized directly in the battery cell. Because the cell was mostly in aqueous form, it provided better flow of ions; increased surface contact with the electrodes; enhancing electrical driving force at the terminals and therefore higher voltage generated. All humic acid batteries registered initial voltage of over 3V. Due to minimal resistance, the voltage of humic acid batteries dropped rapidly until equilibrium at 6 hours (figure 4B). This improved reaction however consumed the electrodes faster. The half-cell reaction at the anode can be presented as:

$$Zn + HSO_4^- \rightarrow ZnSO_4 + H^+ + 2e^- \quad (1)$$

IOP Conf. Series: Materials Science and Engineering **206** (2017) 012067 doi:10.1088/1757-899X/206/1/012067

This process is known as discharging. The negative charged sulphuric ions were dissociated from H_2SO_4 used as one of the additives. Under an electric field, the negative ions moved to the negative terminal and consumed in a reaction. The H^+ protons drifted away in the electrolyte while the electrons moved to cathode. These flows of ions and electrons are attributed to an electrical potential existed in the electrolyte [24].

The other half-cell reaction at cathode or charging process can be depicted as:

$$2PbO + 3H^+ + HSO_4^- + 2e^- \rightarrow Pb_2SO_4 + 2H_2O \qquad (2)$$

The electrons deposited at the terminal causing the cation to reduce. The full-cell reaction can be expressed as:

$$Zn + 2PbO + 2H_2SO_4 \rightarrow Pb_2SO_4 + ZnSO_4 + 2H_2O \qquad (3)$$

The added sulphuric acid provided a mean of temporary recharging of the vermibatteries until the ions were depleted. These cycles of charging-discharging have resulted in fluctuations of voltage in the cells especially at the earlier hours. The trend is more visible with vermicompost batteries due to the internal resistance provided by the solid compost. While for humic acid batteries, the bisulfate ions together with humic acid ions were much freer to flow and react.

The highest voltages generated for vermibattery of vermicompost and humic acid were recorded for batch C at 2.9V and 3.1V respectively. Subtracting these values with the corresponding control value (2.66V) resulted in 0.24V and 0.44V of net voltage generated. Thus, it is proven that compost-derived humic acid can generate electricity, albeit a small amount.

Overall, the voltage generated from vermicompost and humic acid of batch C is marginally higher compared to the rest due to its higher humic acid content. In terms of electrolyte, humic acid can generate higher voltage but vermicompost can sustain the voltage longer. In choosing a suitable compost recipe for vermibattery construction, ones should realize the trade-off of maximum voltage vs. durability as well as the compost quality.

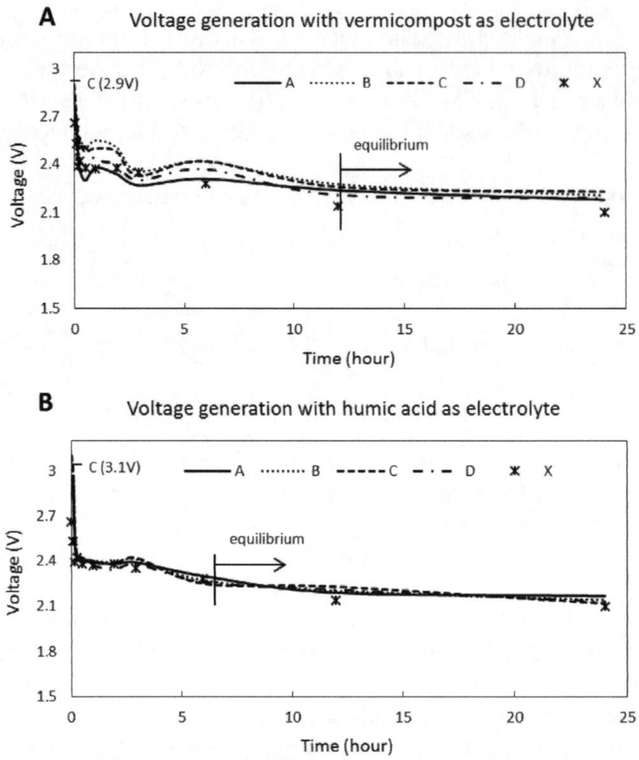

Figure 4. Voltage generation of vermibatteries over time for A) vermicompost or B) humic acid as the electrolyte, with 'X' denotes the control.

4. Conclusion

Composting of organics is highly dependent on the C/N ratio of the raw materials with compost of high N source degraded faster. Balanced C/N ratio between initial and final were achieved at batch B and C, corresponding to C/N ratio of 25 and 30. This behaviour suggests an even decomposition of N and C sources by the worms and bacteria. Humic acid has been proven to generate a small amount of electricity with the highest net voltage production of 0.44V per 0.35g of humic acid for batch C. Vermicompost can also be used to generate comparable electricity to humic acid with an added advantage of better power durability. Batch C gave the overall ideal results in terms of compost quality, humic acid yield and voltage generation.

With the organic mass reduction of up to 35 wt%, coupled with the potential for quality compost production and prospective application as green batteries could make composting an ideal alternative than landfilling when it comes to green waste management. Future work would look into composting formulations to increase the voltage generation, experimenting with other worm species and bacteria as well as scaling-up the process.

References

[1] Cook-Chennault K, Thambi N and Sastry A 2008. *Smart Mater. Struct.* 17 043001.
[2] PSI 2009. http://www.productstewardship.us/page/Batteries.
[3] EPA 1996. 104–142, 104th Congress. E P Agency: US EPA.
[4] Gonsalves A 2007. *Business & Finance.* InformationWeek (California: UBM).
[5] Paul S, Sharma T, Saikia D, Saikia P, Borah D and Baruah M 2015. *International Journal of Plant & Soil Science* 6 218-28.

[6] Bohn H L, Strawn D G and O'Connor G A 2015. *Soil Chemistry.*
[7] Mehta C M, Palni U, Franke-Whittle I H and Sharma A K 2014. *Waste Manage.* 34 607-22
[8] Advanced Composting Technologies I. http://advancedcomposting.com/the-basics-of-composting/.
[9] Epstein E 1996. CRC press.
[10] Chen Y, Chefetz B and Hadar Y 1996. Dordrecht: Springer Netherlands 382-93.
[11] Aso S and Sakai I 1963. *Soil Sci. Plant Nutr.* 9 1-7.
[12] Karim R, Ghani N and Nasari N 2011. *World Academy of Science, Engineering and Technology, International Journal of Electrical, Computer, Energetic, Electronic and Communication Engineering* 5 1356-61.
[13] AIChE 2015 Chem-E-Car Competition. New York. http://www.aiche.org/community/students/chem-e-car.
[14] Richard T and Trautmann N 1996 C/N Ratio. Cornell Waste Management Institute, New York. http://compost.css.cornell.edu/chemistry.html.
[15] Smith M A and Friend D 2016. University of Illinois Extension. http://web.extension.illinois.edu/homecompost/materials.cfm
[16] Dickson N R and T Kozlowski R 1991. Cornell Waste Management Institute. Northeast Regional Agricultural Engineering Service.
[17] Campitelli P and Ceppi S 2008. *Chemometr Intell Lab Syst.* 90 64-71.
[18] Doan T T, Ngo P T, Rumpel C, Nguyen B V and Jouquet P 2013. *Sci Hort.* 160 148-54.
[19] Kalamdhad A S and Kazmi A A 2009. *Chemosphere* 74 1327-34.
[20] Sundberg C 2005. Faculty of Natural Resources and Agricultural Sciences Department of Biometry and Engineering, Uppsala.
[21] Bishop P and Godfrey C 1983. *J Zhejiang Univ Sci B.* 11 (7):497-505.
[22] Aso S and Sakai I 1963. *Soil Sci. Plant Nutr.* 9 1-7.
[23] Das D, Bhattacharyya P, Ghosh B C and Banik P 2016. *Ecol. Eng.* 86 154-61.
[24] Denker J S 2004. https://www.av8n.com/physics/lead-acid.htm.

IOP Conf. Series: Materials Science and Engineering **206** (2017) 012068 doi:10.1088/1757-899X/206/1/012068

Synthesis of asymmetric polyetherimide membrane for CO_2/N_2 separation

A L Ahmad[1,*], Y O Salaudeen[1] and Z.A. Jawad[2]

[1]School of Chemical Engineering, Universiti Sains Malaysia, Engineering Campus 14300, Nibong Tebal, Pulau Pinang, Malaysia.
[2]Faculty of Engineering and Science, Department of Chemical Engineering, Curtin University Malaysia, CDT 250, Miri 98009, Sarawak, Malaysia.

E-mail: chlatif@usm.my

Abstract. Large emission of carbon dioxide (CO_2) to the environment requires mitigation to avoid unbearable consequences on global climate change. The CO_2 emissions generated by fossil fuel combustion within the power and industrial sectors need to be quickly curbed. The gas emission can be abated using membrane technology; this is one of the most promising approaches for selective separation of CO_2/N_2. The purpose of the study is to synthesis an asymmetric polyetherimide (PEI) membrane and to establish its morphological characteristics for CO_2/N_2 separation. The PEI flat-sheet asymmetric membrane was fabricated using phase inversion with N-methyl-2-pyrrolidone (NMP) as solvent and water-isopropanol as a coagulant. Particularly, polymer concentration of 20, 25, and 30 wt. % were studied. In addition, the structure and morphology of the produced membrane were observed using scanning electron microscopy (SEM). Importantly, results showed that the membrane with high PEI concentration of 30 wt. % yield an optimal selectivity of 10.7 for CO_2/Nitrogen (N_2) separation at 1 bar and 25 °C for pure gas, aided by the membrane surface morphology. The dense skin present was as a result of non-solvent (water) while isopropanol generates a porous sponge structure. This appreciable separation performance makes the PEI asymmetric membrane an attractive alternative for CO_2/N_2 separation.

1. Introduction

The complications of global warming and climate change are today being generally discussed in terms of carbon dioxide (CO_2) emission by human activities [1]. Most of the world's primary energy requirement is supplied by fossil fuels, which results in rising emission of greenhouse gasses (GHGs) and related concerns over global warming and climate change [2]. In addition, different concentration of CO_2 in flue gasses depends on the fuel such as coal (12-15 mol-% CO_2), natural gas (3-4 mol-% CO_2), and oil refining (8-9 mol-% CO_2) which must be removed. Typically, CO_2 is the primary anthropogenic greenhouse gas, accounting for 77% of the human contribution to the greenhouse effect in the recent decade. Therefore, to remove CO_2 from the gas streams, various techniques have been improved such as chemical and physical absorption, solid adsorption, cryogenic distillation, membrane separation [3]. Finding a proper separation technology for the removal of CO_2 from different gas streams will drastically lead to the reduction of emissions. Although CO_2 production contributes to global warming, it is also considered as a raw material in many industries. Global CO_2 consumption, for example, has reached approximately 20 million tons per year, which is strictly dedicated to the so-called merchant market, food, beverage, chemical and other industries [4]. Nevertheless, either for storage or for its utilization, CO_2 must be separated. At present, the most widely used CO_2 separation processes consist of reversible chemical and physical absorption. Membrane processes are not used very much; however, they are attractive because of their simplicity and energy efficiency [5]. Actually, there are not yet suitable technologies or technical solutions which can effectively fight against the CO_2 increase in the atmosphere. In the membrane cluster, it is deduced that membrane technology capable of operating under a wide variety of conditions that have a considerable impact on the separation and purification of CO_2 [6–8]. As known, membranes have become an established technology for CO_2 removal since 1981 [9]. However, in order to reduce the overall environmental impacts and cost of CO_2 capture, the priority is given to the technologies with improved CO_2 removal efficiency. In addition, membrane technology is one of the promising alternatives for CO_2 capture, which has been used in different processes such as membrane gas separation, membrane gas absorption, and facilitated transport membrane [10].

Currently, the commercially viable membranes used for CO_2 removal mostly are polymer based. Commercially available polyetherimide (PEI) has several important advantages as a membrane material. This polymer has good chemical and thermal stability. The studies on gas permeation of the PEI dense films reveal that PEI exhibits impressively high selectivity for many important gas pairs especially He and N_2 separation from other gasses with a separation factor of 16.6 [11]. Shamsabadi et al. [12] prepared flat sheet asymmetric polyetherimide membrane for the separation of H_2 from CH_4 with an ideal selectivity of 27.8. Meanwhile, Saedi et al. [13] fabricated flat sheet asymmetric polyethersulfone membrane for separation of CO_2 from CH_4 using PEI as a polymeric additive with a CO_2 permeance of 11.1 at 5 bars. Ren et al. [14] prepared the polyetherimide flat sheet membrane with N-methyl-2-pyrrolidone (NMP)/non-solvent system, and they investigated the influence of the various non-solvents on the membrane morphology. Kurdi and Tremblay [15] produce defect-free asymmetric flat sheet membrane for separation of oxygen from nitrogen using $LiNO_3$ and isopropanol as dual bath coagulation method. Simons et al. [16] investigated CO_2 and CH_4 sorption and transport behavior using the PEI membrane with the CO_2 permeability of 1.4 at 5 bars. lower concentration of 10 and 15 wt. % of the polymer solution results in a porous membrane, which is not suitable for gas separation. This is due to loosely packed structure and weak mechanical strength [17]. In summary, the preparation of the polymer membranes with different concentrations is a method for optimizing the effect of the polymer concentration. Thus, in this research, asymmetric PEI membranes were synthesized with different concentration of the polymer solution by phase inversion method for CO_2/N_2 separation. Also, the performance test, the morphological characteristic of the membrane was investigated.

2. Experimental

2.1. Materials

Polyetherimide (PEI, Ultem®) was purchased from General Electric plastic (Malaysia) where figure1 represents the molecular structure of the used polyetherimide [15]. Table 1 shows the physical properties of the PEI. Also, Isopropanol, anhydrous 1-methyl-2-pyrrolidinone (NMP) (EMPLURA1, 99.5%, water < 0.1%) were supplied from Merck (Malaysia). In addition, the pure gasses including CO_2 (99.9%) and

N_2 (99.9%) were supplied from Technical Gas Services (Malaysia), where these gasses were used in the permeation measurement experiments.

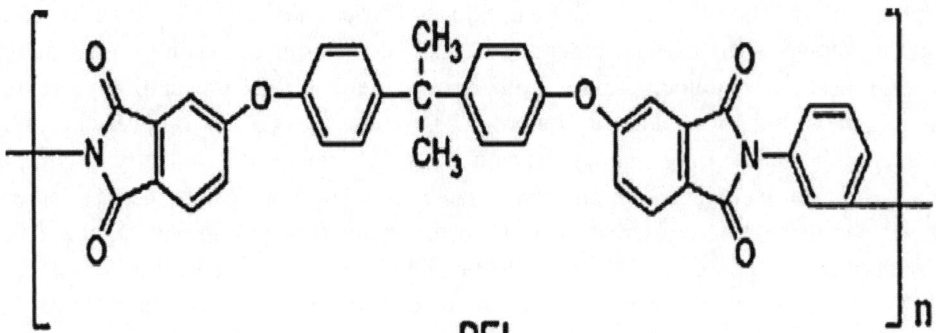

Figure 1. The molecular structure of used PEI where n represents the degree of polymerization.

Table 1. The Physical properties of PEI [12].

Properties	Units	Values
Molecular weight	kg $Kmol^{-1}$	30,000
Density	kg/m^3	1270
Water absorption	mg	20/41
After 24h/996h immersion in water of 23˚C		
Melting point	˚C	219
Glass transition temperature	˚C	206
Thermal conductivity at 23˚C	W/ (m.K)	0.22
Specific heat	$kJ.kg^{-1}.K^{-1}$	2.0
Upper working temperature	˚C	170-200
Elongation at break	%	15
Compressive strength	MPa	140
Tensile strength	MPa	85

2.2. Preparation of asymmetric flat sheet membranes

PEI polymer was dried in a vacuum oven at 110 ˚C overnight in order to remove any trace of adsorbed humidity. The desired polymer solution (w/w) was prepared by dissolving a certain weight of PEI in a pre-determined weight of NMP in a closed glass jar. Table 2 shows the composition of the membranes prepared. A stirring speed of 400 rpm at 60 ˚C was used for efficient mixing for 24 h until a clear yellowish uniform viscous solution was achieved and left overnight. Then the solution was degassed for 30 min in order to remove the air bubbles from the solution. After the degassing process, the polymer solution was cast on a glass plate with a casting knife of 300 μm by a motorized film applicator (Elcometer 4340, Dutech Instruments Malaysia). The cast film along with the glass plate was immediately immersed in a water–isopropanol (4:1) bath to accomplish the phase inversion for 5 min, followed by immersed the membrane in distilled water for 24 h. Then, the membrane was placed in-between two dried glass filled with sheets of filter paper and dried for 24 h at room temperature. Finally, the PEI membrane was dried in an oven at 70±2 ˚C for 12 h for complete removing of the solvents and adsorbed humidity

Table 2. The composition and viscosity of polymer solution.

Solution No	Dope composition (wt. %)		Dope viscosity
	PEI	NMP	
PEI-1	20	80	2200
PEI-2	25	75	8150
PEI-3	30	70	19120

2.3. Gas permeation test

A gas permeation experiment was performed using an assembled setup as shown in figure 2. A cross-flow membrane cell made from stainless steel was used to perform the experiments. The cell consists of two detachable circular compartments clamped together by six screws. The upper part was connected to the feed and the retentate streams. The permeate streams were exhausted from the lower part. Circular membrane discs an effective permeation area of 7.0 cm^2 were used. The pressure-normalized fluxes of the fabricated flat sheet membranes were measured for pure CO_2 and N_2 at 25 °C and 1 bar. The feed gas was supplied to the shell side of the module. The standard permeate volumetric flow rate at atmospheric conditions was measured by a soap bubble flow meter. Each set of data represents an average of 6 replicates. Pressure-normalized flux, called permeance through the membranes was calculated using equation [1] [18]:

$$\left(\frac{p}{l}\right) = \frac{Q}{A\Delta P} \quad (1)$$

where Q is the measured volumetric flow rate (at standard pressure and temperature) of the permeated gas (cm^3 (STP)/s), p is permeance, l is membrane skin layer thickness (cm), A is the effective membrane area (cm^2) and ΔP is the pressure difference across the membrane (cmHg). The common unit of permeance is GPU where GPU is equal to 10^{-6} cm^3 (STP)/cm^2 s cmHg [19]. Also, the ideal gas separation factor (α) can be calculated using equation (2)[20]:

$$\alpha_{CO_2/N_2} = \frac{PCO_2/l}{PN_2/l} \quad (2)$$

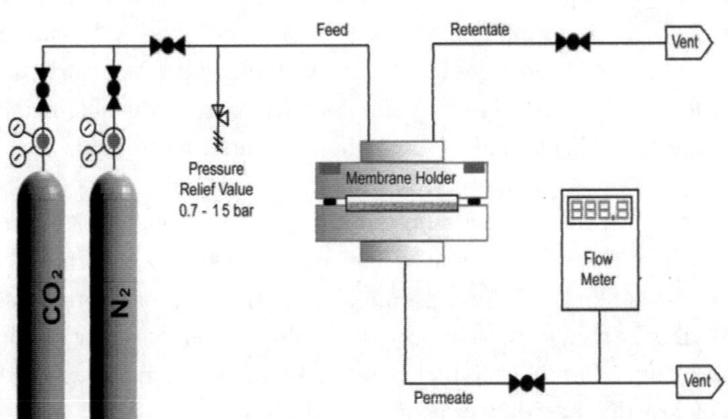

Figure 2. Schematic diagram of gas permeation set up.

2.4. Characterization of flat sheet polyetherimide membrane

The morphology of fabricated flat sheet membranes was investigated by scanning electron microscope (SEM) (TM 3000 Hitachi, High-Tech TOKYO) with an accelerating voltage of 25.0 kV. Prior to scanning, small pieces of the membrane samples were immersed into liquid nitrogen carefully for a minimum of 120 sec. Then, it fractured to obtain a clear cross section. After drying, the fractured fibers were placed on a disc for sputtering with a thin film of platinum before testing. Where, the SEM allows a clear magnification of the overall structure of the membranes; the top surface, the cross section, and the bottom surface can all be observed easily.

3. Results and discussion

3.1. Effect of polymer concentration on the gas separation

SEM images taken of the surface in figure 3 of the membranes, one can see that when 30 wt. % polymer concentrations were used at the dope; the surface of the membrane was smooth with no clear pores. When 25 wt. % was used at dope, pinholes appeared sparsely on the membrane surface, while the use of 20 wt. % polymer concentrations at dope caused the appearance of a large amount of pinholes on the surface. The gas permeance and the selectivities for CO_2/N_2 of the membranes prepared by using 20, 25, and 30 wt. % are shown in table 3. By comparing the data depicted in Table 3, reveals that the flat sheet membranes of PEI-1 and PEI-2 prepared from a lower polymer concentration (20 and 25 wt. %) exhibit much higher permeability but poorer selectivity compared to PEI-3 which was prepared from higher polymer concentration of 30 wt. % [11] . Meanwhile, the CO_2 permeance in PEI-1 and PEI-2 was about 306 and 70 times respectively greater than PEI-3.

As it can be seen from figure 4a, (PEI-1) has a dense outer skin layer with a tiny finger-like layer beneath, these finger-like macrovoids developed to the sublayer in cavity shape and this is surrounded by traces of sponge-like structure. These irregular finger-like macrovoids responsible for the high permeation achieved in the result [12]. Meanwhile, the finger-like structure at the skin layer of the figure 4b (PEI-2) is short and open and the development of sponge-like structure started from the skin layer to the sublayer with obvious pores at this sublayer. The sponge-like structure at the skin layer was responsible for the reduction of the gas flux as presented in table 3 [17]. However, increasing the polymer concentration in the casting solution as in figure 4c (PEI-3) caused changes in thickness, overall membrane porosity and gas permeability. It was observed that the increase in the polymer concentration results in a dense selective layer, a sponge-like structure with obvious pores at the sublayer of the membrane. As shown in the cross-sectional image, the membrane prepared with higher polymer concentration exhibited no finger-like macro void and a less obvious porous substructure. This means, the volume fraction of polymer increases and consequently, a lower porosity is obtained [12]. As a result of fast phase separation at surface layer and slower phase separation at sublayer, an asymmetric membrane with dense and thick skin layer supported by a close cell sublayer was produced by dope of higher polymer concentration. During the immersion process, polymer concentration at the membrane surface which in contact with non-solvent increased due to non-solvent/ solvent exchange process. During rapid exchange process, a skin layer was produced near the membrane non-solvent interface. Therefore, increased of polymer content would expect to support higher polymer precipitation; hence, greater polymer stress was produced in the polymer matrix. As a result, smaller pores were formed at the membrane surface which drastically reduced the gas permeability.

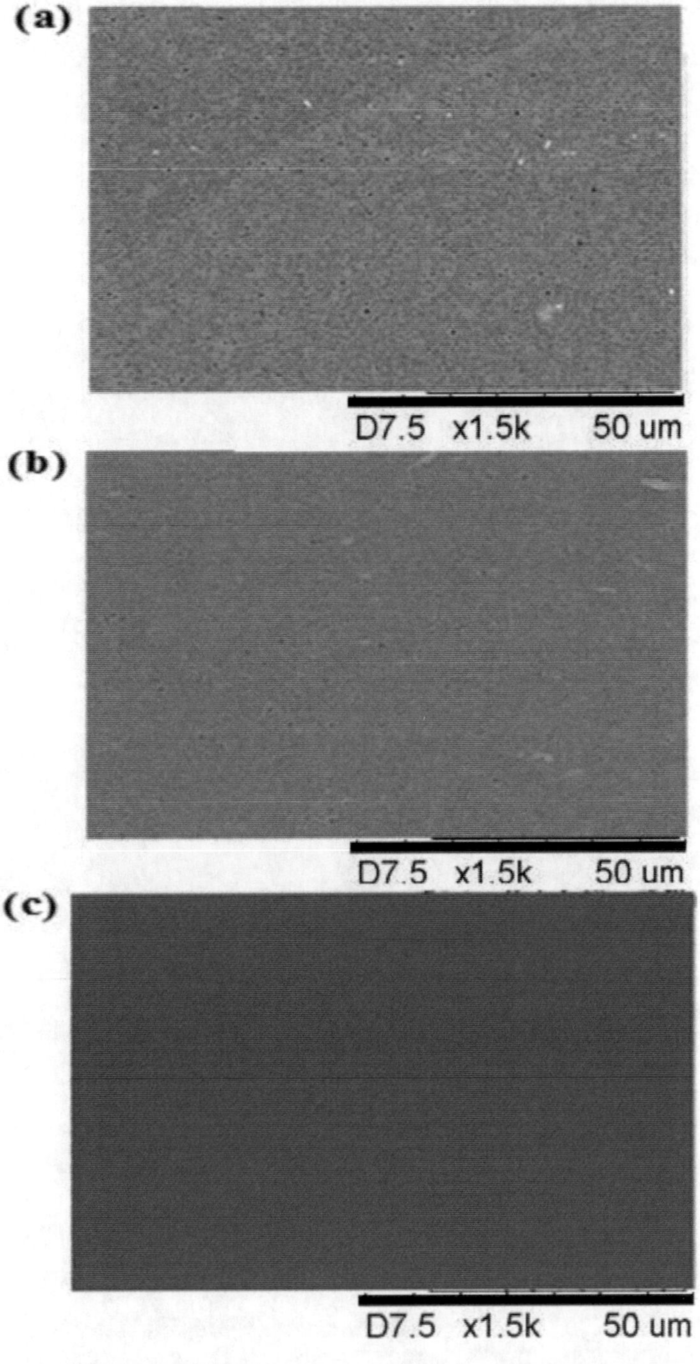

Figure 3. Cross sectional SEM of membrane (a) PEI-1 (b) PEI-2 (c) PEI-3.

IOP Conf. Series: Materials Science and Engineering **206** (2017) 012068 doi:10.1088/1757-899X/206/1/012068

Table 3. CO_2 and N_2 Permeance (GPU) and ideal selectivity at 1 bar for various polymer concentrations.

Samples	CO_2 Permeance (GPU)	N_2 Permeance (GPU)	CO_2/N_2 Selectivity
PEI-1	1942.76	842.53	2.3
PEI-2	443.75	158.7	2.8
PEI-3	6.35	0.59	10.76

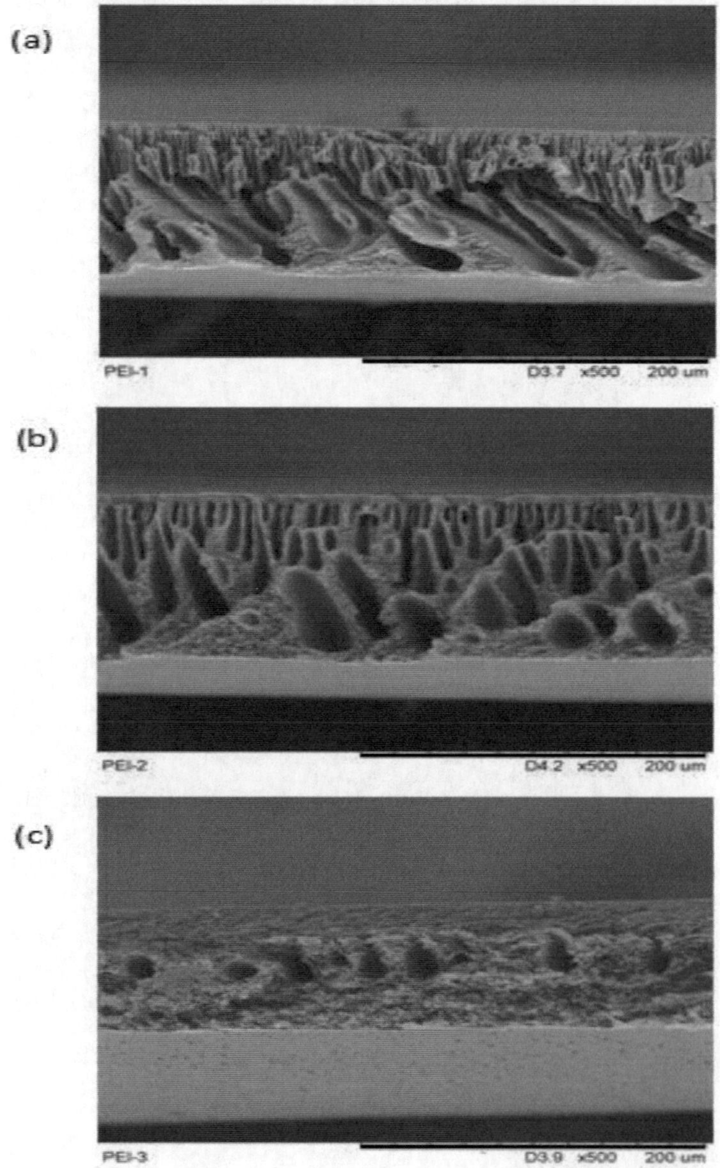

Figure 4. The SEM images of top surface of the membrane cast (a) PEI-1 (b) PEI-2 (c) PEI-3.

IOP Conf. Series: Materials Science and Engineering **206** (2017) 012068 doi:10.1088/1757-899X/206/1/012068

3.2. Determination of critical concentration

As shown in figure 5, viscosity data implies a dramatic slope change in a range of polymer concentration between 20 wt. % and 30 wt. %. Thus, the intersection of the tangents at these two concentrations renders the critical concentration of about 23 wt. %, where a significant change in the degree of chain entanglement occurs [21,22]. Therefore, the membranes which are cast from solutions with polymer concentrations higher than 23 wt. % have a higher entangle conformation, tighter intermolecular chain displacement, and less defective outer skin layers than those are casted from the solution with the polymer lower than 23 wt. % [18]. On the other hand, membrane casted below this critical point has a loosely packed chain thus are able to move freely and allow the non-solvent to penetrate via diffusion mechanism to form microvoid on the membrane surface [17]. Thus, the results showed that the solution viscosity increased as the polymer concentration of the solution was increased.

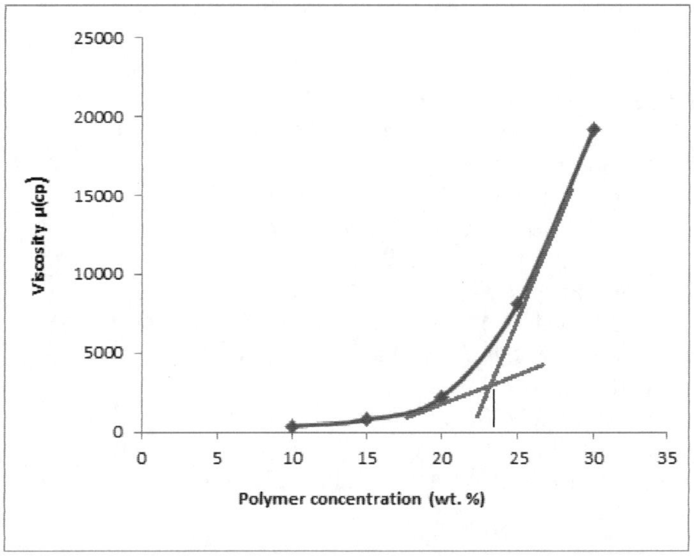

Figure 5. Dope viscosities at 30 °C versus polymer concentration.

4. Conclusion

PEI flat sheet membranes were fabricated via phase inversion process. Isopropanol and water were used as a non-solvent additive in the coagulation bath to improve the phase inversion rate and provide a dense skin-selective layer with a porous sponge-like structure at the sub-layer. In order to eliminate the sub-layer porosity and minimize the gas permeance, the polymer concentration of the solution was increased. Permeation test was conducted for CO_2 and N_2 gas and the considerable separation factor was achieved at 30 wt. % polymer concentration. The critical point diagram reveals that there can only be less defective outer skin layer with a polymer higher than 23 wt. %. The SEM examination exhibits typical morphology for the asymmetric membrane.

References

[1] Lahijani P, Zainal Z A, Mohammadi M and Mohamed A R 2015 *Renew. and Sustain. Energy Rev.* **41** 615–32

[2] Leung D Y C, Caramanna G and Maroto-Valer M M 2014 *Renew. and Sustain. Energy Rev.* **39** 426–43

[3] Songolzadeh M, Soleimani M, Ravanchi T M and Songolzadeh R 2014 *Sci. World J.* **2014** 34

[4] Kargari A and Takht Ravanchi M 2012 *Greenh. Gases - Capturing, Util. Reduct.* **1** 3–30

[5] Abu-Zahra M R M, Feron P H M, Jansens P J and Goetheer E L V 2009 *Int. J. Hydrogen Energy* **34** 3992–4004

[6] Czyperek M, Zapp P, Bouwmeester H J M, Modigell M, Peinemann K V., Voigt I, Meulenberg W A, Singheiser L and Stöver D 2009 *Energy Procedia* **1** 303–10

[7] Shao L, Low B T, Chung T-S and Greenberg A R 2009 *J. Memb. Sci.* **327** 18–31

[8] Baker R W, Lokhandwala K, Baker R W and Lokhandwala K 2008 *J. Ind. Eng. Chem.* **47** 2109–21

[9] Dortmundt D and Doshi K 1999 *Recent Developments in CO_2 Removal Membrane Technology Uop Llc p* 1–32

[10] Brunetti A, Scura F, Barbieri G and Drioli E 2010 *J. Memb. Sci.* **359** 115–25

[11] Wang D, Li K and Teo W K 1998 *J. Memb. Sci.* **138** 193–201

[12] Shamsabadi A A, Kargari A, Babaheidari M B and Laki S 2013 *J. Ind. Eng. Chem.* **19** 1680–8

[13] Saedi S, Madaeni S S and Shamsabadi A A 2014 *Chem. Eng. Res. Des.* **92** 2431–8

[14] Ren J, Zhou J and Deng M 2010 *Sep. Purif. Technol.* **74** 119–29

[15] Kurdi J and Tremblay A Y 1999 *J. Appl. Polym. Sci.* **73** 1471–82

[16] Simons K, Nijmeijer K, Sala J G, van der Werf H, Benes N E, Dingemans T J and Wessling M 2010 *J. Appl. Polym. Sci.* **51** 3907–17

[17] Jamil A, Ching O P and Shariff A M 2016 *ARPN J. Eng. Appl. Sci.* **11** 1687–91

[18] Ismail A F and Yean L P 2003 *Sep. Purif. Technol.* **33** 442–51

[19] Powell C E and Qiao G G 2006 *J. Memb. Sci.* **279** 1–49

[20] Adewole J K, Ahmad A L, Ismail S and Leo C P 2013 *Int. J. Greenh. Gas Control* **17** 46–65

[21] Chung T-S and Kafchinski E R 1997 *J. Appl. Polym. Sci.* **65** 1555–69

[22] Chung T S, Teoh S K and Hu X 1997 *J. Memb. Sci.* **133** 161–75

Acknowledgement
The author would like to acknowledge and thank the research grant 203.PJKIMIA.6071334 provided by Universiti Sains Malaysia that made this study successful.

IOP Conf. Series: Materials Science and Engineering **206** (2017) 012069 doi:10.1088/1757-899X/206/1/012069

Mathematical modelling and simulation on the adsorption of Hydrogen Sulfide (H_2S) gas

N N Zulkefli[1], M S Masdar[1,2,3*], W R W Isahak[1,2], J Jahim[1,2,3], E H Majlan[2], S A M Rejab[4], C C Lye[4]

[1] Department of Chemical and Process Engineering, Faculty of Engineering and Built Environment, Universiti Kebangsaan Malaysia, 43600 UKM Bangi, Selangor, Malaysia
[2] Research Centre for Sustainable Process Technology (CESPRO), Faculty of Engineering and Built Environment, Universiti Kebangsaan Malaysia, 43600 UKM Bangi, Selangor, Malaysia
[3] Fuel Cell Institute, Universiti Kebangsaan Malaysia, 43600 UKM Bangi, Selangor, Malaysia
[4] Sime Darby Research Sdn. Bhd., Jalan Pulau Carey, 42960 Pulau Carey, Selangor, Malaysia

E-mail: shahbud@eng.ukm.my

Abstract. Hydrogen sulfide, H_2S, a pollutant in biofuel gas, i.e., biohydrogen and biomethane, is produced at concentrations ranging from 100 ppm to 10,000 ppm and is recommended to be removed at the early stage of gas purification because it is known as a problematic compound. In this study, adsorption technologies show a promising technique to remove H_2S from biofuel gas, which mainly depends on the operating parameters and adsorbent ability. In this study, the development of the models is important to investigate the fundamentals of H_2S adsorption mechanism. The fitted mathematics model was performed by considering several assumptions made for fixed-bed adsorption, leading to the determination of the breakthrough curve by solving a set of partial differential equations (PDEs). The operating parameters were as follows: varied inlet concentration at 1000 ppm to 10,000 ppm, flow rate at 0.2 L/min to 0.6 L/min, length bed used at 10 cm to 30 cm, and pressure at 1.5 atm to 5 atm. The adsorption performance was also studied by using commercial activated carbon such as palm kernel shell (PKS-AC), coconut shell activated carbon (coconut shell-AC), and zeolite ZSM-5. To support the effectiveness of the mathematical models, the adsorption test was performed by loading the adsorbent into the fixed-bed adsorption column at an overall diameter of 6 cm and height of 30 cm. The system operated under room temperature, H_2S inlet concentration of 1000 ppm, and varying flow rate as in the modelling for PKS-AC. As a result, in the modelling study, the inlet concentration effect was highest in adsorption capacity, breakthrough time, and exhaustion time. However, the increase of flow rate and length bed used only affected the breakthrough and exhaustion times but not adsorption capacity. The total pressure used did not affect adsorption performance. Coconut shell-AC shows longer exhaustion time compared with other adsorbents due to the less frequent changes of adsorbent. In the experimental study, the 1000 ppm inlet concentration shows the highest flow rate effect on the adsorption performance, which, at 0.2 L/min, took almost 23 h to achieve 30 ppm compared with 0.6 L/min, which only took 13 min to exhaust the same outlet concentration. Hence, the adsorption system with the right choice of operational parameters, adsorbent, and fitted mathematical models can optimize the adsorption efficiency, adsorption capacity, breakthrough time, and exhaustion time.

1. Introduction

Biogas is a gas mixture that is formed by various types of microorganisms through the degradation of organic matter under anaerobic conditions. Anaerobic digestion, a well-known process of renewable energy production, is the breakdown and conversion of biomass to biogas by microorganisms. The biogas product is composed of major and minor components. Previous studies found that the major products are Methane (60% to 70%) and Carbon Dioxide (30% to 40%) [1,2,3,4], whereas the minor products consist of Hydrogen Sulfide (H_2S), Ammonia (NH_3), Hydrogen (H_2), Nitrogen (N_2), Carbon Monoxide (CO), saturated or halogenated carbohydrates, Oxygen (O_2), and water (H_2O). The average composition of biogas products as stated in **table 1** is from POME, dairy manure, and waste water treatments. [2, 3, 5, 6].

Table 1. Components and composition of biogas production [4].

Components	Composition
CO_2	30–40
CH_4	60–70
H_2S	0.15–0.3
NH_3	<1
N_2	0–2
CO	<0.6
O_2	0–1
H_2O	5–10

The purification of biogas also attracted considerable attention from many countries in terms of generating renewable energy. As an example, the purification of biogas up to 99.99% of H_2 is needed for the fuel cell system known as polymer electrolyte membrane fuel cell (PEMFC). PEMFC requires below 1 ppm H_2S concentration and is an excellent fuel supply for the generation of energy or electricity.

H_2S are problematic compounds that act as a strong poison for fuel cells, reform catalysts, and also affect the mechanical wear of equipment and the operation [7]. H_2S combustion can also lead to sulfur dioxide emissions, which have harmful environmental effects. Moreover, biogas-produced H_2S leads to the inefficiency of biogas production as it corrodes equipment even at low concentrations. Therefore, early removal of H_2S is recommended in the biogas purification process to upgrade the raw biogas quality, avoid any cost increment, and provide better environment. However, to remove H_2S gas, an acceptable concentration level in the range of 0 ppm to 5 ppm for boiler feed water or industrial process needs to be met [8, 9, 10, 11].

Besides, the removal of H_2S can be achieved by several technologies, namely, biological, absorption, adsorption, and membrane technologies. To maximize the H_2S removal efficiency, the appropriate method should be chosen. In this study, adsorption technology via microporous material (activated carbon) was chosen as the more appropriate and economic method compared with other technologies.

The adsorption technique was reviewed for further study through the performance of mathematical models, as experimental study consumes longer time and higher cost. The mathematical models were performed as necessary to establish the optimization of the design and operation conditions. The ideal model should be convenient, be able to give an exact estimation of the breakthrough behaviour, and evaluate the adsorption effect of each variable. Generally, the modelling of dynamic adsorption consists of macroscopic mass conservation, uptake rate, and isotherm equations. As the mathematical model developed, the performance of fixed-bed adsorption modelling was further studied.

In this modelling, the prediction of the breakthrough method can be used for the determination of fixed-bed adsorption and measuring the full breakthrough curve and equilibrium adsorbed-phase concentrations by using the material balance. This knowledge is important to design and optimize industrial units of adsorbers [12,13]. This method is an early prediction for the evaluation of the adsorption efficiency and adsorption capacity of adsorbents. The adsorption rate also can be determined via time and concentration of H_2S [14].

IOP Conf. Series: Materials Science and Engineering **206** (2017) 012069 doi:10.1088/1757-899X/206/1/012069

Some researchers investigated adsorption performance via several types of adsorbents, operating parameters, and simulation methods. A study by Choo et al. (2013) used impregnated coconut shell activated carbon (coconut shell-AC) in scrubbing units to remove H_2S gas. However, this study had simulated mixing three gases. A study by Lau et al. (2013) investigated the optimization study by using statistical analysis. The adsorbent used was palm shell activated carbon. Melo et al. (2006) discussed the performance of different types of adsorbents (Zinox and Zeolite) to remove H_2S from natural gas.

Based on the information discussed above, the researchers focus on the prediction of breakthrough data, obtained from general rate models by solving the partial differential equation (PDEs) to optimize the parametric performance via the efficiency, adsorption capacity, effect of pressure, and types of adsorbent used toward H_2S removal. The experimental method was tested using H_2S /N_2 gas and a single adsorber column at the same operating parameters used in modelling. Hence, the data of experimental and modelling results were validated.

2. Adsorption principles

Adsorption is a phenomenon of attracting, retaining, and accumulating any substance which results in the higher concentration of molecular species on the surface of another substance as compared with that in the bulk. The solid surface that is exposed toward the gas or liquid will concentrate on the surface. The adsorption process, also known as one of the methods for the removal of undesirable compounds in gas or liquid, penetrates into the pores of the adsorbent and not into the lattice itself [17]. Theoretically, the substance that adsorbed on the surface is called the adsorbate, whereas the substance that is used to adsorb is called the adsorbent.

Adsorption has two types: physical (physisorption) and chemical adsorption (chemisorption). Physical adsorption is a phenomenon of van der Waals attraction between adsorbate and adsorbent. This attraction is well known as weak and reversible since the molecules (adsorbate) are free to move on the surface. This process requires low temperature. Chemisorption requires an attraction between adsorbent and adsorbate through chemical bond at higher temperature.

However, adsorption is suitable for non-combustible gas as it possesses valuable recovery of pollutants and contains very dilute concentrations of pollutant [18]. The adsorption process is important for dehumidifying gas, removing odours or pollutants from the stream, or recovering valuable solvent vapours from the stream.

In the wastewater treatment industry, adsorption via activated carbon is widely used because of the low solubility of contaminant in the waste, as the contaminant has greater affinity for carbon than for the waste or a combination of both. Basic principles of carbon adsorption required mass transfer and adsorption of a molecule from a liquid or gas into solid surface. Hence, commercial microporous material (activated carbon) is produced as extremely porous carbon particles with large surface area, capable of holding adsorbate molecules.

Microporous materials have been tested for applications in gas separation because of their high surface area, micropore volume, and suitable pore size distribution [19]. The variety of microporous materials has also been studied and developed for physical adsorption of gas mixtures [20, 21, 22]. Moreover, the availability of microporous materials is easily obtained as it has been produced in the industrial scale, together with the wide availability of carbon sources. The use of microporous materials is cost-efficient technology, particularly in separation systems, such as pressure swing adsorption (PSA). PSA is commonly accepted to be the most attractive, efficient, and affordable separation method because of its simple control, low operating and capital investment costs, and higher energy efficiency [22, 23, 24].

The fixed-bed adsorption models were investigated with several models that were derived from different assumptions suitable only for limited situations. The models that are widely used are general rate models; linear driving force; wave propagation theory; constant pattern theory; Clark, Thomas, Bohart-Adams, and bed-depth service time Yoon-Nelson, Wang, Zhang, and Cheng models; Wolborska; and the modified dose-response model [25, 26, 27, 28, 29]. Moreover, the ideal model should predict the breakthrough curve, including the breakthrough time, exhaustion time, breakthrough concentration, and exhaustion concentration, as mentioned in **figure 1** below:

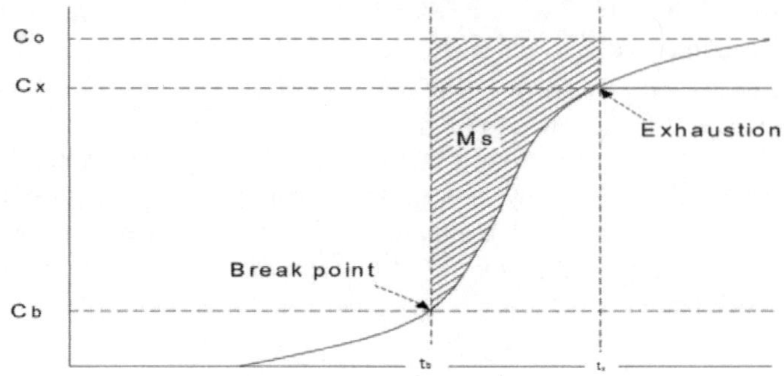

Figure 1. Breakthrough curve.

Where

t_b = Breakthrough time
t_x = Exhaustion time
C_b = Breakthrough concentration
C_x = Exhaustion concentration
C_0 = Inlet concentration
M_s = Adsorption zone from the break point to exhaustion

3. Mathematical model

The mathematical models focused on the calculation of modelling for the fixed-bed adsorption. The breakthrough curve prediction can be determined via solving a set of partial differential equations.

3.1. Fixed-bed adsorption models

As fixed-bed adsorber is commonly used in PSA, the adsorber undergoes four stages, which are pressurization, adsorption, depressurization, and purge. Moreover, the following main assumptions are made for performing the mathematical models: i) ideal gas behavior; ii) axially dispersed plug flow; iii) uniform cross-sectional void fraction; iv) mass and heat transfer are negligible; v) isothermal adsorption; vi) thermal equilibrium occurs between gas and particles; and vii) two component gas, which are adsorb gas (H_2S) and carrier gas (N_2) [31, 32, 33].

According to the aforementioned assumptions, the mathematical model equations for general rate models with length, $L = \partial Z$ may be written as follows. The **figure 2** shows the schematic of adsorption column [34].

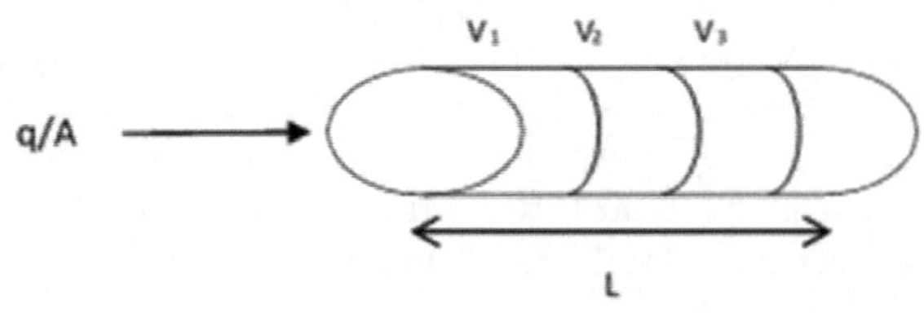

Figure 2. Schematic of adsorption

Then, the general component material balance is given as follows [35, 36]:

$$-D_L \frac{\partial^2 C}{\partial Z^2} + \frac{\partial}{\partial Z}(vC) + \frac{\partial C}{\partial Z} + \left(\frac{1-\epsilon}{\epsilon}\right)\frac{\partial q}{\partial t} = 0 \quad (1)$$

Where
C = Outlet concentration
∈ = Bed void fraction
v = Interstitial molar velocity
Z = Spatial coordinate
D = Effective axial dispersion coefficient
t = Time variable
q = Adsorption capacity

For the determination of the adsorption capacity, the initial and boundary conditions were considered as important parts. By integrating Equation (2) into Equation (3), adsorbed mass (Mads) and adsorption capacity, q (mg/g), can be revealed. This equation is used in the adsorption and desorption (purge) stage. Hence, the initial and boundary conditions are stated below [16, 37]:
Initial condition,

$$c(t,z) = 0 \; for \; t = 0 \tag{2}$$

Boundary condition,

$$C(t,z) = C \; for \; z = 0 \tag{3}$$

$$\frac{\partial C_{(t,L)}}{\partial Z} = 0 \; for \; z = L \tag{4}$$

$$\frac{\partial q}{\partial t} = k_f(C - C^*) \tag{5}$$

$$M_{ads} = QC_0\int_0^t \left(1 - \frac{C}{C_0}\right)dt \tag{6}$$

$$q = \frac{C_0 - C}{M} V \tag{7}$$

Where C_0 and C are the initial and outlet concentrations (mg/L); M is the adsorbent dosage (g); Q is the flow0 rate and V is the volume of the column [34]. Then, the pressurized–depressurized stages need to obey the ideal gas law ($PV= nRT$), and the general material balance is performed as Equation 8 below [33]:

$$\frac{\partial v}{\partial Z} = \frac{1}{p}\frac{\partial p}{\partial t} - E_g\frac{\partial q}{\partial t} \tag{8}$$

Where,

$$E_g = \frac{R_g T_F}{P}\left(\frac{1-\epsilon}{\epsilon}\right) \tag{9}$$

Where P is total pressure; R_g is the gas law constant; T_f is temperature of system; ϵ is porosity; and E_g is concentration gas in systems.

4. Results and discussions

These results are based on the optimization of inlet concentration, flow rate, pressure, length of bed used, and difference of adsorbent through theoretical study. The inlet concentrations varied from 1000 ppm to 10,000 ppm, whereas the gas flow rate and pressure were between 0.1 and 1.0 L/min and 1.5 and 5.0 atm. The H_2S inlet concentration, flow rate, pressure, and length of bed used were discussed based on palm kernel shell activated carbon (PKS-AC), whereas the performance of adsorbent was discussed based on zeolites ZSM-5, coconut shell, and palm kernel shell. The specification of adsorber column from the actual laboratory scale and the data of pore volume and Brunauer–Emmett–Teller (BET) surface area were determined via characterization test. The breakthrough time was predicted as the first outlet concentration came out. Moreover, this study was extended to the experimental study at flow rate (0.2 L/min to 0.6 L/min), constant inlet concentration (1000 ppm) and room temperature. The constant parameters used are stated in **tables 2** and **table 3** [38]

Table 2. Specification of adsorber unit.

Length of adsorber (M)	0.30
Diameter of adsorber (M)	0.06

Table 3. Adsorbent properties.

Parameters	Zeolite ZSM-5	Coconut shell-AC	PKS-AC
Bulk density, kg/m^3	680	500	295
Pore volume, cm^3	0.001	0.413	0.007
BET Surface area, m^2/g	0.064	863.043	6.156

4.1. Effect of the inlet concentration

The optimization data were investigated using PKS-AC. By solving a set of partial differential equations, a breakthrough curve was performed to determine the adsorption capacity, adsorption efficiency, and H_2S adsorbed. In this case, the gas flow rate at 0.2 L/min and the H_2S adsorbed were analyzed at different inlet concentrations, and the result is shown in **figure 3** below:

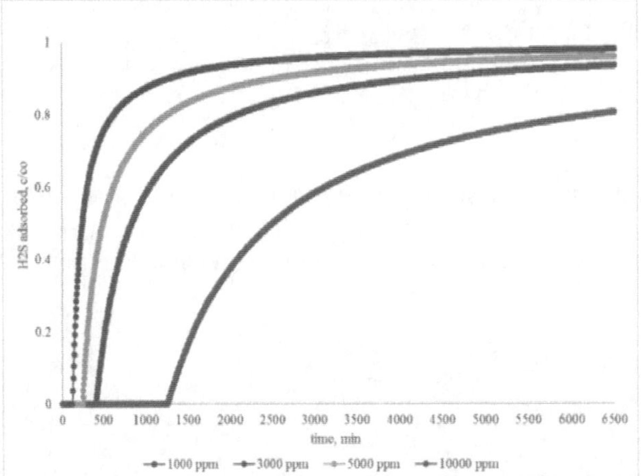

Figure 3. The H_2S adsorbed at different concentrations.

As observed between the different inlet concentrations, the higher the inlet concentration, the faster the exhaustion of the adsorbent bed. The expected results are stated in **table 4**, referring to the phenomenon that the binding site was occupied rapidly and the column was saturated faster compared with the lower inlet concentration [39]. The adsorption capacity was also expected to increase because of the increasing inlet concentration:

Table 4. Effect of H_2S inlet concentration on H_2S adsorption.

Inlet concentration, ppm	Breakthrough time, min	Exhaustion time, min	Adsorption capacity, mg H_2S/g
1000	1255	12,515	3.393
3000	420	4175	10.22

5000	255	2505	16.965
10,000	130	1255	33.929

The increase of the H_2S inlet concentration will increase the H_2S loading, leading to the higher adsorption capacity [14]. Besides, the transition zone in the fixed bed lengthens at lower inlet H_2S concentrations because of the lower apparent reaction rate between H_2S and the adsorbent [40]. Furthermore, as the H_2S concentration decreases, a longer bed or more active sites, is needed to consume a given amount of H_2S within the same period.

The inlet concentration increased from 1000 ppm to 10,000 ppm, and the results show that the increments of inlet concentration affected the H_2S adsorption. The curve pattern is S-shaped. Similar trends were also already proven by several researchers regarding the adsorption study, such as in the removal of H_2S from methane in packed bed, removal of Cu(ii) cations from waste water in fixed bed column, H_2S removal with an iron oxide based CG-4 adsorbent in a fixed-bed reactor, biosorption of Cr(VI) by the thermally activated weed Salvina cucullata, biosorption of acid blue 15 using fresh water Azolla filiculoides, adsorption of Pb (II) cations onto activated tea waste, and immobilized Pinus sylvestris sawdust [41, 42, 43, 44, 45].

4.2. Effect of the gas flow rate

In the following cases, the gas flow rate is shown to play an important role in achieving breakthrough and exhaustion times. The increase of inlet gas flow rate is expected to be faster to meet the breakthrough time. At the inlet concentration of 1000 ppm, the inlet gas flow rate varies at 0.2, 0.4, 0.6, 0.8, and 1.0 L/min.

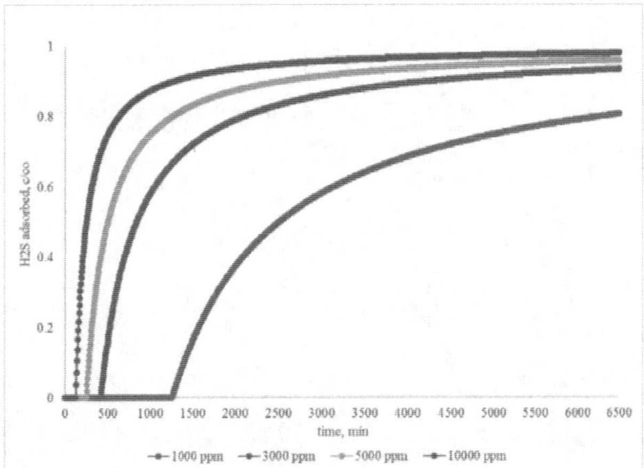

Figure 4. The H_2S adsorbed at different concentrations.

The lower gas flow rate is shown to possess a longer period to saturate and exhaust. The results showed a slight change in adsorption capacity, as shown in **table 5**.

Table 5. Effect of gas flow rate on H_2S adsorption.

Flow rate, L/min	Breakthrough time, min	Exhaustion time, min	Adsorption capacity, mg H_2S/g
0.2	1255	12,515	3.393
0.4	630	6260	3.393
0.6	420	4175	3.407

0.8	315	3130	3.42
1.0	255	2505	3.393

In the study by Natalie (2012), the model predicted that with increasing volumetric flow rate, breakthrough occurs earlier and adsorption capacity remains the same. As the model is not very sensitive to changes in superficial velocity/flow rate, this indicates that the adsorption of H_2S is not limited by kinetics. Practically, the adsorption capacity usually depends on the initial concentration, which provides the necessary driving force to overcome the mass transfer resistances between the aqueous and solid phases. The sorbate interaction increases in terms of vacant sorption site and surface functional group of sorbents when the initial concentration increases [46].

The H_2S is expected to be adsorbed more by increasing the gas flow rate. The fixed-bed column was expected to saturate quickly, as shown in Fig. 4. This results in a higher gas flow rate effect for the mass transfer zone to achieve the exit column faster [47, 48]. Moreover, by increasing the volumetric flow rate, breakthrough occurs earlier, whereas adsorption capacity remains the same. As the rate of introduction of contaminants increases, the active sites are blocked off quicker. These results are similar to the study by Natalie (2012) on the effect of flow rate to the adsorption of H_2S by activated carbon.

4.3. Effect of different types of adsorbent

Three types of adsorbent are used in this research, zeolite, coconut shell-AC, and PKS-AC. All of the adsorbents have different properties, which may lead to different expected results on adsorption efficiency and capacity.

By solving the partial differential equation via varying the bulk density of zeolite, coconut shell-AC, and PKS-AC, the breakthrough curve behaved as shown in **figure 6** by using the constant parameters stated in **table 3**.

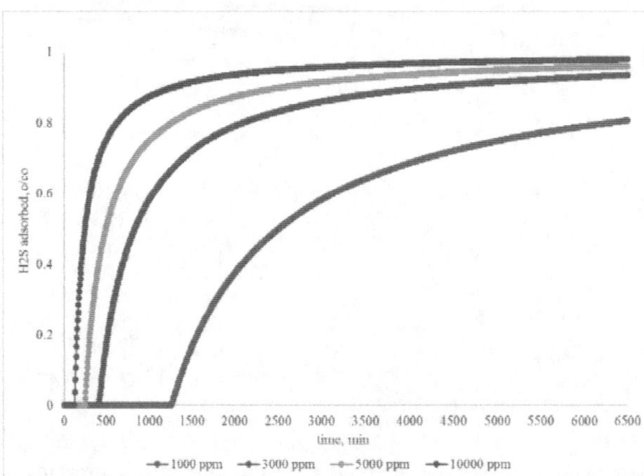

Figure 5. The H_2S adsorbed at different adsorbents.

At 1000 ppm and 0.2 L/min of H_2S gas, the PKS-AC had shorter exhaustion time, which is 12,515 min, compared with zeolite ZSM-5 (28,840 min) and coconut shell-AC (21,210 min). However, increasing either inlet concentration or gas flow rate also affects the period of exhaustion. Hence, breakthrough time, exhaustion time, and adsorption capacity are summarized in **table 6**.

Table 6. Effect of adsorbent types on H_2S adsorption.

Adsorbent types	Breakthrough time, min	Exhaustion time, min	Adsorption capacity, mg H_2S/g

IOP Conf. Series: Materials Science and Engineering **206** (2017) 012069 doi:10.1088/1757-899X/206/1/012069

Zeolite	2885	28,840	1.473
Coconut shell-AC	2125	21,210	2.0
PKS-AC	1255	12,515	3.393

The bulk densities are related to the bed porosity. The adsorption capacity of the column is expected to decrease as porosity increases. This behaviour is expected as less activated carbon is present in more porous beds [49]. Moreover, for higher surface area materials, the higher the concentrations, the higher the adsorption capacity was. As stated by Mangun et al. (1997), the small pores have a higher surface area, which has a higher potential overlap to bind the adsorbate molecule more tightly. The data from BET surface area proved that the larger surface area affected the capability of activated carbon to capture H_2S gas. Therefore, coconut shell-AC is more desirable compared with others because it required a longer period for completing the adsorption process.

4.4. Effect of pressure
A study on the effect of pressure in the adsorption via total operating pressure uses Equations (8) and (9). **Figure 6** shows the breakthrough curve for various total pressures at constant inlet concentration. The total pressure used are 1.5, 2, 3, 4, 5 atm, at an inlet concentration of 1000 ppm. The breakthrough curves show no obvious change when the total pressure was changed.

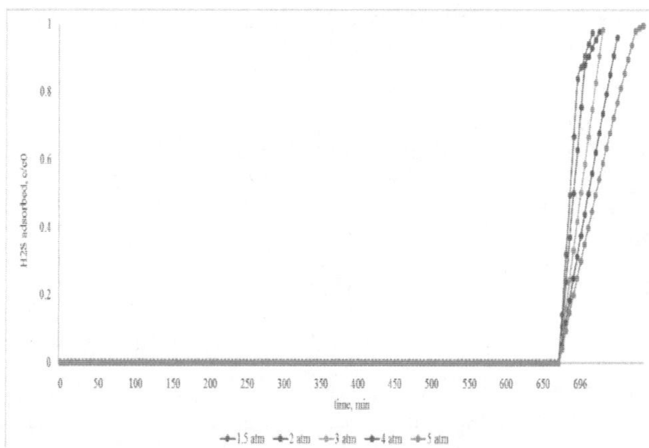

Figure 6. The H_2S adsorbed at different pressure.

As shown in **figure 6**, no changes in total pressure were detected. The total pressure in the range of 1.5 atm to 5 atm shows adsorption capacity at 3.13 mg/g. This result shows a similar trend with a previous study by Wang (2008), wherein the total pressure change in the gas phase does not affect the breakthrough curve characteristics when the other operating conditions are kept constant. This phenomenon is based on a previous study by Garcia-Labiano (2004), which applied the grain model to predict the direct sulfidation of half-calcined $CaCO3 \cdot MgO$ by H_2S. They found that the pre-exponential factor of the Arrhenius-type reaction rate constant decreases when total pressure is higher but the activation energy does not change with total pressure.

4.5. Effect of bed length
The breakthrough curve of the different bed lengths applied in the range of 10 cm to 30 cm was related to the mass of adsorbent and height of the bed used. The H_2S inlet concentration (1000 ppm) and flow rate (0.2 L/min) are kept constant (as shown in **figure 7**).

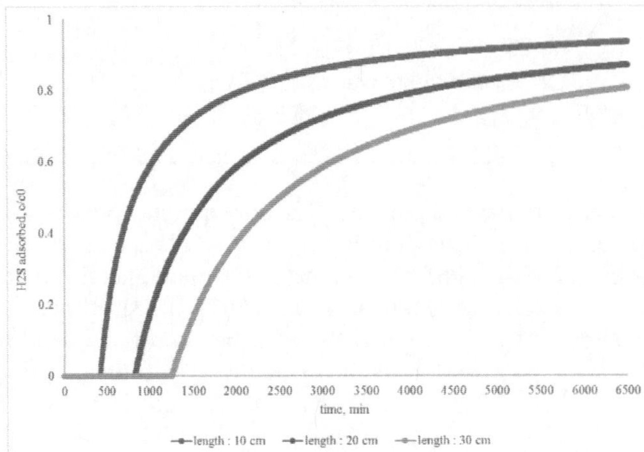

Figure 7. The H_2S adsorbed at different bed length.

By decreasing the bed length, the breakthrough time was shortened and vice versa. This phenomenon is due to the lesser amount of activated carbon used in the adsorber column with shorter bed length. The lesser carbon used refers to the lower active site of palm kernel shell to capture the H_2S gas. The use of a 30-cm bed length resulted in a longer breakthrough time (1255 min) compared with 10 cm (420 min). Hence, more carbon and greater bed length are required for effective removal [49]. **Table 7** shows the relationship between different bed volumes and breakthrough time, exhaustion time, and adsorption capacity at 0.9 of H_2S adsorbed.

Table 7. Effect of different bed lengths on H_2S adsorption.

Length bed used, cm	Breakthrough time, min	Exhaustion time, min	Adsorption capacity, mg H_2S/g
10	420	4525	3.407
20	835	6980	3.407
30	1255	6985	3.393

However, in the context of H_2S removal capacity, the longer the bed length, the more exhaustion time needed. As stated by Andrea (2010), as in the previous study, the decreasing removal capacity is due to a decrease in the contact time and vice versa, which is related to the adsorption reaction time. A previous study shows similar trends in the removal of H_2S gas through anaerobic digestion by-product activated carbon and carbonaceous activated adsorbent [39, 49]. However, the adsorption capacity is constant even if the bed length varied. This happens because of the greater dependence of adsorption capacity on the different concentrations compared with the bed length used.

4.6. Experimental results

The experiment was set up at flow rates of 0.2 L/min to 0.6 L/min, inlet concentration at 1000 ppm, and operation at room temperature. A total of 250 g of PKS-AC was loaded into a 0.848 L adsorber column. The column specification is the same as that used in the modelling study. The outlet concentration was fixed at 30 ppm and is considered the limit for the exhaustion of H_2S gas.

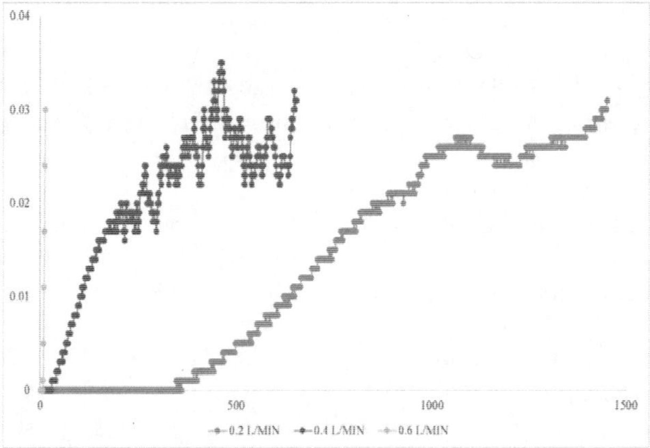

Figure 8. H_2S adsorbed at difference flow rate.

As shown in **figure 8**, the 0.2 L/min flow rate had longer breakthrough and exhaustion times compared to 0.4 and 0.6 L/min. The adsorption capacity is the same due to the different flow rates.

Table 8. Effect of gas flow rate on H_2S adsorption.

Flow rate, L/min	Breakthrough time, min	Exhaustion time, min	Adsorption capacity, mg H_2S/g
0.2	363	1450	3.291
0.4	30	650	3.291
0.6	8	13	3.291

However, the modelling and experimental results show a slight difference on the adsorption capacity, which are 3.4 mg H_2S/g and 3.291 mg H_2S/g, respectively. Basing on these results, the researchers recommended the application of different types of adsorbents, concentrations, and flow rates to capture H_2S particles and purify biogas. By fixing the outlet gas at 30 ppm, the adsorption time can be estimated and controlled via a modelling study, which calculated the exhaustion time. This study also revealed the expected time to stop the adsorption process and change the adsorbents. Hence, the use of H_2S gas and adsorbent can be controlled in experimental studies.

5. Conclusions

The mathematical models are used to predict the breakthrough curve, adsorption capacity, and saturation time earlier, which may be further validated by the experimental study. The theoretical study shows that by using different types of adsorbent. By increasing the inlet concentration, the adsorption capacity, breakthrough time, and exhaustion time were affected. The highest flow rate and bed length used only affected the breakthrough and exhaustion times, but not adsorption capacity. However, varying the total pressure does not affect the adsorption performance. Hence, coconut shell-AC is concluded to be the best adsorbent to use. Moreover, the optimum operating parameters such as flow rate (0.2-0.4 L/min) and 30 cm of length bed used were best to operate the adsorption process at the optimum exhaustion time. These due to the less frequent changes of adsorbent.

6. Acknowledgements

This study was partially supported by a grant from Universiti Kebangsaan Malaysia with the code grant GUP-2014-069 and UKM–Sime Darby Research Project under code KK-2014-013.

IOP Conf. Series: Materials Science and Engineering **206** (2017) 012069 doi:10.1088/1757-899X/206/1/012069

References

[1] Anhuradha S, Vijayagopal V, Radha P and Ramanujam 2007 *CLEAN-Soil Air Water* **35** 197.
[2] Pellerin R L 1987 *Operation and Performance of Biogas-Fueled Cogeneration Systems Energy in Agriculture* **6** 295.
[3] Osorio, F and Torres J C 2009 *Renewable Energy* **34** 2164.
[4] Wellinger A 2011 January *Biogas Upgrading and Utilisation. Retrieved from [internet] IEA Bioenergy Task 24 Energy from Biological Conversion of Organic Waste:* http://www.biogasmax.eu/media/biogas_upgrading_and_utilisation_018031200_1011_24042007.pdf
[5] Wheeler 2000 *Biogas Upgrading and Utilisation IEA Bioenergy Task 24 International Energy Association* (Paris: France)
[6] *Sarawakenergy research & development biomass energy, palm oil mill effluent* http://www.sarawakenergy.com.my/index.php/r-d/biomass-energy/palm-oil-mill-effluent
[7] McCarthy 1998 *Nutzung von Biogas: Probleme und Lösungen für Spurenbestandteile gwf Wasser - Abwasser* **139**(4) 204.
[8] Ministry of Health Malaysia 2004 *National Standard for Drinking Water Quality Engineering Services Division* (MOH Malaysia, Malaysia)
[9] USEPA 2011 *Drinking Water Standards and Health Advisories Environmental Protection Agency* (U.S:Washington)
[10] World Health Organization 2006 *Guidelines for Drinking-Water Quality, First Addendum to Third Edition World Health Organization (WHO)* (Geneva: Switzerland)
[11] Jami M S, Amosa M K, Alkhatib M F R, Jimat D N and Muyibi S A 2013 *Boiler-Feed and Process Water Reclamation from Biotreated Palm Oil Mill Effluent (BPOME): A Developmental Review Chemical and Biochemical Engineering Quarterly* **27** 477.
[12] Clarkson C R and Bustin R M 2000 *Int. J. of Coal Geology* **42** 241.
[13] Yu H, Zhou L, Cheng J and Hu Q 2008 *Int. J. Coal Geol* **73** 115.
[14] Andrea M S 2010 *A Novel and Cost-effective Hydrogen Sulfide Removal Technology using Tire Derived Rubber Particles Ames Iowa: Master Thesis (Iowa State University)*
[15] Choo H S, Lau L C, Mohamed A R and Lee K T 2013 *J. Eng. Sci. Technol.* **8**(6) 741.
[16] Melo, Sobrinho and Santos 2015 *Biogas Purification: Removal of CO2 and H_2S by Process Absorption Chemistry International Conference & Expo on Separation Techniques Separation Technologies* (San Francisco: USA)
[17] Davis M and Cornwell D 2008 *Introduction to Environmental Engineering. Boston: MA: McGraw-Hill*
[18] Wark K 1998 *Air pollution: Its Origin and Control Menlo Park: CA: Addison Wesley Longman*
[19] Dantas T, Luna F, Silva Jr, Torres A, Azevedo, Rodrigues A E and Moreira R F 2011a *Braz. J. Chem. Eng.* **28(3)** 533.
[20] Dantas, T, Luna F, Silva Jr, Azevedo, Grande C A, Rodrigues A E and Moreira R F 2011b *Chem. Eng. J.* **169** 11.
[21] Belmabkhout Y and Sayari A 2009 *Chem. Eng. Sci.* **64** 3729.
[22] Bae Y, Mulfort K L, Frost H, Ryan P, Punnathanam S, Broadbelt L J, Hupp J T and Snurr R Q 2008 *Langmuir* **24** 8592.
[23] Bastin L, Bárcia P S, Hurtado E J, Silva J A C, Rodrigues A E and Chen B 2008 *J. Phys. Chem.* **112** 1571.
[24] Finsy V, Ma L, Alaerts L, De Vos, D E, Baron G V and Denayer J F M 2009 *Mat* **120** 221.
[25] Glueckauf E 1955 *Transactions of the Faraday Society* **51** 1540. [Doi: 10.1039/tf9555101540]
[26] Helfferich F G and Klein G 1970 *Multicomponent Chromatography: Theory of Interference M. Dekker (New York)*
[27] Chern J M and Chien Y W 2002 *Water Research* **36(3)** 647. [Doi: 10.1016/S0043- 1354(01)00258-5]

[28] Clark R M 1987 *Environ. Sci. Technol.* **21(6)** 573. [Doi: 10. 1021/es00160a008]
[29] Yoon Y H and James H N 1984 *The American Industrial Hygiene Association Journal* **45(8)** 509. [Doi: 10.1080/152986684914 00197]
[30] Wang Y H, Lin S H and Juang R S 2003 *Journal of Hazardous Materials* **102(2-3)** 291. [Doi: 10.1016/S0304-3894(03)00218-8]
[31] Crusz P, Santos J C, Magalhaes F D and Mendes A 2005 *Computers and Chem. Eng.* **30** 83.
[32] Crusz P, Santos J C, Magalhaes F D and Mendes A 2003 *Chem. Eng. Sci.* **58** 3143.
[33] Edy H 2006 *Penulenan Gas Hidrogen Menggunakan Sistem Jerapan Buaian Tekanan Terpadat untuk Sel Bahan Api Tesis Doktor Falsafah (Universiti Kebangsaan Malaysia)*
[34] Zahangir A 2005 *J. Environ. Sci.* **19** 103.
[35] Crittenden J C and Weber W J 1978 *Journal of the Environmental Engineering Division* **104(2)** 185.
[36] Crittenden J C, Hutzler N J, Geyer D G, Oravitz J L and Friedman G 1986 *Water Resources Research* **22(3)** 271. [doi:10.1029/aic.690330819]
[37] Wakao N and Funazkri T 1978 *Chem. Eng. Sci* **33** 1375.
[38] Meteku and Benjamin 2013 *Production of Activated Carbon from Palm Kernel Shell for Gold Adsorption using Leachates from Cocoa Husk Ash (Crude Potash) as Activating Agent. Master Thesis of Science in Chemical Engineering* (Kwame Nkrumah University of Science and Technology, Kumasi, Ghana)
[39] Chowndhury Z, Sharifaha and Sarifuddin 2015 *Bioresources* **10(1)** 732.
[40] Adanze J A, Abad F, Garcia-Labiano L F, de Diego and Gayan P 2005 *Fuel* **84** 533-542.
[41] Baral S S, Das N, Ramulu T S, Sahoo S K, Das S N and Chaudhury G R 2009 *J. Hazard. Mater* **161** 1427. [DOI: 10.1016/j.jhazmat.2008.04.127]
[42] Wang D M 2008 *Breakthrough Behavior of H_2S Removal with an Iron Oxide Based CG-4 Adsorbent in a Fixed-Bed Reactor Master Thesis of Department of Chemical Engineering University of Saskatchewan Saskatoon SK* [retrieved from https://ecommons.usask.ca.]
[43] Padmesh T V N, Vijayaranhavan K, Sekaran G and Velan M 2006 *Dyes Pigments* **71** 77. [DOI: 10.1016/j.dyepig.2005.06.139]
[44] Mondal M K 2009 *J. Environ. Manage.* **90** 3266. [DOI: 10.1016/j.jenvman.2009.05.025]
[45] Taty-Costodes V C, Fauduet H, Porte C and Ho Y S 2005 *J. Hazard. Mater.* **123** 135. [DOI: 10.1016/j.jhazmat.2005.03.032]
[46] Amosa M K 2015 *Process optimization of manganese and H_2S removals from palm oil mill effluent (POME) using enhanced empty fruit bunch (EFB)-based powdered activated carbon (PAC) produced from pyrolysis NRF/DST Chair in Sustainable Process Engineering, School of Chemical and Metallurgical Engineering* [University of the Witwatersrand, South Africa (Accepted manuscript). Retrieved from http://dx.doi.org/]
[47] Amosa M K, Alkhatib M F R, Jami M S, Jimat D N, Owolabi R U and Muyibi S A 2014a *Advances in Environmental Biology* **8** 613. (Engineering, University of the Witwatersrand, 1 Jan Smuts Avenue, Private Bag 3, WITS 2050, Johannesburg, South Africa)
[48] Vinodhini V and Das N 2010 *Desalination* **264** 9. (DOI: 10.1016/j.desal.2010.06.073)
[49] Natalie Ho 2012 *Modeling Hydrogen Sulfide Adsorption by Activated Carbon made from Anaerobic Digestion By-Product, master thesis* (Department of Chemical Engineering and Applied Chemistry University of Toronto. Retrieved from https://tspace.library.utoronto.ca)
[50] Garcia-Labiano F 2004 *Energy & Fuels* **18** 761.

IOP Conf. Series: Materials Science and Engineering **206** (2017) 012070 doi:10.1088/1757-899X/206/1/012070

Purification of bioactive phenolics from *Phanerochaete chysosporium* biomass extract on selected macroporous resins

Z M Idris[1], M I H M Dzahir[2], P Jamal[3], A A Barkat[3] and R L W Xian[1]

[1] Department of Chemical Engineering Technology, Faculty of Engineering Technology, Universiti Malaysia Perlis, Kampus UniCITI Alam, Sg. Chuchuh, 02100 Padang Besar, Perlis, Malaysia
[2] School of Bioprocess Engineering, Kompleks Pusat Pengajian Jejawi 3, Universiti Malaysia Perlis, 02600 Arau, Perlis, Malaysia
[3] Department of Biotechnology Engineering, Faculty of Engineering, International Islamic University Malaysia (IIUM), P.O. BOX 10, 50728 Kuala Lumpur, Malaysia

E-mail: zulkarnainidris@unimap.edu.my

Abstract. In this study, two different types of macroporous resins known as XAD-7HP and HP-20 were evaluated for the adsorption and desorption properties against bioactive phenolics extracted from Phanerochaete chrysosporium. From the previous static sorption studies, it was found that the adsorption capacity for both resins had has no significant difference. Then, the kinetic adsorption data were analyzed with both pseudo-first-order and pseudo-second-order equations and the later performed better. The adsorption isotherm data were fitted well by both Langmuir and Freundlich models. Meanwhile in desorption study, HP-20 and XAD-7HP gave 90.52% and 88.28% recoveries, respectively. Considering the desorption results of the macroporous resins, HP-20 and XAD-7HP were packed in chromatography column to further purify the phenolics. For dynamic adsorption, breakthrough capacity of HP-20 (0.522) was found to be higher than XAD-7HP (0.131). Different ethanol concentrations (30% to 50% (v/v)) were investigated at fixed flowrate (1 ml/min) on phenolics recovery from both types of resins. The highest recovery of bioactive phenolics was 94.3% using XAD-7HP resins at 50% (v/v) of ethanol. Only 77.1% of bioactive phenolics were recovered using HP-20 resin at the same experimental conditions. The purified extract subsequently was analyzed using HPLC. The results showed that three phenolics (gallic acid 3,4-dihydroxybenzoic acid and 4-hydroxybenzoic acid) were identified with higher concentrations as compared to non-purified extract. Finally, the purified extract was tested for scavenging activity against DPPH, and it showed that the activity increased significantly to 90.80% from 59.94% in non-purified extract.

1. Introduction

Phenolic compounds are known bioactive constituents from plants and fungi which have remarkable potential to reduce the risk of some health problems due to their antioxidant, antimutagenic, anti-inflammatory and antimicrobial properties [1]. Due to their ability to promote benefits for human health, they are object of great interest of pharmaceutical and food industries. However, crude phenolic compounds extracted from plant or fungi are not suitable for their applications in the pharmaceutical and food industries due to high levels of impurities as well as low concentration of bioactive compounds in the extracts. Therefore, it is essential to separate and purify the responsible

IOP Conf. Series: Materials Science and Engineering **206** (2017) 012070 doi:10.1088/1757-899X/206/1/012070

bioactive compounds (those with the observed biological activities) for in depth pharmacological research and effective applications in medical practice and dietry supplements.

Separation and purification methods of bioactive compounds have to be cost effective, scalable, economical feasible and also can minimize the oxidation, degradation or polymerization of the valuable compounds [2]. Adsorption method has been recognized as the most efficient way of purifying phenolic compounds [3]. Different types of resins have been proposed to purify phenolic compounds such as activated carbon, silica gel, polymeric, cation-exchange and macroporous resins. Activated carbon resins are superior to polymeric resins for phenol and flavonoid adsorption [4][5] but activated carbon resins are difficult to regenerate due to irreversible adsorption. Silica gel resins otherwise suffer low recovery of phenolic compounds because of poor adsorption capacity [6].

At present, macroporous resins have been used widely for the separation and purification of many phenolic compunds including licorice flavonoids [7], scutellarin [8], chlorogenic acid [9], genistein and apigenin [10]. Crude solvent extracts subjected to adsorption-desorption onto macroporous resins showed improvement in their antioxidant [11], immunosuppressive [12] and anti-inflammatory activities [7]. Macroporous resins are durable non-polar and polar polymers. The macroporous adsorption resins exhibit higher adsorption capacity not only because of their similar polarity with the target compounds, but also because of their higher surface area and larger average pore diameter. Because of their characteristics of fast adsorption rate, strong adsorption capacity and ease elution, macroporous resins are gradually regarded as favourable materials for separation and purification of bioactive compounds [13].

Phanerochaete chrysosporium is a filamentous basidiomycete white rot fungus that participates in the degradation process of complex woody materials. This fungus is the subject of many investigations due to its ability to mineralize lignin and other related molecules. It has great potential in many biotechnological applications including bioprotein production and bioremediation. To the best our knowledge, there is no report of the production and recovery of bioactive phenolic compounds from the mycelia biomass of this basidiomycete. Owing to the increasing interest in new natural sources of antioxidant and antimicrobial compounds, this is the first study on *Phanerochaete chrysosporium* mycelia biomass as potential source of bioactive compounds.

2. Materials and methods

2.1. Chemicals and reagents

The chemicals used were Folin-Ciocalteu reagent, sodium carbonate (Na_2CO_3), 95% ethanol, distilled water, sodium hydroxide (NaOH), hydrochloric acid (HCl), potato dextrose agar, molasses, manganese II sulfate ($MnSO_4$), potassium dihydrogen phosphate (KH_2PO_4), di-potassium phosphate (K_2HPO_4), ammonium nitrate (NH_4NO_3), yeast extract, gallic acid, 2,2-diphenyl-1-picrylhydrazyl (DPPH). All solvents for HPLC must be HPLC grade.

2.2. Resins pretreatment

Both macroporous resins (XAD-7HP and HP-20) were treated with 95% ethanol for 24 h and washed several times with deionized water before soaking in 5% (w/v) NaOH solution for 3 h. Then, the resins were washed using distilled water and soaked with 5% (w/v) HCl solution for another 3 h. Finally, the resins were washed with the distilled water thoroughly before pretreated with the ethanol for 12 h at 40 °C before usage.

2.3. Preparation of fungus

The fungal strain, *Phanerochaete chrysosprium* was obtained from the cell bank of Department of Chemical Engineering Technology, University Malaysia Perlis (UniMAP). The fungus was punched into round shapes with dimension of 5mm x 5mm using a cork borer from the petri dish and transferred onto a new Potato-Dextrose Agar (PDA) plate. The strain was incubated at 32°C for 5 days.

2.4. Preparation of inoculum and fermentation
The media used for cultivation of *Phanerochaete chrysosprium* consisted of 0.2% (w/v) of manganese II sulfate ($MnSO_4$), 0.2% (w/v) of potassium dihydrogen phosphate (KH_2PO_4), 1% (w/v) of ammonia nitrate (NH_4NO_3), and distilled water. The pH of the media was adjusted to pH 6. Then, the media was autoclaved at 121°C for 30 minutes. The spore suspension was harvested by washing the fungal plate with 20 ml of sterilized distilled water. About 2% (w/v) of spore suspension prepared was added into the culture media. The fermentation process was done in the incubator shaker at 30°C for 6 days and at 150 rpm to produce mycelial pellets.

2.5. Preparation of powdered biomass
Mycelial pellets were collected and filtered out by using Whatman No.1 filter paper and dried inside the oven at 40°C for overnight. The pellets were then scraped off from the aluminum foil and were grinded into fine powder using pestle and mortar. The powder was weighed in order to obtain the mass of the pellets produced.

2.6. Determination of total phenolic content
The total phenolic content (TPC) was determined using suggested method [14]. About 0.03 mL of sample (extract or standard) was added with 2.37 mL distilled water and 0.15 mL Folin-Ciocalteu reagent in 15 mL test tube. The mixture was vortexed for 1 min. Then, 0.45 mL of 20% (w/v) saturated sodium carbonate was added into the mixture. The mixture was incubated in water bath at 40°C for 30 min after vortexed for 1 min. The absorbance reading was measured at 750 nm. The total phenolic content was determined using the gallic acid standard curve and the result was expressed as mg of gallic acid equivalents per liter of sample (GAE $mg.L^{-1}$).

2.7. Aqueous two phase extraction
The phenolics were recovered using aqueous two-phase extraction system. The fine powder biomass was added into the solution contained ratio of 50% of 80 % (w/v) ethanol, 20% of 40% (w/v) salt (KH_2PO_4) and distilled water. The total mass of 14 g was used which comprised of 7 g of ethanol, 2.8 g of the salt and 0.028 g of the biomass powder and the distilled water. The extraction process was done in the centrifuge tube. The tube was centrifuged (3000 rpm, 25°C) for 10 minutes. Then, the tube was left for 2 hours at the room temperature. The ethanol phase was extracted out and placed in the oven at 40 °C overnight. The dried extract from this phase was used for the following study.

2.8. Dynamic adsorption
The glass columns (1 cm x 20 cm) was used in the dynamic adsorption and desorption experiments, which were wet packed with wet adsorbent HP-20 and XAD-7HP (2.5 g, dry weight) respectively at a BV (bed volume) of 10.0 ml. The dynamic breakthrough experiments was carried out at 298 K. Feed concentration was set constant throughout the experiment at 0.02 GAE mg/ml. Flow rate was fixed constant at elution of 1 mL/min. Samples elute were then collected for TPC detection in effluent [15][16]. A breakthrough curve was plotted and important points are calculated and analyze using Equation (1) till (4).

Total shaded area of breakthrough curve (t_t):

$$t_t = \int_0^{\infty} \left(1 - \frac{c}{c_0}\right) dt \tag{1}$$

t_u time equivalent to the usable capacity or the time at which the effluent concentration reaches its maximum permissible level, t_u is very close to t_b:

$$t_u = \int_0^{t_b} \left(1 - \frac{c}{c_0}\right) dt \tag{2}$$

where t_b is the break-point time.

IOP Conf. Series: Materials Science and Engineering **206** (2017) 012070 doi:10.1088/1757-899X/206/1/012070

Length of bed used up to break point (H_B):

$$H_B = \frac{t_u}{t_t} H_T \quad (3)$$

Length of unused bed (H_{UNB}):

$$H_{UNB} = (1 - \frac{t_u}{t_t}) H_T \quad (4)$$

where H_T is the total bed length.

2.9. Dynamic desorption

The column saturated with sample extract was washed with 2.0 BV of deionized water, and then eluted using different ethanol–water ratios (30:70 (v/v), 40:60 (v/v) and 50:50 (v/v)) at constant elution rate of 1mL/min. Eluents were collected for TPC determination [15][16].

2.10. Antioxidant activity

About 0.05 mL of extract sample was added with 0.15 mL of distilled water and 0.2 mL methanol. Then, 0.05 mL of 1mM DPPH in methanol solution was added to the mixture. The mixture was then incubated at 37°C for 30 minutes. The absorbance was measured at 517 nm. Blank and control used for this method was DPPH solution without the sample extract and absolute ethanol respectively. The scavenging activity will be expressed as percentage of inhibition by using Equation (5):
Scavenging activity (%) = [(ΔA517of control - ΔA517of sample) / ΔA517of control) ×100] (5)

2.11. Identification and quantification of phenolics by using HPLC

The HPLC analysis was done according to the suggested method [17]. For quantification analysis, the standard curve of standard phenolic acids (3,4-dihydroxybenzoic acid, gallic acid and 4-hydroxybenzoic acid) was constructed from their concentration (0.5 mg/mL, 0.05 mg/mL, 0.005 mg/mL and 0.0005 mg/mL). The temperature of the column and the flow rate were set at 30 °C and 1.2 mL/min, respectively. The wavelength and the injection volume were fixed at 280 nm and 50 µL, respectively. The chromatographic separation was performed using C18 column. An isocratic solvent (82:18) of system of A (1.25% glacial acetic acid in water) and B (absolute methanol) were used as the mobile phase. The running time was 30 minutes for each injection. Prior to analysis, the sample was prepared by mixing with 1.8 mL of 1.25% glacial acetic acid in water and then was filtered by using 0.45µm prior to HPLC analysis.

2.12. Statistical analysis

All the measurements were done in triplicates. Analysis of variance (ANOVA) was used to determine the significant differences between the means ± standard deviations of the triplicates.

3. Results and discussion

3.1. Dynamic adsorption

The dynamic breakthrough curves of both resin XAD-7HP and HP-20 were performed based on the volume of injection liquid and the ratio of the outlet to inlet concentration of TPC (Cf/Co). For both resin, dynamic adsorption was proceed under the condition of constant flow rate of 1ml/min, initial concentration TPC of 0.02 GAE mg/mL crude extract and total bed volume of 10 BV packed in column (cross sectional area: 0.7854cm2, height: 12.7cm, mass of adsorbent: 2 g).

From figure 1, both resins have different breakthrough point and XAD-7HP reach breakthrough point earlier than HP-20. This may due to high bed height of HP-20 that creates a longer distance for mass transfer zone to reach outlet of column. Higher concentration of inlet will also shift breakthrough point to the left due to the fact that high feed concentration provides higher concentration gradient causing a higher mass transfer or diffusion coefficient [18]. Thus, breakthrough point of HP-20 can be shift earlier by reducing bed height or increasing feed concentration.

The XAD-7HP resin shows lower maximum C/C_o value than HP-20. This may due to low flow rate of loading samples. Comparatively, at high flow rate, breakthrough point will be shifted earlier. This condition encountered by [19] where at high flow rate, time period to allow adsorption equilibrium to reach maximum is not sufficient, thus adsorbate elute out of column before equilibrium occurs. This will produce a higher value of C/C_o.

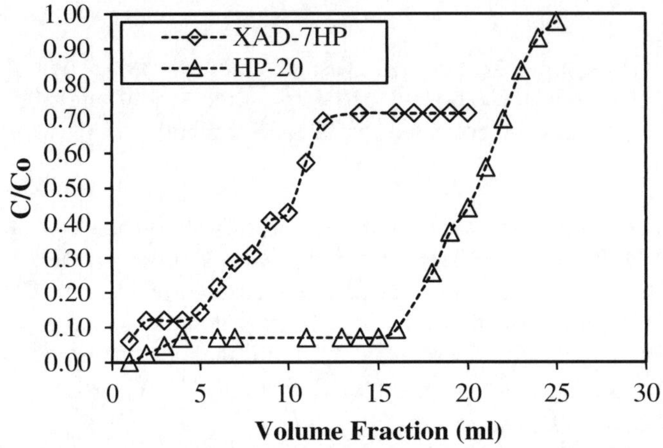

Figure 1. Breakthrough curves for both XAD-7HP and HP-20 resins.

From table 1, breakthrough capacity of HP-20 is much higher than XAD-7HP, which is 0.522 and 0.131 for respective resin. This shows that HP-20 has better ability to absorb phenolic compound upon breakpoint of column. This may assume that due to nature of HP-20 which is weak polar absorbent, which favours non-polar phenolic compound from sample extract. As for XAD-7HP, it shows a much lower breakthrough capacity compare to HP-20. This may due to low feed concentration where it provides low diffusion coefficient affecting the affinity of absorbate towards absorbent.

Table 1. Parameters of fixed bed adsorbents

Adsorbent	t_t (min)	t_u (min)	t_u/t_t	H_B (cm)	H_{UNB} (cm)
XAD-7HP	28.00	3.67	0.131	1.66	11.04
HP-20	29.17	15.24	0.522	6.64	6.07

Besides that, height of unused bed column of HP-20 packed bed is lower than that of XAD-7HP with the value of 6.605cm and 11.037cm respectively. Thus, HP-20 adsorbent-adsorbate system show higher column utilization due to its low solubility in solvent and high affinity towards resin. Therefore, in this experiment, HP-20 resin shows better ability in dynamic adsorption as compared to XAD-7HP.

3.2. Dynamic desorption

For dynamic desorption, column exhausted with sample extract was eluted with different ratios of ethanol-water solution at constant flow rate of 1mL/min. Each eluted fraction was collected to determine the total phenolic content and the results were plotted into desorption curves as shown in figure 2 and figure 3, respectively.

From figure 2, when ethanol concentration in eluent was reduced, elution time increased and serious peak broadening occurred [16]. Therefore, eluent suitable for XAD-7HP resin for desorption of phenolics was at 50:50 (v/v) of ethanol solution.

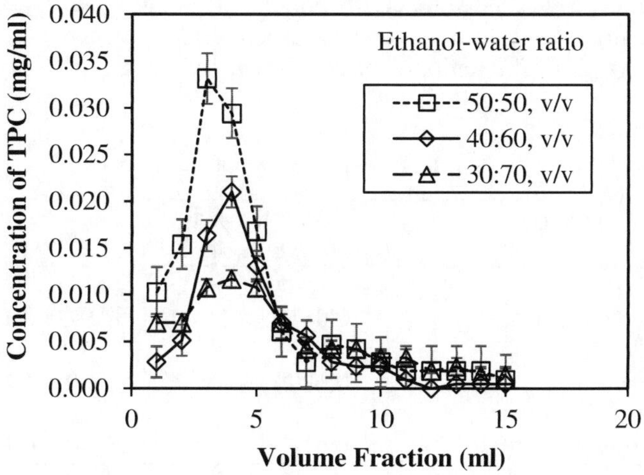

Figure 2. Desorption curve on XAD-7HP resin. Operating condition: Flow rate at 1ml/min; Feed concentration at 0.02 GAE mg.mL-1.

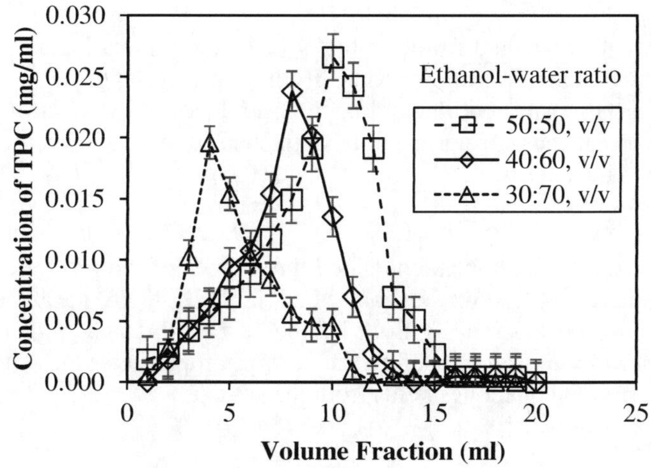

Figure 3. Desorption curve on HP-20 resin. Operating condition: Flow rate at 1ml/min; Feed concentration at 0.02 GAE mg.mL-1.

At this condition, column eluted the highest concentration of phenolic acid within a short elution period. However, the HP-20 resin desorption curve as shown in figure 3 shows that higher ethanol concentration in eluent produced broader peak followed by longer elution time. Therefore, for HP-20 resin, the optimum condition for desorption of phenolics was 30:70 (v/v) of ethanol solution.

Comparatively, both resins preferred different range of ethanol-water concentration in eluent. As XAD-7HP resin preferred higher ethanol concentration in eluent for desorption whereas HP-20 preferred lower ethanol concentration. This may due to XAD-7HP is a polar resin, to allow best desorption, a non-polar eluent is more prefer to reduce polar-polar interaction between adsorbent and eluent.

3.3. Antioxidant analysis

For antioxidant activity, DPPH assay is used to determine the scavenging potential of the antioxidant. The antioxidant activity related to the capability of the phenolic acids in donating the hydrogen and its electrons. The scavenging activity can be calculated by using the (equation 3). Purification from HP-20 has the highest value of the antioxidant activity which was 90.80 ± 0.22% as shown in table 2. The antioxidant activity usually related to the chemical structure of polyphenols, for example number of the hydroxyl group (OH) [20]. The antioxidant activity increased significantly after the purification process.

Table 2. Scavenging activity

Sample	Phenolics acids (GAE mg/mL)	Scavenging Activity (%)
Non-purified extract	0.0512	59.94 ± 0.16
Purified extract (XAD-7HP)	0.0279	80.49 ± 1.34
Purified extract (HP-20)	0.026	90.80 ± 0.22

From the result, it showed that XAD-7HP has the higher concentration of the phenolics acid compared to the HP-20. However, comparing with the scavenging activity, HP-20 has higher activity compared to the XAD-7HP. This may due structure of the phenolics acids as most of the phenolics that bind with the HP-20 contained the hydroxyl group (OH). Moreover, high value of DPPH scavenging activity may also resulted from the bulky or electron-donating substituents on the phenol ring, the meta-position of the electron-donating groups or meta-or ortho-position of hydrophobic groups and finally the ortho-position of the hydrogen-bond donor or the electron-donating groups [21]. Hence, it can be shown that most of the phenolics attached with HP-20 may arrange in the position that gave higher antioxidant activity.

3.4. HPLC analysis

From the HPLC analysis, all the samples contained three types of the phenolics acids; Gallic acid, 3, 4-dihydroxybenzoic acid and 4-hydroxybenzoic acid. Table 3 shows the different of concentration before and after the adsorption process. Both of the resins obtained higher concentration of the phenolics acids compared to the sample before the adsorption process, thus, showing that both of the resins can be used to purify the phenolics acids from the sample.

Table 3. Concentration of phenolics

Sample	Phenolic acids	Concentration (mg/mL)
Non-purified extract	Gallic Acid	0.0011
	3,4-Dihydroxybenzoic acid	0.0011
	4-Hydroxybenzoic acid	0.0031
Purified extract (XAD-7HP)	Gallic Acid	0.0073
	3,4-Dihydroxybenzoic Acid	0.0061
	4-Hydroxybenzoic acid	0.0095
Purified extract (HP-20)	Gallic Acid	0.0556
	3,4-Dihydroxybenzoic Acid	0.0493
	4-Hydroxybenzoic acid	0.0701

4. Conclusion

For dynamic adsorption, breakthrough capacity of HP-20 (0.522) was found to be higher than XAD-7HP (0.131). Different ethanol concentrations (30% to 50% (v/v)) were investigated at fixed flowrate (1 ml/min) on phenolics recovery from both types of resins. The highest recovery of bioactive phenolics was 94.3% using XAD-7HP resins at 50% (v/v) of ethanol. Only 77.1% of bioactive phenolics were recovered using HP-20 resin at the same experimental conditions. The purified extract subsequently was analyzed using HPLC. The results showed that three phenolics (gallic acid 3,4-dihydroxybenzoic acid and 4-hydroxybenzoic acid) were identified with higher concentrations as compared to non-purified extract. Finally, the purified extract was tested for scavenging activity against DPPH, and it showed that the activity increased significantly to 90.80% from 59.94% in non-purified extract.

References

[1] Faria R L, Cardoso L M, Akisue G, Pereira C A, Junqueira J C, Jorge A O and Santos Júnior P V 2011 *J. Appl. Oral. Sci.* **19** 476-82.
[2] Zhao H, Dong J, Lu J, Li Y and Shan L 2006 *J. Agr. Food Chem.* **54** 7277-86.
[3] Soto M L, Conde E, González-López N, Conde M J, Moure A, Sineiro J, Falqué E, Domínguez H, Núñez M J and Parajó J C 2012 *Molecules* **17** 3008-24.
[4] Otero M., Zabkova M, Grande C A and Rodrigues A E 2005 *Ind. Eng. Chem. Res.* **44** 927–36.
[5] Aehle E, Raynaud-Le Grandic S, Ralainirina R, Baltora-Rosset S, Mesnard F, Prouillet C, Maziere J C and Fliniaux M A 2004 *Food Chem.* **86** 579–85.
[6] Jampani C, Naik A and Raghavarao K S M S 2014 *Sep. Purif. Technol* **125** 170-8.
[7] Fu B Q, Liu J, Li H, Li L, Lee F S C and Wang X R 2005 *J. Chromatogr. A* **1089** 18–24.
[8] Gao M, Huang W and Liu C Z 2007 *J. Chromatogr. B* **858** 22–6.
[9] Zhang B, Yang R, Zhao Y and Liu C Z 2008 *J. Chromatogr. B* **867** 253–8.
[10] Liu W, Zhang S, Zu Y G, Fu Y J, Ma W, Zhang D Y, et al 2010 *Bioresource Technol.* **101** 4667–75.
[11] Altiok E, Baycin D, Bayraktar O and Ulku S 2008 *Sep. Purif. Technol* **62** 342–8.
[12] Yin Y, Gong F Y, Wu X X, Sun Y, Li Y H, Chen T and Xu Q 2008 *J. Ethnopharmacol* **120** 1–6.
[13] Du H, Wang H, Yu J, Liang C, Ye W and Li P 2012 *Ind. Eng. Chem. Res.* **51** 7349-54.
[14] Jamal P, Idris Z M, and Alam M Z 2011 Food Chem. **124** 1595–1602.
[15] Lin L, Zhao H, Dong Y, Yang B, and Zhao M 2012 *Food Chem.* **130** 417–24.
[16] Wang Y J, Wu Y F, Xue F, Wu Z X, Xue Y P, Zheng Y G and Shen Y C 2012 *J. Chromatogr. B* **895-896** 146–153.
[17] Wu S, Wang Y, Gong G, Li F, Ren H, and Liu Y 2014 *Food Bioprod. Process* **93** 148–155.
[18] Chowdhury Z Z, Zain S M, Rashid A K, Ra R and Khalid K 2013 *J. Chem.*
[19] Malkoc E, Nuhoglu Y, and Abali Y 2006 *Chem. Eng. J.* **119** 61–8.

[20] Montoro P, Braca A, Pizza C and N D T 2005 *Food Chem.* **92** 349-55.
[21] Jing P, Zhao S J, Jian W J, Qian B J, Dong Y and Pang J 2012 *J. Mol.* **17** 12910-24.

IOP Conf. Series: Materials Science and Engineering **206** (2017) 012071 doi:10.1088/1757-899X/206/1/012071

Ionic liquid-impregnated activated carbon for biohydrogen purification in an adsorption unit

N Y Yusuf[1], M S Masdar[1,2,3], W N R W Isahak[1], D Nordin[1,2], T Husaini[3], E H Majlan[3], S A M Rejab[4] and C L Chew[4]

[1] Department of Chemical and Process Engineering, Faculty of Engineering and Built Environment, 43600, UKM Bangi, Selangor, Malaysia
[2] Research Centre for Sustainable Process Technology (CESPRO), Faculty of Engineering & Built Environment, 43600, UKM Bangi, Selangor, Malaysia
[3] Fuel Cell Institute, Universiti Kebangsaan Malaysia, 43600, UKM Bangi, Selangor, Malaysia
[4] Sime Darby Research Sdn. Bhd., Jalan Pulau Carey, 42960 Pulau Carey, Selangor, Malaysia

E-email: shahbud@ukm.edu.my

Abstract. Biological methods for hydrogen production (biohydrogen) are known as energy intensive and can be operated at ambient temperature and pressure; however, consecutive productions such as purification and separation processes still remain challenging in the industry. Various techniques are used to purify and separate hydrogen. These techniques include the use of sorbents/solvents, membranes and cryogenic distillation. In this study, carbon dioxide (CO_2) was purified and separated from biohydrogen to produce high purity hydrogen gas. CO_2 capture was studied using the activated carbon (AC) modified with the ionic liquid (IL) choline chloride as adsorbent. The physical and chemical properties of the adsorbents were characterized through XRD, FTIR, SEM-EDX, TGA, and BET analyses. The effects of IL loading, flow rate, temperature, and gas mixture were also investigated based on the absorption and desorption of CO_2. The CO_2 level in the biohydrogen composition was analyzed using a CO_2 gas analyzer. The SEM image indicated that the IL homogeneously covered the AC surface. High IL dispersion inlet enhanced the capability of the adsorbent to capture CO_2 gas. The thermal stability and presence of the functionalized group of ILs on AC were analyzed by TGA and FTIR techniques, respectively. CO_2 adsorption experiments were conducted using a 1 L adsorber unit. Hence, adsorption technologies exhibit potential for biohydrogen purification and mainly affected by adsorbent ability and operating parameters. This research presents an improved biohydrogen technique based on adsorption technology with novel adsorbents. Two different types of commercial CO_2 adsorbents were used in the experiment. Results show that the IL/AC exhibited properties suitable for CO_2 adsorption. The IL/AC sample presented a high CO_2 uptake of 30 wt. % IL when treated at 30 °C for 6 h under a flow rate of 1 L/min. The presence of IL increased the selectivity of CO_2 removal during the adsorption process. This IL/AC can be regenerated for several times without any significant loss in the performance.

1. Introduction

Hydrogen is a renewable energy resource that plays a key role in sustainable economic growth because of its carbon-neutral characteristics. Scholars have focused on improving hydrogen and biological production methods. Biohydrogen production is important because it utilizes renewable energy resources and is operated at ambient temperature and atmospheric pressure [1]. Biohydrogen is used in highly energy efficient fuel cells or internal combustion engines [2]. Biohydrogen production requires purified hydrogen. Therefore, hydrogen should be separated from different gaseous by-products, mainly carbon dioxide (CO_2) formed during fermentation.

Studies have investigated the separation of CO_2 from other gasses [3, 4]. Gas mixtures, such as CO_2 and methane (CH_4), are commonly separated through cryogenic distillation, chemical absorption, adsorption, and membrane technology [5-7]. Adsorption is an alternative method for amine scrubbing and membrane technologies, where adsorption capacity is mainly affected by large specific surface area, adsorbent type, and low regenerated heat [8, 9]. Adsorption techniques feature low operational requirements, ease of control, and high efficiency [5].

Adsorption using adsorbents, such as activated carbon (AC), has gained increasing research attention. AC can be used to separates gas mixtures, such as CO_2 and CH_4 [5, 10]. In a previous study, AC derived from carbonized coconut shell char was obtained through CO_2 activation at a carbonization temperature of 600 °C for 2 h [11]. Results showed that the effects of activation temperature and time and flow rates of CO_2 affect the surface area, total and microspore volumes, and yield of AC. The adsorption capacity or uptake of AC is highly influenced by preparation conditions.

Chemical modifications are mainly used to minimize the production cost, increase the removal efficiency of CO_2, and enhance the selectivity of the adsorbent. Chemicals containing nitrogen groups are suitable for modifying the CO_2 adsorbent because of the Lewis acid–base interaction [12]. In the study of Abbott et al. in 2001, a range of quaternary ammonium salts were heated with $ZnCl_2$ and the freezing points of the resulting liquids were measured; the lowest melting point (23 °C - 25 °C) was obtained when choline chloride was used as ammonium salt [13]. This IL was used as CO_2 adsorbent in physical or chemical adsorption. In another study, room temperature ionic liquid (RTIL) was used as substitute for amines because of its properties, such as low vapor pressure, wide liquid range, thermal resistance, and tenability [14].

In this study, an adsorbent material was prepared using AC impregnated with IL to increase the CO_2 adsorption efficiency. These materials were impregnated with IL for gas separation and purification; IL has become a well-known topic in recent separation and purification research because of its unique properties [15]. The developed adsorbent was characterized in terms of surface morphology and element composition (scanning electron microscopy, SEM-EDX), functional group content (Fourier transform infrared spectroscopy, FTIR), crystallinity (X-ray diffraction, XRD), stability and degradation (thermal gravimetric analyzer, TGA), and surface properties (Brunauer–Emmett–Teller, BET). The effects of several parameters, such as flow rate, pressure, temperature, and adsorbent types, on CO_2 adsorption were also investigated. The CO_2 adsorption capacity was compared between the fabricated and fresh (commercial) AC.

2. Materials and methods

2.1. Production of ionic liquid (IL)

The chemicals used were of analytical reagent (AR) grade. Choline chloride ($C_5H_{14}ClNO$) was purchased from Sigma Aldrich Company (USA). Zinc chloride ($ZnCl_2$) was obtained from Friedemann Schmidt Chemical Company (WA). The IL (choline chloride, $2ZnCl_5^-$) was prepared using the process shown in Equation 1:

$$\text{Choline chloride} + 2ZnCl \rightarrow \text{choline chloride. } 2ZnCl5^- \qquad (1)$$

IL was prepared following the standard procedures [16]. IL solution with a ratio of weight percentage of 1:2 was produced by stirring the mixture at 60 °C for 3 h.

2.2. Impregnated AC

AC derived from coconut shells was provided by Afigen Sdn Bhd. (Malaysia). Fresh materials were cleaned and dried at 100 °C for 12 h. The materials were then sieved to obtain > 3 mm particle sizes for adsorption in the column unit. The adsorbents were modified using fresh AC and IL through impregnation methods; this method is commonly known as wet impregnation. Fresh AC and its surface properties were modified by using 30 wt. % IL to determine the effects of IL loading on the physical properties and adsorption performance of AC. The modified adsorbents were wetted in solid–liquid ratio (7:3) for 5 min, heated at 60 °C for 7 h, and dried.

2.3. Characterization of adsorbents

Fresh and impregnated AC was characterized through several techniques including SEM-EDX, FTIR, XRD, TGA, and BET surface area analyses.

2.3.1. Surface morphology and element composition. SEM type ZEISS Supra VP55 was used to examine the pore structure, topography, and composition of the samples. The SEM instrument was equipped with a secondary electron detector to visualize the crystal shape, surface morphology, and structure of particles or agglomerates. Product surface characteristics were also investigated through SEM with detectors for secondary and backscatter electrons. The backscatter detector is designed to analyze the differences in molecular density. The backscatter detector allows the visualization of differences in compositional density in a granule or tablet.

The size range of the sample is ~0.1 μm to few millimeters. The dried adsorbent was coated with a conducting material (e.g., Au) to reflect electrons (sputter coating) prior to the SEM analysis. SEM analysis was performed to determine the porosity and morphology of the adsorbent materials, EDX analysis was used to determine elements inside the adsorbent. This method can be used to observe the specimen in high vacuum or low vacuum, wet conditions, and at cryogenic or high temperatures.

2.3.2. Functional group and crystallinity. IL and the adsorbent AC (fresh and impregnated) were characterized by FTIR spectrometry and recorded with infrared spectrometer instrument-type Buck Scientific M-500 with KBr as pellet at a frequency range of 4000 cm^{-1} to 600 cm^{-1}. The infrared spectra of the adsorbent samples were recorded using FTIR spectrometer (Perkin Elmer, UK). Attenuated total reflection (ATR) method was used for sample preparation. This FTIR analysis generated absorption or transmission spectra as a function of wave number, ranging from 4000 cm^{-1} to 500 cm^{-1}. The sample was weighed (1–5 mg) at ambient temperature by using Spectrum One with SPECAC Golden Gate ATR (Perkin Elmer).

XRD analysis was performed for fingerprint characterization of the crystallinity and structures of the samples. This method was used to identify the shape and type of crystals produced when using AC. XRD was conducted to identify and characterize the crystal structure and size. This method analyzes and determines the distribution of crystal orientation in the thin film crystal sample.

Crystallinity analysis was performed in 1 g of the sample through XRD using Bruker AXS D8 Advance with X-ray radiation source of Cu Ka (40 kV, 40 mA) to record the 2u diffraction angle from 108 to 808 at a specific wavelength (l = 0.154 nm). Based on Bragg's law, the XRD powder (~10 mg) pattern showed the intensity of the diffracted beams as a function of 2-theta (2θ) at temperatures ranging from 5 °C to 80 °C and 25 °C. This analysis was conducted using XRD Bruker AXS D8 Advance with a working voltage of 40 kV and current at 40 Ma, respectively.

2.3.3. Stability/degradation and surface properties. A thermal gravimetric or thermogravimetric analyzer (TGA) was used to measure the amount of weight changes in a material, either as a function of increasing temperature or time, in an atmosphere of nitrogen (N_2) or air. A sample pan that is supported by a precision balance is used. The pan is placed in a furnace and heated or cooled during the experiment, and the sample mass is monitored. The balance obtains weight changes up to 1 μg. In this experiment, 5 – 10 mg of the sample is used for characterization by using N_2 gas carrier. TA-

60WS thermal analysis system from Shimadzu Corporation is used for temperatures up to 1100 °C. The percentage of weight loss over temperature was recorded.

The surface properties were prepared and characterized using Micrometrics ASAP 2020 Version 4.02 in N_2 gas at 77 K. In this analysis, the principle of physical adsorption was used to obtain adsorption and desorption isotherms in the sample. The surface area is calculated by BET (Brunauer–Emmett–Teller) method. Based on BET results, the surface properties of AC adsorbent are shown. The BET surface area, micropore volume and pore size were also calculated from the isotherms. Prior to the analyses, the samples were degassed at 350 °C for 4 h for N_2 adsorption to remove all the volatile compounds and obtain the adsorption isotherm of each sample.

2.4. Adsorption–desorption study for fresh and impregnated AC

The CO_2 separation from gas mixtures of CO_2/H_2 gases was performed in a lab scale experimental single column unit of pressure swing adsorption (PSA) by feeding mixture gases of 50/50 mol % CO_2/H_2 to a single adsorption column unit as shown in figure 1. The column dimensions with a 4.8 cm internal diameter, ID, and 33 cm height were used. The column was connected to 3 mm (1/8 in) stainless steel tubing. The flow rate of feed gases was monitored by gas flow controllers (flow meter). The concentration of product stream was detected by CO_2 analyzer and read by HOBO ware software system in the computer. This procedure of experiment was repeated using the different adsorbent as propose with different parameters.

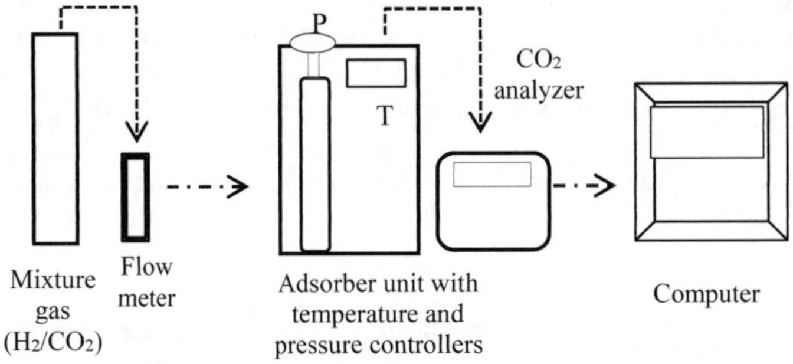

Figure 1. Schematic of the adsorption–desorption system.

In the adsorption–desorption study, fresh and impregnated AC was observed using the same experimental setup throughout the project. For the desorption process, air (3 L/min) was used as carrier gas and supplied through the adsorber unit to eliminate CO_2 from the adsorbent materials. Different flow rates (0.1 L/min, 0.5 L/min, and 1.0 L/min) were used to investigate the performance of fresh and impregnated AC as the adsorbent by passing gas streams containing the CO_2/H_2 mixture through the adsorption column unit. The capacity of adsorption CO_2 was observed using HOBO ware system in a computer connected to the CO_2 analyzer.

3. Results and discussion

3.1. Scanning electron microscopy (SEM-EDX)

The surface micrograph of adsorbents was studied using SEM. This instrument has a common tungsten electron gun with an operational volt ranging 0.1 to 3.0 kV. SEM results show the microscope photos of the compound at different magnifications. Thus, the surface morphology of the chemically modified adsorbents is shown in figure 2.

IOP Conf. Series: Materials Science and Engineering **206** (2017) 012071 doi:10.1088/1757-899X/206/1/012071

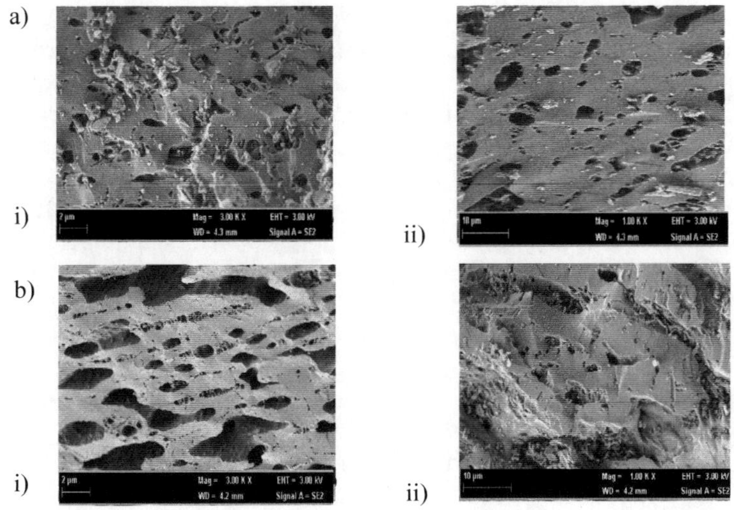

Figure 2. Surface morphologies of the adsorbents a) SEM photograph of fresh ACs: i) Mag: 3.00 K X (2 μ) and ii) Mag: 1.00 K X (10 μ) and b) impregnated ACs: i) Mag: 3.00 K X (2 μ) and ii) Mag: 1.00 K X (10 μ).

As shown in figure 2, the AC show bigger pores in materials that are expected to have a better CO_2 adsorption compared to other types of adsorbent materials. However, for impregnated AC, the surface layer has been fully covered by IL, making surfaces smoother. Based on the adsorption image, the pores in the sample IL was used to improve the adsorption performance and capacity over time. This result is supported by the adsorption performance unit that improved the AC surface and adsorption properties.

EDX was used to confirm the presence of elements on the surface of fresh and impregnated AC. Aside from the presence of pore in the mass percentage of IL loading, the existence of elements O and C was also observed in the physical morphology of AC. As shown in table 1, O and C elements are more abundant in impregnated AC compared with fresh AC because IL causes the increase in element percentage.

The physical adsorption is caused by van der Waals and electrostatic forces between adsorbate molecules and atoms at the adsorbent surface. A large specific surface area is important for a large adsorption capacity; however, a large internal surface area in a limited volume produces a large number of small-sized pores between adsorption surfaces [17].

The increase in physical affinity causes the increase in adsorption capacity. Numerous works about the separation of CO_2 show a high adsorption capacity at atmospheric pressure. Moreover, the advantages of CO_2 separation, such as hydrophobicity, affordability, and low energy requirement, are also discussed.

Table 1. Elements of fresh and impregnated AC.

Materials	Elements	Wt (%)	Σ
Fresh AC	-O	69.0	2.4
	-C	25.5	2.4
	-K	5.5	0.9
Impregnated AC	-O	72.7	0.2
	-C	27.3	0.2

3.2. Fourier transform infrared (FTIR) and spectroscopy and x - ray diffraction (XRD)

In Fourier transform infrared (FTIR) spectroscopy, infrared radiation is passed through a sample. The resulting spectrum represents the molecular absorption, creating a molecular fingerprint of the sample. FTIR was used to provide information about the identification of API and impurities and quantification of components in mixtures, e.g., the determination of amorphous material content in a crystalline sample (percent crystallinity). This method was also used to examine changes in the solid state because of interactions with excipients, i.e., compatibility studies in order to choose the most stable excipients for the final product. The region from 1500 to 400 wave numbers is referred to as the fingerprint region. Absorption bands in this region are generally caused by the intramolecular phenomenon and highly specific to each material.

Up to 0.5 mg of adsorbent sample was used for all tests. The FTIR spectra of the IL and AC are shown in figure 3. IL (figure 3 a) has the most complicated and clear spectrum. A broad adsorption peak was also observed at 3300 cm^{-1} to 3700 cm^{-1} attributed to O–H and N–H (IL case) stretch functional group; this result indicates the presence of bonded hydroxide in all the sample measurements. During infrared radiation, absorbed IR radiation excites molecules into a higher vibrational state. The wavelength of light absorbed by a particular molecule is a function of the energy difference between the at-rest and excited vibrational states. The wavelengths that are absorbed by the sample are characterized by their molecular structure.

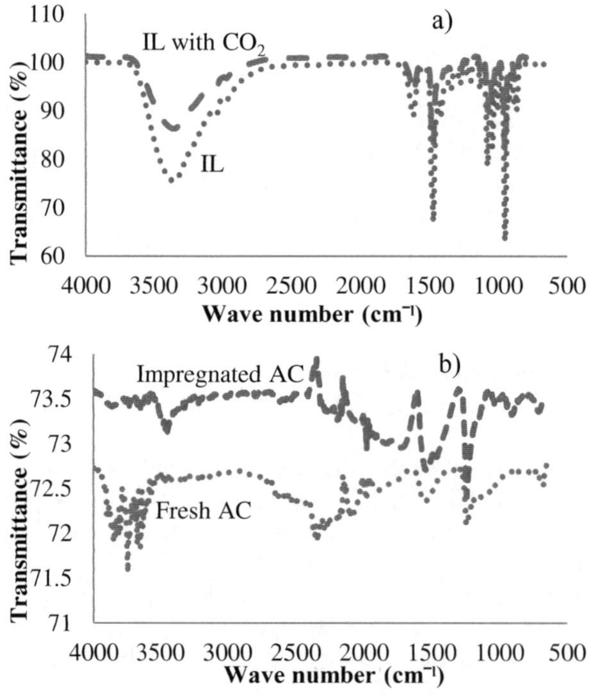

Figure 3. Functional group of IL and AC: a) Comparison between IL and IL injected with CO_2 gas and b) comparison of fresh and impregnated AC.

The O–H stretching absorption of the hydroxyl group is sensitive to hydrogen bonding. In the gas phase and in a diluted CCl_4 solution (0.01 M), small- to moderate-sized alcohols exhibit a sharp absorption at the 3620 cm^{-1} to 3670 cm^{-1} region. In a more concentrated solution or in a pure liquid, a hydrogen bond between the hydroxyl groups occurs, and this bond lowers and broadens the stretching frequencies of the participating O–H bonds.

In the IL region (figure 3 a), N–H bend (primary amines only) comes from 1622 cm^{-1}. The bands at 1474 cm^{-1} show a C–H bend or scissoring stretch. The bands at 1500 cm^{-1} to 1600 cm^{-1} of impregnated AC corresponded to the C=O functional group. The IL sample also shows two important adsorption peaks from 1004 cm^{-1} to 1100 cm^{-1} that refers to the C–O stretching functional group. However, fresh and impregnated AC shows C–N stretching (aliphatic amines) from 1250 cm^{-1} to 1020 cm^{-1}. The C–N stretching vibration of aliphatic amines is observed as medium or weak bands in the region 1250 cm-1 to 1020 cm^{-1}. In aromatic amines, the band is usually strong and at the region from 1335 cm^{-1} to 1250 cm^{-1}.

For impregnated AC (figure 3 b), during the carbonization and activation process, the adsorption peaks of the functional groups disappeared. This phenomenon is caused by the raw material spectrum that vaporized as volatile materials when the sample was heated [18]. The activation process was successfully proven. To confirm this phenomenon, the fresh AC spectrum was used as the reference sample. As a result, the trend of the impregnated AC was almost similar with the fresh AC. Thus, the prepared material of impregnated AC was successfully converted without any carbon loss.

XRD analysis is based on the constructive interference of monochromatic X-rays and a crystalline sample. X-rays are generated by a cathode ray tube, filtered to produce monochromatic radiation, collimated to concentrate, and directed toward the sample. The interaction of the incident rays with the sample produces constructive interference (and a diffracted ray) when conditions satisfy Bragg's Law ($n\lambda = 2d \sin \theta$). This law relates the wavelength of electromagnetic radiation to the diffraction angle and the lattice spacing in a crystalline sample.

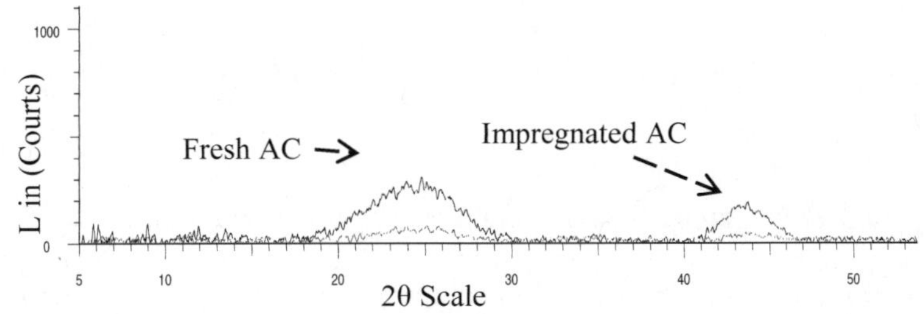

Figure 4. Comparison of the crystallinity of fresh and impregnated AC.

Figure 4 shows the comparison between fresh and impregnated AC. The impregnated AC was prepared by using $ZnCl_2$, which is affordable, and ionic liquid-type choline chloride (ammonia) with a ratio of 30% (w/w). As shown in figure 4, fresh and impregnated AC was in amorphous phase. AC is a non-crystalline material and a type of amorphous carbon that is composed of graphite carbon microcrystals and non-graphite carbon [19].

Results also showed that the intensity peaks increased with increasing the activation temperature but decreases with increasing the impregnation ratio of the materials. The low intensity peaks range from 20° to 40°, which correspond to the graphite lattice. The fresh and impregnated AC samples were observed to determine the characteristic x-ray diffraction pattern by providing a unique "fingerprint" of the crystals in the sample. When properly interpreted, by comparing with the standard reference patterns and measurements, this fingerprint allows the identification of crystalline. Thus, further FTIR analysis is needed for quantification of components in mixtures, e.g., the determination of the amorphous material in a crystalline API, as the absorbance is proportional to the content of the given component. Moreover, the percent crystallinity of a sample is determined using FTIR.

3.3. Thermogravimetric analysis (TGA) and physic sorption properties (BET analysis)

TGA is used to show if a compound contains components such as solvents or other loosely bound molecules that might evaporate upon heating. The presence of solvents or hydrates is usually not desirable because this might cause destabilization of the compound. Furthermore, this method is used to distinguish between solvates and "pure" polymorphic forms. The weight changes of an adsorbent, as a function of time, in N_2 atmosphere was measured.

Table 2. Fresh and impregnated AC decomposition before and after adsorption–desorption

Material in adsorption unit		Percentage reduction (%)		
		1st	2nd	3rd
Fresh AC	Before	0.77	2.73	20.27
	After	1.01	4.38	34.32
Impregnated AC	Before	1.19	6.27	43.61
	After	1.98	8.39	40.34

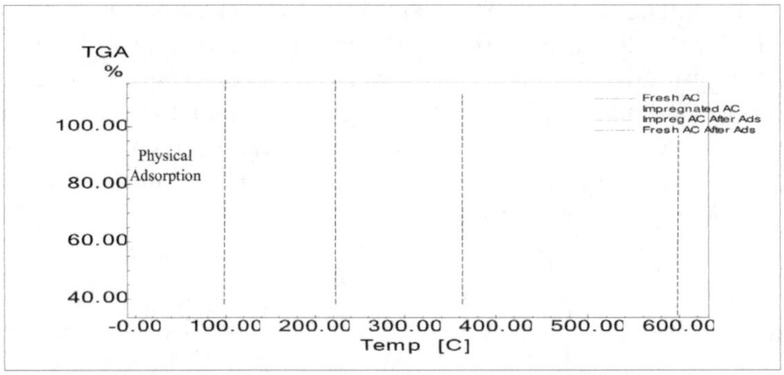

Figure 5. Comparison of weight loss study by TGA method.

The IL (quaternary ammonium salts) starts to decompose at 350 °C, and the high thermal stability of ILs is shown in figure 5. $ZnCl_2$ starts to vaporize and form ZnO at 480 °C. The decomposition of fresh and impregnation AC before and after adsorption is shown in table 2. As shown in figure 5, weight loss for impregnated AC is significant during adsorption compared with before adsorption. This phenomenon is caused by IL surrounding the surface of AC that have higher CO_2 levels compared with fresh AC. These IL show a greater tendency to dissolve CO_2 than other gases, like methane, ethane, oxygen, nitrogen, argon, and carbon monoxide. Therefore, the selective adsorption of CO_2 from flue gas is possible [20, 21].

Biomass product, such as coconut shells, is composed of hemicellulose, cellulose, and lignin. The pyrolysis of coconut shells is attributed to the pyrolysis of these polymers. Nunn et al. (1985) reported that the decomposition of cellulose occurs at 200 °C to 400 °C, whereas lignin decomposes at temperature ranges of 150 °C to 750 °C [22]. Furthermore, the rate of decomposition was low. At temperatures higher than 400 °C, the aromatization process is the most important reaction at low mass and loss rate [23].

Table 3. Comparison of fresh and impregnated AC through BET analysis (N_2).

Material	BET surface area (m^2/g)	Micro-pore area (m^2/g)	Pore volume (cm^3/g)	Micropore volume (cm^3/g)	Pore size (Å)
Fresh AC	863.04	689.79	0.41	0.32	19.14
Impregnated AC	117.49	93.23	0.06	0.04	19.14

BET results showed that the surface area decreased and the pore characteristics changed after chemical modifications. When the pressure is lower, the reaction between AC and CO_2 was slow. Thus, the BET surface area and total and micropore volumes were relatively small. With increasing pressure, the BET surface area and total and micropore volumes were increased because of the higher reaction rate between carbon and CO_2 and higher releasing rate of volatile matter.

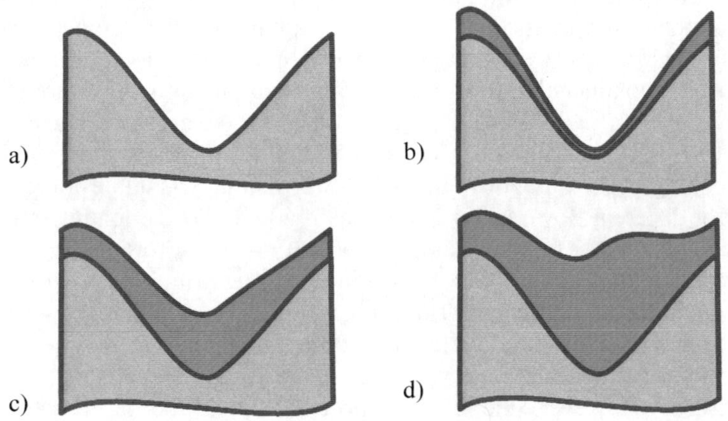

Figure 6. Schematic of the mechanism of pore filling with IL loading (AC: below and IL loading: upper): a) 0 wt. %, b) 10 wt. %, c) 20 wt. %, and d) 30 wt. %.

Table 3 shows the pore volume and micropore area that instantly decreased because of the IL that covered the surface of the adsorbent materials. IL has a greater tendency to cover the entire surface without interfering the area of pore sizes. Based on BET results, as shown in figure 6, 30 wt. % covered mostly the pore, and the capability of impregnated AC to adsorb CO_2 could be affected.

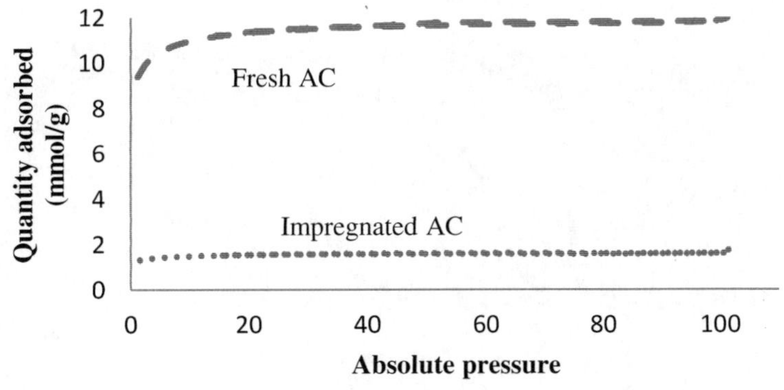

Figure 7. Results of BET analysis for adsorption: isotherm tabular report quantity adsorbed (cm ³/g STP) with absolute pressure (mmHg) for N_2 gas.

As shown in figure 7, 30 wt. % of IL covered all the pores of AC, resulting to a lower adsorption compared. Based on BET results about the physical characteristics of the surface adsorbents, IL covered the surface micropore when the gas carrier flow thorough the measurements. As a result, fresh AC showed higher adsorption results compared with impregnated AC that detect the surface layer of the adsorbent.

4. Adsorption and desorption of CO_2

4.1. Adsorption of fresh and impregnated AC

Active catalysts for this reaction are typically prepared using an impregnation method, and the identification of improvements in methodology preparation produces more active catalysts that retain stability is important. Li et al. (2013) studied the mixture of IL and water that was used to absorb CO_2, and results showed that the aqueous solution of IL was also good at absorbing CO_2 [24].

However, to reduce the cost and viscosity with high capturing effects, a nonionic surfactant was added to the aqueous solution of this IL. The gas mixture of CO_2/H_2 was used throughout the study. The adsorption of CO_2 was conducted in the adsorber single column unit. AC were impregnated by IL at a loading rate of 0, and 30 wt. % of these AC were studied for the adsorption and separation of CO_2 from the flue gas. The adsorption capacity of CO_2 with several flow rates is shown in table 4.

The capacity of loading gas with a flow rate of 0.1 L/min shows stable increments and acts as optimum parameters for different flow rates of gas concentrations. The IL loading rate of the fresh AC was 0 %, whereas the loading rate of impregnated AC was 30 %. Temperature used was 30 °C, and Co is indicated as the initial concentration of CO_2 before the adsorbent (mol/mol), whereas C is shown as the concentration of CO_2 after the adsorbent (mol/mol). figure 8 show that the AC supported by IL increases CO_2 adsorption over time. The flow rate of 1.0 L/min adsorbs faster. The adsorption capacity of impregnated AC in the adsorber unit clearly showed better results that adsorbs nearly 90 % of CO_2 in a for 6 h experiment. However, the cycling process should be performed to compare the performance of the adsorbents between the impregnated AC in the future.

Table 4. Adsorption capacity of fresh and impregnated ACs with different loading and flow rates.

Percentage of CO_2 gas loading (%)	Adsorption of fresh AC with different flow rates (L/min)			Adsorption of impregnated AC with different flow rates (L/min)		
	0.1	0.5	1.0	0.1	0.5	1.0
10	5.47	5.19	5.50	5.46	5.39	5.59
30	30.96	30.41	32.38	30.42	32.37	34.50

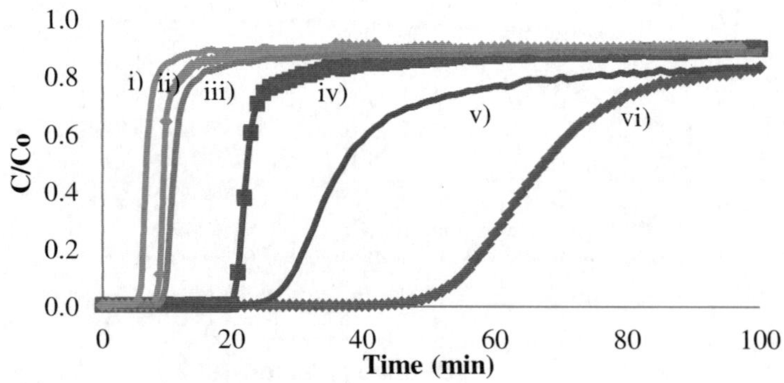

Figure 8. Comparison of CO_2 adsorption behavior of fresh AC on different flow rates of gas: i) 1.0 L/min impregnated AC, ii) 1.0 L/min fresh AC, iii) 0.5 L/min impregnated AC, iv) 0.5 L/min fresh AC, v) 0.1 L/min impregnated AC, and vi) 0.1 L/min fresh AC.

4.2. Desorption of fresh and impregnated AC
Different desorption mechanisms are compared using air with a flow rate of 3.0 L/min as the carrier that has been injected inside the chamber. As shown in figure 9, the impregnated AC was desorbed faster. Desorption of the impregnated AC at a flow rate of 1.0 L/min occurred within a minute. Thus, IL affects the CO_2 adsorption and desorption performance. Moreover, the amino functional IL adsorbs higher amount of CO_2 than other conventional methods [25](Yang & Wang, 2015).

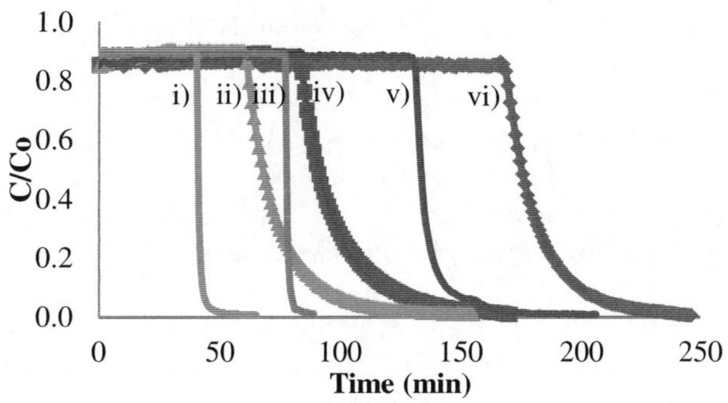

Figure 9. CO_2 desorption behavior of fresh AC at different flow rates of gas: i) 1.0 L/min impregnated AC, ii) 1.0 L/min fresh AC, iii) 0.5 L/min impregnated AC, iv) 0.5 L/min fresh AC, v) 0.1 L/min impregnated AC, and vi) 0.1 L/min fresh AC.

Impregnated AC (covered with IL) show improved adsorption and desorption of CO_2. For a short time period, impregnated AC show a great tendency to capture CO_2; this IL is best described with the cycling performance and affects different loading rates to elucidate the adsorbent performance in future study.

5. Conclusion
This study observed the effects of incorporating IL in AC on CO_2 adsorption. All AC materials (fresh and impregnated) that were characterized by using different analysis methods, such as SEM (morphology of surface), EDX (element composition), TGA (weight loss percentage), physisorption analyzer BET model (surface properties), FTIR (new chemical bond/functional group), and XRD (crystallinity of the adsorbent), were investigated and characterized. IL is for a promising solvent for separations and purifications, specializing in carbon capture process. IL exhibit improved CO_2 adsorption property by using an adsorber in real-time unit. Moreover, the porosity and surface area and IL activation play an important role in the adsorption of flue gas. As a result of adsorption in the actual adsorber unit, the impregnated AC reacts faster and adsorbs larger amounts of CO_2 than those of fresh AC in a short period of time at a high flow rate. To elucidate the adsorption process, further studies should focus on the performance of the adsorber unit. The cycling of adsorbent materials needs to be observed to obtain the optimal performance of the impregnated AC with focus on thermodynamic parameters.

References
[1] Debabrata D, Khanna N and Dasgupta C N 2014 *American J. of Biomass and Bioenergy* **3** 408.
[2] White C M, Steeper R R and Lutz A E 2006 *Int. J. Hydrogen Energy* **31** 1292.
[3] Choi S, Drese J H and John C W 2009 *Chem. Sus. Chem.* **2** 796.
[4] Li J R, Sculley J and Zhou H C 2011 *Chem. Rev.* **112** 869.
[5] McEwen J, Hayman J D and Yazaydin A O 2013 *Chem. Phys.* **412** 72.
[6] Thu K, Kim Y D, Ismil A B, Saha B B and Ng K C 2014 *Appl. Therm. Eng.* **72** 200.

IOP Conf. Series: Materials Science and Engineering **206** (2017) 012071 doi:10.1088/1757-899X/206/1/012071

[7] Su F S, Lu C Y, Chung A J and Liao C H 2014 *Appl. Energy* **113** 706.
[8] Chen S J, Fu Y, Huang Y X, Tao Z C and Zhu M 2016 *App. Energy* **179** 329.
[9] Yan T, Li T X, Wang R Z and Jia R 2015 *Appl. Therm. Eng.* **77** 20.
[10] Yang H, Gong M C and Chen Y Q 2011 *J. Nat. Gas. Chem.* **20** 460.
[11] Shenghui G, Jinhui P, Wei L, Kunbin Y, Libo Z, Shimin Z and Hongying X 2009 *Applied Surface Science* **255** 8443.
[12] Yin C Y, Aroua M K and Daud W M A W 2007 *Separation Purifaction Technology* **52** 403.
[13] Abbott A P, Glen C, David L D, Helen L M, Raymond K R and Vasuki T 2001 *Chem. Commun* 2010–11.
[14] Wang Q, Xia W and Guo W 2013 *Chemistry- An Asian Journal* **8** 1879.
[15] James J and Davis H 2002 *American Chemical Society* 247.
[16] Duan Z, Gu Y and Deng Y 2006 *Catal. Commun.* **7** 651.
[17] Motoyuki S 1990 *Adsorption Engineering* 25 Kodansha LTD Tokyo Japan.
[18] Yacob A 2008 *The Malaysian Journal of Analytical Sciences* **12** 264.
[19] Jiang J 2010 *Chemical Industry Press.*
[20] Jacquemin J, Gomes M F C, Husson P and Majer V 2006 *J. Chem. Thermodynamics* **38** 490.
[21] Anderson J L, Dixon J K and Brennecke J F 2007 *Accounts of Chemical Research* **40** 1208.
[22] Nunn R T, Howard J B, Longwell P J and Peters A W 1985 *Lignin Ind. Eng. Chem. Proc. Des. Dev.* **24** 84.
[23] Fisher T, Hajaligol M, Waymack B and Kellog D 2002 *J. Anal. Appl. Pyrolysis* **62** 331.
[24] Li T, Kong Q and Chen C 2013 *J. Chem. Ind. Eng* 600.
[25] Yang N and Wang R 2015 *J. Serb. Chem. Soc.* **80** 265.

Improvement of CO_2/N_2 separation performance by polymer matrix cellulose acetate butyrate

R J Lee[1], Z A Jawad[1]*, A L Ahmad[2], J Q Ngo[1], H B Chua[1]

[1] School of Engineering and Science, Department of Chemical Engineering, Curtin University Sarawak Malaysia, CDT 250, Miri 98009, Sarawak, Malaysia
[2] School of Chemical Engineering, Engineering Campus, Universities Sains Malaysia, 14300 Nibong Tebal, Penang, Malaysia

E-mail: zeinab.aj@curtin.edu.my

Abstract. With the rapid development of modern civilization, carbon dioxide (CO_2) is produced in large quantities and mainly generated from industrial sectors. The gas emission is the major contributor to global warming. To address this issue, the membrane technology is implemented for the CO_2 removal, due to the energy efficiency and economic advantages presented. Cellulose acetate butyrate (CAB) is selected as the polymeric material, due to the excellent film-forming properties and capable of developing a defect-free layer of neat membrane. This study described the fabrication development of CAB using a wet phase inversion method with different casting conditions. Where the composition of the casting solutions (3-5 wt %) and solvent evaporation time (4-6 min) were determined. The outcomes of these dominant parameters were then used to determine the best CAB membrane for CO_2/Nitrogen (N_2) separation and supported by the characterization i.e. scanning electron micrograph. Gas permeation measurements showed satisfactory performance for CAB membrane fabricated with 5 min evaporation time and 4 wt% polymer composition (M2). Where, its permeance and selectivity are 120.19 GPU and 3.17, respectively. In summary, this study showed a brief outlined of the future direction and perspective of CAB membrane for CO_2/N_2 separation.

1. Introduction

In the past century, carbon dioxide (CO_2) is widely known as the leading contributor to global warming. A total of 35.4 billion tonnes of CO_2 was emitted to the environment annually [1]. With the rising of the greenhouse gasses (GHGs) effect, which increases the atmospheric concentration of CO_2, this had caused lots of concern amongst the researchers in combating this global issue. CO_2 is being classified as the most significant anthropogenic GHGs and partaking around two-third of the GHGs composition [2-4]. The primary source of CO_2 emission is commonly found in the natural gas streams, biogas, and flue gas from fossil fuel combustion [5].

Due to the enormous amount of CO_2 emission back in the mid-19th century, the application of membrane technology had increased remarkably since the first invention in 1981 [6]. The membrane gas separation technology is designed with the purpose to separate individual gas component based on the different permeation rates of each gas component through a thin membrane barrier [7]. The membrane separation performance characteristics are commonly indicated by permeation and selectivity [8]. In an ideal situation, high permeability is preferable while maintaining high selectivity, therefore smaller membrane area and lower driving force is required. Thus, such ideal conditions require less operating cost and hence lower capital cost [9, 10].

There are less than ten types of polymer materials that have been used for at least 90% of the total installed membrane-based gas separation modules, including cellulose acetate butyrate (CAB), cellulose acetate (CA), polyimide, polyamide, polysulfone, polycarbonates, polyethersulfone, and polyphenylene oxide [11]. Based on Basu, Khan [12], the CAB exhibits excellent film-forming properties, the butyryl group in CAB can effectively improve and expand the capacity of the material volume cellulose chain membrane through enlarging the free volume [12]. CAB polymer possess several prominent characteristics including high impact resistance, well-maintained weather resistance, notable chemical resistance with excellent film forming characteristics inherited from acetyl and butyl groups [13].

However, polymeric membrane formation usually involved a series of complex process which depends on a vast variety of parameters [14] such as polymer concentration, and solvent evaporation time [15]. In this regard, Jansen, Macchione [14] proved that the solvent evaporation duration is relatively necessary to determine the membrane gas diffusion and solubility coefficients through sorption measurement [14]. Paulsen, Shojaie [16] investigated the effects of duration of the evaporation step between 10-150 s. They found out that increasing the length of solvent evaporation always tend to damp macro voids formation for CA membrane. The reduction of the microvoids formation eventually improved the overall selectivity of the CA membrane [16]. Also, the polymer concentration plays a key factor in the membrane formation. Where increasing the polymer concentration reduced the microvoids of the membrane due to dense skin formation [17]. The increase in polymer concentration also improved the selectivity performance of the membrane, due to the tighter structural membrane formed [18].

Thus, the aim of the present work is to improve the separation performance of CAB membrane towards CO_2/N_2 by investigating the effect of solvent evaporation time (chloroform) and CAB polymer concentration. Up to date, there is no report has examined this parameter yet. Whereby, both studied dominant parameters will then utilized in the fabrication of mixed matrix membrane (MMM).

2. Methodologies

2.1. Materials
CAB polymer (70,000 Mn) with an acetyl group (12%-15%) was purchased from Sigma-Aldrich (Malaysia). The chemical solutions required for the membrane fabrication were chloroform, isopropyl alcohol (99.6%) and n-hexane (99.8%), which can purchase from Merck (Malaysia).

2.2. CAB membrane preparation
The CAB membrane was fabricated using the phase inversion method. The solution consists of 4 wt% CAB and 96 wt% chloroform, was stirred for 24 hrs [19]. Then, the dope solution was sonicated for 20 min to eliminate the bubble in solution to ensure the surface homogenous of the neat membrane [19]. The dope solution was cast using the automatic film applicator on a glass plate, under the standard operating fume hood. A uniform casting thickness of 250 μm was applied. Allow 5 min solvent evaporation time to evaporate the solvent followed by immersing the membrane film in distilled water (0-22 °C) for 24 hr [20]. The as-spun membrane was immersed in isopropyl alcohol for 1 hr followed by n-hexane for another 1 hr [20]. The resultant membrane (M1) was dry in the oven at 70 °C to eliminate the remaining volatile liquid between two glass plates for 24 hr and stored before use [20].

IOP Conf. Series: Materials Science and Engineering **206** (2017) 012072 doi:10.1088/1757-899X/206/1/012072

2.2.1. Effect of solvent evaporation time. The solution was cast based on the synthesis method described in section 2.2. The studied range was from 4 min to 6 min. The composition of the membrane fabrication process is shown in table 1.

Table 1. The composition of membrane prepared at different solvent evaporation time

Sample Description	CAB (wt%)	Chloroform (wt%)	Casting Thickness (μm)	Evaporation time (min)
M1	4	96	250	4
M2	4	96	250	5
M3	4	96	250	6

2.2.2. Effect of CAB polymer concentration. The solution was cast based on the synthesis method described in section 2.2. The studied range was from 3 wt% to 5 wt%. The composition of the membrane fabrication process is shown in table 2.

Table 2. The composition of membrane prepared at different CAB polymer concentration

Sample Description	CAB (wt%)	Chloroform (wt%)	Casting Thickness (μm)	Evaporation time (min)
M4	3	97	250	5
M2	4	96	250	5
M5	5	95	250	5

2.3. Membrane characterization

2.3.1. Scanning electron microscopy (SEM). The membrane surface morphology and cross-section were observed with a scanning electron microscope (Hitachi TM3000, Tokyo, Japan). Each membrane sample was cut into small pieces and was kept in the cryogenic freezer with plastic petri dish up to -80 °C for 24 hr to give a consistent and clean cut by freezing. The samples were then coated with platinum to prevent the accumulation of static charges at the specimen. Also, the membrane thickness of every sample was calculated based on the frequency count that was observed in these images using Image J software. Approximately 100 measurements were taken to confirm it.

2.4. Gas permeability measurement

Single gas permeation tests were conducted using either pure N_2 or CO_2 gas at room temperature. The kinetic diameter of the N_2 and CO_2 were 3.64 *Å* and 3.3 *Å*, respectively. The schematic diagram of the experimental rig shows in figure 1. For each single gas permeation test, the flow rate of the gas was controlled at 100 ml/min using the mass flow controller (Aalborg AFC26, USA) and gas supplied from compressed gas cylinder tank.

The mass flow controller was connected to a two-channel digital set point/ readout unit (Aalborg 0-200 ml/min, USA). The feed gas pressure for each of the gasses was set from 1-2 bars throughout the experimental investigation. Prior the experiment, leak detection test was conducted to ensure no feed gas escape from the rig pipes. Allowed the feed N_2 gas to flow throughout the rig for 10 min, to purge out the impurities gas inside the pipe. After that, the fabricated membrane was cut into a round disc shape with an effective diameter of 7.065 cm^2 and placed in the membrane permeation cell. Covered and tightened the membrane permeation cell before connecting it back to the streams. The permeate flow rate can be obtained and measured through the volume displacement of the soap bubble flow meters.

IOP Conf. Series: Materials Science and Engineering **206** (2017) 012072 doi:10.1088/1757-899X/206/1/012072

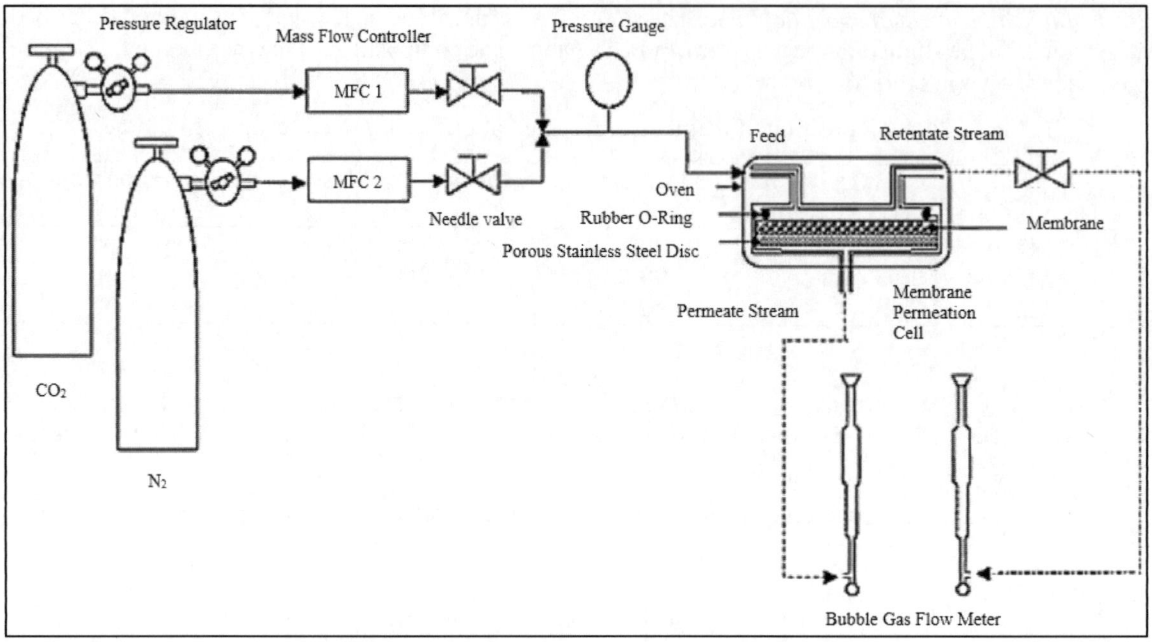

Figure 1. Schematic Diagram of the Experimental Rig

The permeance of the membrane is indicated by (P/ℓ), which can be calculated using equation (1) and the unit of the permeance was expressed in GPU. Whereby the l represent the membrane thickness in cm, Q the volumetric flow rate indicated from flow meter in cm^3/s, A is the effective membrane area in cm^2, and ΔP is the pressure difference in the membrane permeation cell which is expressed in cmHg [22].

$$(P/\ell) = (Q/A\Delta P) \tag{1}$$

Meanwhile, the selectivity of the membrane can be determined using the ideal separation factor (α) as shown in equation (2) [22].

$$\alpha_{CO_2/N_2} = P_{CO_2}/P_{N_2} = [(P/l)_{CO_2}/(P/l)_{N_2}] \tag{2}$$

Whereby, the P_{CO_2} and P_{N_2} represent the permeability of components CO_2 and N_2, respectively [23]. Each membrane specimens were tested at least 4 times to ensure precision of the results generated. The average values together with the standard errors, were reported.

3. Results and discussion

3.1. Effect of solvent evaporation duration

The solvent evaporation period is one of the crucial parameters that manipulates the membrane morphology. SEM was utilized to observe the physical morphology of the CAB, as shown in figure 2. The figure 2a and figure 2c shown smooth surface for M1 with evaporation time of 4 min and M2 with evaporation time of 5 min. On the other hand, the M3 membrane with evaporation time of 6 min is undergoing a transitional phase from smooth surface to porous surface structure. Provided with longer evaporation time, rapid diffusion of the solvent (chloroform) can occur, where the chloroform evaporated from the as-spun membrane before the water immersion [24]. Thus, porous and dense structures were formed.

According to figure 2, M1 (4 min), M2 (5 min), and M3 (6 min) show a thickness of 497.8 μm, 148.2 μm, and 555.7 μm, respectively. This was due to the compact properties of CAB chains in the formation of M2 membrane during the controlled solvent evaporation period (5 min), forming a thin membrane [29]. However, M3 membrane shows thicker membrane of 555.7 μm. This was due to the rapid evaporation of the casting solvent from the membrane due to prolonged evaporation time (6 min), hence causing an expansion in the membrane volume. Moreover, the thicker membrane exerts more flow resistance to the membrane, which typically yields a low separation performance membrane [16, 29].

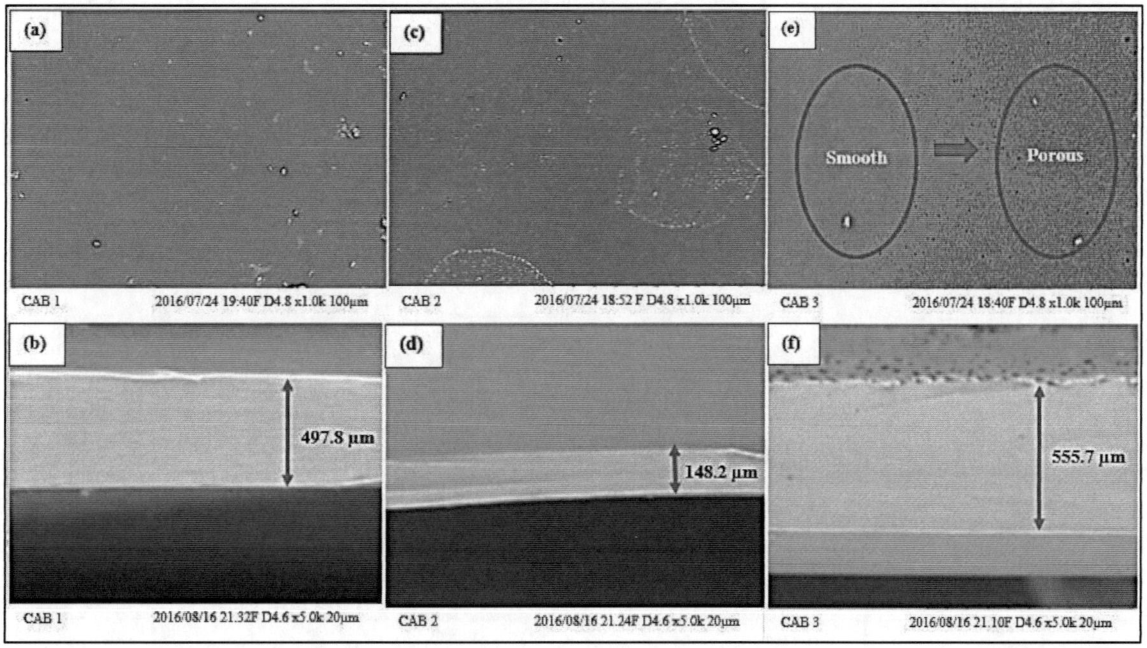

Figure 2. SEM morphologies of surface and cross section of CAB membrane with evaporation time of (a-b) 4 min (M1), (c-d) 5 min (M2), and (e-f) 6 min **(**M3)

According to table 3, M3 with the longest solvent evaporation time up to 6 min yield highest permeance results ranging from 1005-1253 GPU as compared to M1 (107-117 GPU) and M2 (37-120 GPU). This can explain by the arising porous surface structure exhibited in M3 (figure 2e). It was due to prolong the evaporation period of the highly volatile solvent (chloroform), causing a high volume of solvent evaporation outflow during solvent evaporation period [25]. It might allow the gas to escape through the membrane structure quickly. However, the selectivity is significantly reduced due to the high permeance rate causing the trade-off relationship between permeance and selectivity performance. On the other hand, consistent permeance results were observed for both M1 and M2 membranes, and M2 membrane with 5 min solvent evaporation time is preferable due to the better selectivity was shown.

Table 3. Permeability and selectivity performance for M1, M2, and M3

	M1			M2			M3		
Pressure (Bar)	P_{CO2} (GPU)	P_{N2} (GPU)	α	P_{CO2} (GPU)	P_{N2} (GPU)	α	P_{CO2} (GPU)	P_{N2} (GPU)	α
1.0	113.06	116.03	0.97	40.51	119.20	2.94	1010.74	1253.49	1.24
1.5	117.84	107.49	1.10	37.83	116.91	3.09	1005.34	1251.43	1.25
2.0	111.79	109.33	1.02	37.91	120.19	3.17	1006.46	1253.71	1.25

Meanwhile, M2 shows higher CO_2/N_2 separation performance with a selectivity of 2.94-3.17 as compared to M1 (0.97-1.02) and M3 (1.24-1.25) in figure 3. This proved that the CAB membrane achieved the highest selectivity of CO_2 with 6 min solvent evaporation time, due to the smooth surface layer which selectively allows a predetermine amount of CO_2 to pass through the dense membrane which was supported with the solution-diffusion mechanism [26].

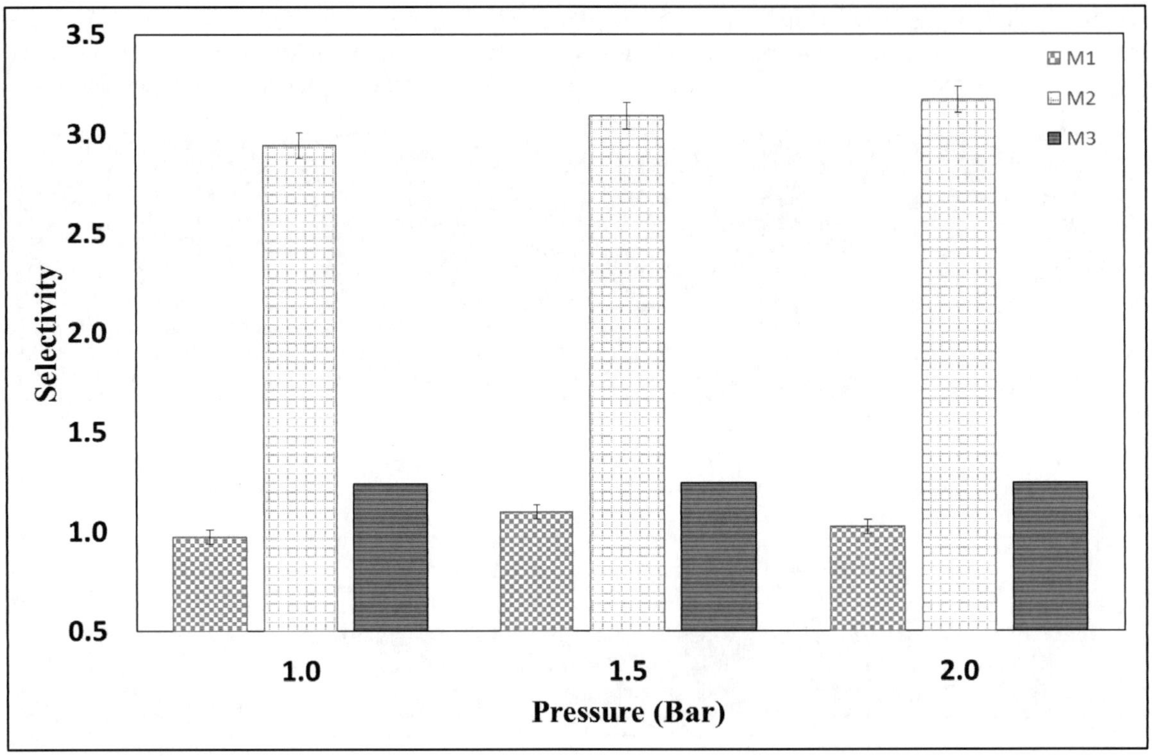

Figure 3. Ideal selectivity of CO_2/N_2 through CAB membrane with evaporation time of 4 min (M1), 5 min (M2), and 6 min (M3).

3.2. Effect of polymer concentration
The polymer loading concentration is another dominant factor that affects the surface morphology of the membrane. Figure 4 demonstrates the top surface and cross-sectional morphology at polymer concentrations of 3 wt% (M4), 4 wt% (M2), and 5 wt% (M5). A porous surface membrane structure was observed for M4 (figure 4a) at 3 wt% of CAB polymer. However, the CAB membrane morphology changes from porous to smooth dense membrane when the polymer concentration increases to 4 wt% (M2) and 5 wt%, as shown in figure 4c and 4e, respectively. This can be explained with the increment in the viscosity of the polymer, hence reducing the fluid of solution, thus causes the delayed demixing in the coagulant bath [18].

Based on table 4, M4 yield highest permeance results ranging from 303-318 GPU as compared to M2 (37-98 GPU) and M5 (50-98 GPU). The possible explanation for M4 membrane having high permeance rate is due to the porous surface layer which allows the gas to escape through a less resistance pathway. As compared to M2 with a thickness of 148.2 µm, M4 and M5 membranes have a thicker membrane of 355.2 µm and 441.2 µm, respectively, as demonstrated in figure 2b and figure 2f. This was due to the effect of a smaller concentration gradient, compensated by a much lower casting solution viscosity [27]. On the other hand, the membrane M4 supposedly should exhibit thin dense membrane. However due to the high solvent composition with low polymer concentration (3 wt%), more solvent evaporated from the membrane resulting in an increase in membrane thickness to 355.2 µm (Asgarkhani et al. 2013). This thicker membrane reduced the separation results of the membranes. Thus, it proved that the effect of the membrane thickness is to exert additional flow resistance to the membrane, therefore less effective membrane yield (Lan et al. 2005).

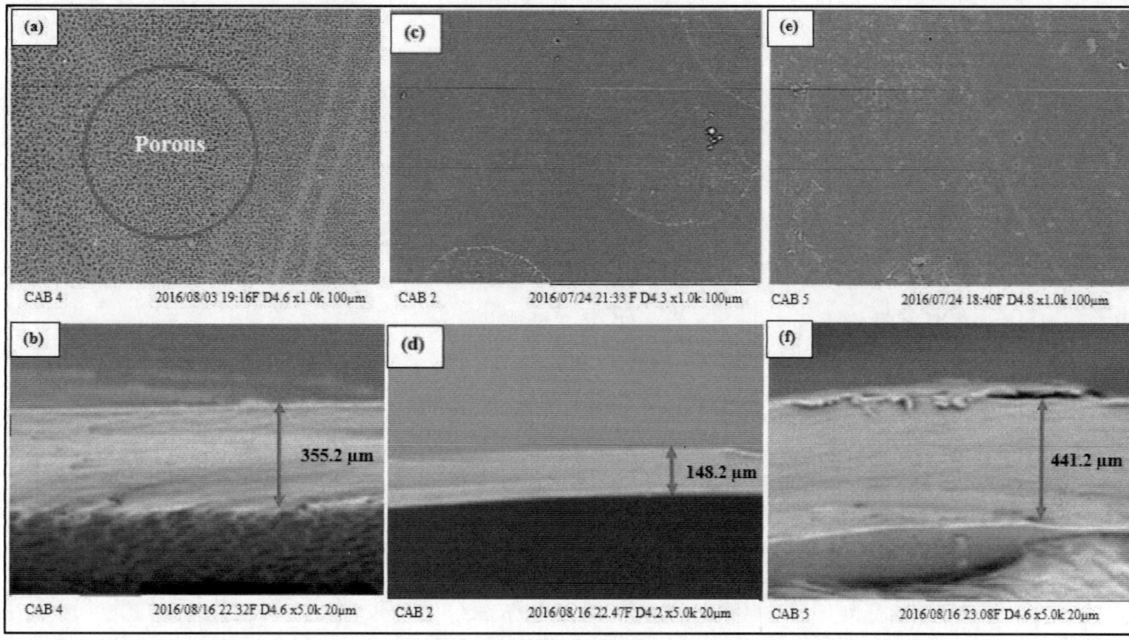

Figure 4. SEM morphologies of surface and cross section of CAB membrane with a polymer concentration of (a-b) 3 wt% (M4), (c-d) 4 wt% (M2), and (e-f) 5 wt% (M5).

The ideal selectivity of M2, M4 and M5 was also investigated as shown in figure 5. The M2 membrane proved to have the best selectivity results amongst the others (M4 and M5). The possible explanation was when low polymer concentration was substituted, the morphology of membrane became less compact with big aperture, resulting in the high flux of membrane while decreasing the diffusion selectivity [27]. Moreover, this causes the structure of membrane become more compact, resulting in less flux, and thus lower the diffusion selectivity [28]. In summary, the M2 membrane with 4 wt% of CAB polymer has the best CO_2/N_2 separation performance.

Table 4. Permeability and selectivity performance for M2, M4, and M5

	M4			M2			M5		
Pressure (Bar)	P_{CO2} (GPU)	P_{N2} (GPU)	α	P_{CO2} (GPU)	P_{N2} (GPU)	α	P_{CO2} (GPU)	P_{N2} (GPU)	α
1.0	317.94	303.43	1.05	40.51	119.20	2.94	98.11	50.28	1.95
1.5	313.05	306.68	1.02	37.83	116.91	3.09	93.63	50.49	1.85
2.0	318.95	304.48	1.05	37.91	120.19	3.17	90.75	50.57	1.79

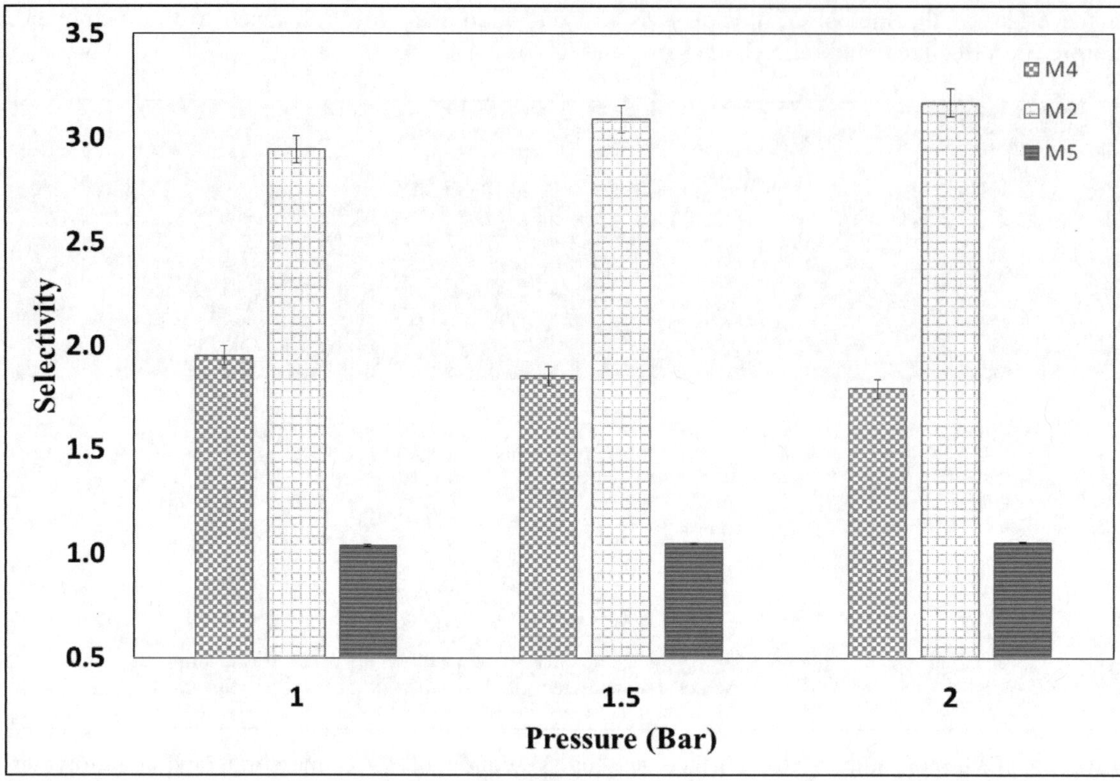

Figure 5. Ideal selectivity of CO_2/N_2 through CAB membrane with polymer concentration 3 wt% (M4), 4 wt% (M2), and 5 wt% (M5).

4. Conclusion

In the present work, it was found that the solvent evaporation duration and concentration of the polymer correlate closely in manipulating the membrane morphology and ultimately affect the CO_2/N_2 separation performance of CAB membrane. The dominant parameters for solvent evaporation time and CAB polymer concentration were 5 min with 4 wt% concentration loading which was represented by M2. Meanwhile, to improve the CO_2/N_2 selectivity of the CAB membrane, it was suggested to incorporate the CAB polymer with inorganic filler such as carbon nanotubes (CNTs) to produce mixed matrix membrane (MMM).

Acknowledgments

The authors would like to appreciate Ministry of Higher Education Malaysia (MOHE) for providing Fundamental Research Grant Scheme (FRGS) (MOHE Ref. No: FRGS/1/2015/TK02/CURTIN/03/1) and Cost Centre: 001048. Also, the authors would also like to thank LRGS USM (Account No: 304/PJKIMIA/6050296/U124), and Curtin Cost Centre: 001047.

References

[1.] Zhang Y et al 2013 *Int. J. of Greenhouse Gas Control*. **12** p 84-107
[2.] Zhang W et al 2008 *Chem. Ind. and Engr. Progr*. **27**(5) p 635
[3.] Nyambura M G et al 2011 *J. of Env. Mgmt*. **92**(3) p 655-64
[4.] Pan S Y, E Chang and P C Chiang 2012 *Aerosol Air Qual Res*. **12**(5) p 770-91
[5.] Zhang Z E et al 2014 *Glob. Nest J*. **16** p 355-74

[6.] Dortmundt D and K Doshi 1999 UOP LLC.
[7.] Kohl A L and R Nielsen 1997 *Gulf Prof Publishing*.
[8.] Ismail A F et al 2009 *Separ. and P. Tech*. **70**(1) p 12-26
[9.] Paradise M and T Goswami 2007 *M & Design*. **28**(5) p 1477-89
[10.] Low B T et al 2013 *J. of Mem. Science*. **431** p 139-55
[11.] Baker R W 2002 *Ind. & Engr. Chem. Research*. **41**(6) p 1393-411
[12.] Basu S et al 2010 *Chem. So.c Reviews*. **39**(2) p 750-68
[13.] Kunthadong P et al 2015 *J. of Polymers and the Env*. **23**(1) p 107-13
[14.] Jansen J C et al 2005 *Polymer*. **46**(25) p 11366-79
[15.] Kesting R E and A Fritzsche 1993 *Wiley-Interscience*.
[16.] Paulsen F G, S S Shojaie and W B Krantz 1994 *J. of Mem. Science*. **91**(3) p 265-82
[17.] Jansen J C et al 2006 *J. of Mem. Science*. **272**(1) p 188-97
[18.] Xie S M et al 2008 *J. of Mem. Science*. **321**(2) p 293-8
[19.] Feng Y, J M Zhang and J Zhang 2015 *ACTA POLYMERICA SINICA*. **12** p 1396-401
[20.] S.Minhas B 1992 *Google Patents*.
[21.] Sadeghi M, M A Semsarzadeh and H Moadel 2009 *J. of Mem. Science*. **331**(1) p 21-30
[22.] Ismail A F et al 2011 *Separ. and P. Tech*. **80**(1) p 20-31
[23.] Baker R W, J Wijmans and Y Huang 2010 *J. of Mem. Science*. **348**(1) p 346-52
[24.] Ong A L et al 2008 *J. of Power Sources*. **183**(1) p 62-8
[25.] Oyama K O N, E L Siqueira and M d Santos 2002 *Brazilian dental J*. **13**(3) p 208-11
[26.] Xing D Y, N Peng and T S Chung 2010 *Ind. & Engr Chem. Research*. **49**(18) p 8761-9
[27.] Yang Y Y, T S Chung and N. Ping Ng 2001 *Biomaterials*. **22**(3) p 231-41
[28.] Mao Y et al 2006 *J. of Mem. Science*. **279**(1–2) p 246-55
[29.] N Widjojo, TS Chung, S Kulprathipanja 2008 *J. of Mem. Science*. **25**(1) p 326-35

IOP Conf. Series: Materials Science and Engineering **206** (2017) 012073 doi:10.1088/1757-899X/206/1/012073

Effective flocculation of *Chlorella vulgaris* using chitosan with zeta potential measurement

Y J Low, S W Lau*

Department of Chemical Engineering, Curtin University Malaysia, CDT 250, 90889 Miri, Sarawak, Malaysia

E-mail: lau.shiew.wei@curtin.edu.my

Abstract. Microalgae are considered as one promising source of third-generation biofuels due to their fast growth rates, potentially higher yield rates and wide ranges of growth conditions. However, the extremely low biomass concentration in microalgae cultures presents a great challenge to the harvesting of microalgae because a large volume of water needs to be removed to obtain dry microalgal cells for the subsequent oil extraction process. In this study, the fresh water microalgae *Chlorella vulgaris* (*C. vulgaris)* was effectively harvested using both low molecular weight (MW) and high MW chitosan flocculants. The flocculation efficiency was evaluated by physical appearance, supernatant absorbance, zeta potential and solids content after centrifugal dewatering. High flocculation efficiency of 98.0-99.0% was achieved at the optimal dosage of 30-40 mg/g with formation of large microalgae flocs. This study suggests that the polymer bridging mechanism was governing the flocculation behaviour of *C. vulgaris* using high MW chitosan. Besides, charge patch neutralisation mechanism prevailed at low MW chitosan where lower dosage was sufficient to reach near-zero zeta potential compared with the high MW chitosan. The amount of chitosan polymer present in the culture may also affect the mechanism of flocculation.

1. Introduction

Harvesting of algal biomass from dilute culture medium is a major bottleneck for economical biofuel production using microalgae as feedstock. It is estimated that the cost of recovery process contributes to 20-30% of the total cost of biomass production, which accounts for a total of about 50% to the final cost of oil production [1]. Conventional harvesting techniques such as filtration, flotation, and centrifugation do not meet the economic requirements for commercialisation of algal biofuels, as compared with the other high-value, low-volume products such as β-carotene derived from microalgae. In particular, the centrifugation method manages to achieve only 50% energy efficiency on the basis of available energy in the recovered biomass relative to the energy input [1]. Therefore, one major challenge to the commercialisation of algal biofuels is to develop a harvesting technique of high efficiency, low energy demand and hence low cost.

Although microalgae are not yet produced at large scale industrial processing, it is believed that advances in research and development of microalgae present vast opportunities for sustainable development of this potential biofuel feedstock within the next 10 to 15 years [2]. Recent studies have proposed several novel microalgae harvesting techniques such as electro-flotation [3], bio-flocculation [4] and membrane separation [5]. It is noteworthy that most studies adopt some form of coagulation and/or flocculation as the sole or supplementary treatment to concentrate and harvest microalgae [3, 6].

IOP Conf. Series: Materials Science and Engineering **206** (2017) 012073 doi:10.1088/1757-899X/206/1/012073

Flocculation is able to concentrate a dilute algal suspension by about 20-100 times into a thick slurry; then centrifugation dewaters the slurry into algal paste of about 25% dry matter content [7].

The effectiveness of chitosan as a flocculant for microalgae harvesting is well known. However, the feasibility of this natural biopolymer for effective microalgae harvesting is yet to be further explored and understood in order to optimise its performance. It is also found that many studies express chitosan dosage in mg/L which confounds the comparison and interpretation of effective flocculant dosage because the biomass concentrations usually differ from one another. This study presents an experimental investigation using low molecular weight (MW) and high MW chitosan flocculants to harvest *C. vulgaris* in order to achieve favourable solids content after centrifugal dewatering. Instead of using the common solvent—acetic acid, hydrochloric acid (HCl) was used as the solvent because chitosan is highly soluble in HCl and the solution is also found to give high separation efficiency [8]. The chitosan dosage is expressed as the weight of flocculant over the weight of biomass solids in the culture, in mg/g, in order to facilitate comparison of performance regardless of its biomass concentration. The study involved two steps of harvesting, with flocculation by chitosan first as the pre-concentrating step, followed by centrifugal dewatering step. Mechanisms involved in the flocculation process were studied and justified based on the evaluation of physical appearance, supernatant absorbance, zeta potential and solids content after dewatering.

2. Experimental procedure

2.1. Cultivation of C. vulgaris

Pure strain of *C. vulgaris* was purchased from Algaetech International Sdn. Bhd., Malaysia. *C. vulgaris* inoculum was cultivated in modified Bold's Basal Medium (BBM) at room temperature, with a photoperiod of 16:8 h light and dark cycle [9], and 15-min aeration every hour during the light cycle. The first batch of cultivation (labelled as Culture A) was cultured for 10 weeks while the second batch (labelled as Culture B) was cultured for 8 weeks, keeping all cultivation conditions unchanged. The modified BBM formulation with each component's concentration in the final medium is given as follows: $NaNO_3$ [25 g/L], $CaCl_2{\cdot}2H_2O$ [2.5 g/L], $MgSO_4{\cdot}7H_2O$ [7.5 g/L], K_2HPO_4 [7.5 g/L], KH_2PO_4 [17.5 g/L], NaCl [2.5 g/L], EDTA [50 g/L], KOH [31 g/L], $FeSO_4{\cdot}7H_2O$ [4.98 g/L], H_2SO_4 [4.98 g/L], H_3BO_3 [11.42 g/L], $ZnO_4{\cdot}7H_2O$ [8.82 g/L], $MnCl_2{\cdot}2H_2O$ [1.44 g/L], $CuSO_4{\cdot}5H_2O$ [1.57 g/L], $Co(NO_3)_2{\cdot}6H_2O$ [0.49 g/L], and 2.5 g sucrose in 1 L of the medium [10]. The final medium was sterilised by autoclave before use.

2.2. Characterisation

After 8–10 weeks of cultivation, mature culture of *C. vulgaris* was sampled for measurement of pH, zeta potential and biomass concentration. Zeta potential of the culture supernatant after 20-min settling was measured by Zetasizer Nano ZS (Malvern) using Laser Doppler Electrophoresis method. To determine the algal biomass concentration, 20-mL microalgae sample was taken from a well-mixed culture suspension and dried in oven at 60°C overnight or until constant weight was obtained. Table 1 shows the characteristics of *C. vulgaris* in both Culture A and Culture B.

Table 1. Characterisation of C. vulgaris.

Culture	A	B
Cultivation period	10 weeks	8 weeks
pH	7.80	7.64
Zeta potential (mV)	-35.2	-28.8
Biomass concentration (g/L)	1.79	2.09

IOP Conf. Series: Materials Science and Engineering **206** (2017) 012073 doi:10.1088/1757-899X/206/1/012073

2.3. Flocculation

Low MW chitosan (50,000–190,000 Da, Aldrich) and high MW chitosan (600,000–800,000 Da, Acros Organics) were prepared as flocculant solutions by dissolving 0.1 g chitosan in 100 mL of 0.1 M hydrochloric acid (HCl) at 40°C under gentle mixing for about an hour or until it was completely dissolved [11]. The solutions were freshly prepared prior to the commencement of experiments to avoid degradation of chitosan polymer in solvent [12]. The flocculant was added into the 100-mL *C. vulgaris* sample at the predetermined dosage, ranging from 0 mg/g to 60 mg/g. The culture medium was rapidly mixed at 500 rpm for 1 min to allow coagulation, followed by slow mixing at 100 rpm for 5 min to promote flocculation. The suspension was then left to settle under quiescent condition for 20 min and the supernatant was sampled at 1-cm depth from the suspension surface for zeta potential measurement [13].

Lambda 25 UV/Visible Spectrometer (Perkin Elmer) was used to measure the absorbance of light by *C. vulgaris*. As shown in figure 1, the characteristic peak wavelength for *C. vulgaris* was found close to 680 nm which was used for the determination of absorbance value as an indirect measurement of cell concentration and flocculation efficiency. The absorbance values were calibrated against cell concentrations for the calculation of flocculation efficiency in equation (1):

$$\text{Flocculation Efficiency (\%)} = 100 \times \left(1 - \frac{C_{supernatant}}{C_{initial}}\right) \tag{1}$$

where $C_{supernatant}$ is the concentration of the algal supernatant with flocculant and $C_{initial}$ the initial concentration of the algal supernatant without flocculant (control).

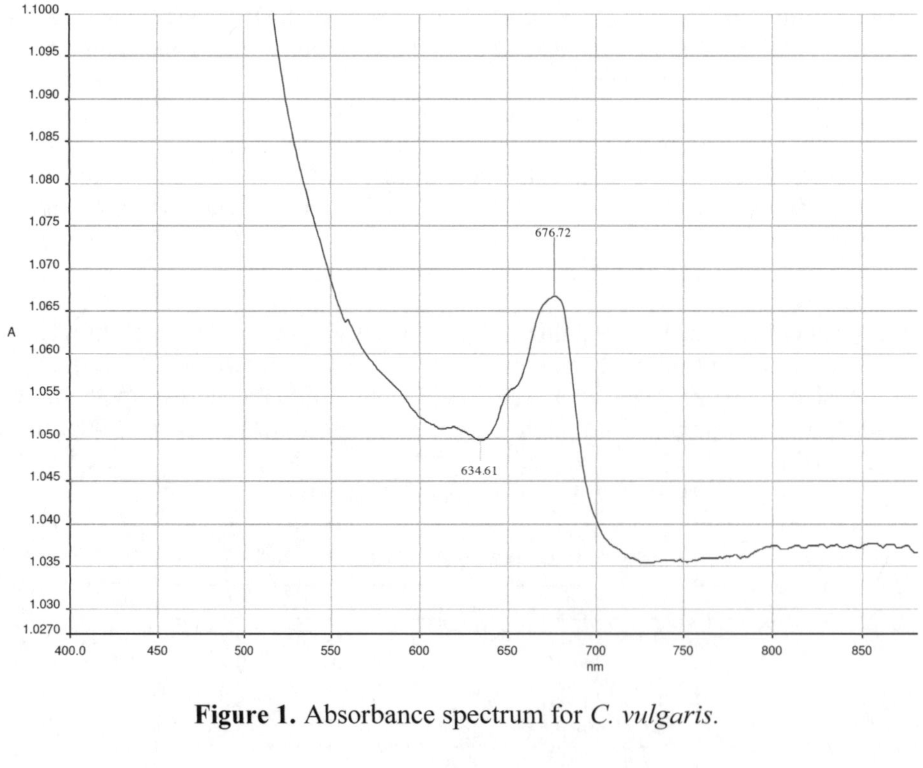

Figure 1. Absorbance spectrum for *C. vulgaris*.

IOP Conf. Series: Materials Science and Engineering **206** (2017) 012073 doi:10.1088/1757-899X/206/1/012073

2.4. Centrifugation

After flocculation, 90% of the supernatant was decanted leaving 10-mL algal flocs which were collected and centrifuged at 2000 rpm for 10 min. The concentrated algal paste was then transferred to a pre-weighted filter paper and dried in the oven at 60°C for 12 hours or until constant weight was obtained. The solids content of the dewatered algal flocs was determined using equation (2), where M_W is the weight of the wet algal paste and filter paper, M_D the weight of the dried algal paste and filter paper, and M_0 the weight of the empty filter paper, all expressed in g.

$$Solids\ Content\ (\%) = \frac{M_D - M_0}{M_W - M_0} \times 100 \tag{2}$$

3. Results

3.1. Effect of molecular weight on flocculation efficiency

From figure 2, it can be seen that high MW chitosan gave high flocculation efficiency of 97% at the dosage of 20 mg/g while low MW chitosan can only achieve 49% at the same dosage in Culture A. Additional 10 g/mg was required to reach optimum flocculation using low MW chitosan. For Culture B, the effect of molecular weight became less significant where both low and high MW chitosans gave high flocculation efficiency of 98-99% at 20 mg/g. However, at lower dosage, high MW chitosan still showed better flocculation of *C. vulgaris* than low MW chitosan. Nevertheless, no visible algal flocs were formed at lower dosage. The microalgae appeared as discrete tiny sediments at the bottom of the suspension which were easily disturbed when external force was exerted. The supernatant was not translucent (or partially translucent) at this stage. This is regarded as a phenomenon of under-dose of chitosan flocculant [7].

3.2. Zeta potential and flocculation efficiency

Flocculation efficiency as a result of charge neutralisation can be examined by correlating zeta potential with the dosage of chitosan flocculant. As the chitosan dosage increased, figure 3 shows the increase of zeta potential from -35.5 mV (Culture A) and -28.2 mV (Culture B) of the control to near-zero value approaching the dosage of 40 mg/g with good flocculation performance, except for Culture B with high MW chitosan. Charge reversal was only noticed for Culture A when flocculated with high MW chitosan at dosage above 50 mg/g. On the other hand, Culture B showed a much gradual increase in zeta potential upon addition of high MW chitosan and no charge reversal was recorded within the range of dosage in this study.

Theoretically, optimal flocculation results from charge neutralisation when zeta potential is near-zero [14]. Upon achieving zero zeta potential, the negative charges on the *C. vulgaris* cell surface are said to be fully neutralised by the cationic amino groups of the chitosan flocculant. Increase in chitosan dosage increases the amount of amino groups and improves the aggregation with formation of larger flocs as observed later in Section 3.3. However, at very high dosage of chitosan, when all available cell surfaces are fully neutralised by the amino groups, repulsion occurs among amino groups especially of long polymer chains, leading to re-stabilisation of flocs and reduced flocculation efficiency [8].

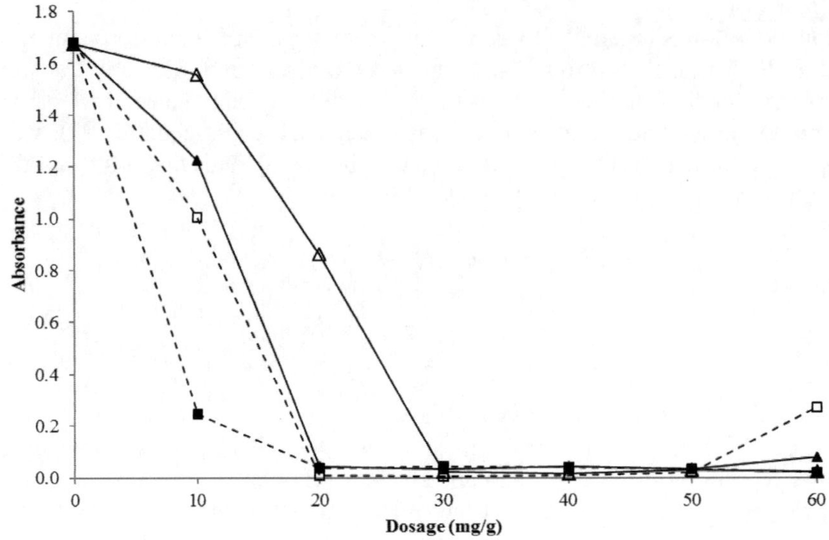

(a)

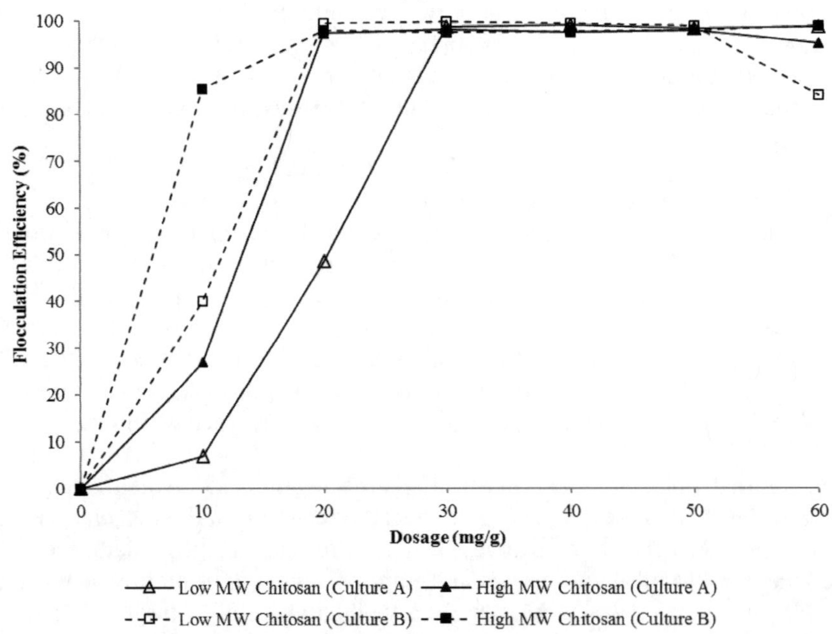

(b)

Figure 2. Effect of molecular weight of chitosan on (a) absorbance measurement, and (b) flocculation efficiency.

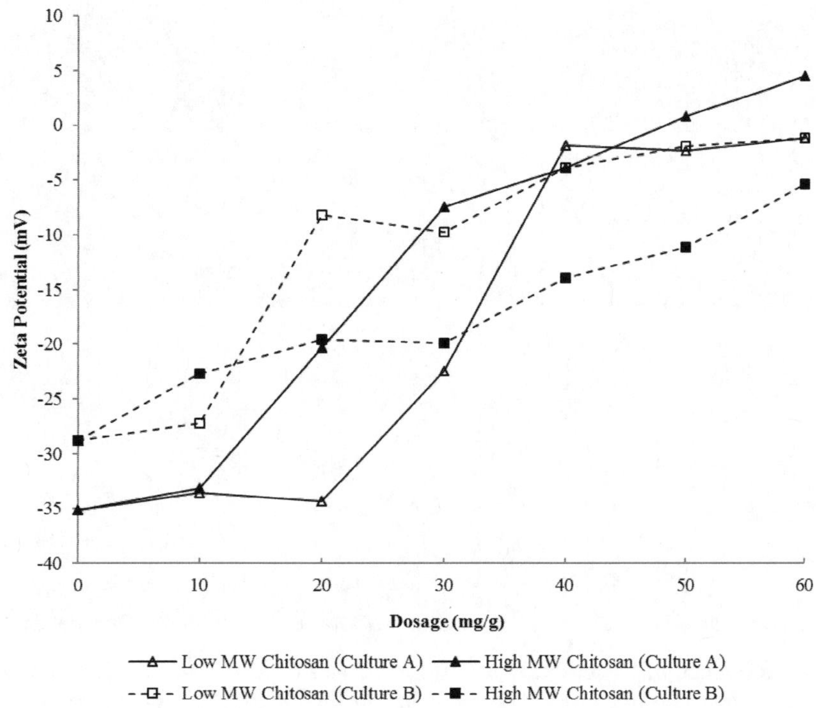

Figure 3. Effect of molecular weight of chitosan on zeta potential values.

3.3. Physical appearance of flocs

It can be seen from figures 4a and 4b that no obvious flocs were formed at the dosage of 10–20 mg/g low MW chitosan in Culture A, which is consistent with the low flocculation efficiency (less than 50%) in figures 2a and 2b. Visible flocs started to form with the addition of 30 mg/g of low MW chitosan and settled fast at the bottom of the beaker after the flocculation mixing was stopped (figure 4c). Although a flocculation efficiency of 98.6% was recorded at this dosage, the respective negative zeta potential value (-22.5 mV) indicated that significant charge repulsion still existed at this point with the formation of smaller flocs. On the other hand, large aggregates and clear supernatant were noticed at the optimum dosage of 40 mg/g with zeta potential approaching zero (i.e. -1.9 mV) (see figure 4d). Overdosing *C. vulgaris* culture at 50–60 mg/g of low MW chitosan did not cause much change to the flocculation efficiency and zeta potential, but the flocs seemed to disintegrate into smaller flocs at 60 mg/g, compared with those formed at 50 mg/g (figures 4e and 4f).

For the flocculation in Culture B, figure 5 shows that at a lower dosage of 20 mg/g low MW chitosan, visible flocs started to form and settled, leaving behind clear supernatant with high flocculation efficiency of 99.4% and zeta potential -8.4 mV which is a quicker response to dosage as compared with Culture A. The colour of *C. vulgaris* cells was found to be lighter green with shorter cultivation period of 8 weeks in Culture B in comparison with Culture A (10 weeks). Besides, the flocs formed in Culture B were generally smaller, even at optimum dosage, unlike the formation of large aggregates in Culture A at optimum dosage. At 60 mg/g, the microalgae cells in Culture B were re-suspended with reduced flocculation efficiency of 83.9% but the zeta potential value remained close to zero (i.e. -1.2 mV). Similar trend of flocculation behaviour was observed for high MW chitosan in Cultures A and B which was not shown here. It is believed that the microalgae culture age has significant effect on the flocculation efficiency and quality. Culture B was harvested at the early stationary growth phase while Culture A was harvested at the late stationary growth phase. More studies are required to understand the effect of culture age on the flocculation behaviour of *C. vulgaris* aided with chitosan flocculant.

IOP Conf. Series: Materials Science and Engineering **206** (2017) 012073 doi:10.1088/1757-899X/206/1/012073

Figure 4. Sequence of flocculation at different doses of low MW chitosan in Culture A: (a) 10 mg/g (-33.6 mV); (b) 20 mg/g (-34.3 mV); (c) 30 mg/g (-22.5 mV); (d) 40 mg/g (-1.9 mV); (e) 50 mg/g (-2.3 mV), and (f) 60 mg/g (-1.2 mV).

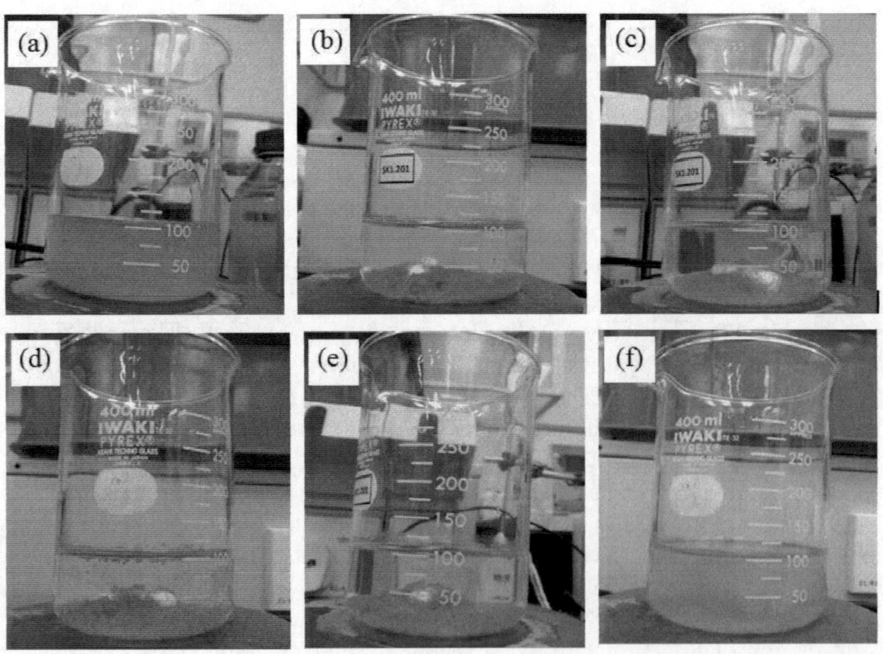

Figure 5. Sequence of flocculation at different doses of low MW chitosan in Culture B: (a) 10 mg/g (-27.2 mV); (b) 20 mg/g (-8.2 mV); (c) 30 mg/g (-9.8 mV); (d) 40 mg/g (-3.9 mV); (e) 50 mg/g (-2.0 mV), and (f) 60 mg/g (-1.2 mV).

In most of the flocculated samples, small amount of algae flocs was found floating on the surface of the medium (see figure 6). However, such phenomenon was more obvious when added with low MW chitosan flocculant. Similar experience was reported by Rashid, Rehman [8] and they attributed it to the formation of fragile flocs with large volume and low weight as a result of intra molecule repulsion among the amino groups of unreacted chitosans. Flotation of microalgae flocs was likely to cause losses of solid mass when the supernatant was decanted.

Figure 6. Flotation of microalgae flocs.

3.4. Solids content after centrifugation dewatering

Flocs formed from various doses of chitosan in Culture B were taken for centrifugal dewatering. Figure 7 shows the concentrated algal paste collected at the bottom of the tube after centrifugation. It can be observed that more biomass solids were captured at high dosage of chitosan.

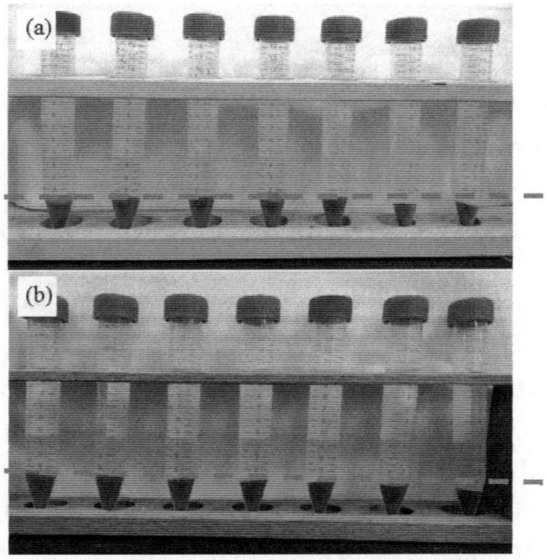

Figure 7. Biomass solids separated from centrifugation after flocculation using (a) low MW chitosan, and (b) high MW chitosan, from the dosage of 60 mg/g to 0 mg/g (left to right).

The solids content of the dewatered flocs was found to increase from 7.2% in the control to 10.8% and 11.8% for low MW and high MW chitosans respectively at the dosage of 10 mg/g (see figure 8). Further increase in flocculant dosage caused the solids content fluctuate between 11.6% and 12.0% for low MW chitosan, and between 10.3% and 11.4% for high MW chitosan. A general impression is that centrifugation effectively separated the algal biomass and chitosan flocculation enhanced the capture of biomass solids. However, a direct relationship between flocculation efficiency and dewatered solids content was not evident from this study, at least not after the centrifugal dewatering. More research effort will be required to better measure the correlation.

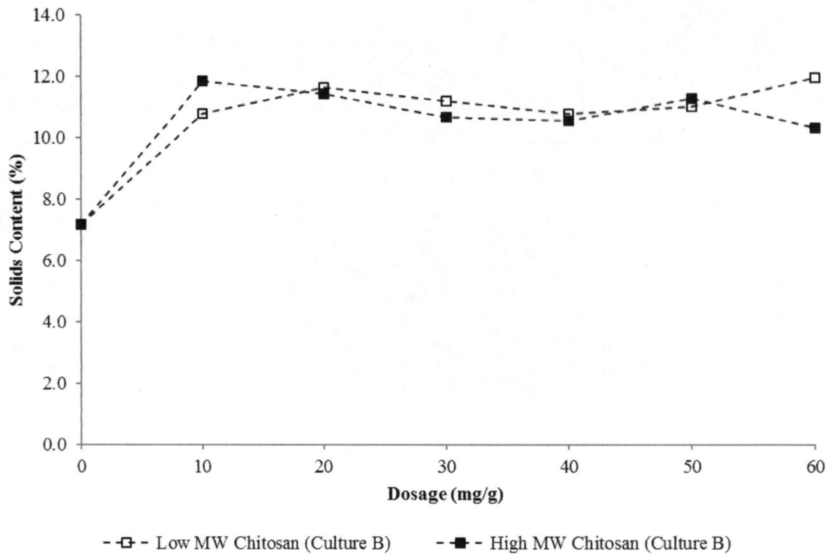

Figure 8. Effect of molecular weight of chitosan on solids content after centrifugal dewatering.

4. Discussion

4.1. Coagulation—charge patch mechanism

In this study, coagulation is the process where *C. vulgaris* cells are destabilised during the rapid mixing step which promotes collision between chitosan flocculant and the microalgae cells. During the coagulation (also known as charge neutralisation or destabilisation) process, the positively charged amino groups of chitosan, which dissociates in acidic medium, attract the negatively charged *C. vulgaris* cells. This results in reduction of electrostatic repulsion among the microalgae cells, and hence reduction in zeta potential as the negative charges on the cell surface are adsorbed or neutralised by the cationic chitosan polymer. The destabilised microalgae cells are attracted to each other and form microflocs. This mechanism is known as the charge patch or electrostatic mechanism [15] in coagulation. Van der Waals' force is responsible in the binding of microalgae cells and destabilising of the suspension [16]. The neutralised cells become capable to "stick" together and form microflocs. However, microflocs are too small to be observed by naked eyes (such as those in figures 4a and 4b).

4.2. Flocculation—polymer bridging mechanism

Formation of microalgae flocs is the main objective of the flocculation step. Ideally, strong flocs of *C. vulgaris* which are not reversible should be obtained, and the layer of clear transparent supernatant can be separated from the solid mass easily which then allows the medium to be recycled and reused. It is suggested that the flocculation process mainly adheres to polymer bridging mechanism. Coagulation process reduces repulsions between the destabilised *C. vulgaris* cells, followed by the flocculation step which initiates the formation of macroflocs. Theoretically, the microalgae flocs will continue to grow until zero zeta potential is achieved. It is evident from this study that large microalgae flocs were formed at zeta potential approaching zero especially in Culture A.

During formation of microalgae flocs, the long chain of chitosan polymer links two or more microflocs together with polymer bridges. The same process may continue even when coagulation is not successful. The polymer chains may bind directly to the microalgae cells and form microflocs, which later grow into macroflocs. According to Blanco, Negro [16], flocculation may produce either soft or hard flocs. Soft flocs are said to be easily break when excessive shear is applied to the flocs, especially when there is turbulence in the medium. Hard flocs formed possess significantly higher strength than

IOP Conf. Series: Materials Science and Engineering **206** (2017) 012073 doi:10.1088/1757-899X/206/1/012073

soft flocs, but will lose their strength over time. According to Norell, Johansson [17], low MW polymer often contributes to formation of soft flocs, where high MW polymer aids the formation of hard flocs. This aspect has not been assessed in this study. Nevertheless, the centrifugal separation of water from the microalgae flocs in this work did not indicate much difference in terms of solids capture for both low MW and high MW chitosans, presuming soft and hard flocs were formed respectively.

Flocculation using high MW chitosan is more prompt to polymer bridging mechanism. Higher MW chitosan possesses longer polymer chains which can bind to more microalgae cells during flocculation process. However, at high dosage, the unbound polymer chains are extended into the bulk solution where the positive charges on the polymer chains contribute to an overall positive zeta potential. This explains the charge reversal which occurred in Culture A when flocculated with high MW chitosan.

4.3. Flocculation—charge patch mechanism

Besides coagulation, charge patch mechanism also promotes the formation of microalgae flocs. During coagulation, destabilised cells are attached together to form the microflocs. For the low MW chitosan in particular, the short polymer chain may adsorb completely onto the cell surface but is not long enough to bind to another cell. It creates positive charges locally on the cell surface and attracts another cell or partially destabilised cell to bind onto itself. However, the microalgae flocs obtained from this mechanism may be smaller in size and are not as strong as those formed from polymer bridging mechanism.

5. Conclusions

Considering both the absorbance measurement for flocculation efficiency as well as zeta potential measurement, the optimal dosage of chitosan for effective flocculation of *C. vulgaris* was found to be 30 mg/g for high MW chitosan and 40 mg/g for low MW chitosan in Culture A. High flocculation efficiency of 98.0-99.0% was achieved at the optimal dosage with formation of large microalgae flocs. Near-zero zeta potential values were attained only at the higher flocculant dosage approaching 40 mg/g (except for Culture B with high MW chitosan) where charge patch neutralisation mechanism is said to be complete. The measurement of zeta potential also revealed that the flocculation of *C. vulgaris* using high MW chitosan required higher dosage to achieve near-zero zeta potential value and charge reversal was observed at increasing dosage. This suggests that the polymer bridging mechanism was governing the flocculation behaviour of *C. vulgaris* using high MW chitosan as the unbound long polymer chains may extend to the bulk microalgal suspension and cause charge reversal. Besides, charge patch neutralisation mechanism was found more significant using low MW chitosan where lower dosage was sufficient to reach near-zero zeta potential compared with the high MW chitosan. The amount of chitosan polymer present in the culture may also affect the mechanism of flocculation. It is proposed that the polymer bridging mechanism prevailed in the flocculation of *C. vulgaris* at lower flocculant dosage since effective flocculation was observed even before charge neutralisation. In addition, the age of microalgae harvested at the early and late stationary growth phase was found to affect the flocculation behaviour of *C. vulgaris*. Although the use of chitosan as a flocculant for microalgae harvesting sounds promising, it may result in microbiological contamination, which affects the downstream processing of microalgal biomass for biofuel production. Future studies should further investigate the influence of the microalgal cell properties or culture conditions on chitosan flocculation, and the respective downstream treatment. Besides, holistic cost evaluation is required to consider both the cost of flocculation as well as the influence on the entire production process.

References

[1]. Xu Y, Purton S, Baganz F 2013. *Bioresour. Technol.* **129** 296.
[2]. Wijffels RH, Barbosa MJ 2010. *Science.* **329** 796.
[3]. Zhou W, Gao L, Cheng W, Chen L, Wang J, Wang H, et al. 2016. *Algal Research.* **18** 7.

[4]. Lananan F, Yunos FHM, Nasir NM, Abu Bakar NS, Lam SS, Jusoh A 2016. *International Biodeterioration & Biodegradation* **113** 391.
[5]. Zhao FC, Chu HQ, Tan XB, Yang LB, Su YM, Zhou XF, et al. 2016. *J. Membr. Sci.* **517** 30.
[6]. Lee D-J, Liao G-Y, Chang Y-R, Chang J-S 2012. *Int. J. Hydrogen Energy* **37** 15643.
[7]. Vandamme D, Foubert I, Muylaert K 2013. *Trends in Biotechnology* **31** 233.
[8]. Rashid N, Rehman MSU, Han J-I 2013. *Chem. Eng. J.* **226** 238.
[9]. Al-Qasmi M, Raut N, Talebi S, Al-Rajhi S, Al-Barwani T, editors. A Review of Effect of Light on Microalgae Growth. World Congress on Engineering 2012; 2012; London.
[10]. Andersen RA 2005. *Algal Culturing Techniques*: Elsevier/Academic Press.
[11]. Divakaran R, Pillai VS 2002. *J. Appl. Phycol.* **14** 419.
[12]. Szymańska E, Winnicka K 2015. *Marine Drugs* **13** 1819.
[13]. Gerde JA, Yao L, Lio J, Wen Z, Wang T 2014. *Algal Research* **3** 30.
[14]. Halverson F, Panzer H 1980. *Kirk-Othmer Encyclopedia of Chemical Technology* **10** 489.
[15]. Wakeman RJ, Tarleton ES 1999. *Filtration: Equipment Selection, Modelling and Process Simulation*: Elsevier Advanced Technology.
[16]. Blanco A, Negro C, Fuente E, Tijero J 2005. *Ind. Eng.Chem. Res.* **44** 9105.
[17]. Norell M, Johansson K, Persson M 1999. *Papermaking chemistry*.

Wax inhibitor based on ethylene vinyl acetate with methyl methacrylate and diethanolamine for crude oil pipeline

S M Anisuzzaman[1,2,*], S Abang[2], A Bono[2], D Krishnaiah[2], R Karali[2], M K Safuan[2]
[1]Energy Research Unit (ERU),
[2]Chemical Engineering Programme, Faculty of Engineering, Universiti Malaysia Sabah, 88400 Kota Kinabalu, Sabah, MALAYSIA.

E-mail: anis_zaman@ums.edu.my, dr.anis.ums@gmail.com

Abstract. Wax precipitation and deposition is one of the most significant flow assurance challenges in the production system of the crude oil. Wax inhibitors are developed as a preventive strategy to avoid an absolute wax deposition. Wax inhibitors are polymers which can be known as pour point depressants as they impede the wax crystals formation, growth, and deposition. In this study three formulations of wax inhibitors were prepared, ethylene vinyl acetate , ethylene vinyl acetate co-methyl methacrylate (EVA co-MMA) and ethylene vinyl acetate co-diethanolamine (EVA co-DEA) and the comparison of their efficiencies in terms of cloud point¸ pour point, performance inhibition efficiency (%PIE) and viscosity were evaluated. The cloud point and pour point for both EVA and EVA co-MMA were similar, 15 °C and 10-5 °C, respectively. Whereas, the cloud point and pour point for EVA co-DEA were better, 10 °C and 10-5 °C respectively. In conclusion, EVA co-DEA had shown the best % PIE (28.42%) which indicates highest percentage reduction of wax deposit as compared to the other two inhibitors.

1. Introduction

Paraffinic crude oil has been a huge historical challenge for operators in the oil and gas industry. The principle problems arise when the equilibrium is disturbed during production or transportation which leads to alteration in pressure, temperature and composition and eventually results in extensive problems such as the precipitation and deposition when the crude oil solidifies on the flow lines of the production system causing the internal diameter to decrease as well as restricts the flow of the fluid. The main composition of crude oil are saturates, asphaltenes, aromatic and resins. Paraffins are the main saturates that contribute to the formation and deposition of wax during the transportation process whereby the solubility decreases as the temperature and operating pressure decrease [1-3]. This phenomena leads to the wax formation and deposition where it tends to create crystalline capable of entrapping liquid crude oil [4,5]. The temperature difference between the pipelines walls and the crude oil driven by molecular diffusion mechanism results in the wax formation and deposition on the surface of the pipelines walls. Wax deposition process can eventually results in significant reduction of the internal diameter of the tubular, casing , perforations, pump strings, rods, pipelines and at worst case causing complete blockage to the flow lines that leads to tremendous profit loss [6,7].

Due to consequential problems faced as a result of wax deposition many treatment methods were developed, such as mechanical, thermal, chemical, and biological treatments that are usually implemented to clean and disperse the deposits based on its nature, location, cost and other influential factors [8]. All these methods are applicable to solve the problem after the wax has been precipitated in the pipeline. In addition to that , there are also several wax prevention techniques such as increasing the operating pressure which promotes the shear rate, insulation the flow lines or even coating the inside of the flow lines with a non-adhering material. Unfortunately the abovementioned techniques come in high cost and usually require partial or complete shutdown of production system in order to be implemented.

Wax prevention techniques have shown an increased interest to the researchers. Many chemical additives were developed to reduce or eliminate this problem. Chemical additives such as wax inhibitors are one of the wax controlling methods that are implemented the industries to help control the deposition and provide flow assurance. Chemical wax prevention methods come in three approaches which are wax dispersants, wax inhibitor (wax crystal modifier) and pour-point depressant (PPD).

Wax inhibitors such as pour point depressants and wax dispersants are used to chemically modify the wax solid structure thus reducing the tendency of the wax crystals to interlock and form three dimensional network growths [9,10]. The main function of wax inhibitors is to impede the paraffin growth. However, these wax inhibitors do not stop the waxing process but it delays the precipitation and hardening process, thus allowing the wax aggregates to be swept away in the oil production steam flow [11].

Even though pour point dispersant and wax dispersants are under the same class that is the chemical additives, but they are different from one another. Pour point dispersant is a wax crystal modifier which functions by co-crystallizing into paraffin structure through van der Waals interactions leaving the polar moiety end tail to create a steric hindrance to interfere with alignment of new incoming wax molecules. Whereas, the wax dispersant functions by absorbing itself into the wax crystal and prevent further sticking of the molecules. Recently, wax dispersants are blended with wax inhibitor to enhance their performance in reducing wax deposition [12-15].

Ethylene vinyl acetate (EVA) is an ester group which can weaken the solid deposition of the wax in the crude oil through van der Waals bonding in which it results in an easy removal of the deposition by sheer force of the flow stream [16]. The influence of the EVA on the crude oil's pour point and cloud point has been investigated by several researchers, 30 wt% of EVA was found to be the most efficient copolymer as a pour point depressant compared to other copolymers [2,17,18]. Methyl methacrylate is the other class of comb polymers commonly used as wax inhibitors. Methyl methacrylate (MMA) is made up from acrylic acid monomer whereas the ester group are made up from long chain of alcohols. EVA and MMA are saturated compounds and the refractive index between both these wax inhibitors is very close ,making them easily blended together [19]. Diethanolamine (DEA) comprises of hydroxyl groups and also contains wax-like portions that allow the molecules to incorporate into the wax deposition structure. The aim of this study was to evaluate the performance of two pour point dispersants (MMA, and DEA) when blended with EVA.

2. Experimental methods

2.1. Preparation of ethylene vinyl acetate co-methyl methacrylate (EVA co-MMA) and co-diethanolamine (EVA co-DEA) wax inhibitor

EVA (40 wt% melt) & DEA (reagent grade $\geq$ 98.0%) was purchased from Sigma Aldrich, USA, MMA ($\geq$ 99.5%) & cyclohexane ($\geq$ 99.5%) was purchased from Merck KGaA, Germany. All these reagents were supplied by Merck Sdn. Bhd, Selangor, Malaysia. Cyclohexane was used as solvent for EVA. For the preparation of the inhibitor, the reaction was conducted on a hot plate at a temperature of about 50-60°C with continuous stirring using magnetic stirrer. A hot plate was used instead of an oven to allow space to work while preparing the chemicals. The individual chemicals, EVA, MMA, DEA and cyclohexane were measured separately of its respective volume and weightage in accordance to the manipulated percentage composition. EVA was measured in grams of mass, meanwhile MMA, DEA and cyclohexane were measure in mL.

The EVA co-DEA and EVA co-MMA was prepared by using a blend polymerization method. For both of the inhibitor formulations 1 g of (30 wt%) EVA was measured using electronic mass balance, and then added to 10 mL of cyclohexane using a micropipette. For the first formulation 5 mL of MMA was added, meanwhile for the second formulation 5 mL of DEA was added to the mixture. Finally the two mixtures were heated in the measuring cylinder at 50-60 °C to melt the copolymer. When the inhibitors were completely melted, the crude oil was poured into the measuring cylinder. In order to obtain the cloud and pour point of the crude oil, cooled down the crude oil with the wax inhibitor from 50 °C to 0 °C with decrement of temperature by 3 °C for every 2 to 3 hours. The experiment was carried out for different amount of the EVA with the fixed amount of DEA and MMA. This is to observe the effect of EVA mass on the viscosity of the crude oil.

2.2. Determination of pour point and cloud point temperature

Pour point and cloud point temperature of both treat and untreated crude oil samples were determined using ASTM D97 method and ASTM D2500 method respectively [20-22,16,17]. To find the pour point ,at which the crude oil fluid becomes semi solid and losses its flow characteristics ,100 mL of the crude oil sample was heated to 50 °C in an oven and kept overnight to ensure that the wax crystals have completely dissolved [22]. The 4 samples were placed in a pour point test jar which was closed tightly by the cork carrying the thermometer in order to monitor the temperature drop .After the sample was heated to 50 °C using the water bath, the water bath temperature was decreased gradually and observation of the temperature decrease was noted. Every 3 °C temperature drop the movement of fluid was checked until the crude oil showed no movement when being held in the horizontal position for exactly 5 seconds. The temperature at which the fluid stops flowing is recorded from the thermometer. That temperature represents the pour point of the fluid. Cloud point measurement was done using similar procedure of pour point except the cloud point temperature is recorded when smallest cluster of hydrocarbon crystals was first observed then the temperature on the thermometer was recorded.

2.3. Determination of viscosity

Viscometer was used to evaluate the viscosity of the crude oil samples. The viscometer used is the Brookfield Programmable Viscometer DV-III + Rheometer. The standard settings of the equipment shown in table1.

Table 1. Standard settings of viscometer

Spindle size	7
Rotational speed	100 rpm
Units	cgs
Reading	Viscosity (cP) Torque (%)

The viscometer was left to warm up for about 10 minutes after start up, and then, auto-zeroed without the spindle and at a levelled position. Each set of samples was tested of its viscosity from 50 °C to 5 °C at every 5 °C intervals. Firstly, the control experiment is conducted which is the blank crude oil at the absence of inhibitor. The crude oil was removed from the oven and allowed to cool down to 50 °C, then measure of the viscosity using the viscometer. The temperature of the sample was observed and controlled using a temperature probe. Then, the sample was allowed to cool down to 45 °C before the viscosity is measured again. The step is repeated at every 5°C interval until the sample temperature drops to 5 °C. To cool the sample below the room temperature, the sample is immersed into an iced water bath. Care was taken to avoid upsetting the spindle. The procedure was repeated for each of the other samples that contain the other inhibitor. Before starting a new sample each time, the spindle and temperature probe is lightly cleaned to avoid contamination of the samples. Again, the spindle is handled with care in order to maintain the calibration.

2.4. Determination paraffin inhibitor efficiency (PIE)

After the wax precipitation has been observed from the crude oil sample at the first sight, decreased the temperature from 10 to 5 °C for 8 hours, and ensured that the crude oil continuous to flow by stirring the sample for every 10-15 minutes. After that, the specimen was heated to increase its temperature so that the wax from the crude oil sample was separated. The wax out from the crude oil was filtered by using filter paper and dried at room temperature for 1 day. The mass of the crude oil was weighed by using the electronic balance and the paraffin inhibition efficiency is determined by using the formula [16,23]:

$$PIE(\%) = \frac{W_f - W_t}{W_f} x\ 100$$

where W_f is the reference amount of wax deposit without chemical treatment in gram (g) and $W_t$is the amount of paraffin deposit with chemical treatment in gram (g).

3. Results and discussion

3.1. Effects of the inhibitors on the pour point and cloud point of the crude oil

Pour point and cloud point of samples were measured by following the ASTM procedures (ASTM D-97 and ASTM D-2500) respectively. These two tests were crucial to evaluate and compare the efficiency of the inhibitor formulations proposed to inhabit the wax formation of the crude oil. Table 2 clearly shows that the untreated crude oil solidified at the temperature of 35 °C. Meanwhile, crude oil treated with EVA , EVA co-MMA and EVA co-DEA solidified at temperature of 15 °C, 15 °C, 10 °C respectively, which confirms the efficiency of the these inhibitor formulations to reduce the temperature of wax formation. This proved the statement that reported by Vieira *et al* [24] whereby it was stated that the WAT is related to the molar mass and structure of the chain in which the waxes that are more linear and have higher molar mass crystallized at higher temperature, meanwhile less linear waxes with lower molar mass crystallized at lower temperature [25]. The molar weight of DEA, MMA and EVA is 105.14 g/mol, 100.12 g/mol and 114.14 g/mol respectively. Since

MMA has the lowest molar weight, it had crystallised at a lower temperature followed by DEA and then EVA. But, since the structure of DEA is less linear compared to MMA, causes the DEA to be crystallised at lower temperature compared to the MMA. The difference in the molar mass between MMA and DEA is quite small making it possible.

Table 2. The effect the inhibitors on the wax precipitation on the crude oil in terms of temperature

Temperature, °C	Crude oil only	Crude oil + EVA	Crude oil + EVA + MMA	Crude oil + EVA + DEA
50	X	X	X	X
47	X	X	X	X
44	X	X	X	X
41	X	X	X	X
38	X	X	X	X
35	X	X	X	X
32	/	X	X	X
30	/	X	X	X
25	/	X	X	X
20	/	X	X	X
15	/	/	/	X
10	/	/	/	/
5	/	/	/	/

whereby, X sign indicates that there is no precipitation of wax and / Sign indicates that there is precipitation of wax.

3.2. Effect of temperature on viscosity for the crude oil with and without wax inhibitor
Figure 1 shows the relationship of between viscosity and temperature for the 4 samples. It can be observed that the viscosity of the crude oil is has a steadily decreased with the temperature decreased whether it is mixed with wax inhibitor or not. The untreated crude oil showed the highest viscosity value at 5 °C that is 1460 cP followed by the treated crude oil with EVA co-MMA, EVA co-DEA and EVA. The viscosity of the treated crude and untreated oil does not have any significant differences between them at low temperature, however it can be observed that the viscosity decline of the samples that contained EVA and EVA co-DEA was very clear compared to the other samples. Therefore, it can be concluded that the sample with EVA co-DEA wax inhibitor shoes best response compared to the other three samples of untreated crude oil, crude with EVA inhibitor and crude with EVA co-MMA inhibitor.

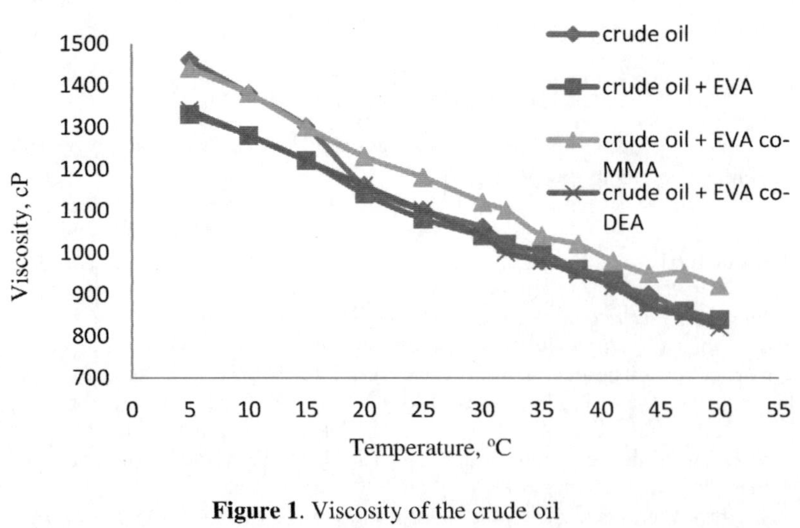

Figure 1. Viscosity of the crude oil

Figure 1 also shows that the crude oil with EVA co-MMA has the highest viscosity compared to the others in the range of 50 °C to 10 °C. The reason for this is that crude oil at high temperature behaves as a Newtonian fluid showing viscosity-temperature dependence. As the temperature of the crude oil falls below the WAT, precipitation of the wax crystals will rapidly increase the viscosity entering into the non-Newtonian fluid region. The yield stress of a wax-solvent mixture is believed to be a function of wax composition, and thermal and shear histories. A typical waxy crude oil would normally exhibit a pseudo-plastic with thixotropic properties [26]. The untreated crude oil is found to behave similarly to pseudo plastic fluid, wherein with increasing shear rate, the viscosity tends to decrease.

Figure 2 (a-d) shows the changes of the viscosity for the sample of the crude oils at different temperature with and without the wax inhibitor. The most significant value of viscosity was when the temperature decreased from 20 °C to 15 °C for the untreated crude oil sample, crude oil with EVA, and crude oil with EVA co-DEA. The viscosity changes for these three types of samples are 140, 80 and 80 cP respectively. Whereas, the highest viscosity recoded of the crude oil with EVA co-MMA was 80 cP, when the temperature was reduced from 15 °C to 10 °C. From the trend of figure 2, it can be observed that at low temperature the viscosity has changed for both treated and untreated crude oil the viscosity at low temperature was higher compared to the viscosity at high temperature. This is because, the wax inhibitor reduces the change in viscosity of the crude oil below 25 °C and 23 °C since adding a wax inhibitor will change the fluid properties of the of crude oil [26-28].

Figure 2 (a-d) also shows that the crude oil with EVA has the lowest change of viscosity as the temperature is dropped from 25 °C to 5 °C, followed by the crude oil with EVA co-DEA, crude oil with EVA co-MMA. The highest change in the viscosity is observed when the temperature is dropped without inhibitor.

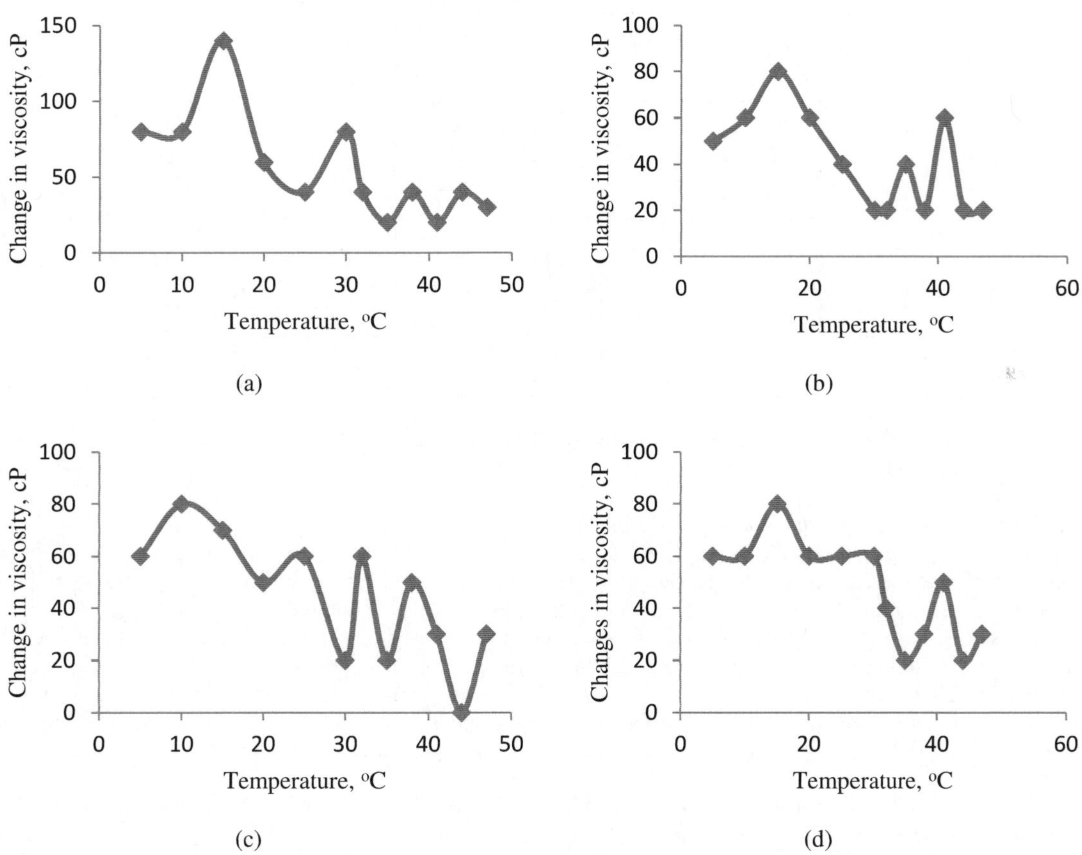

Figure 2. Change of viscosity at different temperatures (a) crude oil (b) crude oil with EVA (c) crude oil with EVA co-MMA (d) the crude oil with EVA co-DEA

3.3. Determination of the crude oil pour point

Table 3 shows the pour point temperature of the crude oil with and without wax inhibitor. From table 3, it can be concluded that at temperature 5-10 °C, the crude oil, crude oil with EVA, and crude oil with EVA co-MMA tend to harden or stop flowing. Whereas, for the crude oil with EVA co-DEA, it stops flowing at a lower temperature than the other at 0-5 °C. These values have confirmed that different types of inhibitor enhances pour point property of crude oil.

IOP Conf. Series: Materials Science and Engineering **206** (2017) 012074 doi:10.1088/1757-899X/206/1/012074

There was no significant difference in the pour point for the untreated crude oil, crude oil treated with EVA and crude oil treated with EVA co-MMA, this might be due the small amount of crude oil sample used. However, the crude oil sample treated with EVA co-DEA has a slightly lower pour point compared to the other samples , this is mainly because at lower temperature the crude oil with EVA co-DEA still have low viscosity and it may be due to the characteristic of EVA co-DEA in which it has structure similar to the nucleation site and co-crystallizing with the paraffin waxes ,therefore preventing the formation of waxes which increase the viscosity of the oil [29,30]. It can be concluded that EVA co-DEA wax inhibitor has proven to have better inhibition characteristics compared to the others. In addition to that ,it has shown the strongest Van Der Waals interaction between wax and the functional group in the wax inhibitor that causes the decrease in the WAT and increase of wax solubility.

Table 3. Pour point temperature

Temperature, °C	Crude oil only	Crude oil + EVA	Crude oil + EVA co-MMA	Crude oil + EVA co-DEA
50	/	/	/	/
47	/	/	/	/
44	/	/	/	/
41	/	/	/	/
38	/	/	/	/
35	/	/	/	/
32	/	/	/	/
30	/	/	/	/
25	/	/	/	/
20	/	/	/	/
15	/	/	/	/
10	X	X	X	/
5	X	X	X	X

whereby, X sign indicates that the crude oil stopped flowing when diverted horizontally for 3 to 5 second, meanwhile / sign indicates that the crude oil still can still flow.

3.4. Effect of EVA mass on the viscosity of the crude oil

In this section the EVA mass was increased to evaluate its effect on the viscosity of the crude oil as the temperature is decreased. Table 4 below illustrates the results. It was observed that increasing the amount of the EVA in the wax inhibitor formulation results in increased viscosity of crude oil. As the viscosity of the crude oil increases with the increase of EVA mass the cloud point of the crude oil remains unchanged, this proves that increasing wax inhibitor dose doesn't affect the wax precipitation temperature. Even if the amount of wax inhibitor is higher than it supposed to be, it will not affect the temperature of the crude oil to form a wax since the wax precipitation will occur when it reached certain temperature. In other words, the amount of the wax inhibitor will have no effect on the WAT of the crude oil except if the amount of the crude oil is increased drastically.

Table 4 shows that the crude oil with EVA co-MMA inhibitor still has the highest viscosity compared to the other inhibitor. The trend figure 2 (a-d) shows that the increasing in viscosity is also affected by the increasing in the concentration of the wax inhibitor and decreasing in temperature. As the viscosity increases the pour point of the crude oil also increases, however table 3 shows that the pour point does not increase as the viscosity of the crude oil sample is increased. This may be due to the concentration of the crude oil does not exceed the point where it changes pour point temperature to stop flow at high temperature.

Table 4. The effect of EVA mass on the viscosity of the crude oil

Temperature, °C	Viscosity, cP					
	Crude oil + EVA (3 g)	Crude oil + EVA (5 g)	Crude oil + EVA (3 g) + MMA	Crude oil + EVA (5 g) + MMA	Crude oil + EVA (3 g) + DEA	Crude oil + EVA (5 g) + DEA
50	860 ± 10	920 ± 12	940 ± 12	1000 ± 15	880 ± 10	900 ± 10
47	900 ± 10	960 ± 12	980 ± 12	1020 ± 15	900 ± 10	940 ± 10
44	920 ± 12	1000 ± 15	1000 ± 15	1040 ± 15	960 ± 10	980 ± 10
41	960 ± 12	1040 ± 15	1040 ± 15	1080 ± 15	1000 ± 15	1040 ± 15
38	1000 ± 15	1080 ± 15	1080 ± 15	1120 ± 18	1020 ± 15	1080 ± 15
35	1040 ±15	1140 ± 18	1120 ± 18	1180 ± 18	1060 ± 15	1140 ± 18
32	1080 ± 15	1180 ± 18	1160 ± 18	1240 ± 18	1120 ± 18	1200 ± 18
30	1140 ± 18	1200 ± 18	1220 ± 18	1300 ± 20	1180 ± 18	1260 ± 18
25	1200 ± 18	1240 ± 18	1300 ± 20	1340 ± 20	1240 ±18	1300 ± 20
20	1260 ± 18	1300 ± 20	1360 ± 20	1420 ± 20	1300 ± 20	1340 ± 20
15	1320 ± 20	1380 ± 20	1440 ± 22	1500 ± 22	1360 ± 20	1400 ± 20
10	1380 ± 20	1460 ± 22	1500 ± 22	1560 ± 22	1440 ± 22	1460 ± 22
5	1420 ± 20	1540 ± 22	1580 ± 22	1620 ± 25	1480 ± 22	1500± 22

3.5. Determination of the performance inhibitor efficiency

Table 5 shows the performance inhibition efficiency (%PIE) of EVA, EVA co-MMA and EVA co-DEA. From table 5 it can be concluded that the EVA co-DEA has the highest %PIE of 28.42% compared to the others.

Table 5. The results of performance inhibitor efficiency

Performance inhibitor efficiency (%PIE)		
EVA	EVA co-MMA	EVA co-DEA
25.32 ± 0.60	24.83 ± 0.50	28.42 ± 0.80

4. Conclusion

EVA co-DEA inhibitor has proven to be the best inhibitor that reduces the wax formation based on paraffin inhibitor efficiency value, cloud point, and pour point that is 28.42% at 10 °C and 5 °C respectively. This study explains the effected of the reduced molar mass of the inhibitor and its ability of reducing the pour point temperature of the crude oil. This can be explained in terms of the changes of the crude oil characteristic from the non-Newtonian to a Newtonian as the temperature decreases. Furthermore this study has also proven that the EVA can act as the main wax inhibitor to be mixed with other wax dispersant to form higher quality wax inhibitor however, not all wax inhibitors are suitable for such reaction. Therefore, it can be concluded that EVA co-DEA is better wax inhibitors compared to the EVA and EVA co-MMA. Meanwhile, EVA co-MMA has shown a lower PIE value and higher viscosity of crude oil compared to EVA that proves that, not all mixture of wax inhibitor with wax dispersant are able to enhance the performance of a wax inhibition compared to the primary wax inhibitor.

Acknowledgements

The authors would like to express greatest appreciation to the Centre of Research and Innovation (PPI), Universiti Malaysia Sabah (UMS) for the support to this project (SGK0006-TK-2015).

References

[1] Anisuzzaman S M *et al* 2015 Proc. *Int. conf. on Mech. Eng. Adv. Mater. (*Kota Kinabalu: Malaysia) 248.
[2] Machado A L C *et al* 2001 *J. Petrol. Sci. Eng.* **32** 159.
[3] Wei B 2015 *J. Petrol. Explor. Prod. Technol.* **5** 391.
[4] Jorda R M 1966 *J. Petrol. Sci. Eng.* **18** 1605.
[5] Towler B F *et al* 2007 *SPE Annual Tech. Conf. Exhib.* (California: U.S.A.).
[6] Abdumula M F 2005 *Ph. D Thesis, University of Miskolc,* Hungary-Miskolc.
[7] Kuzmic A R *et al* 2008 *Fuel* **87** 2943.
[8] Gupta A and Sircar A 2016 *IJLTEMAS* **5** 18.
[9] Pedersen K S and Rønningsen H P 2003 *Energ Fuel* **17** 321.

IOP Conf. Series: Materials Science and Engineering **206** (2017) 012074 doi:10.1088/1757-899X/206/1/012074

[10 Elhaddad E *et al* 2015 *Hemijska Industrija* **69** 269 .
[11] Tinsley J F *et al* 2009 *Energ Fuel* **23** 2065.
[12] Marie E *et al* 2005 **290** 406.
[13] Alcazar-Vara L A and Buenrostro-Gonzalez E 2011 *Fuel Process Technol.* **92** 2366.
[14] Maheshwari A S and Chellani J G 2012 *Fuel* **98** 55.
[15] Behbahani T J *et al* 2015 *Petroleum* **1** 223.
[16] Ridzuan N *et al* 2012 *Oriental J. Chem.* **31** 305.
[17] Ghosh P and Das M 2014 *J. Petrol. Sci. Eng.* **119** 79.
[18] Yang F *et al* 2009 *J. China Univ. Petrol.* **33** 108.
[19] Poomalai P *et al* 2007 *J. Appl. Polym. Sci.* **106** 684.
[20] ASTM - American Society for Testing and Materials 1993 D97-93 - Standard test method for pour point of petroleum products.
[21] ASTM - American Society for Testing and Materials 1998 D2500-98 - Standard cloud-point test method
[22] Annual Book of ASTM Standard 2005 American Society for Testing and Materials, Philadelphia, USA.
[23] Bello O *et al* 2006 *Petrol. Sci. Technol.* **24** 195.
[24] Vieira L *et al* 2008 *Chem. Chem. Technol.* **2** 211.
[25] Neto A A D *et al* 2009 *Braz. J. Petrol. Gas* **3** 149.
[26] Hoffmann R and Amundsen L 2013 *J. Petrol. Sci. Eng.* **107** 12.
[27] Ghosh P *et al* 2009 *Res. J. Chem. Environ.* **1** 17.
[28] Venkatesan R *et al* 2003 *Energ. Fuel* **17** 1630.
[29] Behbahani J T *et al* 2008 *Fuel Process. Technol.* **89** 973.
[30] Al-Sabagh A M *et al* 2013 *Egypt. J. Petrol.* **22** 381.

IOP Conf. Series: Materials Science and Engineering **206** (2017) 012075 doi:10.1088/1757-899X/206/1/012075

Adsorption of Cu(II) Ions in Aqueous Solutions by HCl Activated Carbon of Oil Palm

A Muslim*, Y Syamsuddin, A Salamun, Abubakar, D Ramadhan, D Peiono

Process Technology Laboratory, Department of Chemical Engineering Faculty of Engineering, Syiah Kuala University, Banda Aceh 23111, Indonesia

E-mail: abrar.muslim@che.unsyiah.ac.id

Abstract. Activated carbon was prepared from oil palm empty fruit bunch (OPEFB) by pyrolysis at 873.15 K in a furnace and chemical activation using 0.01 M HCl. Fourier Transform Infrared Spectroscopy, Scanning Electron Microscopy and BET (Brunauer, Emmett and Teller) surface area analyses were taken into account to investigate the chemical functional group, to characterise the surface morphology and to determine total surface area the OPEFB AC, respectively. Experiments in batch mode were conducted to investigate Cu(II) adsorption capacity by the OPEFB AC whereas the system consisted of 1 g the OPEFB AC in 100 mL Cu(II) aqueous solution with initial concentration in the range of 10-70 mg/L, magnetic stirring at 75 rpm, room temperature of 300.15 K (± 2 K), at 1 atm and neutral pH over contact time in the range of 0-150 min. As the result, Cu(II) adsorption capacity increased exponentially over contact time and initial concentration. The Cu(II) adsorption kinetics followed the pseudo second order kinetics with the correlation coefficients (R^2), kinetics rate constant and equilibrium adsorption capacity being 0.98, 4.81 mg/g and 0.15/min, respectively for initial Cu(II) concentration being 58.71 mg/L. In addition, Cu(II) adsorption isotherm followed the Langmuir equation with the R^2 value, the mono-layer and over-all adsorption capacity being 0.99, 5.92 mg/g and 0.17 L/mg, respectively.

1. Introduction

Heavy metals are hazardous and non-biodegradable pollutants in the environment and can be accumulated in the human body. Waste water consisting of heavy metals can be released to the environment from industrial processes such as paper and rubber, mining, petroleum refining, electroplating and smelting [1], [2], [3], [4], [5]. Cu(II) ion is one of the most hazardous ions of heavy metals contained in the wastewater from mining, chemical and electrical industries [6]. It can bio-accumulate in the human body tissues leading to organ cancer and disorder of biological system of human body [7]. Therefore, removing copper from Cu(II) ions-contaminated wastewater is needed before releasing to the environment.

Membrane filtration, electrolysis, chemical precipitation and adsorption using activated carbon have proposed to remove heavy metal ions from inorganic wastewater, and adsorption is one of effective and promoting processes to apply [8], [9]. Wastes from agricultural industries have been investigated to prepare as alternative raw material of adsorbents because it is more economically and environmentally applied compared to others. Utilisation of adsorbents from agricultural wastes which were coarse tea, green tea, Brazil nut shells, coffee bean husks and areca catechu shell, had been investigated for the adsorption of Cu(II) ions from aqueous solutions [10], [11], [12], [13].

Interestingly, oil palm empty fruit bunch (OPEFB) has been proposed as raw material of activated carbon using KOH activator for adsorption of malachite green [14].

The OPEFB as one of waste types from palm oil production is becoming an attractive renewable material for activated carbon because the world palm oil production was projected to sustained and impressive growth in the period of 2016–2020 [15] with the average annual world production of palm oil predicted to be more than 30 million tonnes consisting of approximately 15 million tonnes from Malaysia and the same amount from Indonesia. Indonesia was projected becoming the world's largest exporter and producer of palm oil taking over Malaysia [16]. Hence, it is necessary to investigate the utilisation of carbon active from the OPEFB for adsorption of heavy metal including Cu(II) to answer the global consumption of activated carbon which has been increasing and was forecasted to reach approximately 1.733 million tons by the next two years [17].

This study aimed to prepare activated carbon from the OPEFB for adsorption of Cu(II) ions from aqueous solutions. The chemical functional groups and surface morphology of the OPEFB, carbon and activated carbon were analysed using Fourier transform infrared spectroscopy (FTIR) and scanning electron microscopy (SEM), respectively. This preliminary study was performing by conducting experiments in batch mode at room temperature and neutral pH to obtain Cu(II) adsorption isotherms and kinetics parameters.

2. Materials and method

2.1. Activated carbon from the OPEFB

Preparation of the OPEFB activated carbon (OPEFB AC) and the adsorption experiments in batch mode were done in the Process Technology Laboratory at Department of Chemical Engineering, Syiah Kuala University. An atomic absorption spectrometer (AAS) (Shimadzu AA 6300, made in Japan) in Chemical Analysis Laboratory at the Faculty of Mathematics and Science, Syiah Kuala University was used to obtain Cu(II) ions concentration in the aqueous solution. Nova Station C (Quantachrome Instruments version 11.0, made in USA) which was available in Instrument Laboratory (ITB, Bandung) was used to obtain the BET surface area, pore size and total pore volume of the OPEFB AC.

The OPEFB waste were collected from palm oil processing plant of PT. Mopoli Raya, West Aceh. The OPEFB of 1 kg was washed wholly using tap water and distilled water at the end. It was dried under the sun for 7 days at average temperature of 304.15 K (± 2 K) and relative humidity of 12.5%. The dried OPEFB were crushed to powder using a ball mill. Pyrolysis of the OPEFB was performed using the same method presented in previous study [18] by coating the dried OPEFB with at least four layers of aluminum foil in a Nabertherm furnace (More Than Heat 30-3000 ^{o}C, made in Germany) at 873.15 K for 1 h. Then, the OPEFB carbon of 500 g was activated using 0.01 M HCl (Gatt-Koller) in 1000-mL beaker without stirring it for 2 h. The OPEFB activated carbon was washed and decanted many time with distilled water until reaching pH 7 (± 0.5 K) to remove the excess acid, and it was then filtered using vacuum filter. To remove the excess water, the OPEFB activated carbon was dried in a Memmert oven (NN-ST342M, made in Germany) at 373.15 K until reaching a constant weight. It was sieved to obtain a uniform size of 100 mesh. The OPEFB activated carbon was stored in sealed plastic bottle to use within 2 weeks for experiments.

2.2. Characterization of the OPEFB

Chemical functional groups on the OPEFB powder and activated carbon were identified using the FTIR. The FTIR spectra were obtained between 400 and 4000 cm^{-1} of wavenumber using a Spectrophotometer (Shimadzu IR Prestige 21, made in Japan) which is available in the Analysis Instrument Laboratory of Chemical Engineering Department of Syiah Kuala University, and KBr pellets of 0.1% sample were used to obtain the transmission spectra of the samples. Meanwhile, the physical morphology of the OPEFB powder, carbon and activated carbon surface were identified using a HITACHI TM3000 scanning electron microscopy (made in Japan) with an accelerating voltage of 500VA (1 phase of 50/60Hz) which is avalaible at Mechanical Engineering Department of Syiah Kuala Unversity. All the samples were dried in an oven at 293.15 K for about 10 min prior to the SEM analysis. Prior to measurement of the BET surface area, pore size and total pore volume of the OPEFB, the OPEFB AC sample of 0.039 was degassed under relative pressure in range of 0-1 at 573.15 K for 3 h.

2.3. Cu(II) aqueous solution
The Cu(II) aqueous solution of 1000 mg/L was prepared by dissolving 3.97 g (±0.001) $CuSO_4.5H_2O$ (99% pure from Aldrich) in 1000 mL distilled water in 1000-mL Erlenmeyer flask. The sample of 5 mL was taken from stock solution and analyzed using Atomic Absorption Spectrophotometer (AAS) to obtain the Cu(II) concentration. To prepare all the predetermined concentration of aqueous solution for adsorption experiments, the stock solution was taken using a variable volume pipette, and it was diluted with distilled water to obtain the predetermined Cu(II) concentration calculated based on general dilution formula.

2.4. Adsorption kinetics and isotherm experiments
To obtain an equilibrium time for the Cu(II) ions adsorption on the OPEFB activated carbon, a Cu(II) adsorption experiment of was conducted in batch mode based on the procedure proposed in the literature [14]. The OPEFB activated carbon of 1 g was placed into a 200-mL erlenmeyer flask, and it was mixed with 100 mL of Cu(II) aqueous solution. The set initial concentration was 50 mg/L but based on the AAS reading, it was 49.99 mg/L. The mixture set at neutral pH was stirred using a magnetic stirrer at 75 rpm, room temperature of 300.15 K (± 2 K) and 1 atm. The sample of 1 mL was taken using a variable volume pipette at the predetermined contact time of 0, 15, 30, 45, 60, 75, 90, 105 and 120 min and stirring was stopped for 1 min before sampling. Each sample was added to 5 mL of distilled water in a separated 10-mL vial for the AAS analysis. To determine the real concentration of samples, the dilution factor was taken into account to the concentration calculated from the AAS reading. Based on this experiment, the Cu(II) adsorption reached equilibrium contact time at 120 min. The experiment was repeated for lower Cu(II) initial concentration of 60 mg/L (58.71 mg/L based on the AAS reading) to investigate the effect of initial Cu(II) concentration on adsorption kinetics parameters.

Experiments of adsorption isotherm were conducted at neutral pH, 75 rpm stirring, room temperature of 300.15 K (± 2 K) and 1 atm. The experiment procedure of adsorption kinetics mentioned above was used for adsorption isotherm experiments with the predetermined Cu(II) initial concentration varied in the range of 10-70 mg/L. The samples of 1 mL for the AAS analysis were taken at the contact time of 0 and 120 min.

3. Results and discussion
3.1. Chemical functional groups
The chemical functional groups of the OPEFB powder and the OPEFB AC was obtained using the FTIR spectra in 400-4000 cm^{-1} range, as can be observed in figure 1. As can be seen in figure 1, there would be seven major absorption bands obtained from the spectra of the OPEFB powder and the OPEFB AC. The first band would be at approximately 3500-3700 cm^{-1} of wavenumber with a peak at approximately 3614.61 cm^{-1}, indicating the O–H stretching group of phenol and alcohol [19]. The second band is the C–H stretching with a wide band at approximately 2700-2980 cm^{-1} and 2 intense bands at approximately 2926.01 and 2858.51 cm^{-1}, presenting aldehyde. The third band is in 1500-1700 cm^{-1} range with the peaks at approximately 1687.71 and 1550.77 cm^{-1} that is assigned the C=C stretching of alkene [20]. The fourth band is the C–H symmetrical and asymmetrical bending with the band at approximately 1317-1450 cm^{-1} [21] and the peak at approximately 1456.26 cm^{-1}. Other bands are for the C–O stretching of aromatic ether and ester in 1240-1284 cm^{-1} range with the peak at approximately 1269.16 cm^{-1}, and for the C–O stretching of carboxylic acids esters in 1000-1260 cm^{-1} range [21] with the peak at approximately 1120.64 cm^{-1}. The last band is a weak band which is associated to the C–C stretching at 400-800 cm^{-1} with the intense bands at approximately 758.02 and 563.21 cm^{-1}.

As shown by the FTIR spectra in figure 1, it can be noticed that pyrolysis at 873.15 K for 1 h followed 0.01 M HCl chemical activation decreased the O–H stretching in the OPEFB powder as indicated by the incline in the transmittance of from 85.24 to 80.78%. It means that phenol and alcohol were removed by approximately 5.23% during the utilisation of the OPEFB AC from the OPEFB powder. The utilisation can break the C–H stretching in the OPEFB which is shown by the decrease in the transmittance of the first and second peak from approximately 83.14 to 76.49% and from approximately 84.42 to 78.7%, respectively. As the result, aldehyde in the OPEFB was removed by approximately 14.77%. A higher amount of decrease in chemical functional groups was occurred in the OPEFB powder whereas the C=C stretching transmittance dropped from approximately from 78.91 to 70.73% and from 77.61 to 70.07% for the first and second intense bands, respectively. It indicated that alkene from the C=C stretching was released by approximately 20.08% during the preparation of the OPEFB AC from the OPEFB powder. In addition, alkene from the C–H symmetrical and asymmetrical bending was also released, and it was approximately 11.12% fall shown by the fall in the peak transmittance being falling from 79.02 to 70.23%, as can be seen in figure 1.

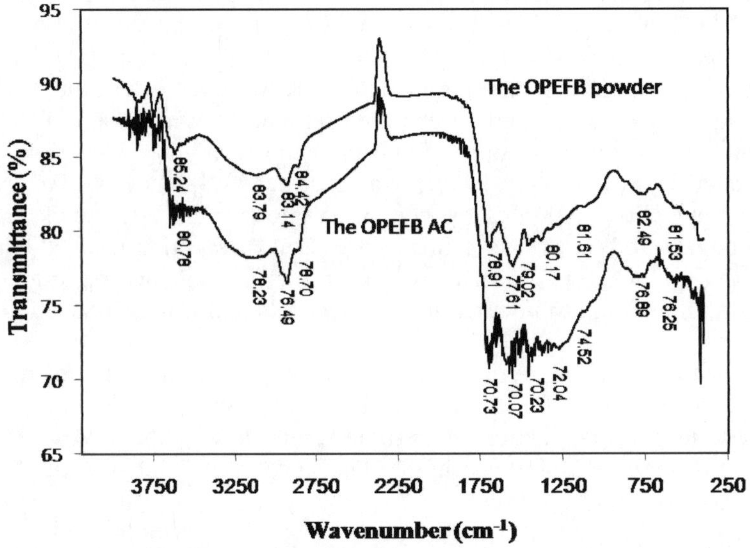

Figure 1. The FTIR spectra of the OPEFB powder and OPEFB AC.

3.2. Surface morphology of the OPEFB activated Carbon

Figures 2(a) and (b) show the SEM micrographs (2000X) of the surface morphology of the OPEFB powder and the OPEFB AC, respectively. A considerable amount of ether and ester in the C–O stretching was removed from the OPEFB powder during the utilisation of the OPEFB AC, and it was by 18.83% indicated by the decrease in the peak transmittance of the C–O stretching of aromatic ether and ester from 80.17 to 72.04% and the decrease in the peak transmittance of the C–O stretching of carboxylic acids esters from 81.61 to 74.52%. Moreover, the last band peak associated to the C–C stretching also reduced by 15.47% from 84.49 to 76.89% for the first peak and from 81.53 to 76.25% for the second peak. Overall, the utilisation of the OPEFB AC from the OPEFB powder reduces the chemical functional groups such as phenol, alcohol, aldehyde, alkene, ether, ester and internal carbon stretching leading to the formation of pores and active site on the OPEFB AC. However, from the transmittance value of 74.64% on average, there could be less pores and active site formed on the OPEFB AC for the adsorption of Cu(II) ions.

IOP Conf. Series: Materials Science and Engineering **206** (2017) 012075 doi:10.1088/1757-899X/206/1/012075

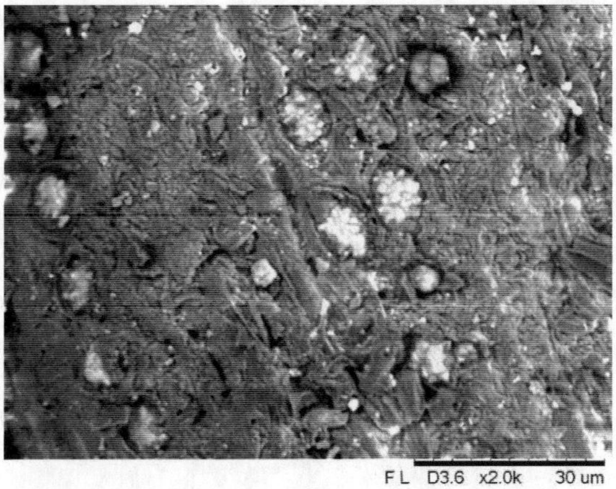

Figure 2(a). The SEM micrographs of the OPEFB

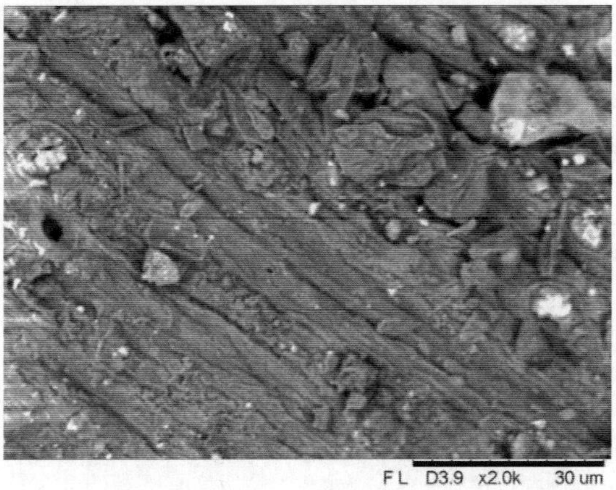

Figure 2(b). The SEM micrographs of the OPEFB AC

As shown in figure 2(a), there are almost regular pores on the OPEFB powder surface, and round materials like flower are in the pores. It seems that the round materials indicate the chemical functional groups of the OPEFB powder, as also highlighted in the previous study [13]. Meanwhile, the round materials of volatile species are not absent at all on the OPEFB AC, as can be seen in figure 2(b). Low concentration of HCl (0.01 M) could be the reason for this. Interesting to note that the pores size and the amount of pores on the OPEFB AC is still small. It could be because the HCl activator concentration being used was still small (0.01 M) to break the C–C stretching and to revome all the volatile matters in the OPEFB powder, as also shown by the overall transmittance in figure 1. Less pores of the OPEFB AC may be formed leading to less surface area, pore size and total pore volume of the OPEFB AC. This expectation was also in line with the results of BET analysis whereas the total surface area, pore volume and pore diameter of the OPEFB AC was 438 m^2/g, and 9.431e-03 cc/g, respectively with the pore diameter being 2.638e+02 A.

3.3. Effect of contact time on Cu(II) adsorption capacity

A batch mode experiment of Cu(II) adsorption was conducted to investigate the effect of contact time on Cu(II) adsorption capacity by the OPEFB AC. The experimental condition shown by the figure 3 legend whereas the system consisted of 50 mg/L Cu(II) initial concentration and 1 g the OPEFB AC.

As the result, Cu(II) adsorption capacity by the OPEFB AC sharply increased during 30-min contact time from 0 to 3.66 mg/g. It moderately inclined to 4.08, 4.25, 4.41 and 4.58 mg/g at 60, 90, 120 and 140-min contact time, respectively. It is clear to see in figure 3 that it reached equilibrium contact time of 140-min because it did not change to much when the contact time increased to 150-min, which was 4.59 mg/g. Overall, Cu(II) adsorption capacity by the OPEFB increased exponentially over contact time until reaching a maximum adsorption capacity, and the same trends also presented in the previous studies [13], [21], [22], [23].

3.4. Effect of contact time on Cu(II) adsorption capacity

The experiments with predetermined initial concentration of Cu(II) ions varied from 10 to 70 mg/L were performed to investigate the effect of initial Cu(II) concentration on the adsorption capacity by the OPEFB AC. As the result, Cu(II) the adsorption capacity by the OPEFB AC considerably increased from 0.87 to 2.26, 3.26, 4.59 and 4.83 mg/g for the increase in the initial Cu(II) concentration in solution from 10 to 30, 40, 50 and 60, respectively, as can be observed in figure 4. However, it fell a bit to 4.62 mg/g when the the initial Cu(II) concentration decreased to 70 mg/g. Overall, Cu(II) adsorption capacity by the OPEFB AC increased exponentially for the increasing initial Cu(II) concentration in solution from 10 to 70 mg/L even if it was a bit fall at 70 mg/L. The same trend also reported in the previous studies [13], [24].

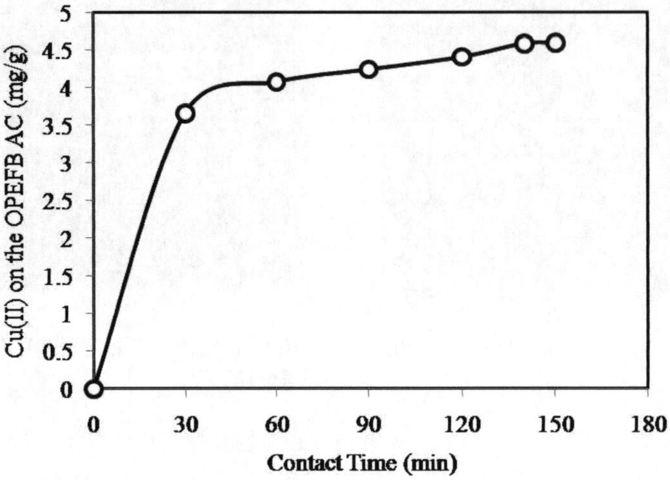

Figure 3. Contact time versus Cu(II) adsorption capacity of the OPEFB AC. Experimental condition: 100 mL aqueous solution of 50 mg/L initial Cu(II) concentration, 1 g the OPEFB AC, 75 rpm magnetic stirring, room temperature of 300.15 K (± 2 K), neutral pH and 1 atm.

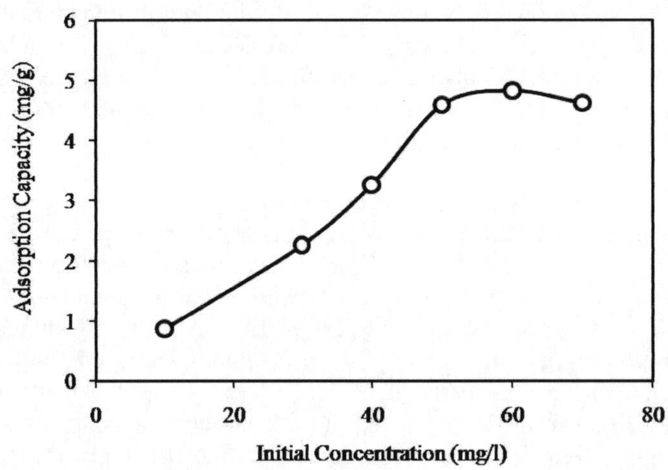

Figure 4. Effect of initial Cu(II) concentration in solution on the adsorption capacity by the OPEFB AC. Experimental condition: 100 mL aqueous solution of Cu(II) ions with initial concentration varied from 10 to 70 mg/L, 1 g the OPEFB AC, 75 rpm magnetic stirring, room temperature of 300.15 K (± 2 K), neutral pH and 1 atm with sampling time of 150-min.

3.5. Cu(II) adsorption kinetics by the OPEFB AC

Adsorption kinetics of Cu(II) by the OPEFB AC represents the amount of Cu(II) adsorbed by the OPEFB AC over time. It can be examined by obtaining the deviation of Cu(II) concentration in solution over contact time, and it can present the rate constant of Cu(II) adsorption on the OPEFB AC.

The common equations are presented by equations (1) and (2) for the pseudo first order [25] and the pseudo second order models [26], respectively:

$$\frac{dq_t}{dt} = k_L(q_e - q_t) \quad (1)$$

$$\frac{dq_t}{dt} = k_H(q_e - q_t)^2 \quad (2)$$

where q_t (mg/g) is the Cu(II) adsorption capacity by the OPEFB AC at the time of t (min), q_e (mg/g) denotes the equilibrium adsorption capacity, k_L represents the rate constant of pseudo first order model (min^{-1}), and k_H is the rate constant pseudo second order model (g/mg.min). Linearizing equations (1) and (2) by integration of equations (1) and (2) with the boudary conditions q_t = 0 mg/g (t = 0 s) and q_t = q mg/g (t = t s), results in equations (3) and (4), respectively [13]:

$$\log(q_e - q_t) = \log q_e - \left(\frac{k_L t}{2.303}\right) \quad (3)$$

$$\frac{t}{q_t} = \frac{1}{k_H q_e^2} + \frac{t}{q_e} \quad (4)$$

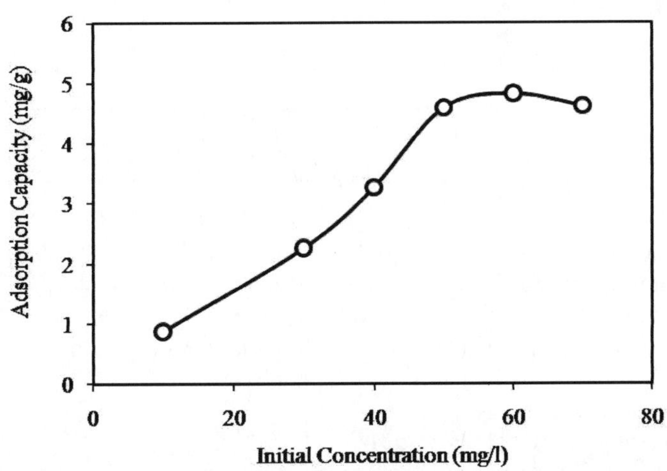

Figure 5. Pseudo first order kinetics of Cu(II) adsorption on the OPEFB AC. Experimental condition: 100 mL aqueous solution of Cu(II) ions with initial concentration varied being 49.99 and 50 mg/L, 1 g the OPEFB AC, 75 rpm magnetic stirring, room temperature of 300.15 K (± 2 K), neutral pH, 1 atm and contact time of 0, 30, 60, 90, 120, 140 and 150 min.

The pseudo first order and second order models were plotted using the data shown in figure 5 and the data from the second experiment of Cu(II) adsorption kinetics with Cu(II) initial concentration being 58.71 mg/L. The Cu(II) adsorption capacity obtained at the contact time of 0, 30, 60, 90, 120, 140 and 150 min was 0, 2.89, 4.09, 4.29, 4.39, 4.47 and 4.65 mg/L, respectively.

The results of Cu(II) adsorption kinetics are shown in figures 5 and 6. As clearly shown by the legends in figures 5 and 6, the Cu(II) adsorption kinetics on the OPEFB activated carbon follows the pseudo second order kinetics, which was justified by the correlation coefficients, R^2 for the pseudo second order kinetics (0.99 and 0.98) being much higher than the pseudo first order kinetics (0.292 and 0.350).

The rate constants of pseudo second order kinetics, k_L and the equilibrium adsorption capacity, q_e determined was 0.19 /min and 4.69 mg/g, respectively for 49.99 mg/L initial Cu(II) concentration, and it was 0.15 /min and 4.81 mg/g, respectively for 58.71 mg/L initial Cu(II) concentration. It means that increasing initial Cu(II) concentration in solution from 49.99 to 58.71 mg/L lifted up the equilibrium adsorption capacity by 2.4%. The increase in the equilibrium adsorption capacity is reasonable because all the active sites on the OPEFB AC could not be completely occupied Cu(II) ions at 50 m/L initial Cu(II) concentration due to the amount of Cu(II) ions could be still less than the avalaible active sites on the OPEFB AC. Meanwhile, the increase in the adsorption kinetics rate constant was caused by the higher driving force to transport Cu(II) ions in solution onto the OPEFB AC overtime, whereas it was caused by the higher deviation between mass of Cu(II) ions in solution and on the OPEFB AC when the higher initial Cu(II) concentration was applied. In other words, the higher initial Cu(II) concentration, the faster adsorption of Cu(II) ions onto the OPEFB AC occurred.

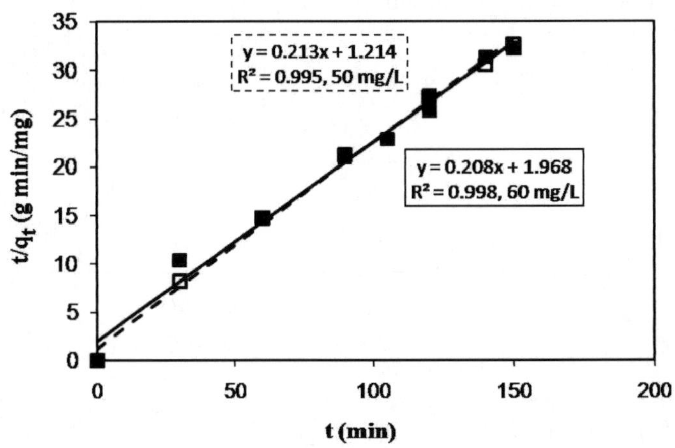

Figure 6. Pseudo second order kinetics of Cu(II) adsorption on the OPEFB AC. Experimental condition: 100 mL aqueous solution of Cu(II) ions with initial concentration varied being 49.99 and 50 mg/L, 1 g the OPEFB AC, 75 rpm magnetic stirring, room temperature of 300.15 K (± 2 K), neutral pH, 1 atm and contact time of 0, 30, 60, 90, 120, 140 and 150 min.

3.6. Cu(II) adsorption isotherm by the OPEFB AC
The adsorption isotherm equations of Langmuir and Freundlich were used to determine the Cu(II) adsorption isotherm by the OPEFB AC. The data presented in figure.6 was fitted to the Langmuir equation [27] in linearized form [13]:

$$\frac{C_e}{q_e} = \frac{1}{q_m K_L} + \frac{1}{q_m} C_e \tag{5}$$

where C_e (mg/L) represents as the equilibrium Cu(II) concentration in solution, q_m (mg/g) denotes as the mono-layer adsorption capacity of Langmuir, and K_L (L/mg) is the over-all adsorption capacity of Langmuir. Plotting C_e/q_e over C_e yields a straight line with the intercept, $1/(q_m K_L)$ and the slope, $1/ q_m$ to determine the parameters value. The Langmuir plot is shown in figure 7. The Freundlich equation [28] is linearized [13] and presented as:

$$\log q_e = \frac{1}{n} \log C_e + \log K_F \tag{6}$$

where K_F (L/mg) represents the over-all adsorption capacity of Freundlich, $1/n$ is the Freundlich adsorption intensity. By plotting log q_e over log C_e yields a straight line with the slope, $1/n$ and the intercept, log K_F to determine the parameters value. The Freundlich plot is shown in figure 8.

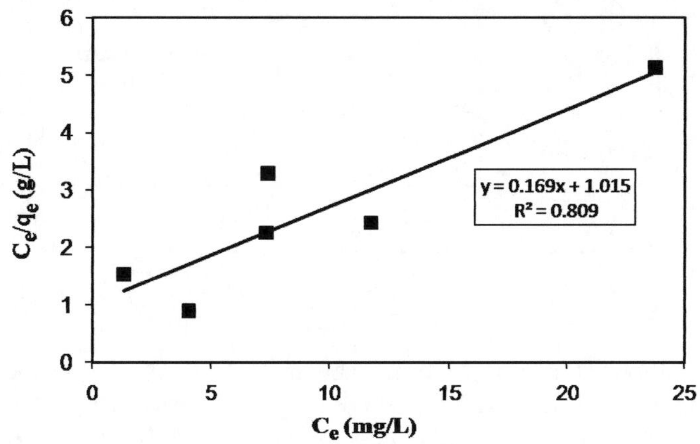

Figure 7. Langmuir isotherm for Cu(II) adsorption by the OPEFB AC. Experimental condition: 100 mL aqueous solution of Cu(II) ions with initial concentration varied from 10 to 70 mg/L, 1 g the OPEFB AC, 75 rpm magnetic stirring, room temperature of 300.15 K (± 2 K), neutral pH and 1 atm with sampling time of 150-min.

As can be seen in figures 7 and 8, the R^2 value on the Cu(II) adsorption isotherm for the Langmuir and Freundlich equations is approximately 0.809 and 0.629, respectively. The mono-layer adsorption capacity and the over-all adsorption capacity of Langmuir obtained was 5.917 mg/g and 0.166 L/g, respectively. Meanwhile, the parameters in the Freundlich model were calculated, and the over-all adsorption capacity and adsorption intensity of Freundlich obtained was 1.006 L/mg and 1.852 respectively. Overall, the Cu(II) adsorption isotherm by the OPEFB AC follows the Langmuir equation, it is favourable adsorption of Cu(II) ions on the OPEFB AC, and each active site of the OPEFB AC adsorb only one molecule of Cu(II) ion to form a monolayer on the OPEFB AC surface.

As highlighted in figures 7 and 8, the correlation coefficients (R^2) of the Cu(II) adsorption isotherm linear plot is approximately 0.95 and 0.57 for the Langmuir and Freundlich models respectively. The parameters based on the Langmuir equation were obtained, and the q_m and K_L values were 5.92 mg/g and 0.17 L/mg, respectively. Meanwhile, the parameters based on the Freundlich equation were calculated, and the K_F and n values were worked out to be 1.01 L/mg and 1.85, respectively. Overall, the adsorption isotherm of Cu(II) on the OPEFB activated carbon follows the Langmuir model. The Cu(II) adsorption capacity by the OPEFB AC is less than the ones by green tea (6.37 mg/g), coarse tea (7.36 mg/g), Cassava peel (11.35 mg/g) and Brazil nut shells (19.45 mg/g) [10], [11]. Therefore, further studies migh be needed to investigate a new method of utilisation activated carbon from the OPEFB to get higher Cu(II) adsorption capacity.

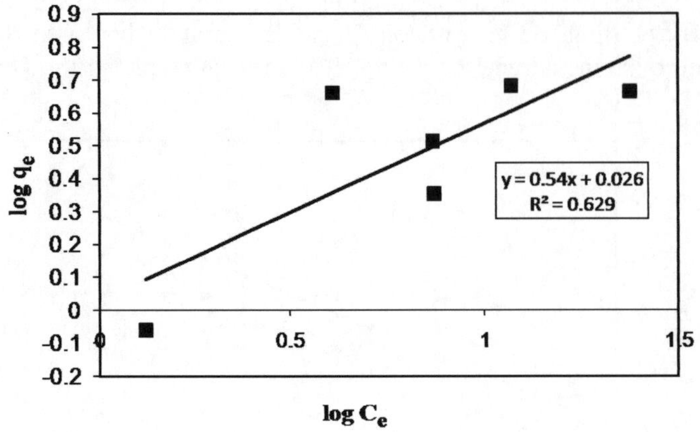

Figure 8. Freundlich isotherm for Cu(II) adsorption by the OPEFB AC. Experimental condition: 100 mL aqueous solution of Cu(II) ions with initial concentration varied from 10 to 70 mg/L, 1 g the OPEFB AC, 75 rpm magnetic stirring, room temperature of 300.15 K (± 2 K), neutral pH and 1 atm with sampling time of 150-min.

4. Conclusions

Oil palm empty fruit bunch (OPEFB) was used to prepare activated carbon by physical activation (pyrolysis at 873.15 K in a furnace for 1 h) and chemical activation (0.01 M HCl for 2 h). Based on the FTIR analysis, it was found that the chemical functional groups such as phenol, alcohol, aldehyde, alkene, ether, ester were removed during the activation leading to the the formation of pores on the OPEFB AC, as also shown by the SEM analysis. The FTIR and SEM analyses was also in line with the results of BET analysis. Experiments in batch mode consisting of 1 g the OPEFB AC in 100 mL Cu(II) aqueous solution with the initial concentration in the range of 10-70 mg/L, magnetic stirring at 75 rpm, room temperature of 300.15 K (± 2 K), at 1 atm and neutral pH over the contact time in the range of 0-150 min were performed to investigate the effect of contact time and initial Cu(II) concentration in solution on Cu(II) adsorption capacity by the OPEFB AC. As the result, Cu(II) adsorption capacity over contact time and initial concentration was typecally exponential trends.

The result of adsorption kinetics study with 58.71 mg/L initial Cu(II) concentration showed that Cu(II) adsorption onto the OPEFB AC followed the pseudo-second-order kinetics with the correlation coefficients (R^2), kinetics rate constant and equilibrium adsorption capacity being 0.99, 4.831 mg/g and 0.21/min, respectively for. Moreover, Langmuir equation was reliable to present Cu(II) adsorption onto the OPEFB AC with the R^2 value being 0.99, and the mono-layer and over-all adsorption capacity was determined to 5.92 mg/g and 0.17 L/mg, respectively. Further studies migh be needed to investigate a new utilisation method to produce the OPEFB AC with higher Cu(II) adsorption capacity.

References

[1] Hawkes S J 1997 *J. Chem. Educ.* **74** 11 1374–80
[2] Srivastava N K and Majumder B C 2008 *J. Hazard. Mat.* **151** 1 1–8
[3] Bala M, Shehu R A and Lawal M 2008 *Bayero J. Pure Appl. Sci.*, **1** 1 6–38
[4] Yan-Biao G, Hong F, Chong C, Chong-Jian J, Fan X and Ying L 2013 *Pol. J. Environ. Stud.* **22** 5 1357–62
[5] Dimple L 2014 *Int. J. Environ. Res. Dev.* **4** 1 41–8
[6] Minamisawa M, Minamisawa H, Yoshioda S and Taki N 2004 *J. Agri. Food Chem.* **52** 18

5606–15
[7] Theophanides T and Anastassopoulou J 2002 *Crit. Rev. Oncol. Hematol.* **42** 1 57–64
[8] Eccles H 1999 *Trends Biotechnol.* **17** 12 462–5
[9] Leung W C, Wong M F, Chua H, Lo W and Leung C K 2000 *Water Sci. Technol.* **41** 12 233–40
[10] Basso M C, Cerrell E G and Cukierman A L 2002 *Ind. Eng. Chem. Res.* **41** 15 3580–85.
[11] Moreno P, Carlos J and Giraldo L 2010 *J. Anal. Appl. Pyrol.* **87** 2 188–93
[12] Baquero M C, Giraldo L, Moreno J C, Suárez-García F, Martínez-Alonso A and Tascón J M D 2003 *J. Anal. Appl. Pyrol.* **70** 2 779–84
[13] Muslim A, Zulfian, Ismayanda M H, Devrina E and Fahmi H 2015. *J. Eng. Sci. Technol.* **10** 12 1654–66
[14] Hidayu A R, Mohamad N F, Matali S and Sharifah A S A K 2013. *Procedia Eng.* **68**, 379–84
[15] Teoh C H 2000 *Land Use and the Oil Palm Industry in Malaysia.* A bridged report produced for the WWF forest information system database under project MY 0057 – Policy assessment of Malaysia conservation issues, Kinabatangan, November 2000
[16] World Growth 2009 *World Growth Palm Oil Green Development Campaign: Palm Oil — The Sustainable Oil.* A Report by World Growth September 2009. Available online at http://www.worldgrowth.org/assets/files/Palm_Oil.pdf, (accessed 7 December 2015)
[17] ResearchInChina 2015 China activated carbon industry report, 2014–2017
[18] Muslim A in press Australian pine cones-based activated carbon for adsorption of copper in aqueous solution *J. Eng. Sci. Technol.*
[19] Yang T and Lua A 2003 *J. Colloid Interface Sci.* **267** 2 408–17
[20] Silverstein R M, Bassler G C and Morrill T C 1981 *Spectrometric Identification of Organic Compounds* 4th ed. (New York: Wiley–Interscience)
[21] Hesas R H, Niya A A, Daud, W M A W and Sahu J N 2003 *Bio Resources* **8** 2 2950–66
[22] Wong K K, Lee C K, Low K S and Haron M J 2003 *Chemosphere* **50** 23–8
[23] Mengistie A A, Siva R T, Prasada R A V and Singanan M 2008 *Bull. Chem. Soc. Ethiop.* **22** 3 349–60
[24] Tseng R L and Tseng S K 2005 *J. Colloid Interface Sci.* **287** 2 428–37
[25] Lagergren S 1989 *K. Sven. Vetensk. Akad. Handl.* **24** 4 1–39
[26] Ho Y S, Wase D A J and Forster C F 1996 *Environ. Technol.* **17** 71–7
[27] Langmuir I 1918 *J. Am. Chem. Soc.* **40** 9 1361–1403
[28] Freundlich H 1906. *J. Phys. Chem.* **57** 384–410

IOP Conf. Series: Materials Science and Engineering **206** (2017) 012076 doi:10.1088/1757-899X/206/1/012076

Effect of dilution and operating parameters on ammonia removal from scheduled waste landfill leachate in a lab-scale ammonia stripping reactor

N M L Hanira [1], C M Hasfalina[1], M Rashid [2], C A Luqman[3], A M Abdullah[4]

[1]Department of Biological and Agricultural Engineering, Universiti Putra Malaysia, 43400 Serdang, Selangor, Malaysia
[2]Air Resources Research Laboratory, Malaysia-Japan International Institute of Technology, 54100 UTM Kuala Lumpur, Malaysia
[3]Department of Chemical and Environmental Engineering, Universiti Putra Malaysia, 43400 Serdang, Selangor, Malaysia
[4]Faculty of Environmental Studies, Universiti Putra Malaysia,43400 Serdang, Selangor, Malaysia

E-mail: nurulhaniramatlazim@yahoo.com

Abstract. A lab-scale ammonia stripping reactor was used to treat raw and diluted (1:1) scheduled waste landfill (SWL) leachate containing ammonia-nitrogen (NH_3-N). Operating parameters such as air-liquid ratio, hydrated lime [$Ca(OH)_2$] dosage, types of packing materials and packing heights were investigated with central composite design (CCD) of response surface methodology (RSM) was used to optimize the parameters affecting NH_3-N removal from the leachate. The percentage removal on turbidity, colour and phosphate were also evaluated in this study. It was observed that the optimal conditions obtained from desirable response (NH_3-N removal) for raw leachate were predicted at air–liquid ratio of 70, $Ca(OH)_2$ dosage of 5 gL^{-1}, packing height of 60 cm and types of packing material was number 3 (non-woven polyester) while for diluted leachate these were 70, 6 gL^{-1}, 60 cm and Type 3 (non-woven polyester), respectively. Quadratic RSM predicted the maximum NH_3-N removal to be 78% for raw leachate and 81% for diluted leachate at these optimal conditions concurred with the experiment which successfully removed 76% and 80% of NH_3-N, respectively. However, higher removal efficiencies of turbidity (97%), colour (88%) and phosphate (93%) was observed in the treatment with diluted leachate compared to raw leachate merely up to 55%, 34% and 49%, respectively. The finding showed that the difference in the removal of NH_3-N in diluted and raw SWL leachate was insignificant. However, turbidity, colour and phosphate showed a significant reduction in the diluted leachate during the treatment. The study suggests that the dilution of SWL leachate does not present a significant effect on the removal of ammonia in the stripping reactor.

1. Introduction

Landfilling is a favorable method for waste disposal due to its low cost, ease of operation and implementation [1]. However, the method generates toxic and highly polluted leachate that contaminates the soil, surface water, groundwater, and also contribute to air pollution [2].

Leachate composition varies widely and among which ammonia-nitrogen (NH_3-N) is known for its toxicity effect, especially when it is exist in high concentration. Generally, in old landfill leachate, the organic matter is well stabilized [3], hence the treatment of leachate with high concentration of NH_3-N by biological systems is ineffective for such effluents and could be failed. In addition, biological treatment required chemical additives to provide alkaline condition for nitrification processes.

Chemical and physical methods are usually applied prior biological method. The major chemical treatment used for leachate treatment includes coagulation-flocculation, chemical or electrochemical oxidation [4, 5]. Meanwhile, the major physical methods used for landfill leachate treatment are adsorption, membrane filtration, sedimentation and air stripping [6]. Aerobic, anaerobic and anoxic processes are the major biological methods used in combination for treating landfill leachate [7]. However, air stripping is favored over other methods for ammonia removal due to few advantages such as low treatment cost, ability of volatile compounds stripping, and degasification of ammonia at high level concentrations [8, 9].

Air stripping is a mass transfer operation involving the transfer of dissolved substances in water from liquid phase to gas phase by physical action. In air stripping process, volatile organic chemicals is removed from aqueous solution by providing maximum contact between the liquid and air in a specifically designed column. An equilibrium state is reached when these two phases are brought into contact. The off-gas or treated air could be released to the atmosphere or being recovered from the air into a strong acid such as sulphuric acid as subsequent treatment [10, 11]. The amount of ammonia-nitrogen (NH_3-N) removed from a liquid in ammonia stripping reactor systems depend on several factors such as NH_3-N concentration, pH, temperature, air-to-liquid ratio, water and ambient air temperature, stripping column high, contact time /stripping time, characteristics of the volatile material (partial pressure, Henry's constant, gas transfer resistance, etc.), turbulence in gases and liquid phases, types of packing materials and the surface area-to-volume ratio [9, 11, 12].

To date, very limited data available on the treatment of scheduled waste leachate is available [13, 14, 15]. The main objective of this work was to investigate the effects of dilution on NH_3-N removal efficiency and to compare the effectiveness of raw and diluted leachate treatment by using lab-scale ammonia stripping reactor. The RSM was used for the optimal experimental design of ammonia stripping process. Both raw and diluted leachate was subjected to aerobic chemical and physical treatment by ammonia stripping reactor at different air-liquid ratio, hydrated lime [$Ca(OH)_2$] dosage, types of packing materials and packing heights. Design-Expert software was used in this study to produce 3D surface responses of quadratic model and to evaluate the interactive relationships between independent variables and response. For each 3D surface plot, one variable was kept constant, while the other two were varied within the experimental ranges.

2. Materials and method

2.1. Leachate sampling and characterization

Leachate samples were taken from a scheduled waste landfill facility. The collection and preservation of sample was done based on Standard Methods for the Examination of Water and Wastewater [16]. The sample was characterized immediately in the laboratory. Table 1 presents the characteristics of raw and diluted leachate samples of the Scheduled Waste Landfill (SWL).

Table 1. Characteristics of raw and diluted leachate from SWL

Parameter	Units	Raw Leachate		Diluted Leachate	
		Range	Mean	Range	Mean
pH	-	9.25-9.36	9.31	9.12-9.31	9.18
BOD_5	mgL^{-1}	1610 - 2670	2053	NA	NA
COD	mgL^{-1}	3430 - 4610	4127	1283-1650	1319
TSS	mgL^{-1}	0 - 870	313	NA	NA
Phosphate	mgL^{-1}	61-71.4	65	61-65.5	62.8
Colour	Pt-Co	2654-2890	2765	1988-2220	2126
Total Iron	mgL^{-1}	0.24-0.41	0.29	0.23-0.37	0.28
Turbidity	NTU	206-212	209	52.7-59.4	55.1
NH_3-N	mgL^{-1}	1790 - 2570	2267	1020-1587	1267

2.2 Ammonia stripping

A 2.5 L diluted (diluted 1:1 with water) and raw leachates were used in each experiment. Hydrated lime ($Ca(OH)_2$) was applied to leachate prior ammonia stripping at different dosage to enhance ammonia and other pollutants removal [13, 14, 15]. The stripping reactor consisted of a 1.20 m (H) x 0.5m (D) column with compressed air introduced at the bottom. Leachate flow rate was maintained at 0.1 L min^{-1}, while the air flow rate varied at 2, 5 and 8 L min^{-1}. The stripping reactor was operated on a batch mode. Each experiment was carried out for 12 hours with sample withdrawn and analysed on hourly basis. Figure 1 shows the schematic diagram of an ammonia stripping reactor where the leachate was carried up to the top of the column using a dosing pump with perforated PVC distribution plate.

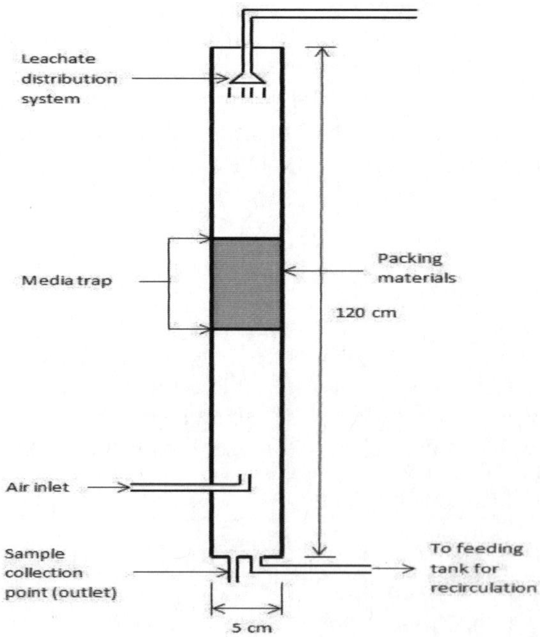

Figure 1. The schematic diagram of ammonia stripping reactor.

IOP Conf. Series: Materials Science and Engineering **206** (2017) 012076 doi:10.1088/1757-899X/206/1/012076

2.3 Experimental Design and Mathematical Modelling

The Design Expert Software (version 7.0) was used for the statistical design of experiments and data analysis. The central composite design (CCD) of response surface methodology (RSM) was applied to optimize the four operating parameters and their ranges; air-liquid ratio of 20, 60 and 80, $Ca(OH)_2$ dosage of 4, 5 and 6 gL^{-1}, with three different types of packing materials (polyurethane foam, polyurethane nylon and non-woven polyester) and packing height of 20, 40 and 60 cm.

The CCD for 30 factorial designs with six additional experimental trials was used as the replicates of the central point to get a good estimate of the experimental error. The coded values for air-liquid ratio (X_1), $Ca(OH)_2$ dosage (X_2), types of packing materials (X_3), and packing height (X_4), were considered at three levels, assigned as -1 (minimum), 0 (central), and +1 (maximum), respectively. Experimental results are shown as removal efficiency (%) of NH_3-N. Analysis of variance (ANOVA) was used for graphical analyses of the data to attain the interaction between the process variables and the responses. The interactive effects of the independent variables on the dependent variables were illustrated by three- and two-dimensional contour plots. From the 3D plots, the simultaneous interaction of the two factors on the responses was studied. The optimum region was also identified based on the main parameters in the overlay plot.

3. Results and discussion

3.1 ANOVA

The process performance of ammonia stripping was evaluated by analysing the experimental results of all the terms of coded factors were represented by the following Equation 1;

$$Y_{NH3\text{-}N} = 55.35 + 8.57\,X_1 - 2.29\,X_2 + 10.28\,X_3 + 4.11\,X_4 - 1.22\,X_1X_2 + 0.81\,X_1X_3 + 3.89\,X_1X_4 + 3.73\,X_2X_3 + 3.48\,X_2X_4 - 5.41\,X_3X_4 - 10.03X_1^2 + 4.33\,X_2^2 + 2.70\,X_3^2 + 2.25\,X_4^2 \qquad (1)$$

The optimal conditions for NH_3-N removal efficiency were selected based on responses. The final second-order polynomial regression in obtained at 70, 6 gL^{-1}, 60 cm and type 3 (non-woven polyester) for air–liquid ratio, $Ca(OH)_2$ dosage, packing height and types of packing material, respectively. The adequacy of the response surface models were validated statistically by ANOVA. Table 2 presents the ANOVA of regression parameters of the predicted response surface quadratic model for NH_3-N removal efficiency which showed that the model was significant ($p<0.0001$) for ammonia stripping process. Table 2 demonstrates that all the variables contributing to the model were significant at the 5% confidence level. The optimal points of working conditions and predicted NH_3-N removal efficiencies were in good agreement between the experimental and predicted with the R^2=0.9943.

Table 2. ANOVA for response surface quadratic model for ammonia stripping

	Sum of square	df	Mean square	F-value	Prob>F
Model	5051.24	14	360.80	187.20	< 0.0001
X_1	1323.38	1	1323.38	686.64	< 0.0001
X_2	93.98	1	93.98	48.76	< 0.0001
X_3	1902.01	1	1902.01	986.86	< 0.0001
X_4	304.55	1	304.55	158.02	< 0.0001
X_1^2	260.60	1	260.60	135.21	< 0.0001
X_2^2	48.49	1	48.49	25.16	0.0002
X_3^2	18.83	1	18.83	9.77	0.0069
X_4^2	13.13	1	13.13	6.81	0.0197
$X_1 X_2$	23.99	1	23.99	12.44	0.0030
$X_1 X_3$	10.48	1	10.48	5.44	0.0340
$X_1 X_4$	242.50	1	242.50	125.82	< 0.0001
X_2X_3	222.08	1	222.08	115.23	< 0.0001
X_2X_4	193.42	1	193.42	100.36	< 0.0001
X_3X_4	468.61	1	468.61	243.14	< 0.0001
Residual	28.91	15	1.93		
Lack of fit	23.88	10	2.39	2.38	0.1760
Pure Error	5.03	5			
Cor Total	5080.15	29			
Std. Dev.	1.39		R^2	0.9943	
Mean	54.89		Adj R^2	0.9890	
C.V%	2.53		Pred R^2	0.9707	
Press	148.98		Adeq. precision	58.972	

3.2. Effects of operational parameters on removal efficiency

Figure 2 to Figure 7 shows the effect of air–liquid ratio, $Ca(OH)_2$ dosage, packing height and types of packing material on NH_3-N removal efficiency for diluted leachate. NH_3-N removal increased with a rise in air–liquid ratio and packing height. The response surface of NH^3-N removal illustrated that the higher the air-liquid ratio, the higher NH_3-N removal efficiency (Figures 2, 3 and 4). It was observed the maximum NH_3-N removal was at air-liquid ratio 80 and packing height of 60 cm. Lin et al. [17] observed higher ammonia removal efficiency with the higher air-liquid ratio of stripping process. This is because higher air to liquid ratio promotes the transfer of ammonia into the air phase [9, 18].

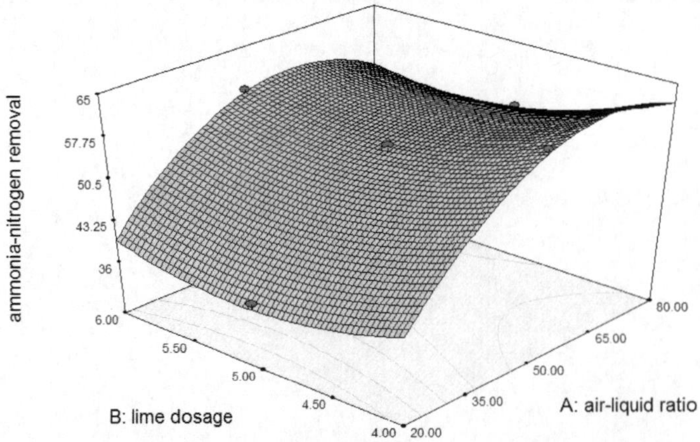

Figure 2. Design-expert plot; response surface plot for NH_3-N removal at different air-liquid ratio and lime dosage.

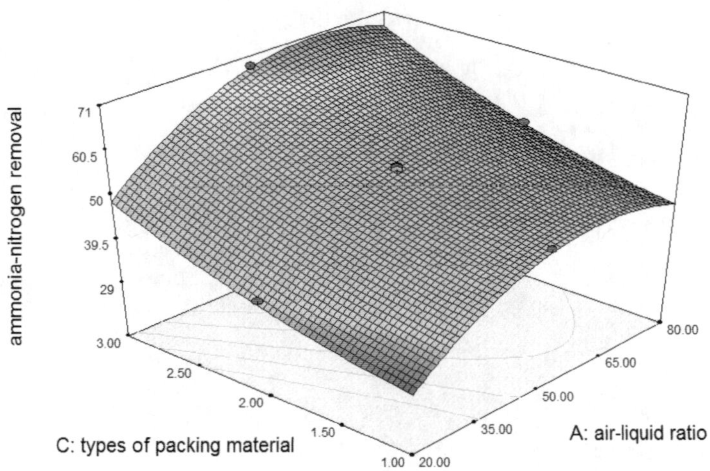

Figure 3. Design-expert plot; response surface plot for NH_3-N removal at different air-liquid ratio and types of packing material.

However, the effect of packing height and types of packing material was more significant than air-liquid ratio and $Ca(OH)_2$ dosage for NH_3-N removal in this study. An increased in packing height of stripping reactor results in higher ammonia removal efficiency regardless of packing materials due to increase of contact time between air and liquid [9]. In this study, the maximum packing height (60 cm) showed higher ammonia removal efficiency for diluted leachate (Figures 5 and 6). Alam and Hossain [9] also investigated few types of packing materials including plastic ring, coal, stone chips, and wood chips in which plastic ring was found to be the most efficient for removal of ammonia, with 91.60% removal efficiency. The quality of packing materials had significant effect for stripping column efficiency. Light, easily produced and high surface area of packing materials gave the best performance for ammonia removal in ammonia stripping column. For this study the highest removal efficiency was obtained when packing material Type 3 (non-woven polyester) was used. Experiments with raw leachate also favour packing material Type 3 (non-woven polyester) resulted the highest NH_3-N removal efficiency.

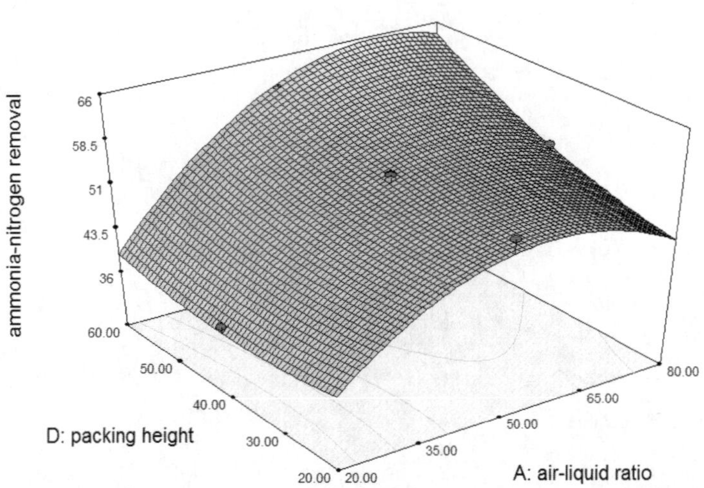

Figure 4. Design-expert plot; response surface plot for NH_3-N removal at different air-liquid ratio and packing height.

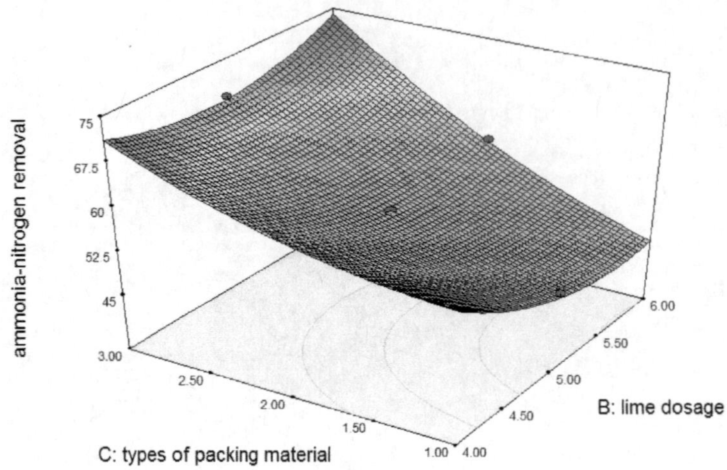

Figure 5. Design-expert plot; response surface plot for NH_3-N removal at different types of packing materials and lime dosage.

Furthermore, NH_3-N removal was higher when the hydrated lime ($Ca(OH)_2$) dosage was at 6 gL^{-1} for diluted leachate treatment. A high pH value by the addition of alkaline agents such as lime and sodium hydroxide is crucial to obtain higher ammonia removal through the displacement of ammonium/ammonia equilibrium. In wastewater, ammonium ions (NH_4^+) exist in equilibrium with gaseous ammonia (NH_3) as in Equation 2 [15, 19]. Hydroxide ion (OH^-) from $Ca(OH)_2$ reacts with ammonium ion (NH_4^+) in the leachate to form ammonia gas (NH_3). Most ammonium ions (NH_4^+) are in the form of NH_3 at pH 10.5 to 12. Thus, at higher pH, there is greater proportion of nitrogen in the form of gaseous ammonia and more ammonia is stripped off from the liquid. For raw leachate, less $Ca(OH)_2$ dosage was used in the study. The pH of raw leachate was 9.31, whilst 9.18 of diluted leachate. Thus, at lower pH, diluted leachate demands for more OH^- ions to transform NH_4^+ ion to NH_3. Therefore, more alkaline chemicals are needed to aid ammonia removal in diluted leachate. This theory is in good agreement with the results of this study.

$$\underset{\text{ammonium ion}}{NH_4^+} + \underset{\text{hydroxide ion}}{OH^-} \rightleftharpoons \underset{\text{ammonia gas}}{NH_3} + \underset{\text{water}}{H_2O} \quad (2)$$

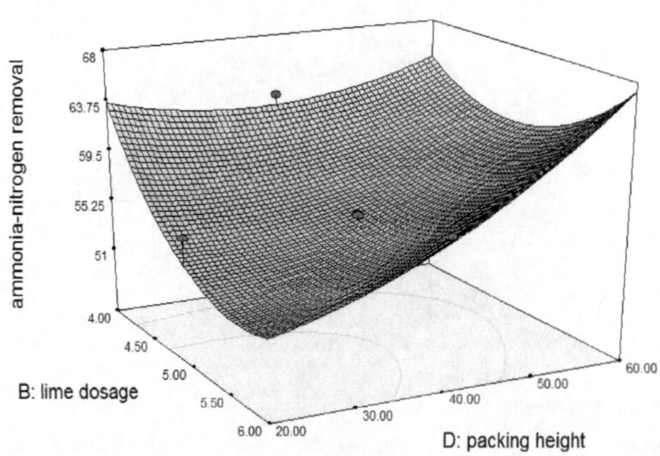

Figure 6. Design-expert plot; response surface plot for NH_3-N removal at different types of packing heights and lime dosage.

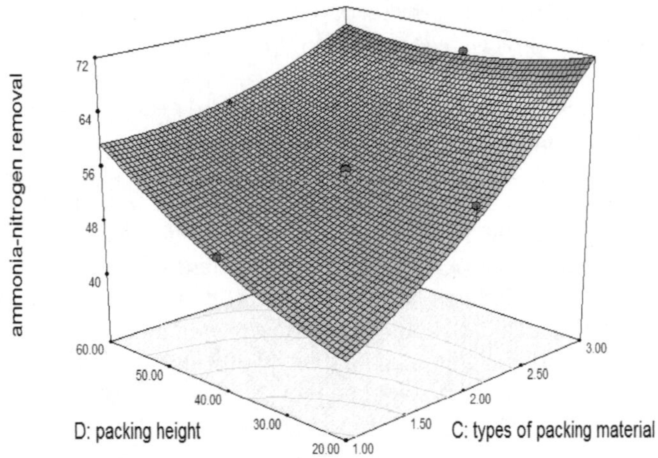

Figure 7. Design-expert plot; response surface plot for NH_3-N removal at different types of packing materials and packing height.

3.3 Process optimization and comparison with raw leachate treatment

The second order polynomial model generated from RSM was used to confirm the validity of the statistical experimental strategies at optimum conditions for air-liquid ratio, $Ca(OH)_2$ dosage, packing height and types of packing material. When the model was developed from experimental results and checked for adequacy, optimization criteria can be selected to find out the most optimum operating conditions. Based on Table 3, RSM results suggested that the optimal conditions for maximum removal of NH_3-N from raw leachate treatment were at air-liquid ratio of 70, $Ca(OH)_2$ dosage of 5 gL^{-1}, packing height of 60 cm and types of packing material 3, while for diluted leachate treatment, RSM results suggest the optimal conditions should be at air-liquid ratio of 70, $Ca(OH)_2$ dosage of 6 gL^{-1}, packing height of 60 cm and types of packing material 3 (non-woven polyester). Both raw and diluted leachate treatment suggest the same air-liquid ratio (70), packing height (60 cm) and types of packing material (Type 3: non-woven polyester), except for $Ca(OH)_2$ dosage, in which diluted leachate required more dosage compared to raw leachate treatment. It was estimated that based on optimum operating conditions, the maximum value for NH_3-N removal via air stripping was estimated as 78% and 81% for raw leachate treatment and diluted leachate treatment, respectively, after 8 hours. To validate this, verification experiments were carried out thrice based on predicted RSM. The selected combinations of the four variables resulted in 76% NH_3-N removal for raw leachate treatment and 80% NH_3-N removal after 8 hours. Prolonged treatment up to 12 hours successfully obtained 88% of NH_3-N removal efficiency and 92% for raw and diluted leachate treatment, respectively. In addition, for raw leachate treatment, 55%, 38% and 49% of turbidity, colour and P-reactive were successfully removed, respectively. For diluted leachate, the removal efficiency for turbidity, colour and P-reactive were 98%, 89% and 98%, respectively, after 12 hours.

The results obtained in this study were relatively similar to other studies, despite different type of sample used. Campos et al. [20] treated the same amount of leachate samples (2.5 L) and after 7 hours of treatment, 91% of ammonia was successfully removed. Nevertheless, high temperature of 60 °C was required to enhance the ammonia removal in their study. In the long term, the requirement of high temperature significantly increases the usage of electricity, which consequently increase the treatment cost. Therefore, low-cost and simple mechanical or air stripping process is chosen over other methods. In addition, previous studies by Cheung et al. [21] documented high ammonia removal efficiency about 93% and >99.5% for 1,600 (24 h). Meanwhile, 88% of NH_3-N was removed from raw leachate in this study which accounted about 1767 mg L^{-1} (influent = 2019 mg L^{-1}, effluent= 252 mg L^{-1}) within 12 hours. As for diluted leachate, about 98% of NH_3-N removal efficiency was obtained within 12 hours, which accounted about 1182 mgL^{-1} (influent = 1280 mg L^{-1}, effluent= 98 mg L^{-1}). In terms of NH_3-N concentration removal, direct treatment using raw leachate was better as higher NH_3-N concentration can be reduced with less dosage required.

The treatment using ammonia stripping reactor in this study seems promising as high strength wastewater with larger volume and high NH_3-N concentration varied from 1020 – 2570 mg L^{-1} could be reduced within a short time (12 hours). Compared to Marttinen et al. [6] study, 24 hours was required to treat smaller volume and low-strength leachate (74–220 mg L^{-1}), with relatively similar ammonia removal efficiency. Since the final effluent for raw and diluted leachate were 252 and 98 mg L^{-1}, respectively, and still not meet the permissible limit standard of NH_3-N (20 mg L^{-1}) and pH 9 as stated in Environmental Quality (Industrial Effluent or Mixed Effluent) Regulations 2009, Fifth Schedule (Standard B) [22], the effluent should be further treated with biological method. Biological method can take place efficiently as the concentration of final effluent is low and suitable for biological methods to take place until the discharged requirement was achieved.

4. Conclusion

In this study, ammonia stripping for scheduled waste landfill leachate treatment with and without dilution was investigated. Ammonia stripping process optimization concentrated on the influence of operating parameters such as air-liquid ratio, $Ca(OH)_2$ dosage, packing materials and packing height using CCD of RSM. All interactions among the variables were evaluated by RSM. The multiple correlation of determination R^2 was found to be 0.9943, which was near to 1 showing that the actual data was well fitted with predicted data. The optimum results obtained from the model indicated that air liquid ratio of 70, $Ca(OH)_2$ dosage of 6 gL^{-1}, packing height of 60 cm and types of packing material type 3 (non-woven polyester) were required to achieve 88% and 92% of NH_3-N removal efficiency for contact time of 8 hours and 12 hours, respectively. According to the obtained result, diluted leachate required more $Ca(OH)_2$ dosage (6 gL^{-1}) compared to raw leachate which required only 5 gL^{-1}. For other pollutants removal, treatment with diluted leachate is far better as the removal efficiency is much higher when diluted leachate was used compared to raw leachate. It can be concluded from the study that using diluted leachate for ammonia stripping reactor is better to achieved higher NH_3-N, turbidity, colour and P-reactive removal efficiency. Nevertheless, the optimum conditions for efficient NH_3-N removal and operating parameters that would make the technology economically viable should be further studied.

Table 3. The optimum parameters and NH_3-N, turbidity, colour and P-reactive removal efficiency for raw and diluted leachate treatment

Types of leachate treatment	$Ca(OH)_2$ dosage (g L^{-1})	Average initial concentration (mg L^{-1})	Average final concentration (mg L^{-1})	RSM (%)		Turbidity removal efficiency (%)	Colour removal efficiency (%)	P-reactive removal efficiency (%)
				Actual	Predicted			
Raw leachate	5	2019	493	76 (8 hour)	78	50	34	46
		2019	252	88 (12 hour)		55	38	49
Diluted Leachate	6	1280	262	80 (8 hour)	81	97	88	93
		1280	98	92 (12 hour)		98	89	98

Acknowledgements

The authors wish to acknowledge their gratitude for the financial supports from Universiti Putra Malaysia (UPM Putra Grant, Vot No.: 9393400), Ministry of Higher Education Malaysia for IPB grant (Vot No.: 9442500) and MyBrain15-MyPhD. The acknowledgements also go to the waste management centre for providing the leachate sample in this study.

References

[1] Ngoc U N and Schnitzer H 2009 *Waste Manage.* **29** 1982–1995
[2] Mohd Dinie M S and Mashitah M D 2013 *J. Teknol.* **62** (1) 95–101
[3] Baig S, Coulomb I, Courant P and Liechti P 1999 *Ozone Sci. Eng.* **21**(1) 1-22
[4] Chiang L, Chang J and Chung C 2001 *Environ. Eng. Sci.* **18**(6) 369-378
[5] Ahn D H, Yun-Chul C and Won-Seok C 2002 *J. Environ. Sci. Health A Tox. Hazard. Subst. Environ. Eng.* **37**(2) 163-173
[6] Marttinen S K, Kettunen R H, Sormunen K M, Soimasuo R M and Rintala J A 2002 *Chemosphere* **46** 851-858
[7] Im J, Woo H, Choi M, Han K and Kim C 2001 *Water Res.* **35** (10) 2403-2410 http://dx.doi.org/10.1016/S0043-1354(00)00519-4
[8] Budzianowski W and Koziol A 2005 *Chem. Eng. Res. Des.* **83** 196–204
[9] Alam R and Hossain M D 2009 *J. Water Resource Prot.* **1** (3) 210-215 doi: 10.4236/jwarp.2009.13026
[10] Wang G, Mu Y and Yu H Q 2005 *Biochem. Eng. J.* **23** 175-184
[11] Abdullahi M E, Hassan M A A, Noor Z Z and Raja Ibrahim R K 2014 *Rev. Chem. Eng.* **30**(5) 431–451
[12] Hossini H, Rezaee A, Ayati B and Mahvi A H 2016 *Health Scope* **5** (1) e26479 doi: 10.17795/jhealthscope-26479
[13] Sani A, Rashid M, Nurul Hanira M L and Hasfalina C M 2014 *J. Teknol.* **68** (5) 25-28
[14] Hanira, N M L, Hasfalina C M, Sani A and Rashid M 2015 *AIP Conf. Proc.* **1660** 090010-1–090010-8 doi: 10.1063/1.4915854
[15] Hanira, N M L, Hasfalina C M, Sani A and Rashid M 2017 *J. Teknol.* **79** (1) 107–118
[16] APHA 2005 *Standard Methods for the Examination of Water and Wastewater* American Public Health Association
[17] Lin M, Zhao Z, Cui F, Wang Y and Xia S 2012 *Desalin. Water Treat.* **40** 215-223
[18] Montgomery J M 1985 *Water treatment principles & design* Consulting Engineers Inc. pp. 237–261 John Wiley & Sons, New York
[19] Muniz C and Roberts P V 1989 *Water Res.* **23** (5) 589-601
[20] Campos J C, Moura D, Costa A P, Yokoyama L, Araujo F V F, Cammarota M C and Cardillo L 2013 *J.* Environ. *Sci. Health A* **48** 1105–1113
[21] Cheung K C, Chu L M and Wong M H 1997 *Water, Air Soil Poll.* **94** 209-221
[22] Laws of Malaysia *Environmental Quality Act and Regulations (20th ed.)* 2010 Kuala Lumpur: MDC Publishers Sdn. Bhd.

IOP Conf. Series: Materials Science and Engineering **206** (2017) 012077 doi:10.1088/1757-899X/206/1/012077

Ammonia nitrogen removal from aqueous solution by local agricultural wastes

I Azreen, Y Lija, and A Y Zahrim

Faculty of Engineering, Universiti Malaysia Sabah, Jalan UMS, 88400, Kota Kinabalu, Sabah, Malaysia

E-mail: azreen@ums.edu.my

Abstract. Excess ammonia nitrogen in the waterways causes serious distortion to environment such as eutrophication and toxicity to aquatic organisms. Ammonia nitrogen removal from synthetic solution was investigated by using 40 local agricultural wastes as potential low cost adsorbent. Some of the adsorbent were able to remove ammonia nitrogen with adsorption capacity ranging from 0.58 mg/g to 3.58 mg/g. The highest adsorption capacity was recorded by Langsat peels with 3.58 mg/g followed by Jackfruit seeds and Moringa peels with 3.37 mg/g and 2.64 mg/g respectively. This experimental results show that the agricultural wastes can be utilized as biosorbent for ammonia nitrogen removal. The effect of initial ammonia nitrogen concentration, pH and stirring rate on the adsorption process were studied in batch experiment. The adsorption capacity reached maximum value at pH 7 with initial concentration of 500 mg/L and the removal rate decreased as stirring rate was applied.

1. Introduction

Nitrogen is a vital nutrient for living organism but excess of it in waterways will result depletion of dissolved oxygen, eutrophication and toxicity to fish and other aquatic organisms [1]. Ammonia nitrogen (NH_3-N) is one of the most common and toxic species of nitrogen and contamination of ammonia nitrogen in waterways have become one of the biggest environmental challenges recently. This pollutant usually found in municipal sewage, domestic and industrial wastewater or decomposed from organic N compounds in those wastewater and wastes [2]. The ammonia nitrogen discharged to the environment is strictly regulated and the maximum limit for ammonia nitrogen discharged onto waterways is 5 mg/L according to the Department of Environment Malaysia [3].

Various treatment methods of ammonia nitrogen have been developed such as chemical precipitation, ion exchange, adsorption, air stripping, biological nitrification and de-nitrification. Adsorption is one of the most attractive methods which are easy, effective and economical feasible [4] compared to others as the conventional methods are high in cost and complicated operation [5].

The utilization of agricultural biomass as biosorbent has been studied extensively as it offers economic and environmental friendly. Agricultural waste is convenient for experimental purpose as they are cheap, abundance and unlimited [6]. Besides, low cost adsorbent also can reduce the production of solid waste in Malaysia and they can be used as composting material and fertilizer after the adsorption process compared to conventional adsorbent such as activated carbon and zeolites [7].

The effectiveness of biosorbent in removing variety of pollutants such as dyes [8, 9], heavy metals [10], and phenolic compound [11] has been reported by many researchers. However, the availability

IOP Conf. Series: Materials Science and Engineering **206** (2017) 012077 doi:10.1088/1757-899X/206/1/012077

on the information of ammonia nitrogen removal by biosorbent is still limited. Hence, this study was intended to study the ability of 40 potential local biosorbent for adsorption of ammonia nitrogen. Biosorbent with highest adsorption capacity then was analyzed further for the effect of initial concentration of ammonia nitrogen solution, pH and stirring rate on the adsorption process.

2. Materials and methods

2.1. Preparation of adsorbents

A total of 40 local biomasses were collected from area of Kota Kinabalu, Sabah. The samples were washed thoroughly with tap water to remove dust and impurities. Then, the samples were cut into small pieces and washed again with distilled water. Next, the samples were allowed to dry in an oven at 40°C for 48 hours. The dried materials were crushed and grounded to desired size (1-2 mm).

2.2. Preparation of stock solution

Synthetic ammonium (NH_{4+}) solutions were used throughout the adsorption experiments. Ammonium chloride salt (NH_4Cl) with calculated weight was dissolved in 1L of distilled water for desired concentration. The initial pH value of solution was adjusted to 7.0 using dilute solution of 0.1 M of sodium hydroxide (NaOH).

2.3. Screening experiment

The screening processes of each 40 samples were conducted by mixing 20 g of adsorbent in 500 mg/L of synthetic solution for 40 minutes of contact time. The final concentration of NH_3-N solution was determined by using UV-Vis spectrophotometer and adsorption capacity can be calculated.

2.4. Characterization of adsorbent

The characterization of adsorbent was assessed by zeta potential test. The zeta potentials of the adsorbent were determined at different pH value ranged from pH 3 to pH 10. This experiment was conducted in triplicate by using Malvern Instruments Zetasizer Nano-ZS and the results averaged.

2.5. Batch adsorption experiment

Batch adsorption experiments consisted of analysing the respective effect of initial stock solution concentration, pH and stirring rate on the adsorption of NH_3-N from aqueous solution. The sample with highest adsorption capacity obtained from screening experiments was selected for equilibrium characteristics study of NH_3-N adsorption.

The effect of initial ammonia nitrogen concentration was performed by using concentration range from 5 to 1000mg/L. Effect of initial pH was assessed by varying pH range from pH 3 to pH 11. Hydrochloric acid (HCl) and Sodium Hydroxide (NaOH) solutions were used for pH adjustment and the volume added for pH adjustment must be not exceeding 1% of the solution volume. The effect of stirring rate on ammonia nitrogen adsorption was assessed by using stirring rate range from 5 to 200 rpm. These experiments were conducted at room temperature and mass of samples and contact time were fixed to 20 g and 40 minutes respectively. Each experiment was made at least in two runs and result is averaged.

The adsorption capacity, q (mg/g) and removal rate (%) of NH_3-N were calculated for screening and batch adsorption experiment. Both equations are shown below respectively:

$$q = \frac{(C_o - C_t)\,V}{M} \quad (1)$$

$$\text{Removal rate (\%)} = \frac{C_o - C_t}{C_o} \times 100 \quad (2)$$

IOP Conf. Series: Materials Science and Engineering **206** (2017) 012077 doi:10.1088/1757-899X/206/1/012077

where q (mg/g) is the adsorption capacity, C_o (mg/L) and C_t (mg/L) are respectively the initial and at time t ammonia nitrogen concentrations, V is the volume of aqueous solution (L) and M is the mass of sample used as adsorbent (g).

3. Results and discussion

3.1. Screening experiment

The results shows that 18 out of 40 samples were able to adsorb ammonia nitrogen from the aqueous solution and the rest of the samples show negative values. The negative values may be due to ammonia nitrogen adsorbed from solution was less than the ammonia nitrogen released in these samples [12]. Table 1 shows the adsorption performances of 40 samples. Langsat peels (LP) show the highest adsorption capacity with 3.58 mg/g and followed by Jackfruit seeds (JS) and Moringa Oleifera peels (MOP) with adsorption capacity of 3.37 mg/g and 2.64 mg/g respectively.

Screening experiment conducted in this study suggested that some of the local agricultural wastes have potential to be used as adsorbent for removal of NH_{4+} ions. Agricultural biomass is an organic compound comprised of cellulose, hemicelluloses, lignin, lipids, proteins, hydrocarbons, starch and the samples may contain polar functional groups, such as alcohols, aldehydes, ketones, carboxyl, phenol and ether groups. These constituents have the ability to bind pollutants to some extent [13] thus making the adsorption of ammonia nitrogen is possible on the surface of agricultural biomass. Despite having the highest adsorption capacity, JS was selected over LP for further adsorption equilibrium characteristics study due to its availability throughout the years compared to LP and the adsorption capacity of JS is not much difference with LP. Previous studies proved that JS can be utilized as adsorbent for pollutant removal such as dye [14] and heavy metal [15].

Jackfruit seed is high in starch content with approximately 93 to 95% [16] and the starch consists of two macromolecules: amylose (20-30%) and amylopectin (70-80%), which are associated with each other by hydrogen bonds [17]. Amylose is a linear polymer consisting mainly α-1, 4 linked D-glucose units and Amylopectin consists of branched polymer comprised of α-1, 4 linked glucose units with α-1, 6 branching links [18]. The structure and physicochemical properties of starch are depending on these macromolecules content and it can affect the sorption performance of starch [19]. The ability of starch to adsorb NH_{4+} ions may arises from the electrostatic force of attraction between the anion from the sample surface and the positively charged NH_{4+} ions [20].

The characterization of adsorbents used in this experiment was studied by zeta potential test. The magnitude and sign of surface charge of a solid immersed in a water solution are related to zeta potential that occurs at the slipping plane of the solid. Thus, this test is used for measuring the potential of solid at the shear plane to characterize its electro-kinetic behaviour [21]. Zeta potentials of 40 agricultural biomasses at pH 7 were determined and presented in Table 1. The zeta potential properties of JS over pH range of 3-10 were further analysed and results depicted in Figure 1.

The adsorbent exhibits negative value of zeta potential over the entire range of pH value and become more negative as the pH increases. This condition is favourable for ammonium adsorption due to the attraction between the negatively charged adsorbent and the positively charged adsorbate species [13]. This result is in accordance with previous studies that report on the dependency of zeta potential over pH value [22-24].

IOP Conf. Series: Materials Science and Engineering **206** (2017) 012077 doi:10.1088/1757-899X/206/1/012077

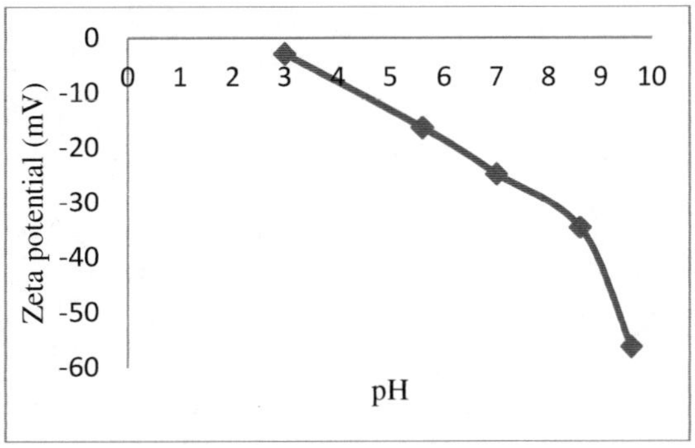

Figure 1. Zeta potential against pH value for Jackfruit seeds

Table 1. Adsorption capacity and removal rate of 40 local biosorbent.

Waste	Scientific Name	Removal rate (%)	Adsorption capacity (mg/g)	Zeta potential (mV)	Season (month)
Banana peel	Musa Acuminata	8.22	1.51	-27.4	Jan-Dec
Banana comb	Musa Acuminata	7.03	0.88	-51.37	Jan-Dec
Papaya peel	Carica Papaya	-4.96	-0.81	-32.4	Jan-Dec
Papaya seed	Carica Papaya	-17.3	-3.03	-31.6	Jan-Dec
Pomelo peel	Citrus Maxima	13.95	1.74	-29.0	Aug-Oct /Jan-March
Pineapple peel	Ananas Comosus	-39.07	-4.88	-12.4	Jan-Dec
Pineapple stem	Ananas Comosus	2.93	1.07	-14.0	Jan-Dec
Langsat (Lanzones) peel	Lansium Domesticum	28.67	3.59	-36.3	Jun-July / Nov -Feb
Langsat (Lanzones) seed	Lansium Domesticum	5.36	0.79	-48.3	Jun-July / Nov -Feb
Soursop peel	Annona Muricata	-1.96	-0.25	-22.4	March- April / June-July
Coconut coir	Cocos Nucifera	12.73	1.59	-37.9	Jan-Dec
Sugarcane bagasse	Saccharum Officinarum	3.67	1.59	-20.3	Jan-May
Corn (silk)	Zea Mays	-3.74	-4.68	-37.87	Jan-Dec
Corn (shank)	Zea Mays	-7.56	-0.95	-22.75	Jan- Dec
Corn comb	Zea Mays	-10.45	-1.56	-23.7	Jan-Dec
Corn husk leaf	Zea Mays	5.55	0.69	-20.15	Jan- Dec
Passion fruit	Passiflora Edulis	6.34	0.79	-35.1	Nov-April
Cassava peel	Manihot Esculenta	5.81	0.73	-40.11	Jan-Dec
Banana false stem (Pseudostem)	Musa Acuminata	-3.14	-0.39	-32.6	Jan-Dec
Sapodilla peel	Manilkara Zapota	-73.16	-9.32	-11.9	Jan-Dec
Limau madu (Citrus) peel	Citrus Suhuiensis	15.59	1.95	-38.05	March-April / Oct-Nov
Jackfruit peel	Artocarpus Heterophyllus	16.26	2.03	-18.03	Jan- Dec
Jackfruit seed	Artocarpus Heterophyllus	26.96	3.3702	-24.8	Jan- Dec

Areca palm	Areca Catechu	-4.32	-0.54	-31.9	Jan- Dec
Coconut mesocarp	Cocs Nucifera	-72.0	-10.08	-29.7	Jan -Dec
Honey Dew peel	Cucumis melo var. inodorus	-5.79	-0.72	-27.2	Jan- Dec
Purple Yam peel	Dioscorea Alata	-43.05	-5.38	-47.1	Jan-Dec
Mango seed	Mangifera Indica	-78.65	-9.78	-42.07	Jan-Dec
Mango peel	Mangifera Indica	-3.74	-0.47	-26.23	Jan-Dec
Sweet potato peel	Ipomoea Batatas	6.97	0.87	-19.53	Jan-Dec
Cat's Eye (Mata Kucing) peel	Euphoria Malaiense	-52.93	-6.62	-27.1	Jan-Feb
Mata Kucing (Cat's Eye) seed	Euphoria Malaiense	-2.37	-0.36	-28.4	Jan-Feb
Moringa peel (young)	Moringa Oleifera	-76.33	-9.54	-42.33	Jan-Dec
Moringa seed (young)	Moringa Oleifera	-6.99	-0.87	-28.37	Jan-Dec
Moringa peel (mature)	Moringa Oleifera	21.09	2.64	-44.07	Jan-Dec
Moringa seed (mature)	Moringa Oleifera	-7.35	-0.92	-9.47	Jan-Dec
Breadnut stem	Artocarpus Camansi	-4.32	-0.64	-25.8	Jan-Dec
Breadnut peel	Artocarpus Camansi	10.97	1.37	-35.2	Jan-Dec
Bambangan peel	Mangifera Pajang	4.67	0.58	-21.8	May-Nov
Bitter bean (Petai) peel	Parkia Speciosa	-1.43	-0.18	-50.5	March- April/ Aug-Sept

3.2. Effect of initial stock solution concentration

The effect of initial NH_3-N concentration on adsorption process by JS was studied by varying the concentration in a range of 5 to 1000mg/L. The concentration range used in this study is based on typical range of NH_3-N concentration from landfill leachate [25] and industrial wastewater [26]. The experiments were carried out at room temperature with adsorbent dosage of 20 g and 40 minutes of contact time. The variation in the sorption of NH_3-N as a function of concentration is depicted in Figure 2.

Increasing the initial stock solution concentration lead to an increase in the ammonium (NH_{4+}) ions uptake by JS, however the adsorption capacity decreased at 1000 mg/L of solution. The increase in NH_{4+} ions uptake may be due to the increasing driving force of the concentration gradient as the initial concentration increased [7]. As the initial concentration increased to 1000mg/L, the adsorption capacity decreased to 1.34 mg/g. This behaviour can be explained due to the insufficient surface sites of an adsorbent. Since the concentration of solution increased, more sites on the surface of JS were taken up [27].

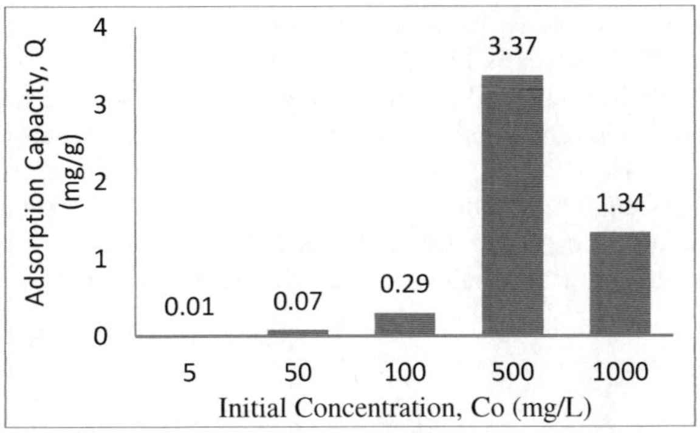

Figure 2. Effect of initial NH_3-N concentration solution on the adsorption capacity of Jackfruit seeds.

3.3. Effect of initial pH

The initial pH solution is one of the most critical controlling parameters in adsorption processes. The effect of initial pH on the adsorption process was observed by taking the common pH values of wastewater where the value lies in range of 3 to 11. The adsorption capacity of ammonia nitrogen removal at different pH is shown in Figure 3 for the initial solution concentration of 500 mg/L. The adsorption capacity of JS on the NH_3-N removal process shows negative value at pH 3, increased to 0.69 mg/g at pH 5 and reached maximum adsorption capacity (3.37 mg/g) at pH 7. The adsorption values decreased to 0.44 mg/g as the pH value increased to pH 9 and reached negative value at pH 11.

The negative values of adsorption capacity obtained are due to amount of ammonium released more than amount adsorbed. Low adsorption capacity obtained at lower pH is mainly attributed to the facts that more hydrogen (H^+) ions is present in the aqueous solution thus making H^+ ions compete effectively with NH_4^+ cations [28]. Increasing the pH value to pH 7 seems favorable for the adsorption process as the adsorption capacity increased along with it. This might be due to the fact that the fibers chains become negatively charged at higher pH and thus enhance the replacement of exchangeable cations present at the exchangeable sites in the aqueous solution through attraction forces [5]. At higher pH, NH_4^+ is neutralized by hydroxyl (OH^-) ion and OH^- compete for NH_4^+ with the active sites on the surface of the adsorbent thus lower the adsorption capacity [29].

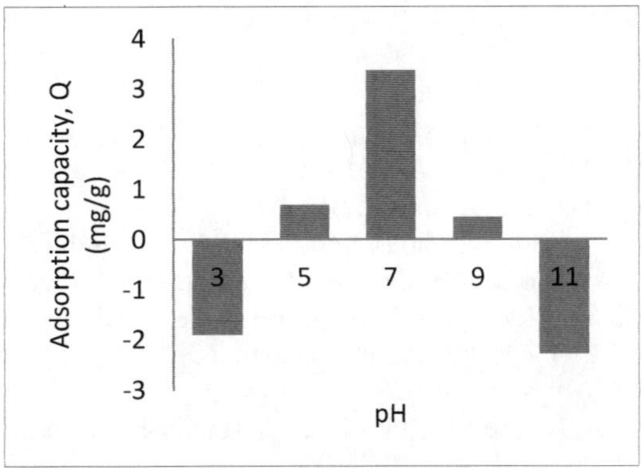

Figure 3. Effect of initial pH on the adsorption capacity of Jackfruit seeds.

3.4. Effect of stirring rate

Figure 4 shows the adsorption capacity of jackfruit seed at different stirring rate range from 0 to 200 rpm with contact time of 40 minutes. The effect of stirring rate at static conditions is important to be investigated since it will give a better understanding about nutrients adsorption capacity from wastewater onto lignocellulosic fibers [30, 31]. The ability of adsorbent to remove pollutant in moving or fast-flowing water can be evaluated as the stirring speed increases. Maximum adsorption occurred when no stirring rate was applied to the solution where the adsorption capacity reached 3.37 mg/g. The adsorption capacity of JS slowly decreased as the stirring rate increased and reached negative value at highest stirring rate applied. This reduction is related to the fact that rotation caused the ammonium ions which bind to the adsorbent to be released again and ammonia nitrogen molecules does not have enough time to come into contact with the sorbent at higher speeds [32].

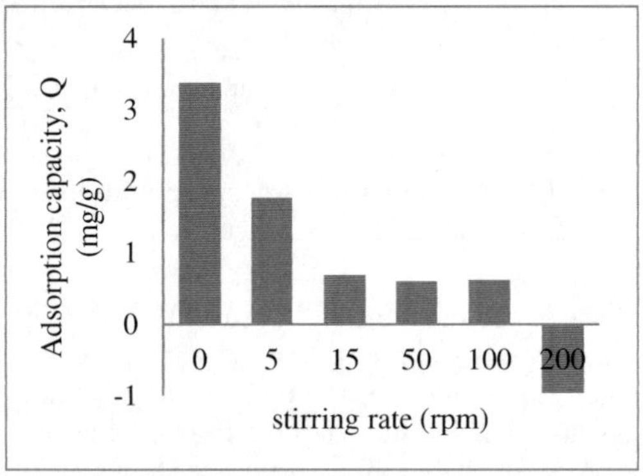

Figure 4. Effect of stirring rate on adsorption capacity of Jackfruit seeds.

4. Conclusions

The ability of 40 agricultural wastes as potential low cost adsorbent for ammonia nitrogen removal was tested in this study. The experimental findings suggested that 18 out of 40 samples were able to remove ammonia nitrogen with highest adsorption capacity of 3.58 mg/g. Jackfruit seed was chosen for further equilibrium characteristic analysis due to its availability and high adsorption capacity. Maximum adsorption occurred at pH 7 with initial ammonia nitrogen concentration of 500 mg/L.

Acknowledgments

This study was financially supported by Ministry of Higher Education Malaysia and University Malaysia Sabah under research Project Grant FRG0420-TK-1/2015.

References

[1] Ugurlu M and Karaoglu M H 2011 *Microporous and Mesoporous Mater.* **139** 173-8
[2] Farhad J M M 2015 *J. Environ. Chem. Eng.* 1-17
[3] Department of Environment Malaysia 2010 1-82
[4] Ibrahim A, Yusof L, Beddu N, Galasin N, Lee P, Lee R and Zahrim A 2016 *IOP Conf. Series: Earth Environ. Sci.* vol 36 (Sabah: IOP Publishing) p 012020
[5] Wahab M A, Jellali S and Jedidi N 2010 *Bioresource Technol.* **101** 8606-15
[6] Zahrim A, Lija Y, Ricky L and Azreen I 2016 *IOP Conf. Series: Earth Environ. Sci.* vol 36 (Sabah: IOP Publishing) p 012028
[7] Jellali S, Wahab M A, Anane M, Riahi K and Jedidi N 2011 *Desalination* **270** 40-9
[8] Crini G 2006 *Bioresource Technol.* **97** 1061-85
[9] Hameed B H 2009 *J. Hazard. Mater.* **161** 753-9
[10] Feng N, Guo X and Liang S 2010 *Trans. Nonferrous Met. Soc. China* **20** 146- 52
[11] Vimala R and Grace A N 2013 *Int. J. Pharm. Bio. Sci.* **4** 263-73
[12] Liu H, Dong Y, Liu Y and Wang H 2010 *J. Hazard. Mater.* **178** 1132-6
[13] Liu H, Dong Y, Wang H and Liu Y 2010 *Desalination* **263** 70-5

[14] Karmaker S, Uddin M N, Ichikawa H, Fukumori Y and Saha T K 2015 *J. Environ. Chem. Eng.* **3** 583-92
[15] Kannan N and Veemaraj T 2010 *Electron. J. Environ. Agric. Food Chem.* **9**
[16] Kooh M R R, Khairud M D and Lim L B L 2016 *J. Environ. Biotechnol. Res.* **4** 7-16
[17] Madruga M S, Albuquerque F S M, Silva I R A, Amaral D S, Magnani M and Neto V Q 2014 *Food Chem.* **143** 440-5
[18] Dehabadi L and Wilson L D 2014 *Carbohydr. Polym.* **113** 471-9
[19] Copeland L, Blazek J, Salman H and Tang M C 2009 *Food Hydrocolloids* **23** 1527-34
[20] Wahab M A, Boubakri H, Jellali S and Jedidi N 2012 *J. Hazard. Mater.* **241** 101-9
[21] Alkan M, Tekin G and Namli H 2005 *Microporous and Mesoporous Mater.* **84** 75-83
[22] Gu L, Xu J, Lv L, Liu B, Zhang H, Yu X and Luo Z 2011 *Desalination* **269** 206-13
[23] Boonamnuayvitaya V, Chaiya C, Tanthapanichakoon W and Jarudilokkul S 2004 *Sep. Purif. Technol.* **35** 11-22
[24] Chang S-H, Lin H-T V, Wu G-J and Tsai G J 2015 *Carbohydr. Polym.* **134** 74-81
[25] Salleh N F D M and Hamid K H K 2013 *Malaysian J. Anal. Sci.* **17** 24-9
[26] Mohammadi M, Man H C, Hassan M A and Yee P L 2010 *Afr. J. Biotechnol.* **9** 6233-43
[27] Hussain S, Aziz H A, Isa M H, Adlan M N and Assari F A H 2006 *Bioresource Technol.* **98**
[28] Wang M, Liao L, Zhang X, Li Z, Xia Z and Cao W 2011 *Clays and Clay Miner.* **59** 459-65
[29] Zheng Y, Liu Y and Wang A 2011 *Chem. Eng. J.* **171** 1201-8
[30] Yaser A Z, Abd R R and Kalil M S 2007 *Pakistan J. Biol. Sci. PJBS* **10** 4473-8
[31] Zahrim A and Asis T 2010 *J. Inst. Eng. Malaysia* **71** 11-7
[32] Fauzia S, Furqani F, Zein* R and Munaf E 2015 *J. Chem. Pharm. Res.***7** 573-82

IOP Conf. Series: Materials Science and Engineering **206** (2017) 012078 doi:10.1088/1757-899X/206/1/012078

Estimation of Melanoidin concentration in palm oil mill effluent ponding system and its treatment using Calcium Lactate

I Azreen, A Y Zahrim, S H Chong and S W Ng

Chemical Engineering Department, Faculty of Engineering, Universiti Malaysia Sabah, Jalan UMS, 88400 Kota Kinabalu, Sabah

E-mail: azreen@ums.edu.my

Abstract. Colour removal from wastewater is among the major challenge in water and wastewater treatment. Among others, melanoidin could be the source of colour in wastewater. In this study, the estimation of melanoidin concentration in conventional palm oil mill effluent (POME) ponding system was investigated. Melanoidin was analyzed by detecting its absorption using double beam UV-Vis spectrophotometer. For melanoidin, the maximum absorption is 330nm. From the analysis, the melanoidin concentration decrease from anaerobic pond 1 to anaerobic pond 3 and slightly increase in anaerobic pond 4 and aerobic pond 1. After that, the melanoidin concentration decreased from aerobic pond 1 to final discharge. It is estimated that the anaerobic pond 1 had the highest melanoidin concentration which was 87.3 mg/L. Finally, the effectiveness of melanoidin removal using a coagulation/flocculation process was also studied. Calcium lactate was used as a coagulant and low molecular weight anionic polyacrylamide was used as a coagulant aid. The jar test experiment was carried out by using 0.3g/L calcium lactate solution and dosage of anionic polyacrylamide was altered in order to find out the best melanoidin removal. Experiments carried out by using sedimentation time of 20 minutes showed that the highest percentage removal of melanoidin was 80.93% at the dosage of 0.3g/L of calcium lactate without any anionic polyacrylamide being added. This result concluded that the addition of anionic polyacrylamide as coagulant aids is not significant when compared to the use of calcium lactate only.

1. Introduction

Melanoidins are dark brown to black coloured natural condensation products which are produced by non-enzymatic browning reactions called Maillard reactions [1]. Maillard reaction is a type of non-enzymatic browning reaction that formed between amino compounds and carbohydrates [2]. The Maillard reaction products are a particularly complex mix of different compounds of different molecular weight. They include a lot of functional groups such as aldehydes, ketones, dicarbonyls, acryl amides and heterocyclic amines and all of that will contribute to flavor, melanoidins and advanced glycation end products (AGEs), which are a polymeric product formed at the advanced steps of Maillard reaction [3]. Naturally, melanoidins are widely distributed in food and drinks. Besides that, melanoidins are also widely discharged in large amount by various agro-based industries especially from cane molasses based distilleries and fermentation industries as environmental pollutants [1]. Melanoidin is a main colour pollutant from molasses-based fermentation plants, alcohol distilleries, spent wash and sugar-based factories [4].

Melanoidin in the treated effluent brings threats. The dark brown characteristics can prevent the sunlight penetration to the water [2]. This will reduce the photosynthetic activity of the plants in the water and reduce the dissolved oxygen level of surface waters. This will then greatly affect the ecosystem in water. Besides, melanoidins also have such characteristics that it is a water-soluble recalcitrant colouring compound, which are highly resistant to the microbial attack. Therefore, conventional biological process and activated sludge treatment process are unable to treat these melanoidins containing wastewater. Therefore, such wastewaters require a proper pretreatment before it can be safely disposed into the environment [1].

Coagulation and flocculation is a vital process in water surface treatment which includes pre-treatment, post-treatment and main treatment. However, flocculants play an important role in affecting the flocculation effect during the process. There are a lot of flocculants commercially available in the market. However, some of them are believed to bring a lot of negative impacts to the environment and also human health [5].

Malaysia is one of the largest producers of palm oil in the world. Therefore, there is large amount of palm oil mill effluent (POME) being produced annually [6]. Palm oil mill effluent (POME) is a voluminous liquid waste. It has high biochemical oxygen demand (BOD) which normally discharged at 75-85°C. It is a colloidal dispersion of organics with an unpleasant odor. POME is a liquid waste that will bring harm to the environment. POME has total solids content of 5-7% of which more than half is dissolved solids and about the other half is mixture of various forms of organic and inorganic suspended solids [7].

Palm oil mill effluent contains amino acids, a mixture of carbohydrates such as hemicelluloses to simple sugars, organelles, free organic acids, nitrogenous constituents and inorganic nutrients such as sodium, potassium, magnesium, calcium and short fibers [8]. In 2009, Zahrim et al. has hypothesized that the anaerobic digested POME may contain melanoidin due to the existence of amino acid and carbohydrate compound in POME. Melanoidin which contribute to the colour of the POME should be treated before it is discharged into the river or seawater. By doing this, the ecosystem of the aquatic life can be preserved and protected.

There has been research done on the colour treatment of the melanoidins. Many types of conventional coagulants have been used to treat these melanoidins such as aluminium sulfate, ferric sulfate, aluminium chloride and ferric chloride and alum [9, 10]. In this research, calcium lactate is used instead of these kinds of conventional coagulant. This is because with the use of inorganic salts like aluminium chloride or aluminium sulfate, it can cause harm to human health by contributing to Alzheimer disease and produce a large scale of the sludge [5]. It is believed that calcium lactate with the combination of anionic polyacrylamide can be used to remove the melanoidin. The aims of this research are to estimate melanoidin concentration in palm oil mill effluent (POME) and analyze the usage of calcium lactate as a coagulant together with the help of anionic polyacrylamide (A-PAM) as coagulant aid in the coagulation/flocculation process of aerobic POME.

2. Experimental

2.1. Experimental materials

Raw material collection: Samples of palm oil mill effluent were collected from a palm oil mill in Beaufort, Sabah. The POME ponding system contained a cooling/grid pond, four stages of anaerobic ponds and four stages of aerobic ponds as shown in Figure 1. The collection bottles were rinsed with distilled water before samples were collected. Palm oil mill effluent samples were then stored less than 4°C in the refrigerator.

Standard melanoidin preparation: Standard solution of melanoidin was prepared by mixing 1M of glucose (Fluka) and 0.5M of glycine (Merck). After several processes of dilution and heating, the mixture will become hard solid objects. The solid will be left in the lab for drying for a day. The dried melanoidin was removed from the beaker and crushed into powder form.

IOP Conf. Series: Materials Science and Engineering **206** (2017) 012078 doi:10.1088/1757-899X/206/1/012078

Jar test: The coagulant that will be used is calcium lactate powder with Merck brand while the flocculent that will be used is low molecular weight of anionic polyacrylamide. The anionic polyacrylamide is supplied by Guang Xi Nanning Bonglin Business.

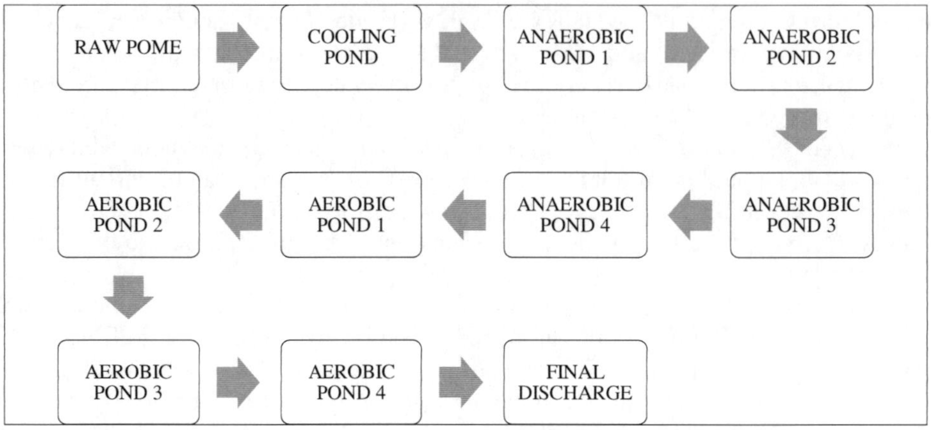

Figure 1. Flowchart of ponding treatment system of POME

2.2. Jar test

Melanoidin concentration in each samples taken from four anaerobic ponds, three aerobic ponds and final discharge were first estimated before the jar test was performed. The coagulation procedure started with adding pre-determined amount of calcium lactate stock solution into the POME. The mixture of POME and calcium lactate solution went through rapid mixing (258 rpm) for 3 minutes, followed by slow mixing (39 rpm) for 10 minutes. The stirrer will not be stopped during the changing of mixing speed. After that, low molecular mass of anionic polyacrylamide (if required) was added and the stirrer was stopped after 10 minutes passed. The total slow mixing period was 20 minutes. Then, the POME was allowed to settle for 20 minutes. The jar test was carried out for three times. For each run of test, the dosage of calcium lactate was made constant at 0.3 g/L while varying the dosage of anionic polyacrylamide at 0 ml, 0.025 ml, 0.05 ml, 0.25 ml, 0.5 ml and 1.0 ml.

2.3. UV-Vis test

After measuring the pH, a small portion of each of the melanoidin stock solution was transferred into the cuvette. The sample was placed in the UV-Vis beam and a graph of the transmittance or absorbance versus the wavelength was obtained. Alternatively, samples were prepared in known concentrations and their absorbance at the specific wavelength was read by the UV-Vis spectrophotometer. A standard graph for the melanoidin was made after the absorbance were obtained for the five different concentration of the melanoidin with pH 5, pH6, pH8 and pH9 as shown in Figure 2.

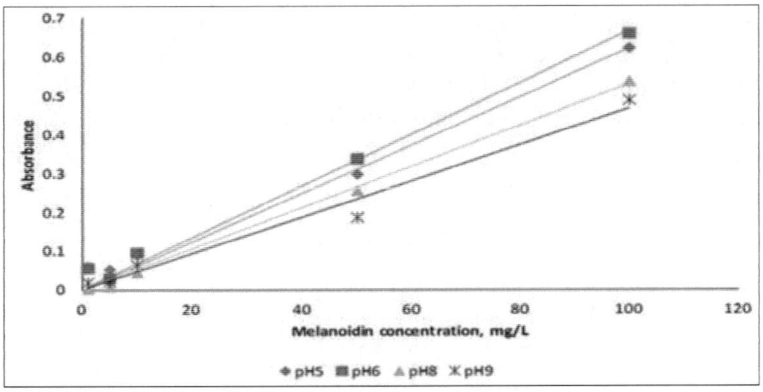

Figure 2. Standard graph of Melanoidin

3. Results and discussion

3.1. Melanoidin analysis

It can be seen from Figure 3 that the melanoidin concentration decreases from anaerobic pond 1 to anaerobic pond 3 and slightly increases in anaerobic pond 4 and aerobic pond 1. After that, the melanoidin concentration decreased from aerobic pond 1 to final discharge. It also can be seen that the anaerobic pond 1 had the highest melanoidin concentration which was 87.3 mg/L. This might be since the effluent from anaerobic pond has not undergone treatment yet. This was due to the higher colour intensity in the influent [11]. Therefore, it has the highest melanoidin concentration.

The melanoidin concentration is decreasing from the anaerobic pond 1 to pond 3. This indicated that the colour treatment efficiency is increasing from anaerobic pond 1 to pond 3. The sudden decrease of the melanoidin concentration from anaerobic pond 1 to pond 2 because of the lower pH operation in biological treatment in the POME (pH5) since melanoidin decolourisation had maximum decolourisation activity in optimum pH range of 5.0-6.0.12-13. During this pH range, the cleavage of the ethylinic C=C and azomethine C=N linkages might occurred where the presence of conjugated C=C and C=N bonds impart colour to the melanoidin [4]. Each microorganism had a specific pH for their microbial activity. All enzymes were protein in nature, therefore some proteins will denature at the unfavourable pH. This will inhibit the activity of the enzymes as well as its production [12]. It explained why the melanoidin decolourisation was so high from the anaerobic pond 1 to anaerobic pond 2.

However, the melanoidin concentration increased from anaerobic pond 4 to the aerobic pond 1. This phenomenon was due to the re-polymerization of colouring compounds after anaerobic treatment. For a palm oil mill, there was no chemical addition during the processing operation. Therefore, the colour from the palm oil mill effluent appeared from plant constituents such as lignin and phenolic while the colour of the sugar processing effluent were contributed by phenolic (tannic and humic acids) from the feedstock and melanoidins from Maillard reaction of sugars (carbohydrate) with proteins [13].

Aerobic treatment on the melanoidin was quite efficient from the aerobic pond 1 to the final discharge due to the decrease of the melanoidin concentration. However, the removal of melanoidin concentration was not very significant. This was due to melanoidin widely distributed in nature and was not readily susceptible to microbial degradation [14].

3.2. Melanoidin treatment

The standard melanoidin graph as shown in Figure 2 was used to determine the concentration of melanoidin at specific pH. Table 1 showed melanoidin concentration with varies anionic polyacrylamide concentration as performed via coagulation flocculation experiment.

Table 1. Melanoidin concentration with variation of anionic polyacrylamide concentration.

Calcium lactate concentration (g/L)	Anionic polyacrylamide concentration (mg/L)	Melanoidin concentration (mg/L)
0	0	46.98 (12.7573)
0.3	0	9.46 (6.1360)
0.3	0.05	19.79 (12.8740)
0.3	0.1	15.32 (1.8429)
0.3	0.5	26.98 (0.6343)
0.3	1	21.80 (0.8263)
0.3	2	19.02 (0.4417)

* ( ) value indicates the standard deviation of the melanoidin concentration

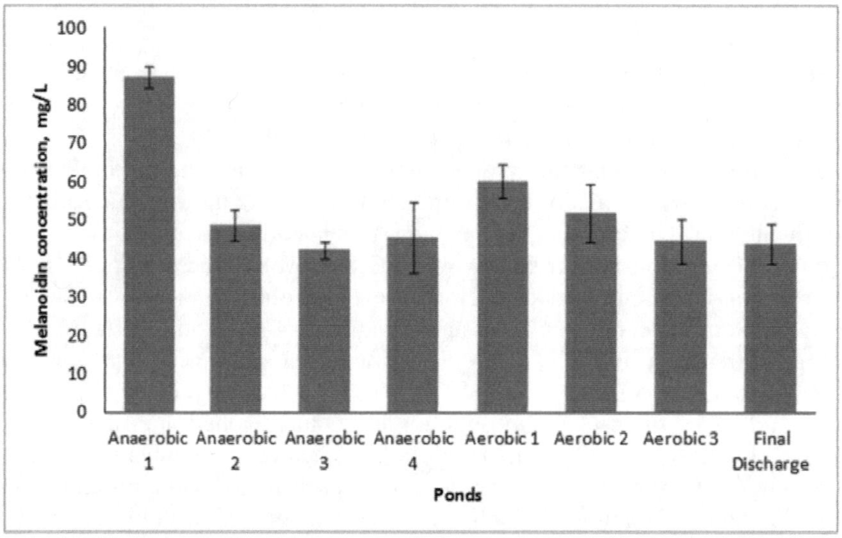

Figure 3. Melanoidin concentration against different ponds

The coagulation of the melanoidin colloid was done by the calcium lactate to reduce or neutralize the electrical repulsion between the colloids and causing them to aggregate to form microfloc. From Figure 4, it can be seen that the percentage removal of melanoidin is showing a decreasing trend from 0 mg/L to 2 mg/L of concentration of anionic polyacrylamide. It can also be seen that the highest percentage removal of melanoidin was 80.93% at the dosage of 0.3g/L of calcium lactate without any anionic polyacrylamide added. This phenomenon may be due to the formation of larger aggregates which were too hard for them to settle at the stated polymer doses [15]. This was because the sudden add-on of the anionic polyacrylamide at 0.05mg/L from none before and gradually increases of the polymer dosages from 0.1mg/L to 0.5mg/L might gave them some difficulty to settle down immediately. However, the density of the aggregates increased as the polymer doses increases [15]. This can be used to explain the better removal obtained for the polymer doses that are higher than 0.05 and 0.5mg/L.

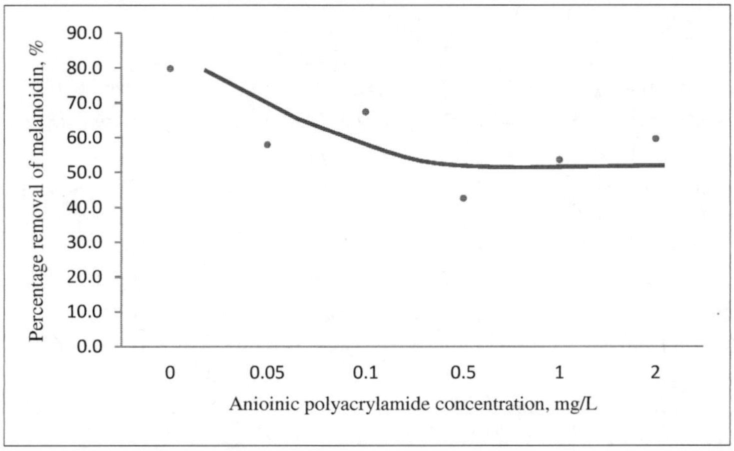

Figure 4. Graph of percentage removal of melanoidin concentration against anionic polyacrylamide concentration

The results show that there is only minor improvement in the removal of melanoidin by using anionic polyacrylamide (flocculants) as compared to the removal of melanoidin by using only calcium lactate (coagulant). This might be due to the molecular weight and charge density of the A-PAM led to the random coil rather than an extended conformation [9].

However, the melanoidin removal percentage was increasing at 0.5mg/L dosage of anionic polyacrylamide from 40.20% to 57.84% with the dosage of 2mg/L. This implied that the removal of melanoidin colloids increased with the increased in dosage of the anionic polyacrylamide. This was most probably because of the enlargement of flocs occurred. Enlargement of flocs can only occurred when there was enough adsorption affinity between the polymers and the flocs surfaces [15]. Polymers adsorbed onto the melanoidin colloid surface with the adequate amount of polymers. Besides that, the adequate amount of anionic polyacrylamide also produced sufficient long loops that can be attached to the surface of the melanoidin colloid that required for bridging flocculation [16]. By looking at the results obtained in this study, A-PAM may have disturbed the formation of flocs. Therefore, the main mechanism is believed to be charge neutralization-precipitation.

4. Conclusions

Treatment of aerobic palm oil mill effluent using combination of the calcium lactate and anionic polyacrylamide has been studied. It can be concluded that anionic polyacrylamide do not significantly assists the overall performance of the removal of melanoidin from palm oil mill effluent. Anionic polyacrylamide only showed some minor effect on the removal of melanoidin which is dark brown to black coloured natural condensation products of sugar and amino acid produced by non-enzymatic browning reactions called Maillard reaction. There are some further study that can be carried out on the POME such as a combined process that involving coagulation/flocculation and membrane filtration so that a complete removal can be achieved.

Acknowledgment

This study was financially supported by Ministry of Higher Education Malaysia and University Malaysia Sabah under research Project Grant FRG0420-TK-1/2015.

References

[1] Chandra R, Bharagava R N and Rai V 2008 *Bioresour. Technol.* **99** 4648-60

[2] Liang Z, Wang Y, Zhou Y and Liu H 2009 *Chem. Eng. J.* **152** 88-94
[3] Wang H Y, Qian H and Yao W R 2011 *Food Chem.* **128** 573-84
[4] Naik N, Jagadeesh K and Noolvi M 2010 *Iranica J Energ. Environ.* **1** 347-51
[5] Li J, Jiao S, Zhong L, Pan J and Ma Q 2013 *Colloids and Surf. A: Physicochemical and Eng. Aspects* **428** 100-10
[6] Bello M, Nourouzi M, Abdullah L C, Choong T S, Koay Y and Keshani S 2013 *J. Hazard. Mater.* **262** 106-13
[7] Bhatia S, Othman Z and Ahmad A L 2007 *J. Hazard. Mater.* **145** 120-6
[8] Kamal S A, Jahim J M, Anuar N, Hassan O, Daud W R W, Mansor M F and Rashid S S 2011 *Int. J. Adv. Sci., Eng. Inf. Technol.* **2** 325-31
[9] Liang Z, Wang Y, Zhou Y, Liu H and Wu Z 2009 *Sep. Purif. Technol.* **68** 382-9
[10] Dwyer J, Griffiths P and Lant P 2009 *Water res.* **43** 553-61
[11] Shahata A, Omata T and Urase T 2013 *J. Water and Environ. Technol.* **11** 539-46
[12] Tiwari S, Gaur R and Singh R 2012 *BMC Biotechnol.* **12** 1
[13] Zahrim A, Rachel F, Menaka S, Su S, Melvin F and Chan E 2009 *World Apply Sci. J.* **5** 126-9
[14] Gonzalez T, Terron M C, Yague S, Junca H, Carbajo J M, Zapico E J, Silva R, Arana-Cuenca A, Tellez A and Gonzalez A E 2008 *Res. Microbiology* **159** 103-9
[15] Zahrim A, Tizaoui C and Hilal N 2010 *J. Hazard. Mater.* **182** 624-30
[16] Fleer G and Lyklema J 1976 *J. Colloid Interface Sci.* **55** 228-38

Ammonia removal via microbial fuel cell (MFC) dynamic reactor

I Alabiad, U F M Ali*, I A Zakarya, N Ibrahim, R W Radzi, N Z Zulkurnai and N H Azmi

School of Environmental Engineering, Universiti Malaysia Perlis, Kompleks Pusat Pengajian Jejawi 3, 02600 Arau, Perlis, Malaysia

E-mail: umifazara@unimap.edu.my

Abstract. Landfill leachate is generally known as high-strength wastewater that is difficult to handle and contains dissolved extracts and suspended matter. Microbial fuel cells (MFCs) were designed to treat landfill leachate while continuously producing power (voltage output). Three different anodes were tested in MFC reactors: carbon black, activated carbon, and zinc electrodes. Movements in the MFC reactor during treatment were also a key factor for testing. Results showed a difference in ammonia levels in the three anodes used. The study compared the efficiency of static and dynamic modes of MFC in removing ammonia. Continual leachate movement in the reactor could increase the rate of removal of the ammonia components. The setup provided a viable condition for maximum removal because the reactor movement caused the sludge to disintegrate, which allowed ammonia to separate easily from the parent leachate. Ammonia removal also resulted from the transfer of ammonium through the membrane or from ammonia loss. Constant exchange of ionic content benefited the MFC performance by increasing power production and decreasing internal electrode material resistance. This paper presents the results of the analyses of leachate treatment from the solid waste landfill located in Padang Siding Landfill, Perlis. The performance of ammonia removal was enhanced using different types of electrodes. In both modes, activated carbon performed better than black carbon and zinc. The respective percentages of ammonia removal for activated carbon of dynamic over static were 96.6%, 66.6%, and 92.8% for activated carbon, zinc, and black carbon. The results provide further information on the possibility of using MFCs in landfill leachate treatment systems.

1. Introduction

Th Leachate is a major challenge that threatens environmental quality in Malaysia. Leachate results from the production of large amounts of solid waste, which comprises unwanted matter produced by human activity [1, 2, 4, 5]. Population growth in urban areas in Malaysia has increased the generation of solid waste. Malaysia is one of the most economically successful countries owing to its stable political conditions (Solid Waste Management and Public Management Cleansing Corporation, Composition Solid Waste Management and Public Management Cleansing Corporation 2011[2, 4, 5-7]. Rapid economic transition and urban population increase reflect the escalation of solid waste generation [1, 8-11]. The Solid Waste Management and Public Management Cleansing Corporation, a general policy for solid waste management, was implemented in 2011. The urban population in Malaysia has increased by over 50% in the last few decades [10-15, 17]. This population growth has caused serious concerns in solid waste production (Composition Solid Waste Management and Public Management Cleansing Corporation 2011 [2, 4, 5-7]. The number of cities in peninsular Malaysia increased by 400% from 1957 to 2000 [14, 18-22]. Urbanization and industrialization in Malaysia have changed the characteristics of solid waste [22-24]. Thus, effective solid waste management practices must be updated to suit the current waste quantity. Landfilling is the main disposal method in Malaysia. Malaysia, however, is facing serious landfill problems, such as, insufficient capacity, overflow of landfill sites, serious odour and lack of ventilation, and lack of leachate treatment facilities. Therefore, ammonia removal using the proposed method will improve the treatment of landfill waste and reduce the total cost of handling waste in Malaysia.

Several approaches of leachate treatment have been reported such as aerobic biological treatment, anaerobic biological treatment, physiochemical treatment, coagulation and advanced techniques including carbon adsorption and ion exchange. However, most of the approaches have one or more disadvantages, until recently, micro fuel cell (MFC) received significant attention as they enable the use of microorganism as promoting agent or catalyst to convert chemical energy of the electro donors into electricity. In addition to that, MFC have shown capabilities in various treatment of ammonia and ammonia related chemicals [22-26]. Therefore, the present work focused on the ammonia removal from the leachate existing in a landfill located in Padang Siding Landfill, Perlis. The performance of ammonia removal using different types of electrodes at static and dynamic modes of MFC will be analysed.

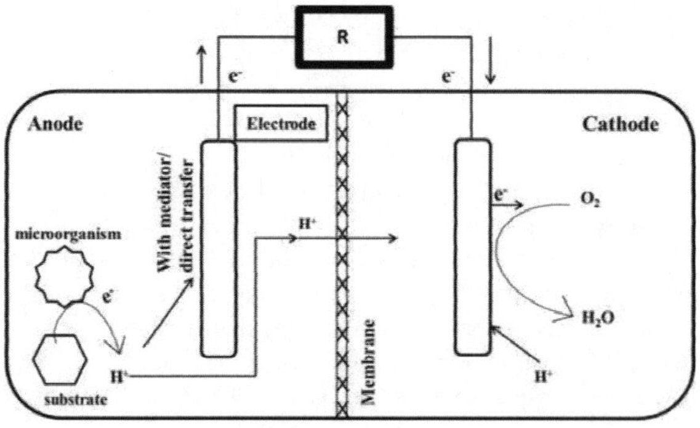

Figure 1. Schematic diagram of The MFC

2. Methods

A two-chamber microbial fuel cell (MFC) as shown in figure 1 is required to perform the bio-electrochemical reactions. Square, circular, and large-scale MFC designs are required for characterization, and a pH meter is required to determine pH levels. The sample was collected from Padang Siding Landfill by determining the physical and chemical parameters of the target ammonia. It was taken with proper precautions to prevent contamination and protected from rainfall. The leachate was collected from the landfill and sun-dried prior to laboratory testing. The temperature during sampling was 35 °C. The leachate sample was collected from the solid waste landfill pond at the back of the landfill. A total 20 L of leachate was manually collected and placed in plastic containers. After collecting the sample, it was placed in a controlled environment. The sample was immediately transported to the laboratory, characterized, and cooled at 4 °C to minimize biological and chemical reactions [22-25]

IOP Conf. Series: Materials Science and Engineering **206** (2017) 012079 doi:10.1088/1757-899X/206/1/012079

The next step is the MFC construction. The MFC is constructed from Perspex, with valves, wire, three material types and electrodes, a proton exchange membrane, and a silicon sealant that can produce electricity and harbor bacteria. The electrode and proton exchange membrane are encased in a Plexiglas compartment that is designed with valves and wires connected to the electrode and rheostat (R). A sealant is then used to ensure the compartment is airtight. A flask (1 L capacity) is used as the anode chamber, and a cylinder of Perspex (125 mL capacity) with two end pieces, one of them with a hole, is fixed to the anode chamber. Both chambers are physically separated by a proton exchange membrane with an interchange surface of 7 cm^2. Each chamber contains a zinc, activated carbon, or carbon black electrode. The total accessible geometrical surface of the electrodes is intended to be 6.5 cm^2. The electrodes are attached to the external system by a copper wire with all exposed surfaces sealed with non-conducting silicone. The MFC are operated in two different modes of electrode; static and dynamic. In the dynamic mode, the leachate is continuously stirred whilst the leachate is in quiescent condition at static mode.

3. Results and Discussion

The oxygen required for the initial starting of the device is obtained from the leachate itself through the nitrification process and the cathodic electrode oxygen reduction reaction, which relies on ion diffusion through the membrane cathode. However, the dynamic feeding medium may bring some dissolved oxygen to the system in addition to the oxygen already present in the leachate. The MFC is constructed with three configurations based on the anode electrode [5, 7]. Table 1 shows the results obtained using different anodes at a dynamic system which have almost linear removal over 4 weeks. The maximum amount of ammonia removal was by activated carbon (AC) electrode at 96.6% at week 4.

Table 1. Comparison of the rate of ammonia removal by different anode materials.

Anode Material / Day	Week 1 % of Removal	Week 2 % of Removal	Week 3 % of Removal	Week 4 % of Removal
Zinc	20.0	53.3	56.3	66.3
AC	33.3	55.5	83.3	96.6
CB	32.3	54.9	80.3	92.8

Three MFC electrodes were evaluated in terms of their efficiency in removing ammonia. The process was dynamic for liquor sample cycling from the two tanks. The waste was recycled again by pump. All three operations were conducted under dynamic mode conditions. After the process, the MFC required approximately 30 days to achieve an optimal removal rate. This process was possible when the bacterial community found the necessary conditions to enrich and allow the MFC to operate for this long period. Figure 2 shows the ammonia removal by the zinc electrode dynamic system [2, 16, 20-24].

A linear reduction in ammonia occurred with the passage of time. The maximum amount of reduction after one week was approximately 20%. The amount of reduction varied with different types of materials. The removal rate of activated carbon was higher than those of zinc and carbon black. The dynamic process of the activated carbon sample produced a higher removal potential than any other anode material. This result may be due to the fact that activated carbon has low internal resistance resulting from the porosity of the materials that supports mass transfer. Generally, the rate of ammonia removal depends on factors such as mass transfer and internal resistance. The low internal resistance of activated carbon allows for increased mass transfer.

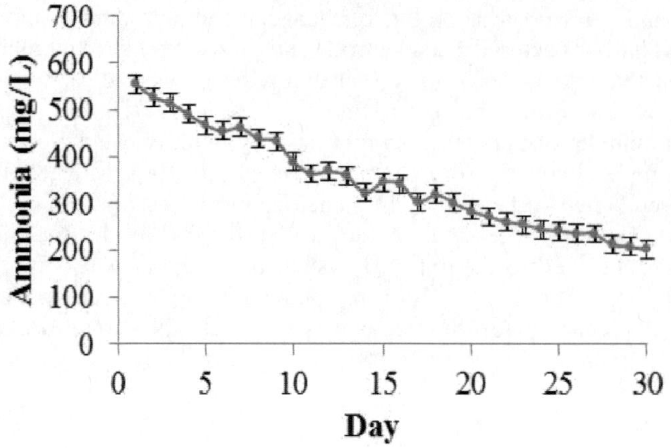

Figure 2. Ammonia removal by zinc electrode dynamic system.

Ammonia removal increased linearly throughout the experiment in the case of activated carbon (figure 3), which also produced a stable voltage ranging from 5 mV to 10 mV. However, the zinc electrode produced a slightly higher voltage than the two carbon materials, because electrochemically active bacteria were more dominant in zinc than in carbon materials (Table 2). Zinc produces low voltage output and low ammonia removal rate because zinc materials possess higher resistance than do carbon materials.

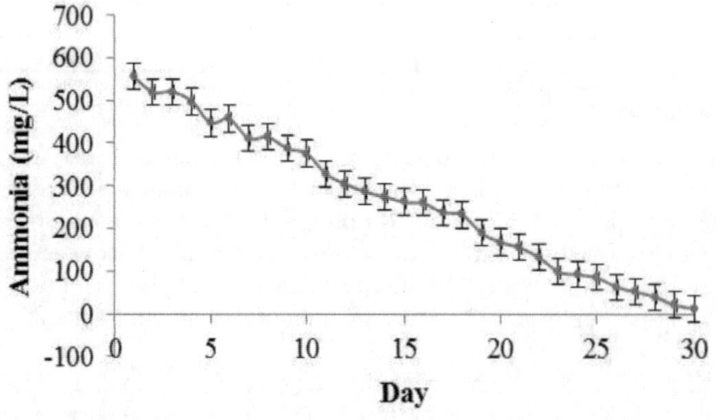

Figure 3. Ammonia removal by activated carbon dynamic system.

Figure 4 shows the results of carbon black anode materials. The data for this process were also monitored for 30 days before and after the MFC process. The ammonia removal percentages were 32.3%, 55.6%, 80.3%, and 92.8% for the first, second, third, and fourth weeks, respectively. Although zinc can generate high power density (voltage), its ability to remove ammonia was lower than that of carbon materials. The removal rates of both activated carbon and black carbon were high. However, activated carbon had removal rates of 33.3%, 55.5%, 83.3%, and 96.3%. The effluent ammonia concentration was 600 mg/L in all cases.

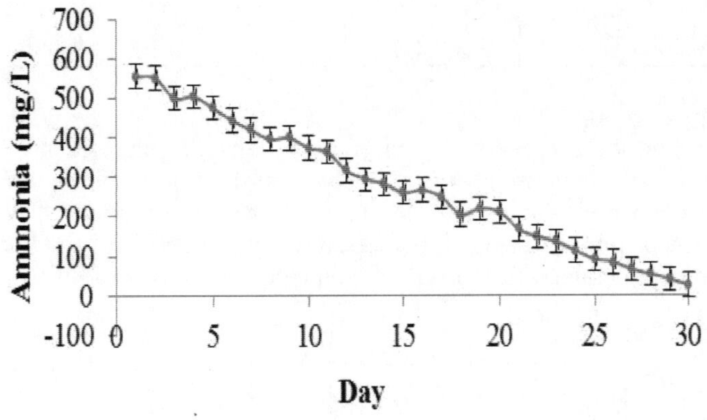

Figure 4. Ammonia removal by carbon black dynamic system.

As stated earlier, the MFC can simultaneously remove ammonia and generate electricity. This ability has made MFCs a viable approach to treat water and leachate. The highest rate of voltage generation in this study was achieved when the MFC was operated with zinc anode material (0.3 V), followed by carbon materials (5–10 mV). However, figure 5 shows that activated carbon provides the highest and most consistent rates of ammonia removal. The highest efficiency of ammonia removal was achieved by activated carbon (96.3%), followed by carbon black (92.6%) and zinc (66.6%). Activated carbon is the most efficient among the three types of anode materials.

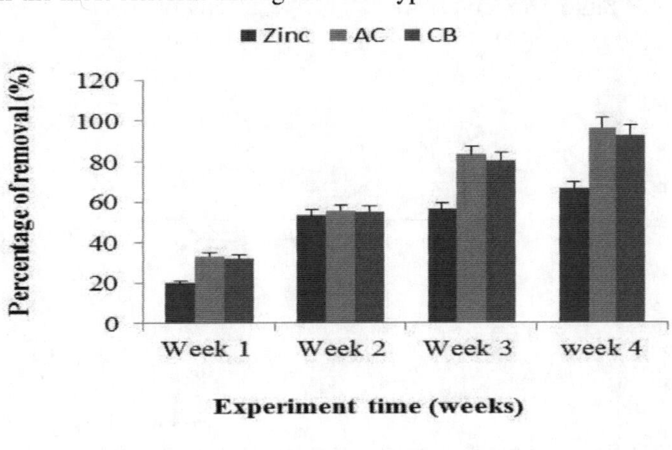

Figure 5. Comparison of rate of ammonia removal of three anode materials.

Table 2 shows the power generated by different materials. Zinc electrode achieved the highest power generation (3.60 mW), followed by activated carbon electrode (0.75 mW). The lowest power generated for activated carbon (0.4 mW) resulted from the passive nature of activated carbon in supporting electron transfer.

Table 2. Comparison of power generation by three anode materials

Electrode	Voltage (mV)	Current (A)	Power (mW)
Activated carbon	5.00	0.15	0.75
Zinc	9.00	0.40	3.60

Carbon black	4.00	0.10	0.40

As mentioned earlier, the use of MFCs can produce voltage. Figure 6 and 7 show the voltage values for the dynamic and static mode, respectively from the start until the end of treatment. The highest voltage obtained from zinc electrode was 0.761 mV, and the lowest was 0.492 mV. The highest voltage obtained from activated carbon electrode was 0.260 mV, and the lowest was 0.038 mV. The highest voltage obtained from black carbon electrode was 0.749 mV, and the lowest was 0.372 mV.

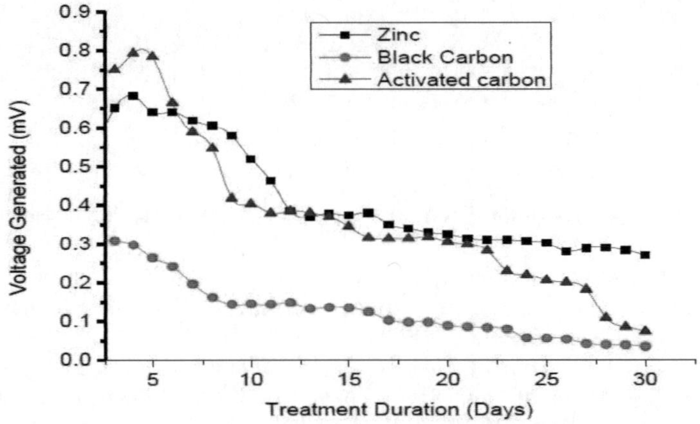

Figure 6. Comparison of voltage generation by three anode materials for the dynamic mode.

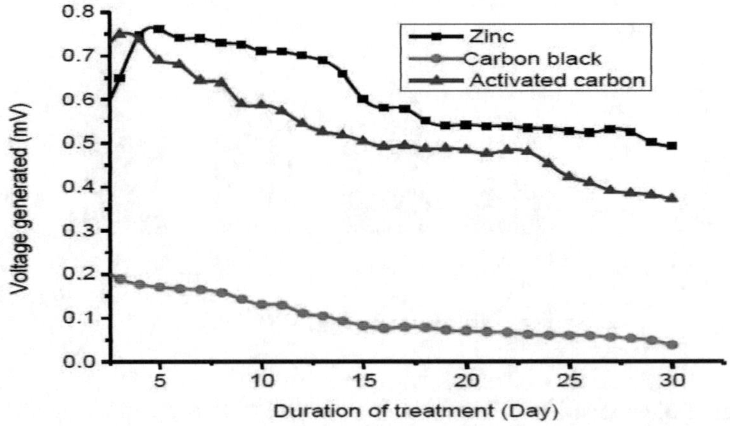

Figure 7. Comparison of voltage generation by three anode materials for the static mode.

4. Conclusion

This study demonstrated the process of ammonia removal by using an MFC with three different electrodes: zinc, activated carbon, and carbon black. The MFC shows the potential to use the bacterial community in mixed cultures of leachate and thus treat waste and produce electricity. The efficiency of anode materials to remove ammonia was successfully compared. Carbon materials were more efficient in removing ammonia than zinc. Over a period of 30 days, the removal rates of ammonia were 96.3% for activated carbon and 66.6% for zinc. However, zinc is a more appropriate choice for power generation. Therefore, MFCs are a promising tool in waste treatment, wastewater treatment, and electricity generation.

Acknowledgement
This project was supported by the School of Environmental Studies, University Malaysia Perlis (UniMAP)

References

[1] Adhikari B, Dahal K R and Khanal S N 2014 A Review of Factors Affecting the Composition of Municipal Solid Waste Landfill Leachate *Int. J. Eng. Sci. Innov. Technol.* **3**(5) 273–281
[2] Ghasimi S M D, Idris A, Ahmadun F R, Tey B T and Chuah T G 2008 Batch Anaerobic Treatment of Fresh Leachate from Transfer Station *J. Eng. Sci. Technol.* **3**(3) 256–264
[3] Diamadopoulos E 1994 Characterization and Treatment of Recirculation Stabilized Leachate *Wat. Res.*, **28**(12) 2439-2445
[4] Evren A T, Pelin C and Baris C 2013 Bio-Electrochemical Post-Treatment of Anaerobically Treated Landfill Leachate *Bioresource Technol.* **128** 266–272
[5] Ganesh K and Jambeck J R, 2013 Treatment of Landfill Leachate Using Microbial Fuel Cells: Alternative Anodes and Semi-Continuous Operation *Bioresour. Technol.* **139** 383–387
[6] Karmakar S, Kundu K and Kundu S 2010 Design and Development of Microbial Fuel Cell Current Research *Technology and Education Topics in Applied Microbiology and Microbial Biotechnology* 1029-1034
[7] Puig S 2011 Microbial Fuel Cell Application in Landfill Leachate Treatment *J. Hazard. Mater.***185**(2-3) 763–767
[8] Qixing W L , Zhou and Tao H 2010 Removal of Organic Matter from Landfill Leachate by Advanced Oxidation Processes: A Review *Int. J Chem.Eng.* 1-10
[9] Raghab S M, Abd El Meguid A M and Hegazi H A 2013 Treatment of leachate from municipal solid waste landfill *HBRC J.* **9**(2) 187–192
[10] Rahimnejad M, Adhami A, Darvari S, Zirepour A and Oh S E 2015 Microbial fuel cell as new technology for bioelectricity generation: A review *Alexandria Eng. J.* **54(3)** 745–756
[11] Kettunen R H, Hoilijoki T H and Rintala J A 2009 Anaerobic and sequential anaerobic-aerobic treatments of industrial and municipal landfill leachate at low temperatures *Bioresour. Technol.* **58** 40-41
[12] Ringeisen B R, Henderson E, Wu P K, Pietron J, Ray R, Little B and Jones-Meehan J M. 2006. High power density from a miniature microbial fuel cell using Shewanella oneidensis DSP10 *Environ. Sci Techno.* **40**(80) 2629-2634
[13] Sani A 2014 The influence of pH on the removal of ammonia from a scheduled waste landfill leachate *J. Tech. (Sci. Eng.)* **68(5)** 25–28
[14] Seow- Wee 2012 New Perspective of Integrated Solid Waste Management in Malaysia *Proceeding 3rd International Conference on Human Habitat & Environment in the Malay World* 19-20 Jun 2012 UKM Bangi
[15] Solid Waste Management and Public Management Cleansing Corporation 2011 Implementation of Solid Waste and Public Cleansing Management Act (Act 672), Annual Report 2011 Chapter 4 Kuala Lumpur: Solid Waste and Public Cleansing Management Corporation
[16] Tong, Zhnag B, Liu Y, S., Zheng M, Zhao Y, Tian C and Feng C 2014 Enhancement of bacterial denitrification for nitrate removal in groundwater with electrical stimulation from microbial fuel cells *J. Power Sour.* **268** 423-429
[17] Watanabe K and Biosci J 2006 Recent developments in microbial fuel cell technologies for sustainable bioenergy *J. Biosci. Bioeng.* **106** 528-536
[18] Woo J H, Im H J M W, Choi K B, Han C W and Kim 2001 Simultaneous organic and nitrogen removal from municipal landfill leachate using an anaerobic-aerobic system *Water Res.* **35** 2403-2410

IOP Conf. Series: Materials Science and Engineering **206** (2017) 012079 doi:10.1088/1757-899X/206/1/012079

[19] Wu Y, Zhou S, Qin F, Ye X and Zheng K 2010 Modelling physical and oxidative removal properties of Fenton process for treatment of landfill leachate using response surface methodology (RSM) *J. Hazard. Mat.* **180**(1-3) 456–465

[20] Zhang Q Q 2013 Investigation on characteristics of leachate and concentrated leachate in the landfill leachate treatment plants *Waste Manage.* **33**(11) 2277–2286

[21] Zhou W, Qin Y S, Zheng F K and Ye X 2010 Modelling the oxidation kinetics of Fenton's process on the degradation of humid acid *J. Hazard. Mat.* **179**(1-3), 533– 539

[22] Jayesh M S, Enrico M and Prakash C G 2014 Treatment of domestic and distillery wastewater in high surface microbial fuel cells *Int. J. Hydrogen Energ.* **39** 21819-21827

[23] Guodong Z, Yan J and Duu-Jong L 2015 A lab-scale anoxic/oxic-bioelectrochemical reactor for leachate Treatments *Bioresour. Technol.* **186** 97–105

[24] Zhao G D, Zhang Q L, Jiao Y, Wang K, Lee D J and Ren NQ 2012 Biocathode microbial fuel cell for efficient electricity recovery from dairy manure *Biosens. Bioelectron.* **31** 537–543

[25] Jiao Z G D, Zhao Q L Y, Zhang J N, Jiang J Q, Ren N Q and Kim B H 2011 Improved performance of microbial fuel cell using combination biocathode of graphite fiber brush and graphite granules *J. Power Sour.* **196** 6036–6041

[26] Wang Z F W, Zhang X and Tao G, 2011 Electricity generation in a membraneless microbial fuel cell with down-flow feeding onto the cathode *Bioresour. Technol.* **102** 7324–7328

IOP Conf. Series: Materials Science and Engineering **206** (2017) 012080 doi:10.1088/1757-899X/206/1/012080

Performance Study of Ceramic Filter Module in Recirculated Aquaculture System (RAS)

L Y Ng[1], C Y Ng[2]

[1]Centre for Environmental and Green Technology, Department of Chemical Engineering, Lee Kong Chian Faculty of Engineering and Science, Universiti Tunku Abdul Rahman, Jalan Sungai Long, Bandar Sungai Long, Cheras 43000 Kajang, Selangor, Malaysia.
[2]Department of Chemical & Petroleum Engineering, Faculty of Engineering, Technology and Built Environment, UCSI University (North Wing), Lot 12734, Jalan Choo Lip Kung, Taman Taynton View, 56000 Cheras, Kuala Lumpur, Malaysia.

E-mail: ngcy@ucsiuniversity.edu.my

Abstract. The growth of world population has led to significant increase in seafood demand over the world. Aquaculture has been widely accepted by many countries to increase the seafood production owing to the decline of natural seafood resources. The aquaculture productivity, however, is directly linked to the pond water quality. In this study, attempts were made to employ ceramic micro-filter to improve the pond water quality through filtration processes. There were two batches of filtration processes, short term (1 hour) and long term (48 hours). Significant improvements on real pond water quality were recorded through the short term microfiltration process, which reduced turbidity (96%), total suspended solids (TSS) (80%), biochemical oxygen demand (BOD) (72%), chemical oxygen demand (COD) (55%), ammonia (60%), nitrate (96%) and phosphorus (83%). The long term filtration process also showed high efficiency in the removal of solid particle and organic matters. The results showed that all of the parameters were successfully reduced to acceptable ranges (turbidity$<$80 NTU, TSS$<$400 mg/L, BOD$<$5 mg/L, COD$<$70 mg/L, phosphate$<$3 mg/L and ammonia$<$0.05 mg/L) for fish culturing activity. Based on current study, there was a drastic increase in nitrate content after 24 hours due to the nitrification process by regenerated bacteria in the filtered pond water. Current study showed that the microfiltration using ceramic micro-filter has high potential to be used in recirculating aquaculture system throughout the aquaculture activities in order to maintain the pond water quality, thus, increase the survival rate of cultured species.

1. Introduction

Aquaculture is important in providing food and generating income in many developing countries such as Thailand, India, China and Malaysia. Continuous growth of world population makes food scarcity a critical issue around the world. Malaysia government has foresighted the future prospect of aquaculture, thus, heavily promoting the aquaculture activities locally [1]. Fish farming, thus, has been blooming since a decade ago in Malaysia. The main challenge that farmers are facing now, however, is proper water quality control. To mitigate this problem, several methods and system have been

IOP Conf. Series: Materials Science and Engineering **206** (2017) 012080 doi:10.1088/1757-899X/206/1/012080

introduced and attempted such as sedimentation control, probiotic treatment and treating the effluent by microalgae [2].

Filtration can be used to retain suspended solid matter from a liquid by passing the fluid through the pores of filter medium. The presence of suspended solids in the pond water is normally resulting from fish waste and uneaten food, which contribute to a portion of the oxygen demand and toxic ammonia in the farming system [2]. In addition, filtration process has low energy consumption, easy to be set up and economically feasible to be integrated into several separation processes [3]. Therefore, filtration could be a highly potential method to mitigate the water contamination problem in fish farming although it is still not widely adopted by aquaculture industry.

Water quality in the fish farming is one of the most complex solution as it consists of many controllable and un-controllable parameters. A poorly managed pond water is prone to the outbreak of diseases because of the blooming of bacterial and/or viruses in the water. One of the most common diseases in aquaculture farms is White Spot Syndrome Virus (WSSV) disease. Poorly managed pond water will encourage the replication of WSSV, which consequently lead to higher mortality rate during the farming cycle [4,5]. Parameters in aquaculture that require intensive monitoring and controlling including pond water pH, total suspended solids, salinity, ammonia, ammonium, nitrate, nitrite, heavy metals, temperature and dissolved oxygen content. The survival rate of cultured species can be increased if the parameters listed above are well-controlled within the acceptable limit. Filtration system using ceramic module, therefore, is proposed in this study to be integrated into the pond water system in order to improve and maintain the pond water quality.

Table 1. Optimum water parameters for freshwater aquaculture activity (catfish) [6]

Parameter	Optimum range
pH	pH 6 - 9
Turbidty	< 80 NTU
Total suspended solids (TSS)	< 500 mg/L
COD	< 70 mg/L
BOD	< 5 mg/L
Ammonia	< 0.05 mg/L
Nitrate	< 1 mg/L
Phosphate	< 3 mg/L

2. Materials and methods

Pond water was collected from a freshwater catfish pond during 60 culturing days and it is located at Hulu Langat province, Selangor, Malaysia. The water sample was stored in 2.5 L polyethylene bottles and kept at 4°C prior to the experiments and water analyses. The pond water pH was collected and measured using Eutech pH700 (Thermo Fisher Scientific Inc. MA, USA). The Doulton ceramic filter (Southfield, United State) with an average pore size of 0.6 μm was selected to be used as the filter module in this study. The chemicals and reagents such as ammonia salicylate reagent powder pillows, nitra Ver5 nitrate reagent powder pillows, molydovanadate reagent, COD digestion solution and copper and iron standard solution were purchased from Hach (Colorado, United State).

The tested parameters were turbidity, total suspended solids (TSS), chemical oxygen demand (COD), biochemical oxygen demand (BOD), total ammonia nitrogen (TAN), nitrate ($NO^{3-}N$), phosphate ($PO4^{3-}$) and heavy metal (Fe, Cu). All of the parameters were analyzed using Hach DR3900 spectrophotometer and following standard procedures (U. S. Environmental Protection Agency 1983). In the COD test, samples need to go through the digestion process using Hach DRB 200 digital reactor prior to the concentration reading by Hach DR3900 spectrophotometer. Perkin Elmer Analyst 400 Atomic Absorption Spectrometer (Ohio, United State) was used to analyze the heavy metal in all the samples.

IOP Conf. Series: Materials Science and Engineering **206** (2017) 012080 doi:10.1088/1757-899X/206/1/012080

2.1. Calculation
The flux (L/m^2.min) of permeate can be calculated using the Equation (1):

$$Flux, J = \frac{V}{A.\Delta t} \quad (1)$$

where V is the volume of permeate collected (in unit L), A is surface area of filter (in unit m^2) and Δt is interval of time (minute). Normalized flux is a ratio of permeate flux to initial pure water flux (J/Jo), as shown in Equation (2), at similar operating pressure and temperature.

$$Normalize\ flux = \frac{Permeate\ flux}{Initial\ water\ flux} = \frac{J}{J_o} \quad (2)$$

Rejection of the tested parameters can be expressed in terms of percentage as shown in Equation (3):

$$Rejection, Rej\ (\%) = \left(1 - \frac{Cp}{Cf}\right) x100\% \quad (3)$$

where Cp and Cf represent the concentration of concerned components in permeate and feed streams, respectively.

2.2. Experimental setup
A recirculating aquaculture system (RAS) was set-up using a filtration module as shown in figure.1 The RAS was conducted at short and long term in this study. Filtration process using pure water was conducted for 15 minutes before the filtration of pond water samples. This is to ensure that the filter module is in good condition and also for the filter wetting purpose. The operating conditions for each of the experimental run are shown in table 2. A water pump (BLDC Pump DC-30A, 12V, Guangzhuo, China) was used to supply the desired pressure throughout this study. Permeated water sample was collected and recorded for an interval of 1 minute over an hour of filtration period for water flux calculation. The collected permeate was stored at 4°C prior to the water analyses.

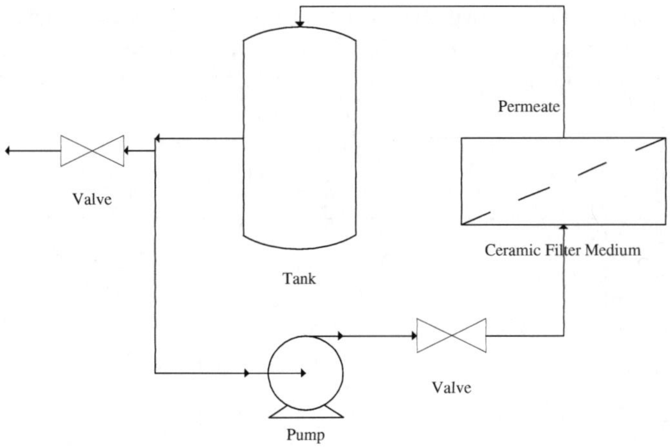

Figure 1. The schematic diagram of filtration system

Table 2. Parameters of each experimental run

Experiment set	Abbreviation	Filtration Time	Pressure (bar)	Temperature (^{o}C)
Short term (batch 1)	S.T1	1 hour	1	25
Short term (batch 2)	S.T2	1 hour	1.5	25
Long term	L.T	48 hours	1	25

For the long term filtration process, the permeate was returned to the pond water tank continuously for a duration of 48 hour. The pressure and temperature applied was same as the short term filtration set (S.T1). At the end of long term filtration process, the sample was sent for water analyses. The produced flux for all batches of filtration were evaluated and discussed. All the tested parameters for all batches were also compared and discussed.

3. Results and discussion

3.1 Flux decline

Figure.2 and figure.3 shows the normalized flux decline curves for experimental set S.T1 and S.T2, respectively. Similar normalized flux decline curves were observed for both batches of short term filtration processes. Normalized flux reduction, as much as 97%, were achieved for both S.T1 (6 min) and S.T2 (5 min) within a very short period of filtration time. The pond water sample is a complex mixture which consists of various components with a broad range of particle sizes and different properties. These factors may affect the performance of used ceramic filter. Ceramic filter with high porosity is highly susceptible to fouling during the starting of filtration process owing to the concentration of foulants onto the pores present on the ceramic filter, which led to a drastic decline of solution flux [7; 8]. This drastic flux decline can be observed in the normalized flux decline curves as plotted (figure.2 and figure.3).

Ceramic filter with an average pore size of 0.6μm encouraged the fouling phenomenon to be observed on the ceramic filter owing to a broad distribution of particle sizes in the pond water sample [9]. Ceramic filter used in this study can be categorized as microfiltration which restrict the passage of micro-sized or larger constituents through the filter pores. For instance, the constituents including TSS, dissolved solids, natural organic matter (NOM), inorganic matter, bacteria, proteins traces from uneaten food and feces and solid particles [10]. The presence of these components increased the fouling rate of ceramic filter module, thus, reduced the volume of permeate. Fouling mechanisms such as pore blocking and cake layer formation are the possible explanations for the reduced permeate as can be observed in this study. This, however, requires further experimental data in the future to verify the fouling mechanism observed on the ceramic filter. After 5 minutes of filtration, steady-state flux was achieved. This condition normally occurs when a cake layer with certain thickness covered and compacted on the surface of a filter [11; 12; 13]. Cake formation, thus, could be formed during these short term filtration processes using the ceramic filter in this study.

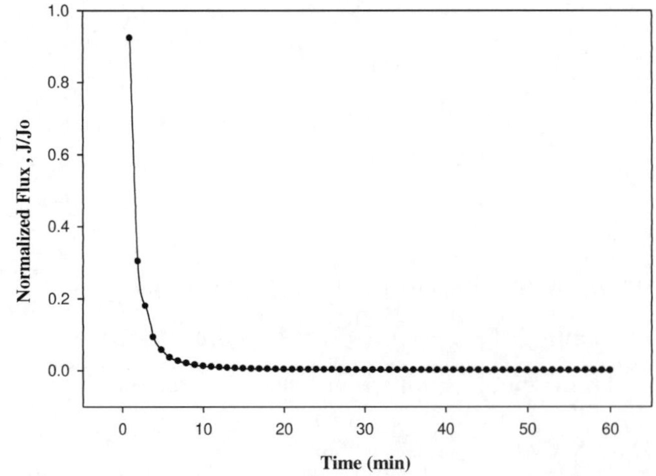

Figure 2. Flux decline of pond water for filtration set S.T1

IOP Conf. Series: Materials Science and Engineering **206** (2017) 012080 doi:10.1088/1757-899X/206/1/012080

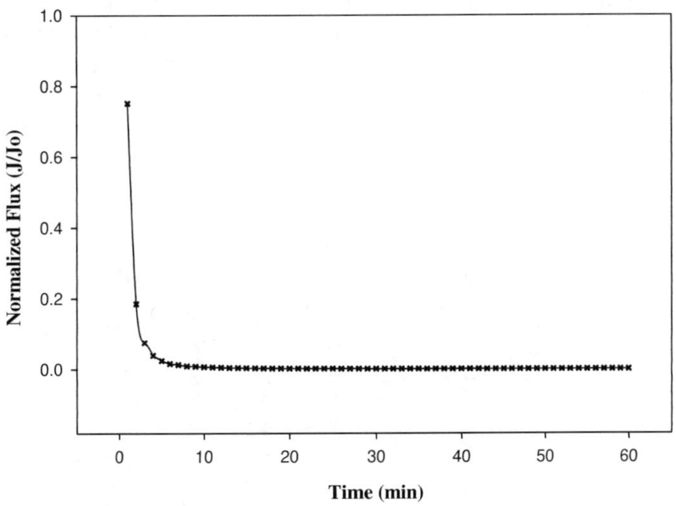

Figure 3. Flux decline of pond water for filtration set S.T2

3.2 Sample analyses

Performance and efficiency of ceramic filter in fish pond water treatment was evaluated through the water analyses. This study analyzed the water samples based on several parameters including pH, turbidity, TSS, BOD, COD, ammonia, nitrate, phosphate and heavy metals. The feed stream and permeate samples collected from the filtration processes (T.S1, T.S2 and L.T) were sent for water analyses. Range of water sample pH 5.0 – 6.50 was recorded for all the feed and permeate streams of S.T1 and S.T2. The results obtained is reasonable owing to the fact that a healthy freshwater fish culturing activity has to be conducted at pH values near to the neutral condition [14]. Extremely acidic or alkaline conditions are not suitable for freshwater fish farming, which might lead to lower survival rate of the cultured fish. Water pH, therefore, is one of the important parameters that need to be continuously monitored in aquaculture industry. Close monitoring of the pond water pH is one of the important practices that normally employed to minimize the stress in cultured species, thus, diseases and viruses infection [15].

The pond water or feed stream sample of each batch of filtration collected was sent for water analyses based on the parameters as mentioned earlier. Table 3 shows the water analysis results in feed streams for all filtration batches. Based on the analysis results, no heavy metals (copper and iron) elements were present in the pond water sample collected.

Table 3. Water analyses of feed stream for all batches of filtration process

Parameters	Value
pH	6.61 ± 0.00
Turbidity (NTU)	62.55 ± 1.35
TSS (mg/L)	1072.00 ± 0.30
BOD (mg/L)	4.71 ± 0.31
COD (mg/L)	128.67 ± 1.52
Ammonia (mg/L)	0.47 ± 0.05
Nitrate (mg/L)	1.15 ± 0.05
Phosphate (mg/L)	5.10 ± 0.10
Copper (mg/L)	n.a
Iron (mg/L)	n.a

Figure.4 illustrates the rejection percentages for all water parameters in S.T1 and S.T2 after the filtration process using ceramic filter. Turbidity and nitrate achieved higher rejection rate for both short term filtration processes in comparison to other parameters. Turbidity analysis showed 96.35 ± 0.08% and 97.24 ± 0.07% rejections for filtration set S.T1 and S.T2, respectively. For nitrate removal study, 94.50 ± 0.83% and 95.89 ± 0.84% retentions were recorded during the filtration set S.T1 and S.T2, respectively.

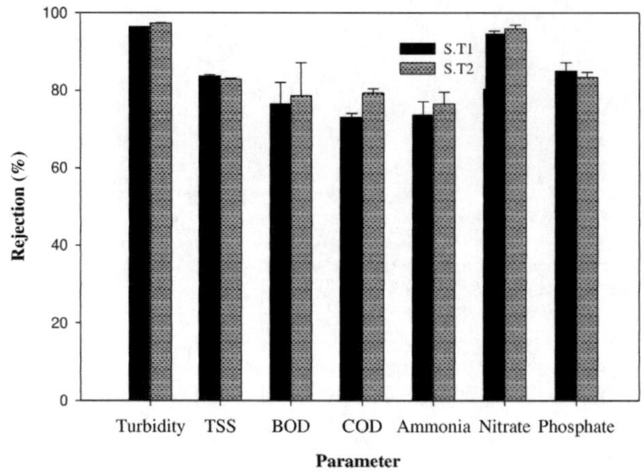

Figure 4. Rejection percentages for all water parameters in S.T1 and S.T2 after filtration process using ceramic filter

Above 80% of rejections were observed for TSS and phosphate for both S.T1 and S.T2 filtration sets. On the other hand, 73 – 80% of BOD, COD and ammonia were retained by ceramic filter during the short term filtration processes, S.T1 and S.T2. Pond water comprises various particles or compounds, which own broad ranges of particle sizes and different characteristics. Combination of several factors have contributed to the rejection results during the filtration processes using ceramic filter. Besides particle size distribution factor, ionic organic matters would attach to or combine with oppositely charged compounds forming larger particles that were restricted from passing through the ceramic filter [16]. The ionic compounds that were investigated in this study including ammonia ions, nitrate, phosphate and so forth. These particles are possible to be retained on the filter surface or deposited on the filter pore wall if their sizes are similar to or larger than the ceramic filter pores [9]. In this study, rejection of all of the water quality parameters were highly encouraging in which most of their rejections were above 70%.

Visualization of the water quality difference can be done based on figure.5, which illustrates the water quality after the filtration process using ceramic filter. The improvement can be verified based on the water color, odor and turbidity content. After the filtration treatment, turbidity and color of the water are clearly reduced. This observation, coincidently, can be confirmed by the results in figure.4, which shows that most of the foulants have been successfully retained by ceramic filter.

Figure 5. Quality of pond water before and after treatment using ceramic filter module

3.3 Long term filtration

The operating condition for long term filtration process was fixed as stated in table 2. The initial permeate flux for the long term filtration was recorded as 1101.34 L/m^2.hr, while the permeate flux after 48 hours was reduced to 0.5580 L/m^2.hr. The reduction of permeate flux was 99.95 %. This revealed that a tremendous fouling had occurred during the long term filtration. This observation is reasonable as the filtration process in this study was carried out using a dead-end mode, which retain the water contaminants, thus, the build-up of the foulants within the ceramic module. Cake formation could be one of the fouling mechanism on the surface of ceramic filter after 48 hours of filtration process. Similar result was also reported by several researchers who performed the long term filtration in their study [10; 17; 18]. Pore blocking mechanism, however, could be the contributor to the drastic reduction in the permeate flux after 48 hours [19].

Figure. 6 shows the concentration of turbidity, TSS, COD and BOD from initial (0 hour) until the end (48 hours) of long term filtration process using ceramic filter medium. The water parameters were determined at 0, 6, 12, 24 and 48 hours of filtration. Concentration of TSS had dropped drastically from the initial until 6 hours of filtration, which was followed by gradual decline up to the end of the filtration. However, other parameters (turbidity, COD and BOD) had slowly declined from the beginning until the final stage of long term filtration process. This outcome disclosed that the ceramic filter medium is able to effectively remove the TSS from aquaculture pond water throughout the filtration process [8; 20]. Most of the suspended solids consist of inorganic matters and their particle sizes can reach up to 2 microns [20]. TSS particle sizes are much bigger in comparison to ceramic filter pores. Most of the TSS, thus, could be retained by the ceramic filter module. Turbidity measurement is an indicator of water quality based on the water clarity and estimated TSS in water [8]. In this study, turbidity of pond water reduced during the long term filtration along with reduced TSS content. Concentration of TSS, turbidity, COD and BOD parameters in this study were successfully reduced to the optimum water parameter ranges as shown in Table 1 [21].

The concentration of ammonia, nitrate and phosphate at various time intervals (at 0, 6, 12, 24 and 48 hours) during the long term filtration study were plotted in figure.7. A declining trend was shown by ammonia and phosphate parameters throughout the long term filtration. Nitrate analysis, however, showed different trend when compared to ammonia and phosphate contents for several intervals between 0 and 48 hours of the long term filtration process. This distinctive result is logical owing to the nitrification process in the water sample. In this study, ceramic filter medium with 0.6 μm average pore size is permeable to microorganism and bacteria [16; 4]. The presence of nitrifying bacteria converts the ammonia in permeate streams into nitrate through nitrification process when sufficient time was provided during the long term filtration using ceramic filter [3]. Consequently, the content of nitrate in pond water sample was increased after 12 hours of filtration. The decline of nitrate in the beginning of long term filtration process can be explained by the retention of nitrate by ceramic filter, which was verified through the short term filtrations that were conducted in this study. The nitrification process by bacteria in the initial stage of the long term filtration, in addition, may not be initiated as the growth of bacteria require longer duration before the nitrification process can become significant [15; 17]. After the long term filtration of pond water sample using ceramic filter, concentration of ammonia and phosphate were still within the optimum range for aquaculture activity. At 48 hours, nitrate concentration was 1.2 mg/L, which was slightly higher than the optimum limit ($<$ 1 mg/L). Continuous and sequential filtration system using a combination of ceramic filter media and ultrafiltration membrane may be suggested in the future to remove the excess nitrate content to further improve the pond water quality [22; 9]. Possibility of disease or virus infection on cultured fish can be further minimized thus higher survival rate by reducing the stress induced on the cultured species when the pond water quality can be properly managed through the ceramic filter.

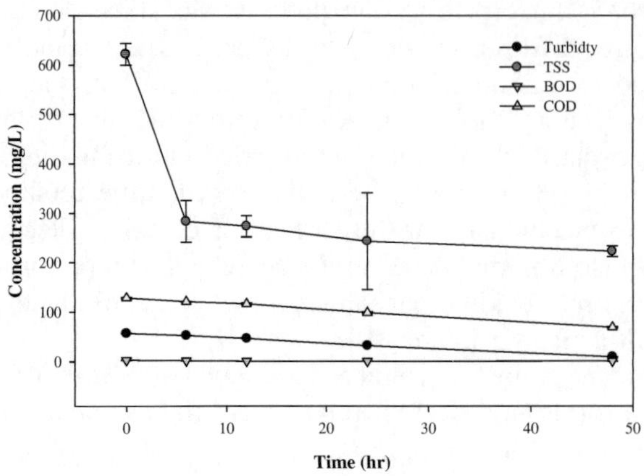

Figure 6. Interval results of turbidity, TSS, BOD and COD analyses between 0 and 48 hours of long term filtration process using ceramic filter.

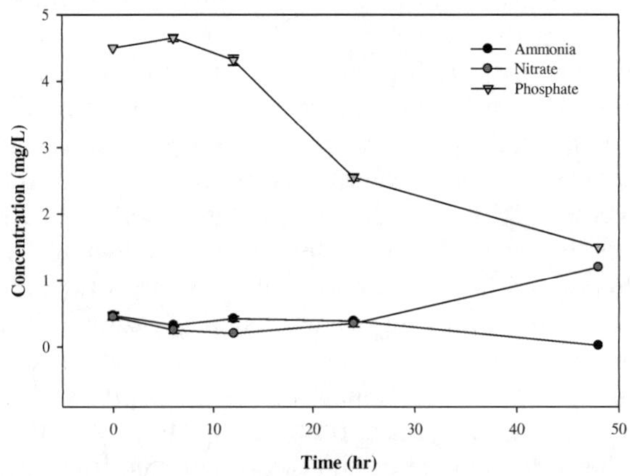

Figure 7. Interval results of ammonia, nitrate and phosphate analyses between 0 and 48 hours of long term filtration process using ceramic filter.

Figure. 8 illustrates the condition of ceramic filter medium used in this study before and after the filtration process using pond water. There was a significant distinction between fresh and used ceramic filter. Based on the observation, cake formation on the ceramic filter was one of the main mechanisms that contributed to the tremendous flux reduction as shown in figure.2 and figure.3. A thick layer of greenish cake consists of various kinds of foulants was coated on outer surface layer of the ceramic filter used during the study. Fish pond water, which consists of plenty of microalgae or fungus, acts as a medium to absorb CO_2 while provide O_2 to aquatic organisms through photosynthesis [23]. This is an important process to cultured-fish in ensuring a continuous supply of oxygen. Pond water is rich with various nutrients that encourages the growth of microalgae. The presence of microalgae in fish pond water is important in decomposing the organic matter in sediment while keeping the pond water clean [24]. Microalgae culturing could be one of the method to be used to manage and maintain the high quality of culture pond water [2; 25]. Excessive bloom of microalgae, however, can lead to high consumption of the pond water nutrients. When the nutrients in the pond water are depleted, large amount of dying microalgae and their decompositions may compete with the cultured fish for oxygen supply, which may reduce the fish survival rate. Relationship between the filtration flow rate and microalgae growth rate may need to be taken into consideration during a fish pond water filtration in order to maintain the presence of nutrients and quality of the pond water.

IOP Conf. Series: Materials Science and Engineering **206** (2017) 012080 doi:10.1088/1757-899X/206/1/012080

(a) (b)

Figure 8. Illustrations of ceramic filter (a) before and (b) after filtration process using fish farming pond water

Ceramic filter has been selected in this study as it can resist the bacterial attack better than polymeric materials. To study its feasibility in terms of economic perspective and performance reliability, ceramic filter in this work was cleaned after the long term filtration study. Clean water was used as a washing agent. 76.43% of flux recovery was successfully recorded by applying the washing on used ceramic filter medium after long term filtration (initial pure water flux was 1101.34 L/m^2.hr and water flux after cleaning was 841.75 L/m^2.hr). High flux recovery revealed that a great potential use of ceramic filter in aquaculture in which its performance can be further improved through periodical backwash and conducting the ceramic filter in a scale-up cross flow filtration system during a fish farming period [10; 26]. Implementation of a such filtration system, together with a periodical backwash step in aquaculture industry, can further minimize the water management cost and other chemical consumptions [27]. Based on its high flux recovery and good retention capability of undesired solutes, ceramic microfilter can be suggested to be integrated into a recirculating aquaculture system.

4. Conclusion

Performance of ceramic filter has been successfully evaluated in terms of flux decline and retention capability towards different water parameters such as turbidity, TSS, BOD, COD, ammonia, nitrate and phosphate using real pond water sample from a catfish farm. Deposition of the foulants onto the ceramic filter was one of the main factors that contributed to the flux reduction during the filtration of pond water. Pond water consists of various components that possess broad particle sizes and different properties. These components, thus, contributed to the high flux decline profiles as observed in this study. The mechanisms of fouling that dominated the filtration of pond water could be pore blockage and/or cake formation. Images of ceramic filter, before and after the filtration process, successfully demonstrated the retention of the foulants from the pond water.

All of the tested water parameters showed significant rejection after short term filtrations (S.T1 and S.T2), whereby the highest rejection was towards turbidity and nitrate at 97.84% and 95.89%, respectively. Meanwhile, rejection percentage for TSS, BOD, COD, phosphate and ammonia were between 73 to 80%. Based on the results, pond water quality has been successfully improved using ceramic filter module during the short term filtration processes. Significant results based on the water parameters also have been demonstrated during the long term filtration process. All of the tested parameters were successfully reduced to the optimum conditions for freshwater fish farming activity, excluding nitrate content. This could be explained by the nitrification of ammonia by the nitrifying bacteria after 12 hours. Sequential filtration system, by integrating microfiltration and ultrafiltration, is suggested for future investigation to further reduce the nitrate and microorganism content for pond water quality improvement.

Flux recovery, as high as 76.43%, was successfully obtained after the washing process done on the used ceramic filter. This result implies that the ceramic filter could be periodically backwashed in order to prolong its lifespan for continuous application in fish farming. A cross-flow filtration system can be suggested in future to maintain the quality of pond water while minimizing the build-up of the foulants on the ceramic filter. Chemically stable, good resistance towards bacterial-attack, recoverable water flux and effectiveness in retention capability towards various contaminants have suggested the ceramic filter to be further investigated in the future. This is as part of the effort to integrate the microfiltration into an aquaculture water management system for a sustainable development in that industry.

Acknowledgement
Authors would like to gratefully acknowledge UCSI University for Pioneer Scientist Incentive Fund as the main financial support.

5. References

[1] Awangku H. B. P. B A. 2010 *Marine Sci. & Aquac.* 2000 1–9
[2] Amir N, Muki S and David B E 2000 *Aquac.* **186** (3–4) 279–29
[3] Michael M and Charles C 2000 *Inst. Food Agric. Sci.* 1–4
[4] Flegel T W and Alday-Sanz V 1998 *J. Appl. Ichthyol.* **14** 269–273
[5] Xixian X, Hongyan L, Limei X and Feng Y 2005 *Virus Res.* **108** (1–2) 63–67
[6] Dieter K, Friederike K, Sebastian M, Tom L, Mikael A, Derek G and Annie D 2011 *Angew. Chemie - Int. Ed.* **50** (24) 5438–5466
[7] Mehrdad E, Daniel W, Kikavous S A, Larisa E, Laura P, Peter M, Peter B and Peter C 2010 *Desalination* **250** (3) 991–996
[8] Tewari P K, Singh R K, Batra V S and Balakrishnan M 2010 *Sep. Purif. Technol.***71**(2) 200–204
[9] Benjamin M., Annegret D., Johannes F., Christos A., and Thomas G., 2011. *Sep. Purif. Technol.* **81** (1), 77–87.
[10] In-Soung C and Su-Na K 2005 *Process Biochem.* **40** (3–4) 1307–1314
[11] Behnaz R, Abdolreza A and Mahdi F 2012 *Food Bioprod. Process.* **90**(4) 841-848
[12] Blanpain-Avet P, Fillaudeau L and Lalande M 1991 *Food Bioprod. Process.* **77**(2) 75-89
[13] Lalande M and Blanpain P 1997 *Filt. Separat.* **34**(10) 1065-1069
[14] Fabiano D S S, Altair B M, Marcia C B, Sonia M N G and Maria J S Y 2008 *Ecol. Indic.* **8** (5) 476–484
[15] Shi-Yang Z, Gu L, Hui-Bi W, Xing-Guo L, Yan-Hong Y, Ling T and Huang L 2011 *Aquac. Eng.* **45**(3) 93-102
[16] Abdul W M, Ng C Y, Lim Y P and Ng G H 2012 *Food Biopro. Technol.* **5**(4) 1143-1156
[17] Rosenberger S, Laabs C, Lesjean B, Gnirss R, Amy G, Jekel M and Schrotter J C 2006 *Water Res.* **40** (4) 710–720
[18] Asaadi M and White D A 1992 *Chem. Eng. J.* **48**(1) 11–16
[19] Glen B, Dan L and Ralf K 2006 *J. Memb. Sci.* **277** (1-2) 75–84
[20] Guo Q 2006 *Correlation of Total Suspended Solids (TSS) and Suspended Sediment Concentration (SSC) Test Methods.* Final Report, New Jersey Dep. Environ. Prot.
[21] Michael B N, James H, Tidwell L R D and Methil K 2011 *Freshwater Prawns: Biology and Farming*. John Wiley & Sons
[22] Shuji N and Alan S M 1992 *J. Membr. Sci.* **69**(3) 189–211
[23] Qunlan Z, Kangmin L, Xie J and Liu B 2009 *Bioresour. Technol.* **100** (16) 3780–3786
[24] David J W M 1997 *Aquaculture* **151**(1-4) 333–349
[25] Muki S, Amir N, Dan M P and Hillel G 1993 *Aquaculture* **117** (1-2) 115-128
[26] Salem A and Abdul W M 2014 *J. Water Process Eng.* **4** 107–133
[27] Xing Z, Mathias E, Peter M H and Martin J 2010 *Water Res.* **44** (18) 5212–5221

Column performance of carbon nanotube packed bed for methylene blue and orange red dye removal from waste water

G K Gill [1], N M Mubarak[2*], S Nizamuddin [3] H S Al-Salim[1], J N Sahu[4]

[1]Department of Chemical and Petroleum Engineering, Faculty of Engineering, UCSI University, Kuala Lumpur-56000, Malaysia
[2]Department of Chemical Engineering, Faculty of Engineering and Science, Curtin University, 98009 Sarawak, Malaysia
[3]Department of Chemical engineering, Scholl of engineering RMIT University Melbourne, Australia 3001
[4]Petroleum and Chemical Engineering Programme Area, Faculty of Engineering, Institut Teknologi Brunei, Tungku Gadong, P.O. Box 2909, Brunei Darussalam

Email: mubarak.mujawar@curtin.edu.my; mubarak.yaseen@gmail.com

Abstract: Environmental issues have always been a major issue among human kind for the past decades. As the time passes by, the technology field has grown and has helped a lot in order to reduce these environmental issues. Industries such as metal plating facilities, mining operations and batteries production are a few examples that involves in the environmental issues. Carbon nanotube is proven to possess excellent adsorption capacity for the removal of methylene blue and orange red dyes. The effect of process parameters such as pH and contact time was investigated The results revealed that optimized conditions for the highest removal for methylene blue (MB) (97%) and orange red (94%) are at pH 10, CNTs dosage of 1 grams, and 15 minutes for each dyes removal respectively. The equilibrium adsorption data obtained was best fit to Freundlich model, while kinetic data can be characterized by the pseudo second-order rate kinetics.

1. Introduction

Water is an important factor for a wide range of activities. The main source of contamination in water bodies is waste water run-off. Various industries discharge effluents that contain pollutants, which must be removed before it is released to the environment [1, 2]. Industrialization, agriculture practices and ecology changes are impacting the condition of water resources [3, 4]. Industries such as paper, pulp mills, paint, pharmaceutical, cosmetics, photographic, textile, leather, food, plastics and printing use huge amounts of dyes every year [5-7]. Large number of organic, inorganic and biological pollutants including pesticides, fertilizers, heavy metals, dyes, microbes etc. has been identified as water contaminants out of which some have serious side effects and few are believed to be dangerous and carcinogenic [3, 8-11]. World Bank appraises that material coloring and treatment contribute up to 17-20 percent of total industrial water pollution. Various techniques are available for the removal of heavy metal and dyes such as coagulation, adsorption, oxidation, chemical precipitation, membrane filtration, solvent extraction, electrochemical treatment, liquid-liquid extraction, reverse osmosis, photocatalytic degradation [12-14]. Adsorption is considered the most effective because it offers flexibility in operation and design; it also has a reversible nature as adsorbents can be regenerated by suitable

IOP Conf. Series: Materials Science and Engineering **206** (2017) 012081 doi:10.1088/1757-899X/206/1/012081

desorption processes for multiple use [15, 16]. Since discovery of carbon nanotubes (CNTs) is a new member in the carbon family [17]. CNTs have attracted researchers' interest as a type of adsorbent and offer an attractive option for the removal of organic and inorganic contaminates from water [18-21]. In addition CNTs is used for adsorption because of its unique properties; chemical and physical properties, optoelectronic, semiconductor, electrical, mechanical, structural properties, large specific surface area and extremely high thermal conductivity [22-24]. Hence, CNTs is most promising candidate especially in the field of remediation of toxic heavy metals, dyes contaminated industrial waste water [25, 26].

In this research, the comparative study on the removal of methylene blue (MB) and orange -red using batch adsorption study. The effect of the process parameters such as pH, contact time was investigated. Furthermore the kinetic and isotherm studies for the removal of both dyes were discussed.

2. Materials and method

2.1 Raw Materials

Microwave assisted CNTs were obtained from our previous [27]. The dyes used in this research are methylene blue and orange red obtained from Merck. Preparation of stock solution

2.2 Preparation of Stock Solution

Analytical grade of methylene blue (MB) and orange- red standards which were obtained from Merck were used to prepare stock solutions containing 1000 mg/L MB and orange -red which were further diluted with distilled water to obtain the required concentrations. In this research, the initial concentration of both MB and orange red dyes were set to 2 mg/L and prepared stock solution was used to batch adsorption experiments

2.3 Adsorption Experiment

The batch adsorption experiment was carried out using prepared stock solution with 2mg/L of both initial dyes concentration. The lower and upper values of each parameter (pH, and contact time) were chosen based on previous study on the removal of MB and orange red. The effect of process parameters such as pH (2 to 11) and contact time (10-50) using single factor optimization. The height of bed height was fixed 2cm (1 gram of CNTs). The experiments were carried out at constant room temperature. The pH values of the dye stock solutions were altered by using either 1.0 M of hydrochloric acid or 1.0 M of sodium hydroxide, depending on the value of desired pH. For the removal of MB and orange-red with fixed height (2 cm) which is used in this research, its adsorption capacity at time t was determined by using (1). As for the adsorption equilibrium of the dye, it was determined based on (2).

$$q_t = \frac{(C_O - C_t)V}{m} \quad \textbf{(1)}$$

$$q_e = \frac{(C_O - C_e)V}{m} \quad \textbf{(2)}$$

Where Co represents the initial concentration of the dye solution (mg/L), Ct is the concentration of dye solutions at time t (mg/L), Ce is the concentration of dye at equilibrium, V is the volume of dye solution (L) and m is the weight of magnetic biochar adsorbent used (g).

2.4 Characterization of Dye Removal

The physical properties, structure, and morphology of microwave assisted CNTs were reported earlier. Fourier transform infrared (FTIR) spectroscope (Bruker, USA) was used to analyse the before and after adsorption of dye on the surface of CNTs for the determination of the surface functional groups. All pH measurements were carried out using Mettler Toledo type (MP220 model, USA). The removal of dyes concentrations were measured by using the UV-Vis Spectro (U-29010 double beam, Perkin Elmer)

3. Results and discussion

3.1.1. Effect of pH on removal of MB and orange red

The pH factor is an important factor for the uptake of MB and orange red because the pH of a medium manipulates the magnitude of the electrostatic charges which are imparted by the ionized dye molecules [28, 29]. The pH factor causes the chemical characteristics of both adsorbent and adsorbate [30]. Hence, the rate of adsorption varies greatly in different mediums with different pH values. Fig 1(a-b) shows the effect of pH on removal of MB and orange-red respectively. From the observation as pH increase from pH2 to pH 10 increasing rapidly for both dye removal [31]. This is due to the pH increases, the number of H^+ ions decreases, therefore making the process more favourable. Further increase in pH 11 decrease in removal of both dyes was observed probably due to the presence of excess of OH ions competing with the dye anions for the adsorption sites [32-34]. In addition, an increase in pH will cause the density of the surface charge to decreases and this causes a lowering in the electrostatic repulsion between the surface of the CNT and the charge of MB and orange-red therefore the rate of adsorption increases. [35, 36]. Hence pH plays a vital role in adsorption of dye removal.

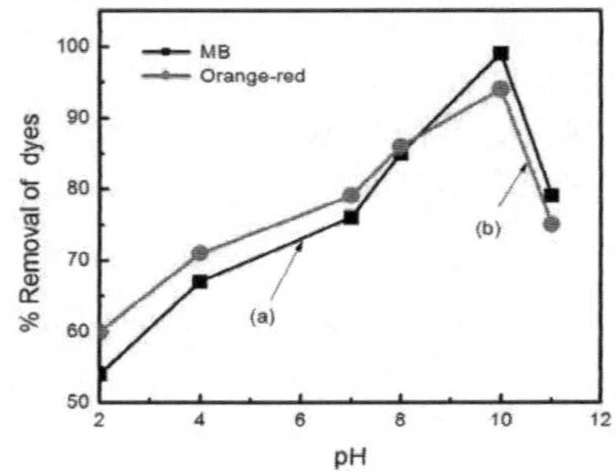

Figure 1. Effect of pH on (a) MB and (b) orange-red removal

3.2 Effect of contact time on MB and Organe-red

Figure 2 (a-b) shows the effect of contact time on removal of MB and orange-G respectively. It was observed as contact time increases from 10 min to 15 min rapid increase in removal of MB (99%) and orange-G (94 %) was observed. This is due to adsorption sites are vacant initially and the concentration gradient of solute is very high at the beginning of contact time. Further increase the contact time decreases percentage of removal of dyes was observed. The initial rapid rate of adsorption may be due to the availability of the positively charged surface of the adsorbent for anionic MB and orange-red species present in the solution. Occupation of the remaining vacant sites did not take place due to the high repulsive forces between the molecules of the dye at the adsorbent surfaces and in the bulk solution [37].

IOP Conf. Series: Materials Science and Engineering **206** (2017) 012081 doi:10.1088/1757-899X/206/1/012081

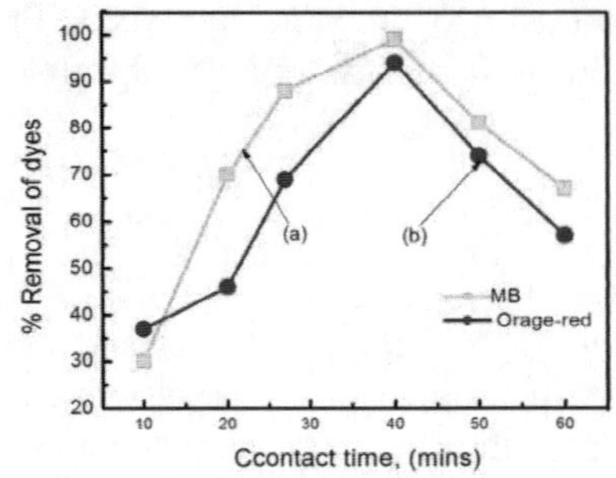

Figure 2. Effect of contact time on MB and orange-red removal.

3.3 Adsorption kinetic studies of MB and orange-red

The experiments for kinetic study were carried out for 2h minutes where three different pH values and initial concentration of the dye stock solution were used for the runs. From these three different beakers with different pH values of pH 4, pH 7 and pH 10 with initial concentration of 2 mg/L of MB and orange red respectively. The dye solution was measured every 5 minutes where later on the concentration at that specific time was checked by using the UV-Vis Spectrophotometer. Figures 3(a) and 3(b) are graphs of amount of dye adsorbed at time t, qt against the contact time for MB and orange-red respectively. Adsorption of high initial concentration (2mg/L) results in an increase in the adsorption uptake due to increase in driving force leading to higher adsorption rate [14]. From the observation the rapid equilibrium was reached both dyes at 15 miniatures and completed equilibrium was reached at 60 minutes. No significant additional adsorption was observed after 60 minutes. The contact time was thus maintained for 2 hr to ensure that the equilibrium could be achieved. These results was good agreement with others researchers [13, 31].

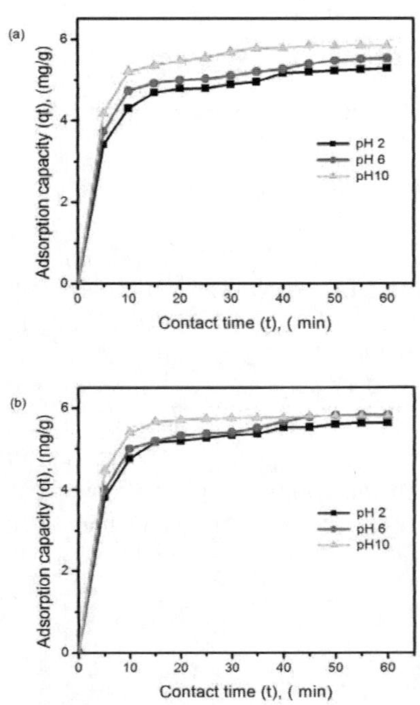

Figure 3 (a-b). The effect of contact time on adsorption capacity (a) MB, (b) orange-.red

3.3.1 Adsorption kinetic study of MB and orange- red

In this research, the equilibrium data for the adsorption of the two dyes with varying concentrations of were fitted to two important equations; Langmuir equation and the Freundlich equation. The Langmuir isotherm could give a good relation between the solid phase concentration at equilibrium, q_e and equilibrium concentration of the dye solution, C_e, as shown in (3).

$$\frac{C_e}{q_e} = \frac{1}{k_L q_m} + \frac{C_e}{q_m} \tag{3}$$

Where C_e (mg/L) is the equilibrium concentration of dye solution, q_e is the amount of dye adsorbed at equilibrium (mg/g), K_L is the adsorption equilibrium constant (L/mg) and q_m is a constant and it reflects a complete monolayer (mg/g). K_L is a constant which links the affinity of the binding sites and q_m represents the maximum adsorption capacity of the magnetic biochar as adsorbent.

Fig. 4(a-b) are the linearized Langmuir models of C_e/q_e against C_e for isotherm studies for the removal of methylene blue and orange red dyes respectively. From these two figures, the values of K_L and q_m can be obtained. According to Fig. 4, the calculated value of q_m was found to be 45.66 mg/g and K_L to have a value of 0.17755 L/mg and for Fig. 4(b), the q_m value was calculated to be 31.26 mg/g and K_L is 0.384 L/mg. The Langmuir isotherm produces good fits over the range of concentration for the removal of these two dye solutions with R^2 values of 0.89 and 0.98 for methylene blue and orange red respectively. Another model of equation to be studied is the Freundlich model. Equation (4) shows the Freundlich equation for heterogeneous surface energy systems.

$$\ln q_e = \ln K_F + \frac{1}{n} \ln C_e \tag{4}$$

Where K_F and n are both the Freundlich constants that will be determined from the plot of ln q_e against ln C_e. K_F and 1/n are interrelated to both the sorption capacity as well as the intensity of the sorption of the system. Besides, from the magnitude of the 1/n term, one can determine the favourability of the sorbent/adsorbate systems [8]. Fig. 5(a-b) are plots of the Freundlich adsorption isotherm for the removal of methylene blue and orange red dyes respectively. From observation it is found that the relationships between the uptake of dyes, q_e (mg/g) and the residual dye equilibrium concentration, C_e (mg/L) based on (6) were shown. Figure 5(a), the value of n is 1.462 and K_F to have a value of 7.422 (mg/g) (mg/L) $^{1/n}$ and the R^2 value is 0.9310. As for the removal of orange red based on Fig.5 (b), the K_F value is 1.462 (mg/g) (mg/L)$^{1/n}$ and n is 1.83. The R^2 value for the removal of this dye was found to be 0.96 and 0.989 for MB and orange red. Table 1 shows isotherm constant for both Langmuir and Freundlich isotherm.

Table 1. Isotherm parameters for dyes removal

Adsorbent	Langmuir Isotherm			Freundlich Isotherm		
	qm (mg/g)	KL (L/mg)	R^2	n	KF (L/mg)	R^2
MB	45.66	0.1775	0.8996	1.462	7.4224	0.9653
Orange-red	31.26	0.3844	0.98	1.832	9.148	0.998

IOP Conf. Series: Materials Science and Engineering **206** (2017) 012081 doi:10.1088/1757-899X/206/1/012081

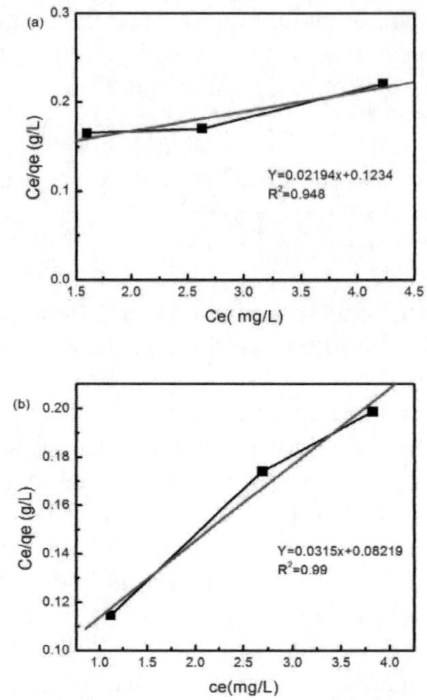

Figure 4. a) Langmuir adsoption MB and b) orange- red

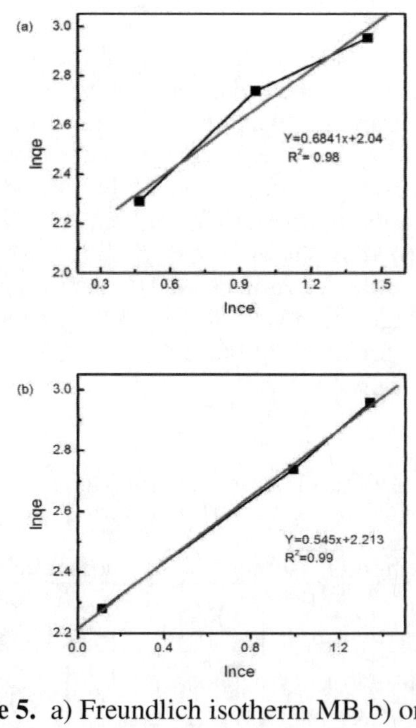

Figure 5. a) Freundlich isotherm MB b) orange- red.

The kinetic studies for the pseudo-first order and pseudo-second order were also carried out in this research. The purpose of carrying out this study is to develop a better understanding on the dynamics of adsorption process. The linearized equations 5 and 6 are both are used to study the pseudo-first order and second order models respectively.

$$\log(q_e - q_t) = \log q_e - (\frac{k_1}{2.303})t \quad (5)$$

$$\frac{t}{q_t} = \frac{1}{K_2.q^2} + \frac{1}{q_e} \quad (6)$$

where qe represents the amount of dye being adsorbed onto the adsorbent at equilibrium (mg/g), qt represents the amount of dye being adsorbed onto the adsorbent at time t (mg/g), K_1 (min-1) is the rate constant for pseudo-first order adsorption and K_2 (g mg-1 min-1) represents the rate constant for the pseudo-second order adsorption. . Besides that, according to the calculated values on Tables 2 and 3, the obtained R^2 values for pseudo-second order that are approaching unity are way higher than that of pseudo-first order. Also, the values of q_e and K_2 for pseudo-second order are proven to be consistent with the pH values

Table 2. Pseudo-first-order and pseudo-second-order kinetics for adsorption of MB and orange red

	MB			Orange Red		
pH	q_e (mg/g)	K_1 (min^{-1})	R^2	q_e (mg/g)	K_1 (min^{-1})	R^2
2	41.5	0.0001	0.7006	27.48	0.0001	0.851
6	41.21	0.0001	0.7216	27.98	0.0001	0.794
10	40.65	0.0606	0.5574	28.22	0.0001	0.896

Table 3. Pseudo-second-order kinetics for adsorption of MB and orange red

	MB			Orange red		
pH	q_e (mg/g)	K_2 (g/mg min)	R^2	q_e (mg/g)	K_2 (g/mg min)	R^2
2	5.48	0.0180	0.998	5.82	0.0246	0.999
6	5.71	0.0183	0.998	5.31	00264	0.999
10	5.96	0.0311	0.999	5.12	0.0223	0.999

3.4. Characterization of dyes Removal

3.4.1 FTIR analysis of dyes

Figure 6 (a-c) shows the FTIR image for the surface functional group of the raw CNT, MB and orange-red. The results indicate that the functional groups have been successfully impregnated on the surface of the adsorbent. The main bonds that represent the vibration of functional groups on the surface of CNTs. Figure 6 (a) shows less functional on the surface of CNTs was observed after adsorption of MB and orange-red as shown in figures 6(b-c). The peak at approximately 3500 cm^{-1}, shows characteristic of an H bonded O–H stretch on the CNT. Peak at 700-610 cm^{-1} represents alkynes group with –C≡C–H: C–H bond in CNT. A sharp peak at approximately 1600-1750 cm^{-1} corresponding to C═O stretching vibration of the carboxylic acid on the CNT. A strong peak appears at 2850 and 3000 cm^{-1} due to the C–H stretching in the adsorbent. The spectra of both dyes displays all characteristic peaks of the bonds in dyes, namely 1150 cm^{-1} due to C–H stretching vibrations in aromatic ring, 1650 cm^{-1} due to C=C stretching vibrations in aromatic nuclei, 2900 cm^{-1} due to C–H stretching vibrations of the crystal violet and a broad peak at 3400 cm^{-1} due to the presence of O–H bonds of the water molecules adsorbed on both dyes. The spectra of Copper metal display the peaks of bonds, 3200-3500cm^{-1} O–H stretch, 1680–1640cm^{-1} –C=C– stretch for alkenes functional group and C-H bond at 900-675cm^{-1} having aromatic functional group. These results was good agreement with other researchers [29, 38].

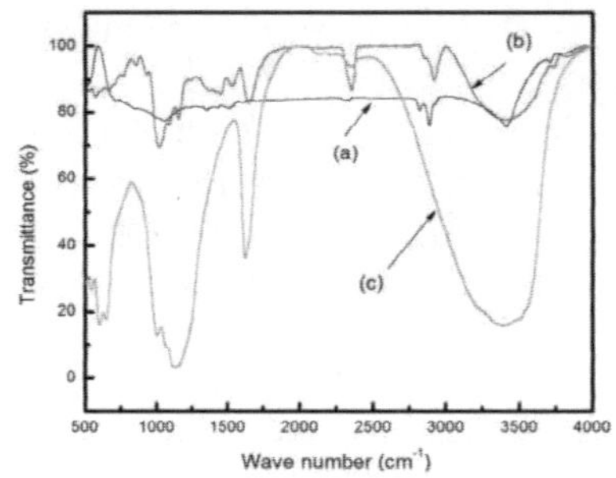

Figure 6. FTIR Images of a) Raw CNTs, b) MB C) orange –red

3.4.2 FESEM Analysis of dye

The CNT used for this research were analysed by using the Field Emission Scanning Electron Microscopy (FESEM) where Figure 7 (a-b) shows the morphology of the CNT. Figure 7 (a) shows the raw image of CNT before adsorption whereas Figure 7(b) shows after adsorption of dyes onto CNT with magnification scale of 1 μm. As it can be observed from Figure 7 (a), it is clearly shown that the surface texture of this raw CNT has a rather smooth surface and few pores can be seen on the surface. Figure 7 (b) shows the tubes after adsorption they open up as some functional groups on the surface of CNTs during adsorption therefore the surface is less smooth compared with the raw CNT. Similar observation was observed in previous researchers [39, 40]

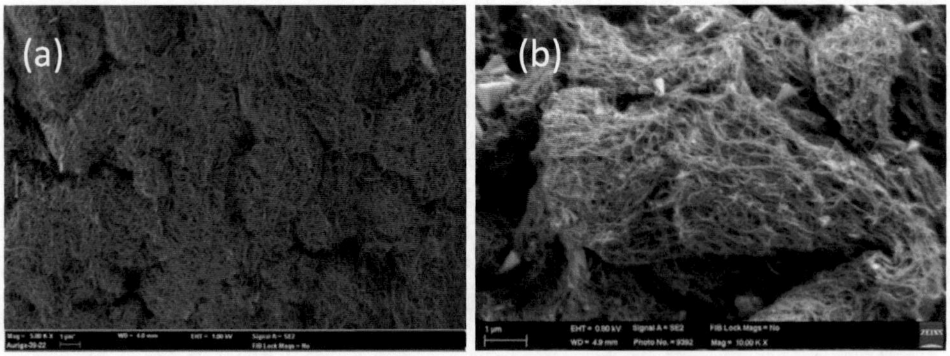

Figure 7. FESEM images of (a) raw CNT and (b) CNT after adsorption from dye.

IOP Conf. Series: Materials Science and Engineering **206** (2017) 012081 doi:10.1088/1757-899X/206/1/012081

4. Conclusion

Carbon nanotube is proven to possess excellent adsorption capacity for the removal of MB and orange red dyes. The process parameters for the maximum adsorption of MB and orange red dyes were pH 10, adsorbent dosage of 1.00 gram, and contact time of 15 minutes respectively. The equilibrium data for this research fitted very well in both Langmuir and Freundlich isotherms but was best described by the Langmuir model with a maximum adsorption capacity of 45.66 mg/g for methylene blue and 31.6 mg/g for MB and orange red respectively. As for the order of the reaction, the rates of adsorption for both dyes were found to comply with the pseudo-second order kinetic model.

References

[1] Arzani K, Ashtiani B G and Kashi A H A 2012 *无机材料学报* **12050**
[2] Muyibi S A, Ambali A R and Eissa G S 2008 *Water Policy* **10** pp193-206
[3] Kumar S, Bhanjana G, Dilbaghi N and Umar A 2014 *J. Nanosci. Nanotechnol.* **14** pp7054-7059
[4] Wang S, Boyjoo Y, Choueib A and Zhu Z H 2005 *Water Res.* **39** pp129-138
[5] Aysu T and Küçük M 2015 *Int. J. Environ. Sci. Technol.* **12** pp2273-2284
[6] Namasivayam C and Arasi D J S E 1997 *Chemosphere* **34** pp401-417
[7] Namasivayam C, Yamuna R T and Arasi D J S E 2002 *Sci. Technol. Sep.* **37** pp2421-2431
[8] Mubarak N, Sahu J, Abdullah E and Jayakumar N 2014 *Sep. Purif. Rev.* **43** pp311
[9] Ruthiraan M, Mubarak N M, Thines R K, Abdullah E C, Sahu J N and Jayakumar N S 2015 *Korean J. Chem. Eng.* **32** pp446-457
[10] Dehghani MH, Mostofi M, Alimohammadi M, McKay G, Yetilmezsoy K, Albadarin AB 2016 *J Ind Eng Chem* **35** pp63-74.
[11] Mubarak NM, Fo YT, Al-Salim HS, Sahu JN, Abdullah EC, Nizamuddin S 2015 *Int J* Nano **14** *1550009.*
[12] Kant R 2012 *Natural science* **4** 22.
[13] Madrakian T, Afkhami A, Ahmadi M and Bagheri H 2011 *J. Hazard. Mater.* **196** pp109-114
[14] Gong J-L, Wang B, Zeng G-M, Yang C-P, Niu C-G and Niu Q-Y 2009 *J. Hazard. Mater.* **164** *pp1517*
[15] Shahryari Z, Goharrizi A S and Azadi M 2010
[16] Sabna V, Thampi SG and Chandrakaran S 2015 *Ecotoxicol Environ. Saf.*
[17] Iijima S 1991 *Nature* **354** pp56-58
[18] Dehghani M H, Mostofi M, Alimohammadi M, McKay G, Yetilmezsoy K and Albadarin AB 2016 *J. Ind. Eng. Chem.* **35** pp63-74.
[19] Onundi Y B, Mamun A, Al Khatib M, Al Saadi M, Suleyman A. 2011 *Int. J. Env. Sci. Technol.* **8** pp799-*806*
[20] AlSaadi M A, Al Mamun A, Alam M Z, Amosa M K and Atieh M A 2016 *Nano* **11** 1650011
[21] Yu J-G, Zhao X-H, Yang H, Chen X-H, Yang Q and Yu L-Y 2014 *Sci. Total Environ.* **482** pp241-251
[22] Ajayan P M 1999 *Chem. Rev.* **99** pp1787
[23] Terrones M 2003 *Annual Rev. Mater. Res.* **33** pp419
[24] Dai L and Mau A W H 2001 *Adv. Mater.* **13** pp899
[25] Rao GP, Lu C and Su F 2007 *Sep. Purif. Technol.* **58** pp224
[26] Yit Thai O, Abdul Latif A, Sharif Hussein Sharif Z and Soon Huat T 2010 *Brazilian J. Chem. Eng.* **27**
[27] Mubarak N, Sahu J, Abdullah E, Jayakumar N and Ganesan P 2014 *Diam. Relat. Mater.* **48** pp52-59
[28] Tabrizi N and Yavari M 2015 *Chem. Eng. Res. Des.* **94** pp516-523
[29] Ghaedi M, Hajati S, Zaree M, Shajaripour Y, Asfaram A and Purkait M 2015 *Adv. Powder Technol.* **26** pp1087-1093
[30] Yasin Y, Hussein M Z and Ahmad F H 2007 *The Malaysian J. Analyt. Sci.* **11** pp400-406
[31] Ai L, Zhang C, Liao F, Wang Y, Li M and Meng L 2011 *J. Hazard. Mater.* **198** pp282-290
[32] Wang S, Boyjoo Y, Choueib A and Zhu Z 2005 *Water Res.* **39** pp129
[33] Ma J, Yu F, Zhou L, Jin L, Yang M and Luan J 2012 *ACS Appl. Mater. Interfaces* **4** pp5749-5760
[34] Gupta V K, Kumar R, Nayak A, Saleh T A and Barakat M 2013 *Adv. Colloid Interface Sci.* **193** pp24-34
[35] Gupta V and Nayak A 2012 *Chem. Eng. J.* **180** pp81-90
[36] Ngah W W, Teong L and Hanafiah M 2011 *Carbohydr. Polym.* **83** pp1446-1456
[37] Belay K and Hayelom A. 2014
[38] Sadegh H, Shahryari-ghoshekandi R, Masjedi A, Mahmoodi Z and Kazemi M 2016 *In.t J. Nano Dimension* **7** pp109-120

[39] Păcurariu C, Paşka O, Ianoş R, Muntean SG 2016 *Clean Technologies and Environmental Policy* **18** pp705-715.

[40] Yao Y, Bing H, Feifei X, Xiaofeng C 2011 *Chemical Engineering Journal* **170** pp82-89.

IOP Conf. Series: Materials Science and Engineering **206** (2017) 012082 doi:10.1088/1757-899X/206/1/012082

Removal of oil and grease from automobile garage wastewater using electrocoagulation

A M Manilal*, M G Harinarayanan Nampoothiri, P A Soloman

Department of Chemical Engineering, Government Engineering College Trichur, Thrissur, Kerala- 680 009, India

E-mail: manilalam@gectcr.ac.in

Abstract. Wastewater from automobile garages and workshops is an important contributor to the water pollution. Oil and grease is one of the major content of wastewater from vehicle garages. Wastewater from a public transport depot at Thrissur district in Kerala, India was collected for the study. A batch reactor has been devised to assess the efficacy of electrocoagulation in removing oil and grease from the wastewater. Aluminium and iron were tested as the anode material with stainless steel as cathode. Experiments were conducted to investigate the effect of various operating parameters such as current density, pH, time and salt concentration on oil and grease removal. The results shown that aluminium is superior to iron in removing the oil and grease from the wastewater. The reactor with aluminium as anode was able to remove 90.8 % of the oil and grease at a current density of 0.6 A/dm^2 in 15 minutes. The calculated specific energy consumption is also less for aluminium in comparison with iron.

1. Introduction

Environmental pollution is a potential threat to living conditions in the earth. Overpopulation has resulted in the larger utilization of natural resources which paved the way for the increase of pollution in the environment. The conservation of the natural sources such as air, water etc. is the responsibility of each human being for their next generations. Water conservation is one area which requires a significant attention. It is reported that in underdeveloped countries about 80% of the water used for irrigation contains pollutants to a level more than their limits [1]. In day to day life water is polluted in many ways. Industries are one of the major sources of water pollution. These include industries like oil and gas production units, oil refineries, edible oil industries, textile industries to name a few [2]. Industries pollute the water with different contaminants. Water from oil and gas industries, petroleum and petrochemical industries contains large amount of oil [3]. One of the major contributor of pollution from such industries is the oil and grease generated and used in their daily operations. Similarly the events like oil spills have affected the balance of an entire life in maritime fields. Kuwaiti oil fire was one of the instances of an oil spill that caused a discharge of 42 Billion US Gallons of oil to the sea; and affected the sea life adversely by contaminating the flora and fauna. Thus the impact of oil content in water is of high significance and separation of oil from water hence become highly important. Another important area where large quantity of oily wastewater generated is the automobile service stations. The automobile service stations use large amount of water for washing and maintenance activities. Thus the possibility of mixing of water and oil particles is unavoidable in the units. Water discharging from the automobile workshops and garages causes water pollution in

IOP Conf. Series: Materials Science and Engineering **206** (2017) 012082 doi:10.1088/1757-899X/206/1/012082

urban areas [4]. Water is consumed in large amounts (1000-25000 liters) in these service stations. Water treatment is unnoticed in most of these vehicle garages and workshop.

There are several methods available for oil removal from effluent, such as physical treatment, chemical treatment, biological treatment, membrane treatment etc. [3]. Physical treatment requires longer time, and has lower removal efficiency. Moreover they cannot remove colloidal and dissolved matters effectively. Chemical treatments include chemical precipitation, chemical oxidation, photo catalytic treatment, ozone treatment etc. These processes have widely used in industrial plants. But these methods need further treatment mechanisms for the removal of large amounts of sludge and require the help of trained operators. Biological treatments and membrane treatments also demand for skilled labor and overall efficiencies of treatment plants are also a cause of concern.

Electrocoagulation technique (EC) is one of the modern technologies in wastewater treatment. It is the combination of chemical and physical treatments [5]. It is an electrochemical method which applies the basic precepts of traditional water treatment methods, coagulation and flocculation [6]. In contrast to conventional coagulation method in which particles known as coagulants are added to aggregate the pollutant, EC uses electrodes to release coagulants. Electrocoagulation has been under research for removal of pollutants from wastewater. Electrocoagulation has the advantages like low sludge production, high process efficiency, and the effluent need no further treatment. The viability of electrocoagulation is being investigated in several studies which includes the treatment of wastewater including dairy, dye and textile, heavy metals containing wastewaters and wastewaters from the pulp and paper mill industry treatment [7-9]. Hence the possibility of electrocoagulation treatment for smaller water treatment units like automobile garages is investigated.

A method to treat the wastewater using electrocoagulation is proposed for the wastewater from vehicle garages. The used water from Kerala State Road Transport Corporation (KSRTC) Thrissur depot is taken for analysis. KSRTC depot Thrissur uses large quantities of water (5000 liters per day) and large quantity of oil contained wastewater is brought forth. Treatment of this wastewater has large scope because recycling and reuse of the same water can save large amounts of water.

The objective of the present research is to examine the feasibility of electrocoagulation in treating automobile garage wastewater and to determine the optimum operational conditions.

2. Mechanism of electro-coagulation

Electrocoagulation (EC) is a complex process involving many chemical and physical phenomena. It is carried out in an electrolytic cell/ reactor where a DC voltage is applied across the electrodes. The anode used is a sacrificial electrode, it provide ions into the wastewater stream to be treated. The ions which are the major contributor for coagulation are produced 'in situ' by the dissolution of anode. Broadly, the phenomenon of EC involves three successive stages: (i) electrolytic oxidation of the 'sacrificial electrode' and thus the formation of coagulants, (ii) destabilization of the contaminants and breaking of emulsions and (iii) aggregation of the destabilized phases to form flocs [10]. The coagulants remove the dissolved organic part by precipitation and suspended fines by adsorption and density separation.

In the present research two different anode materials were studied to investigate its capability in oil removal process. Mild steel and aluminium were the materials selected as the anode and stainless steel is used as the cathode. The electrochemical reactions with iron as the anode material are presented in the following equations [11]:

Anode:

$$Fe \rightarrow Fe^{3+} + 3e^- \quad E^0 = -0.04\,V \tag{1}$$

$$Fe \rightarrow Fe^{2+} + 2e^- \quad E^0 = -0.44\,V \tag{2}$$

$$Fe^{2+} \rightarrow Fe^{3+} + e^- \quad E^0 = -0.77\,V \tag{3}$$

IOP Conf. Series: Materials Science and Engineering **206** (2017) 012082 doi:10.1088/1757-899X/206/1/012082

Cathode:

$$2H_3O + 2e^- \rightarrow H_2 + 2H_2O \qquad pH > 7 \tag{4}$$

$$2H_2O + 2e^- \rightarrow 2OH^- + H_2 \qquad pH > 7 \tag{5}$$

Bulk:

$$4Fe^{2+} + 2H_2O + O_2 \rightarrow 4Fe^{3+} + 4OH^- \quad pH > 7 \tag{6}$$

$$4Fe^{2+} + 4H^+ + O_2 \rightarrow 4Fe^{3+} + 2H_2O \quad pH > 7 \tag{7}$$

The generated Fe^{3+} (aq) ions will immediately undergo further spontaneous reactions to produce corresponding hydroxides and/or polyhydroxides [12]. The metal hydroxide flocs formed can ensure the removal of the organic matter present in the solution predominantly by precipitation and/or adsorption mechanism and subsequent density separation. The hydrogen gas generated as a result of the cathode reaction can assist the flocculated particles to float to the surface. The monomeric forms of metal hydroxides can get polymerized giving various forms of ferric hydroxo complexes viz., $Fe(H_2O)_6^{3+}$, $Fe(H_2O)_5(OH)^{2+}$, $Fe(H_2O)_4(OH)^{2+}$, $Fe_2(H_2O)_8(OH)_2^{4+}$, $Fe_2(H_2O)_6(OH)_2^{4+}$, $Fe(OH)_4^-$, etc. The polyhydroxide complex (Fe $(OH)_n$) formed remains in the aqueous stream as a gelatinous suspension, which can remove the pollutants from wastewater either by complexation or by electrostatic attraction, followed by coagulation [13].

When aluminium is used as the electrode material the reactions are as follows:

Anode:

$$Al \rightarrow Al^{3+} + 3e^- \tag{8}$$

Cathode:

$$3H_2O + 3e^- \rightarrow \frac{3}{2}H_2 + 3OH^- \tag{9}$$

Bulk:

$$Al + 3H_2O \rightarrow Al(OH)_3 + 3H^+ \tag{10}$$

$$nAl(OH)_3 \rightarrow Al_n(OH)_{3n} \tag{11}$$

The aluminium hydroxides formed in the above reactions are insoluble and will precipitate along with the suspended or colloidal matter present in the wastewater. All the other mechanism specified above *viz.* coagulation, adsorption, floatation etc. will contribute to the pollutant removal process.

The amount of metal ions (m_{metal}, g) generated can be calculated from Faraday's law (Eq. (12)). The current density (i, Am^{-2}) (Eq. (13)) can be calculated as the current in the EC cell (I, A) per effective anode area below the water surface (A_{eff}, m^2).

$$m_{metal} = \frac{ItM}{nF} \tag{12}$$

$$i = \frac{I}{A_{eff}} \tag{13}$$

Where t is the treatment time of the EC process (s), M the molar mass of the metal concerned (g mol^{-1}), n the number of electrons in the oxidation reaction and F Faraday's constant (96,500 C mol^{-1}). It has been experimentally observed that the generation of metal ions has been higher than the value calculated by Faraday's law. This phenomenon has been referred as superfaradaic efficiencies.

IOP Conf. Series: Materials Science and Engineering **206** (2017) 012082 doi:10.1088/1757-899X/206/1/012082

The method by which the Fe ion affects the impurities removal is generally explained in terms of two distinct mechanisms: one is the charge neutralization of negatively charged colloids by cations and the other is incorporation of impurities in the amorphous (mono and/or poly) hydroxide precipitate as sweep flocculation. Other contributing factors influencing the efficiency of the EC process are: electrode materials, current density and treatment time, composition of wastewater and the type of pollutants to be removed, conductivity, electrode gap, passivation of the anode, temperature, solution chemistry, etc.

3. Materials and methods

3.1. Wastewater Characteristics

The samples were collected from the workshop of the KSRTC Thrissur depot. Around 5000 litres of water per day is used for different purposes in the KSRTC workshop. The samples were analyzed for its chemical oxygen demand (COD) and oil & grease by fluorescence spectrometry. F-4660 FL, HITACHI Fluorescent spectrometer is used for the analysis. The fluorescence peak is held between 300-340 nm. The excitation wavelength of oil is 337 nm [14]. The other parameters estimated were given below.

Table 1. Wastewater characteristics.

Parameter	Values
COD	1450 mg/L
pH	6.5
Fluorescence analysis for oil	1260 mg/L
Total Solids	363 mg/L
Total Dissolved solids	225 mg/L
Total Suspended Solids	138 mg/L
Density	981g/cm^3

3.2. Electrocoagulation Set Up

Experimental set up for electrocoagulation is shown in figure 1. A reactor with a capacity of 2 litres and having dimensions 20cm×10cm×10cm was selected. The reactor was made of plastic. Two different anodes were used for the study *viz.* mild steel and aluminium. The cathode was made up of

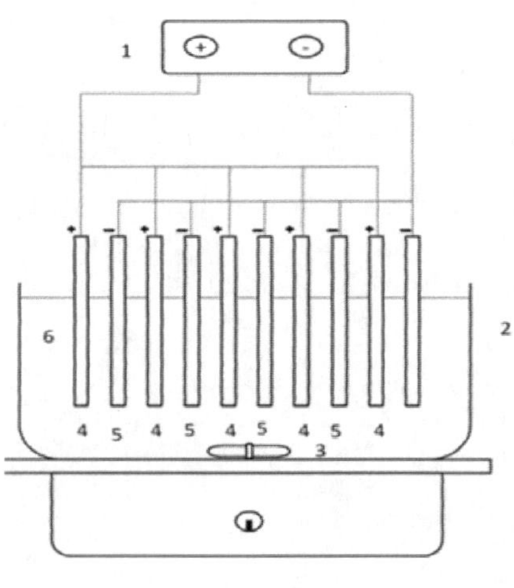

1. Power supply unit
2. Reactor
3. Magnetic pellet
4. Anodes
5. Cathodes (stainless steel)
6. Wastewater to be treated

Figure 1. Schematic of EC reactor

stainless steel and are aligned vertically and parallel to each other with an inter electrode gap of 1 cm. Five pairs of electrodes, of active area (8 cm×4 cm) 40cm^2 each, were used. The electrode plates were cleaned manually by abrasion with sand paper and by the treatment with 10 % hydrochloric acid followed by washing with distilled water prior to every run. All experiments were performed at a room temperature of 30 °C.

Total electrode area for 10 electrodes was 400 cm^2. Number of electrodes were selected and arranged with minimum gap with the objective for efficient EC operation. A DC power supply was used to provide current and potential across the electrodes. APLABS regulated DC power supply was used for the purpose. The input voltage specification was 0-230 V AC, and output voltage was 0-30 V DC .The current output of the power supply specified at 2 A. The power supply had short circuit protection which helped in protecting it from probable contacts with electrodes. A magnetic stirrer was provided in the reactor with varying rotational speeds.

4. Results and discussions

4.1. Calibration of Fluorescence Spectrometry

Fluorescence spectrometer was used for finding the oil content from the fluorescence values of different emission wavelengths. The excitation wavelength of oil is found to be 337 nm. Fluorescence spectrometer generates the spectrum. The emission wavelength is given in the range of 300-500nm. Figure 2 shows the fluorescence emission spectra obtained for a sample.

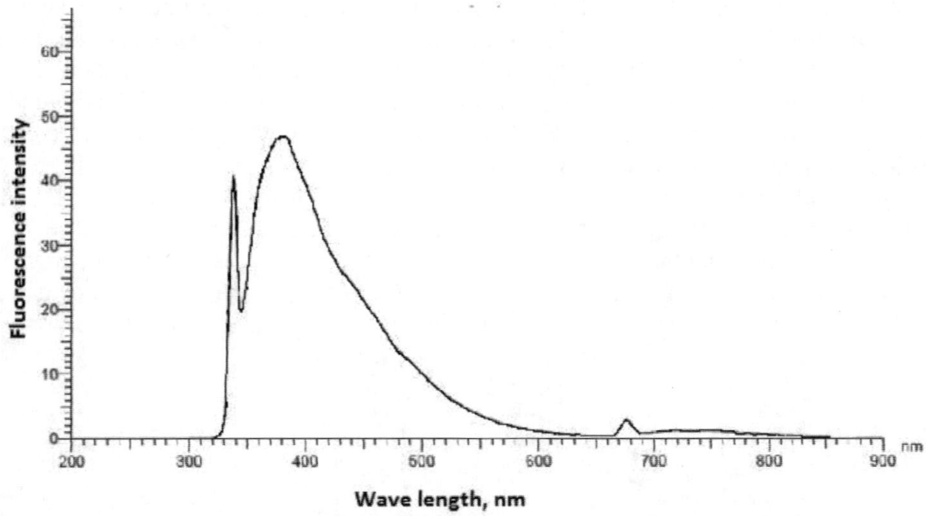

Figure 2. Fluorescence emission spectrum for oil in water emulsion

Fluorescence was calibrated in terms of concentration, ppm. The calibration was done with known concentrations. The calibration curve is shown in figure 3.

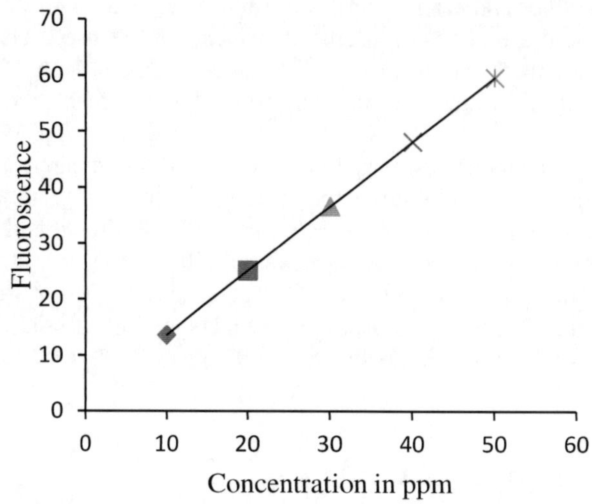

Figure 3. Calibration curve for fluorescence analysis

4.2. Effect of current density on electrocoagulation

Current density is one parameter which can be varied externally .The term current density is the current loaded per unit area. It was observed that there was a direct relationship between current density and the percentage oil removal. The oil removal increases with respect to an increase in the current density. Figure 4 shown below gives the effect of current density on percentage oil removal for the EC using Al electrode and Fe electrode. This is very much in accordance with the Faraday's law of electrolysis. It is evident from equation (12) that the release of anode material in an electrochemical process is directly proportional to the current density.

The consumption of electrode material with the current density were calculated. This is again strongly influenced by operating conditions and wastewater characteristics. Figure 5 shows the influence of current density on the amount of anode material released. As clearly indicated by equation (12) there exists a linear relationship between the current density and the amount of anode material released. Since the equivalent weight of iron is greater than the equivalent weight of aluminium, as per

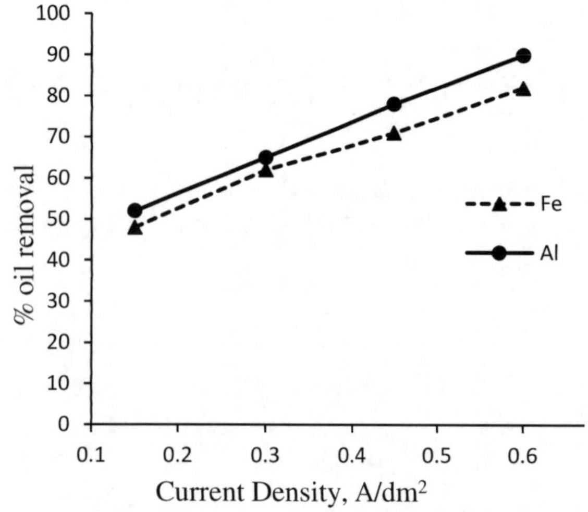

Figure 4. Effect of current density on oil removal (pH 7.5 and EC time 15 min).

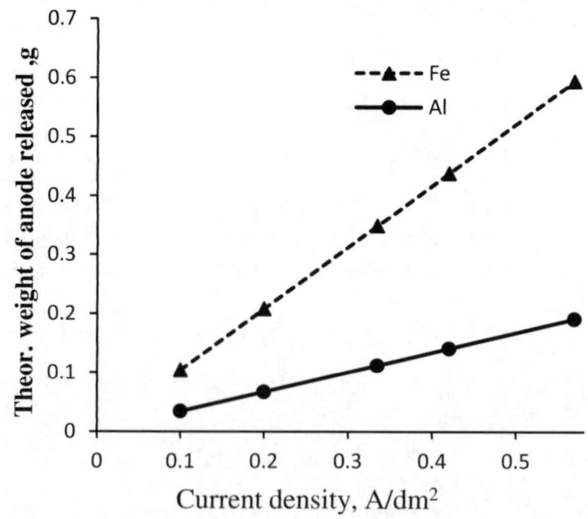

Figure 5. Theoretical weight of anode material released with current density (pH 7.5 and EC time 15 min).

Faradays law the weight of the anode material released is more for iron than aluminium for the same amount of charge passed. In few cases the actual electrode material consumption went beyond the theoretical value, a phenomenon reported as superfaradic efficiency.

4.3. *Effect of pH on electrocoagulation*

Initial pH has significant role in the performance of electrochemical process [15]. Its effect has been examined by adjusting the pH of the wastewater to the desired value for each experiment by using sodium hydroxide or sulphuric acid. Figure 6 depicts the influence of wastewater pH in percentage oil removal. Oil removal was observed to be varying between pH from 4 to 10. In both cases the maximum removal was observed at near neutral pH.

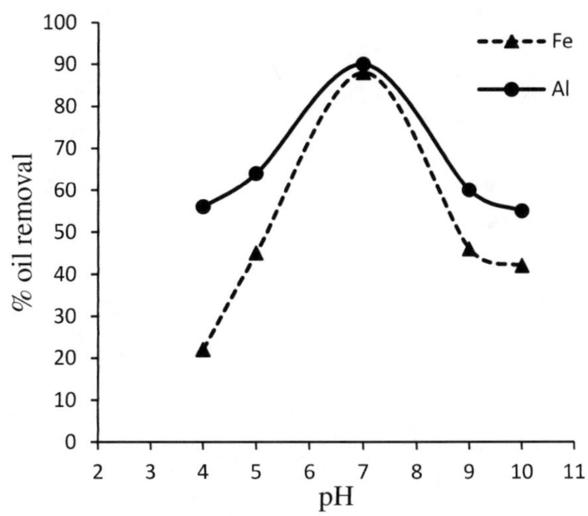

Figure 6. Effect of influent pH on oil removal (current density 0.57 A/dm^2 and EC time 15 min).

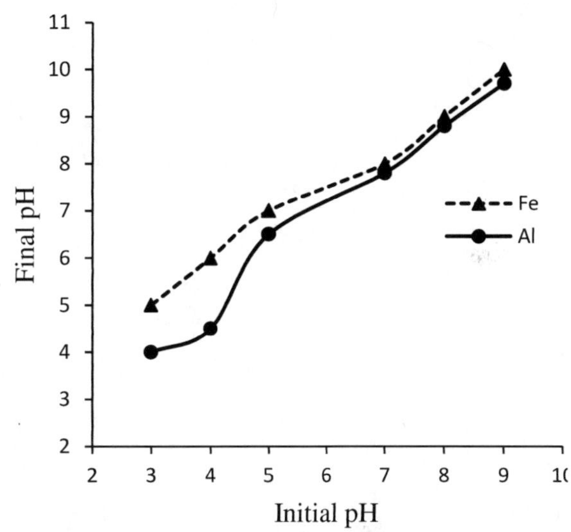

Figure 7. pH change after electrocoagulation (current density 0.57 A/dm^2 and EC time 15 min).

Such a radical change in the percentage removal for different pH may be due to the different chemical reactions happening inside the EC reactor for different pH's. Compounds like $Fe(OH)_2$,$Fe(OH)_3$, etc in the case of mild steel electrode and $Al(OH)^{2+}$,$Al(OH)_2{}^{+}$,$Al(OH)_2{}^{4+}$,$Al(OH)_3$ etc. in the case of aluminium electrodes were reported to be formed at different pH ranges [16]. The removal efficiencies remained much unchanged at pH less than 4 and more than 10.

This is because compounds like $Al(OH)^{4-}$ is forming at higher H. It shows amphoteric behavior so that less adsorption of pollutant occurs. At lower pH generation of Al^{3+} is more which also reduces the efficiency. Intial pH is an important parameter for EC using mild steel electrodes also. The reactions of mild steel electrodes were different at different pH values. More hydroxide formation can be explained as a reason for higher removal efficiencies around pH 7.The solubility of pollutants were less at neutral pH which aided in the adsorption at coagulant surfaces.

The final pH of the wastewater sample was found to increase during the EC process. When the initial influent pH was low this increase was found marginal. Towards the higher influent pH there was not much increase in the effluent pH. The results are shown in Fig. 7. The increase in pH may be due to the release of OH^- ion during the production of hydrogen gas at cathode.In addition, if some anions such as, Cl^-,SO_4^{2-},$HSiO_4^-$,NO_3^- are present in the influent wastewater then they can exchange the OH^- ions in $Al(OH)_3$ to release OH^- and this will result in pH increase.

IOP Conf. Series: Materials Science and Engineering **206** (2017) 012082 doi:10.1088/1757-899X/206/1/012082

The reduction in pH change at high influent pH is attributed to the reaction of $Al(OH)_3$ with OH^- ions at very high pH to form $Al(OH)_4^-$, leading to a reduction in pH. Moreover oxygen evolution reaction also leads to pH increase.

4.4. Effect of salt concentration on EC

Electrochemical processes depends on the conductivity of the electrolyte. Hence NaCl was added to the influent wastewater to improve its conductivity. By increasing NaCl concentration the conductivity was increased and this in turn increased the current passing through the electrochemical cell and thereby increasing the current density. The effect on percentage removal for different NaCl concentrations is shown in figure 8.

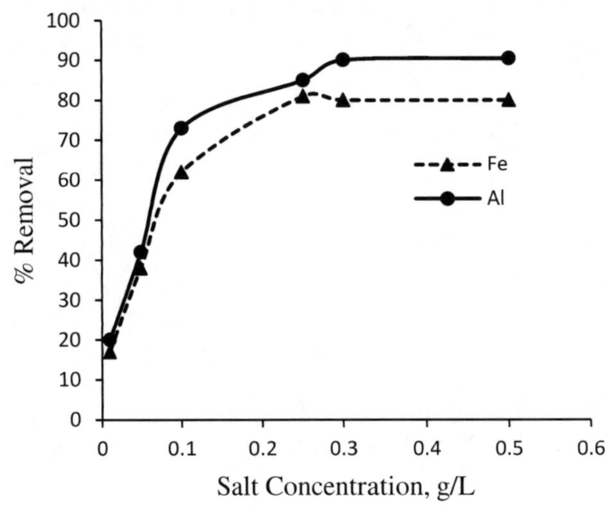

Figure 8. Effect of salt concentration on oil removal (current density 0.57 A/dm^2 and EC time 15 min).

The oil removal was increased as the salt concentration increased and reached the maximum removal of 81% for mild steel electrode at around a salt concentration of 0.25 g/L and 90.4% for the aluminium electrode at around a salt concentration of 0.3 g/L. Addition of NaCl has additional advantages along with increase in conductivity that the chloride ions can aid disinfection of the wastewater.

4.5. Effect of time on oil removal by EC

The time given for EC is an important parameter which influences the oil removal. The influence of time is examined by conducting the experiments for different times starting from 3 minutes to 20 minutes, keeping the other parameters fixed at the following values *viz.* current density 0.5 A/dm^2, salt composition 0.5 g/L, and initial pH 7. The results obtained are shown in figure 9. After 15 minutes it was observed that the percentage removal was attaining a maximum around 90% for the aluminium electrode and 85% for the mild steel electrode.

In the EC process the anode materials produces metal ions during electrolysis. As discussed earlier this metal ions are the primary motive for the destabilization and the resulting removal of the colloidal matter present in the wastewater. If the time provided for the EC was low then the metal ion dosage was not enough to destabilize the colloidal matter and hence resulted in low percentage oil removal. Furthermore the higher value of percentage removal was attributed to the overall progression of electrolytic process comprising electro coagulation, electrolytic oxidation and the floatation.

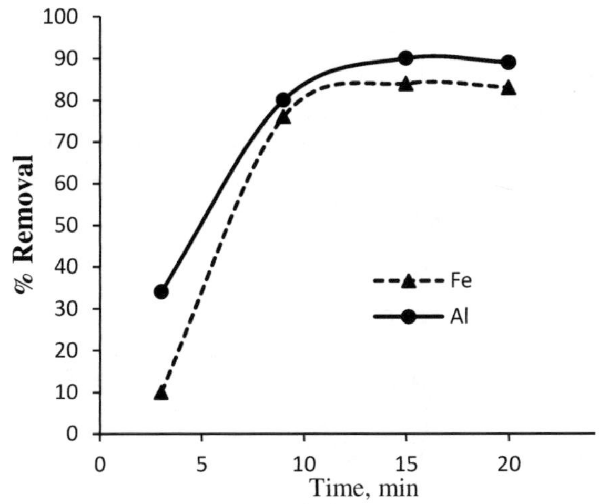

Figure 9. Effect of time on oil removal (current density 0.57 A/dm^2 and pH 7.5).

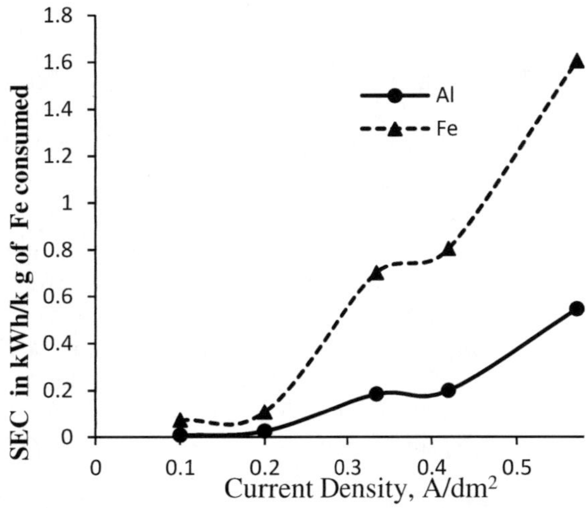

Figure 10. Effect of current density in SEC.

4.6. Specific Energy Consumption (SEC)
The study can be made more meaningful by estimating the energy consumptions for the process. This was examined in detail by estimating the specific energy consumption (SEC) based on the amount of anode dissolved (SEC in kWh/(kg of Fe), using the following equation:

$$\text{Specific Energy Consumptio n} = \frac{\text{Energy Consumed}}{\text{kg of Fe consumed}} \tag{14}$$

The figure 10 shows the SEC calculated against different current densities. It is evident from the figure that the SEC for the EC process is fairly less for aluminium anode in comparison with iron anode for the same current density. This information can be very much useful in making decisions on the selection of anode material for EC.

5. Conclusions

The viability of EC process for the removal of oil and grease from an automobile garage was investigated. The experiments were done with two different anode materials *viz.* mild steel and aluminium. In both cases stainless steel was used as the cathode. The influence of various operating parameters such as current density, pH, salt concentration and time were examined. Effective removal of the oil and grease from the wastewater were observed during the study. Compared to mild steel anode aluminium anode has given better removal. The optimum operating conditions for the given EC system with aluminium anode were estimated as current density 0.6 A/dm^2, pH 7, EC time 15 min and a salt composition of 0.3 g/L. The same for the EC system with mild steel anode were current density 0.57 A/dm^2, pH 7, EC time 17 min and a salt composition of 0.25 g/L. The calculated SEC is also in favour of EC with aluminium anode as its SEC is far below than that of iron anode. Based on the results obtained from the present experiments it may be concluded that EC with aluminium anode and stainless steel cathode may be considered as a good treatment option for treating automobile garage wastewater. Also among these two anode materials used aluminium stands superior to iron in performance.

References

[1] Rosenstock L 2003 *Environ. Health Persp.* **111**(7) A376-A377
[2] Kromm D E 1973 *Annals Assoc. American Geo.* **63**(2) 208-217
[3] Fakhru'l-Razi A, Pendashteh A, Abdullad L C, Biak D R A, Madaeni S S and Abidin Z Z 2009 *J. Hazard. Mater.* **170** 530-551
[4] Mara D and Cairncross S 1989 *World Health Organization* Geneva 187
[5] Mollah M A, Schennac R, Parga J R and Cocke D L 2001 *J. Hazard. Mater.* **84** 29–41
[6] Debabrata Mazumder and Somnath Mukherjee 2011 *Int. J. Environ. Sci. Dev.* **2** (1) 64-69
[7] Chen X, Chen Guohua and Po Lock Yue 2000 *Sep. Purif. Technol.* **19** 65–76
[8] Isa M H, Ezechi E H, Ahmed Z, Magram S F and Kutty S R M 2014 *Water Res.* **51** 113 - *123*
[9] Soloman P A, Basha C A, Velan M, Balasubramanian N and Marimuthu P 2009 *Sep. Purif. Technol.* **69** 109–117
[10] Mollah M Y A, Morkovsky P, Gomes J A G, Kesmez M, Parga J and Cocke D L 2004 *J. Hazard. Mater.* **114** 199–210
[11] Mollah M Y A, Robert Schennach, Jose R Parga and David L Clarke, *J. Hazard. Mater.* **B84** 29-41
[12] Xu X and Zhu X 2004 *Chemosphere* **56** 889–894
[13] Chen X, Chen G and Yue P L 2000 *Sep. Purif. Technol.* **19** 65–76
[14] Donald C Gordan Jr, Kaizer P D and Jacqueline Dale 1974 *Marine Chemistry* **2** 251-261
[15] Zhao S , Huang G , Cheng G, Wangc Y and Fua H 2014 *Desalination* **344** 454–462
[16] Yavuz Y, Koparal A S and Ogutveren U B 2010 *Desalination* **258** 201–205

Preparation of PVDF/SiO_2 composite membrane for salty oil emulsion separation: Physicochemical properties changes and its impact on fouling propensity

H P Ngang*, A L Ahmad, S C Low and B S Ooi*

School of Chemical Engineering, Engineering Campus, Universiti Sains Malaysia, Seri Ampangan 14300, Nibong Tebal, Pulau Pinang, Malaysia

Email: hueyping0404@gmail.com (H P Ngang), chobs@usm.my (B S Ooi)

Abstract. In this study, polyvinylidene fluoride (PVDF)/silica (SiO_2) composite membranes were prepared by diffusion induced phase separation through direct blending method. The roles of SiO_2 particles concentration on membrane physicochemical properties were evaluated through oil emulsion separation under high ionic strength environment whereby hydrophobic interaction is prevalent. Membranes were characterized using field emission scanning electron microscope (FESEM), atomic force microscopy (AFM), contact angle measurement, membrane porosity and pore size distribution. It was expected that by adding the monodispersed SiO_2, it will render the membrane with hydrophilic characteristic. However, it is concomitantly changing the physical properties of the membrane. Addition of SiO_2 caused the changes to the physicochemical properties of the composite membrane and its effects on the fouling propensity were evaluated. It was found that the mean pore size of the membranes increased with the increase of SiO_2 concentration. The addition of hydrophilic SiO_2 had accelerated the precipitation of the membrane dope solution resulting in changes of membrane cross section morphology. FESEM images showed the membrane cross-section morphology of PVDF/SiO_2 composite membrane had gradually changed from finger-like to macrovoid-like structure with the increased of SiO_2 concentration. The hydrophilicity of the PVDF/SiO_2 composite membrane was enhanced which is a desired property for water purification. However, the changes in physical properties (pore size, porosity, and surface roughness) had played more dominant role in the oil emulsion fouling behaviour rather than hydrophilicity enhancement. Due to the salting out effect under high ionic strength environment, hydrophobic interaction played an important role in the oil adsorption. The increment in membrane pore size, porosity, and surface roughness after incorporation of SiO_2 particles had encountered more serious relative flux reduction and lower flux recovery ratio.

1. Introduction

Produced-water is the largest waste stream generated in oil and gas industry [1]. The global produced-water production was estimated to be ~ 250 million barrels per day in year 2007 [2]. It contains mixture of organic and inorganic compounds including dissolved minerals, dispersed oil compound, suspended particles, chemical compound such as corrosion inhibitor, dissolved gases, and surfactant [2]. Discharging produced-water without treating it can pollute surface and underground water soil, which leads to serious environmental problems. The maximum total oil and grease concentration in discharge waste was set by environmental regulation to be 10-15 mg/L. However, major oil fields industries

IOP Conf. Series: Materials Science and Engineering **206** (2017) 012083 doi:10.1088/1757-899X/206/1/012083

generate produced water with concentration range between 100 and 1000 mg/L or higher depending on the nature of the crude oil [3, 4].

Many researchers have focused on treating oily saline produced-water in order to meet environmental regulations as well as reuse and recycling of produced-water. Produced-water is conventionally treated through various physical, chemical, and biological methods which include gravity separation and skimming, air flotation, centrifugation, fibrous/packed bed coalescence, de-emulsification and flocculation, which have the intrinsic disadvantages such as low efficiency, high operating cost as well as corrosion problems [5-7]. Most of the conventional treating method cannot effectively treat stable oil emulsion (size < 20 μm) [3, 8]. Therefore, it is appropriate to incorporate membrane technology to refine the final effluent. The use of membrane technology offers a potential solution to the problems of micro or submicron sized of oily wastewater. Ultrafiltration membrane was considered a promising method for treating oily wastewater due to it suitable pore size (0.002-0.05 μm), as compared to common oil droplet size in oily wastewater (0.1-10 μm), which has the capability to remove emulsified oil droplet without any de-emulsification processes [9].

Polyvinylidene fluoride (PVDF) has been widely used for ultrafiltration membrane due to its outstanding thermal stability, chemical resistant, as well as high mechanical strength. However, its low surface energy and hydrophobicity often leads to severe fouling problem and decline in permeability which had limited their application in water treatment process [10, 11]. Therefore, it is necessary to modify the PVDF membrane to improve its membrane hydrophilicity. Membrane surfaces modification by altering their hydrophilicity, surface charge and morphology, are some of the practical ways to reduce membrane fouling propensity [12]. It is well known that nanoparticles can be used to improve the capabilities of polymers in certain extent due to their small size and large surface to volume ratio. Thus, organic-inorganic composite membranes have become an emerging technology for improving their membrane performances, at the same time it can acquires benefit from both organic and inorganic materials such as superior hydrophilicity, permeability and selectivity [13].

Nanoparticles that have been used to modify organic membranes include titanium dioxide (TiO_2), alumina (Al_2O_3), silica (SiO_2), etc [14-18]. Among different inorganic materials used, silica (SiO_2) has received most attention owing to its convenient operation, mild reactivity, good thermal and mechanical stability, and well known chemical properties [19]. These studies showed that the blending of the hydrophilic inorganic materials into membrane matrix increased the surface hydrophilicity and reduce potential adsorption of foulant. However, it is known that the addition of inorganic components as additive to a polymer dopes solution did play an important role in affect the formation of membrane structure. The final membrane morphology varies greatly in term of membrane pore size, porosity, hydrophilicity as well as membrane surface roughness which are the controlling factors for membrane fouling.

The aim of this paper is to study the role of SiO_2 particles towards membrane physicochemical properties changes and its fouling evaluation using high salinity oil emulsion as the foulant. The physicochemical properties of prepared membranes were expected to change during high salinity oil emulsion separation due to exclusion of charge repulsion and salting out effect, which could indirectly affect the fouling propensity.

2. Experimental

2.1. Materials

Tetraethylorthosilicate (TEOS, 99.999 %), sea salts, and Triton X-100 were purchased from Sigma Aldrich. Ethanol, ammonium solution 25 %, and Dimethylformamide (DMF) were purchased from Merck, Germany. Polyvinylidene fluoride (Solef® PVDF) was supplied by Solvay Solexis, France. Crude oil was supplied by Petronas Malaysia. PVDF was dried in oven at 70 °C overnight prior to use, while other organic chemicals were obtained in reagent grade and used as received.

IOP Conf. Series: Materials Science and Engineering **206** (2017) 012083 doi:10.1088/1757-899X/206/1/012083

2.2. Preparation of silica particles

Monodispersed silica particles were synthesized using Stöber method [20]. Firstly, 200 mL of ethanol, 22 mL of deionized water and 12 mL of ammonium solution 25 % were added together into 500 mL three neck flask and stirred using magnetic stirrer at 350 rpm. Next, 12 mL of TEOS was added to the mixture at 30 °C. The reaction was continued for 24 h. The produced silica particles were purified by three cycles of centrifugation and re-suspension in deionized water. The SiO_2 particles were then re-dispersed in DMF solvent and the particle concentration was determined using gravimetric method prior to use.

2.3. Membrane Preparation

The predetermined amount of SiO_2 particles suspension was dispersed in DMF solvent under sonication for 15 min. 18 wt.% of PVDF powder was dissolved into the SiO_2 particles suspension and stirred at 65 °C for 4 h. The solution was left to be stirred overnight at 40 °C to form a homogenous solution. The final solution was then subjected to degassing for 10 min to remove all the trapped bubbles and let it stand for 30 min prior to membrane casting. The solution was cast on the tightly woven polyester sheet using an automatic film applicator (Elcometer 4340, E.U.). The membrane casting nominal thickness was fixed at 200 μm and cast at constant speed of 0.06 cm/s. It was then immediately immersed into the water bath of distilled water (25 °C) and leave it overnight in order to remove the DMF solvent residual. The detail of membrane synthesis parameter used was summarized in Table 1.

Table 1. Detail of the membrane synthesis parameter.

Membrane	PVDF (wt.%)	DMF (wt.%)	SiO_2 (wt.%)
M18-0	18	82.0	0.0
M18-0.1	18	81.9	0.1
M18-0.5	18	81.5	0.5
M18-1.0	18	81.0	1.0
M18-2.0	18	80.0	2.0

2.4. Preparation of oil emulsion

Oil emulsion was freshly prepared before every filtration test. Firstly, saline solution was prepared by dissolving 32 g of sea salts in 1 L of distilled water under stirred condition. Subsequently, 50 mg of crude oil was added into the prepared salty solution to obtain 50 mg/L high salinity oil emulsion. Surfactant (Triton X-100) was then added into the solutions to stabilize the oil emulsion. The prepared oil emulsion was then dispersed using a blender (Khind BL-1515, power = 300 W) for 2 min. Finally, the dispersed solution was sonicated for 5 min to remove any air bubble which might exist during the dispersing process.

2.5. Characterizations

2.5.1. Field emission scanning electron microscope. The membrane surface and cross sectional morphology were observed using Field Emission Scanning Electron Microscope (FESEM, Quanta Feg 450, FEI, USA) at an accelerating voltage of 5 kV. The membranes were immersed into liquid nitrogen and cryogenically cracked for membrane cross-sectional observation to prevent membrane structure rupture.

2.5.2. Pore size distribution. Capillary Flow Porometer, Porolux 1000 (Benelux Scientific, Belgium) was used to determine the pore size distributions of the prepared membranes. Perfluoroethers was used as a pore wetting agent. Three membrane samples (diameter of 20 mm) were characterized and analysed using the LabView software to minimize the experimental error.

2.5.3. Contact angle measurement. Water contact angle instrument (Rame-Hart Model 300 Advanced Goniometer) based on sessile drop methods was employed to measure the membrane surface wettability. The acquired images were analysed using DROPimage software to obtain the contact angles.

2.5.4. Atomic force microscopy. The surface topography and roughness of the membranes in the dry state were analysed by using atomic force microscope, AFM (Park Scientific, Korea, XE-100). The membrane areas of 20 μm × 20 μm were scanned by non-contact mode.

2.5.5. Dynamic light scattering analysis. A dynamic light scattering apparatus (Malvern Zetasizer Nano ZS90) was used to measure the size of the oil droplets in oil emulsion with refractive index of 1.47 and 1.335 for the crude oil and dispersant (sea salt solution), respectively. The refractive index of 1.46 (SiO_2) and 1.428 (DMF) were applied to measure the hydrodynamic diameter (D_h) of SiO_2 particles in DMF solvent.

2.5.6. Membrane porosity. The membrane porosity (ε_p) represents the total volume of the pores divided by the total volume of the porous membrane. Five pieces of square-sized membranes with an area of 6.25 cm^2 each were dried in an oven at 60 °C until a constant weight was observed. Then, the membrane dry weight was measured. The dried membranes were then immersed into 2-butanol and degassed for 30 s at room temperature to prevent air trapped in the membrane internal pores and let in room temperature for 2 h. After that, the membrane surface was dried using filter paper and weighed immediately to avoid evaporation of 2-butanol from membranes internal pores. The membrane porosity was calculated using equation (1),

$$\varepsilon_p = (\frac{w_1 - w_2}{\rho_b})/(\frac{w_1 - w_2}{\rho_b} + \frac{w_2}{\rho_p}) \tag{1}$$

where ε_p is the membrane porosity (%), w_1 is the weight of the wet membrane (g), w_2 is the weight of the dry membrane (g), ρ_p is the specific gravity of the PVDF polymer (1.78 g/cm^3) and ρ_b is the specify gravity of 2-butanol (0.81 g/cm^3). 2-butanol was chosen as a wetting liquid because it did not swell the PVDF membrane and could wet the hydrophobic PVDF membrane well to ensure complete pores filling.

2.6. Membrane permeation test

The membrane performance in high salinity oil emulsion separation was evaluated using cross-flow ultrafiltration (UF) unit (Figure 1). The membrane chamber (Sterlitech, CF042) has an effective area of 0.0042 m^2. The pure water and oil emulsion were re-circulated at a constant flow rate of 0.4 L/min using a peristaltic pump (Flex-Pro A4 Series). The membranes were subjected to 30 min compression at transmembrane pressure (TMP) of 1.5 ± 0.1 bar prior to each filtration process. After the membrane compression process, the pure water permeation was carried out to measure the membrane pure water flux (J_1) by using equation (2). The permeate was collected at a time interval of 1 min and the readings were directly recorded using an electronic balance which was connected to a data acquisition system (AND Super Hybrid Sensor, Model: Fx-3000i, A&D Company, Limited). One litre of 50 mg/L oil emulsion was first prepared in the feed tank. The feed solution was then continuously pumped and recirculated through the cross-flow filtration cell. Feed pressure was monitored continually throughout the experiments. After 1 h of oil emulsion filtration, the membranes were rinsed by circulating pure water (25 ± 1 °C) for 10 min to remove loosely bound oil droplet. The pure water flux (J_2) ($L/m^2{\cdot}h$) at 25 ± 1 °C was recorded and the flux recovery ratios (FRR) was calculated based on equation (2) and equation (3). The relative flux reduction (RFR) was applied to examine the antifouling properties by using equation (4).

$$J = \frac{V}{A \cdot t} \tag{2}$$

IOP Conf. Series: Materials Science and Engineering **206** (2017) 012083 doi:10.1088/1757-899X/206/1/012083

$$FRR = \frac{J_2}{J_1} \times 100\% \tag{3}$$

$$RFR = (1 - \frac{J_p}{J_1}) \times 100\% \tag{4}$$

where V (L) is the volume of water permeate, A (m^2) is the membrane area, t (h) is time taken for permeate collected, J_1 is the initial pure water flux, J_2 is the pure water flux after 10 min crossflushing, and J_p is oil emulsion permeate flux. All the presented results are average data obtained from three measurements. Membranes with higher value of FRR with lower value of RFR indicate better antifouling properties.

The oil concentration in the permeate samples were determined using UV-visible spectrophotometer (UV mini-1240, Shimadzu) at maximum absorbance of wavelength 220 nm. The rejection of the solute, R (%) was calculated based on equation (5):

$$R = (1 - \frac{c_p}{c_f}) \times 100\% \tag{5}$$

where c_p is the oil concentration in the permeate (mg/L) and c_f is the initial concentration of the oil in the feed (mg/L).

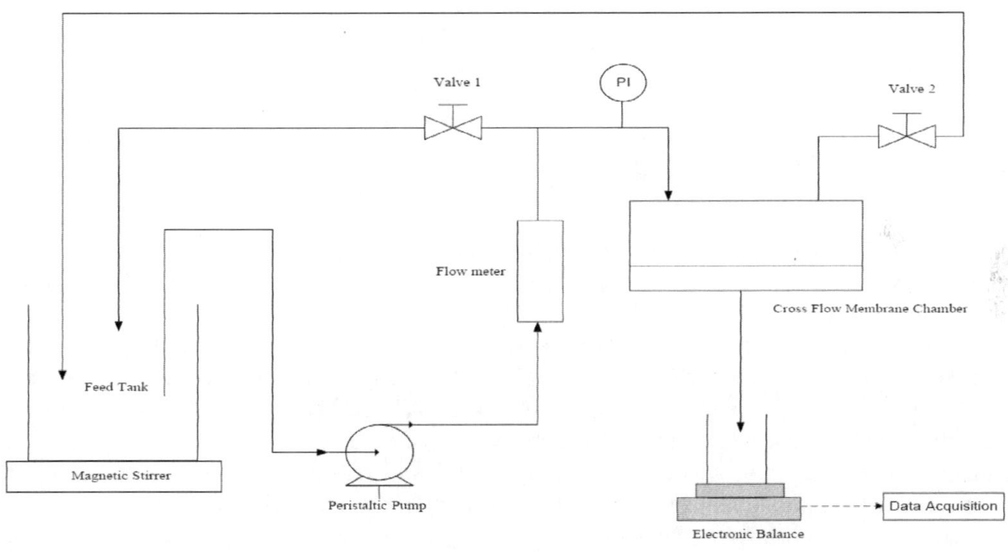

Figure 1. Schematic diagram of cross flow ultrafiltration unit.

3. Results and discussion

3.1. Characterization of oil emulsion and SiO_2 particles

The hydrodynamic diameter (*Dh*) of the oil emulsion was calculated based on Brownion motion of the oil droplet. Figure 2 shows the measurement of intensity based oil droplet hydrodynamic diameter distribution. The oil droplet size distribution for 50 mg/L oil emulsion was found in the range of 38 – 900 nm with an average oil droplet diameter of 153 ± 4.83 nm. The bigger oil droplet size compare to membrane mean pore size (40 – 50 nm) was used as foulant to minimize direct pore blocking effect.

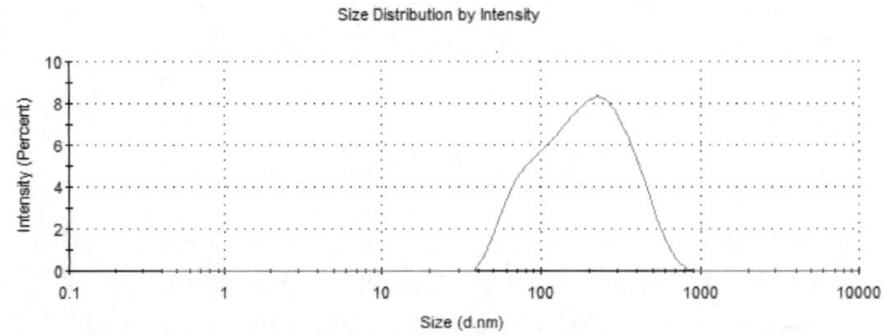

Figure 2. Size distributions of oil droplets in oil emulsion (oil concentration = 50 mg/L).

The hydrodynamic size distribution of SiO_2 particles in DMF solvent was characterized using dynamic light scattering prior to use. Figure 3 shows the SiO_2 particle size distribution in DMF solvent. The result shows average SiO_2 particle size with narrow distribution at 248.6 nm with 0.018 polydispersity index.

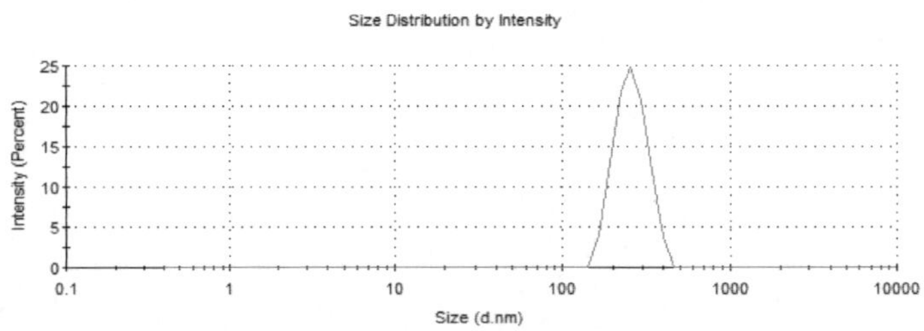

Figure 3. Size distributions of SiO_2 particles in DMF solvent.

3.2. Physicochemical properties of the PVDF/SiO_2 composite membrane

Surface and cross-sectional images of membrane prepared from PVDF/SiO_2/DMF systems with different contents of SiO_2 additive are shown in Figure 4. The SEM images show that all prepared membranes were typical highly porous surface and combination of finger-like and sponge-like cross-section. It can be seen that the membrane surface pore sizes as well as surface porosity had increased with the increase of SiO_2 concentration as shown in Figure 4 (left). The SiO_2 particles dispersed well on the membrane surface and the support structure without observable particle aggregation. The phenomenon of pore widening was observed due to two fold reasons. Addition of hydrophilic SiO_2 particles increased the non-solvent influx and cause instantaneous demixing and pore enlargement. In addition, the different in shrinkage rate between polymer and SiO_2 during membrane drying process had further enhanced the pore rupture and enlargement.

Figure 4 (right) shows the SEM cross-sectional images of membranes prepared with different SiO_2 concentration added. Generally, non-solvent (water) inflow and solvent outflow (DMF) will be changed accordingly with the addition of additives into polymer dopes solution. The addition of SiO_2 particles increases the instability and viscosity of the polymer dope solution, caused instantaneous demixing during phase inversion process, resulting in macrovoid-like structure at higher SiO_2 concentration. The hydrophilic nature of SiO_2 particle (high affinity of SiO_2 towards water) enhances the kinetic of water influx during the phase inversion process. The diameter of finger-like structure became larger and irregular in shape after addition of 2 wt.% of SiO_2 particles. Apart from thermodynamics aspects, the membrane formation process is also influenced by the kinetic aspects during phase inversion process. In term of kinetic aspects, increasing SiO_2 concentration will enhance the viscosity of the polymer dopes solution. Higher viscosity will cause a slower diffusion rate of solvent and non-solvent exchanges, resulting macrovoid formation in the membrane structure [21, 22].

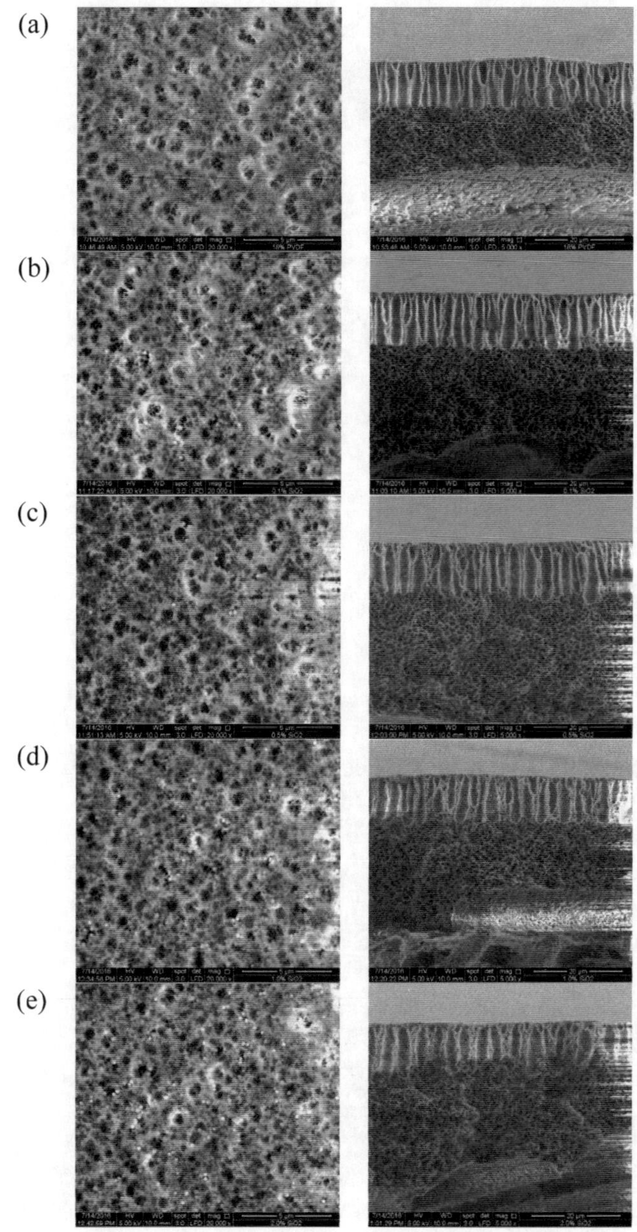

Figure 4. FESEM images of membrane surface structure (left) and cross sectional structure (right) for (a) M18-0, (b) M18-0.1, (c) M18-0.5, (d) M18-1.0, and (e) M18-2.0.

The surface morphology of the prepared membranes was further analysed using AFM as shown in Figure 5. The surface roughness values (R_q) was tabulated in Table 2. The pure PVDF membrane exhibits the R_q value of 63.128 nm, which is lowest among all prepared membranes. These images indicated that the membranes surface micro-roughness increased structurally with the increased of SiO_2 contents which might relate to the higher surface porosity as seen in previous FESEM surface morphology (Figure 4 (left)). Results in Table 2 show that the *Rq* had increased from 63.128 nm to 87.430 nm with the increase of SiO_2 concentration from 0 wt.% to 2 wt.%. The surface roughness increment was governed by the SiO_2 particles itself rather than the polymer nodular structure.

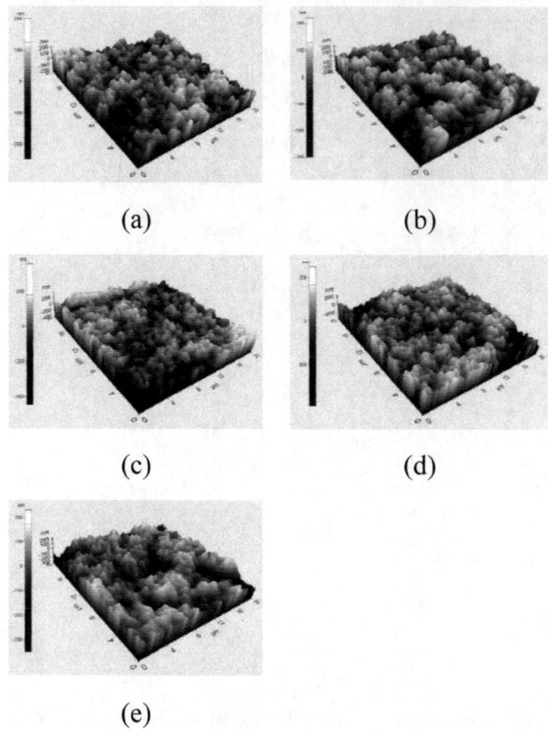

(a) (b) (c) (d) (e)

Figure 5. AFM images of PVDF membrane surface for (a) M18-0, (b) M18-0.1, (c) M18-0.5, (d) M18-1.0, and (e) M18-2.0.

Table 2. The effect of SiO_2 concentration on membrane physical properties.

Membrane	R_q (nm)	Porosity (%)	Pore Size (nm)
M18-0	63.128	64.17 ± 0.06	39.81 ± 6.80
M18-0.1	65.162	64.73 ± 0.13	40.06 ± 8.44
M18-0.5	74.874	65.39 ± 0.33	44.87 ± 7.29
M18-1.0	85.132	65.90 ± 0.28	47.97 ± 5.75
M18-2.0	87.430	66.28 ± 0.19	50.78 ± 8.08

The PVDF/SiO_2 composite membranes were then characterized in term of surface wettability. Theoretically, surface wettability (contact angle) is affected by membrane surface porosity, roughness, and its hydrophilicity. It was expected that the membrane prepared at higher concentration of hydrophilic additives will possess lower contact angle. Figure 6 shows that at low SiO_2 content, the hydrophilic nature of the particles slightly reduced the surface contact angle. However, at higher particle concentration (2 wt.%), concomitant increased of porosity induced lower surface energy and gave higher contact angles.

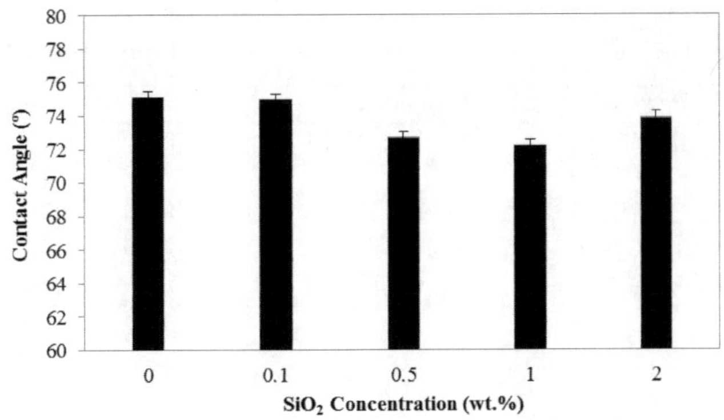

Figure 6. Effect of SiO_2 concentration on membrane contact angle.

3.3. Permeability and antifouling properties of PVDF/SiO_2 composite membranes for salty oil emulsion separation

The prepared membranes were evaluated in terms of membrane permeability, oil emulsion rejection, relative flux reduction (RFR) and flux recovery ratio (FRR) using cross-flow ultrafiltration experiments. Table 3 shows the effect of SiO_2 concentration on membrane performance. The membrane pure water fluxes increased with the increase of SiO_2 concentration from 57 to 93.86 L/m^2·h for M18-0 to M18-2.0, with the mean pore size from 39.81 to 50.78 nm, respectively. Theoretically, pore enlargement and hydrophilicity improvement are considered as main reason for membrane flux enhancement. However, in this study, the role of pore size enlargement seems to be more dominant than membrane hydrophilicity as can be seen from the highest contact angle of M18-2.0 which still having high flux. The addition of SiO_2 particles gave a positive effect on membrane pure water flux due to the pore size enlargement which allow higher passage of water permeate through the membrane. This result was also in accordance to the results of porous FESEM surface morphology for PVDF/SiO_2 composite membrane (Figure 4 (left)) as discussed earlier.

The oil rejection performance data was demonstrated in Table 3. It was found that all prepared membranes showed very promising oil rejection with 99 % of oil removal achieved. It is due to the bigger mean oil droplet size (153 nm) as compared to prepared membranes mean pore size (39.81 - 50.78 nm). The prepared membrane pores were relatively smaller which enable it to retain bigger oil droplet in high salinity oil emulsion from passing through.

The relative flux reduction (RFR) of oil emulsion was calculated and compared to evaluate the membrane fouling behaviour. Membrane with lower RFR value possesses high antifouling properties. Table 3 shows the RFR values for PVDF membrane with different SiO_2 concentration added. It is interesting to know that the RFR value of the prepared membrane is increased with the increased of SiO_2 concentration, which indicated that the oil droplet fouling behaviour was getting serious after introducing SiO_2 particles into the polymer matrix. The oil droplets start to deposit or adsorb onto the membrane surface by hydrophobic interaction. Figure 7 shows drastic reduction of relative flux for membrane M18-2.0 (0.42) compared to M18-0 (0.65) once the oil emulsion was introduced into the system. More serious oil permeate flux reduction was observed in PVDF/SiO_2 composite membrane as compared to pure PVDF membrane especially at high SiO_2 concentration (M18-2.0), which is due to the higher membrane porosity, pore size, and surface roughness as tabulated in Table 2. The chemical property of hydrophilic PVDF/SiO_2 membranes deteriorate due to the exclusion of charge repulsion under high salinity oil emulsion separation. The presence of salt ions in oil emulsion could disrupt the SiO_2 hydration layer, thus losing its hydrophilicity (Hofmeister effect) [23-25]. In this case, the membrane physical properties (surface roughness, porosity and pore size) are playing more dominating role in oil emulsion fouling.

Table 3. The effect of SiO_2 concentration on membrane performance

Membrane	PWF (L/m^2·h)	Rejection (%)	RFR (%)	FRR (%)
M18-0	57.00 ± 2.53	99.35 ± 0.13	81.45 ± 1.07	57.89 ± 1.20
M18-0.1	59.57 ± 1.27	99.19 ± 0.17	89.92 ± 1.17	48.92 ± 1.21
M18-0.5	80.71 ± 3.15	99.08 ± 0.20	93.27 ± 1.26	41.24 ± 1.39
M18-1.0	87.14 ± 2.90	99.09 ± 0.21	93.30 ± 1.20	40.31 ± 1.20
M18-2.0	93.86 ± 4.36	98.99 ± 0.24	93.76 ± 1.40	33.64 ± 1.47

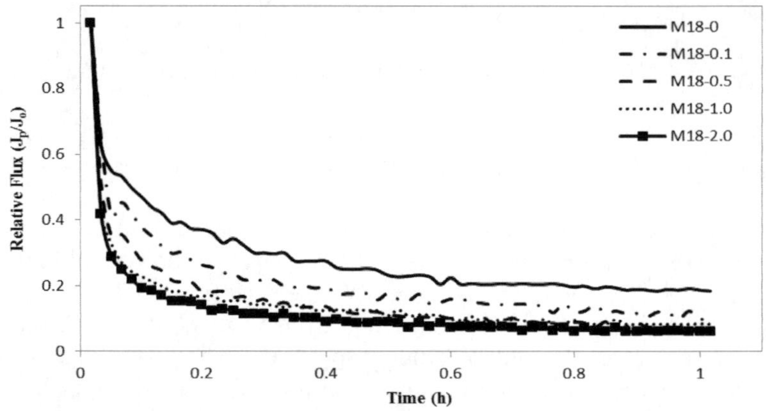

Figure 7. Effect of SiO_2 concentration on relative flux profile during oil emulsion separation.

The regeneration or reusability of the membrane is always a critical concern in membrane technology. The membrane (after oil emulsion separation) were washed thoroughly by circulating the fresh distilled water for 10 min, and the cleaned membrane's water flux was measured again in order to calculate the flux recovery ratio (FRR). The membrane FRR has direct relation with membrane physical properties including membrane pore size, porosity and surface roughness, as discussed earlier. Membrane with bigger pore size and porosity was expected to have difficulties in membranes cleaning, resulting in lower values of flux recovery ratio. Table 3 reveals that the FRR values were decreased with the increased of SiO_2 concentration. This phenomenon was due to the higher pore size and surface roughness at higher SiO_2 concentration which can easily trap the foulant in the valley as well as inside the pore, thus resulting in difficulty to wash away via crossflushing.

4. Conclusion

PVDF/SiO_2 composite membranes were synthesized by using diffusion induced phase separation technique for oil emulsion separation. The influence of SiO_2 concentration on membrane properties and fouling evaluation were studied. FESEM results show that addition of hydrophilic SiO_2 particles had increased the non-solvent (water) inflow thus caused instantaneous demixing to occur during phase inversion process, resulting in gradual change of membrane cross-sectional structure from finger-like to macrovoid-like structure. The results showed that the pore size, porosity and surface roughness of the PVDF membranes were increased with the addition of SiO_2 particles. Although the addition of SiO_2 particles increases the surface wettability, however the effect of microstructure roughness, pore size, and porosity increment has compensated the hydrophilic characteristic of SiO_2 particles during high salinity oil emulsion separation. The RFR value of the prepared membrane increased with the increase of SiO_2 concentration, indicated the oil droplet fouling behaviour getting serious after introducing of

SiO_2 particles into the polymer matrix. The membrane physical properties play an important role in the oil emulsion fouling behaviour rather than its hydrophilicity enhancement. It is because the high salinity oil emulsion resulted in exclusion of charge repulsion and at the same time disrupts the hydration layer of SiO_2 particles, thus losing its hydrophilicity properties. Furthermore, the results show low flux recovery ratio for PVDF/SiO_2 composite membranes compared to pure PVDF membrane. It is due to the higher pore size and surface roughness which can easily trap the foulant within the valley and pores thus exert higher difficulty in cleaning the membrane via crossflushing.

Acknowledgements
The authors are grateful for the financial supports from Universiti Sains Malaysia (RU Grant) (1001/PJKIMIA/814210) and Membrane Science and Technology Cluster (1001/PSF/8610011). In addition, the first author also thanks the Ministry of Higher Education (MyPhD) for financial support.

References
[1] Munirasu S, Haija M A and Banat F 2016 *Process Saf. Environ.* **100** 183-202.
[2] Fakhru'l-Razi A, Pendashteh A, Abdullah L C, Biak D R A, Madaeni S S and Abidin Z Z 2009 *J. Hazard. Mater.* **170** 530-51.
[3] Chakrabarty B, Ghoshal A K and Purkait M K 2010 *Chem. Eng. J.* **165** 447-56.
[4] Tummons E N, Tarabara V V, Chew Jia W and Fane A G 2016 *J. Membr. Sci.* **500** 211-24.
[5] da Silva S S, Chiavone-Filho O, de Barros Neto E L and Foletto E L 2015 *Desalin. Water Treat.* **56** 1802-08.
[6] Krebs T, Schroën C G P H and Boom R M 2012 *Chem. Eng. Sci.* **71** 118-25.
[7] Chen X, Lu H, Jiang W, Chu L-Y and Liang B 2010 *Ind. Eng. Chem. Res.* **49** 9279-88.
[8] Chen W, Peng J, Su Y, Zheng L, Wang L and Jiang Z 2009 *Sep. Purif. Technol.* **66** 591-97.
[9] Chen W, Su Y, Zheng L, Wang L and Jiang Z 2009 *J. Membr. Sci.* **337** 98-105.
[10] Kang G-d and Cao Y-m 2014 *J. Membr. Sci.* **463** 145-65.
[11] Zuo D-y, Xu Y-y, Xu W-l and Zou H-t 2008 *Chinese J. Polym. Sci.* **26** 405.
[12] Rana D and Matsuura T 2010 *Chem. Rev.* **110** 2448-71.
[13] Cong H, Radosz M, Towler B F and Shen Y 2007 *Sep. Purif. Technol.* **55** 281-91.
[14] Ngang H P, Ooi B S, Ahmad A L and Lai S O 2012 *Chem. Eng. J.* **197** 359-67.
[15] Razmjou A, Mansouri J and Chen V 2011 *J. Membr. Sci.* **378** 73-84.
[16] Yan L, Li Y S and Xiang C B 2005 *Polymer* **46** 7701-06.
[17] Huang X, Zhang J, Wang W, Liu Y, Zhang Z, Li L and Fan W 2015 *RSC Adv.* **5** 18258-66.
[18] Ahmad A L, Majid M A and Ooi B S 2011 *Desalination* **268** 266-69.
[19] Ogoshi T and Chujo Y 2005 *J. Polym. Sci. Poly. Chem.* **43** 3543-50.
[20] Stöber W, Fink A and Bohn E 1968 *J. Colloid Interf. Sci.* **26** 62-69.
[21] Zheng Q-Z, Wang P, Yang Y-N and Cui D-J 2006 *J. Membr. Sci.* **286** 7-11.
[22] Pagidi A, Saranya R, Arthanareeswaran G, Ismail A F and Matsuura T 2014 *Desalination* **344** 280-88.
[23] Jhon Y K, Bhat R R, Jeong C, Rojas O J, Szleifer I and Genzer J 2006 *Macromol. Rapid Comm.* **27** 697-701.
[24] Schild H G and Tirrell D A 1990 *J. Phys. Chem.* **94** 4352-56.
[25] Von Hippel P H and Schleich T 1969 *Accounts Chem. Res.* **2** 257-65.

IOP Conf. Series: Materials Science and Engineering **206** (2017) 012084 doi:10.1088/1757-899X/206/1/012084

Phytoremediation Capabilities of *Spirodela polyrhiza* and *Salvinia molesta* in Fish Farm Wastewater: A Preliminary Study

Y S Ng, N I S Samsudin, D J C Chan*

School of Chemical Engineering, Universiti Sains Malaysia, 14300 Nibong Tebal, Penang, Malaysia

Email: chderekchan@usm.my (D J C Chan)

Abstract. Fish farm wastewater needs to be treated as it contains considerably high loading of suspended solids and dissolved nutrients from accumulation of by-products e.g. fish excretions and uneaten feed. In this study, macrophytes, namely *Spirodela polyrhiza* and *Salvinia molesta* were examined for their phytoremediation efficiency in treating fish farm wastewater in a raceway pond rig. It was carried out indoor for 14 days under controlled environment. Water samples was collected once every 2 days for analysis of NO_3^-- N, PO_4^{3-}, NH_3-N, COD, turbidity, MLVSS and pH. The results showed that there was decrement of phosphate in fish farm wastewater using either *S. polyrhiza* or *S. molesta*. Interestingly, *S. polyrhiza* was found to be more efficient in phosphate uptake as it removed 72% phosphate at day 4 and up to 95% in the end of the experiment whereas 72% phosphate removal was only achieved by *S. molesta* at day 10. Similar ammonia decrement was observed for both plants and most of the ammonia were not detected in the wastewater by day 10 for *S. polyrhiza*, while by day 8 for *S. molesta*. Nitrate showed increment for both plants which could be due to nitrification. Both plants achieved highest COD removal on day 12, whereby 68% for *S. polyrhiza* and 63% for *S. molesta*. They were able to reduce turbidity and total suspended solids (TSS) to very low level and significantly increase clarity of wastewater. *S. polyrhiza* reduced up to 96% of initial turbidity value and 86% of TSS. 82% reduction of initial turbidity and 79% TSS decrement were observed for *S. molesta*. pH fluctuations were minimum for both plants, with a range between 7.62 to 7.77. Both plants demonstrated biomass increment for fresh weight in which 84% for *S. polyrhiza* while 85% for *S. molesta*. This study proved that the macrophytes were able to treat fish farm wastewater by significantly removing phosphate, ammonia, turbidity and TSS. It aids in minimizing pollutants released to receiving waters and producing biomass which can be utilized for agriculture sector.

1. Introduction

According to The State of World Fisheries and Aquaculture 2016, world total fish production had reached 167.2 million tonnes in 2014 while aquaculture production alone accounted for about 44% of the total fish production. Provided that aquaculture comprised only 7 percent of fish for human consumption in 1974, this share had rose up to 26 percent in 1994 and 39 percent in 2004 [1]. It is not surprisingly that the figure will soon overtake the wild-caught fish production after 2014. Therefore, aquaculture would play a major role in world fish production now and future to ensure food security and nutrition to ever-growing human population. Malaysia was listed as one of the top 25 major

IOP Conf. Series: Materials Science and Engineering **206** (2017) 012084 doi:10.1088/1757-899X/206/1/012084

aquaculture producers in the world with total production of 521.0 thousand tonnes in 2014, ranked 15th among the countries [1]. Its inland aquaculture covered an area of about 794.2 thousand hectares [2]. However, improper management of the aquaculture site in term of effluent discharge would bring harm to the nearby water resources and environment.

In an enclosed, intensive inland aquaculture, the water used to culture the fish are generally easier to be concentrated with suspended solids and dissolved nutrients due to accumulation of by-products eg. fish excretions and uneaten feed [3]. In order to maintain the health and welfare of the fish, water exchange need to be done regularly [4]. However, this effluent is normally either directly discharged into the nearby waterways or undergone sedimentation pond before released. Sedimentation may help reduce suspended solids, but not to remove dissolved nutrient, so eventually fish farm wastewater still poses risk of harming the receiving water. This phenomenon is attributed to rural farmers who are characterised as low capital cultivator, making advanced treatment system is too expensive for them to be installed and operated whereas no clear provision made with regard to local aquaculture effluents [5] also cause no further treatment of the effluents since the issue is not prioritised. Therefore, an affordable, efficient yet easy to implement treatment system for the fish farm wastewater is needed to ensure successful of the system. The system will give the farmer a short in the arm if it can generate valuable products or side income.

Phytoremediation is identified to be a treatment system which fulfilling those criteria. It is relatively low cost to maintain since it is solar-driven and no complex containment system needed. It is cheaper than conventional treatment methods that rely on electricity, pumping, aeration or chemicals additions and usually need large concrete or steel vessels. It is also the least harmful method as it uses naturally occurring organisms and preserves environment in more natural way, and it is some more aesthetically pleasing. Significant amounts of pollutants eg. phosphate, ammonia and etc. can be efficiently reduced by phytoremediation as shown in the available literature. Jesus et al. [6] demonstrated that phytoremediation of freshwater aquaculture using *Typha latifolia* in constructed wetland was able to reduce 61%, 78% and 98% of NH_4, NO_2 and PO_4 respectively. In Effendi et al. [7] study, they showed that ammonia, nitrate and orthophosphate can be reduced until 84.6%, 34.8% and 44.4% in freshwater crayfish culture wastewater by applying water spinach aquaponic system. Another phytoremediation work by Olguin et al. [8] indicated that *Salvinia minima* managed to decrease TKN, NH_4-N and P-PO_4 up to 97%, 99% and 88% respectively in anaerobic effluents of coffee wastewater. When the phytoremediation system is coupled with sedimentation pond, it will aid in removing the dissolved nutrients in the effluent as well the suspended solids.

The macrophytes are plants that have their own potential uses. Traditionally, *Wolffia arrhiza* has been eaten in Myanmar, Laos, and northern Thailand [9]. King et al. [10] showed that inclusion of *Salvinia molesta* in commercial fish feed diet will have higher fish weight on Nile tilapia (*Oreochromis niloticus*) compared to feeding with commercial feed alone and significant effect might be observed if feeding period is prolonged. The extracted lipid of *S. molesta* can be processed into biodiesel [11] and its biomass has the potential to be converted into organic fertilizer via vermiremediation [12]. Similarly, *Spirodela polyrhiza* can be promising substrate for biohydrogen production [13] and also be included in fish meals [14]. Hence, the aquaculture farmers can earn extra income out from the valuable plant stock harvested besides being applied to remediate the fish farm wastewater.

By taking the context mentioned earlier into consideration, the aim of this study is to examine the the water quality of fish farm wastewater after phytoremediation and its effect toward biomass of macrophytes. Since *S. molesta* and *S. polyrhiza* is the macrophytes species that locally available, they were employed for the study and their respective performances were being compared.

2. Materials and methods

2.1. Characterisation of fish farm wastewater

The fish farm wastewater was collected from a catfish pond near Sungai Udang, Nibong Tebal. It was then analysed and characterised according to standard protocol as stated in subsection 2.4. The water quality parameters taken for analysis included nitrate (NO_3^--N), phosphate (PO_4^{3-}), ammonia (NH_3-N), COD, turbidity, MLVSS and pH. The wastewater was later stored in cold room at around 4°C before used.

2.2. Plant stock establishment

Axenic *S. molesta* and *S. polyrhiza* were obtained from USM SCE's laboratory after previous successful isolation of the plants. The aquatic plants were subcultured and maintained in liquid Hoagland medium with 15g/L sucrose to provide needed stock for the study. All culture media were adjusted to pH 5.8 using NaOH before autoclaving at 121°C for 15 min. The explants of *S. molesta* were cut from the sterilized plant by stem splitting while healthy green fronds of *S. polyrhiza* were selected, for inoculation in the medium. Each 500ml Erlenmeyer flask contained 200ml liquid medium with five explants. All cultures were then incubated at 25±1°C under fluorescent tubes (1000 lux) with a 24h-photoperiod for 15 days.

2.3. Lab-scale phytoremediation on fish farm wastewater

This experiment was conducted to investigate the water quality of fish farm wastewater after phytoremediation and its effect toward biomass of macrophytes. *S. molesta* and *S. polyrhiza* were the floating macrophytes employed for phytoremediation and their treatment performances were compared. This study was carried out in bench scale raceway pond rig with its configuration as shown in [15] with dimensions of 50cm x 25cm x 9cm, inside the laboratory under controlled environment. 12 L of fish farm wastewater was poured into the storage tank and submerged pump was started to fill up and circulate the wastewater in the raceway pond rig system Healthy macrophytes were selected from axenic plant stock and undergone thorough rinsing and washing with water, so that the unutilised inorganic nutrients and sucrose remained on plant surfaces can be removed to avoid them from affecting the result of study. They were later dried by blotting with cloth and being weighed for 77 g before placing evenly on the surface of raceway pond. The circulation flow rate of wastewater in the rig was then maintained at 50 ml/min and the rig was closely monitored until equilibrium was achieved in the system whereby water level in storage tank reached steady state. The level was being marked. The phytoremediation study was carried out at 25±1°C under fluorescent tubes (1000 lux) with a 24h-photoperiod for 14 days. 100ml of water sample was collected each of every 2 days starting day 0 towards the end of experiment. The water level in the storage tank was ensured to maintain at the initial marked level before collection. Distilled water was added to the storage tank if the water level dropped below the marked level, which was due to evaporation. All water samples were contained with centrifuge tubes and kept in cold room at around 4°C before analysis. The water samples were tested for nitrate (NO_3^--N), phosphate (PO_4^{3-}), ammonia (NH_3-N), COD, turbidity, MLVSS and pH to evaluate the changes of water quality during phytoremediation period. At day 14, the macrophytes were harvested and dried carefully as the same way during preparation before being weighed for its final fresh weight.

2.4. Analytical analysis

2.4.1. Determination of nitrate and phosphate concentration for water samples.

The nitrate was determined by Cadmium Reduction Method (HACH method 8039) using NitraVer®5 Nitrate Reagent Powder Pillows by HACH DR2800 spectrophotometer at 500nm with allowable detection of high range (0.3-30.0mg/l NO_3^--N). The phosphate was determined by Ascorbic Acid Method (HACH method 8048) using PhosVer® 3 Phosphate Reagent Powder Pillows by HACH

DR2800 spectrophotometer at 880nm with detection range of (0.02-2.50mg/l PO_4^{3-}). This phosphate determination is in accordance to USEPA method 365.2 and Standard Method 4500-P-E for wastewater. Each method consumed 10ml of wastewater samples.

2.4.2. Determination of ammonia concentration and COD for water samples.
The ammonia was determined by Salicylate Method (Lovibond method 66) using VARIO Am tube test Reagent, Set HR, F5 (VARIO Ammonia Salicylate, F5 and VARIO Ammonia Cyanurate, F5 powder packs as well as VARIO Am Diluent Reagent High Range reaction tube) by LOVIBOND Maxidirect MD600 photometer at 660nm with detectable range of (0-50mg/l NH3-N). This method consumes 0.1ml water sample. The COD was determined by Dichromate/H2SO4 Method (Lovibond method 131) using COD VARIO 0 - 1500 mg/l tube test Reagent by LOVIBOND Maxidirect MD600 photometer at 610nm with detectable range of (0-1500mg/l COD/CSB). This method complies with Standard Methods for the Examination of Water and Wastewater.

2.4.3. Determination of turbidity of water samples.
The water samples were well-mixed by shaking the centrifuge tube vigorously. A clean cuvette was filled with water sample until the level mark. The bubbles in water samples were removed by placing cuvette in the degaser. The cuvette was wiped with lint-free tissue before inserting into measurement cell to measure its turbidity value. The turbidity value was determined by HANNA HI 93703 microprocessor turbidity meter peaking at 890nm with range of 0-1000NTU. The turbidity measurement was performed according to the ISO 7027 International Standard.

2.4.4. Determination of MLVSS.
Well-mixed water samples were prepared by shaking the centrifuge tube vigorously. The 30ml samples were filtered through a 47mm diameter weighed Whatman™ glass microfiber filters with mini air pump and the residues retained were dried in an oven to a constant weight at 105°C for 1 hour. Increase in weight of filter represents the total suspended solids. The filters with the residues were ignited in muffle furnace to constant weight at 550°C for 20 minutes. The remaining solids represent the fixed solids while weight lost on ignition was the volatile solids. This MLVSS test was carried out based on APHA 2540D and APHA 2540E.

2.4.5. Determination of pH.
pH measurement was performed using Hanna Edge® pH meter HI-2020. The water sample was continuously stirred with pH probe in order to obtain an accurate pH reading.

3. Results and discussion

3.1. Lab-scale phytoremediation on fish farm wastewater
This experiment was carried out to determine the water quality of fish farm wastewater during 14 days of phytoremediation and its effect toward biomass of macrophytes namely *S. molesta* and *S. polyrhiza.* Specifically, nutrient uptake by both macrophytes such as nitrate, phosphate and ammonia were accessed while COD level, turbidity, pH and MLVSS were also checked. Biomass in term fresh weight was being determined in the end of experiment. Table 1 shows the characteristics of fish farm effluent used for the experiment.

IOP Conf. Series: Materials Science and Engineering **206** (2017) 012084 doi:10.1088/1757-899X/206/1/012084

Table 1. Characteristics of fish farm wastewater.

Parameter	Units	Value
Nitrate (NO_3^- -N)	mg/L	-
Phosphate (PO_4^{3-})	mg/L	16.9
Ammonia (NH_3 - N)	mg/L	22.6
Chemical oxygen demand (COD)	mg/L	439
pH		7.76
Turbidity	NTU	315.3
Total Suspended Solids	mg/L	500.7
Volatile Solids	mg/L	334.0
Fixed Solids	mg/L	166.7

3.2. Nutrient uptake of the macrophytes

Nitrate, ammonia and phosphate are among the important nutrients which taken up by the aquatic plants from water body to sustain its growth.

Based on Figure 1, throughout the 14 days of experiment, the concentration of nitrate in fish farm wastewater for *S. molesta* macropyhtes was increasing gradually up to day 14. For *S. polyrhiza*, the nitrate concentration having similar trend with *S. molesta* macropyhtes. The reason both macropyhtes showed a higher nitrate level until day 14, was because nitrification process took place in phytoremediation system. Nitrification is a microbial process by which reduced nitrogen compounds (primarily ammonia) are sequentially oxidized to nitrite and nitrate. It occurred when the ammonia present was oxidized to nitrite through a bacteria group called Ammonia-Oxidising Bacteria (AOB) while Ammonia-Oxidising Archaea (AOA) oxidized ammonia to nitrite and produce nitrate [16]. For phytoremediation system, the wastewater was exposed to the direct light from lamp. Hence, great amount of oxygen was released by the macrophytes and filamentous algae in the wastewater through photosynthesis. This in turn promoted nitrification process since abundant oxygen was dissolved in the water column [17], leading to increasing nitrate concentration in the wastewater. The reason the plants did not remove nitrate was due to preference of plants in taking up ammonia over nitrate [18].

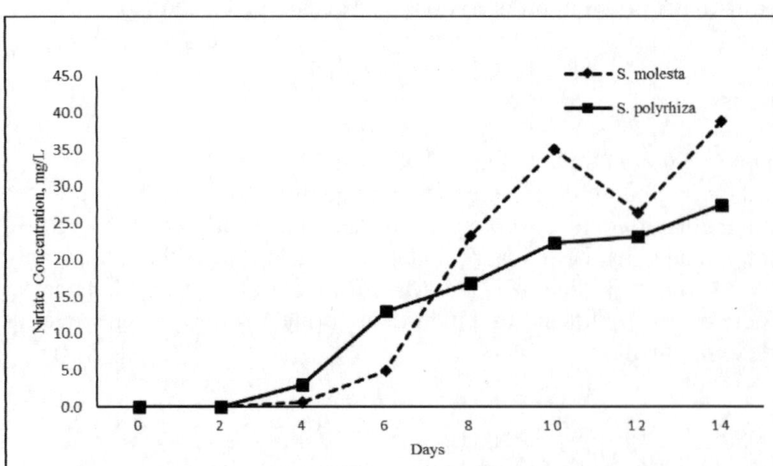

Figure 1. Concentration profile of nitrate in fish farm wastewater grown with *S. molesta* and *S. polyrhiza* species.

From Figure 2, it showed that there was decrement of phosphate in fish farm wastewater using either *S. polyrhiza* or *S. molesta*. Interestingly, *S. polyrhiza* was found to be more efficient in phosphate uptake as it removed 72% phosphate at day 4 and up to 95% in the end whereas 72% phosphate removal was only achieved by S. molesta at day 10. This indicated that *S. polyrhiza* removes phosphate more efficiently and in higher rate than *S. molesta* in the fish farm wastewater. This huge phosphate decrement trend was also shown in [15]. The phosphate can be taken up by plants roots, absorption through leaves and shoots in aquatic plants [17]. The concentration of phosphate or specifically called orthophosphate decreased throughout the experiment by plant uptake, was due to continuous need of metabolic activity by plant itself during growth which requires phosphorus for building ATP, ATP as well as other plant structural component. This phenomenon was also shown in the work of Ghaly et al. [19] in which the wastewater effluent contained less amount of phosphate at the end of the run and the plants gained weight significantly after phytoremediation.

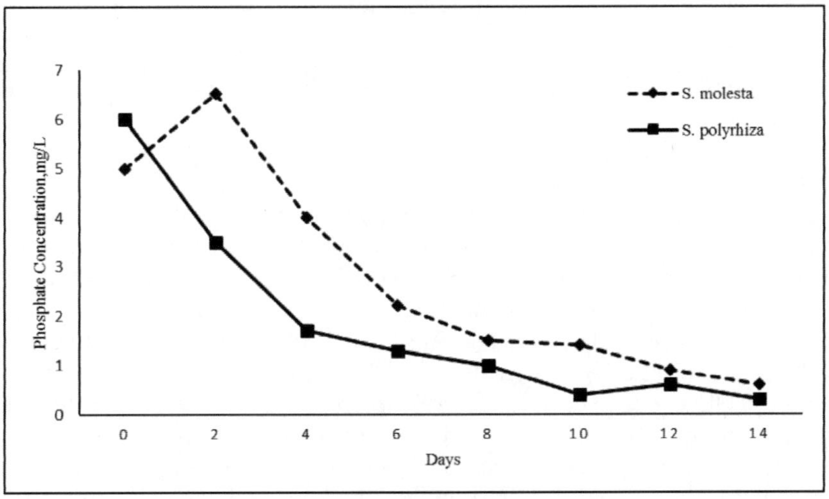

Figure 2. Concentration profile of phosphate in fish farm wastewater grown with *S. molesta* and *S. polyrhiza* species.

Figure 3 showed the trend for ammonia concentration for *S. molesta* and *S. polyrhiza* macropyhtes during 14 days. Similar ammonia decrement was observed for both plants and most of the ammonia were not detected in the wastewater by day 10 for *S. polyrhiza*, while by day 8 for *S. molesta*. The reason the ammonia level dropped significantly after going through phytoremediation was because of both direct plant uptake and the consumption of ammonia in nitrification [20]. Furthermore, similar study was conducted by Azim et al. [21] showed that ammonia concentration in a pond treated with bamboo was lower and the *periphyton* macropyhtes improved water quality in aquaculture system by nitrification. In addition, enhanced bacterial biofilms on the substrates might reduce ammonia level through nitrification [22].

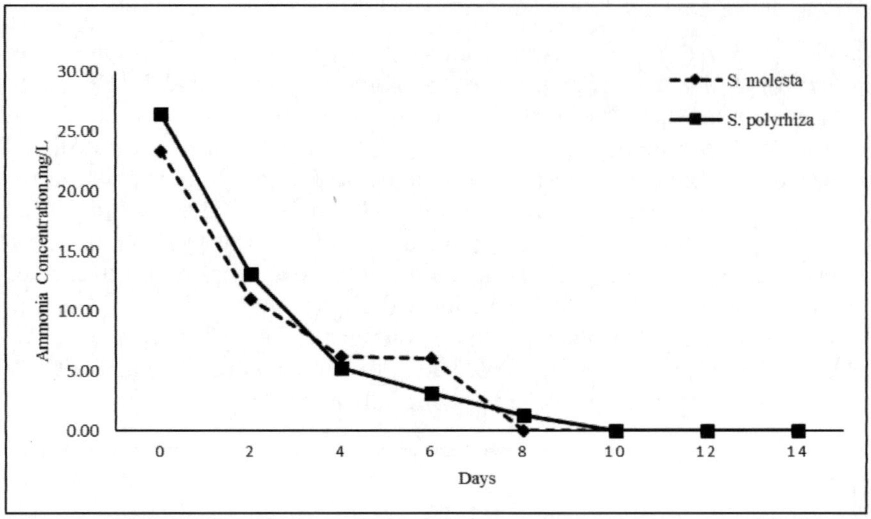

Figure 3. Concentration profile of ammonia in fish farm wastewater grown with *S. molesta* and *S. polyrhiza* species.

3.3. Water quality after phytoremediation

The water quality of fish farm wastewater which undergoes phytoremediation may either deteriorate or improve. Chemical Oxygen Demand (COD), Mixed Liquor Volatile Suspended Solids (MLVSS), turbidity and pH are among the analysis tests used to examine the water quality of the wastewater.

Chemical Oxygen Demand (COD) is a measure of the oxygen required to oxidize soluble and particulate organic matter in the water. It is an important water quality parameter as it provides an index to assess the effect of discharged wastewater will have on the receiving environment. Figure 4 showed the trend for COD concentration for *S. molesta* and *S. polyrhiza* macropyhtes during 14 days of phytoremediation period. Both plants showed reduction in COD level in their respective wastewater. COD level for *S. molesta* decreased from 186 mg/L at day 0 to 75 mg/L at day 14. Meanwhile, COD level for *S. polyrhiza* dropped gradually from 77 mg/L at day 0 to 30 mg/L towards the end of the run. Overall COD reduction for *S. molesta* after 14 days of phytoremediation was about 60% whereas *S. polyrhiza* was able to remove about 61 % COD in the wastewater. Both plants achieved highest COD removal on day 12, whereby 68% for *S. polyrhiza* and 63% for *S. molesta.* The COD decreased significantly during the treatment because the plants were fully developed causing the filtration capacity of the roots towards suspended solids and the absorption of dissolved nutrients increased [19].

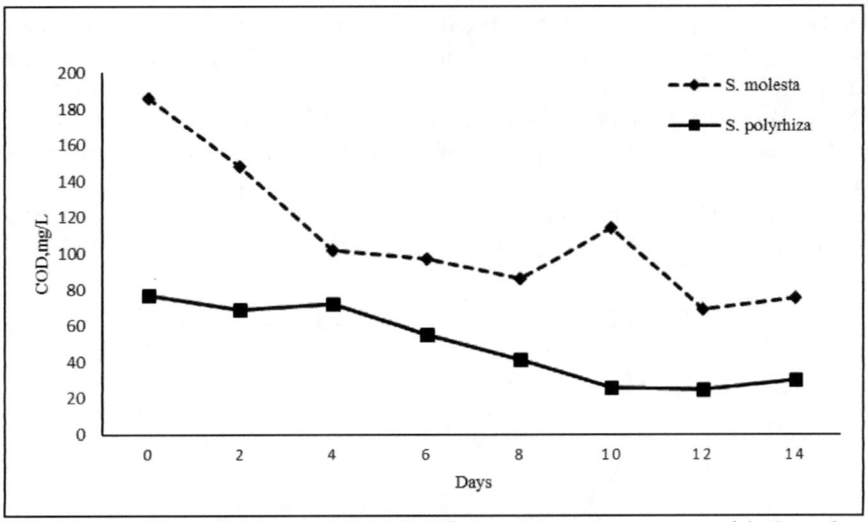

Figure 4. Concentration profile of COD in fish farm wastewater grown with *S. molesta* and *S. polyrhiza* species.

For pH analysis, Figure 5 showed that pH value did not fluctuate much for *S. molesta* macropyhtes with 7.69 at day 0 and 7.67 towards the end of the experiment whereas pH reading for *S. polyrhiza* at day 0 and day 14 is 7.71. Both plants had similar trend in pH value and pH fluctuations were minimum for both plants, with a range between 7.62 to 7.77. Akinbile and Yusoff [23] stated the pH level reduction is due to microbial activities and increase in CO_2 level in photosynthesis. Nitrification occurs over a wide pH range in soil while the optimal pH was estimated to be between pH 6.6 to 8.0 [17]. This may show that nitrification activity in our phytoremediation system is in optimal condition where its pH lain within 6.6 to 8.0, causing increasing nitrate level in the system.

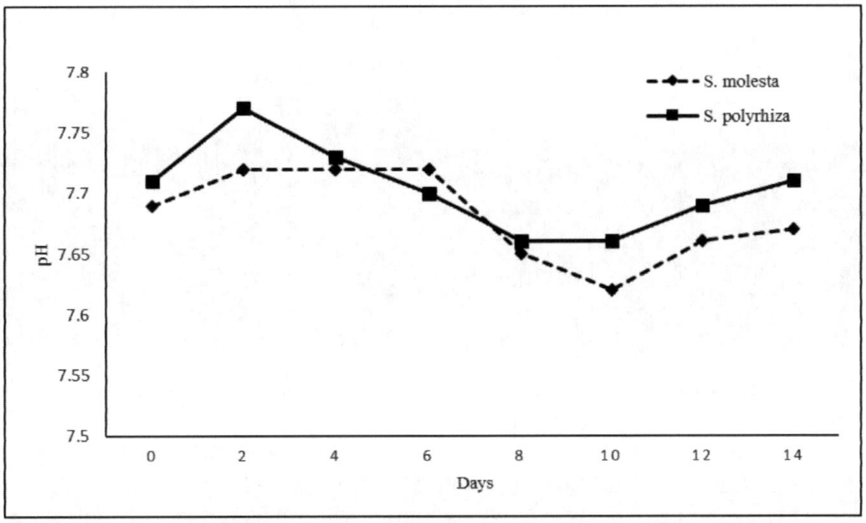

Figure 5. Concentration profile of pH in fish farm wastewater grown with *S. molesta* and *S. polyrhiza* species.

The results showed that significant improvement had taken place in increasing clarity of fish farm wastewater after 14 days of phytoremediation. Obvious change can be observed in the clarity of the fish farm wastewater when the level of turbidity drastically reduced during the first 6 days of experiment. As the results shown in Figure 6, the turbidity reading for *S. molesta* decreased gradually from 298 NTU at day 0 to 54 NTU at the last day of experiment. Same goes to *S. polyrhiza,* the turbidity decreased from 137 NTU to 5 NTU at day 14. 82% and 96% reduction in turbidity were achieved for *S. molesta* and *S. polyrhiza* respectively. It is believed that the microbes and biofilms that grew, suspended in the wastewater column and resided in long hairy extensive roots eg. *S. molesta* as shown in Figure 7 consumed the organic suspended solids for their growth or specifically respiration. The organic suspended solids were metabolized or used up by the cells and being converted to energy, carbon dioxide and water. This would explain the reduction of COD and turbidity level in the fish farm effluent.

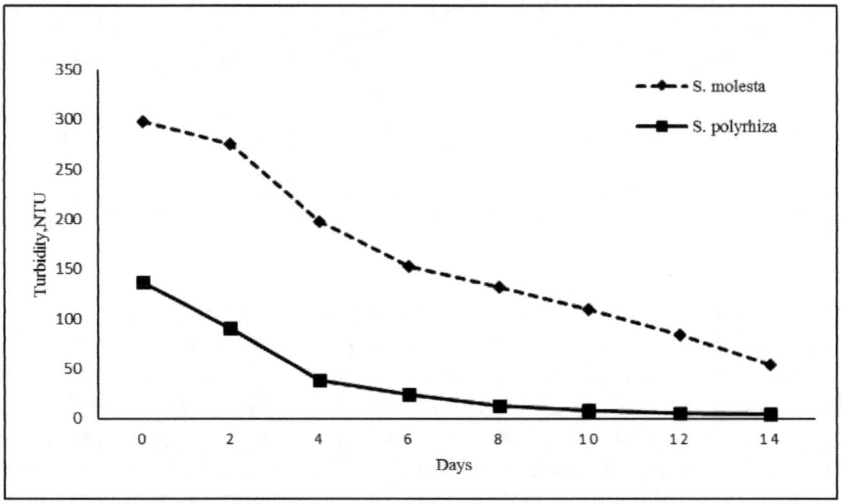

Figure 6. Turbidity profile of fish farm wastewater grown with *S. molesta* and *S. polyrhiza* species.

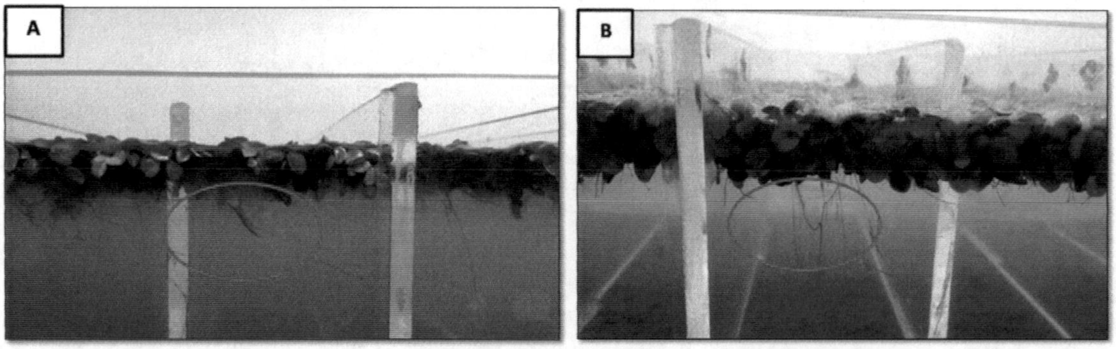

Figure 7. Biofilms or filamentous algae formed and attached to the roots of macropytes (A) *S. molesta* and (B) *S. polyrhiza*.

The red arrow in Figure 7 indicates the location where biofilms or filamentous algae attach to the roots in both macropyhtes system. They can be seen after 4 days. The roots of the plants become the place for bacteria to grow and the hairy roots makes nutrients easy to being absorbed into their roots.

Figure 8 showed the MLVSS result for both macrophytes. It showed that the concentration of total suspended solid (TSS) for both *S. molesta* and *S. polyrhiza* macropyhtes gradually decreased to 40 mg/L and 23 mg/L respectively in the end of the experiment. *S. molesta* reduced 79% of TSS while *S. polyrhiza* decreased 86% of TSS. Meanwhile, *S. molesta* achieved 67% reduction whereas *S. polyrhiza* managed to achieve 84% reduction in term of volatile solids. The concentration of fixed solids also depicted decrement in which 93% of reduction was observed for *S. polyrhiza* and 96% reduction for *S. molesta* in phytoremediation system. The suspended solids concentrations declined during the experimental period due to several reasons. First, the plants were fully developed and the filtration capacity of the roots increased with the growth of the plants. Biofilms or filamentous algae were able to grow on them. This allowed those microbes to multiply and consumed the organic suspended solids causing the TSS to decrease throughout the experiment. Brix [24] stated that the reduction in suspended solids caused by the reduction of water current, which enhance the sedimentation of suspended solids and the filtration of solids by the plant tissue. TSS was thus further reduced in the wastewater. There was a study conducted by Lin et al. [25] in which more than 90% reduction in suspended solids of aquaculture wastewater using subsurface flow wetland planted with reed named *Paspalum vaginatum.*

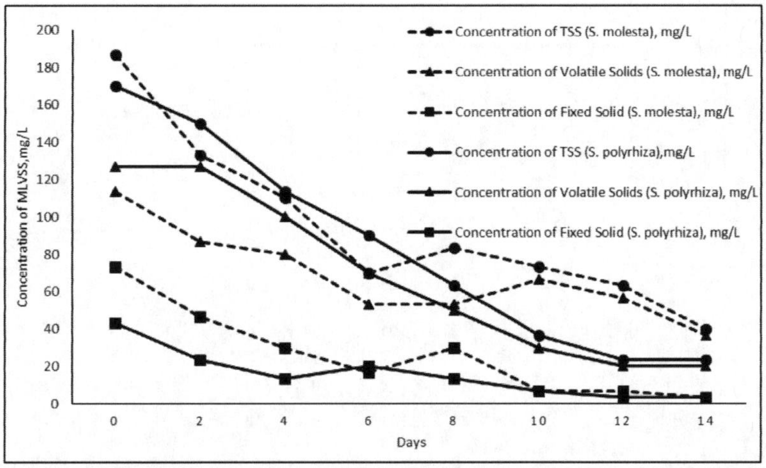

Figure 8. Concentration profile of MLVSS in fish farm wastewater grown with *S. molesta* and *S. polyrhiza* species.

3.4. Effect of phytoremediation towards biomass

After the phytoremediation, the macrophytes were checked for changes of their fresh weight, so that effect of phytoremediation towards biomass can be known.

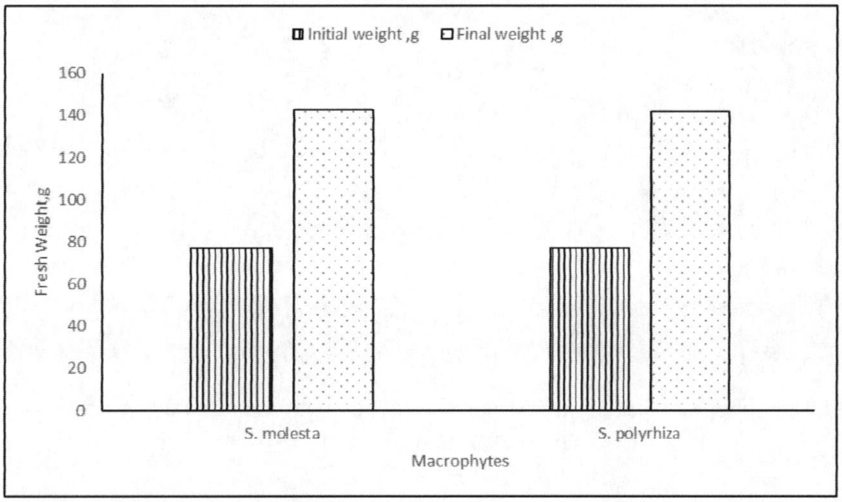

Figure 9. Changes of biomass (fresh weight) of different macropyhtes in phytoremediation system

IOP Conf. Series: Materials Science and Engineering **206** (2017) 012084 doi:10.1088/1757-899X/206/1/012084

As shown in Figure 9, fresh weight of *S. molesta* macropyhtes increased to 142.43g in which 85 % increase in weight was achieved. On the other hand, fresh weight of *S. polyrhiza* macropyhtes rose up to 141.62g where 84 % increase in weight was obtained. Both showed similar increment in biomass. The positive increment of biomass was attributed to the sufficient nutrients available in the fish farm wastewater to support the growth of macrophytes. The nutrients came from inorganic dissolved nutrients as well as mineralization of the organic suspended solids which were originated from uneaten feed and fish excretion. It showed suitability of the effluent in cultivating the plants, which was also shown in [26]. The huge increment of biomass for both species also allowed aquaculture farmers to produce valuable by-products such as fertilisers [12], biofuel [11, 13] and fish feed [10, 14] which can ultimately generate side income. Figure 10 and Figure 11 showed the appearance of both macrophytes on raceway pond at day 0 and day 14.

Figure 10. The appearance of *S. molesta* on the raceway pond at (A) day 0 and (B) day 14 of phytoremediation period

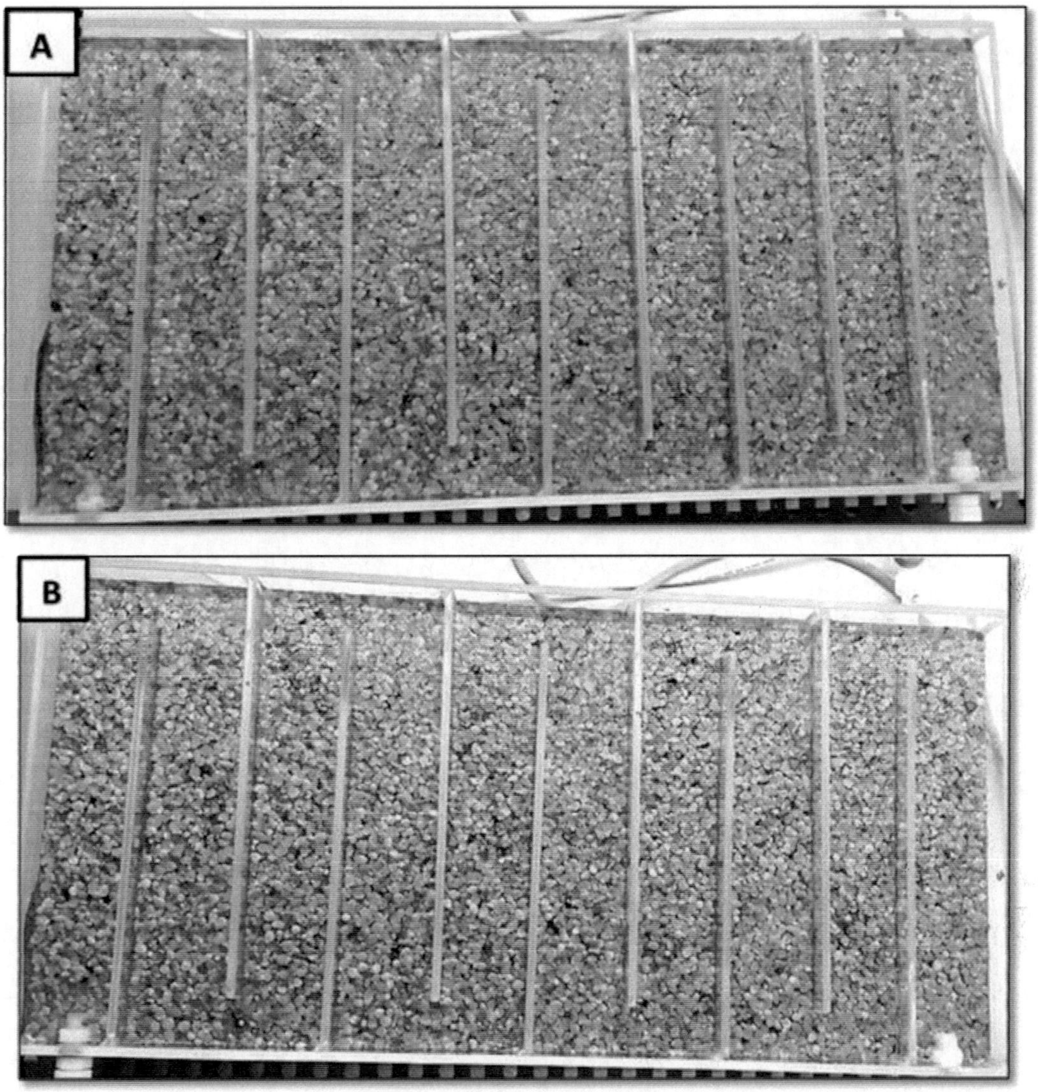

Figure 11. The appearance of *S. polyrhiza* on the raceway pond at (A) day 0 and (B) day 14 of phytoremediation period

4. Conclusions

The macropytes namely *S. molesta* and *S. polyrhiza* showed promising result in treating the fish farm wastewater. It was demonstrated that the phytoremediation system was capable of depurating the dissolved nutrients such as phosphate and ammonia and decreasing the suspended solids resulted from fish culture. Turbidity and COD was also lowered significantly. *S. polyrhiza* was more efficient and better in phosphate uptake as it removed phosphate faster and up to 95% in the end of the experiment whereas *S. molesta* removed only 72% at day 10. Both macrophytes had similar ammonia removal rate and managed to almost all ammonia within 8 to 10 days. This will surely help in minimizing the occurrence of eutrophication in receiving water since the effluent discharged were low in phosphate and ammonia. *S. polyrhiza* and *S. molesta* achieved highest COD removal on day 12 with 68% and 63% respectively. The total suspended solids in the wastewater was reduced drastically after phytoremediation as observed for both *S. polyrhiza* and *S. molesta.* It aids in avoiding the depletion of oxygen in the water course that leads to suffocation and stress of aquatic living things. Turbidity of the wastewater was also significantly reduced in which 96% and 82% decrement shown for *S. polyrhiza* and *S. molesta* respectively. The clarity of fish farm effluent was improved tremendously eventually which allowed light to penetrate the water to ensure healthy living of the organisms beneath and prevent water-borne diseases.

Both macrophytes also exhibited high productivity in biomass in the fish farm wastewater besides their ability in treating the effluent. Biomass increment in fresh weight for *S. polyrhiza* was 84% while 85% was shown for *S. molesta.* The huge biomass harvested can be utilized by the aquaculture farmers to become fish feed, fertilizer or biofuel that in turn gain extra benefit.

Acknowledgements
This work is supported by FRGS Grant (6071271) and Research University Grant (814209). Yin Sim Ng is financially assisted by MyMaster scholarship from Ministry of Higher Education of Malaysia. All authors are affiliated to Membrane Science and Technology cluster USM. N I S Samsudin is thanked for all experimental data collection.

References
[1] FAO 2016 *The State of World Fisheries and Aquaculture 2016 Contributing to food security and nutrition for all.* ed Fisheries and Aquaculture Department (Rome: The Publishing Group in FAO's Office for Corporate Communication) p 200
[2] Department of Fisheries, Malaysia 2014 *Aquaculture Data Table* ed DOF (Malaysia: DOF)
[3] Pfeffer E 1990 *DTW Deutsche tierarztliche Wochenschrift* **97** 273-5
[4] Johansen R, Needham J, Colquhoun D, Poppe T and Smith A 2006 Guidelines for health and welfare monitoring of fish used in research *Lab. Animals* **40** 323-40
[5] FAO 2016 *National Aquaculture Legislation Overview - Malaysia. National Aquaculture Legislation Overview (NALO) Fact Sheets.* ed Skonhoft A (Rome: FAO Fisheries and Aquaculture Department) [cited 12 August 2016]. Available from: http://www.fao.org/fishery/legalframework/nalo_malaysia/en.
[6] Jesus J, Borges M, Calheiros C S and Castro P M 2012
[7] Effendi H, Utomo B A and Darmawangsa G M 2015 *Aquaculture, Aquarium, Conservation &* Legislation-*Int. J. Bioflux Soc. (AACL Bioflux)* **8**
[8] Olguin E, Rodriguez D, Sanchez G, Hernandez E and Ramirez M 2003 *Acta Biotechnol.* **23** 259-70
[9] Bhanthumnavin K and Mcgarry M G 1971 Wolffia arrhiza as a possible source of inexpensive protein.
[10] King C, McIntosh D and Fitzsimmons K 2004 *Giant salvinia (Salvinia molesta) as a partial feed for Nile tilapia (Oreochromis niloticus). Proc. of the 6th Int. Symp. on Tilapia in Aquaculture, Bureau of Fisheries and Aquatic Resources* 2004
[11] Mubarak M, Shaija A and Suchithra T V 2016 *Environ. Sci. Pollut. R* 1-9.
[12] Hussain N, Abbasi T and Abbasi S A 2016 *Ecol. Eng.* **91** 432-40
[13] Xu J and Deshusses M A 2015 *Int. J. Hydrogen Energ.* **40** 7028-36
[14] Cruz-Velásquez Y, Kijora C, Agudelo-Martínez V and Schulz C 2014 *Orinoquia* **18** 229-36
[15] Ng Y S and Chan D J C 2017 *J. WaterProc. Eng.* **15** 107-15
[16] Ward B B 2013 *Reference Module in Earth Systems and Environmental Sciences* ed Scott AE (Amsterdam: Elsevier)
[17] Vymazal J 2007 *Sci.Total Environ.* **380** 48-65
[18] Fang Y Y, Babourina O, Rengel Z, Yang X E and Pu P M 2007*Annals of Botany* **99** 365-70
[19] Ghaly A, Kamal M and Mahmoud N 2005 *Environ. Int.* **31** 1-13

[20] Gloger K, Rakocy J, Conter J, Bailey D, Cole W and Shultz K 1995 *Aquacultural Engineering and* Waste *Management Proc. from the Aquaculture ExpoVIII and Aquaculture in the Mid-Atlantic Conf.* 1995 USA (Washington, DC: Northeast Regional Agricultural Engineering Service) pp 272-300
[21] Azim M E, Wahab M A, van Dam A A and van Rooij J M 2002 *Aquatic Living Resour.* **15** 231-41
[22] Mridula R, Manissery J, Keshavanath P, Shankar K, Nandeesha M and Rajesh K 2003 *Bioresour. Technol.* **87** 263-7
[23] Akinbile C and Yusoff M S 2012, *Int. J. Phytoremediation* **14** 201-11
[24] Brix H 1997 *Water Sci.Ttechnol.* **35** 11-7
[25] Lin Y-F, Jing S-R, Lee D-Y and Wang T-W 2002 *Water Environ.Res.* **74** 136-41
[26] Ng Y S, Lim C R and Chan D J C 2016 *J. Environ. Chem. Eng.* **4**(4, Part B) 4890-6

IOP Conf. Series: Materials Science and Engineering **206** (2017) 012085 doi:10.1088/1757-899X/206/1/012085

Application of Sawdust modified with m-DMDHEU/ Choline chloride for the removal of CrO_4^{2-} and $H_2AsO_4^-$ in water

N T Dang[1], P T Quan, P H P Sang

Faculty of Chemical Engineering – Ho Chi Minh City University of Technology – Vietnam

[1]E-mail: thuongdang85@yahoo.com

Abstract. This article studies the use of *acacia auriculiformis* wood sawdust modified with 4,5-dihydroxy-1,3-bis (methoxymethyl) imidazolidin-2-one (m-DMDHEU) and choline chloride for separating CrO_4^{2-} and $H_2AsO_4^-$ ions in water. NaOH 0.2N/ ethanol 70° solution was used to remove lignin from the raw material, the material was then immersed in m-DMDHEU/choline chloride aqueous solution for 24 hours, after that the material was activated at 140°C for one hour. The ability to adsorb and exchange ions of the material was examined using solutions containing CrO_4^{2-} and $H_2AsO_4^-$ ions in different conditions. The results suggested that the ability to separate CrO_4^{2-} and $H_2AsO_4^-$ ions of the modified material was better than that of anion resin at pH = 7.0; the chromate adsorption capacity was the highest in acidic condition; the presence of arsenate (V) anions had no effect on the ability to remove chromate. Lastly, the modified material was used to treat water samples containing concentrations of arsenic similar to groundwater in several arsenic-contaminated areas of Vietnam.

1. Introduction

From mining activities, tanneries, electronics industries, metal plating, lead-acid battery manufacture, petroleum refining, or textile and dyeing industries, etc. a large amount of heavy metals like Fe, Cu, Pb, Ni, Cd, As, Hg, etc. has been discharged [1,4]. These heavy metal are directly associated with genetic mutation, cancer, as well as many adverse effects on the environment [1,2]. For developing countries like Vietnam, the industrial activities are usually conducted on a small scale or medium scale, thus wastewater treatment can be a difficult task because of the high cost. For that reason, agricultural by-products are being studied for water treatment due to their benefits: low cost, renewability, and they are mostly made of polymers which are easy to modify and have a high potency for adsorption/ion exchange [1-3,5].

Citric acid is commonly used as the modifying agent for cellulose to remove metals in cation form such as Fe, Cu, Pb, and Ni, etc. Compared with previous methods to modify cellulose, citric acid shows many advantages: low cost, simple condition for reaction, citric acid is non-toxic, modified material gives a high performance in ion separation [6-11].

To separate metals in anions such as chromate, arsenate, etc. have just had a few publicated researches on using modified cellulose [12-14]. Notably, Marshall et al. (2005) carried out experiments to modify soybean hulls, sugarcane bagasse, and corn stalks using DMDHEU and choline chloride (CC) as modifying agents. The separation of chromate, arsenate, and selenate ions was significantly enhanced [12].

In this research, sawdust was modified with 4,5-dihydroxy-1,3-bis(methoxymethyl)imdazolidon-2-one (m-DMDHEU) and choline chloride following the mechanism illustrated in Figure 1 [14].

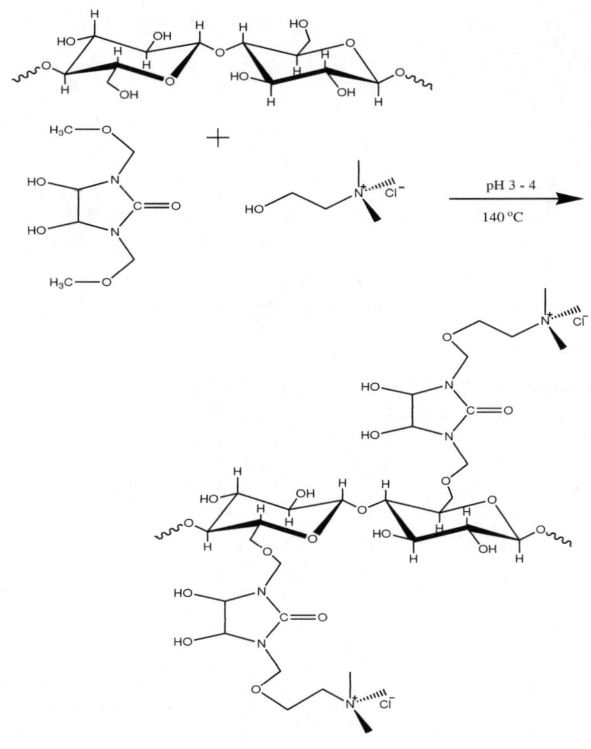

Figure 1. Cellulose modification with m-DMDHEU and choline chloride

After modification, there are $-N(CH_3)_3Cl$ groups on the cellulose chain, these groups play the main role in the ability to exchange ions of the modified material.

$$2R\text{-}N(CH_3)_3Cl + CrO_4^{2-} \longrightarrow (R\text{-}N(CH_3)_3)_2CrO_4 + 2Cl^-$$
$$R\text{-}N(CH_3)_3Cl + K^+ + CrO_4^{2-} \longrightarrow (R\text{-}N(CH_3)_3)KCrO_4 + Cl^-$$

2. Materials and methods

2.1 Materials

The chemicals used in the experiments are: m-DMDHEU 30% produced by BASF; choline chloride 98% made in India; K_2CrO_4; KH_2AsO_4; anion resin GA13 made in India; and distilled water. The desired solutions were prepared by mixing the chemicals with distilled water without any further treatment. Sawdust from acacia auriculiformis wood was used as the base material.

2.2 Process of sawdust modification

Sawdust was immersed in m-DMDHEU 5% and choline chloride 10% solution at a set pH (pH = 3÷4) for 24 hours so that the solution could infiltrate deep between the fibers; after 24 hours the material was filtered out, dried at room condition until its moisture content was lower than 20%; after that the material was activated at 140°C for 1 hour. The activated material was then rinsed thoroughly with distilled water, dried at room condition and stored in plastic bags.

2.3 Equipment

Measuring of the solid materials was done on SARTORIUS CPA2245 analytical scale and moisture content was determined with SARTORIUS MA35 moisture analyzer at the Laboratory of Organic Chemical Engineering, University of Technology, HCMC; chromate concentration was measured with SHIMADZU AA-6300 atomic absorption spectrophotometer and arsenate concentration was measured with Agilent 7500 ICP-MS system at the Laboratory of Analytical Chemistry, University of

IOP Conf. Series: Materials Science and Engineering **206** (2017) 012085 doi:10.1088/1757-899X/206/1/012085

Science, HCMC; FT-IR spectra of the material before and after modification were obtained at the Key Laboratory of Chemical Engineering and Petroleum Processing, University of Technology, HCMC.

3. Results and discussion

3.1 Effect of activation time

The sawdust was processed with procedure similar to that in section 2.2, the effect of activation time (ranging from 0.5 hours to 3 hours) was evaluated. After being rinsed with distilled water, the material was dried at room condition, then the material was weighed to determine the increase in the mass of material, the capacity of the material to exchange ions was also examined using 200 ml of solution containing chromate 200 ppm at pH = 7.6 and room temperature (28±5°C). The results are presented in Table 1.

Table 1. Effect of activation time on mass increase and ion-exchange capacity

Time (hour)	Dry matter (g)	Dry matter after activation (g)	Mass increase (g/g)	mmol CrO_4^{2-} /g of dry matter
0	0.9476	0.9666	0.02	0.012
0.5	0.9517	1.0278	0.08	0.071
1	0.9118	1.0851	0.19	0.101
1.5	0.9143	1.0423	0.14	0.074
2	0.9412	1.0542	0.12	0.065
3	0.9216	1.0322	0.12	0.054

There was a significant increase in mass and capacity after activation. At activation time of 0 hour the increase of mass was unnoticeable and the ion-exchange capacity was little. It can be explained as there was no reaction between m-DMDHEU/CC and sawdust at 0 hour, the chromate ions were merely adsorbed on to the surface of the material.

Maximum increase of mass (0.19 g/g) and maximum capacity (0.101 mmol/g) was achieved at 1 hour. A conclusion can be made that after 1 hour, there was no more reaction between CC and sawdust through m-DMDHEU linkage. By combining the results, the chosen optimum for activation time was 1 hour.

3.2 Effect of liquor ratio

The sawdust was processed with procedure similar to that in section 2.2, with liquor ratio ranging from 1:20 to 1:100. The results are presented in Table 2.

Table 2. Effect of liquor ratio on mass increase and ion-exchange capacity

Liquor ratio	Dry matter (g)	Dry matter after activation (g)	Mass increase (g/g)	mmol CrO_4^{2-} /g of dry matter
1:20	1.1612	1.3360	0.15	0.081
1:60	1.1684	1.3904	0.19	0.103
1:100	1.0896	1.2857	0.18	0.101

The mass increase and ion-exchange capacity of the material rose when the liquor ratio was raised from 1:20 to 1:60, and remained approximately the same when liquor ratio was raised from 1:60 to 1:100. This can be explained as when the liquor ratio is low, the diffusion of the substances into the fibers is hindered, resulting in a low impregnation efficiency.

3.3 FT-IR spectra of sawdust

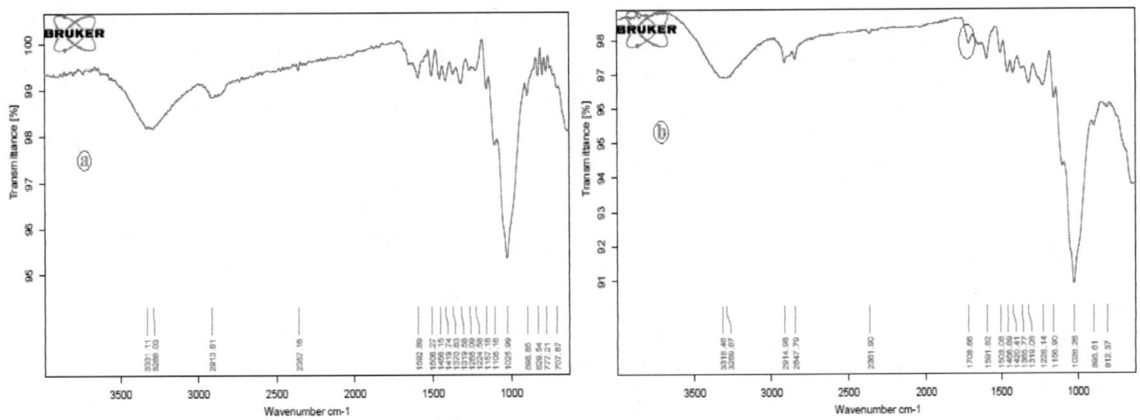

Figure 2. FT-IR spectra of sawdust before (a) and after (b) modification.

Compared to the spectrum of sawdust before modification, there is an additional peak at wavenumber 1708 cm^{-1} in the spectrum of modified sawdust, this indicates the presence of C=O group in the N-CO-N urea group. This indicates that there was reaction between m-DMDHEU, choline chloride, and sawdust.

3.4 Effect of pH on ion-exchange capacity

Three tests were carried out using a solution containing chromate 200 ppm, 200 ml of solution was used for each test, and each test was conducted at a different pH value. The results are presented in Table 3.

Table 3. Ion-exchange capacity at different pH values

pH	Condition	Dry matter (g)	mmol CrO_4^{2-}/g of dry matter
3.2	Acidic	1.07	0.161
6.8	Neutral	1.05	0.103
10.3	Basic	1.05	0.078

The results showed that in acidic condition, the ion-exchange capacity was the best. In basic condition, the capacity of the material dropped significantly, the reason is the presence of OH- ions, they occupied the ion exchanging sites and prevented contact with chromate ions, leading to a drop in ion-exchange capacity of the material. Meanwhile in acidic condition, chromate ions form dichromate ions ($Cr_2O_7^{2-}$), which have the same electrical charge as chromate ions (CrO_4^{2-}), but greater molecular mass, and double the amount of chromium atoms; in other words, with the same number of ion exchanging sites, the material can separate more chromate ions from water in acidic condition.

3.5 Evaluation of the ability to separate chromate and arsenate ions in a packed column

The solution containing CrO_4^{2-} and $H_2AsO_4^{-}$ ions at pH = 7.0 was used to assess and compare the efficiency of the modified sawdust with high-quality commercial anion resin. The columned used in the experiment had a diameter of 30 mm; the experiment was carried out at room temperature (28 ± 2 °C). The results are presented in Table 4.

Table 4. Efficiency of ion separation process in packed column

Target anion		CrO_4^{2-}	$H_2AsO_4^-$
Modified sawdust (g), (dry matter)		10	
Flow rate (ml/min)		5	
Total volume (ml)		200	
Concentration (ppm)		100	50
After using sawdust	Concentration (ppm)	0.87	0,04
	Efficiency (%)	99.1	99.9
After using anion resin	Concentration (ppm)	2.0	1.2
	Efficiency (%)	98.0	97.6

As shown in the results, the presence of arsenate ions did not interfere with the ability of the material to separate chromate ions. The efficiency of the ion exchanging process was almost ideal, yielding 99.1 % for chromate ions and 99.9 % for arsenate ions, meanwhile the values obtained with anion resin were 98.0 % and 97.6 %, respectively.

3.6 Treating water containing arsenic in a packed column

The modified sawdust was used to treat water with a high arsenic concentration similar to groundwater from tube wells in arsenic-contaminated areas in Vietnam. The arsenic concentration of the prepared water sample was 2500 ppb, which is 50 times higher than permitted in Vietnam National Technical Regulation QCVN 02:2009/BYT. The columned used in the experiment had a diameter of 30 mm; the experiment was carried out at room temperature (28 ± 2 °C); the flow rate was 5 ml/min; pH of the water sample was at 7. The results are presented in Table 5.

As seen in Table 5, the efficiency of the treatment was 98.2%. Furthermore, the concentration of arsenic after treatment was 44 ppb; this is below the permitted concentration.

Table 5. Efficiency of water treatment

Target element	As
Modified sawdust (g), (Dry matter)	5
Flow rate (ml/min)	5

IOP Conf. Series: Materials Science and Engineering **206** (2017) 012085 doi:10.1088/1757-899X/206/1/012085

Total volume (ml)	100
Concentration (ppb)	2500
Concentration after treatment (ppb)	44
Efficiency (%)	98,2
Permitted concentration (ppb)	50

4. Conclusion

From the results of this study, it can be concluded that there was a significant increase in the ability of sawdust to adsorb and exchange chromate ions after being modified with m-DMDHEU and choline chloride. Compared with commercial anion resin, the performance in separating chromate and arsenate ions of modified sawdust was better at pH=7.0. Modified sawdust can be applied for treating contaminated groundwater with arsenic concentration 50 times higher than permitted in Vietnam National Technical Regulations. In conclusion, m-DMDHEU was proven to be an effective agent to link choline chloride to cellulose chain; this suggested a new approach to produce lignocellulosic anion exchangers from other agricultural by-products to separate heavy metal anions in aqueous solution [14].

References

[1] Babarinde N A A, Babalola J O and Sanni R A 2006 *Int. J. Phys. Sci.* **1**(1) pp. 023-026.
[2] Han J S 1999 *2nd Inter-Regional Conf. on Environment-Water* **99**
[3] Igwe J C, Nwokennaya E C and Abia A A 2005 *African J. Biotechnol.* **4**(10) pp. 1109-1112.
[4] Patterson J W 1985, *Industrial Wastewater Treatment Technology, Second Edition, Jenson* book *Inc.*
[5] Marshall W E, Chatters A Z, Wartelle L H, McAloon A 2001, *Ind. Crops Prod.* **14** pp. 191-199
[6] Mc Sweeny J D, Rowell R M, Min S H 2006 *J. Nat. Fibers* **3**(1) pp. 43-58
[7] Marshall W E, Wartelle L H, Boler D E, Johns M M, Toles C A 1999 *Bioresour. Technol.* **69**(3) pp. 263-268
[8] Khan N A, Ibrahim S and Subramaniam P 2004 *Malaysian J. Sci.* **23** pp. 43-51
[9] L T Hung, P T Quan, N M Tam, N X Thom 2008, *J Sci. & Technology Development* (ISSN 1859-0218), **11**(8), pp. 5-12.
[10] Quan P T, N T Dang N T and Danh T T 2010 *J. Chem.* **48-4C** pp.490-495
[11] Quan P T, Dang N T and Hue C M 2012 *J. Sci. Technol.* **50**(3A) pp. 265-270
[12] Marshall W E and Wartelle L H 2005 *Water Res.* **39** pp. 2869–2876.
[13] Bajpai S K and Rohit V K 2009 *J. Environ. Protect. Sci.* **3** pp. 23-33
[14] Dang N T, Quan P T and Trung T C 2016 *J. Sci. Technol. Dev.* **19**(K5) pp.62-68

IOP Conf. Series: Materials Science and Engineering **206** (2017) 012086 doi:10.1088/1757-899X/206/1/012086

A Review on overboard CEOR discharged produced water treatment and remediation

H Rawindran[1,2], S Krishnan[1,2], C M Sinnathambi*[1,2]

1. Centre of Research Enhanced Oil Recovery (COREOR), Universiti Teknologi Petronas (UTP), Seri Iskandar, 32610 Tronoh, Perak, Malaysia
2. Department of Fundamental and Applied Sciences, Universiti Teknologi Petronas (UTP), Seri Iskandar, 32610 Tronoh, Perak, Malaysia

E-mail: chandro@utp.edu.my

Abstract. Produced water is a waste by-product generated during oil and gas recovery operations. It contains the mixture of organic and inorganic compounds. Produced water management is a challenge faced by the petroleum practitioners worldwide. Build-up of chemical wastes from produced water causes huge footprint, which results in high CapEx and OpEx. Different technologies are practiced by various practitioners to treat the produced waste water. However, the constituents removed by each technology and the degree of organic compound removal has to be considered to identify the potential and effective treatment technologies for offshore industrial applications. Current produced water technologies and their successful applications have advantages and disadvantages and can be ranked on the basis of several factors, such as their discharge limit into water bodies, reinjection in producing well, or for any miscellaneous beneficial use. This paper attempts to provide a review of existing physical and chemical treatment technologies used for management of produced water. Based on our analysis, suitable methods will be recommended for offshore waste water treatment technologies.

1. Introduction

Oil production is separated into three phases; primary process, which involves the usage of artificial lift for oil recovery. Secondary process utilizes water alternating gas injection (WAG) method. These two recovery methods leave up to 70% oil in the well. Tertiary oil recovery method plays its role to recover the conventional primary and secondary oil recovery phase could not produce. There are several EOR techniques available in the industry; which includes, Chemical (C_{EOR}), Thermal (T_{EOR}), and miscible displacement process [1].

Oil recovery method has become highly effective and technological, therefore there are always renewed interest in enhanced oil recovery (EOR), specifically chemical enhanced oil recovery (C¬EOR). C¬EOR has the potential to unlock million barrels of unrecoverable oil from offshore reservoirs. Nevertheless, several technical and regulatory limitations has to be met before these techniques are implemented by the industry.

During the oil production, produced water that comes along with the recovered oil has to go through a series of treatment before it is discharged to the water bodies. Produced water is one of the largest by-product stream associated with oil and gas production. Therefore, driven by the urgent need to develop cost-effective produced-water management for the best interest of society, a lot of conventional as well as new treatment options has been introduced. This is particularly applicable to the treatment of produced water in offshore EOR applications, such as CEOR [2].Treatment of produced water can lead to value through incremental oil recovery from water flood projects and by incurring a lower cost for disposal via injection of clean water [3].

IOP Conf. Series: Materials Science and Engineering **206** (2017) 012086 doi:10.1088/1757-899X/206/1/012086

1.1 Produced Water

Oil is the primary energy sources worldwide. In offshore platform, there is naturally trapped water underground called formation water. During oil recovery, this water along with the injected chemicals will flow through the well. The water that comes up to the surface along with the oil and gas is called the produced water.

Produced water originated from the formation water that was in contact with hydrocarbon-bearing formations. At the surface, hydrocarbons are separated from the produced fluid or produced water [4]. Therefore, it contains some of the chemical characteristics of the formations as well as the hydrocarbons. It may also include water from the reservoir, water previously injected into the formation, and residuals of those chemicals added during the production process [6].

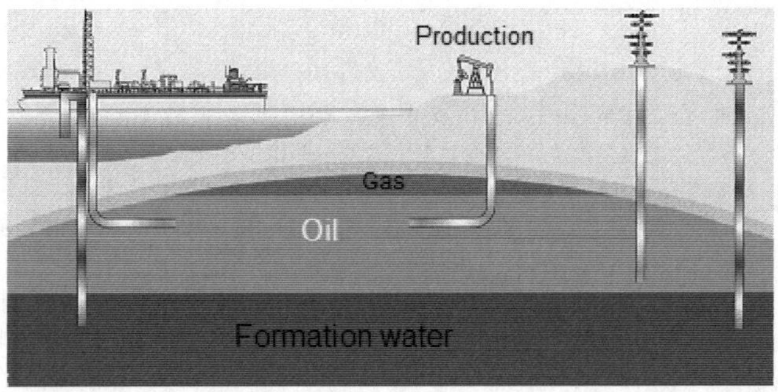

Figure 1. Illustration on the overview of oil and gas platform [5]

1.2 Characteristics of Produced Water

Produced water can be characterized as saline water [7]. Generally, the constituents and characteristics of chemical substances in the produced water are interrelated to the geological characteristics of their respective reservoirs. The quality of produced water varies among the wells and fields. Major constituents of produced water are; salt content (Degree of salt present is expressed as salinity, in terms of total dissolved solids (TDS) or electrical conductivity), oil and grease (this is measured by an analytical test that measures the group that organic compounds belong to), natural inorganic and organic compounds (Ca, Mg, Ba and S are some of the chemicals that causes hardness and scaling) [8]. Other than these, components such as dispersed oil, aromatic hydrocarbons, alkylphenols (AP), heavy metals, and naturally occurring radioactive material (NORM) are some of the chemical additives that are used during drilling, fracturing and well operation which may hold toxic properties that exist in the produced water and serve as environmental threat [9].

2. Literature Review

2.1 Treatment Technologies of Produced Water

The offshore oil and gas industry generates hundreds or thousands litres of produced water from offshore oil production facilities in daily basis. Most of them are discharged into the ocean [10]. Operational discharges of produced water from offshore oil and gas platforms are a continuous source of contaminants to the ecosystems. Often disposal is chosen as an option to manage the produced water, as that is the cost effective method. But, prior to the disposal, many researches as well as technologies has been developed. This is to ensure that the environmental regulations are met.

Regulators usually specify the standards of pollutant discharges to offshore operators based on the best available technology (BAT). This BAT for the treatment of produced water has to be efficient in removing most free oil before discharged into the water bodies. Therefore, the oil that remains in the produced water is discharged into the ocean in dispersed and dissolved forms [10].

2.2 Physical Treatment of Produced Water
To meet up with these objectives, the regulators have developed treatment options for produced water management by physical or chemical means [11]. Treatment of produced water has the potential to be a harmless and valuable product rather than a waste. Some of these options are reviewed in the next section.

The treatment goals are essentially the same, i.e.; for reuse, reinjection or disposal, although the level of organic pollutant removal required for reuse can be significantly higher, depending on the quality of the produced water. Therefore, some of the treatment technologies employed are:

2.2.1 Physical Adsorption
Adsorption process is the physical adhesion of the contaminant chemicals onto a solid surface [12]. Adsorption process can be carried out by utilizing zeolites, organoclays, activated alumina and activated carbon. These adsorbents are potential of removing chemicals such as iron, manganese, total organic carbon, BTEX compounds, heavy metals, as well as oil from produced water. Organic compounds of produced water therefore adhere to porous media of carbon surfaces. After several experiments, the wet air oxidation process can regenerate activated carbon. This activated carbon can remove soluble BTEX which differs from organoclay that can remove insoluble free hydrocarbons. Organoclay is a mixture of sodium montmorillonite clay with a cationic quaternary amine salt. When organoclay is used in conjunction with activated carbon, hydrocarbon concentration falls below water quality standards. Copolymer beads are prepared based on methylmethacrylate (MMA) and divinylbenzene (DVB) by suspension polymerization. These copolymers can reduce oil content of produced water to around 85 %[14].

Zeolites are commonly used as ion-exchange resins. This hydrophobic zeolite pellets are usually in a fixed bed to adsorb dissolved organic compounds in produced water. At the same time, another study conducted by Mitchell et al, showed that a resin-filled column can be used to eliminate soluble organic compounds. The similarity of these two processes is, acid backwash and solvents can regenerate the adsorbers [4].

Adsorbents are capable of being exerted with large concentrations of organics, which makes this process to serve best as a polishing step rather than as a primary treatment process. This process works on the aid of gravity and does not require any energy, except for the backwash process [13].

Media usage rate is known as one of the main operational costs for adsorptive processes. When all active site of the adsorptive material have been consumed, the material must either be regenerated or disposed. Regenerating the materials will result in a liquid waste for disposal. Solid waste disposal is necessary when the material needs to be replaced entirely.

2.2.2 Hydrocyclone
The function of hydrocyclones is to separate solids from liquids according to the density of the material. Hydrocyclones usually has a conical base and cylindrical section at the top portion, where the liquid is fed tangentially (Figure 2).

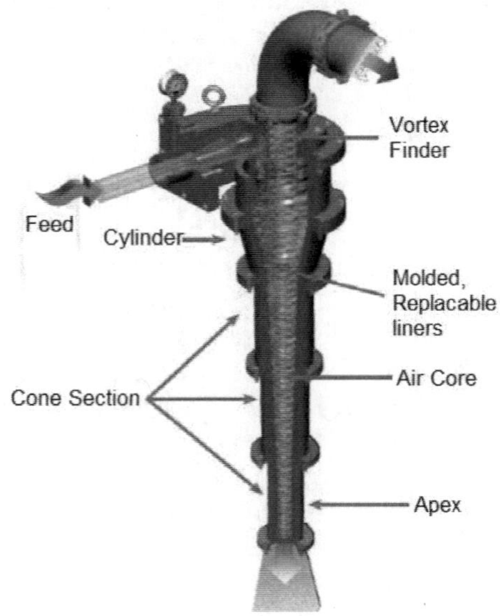

Figure 2. Overview of hydrocyclone structure [10].

The utilization of hydrocyclones do not require any pre- or post-treatment nor any chemicals or energy. There will be no energy consumption with this equipment usage, as long as the plant setup needs to deliver water to the hydrocyclone through forwarding pump. Depending on the size and configuration of the hydrocyclone, a large pressure drop can occur across the hydrocyclone. Hydrocyclone generates waste in the slurry form of concentrated solids. This is the product from this equipment that requires disposal [13].

Compared to the conventional produced water treatment system, hydrocylones offers offshore operators many advantages. This includes its benefits of being light weight, compact size, easy and reliable operation, and low maintenance, complete insensitivity to platform motions, low utility requirements and highly predictable capacity. When water production increases, hydrocyclones can be added without extensive piping modifications or additional costly construction. All these features are present at lower costs for a non-revenue producing function, and provides a cost effective alternative for meeting stringent environmental standards [5]. However, the disadvantage of using a hydrocyclone is that, this equipment has low efficiency and it is unable to remove dissolved components [16].

2.2.3 Evaporation method

Evaporation does not require any kind of chemical treatment that prevents the risk of secondary sludge handling. This process also does not require highly skilled labour. On the contrary, the requirement of energy for evaporation process is very high especially in the offshore platform, which could lead to increase in operating cost. The energy consumption could be brought down by reusing hot vapor to heat the fresh feed [15].

There are several types of evaporation method present for produced water treatment; such as, vertical tube, falling film, and vapour compression evaporation. These methods are effective because they eliminate physical and chemical treatments. Hence, no chemical sludge is produced, causing costs of waste and life cycle to be lowered [4].

A study conducted by Heins et.al, states that evaporation process by generating steam can be carried out at any heavy oil production to recover up to 98% of the produced water as high quality distillate (<10 ppm [mg/L] non-volatile inorganic TDS) for reuse or reinjection [15]. By using this

IOP Conf. Series: Materials Science and Engineering **206** (2017) 012086 doi:10.1088/1757-899X/206/1/012086

approach, the overall life cycle costs were found to be lesser for an evaporative produced water treatment system compared to the conventional approach [16].

2.3 Chemical Treatment of Produced Water

2.3.1 Demulsifiers
Demulsifiers functions to improve the oil and produced water separation. Depending on the respective demulsifier properties, the chemicals will partition between the oil and the water in the process. The reaction products and residual chemicals from those that partition to the water phase will be released to the water bodies with the produced water discharge. They reduce the oil-water interfacial tension. Demulsifiers are surface active agents that would disrupt the effects of surfactants. But a number of solids like silts, iron sulphide and paraffin, etc., present in the crude oil complicate the process [15].

2.3.2 Advanced Oxidation Process
Advanced oxidation process (AOP) involves the generation of highly reactive hydroxyl radicle (HO•) and this have emerged as one of the important and promising technology to reduce chemical contaminants and toxicity. The degraded effluent can be reintroduced to receiving streams and/or to conventional sewage streams. One class of AOP utilizes photo-activated reactions for degradation of surfactant. These reactions are characterized by the free radical mechanism initiated by the photon interaction with a specific amount of energy with a catalyst (photocatalytic oxidation). *Madjene et al.,* stated that photocatalytic oxidation reactions tend to completely mineralize organic compound to carbon dioxide, water vapor and inorganic substance by UV light (Figure 3) [18].

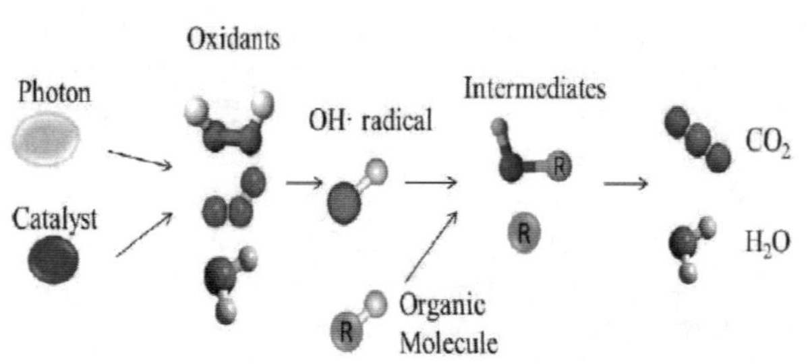

Figure 3. Degradation process using AOP process [17]

Therefore, attack by (HO•) leads to mineralization of the organic compounds and AOP results in reducing the contaminants concentration from several hundred ppm to less than 5ppb. The strength of the free radical depends on its relative oxidation power (ROP). Various oxidizing species are available for this application. Table 1 listed the various relative ROP.

IOP Conf. Series: Materials Science and Engineering **206** (2017) 012086 doi:10.1088/1757-899X/206/1/012086

Table 1. Relative oxidation power of oxidation species [18]

Oxidizing Species	Relative Oxidation Power
Chlorine	1.00
Hypochlorous Acid	1.10
Permanganate	1.24
Hydrogen Peroxide	1.31
Ozone	1.52
Atomic Oxygen	1.78
Hydroxyl Radical	2.05
Titania^{+}	2.35

AOP's offer many ways for hydroxyl generation using non-photochemical or photochemical (with UV radiation) methods. The UV radiation acts as a source of energy to activate and enhance the production of hydroxyl radicals from various sources like water. Unfortunately HO● is non-selective but powerful. Targeting the surfactant using activated HO● makes AOP technique a better and favourable option.

Figure 4, summaries the different AOP technologies, and the reactive species produced for the degradation of organic pollutants.

AOP	Reactive species
Ozone treatment: O_3	$^{\bullet}OH$, $HO_2^{\bullet}$, $HO_3^{\bullet}$, $O_2^{-\bullet}$, $O_3^{-\bullet}$
O_3/H_2O_2	$^{\bullet}OH$, $O_2^{-\bullet}$, $O_3^{-\bullet}$
Fenton processes: H_2O_2/Fe^{2+}	$^{\bullet}OH$, $HO_2^{\bullet}$
Photo-Fenton processes	$^{\bullet}OH$
UV/O_3, UV/H_2O_2 and $UV/O_3/H_2O_2$	$^{\bullet}OH$, $HO_2^{\bullet}/O_2^{-\bullet}$, $O_3^{-\bullet}$
V–UV ($\lambda < 190$ nm)	$^{\bullet}OH$, H^+, e^-_{aq}
Photocatalytic treatment: UV/Vis light using TiO_2, ZnO, etc. as catalysts	$^{\bullet}OH$, h^+, $O_2^{-\bullet}$, e–, $^1O_2$, $HO_2^{\bullet}$, $HOO^{\bullet}$
Ultrasonic treatment	$^{\bullet}OH$, $^{\bullet}H$
γ-Radiolysis	$^{\bullet}OH$, $^{\bullet}H$, e^-_{aq}

Figure 4. Reactive Species for Different AOP's [18]

2.3.3 Chemical precipitation

Suspended solids and colloidal particles can be removed by the means of coagulation and flocculation. There are few coagulants, such as modified hot lime and FMA (a mixed metal polymer), calcite and ferric ions used as coagulant to treat produced water. In a study conducted by Fakhrul et. al, modified hot lime process produced water containing 2000 ppm hardness, 500 ppm sulfides, 10,000 ppm TDS, and 200 ppm oil could be successfully converted to steam generator-quality feed-water[15].

In another study on treatment of offshore produced water, an oxidant, ferric ions, and few flocculants were used to remove hydrocarbons, arsenic, and mercury. The oxidation–reduction potential of the wastewater was controlled by oxidant addition to allow the required arsenic oxidation to occur while maintaining the mercury in elemental form. Results showed that effluent streams had less than about 10 parts per billion (ppb) of mercury (calculated as Hg), less than about 250 ppb, and preferably less than 100 ppb of arsenic (calculated as As), and less than about 40 ppm of TPH. The major drawback of this process is its ineffectiveness in treating for dissolved components and increased concentration of metals in the sludge formed.

2.5 Comparative studies of physical and chemical treatment options.

Conventionally, physical treatment has been used as produced water treatment option. However, the physical technique does not degrade the pollutants, but only isolate or traps them [19]. Commonly, physical processes of treatment are considered to be less environmentally friendly, costly and capable of generating large volumes of chemical sludge. These technique usually come with many limitations, especially in offshore operation. The major problem is that, the treatment of high volume effluent requires high residence time for separation. Another problem is the build-up of chemical waste or sludge, which is environmental hazard. This is unfavourable for offshore operation as it incurs large storage, huge footprint resulting in high operating and capital costs together with its environmental unfriendly by-products. As a consequence, chemical techniques are found to be suitable for destruction of organic pollutant (mineralization) and disposed as safe by-product as a result protecting the environment.

3. Conclusion

In the treatment of produced water, it is necessary to ensure that the level of contaminants/ organic compounds present in the water are within the acceptable range set by the regulators. Achieving the various treatment goals requires the use of multiple treatment technologies, including physical and chemical treatment processes. Current produced water technologies and their successful applications have advantages and disadvantages and can be ranked on the basis of their efficiency in eliminating the organic pollutant in produced water. Therefore, from the literature reviews done, the suitable technology for produced water treatment will be the combination of both physical and chemical process. This will aid in the trapping as well as the degradation of contaminants in produced water.

Acknowledgement

The authors would like to thank Universiti Teknologi PETRONAS (UTP) and the Enhanced Oil Recovery, mission oriented research division for their financial support. We would also like to express our sincere and honest gratitude to colleagues from UTP and PETRONAS Research Sdn Bhd (PRSB), and friends who have guided and assisted us during the study.

References

[1] Pwaga S, Iluore C, Hundseth O, Perales F J and Idrees M U 2010 Comparative Study of Different EOR Methods *Department of Petroleum Engineering* Norwegian University of Science & Technology, Trondheim, Norway

[2] Boysen B, Henthorne L, Johnson H and Turner B 2013 *Project System Technologies Oil Gas Facilitie*

[3] Duhon H 2012 *Oil Gas Fac.* **1** (1) 29-30

[4] Fakhru'l-Razi A, Pendashteh A, Abdullah L C, Biak D R A, Madaeni S S, and Abidin Z Z 2009 *J. Hazard. Mater.* **170** (2) 530-551

[5] Westgarth V W S 2014 Produced Water, http://www.vwswestgarth.com/en/markets/offshoreoilandgas/producedwater/

[6] Veil J 2015 U.S. Produced Water Volumes and Management Practices in 2012, Ground Water Protection Council United States.

[7] James P R and Engelhardt F R 1992 Produced Water: Technological/Environmental Issues and Solutions, Pittsburgh, Pennsylvania.

[8] NPC North American Resource Development Study 2011 Management of Produced Water from Oil and Gas Wells.

[9] Neff J, Lee K and De Blois E M, 2011 Produced water: overview of composition, fates, and Produced Water, *Springer*, New York
[10] Fraser G S, Russell J and Von Zharen W M 2006 *J. Marine Ornithology* **34** 147–156
[11] Ebenezer T I and George Z C 2012 *Int. J. Low Carbon Technol.* ***0*** 1-21.
[12] Okiel K, Mona El-Sayedb and El-Kadyc M Y 2011 *Egyptian J. Petrol.* **20** 9–15
[13] Drewes J E, Cath T Y, Xu P, Graydon J, Veil J and Snyder S 2009 An integrated framework for treatment and management of produced water RPSEA Project, 07122-12.
[14] Choi M. S 1990 Hydrocyclone produced water treatment for offshore developments *SPE Annual Technical Conference and Exhibition. Society of Petroleum Engineers*.
[15] Duraisamy R T, Beni A H and Henni A 2013 *State of the art treatment of produced water* INTECH Open Access Publisher.
[16] Heins W and Peterson D 2005, *J. Canadian Petrol. Technol.* 44 01.
[17] Madjene F, .Lamine J, Sadek I, Hafida L and Belkacem B 2013 *J. Sci. Technol.* **3** 10
[18] Bokare A. D and Choi W 2014, *J. Hazard. Mater.* **275** 121-135
[19] Ikehata K and El-Din M G 2004 *Ozone: Sci. Eng.* **26** (4) pp 327-343

IOP Conf. Series: Materials Science and Engineering **206** (2017) 012087 doi:10.1088/1757-899X/206/1/012087

Preliminary Studies on Oleochemical Wastewater Treatment using Submerged Bed Biofilm Reactor (SBBR)

Z Ismail[1], N A N Mahmood[1], U S A Ghafar[1], N A Umor[2], S A F Muhammad[1*],

[1]Department of Bioprocess and Polymer Engineering, Faculty of Chemical and Energy Engineering, Universiti Teknologi Malaysia, 81310 UTM Skudai, Johor Bahru, Johor, Malaysia
[2]Department of Microbiology, Faculty of Applied Science, University Teknologi MARA, 72000 UiTM Kuala Pilah, Negeri Sembilan, Malaysia

E-mail: syed@utm.my

Abstract. Wastewater discharge from the industry into water sources is one of the main reason for water pollution. The oleochemicals industry effluent produces high content of chemical oxygen demand (COD) with value between 6000-20,000 ppm. Effective treatment is required before wastewater effluent is discharged to environment. The aim of the study is to develop submerged bed biofilm reactor (SBBR) with packing materials in the cosmoball® carrier. Water quality such as chemical oxygen demands (COD), turbidity and pH were analysed. The result shows that the initial COD of 6000 ppm was reduced below 200 ppm. The optimum conditions for SBBR were obtained when green sponges used as packing material in cosmoball®; effluent flowrate set at 100 mL/min; 1:1 ratio of cosmoball® volume to reactor volume and 1:1 ratio of active sludge (mixed culture) volume to reactor volume. Turbidity and pH were recorded with 9.0 NTU and 7.0 respectively, which indicated that SBBR is feasible as an alternative for conventional biological treatment in oleochemical industry.

1. Introduction

Oleochemicals industry is one of major industries in Malaysia that at some instance produces effluent and affecting the water quality in the environment. This is due to large consumption of mixture of water and chemicals for processing crude palm oil. Without proper treatment, wastewater effluent produced eventually contributes serious and long lasting consequences to human and life [1]. A oleochemical wastewater plant usually generates around of 200 m^3 of wastewater daily and mainly comes from the processes of degumming, deacidification, neutralization, bleaching, and deodorization [2]. However, current oleochemicals wastewater treatment is an inefficient, consuming higher hydraulic retention time and use abundant resources [3]. It has been reported that, the treatments are consisting of physical, chemical and biological treatment which consume high hydraulic retention time and requires huge space. Furthermore, activated sludge system which is one type of biological treatment in conventional oleochemical wastewater treatment plant has higher operating costs [4].

In compliance to Environmental Quality Act 1974, which consists of the Environmental Quality (Sewage) Regulations 2009 and Environmental Quality (Industrial Effluent) Regulations 2009, the

IOP Conf. Series: Materials Science and Engineering **206** (2017) 012087 doi:10.1088/1757-899X/206/1/012087

water quality index is use to determine the quality of the surface water and the suitability of surface water for irrigation. Based on the designated classifications in the National Water Quality Standards for Malaysia established by DOE [5], the water quality of Chemical Oxygen Demands (COD) should be below 200 ppm before it can be released into the environment.

1.1. Submerged Bed Biofilm Reactor (SBBR)

Submerged biofilms systems including fixed and suspended carriers have been proven highly effective and efficient in removing organic carbon and nitrogen by attached growing biofilms. The system exhibit lots of advantages as compared to semi-batch activated sludge processes. The advantages are stability and long retention time for formation of microorganisms' biofilm that enables removal of pollutant, much less formation of surplus biomass or sludge [6], needs less air for aeration, can be upgraded to fit increased population, effective for BOD and nitrogen removal with no odor release and quick restarting during power failure [7][8].

1.2. Cosmoball® Carrier

Cosmoball® carrier is a new alternative system used for biological wastewater treatment. Cosmoball® carrier can be used for both aerobic and anaerobic biological treatment that made of strong polyethylene plastics which can resist corrosive and hazardous wastewater [7]. Previous studies show that cosmoball® carrier was used effectively in industrial wastewater treatment of textile wastewater which reduced the COD approximately about 60% [9-10]. Similar studies showed 80% reduction of COD for sewage treatment plant (Ekhmaj, 2012). Installation of new treatment system like cosmoball® in submerged bed biofilm reactor (SBBR) is likely to improve biological treatment in oleochemical plant. In this present study, the cosmoball® treatment system that fitted in SBBR is developed and analyzed for its feasibility and effectiveness in the treatment of oleochemical wastewater.

2. Materials and methods

2.1. Sampling of Oleochemical Wastewater

Samples of wastewater was obtained from Natural Oleochemicals (NatOleo) Sdn. Bhd. that located in the Pasir Gudang Industrial Area, Johor, Malaysia. Meanwhile, at the same location the lab scale of the SBBR was installed with continuous flow of wastewater. Table 1, shows the characteristics of wastewater samples taking from wastewater treatment plant in NatOleo.

Table 1. Characteristics of Wastewater

Characteristics	Value
Flowrate, m^3/day	100
Chemical Oxygen Demands, mg/L	10000
Biological Oxygen Demands, mg/L	1000-2000
Temperature, °C	<60
pH	8-10
Suspended Solids, mg/L	200-400
Oil and Grease, mg/L	1000

2.2. Design of Aerobic SBBR Lab Scale

The SBBR was made from Perspex glass with a wall thickness of 5.0 mm. Perspex glass was used because easy to handle, low cost and transparent. Each SBBR tanks has dimensions of 30.0 cm length,

30.0 cm width and 45.0 cm of height. Three SBBR tanks were connected serially using tubes that will allow wastewater to flow continuously.

2.3. Experimental Procedure

The SBBR tank system was labelled as SBBR1, SBBR2 and SBBR3 as shown in Figure 1. The system was designed for maximum reaction and contact area between wastewater and the carrier that will improve the degradation rate of the contaminant content in the wastewater

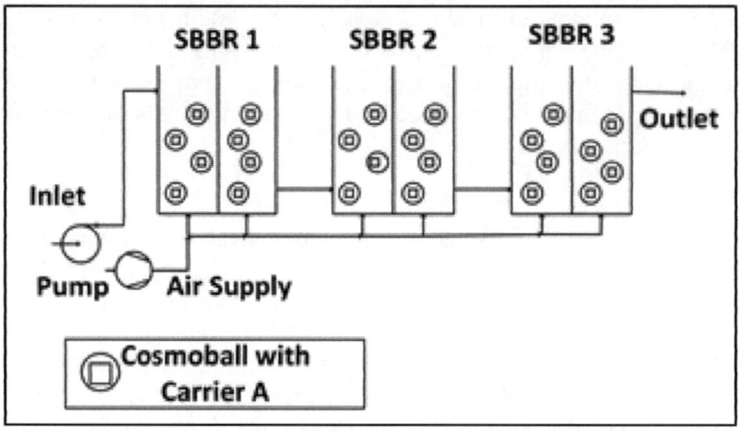

Figure 1. Experimental Setup of the SBBR in the lab scale

Carrier A as shown in Figure 2 was added inside the cosmoball® carrier to test whether such packing material will improve the quality of Palm Oil Refinery (POR) wastewater. Carrier A is rectangular sponge pads with rough fabric surface and non-scratches material. Carrier A has the size of 14.7cm of length, 9.7cm of width and 0.7cm of thickness. Filling ratio of the cosmoball® inside the SBBR tank was 50% by volume of the tank. Previously, it was observed that 50% of the carrier filling ratio was the optimum condition for this type of treatment [11]. The tank was inoculated with the activated sludge that was taken from the biological clarifier at the NatOleo. The ratio of the inoculated activated sludge to reactor volume is 1:1. The experiment was operated with flowrates of 100 mL/min for the wastewater effluent. Sample of the wastewater was taken daily (every 24 hours) at 0800 hour for the effluent of first SBBR1 and effluent of the third SBBR3. The water quality was analysed daily. The SBBR tanks were cleaned once a week before run the new set of experiment.

Figure 2. Carrier A pack in the Cosmoball®

2.4. Analytical Methods

Wastewater effluent in the SBBR1 and SBBR3 were analysed according to the Standard Methods for Examination of Water and Wastewater [12]. The parameters that were analysed; pH, COD and

Turbidity. The analysis was done trice for each set of experiment to get the average water quality results.

3. Results and discussions

3.1. Treatment Efficiency

As shown in Figure 3, COD effluent after the treatment with and without carrier. COD is a measure of oxygen equivalent of the organic matter in a water sample that is susceptible to oxidation by a strong chemical oxidant [13]. The initial COD of the oleochemical wastewater after chemical treatment was 6000 ppm. However, by combining carrier A with the flowrate of the 100 mL/min and 50% of sludge volume, the system was capable to reduce the COD effluent to 153 ppm which is about 97% of COD reduction for the first day. The usage of cosmoball® only shows the COD reduction from 6000 ppm to 1700 ppm which is only 71.5% COD reduction. This indicates that large surface area of cosmoball® with carrier A as the packing material helps to increase the degradation rate of the organic matter in the wastewater and similar observation was documented which used K1 packing material for combination of hybrid MBR-MBBR where the wastewater had 90% of COD reduction [14]. The present cosmoball® packed with carrier A was more efficient as previous researcher that succeeded to reduce 50% to 90% of COD [15-17].

Turbidity in the wastewater indicates the amount of suspended materials contained in the wastewater. Too high-suspended solid will eventually clog and increase the turbidity value [18]. However, most of the suspended solids in the oleochemicals wastewater had been removed during the pre-treatment like physical treatment that use to separate solids from the water. Figure 4, shows the turbidity percentage reduction for both cosmoball® and combination of cosmoball® and carrier A from the initial turbidity after chemical treatment. It was observed that by using the combination of the cosmoball® with carrier A and cosmoball® only, both succeed in 92% and 83% of the turbidity reduction for each set. It was hypothesized that the cosmoball® with carrier A not only acted as the home for attached growth bacteria for the degradation of the contaminant in the wastewater but also acted as the filter to the suspended solids [19]. The uniqueness design of the cosmoball® with carrier A make it less prone clogging as the 85% excess of void spaces provided [7]. So, the combination of the cosmoball® and carrier A can help for better COD and turbidity reduction in the wastewater treatment. Table 2 and 3 shows the COD and turbidity analysis of mean and standard deviation for both sets of experiments.

pH is also one of the factors that indicates the water quality and measures the acidity or alkalinity of the water. Figure 5 shows the pH for both set of experiments which is combination of the cosmoball® and carrier A and cosmoball® only. Both sets succeed in achieved neutral pH of 7 after the treatment. This indicates that mixed culture bacteria succeed in pH adjustment due to the bacteria act in dynamic population. This mixed culture bacteria had a concept of mix well defined microbe with high activity and the risk of contaminated is reduced because the diversity of inoculum can out bet the contaminating organisms [20].

Table 2. COD Analysis

Types and Analysis	Days	Mean (ppm)	Standard deviation
Cosmoball® (COD Analysis)	1	2000	14.14
	2	1880	28.28
	3	1800	21.21
	4	1750	28.28
	5	1710	21.21
	6	1700	21.21
	7	1700	21.21

Cosmoball® with Carrier A (COD Analysis)	1	185	2.82
	2	230	4.24
	3	250	14.14
	4	200	14.14
	5	153	2.82
	6	160	4.24
	7	155	1.41

Table 3. Turbidity Analysis

Types and Analysis	Days	Mean(ntu)	Standard deviation
Cosmoball® (Turbidity Analysis)	1	90	0
	2	82	1.41
	3	65	2.82
	4	34	1.41
	5	30	0
	6	25	1.41
	7	17	0
Cosmoball® with Carrier A (Turbidity Analysis)	1	88	2.82
	2	58	0
	3	23	4.24
	4	24	2.82
	5	23	0
	6	17	2.82
	7	9	1.41

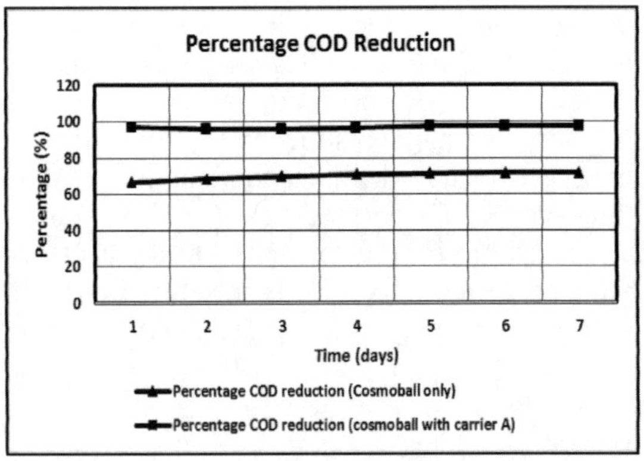

Figure 3. Percentage COD Reduction

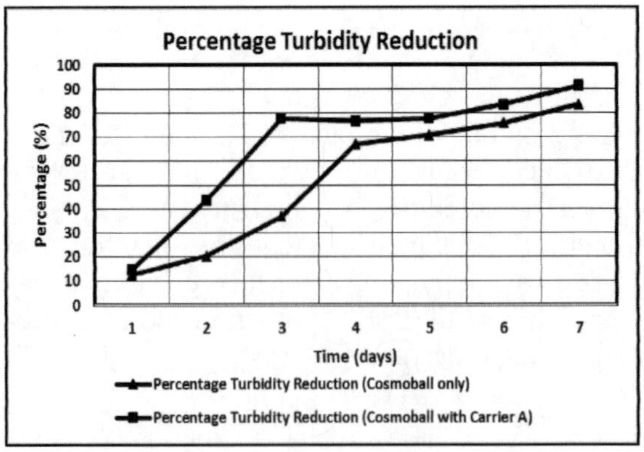

Figure 4. Percentage Turbidity Reduction

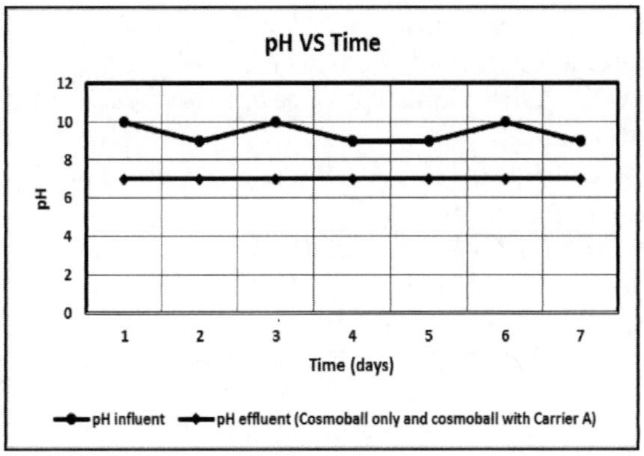

Figure 5. pH of Wastewater

4. Conclusion

The SBBR wastewater treatment system was demonstrated for its ability to treat oleochemical wastewater treatment with high reduction of COD with 97% reductions, the turbidity 92% reductions and neutralized the pH of the wastewater. The combination of cosmoball® and carrier A has been proven to be the best condition for efficient treatment of oleochemical wastewater.

Acknowledgement

The authors are grateful for financial support from the Universiti Teknologi Malaysia (UTM) and Malaysian Ministry of Higher Education by providing Research University Grant (RUG) Tier 1.0 PY/2014/03724. The authors graciously thank Ya'akob Sabudin from the Department of Bioprocess & Polymer Engineering and Halilah Yusof from NatOleo for providing technical support.

References

[1] Abbasi M, Sebzari M R, Abadi S R H, Mahmood M T 2013 *Desalin. Water. Treat.* **51**(13-15) pp.2543-2553
[2] Rajkumar K, Muthukumar M and Sivakumar R 2010 *Resour. Conserv. Recyc.* **54**(10), pp.752-758
[3] Pandey R A, Sanyal P B, Chattopadhyay N and Kaul S N 2003 *Resour. Conserv. Recyc.* **37**(2), pp.101-117.
[4] Wei Y, Van Houten R T, Borger A R, Eikelboom D H and Fan Y 2003 *Water Res.* **37**(18), pp.4453-4467.
[5] Kaur R, Wani S P, Singh, A K and Lal K 2012 May. In *National Report presented at the 2 nd regional workshop on Safe Use of Wastewater in Agriculture*.
[6] El-Shafai S A and Zahid W M 2013 *Bioresour. Technol.* **143** pp.476–482
[7] Ekhmaj K M S I 2012 **6**(2) pp.268–271
[8] Gómez-Villalba B, Calvo C, Vilchez R, González-López J and Rodelas B 2006 *Appl. Microbiol. Biotechnol. 72*(2) pp.393-400
[9] Ahmed M, Idris A and Adam A 2005 *Suranaree J. Sci. Technol.* **12**(4), pp.286-295
[10] Ahmed M A H D I, Idris A Z N I and Adam A O F A H 2007 *J. Eng. Sci.Technol.(JESTEC)* **2**(1) pp.55-69
[11] Gu Q, Sun T, Wu G, Li M and Qiu W 2014 *Bioresour. Technol.* **166** pp.72-78
[12] American Public Health Association, American Water Works Association & Water Environment Federation, 2012. Standard Methods for the Examination of Water and

Wastewater. Standard Methods (102) p.741
[13] Stokes T L 2008 Second Edition *Annals of Surgery* **142**(6) p.1001
[14] Leyva-Díaz J C, Martín-Pascual J, González-López J, Hontoria E and Poyatos J M 2013 *Chem. Eng. Sci.* **104** pp.808-816
[15] Jahren S J, Rintala J A and Ødegaard H 2002 *Water Res.* ***36***(4) pp.1067-1075
[16] Martín-Pascual J, López-López C, Cerdá A, González-López J. Hontoria E and Poyatos J M 2012 *Water Air Soil Pollut.* **223**(4) pp.1699-1712
[17] Leyva-Díaz J C, Martín-Pascual J, Muñío M M, González-López J, Hontoria E and Poyatos J M 2014 *Ecol. Eng.* **70** pp.227-234
[18] Hannouche A, Ghassan C, Ruban G, Tassin B and Lemaire B 2011 *Water Sci. Technol.* **64** (12) pp.2445-52
[19] Hussain S A, Tan H T and Idris A 2010 *J. Appl. Sci.(Faisalabad)* **10**(24) pp.3384-3387
[20] Kalia V C, Prakash J and Koul S 2016 *Indian J. Microbiol.* **56**(2) pp.113-125

IOP Conf. Series: Materials Science and Engineering **206** (2017) 012088 doi:10.1088/1757-899X/206/1/012088

Treatment of azo dye Acid Orange 52 using ozonation and completed-mixed activated sludge process

C Z A Abidin[1], Fahmi, S A Ong, R Ahmad and S N Sabri

[1]School of Environmental Engineering, Universiti Malaysia Perlis (UniMAP), Kompleks Pusat Pengajian Jejawi 3, 02600 Arau, Perlis, Malaysia

E-mail: zulzikrami@unimap.edu.my

Abstract. In this study, the characteristic of colour and COD removal of azo dye Acid Orange 52 (AO52) by ozonation, in combination with complete-mixed activated sludge process (CMAS) was evaluated. The experimentation was arranged in two phases: during the first one, only ozonation was performed, while, during the second phase, it was integrated with CMAS. The performance of colour and COD concentration of AO52 with and without CMAS treatment, is compared and evaluated. From the results, it is obvious that high decolourization from the start of CMAS was contributed from the pre-treatments. The colour removal was due to the fact that ozonation able to cleave the azo bonds that represent colour. Thus, CMAS without pre-treatment are unable to decolourize the dyes sufficiently. 59.6% COD was removed from the first-stage, while merely 9.8% COD fraction removed from the subsequence second-stage CMAS. It is suggested that the rapid COD removal without ozonation are due to activated sludge adsorption processes. The decreased of mixed liquor suspended solids (MLSS) affected the CMAS performances, as the biomass decreased due to lack of nutrient for activated sludge microorganisms to multiply. Results from pre-ozonation alone contributed more than 50% of total COD removal, which indicated that at higher ozone dosage, tend to mineralize azo dye. Thus, ozonation not oxidized the dye though complete mineralization that produce carbon dioxide and water. However, it is a potential process for enhancing colour removal and biodegradability of dye-containing wastewater, once the appropriate ozonation time is determined. Therefore, the role of ozonation seems to break down the dye molecules and created ozonation by-product that is easily biodegraded in the subsequent biological treatment.

1. Introduction

The presence of colour and its causative compounds has always been undesirable in water used for either industrial or domestic needs. Different colouring agents like dyes, inorganic pigments, tannins, lignins that usually impart colour. Amongst complex industrial wastewater with various types of colouring agents, dye wastes are predominant. Even though the concentration of dyes in wastewater is low, it often receives attention due to strong colour visible even at very low dye concentrations [1]. The azo dyes constitute the largest and the most important class of commercial dyes that is widely used in textile, plastic, leather, and paper industries as additives. The main environmental concern is related to the presence of colour and chemical load it carries, that imparts light penetration and compromises the ecosystems in the receiving water [2,3]. Without sufficient treatment, these dyes are stable and persist in the environment for an extended period of time.

IOP Conf. Series: Materials Science and Engineering **206** (2017) 012088 doi:10.1088/1757-899X/206/1/012088

Due to the complexity and diversity of the dyes employed in the industries, it has become rather difficult to find an effective and economic treatment for eliminating the wide diversity of industrial wastewater. Conventional physical and chemical methods such as coagulation, activated carbon adsorption, ion-exchange or ultra-filtration and reverse osmosis can be used efficiently to remove industrial wastewater containing dye. However, these processes are considered as non-destructive because it merely transfers the dye from liquid to solid wastes, and expensive operation is required for the post-treatment of solid wastes. Therefore, how to decolourize and degrade dyes without causing secondary pollutant has become a research emphasis nowadays. Thus, chemical method to oxidize organic materials by oxidizing agents, such as ozone, H_2O_2, UV light or combination of such oxidants that is known as Advanced Oxidation Processes (AOPs) has become a promising process. Nevertheless, it cannot satisfy the environmental discharge standard by itself alone and the cost is fairly high. The combination of ozonation and biodegradation seems to be promising unit processes to remove residual colour and COD of wastewater containing dye. The reaction between the oxidizing agent with dye in an aqueous environment lead to the decrease in the aromaticity and molecular weight, which eventually result in an increase in biodegradability and colour removal of dye. The biodegradable compound produced during ozonation would be removed by the following biodegradation. Therefore, the biodegradability enhancement is considered to be the essential factor that determines the performance of ozonation-biological treatment process.

The combined treatment processes have been suggested to treat dye-containing wastewater, including UV/O_3 or UV/H_2O_2 [4,5], followed by biological treatment. Several other researchers also have investigated the application of partial ozonation or AOPs as pre-treatment steps, followed by subsequence biological treatments for treating the dye-containing wastewaters [6,7]. Guelli Ulson de Souza *et al.* [4] stated that ozonation as a pre-treatment for combined chemical-biological treatment is a potential process for enhanced colour removal and biodegradability of wastewaters containing azo dyes Remazol Black B, once the appropriate ozonation period is determined. In addition, the measurement of combined treatment efficiency mainly depends on the purpose of treatment, but also involves the optimization of each chemical oxidation and biological steps [8].

Our previous study reported that ozonation effectively decolourized azo dye Reactive Red 120, even with lower ozone dosage [9]. However, significant COD removal was only observed within the higher range of ozone dosages. Moreover, ozonation transforms the functional groups in azo dye to produce more biodegradable byproducts, which is easily removed by biological treatment. In this study, colour and COD removal of Acid Orange 52 (AO52) were evaluated to be applying for wastewater treatment containing azo dye. The experimentation was arranged in two main phases. The first phase was aimed to evaluate the oxidation process to achieve complete colour removal of dye and partial cleavage of aromatic amines to make them easily biodegradable. Further, degradation can be achieved in the second phase by treating effluent from ozonation by biological treatment.

2. Materials and methods

2.1. Materials

The AO52 was of analytical grade obtained from Acros Organics, and used without further purification. It has a simple molecular structure, slow decomposition rate in light, relatively high stability and toxicity, complex structure and non-biodegradability is widely used in the textile, chemical and paper industries [10–12]. The chemical structures according to the manufacturer are shown in Figure 1. The molecular formula and its molecular weights are $C_{14}H_{14}N_3NaO_3S$, and 327.32 g/mol, respectively. The pH of aqueous solutions was adjusted using 0.1N NaOH to raise the pH or 0.1N HCl to lower the pH upon decolourization and degradation. All solutions were prepared using ultrapure water.

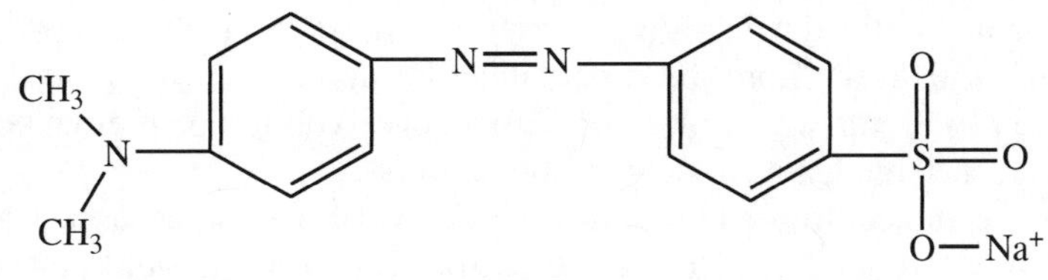

Figure 1. Chemical structure of Acid Orange 52.

2.2. Phase 1: Pre-ozonation treatments

The experimental setup shown in Figure 2 consists of an oxygen cylinder, ozone generator, glass reactor and KI reactor. Ozone was generated from the A2ZZ-3G ozone generator utilizing pure oxygen of 99.5% concentration, introduced at a rate of 3 L/min. Ozone applications were semi-batch reactions carried out in a cylindrical glass reactor of the 2,000 mL working volume. The dye solution with the concentration of 100 mg/L was added into the glass reactor. Constant ozone was supplied from the bottom of the reactor through a diffuser at a dosage of 10.2 mg/min. Then, followed by 5 min aeration to remove residual ozone. All connections from the ozone generator to the reactor were made through Teflon tubing. The excess ozone, leaving the reactor was decomposed through 2,000 mL flasks filled with 2% KI solution. The experiments were conducted at room temperature, and samples were withdrawn at definite time intervals for determination of pH, colour, COD and UV-Vis spectra analysis.

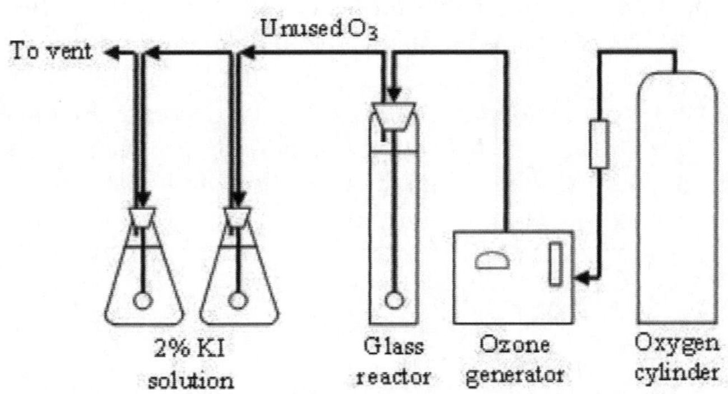

Figure 2. Schematic diagram of the bench-scale reactor system.

2.3. Phase 2: Biological treatment

The biological treatment was performed by an aerobic complete-mix activated sludge (CMAS) reactor, consists of a 0.15 m deep cylindrical glass with an internal diameter of 0.10 m. The aerobic activated sludge as seed for CMAS reactor was collected from the biological treatment system of a glove manufacturing plant, located in Perlis, Malaysia. The bioreactor had an operating liquid volume of 1,000 mL and equipped with an air diffuser and stirrer. The air was bubbled at the bottom of the reactor by air diffuser. The agitation is maintained during the whole experiment in order to mix the medium and avoid concentration gradients within the bioreactor. The DO concentration in the medium was monitored daily to ensure oxygen concentrations between 6-8 mg/L. The overall experiments were conducted at room temperature.

The bioreactor operated in batch mode, with 24 h treatment cycles that consist of four consecutive phases: filling, reaction, settling and drawing. It was previously seeded by 500 mL of aerobic activated sludge. Then, at the beginning of the filling cycle, the reactor was filled with 500 mL of pretreated dye influent. Subsequently, aeration and agitation were switched on to begin the reaction phase to allow the biological removal of biodegradable pollutants. Once completed, both aeration and agitation were stopped to let the biomass settle down. Finally, after 1 h of settling phase, approximately 300 mL volume of samples was withdrawn from the supernatant during the drawing phase, without disturbing the settled sludge. The bioreactor was replaced with an equal volume of pretreated dye samples, and aeration-agitation was turned on again for new treatment cycles.

The dye samples were subject to pre-ozonation, with 0, 5, 10, 15 and 20 min contact times, and left for approximately 24 h before feeding into the bioreactor. Therefore, possible interferences from any quantity of residual DO in the biological treatment were eliminated. Samples from each reactor were collected daily (until day 30) and then filtered through a 0.45 μm glass microfibre filter (Fisherbrand, 25 mm) prior to analyses to remove the biomass.

2.4. Analytical methods

The ozone concentration in feed gas was determined by the KI-starch titration method [13]. The decolourization and degradation of azo dye were determined by measurements of absorbance at wavelengths ranging from 200 to 800 nm by UV-Vis spectrophotometer (Hitachi U-2810). The maximum visible region (λ_{max} 465 nm) was employed as a base for characterization of decolourization for AO52, using a 10 mm quartz cell. The pH was measured by Hanna Instruments HI223 pH meter, and COD determined in accordance with the closed reflux, colorimetric method [13].

3. Results and discussions

3.1. Effect of pH

The reaction pH is an important parameter that influences the formation of hydroxyl radicals during ozonation. Figure 3 showed the pH of AO52 as a function of contact time. Dye solution decreases rapidly as the early stage of the experiment, dropping from the initial solution of pH 7.0 to about 3.5 after 5 min (when almost decolourization occurred). Then, the decrease continued until a steady state was reached. The drop of pH showed the formation of acidic by-products, such as H_2SO_4 originating from the sulfonate groups and lesser extent, from the formation of aliphatic acids and NH_4^+ [14].

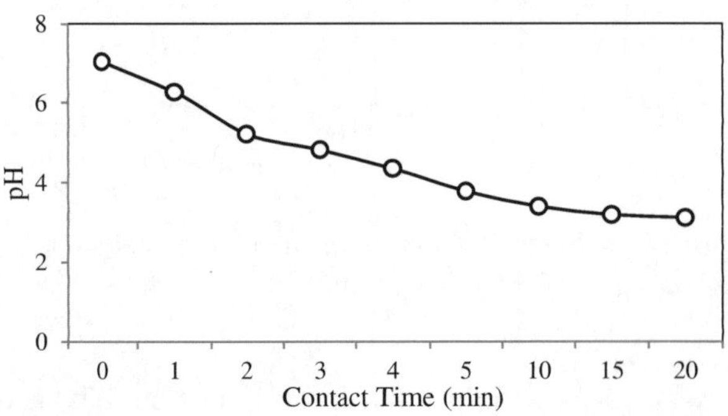

Figure 3. pH evolution in relation to ozonation contact time.

3.2. Effect of pre-ozonation on colour and COD removal

The effect of pre-ozonation on the UV-vis spectra of AO52 after 20 min of treatment was shown in Fig. 3. The peak was slowly collapsed with the increase of contact time, and finally no peak was observed during 20 min of ozonation. The spectra of initial solution represent the UV band characteristics of -N=N- groups (465 nm) and related it to the benzene rings (240 nm) bonded to the -N=N- group [15–17]. During the experiment, the intensity of absorption at 465 nm declined extremely rapidly in comparison to benzene rings, as reaction time proceeded. This resulted in the decolourization of the dye sample. The colour removal percentage reached 100% during 10 min of contact times.

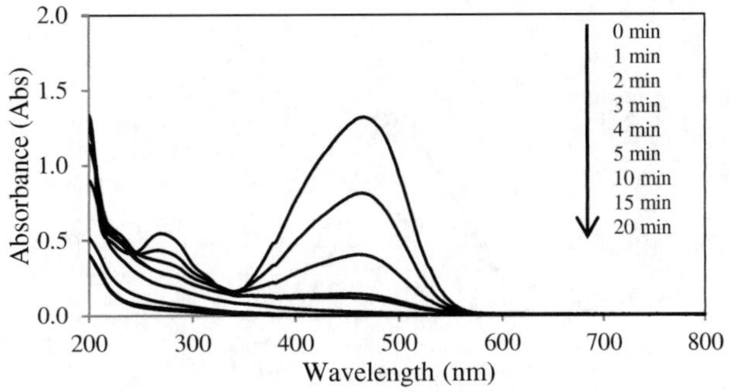

Figure 4. UV-Vis absorbance spectra in relation to ozonation time (100 mg/L, pH 7).

Figure 5 showed the COD concentration for AO52 after ozonation treatment. For a given time, an increase in contact time resulted in an increase in COD removal efficiency. Initial COD value is 128 mg/L, and decrease to 43 mg/L, which indicates partial mineralization of dyes. Continuous increase of reaction time had not increased the COD removal quickly. As the oxidation continued, recalcitrant organic compounds that could be oxidized became less available and that the remaining inert organic compounds after ozonation were difficult to break down.

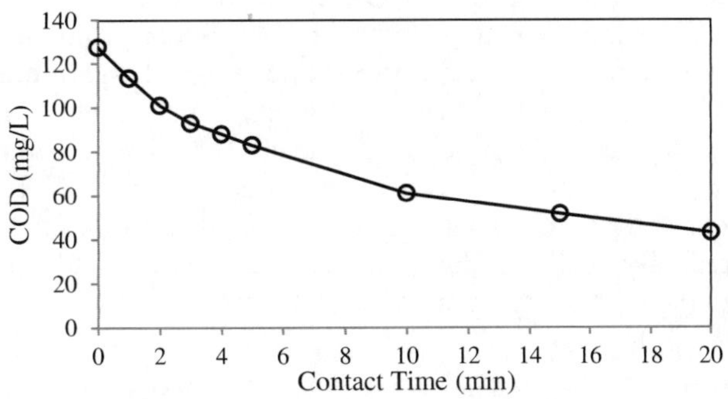

Figure 5. COD concentration in relation to ozonation time (100 mg/L, pH 7).

3.3. Effect of pre-ozonation on CMAS treatment and for colour and COD removal

Figure 6 shows the wavelength of AO52 indicated by UV-Vis Spectrophotometer, submitted to different pre-ozonation contact time and CMAS treatment (between day 0 and 30). It is obvious that 0 min (without ozonation pre-treatment) curves of day 0 and 30 almost overlapped suggesting that the CMAS treatment alone hardly affects the decolourization and degradation. However, the specified wavelength peaks of dyes with ozonation (5, 10, 15 and 20 min ozonation) pre-treatment was disappeared, which indicated that the azo group of the synthetic dyes were transformed by the ozonation-CMAS treatment. Nevertheless, it seemed possible to obtain bio-treatable solutions once the colour had disappeared. The biological process is more efficient at mineralizing the by-products of the ozonation, but less efficient in colour reduction than the ozonation process itself [4].

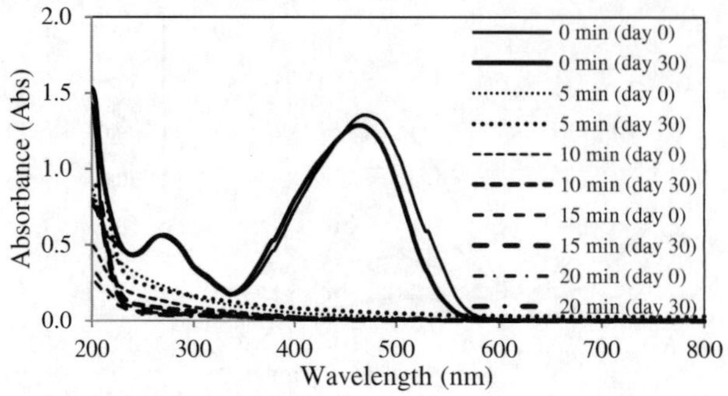

Figure 6. UV-Vis absorbance spectra of pre-ozonation and CMAS treatment (100 mg/L, pH 7).

From the data in Figure 7, it is apparent that concentrations gradually decrease from the initial concentration of 84, 62, 52 and 44 mg/L for 5, 10, 15 and 20 min ozonation, respectively. However, after day 30 of biological treatment, merely pre-ozonation of 10, 15 and 20 min shows a significant trend of reduction. Even though 20 min seems to have higher removal compared to a 15 min, it is suggested that 15 min is economically desirable due to the low ozonation required for adequate mineralization of AO52. These results imply that subsequent biological treatment after ozonation for 15 min was effective enough for the reduction of COD concentration.

The products obtained from the cleavage of the azo bonds are aromatic amines, which further mineralization of these compounds has been reported under aerobic conditions by Fongsatitkul *et al.* [19]. Since no external carbon source was added into the treatment systems, all of the COD (organic matter) was contributed by the dyes. In general, the results could be attributed to adsorption, bio-degradation, or simultaneous adsorption and bio-degradation processes. The basis of adsorption phenomenon and the extent can be co-related to increasing adsorption sites to increase in reaction time. The removals are gradually declining due to the exhaustion of the adsorbent site. Moreover, the biomass decreased affected the CMAS treatment for COD removals.

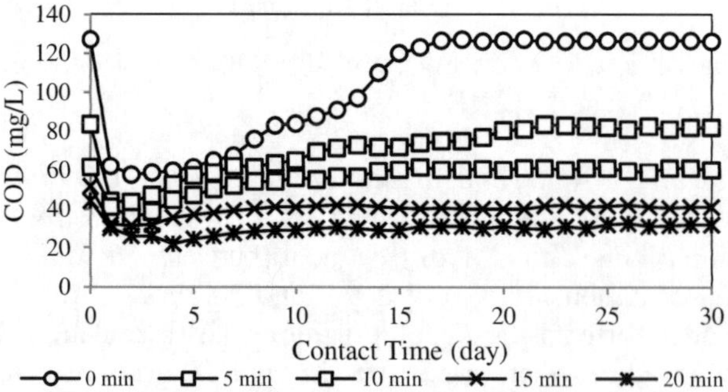

Figure 7. COD concentration in relation to ozonation time and CMAS treatment (100 mg/L, pH 7).

Results from pre-ozonation alone contributed more than 50% of total COD removal, which indicated that at higher ozone dose, tend to mineralize azo dye. On the other hand, at a medium ozone dose (5 to 10 min), the role of ozonation seems to breakdown the azo dye molecule and created ozonation product that is easily biodegraded in biological treatment. Therefore, ozone was not oxidized azo dye though complete mineralization to produce CO_2 and H_2O. Khadhraoui *et al.* [14] mentioned that ozonation of dyes usually leads to small organic molecular fragments, such as acetic acids, epoxides, aldehydes, and ketones that leads to a residual COD. Therefore, COD resulting from these small new formed molecules can be further degraded by a biological process.

4. Conclusion

In this study, the characteristic of colour and COD removal of AO52 azo dye by pre-ozonation followed by CMAS treatment was evaluated for applying in azo dye industrial effluent treatment. The drop of pH indicated the formation of acidic by-products. Based on the results, the biological treatment is more efficient at mineralizing the by-products of the ozonation, but less efficient in colour reduction than the ozonation process itself. Even though the combine treatment was not efficient enough to completely oxidize the contaminants, it contributes to higher dye mineralization in comparison to ozonation by itself. Ozonation of dyes usually leads to small organic molecular fragments that lead to a residual COD. Therefore, COD resulting from these small new formed molecules can be further degraded by biological treatment. Therefore, it is recommended that ozonation should be applied at medium ozone dosages as a pre-treatment for combined ozonation-biological treatment for azo dye removal.

Acknowledgement

Authors wishing to acknowledge Ministry of Higher Education (MOHE), Malaysia for the financial support (FRGS 9003-00513).

References

[1] Abidin C Z A, Muhammad Ridwan F, Ong S A, Mohd Makhtar S N N and Rahmat N R 2015 *Sci. Asia* **41** 49–54
[2] Tomei M C, Pascual J S and Angelucci D M 2016, *J. Clean. Prod.* **129** 468–77
[3] Franca R D G, Vieira A, Mata A M T, Carvalho G S, Pinheiro H M and Lourenco N D 2015 *Water Res.* **85** 327–36
[4] Guelli Ulson de Souza S M A, Santos Bonilla K A and Ulson de Souza A A 2010 *J. Hazard. Mater.* **179** 35–42
[5] Ledakowicz S, Solecka M and Zylla R 2001 *J Biotechnol* **89** 175–84
[6] Qi L, Wang X and Xu Q 2011 *Desalination* **270** 264–68
[7] Gokcen F and Ozbelge T A 2006 *Chem. Eng. J* **123** 109–15
[8] Oller I, Malato S and Sanchez-Perez J A 2011 *Sci. Total Environ.* **409** 4141–66
[9] Fahmi, Abidin C Z A and Rahmat N R 2010 *Int. J. Environ. Sci. Dev.* 1 **2** 193–8
[10] Liu J, Ma S and Zang L 2013 *Appl. Surf. Sci.* **265** 393–8
[11] Zhao Y, Chu J, Li S H, Chen Y, Sheng G P, Chen Y P, Li W W, Liu G, Tian Y C, Xiong Y and Yu H Q 2011 *Chem. Eng. J.* **170** 440–4
[12] Chen L Y 2000 *Water Res.* 34 **3** 974–82
[13] APHA-AWWA-WEF 2005 *Standard Methods for the Examination of Water and Wastewater* 21st ed. (Washington DC: APHA)
[14] Khadhraoui M, Trabelsi H, Ksibi M, Bouguerra S and Elleuch B 2009 *J. Hazard. Mater.* 161 **2-3** 974–81
[15] Wang A, Qu J, Liu H and Ge J *Chemosphere* **55** 1189–96
[16] Feng W, Nansheng D and Helin H 2000 *Chemosphere* **41** 1233–38
[17] Solozhenko E G, Soboleva N M and Goncharuk V V 1995 *Water Res.* 29 **9** 2206–10
[18] Fongsatitkul P, Elefsiniotis P, Yamasmit A and Yamasmit N 2004 *Biochem. Eng. J.* **21** 213–20

IOP Conf. Series: Materials Science and Engineering **206** (2017) 012089 doi:10.1088/1757-899X/206/1/012089

Comparison of various advanced oxidation processes used in remediation of industrial wastewater laden with recalcitrant pollutants

S Krishnan[1,2], H Rawindran[1,2], C M Sinnathambi[1,2 *], J W Lim[2]
[1]*Centre of Research Enhanced Oil Recovery (COREOR), Universiti Teknologi PETRONAS, 32610 Seri Iskandar, Perak, Malaysia*
[2]Department of Fundamental and Applied Sciences, Universiti Teknologi PETRONAS, 32610 Seri Iskandar, Perak, Malaysia

*E-mail: chandro@utp.edu.my (Chandra Mohan Sinnathambi)

Abstract. Due to the scarcity of water, it has become a necessity to improve the quality of wastewater that is discharged into the environment. Conventional wastewater treatment can be either a physical, chemical, and/or biological processes, or in some cases a combination of these operations. The main purpose of wastewater treatment is to eliminate nutrients, solids, and organic compounds from effluents. Current wastewater treatment technologies are deemed ineffective in the complete removal of pollutants, particularly organic matter. In many cases, these organic compounds are resistant to conventional treatment methods, thus creating the necessity for tertiary treatment. Advanced oxidation process (AOP), constitutes as a promising treatment technology for the management of wastewater. AOPs are characterised by a common chemical feature, where they utilize the highly reactive hydroxyl radicals for achieving complete mineralization of the organic pollutants into carbon dioxide and water. This paper delineates advanced oxidation processes currently used for the remediation of water and wastewater. It also provides the cost estimation of installing and running an AOP system. The costs are separated into three categories: capital, operational, and operating & maintenance.

1. Introduction

Wastewater is generally categorized by properties like total dissolved solids (TDS), turbidity, chemical oxygen demand (COD), biological oxygen demand (BOD), dissolved oxygen, hardness, pH, and colour. These characteristics vary between industries and also within each industry. Dissolved oxygen content in water is reduced by the presence of organic compounds. Therefore, disposing these wastewaters, without treatment, to water sources can be harmful to aquatic life.

Many types of conventional treatment methods can be used for wastewater treatment. These methods can be either physical, chemical, or biological processes, or in some cases a combination of these operations. The main purpose of wastewater treatment is to eliminate solids, organic matter and, in some cases nutrients from wastewater.

Physical treatment involves isolation or separation of the waste material from the mainstream. In this method, little or no degradation of the waste is involved, such as in coagulation, filtration, etc. On the other hand, for biological treatment, the main process involves the use of microbes to feed on the organic waste. In this case, special care like pH and aeration, should be monitored to sustain the microbes' activities.

However, there are some contaminants found in wastewater that are recalcitrant to several of the generally applied physical and/or chemical treatment processes. Chemical oxidation processes can be used to expand the current treatment processes, by transformation processes that can be used to destroy the recalcitrant compounds through oxidation and reduction reactions. The objective of this review is to investigate advanced oxidation processes that have been used for treatment of various types of contaminants, to determine the best methods for wastewater generated from different industrial processes, as well as provide a cost estimation of running an advanced oxidation system.

2. Advanced Oxidation Process

Advanced Oxidation Process (AOP) is a chemical treatment method that has been growing in the wastewater management industry. It is a highly suggested method for the removal of problematic organic matter. The basic principle of AOP involves the production of hydroxyl radicals (HO•), which can be generated from hydrogen peroxide (H2O2), ozone, photo-catalysis, or oxidants in combination with using ultraviolet (UV) radiation. In some cases, two or more radical generators are used in combination. However, it is the HO• that is mainly responsible for the degradation of organic compounds.

The HO• is an unselective strong chemical oxidant. Once produced, it attacks nearly all organic complexes. Therefore, attack by the HO• leads to a complete breakdown of the organic compound, and as a result, AOPs diminish the concentration of the pollutant from a few hundred ppm to less than 5ppb [1]. Studies show that with AOP, organic chemicals disintegrate and become smaller and easily biodegradable. The HO• takes away a hydrogen atom from an organic compound (R–H) and causes the formation of an organic radical (•R) as in Equation (1). This radical then goes through a succession of chemical reactions to form several products and by-products [2]. Theoretically, AOPs should fully mineralize organic compounds to carbon dioxide and water as illustrated in figure 1.

$$R - H + HO\cdot \rightarrow H_2O + \cdot R \quad (1)$$

Advanced oxidation processes use vigorous oxidants, such as hydrogen peroxide, ozone, or Fenton's reagent to generate highly reactive intermediates. In some cases, these reagents are used in the presence of ultraviolet light, which enhances the oxidation process.

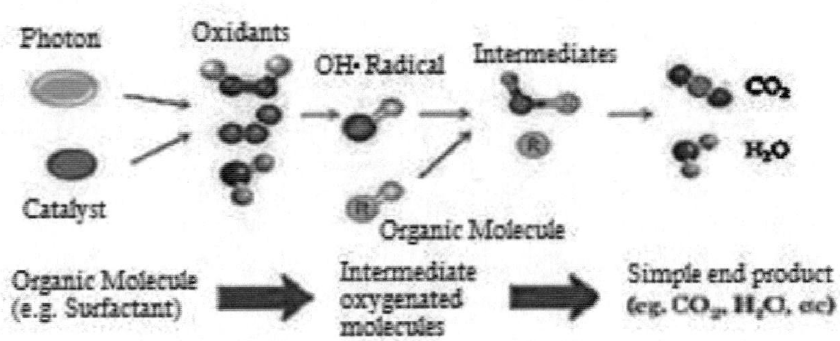

Figure 1. Principle of advanced oxidation processes [3]

The advantages of the AOPs are due to the fast reaction rates and non-selective oxidation, allowing the simultaneous treatment of multiple contaminants. Table 1 shows the relative oxidation power of the different oxidizing species, and it can be established that the oxidation power of HO• is relatively higher.

IOP Conf. Series: Materials Science and Engineering **206** (2017) 012089 doi:10.1088/1757-899X/206/1/012089

Table 1. Relative oxidation power [4]

Oxidizing Groups	Relative Oxidation Power
Chlorine	1.00
Hypochlorous Acid	1.10
Permanganate	1.24
Hydrogen Peroxide	1.31
Ozone	1.52
Atomic Oxygen	1.78
Hydroxyl Radical	2.05
TiO2+	2.35

Generally, the rate constant for the reaction of the contaminant with HO• determines the rate of destruction of a contaminant. Table 2 shows the rate constants of destruction of various compounds with ozone and hydroxyl radicals. It clearly indicates that the reaction rates are much faster with the hydroxyl radical compared to ozone.

Table 2. Reaction rate constants of ozone vs. hydroxyl radical [4]

Compound	O_3 ($M^{-1}s^{-1}$)	HO•($M^{-1}s^{-1}$)
Chlorinated alkenes	10^3–10^4	10^9–10^{11}
Phenols	10^3	10^9–10^{10}
N-containing organics	10–10^2	10^8–10^{10}
Aromatics	1–10^2	10^8–10^{10}
Ketones	1	10^9–10^{10}
Alcohols	10^{-2}–1	10^8–10^9

There are a few different ways by which AOPs can generate hydroxyl and other radicals [5]. Table 3 summarizes the various AOPs that have been studied for the removal of taste and odour compounds, along with the reactive species formed during each process. By studying Table 3 below, it clearly indicates that the highly reactive HO• is the main driver for all AOPs, and the effectiveness of the method depends on the efficiency by which HO• is produced. From the many AOP systems, ozonation, UV/ozone, UV/hydrogen peroxide, and photocatalysis are most frequently studied, and utilized for many applications.

Table 3. Common AOPs and the reactive species produced in each process [5]

AOP	Reactive Species
Ozone treatment: O_3	HO•, HO_2• , HO_3• , O_2^-•, O_3^-•
O_3/H_2O_2	HO•, O_2^-•, O_3^-•
Fenton processes: H_2O_2/Fe^{2+}	HO•, HO_2•

IOP Conf. Series: Materials Science and Engineering **206** (2017) 012089 doi:10.1088/1757-899X/206/1/012089

Photo-Fenton processes	HO•
UV/O_3,UV/H_2O_2, UV/O_3/H_2O_2	HO•, HO_2•/ O_2^-• , O_3^-•
V-UV (λ < 190 nm)	HO•, H^+, e^-
Photocatalytic treatment: UV/Vis light using catalysts	HO•, h^+, O_2^-, e-, $^1O_2$, HO_2•, HOO•
Ultrasonic treatment	HO•, $\dot{H}$
γ-Radiolysis	HO•, $\dot{H}$, e^-

2.1. Ozonation

The formation of radicals such as the hydroxyl radical (HO•) and superoxide radical ($O2^-$•) occurs when ozone is added to water, through a complex series of reactions. The disintegration rate of ozone in water is better at higher pH levels [4]. A combination of reactions with molecular ozone and reactions with HO• causes the oxidation of organic species.

$$3O3 + OH^- + H^+ \rightarrow 2HO\bullet + 4O2 \quad (2)$$

The decomposition cycle of ozone can be enhanced by the addition of hydrogen peroxide to result in the formation of HO• [6]. Hydrogen peroxide partially dissociates in water to produce the $HO2^-$. This hydroperoxide ion reacts with the ozone rapidly producing the HO•. In summary, the combination of reactions leads to the following:

$$2O3 + H2O2 \rightarrow 2HO\bullet + 3O2 \quad (3)$$

Contact time, alkalinity of water, and the dose of ozone determine the performance of the process. By introducing the dose of hydrogen peroxide once the oxidation of highly reactive substances with ozone takes place, it enhances the degradation achieved. The process of ozonation by H2O2/O3 systems seem to be the most established AOP in remediation for water as compared to other AOPs. Thus, there is a field-proven history of operation in the implementation of H2O2/O3 systems.

The second way to speed up ozonation is by the use of homogenous or heterogeneous catalysts. Studies have been carried out with various metal oxides and metal ions, and have shown significant changes in decomposition, however only in some cases. The concentration of ozone in the inlet gas and the ozonation time has been considered as predominant factors to the enhancement of the degradation rate. Oxidation studies have been carried out on various organic contaminants using Fe2O3, MnO2, TiO2–Me, Fe^{2+}, Fe^{3+}, and Mn^{2+}. The ozone/catalyst system appears to be more effective for the reduction of chemical oxygen demand (COD) and total organic carbon (TOC), than oxidation with ozone alone at higher pH values [7]. In experiments carried out by Horáková *et al.* [8], it was observed that decomposition of the chemicals is effectively improved by the synergistic effect of photo catalytic reactions occurring on TiO2-photocatalyst.

Ozonation can also be enhanced when ozone decomposes by readily absorbing UV radiation (λ = 254nm), which generates H2O2 as an intermediate, and then decomposes to HO•. The reactions indicated in equations 4 and 5 illustrate the photolysis of ozone to generate hydrogen peroxide, which results in the production of the highly reactive HO• [9]. There are however, several oxidation mechanisms that result in the destruction of organic contaminants, but HO• is the predominant removal

mechanism.

$$O3 + H2O + h\upsilon \rightarrow O2 + H2O2 \text{ at } (h\upsilon: \lambda < 300 \text{ nm}) \quad (4)$$

$$2\ O3 + H2O2 \rightarrow 2\ HO\bullet + 3\ O2 \quad (5)$$

This method comprises all the organic degradation mechanisms through O3/UV, H2O2/O3 and H2O2/UV. It is more effective for the destruction of organic substances than using O3 or UV alone, and it is a more efficient way of generating HO• as compared to the UV/H2O2 method at equal oxidant concentrations [4].

Nevertheless, this process is cost and energy intensive [9]. Furthermore, ozone methods have a potential for the formation of bromate. Interfering compounds can affect the absorption of UV radiation, thus reducing the efficiency of the system. However, this problem can be mitigated by raising the ratio of peroxide to ozone, reducing the pH, or making the concentration of other radical scavengers higher [9]. Turbidity also affects the amount of UV radiation absorbed by the ozone. The energy costs for ozone generation is the main operation expenditure for this process. The presence of bicarbonate or carbonate may act as scavengers of the HO• in natural systems is another drawback of these processes [4].

The use of AOP with ozonation has more industrial applications compared to all other AOPs in water treatment. Ozone technology has been used to manage contamination in wastewater from electroplating wastes, recycling, electronic chip manufacturing, marine aquaria, textiles, and petroleum refineries. Ozonation has also been effectively applied to the food industry, to treat wastewater from distilleries, olive mill, meat industry, and molasses [10]. There are recent applications for ozone in treating landfill leachates, rubber additive wastewaters, and detergents in municipal treatment plants [11]. An industrial scale UV/ozonation system has been set-up in Oklahoma to handle refractory organics and metal complexed cyanides [4]. Cadillac Motor Car Division in Detroit, started using ozone treatment since 1978. They have managed to maintain the total cyanide levels in the effluent below 1 mg/L [4]. Different types of ozonation systems have been studied in their ability to reduce the harmful of discharge from the pulp and paper manufacturing [12].

In the cost estimation shown by Kommineni *et al.* [9], the vendor provided estimates for their H2O2/O3 system per 3.79 cubic meters of water treated. These costs varied between \$0.35 for 1380m^3/h, at influent MTBE concentration of 20 μg/L and \$3.62 for 13.8m^3/h, at initial Methyl tert-butyl ether (MTBE) concentration of 2,000 μg/L. In their study the capital costs for this system were higher, while the operating & maintenance (O&M) costs are suggestively lower.

2.2. AOP using hydrogen peroxide

Hydroxyl radicals are formed by the photolytic dissociation of hydrogen peroxide in water by UV irradiation at a wavelength of 254nm. This formation is indicated in Equation 6 [13]. Medium Pressure UV lamps produce a higher radical generation.

$$H2O2 + h\upsilon \rightarrow 2\ HO\bullet \quad (h\upsilon : \lambda < 300\text{ nm}) \qquad (6)$$

Andreozzi *et al.* [6] found that the rate of photolysis of aqueous H2O2 to be pH dependent, and it increases when more alkaline conditions are used. As with the UV/Ozone system, interfering compounds and turbidity can absorb the UV light, thus reducing the efficiency of the system. Hydrogen peroxide can be introduced into the system as a single dose or at many points in the system [9]. Each water source requires optimization based on lab scale testing, in order to ascertain the ideal dose of hydrogen peroxide. If there is excess H2O2, it reacts with other contaminants (e.g. elemental sulphur) producing complex, oxidizable materials. Vaferi *et al.* [13] concluded that this system can be used to appropriate design, and for scaling up of industrial batch reactors for wastewater treatment.

UV/H2O2 system is seen as a beneficial method for AOP, as it has high reaction rates, and is flexible in the design of reactors, which leads to a lesser footprint in the treatment plant. The use of ultraviolet lamps also provide disinfection of the water. Akin *et al.* [14] suggested that the polyethoxy chain of the surfactant used was more susceptible to degradation in the H2O2/UV treatment process. In their comparisons between H2O2/UV and Photo-Fenton processes, they concluded that H2O2/UV treatment was more efficient than the Photo-Fenton process in the overall TOC removal efficiencies of the nonylphenol polyethoxylate surfactant. This is supported in studies conducted by Antonopoulou [5], who reported that the UV/H2O2 process was proved effective to reduce odorous aldehyde concentrations. They further state that the findings showed that direct UV photolysis is the main mechanism involved in this degradation process.

This method for AOP treatment is cheaper because of the lower cost and easy availability of hydrogen peroxide [9]. However, this system has some drawbacks. H2O2 has poor UV light absorption characteristics, thus this can be considered as wasting most of the light input. The formation of bromate for this system can hinder this process, but by lowering the pH, and increasing the peroxide to ozone ratio can reduce this problem.

Most studies conducted on this method stress the importance of optimization of the parameters (dose and concentration of H2O2, concentration of contaminant, retention time, UV intensity, etc.) [2, 5, 13 - 15]. It has been concluded in a number of studies that the removal efficiency of organic contaminants is above 95%, however, the COD reduction increases to a certain extent and then declines after a certain point indicating the reactions of hydroxyl radicals with other inorganic species to form oxidizable materials [16 -17].

Phenolic complexes that are usually found in olive mill wastewater have been treated using UV/H2O2 method. UV/H2O2 has also been successfully used for removal of dyes from textile industry effluents. Various studies on textile effluent treatment show that increase in UV dosage and H2O2 concentration increases the rate of decolorization. Furthermore, there are a number of H2O2/UV industrial applications for remediation of drinking water. For example, Calgon Carbon runs two industrial AOP applications for drinking water treatment. This method is the only AOP that is used in drinking water remediation [4]

Calgon Carbon Corporation provided cost estimates for the UV/H2O2 system which is discussed in a study provided by Kommineni [9]. The cost shown per $3.79m^3$ of treated water ranged from \$0.32 for a flow rate of $1380m^3/h$, at 20 µg/L pollutant concentration, and \$4.11 at $13.8m^3/h$ with 2,000 µg/L pollutant concentration. Their system had the lowest capital costs, but O&M expenditures were more due to higher energy utilisation [9].

2.3. AOP using Fenton's Reagent, Fenton-Like and Photo Fenton Processes

Radicals are produced when iron (II) reacts with hydrogen peroxide, where Fe^{2+} ion acts as a catalyst for this reaction. Fenton's Process is a simple method to produce HO• without the need for neither special apparatus nor chemicals, and takes place at ambient temperature and pressure. This method is an attractive way for oxidation, as hydrogen peroxide and iron salts are easily available, easy to handle and environmentally safe. The destruction of organic compounds takes places by reacting with the HO•. Equation 7 shows how the HO• are formed [6]:

$$Fe^{2+} + H2O2 \rightarrow Fe^{3+} + OH^- + HO\bullet \qquad (7)$$

The rate of degradation of organic pollutants with Fenton–Fenton like reagents is strongly accelerated by irradiation with UV-VIS light at wavelengths greater than 300nm. The photolysis of Fe^{3+} complexes allows Fe^{2+} to be regenerated. The occurrence of Fenton reactions in the presence of hydrogen peroxide is shown in equation 8 [15].

$$Fe(OH)2+ h\nu \rightarrow Fe^{2+} + HO\bullet + OH^- \qquad (8)$$

This process, like the other AOPs, is initiated by the hydroxyl radical for the degradation of organic compounds, mainly by oxidation reactions. However, the Fenton process is strict and requires pH control [2, 14], so as to prevent precipitation of the iron. For this method, the reactor must be constructed so as to allow proper mixing of the Fe(II) and H2O2. This allows for optimum hydroxyl radical formation, and degradation of the contaminants. The extraction of iron from the effluent water could possibly raise the costs of using this system.

For Fenton-like processes, elements with several redox states such as chromium, copper, and can be used to decompose H2O2 into HO• through conventional ways [2]. However, the Fenton-like system, each non-ferrous catalyst could counterbalance the practical gains of better catalyst stability and working at neutral pH. Therefore, the activation process for H2O2 is specific to the nature of the catalyst and is influenced by its composition.

Toxic compounds such as phenols and herbicides in wastewater can be destroyed by this method [6]. However, the low pH value that is required for this system to keep the iron in solution increases the maintenance and operation costs. Advanced oxidation pre-treatment using Fenton reagent has been found to be very effective at enhancing the biodegradability of wastewater containing surfactants The Fenton, photo-Fenton and Fenton-Like methods are popular methods for AOP due to their flexibility, simplicity, and integration into existing water remediation processes such as coagulation.

Furthermore, this method has a rapid reaction between iron and H2O2, which then generates HO• in the shortest time compared to other AOPs [2].

The optimum amount of H2O2 for this process needs to be determined since any excess H2O2 reacts with other inorganic matter, affecting the COD reduction. This method has high elimination for the degradation of linear alkylbenzene sulfonates (LAS), and alkyl benzene sulfonate (ABS) surfactants.

However, there is one major drawback of using the Fenton process, which is the production of sludge that contains iron hydroxide as a byproduct. Large amounts of flocs of various sizes have been observed during a number of studies conducted [2, 18]. The disposal of the sludge formed has to be included when estimating the cost of this process. Nevertheless, chemical coagulation has been found as an effective method for the removal of the flocs, as well as COD [16, 19].

Until now, Fenton and photo Fenton processes have been used to treat effluents from dye and chemical manufacturing, pulp bleaching, and agricultural processing. Additionally, Fenton pretreatment can be used to enhance biological wastewater treatment. Since this process is deemed as easy-to-handle, Fenton's reaction has demonstrated to be more efficient in terms of operating expenses for the treatment of toxic and food industrial wastewater [20].

In a study of the degradation of phenol from aqueous solutions by photo-Fenton process by Alam *et al.* [21] optimum economic condition was found by using 30 mmol/L of H2O2 and 0.5 g/L of FeSO4·7H2O for irradiation time of 60 min. The total estimated costs with maximum degradation were found to be \$2.88 US/m^3.

2.4. AOP using titanium dioxide (TiO2)

Heterogeneous photocatalysis using titanium dioxide (TiO2) is an efficient advanced oxidation method in industrial effluent treatment process. When the TiO2 is illuminated by UV, it generates a conduction band of electrons (ecb) and valence band holes (h^+) (Eq. 9-11). These band electrons interact with surface adsorbed oxygen to produce superoxide radical anions. The hydroxyl radicals are produced when the band holes interact with water.

$$\text{TiO2} + h\nu \rightarrow \text{ecb} + h^+ \quad (9)$$

$$\text{ecb} + \text{O2} \rightarrow \text{O2}^{-}\bullet \quad (10)$$

$$h^+ + \text{H2O} \rightarrow H^+ + \text{HO}\bullet \quad (11)$$

It has been shown that many organic complexes can be degraded by oxidation in the presence of UV/TiO2 [20, 22, 23] . As with all the other processes in AOP, the degradation of organic compounds takes place by reacting with the hydroxyl radicals.

Furthermore, TiO2 nanoparticles are deemed suitable for wastewater treatment, because they are non-toxic, chemically and biologically inert, and inexpensive [4]. In addition, TiO2 has a comparatively high oxidative power as compared to other oxidizing species [22]. Suspended TiO2particles were used in most studies that pertain to the photocatalytic oxidation of contaminants [23 – 25] . However, the recycling and recovery of these nanoparticles become cumbersome and expensive, thus making it a disadvantage of using suspended systems. This results in the need for composing new methods for using immobilized TiO2 to create systems with an immobilized active phase [22, 26 - 27]. Therefore, different techniques to create immobilized TiO2 have been studied using white plaster cement for the photodegradation of surfactants present in produced water [27].

Since photocatalytic oxidation is considered as an emerging technology, there aren't many industrial applications of the process. However, several studies show the photocatalytic process has been successfully used in the treatment of effluents from winery and distillery, olive mill, dairy industry, molasses, candy and sugar industry, fresh-cut vegetable industry, etc [20]. Photocatalysis was found to be effective for reducing counts of bacteria, molds, and yeasts.

Hydroxyl Systems Incorporated stipulated the costs for TiO2/UV AOP system, in the study conducted by Kommineni *et al.* [9]. The estimated costs per 3.79m^3 of treated water range between

IOP Conf. Series: Materials Science and Engineering **206** (2017) 012089 doi:10.1088/1757-899X/206/1/012089

\$1.01 for a flow 1380m^3/h at 20 µg/L contaminant concentration and \$5.17 for 13.8m^3/h at 2,000 µg/L. Compared to the other processes, this system appears to be the least cost effective.

3. Cost estimation

The expenses for AOPs greatly rely on the quality of the effluent to be treated and the aim of the treatment. Hence, there are very few reports on the cost estimation of the different types of AOPs. According to some of the studies that include cost analysis, it is recommended to carry out pilot testing to ascertain site specific costs.

It was found that the main factors influencing the prices were the removal efficiency and flow-rate, not owing to the concentration of the influent [4, 9, 18, 28]. The operating and management costs, as well as the capital costs, increased considerably with higher removal efficiency [4, 9]. Location of the plant, the time requirement for treatment, environmental trepidations and community influences should also be taken into account when evaluating the cost.

For this cost estimation, we have followed Kommineni's [9] cost evaluation guidelines. The cost estimation has been adjusted to the current year, with minimum 3% inflation rate. Out of the four cost estimations shown in Kommineni's study, Applied Process Technology provided the most cost effective rates for Capital expenditure (CAPEX) and operating expenditure (OPEX) combined for each process. Their H2O2/O3 system per 3.79m^3 of water, costs between \$0.35 and \$3.62 for 1380m^3/h (20 µg/L), and 13.8m^3/h (2000 µg/L) respectively to be treated.

In their study, Kommineni compared costs of the various AOPs based on the cost estimates. Each study was provided with a number of treatment conditions. The budget included prices for equipment, electricity, chemical quantities and spare parts. The budgets were prepared for given influent water characteristics, as shown in Table 4.

Table 4. Influent water characteristics [9]

Parameters	**Value**
Hardness (mg/L as CaCO3)	200
Alkalinity (mg/L as CaCO3)	250
Bromide	ND
Iron (mg/L)	<1
pH	7.0
Temperature (°F)	65
TDS (mg/L)	500
Nitrate (mg/L as NO3)	25
(mg/L as N)	5

The costs were divided into three categories, which are capital, operational, and O&M. The investment costs contain the full treatment system and its setting up. Under the capital cost, each of the installing costs were calculated. Valves, Piping, and electrical work were approximated at 30%, site work at 10%, engineering at 15% and contractors at 15% of the equipment costs. In addition, 20% of the overall cost was included as a contingency. Amortization of the annual capital cost was based on a 30-year period, at a discount rate of 7%. The full illustration of the calculation is shown in Table 5.

IOP Conf. Series: Materials Science and Engineering **206** (2017) 012089 doi:10.1088/1757-899X/206/1/012089

Table 5: Cost estimation for AOP Unit [9]

Item	**Costs (USD)**
Advanced oxidation unit	1,200,000
Piping, valves, electrical (30%)	360,000
Site work (10%)	120,000
Contractor O&P (15%)	252,000
Engineering (15%)	289,800
Contingency (20%)	444,360
TOTAL CAPITAL	2,666,160
Amortized capital	214,864
Annual O&M	207,507
Total annual cost	422,371
Total cost per 1,000 gallons treated	1.34

For the annual OPEX, replacement expenses are based on the estimate of 1.5% capital cost. The labour costs will vary for different countries. Analytical costs are based on the number of sampling conducted per week. The chemical costs will be based on dosages. Power is based on estimated consumption and price per unit for the country of operation.

In a study conducted by Mahmuni et. al [28], a method to estimate treatment costs of various AOPs was developed. Kinetic data of pollution removal from various literature were used for their estimation. The rate constants for the kinetic data were then used to estimate the time it takes for degradation of the pollutant by 90% of its original concentration. This time was taken to be the residence time of the reactor. The residence time was then multiplied with the design flowrate to calculate the reactor capacity. Total energy requirement in the AOP reactor was calculated, and this used to find the estimated cost of the AOP reactor.

4. Conclusions and ways forward

Advanced oxidation process is gaining popularity in the wastewater treatment industry. The main focus of this process is the hydroxyl radical, which once generated aggressively attacks virtually all organic compounds. Thus far, photocatalytic (TiO2/UV) process, ozonation, H2O2/UV process and Fenton's reactions have been studied and extensively used for the removal of recalcitrant organic chemicals thus reducing the COD, TOC, dyes, and phenolic compounds that are commonly found in industrial and municipal wastewater. Major factors affecting these processes are the initial concentration of the pollutant, the quantity of oxidizing agents and catalysts, light intensity, irradiation time and the nature of the wastewater's solution (pH, presence of solids and other ions). It is pertinent to conduct experimental studies in order to develop a method suitable for the specific wastewater. It has been established that it is necessary to estimate the capital costs, and overhead and management costs by conducting pilot studies, since pilot plant studies are better capable of providing closer conditions to estimate accurate costs. Based on the limited reviews, H2O2/O3 and H2O2/UV appear to be the two most promising AOP systems, and they are economically practical.

Few AOPs have been examined in detail under controlled experimental conditions, for many it is uncertain what the exact chemical mechanisms are, and few field tests have been carried out in such a manner that the entire process could be critically examined. Hence, process optimization should be carried out for specific wastewater, to ensure no harmful by-products are generated, and residual reagents are limited in the effluent from the AOP system. At the same time the knowledge about the exact mechanisms of AOPs is still incomplete. Furthermore, one should clearly see the importance of AOPs as a technological tool for environmental management, and they must be developed from this firm scientific and engineering basis.

Acknowledgments
The authors would like to thank Universiti Teknologi PETRONAS (UTP) and the Enhanced Oil Recovery, Mission Oriented Research (MOR-EOR) division for their financial support. We would also like to express our sincere and honest gratitude to colleagues from UTP and PETRONAS Research Sdn Bhd (PRSB), and friends who have guided and assisted us during the study.

References

[1] Mohajerani M, Mehrvar M and Ein-Mozaffari F 2009 An Overview Of The Integration Of Advanced Oxidation Technologies And Other Processes For Water And Wastewater Treatment *Int. J. Eng.(IJE).* **3 (2)** 120 – 46

[2] Bokare A D and Choi W 2014 Review of iron-free Fenton-like systems for activating H2O2 in advanced oxidation processes *J Hazard.Mat.* **275** 121-35

[3] Yao H 2013 Application of advanced oxidation processes for treatment of air from livestock buildings and industrial facilities Department of Engineering, Aarhus University. Denmark. 36 pp. - *Technical report BCE -TR-8*

[4] Munter R 2001 Advanced Oxidation Processes – Current Status And Prospects. *Proc. Estonian Acad. Sci. Chem.* **50** *(***2***)* 59–80

[5] Antonopoulou M, Evgenidou, E, Lambropoulou D and Konstantinou I 2014 A review on advanced oxidation processes for the removal of taste and odor compounds from aqueous media *Water Res.* **53** 215-34

[6] Andreozzi R, Caprio V, Insola A and Marotta R 1999 Advanced oxidation processes (AOP) for water purification and recovery *Cat. Today* **53** 51–59

[7] Cortes S, Sarasa J, Ormad P, Gracia R and Ovelleiro J 1998 Comparative efficiency of the systems O3/high pH and O3/CAT for the oxidation of chlorobenzenes in water *In Proc. Int. Reg. Conf. Ozonation and AOPs in Water Treatment, Poitiers, France*, 14-1–15-1.

[8] Horáková M, Klementová Š, Kříž P, Balakrishna S K, Špatenka P, Golovko O, Hájková P and Exnar P 2014 The synergistic effect of advanced oxidation processes to eliminate resistant chemical compounds *Surf. Coat. Technol.* **241** 154-58

[9] Kommineni S, Zoeckler J, Stocking A, Liang P S, Flores A, Rodriguez R, Tom Browne, Roberts PE R and Brown A, 2000, 3.0 Advanced Oxidation Processes. *Center for Groundwater Restoration and Protection National Water Research Institute*

[10] Krzemińska D, Neczaj E and Borowski G 2015 Advanced oxidation processes for food industrial wastewater decontamination *J. Ecol. Eng.* **16(2)** 61-71

[11] Rice R G 1996 Applications of ozone for industrial wastewater treatment—a review *Ozone: Sci. Eng.* **18(6)** 477-515

[12] Friedrich M 1995 Actual development of ozone technology, explained with examples for treatment of chemical wastewater with ozone. *In Proc. 12th Ozone World Congress of IOA*, 15–18 May, 1995, Lille, France, **2**, 469–476

[13] Vaferi B, Bahmani M, Keshavarz P and Mowla D 2014 Experimental and theoretical analysis of the UV/H2O2 advanced oxidation processes treating aromatic hydrocarbons and MTBE from contaminated synthetic wastewaters *J. Environ. Chem. Eng.* **2 (3)** 1252–126

[14] Akin K, Idil A-A, Miray B, Gul O and Buket A 2014 H2O2/UV-C and Photo-Fenton treatment of a nonylphenol polyethoxylate in synthetic freshwater: Follow-up of degradation products, acute toxicity and genotoxicity *Chem. Eng. J.* **241** 43–51
[15] Karci A, Arslan-Alaton I, Bekbolet M, Ozhan G and Alpertunga B 2014 H2O2/UV-C and Photo-Fenton treatment of a nonylphenol polyethoxylate in synthetic freshwater: Follow-up of degradation products, acute toxicity and genotoxicity *Chem. Eng. J.* **241** 43–51
[16] Ikehata K and El-Din M G 2004 Degradation of recalcitrant surfactants in wastewater by ozonation and advanced oxidation processes: a review *Ozone: Sci.Eng.* **26(4)** 327-43
[17] Paphane B D and Ramirez L L 2013 Chemical Pre-Treatment of Anionic Surfactants Contaminated Waste Water at Enaspol AS using H2O2/UV Light Waste Water Pre-Treatment Method *J Environ Anal Toxicol*. **3(181)** 2161-0525
[18] Canizares P, Paz R, Sáez C and Rodrigo M A 2009 Costs of the electrochemical oxidation of wastewaters: a comparison with ozonation and Fenton oxidation processes *J. Env. Manage.* **90(1)** 410-20
[19] Lin S H and Leu H G, 1999, Operating characteristics and kinetic studies of surfactant wastewater treatment by Fenton oxidation. *Water Res.* **33(7)** 1735-41
[20] Heponiemi A and Lassi U 2012 Advanced Oxidation Processes in Food Industry Wastewater Treatment-A Review *INTECH Open Access Publisher*.
[21] Alam M G, Tawfik A and Ookawara S 2014 Investigation of optimum conditions and costs estimation for degradation of phenol by solar photo-Fenton process *Appl. Water Sci.* 1-8
[22] Akpan U G and Hameed B H 2009 Parameters affecting the photocatalytic degradation of dyes using TiO2-based photocatalysts: a review *J. Hazard. Mat.* **170(2)** 520-529
[23] Al-Dawery S K 2013 Photo-Catalyst Degradation Of Tartrazine Compound In Wastewater Using TiO2 And UV Light *J. Eng. Sci. Technol.* **8(6)**, 683-91
[24] Czech B and Curie-Sklodowska M 2012 Advanced photooxidation of surfactants in wastewater.*CHEMIK 2012* **66** 1314-1325
[25] da Silva S W, Klauck C R, Siqueira M A and Bernardes A M 2015 Degradation of the commercial surfactant nonylphenol ethoxylate by advanced oxidation processes *J. Hazard. Mat.* **282** 241-48
[26] Munusamy S, Sai Laxmi Aparna R and Gunneswara Subramanya Vara Prasad R 2013 Photocatalytic effect of TiO2 and the effect of dopants on degradation of brilliant green *Sustain Chem Process* **1(1)**, 1-8
[27] Sopyan I, Hafizah N and Jamal P 2011 Immobilization of TiO2 with cement: Photocatalytic degradation of phenol and its kinetic studies *Indian J. Chem. Technol.* **18(4)** 263-70
[28] Mahamuni N N and Adewuyi Y G 2010 Advanced oxidation processes (AOPs) involving ultrasound for waste water treatment: a review with emphasis on cost estimation *Ultrason. Sonochem.* **17(6)** 990-1003

IOP Conf. Series: Materials Science and Engineering **206** (2017) 012090 doi:10.1088/1757-899X/206/1/012090

Removal of Methylene Blue Dye from Aqueous Solution Using a Newly Synthesized TiO_2-SiO_2 Photocatalyst in the Presence of Active Chlorine Species

C G Joseph and L Elilarasi

Sonophotochemistry Research Group, Faculty of Science and Natural Resources, University Malaysia Sabah, Kota Kinabalu, Sabah, Malaysia

E-mail: collin@ums.edu.my

Abstract. Industrialization and urbanization demand high amount of water consumption, which contributes to their polluted condition. Thus, there is a need to develop a sustainable wastewater remediation technique in order to provide sustainable use of clean water for future generations without ramifications to the economic sectors. The newly synthesized TiO_2-SiO_2 photocatalyst was used to remediate Methylene Blue contaminated aqueous solution in the presence of active chlorine species. The doping of SiO_2 into TiO_2 enhanced the removal rate of Methylene Blue dye from the solution by increasing the surface area, thermal stability and surface acidity of the TiO_2. The active chlorine species further enhanced the removal rate of Methylene Blue dye from the solution by contributing more reactive species, chlorine radicals, which broke down the dye molecules. The experiments were conducted via Taguchi analysis. The findings show that combining TiO_2, SiO_2 and active chlorine species enhanced the removal percentage of Methylene Blue dye compared to using TiO_2 alone by 70%. About 70% of 50ppm Methylene Blue was degraded by 1 g of TiO_2-SiO_2 in the presence of 0.3 ppm $Ca(OCl)_2$ under 9 Watts solar irradiation within 3 hours. The enhanced dye removal method brings photocatalysis a step closer to sustainable wastewater remediation methods.

1. Introduction

Sunlight or solar energy is a well known renewable energy and largest source of carbon neutral energy [1]. Approximately 95% of UV-A and 5% of UV-B reaches the Earth's surface since ozone and oxygen absorb majority UV-B and UV-C [2]. The UV wavelength (100-400 nm) consists of UV-C (100-280 nm), UV-B (280-315 nm) as well as (315-400 nm) [2-4]. The titanium oxide, TiO_2, is a well known photocatalyst, utilized widely in numerous catalyzed reactions including photocatalysis, oxidation, Fischer-Tropsch reaction [5] delivery of drug, chemical sensors as well as photovoltaic cells [6]. The SiO_2 addition into TiO_2 enhances TiO_2 suspension in aqueous solution, enhances the adsorption of pollutants [7], induce high mechanical and thermal stability [5].

In previous research done by our team, 10 ppm of Methylene Blue (MB) was successfully removed from simulated wastewater using Solar/HOCl treatment method. Active chlorine species produced due to addition of HOCl was utilized as chemical oxidant due to its simplicity, strong oxidation and disinfection capability and low cost. Optimum conditions required for remediation of MB using 9 Watts solar irradiation was at pH 4, lower HOCl dosage as well as high initial concentration of MB. In order to further enhance the removal efficiency of MB from simulated wastewater using solar irradiation, in this we study we included photocatalyst, TiO_2-SiO_2.

In this study, commercial P-25 TiO_2 doped with SiO_2 was synthesized via sol gel method in order to increase the surface area and acid properties [8]. MB dye was utilized as a model dye to analyze the photo activity of TiO_2-SiO_2. The Taguchi design was employed to study the influence of different parameters on removal efficiency of MB [9]. Taguchi design was employed because compared to full factorial design, the number of experiments required for Taguchi method is lowered [9].

2. Materials and Methods

2.1. Preparation of TiO_2-SiO_2

Commercial TiO_2 (Degussa P25) and TEOS precursor were used to prepare TiO_2-SiO_2 photocatalyst. SiO_2 sol was prepared by addition of 22.2 mL of TEOS, to a mixture of 29.2 mL ethanol, 7.2 mL distilled water and 0.4 mL of 3.6% of hydrochloric acid followed by stirring at 90°C in air for 1 hour. 0.2g of Degussa TiO_2 was added to a mixture of SiO_2 sol which was subsequently diluted with absolute ethanol in order to yield a final volume of 50 mL. The resulting sol was dried in air at 80°C for 24 h [10]. Dried TiO_2-SiO_2 were stored in small clean screw cap vials for subsequent experiments [11]. The synthesized TiO_2-SiO_2 was later irradiated with UV-B, UV-C and Solar lamps in order to facilitate the formation of Ti-O-Si bond. Calcium hypochlorite ($Ca(OCl)_2$) salt was utilized to prepare stock aqueous chlorine solutions and were filtered through a 0.45 μm filter [12]. All free chlorine solutions were prepared by dilution of the calcium hypochlorite stock solution with distilled water in glassware protected from light to minimize free chlorine photodecay [13].

2.2. Photo-activation of TiO_2-SiO_2

Photoactivation of TiO_2-SiO_2 was conducted inside a beaker placed on a hotplate to enable constant stirring of the solution using a magnetic stirrer. Three different types of lamps emitting UV-C, UV-B and Solar were utilized to activate the TiO_2-SiO_2. The height of UV lamps were adjusted so that the height of the MB solution is equal or more than midpoint of the lamps for maximum exposure [11]. The photocatalyst was added into MB solution and stirred using magnetic stirrer for 40 min in the dark to achieve adsorption equilibrium and uniformity of the reaction system [14]. All the experiments were conducted using batch photo-reactor [15]. 250 mL of simulated Methylene Blue wastewater (10, 30 or 50 ppm) was prepared and TiO_2-SiO_2 (0.5, 1.0 or 1.5 g) was added. After 40 min of adsorption, 50 mL of prepared HOCl (0.1, 0.2 or 0.3 ppm) solution was added. 9 Watts of Solar lamp was used to irradiate the solution. At every 10 min intervals, the dispersion was extracted and filtered to separate the photocatalyst particles. Any changes in the concentration of MB after the experiments were tracked using a Perkin Elmer Lambda 25 UV-Vis spectrophotometer [16]. The removal efficiency of MB was calculated from the peak intensity of the characteristic absorption band of MB at 665 nm. The removal efficiency of MB was calculated using [17]:

$$Removal\ Efficiency\ (\%) = \frac{(C_o - C_t)}{C} \times 100 \quad (1)$$

In which C_0 (mg L^{-1}) and C_t (mg L^{-1}) represent the initial concentration of dye and concentration at a particular time, respectively.

2.3. Taguchi Design

First, the outcome characteristic (removal efficiency of MB dye) was identified, and process parameters that are to be investigated (effect of different wavelength irradiated TiO_2-SiO_2, HOCl dosage, TiO_2-SiO_2 dosage as well as initial concentration of MB) are selected [18-19]. Second, the amount of levels of each parameters are selected. Next, suitable orthogonal array (OA) was selected and the parameters were assigned to the orthogonal array. Forth, experiments were conducted based on the arrangement of the orthogonal array. The experiments were conducted twice and average value was determined and used for the analysis. Fifth, the analysis on the obtained data were conducted and analyzed using analysis of variance (ANOVA). In this study, four parameters were selected to study removal efficiency of MB using Taguchi design. Taguchi orthogonal array L9 (3^4) with three levels for each factor [20] was utilized in order to analyze the impact of four factors (effect of irradiated TiO_2-SiO_2, HOCl dosage, TiO_2-SiO_2 dosage as well as MB concentration) on the removal efficiency of Methylene Blue dye. It should be noted that in the orthogonal arrays, for each level of any factor, each level in the factor occur equal number of times in which gives a balanced design [9]. Following are the different levels for each factor, (1) Different wavelength utilized to irradiate TiO_2-SiO_2 (UV-C/ 100-280 nm, UV-B/ 280-315 nm, & solar/ 280-700 nm), (2) HOCl dosage (0.1, 0.2 & 0.3 ppm) (3) TiO_2-SiO_2 dosage (0.5, 1.0 & 1.5 g) and lastly initial concentration of MB (10, 30 & 50 ppm). UV-A (315-400 nm) was not included in the study since the natural sunlight consists of UV-A [21]. The experiments were conducted twice for each experiment to ensure accuracy and the average response is shown in the table [19]. The influence of the parameters on MB removal efficiency were investigated through analysis of variance (ANOVA) with 95% confidence level [9, 22].

3. Results & Discussion

3.1. Effects of parameters on the removal efficiency of MB dye

No significant change in MB concentration was determined under solar irradiation without the presence of TiO_2-SiO_2 [23]. Following Table 1 and Table 2 show selected factors and their levels (Coded and Actual). All the Taguchi experiments showed removal of MB from aqueous solution with removal efficiency ranging from 82.6-99.3%. From the analysis of UV-Vis absorbance peaks of Methylene Blue, the characteristic peaks of MB appears at 665 nm, 293 nm and 246 nm. The peak at 665 nm represents dimenthylamino group / thiazine (the chromophore), as well as peaks at 293 nm and 246 nm represent the benzene rings in the MB molecule [24-25]. At the end of all 9 experiments, the dominant peak at 665 nm decreases due N-demethylation and deamination of MB, while the peaks at 293 nm and 256 nm decrease due to cracking of benzene ring [24].

As shown in the Fig. 1-4, highest removal efficiency of Methylene Blue was achieved when lower wavelength (UV-C/ 100-280 nm) was used to irradiate TiO_2-SiO_2 (UV-C), lower HOCl dosage, lower concentration of Methylene Blue as well as higher amount of TiO_2-SiO_2 dosage.

Table 1. Selected factors and their levels (actual)

Factors	Symbol	Level		
		1	2	3
Different wavelength irradiated TiO_2-SiO_2 (Factor A)	-	UV-C irradiated TiO_2-SiO_2	UV-B irradiated TiO_2-SiO_2	Solar irradiated TiO_2-SiO_2
HOCl dosage (Factor B)	ppm	0.1	0.2	0.3
TiO_2-SiO_2 dosage (Factor C)	g	0.5	1.0	1.5
Conc. of MB (Factor D)	ppm	10	30	50

Table 2. Taguchi L_9 (3^4) orthogonal array of selected factors (coded)

Standard	Run Order	Factor A	Factor B	Factor C	Factor D	Removal efficiency (Average)
2	1	1	2	2	2	97.2
5	2	2	2	3	1	99.3
6	3	2	3	1	2	82.6
8	4	3	2	1	3	85.3
4	5	2	1	2	3	92.7
3	6	1	3	3	3	95.9
1	7	1	1	1	1	97.5
7	8	3	1	3	2	97.0
9	9	3	3	2	1	95.1

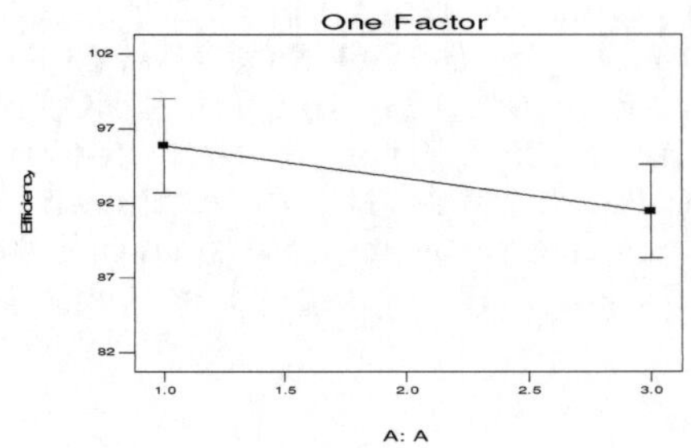

Figure 1. Factor A

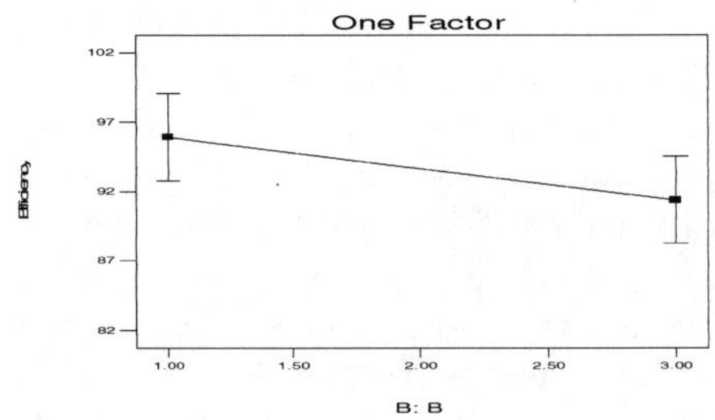

Figure 2. Factor B

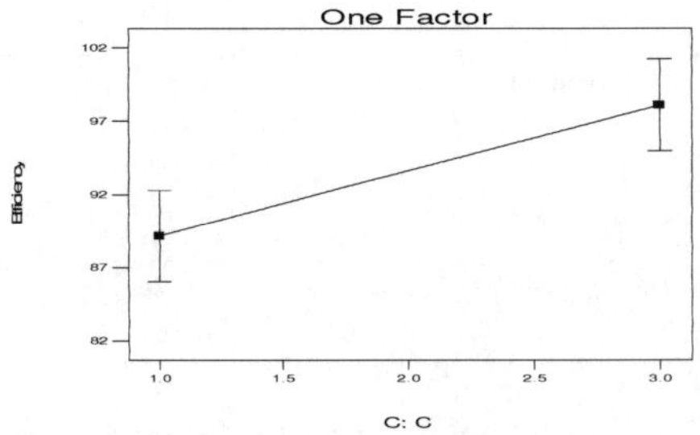

Figure 3. Factor C

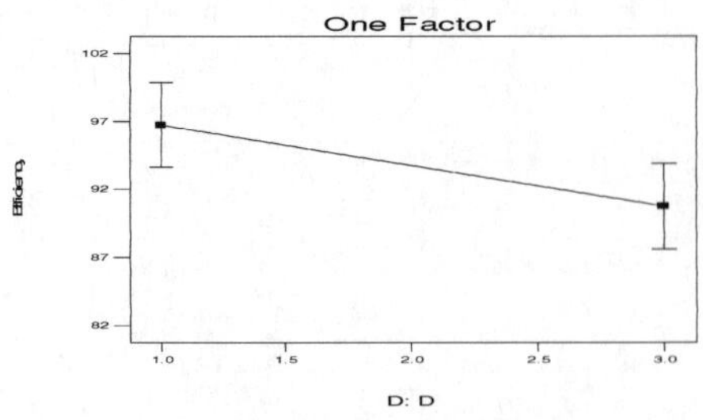

Figure 4. Factor D

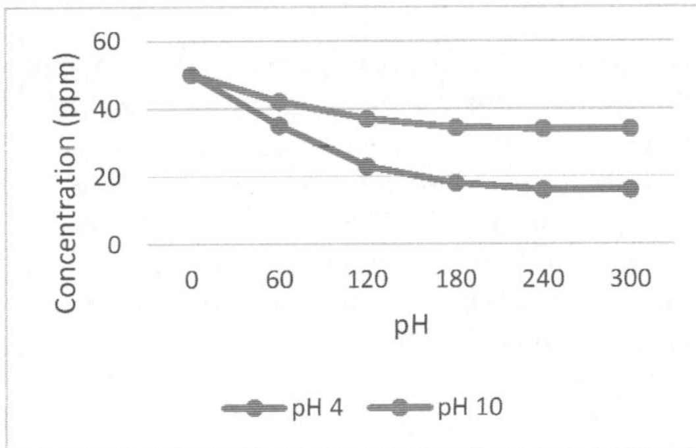

Figure 5. Concentration vs pH (50 ppm of Methylene Blue was degraded by 1 g of TiO_2-SiO_2 in the presence of 0.3 ppm $Ca(OCl)_2$ under 9 Watts solar irradiation within 3 hours)

Table 3. Analysis of variance

Source	Sum of Squares	Df	Mean Square	F value	p-value Prob>F
Model	233.57	4	58.39	6.33	0.0507
A	29.04	1	29.04	3.15	0.1506
B	30.83	1	30.83	3.34	0.1415
C	119.71	1	119.71	12.98	**0.0227**
D	54.00	1	54.00	5.86	0.0728

This shows that the newly synthesized TiO_2-SiO_2 has good photo degradation ability in relative to low MB concentration [24, 26]. At low initial concentration of MB, maximum amount of dye molecules adsorbed on TiO_2-SiO_2 dosage leading to high adsorption efficiency [26]. However, at high concentration of MB, lower adsorption efficiency was reported due to saturation of active sites on the adsorbent's surface [26].

The maximum adsorption percentage was achieved at 1.5 g of TiO_2-SiO_2 because more active sites are available as the dosage increased from 0.5 to 1.5 g. If the dosage is further increased, it may cause aggregation of adsorption sites and increased diffusion path length which may decrease the availability of total available surface area of adsorbent [26]. From the Fig. 2, it shows that increasing the HOCl dosage decreases the removal efficiency of MB from simulated wastewater. This may due to pH of the solution which was alkaline (pH 10). At pH < 3, Cl_2 is the main chlorine species HOCl in the region 3≤ pH ≤7, and OCl^- at pH> 7 in which among these, HOCl is the strongest oxidant and the predominant species at pH 4 [27]. Hence, another experiment was conducted to investigate the effect of HOCl at pH 4 as shown in Graph 1. About 70% of 50ppm Methylene Blue was degraded by 1 g of TiO_2-SiO_2 in the presence of 0.3 ppm $Ca(OCl)_2$ under 9 Watts solar irradiation within 3 hours for pH 4 while 32% for pH 10. The effectiveness of 254 nm radiation (UV-C) was attributed to the shorter penetration capability of the higher energy photons [28]. Compared to UV-A, UV-B and solar irradiation, UV-C contributed the most to chemical or morphological modification on the surface of TiO_2-SiO_2 by creating uniformly distributed pores and formation of large amount of Ti-O-Si bonds. The Table 3 consists of "p" values that is used to investigate the statistical significance of each parameter [19, 29]. It was found that only factor C (TiO_2-SiO_2 dosage) exhibit significant influence on removal efficiency of MB from simulated wastewater.

IOP Conf. Series: Materials Science and Engineering **206** (2017) 012090 doi:10.1088/1757-899X/206/1/012090

3.2. Mechanism of TiO_2-SiO_2

It was reported that TiO_2 either successfully or partially incorporated with amorphous SiO_2 through Si^{4+} entering the TiO_2 lattice fringes [30]. This is due to smaller size of Si^{4+} ionic radii (0.039 nm) compared to TiO_2 (0.068 nm) which enables Si^{4+} ions enter into crystal lattice of TiO_2 as a substitute for Ti^{4+} [30]. In the TiO_2-SiO_2, the SiO_2 acts as an excellent adsorbent due to enormous amount of acid sites as well as hydroxyl groups on its amorphous surface [31]. The SiO_2 on the surface of TiO_2 acts as favored adsorption sites for dye molecules, thus bringing the pollutant molecules within close proximity to high $^{\bullet}OH$ concentration leading to enhanced photo degradation rate [10, 32]. In addition, there is a possibility of other ROS (e.g., superoxide anion) production enhanced due to surface modification using SiO_2 [32].

The mechanism of TiO_2-SiO_2's superhydrophilicity is still not explained well. Three possible explanations for this enhanced property are; 1) enhanced surface roughness cause an increase in hydrophilicity, 2) Changes in the surface structure (either chemically or morphologically) caused conversion of Ti^{4+} sites to Ti^{3+} sites via conduction band electron trapping which is favors the dissociative water adsorption, 3) increased surface acidity of Si-O-Ti bonds at the interfaces lead huge amount of hydroxyl groups on the surface [10]. The possible interaction that occur during adsorption of MB dye molecules onto adsorbents surface are the electrostatic interaction between negatively charged surface of the adsorbent and positively charged active site of dye molecule; dipole-dipole H-bonding interaction between –OH groups present on the surface of the adsorbent and the electronegative residue (N lone pair) in MB^+ molecules; as well as Yoshida H- bonding interaction between silanol –OH groups, large number of –OH groups present on the TiO_2 surface and the active sites of MB^+ dye (aromatic ring) [26]. The existence of silanol groups on the surface of SiOH and these groups are able to interact with MB molecules. At low pH values, the H^+ and MB^+ compete with each other for adsorption sites and there is repulsive force between the positive charged TiO_2-SiO_2 and MB^+. However with increasing pH, the silanol groups on the surface increases which then interact with MB^+. Moreover, the amount of negatively charged adsorption sites increases at alkaline condition, which favors the cationic MB adsorption [33].

$$-SiOH + H^+ \rightarrow -SiOH_2^+ \quad (2)$$
$$-SiOH + MB^+ \rightarrow -Si\text{-}O\text{-}MB + H^+ \quad (3)$$
$$-SiOH + OH^- \rightarrow -SiO^- + H_2O \quad (4)$$
$$-SiO^- + MB^+ \rightarrow -Si\text{-}O\text{-}MB \quad (5)$$

Meso-macroporous TiO_2-SiO_2 film exhibited higher photo degradation compared to silica-etched TiO_2 and dense TiO_2 films. The higher degradation rate is attributed to meso-macroporous structure and TiO_2-SiO_2 hybridization [6]. The doping of SiO_2 onto TiO_2 enhances the photo degradation efficiency of MB dye molecules due to the SiO_2's ability to adsorb the dye molecules in the vicinity of TiO_2 photo active centre as well as its ability to block the recombination of photo generated holes and electrons from TiO_2. The pores offer larger surface area with greater possibility of target pollutant attack by a higher number of oxidizing agents such as positive holes h^+_{VB} and •OH formed between holes and HO^- groups on the surfaces of TiO_2 or H_2O as well as reducing agents such as electrons, e^-_{CB} and the superoxide radical anion, $O_2^{\bullet-}$ formed due to the reaction of O_2 adsorbed on the TiO_2 surface or dissolved in H_2O and photogenerated electron [6].

Two degradation pathway of MB dye molecule under solar light irradiation [23];

i) Self-sensitized reaction: Direct photon absorption by MB, excitation of MB molecule, injection of electron from MB into conduction band of TiO_2, electron reacts with O_2 to produce super oxide radical which oxidize MB molecules.

ii) Excitation of TiO_2: photon absorption by TiO_2, electron excitation from valence band into conduction band to create e^-/h^+ pairs. The electron reacts with O_2 to produce super oxide radical and holes oxidize H_2O to produce OH radicals which oxidize the MB molecules.

4. Conclusion

The TiO_2-SiO_2 acts as an excellent adsorbent and photocatalyst which can work effectively under natural sunlight. The enhanced photocatalytic activity of UV irradiated TiO_2-SiO_2 is due to the narrowed bandgap aided by production of many $HO^{\bullet}$ free radicals, larger surface area as well as more active sites for dye adsorption of TiO_2-SiO_2. Combination of crystalline nature, high amount of hydroxyl groups and porosity are the main contributors for high photocatalytic activity by TiO_2-SiO_2. Compared with non-irradiated TiO_2-SiO_2, irradiated TiO_2-SiO_2 exhibited higher removal efficiency of MB. Taguchi analysis showed that removal efficiency of MB decreases with increasing irradiated TiO_2-SiO_2, HOCl dosage and initial concentration of MB. However, removal efficiency of MB increases with increasing TiO_2-SiO_2 dosage. From the ANOVA analysis, TiO_2-SiO_2 dosage significantly influences the removal efficiency of MB.

References

[1] Kibombo H S, Peng, R, Rasalingam S and Koodali R T 2012 *Cat. Sci. & Technol.* **2** 1737-1766
[2] Schuch A P, Garcia C C M, Makita K and Menck C F M *Photochem. & Photobiological Sci.* **12** 1259-1272
[3] Mallet, J D and Rochette P J 2013 *Photochem. & Photobiological Sci.* **12** 1310-1318
[4] Caputo F, De Nicola M, Sienkiewicz A, Giovanetti A, Bejarano I, Licoccia S and Ghibelli L 2015 *Nanoscale.* **7** 15643-15656
[5] Wei Q, Li Y, Zhang T, Tao X, Zhou Y, Chung K and Xu C *Energy & Fuels.* **28** 7343-7351
[6] Jeon H, Lee C S, Patel R and Kim J H 2015 *ACS app. Materi. & interfaces.* **7** 7767-7775
[7] Bo Z, Eaton T R, Gallagher J R, Canlas C P, Miller J T and Notestein J M *Chem. of Mater.* **27** 1269-1277
[8] Chen L C and Huang C M 2004 *Ind. & eng. Chem. research.* **43** 6446-6452
[9] Engin A B, Özdemir Ö, Turan M and Turan A Z 2008 *J. of hazard. mater.* **159** 348-353
[10] Fateh R, Dillert R and Bahnemann D 2013 *Langmuir.* **29** 3730-3739
[11] Bitter J L, Yang J, Milani S B, Jafvert C T and Fairbrother D H 2014 *Environ. Sci.: Nano.* **1** 324-337
[12] Rao B, Estrada N, McGee S, Mangold J, Gu B and Jackson W A 2012 *Environ. sci. & technol.* **46** 11635-11643
[13] Powell J, Luh J and Coronell O 2014 *Environ. sci. & technol.* **48** 2741-2749
[14] Kang C, Jing L, Guo T, Cui H, Zhou J and Fu H 2008 *The J. of Physic. Chem. C.* **113** 1006-1013
[15] Joseph C G, Puma G L, Bono A, Taufiq-Yap Y H and Krishnaiah D 2011 *Desalination.* **276** 303-309
[16] Noimark S, Dunnill C W, Kay C W, Perni S, Prokopovich P, Ismail S and Parkin I P 2012 *J. of Mater. Chem.* **22** 15388-15396
[17] Banerjee S, Gautam R K, Jaiswal A, Chattopadhyaya M C and Sharma Y C 2015 *RSC Adv.* **5** 14425-14440
[18] Srivastava V C, Mall I D and Mishra I M 2007 *Ind. & eng. Chem. research.* **46** 5697-5706
[19] Devadi M A H, Krishna M Murthy H N and Sathyanarayana B S 2014 *Procedia Materi. Sci.* **5** 612-621
[20] Gönder Z B, Arayici S and Barlas H 2012 *Ind. & Eng. Chem. Research.* **51** 6184-6195
[21] Liu Y, Dong S, Kuhlenschmidt M S, Kuhlenschmidt T B, Drnevich J and Nguyen T H 2015. *Environ. Sci.: Water Research & Technol.* **1** 188-198
[22] Taheri M, Moghaddam M R A and Arami M 2015 *Environ. Progress & Sustainable Energy.* **34** 1568-1575
[23] Kokane S B, Sartale S D, Betty C A and Sasikala R 2014 *RSC Adv.* **4** 55539-55547
[24] Duan T, Chen Y, Wen Q and Duan Y 2014 *RSC Adv.* **4** 57463-57475
[25] Lau Y Y, Wong Y S, Teng T T, Morad N, Rafatullah M and Ong S A 2015 *RSC Adv.* **5** 34206-34215
[26] Ghorai S, Sarkar A, Raoufi M, Panda A B, Schönherr H and Pal S 2014 *ACS app. mater. & interfaces.* **6** 4766-4777

IOP Conf. Series: Materials Science and Engineering **206** (2017) 012090 doi:10.1088/1757-899X/206/1/012090

[27] Barrera-Díaz C, Roa-Morales G, Avila-Córdoba L, Pavón-Silva T and Bilyeu B 2006, *Ind. & Eng. Chem Research.* **45** 34-38
[28] Bayarri B, Abellán M N, Giménez J and Esplugas S 2007 *Cat. Today.* **129** 231-239
[29] Giuca O and Grozescu I 2011 *Chem. Bull. "Politehnica" Univ. Timisoara, Romania, Ser. Chem. Environ. Eng.* **56** 105-114
[30] Yang P, Tang Q and He B 2015 *RSC Adv.* **5** 46260-46266
[31] Ajmal A, Majeed I, Malik R N, Idriss H and Nadeem M A 2014 *RSC Adv.* **4** 37003-37026
[32] Liga M V, Maguire-Boyle S J, Jafry H R, Barron A R and Li Q 2013 *Environ Sci Technol* **47** 6463-6470
[33] Zhang L, Xing Z, Zhang H, Li Z, Wu X, Zhang X and Zhou W 2016 *App. Cat. B: Environ.* **180** 521-529

IOP Conf. Series: Materials Science and Engineering **206** (2017) 012091 doi:10.1088/1757-899X/206/1/012091

A Survey on the Usage of Biomass Wastes from Palm Oil Mills on Sustainable Development of Oil Palm Plantations in Sarawak

K Y Phang, S W Lau[1]

Department of Chemical Engineering, Curtin University, Malaysia, CDT 250, Miri 98009, Sarawak, Malaysia

E-mail:[1]lau.shiew.wei@curtin.edu.my

Abstract. As one of the world's largest palm oil producers and exporters, Malaysia is committed to sustainable management of this industry to address the emerging environmental challenges. This descriptive study aims to evaluate the oil palm planters' opinions regarding the usage of biomass wastes from palm oil mills and its impact on sustainable development of oil palm plantations in Sarawak. 253 planters across Sarawak were approached for their opinions about the usage of empty fruit bunch (EFB), palm oil mill effluent (POME), mesocarp fibre (MF), and palm kernel shell (PKS). This study revealed that the planters had generally higher agreement on the beneficial application of EFB and POME in oil palm plantations. This could be seen from the higher means of agreement rating of 3.64 – 4.22 for EFB and POME, compared with the rating of 3.19 – 3.41 for MF and PKS in the 5-point Likert scale (with 5 being the strongest agreement). Besides, 94.7 percent of the planters' companies were found to comply with the Environmental Impact Assessment (EIA) requirements where nearly 38 percent carried out the EIA practice twice a year. Therefore high means of agreement were correlated to the compliance of environmental regulations, recording a Likert rating of 3.89 to 4.31. Lastly, the usage of EFB and POME also gained higher Likert scale point of 3.76 to 4.17 against MF and PKS of 3.34 to 3.49 in the evaluation of the impact of sustainability in oil palm plantations. The planters agreed that the usage of EFB and POME has reduced the environmental impact and improved the sustainable development, and its application has been improved and increased by research and development. However the planters were uncertain of the impact of usage of biomass wastes with respect to the contribution to social responsibility and company image in terms of transparency in waste management.

1. Introduction

The oil palm planted area in the state of Sarawak, Malaysia is expanding rapidly, with its current region of approximately 1.4 million hectares of oil palm land bank, including 0.5 million hectares on peat [1, 2]. Due to the limitation of fertile soil profile, Sarawak planters have established valuable experience and knowledge over decades for oil palm cultivation on peat [3]. With the increase of plantations and the maturing palm age profile, the number of palm oil mills grows almost linearly in Sarawak in order to match the production of fresh fruit bunches (FFB). At the same time, the greatest challenge faced by this industry is to strike a balance between the environment and the social economy of the country.

Sarawak Energy Berhad (SEB), the state's energy provider, projects an increase of FFB production from 9.3 million ton/year in 2008 to 21.0–25.5 million ton/year in 2015 [4]. With such a tremendous capacity of FFB production, processed by 73 palm oil mills across Sarawak, a systematic, effective and efficient waste management system is indispensable especially for the palm biomass wastes for sustainable development of the industry.

[1] Corresponding author

The biomass residues from the palm oil industry in Sarawak have proven to have great potential to achieve industry self-reliance in fuel by using the waste to produce high pressure steam for turbines in power generation. This reduces the use of fossil fuel and subsequently minimise the emission of greenhouse gases (GHGs). It also creates more job opportunities especially in renewable energy industry in the state. Trapping and converting bio-methane gas emitted from the POME treatment pond has also successfully contributed to clean energy supply and reduction of GHG emission [5, 6]. Beside its application in palm oil mills, the composted POME and EFB can be applied as organic fertilisers in oil palm plantations in order to reduce the application of inorganic fertilisers, increase the oil palm yield, and improve the physical and chemical properties of soil by increasing the organic matter (OM) content. Bunch ash (BA) produced from the complete combustion of EFB in furnaces is also widely used as alkaline and potassium sources for nutrient recycle, especially for the plantations on peat area [5, 7].

This study aims to study the oil palm planters' opinions regarding the usage of the palm oil mill biomass wastes and to evaluate the impact of their usages on sustainable development in oil palm plantations in Sarawak. This study also provides illustration regarding the compliance of government policy and environmental regulations, especially the Environmental Quality Act (EQA) 1974 and Environmental Quality (Prescribed Premises) (Crude Palm Oil) Regulations (1977) which influence the growth of palm oil industry toward sustainable development.

2. Research background and framework

Non-governmental organisations (NGOs) have constantly criticised the development of palm oil industry, particularly in Malaysia and Indonesia because both countries have supplied almost 90% of the global palm oil demand. Misleading studies have always linked the palm oil industry to several environmental issues such as deforestation, species extinction, pollution, climate change, and forest fire brought by burning during land clearing activity in the preliminary stage of oil palm planting, for example, the Kalimantan forest fire during 1997 – 1998 [8]. In Malaysia, biomass wastes generated from palm oil industry account for 85.5% of total biomass production in Malaysia. This gigantic amount of biomass wastes will trigger a tremendous impact to the environment if they are not handled in an appropriate manner [9].

To address the severe sustainability challenges of palm oil industry, the Roundtable on Sustainable Palm Oil (RSPO) was established in 2004 to answer to the global call on sustainable palm oil production [10, 11]. To obtain the RSPO certification, the respective production process is subject to the rigorous verification of RSPO Principles and Criteria in order to fulfil the regulations, social benefit, environmentally friendliness, as well as economic viability, which broadly cover various aspects such as transparency, compliance of regulations and laws, commitment to long term economic and financial viability, best practices by growers and millers, sustainable environment, and continuous improvement [11, 12]. The produced palm oil from the certified growers and millers of RSPO is known as Certified Sustainable Palm Oil (CSPO). Malaysia and Indonesia are the world CSPO leading players which have contributed 90% or approximately 5.4 million tons out of the total 6 million tons of CSPO in 2012, with the remaining contributed by countries like Papua New Guinea, Brazil, Colombia and others. The total CSPO produced has achieved 12.53 million tons, being 20% of the total palm oil produced globally in March 2015, as reported by RSPO [11]. This implies that the standard sustainable management of biomass wastes from palm oil industry has been in place for some times.

In Malaysia, several practices have been adopted by the oil palm industry to enhance the sustainability of the industry in terms of environmental regulations, agriculture practices, codes of practices, and conservation efforts. One of the greatest challenges for palm oil millers toward sustainable development is the handling of the biomass wastes in compliance to environmental regulations, such as the EQA 1974 [13]. Composting technique is technologically simple and relatively inexpensive in treating the biomass wastes which is widely carried out by most oil palm plantations in Sarawak. Besides, trapping and converting methane gas emitted from the POME treatment pond to generate electricity power is another potential beneficial use of the biomass wastes in Sarawak. Evidently, a recent biomass power plant project launched in Mukah, Sarawak in November 2015 has projected a production capacity of 90 tons per hour or equivalent to a reduction of 60 thousand tons carbon dioxide annually. It is undeniable that the biomass wastes from palm oil mills have great potential in various aspects, which bring about a win-win situation toward achieving sustainable development for the industry.

3. Research methodology

3.1. Research approach and design

This is a descriptive study where questionnaires were administrated to the oil palm planters across Sarawak to collect the planters' opinion regarding the usage of biomass wastes from palm oil mills and its impact on sustainable development of oil palm plantations. The study was conducted on 20 percent of 1,500 palm oil planters in Sarawak, mainly from the northern, central and southern regions of Sarawak, with membership or non-membership of Incorporated Society of Planters (ISP), a professional body representing the Planters of Malaysia and other territories.

The hypotheses generally align to correlation between sustainable development and usage of the major biomass wastes from the millers. The major biomass wastes under study were EFB and POME which typically account for 30-32 percent and 57.4 percent respectively, while MF and PKS normally contribute to 18 percent and <10 percent respectively [14, 15]. As such, the hypotheses are listed as below:

Hypothesis 1: Usage of EFB and POME from palm oil mill influences sustainable development in oil palm plantations.

Hypothesis 2: Usage of MF and PKS from palm oil mill influences sustainable development in oil palm plantations.

3.2. Data collection and analysis

This study proposed a stratified random sampling with sampling population based on the area of plantations, i.e. small (below 3,000 hectares), medium (3,000 to 5,000 hectares) and corporate (more than 5,000 hectares) plantations in Sarawak. The primary data were collected based on the oil palm planters' experience in operation and management of oil palm estates or plantations. The respondents were given self-administered questionnaires. The first part of the questionnaire was designed to understand the oil palm planters' employment background such as the present working region, professional membership, work responsibility, company status, work experience, palm oil mill facilities and compliance of environmental assessment. The second part of the questionnaire collected information related to the planters' experience in utilisation, application and disposal of biomass wastes from palm oil mills in their oil palm estates or plantations. Lastly, in the third part of the questionnaire, the respondents were surveyed for three attributes which were the benefits and advantages derived from the usage of biomass wastes, the enforcement of environmental regulations, and the contribution of biomass waste usage to five sustainable criteria. The respondents were required to grade their agreement using the 5-point Likert scale, ranging from 1 being "strongly disagree" to 5 "strongly agree". Oral explanation was provided by the researcher to help respondents to understand the questions. The sampling via questionnaire survey took approximately three months to complete, from November 2015 to January 2016. The data from the questionnaires were interpreted and analysed using the statistical analysis tool, SPSS (originally abbreviated for Statistical Package for the Social Sciences) software. For the estimate of reliability, the Cronbach's Alpha was applied to assess the internal consistency.

4. Findings and discussion

4.1. Background of the respondents

Approximately 300 respondents were met for the survey and 253 persons responded accordingly, which represented 84% of the targeted sampling size. From the frequency analysis as illustrated in figure 1, the respondents from the smallholder planters occupied 56.5 percent out of 253 respondents. The medium-size and the corporate planters corresponded to 25.7 percent and 17.8 percent respectively. As the majority respondents, the smallholder planters were mostly field conductors, supervisors and assistant managers working for the associated and subsidiary companies and divisions of corporate plantations.

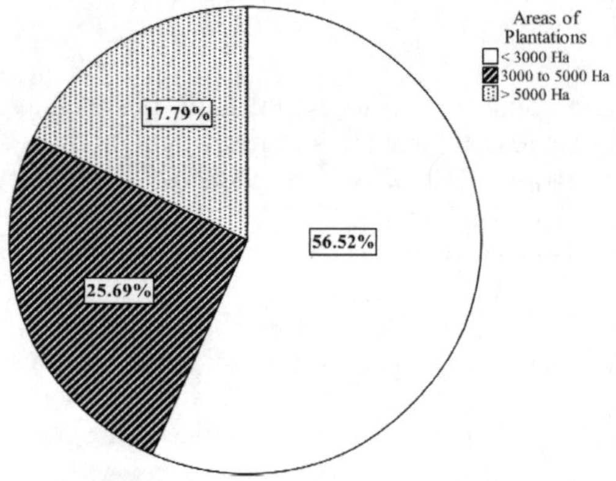

Figure 1. Characteristic of the respondents based on the areas of oil palm plantations in Sarawak.

53.4 percent of the respondents came from northern Sarawak, such as Bintulu, Tatau, Miri, Suai, Limbang and Lawas. Respondents from the central region (Sarikei, Bintangor, Sibu, Daro and Mukah) and southern region (Kuching, Sri Aman, Betong, and Pusa) of Sarawak are relatively small, which made up 28.9 percent and 17.8 percent respectively. The availability and accessibility of the oil palm planters in northern Sarawak was the major reason of high response rate in this region.

The oil palm planters who took part in this survey study were represented by the ratio of the ISP members to non-ISP members of four to six (4:6). There is no special requirement to be an ISP member. Most of the palm oil machinery suppliers, mill engineers, and materials suppliers were engaged with the ISP membership by subscription of an annual membership fee of RM 170 for the purpose of business trading with the plantation directors and planters during the planters' conferences and other related business activities. Number of respondents with ISP membership is 21 percent lower than those of nonISP membership. This is because the ISP members are mostly holding managerial and directorship role and they prefer to appoint their non-ISP subordinates (mainly the smallholder planters) to attend to the questionnaire study. However the subscription of ISP membership would not affect the oil palm planters' experiences and professionalism in answering the questionnaire survey of this study.

The reliability analysis was determined by the Cronbach's Alpha index. Table 1 shows that this study has achieved significant and approachable Cronbach Alpha index at 0.870. Therefore, this study has sufficient reliability. The use of simple English in the survey questionnaire and the design of questions which are closely related to the planters' experiences and knowledge regarding the usage of palm oil mill biomass have contributed to the reliability of this study. Additional explanation was provided to the respondents during the survey which may help to boost the Cronbach's Alpha figure and hence increase the reliability of this study.

Table 1. The reliability analysis of Cronbach's Alpha.

Cronbach's Alpha	Cronbach's Alpha Based on Standardized Items	N of Items
0.870	0.874	24

IOP Conf. Series: Materials Science and Engineering **206** (2017) 012091 doi:10.1088/1757-899X/206/1/012091

4.2. Descriptive statistics

Table 2 summarises the oil palm planters' opinions as means of agreement rating based on the 5-point Likert scale (from 1: strongly disagree, to 5: strongly agree) concerning the following three main parts:

Part A: Usage of biomass wastes from palm oil mills in oil palm plantations

Part B: Government policy regarding the usage of palm oil mill biomass wastes

Part C: Impact of the usage of palm oil mill biomass wastes on sustainable development

Part A is designed to capture planters' opinions on the several benefits obtained from the usage of the palm oil mill biomass wastes. The common benefits listed are: a) increase crop yields, b) increase nutrient uptake, c) reduce operational cost, d) good commercial value, e) reduce the usage of inorganic fertiliser, and f) good nutrient value. The planters indicated higher means of agreement of 3.64 to 4.22 for the beneficial usage of EFB and POME, whereas the usage of MF and PKS recorded a neutral range of agreement of 3.14 to 3.41. Among the six benefits rated for EFB and POME, the planters held a more neutral opinion for their commercial values, which recorded the lowest mean of 3.64 compared with the other benefits as shown in table 2. In contrast, the commercial values of MF and PKS were more recognised by the planters with the highest mean of agreement of 3.64 among the other benefits for these two biomass wastes.

Part B of the questionnaire collects the oil palm planters' opinions about the role of government policy especially the EQA 1974 and Environmental Quality (Prescribed Premises) (Crude Palm Oil) Regulations (1977) whereby the EIA requirements are embedded within the framework of the EQA 1974 and are enforced by the state's Natural Resources and Environment Board (NREB) to enable the state government to promote sustainable management of natural resources within Sarawak. Table 2 indicates that the planters strongly agreed that NREB plays an important role in enforcing environmental regulations and policies in Sarawak with a rating of 4.31. They also agreed that the practice of EIA under EQA 1974 is important concerning the usage of biomass wastes with a rating of 4.07.

Part C is intended to study the planters' opinions on the impact of the usage of biomass wastes from palm oil mills on sustainable development in five criteria, namely: a) reduce environmental impact, b) improvement by research and development, c) improve sustainable development in palm oil industry, d) increase social responsibility, and e) improve company image in terms of transparency in waste management. Again, the planters indicated higher rating of agreement for the usage of EFB and POME in contribution to sustainable development with means of rating from 3.76 to 4.17 for the aforementioned six criteria. For the usage of MF and PKS, although the planters recognised the impact of the usage of these two biomass wastes to sustainable development with level of agreement rated at 3.70, they showed lower agreement for the impact in reducing environmental impact (3.34), increasing social responsibility (3.48), and improving company image (3.49). Comparison of means analysis on the impact of the usage of EFB and POME as well as MF and PKS for these three parts of questionnaire survey is shown in figure 2.

Table 2. Descriptive analysis of the oil palm planters' opinions on the usage of biomass wastes from palm oil mills in Sarawak.

Descriptions	N	Minimum	Maximum	Mean	Variance	Std. Deviation
Part A. The opinions of the usage of biomass waste from palm oil mills.						
I) EFB & POME						
a. Increase Crop Yields	253	0	5	4.22	0.671	0.451
b. Increase Nutrient Uptake	253	2	5	4.15	0.695	0.482

IOP Conf. Series: Materials Science and Engineering **206** (2017) 012091 doi:10.1088/1757-899X/206/1/012091

c. Reduce Operational Cost	253	0	5	3.87	0.925	0.855
d. Good Commercial Value	253	1	5	3.64	0.923	0.851
e. Reduce Inorganic Fertiliser	253	0	5	3.76	0.899	0.809
f. Good Nutrient Value	252	2	5	3.98	0.730	0.534
II) MF & PKS						
a. Increase Crop Yields	253	1	5	3.23	0.862	0.743
b. Increase Nutrient Uptake	253	1	5	3.19	0.887	0.787
c. Reduce Operational Cost	253	0	5	3.14	0.897	0.805
d. Good Commercial Value	253	1	5	3.41	0.857	0.735
e. Reduce Inorganic Fertiliser	253	0	5	3.22	0.903	0.816
f. Good Nutrient Value	253	1	5	3.27	0.831	0.691

Part B. The government policy regarding the usage of the palm oil mill biomass wastes.

a. EQA 1974 (EIA)	253	2	5	4.07	0.731	0.535
b. Regulations 1977	253	2	5	3.89	0.812	0.659
c. NREB	252	2	5	4.31	0.813	0.661

Part C. The impact of the usage of biomass wastes on sustainable development.

I) EFB & POME						
a. Reduce Environmental Impact	253	2	5	4.17	0.723	0.523
b. Sustainable Development	253	2	5	4.07	0.607	0.368
c. Increase Social Responsibility	253	2	5	3.76	0.746	0.557
d. Improve Company Image	253	1	5	3.76	0.807	0.652
II) MF & PKS						
a. Reduce Environmental Impact	253	0	5	3.34	1.104	1.219
b. Sustainable Development	253	1	5	3.70	0.765	0.586
c. Increase Social Responsibility	253	1	5	3.48	0.738	0.544
d. Improve Company Image	253	1	5	3.49	0.800	0.640
III) All Biomass Wastes						
e. Continuous Improvement by Research and Development (R&D)	253	2	5	4.04	0.645	0.415

IOP Conf. Series: Materials Science and Engineering **206** (2017) 012091 doi:10.1088/1757-899X/206/1/012091

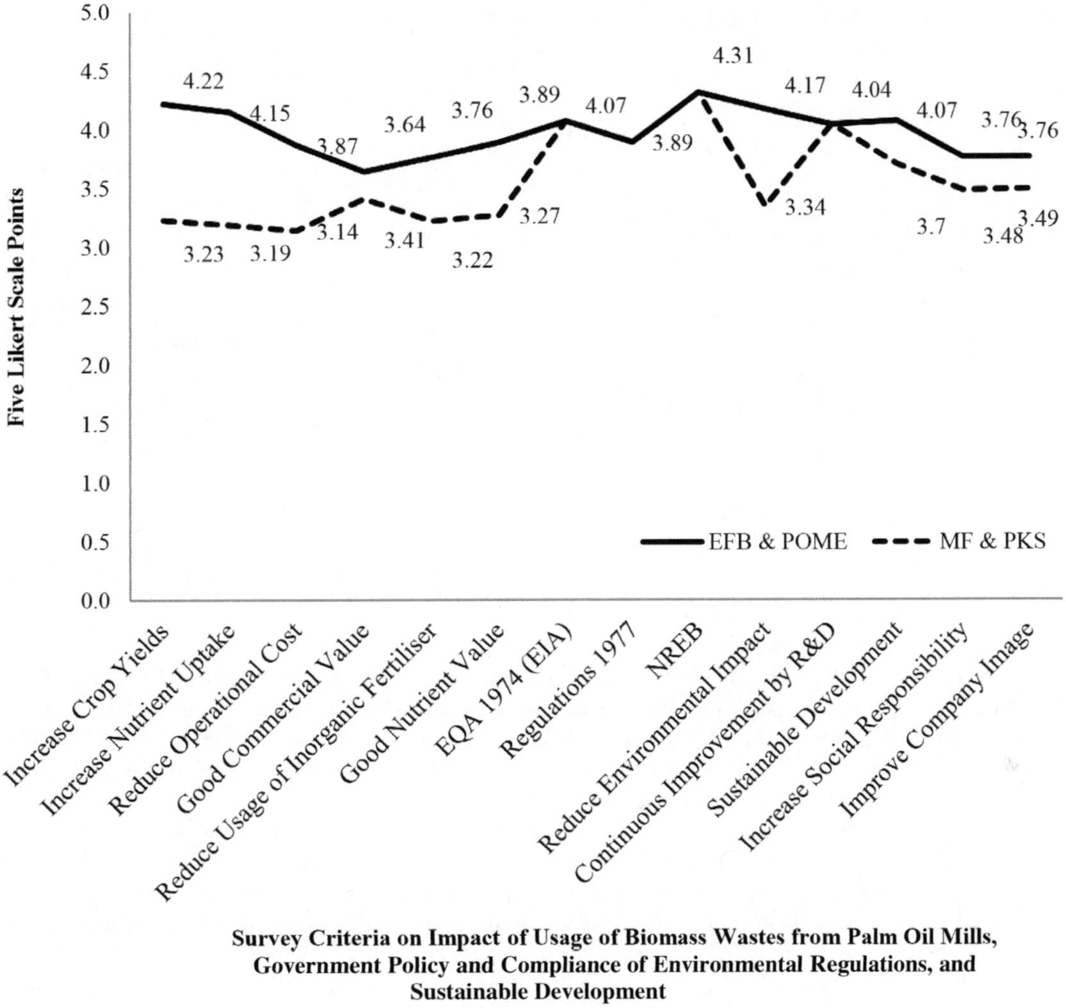

Figure 2. Comparison of means analysis on impact of the usage of biomass wastes from palm oil mills on sustainable development in oil palm plantations in Sarawak.

4.3. Usage of biomass wastes from palm oil mills in oil palm plantations

4.3.1. Increase crop yields

As shown in table 3, none of the respondents disagreed on the impact of the usage of EFB and POME to increase crop yields. 227 out of 253 respondents (89.7%) agreed on the benefit of MF and PKS in crop yields. Among them, 85 planters rated strong agreement. On the other hand, 46 out of 253 respondents (18.2%) disagreed on the benefit of MF and PKS in crop yields while 43.1 percent held neutral opinions. The positive responses from the planters on the usage of EFB and POME indicate that these two biomass wastes are well recognised for their applications in the fields. MF and PKS are mainly used by millers as feedstock of bio-fuel for boiler station and engine room instead of field application.

Table 3. Cross tabulation analysis of the impact to increase crop yields.

Particulars		Increase Crop Nutrients Uptake by Usage of EFB & POME					Total
		Strongly Disagreed	Disagreed	Neutral	Agreed	Strongly Agreed	
Area of Plantations	< 3,000 Ha		1	21	78	43	143
	3,000 to 5,000 Ha		0	8	34	23	65
	> 5,000 Ha		1	10	20	14	45
Total			2	39	132	80	253
Particulars		**Increase Crop Nutrients Uptake by Usage of MF & PKS**					Total
		Strongly Disagreed	Disagreed	Neutral	Agreed	Strongly Agreed	
Area of Plantations	<3,000 Ha	3	24	66	43	7	143
	3,000 to 5,000 Ha	0	8	31	19	7	65
	>5,000 Ha	5	8	21	9	2	45
Total		8	40	118	71	16	253

4.3.2. Increase nutrient uptake

Similarly, table 4 shows that 83.8 percent of the planters agreed on the benefit of EFB and POME in increasing the nutrient uptake of the crops in oil palm plantations. Razak, Shylaja [5] and Yusoff [16] reported that the application of EFB and POME in the fields directly increased the soil organic matter and soil condition, especially when coupled with the use of bunch ash (produced when EFB is fully incinerated) in peat areas to reduce the soil acidity.

65.6 percent of the planters either disagreed or gave neutral views towards the usage of MF and PKS in oil palm estates to increase crop nutrient uptake. The remaining agreed on the increase of nutrient uptake by the usage of MF and PKS which is likely attributed to their experience in using MF as raw material to compose with decanter cake in EFB composting plant. The mixed compost is then applied in oil palm fields to increase nutrient uptake.

4.3.3. Reduce operational cost

It can be seen from table 5 that majority of the oil palm planters in Sarawak (72.7%) agreed on the reduction of operational cost by using EFB and POME in oil palm plantations while 17.8 percent were neutral about this impact and 9.1 percent disagreed. The dissent may arise from the concern of higher labour cost required to apply EFB as mulching in the estates than to apply the common fertilisers. Higher disagreement (21.3%) was recorded for the usage of MF and PKS because of less exposure of the usage of these two biomass wastes.

Table 4. Cross tabulation analysis of the impact to increase crop nutrient uptake.

Particulars		Increase Crop Nutrients Uptake by Usage of EFB & POME					Total
		Strongly Disagreed	Disagreed	Neutral	Agreed	Strongly Agreed	
Area of Plantations	<3,000 Ha		1	21	78	43	143
	3,000 to 5,000 Ha		0	8	34	23	65
	>5,000 Ha		1	10	20	14	45
Total			2	39	132	80	253
Particulars		**Increase Crop Nutrients Uptake by Usage of MF & PKS**					Total
		Strongly Disagreed	Disagreed	Neutral	Agreed	Strongly Agreed	
Area of Plantations	<3,000 Ha	3	24	66	43	7	143
	3,000 to 5,000 Ha	0	8	31	19	7	65
	>5,000 Ha	5	8	21	9	2	45
Total		8	40	118	71	16	253

Table 5. Cross tabulation analysis of the impact to reduce operational cost.

Particulars		Reduce Operational Cost by Usage of EFB & POME						Total
		0	Strongly Disagreed	Disagreed	Neutral	Agreed	Strongly Agreed	
Area of Plantations	<3,000 Ha	0	0	17	22	69	35	143
	3,000 to 5,000 Ha	0	0	3	15	32	15	65
	>5,000 Ha	1	1	2	8	21	12	45
Total		1	1	22	45	122	62	253
Particulars		**Reduce Operational Cost by Usage of MF & PKS**						Total
		0	Strongly Disagreed	Disagreed	Neutral	Agreed	Strongly Agreed	
Area of Plantations	<3,000 Ha	1	1	30	66	39	6	143
	3,000 to 5,000 Ha	0	0	10	30	19	6	65
	>5,000 Ha	0	5	8	17	13	2	45
Total		1	6	48	113	71	14	253

4.3.4. Good commercial value

Research suggests that biomass wastes from palm oil mills have good commercial value in palm oil supply chain to enhance its economic viability [17]. This is supported by the positive agreement of 157 planters (62.7%) on EFB and POME, and 117 planters (46.2%) on MF and PKS, out of 253 planters surveyed in this study (see table 6). These planters mainly came from the companies which own palm oil mill facilities. These companies trade their EFB and POME to the outsiders, especially the nearby smallholder planters for mulching purpose. When the biomass wastes were circulated for internal use or for own plantations, the respective planters inclined to disagree or hold neutral opinion about the commercial value of these wastes. For example, the Tradewinds group of plantations in Sarawak stipulates that the palm oil mill operations would subsidise RM 5 per metric ton for the EFB being used by their own estates to encourage reuse, reapplication and disposal of these organic wastes, in order to address the environmental impact. In fact, MF and PKS may have greater commercial value than EFB and POME as the former is potential exported commodity to be processed into activated carbon and biofuel. Some local private limited millers pelletise MF and PKS for export market in order to diversify their business activities and to increase revenue.

Table 6. Cross tabulation analysis of the good commercial value.

Particulars		Commercial Values by Usage of EFB & POME					Total
		Strongly Disagreed	Disagreed	Neutral	Agreed	Strongly Agreed	
Area of Plantations	<3,000 Ha	0	17	42	69	15	143
	3,000 to 5,000 Ha	0	7	10	34	14	65
	>5,000 Ha	2	8	10	14	11	45
Total		2	32	62	117	40	253
Particulars		**Commercial Values by Usage of MF & PKS**					Total
		Strongly Disagreed	Disagreed	Neutral	Agreed	Strongly Agreed	
Area of Plantations	<3,000 Ha	0	20	58	52	13	143
	3,000 to 5,000 Ha	0	5	29	26	5	65
	>5,000 Ha	3	5	16	16	5	45
Total		3	30	94	94	23	253

4.3.5. Reduce the usage of inorganic fertiliser

Table 7 reveals that 66.4 percent of the planters agreed that the use of EFB and POME in field application has substantially reduced the usage of inorganic fertiliser whilst only 37.5 percent agreed on the beneficial use of MF and PKS in this respect. It is well known that the bunch ash from EFB is a great alternate source of alkaline and potassium to replace the chemical fertiliser [18]. However the usage of MF and PKS in field application is uncommon practice for most planters in this survey.

4.3.6. Good nutrient value

Similarly, 75.5 percent of the planters recognised the nutrient value of EFB and POME while only 36.0 percent agreed on the nutrient value of MF and PKS as shown in table 8. As mentioned earlier, most planters only explored the usage of EFB and POME in oil palm estates. Hence they were less aware of the beneficial properties of MF and PKS. This also explains why most researches concentrate on the exploration of useful chemical characteristics of POME, EFB and bunch ash, instead of MF and PKS [19].

Table 7. Cross tabulation analysis of the impact to reduce the usage of inorganic fertiliser.

Particulars		Reduce in the Usage of Inorganic Fertilizers by Usage of EFB & POME						Total
		0	Strongly Disagreed	Disagreed	Neutral	Agreed	Strongly Agreed	
Area of Plantations	< 3,000 Ha	0	1	16	42	65	19	143
	3,000 to 5,000 Ha	0	0	1	13	35	16	65
	> 5,000 Ha	1	0	3	8	18	15	45
Total		1	1	20	63	118	50	253
Particulars		**Reduce in the Usage of Inorganic Fertilizers by Usage of MF & PKS**						**Total**
		0	Strongly Disagreed	Disagreed	Neutral	Agreed	Strongly Agreed	
Area of Plantations	<3,000 Ha	1	3	22	65	48	4	143
	3,000 to 5,000 Ha	0	0	8	33	16	8	65
	>5,000 Ha	0	4	8	14	15	4	45
Total		1	7	38	112	79	16	253

Table 8. Cross tabulation analysis of the impact of good nutrient value.

Particulars		Good Nutrients Value by Usage of EFB & POME					Total
		Strongly Disagreed	Disagreed	Neutral	Agreed	Strongly Agreed	
Area of Plantations	< 3,000 Ha		3	34	73	33	143
	3,000 to 5,000 Ha		0	14	34	16	64
	> 5,000 Ha		1	9	23	12	45
Total			4	27	130	61	252
Particulars		**Good Nutrients Value by Usage of of MF & PKS**					**Total**
		Strongly Disagreed	Disagreed	Neutral	Agreed	Strongly Agreed	
Area of Plantations	<3,000 Ha	0	19	75	40	9	143
	3,000 to 5,000 Ha	2	6	33	18	6	65
	>5,000 Ha	2	7	18	15	3	45
Total		4	32	126	73	18	253

4.4. Government Policy and Compliance to Environmental Regulations

Out of 246 planters whose companies own palm oil mill facilities in this study, 94.7 percent have complied with the EQA 1974 to carry out the EIA practice. The remaining 5.3 percent are less sensitive to EIA requirements in which their business operations are under conservative proprietorship and joint venture with the local natives who are less aware of the environmental regulations.

Figure 3 illustrates that palm oil mills with capacity of 60 to 90 metric tons per hour (Mts/hr) contributed most to the frequency of EIA practices. Among the 129 mills of this capacity, 40 mills carried out EIA once a year, 47 did it twice a year, and 42 practised quarterly EIA. In terms of area of plantations, small planters (below 3,000 hectares) contributed significantly to the practice, with 46 who practised EIA once a year, 40 half yearly, and 38 on a quarterly basis (see figure 4).

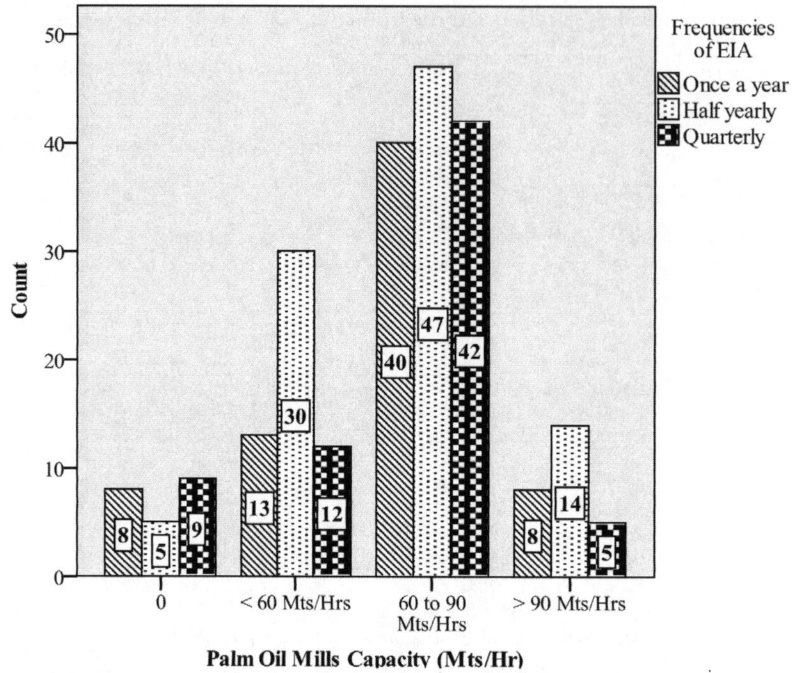

Figure 3. Frequency of EIA practice versus palm oil mill capacity.

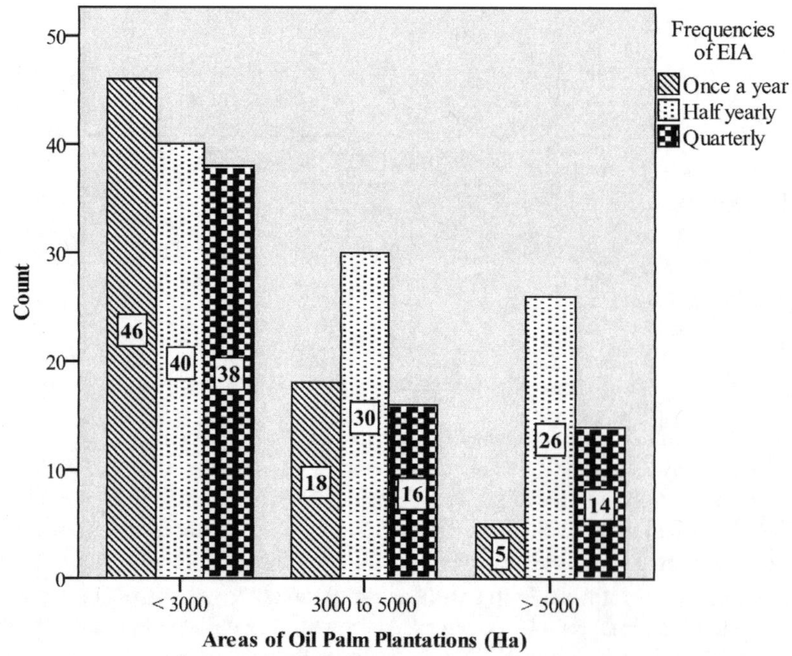

Figure 4. Frequency of EIA practice versus area of oil palm plantations.

From table 9, the correlation index (sig. 2-tailed) between palm oil mill capacity and frequency of EIA practice is 0.646, which indicates that there is a positive relationship between palm oil mill capacity and frequency of EIA practice. This is due to the fact that greater palm oil mill production capacity directly generates more biomass wastes, and hence higher requirement of EIA practice.

Table 9. Correlation analysis between area of oil palm plantations versus palm oil mill capacity and frequency of EIA practice.

Particulars		Plantations Hectares	Palm Oil Mills' Capacity	Frequency of EIA Practice
Area of Plantations	Pearson Correlation	1	0.190**	0.119
	Sig. (2-tailed)		0.003	0.069
	Sum of Squares and Cross-products	150.040	30.337	16.661
	Covariance	0.595	0.124	0.072
	N	253	246	233
Palm Oil Mills' Capacity	Pearson Correlation	0.190**	1	-0.030
	Sig. (2-tailed)	0.003		**0.646**
	Sum of Squares and Cross-products	30.337	173.337	-4.309
	Covariance	0.124	0.707	-.019
	N	246	246	233
Frequency of EIA Practice	Pearson Correlation	0.119	-0.030	1
	Sig. (2-tailed)	0.069	**0.646**	
	Sum of Squares and Cross-products	16.661	-4.309	136.996
	Covariance	0.072	-0.019	0.590
	N	233	233	233

**. Correlation is significant at the 0.01 level (2-tailed).

As discussed in Section 4.2, the planters agreed that the practice of EIA under EQA 1974 plays an important role to reduce, reuse and dispose the biomass wastes from palm oil mills with a rating of 4.07. The agreement on Environmental Quality (Prescribed Premises) (Crude Palm Oil) Regulations (1977) is slightly lower, recording a mean of agreement of 3.89. This may be due to the nature of Regulations 1977 which mainly regulate the palm oil millers in compliance with the proper treatment of biomass wastes produced from the mills, hence giving less impact to the oil palm planters.

Meanwhile, the role of NREB was strongly supported by the high mean of agreement of 4.31 from the planters as stated earlier (see table 2). This may be interpreted as a significant responsibility of NREB acknowledged by the planters to ensure the palm oil players and growers adhere to the environmental regulations.

4.5. Evaluation of the impact of palm oil mill biomass wastes usage on sustainable development
Respondents in this study generally agreed that the usage of biomass wastes especially EFB and POME was able to reduce the environmental impact from palm oil industry because it helps to reduce the amount of biomass residues need to be disposed from palm oil mills and subsequently mitigate the possible contamination to the environment. Besides, EFB and POME can be used as organic fertiliser, soil conditioner as well as alkaline and potassium sources to partially replace the usage of chemical fertiliser which further contribute to sustainable development.

On the other hand, the planters in this survey were less convinced of the impact of biomass waste usage to increase social responsibility and to improve company image in terms of transparency in waste management (means of agreement between 3.48 and 3.76). Social responsibility may be demonstrated by mutual distribution of benefits from the proper management of source of biomass wastes from palm oil mills. For example, the palm oil mills may use the appropriate biomass residues such as MF and PKS to generate electricity which can be channelled to the nearby community as a form of social responsibility. Most palm oil mills in Sarawak use the power supply generated from the biomass wastes for worker quarters, daily mill and plantation operations instead of supplying it to the native communities who are living around. There are many impeding factors such as the limited capacity of power convertor, partly due to the company budget, and the concern of maintenance cost.

Besides, high transparency in the documentations, recordings and information sharing concerning the palm oil production processes to all stakeholders may boost the corporate image, especially in handling the biomass wastes from palm oil mills. This may include the involvement of external and internal audits in the administrative and operational processes. Several corporate plantations visited by the researcher during the process of sampling displayed EIA certificate issued by NREB at their respective administrative departments which evidently projected a healthy public image to the visitors. Although the respondents were uncertain about the impact brought to the company image by practising transparent management, more work can be done to reinforce this aspect as part of sustainable development in this industry.

Lastly, the planters agreed (with a rating of 4.04 in table 2) that research and development has continuously improved and increased the usage of biomass wastes in oil palm plantations. Previous studies on the nutrient value of EFB and POME supported the usage of these biomass wastes in oil palm plantations as soil conditioner to improve the soil organic matter, to raise soil pH, and to replace chemical fertiliser to some extent [3]. Indeed there exists a better understanding of the usage of palm oil mill biomass wastes through research and development.

5. Conclusion

In this study, the respondents rated their agreement from 3.64 to 4.22 in the Likert scale (with 5 being the strongest agreement) for several benefits brought about by the usage of EFB and POME in oil palm plantations. In general, they agreed that the usage of EFB and POME can increase crop yields, increase nutrient uptake, reduce operational cost, reduce the usage of inorganic fertiliser, and has good nutrient value. However they were more neutral about the commercial value of these two biomass wastes. In contrast, the usage of MF and PKS received lower acceptance with a neutral range of agreement of 3.14 to 3.41 for all the benefits surveyed. Besides, 94.7 percent of the planters' companies were found to comply with the EIA requirements. The planters agreed to the significant role performed by NREB to enforce the EIA practice under EQA 1974 as well as compliance of Environmental Quality Regulations 1977, recording a Likert rating of 3.89 to 4.31.

Lastly, the planters agreed that the usage of EFB and POME has reduced the environmental impact and improved the sustainable development, and they recognised the importance of research and development to improve and increase the usage of biomass wastes. However the planters were uncertain of the impact of usage of biomass wastes with respect to the contribution to social responsibility and company image in terms of transparency in waste management. The usage of EFB and POME gained higher Likert scale point of 3.76 to 4.17 against MF and PKS of 3.34 to 3.49 in the evaluation of the impact of sustainability in oil palm plantations. Therefore the researchers accepted Hypothesis 1 in this study which states that the usage of EFB and POME from palm oil mill influences sustainable development in oil palm plantations.

References

[1] Paramananthan S 2013 Peat Soil of Malaysia: Their Extent, Characteristics, Mapping, and Classification. *The Planter* **89 (1051)**
[2] Lim How Pim. 2 mil ha to be planted with oil palm by 2020. The Borneo Post. 2016.
[3] Tang M K, Shylaja V N, Iswari N. Zero Waste Management- Sime Darby Plantation's Experience.

In: E Pushparajah, Patrick H C Ng, Sharifah Shahrul Rabiah Syed Alwee, editors. 12th ISP National Seminar 2016; Kuching, Sarawak: The Incorporated Society of Planters; 2016. p. 15560.

[4] Sarawak Energy Berhad. Biomass: Introduction of Biomass As Energy Resource 2013 [updated 21 October 2015]. Available from: http://www.sarawakenergy.com.my/index.php/rd/biomassenergy.

[5] Razak, Shylaja, Johari 2014 Utilization of Palm Oil Processing Wastes. *The Planter* **90 (1055)** pp 108-13.

[6] Anonymous. Production of Crude Palm Oil for the Month of December 2014: Malaysia Palm Oil Board; 2015 [22nd June 2015]. Available from: http://bepi.mpob.gov.my/index.php/statistics/production/125-production-2014/659-productionof-crude-oil-palm-2014.html.

[7] Donough CR, Witt C, Fairhurst TH, Griffiths W, Agfroerer-Kerstan. Concept and Implementation off Best Management Practice for Maximum Economic Yield in Oil Palm Plantations. International Planters Conference, Higher Productivity and Efficient Practices for Sustainable Plantation Agriculture: The Incorporated Society of Planters; 2006.

[8] Rao V 2006 A Supply Chain for Sustainable Palm Oil? *Higher Productivity and Efficient Practices for Sustainable Plantation Agriculture* **1** pp 1-16.

[9] Obidzinski K, Andriani R, Komanidin H, Andrianto A 2012 Environmental and social impacts of oil palm plantations and their implications for biofuel production in Indonesia. *Ecology and Soc.* **17** pp 25.

[10] Laurance WF, Koh LP, Butler R, Sodhi NS, Bradshaw CJ, Neidel JD, et al. 2010 Improving the performance of the roundtable on sustainable palm oil for nature conservation. *Conservation Biology* **24** pp 377-81.

[11] RSPO. How RSPO Certification Work Kuala Lumpur, Malaysia: Roundtable On Sustainable Palm Oil 2015 [updated 30th June 2015]. Available from: http://www.rspo.org/certification/howrspo-certification-works.

[12] Andrew N, Siew LS 2006 From Inseption to Action: RSPO as a Global Multi-stakeholder Platform for Sustainability in Palm Oil. *Higher Productivity and Efficient Practices for Sustainable Plantation Agriculture* **1** pp 3-24.

[13] Ahmad A, Chan C 2009 Sustainability of palm oil industries: an innovative treatment via membrane technology. *J. of Appl. Sci.* **9** pp 3074-9.

[14] Shuit SH, Tan KT, Lee KT, Kamaruddin A 2009 Oil palm biomass as a sustainable energy source: a Malaysian case study. *Energy* **34** pp 1225-35.

[15] Ng WPQ, Lam HL, Ng FY, Kamal M, Lim JHE 2012 Waste-to-wealth: green potential from palm biomass in Malaysia. *J.l of Cleaner Prod.* **34** pp 57-65.

[16] Yusoff S 2006 Renewable energy from palm oil–innovation on effective utilization of waste. *J. of Cleaner Prod.* **14** pp 87-93.

[17] Lam MK, Lee KT 2011 Renewable and sustainable bioenergies production from palm oil mill effluent (POME): win–win strategies toward better environmental protection. *Biotechnology Advances* **29** pp 124-41.

[18] Etukudo, Mbosowo. M., Roberts E, M. I. and Omeje, F I 2014 Early Seedling Growth and Development Response os Zea Mays. L. To Extract from Palm Bunch Ash (PBAE) Elaeis Guineensis Jacq *Int. J. of Recent Scientific Res.* **5** pp 1307-10.

[19] Singh R, Ibrahim MH, Esa N, Iliyana M 2010 Composting of waste from palm oil mill: a sustainable waste management practice. *Reviews in Environmental Sci. and Bio/Technol.* **9** pp 331-44.

IOP Conf. Series: Materials Science and Engineering **206** (2017) 012092 doi:10.1088/1757-899X/206/1/012092

Effect of Catalyst Loading on Photocatalytic Degradation of Phenol by Using N, S Co-doped TiO_2

N N Yunus[1], F Hamzah[2], M S So'aib[1] and J Krishnan[3]
[1] Faculty of Chemical Engineering, Universiti Teknologi MARA, 40450 Shah Alam, Selangor, Malaysia.
[2] Biocatalysis & Biobased Material Research Group, Green Technology and Sustainable Development Research Community, Faculty of Chemical Engineering, Universiti Teknologi MARA (UiTM), 40450 Shah Alam, Selangor, Malaysia
[3] SSN School of Engineering, Old Mahabalipuram Road, Kalavakkam – 603110, Tamil Nadu, India

E-mail: fazlena@salam.uitm.edu.my

Abstract. The study on the effect of catalyst loading of photocatalytic degradation of phenol by using N, S co-doped TiO_2 was investigated. The precursor of titania was Titanium (IV) isopropoxide (TTIP), while the sources of Nitrogen and Sulfur were ammonium nitrate and thiourea respectively. The photocatalyst were prepared by using dopant concentration at 1% of both Nitrogen and Sulphur that were prepared via sol-gel method. The photocatalyst were tested by different catalyst loading which were 1 g/L, 2g/L and 3 g/L. The gel obtained from the mixing process was dried and calcined at 600 °C. The performance of the photocatalyst were tested by using phenol as a model pollutant. The mixture of photocatalyst and pollutant was left under visible light for five hours for irradiation time. The experiment showed that catalyst loading of 3 g/L able to fully degrade phenol while 1 g/L and 2 g/L of photocatalyst degraded phenol at 69.9% and 96.2% respectively.

1. Introduction

Growing numbers of chemical industrial can produce tonnes of phenolic compounds in their effluent. Some of the examples are production of resins, coal conversion, petrochemical; and pharmaceutical industries [1,2]. The toxicity of phenol can affect aquatic microorganisms and can be lethal to fish at concentrations of 5 to 25 mg/L. Meanwhile, exposure towards human can cause damage towards internal organs. Hence, the World Health Organization (WHO) regulates the phenol concentration must not exceed 1 μg/L in drinking waters [2,3].

There are many treatments available to remove phenol completely by physical, biological and chemical means. One of the popular method of physical treatment is physical adsorption via activated carbons due to the effectiveness and low in cost [4]. The activated carbons can be made up of various materials such as wood, coconut shells, coal, date-pit, tamarind bean etc.[1,5,6]. Meanwhile, biological treatment of phenol is utilizing microorganisms from many genera which includes *Pseudomonas, Bacillus, Aspergillus* and others. The microorganisms are used in forms of pure, mixed culture and immobilized cells [7]. Other than that, the chemical treatment of phenol is via chemical oxidations by

IOP Conf. Series: Materials Science and Engineering **206** (2017) 012092 doi:10.1088/1757-899X/206/1/012092

using chlorine, ozone, hydrogen peroxide and potassium permanganate and chlorine dioxide are used to treat phenol [8].

Ozone (O_3) and hydrogen peroxide (H_2O_2) are capable in oxidizing inorganics to their higher oxidation while at the same time organic compounds are being oxidized into carbon dioxide and water. Since both methods have lower reaction rate if being used alone without ultraviolet (UV) radiation, the development of Advanced Oxidation Processes (AOPs) was applied to cater the problem. The UV rays were used with ozone or hydrogen peroxide to generate hydroxyl radicals which will cause the complete elimination of organic chemicals [9,10]. Some examples of AOPs in phenol treatment are photocatalytic degradation, electrochemical oxidation, photo-Fenton and sono-Fenton etc. [9]. In addition, photocatalytic degradation by titanium dioxide (TiO_2) and zinc oxide (ZnO) show promising results when it comes to phenol treatment. The usage of titania is very common because of the high efficiency, inexpensive and non-toxic [11].

According to Ahmed, Rasul, Martens, Brown, & Hashib (2010); photocatalytic activity of titania are reliant on surface and structural properties of the semiconductor in which including the surface area, crystal composition, particle size distribution, band gap and others. These factors have a huge impact on the adsorption behaviour during degradation of pollutants. Titanium dioxide photocatalyst are proven many times in treatment of organic pollutants despite its high band gap energy (3.2 eV) which requires UV light for photoactivity. By introducing other elements such as Nitrogen, Carbon, Sulphur and Ferum as dopant, the band gap energy can be reduced [13]. The photocatalytic degradation efficiency is highly dependent on the catalyst loading. The optimum catalyst loading can vary from 0.6 g/L of photocatalyst up to 2.5 g/L based on review by Ahmed et al. (2010)[12]. Past researches showed that the rate increased proportionally with catalyst loading but started to decrease at excessively high values as light scattering and screening effects tend to become lower. At high catalyst loading, the surface area of the photocatalyst tend to become lower because of agglomeration (particle-particle interaction) in which may affect the light absorption.

In this study, the effect of catalyst loading of photocatalytic degradation of phenol by using N, S co-doped TiO_2 was investigated. The photocatalyst was synthesised by using sol-gel method. The photocatalyst was characterized with XRD and FE-SEM and the degradation efficiency was tested at catalyst loading of 1 g/L, 2 g/L and 3 g/L under visible light.

2. Methodology

2.1. Preparation of Photocatalyst via Sol-gel Method

The N, S co-doped TiO_2 photocatalyst was prepared by using Titanium (IV) isopropoxide (TTIP) as the titania precursor while using ammonium nitrate (NH_4NO_3) and thiourea ($CS(NH_2)_2$) were used as nitrogen and sulphur sources respectively. Ethanol (C_2H_5OH) and analytical reagent (AR) grade of acetic acid (CH_3COOH) were used in the present study as solvents. The amount of NH_4NO_3 and thiourea were weighed based on the dopant concentration which were 1% nitrogen and 1% sulphur. The dopants were added with 180 mL deionized water and 20 mL acetic acid and labelled as Solution A. On the other hand, 60 mL TTIP were dissolved in 200 mL ethanol and labelled as Solution B. The Solution A was added to Solution B in a dropwise manner with constant stirring for an hour. Later, the solution was left stirred at 200 rpm by using magnetic stirrer for another 2 hours until gel was formed. The gel was covered with aluminium foil and left 24 hours for aging phase. The gel was dried in oven at 100 °C until fully dried. The dried photocatalyst were ground and calcined at 600 °C for 3 hours at 5 °C/min as the heating rate.

2.2. Characterization of Photocatalyst

The photocatalyst were characterized by using X-ray Diffractometer (XRD model: Rigaku Ultime IV) with a Cu Kα radiation. The scanning rate of XRD was 5°/min in a 2θ range of 20 – 80° to analyse the phases of the crystals. The morphologies of the photocatalyst were observed by using Field Emission-Scanning Electron Microscope (FE-SEM, model DSM 982 Gemini Supra 40VP).

2.3. Photocatalytic Degradation Study

The photocatalytic degradation was carried out by using phenol prepared with the starting concentration of 15 mg/L and volume of 100 mL. Later, 1g/L of photocatalyst was weighed and added to the solution. The solution was placed inside a photoreactor equipped with 24W Philips Compact Fluorescent Lamp (CFL) at the height of 6 cm from the beaker which acted as a visible-light source. The solution was stirred in the dark for 30 minutes to ensure in equilibrium state. The solution was left stirred for five hours and 5 mL sample was collected in every 30 minutes interval. The collected samples were centrifuged for 10 minutes at 10,000 rpm to separate phenol and the photocatalyst. The concentrations of phenol were recorded to determine the degradation rate. The steps were repeated for 2 g/L and 3 g/L of catalyst loading.

The quantifications of phenol were analysed by using rapid colorimetric estimation by Martin (1949). The reagent will turn from yellow to reddish colour in the presence of phenol. The solutions were left for 15 minutes followed by absorbance measurement at wavelength of 510 nm. The degradation efficiencies of phenol were calculated by using Equation (1).

$$\eta = \frac{C_o - C_t}{C_o} \times 100\% \tag{1}$$

where:

η	: phenol degradation efficiency
C_o	: initial concentration of phenol
C_t	: concentration of phenol at specific time

3. Results and discussions

3.1. Characterization of Photocatalyst

The X-ray diffraction (XRD) pattern for the N, S-TiO_2 photocatalyst is shown in Figure 1. The XRD pattern is used to determine the phase constitution and the crystallite size of the photocatalyst [15]. From the result, characteristic peak at $2\theta = 25.356^o$ was obtained that represent anatase phase (JCPDS file number = 01-089-4921). There are no trace of rutile and brookite phases which is favourable because anatase has better photocatalytic activity than other phases [16,17].

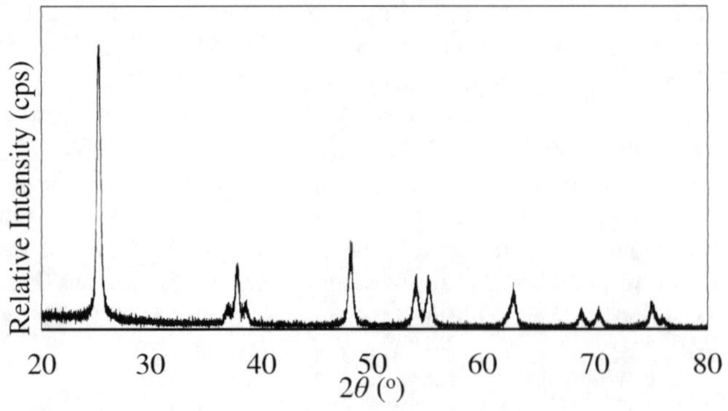

Figure 1. XRD patterns of 1.0% N, S-TiO_2 (fitting well with JCPDS file number 01-089-4921)

The surface morphology of the photocatalyst by FE-SEM can be referred in Fig. 2. The figure shows spherical particles, unevenly sized and agglomeration occur at certain area. The agglomeration can occur due to the formation of solid bridge between surface hydroxyl groups during drying process or combination of particles after being exposed to high temperature while calcination [18,19]. Agglomerated particles is undesirable because of reduction the surface area of the photocatalyst and eventually affecting the degradation performance of phenol [17].

Figure 2. Surface morphology of N, S-TiO_2 photocatalyst depicted by FE-SEM.

3.2. Effect of Catalyst Loading on Photocatalytic Degradation of Phenol

The effect of catalyst loading on photocatalytic degradation of phenol were studied by using the amount of catalyst of 1 g/L, 2 g/L and 3 g/L. The photocatalytic degradation of phenol against time is shown Figure 3. From the figure, the concentrations of phenol declined gradually but increased at some point. The increased of concentrations of phenol might be because of the separation process between phenol and photocatalyst. During separation via centrifuge, the residue of the photocatalyst powder may still left inside the sample and affect the light penetration during quantification of phenol.

The fastest degradation of phenol at 3 g/L in which took only 4.5 hours to complete and had the steepest slope. On the other hand, the other catalyst loading were unable to degrade within the given irradiation time. By using Equation (1), 3 g/L of photocatalyst loading shown the highest photodegradation performance which is 100.0%. Whereas 1 g/L and 2g/L of photocatalyst loading resulting in 69.9% and 96.2% degradation of phenol respectively.

The results suggest that the degradation performance increase proportionally with the catalyst loading which is comparable to a review by Ahmed et al. (2010) which state the optimum loading of phenol by using titania photocatalyst was 2.0 g/L using UV light as the light source. Therefore, it can be surmised that doping Nitrogen and Sulphur to TiO_2 at the highest loading (3 g/L) corresponds to highest phenol degradation rate as energy band gap of TiO_2 is reduced allowing visible light as the light source. The best catalyst loading for N, S-TiO_2 is 3 g/L. However, at higher catalyst loading, the degradation efficiency can be expected to decrease because of agglomeration and thus lower the photocatalytic activity.

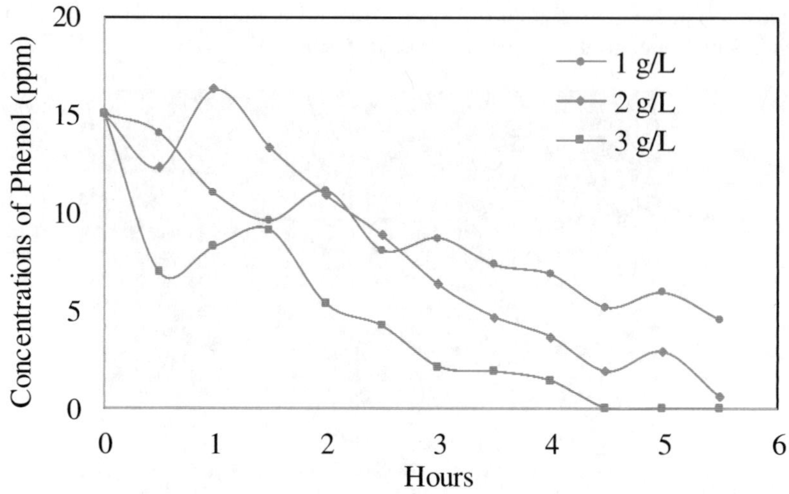

Figure 3. The effect of catalyst loading on photodegradation of phenol

4. Conclusions

N, S-TiO_2 photocatalyst was prepared via sol-gel method at dopant concentration of 1%. The characterization of photocatalyst shows anatase phase transformation of TiO_2. From FE-SEM results, the surface of the photocatalyst was unevenly sized and agglomeration occur at certain areas. The photocatalyst was tested for effect of catalyst loading on degradation of phenol. The catalyst loading of 3 g/L able to fully degrade phenol and taken the least time to complete the degradation.

Acknowledgement
The work was fully funded by Ministry of Education Malaysia for research funding through FRGS GRANT (600-RMI/FRGS 5/3 (96 /2013)). Authors also would like to thank Institute of Graduate Studies (IPSIS), UiTM for providing facilities needed for the study.

References

[1] Al-jiboury K F C 2013 Adsorption of Phenol from Industrial Wastewater using Commercial Powdered Activated Carbon
[2] Al-Khalid T and El-Naas M H 2012 *Crit. Rev. Environ. Sci. Technol.* **42** 1631–90
[3] Banerjee A and Ghoshal A K 2011 *Int. Biodeterior. Biodegradation* **65** 1052–60
[4] Dakhil I H 2013 Removal Of Phenol From Industrial Wastewater Using Sawdust **3** 25–31
[5] El-Naas M H, Al-Zuhair S and Alhaija M A 2010 *Chem. Eng. J.* **162** 997–1005
[6] Kulkarni, Sunil J and Kaware J P 2013 *Int. J. Sci. Res. Publ.* **3** 1–5
[7] Basha K M, Rajendran A and Thangavelu V 2010 Recent advances in the Biodegradation of Phenol : A review **1** 219–34
[8] Diya'Uddeen B H, Daud W M A W and Abdul Aziz A R 2011 *Process Saf. Environ. Prot.* **89** 95–105
[9] Babuponnusami A and Muthukumar K 2011 *Clean - Soil, Air, Water* **39** 142–7
[10] Oller I, Malato S and Sánchez-Pérez J A 2011 *Sci. Total Environ.* **409** 4141–66
[11] Velasco L F, Parra J B and Ania C O 2010 *Appl. Surf. Sci.* **256** 5254–8
[12] Ahmed S, Rasul M G, Martens W N, Brown R and Hashib M A 2010 *Desalination* **261** 3–18

[13] Sikirman A, Krishnan J and Mohamad E N 2014 *Appl. Mech. Mater.* **661** 34–8
[14] Martin R W 1949 Rapid Colorimetric Estimation of Phenol 1419–20
[15] Yu C, Cai D, Yang K, Yu J C, Zhou Y and Fan C 2010 *J. Phys. Chem. Solids* **71** 1337–43
[16] Rahimi R, Safalou Moghaddam S and Rabbani M 2012 *J. Sol-Gel Sci. Technol.* **64** 17–26
[17] Hamadanian M, Reisi-Vanani a. and Majedi a. 2010 *J. Iran. Chem. Soc.* **7** S52–8
[18] Sikirman A, Krishnan J, Jai J and Senusi F 2014 *Adv. Mater. Res.* **894** 245–9
[19] Krishnan J, Mohamad E N and Hadi A 2014 *Adv. Mater. Res.* **661** 63–7

IOP Conf. Series: Materials Science and Engineering **206** (2017) 012093 doi:10.1088/1757-899X/206/1/012093

Electrochemical Recovery of Gold from Waste Electric and Electronic Equipment Using Circulating Particulate Bed Reactor (CPBE)

T Ravinder, U F M Ali*, F M Ridwan, N Ibrahim, N H Azmi
School of Environmental Engineering, Universiti Malaysia Perlis, Kompleks Pusat Pengajian Jejawi 3, 02600 Arau, Perlis, Malaysia

Email: umifazara@unimap.edu.my

Abstract. The utilization of electrochemical process recovery involving low reactant concentrations of metal requires electrodes with high mass transport rates and specific surface areas. This is essential to increase cross-sectional current densities whilst optimizing the capital and operating costs. Experimental results demonstrated that Circulating Particulate Bed Reactor (CPBE) is suitable for the recovery of low concentrations of gold from aqueous chloride solution containing $AuCl_4^-$ and $AuCl_2^-$ of less than 0.5 mol m^{-3}($< 10^2$ g m^{-3}). Elemental gold was successfully obtained on 0.5-1 mm gr particles in an electrochemical reactor incorporating a cation- permeable membrane and operated in bath recycle mode. Depletion to concentration $< 5 \times 10^{-3}$ mol m^{-3} (< 1 g m^{-3}) appeared to be mass transport controlled at an applied potential of +0.20 V (SCE), specific electrical energy consumption (SEEC) of ca. 800-1300 kWh h (tonne Au)$^{-1}$ for cell voltages (U) of 2.0-3.0 V, and fractional current efficiencies of ca. 0.95. However, atomic absorption and UV spectrophotometry established that as the ([$AuCl_4^-$+ [$AuCl_2^-$]) concentration decayed, the [$AuCl_4^-$]:[$AuCl_2^-$] molar ratio changed. A multi-step mechanism for reduction of $AuCl_4^-$ ions explained this behavior in terms of changing overpotentials for $AuCl_4^-$ and $AuCl_2^-$ reduction as total dissolved gold concentration decreased. In addition, SEM images confirmed that adherent and coherent Au deposits were achieved with CPBE for Au deposition under mass transport control at 0.20 V (SCE).

1. Introduction

The production of electrical and electronic equipment (EEE) is known as one of the fastest growing businesses in the world [1]. New applications of EEE are increasing significantly at the same time as the average replacement process accelerating. The rapid development has resulted in an increase of end of life or waste electric and electronic equipment (WEEE), which can be a valuable source of secondary raw materials if treated appropriately [2].

Technologies that are able to combine metal removal with metal recovery are becoming increasingly important. This is because they are able to convert wastes from an economic liability to a saleable resource, which results in the production of little or no waste, and simultaneously protects the environment.

Current recycling processes are mainly pyrometallurgical, the WEEE being smelted at high temperature which consists of the thermal treatment of minerals and metallurgical ores and concentrates to bring about physical and chemical transformations in the materials to enable recovery of valuable metals [2]. The cost effectiveness of pyrolytic recycling for WEEE, apart from those with relatively high precious metal content, coupled with increasing ecological concerns has cast doubt over the long-term viability of this methodology and has provided impetus for the development of more sustainable approaches embracing mechanical and hydrometallurgical technologies. Although majority of metals can be recovered by pyrometallurgical processes, the organic yields are combusted and subsequently produce noxious gasses that are harmful to the environment [3]. Hydrometallurgical processes allow a more environmentally benign and energy efficient route which requires two stages of process to metal reclamation from shredded WEEE; metal dissolution into an aqueous media and followed by recovery of metal ions.

A number of papers [4-8], including a more recent one [9], has been written on the electrodeposition of gold particularly from chloride complexes. However, in each case, it is from a solution at different concentration several orders of magnitude greater than would be expected from gold concentration in WEEE leachate which was around 0.05 mol dm^{-3}. Nicol [5] have studied the gold electrodeposition in acidic chloride media and found gold deposition to be mass transport controlled and the limiting current proportional to the Au^{III} concentration. Nonetheless, they were able to detect Au^{I} intermediates by oxidation at a Pt ring electrode. Harrison and Thomson [6 also studied gold electrodeposition from chloride electrolytes using rotating disc electrode but unfortunately neglected the presence of Au^{I} species. Diaz [8] studied the effects of Au^{I}: Au^{II}molar ratio, pH and Cl^- concentration on the voltammetric behavior of a Pt electrode to enable the kinetics and mechanism of gold dissolution and deposition to be determined. According to their investigations, at +0.80 V (SCE), Au dissolved exclusively as $AuCl_2^-$ ions, the proportions of $AuCl_4^-$ ions increasing at higher potentials.

Removing gold from chloride solution specifically produced from WEEE was further investigated by Cheng [9] using a packed bed cathode operated in batch recycle mode with a stirred tank reservoir. The packed bed cathode of 2 -3 mm graphite particles was able to deplete ($[AuCl_4^-]$+ $[AuCl_2^-]$) in 1 kmol m^{-3}aqueous HCl from 0.5 mol m^{-3} (10^2g m^{-3}) to < 5 x 10 mol m^{-3}(<1 g m^{-3}), essentially under mass transport control, with current efficiencies of <4F (mol Au)$^{-1}$. The investigation established that as the $[AuCl_2^-]$:$[AuCl_4^-]$ molar ratio decayed from an initial value of ca. 0.3 to < 0.05 over the first ca. 500s which then increased passing through a maximum after 500-1500 s depending on the flow rate. In addition, a multi-step mechanism for reduction of $AuCl_4^-$ ions involving adsorbed $Au^{II}$and Au^{I}species was proposed to explain this behavior in terms of changing overpotentials for $AuCl_4^-$ and $AuCl_2^-$ reduction as the total dissolved gold concentration decreased and changing $[AuCl_2^-]$: $[AuCl_4^-]$ molar ratios.

For a low concentration of metal ion in waste electronics such as gold, platinum and palladium, a three-dimensional cathode, with a high surface area per unit volume and high mass rate coefficients, is the best option to be used for metal recovery. Since the concentrations of gold in leachate solutions from WEEE are likely to be below 1 mol m^{-3}, a CPBE reactor is attractive for the deposition of gold. In this paper, the cathode potential, charge yields, specific electrical energy consumptions, and deposit morphologies of gold electrodeposition in CPBE will be determined.

2. Procedure

2.1. Experimental Setup

Figure 1 shows arrangement of the components in an individual unit cell, each contained by a Perspex frame (240 mm (h) x 145 mm (w)x 50 mm (t)), separated by a 2 mm thick rubber gasket to prevent short circuit by contacts between cathode, membrane and anode, while allowing electrolyte, and hence ionic current, to pass through. The electrolyte flowed from the bottom to the top of reactor and even distribution of the electrolyte. The unit cell were compressed together by two stainless steel plates on each side of the frame to ensure good electrical contacts between the Ti-feeder cathode and the graphite particles electrodes. The anode was a mesh Ti/Ta_2O_5-IrO_2 while the particulates bed cathode consisted of 40 cm^3 of graphite particles with an averaged diameter of 0.5- 1.0 mm and a total surface area of c.a. 0.2 m^2.

The applied feeder cathode potential was controlled with respect to a saturated calomel reference electrode (SCE.) by a computer-controlled AUTOLAB PGSTAT30 potentiostat/galvanostat (coupled with a 10A current booster), which also measured the current through the reactor. The catholyte and

IOP Conf. Series: Materials Science and Engineering **206** (2017) 012093 doi:10.1088/1757-899X/206/1/012093

anolyte compartment were separated by a Nafion-N424 membrane to prevent partially reduced metals ions from the cathode transferred to the anode, re-oxidised subsequently and reducing charge efficiency. The catholyte was 4 kmol m^{-3} HCl + 1 kmol m^{-3} NaCl solutions containing 0.5 mol $AuCl_4^-$ m^{-3} and the anolyte was 1 kmol m^{-3} H_2SO_4. Both catholyte and anolyte were operating in batch recycle mode with separate reservoirs. The catholyte flow rate was measured by a flow meter (Omega FTB603) and logged by data acquisition device (NI DAQPad-6015, National Instruments) and software (LabView^{TM}).

The reason for the flow of the anolyte was to aid the removal of gas bubbles evolved from the anode, since build-up of bubbles would reduce the effective conductivity of the anolyte and increase the cell voltage. A schematic of the flow circuit is shown below. The piping arrangement used was George Fischer PVDF piping. Two 12 V DC Totton pumps were used to pump the fluids from their reservoirs through the anode and cathode chambers and back to the reservoirs again. Two 10 L glass reservoirs were used as catholyte and anolyte reservoirs. The schematic diagram of electrochemical reactor flow circuit is shown in figure 2.

2.2 Characteristics of samples

Samples of catholyte were taken from the reservoir at fixed time intervals. Each sample was diluted at a ratio of 1:50 in 0.5 M HCl before metal ions concentration was analyzed by ICP (Inductively Coupled Plasma – Perkin Elmer Instruments Optical Emission Spectrometer Model Optima 2000 DV). In addition, the morphology of metal deposits on pre-metallised graphite particulates obtained under different potentials had been studied using SEM (Scanning Electrons Microscopy, SEM – JEOL 5610). The effects of applied feeder cathode potential and electrolyte flow rate on the Cu recovery rates and charge efficiency had also been investigated.

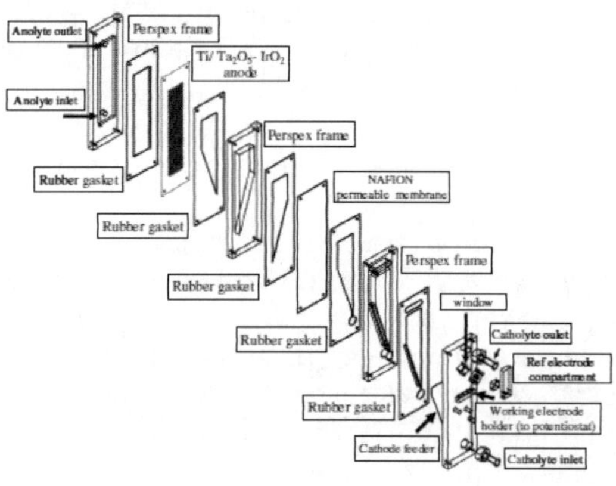

Figure 1. Arrangements of components in a cell

IOP Conf. Series: Materials Science and Engineering **206** (2017) 012093 doi:10.1088/1757-899X/206/1/012093

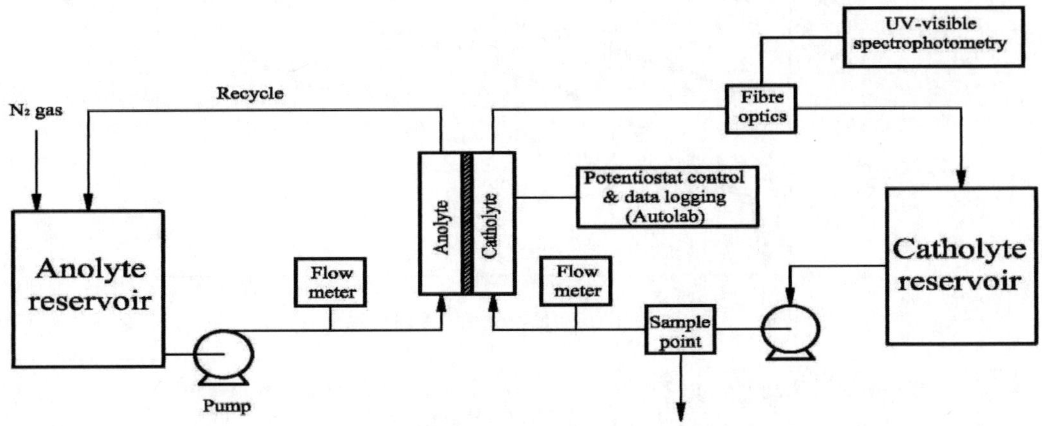

Figure 2. Schematic diagram of the experimental setup of the circulating particulate bed cathode electrochemical reactor (CPBE).

3. Results and discussion

3.1 Electrode potential analyses

Figure 3 shows the results of the negative-going linear potential sweep from +0.60 V (SCE) to 0 V (SCE) for the reduction of $AuCl_4^-$ ions in the CPBE reactor, at a flow rate of 0.25 dm^3 min^{-1}. The current was noisier than for a planar rotating disc, due to the small potential drop in the 3D electrode and the noise due to the variation in number of particles in contact with the feeder at any time. However, it was possible to see that the reduction of gold is mass transport controlled at +0.20 V (SCE). The mass transport controlled current when the particles were circulating, was about half the initial current with no flow, suggesting that about 50 % of the particles were active at any one time. While it would have been possible to deposit gold as a packed bed, if the particles were not moving, there was the risk of forming bridging deposits and poor deposit morphologies. With no flow to replenish reactants, the gold would also have been depleted and the reaction rate would have decreased quickly in a packed bed reactor, and so overall the gold deposition rate was unlikely to be faster for the CPBE reactor. The reactions that take place at the cathode are as follows:

$$AuCl_2^- + e^- \rightleftarrows Au + 2Cl^- \quad (1)$$

$$E_{AuCl_2^-/Au}(\text{SHE})/\text{V} = 1.152 + 0.0296\log\left(AuCl_2^-\right) - 0.118\log\left(Cl^-\right) \quad (2)$$

$$AuCl_4^- + 3e^- \rightleftarrows Au + 4Cl^- \quad (3)$$

$$\mathrm{E_{AuCl_4^-/Au}(SHE)/\ V = 1.0006 + 0.0197\ log(AuCl_4^-) - 0.0789\ log\ (Cl^-)} \quad (4)$$

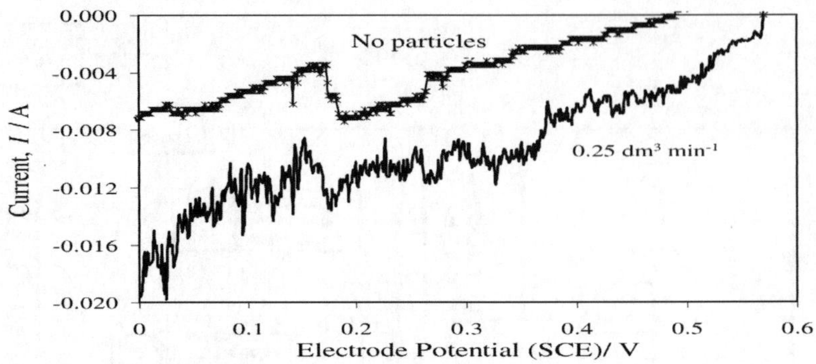

Figure 3. Steady state current-potential behavior for reduction on 0.5 mol $AuCl_4^-$ m^{-3} in CPBE reactor.

During the electrodeposition process of gold onto the unseeded graphite particles, the Au^{III} concentration was calculated from the solution absorption, measured by UV-Vis spectrophotometry through fibre optics apparatus, using the absorptivity for Au^{III} concentrations (λ_{max} = 312 nm, molar absorptivity, ε_{312} = 662.1 mol^{-1} m^2), so enabling Au^{III} / Au^{I} molar ratios to be determined. The total of gold concentration of each sample was also analyzed using ICP (Inductively Coupled Plasma – Perkin Elmer Instruments Optical Emission Spectrometer Model Optima 2000 DV). Figure 4 depicts the reading of the absorbance of solution during gold electrodeposition in CPBE for ca. 3 hours. The peak at ca. 225 nm was contributed by the concentration of aqueous chlorine in the sample.

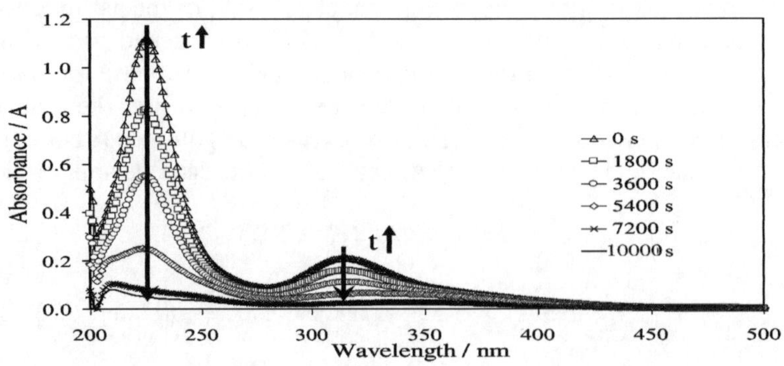

Figure 4. Absorbance of solution during gold electrodeposition in CPBE at +0.20V (SCE); path length 1 mm measured via fibre optics, at time, t.

As can be seen figure 5, and as was expected from the kinetic measurements at the planar RDE, the reduction of $AuCl_4^-$ ions was faster than that of $AuCl_2^-$ ions on carbon due to the additional overpotential driving force required for the nucleation of a new phase. This resulted in an initial accumulation of $AuCl_2^-$ ions during the first electrodeposition process onto fresh graphite

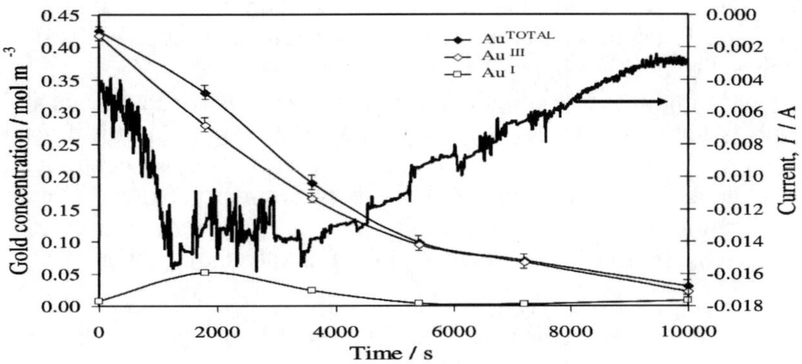

Figure 5. Current and gold concentration with time during electrodeposition on virgin graphite particles in CPBE reactor at +0.20 V (SCE); flow rate= 0.25 dm^3 min^{-1}

Figure 6 shows images of Au deposits in the CPBE (a) on the feeder Ti cathode and (b) on the graphite particulates in the catholyte compartment. Even Au deposition was observed on the feeder cathode unlike in the packed bed of graphite particulates. On the feeder cathode, the majority of Au deposition occurred along the area corresponding to the draft tube as well as on the feeder cathode area corresponding to the particulates packed bed. Since the voidage in the draft tube and above the packed bed was higher with respect to that in the packed bed, the effective conductivity of electrolyte in those areas was higher, and hence became the preferred paths for the current flow. In order to ensure maximum amount of current passing through the particulates packed bed and to fully utilise the large surface area provided by the graphite particulates to ensure high Au removal rates, the feeder cathode areas corresponding to the draft tube and above the packed bed were insulated (lacquered) with acrylic to block the current routes by-passing the packed bed [10].

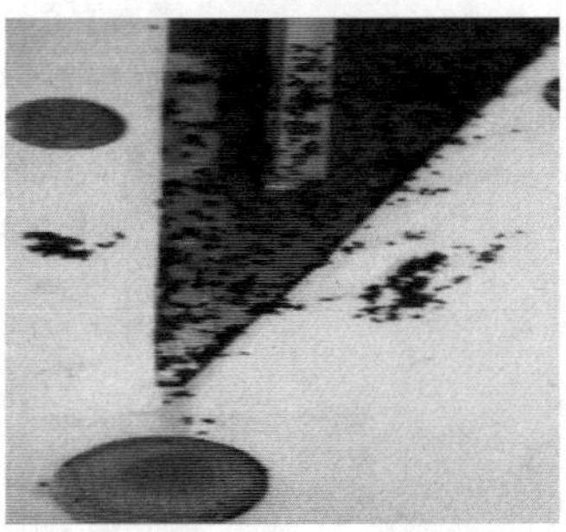

Figure 6. (a)

Figure 6. (b)

Figure 6. Images of gold deposits in the CPBE: (a) on the feeder cathode electrode and (b) on the Au particulates inside the catholyte compartment.

The peak in the magnitude of the (negative) reduction current after 3500 s was as a result of the interfacial area increasing with time until the growing nuclei coalesced, while the concentration of dissolved gold in solution decreased. Having metallised the graphite particles, in further electrodeposition experiments at +0.20 V (SCE), >99 % of the gold was recovered from solution in just over 2 hours (figure7). The initial current was nearly double that with the bare graphite, even though the initial concentration current was lower (0.33 mol m^{-3} against 0.44 mol m^{-3}), probably because the kinetics of gold reduction are slower on graphite than on gold. Therefore, while the reduction of $AuCl_4^-$ ions to Au on a gold surface is mass transport controlled at +0.20 V (SCE), this is not the case on graphite. The conductivity of the particles would also have increased once coated with gold, so more of the bed would have been electrically active.

It can also be seen in Figure 7 that on gold, the reduction of $AuCl_4^-$ ions occurs at the same rate as the reduction of $AuCl_2^-$ ions. Hence since Au^I does not appear in solution, it is possible to follow the electrowinning of gold, at potentials where the reduction is mass transport controlled online via fibre optics by measuring the Au^{III} concentration with UV-Vis spectrometry. However, online measurement is only applicable for a diluted gold concentration (< 1 mol m^{-3}). The plateau in current of about 4 mA after ca. 8000 s (Figure. 7), is probably due to reduction of dissolved oxygen by:

$$O_2 + 4H^+ + 4e^- \rightarrow 2H_2O \tag{5}$$

The main source of dissolved oxygen was the aqueous H_2SO_4 anolyte, which would have been supersaturated with oxygen and for which Nafion membranes have been reported to have high solubility and diffusion coefficients [11]. For typical Nafion membranes, of 1.78×10^{-5} m thickness, the diffusion coefficient of oxygen was 5×10^{-11} m^2 s^{-1} and the solubility was 15 mol m^{-3} [12]. As oxygen was generated at the anode of the CPBE reactor, the anolyte dissolved oxygen concentration would have increased and because of the diffusion across the membrane, the catholyte dissolved oxygen concentration would have also increased with time. The maximum current at the 0.1 m^2 area graphite cathode, from the reduction of dissolved oxygen, limited by diffusion across a typical Nafion membrane, was predicted to be 3.3 mA. This reduction of dissolved oxygen decreased the charge efficiency for gold electrowinning to ca. 0.90. However, this loss of efficiency will be less of a problem if the leaching and electrochemical reactors are coupled, since the anolyte will be a chloride solution to generate chlorine which has a much lower solubility and diffusion coefficient through Nafion membranes [12].

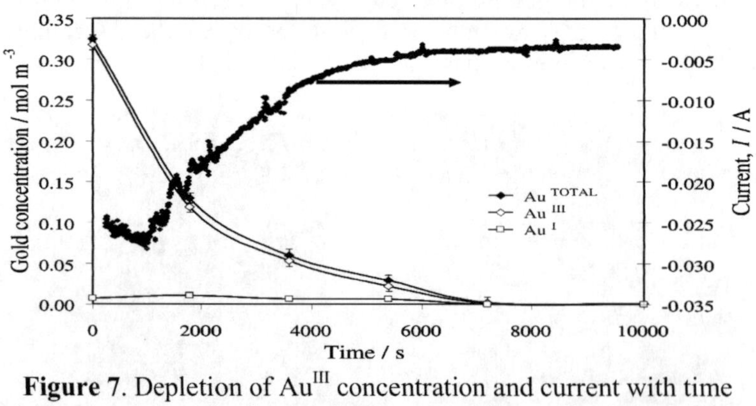

Figure 7. Depletion of Au^{III} concentration and current with time during batch recycle electrowinning on gold metallised particles in CPBE reactor with feeder electrode at +0.20V (SCE); flow rate= 0.25dm^3 min^{-1}.

The concentration decay during the electrodeposition of gold under mass transport control can be approximated to the general form expected for a model steady-state plug flow reactor (PFR), coupled to a CSTR. Hence, for a mass transport controlled reaction, the concentration should decay exponentially with time according to equation 3.6, and as shown experimentally in figure 8. The k_mA_a values were measured as 1.1×10^{-5} m s^{-1} for deposition on virgin graphite particles and 1.3×10^{-5} m^3 s^{-1} for deposition on Au-plated graphite particles.

$$c_t = c_o \exp\left\{-\frac{\text{v}t}{V}\left[1-\exp\left(-\frac{k_m AaL}{\text{v}}\right)\right]\right\} \tag{6}$$

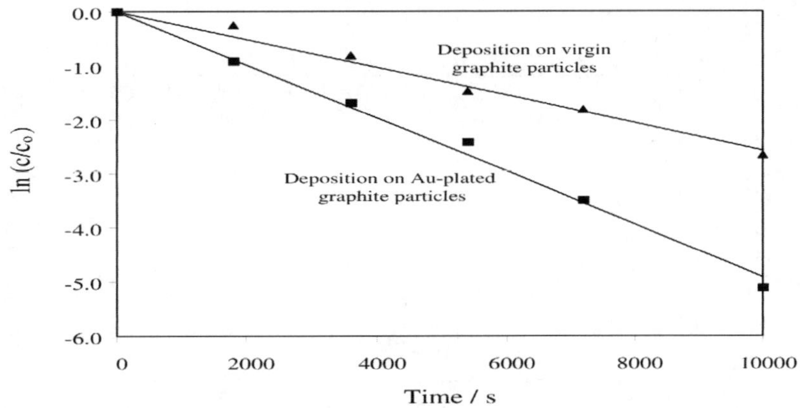

Figure 8. Depletion of dissolved Au^{III} concentration with time during batch recycle reduction in CPBE reactor with feeder at +0.20 V (SCE); flow rate = 0.25 dm^3 min^{-1}

It was considered that increasing the inlet flow rate increased the mass transfer rate within the CPBE and therefore, may increase electrodeposition rates. However, at higher flow rates, the graphite particles were entrained up the draft tube from the bottom of the particle bed faster than they could be replaced by the falling particles from above. Hence, the lower part of the bed was not a descending packed bed in contact with the feeder electrode and was unable to deposit gold at the electrode. Therefore, the overall electrodeposition current at higher flow rates was lower as shown in figure 9.

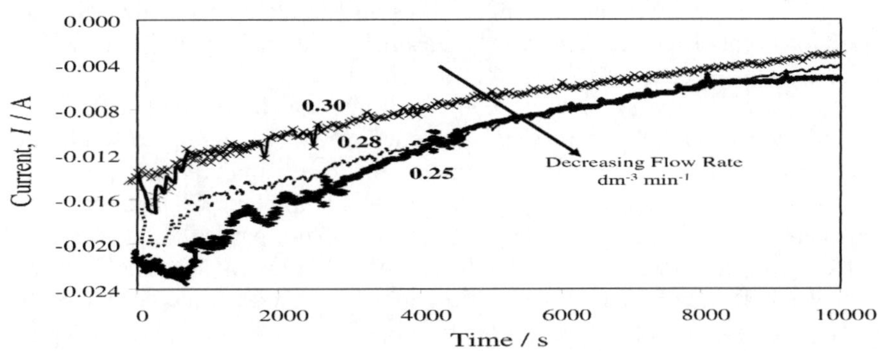

Figure 9. Effects of flow rate on the time dependency of current, for gold deposition in a CPBE reactor at a feeder cathode potential of +0.20 V(SCE) and an initial gold concentration of 0.5 mol m^{-3}.

It has been shown that gold chloride ions can be depleted from solution by electrodeposition, to concentrations that were analytically undetectable. Gold was recovered with high current efficiencies (figure 10) and very low specific electrical energy consumption (equation 3.7) of 800-1300 kW h (tonne Au)$^{-1}$ for cell voltages (U) of 2.0-3.0 V and fractional current efficiencies of ca. 0.95.

$$w \ / \ \text{kW h (tonne } Au)^{-1} = \frac{\nu_e F}{\Phi_i^e} \cdot \frac{U}{3.6 M_{Au}} \qquad (7)$$

IOP Conf. Series: Materials Science and Engineering **206** (2017) 012093 doi:10.1088/1757-899X/206/1/012093

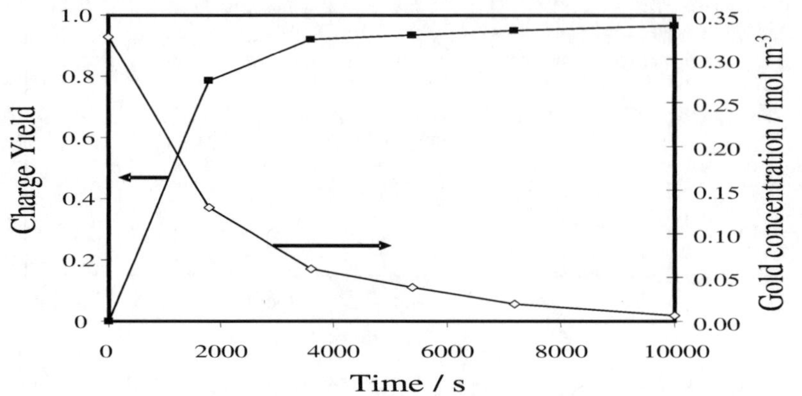

Figure 10. Depletion of dissolved gold concentration with time during batch recycle reduction in CPBE reactor with feeder electrode at +0.20 V (SCE); flow rate = 0.25 dm^3 min^{-1}.

3.2 Deposit morphologies

The morphology of the Au deposits was one of the major factors that determine if the CPBE was feasible for the recovery of Au from the catholyte (leach solution). Au deposits formed from a chloride media tend to be dendritic [4],[5], which break from the electrode's surface easily, re-dissolve in the catholyte and hence, reducing current efficiency. Coherent and adherent Au deposit was essential for maintaining a high current efficiency, achieving a reasonable Au loading on the electrode and enabling the transfer of metals-coated electrode to the electro-refining stage [9].

Figure 11 shows a SEM image of virgin graphite beads which will be used as a cathode in the CPBE. As the density of the graphite particles is recorded ca. 2.5 g cm^{-3}, it is essential to metallise the virgin graphite particles. Thus, a layer of gold of at least 5 μm thick was required to increase the effective density enough to prevent the particles floating in a 5 kmol m^{-3} Cl^- solution.

Figure 12 depicts a SEM image of Au deposits on a graphite particulate in the CPBE, operating mass transport control, at a feeder cathode potential of +0.20 V (SCE). The average diameter of graphite particulates before and after Au deposition, measured under a microscope, was 520 and 530 μm, respectively which are shown in Figure 12. Therefore, the thickness of the Au deposits was ca. 10 μm.

Subsequently, the metallised graphite particulates were used to deposit Au from the actual WEEE leachate. The morphology obtained at +0.20 V (SCE), Au deposited under total mass transport control were adherent to the particulates and depicts a compact cauliflower structure and little voidage. Random samples of Au particulates had been chosen from the packed bed cathode showed similar structure to that of figure 13, which proved that the constant collisions between the Au particulates in the CPBE were able to improve the morphology of the deposits.

IOP Conf. Series: Materials Science and Engineering **206** (2017) 012093 doi:10.1088/1757-899X/206/1/012093

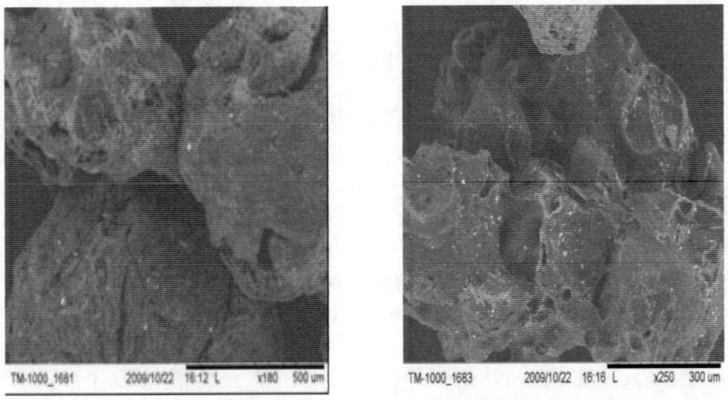

Figure 11. SEM image of virgin graphite particles (0.5- 1 mm) used as cathode in the CPBE.

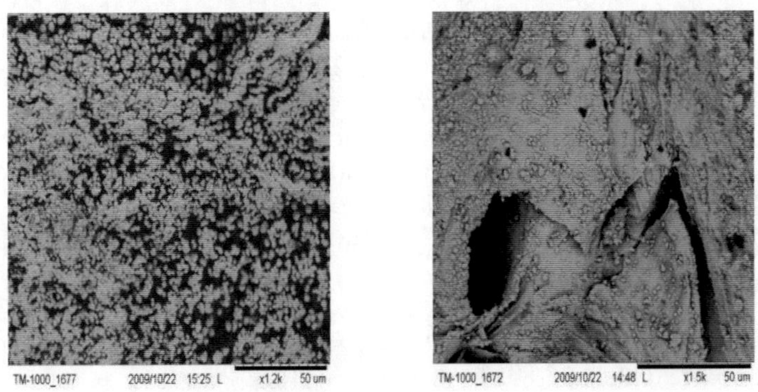

Figure 12. SEM images of 10 μm gold deposits on virgin graphite particles.

Figure 13. SEM image of gold deposits on a Au-pre plated graphite particulate in the CPBE operating at a feeder cathode potential of +0.20 V (SCE).

4. Conclusion

Experimental results demonstrated that the CPBE was suitable for the recovery of Au from the leach solution produced by leaching metals from WEEE. Concentrations <0.5 mol m^{-3} (< 10^2 g m^{-3}) of $AuCl_4^-$ and $AuCl_2^-$ ions in 4 kmol m^{-3} HCl + 1 kmol m^{-3} NaCl in aqueous electrolyte were successfully reduced to elemental gold in CPBE of 0.5-1 mm graphite particles, in an electrochemical reactor incorporating a cation- permeable membrane and operated in bath recycle mode. Depletion to concentrations < 5 x 10^{-3} mol m^{-3} (< 1 g m^{-3}) appeared to be mass transport controlled at an applied potential of +0.20 V (SCE) and specific electrical energy consumption (SEEC) of ca. 800-1300 kW h (tonne Au)$^{-1}$ for cell voltages (U) of 2.0-3.0 V and fractional current efficiencies of ca. 0.95. However, atomic absorption and UV spectrophotometry established that, as the ([$AuCl_4^-$]+ [$AuCl_2^-$]) concentration decayed, the [$AuCl_2^-$]:[$AuCl_4^-$] molar ratio changed. A multi-step mechanism for reduction of $AuCl_4^-$ ions proposed by Cheng et al. explained this behavior in terms of changing overpotentials for $AuCl_4^-$ and $AuCl_2^-$ reduction as total dissolved gold concentrations decreased. In addition, SEM images confirmed that adherent and coherent Au deposits were achieved with CPBE for Au deposition under mass transport control, which is at +0.20 V (SCE).

Acknowledgements
The authors are grateful for financial support provided by the Knowledge Transfer Programme (KTP) (9011-00026), which made this research possible.

References
[1] Babu B R, Parande A K, and Basha C A 2007 *Waste Manage.* **25** 307-318
[2] Ongondoa F O, Williams I D and Cheerretta T J 2011 *Waste Manage.* **31** 714-730
[3] Pilone D and Kelsall G H 2006 *Electrochim. Acta* **51** 3802-3808
[4] Schalch E, Nicol M J, Diggle J W, Charlton B D and Vaessen J P H 1976 The Cathodic Deposition of Gold Report No.1848, National Institution Metallurgy, Mineral Process Chemical Division, Johannesburg, South Africa
[5] Nicol M J 1976 Report No.1848, National Institution Metallurgy, Mineral Process Chemical Division, Johannesburg, South Africa
[6] Harrison J A and Thompson J 1975, *Electroanalyt. Chem.* **59** 273-280
[7] Muir R D, Stett C W A and Priddle C J 1982 *Hydrometallurgy-Res.* Development and Plant Practice AIME Warrendale 825-838
[8] Diaz M A, Kelsall G H and Welham N J 1993. *J. Electroanalyt. Chem.,* **361**
[9] Cheng, C Y, Diaz, M A, Kelsall, G H, and Robson, A C 2006. *ECS Trans* **2** 317-328
[10] Dawson R J 2006 Electrochemical Recovery of Platinum and Palladium from Alumina-Supported Catalysts Ph.D. Thesis, Chemical Engineering (Imperial college London).
[11] Sakai T, Taenaka H, and Torikai E 1986 *J. Electrochem. Soc.* ***133*** 88-92
[12] Davis T A, Genders J D and Pletcher D 1997 A First course in ions permeable membrane Electrochemical Consultancy Romsey, England.

Effect of Recycle Sludge on Anaerobic Digestion of Palm Oil Mill Effluent in A Thermophilic Continuous Digester

Irvan*[1,2], B Trisakti[1,2], Y Tomiuchi[3], U Harahap[4] and H Daimon[5]
[1]Chemical Engineering Department, University of Sumatera Utara, Medan, 20155, Indonesia
[2]Sustainable Energy and Biomaterial Center of Excellence, University of Sumatera Utara, Medan, 20155, Indonesia
[3]Metawater Co., Ltd. 7, Yawata-kaigandori, Ichihara-city, Chiba 290-8511, Japan
[4]Faculty of Pharmacy, University of Sumatera Utara, Medan, 20155, Indonesia
[5]Department of Environmental and Life Sciences, Toyohashi University of Technology, Toyohashi, 441-8580, Japan

E-mail: i_v_a_n_mz@yahoo.com

Abstract. The objective of this research is to maintain short retention time and high degradation of palm oil mill effluent (POME) to biogas by applying recycle sludge. Fresh POME from Rambutan Mill without further treatment was used as feed. Two lab-scale digesters supported from Metawater Co. Ltd. have been applied to treat POME at thermophilic (55°C) condition. Both digesters were operated under intermittent operation mode. Experiments were performed in two methods: with and without recycle sludge. Hydraulic retention time (HRT) of both methods was maintained at 6 days, while sludge retention time (SRT) was maintained at various days. The result showed that by extending SRT in return sludge process where 25% of digested slurry recycled to the digester, improvement of volatile solid (VS) decomposition was obtained around 84% at HRT of 6 days and SRT of 21 days. Then, chemical oxygen demand (COD) removal efficiency could be reached until 85% by using recycle sludge.

1. Introduction

Palm oil manufacturing is one of Indonesia's major industries. Approximately 23.5 million tons of crude palm oil (CPO) was produced in 2012 [1]. It makes Indonesia become one of the countries with largest production of CPO, followed by Malaysia. The amount of the palm oil mill effluent (POME) discharged is estimated to be three times CPO production.

The common POME treatment method utilized by palm oil mills is the application of lagoon systems. POME is treated in a series of anaerobic lagoons [2]. This method is an easy operating system but it has disadvantages such as, requires a long retention time of 20 to 200 days [3], the use of large areas (5 hectares for a palm oil mill with a capacity of 30 tons fresh fruit bunches/hour), bad odor and uncontrolled methane emission from the ponds to the atmosphere. In addition, the cost of dredging and handling the sludge from the lagoons is high because sludge accumulates at the bottom of the lagoons. This accumulation reduces the effective volume of the lagoons.

Nowadays, anaerobic digestion process becomes a popular method to treat POME. This method involves a group of microorganism digesting organic matters to produce methane and carbon dioxide as major components in biogas. Many studies were conducted to convert POME to biogas, even several of them were already implemented in the palm oil mills at pilot or commercial scale [4]. One of the companies that provide technologies for biogas plants was Novaviro Technology Sdn. Bhd. [5]. Even though they have already constructed many biogas plants at large scale, however the process still

IOP Conf. Series: Materials Science and Engineering **206** (2017) 012094 doi:10.1088/1757-899X/206/1/012094

requires relatively long hydraulic retention time (HRT) which is 18.75 days. Therefore, it needs high investment cost for the application [4].

Irvan et al in 2012 conducted POME conversion into biogas using anaerobic microbes in a 2 (two) liters continuous stirred tank reactor (CSTR) at temperature of 55 °C (thermophilic), closed systems, intermittent, and HRT variation. They succeeded in converting POME into biogas at HRT 6 days with quantity and quality of biogas similar to the Novaviro's process, although decomposition of COD was still less than 80%, at HRT 10 days [6]. Therefore, the purpose of this research is to study the effect of recycle sludge in the new methane fermentation system capable of maintaining high speed and high degradation of POME to biogas.

2. Materials and methods

2.1. Characteristics of POME

Fresh POME was collected from a fat pit of palm oil mill's wastewater treatment facility belongs to Rambutan Palm Oil Mill PTPN III, located at North Sumatera, Indonesia. The collected fresh POME was kept in 10 L clean containers and stored at 4°C in refrigerator for further analysis and use. It was used as feed in the digester having characteristic shown in Table 1. While, anaerobic digested wastewater taken from anaerobic pond of the same treatment facility was used as inoculum.

Table 1. Characteristics of fresh POME

Parameters*	Values
pH	3.9 – 4.6
TS	28,000 – 64,700
VS	23,200 – 53,900
COD_{Cr}	41,000 – 99,000
Kj-N	560 – 1,200
Fat	1,700 – 9,200

** Unit for all parameter is mg/L except pH*

In this experiment, supporting materials were added namely $NaHCO_3$, metallic solution of trace metal; $FeCl_2$ (25 mg/L), $NiCl.6H_2O$ (0.5 mg/L), and $CoCl_2.6H_2O$ (0.4 mg/L). The addition of $FeCl_2$ is to minimize H_2S production, and the addition of $NiCl.6H_2O$ and $CoCl_2.6H_2O$ are required for anaerobic microbial metabolism.

2.2. Mass Balance of Recycle Sludge

A continuous stirred tank reactor (CSTR) was employed in this study. In complete mix systems, digested slurry is settled in an external clarifier, and the microbe-rich sludge is recycled back to the digester. Mass balance of recycle sludge in the anaerobic digester was analyzed. Schematic drawing of anaerobic digester with recycle sludge is shown in Figure 1. It identifies which variables are associated with the streams, such as: influent volumetric flow rate Q_{In} (L/day), effluent volumetric flow rate Q_{Out} (L/day), sludge recycle volumetric flow rates QR (L/day), digested slurry volumetric flow rates QD (L/day) influent concentrations C_{In} (gr/L), effluent concentrations C_{Out} (gr/L), sludge recycle concentrations CR (gr/L), digested slurry concentrations CD (gr/L) and the digester volume V (L). As depicted in Figure 1, HRT and solids retention time (SRT) can be calculated by Equation 1 and 2, respectively.

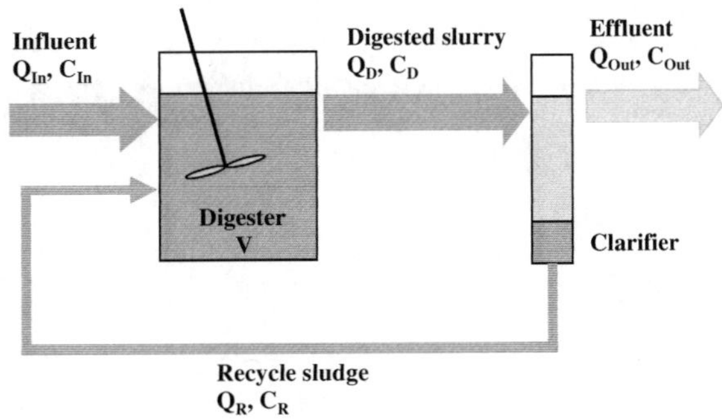

Figure 1. Schematic drawing of anaerobic digester with recycle sludge

$$HRT = \frac{V}{Q_{in}} \tag{1}$$

$$SRT = \frac{V_C}{Q_{Out}C_{Out} - Q_R C_R} \tag{2}$$

2.3. Experimental Method

All experiments were carried out in two transparent jar digesters (EYELA, model MBF 300ME) which each has a total capacity of 2.2 L and a working volume of 2 L. It worked as a well-stirred reactor with a continuous agitation of 250 rpm, and the temperature was maintained at 55 ± 1oC by means of a water jacket. A data logger (KEYENCE, Model NR-250) was connected to computer to enable automatic recording of temperature and pH, provided by censoring equipment attached to digester [7].

Experiments were performed in two methods: with and without recycle sludge. Loading up was carried out based on the increased production of biogas which measured by using a wet gas meter (SHINAGAWA, Model W-NK-0.5B). Hydraulic retention time (HRT) of both methods was maintained at 6 days, while sludge retention time (SRT) was maintained at various days (8 to 22 days). If biogas production raised by 20%, then the loading up was increased by 20% as well, until HRT 6 days was achieved.

Digested slurry was allowed to settle for 6 hours in a 1,000 ml volumetric flask as an external clarifier before 25% of it was recycled to the digester. After that, the recycled sludge was analyzed by chemical oxygen demand (COD), total solid (TS), volatile solid (VS), alkalinity, and pH. Concentrations of H2S and CO2 in the biogas were measured by using a suction gas injector (GASTEC, type GV-100S) and inspection tube (GASTEC, 25 ~ 1600 ppm).

3. Results and discussion

3.1. Biogas Production in Digester

Biogas production depends mainly on operational parameters such as feeding material, pH and temperature. Figure 2 shows the biogas production during the fermentation proses with and without recycle sludge. At a glance if we notice Figure 2, there is no difference between the two graphs. However, in order to achieve HRT 6 days, these two systems are different.

IOP Conf. Series: Materials Science and Engineering **206** (2017) 012094 doi:10.1088/1757-899X/206/1/012094

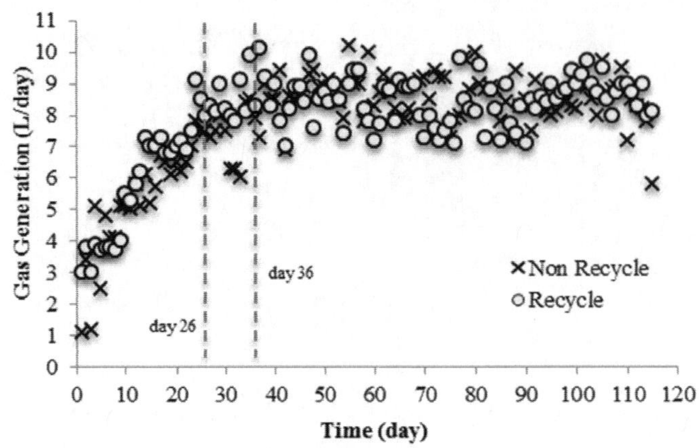

Figure 2. Biogas production of anaerobic digester with and without recycle sludge

For fermentation with recycle sludge, HRT 6 days was achieved at day 26, while for the non-recycle sludge was at day 36, which means there is a difference of 10 days to reach HRT 6 days. This could happen because digester with recycle contained the numbers of bacteria more than digester without recycle, due to the return of sludge every day. Furthermore, stable condition in the recycle system was achieved faster than without recycle. Where, fermentation with recycle, biogas began to increase until it reached a stable condition at day 43 and the production of biogas ranged from 7 to 9 L/day, while for the non- recycle fermentation, stable condition was achieved at day 53 and the production of biogas ranged between 8 to 9 L/day. As it is observed in Figure 2, the recycle did not have a significant effect in the biogas generation, since there were not significant differences between the two graphs.

3.2. Effect of Recycle Sludge on Alkalinity and pH

Alkalinity and pH adjustment within the optimum range is very important for effective anaerobic digestion process [8]. Therefore, additional substances are often inserted into the digester which containing insufficient alkalinity. POME used in this study was acidic with pH ranged 3.9 – 4.6. Thus alkalinity and pH adjustment to within the range of 2,000 – 5,000 mg/L [9] and 6.8 – 7.2 [10] for alkalinity and pH range, respectively, was taken into consideration in this experiment. In the feed preparation of this experiment, 2.5 g/l of sodium bicarbonate ($NaHCO_3$) was added to the influent. The pH in the digester with recycle sludge increased to 7.5 - 7.7 from an initial value of 4.0. While, pH in the digester without recycle sludge increased from 4.0 to 7.7 - 7.5.

Figure 3 shows the changes in alkalinity for fermentation with and without recycle. Alkalinity in the digester with recycle sludge was higher than without recycle. This is due to the sludge which recycled into the digester has high concentration of sodium, whereas the addition of NaHCO3 in the fresh POME was still performed, so that the sodium accumulated in the digester.

IOP Conf. Series: Materials Science and Engineering **206** (2017) 012094 doi:10.1088/1757-899X/206/1/012094

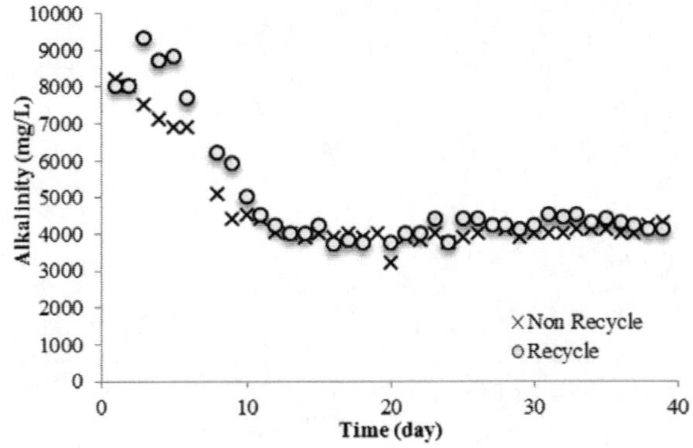

Figure 3. Effect of recycle sludge on alkalinity and pH

3.3. Effect of Recycle Sludge on Total Solid

Total solid (TS), as defined by Standard Methods, is the material residue left in a vessel after evaporation of a sample and its subsequent drying in an oven at 103 to 105°C for one hour [11]. Figure 4 shows the profile of TS inside the digester with and without recycle sludge

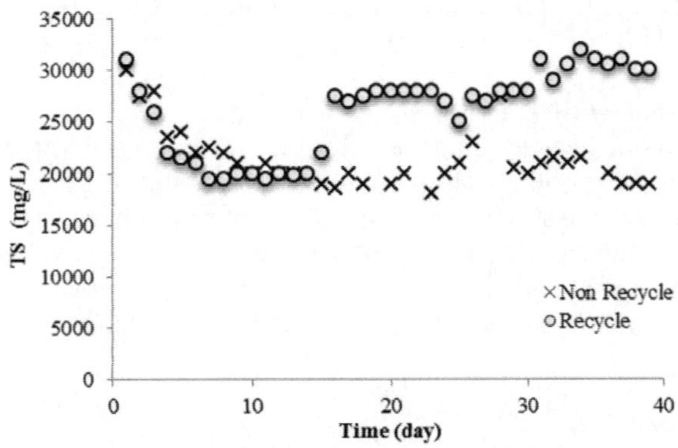

Figure 4. Effect of recycle sludge on total solid

As shown in Figure 4, in the fermentation with recycle sludge, TS decreased from 30,000 mg/L to at about 20,000 mg/L, then relatively constant at 28,000 – 30,000 mg/L. While, without recycle sludge TS decreased from 30,000 mg/L to at about 20,000 mg/L, then relatively constant at that value. TS in the digester with recycle sludge was higher than without recycle, this is because the sludge recycled to the digester came from the bottom of the clarifier are still containing material with high TS.

3.4. Effect of Recycle Sludge on VS Decomposition

Conversion of volatile solids into biogas is affected by SRT rather than HRT. If SRT is low then there is no sufficient time for the bacteria to grow and to replace the bacteria lost in the effluent. Basically, the purpose of recycle time is to increase the SRT, because the residence time of sludge in the system becomes longer [12]. Effect of recycle sludge to decomposition VS is presented in Figure 5.

IOP Conf. Series: Materials Science and Engineering **206** (2017) 012094 doi:10.1088/1757-899X/206/1/012094

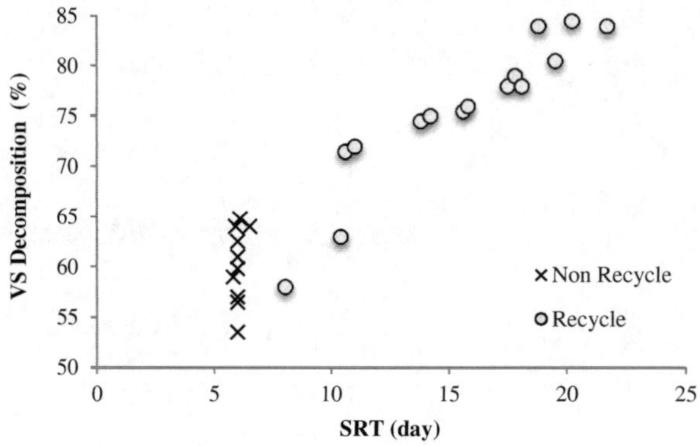

Figure 5. Effect of recycle sludge on VS decomposition

As shown in Figure 5, in the fermentation process without recycle sludge at HRT 6 days, VS decomposition had the highest value of 65% and the lowest value of 53%. While in the fermentation process with recycle sludge, due to the increase of SRT then VS decomposition was also increased. At SRT 21 days, VS decomposition reached maximum value at 84 %, while the lowest value of 58% was achieved at SRT 8 days. From the results obtained, it can be confirmed that volatile solids conversion to biogas is a function of SRT.

3.5. Effect of Recycle Sludge on COD Removal

The advantageous effect of the recycle sludge in terms of COD was observed. Figure 6 shows the effect of recycle sludge on COD removal. At HRT 6 days, fermentation of POME to biogas using recycle sludge, COD removal efficiency was 85%, while using non recycle sludge COD removal efficiency was only 76 %. This proves that the recycle sludge can increase the COD removal in anaerobic fermentation process. This was confirmed by previous researchers who have demonstrated that COD removal efficiency in the digester with recycle sludge was lower than that without recycle sludge [13, 14, 15]

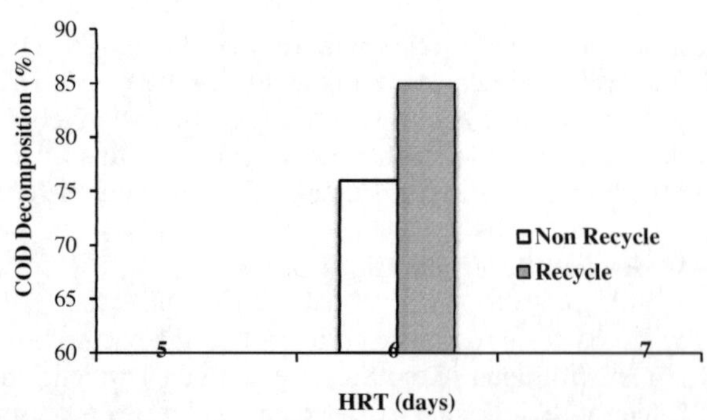

Figure 6. Effect of recycle sludge on COD decomposition

4. Conclusion

This study has demonstrated the feasibility of anaerobic digestion of POME with recycle sludge. During the fermentation process with and without recycle sludge, biogas production was relatively similar. Alkalinity and TS in the digester with recycle was higher than digester without recycle sludge. The same thing was also shown by the decomposition of VS and COD removal efficiency.

Acknowledgment
This research was supported by Hibah Penelitian Unggulan Perguruan Tinggi No: 017/SP2H/LT/DRPM/II/2016, date February 17th, 2016.

References
[1] Rifai N, Syaukat Y, Siregar H and Sa'id EG 2014, *IOSR J. Econ. Finance* **4** 27-39
[2] King LS and Yu LC 2013, *Eur. Int. J. Sci. Tech.* **2** 106-114
[3] Yacob S, Hassan MA, Shirai Y, Wakisaka M and Subash S 2005, *Chemosphere* **59** 1575–1581
[4] Ji CM, Eong PP, Ti TB, Seng CE and Ling CK 2013, *Renew. Sustainable Energy Rev.* **26** 727-726
[5] Tong SL and Jaafar AB 2004 *Energy Smart* **14** 4-11
[6] Irvan, Trisakti B, Wongistani V and Tomiuchi Y 2012, *Int. J. Sci. Eng.* **3** 32-35
[7] Irvan, Trisakti B, Sosanty F and Tomiuchi Y 2016, *Asian J. Chem.* **28** 377-380
[8] Olvera JR and Lopez AL 2012 Biogas production from anaerobic treatment of agro-industrial wastewater, Rijeka: Intech
[9] Metcalf and Eddy, Inc. 2003 *Wastewater engineering: Treatment and reuse*, McGraw-Hill New York
[10] Gerardi MH 2006 *Wastewater bacteria*, Wiley-Interscience New Jersey
[11] American Public Health Association - APHA 2005 *Standard methods for the examination of water and wastewater* 21st edition Washington: Port City Press USA
[12] Burke DA 2001 *Dairy Waste Anaerobic Digestion Handbook*, Environmental Energy Company Olympia
[13] Lee DY, Ebie Y, Xu KQ, Li YY, and Inamori Y 2010, *Bioresour. Technol.* **101** S42-S47
[14] Busu Z, Sulaiman A, Hassan MA, Shirai Y, Aziz SA, Yacob S and Wakisaka M 2010, *Pertanika J. Trop. Agric. Sci.* **33** 27-37
[15] Gnanapragasam G, Senthilkumar M, Arutchelvan V, Sivarajan P and Nagarajan S 2010, *World J. Microbiol. Biotechnol.* **26** 1093-1098

IOP Conf. Series: Materials Science and Engineering **206** (2017) 012095 doi:10.1088/1757-899X/206/1/012095

Removal of Mn(II) from the acid mine wastewaters using coal fired bottom ash

M Mahidin[1,2,3], T N Sulaiman[1], A Muslim[1,2] and A Gani[1,2,3]

[1] Postgraduate Program of Chemical Engineering, Syiah Kuala University, Banda Aceh 23111, Indonesia

[2] Department of Chemical Engineering, Syiah Kuala University, Banda Aceh 23111, Indonesia

[3] Environmental and Natural Resource Research Center, Syiah Kuala University, Banda Aceh 23111, Indonesia

E-mail: mahidin@unsyiah.ac.id

Abstract. Acid mine wastewater (AMW), the wastewater from mining activities which has low pH about 3-5 and contains hazardous heavy metals such as Cu, Fe, Mn, Zn, Pb, etc. Those heavy metals pollution is of prime concern from the environmental view point. Among the heavy metals, Mn occupies the third position in the AMW from one the iron ore mining company in Aceh, Indonesia. In this study, the possibility use of bottom ash from coal fired boiler of steam power plants for the removal of Mn(II) in AMW has been investigated. Experimental has been conducted as follows. Activation of bottom ash was done both by physical and chemical treatments through heating at 270 °C and washing with NaOH activator 0.5 and 1 M. Adsorption test contains two parts observation; preliminary and primary experiments. Preliminary study is addressed to select the best condition of three independent variables i.e.: pH of AMW (3 & 7), bottom ash particle size (40, 60 & 100 mesh) and initial Mn(II) concentrations (100 & 600 mg/l). AMW used was synthetics wastewater. It was found that the best value for NaOH is 1 M, pH is 7, particle size is 100 meshes and initial Mn(II) concentration is 600 mg/l from the adsorption efficiency point of view. The maximum adsorption capacity (q_e) is 63.7 mg/g with the efficiency of 85%.

1. Introduction

One of the heavy metals pollution source is acid mine wastewater (AMW), such as iron core mining. The highest heavy metals contain in the AMW of iron core mining are iron (Fe) and manganese (Mn). Those heavy metals are hazardous and toxic materials. The heavy metals such as Fe and Mn when it accumulated in the human body will attach neural system, blood cell, kidney and bone-calcium content.

In this study, we developed a method for those heavy metals removal in order to solve the environmental problems cause by the AMW. The investigated method for the AMW's metals removal was adsorption, this method is a very usual method used in the wastewater treatment since the method is a lower in cost, without any further effect from hazardous materials and capable also to shift an organic material.

Patil et al. (2016) in their review on the Mn removal technologies both in the water and wastewater concluded that the chemical precipitation (degree of Mn(II) or Mn^{2+} ions recovery 99.5%), oxidation/filtration (degree of Mn(II) shifting 95%) and adsorption (degree of Mn(II) removal up to 99.42%) can effectively be applied for that propose. The most promising technology is membrane filtration, with the degree of Mn(II) rejection almost 100%, however, this technology is still expensive. Others technology such as coagulation, flotation, ion exchange and electro-chemical treatment stand below the four technologies mentioned above.

In Mn(II) absorption by using hydrothermally modified bottom ash so that has a similar structure to zeolite (zeolite-like) and zeolite-like modified by dithizone, thermodynamics Mn(II) adsorption behavior by both adsorbent presented the change in Gibbs free energy is positive (endothermic reaction), which mean non-spontaneous chemical reaction. Meanwhile, its entropy change was negative that mean adsorption process occur spontaneously at low temperature. Optimum conditions of Mn(II) adsorption by zeolite-like were at pH 6, adsorption time 60 min, initial metal ion concentration 20 mg/L and temperature 50°C. Further, optimum conditions for adsorption of Mn(II) by zeolite-like modified were observed at pH 8, time 120 min, temperature 60°C and the same initial metal ion concentration (Putra et al., 2015).

Investigation on adsorption of Fe(II) AMW-contaminated ground water has been conducted by Orakwue et al. (2016) by using coal fly ash, bottom ash and bentonite. The highest adsorption capacity provided by bottom ash (~80 g/g), compare to fly ash and bentonite with adsorption capacity only at ~50 mg/g. Moreover, utilization of bottom ash as adsorbent for removing Cu^{2+} and Ag^{+} in the industrial wastewater has been examined by Kusmiyati et al. (2012). Maximum adsorption capacity occurred at initial metal concentration 25 ppm. Equilibrium state reached after 300 min.

Utilization of bottom ash and fly ash as activated carbon has also been performed by Damayanti et al. (2000). In the study, bottom ash and fly ash activated chemically were applied to reduce Fe and Mn concentrations in the AMW. It was obtained that degree of metal shifting was significant; about 93.6–95.9% for fly ash and about 55.9% for bottom ash.

This paper deal with utilization of bottom ash as an adsorbent for removal the Mn^{2+} ions in the AMW (used synthetic waste); to evaluate the effect of adsorbent particle size, pH of AMW, concentration of NaOH (as activator) and initial Mn^{2+} ions concentration on the adsorption capacity.

2. Experimental procedure

The experimental procedure comprises of materials and apparatus, adsorbent preparation and adsorption study as following.

2.1. Materials and apparatus

Material used in the experiment were $MnSO_4$ [Mn^{2+} ions source], bottom ash from power plant, NaOH, and H_2O. Thus, apparatus needed were magnetic stirrer, pen stirrer, digital balance, 100 ml measuring jug/glass, 250 ml Erlenmeyer, 200 ml beaker glass, 10 ml pipette volume, filter paper, furnace, oven, ball mill, screen, SEM, XRD and AAS.

2.2. Adsorbent preparation

Adsorbent preparation procedure is presented in Figure 1. The chemical treatment was done at two activator (NaOH) concentrations, i.e. 0.5 and 1 M, respectively.

IOP Conf. Series: Materials Science and Engineering **206** (2017) 012095 doi:10.1088/1757-899X/206/1/012095

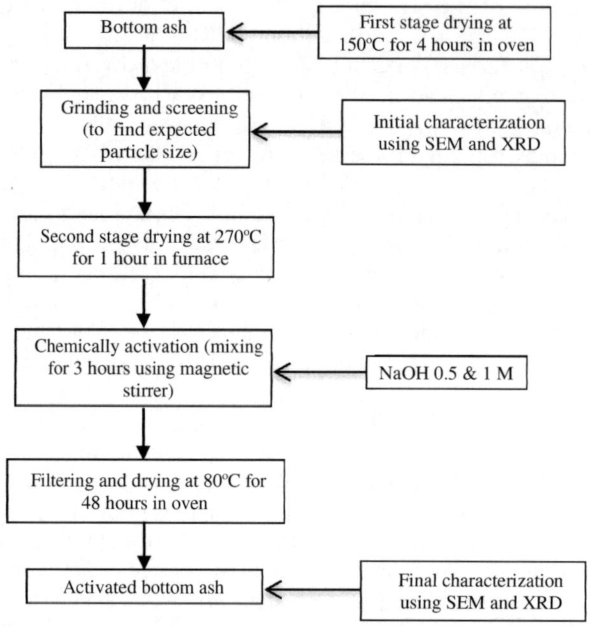

Figure 1. Scheme for adsorbent preparation.

2.3. Adsorption study

Main experimental was done for three independent variables (initial Mn^{2+} ion concentration, pH of synthetic wastewater and bottom ash particle size) as described in Figure 2.

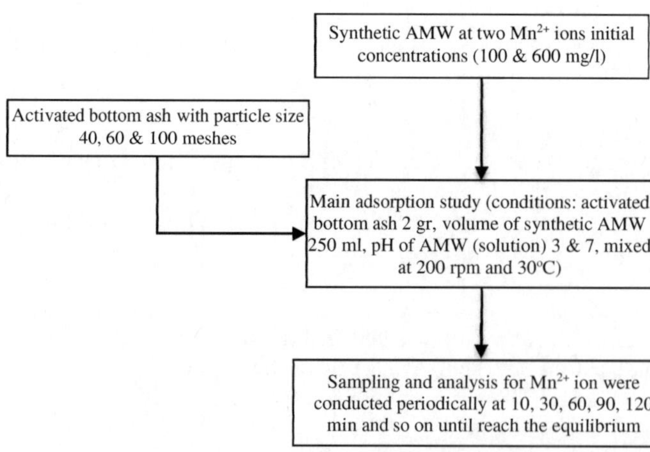

Figure 2. Scheme for Mn(II) ion adsorption experimental.

Sampling and analysis for ion metal concentration were done at 10, 30, 60, 90, 120 min, etc. until reached the equilibrium state or t_e (t_e = equilibrium time). pH of solution was selected at acid condition (3) and neutral one (7), by considering the real condition in mining site. Most the heavy metal mining produce the AMW that has a lower pH, however, in rainy session the AMW become almost neutral since the lagoon is mostly in open site.

IOP Conf. Series: Materials Science and Engineering **206** (2017) 012095 doi:10.1088/1757-899X/206/1/012095

3. Results and discussion

The examination results will be provided in this section and discussion on them as well. Attention will be paid on the morphology and chemical constituents of adsorbent, followed by the discussion on the effect of the adsorbent particle size, initial Mn(II) concentration, pH of AMW and activator concentration on the adsorption capacity and efficiency.

3.1. Morphology of adsorbent

Figure 3 present the Scanning Electron Microscope (SEM) of adsorbent before and after activated at two magnifications 5.000x (diameter 5 μm) and 10.000x (diameter 1 μm) both for the adsorbent particle size 40 and 100 meshes.

It is evident that the adsorbent before activation has small pore size and number of pore as well as many impurities trapped in the pore as presented in Figure 3(a), (b), (e) and (f). Meanwhile, after activation the size and number of pore increase significantly as shown by Figure 3(c), (d), (g), (h), (i) and (j), especially Figure 3(g) and (h). All the figures clearly reveal the surface structure, porous texture and porosity of the adsorbent. The bottom ash particles are in the form of indefinite shape and of a wide range of sizes. As shown in the images, many small holes are seen on the surface of activated bottom ash. Based on these photomicrographs, particles of the bottom ash can approximated as spheres (Gupta et al., 2004). It was also observed effect of the particle size on the cation exchange capacity (CEC) and weight fraction; it was found that decrease in the particle size caused increase of CEC and oppositely decrease of weight fraction (Gupta et al., 2005).

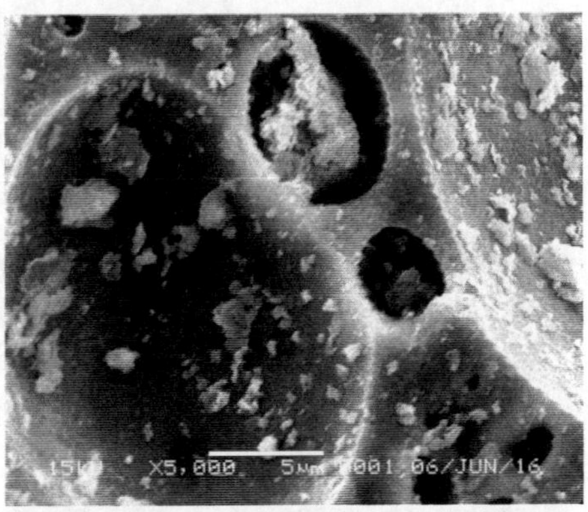

(a) Before activated (particle size 40 mesh; 5,000x)

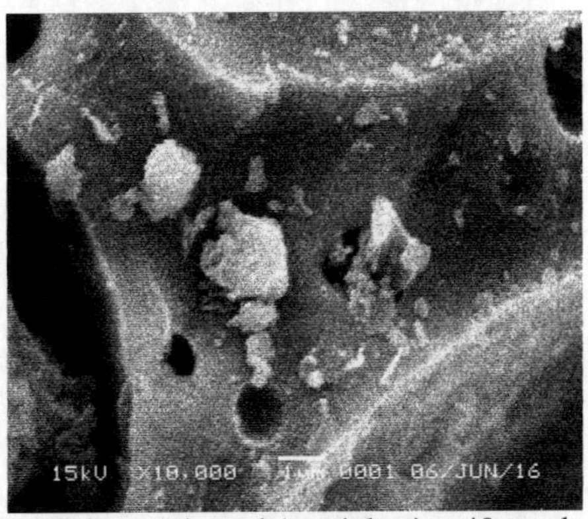

(b) Before activated (particle size 40 mesh; 10,000x)

(c) After activated (particle size 40 mesh and NaOH concentration 1 M; 5,000x)

(d) After activated (particle size 40 mesh and NaOH concentration 1 M; 10.000x)

IOP Conf. Series: Materials Science and Engineering **206** (2017) 012095 doi:10.1088/1757-899X/206/1/012095

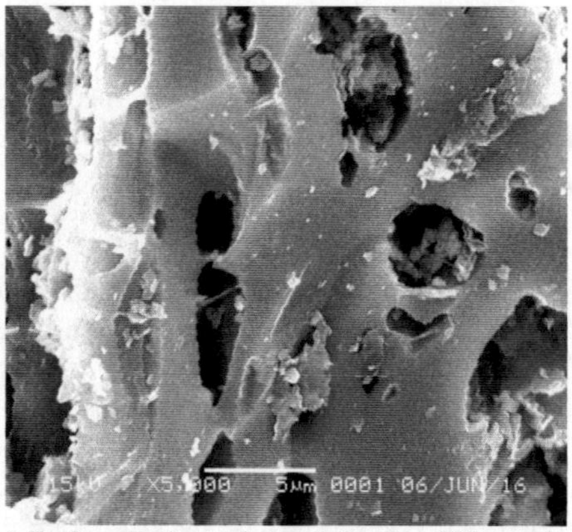

(e) Before activated (particle size 100 mesh; 5,000x)

(f) Before activated (particle size 100 mesh; 10,000x)

(g) After activated (particle size 100 mesh and NaOH concentration 0.5 M; 5,000x)

(h) After activated (particle size 100 mesh and NaOH concentration 0.5 M; 10,000x)

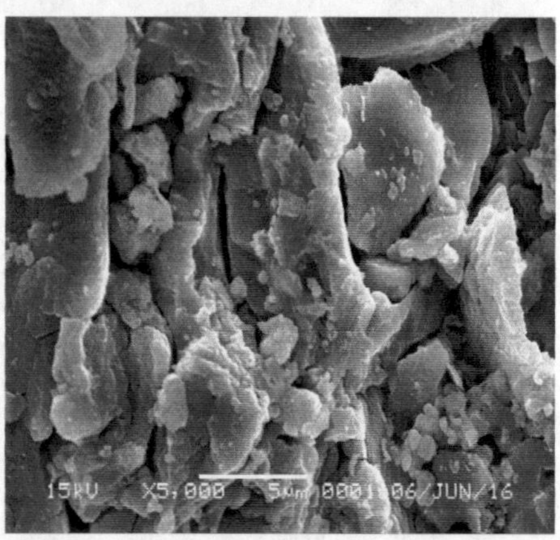

(i) After activated (particle size 100 mesh and NaOH concentration 1 M; 5,000x)

(j) After activated (particle size 100 mesh and NaOH concentration 1 M; 10,000x)

Figure 3. Morphology of bottom ash (40 and 100 meshes) before and after activated.

Bottom ash with particle size 100 mesh before activated has less pore and big pore diameter. Meanwhile, after activated was significantly changed; the number of pore greatly enhanced and the pore diameter significantly reduced. It is predicted that the NaOH activator plays an important role chemically to modify the surface and pore structure of bottom ash.

3.2. Chemical constituents content

Characterization by using the X-ray diffraction (XRD) was performed in order to identify chemical constituent in the adsorbent. Figure 4 shows that several peaks of bottom ash appear at 2θ around 10o-80o, 33 peaks for the raw bottom ash (before activated); for the activated bottom ash with NaOH 0.5 and 1 M the number of peaks are significantly increase to 41 and 40, respectively.

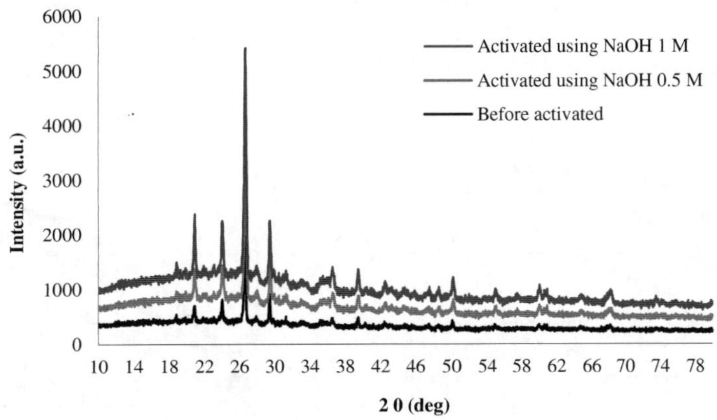

Figure 4. Wavelength of infrared spectra for adsorbent

For the raw bottom ash, the peak at 26.597° corresponds to a trigonal (hexagonal axes) crystal structure of $CaCO_3$, followed by others peaks at 29.380°, an orthorhombic crystal structure of C (graphite) and 23.982°, a trigonal crystal structure of SiO_2 (quartz). Then, the sharp peaks for the activated bottom ash with NaOH 0.5 M occurred at 26.712° described a trigonal crystal structure of SiO_2 (quartz), followed by others peaks at 29.482°, a trigonal crystal structure of $CaCO_3$ (calcite) and 24.093°, also a trigonal crystal structure of F_3Fe (iron fluoride). Finally, the peaks for the activated bottom ash with NaOH 1 M appeared at 26.598° shows a trigonal crystal structure of SiO_2 (quartz), followed by others peaks at 20.853°, a trigonal crystal structure of $CaCO_3$ (calcium carbonate) and 23.974° with a cubic crystal structure of C.

Based on above finding data, it can be postulated that the bottom ash contains the quartz component. In particular for the activated bottom ash, the sharp and higher peaks observed at around 26° correspond to the quartz component, so that the quartz (SiO_2) intensity is the strongest as was investigated by Faridah et al. (2015) and Sutarno et al. (2004). This fact describes that the crystallinity level of the quartz is high. A similar characteristic was exhibited by fly ash which contains silica more than 44% (Sharma et al., 2007).

3.3. Effect of the adsorbent particle size, initial Mn(II) concentration, pH of AMW and activator concentration on the adsorption capacity

Figures 5 to 10 present the effects of the adsorbent particle size, initial Mn(II) ion concentration, pH of AMW and activator concentration on the adsorption capacity. It can be seen that the initial Mn(II) concentration strongly influences the adsorption capacity, whereas the adsorbent particle size, activator concentration and pH of solution (AMW) show the slightly effect only. This investigated fact correspond well with the result observed by Sharma et al. (2007), especially on the role of Mn(II) concentration.

IOP Conf. Series: Materials Science and Engineering **206** (2017) 012095 doi:10.1088/1757-899X/206/1/012095

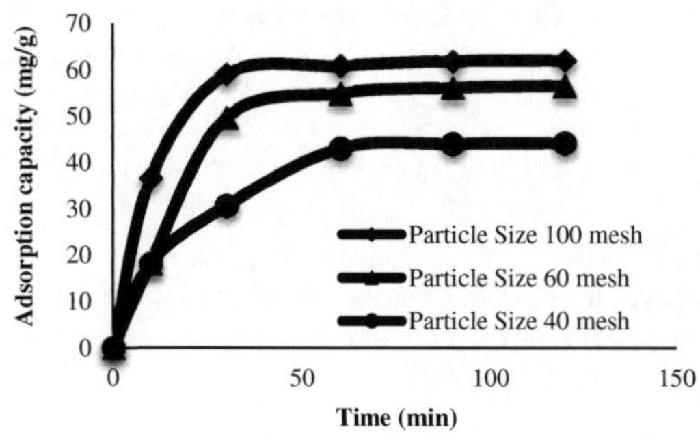

Figure 5. Adsorption capacity [initial Mn(II) ion concentration 600 mg/l; pH 7; NaOH 1 M].

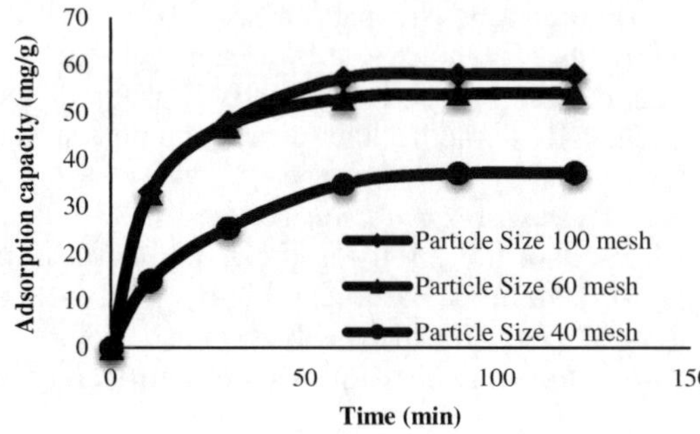

Figure 7. Adsorption capacity [initial Mn(II) ion concentration 600 mg/l; pH 7 & NaOH 0.5 M].

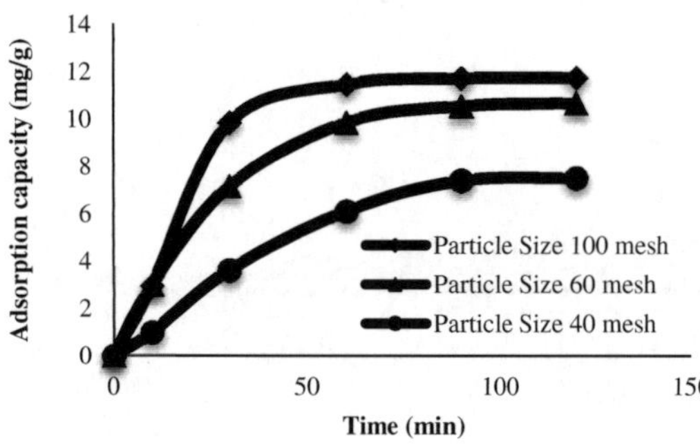

Figure 6. Adsorption capacity [initial Mn(II) ion concentration 100 mg/l; pH 7; NaOH 1 M].

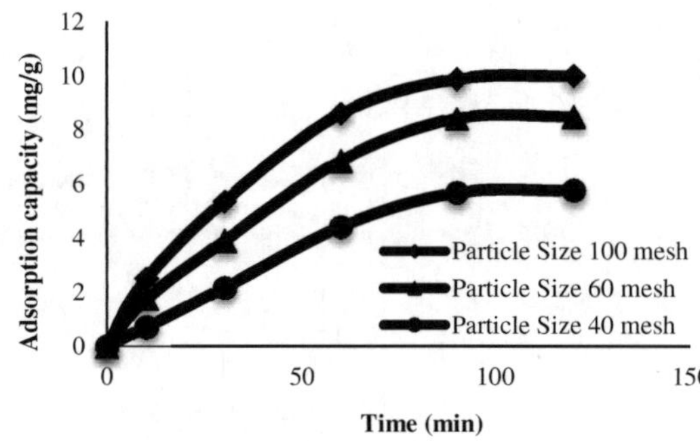

Figure 8. Adsorption capacity [initial Mn(II) ion concentration 100 mg/l; pH 7; NaOH 0.5 M]

It is clearly present by Figures 5 to 10 for the initial Mn^{2+} ions concentration of 600 mg/l provide the adsorption capacity up to 60 mg/g or more, while for the 100 mg/g ones reach only the level of 12 mg/g (max). As expected, at a small particle size the adsorption capacities are higher since it has big number of pore and small pore diameter compare to the large particle size one, consequently has higher surface area to ensure the contact between Mn^{2+} ions and adsorbent. In addition, pH of solution reflects a similar pattern to those parameters even not deeply influence the the adsorption capacity.

IOP Conf. Series: Materials Science and Engineering **206** (2017) 012095 doi:10.1088/1757-899X/206/1/012095

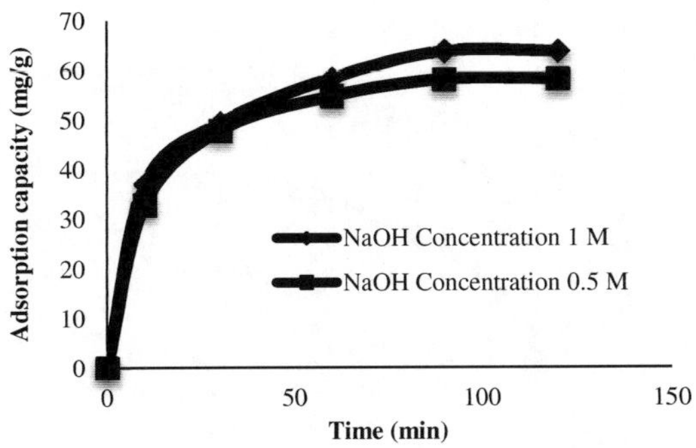

Figure 9. Adsorption capacity [initial Mn(II) ion concentration 600 mg/l; pH 3; particle size 100 mesh].

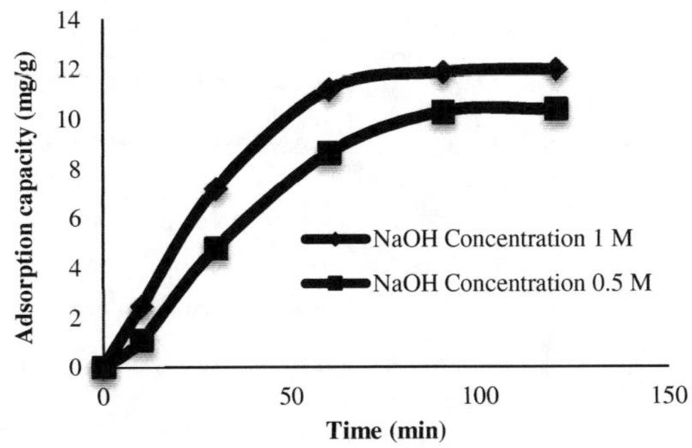

Figure 10. Adsorption capacity [initial Mn(II) ion concentration 100 mg/l; pH 3; particle size 100 mesh].

3.4. Effect of the adsorbent particle size, initial Mn(II) concentration, pH of AMW and activator concentration on the adsorption efficiency

It was found that the removal efficiency to be highly concentration dependent and higher removal efficiency was obtained at low concentrations of Mn(II) (Sharma et al., 2007). Figures 11 to 16 strongly support the previous observed facts. All the figures evidently describe that the higher efficiencies are detected at the lower initial Mn^{2+} ions concentration. The highest efficiency in this study is 85%, it is fantastic compare to the past result up only 74.2% (Sharma et al., 2007). McKay (1982) proposed that higher percentage removal in lower concentration ranges has lot of industrial significance as in most cases the wastewaters have been reported to have lower concentrations of metallic species including Mn(II).

McKay (1982) and McKay and Sweeny (1980) also discusses the effect of adsorbent particle size. Higher external surface area in smaller particle sizes at a constant amount of the adsorbent is the reason for higher removal of Mn(II) at low particle sizes. Increase in removal does not have a 'directly proportional' relationship with particle size. Possible reason for this finding may be the lower free concentration of Mn(II) for smaller adsorbent particles is significantly less than that for the larger ones.

IOP Conf. Series: Materials Science and Engineering **206** (2017) 012095 doi:10.1088/1757-899X/206/1/012095

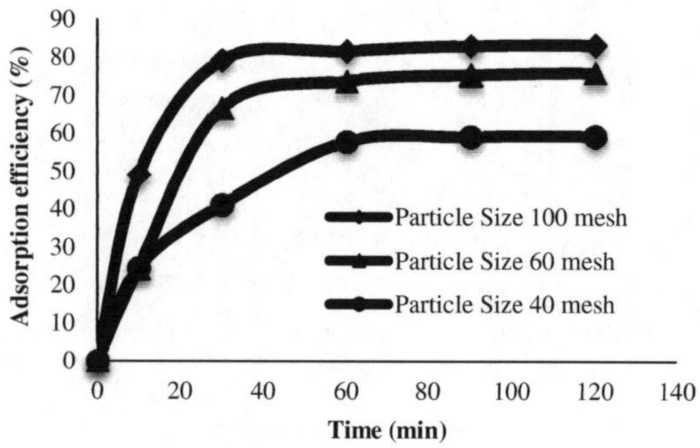

Figure 11. Adsorption efficiency [initial Mn(II) ion concentration 600 mg/l; pH 7; NaOH 1 M].

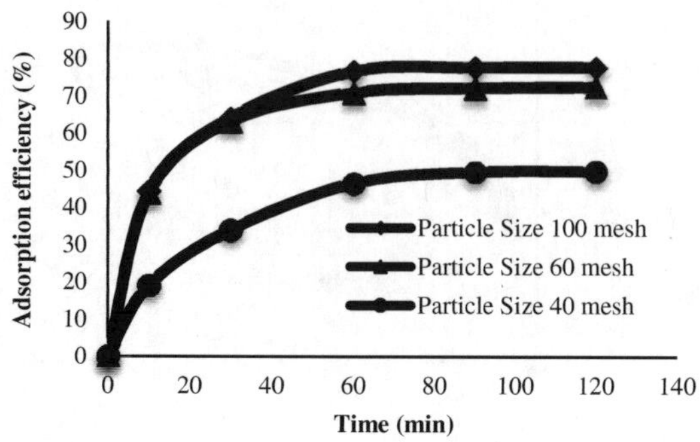

Figure 13. Adsorption efficiency [initial Mn(II) ion concentration 600 mg/l; pH 7; NaOH 0.5 M]

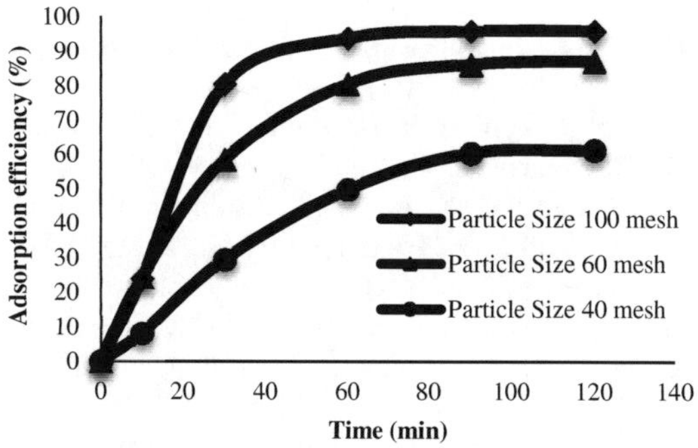

Figure 12. Adsorption efficiency [initial Mn(II) ion concentration 100 mg/l; pH 7; NaOH 1 M].

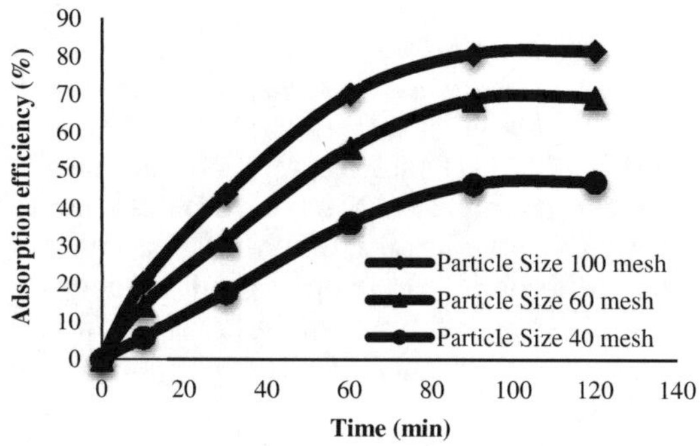

Figure 14. Adsorption efficiency [initial Mn(II) ion concentration 100 mg/l; pH 7; NaOH 0.5 M]

Furthermore, Sharma et al. (2007) and Taffarel and Rubio (2009) explained on pH influence. Sharma et al. eximined effect of pH on removal of Mn(II) by adsorption on fly ash faced that removal was higher in alkaline range of pH. The removal increased from 41.7 to 92.2% by varying the pH of the solution from 3.0 to 8.5. Their experiment was performed at very low Mn(II) concentration of 1.5 mg/l and temperature of 25°C. That data is in good agreement with the present study. Taffarel and Rubio then wrote the medium pH influenced significantly the Mn^{2+} ions adsorption capacity and best results were obtained at pH 6–6.8. In acid medium H_3O^+ are competitive ions, similar to OH ions in alkaline medium. Those investigated data therefore futher strengthen the present observation that the best pH is at 7 (neutral condition).

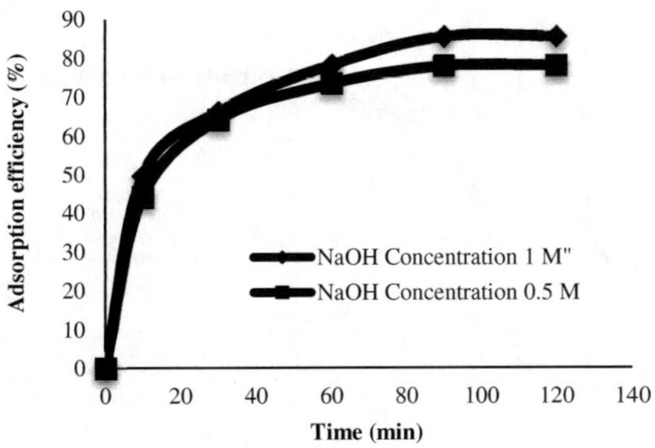

Figure 15. Adsorption efficiency [initial Mn(II) ion concentration 600 mg/l; pH 3; particle size 100 mesh].

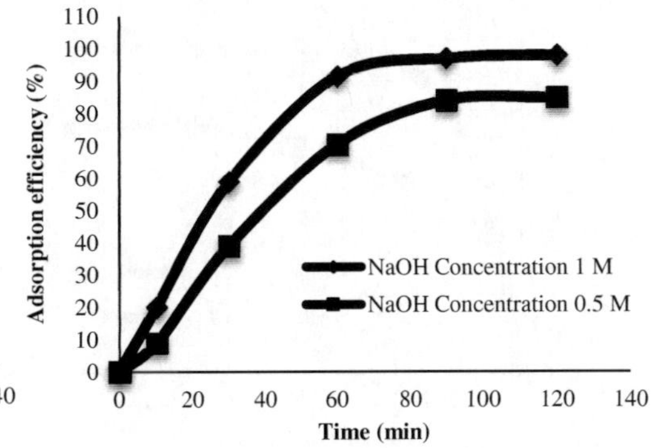

Figure 16. Adsorption efficiency [initial Mn(II) ion concentration 100 mg/l; pH 3; particle size 100 mesh].

3.5. Comparison between pH 3 and 7

The pH of the solution is an important controlling parameter in the adsorption processes (Elliott and Huang, 1981) and metal ion removal usually increases with increasing pH values (Huang and Ostovic, 1978). According to those authors, the pH may affect the ionization degree (species formation) of the adsorbate and the surface property of the adsorbent (Lin and Yang, 2002).

Figures 17 to 20 exhibited the adsorption efficiency as a function of pH. The experiment was conducted at the particle size 100 mesh. Figures 17 and 18 show that the amount of Mn^{2+} ions adsorbed on the adsorbent increases with the increase of the initial pH, but it is not for Figs. 19 and 20. This phenomenon appears to be due to the fact that bottom ash were highly selective for H_3O^+ ions when the H_3O^+ ions concentration was high (Taffarel and Rubio, 2009). At lower pH the H_3O^+ ions compete with Mn^{2+} ions for the exchange sites in adsorbent leading to a low removal of Mn^{2+} ions (Shriver et al., 1990).

The increase in Mn^{2+} ion removal with the increases in pH can be explained on the basis of a decrease in competition between protons and Mn^{2+} cations for the same functional groups and by the decrease in positive surface charge, which results in a lower electrostatic repulsion between the surface and Mn^{2+} ions before ion-exchange. Adsorption results showed an optimum pH value of about 6 (equilibrium pH fluctuated between 6.0 and 6.8) which is also important in the disposal of the treated effluents (required for neutrality of the bodies ''receivers" water or water reuse). The decrease in adsorption at higher pH (above pH 7), is probably due to the formation of insoluble manganese hydroxide $Mn(OH)_2$ (Taffarel and Rubio, 2009).

Unfortunately, the effect of pH in this study cannot be distinguished clearly from all points of view in overall.

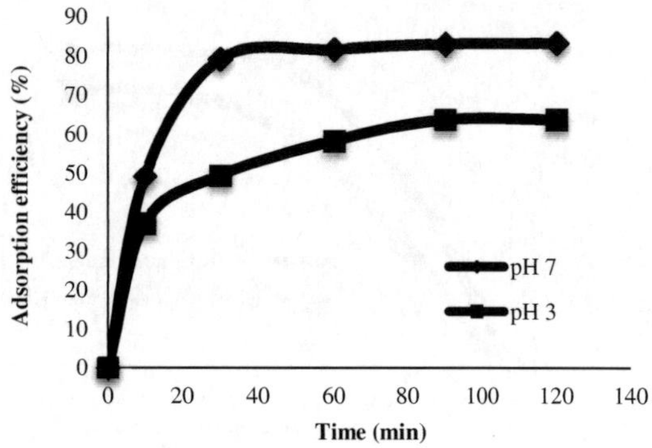

Figure 17. Adsorption efficiency [initial Mn(II) ion concentration 600 mg/l; NaOH 1 M].

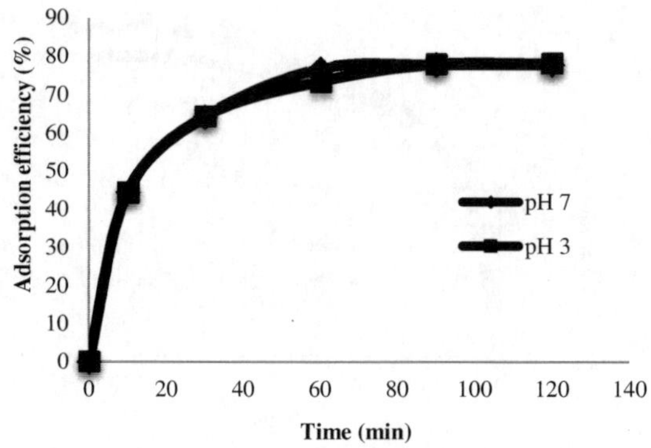

Figure 19. Adsorption efficiency [initial Mn(II) ion concentration 600 mg/l; NaOH 0.5 M].

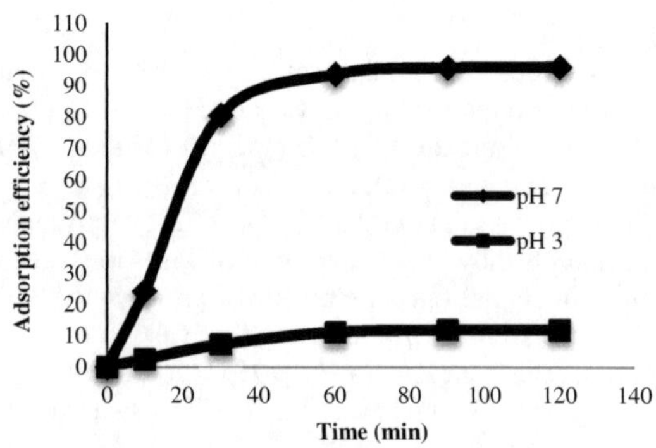

Figure 18. Adsorption efficiency [initial Mn(II) ion concentration 100 mg/l; NaOH 1 M].

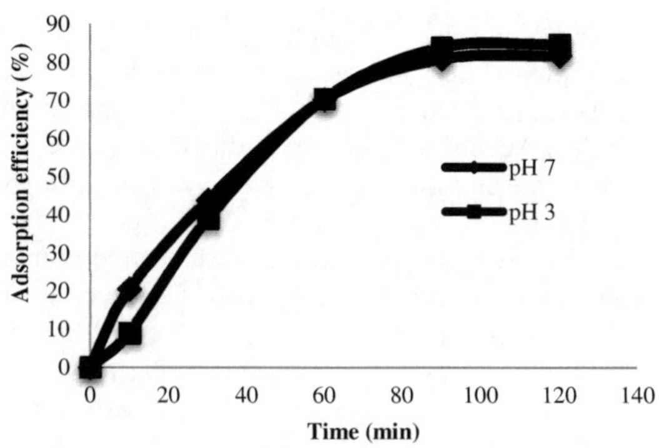

Figure 20. Adsorption efficiency initial Mn(II) ion concentration 100 mg/l; NaOH 0.5 M].

4. Conclusions

It concluded that the coal fired bottom ash can be applied as an adsorbent for the heavy metal removal (in this study was Mn^{2+}). Higher initial Mn^{2+} ions concentration (of 600 mg/l) provide the higher adsorption capacities (up to 60 mg/g or more), however the higher adsorption efficiencies (up to 100%) are found at lower initial Mn^{2+} ions concentration (of 100 mg/l). The best recorded particle size, pH and NaOH concentration were 100 mesh, 7 and 1 M, respectively.

References

[1] Damayanti R., Selinawati T D and Djuarsih 2000 *Pros. Sem. Nas. Kim. VIII* (Univ. Gajah Mada) **2** (Yogyakarta: Indonesia) p 43
[2] Elliott H A and Huang C P 1981 *Water Res.* **15** 849
[3] Faridah A M, Widiastuti N and Prasetyoko D 2012 *Pros. Sem. Nas. Kim. Unesa* (Jurusan Kimia FMIPA Unesa) seri 2012 (Surabaya: Indonesia) p C-198
[4] Gupta V K, Mittal A, Krishnan L and Gajbe V 2004 *Separ. and Purific. Technol.* **40** 87
[5] Gupta V K, Ali I, Saini V K, Gerven T V, Van der Bruggen B and Vandecasteele C 2005 *Ind. Eng. Chem. Res.* **44** 3655
[6] Huang C P and Ostovic F B 1978 *J. Env. Eng. ASCE*, **104** 863
[7] Kusmiyati, Lystanto P A and Pratiwi K 2012 *Reaktor* (in Indonesia) **14** 51
[8] Lin C Y and Yang D H 2002 *Env. Sci.and Health, Part A: Toxic/Hazard. Subst. Env. Eng.* **37**

1509
[9] McKay G 1982 *J. Chem. Technol. Biotechnol*, **32** 759
[10] McKay G and Sweeny A G 1980 *Water Air Soil Pollut.*, **14** 3
[11] Orakwue E O, Asokbunyarat V, Rene E R, Lens P N L and Annachhatre A 2016 *Water Air Soil Pollut.* **227** 1
[12] Patil D S, Chavana S M and Oubagaranadinb J U K 2016 *J. Env. Chem. Eng.* **4** 468
[13] Putra R, Khamidinal and Krisdiyanto D 2015. *Pros. Sem. Nas. Tek. Kim.* (Prodi Tek. Kim. FTI UPN "Veteran") seri *"Kejuangan"* (Yogyakarta: Indonesia) p G3 -1
[14] Sharma, Y C, Uma, Singh S N, Paras and Gode F 2007 *Chem. Eng. J.* **132** 319
[15] Shriver D F, Atkins P W and Langford C H 1990 *Inorg. Chem.* (New York: Freeman)
[16] Sutarno, Arryanto Y and Budhyantoro A 2004 *J. Matematika dan Sains* **9** 285
[17] Taffarel S R and Rubio J 2009 *Minerals Eng.* **22** 336

IOP Conf. Series: Materials Science and Engineering **206** (2017) 012096 doi:10.1088/1757-899X/206/1/012096

Microwave-assisted extraction of lipid from fish waste

M A Rahimi[1], R Omar[1*], S Ethaib[1, 2], M K Siti Mazlina[3], D R Awang Biak[1], and R Nor Aisyah[1]

[1]Department of Chemical & Environmental Engineering, Universiti Putra Malaysia, Serdang, Malaysia

[2]University of Thiqar, Ministry of Higher Education of Iraq, Al-Nasiriya

[3]Department of Food & Process Engineering, Faculty of Engineering, Universiti Putra Malaysia, Serdang, Malaysia

Email: rozitaom@upm.edu.my (R Omar)

Abstract. Processing fish waste for extraction of value added products such as protein, lipid, gelatin, amino acids, collagen and oil has become one of the most intriguing researches due to its valuable properties. In this study the extraction of lipid from sardine fish waste was carried out using microwave-assisted extraction (MAE) and compared with Soxhlets and Hara and Radin methods. A mixture of two organic solvents isopropanol/hexane and distilled water were used for MAE and Hara and Radin methods. Meanwhile, Soxhlet method utilized only hexane as solvent. The results show that the higher yield of lipid 80.5 mg/g was achieved using distilled water in MAE method at 10 min extraction time. Soxhlet extraction method only produced 46.6 mg/g of lipid after 4 hours of extraction time. Lowest yield of lipid was found at 15.8 mg/g using Hara and Radin method. Based on aforementioned results, it can be concluded MAE method is superior compared to the Soxhlet and Hara and Radin methods which make it an attractive route to extract lipid from fish waste.

1. Introduction

Fish is one of protein sources for humans. It has high value of nutrient and rich in not only protein, but also packed with essential fatty acids, minerals and micronutrients. Fish demands in recent years have increased, due to the growing of population worldwide and the demand for fish for functional food. Malaysia is also known as fish eating nation. Therefore, fish demand in years has increased. In year 2008 the fisheries sector produced 1.75 million tonnes of fish with a value of RM 7.5 million in Malaysian Ringgit [1]. Fish also can easily be digested not only as protein, but for its essential fatty acids, mineral and vitamins to Malaysian diet.

During the fish processing, 20-80% amount of fish waste are produced, depending on the type of fish used and usually fish waste are disposed in the ocean. The commercial fish industry is more export-oriented and encompasses of processing of prawns, canning of fish, and the production of surimi and surimi-based products. The export values exceed RM1.7 billion per annum of which frozen shrimps and prawns constitute more than RM 1 billion. In Malaysia, there are 120 commercial and 5,000 small and medium fish processing plants actively operating over the country. Most of these processing plants only utilize only fish flesh and discards the skins, bones, head, scales, fins, offal and viscera. These by-products are good source of essential amino acids and rich in energy, minerals and essential fatty acids, potential to be converted to the functional foods, nutraceutical and pharmaceutical products [2].

IOP Conf. Series: Materials Science and Engineering **206** (2017) 012096 doi:10.1088/1757-899X/206/1/012096

The standard method used to extract lipid in chemical method is solvent extraction. This is the traditional method to extract lipid. Oils are highly soluble in organic solvents such as hexane, ethers, benzene, cyclohexane, acetone and chloroform. Factors to be considered in choosing an organic solvent include preferential solubility of the compound of interest, low boiling point for easy recovery, economics, toxicity, availability, safety and re-usability. Hexane is one of the few solvents with such qualities and used in large scale extraction [3]. There are various forms of lipids in tissue matrix. The simple lipids exist as part of large aggregates in storage and are readily extractable. The complex lipids usually exist as constituent of membranes in close association with protein and polysaccharides. Therefore, the interaction with these compounds is not extracted easily [4].

A Soxhlet extraction method is one of the oldest methods that were originally designed for the extraction of a lipid from solid materials [5]. In recent years, microwave-assisted extraction has been successfully applied for extraction which involves deriving natural compounds from raw plants [6]. Microwave extraction allows organic compounds to be extracted more rapidly, with similar or better yield as compared to conventional extraction methods [7] as a results for unique features of microwave heating such as high heating rate and ease of operation [8]. Lipid investigators [9] have established an alternative method for extraction using low toxicity solvent such as isopropanol/hexane. Therefore this study aims to evaluate the extraction efficiency of liped yield from fish waste by the microwave-assisted extraction method and compared to soxhlet extraction method and Hara and Radin method.

2. Methodology

2.1. *Equipment and apparatus*

The microwave-assisted extraction was carried out in a domestic microwave oven (NN-ST340M, Panasonic, Kadoma, Osaka Prefecture, Japan) at a frequency of 2.45 GHz with maximum output power 800 Watt. Soxhlet apparatus was used for soxhlet extraction. A 250 ml conical flask was used for extraction in Hara and Radin method. All chemicals are of reagent grade.

2.2 Substrate preparation

The fresh sardine fish was purchased from local market and the cut to obtain the head, tail and bones. These parts were chopped to small size using household blender with the adding water to the blender with ratio1:3 (1of water to 3 of wastes). Then, the mixture was filtered to separate solid parts and stored in plastic bag at -18°C until further use.

2.3 Experimental procedure

The overall layout of extraction process of the three extraction methods which include microwave-assisted extraction, Soxhlet, and Hara and Radin methods is described in the flowchart as shown in Figure 1.

IOP Conf. Series: Materials Science and Engineering **206** (2017) 012096 doi:10.1088/1757-899X/206/1/012096

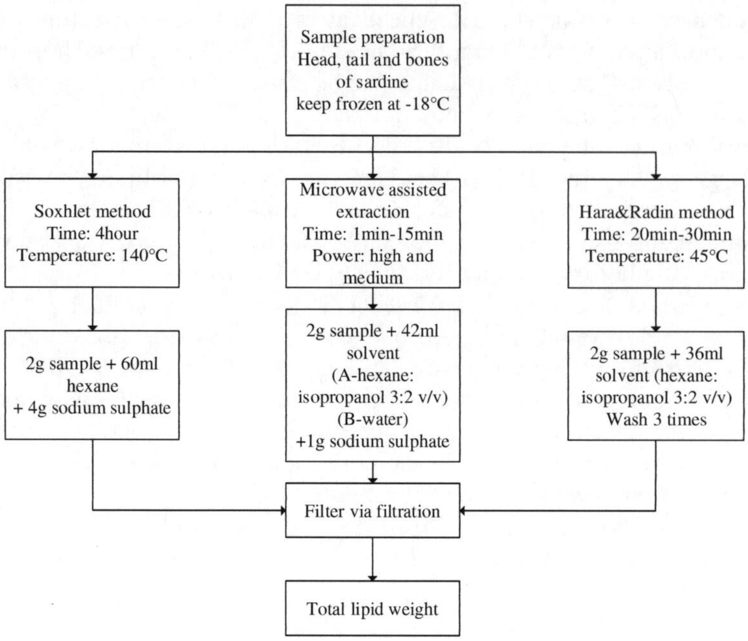

Figure 1. Experimental layout of extraction process

2.3.1 Microwave-assisted extraction method. Microwave-assisted extraction was carried out using distilled water and hexane/isopropanol mixture with ratio 3:2 (v/v) as extraction solvents. 2 g of fish waste was mixed with 1 g sodium sulphate and soaked in 42 ml of the extraction solvent. Then the mixture was subjected to microwave power at 800 Watt. At this power the sample irradiated was for 2, 3, 5, 7 and 10 min when using water as solvent and 1-4 mins when using the organic solvent. After that, mixture was filter using filter paper. 5 ml hexane was added to separate the filtrate mixture. The upper layer was taken and dried. After that, lipid weight was measured.

2.3.2 Soxhlet method. Soxhlet apparatus was used for the extraction of lipid from the fish waste. 2g of fish waste was mixed with 60 ml hexane and 4 g of sodium sulphate was added to the mixture before fixing the Soxhlet apparatus. The temperature of the Soxhlet was set at 140ºC for 4 hour. By the end of the extraction time, the mixture was filtered using filter paper. Then, the filtrate was dried in rotary evaporator at 45ºC and the dried part was weighed to obtain the lipid weight.

2.3.3 Hara and Radin method. In this method distilled water and hexane:isopropanol with ratio 3:2 (v/v) were employed as extraction solvents [9]. 2 g of sample was added to 36 ml solvent and homogenize for 30 seconds. After homogenize, the mixture was filtered into a round bottom flask. The residual on the filter paper was rinsed twice with additional 10 ml hexane which was also filtered into the same round bottom flask. 24 ml aqueous sodium sulfate was added to the filtrate. The aqueous sodium sulphate was prepared by mix 30 ml water and 1 g sodium sulphate. After addition of the aqueous sodium sulphate solution two layers were formed. The lipids were in the upper layer, pipetted out and dried to obtain the lipid weight. Figure 2 shows the layout of the procedure for Hara and Radin method.

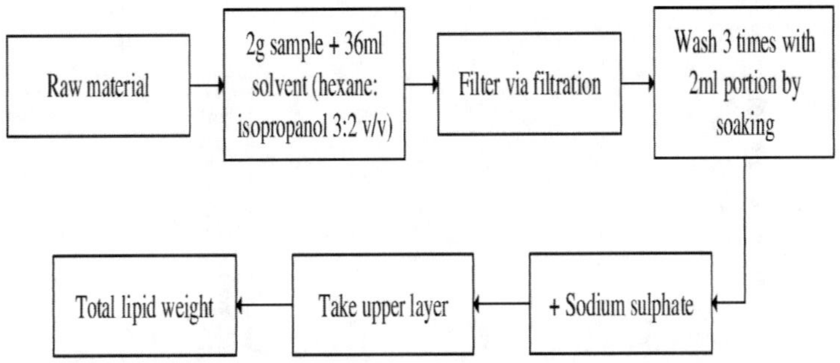

Figure 2. Hara and Radin method procedure

3. Results and Discussion

Figure 3 shows the lipid yield from fish waste when distilled water was used at high microwave power. The results revealed that lipid weight increased from 33.0 mg/g to 80.5 mg/g by increase the treatment time from 2 min to 10 min. The highest yield at 80.5 mg/g resulted from 10 min treatment time can be attributed to the microwave heating which based on the polar characteristic of the treatment solvent [10,11]. Water as solvent has high dielectric constant which can enhance the internal heat generation within the sample that helps to extract out the lipid from fish particle waste. Moreover, extending the treatment time at high power level caused an increase in the heat generation within the sample. The highest yield when using organic solvent as solvent at 4 mins is at 61 mg/ as shown in Figure 4. Increasing treatment time beyond 4 mins only resulted in evaporation of solvent. Increasing the time gave positive result to the lipid yield. In fact, at 4 mins, the yield drastically increased. This could be attributed by the heating effect of the solvent and sample, which could have approach a suitable temperature for extraction. However, the yield is 24% lower than the yield in MAE water extraction albeit at less than half the time. Comparative amount of lipid was extracted in MAE-water as early as 3 mins, possibly due to the fact that water has better dielectric characteristics (dielectric constant = 80 at 20ºC) compared to hexane/isopropanol mixture (dielectric constant=20 at 20ºC) which absorbs the microwave faster and converts to heat better. The higher yield when using water as the solvent is promising as it is a green solvent plus the separation of lipid and water is almost instant.

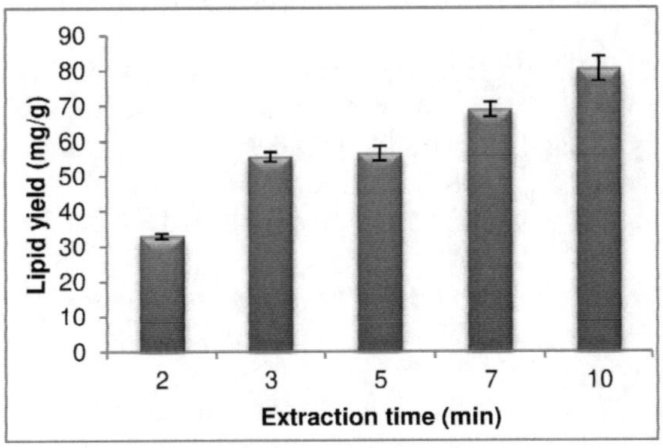

Figure 3. Lipid yield during microwave assisted extraction using distilled water.

For Soxhlet method, in this process the hexane was selected and used as extraction solvent. The results show the lipid yield was 46.6 mg/g after 4 hours extraction time and at 140ºC. The yield is only 58% of yield found during microwave-assisted extraction. In Hara and Radin method, the same solvents with microwave-assisted extraction were employed to compare the microwave effect. Only 15.8 mg/g lipid was recovered when using Hara and Radin method, only 25% of the MAE-organic solvent maximum yield. This shows that higher temperature of solvent is needed to extract lipid out of the cells.

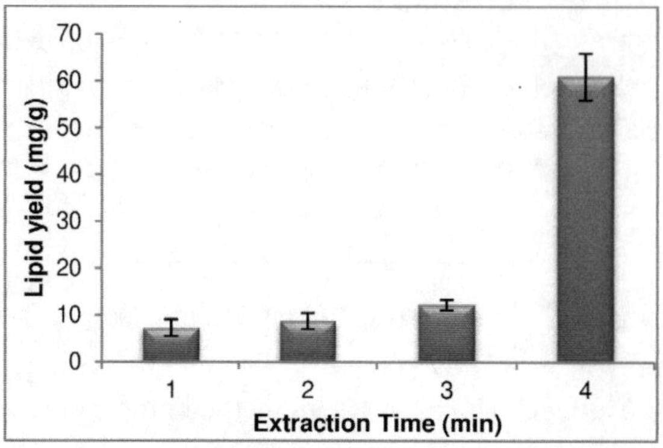

Figure 4. Lipid yield during microwave assisted extraction using hexane: isopropanol

4. Conclusions

The extraction of lipid was done using the three different methods. Among the experimented methods, the microwave- assisted extraction achieved the highest yield of lipid from fish wastes using distilled water as an extraction solvent meanwhile Soxhlet, and Hara and Radin methods yielded 58 % and 20% of the highest yield found using MAE method. These results suggest that the microwave-assisted extraction process can enhance the lipid extraction from fish waste using a green solvent such as water which makes this process an attractive route. In future study it is essential to identify the optimal extraction conditions such as microwave power and extraction time.

References

[1] Tan B P, Ishakm N, Badariah M A 2015 *Food and Fertilizer Technology Center.* http://www.fftc.agnet.org/library.php?func=view&id=20150106111454. Retrieved on 3 April 2015

[2] Tan B P 2014. *Food and Fertilizer Technology Center.* http://www.agnet.org/library.php?func=view&id=20140401154124. Retrieved on 3 April 2015.

[3] Mercer P and Armenta R E 2011 Eur. *J. Lipid Sci. Tech.* **113** (5) 539-547.

[4] Adeoti I A and Hawboldt K 2014 *Biomass & Bioenergy* **63** 330-340.

[5] Harwood L M, Moody C J and Harwood L M 1989 *Experimental organic chemistry: Principles and Practice* (Oxford: Blackwell Scientific).

[6] Ethaib S, Omar R, Mustapa Kamal S M and Awang Biak D R 2015 *J. Eng. Sci. & Technol.* **2** 97-109.

[7] Mandal V, Mohan Y, and Hemalatha S 2007 *Pharmacognosy Rev.* **1** (1) 7-18.

[8] Ethaib S, Omar R, Mustapa Kamal S M, Awang Biak D R and Syafiie 2016, *BioResources* **11** (3) 5687-5702.

[9] Hara A and Radin N S 1978 *Anal. Biochem.* **90** (1) 420-426.

[10] Ethaib S, Omar R, Mustapa Kamal S M, Awang Biak D R, Syafiie, and Harun M Y 2016 *J. Wood Chem. & Technol.* **1** 1-17.

[11] Ethaib S, Omar R, Mustapa Kamal S M, Awang Biak D R, Syafiie, and Harun M Y 2016 *Intl. Food Research J.* **23** 149S-154S.

IOP Conf. Series: Materials Science and Engineering **206** (2017) 012097 doi:10.1088/1757-899X/206/1/012097

Phycocyanin extraction in Spirulina produced using agricultural waste

N Taufiqurrahmi[1], P Religia[2], G Mulyani[1], D Suryana[4], Ichsan[4], F A Tanjung[5], Y Arifin[3],[5]

[1]Chemical Engineering Department, Surya University, Boulevard Gading Serpong, Tangerang 15810, Indonesia

[2]Department of Food Technology, International University Liaison Indonesia, IULI – Eco Campus, The Breeze BSD City, Tangerang Selatan 15345, Indonesia

[3]Department of Food Business Technology, Prasetiya Mulya University, BSD Raya Utama, BSD City, Kabupaten Tangerang 15339, Indonesia

[4] PT Maris Sustainable Indonesia, Jl Tipar Cakung, Pool PPD 5, Cakung, Jakarta Timur 13910, Indonesia

[5]Department of Chemical Engineering, Curtin University Sarawak Campus, CDT 250, Miri 98009, Sarawak, Malaysia

Email: yalun.arifin@prasetiyamulya.ac.id

Abstract. Phycocyanin is a pigment-protein complex synthesized by blue-green microalgae such as Arthrospira (Spirulina) platensis. This pigment is used mainly as natural colouring in food industry. Previous studies have demonstrated the potential health benefits of this natural pigment. The price of phycocyanin is a vital factor that dictates its marketability. The cost of culturing the algae, particularly from the substrate used for growth, is one of the main factors that determine the price of phycocyanin. Another important factor is the growth yield of the algae. In our research, agricultural waste such as charcoal produced from rice husk was utilized for the algae cultivation to replace the synthetic chemicals such as urea and triple superphosphate used the mineral medium. The use of this low cost substrate increases the cell concentration by 60 % during 8 days' cultivation to reach 0.39 g/l. The phycocyanin extraction was performed using water at the different biomass-to-solvent ratio and shaking rates. The phycocyanin concentration and purity (A_{615}/A_{280}) obtained were 1.2 g/l and 0.3. These values are 40 % and 20 % lower than the value obtained from the algae produced using the synthetic chemicals. Further purification produced the extract purity required for food grade. The biomass-solvent ratio does not significantly affect the extract purity; however, the higher shaking rate during extraction reduces the purity. This finding demonstrates the potential of using rice husk as an alternative substrate to cultivate algae for phycocyanin extraction.

1. Introduction

Arthrospira platensis, also known as Spirulina, has received considerable attention as food supplement, feed, and medicinal properties due to its high protein content (up to 70%) and nutritional values [1-3]. This cyanobacterium has a potential for inhibiting cancer in animal cells [4]. Other studies show that Spirulina has demonstrated antiviral effect [5], and stimulated immune system [6].

Spirulina contains a pigment called phycocyanin. It is a protein-pigment complex from the phycobiliprotein (PBP) family characterized by its intense blue color. PBPs build a peripheral accessory light-harvesting complex called phycobilisome (PBS), which is assembled on the surface of the thylacoid membrane. Phycocyanin is a hydrophilic protein that has a fluorescent pigment that divided into three groups: phycocyanin (dark blue), phycoerythrin (red) and allophycocyanin. Its main function is to transfer the excitation energy to the center of reaction where the maximum wavelength of absorption is near to 620 nm. Recently, phycocyanin is used as food coloring that is approved by the Federal Drug Administration (FDA) in the US. This natural pigment has shown antioxidant acitivities both in vivo and in vitro [7]; therefore, it can be considered a nutraceutical compound.

IOP Conf. Series: Materials Science and Engineering **206** (2017) 012097 doi:10.1088/1757-899X/206/1/012097

Phycocyanin harvesting from Spirulina biomass is achieved by extraction. There many factors that affect the result of the extraction, such as the biomass content of phycocyanin, size distribution, solvent, extraction time, temperature, mixing rate, and ratio of biomass to solvent. Previous study shows water is a good solvent for phycocyanin extraction [8].

The cost of phycocyanin production depends on the cost of Spirulina cultivation. The mineral medium used is the largest cost component in the cultivation. Our research is trying to produce phycocyanin from the Spirulina that is cultivated using the medium from agricultural waste. Rice husk biocharcoal (RHB) was selected as one of the main components in the growth medium. Rice husk, which is 20 % of the rice biomass, is one of the major agricultural wastes. Several studies have been done in utilizing RHB, a product of partial combustion of rice husk, for improving terrestrial farming [9-10], but so far no application in cultivating aquatic biomass such as Spirulina.

In our research, the phycocyanin productivity is compared with the result from Spirulina cultivated with a standard mineral medium. The biomass to solvent ratio and shaking speed were varied to obtain the optimum values and to study the effect to the phycocyanin concentration and purity.

2. Materials and methods

2.1. Culture condition

Spirulina was cultivated in a 1-liter medium using 2-liter flask. The standard medium MSI used is developed by Maris Indonesia, which consists of 0.5 g/l $NaHCO_3$, 0.5 g/l NaCl, 0.05 g/l urea, 0.02 g/l triple superphosphate (TSP), and 0.01 g/l cobalamin (vitamin B12).

The rice husk biocharcoal (RHB) medium consists of 0.4 g/l $NaHCO_3$, 0.5 g/l NaCl, 0.01 g/l cobalamin, 0.025 g/l urea and 20 ml/l of RHC extract. The extract was obtained by batch extraction using 8 gram of rice husk charcoal (supplied by Maris Sustainable Indonesia) mixed with 400 ml of water. The extraction was performed at room temperature for 12 hours, followed by a filtration to remove the solid.

Mixing was done by sparging the air using a pump to make sure the cells and nutrients are mixed uniformly throughout the cultivation system. After 8 days' cultivation, the biomasswas harvested using 40-mesh filter paper. The harvested biomass is covered with 10 mm transparent glass, sun-dried for 4 hours and stored at -18°C for cell disruption. Dried biomass was crushed and sieved (160 mesh) to make the size distribution same so the contacts between solvent and biomass will be uniform. The biomass dry weight was measured.

2.2. Phycocyanin Extraction

The extraction method is developed based on the previous publication [8]. Two gram of biomass was suspended in 30, 50, 75, or 100 ml of water in a 250-ml baffled shake flask The flask was placed in incubator shaker (Heidolph Incubator 1000, Schwabach, Germany) that run at 25°C (Figure 1) for 30 hours. Two speeds were used for the shaking: 150 and 200 rpm. Samples were collected and followed by centrifugation at 4°C and 6000 rpm for 10 min. After centrifugation, the supernatant was collected for analysis using UV-Vis spectrophotometer (Biochrom Libra S12, Cambridge, UK) at 652nm, 615nm, and 280nm. Phycocyanin concentration (PC) was defined as:

$$PC=\frac{(OD_{615}-0.474(OD_{652}))}{5.34} \tag{1}$$

where PC is phycocyanin concentration (mg/ml), OD_{615} is optical density of sample at 615nm, and OD_{652} is optical density of sample at 652nm.The purity of phycocyanin was determined by the A_{615}/A_{280} ratio. Absorbance at 615 indicates the phycocyanin concentration and 280 indicates total protein concentration in the solution:

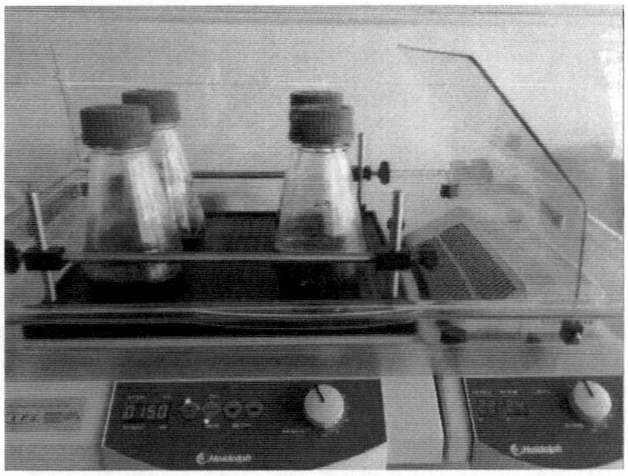

Figure 1. Extraction apparatus

$$EP = \frac{OD_{615}}{OD_{280}} \quad (2)$$

where EP is extract purity, OD_{615} is optical density of sample at 615nm, and OD_{280} is optical density of sample at 280nm.

2.3. Phycocyanin Purification

Phycocyanin was purified using a standard ammonium sulfate precipitation method in fractionation of 0-20%/20-50%. Sample was centrifuged 6000 rpm and 4°C for 10 minutes. The supernatant was poured into a beaker glass under cold condition. Ammonium sulfate was added slowly with stirring at 200 rpm to prevent clumps. After all powder was added, the stirring continued for an hour, followed by centrifugation at 6000 rpm and 4°C for 20 minutes. The pellet was resuspended with 10 mM sodium phosphate buffer pH 7.00.

3. Results and discussion

3.1. Improved biomass production in cultivation using rice husk biocharcoal (RHB) medium

The biomass production of Spirulina in rice husk biocharcoal (RHB) medium is 60 % higher than the result from the mineral medium (Table 1). Rice husk charcoal extract was intended to replace urea and TSP as the source of nitrogen and phosphorus. Rice husk biocharcoal typically contains significant amount of silica, potassium, calcium, and magnesium at 100 - 200 mg/kg and some trace metals [9]. Previous study shows the utilization of RHB increases the growth of water spinach [9] and paddy [10]. The addition of these elements is expected to improve the growth yield and maintain the alkalinity of the medium. Spirulina grows optimally at slightly alkaline pH at 8 – 9. The improvement in the growth yield is beneficial for future biorefinery from this biomass.

Although the biomass concentration is higher, the PC and EP values obtained from the Spirulina grow in RHB are lower than the values obtain when the cyanobacterium grow in mineral medium MSI. In other words, the use of RHB increases the biomass production but not the phycocyanin level. When the biomass to solvent ratio is increase to 2 g: 30 ml, the PC and EP values are similar to the data from MSI medium cultivation. The lower PC and EP values seen in the extract from the cell growing in RHB medium may indicate lower phycocyanin content that normally seen in the cells growing heterotrophically (lower photosynthetic activity) [11].

Table 1. The Spirulina cell biomass concentration in the culture, phycocyanin concentration (PC), and phycocyanin extract purity (EP) from the cultivation using mineral medium (MSI) and rice husk biocharcoal medium (RHB). The data is from the experiment with the shaking speed of 200 rpm after 10-hour extraction.

	Medium (biomass to solvent ratio)		
Parameters	**MSI (2 g: 50 ml)**	**RHB (2 g: 50 ml)**	**RHB (2 g: 30 ml)**
Cell concentration (g/l)	0.24	0.39	0.39
PC (mg/ml)	2.03	1.20	2.25
EP	0.36	0.32	0.31

3.2. Effect of Biomass to Solvent ratio, Extraction Time and Shaking Rate

The phycocyanin concentration (PC) and purity (EP) from the Spirulina grow on RHB medium are shown in Figure 2 to 3. The highest ratio of biomass to solvent (2 g: 30 ml) produced the highest content of phycocyanin extracted as seen in Figure 2. However, the change in the ratio did not affect the extract purity (Figure 3). The same results also observed in a previous research [8] when the mineral Zarrouk medium was used. Still, the higher biomass to solvent ratio has an economic advantage in the downstream processing. The lower amount of solvent used that resulting in a higher PC will reduce the cost for bioseparation.

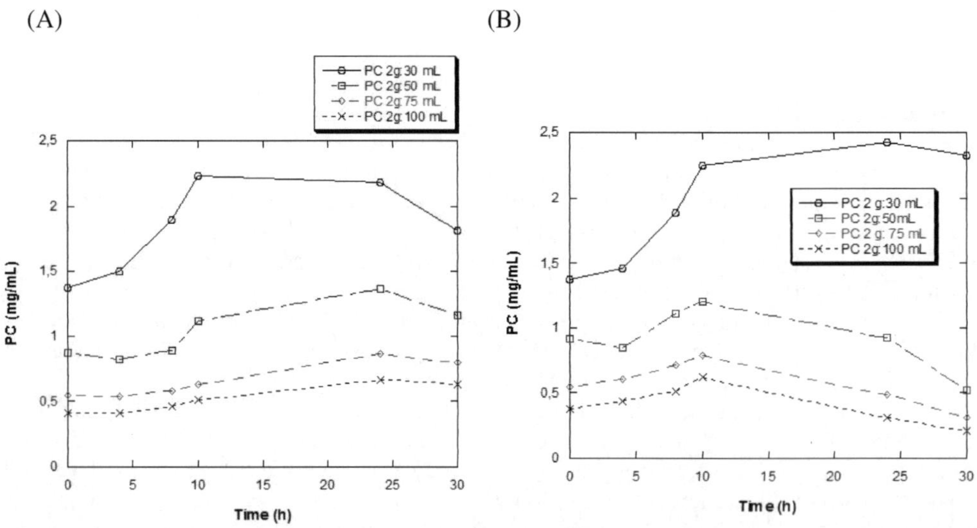

Figure 2. (A) Phycocyanin concentration (PC) from different biomass to solvent ratio on shaking rate 150 rpm (A) and 200 rpm (B)

The PC from all experiments increased for the early hours (<10 hours), but the decreases were generally observed at the final hours of the extraction process. EP values were relatively constant at 0.3 ± 0.06, except when the stirring speed was at 200 rpm. Longer extraction period did not increase PC and EP. Several factors could cause the decrease of PC and EP. The temperature of the suspension was maintained constant; therefore, it is unlikely that the decrease is due to the increase in temperature that may degrade the phycobiliproteins. Previous study shows that phycocyanin is stable at up to 40 °C [12]. We predict that the mechanical impact may be the cause of PC and EP reduction. This needs to be further investigated.

The possibility that the mechanical factor may damage phycocyanin can be further observed when the stirring speed is increased. The significant drops in the PC and EP values were seen at 200 rpm after 10-hour extraction, whereas the values were relatively constant when the extraction was at 150 rpm. Although the higher shaking rate is expected to assist the phycocyanin extraction by improving cell disruption, this strategy is only useful at the early hours of extraction (<10 hours).

Table 2. The phycocyanin concentration (PC), and phycocyanin extract purity (EP) from the cultivation using mineral medium (MSI) and rice husk biocharcoal medium (RHB) after purification using ammonium sulfate. The data is from the experiment with the shaking speed of 200 rpm after 10-hour extraction.

	Medium (biomass to solvent ratio)		
Parameters	**MSI (2 g: 50 ml)**	**RHB (2 g: 50 ml)**	**RHB (2 g: 30 ml)**
PC (mg/ml)	2.42	1.50	2.62
EP	0.71	0.72	0.81

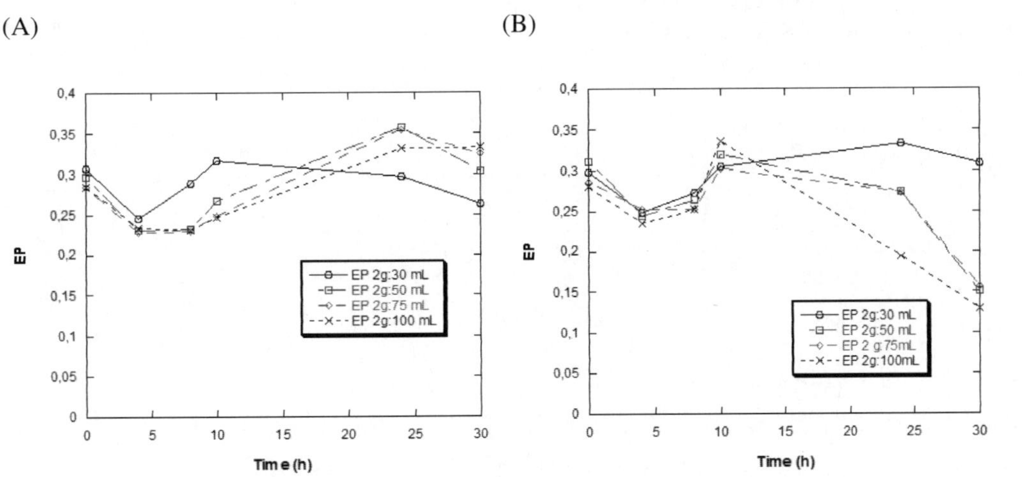

Figure 3. Extract Purity (EP) from different biomass:solvent ratio at shaking rate 150 rpm (A) and 200 rpm (B)

The PC values increase slightly (about 20%) after the purification (Table 2). The EP values increase more than double reaching the value for required for food grade phycocyanin (EP > 0.7) [13]. This result shows that the phycocyanin can be produced from *Spirulina* grow in the medium from agricultural waste such as rice husk.

IOP Conf. Series: Materials Science and Engineering **206** (2017) 012097 doi:10.1088/1757-899X/206/1/012097

4. Conclusion

In this research, we have demonstrated that an agricultural waste such as rice husk can be used to cultivate Spirulina for its phycocyanin extraction. The replacement of urea and triple superphosphate in the mineral medium with rice husk biocharcoal increases the biomass concentration but not the phycocyanin level. The extraction should be performed at high biomass to solvent ratio, high shaking rpm, and at the period of less than ten hours. Longer extraction period will reduce the phycocyanin concentration and purity.

References

[1] Khan Z, Bhadouria P, Bisen P S 2005 *Curr. Pharma. Biotechnol.* **6** 373-379

[2] Simpore J, Kabore F, Zongo F, Dansou D, Bere A, Pignatelli S, Biondi D M, Ruberto G, Musumeci S 2006 *Nutr. J.* **5** 3

[3] Teimouri M, Amirkolaie A K, Yeganeh S *Aquaculture* **396** 14-19

[4] Roy K R, Arunasree K M, Reddy N P, Dheeraj B, Reddy G V, Reddanna, P 2007 *Biotechnol. Appl. Biochem.* **47** 159-167.

[5] Shih S R, Tsai K N, Li Y S, Chueh C C, Chan E C *J. Med. Virol.* **70** 119-125.

[6] Qureshi M A and Ali R A 1996 *Immunopharmacol. Immunotoxicol.* **18** 457-463.

[7] Rojas B F, Juarez J H, Chaverri J P 2014 *J. Funct. Foods.* **11** 375-392.

[8] Silveira S T, Burkert J F M, Costa J A V, Burkert C A V, Kalil S J 2007 *Bioresour. Technol.* **98** 1629-1634.

[9] Milla O V, Rivera E B, Huang W J, Chien C C, Wang Y M 2013 *J. Soil Sci. Plant Nutr.* **13** 251-266

[10] Pratiwi E P A and Shinogi Y 2016 *Paddy. Water. Environ.* 1-12

[11] Marquez F J, Sasaki K, Kakizono T, Nishio N, Nagai S 1993 *J. Ferment. Bioeng.* **76** 408-410

[12] Sarada R, Pillai M G, Ravishankar G A 1999 *Process Biochem.* **34** 795-801.

[13] Silva L A, Kuhn K R, Moraes C C, Burkert C A V, Kalil S J 2009 *J. Braz. Chem. Soc.* **20** 5-12

Kinetics of heavy metal removal in a suspended and immobilized bioreactors

S R M Kutty, E H Ezechi, S G Khaw, C L Lai, M H Isa

Civil and Environmental Engineering Department, Universiti Teknologi PETRONAS, 32610 Bandar Seri Iskandar, Perak, Malaysia.

E-mail: honhenry2k5@yahoo.com

Abstract. The capacity of microorganisms to remove heavy metal from wastewater has been a subject of diverse interest. Whereas some heavy metals are essential for effective microbial activity, some heavy metals could be toxic to the microorganisms at concentrations higher than their minimal inhibitory limit. The kinetics of Zn^{2+} removal from aqueous solution was evaluated in terms of substrate removal rate for two identical suspended and immobilized bioreactors. The suspended growth bioreactor was used as a control system (CS) and contains only biomass. The immobilized bioreactor (IB) contains both biomass and microwave incinerated rice husk ash (MIRHA). The bioreactors were operated at a fixed HRT of 29.1 hours, whereas Zn^{2+} influent concentration was varied in the range of 0.5, 1, 2, 5, 10 and 15 mg/L. At steady state conditions, the results show that Zn^{2+} removal was in the range of 72, 75, 72.5, 68.2, 70.3 and 58.7% for CS, whereas it was in the range of 88, 90, 83, 88.6, 86.2 and 83.7% for IB. The substrate removal rate was found as 1.1856 g/L.d for CS and 4.2693 g/L.d for IB. The results clearly show that Zn^{2+} removal was more favorable in IB, indicating that the performance of the bioreactor was enhanced by the addition of MIRHA.

1. Introduction

Biological wastewater treatment system is an age long treatment technology, associated with numerous benefits such as low capital and operational cost, effective organic matter degradation and biogas production. It has been widely used in various manufacturing applications such as food and beverages, pharmaceuticals, refineries and petrochemical industries as secondary treatment. The success of biological treatment technology has largely dependent on the capability of microorganisms to biodegrade wastewater contaminants into acceptable end products. However, the application of biological treatment technologies for the treatment of wastewater from some industries which utilize heavy metals such as tannery and electroplating are greatly hinder by the inability of microorganisms to biodegrade heavy metal. Whereas microorganisms can easily oxidize organic matter to carbon dioxide and other acceptable end products, it can only alter metal speciation and transform it from one chemical state to another. Heavy metals inhibits microbial activities by blocking the essential functional groups on microorganisms, displace essential metal ions or modify the active form of biological molecules [1]. However, while some heavy metals are toxic to microorganisms even at low concentrations, others are essential for microbial composition and activity.

Microorganisms can resist and tolerate metal toxicity. Depending on the intrinsic and detoxification mechanisms of the microorganisms as well as the environmental factors, some microorganisms can grow in the presence of high metal concentration [2]. The intrinsic mechanisms include the ability of microorganisms to tolerate and cope with metal toxicity whereas detoxification mechanisms is the stimulated microbial direct response to resist metal toxicity [2]. The bacteria resistance to heavy metal toxicity can further arise from non-specific mechanisms such as cellular impermeability or by specific resistance transfer factors [3]. The microbial resistance and tolerance to metals are based on the two

IOP Conf. Series: Materials Science and Engineering **206** (2017) 012098 doi:10.1088/1757-899X/206/1/012098

significant principles of bioaccumulation and biosorption. In both bioaccumulation and biosorption, heavy metal removal is based on the interaction of the metal ions with the functional groups of the microorganisms [2, 4].

The immobilization of microorganisms has demonstrated higher potential for metal accumulation [5]. The advantage of immobilization is that the immobilization material, howbeit non-toxic, can contribute its properties in various forms for metal accumulation. Microorganisms can grow on insoluble inert material either attach to the material surface or remain suspended in the bulk solution. However, it is well known that those microorganisms which attach on the surface of the inert material have faster growth rate than the suspended counterpart as a result of their exposure to available food resources [6].

The knowledge of the substrate removal rate by biological system is necessary in order to evaluate the performance of the bioreactor. It gives information about the state of the microorganisms, level of bioaccumulation and biosorption, the rate at which metal speciation are altered and a basis for process design and scale up. The objective of this study was to compare the performance of a suspended and an immobilized bioreactor on the basis of their substrate removal rate, treating a heavy metal laden aqueous solution.

2. Materials and Methods

The materials and methods used in this study are presented in this section.

2.1. Bioreactor setup

Two rectangular bioreactors of equal size (8.5L) were made using acrylic glass and have a thickness of about 5 mm. An influent port was made towards the bottom of the bioreactors whereas the effluent port was located towards the top of the bioreactor opposite the influent port. A settling unit was created by means of baffle towards the effluent port to prevent wash-out of biomass into the effluent unit. Horizontal air tube diffusers were installed at the bottom of the bioreactors to provide oxygen in upflow pattern. The bioreactors were then started with a seed sludge from a local domestic wastewater plant with a mixed liquor suspended solid (MLSS) concentrations of about 4000 mg/L. The solid retention time (SRT) was fixed at 30 days through sludge wasting.

2.2 Preparation of immobilized material

The immobilized material was prepared from rice husk. Rice husk was purchased from a local retail shop in Malaysia. Rice husk is an abundant agricultural waste materials. It was washed, dried for 2 hours at 105 □C using a conventional oven and carbonized for 2 hours at 800 □C using a Microwave incinerator. Microwave incinerated rice husk ash (MIRHA) was then produced and stored. No additional modification was carried out on MIRHA.

2.3 Wastewater preparation

Synthetic wastewater was used in this study and was prepared by dissolving appropriate amount of an organic substrate (Purina Alpo) in tap water according to the experimental plan. Synthetic wastewater was used in order to provide consistent loading on the bioreactor. A specific volume (0.15 mL) of phosphate buffer was added to the wastewater for pH stability. The wastewater has a C:N:P ratio of about 100:24:3 which meets the basic domestic wastewater nutrient requirement. The synthetic wastewater was characterized and has an average influent COD, BOD_5 and TSS concentrations of 500, 250, and 300 mg/L, respectively. Various Zn^{2+} concentrations was simulated in the wastewater by the addition of appropriate quantity of Zinc chloride. Zn^{2+} concentrations was varied in the range of 0.5, 1, 2, 5, 10 and 15 mg/L, respectively.

2.4 Bioreactor operation

The experiment was conducted at a fixed influent flowrate of 7 L/d, corresponding to a hydraulic retention time (HRT) of 29.1 hours. For the suspended growth bioreactor (CS), the mixed liquor volatile suspended solids (MLVSS) from a local sewage treatment plant was used only. For the immobilized bioreactor (IB), first, the MLVSS from a local sewage treatment plant was used. Then,

IOP Conf. Series: Materials Science and Engineering **206** (2017) 012098 doi:10.1088/1757-899X/206/1/012098

MIRHA was added to the bioreactor (IB) to immobilize the microorganisms. The experiments were conducted in 8 different phases. In phase 1, acclimation of the biomass to the wastewater was carried out in both CS and IB. In phase 2, MIRHA was added to the IB to acclimate with the biomass and initiate the immobilization process and was maintained at 2000 mg/L throughout phase 2 by daily adding MIRHA to IB. The SRT in phases 1 and 2 were controlled through daily recycling and wasting. MIRHA was daily added to IB. In phase 3, MIRHA concentration of 100 mg/L was added daily to IB, taking into consideration the MIRHA wasted daily. Then, 0.5 mg/L of Zn^{2+} was fed to both the CS and IB in phase 3. Sludge wasting was terminated in this phase and only recycling was conducted. In phases 4, 5, 6, 7 and 8, Zn^{2+} concentrations of 1, 2, 5, 10 and 15 mg/L was fed to both reactors, respectively. Zn^{2+} concentration was measured from the effluent port on a daily basis in accordance with Hach colorimetric (Zincon) method.

2.5 Substrate removal rate

The substrate removal rate for both CS and IB was evaluated from the average of the steady state data in each phase using the expression below :

$$\frac{Q(S_o - S_e)}{V} \tag{1}$$

where, Q is the influent flowrate (L/d), S_o and S_e are the influent and effluent Zn^{2+} concentration, V is the reactor volume (L).

A plot of $Q(S_o\text{-}S_e)/V$ vs S_e should give a straight line from which the substrate removal rate (K) is obtained from its slope.

3. Results and discussion

The steady state result for all experimental conditions for CS and IB is summarized and presented in Table 1.

Table 1. Steady state results.

Days	Influent (mg/L)	CS (mg/L)	IB (mg/L)	CS (%)	IB (%)
1-12	Acclimation	-	-		
13-29	MIRHA addition	-	-		
30-47	0.5	0.14	0.06	72	88
48-63	1	0.25	0.10	75	90
64-78	2	0.55	0.34	72.5	83
79-95	5	1.59	0.57	68.2	88.6
96-112	10	2.97	1.38	70.3	86.2
113-125	15	6.2	2.44	58.6	83.7

The obtained results clearly show that IB performance was superior to CS at all experimental phases investigated. After the addition of MIRHA to IB in phase 2 (day 13-29), a Zn^{2+} concentration of 0.5 mg/L was fed to CS and IB in phase 3 (Day 30-47). The steady state result shows that a removal of 72 and 88% was achieved in both CS and IB, respectively. In phase 4 (Day 48-63), the influent Zn^{2}+ concentration was increased to 1 mg/L and the removal efficiency for both CS and IB was 75 and 90%. In phase 5, 6, 7 and 8, the removal efficiency of Zn^{2+} was in the range of 72.5, 68.2, 70.3 and 58.7% for CS and 83, 88.6, 86.2 and 83.7% for IB, respectively. The IB demonstrated high performance in all phases when compared with CS.

The known principle for heavy metal removal by microorganisms include bioaccumulation and biosorption. Bioaccumulation involves several other mechanisms such as adsorption, precipitation, complexation and active intracellular transport. Several other physicochemical factors such as pH and ionic strength also influence the magnitude of bioaccumulation. Accumulation of heavy metals could specifically vary among microorganisms and the potential formation of minerals within the nucleation sites of the microorganism can occur [2]. It is well known that microorganisms are typically small in size and possess a high surface area to volume ratio, which provide a considerable contact area for interaction with metals in activated sludge process. Additionally, some metals can undergo transformation by redox or alkylation processes and form compounds that significantly differ from the original both in mobility and toxicity [2]. It has been reported that the absence of sharp inflections during the titration of bacteria, fungi and yeast suggests that various abundant ligands which have the capacity to bind metals and other elements were present on microorganisms [7]. Additionally, sorption of microorganisms by heavy metals involves several steps depending on the type of microorganisms. The biosorption mechanisms greatly rely on the electrostatic attraction between heavy metals and the reactive groups on the surface of microorganisms which causes these sites to nucleate and accumulate more metals as well as counters ions. The accumulated metals which have nucleated then grow as an aggregate [2]. Thus, the removal of Zn^{2+} by both CS and IB was achieved through bioaccumulation and biosorption microbial mechanisms.

The superior performance of the IB system clearly shows that the addition of MIRHA enhanced the activities of the microorganisms for heavy metal accumulation and biosorption. In a related study, the performance of a bioreactor immobilized with groundwater treatment plant sludge (GWTPS) was compared with a suspended growth bioreactor for Zn^{2+} removal. The authors found that the performance of the GWTPS bioreactor was superior to that of the suspended growth bioreactor [8]. Thus, while it is certain that microorganisms can tolerate heavy metals upto their minimum inhibitory limits, the addition or immobilization of microorganisms with specific materials can further improve heavy metal accumulation

3.1 Substrate removal rate

The substrate removal rate for CS and IB systems are presented in Fig. 1 and Fig. 2, respectively. The results confirm our earlier position that MIRHA enhanced the accumulation of heavy metals in the microorganisms. The substrate removal rate for CS system was found as 1.1856 g/d.L whereas it was 4.2693 g/L.d for the IB system. In a similar study, it was observed that toxic heavy metals effectively accumulated in *Saccharomyces cerevisiae* immobilized in sol-gel matrix. The entrapment of *Saccharomyces cerevisiae* into a sol-gel matrix enhanced the accumulation and overall uptake of some heavy metals investigated [9]. Additionally, *Rhizopus oligosporus* reportedly demonstrated high Cd^{2+} accumulation capacity when immobilized compared to the suspended system (free cells) [10].

The high performance of immobilized systems for heavy metal accumulation has been attributed to various phenomenon. One of such phenomenon is the composition of the growth medium (immobilized material). The composition of the immobilized material could contribute to heavy metal accumulation through the formation of functional groups on their cell surface. For instance, it is reported that cysteine can insert S- and N-Ligands, ammonium N-Ligands, glucose C-Ligands and phosphate p-ligands on *Saccharomyces Cerevisiae*. A higher accumulation of Cu and Pb was observed when the microorganisms were grown in Cystein-rich media. On the other hand, addition of phosphate improved Zn selectivity [11].

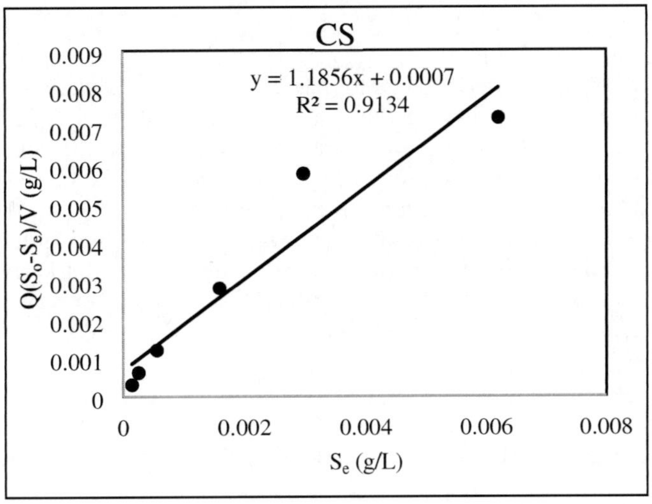

Figure 1. Substrate removal rate for CS.

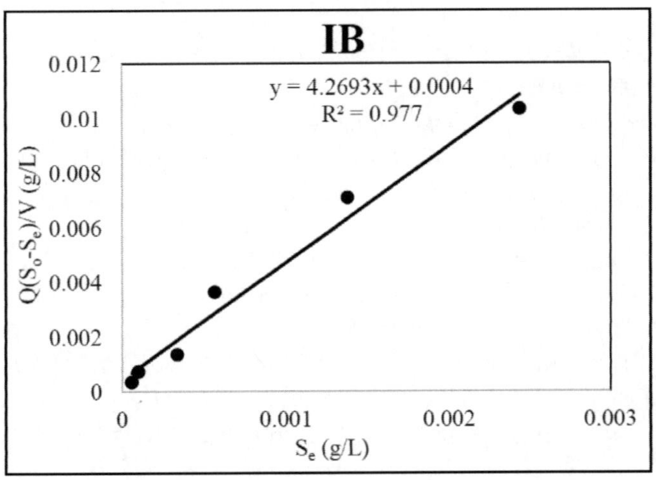

Figure 2. Substrate removal rate for IB.

The basic phenomenon in the enhancement of metal accumulation or uptake by immobilized systems could be through the contribution of the properties of the immobilized material for the resistance of metal toxicity, formation of extra ligands on the surface of the microorganisms that helps in the binding of metal, electrostatic attraction between heavy metals and the reactive groups of the microorganisms, electrostatic attraction between the heavy metals and the reactive groups of the immobilized materials and the nucleation and growth of the formed aggregates on the immobilized material. Additionally, chelation of heavy metals to various components of the media as well as the potential formation of various complexes could reduce the activities (toxicity) of free metals and ultimately hinders the capability of the metals to approach the microbial minimal inhibitory concentration [1]. For instance, it is reported that about 30-40% of mercury was found to form complexes with tryptone and yeast extracts of the components of Luria-Bertani broth (LB) during the growth of *Pseudomonas aeruginosa* [12].

Thus, the substrate removal rate for CS and IB systems were in the ratio of 1:3.6 respectively under the same experimental conditions.

4. Conclusion

The performance of two identical bioreactors operated as suspended growth system (CS) and immobilized bioreactor (IB) were evaluated for the removal of Zn^{2+} at various influent concentrations. MIRHA was used as the immobilizing material. The results show that IB demonstrated superior performance compared to CS. At all the influent Zn^{2+} concentration, IB demonstrated high removal efficiency. Additionally, IB also demonstrated higher substrate removal rate of 4.2693 g/L.d compared to the 1.1856 g/L.d for CS. Thus, the superior performance of the immobilized bioreactor in this study in terms of removal efficiency and substrate removal rate clearly shows that MIRHA, a low cost adsorbent produced from abundant agricultural waste material (Rice husk) can effectively enhanced the microbial tolerance towards heavy metal toxicity. The results also serve as a basis and provide kinetic information for the potential upgrading of the laboratory system into a pilot system for extensive use

Acknowledgment

This study was supported by Universiti Teknologi PETRONAS (UTP). The authors are therefore grateful to UTP.

References

[1] Hassen A., Saidi N., Cherif M., and Boudabous A., 1998 *Bioresource Technol.* **64**, 7-15.
[2] Ledin M. 2000 *Earth Sci. Rev.,* **51**, 1-31.
[3] Sterritt R. and Lester J. 1980 *Sci. Total Environ.* **14**, 5-17.
[4] Lovley D. R and Coates J. D 1997 *Curr. Opin.Biotech,* **8,** 285-289.
[5] Gadd G. M. 2000 *Curr. Opin. Biotech.* **11**, 271-279.
[6] Wang and Li Z.-W. Y. 2014 *Biochem. Eng. J.,* **82**, 134-138.
[7] Chen P. and Ting Y., 1995 *Biotechnol. Lett.* **17**, 107-112.
[8] Kutty S., E. Ezechi, S. Khaw, C. Lai, and M. Isa, 2016 *Desal. Wat. Treat.,* **57**, 28932-28938.
[9] Al-Saraj M., Abdel-Latif M., El-Nahal I., and Baraka R 1999 *J Non-Cryst. Solids.* **248**, 137-140.
[10] Aloysius R., Karim M., and Ariff A., 1999 *World J. Microb. Biot.***15** 571-578.
[11] Engl A. and Kunz , 1995 *J. Chem. Tech. Biot.* **63**, 257-261.
[12] Chang J. S., Hong J., Ogunseitan O. A, and Olson B. H., 1993 *Biotechnol. Progr.* **9**, 526-532.

IOP Conf. Series: Materials Science and Engineering **206** (2017) 012099 doi:10.1088/1757-899X/206/1/012099

Chlorinated phenol removal from aqueous media by tea (*Camellia sinensis*) leaf waste tailored activated carbon

C G Joseph[1*], S M Anisuzzaman[2], W M A W Daud[3], D Krishnaiah[2], K A Ng[1]
[1]Water Research Unit, Faculty of Science & Natural Resources, Universiti Malaysia Sabah, 88400 Kota Kinabalu, Sabah, Malaysia
[2]Chemical Engineering Programme, Faculty of Engineering, Universiti Malaysia Sabah, 88400 Kota Kinabalu, Sabah, Malaysia
[3] Department of Chemical Engineering, Faculty of Engineering, University of Malaya, 50603 Kuala Lumpur, Malaysia.

Email: collin@ums.edu.my

Abstract. In this study, activated carbons (ACs) wereprepared from tea leaves by using a two-stage self-generated atmosphere method. The process was done by semi-carbonizing the precursor at 300 °C for 1 h, followed by the impregnation of the resulting char at 85 °C for 4 h and finally activation at 500 °C for 2 h. The semi-carbonised samples were impregnated with different ratios of zinc chloride ($ZnCl_2$) and their physicochemical effect was studied. The prepared ACs underwent several aspects of both, chemical and physical characterizations, such as the percentage of yield, moisture content, ash content, pH, porosity, adsorption capacity of 2,4-dichlorophenol (2,4-DCP), surface area, porosity, morphology and surface chemistry studies. It was found that sample AC2, with an impregnation ratio of 2:1 was the best AC produced in this study. The maximum Brunauer, Emmett and Teller surface area of AC2 was found to be 695 m^2/g. Langmuir, Freundlich and Temkin isotherm models were used to examine the experimental isotherms while the kinetic data was analyzed using the pseudo-first-order, pseudo-second-order and intraparticle diffusion kinetic models. The 2,4-DCP adsorption isotherm results complied well to the Langmuir isotherm for the equilibrium data while the adsorption kinetic data fitted well to the pseudo-second order model, indicating that chemisorption by valency forces via the sharing (covalent bond) or exchanging of electrons between the AC and the 2,4-DCP molecules were mainly responsible for the adsorption process. From these findings, it is concluded that tea leaves can be used as a low cost precursor for the removal of 2,4-DCP in aqueous medium.

1. Introduction

2,4-Dichlorophenol (2,4-DCP) is one of the constituent of pollutant found in wastewater and sewage. It is used in the manufacture of pesticides, germicides, resins, seed disinfectants, and antiseptics;therefore it can be found in degraded herbicides, saw mills, wood waste incinerators and hazardous waste sites [1]. The main threat of 2,4-DCP is to human health. Of the various water treatments methods have been employed to remove this toxic pollutant using chemical, biological, physicochemical and mechanical methods, adsorption using activated carbon (AC) is one of the most promising solutions due to its inert properties, low cost factor and ease of disposal of the spent carbon [2-7].

The unique characteristics of AC, such as the large surface area, highly developed porosity and high adsorption capacity makes it useful for a wide range of applications. Other than that, AC is suitable for the removal of pollutants because it is resistant to heat, acid, alkali, do not swell or shrink when expose to different pH conditions, and has good mechanical strength [8]. Due to these reasons, AC is commonly regarded as an excellent adsorbent and is more commonly used in environmental applications for removing pollutant.

The preparation of AC is considered expensive as the precursor undergoes physical and chemical treatment but produces a low yield. It requires high energy consumption during production and subsequent thermal treatments for its regeneration while resulting in mass loss during the regeneration process (Peláez-Cid and Teutli-León 2012) [9]. Not only that, most of the commercial AC are coal based or petroleum pitch based which is a non-renewable resource and costly precursor. This limits the production and usage of AC. Other than that, it is reported that the cost of AC is expected to rise as the annual consumption of AC is growing at 9% per annum [10].

Numerous studies were done revolving around the preparation of AC from precursors which are low-cost, renewable organic by-products. Wood and coconut shells have been used to replace coal and charcoal. They have been the major precursors for AC with annual world production of 300,000 MT. The selection of precursor depends on availability, cost and purity. Precursors have to be materials with high carbon content and low ash content for a more promising yield during production (Hernández-Montoya & Bonilla-Petriciolet, 2012). Due to this factor, researchers have studied many agricultural by-products such as pistachio shell, cotton stalks, acorn shell, bagasse and rice husk, corn cob, safflower seed, press cake, elaeagnusangustifolia seeds, barley husk, nutshells, wood sawdust, jackfruit peel, orange peel,etc. for the preparation of AC [10, 12-23].

Tea leave waste is easily available all over the world and it has a great potentialto be used as a precursor to prepare AC [24 & 25]. Tea is the second largest consumed beverage after water with a worldwide per capita consumption of approximately 0.12 L/year [26]. In 2011, Food and Agriculture Organization of the United Nations ranked Malaysia as the 18th largest tea producer in Asia. Tea leaves are lignocellulosic biomass, as the cell wall of tea consists of cellulose and hemicelluloses, lignin, condensed tannins and structural proteins [27]. Tea leaves have low ash content varying from 3-8% of the tea leaf's total weight making it suitable for preparation of AC [26].

Therefore, in this study, tea leaves (Camellia sinensis) were used as the precursor for an alternative sorbent and experiments were conducted to determine its effect on the removal of 2,4-DCP from aqueous solution with the intention of replacing AC for water treatments commercially. The AC was produced using a two-stage self-generated atmosphere method. Zinc chloride ($ZnCl_2$) was used as the dehydrating agent with a fixed ratio of activating agent to precursor used. In the previous method, the precursor would be impregnated with the dehydrating agent prior to carbonisation and activation. However, in this novel method, the precursor was charred before the impregnating process and then activated in a furnace. This gave some very interesting findings which are reported here.

2. Experimental

2.1. Preparation of AC

In this study, Tea leaves (1-2 mm) from Camellia sinensis species were used as a precursor to produce the activated carbon using the two-staged self-generated atmosphere method. Firstly, the tea leaves were soaked and washed with hot water several times to remove the soluble and coloured components. The washed samples were dried in an oven at 110 °C for 24 h. The dried samples were semi-carbonized at 300 °C for 1 h under self-generated atmosphere in a muffle furnace and then left to cool down to room temperature inside the muffle furnace. The semi-carbonized samples were agitated with 200 mL of aqueous solution containing varying $ZnCl_2$ to tea leaf (dry weight) ratio. The impregnation ratios were altered basedon weight; weight of $ZnCl_2$: y from 1:1 to 5:1 producing samples AC1-AC5. The chemical activating agent and semi-carbonized samples were homogeneously mixed at 85 °C for 4 h. After mixing, the mixture was dried in an oven at 110 °C for 24 h. The resulted samples were then activated in a muffle furnace under self-generated atmosphere at 500 °C for 2 h, and allowed tocool to room temperature before removal for further analysis [28].

2.2 Washing Process

Traditional water-washing cannot effectively remove the chemical residues which is deep within the AC as this in turn may reduce the total pore volume. As such, acid-washing procedure was employed to wash away the chemical residue. The activated sample was soaked with a 0.01 M hydrochloric acid (HCl) solution, at 85 °C for 30 min, to eliminate excess $ZnCl_2$. The liquid was filtered away using filter

IOP Conf. Series: Materials Science and Engineering **206** (2017) 012099 doi:10.1088/1757-899X/206/1/012099

paper. The sample was then washed with hot distilled water several times to remove the residual acid. Finally the sample was washed with distilled water. The washed samples were dried in an oven at 110 °C for 24 h to get the final product [29].

2.3 Characterization of AC

Moisture content was determined according to the ASTM D2867–04 standard whereas ash content was determined using the ASTM D2866 – 94 standard. Details for determination of the percentage of yield, moisture content, ash content and pH were described in our previous work (Anisuzzaman et al, 2015a). Surface chemistry of the prepared carbon was analyzed using Fourier Transform Infrared (FTIR) spectroscope (Thermo Nicolet NEXUS 670) analysis to determine the surface functional groups, where the spectra were recorded in the range of 4000-650 cm^{-1}. Scanning Electron Microscope (SEM) (JEOL JSM-5610LV, Japan) at 20 kV was used to observe morphological surface structure of the ACs. The specific surface area and the pore-size distribution were determined using the Brunauer, Emmet and Teller (BET) and Barret, Joyner and Halenda (BJH) methods, respectively. The BET surface area and pore size distribution were determined from nitrogen isotherm at 77.3 K using Quanta chrome autosorb automated gas sorption instrument [2].

2.4 Adsorption capacity of 2,4-DCP

0.5g of AC sample was added into 5 different 1000 mL Erlenmeyer flask with each flask filled with 400 mL of 40 mg/L 2,4-DCP and agitated for 3h at room temperature. The solution was extracted using a syringe and then filtered using a Sterlitech 45 µm syringe filter to make the solution adsorbent free. The solution concentration was determined by using UV-Vis spectroscopy at a wavelength of 285 nm to determine 2,4-DCP uptake. The sample with best adsorption capacity was used for subsequent batch adsorption tests. Adsorption capacity was displayed as adsorption percentage, indicating the amount of 2,4-DCP adsorbed and it was calculated using the formula as below:

$$\% = \frac{c_i - c_t}{c_i} \times 100 \tag{1}$$

where c_i is the Initial concentration of adsorbate solution, mg L^{-1} and c_t is the concentration of adsorbate solution at time *t*, mg L^{-1}.

Batch adsorption tests were carried out with differing adsorbent dosage (0.1, 0.3 and 0.5 g), pH (3, 7 and 9) and initial 2,4-DCP concentration (10, 20, 30 and 40 mg/L).

2.5 Adsorption isotherms

The resulting data was analysed using Langmuir, Freundlich and Temkin isotherm models. Linearized form of the model equations are shown as below:

$$\frac{c_e}{q_e} = \frac{1}{q_m} c_e + \frac{1}{q_m K_L} \tag{2}$$

$$\log q_e = \log K_F + \frac{1}{n} \log C_e \tag{3}$$

$$q_e = \frac{RT}{b} \ln K_t + \frac{RT}{b} \ln c_e \tag{4}$$

where, q_m is the Langmuir constant (mg g^{-1}); K_L is the Langmuir rate constant (L mg^{-1}); K_F is the Freundlich rate constant; K_t is the Temkin rate constant or adsorption potential (L mg^{-1}); *b* is the heat of adsorption (kJ mol^{-1}); *R* is the gas constant; *T* is the absolute temperature (K). A plot of c_e/q_e against c_e; log q_e against log c_e and q_e against lnc_e was done for Langmuir isotherm; Freundlich isotherm and for Temkin isotherm respectively.

2.6 Adsorption kinetics
For adsorption kinetics, pseudo-first-order, pseudo-second-order and intraparticle diffusion model were used to analyse the experimental data. Linearized form of the model equations are shown as below:

$$\log(q_e - q_t) = \log q_e - \frac{k_1}{2.303} t \quad (5)$$

$$\frac{t}{q_t} = \frac{1}{q_e} t + \frac{1}{k_2 q_e^2} \quad (6)$$

$$q_t = k_{id} t^{0.5} + C \quad (7)$$

where, q_e and q_t is the amount of 2,4-DCP adsorbed by AC at equilibrium and time t respectively (mg g^{-1}); t is time (min); k1 is the rate constant for pseudo-first-order (min^{-1}); k2 is the rate constant for pseudo-second-order (g mg^{-1} min^{-1}); k_{id} is the rate constant for intraparticle diffusion (mg g^{-1} $min^{0.5}$); C is the plot intercept. A plot of log ($q_e - q_t$) against t; t/q_t against t and q_t against $t^{0.5}$ was done for pseudo first-order model, pseudo second-order model and intraparticle diffusion model respectively.

3. Results and discussion

3.1 Physical Characterization
The yield percentage, moisture and ash content along with pH of AC1-AC5 are shown in Table 1. It can be seen that the total yield percentage increased in proportional to the increased $ZnCl_2$ impregnation ratio with the exception of AC5, which reduced to 50.015%. The moisture content of ACs also showed a gradual increment when $ZnCl_2$ impregnation ratio increases. Ash content of ACs range from 13.989% to 28.398% with AC2 having the lowest ash content whereas AC5 having the highest ash amount. Ash content does not show any trend with the increasing $ZnCl_2$ impregnation ratio. pH of ACs did not show drastic changes as it stayed between 5.80 and 5.95. When compared to the work done by Gundogdu *et al.* [24] the moisture content showed a better result while the yield and the ash content of this work was less favourable.

Table 1. ACs with their relative yield percentage, moisture, ash and pH

Sample	Yield, (%)	Moisture content (%)	Ash content (%)	pH
AC1	22.568	3.629	28.226	5.85
AC2	30.271	4.537	13.989	5.95
AC3	43.832	4.742	17.938	5.87
AC4	64.327	4.790	15.968	5.83
AC5	50.015	5.477	28.398	5.80

3.2 Surface Functional Group of AC
FTIR spectras of the tea leaves and ACs are shown in Fig. 1 and Fig 2. The band at 3400, 2923, 2851, 2366 and 1628.18 cm^{-1} were contributed by OH stretching vibration in hydroxyl group, asymmetric C-H, symmetric C-H, saturated C-H stretching and aromatic ring vibration (vibration mode of olefinic C=C bonds) respectively. The bands around 1383.44 cm^{-1} and 1436.84 cm^{-1} consists carboxylic-carbonate groups, C=O and nitrate groups, N=O. The band present around 1156-1103 cm^{-1} indicates C-C, C-O-H and C-O-C stretch in primary and secondary alcohol. The bands at 876.14 and 609.15 cm^{-1} were contributed by C-C stretching and -C-C- group respectively. The band at 2923, 2851 and 876.14

cm^{-1} disappeared in the FTIR spectra of ACs but is present in FTIR spectra of the tea leaves. Comparison of tea leaves and AC samples showed that the intensities were stronger in tea leaves and was weakened after the precursor was carbonized and activated. The combined FTIR of AC1-5 is shown in Fig. 2. As shown in Fig. 2 that intensities decrease as impregnation ratios increased.

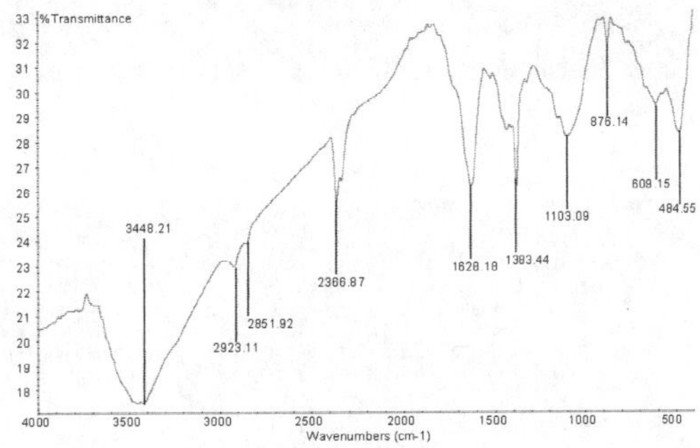

Figure 1. FTIR of tea leaves

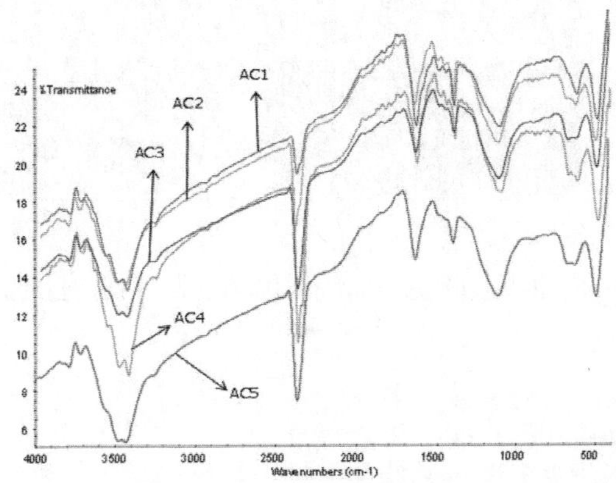

Figure 2. FTIR of combined AC1-5

3.3 Morphological Study of AC

SEM was used to compare the morphology of tea leaves and AC. It was also used to compare the morphology of AC of different impregnation ratio. The micrographs are shown in Fig. 3 (a-f). Before activation, it can be seen that there are plenty fibres on the surface of tea leaves and pores cannot be found on the surface. Other than that, it can also be observed that with each different impregnation ratio, the surface morphology also changes as there are no similarities among the five activated carbon samples. Among the five different AC samples, porosity in AC1 and AC3 were not displayed well, however pore channels still can be observed. Photos of AC2 and AC4 are able show large pores clearly. Micrograph of AC2 with magnification of shows the pores are attached to the pore channels. The pore structures in AC4 closely resemble those of sponge-like structures and the pores have an opening larger than 10μm. In the micrograph of AC5, pores can be seen however the surface looks damaged. In overall, the ACs show some form of irregularities which is probably caused by impurities or the $ZnCl_2$ residue.

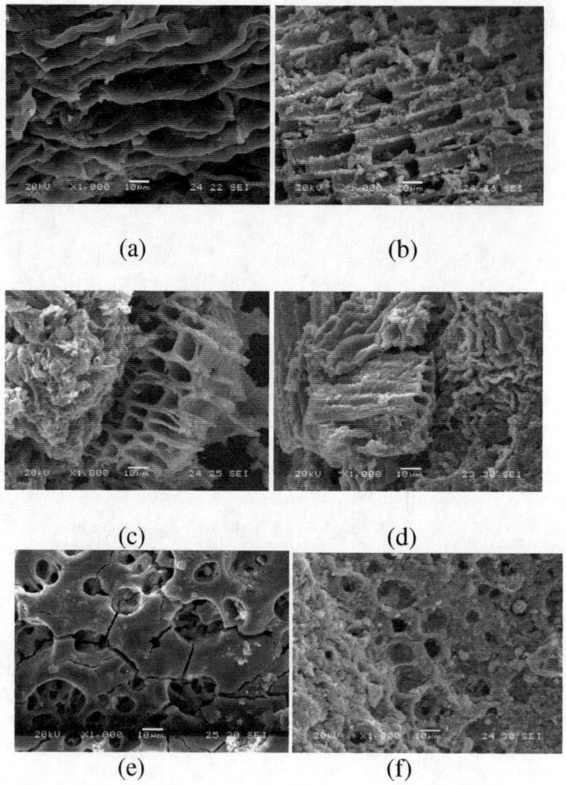

Figure 3. SEM micrographs at magnification of x1000 for (a) tea leaves, (b) AC1, (c) AC2, (d) AC3, (e) AC4 and (f) AC5

3.4 Specific Surface Area and Pore-distribution

Table 2 contains the porous and surface characteristics data of tea leaves. Fig. 4 shows the isotherms from N_2-sorption measurements of the Tea leaves which contain specific information on the porosity of the particles at the temperature of liquid nitrogen. Type I isotherm with no hysteresis loop was observed for the selected adsorbent, AC2. This type of isotherm pattern is identified as a single-molecule adsorption process. The characteristic of this isotherm is demonstrated by a continuous increase in the adsorption volume until the relative pressure reaches and exceeds a certain value. This isotherm corroborate with the calculated Langmuir isotherm model which indicates monolayer coverage with chemisorption properties due to the compliance with the pseudo-second-order reaction kinetics. This type of corroboration is typical for adsorption in microporous solids (Anisuzzaman et al., 2015a).

Table 2. ACs with their relative yield percentage, moisture, ash and pH

Surface area	
Langmuir surface area, m^2/g	$6.95x10^2$
BJH method cumulative adsorption surface area, m^2/g	$6.13x10^2$
BJH method cumulative desorption surface area, m^2/g	$6.52x10^2$
Pore volume	
BJH method cumulative adsorption pore volume, cc/g	$0.22x10^{-2}$
BJH method cumulative desorption pore volume, cc/g	$0.23x10^{-2}$
Pore size	
Average pore diameter, Å	25.70
BJH method adsorption pore diameter (mode), Å	8.82
BJH method desorption pore diameter (mode), Å	9.09

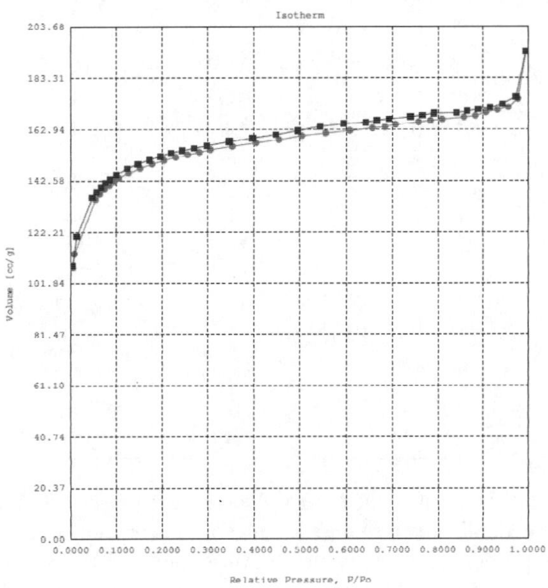

Figure 4. Nitrogen adsorption-desorption isotherm for tea leaves

3.5 Adsorption capacity of the ACs

The adsorption capacity data are presented in Table 3 and it shows that AC2 has the highest adsorption capacity among the different ratio impregnated ACs. AC2 was capable of removing up to 95 % while AC5 gave the lowest adsorption capacity, with only 75 % of the target pollutant removed from the solution. This can be explained by comparing the ash content of AC1 and AC5. AC2 shows a high removal percentage of 2,4-DCP due to the lowest ash content and a fairly low moisture content. Since AC2 gave the best adsorption capacity, it was used to carry out the batch adsorption tests, adsorption

IOP Conf. Series: Materials Science and Engineering **206** (2017) 012099 doi:10.1088/1757-899X/206/1/012099

kinetics and adsorption isotherm analysis. For batch adsorption studies, the adsorption dosage is related to the removal efficiency. As such, it can be seen that using 0.5 g of AC2 effectively removed up to 95% of 2,4-DCP from the solution as compared to 43.5 % and 67% of 0.1 and 0.3g of AC2, respectively. Therefore to effectively remove the pollutant from an aqueous solution, a sufficient amount of AC is required to ensure the optimum adsorption removal. The pH of the solution is one of the important parameters affecting adsorption efficiency. Removal of 2,4-DCP was optimal under acidic condition. At pH 3, the adsorption percentage reached 95.25 % however under neutral condition, the percentage reaches 94.25 % whereas under basic condition, the percentage only reaches 93.50 %. The adsorption percentage only reduced when the solution pH shifted away from acidic to basic medium. It can explained that at high pH, 2,4-DCP anions will be formed, which will repel the negatively charged activated carbon thus reducing the adsorption efficiency. This concludes that in order to remove 2,4-DCP from a solution using AC, the solution pH should be adjusted to acidic condition to allow for optimal removal.

Table 3. Adsorption test result for (a) adsorption capacity, (b) AC dosage, (c) pH and (d) initial 2,4-DCP concentration

Parameters		Adsorption percentage, %
Adsorption capacity	AC1	83.00
	AC2	95.00
	AC3	92.75
	AC4	90.00
	AC5	75.00
Adsorbent dosage, g	0.1	43.50
	0.3	67.00
	0.5	95.00
pH	3	95.25
	7	94.25
	9	93.50
Initial adsorbate concentration, mg L^{-1}	10	98.00
	20	97.50
	30	96.33
	40	95.00

Altering the initial concentration of 2,4-DCP gave varying results. When the solution had low initial concentration, the available surface and pores for adsorption was abundant but at higher initial concentration, these properties became the limiting factor and as a result, the removal percentage decreases. At 10 mg/L, 0.5 g of AC2 can remove the pollutant up to 98 % whereas in 40 mg/L, only 95% was removed. Although the differences are small, it can be concluded that the initial pollutant concentration in the solution has to be evaluated prior to determining the suitable amount of AC to be used in the treatment process.

3.6 Adsorption Isotherms of the ACs

The plot for Langmuir, Freundlich and Temkin are shown in Fig. 5 whereas the data is shown in Table 4. $R_L$for Langmuir isotherm was found to be 0.023; which was in between 0 and 1, indicating that the adsorption was favourable for monolayer adsorption process. The maximum adsorption capacity, q_m was calculated from the plot was capable of adsorbing 44.25 mg of 2,4-DCP per gram of AC2. The value of $1/n$ obtained from Freundlich isotherm is less than 1, indicating that the adsorption was a normal adsorption process while n = 1.7058, indicates that it was a favourable adsorption (multi-layer adsorption process). K_F shows a fairly large value which shows that there was an adsorption intensity during the adsorption process. The heat of adsorption, b obtained from Temkin isotherm equalled to 0.2528 kJ mol^{-1} indicating a physical adsorption process. Compliance with all three isotherms indicated that the adsorbate (2,4-DCP) can bind to more than one surface on the adsorbent. When compared among the three adsorption isotherm models, Freundlich had a lower R^2 of 0.9852 whereas Temkin had a higher R^2 value of 0.9939 but Langmuir has the highest R^2 of 0.997. Although both Freundlich and Temkin have high R^2 value, Langmuir isotherm showed the best fitting. This indicates that the adsorption of 2,4-DCP taking place with the AC was mainly monolayer.

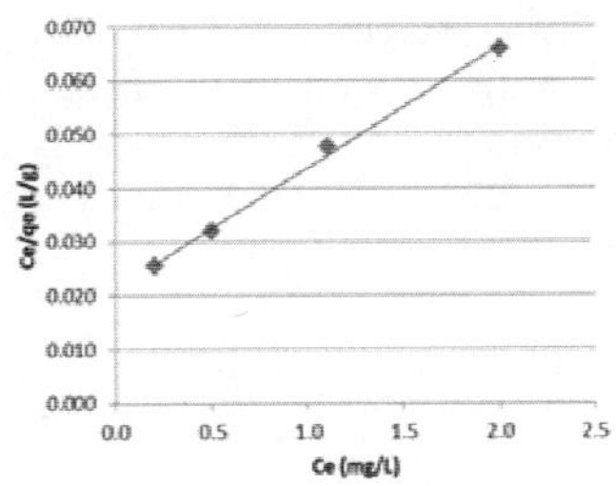

(a)

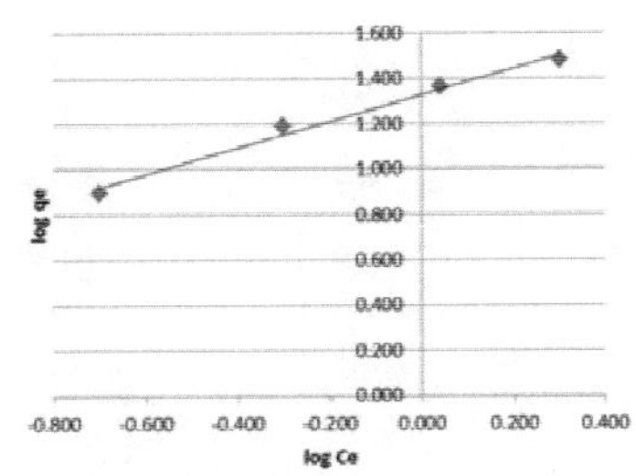

(b)

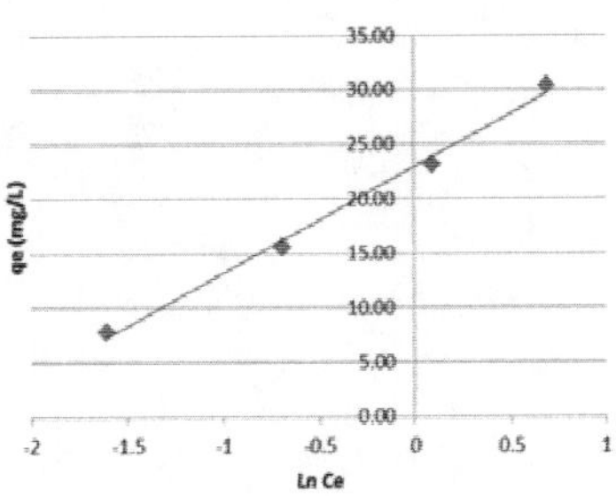

(c)

Figure 5. (a) Langmuir plot, (b) Freundlich plot (c) Temkin plot

Figure 6 (a) shows the uptake rate of 2,4-DCP for 3 h of period. It can be observed that at 90 min onward when adsorption percentage reached 90%, the uptake process was slowed down and stayed around 95 %. The calculated q_e and rate constant for both the pseudo-first-order and pseudo-second-order was calculated from the intercept and slope of the plot.

The plot for pseudo-first-order, pseudo-second-order and intraparticle diffusion is shown in Fig. 6 (b), (c) and (d) respectively whereas the data is shown in Table 5. The q_e from experiment and the q_e

calculated from the pseudo-first-order showed a large difference. The experimental q_e is more similar to the one calculated from pseudo-second-order model. Not only that, the coefficient of correlation shows that the data does not fit well to pseudo-first order model with R^2 of only 0.9681 when compared to pseudo-second-order with a R^2 of 0.9998 which is closer to 1.

For the intraparticle diffusion plot, two slopes were observed. The slope does not pass through the origin, indicating that the adsorption was not controlled solely by intraparticle diffusion. The first slope suggest rapid uptake of the target pollutant by the boundary layer or adsorption on the solid surface. The less steep, second slope suggests adsorption occurs by slow diffusion inside the particles. This is the step where intra-particle diffusion within the pores is the rate-limiting step. Using the second slope as the comparison, the correlation of coefficient, R^2 was 0.9227. This value is lower compared to the R^2 value for the pseudo-second-order. This indicates that the adsorption of 2,4-DCP did not follow the pseudo-first-order reaction kinetic but fitted well to the pseudo-second-order model more. The rate determining step for the adsorption process was most probably a chemisorption process, which was restricted to monolayer adsorption on the surface. In chemisorption, 2,4-DCP reacted with the AC surface to form a chemical bond with the carbon matrix.

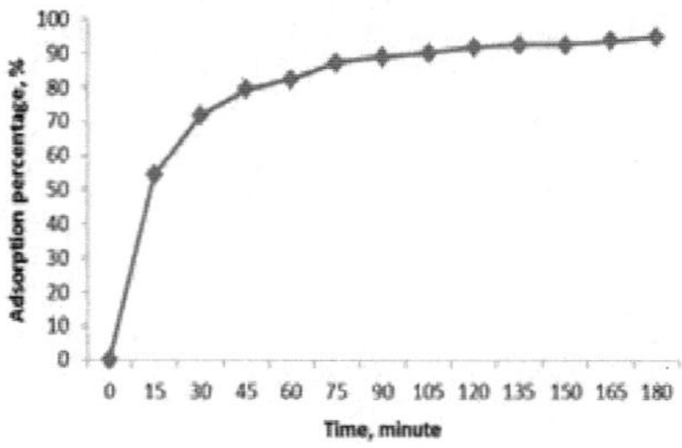

(a)

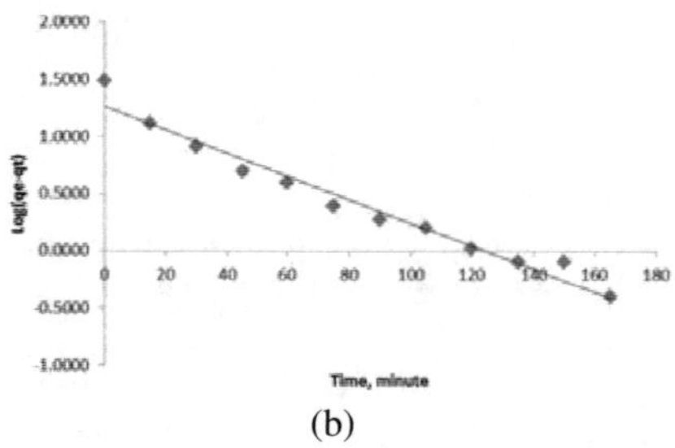

(b)

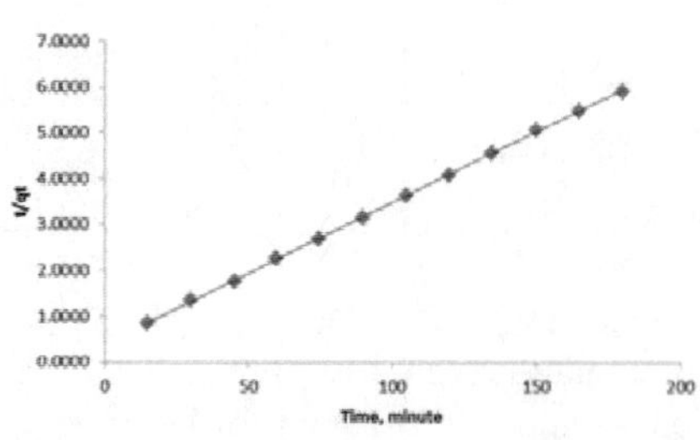

(c)

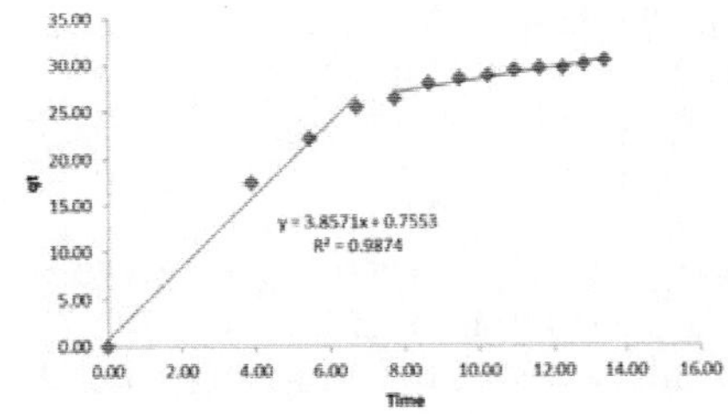

(d)

Figure 6. 2,4-DCP adsorption, (b) pseudo-first-order plot, (c) pseudo-second-order plot and (d) intraparticle diffusion plot

Table 4. Adsorption isotherm data using AC2

Sample	Langmuir model			Freundlich model			Temkin model		
	K_L	q_m	R^2	K_F	n	R^2	K_t	b	R^2
AC2	1.0610	44.25	0.9970	21.3698	1.7058	0.9852	10.3588	0.2528	0.9939

Table 5. Adsorption kinetics data

k_1 (min^{-1})	$q_{e,exp}$ mg g^{-1}	$q_{e,cal}$ mg g^{-1}	R^2	k_2, g mg^{-1} min^{-1}	$q_{e,cal}$ mg g^{-1}	R^2	k_{id} mg g^{-1} $min^{-0.5}$	C mg g^{-1}	R^2
pseudo-first order				pseudo-second order			Intraparticle diffusion		
0.2326	30.4	18.2348	0.9681	3.8297 x10^{-4}	32.4675	0.9998	0.6107	22.354	0.9227

4. Conclusion

In this study, AC was successfully prepared from tea leaves using a two-stage self-generated atmosphere method. The yield percentage ranged from 22.568 to 50.015 % while moisture content showed a low percentage ranging from 3.629 to 5.477 %. On the contrary, ash content ranged from 13.989 to 28.398 % indicating an undesirable result due to the impregnation ratio. The FTIR analysis showed that changes occur when AC was prepared from tea leaves. From the SEM micrographs, morphological changes on the ACs surfaces were observed. Comparison of tea leaves and ACs showed that pores were formed from a non-porous surface. Comparison of the different impregnation ratio AC showed that each impregnation ratio produced a different effect on the activated carbon. The maximum Brunauer, Emmett and Teller (BET) surface area of the best produced AC was found to be 695 m^2/g. Adsorption capacity showed that AC2 had the highest capacity among the other ACs. The theoretical maximum adsorption capacity of AC2 was found to be 44.25 mg g^{-1}. Multi-linearity in intraparticle diffusion plot suggested that the intraparticle diffusion was not the only rate limiting step. Adsorption kinetics showed that pseudo second-order fitted more when compared to pseudo first-order. Experimental results are in good agreement with Langmuir, Freundlich and Temkin isotherm with Langmuir giving the best compliance. The data suggested that adsorption rate was limited by chemisorption and the adsorption process was mostly a monolayer adsorption process.

Acknowledgement
This research was supported by the Research Management Center of Universiti Malaya in collaboration with the Center of Research and Innovation of Universiti Malaysia Sabah (Grant No. GL0111 or UM Project code: CG071-2013). These contributions are gratefully acknowledged.

References
[1] Swaminathan G and Ramanujam T K, 1999 *Bioprocess Eng.* **21** 169
[2] Nowicki P, Kazmierczak J and Robert P 2015 *Powder Technol.* **269** 312
[3] Joseph C G, Bono A, Anisuzzaman S M and Krishnaiah D 2014 *J. Appl. Sci.* **14** 3182
[4] Krishnaiah D, Anisuzzaman S M, Bono A and Sarbatly R 2013 *J. King Saud Univ. Sci.* **25** 251
[5] Anisuzzaman S M, Joseph C G, Daud W M A W, Krishnaiah D and Yee H S 2015, *Int. J. Ind. Chem* **6** 9
[6] Anisuzzaman S M, Joseph C G, Krishnaiah D, Bono A and Ooi L C 2015 Parametric and adsorption kinetic studies of methylene blue removal from textile simulated sample using Durian (*DuriozibethinusMurray*) skin *Manuscript accepted in the journal of Water Sci. Technol.*
[7] Anisuzzaman S M, Joseph C G, Taufiq-Yap Y H, Krishnaiah D and Tay V V 2015. *J. King Saud Univ. Sci.* (in press).
[8] Daud W M A W and Houshamnd A H 2010 *J. Nat. Gas. Chem.* **19** 267
[9] A. Alicia Peláez-Cid and M.M. Margarita Teutli-León 2012. Lignocellulosic Precursors Used in the Elaboration of Activated Carbon, Lignocellulosic Precursors Used in the Synthesis of Activated Carbon - Characterization Techniques and Applications in the Wastewater Treatment, Dr. Virginia Hernández Montoya (Ed.), InTech, DOI: 10.5772/39364.
[10] Özhan A, Ömer S, Küçük M M and Cafer S 2014 *Cellulose.* **21** 2457
[11] Virginia Hernández-Montoya and Adrian Bonilla-Petriciolet 2012, 1stCroatia Vol. 1
[12] Joseph C G and Yii F 2008. *J. Eng. Sci. Technol.* **3** 234
[13] Kim J W, Sohn M H, Kim D S, Sohn S M and Kwon Y S 2001 *J. Hazard. Mater.* **85** 301
[14] Karen J S, Greisly D M and Puello J 2014 *Chem.Eng. Transactions.* **37** 721
[15] Tan I A W, Ahmad A L and Hameed B H 2008 *J. Hazard. Mater* **154** 337–346
[16] Bulut Y and Aydın H 2006 *Desalination.* **194** 259
[17] Gupta V K, Agarwal S and Saleh T.A. 2011 *J. Hazard. Mater.* **185** 17
[18] Gupta V K and Nayak A 2012 *Chem. Eng. J.* **180** 81
[19] Hameed B H and Ahmad A A 2009 *J. Hazard. Mater.* **164** 870
[20] Hartono S B, Ismadji S, Sudaryanto Y and Irawaty W 2005 *J. Ind. Eng. Chem.* **11** 864
[21] Mittal A, Kaur D, Malviya A, Mittal J and Gupta V K 2009 *J. Colloid. Interf.Sci.* **337** 345.
[22] Tham Y J, Latif P A, Abdullah A M, Shamala-Devi A and Taufiq-Yap Y H 2011. *Bioresources Technol.* **102** 724
[23] Al-Qodah Z and Shawabkah R 2009 *Braz. J. Chem. Eng.* **26** 127
[24] Gundogdu A, Duran C, Senturk H B, Soylak M, Ozdes D, Serencam H and Imamoglu M 2012 *J. Chem. Eng. Data.* **57** 2733.
[25] Ahmaruzzaman M and Laxmi G S 2010 *J. Chem. Eng. Data.* **55** 4614
[26] Graham H N 1992 *Prev.Med.* **21** 334
[27] Wankhade A A and Ganvir V N 2013 *Int.Res. J. Environment Sci.* **2** 53
[28] Kennedy L J, Vijaya J J, Kayalvizhi K and Sekaran G 2007 *Chem. Eng. J.* **132** 279
[29] Sahu J N, Acharya J and Meikap B C 2010 *Bioresource Technol.* **101** 1974

IOP Conf. Series: Materials Science and Engineering **206** (2017) 012100 doi:10.1088/1757-899X/206/1/012100

Polymer-based adsorbent for heavy metals removal from aqueous solution

H N M E Mahmud[1*], A K O Huq[1,2] and R Yahya[1]
[1]Department of Chemistry, University of Malaya, 50603, Kuala Lumpur, Malaysia
[2]Department of Food Technology and Nutritional Science, Mawlana Bhashani Science and Technology, University, Santosh, Tangail-1902, Bangladesh

E-mail: ekramul@um.edu.my

Abstract. A novel conducting polymer-based adsorbent, polypyrrole (PPy) fine powder has successfully been prepared as a new adsorbent and utilized in the adsorption of heavy metal ions like arsenic, zinc and cadmium ions from aqueous solution. PPy was chemically synthesized by using $FeCl_3.6H_2O$ as an oxidant. The prepared PPy adsorbent was characterized by Brunauer-Emmet-Teller (BET) surface analysis, field emission scanning electron microscopy (FESEM) and attenuated total reflectance fourier transform infrared ATR-(FTIR) spectroscopy. The adsorption was conducted by varying different parameters such as, contact time, pH and adsorbent dosage. The concentrations of metal ions were measured by inductively coupled plasma mass spectroscopy (ICP-MS). The results show that PPy acts as an effective sorbent for the removal of arsenic, zinc and cadmium ions from aqueous solution. The as-prepared PPy fine powder is easy to prepare and appeared as an effective adsorbent for heavy metal ions particularly arsenic in wastewater treatment.

1. Introduction

Industrial wastewaters containing heavy metal ions such as As(III)/AS(V), Pb(II), Cd(II), Ni(I), Cr(III)/Cr(VI), Zn(II), Cu(II), Hg(I)/Hg(II), Co(II) etc. are directly or indirectly discharged into the streams, lakes, rivers or oceans [1]. Usually, heavy metals cannot be decomposed or biodegraded. Various methods such as chemical precipitation, ion-exchange, membrane filtration, adsorption, electrochemical technologies etc. are applied for the removal of various heavy metal ions from wastewater or aqueous solution [2]. Among them, adsorption is widely used because of low cost, simple design and ease of operation [3].

Some conducting polymer based-adsorbents like polyaniline, polypyrrole (PPy) and their composites have received considerable attention due to their potential application as adsorbents of heavy metal ions such as chromium, lead, zinc, copper, nickel etc. from aqueous solution [2,3].

Polypyrrole based-adsorbents play a major role in removing heavy metal ions due to their ease of synthesis, biocompatibility and redox properties. Since the adsorption efficiency largely depends on the preparation conditions of PPy, a lot of efforts have been made to remove heavy metal ions by polypyrrole conducting polymer prepared by chemical oxidative polymerization of pyrrole in presence of various oxidants at different conditions [4,5].

In this present effort, polypyrrole fine powder, a novel conducting polymer, has been chemically synthesized paying attention to the monomer to oxidant ratios for the removal of arsenic, zinc and cadmium ions from aqueous solution. The effects of different parameters such as adsorbent dosage, contact time and initial pH on the uptake of arsenic, zinc and cadmium ions have been investigated.

2. Materials and methods

2.1. Materials

High purity chemicals were used in all reagents and the stock solution was prepared using deionized water. Freshly distilled pyrrole (Sigma-Aldrich) was used as monomer and ferric chloride hexahydrate (Sigma-Aldrich) was used as an oxidant. HCl and NaOH (Analytical Grade) were used to control the pH of the stock solution.

IOP Conf. Series: Materials Science and Engineering **206** (2017) 012100 doi:10.1088/1757-899X/206/1/012100

2.2. Characterization

The surface morphology was observed by a Field Emission-Scanning Electron Microscope (FE-SEM) [Hitachi, SU 8220] and functional groups were observed by Attenuated Total Reflectance–Fourier Transform Infrared Spectrophotometer (ATR-FTIR) [Perkin-Elmer].

2.3. Polypyrrole synthesis

Polypyrrole adsorbent was prepared by chemical oxidative polymerization of pyrrole monomer using iron (III) chloride hexahydrate as an oxidant. The different mole ratios of monomer to oxidant (1: 0.25, 1: 0.5, 1:1, 1:2 and 1:3) were followed in aqueous solution for complete polymerization. The change of the initial color from green to finally black indicated the formation of PPy. The black mixture was then filtered and washed several times with distilled water to remove excess reactants before drying at 65 ^{o}C for 24h.

2.4. Batch adsorption

The batch adsorption was carried out for known concentration of arsenic, zinc and cadmium ions with the prepared Ppy adsorbents prepared from using 1:1 mole ratio of monomer to oxidant.

3. Results and discussion

Different physical appearances of PPy were observed with different amounts of yields during the synthesis of polypyrrole using different monomer to oxidant ratios. It was observed that the formation of PPy (yield) was increased with increasing the amount of oxidant at the same preparation conditions (Table 1). The higher monomer:oxidant ratio (1:3 and 1:2) gives the higher amount of yield but the monomer:oxidant ratio of 1:1 shows the maximum adsorption efficiency for all the heavy metal ions. Polymerization with higher oxidant ratios produces a very hard and compact polymer with less active site available for the binding interactions with metal ions.

Table 1. Monomer to oxidant ratio of polypyrrole versus physical appearance.

Monomer:oxidant ratio	Yield (g)	Physical Appearance
1:0.25	0.1247	Very thin layer, glazy
1:0.50	0.2986	Irregular thin leaves likes, glazy and shiny
1: 1	0.6549	Irregular flakes, glazy, shiny, fine powder
1:2	1.2172	Thick, muddy likes, not shiny
1:3	1.6831	Very thick, coarse and compact

On the other hand, polymerization with lower monomer to oxidant ratio (1:0.5) and (1:0.25) produces very thin and soft layer powder which appears very difficult to handle. Polymerization with 1:1 ratio produces an irregular flaky, shiny and stable fine powder with balanced amount of oxidant. It provides more active sites availability (nitrogen in amine functional groups of PPy) and therefore, the adsorption of heavy metal ions by polypyrrole adsorbent prepared from 1:1 mole ratio of monomer to oxidant has been reported.

3.1 Effect of adsorbent dosage

Figures 1-3 show the effect of adsorbent dosage on the adsorption of metal ions. It shows that the maximum efficiency of metal adsorption was exhibited by using 0.2-0.25 g of the prepared adsorbent. With the increase in PPy dosage beyond 0.2-0.25 g, the adsorption efficiency remained almost unaltered. More adsorbent dosage inhibits further adsorption of metal ions due to the saturation of the active sites.

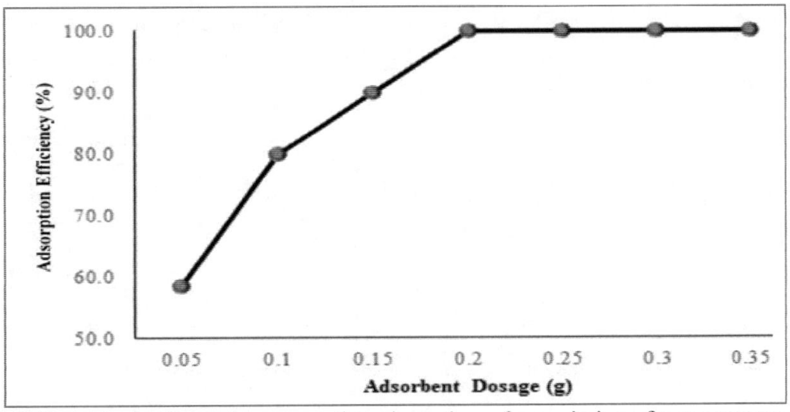

Figure 1. Effect of adsorbent dosage on the adsorption of arsenic ions from aqueous solution.

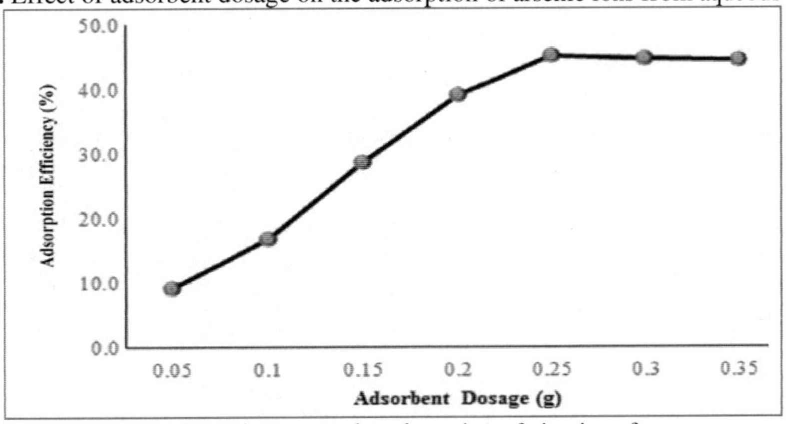

Figure 2. Effect of adsorbent dosage on the adsorption of zinc ions from aqueous solution.

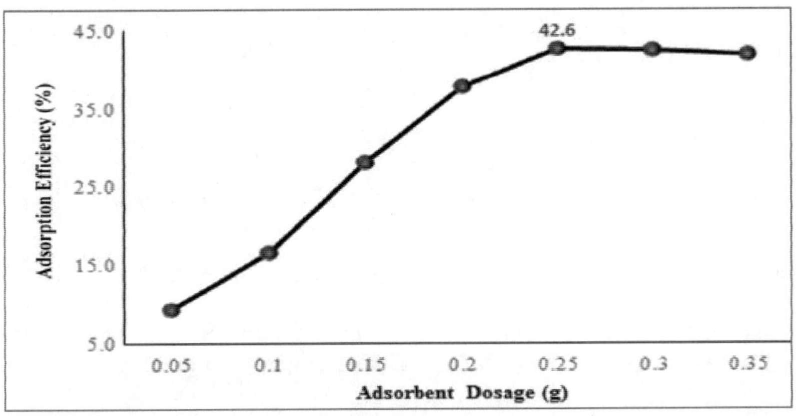

Figure 3. Effect of adsorbent dosage on the adsorption of cadmium ions from aqueous solution.

3.2. Effect of contact time

The adsorption efficiency of arsenic ions on certain dosage (0.2g) was measured as a contact time of 1, 2, 4, 6, 8 and 10 hours keeping all other parameters constant. Figure 4 shows the effect of these contact time and indicated that maximum efficiency was found at a contact time of 6 hours. It was observed that the adsorption efficiency increased very first in first few hours and then slowly increased over the time upto 6 hours and beyond that time adsorption efficiency was steady. This is due to equilibrium that has been achieved at that time.

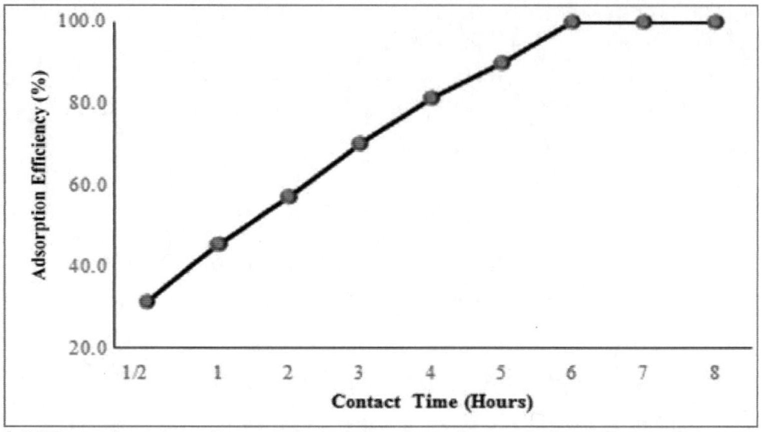

Figure 4. Effect of contact time on the adsorption of arsenic ions from aqueous solution.

The adsorption efficiency of zinc ions on certain dosage (0.25g) of PPy adsorbent was measured at a contact time of 15 minutes interval keeping all other parameters constant. The adsorption of zinc ion is dependent on contact time and the maximum efficiency (45.1%) was found at a contact time of 75 minutes (Figure 5).

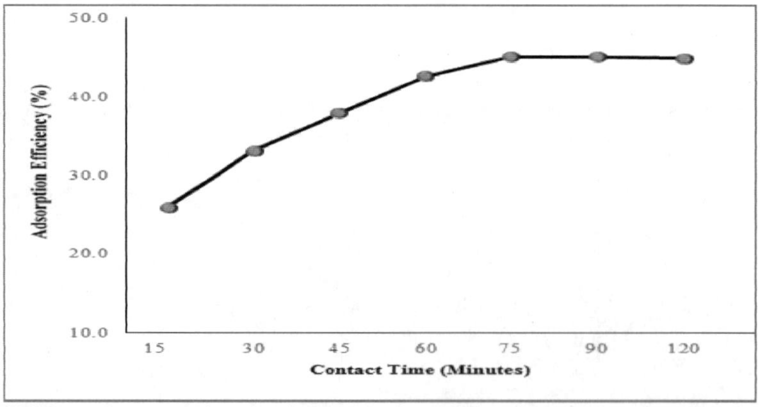

Figure 5. Effect of contact time on the adsorption of zinc ions from aqueous solution.

Figure 6 showed that the adsorption efficiency of cadmium ions was rapid at first 30 minutes and then slowly increased over the time up to 150 minutes and beyond that time adsorption efficiency was steady.

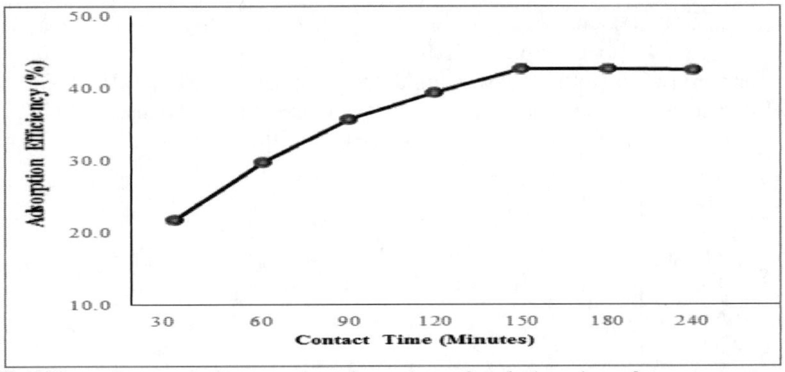

Figure 6. Effect of contact time on the adsorption of cadmium ions from aqueous solution.

3.3. Effect of pH

All heavy metal ions are pH sensitive. In case of arsenic ion adsorption by several polymeric and non-polymeric adsorbents, the optimum pH range is 5-7 [6]. In this study, it was observed that the optimum pH was 6.5 at certain dosage and constant conditions for the removal of arsenic ions from aqueous solution (Figure 7).

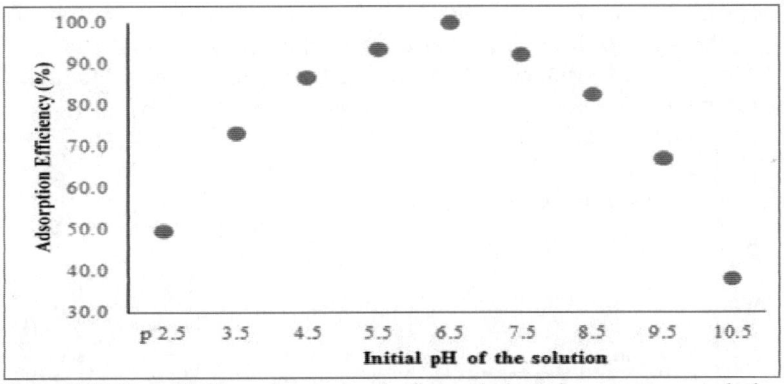

Figure7. Effect of pH on the removal of arsenic ions from aqueous solution.

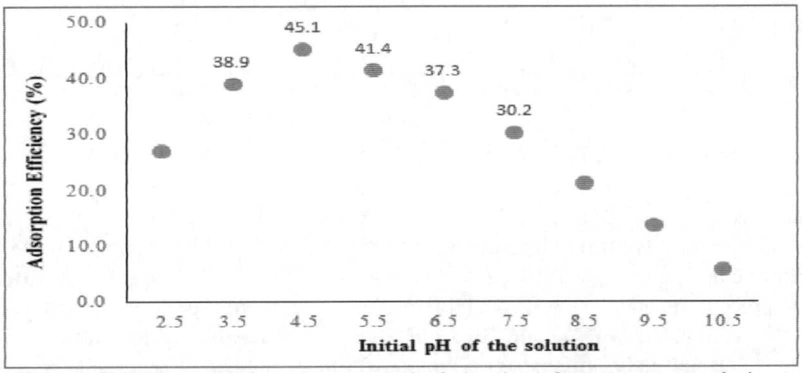

Figure 8. Effect of pH on the removal of zinc ions from aqueous solution.

In the case of zinc ion adsorption from aqueous solution, the optimum range of pH was found 3-7 using several adsorbents [7]. Figure 8 shows the effect of pH on zinc ion sorption by PPy adsorbent.

The adsorption of zinc ions onto PPy was highly pH dependent and the optimum pH was found to be 4.5. The effect of initial pH on cadmium sorption onto PPy adsorbent was studied at pH 2.5-10.5. The adsorption of cadmium ions onto PPy was highly pH dependent and the pH 5.5 has been found as the optimum pH for cadmium ion adsorption (Figure 9). Several researchers showed the optimum range of pH of cadmium ions adsorption was found between 4-7 [8,9].

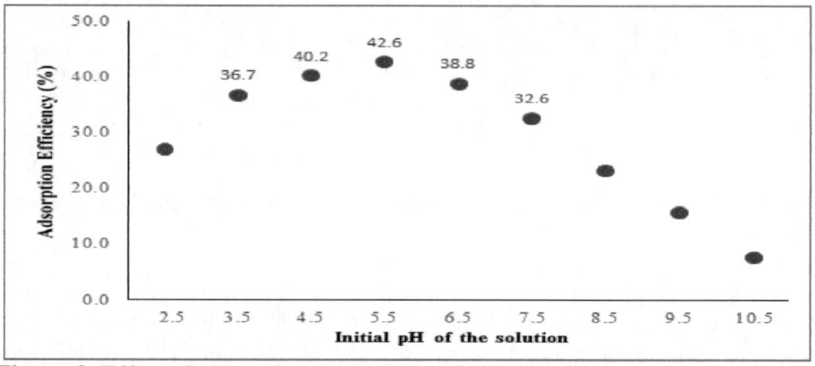

Figure 9. Effect of pH on the removal of cadmium ions from aqueous solution.

3.4. FTIR spectroscopy

Figure10 shows the ATR-FTIR spectra of PPy (1:1) before adsorption (a) and after adsorption (b) of arsenic ions. All peak band positions of PPy before adsorption was shifted to higher wavelength after arsenic ions adsorption.

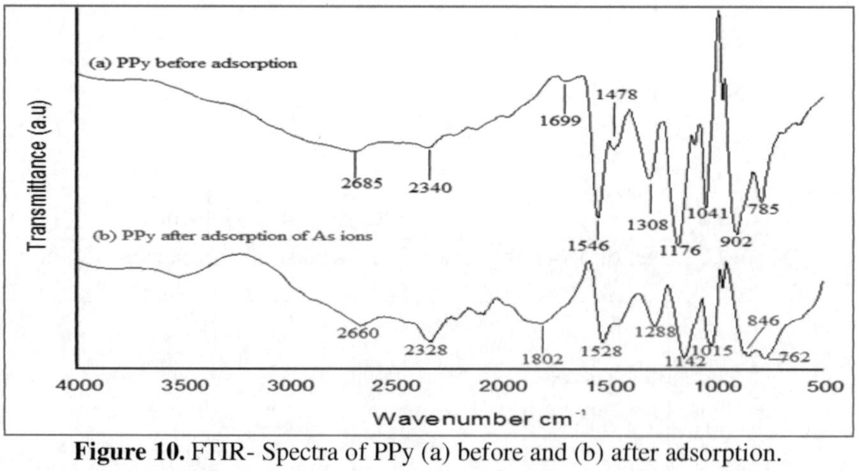

Figure 10. FTIR- Spectra of PPy (a) before and (b) after adsorption.

The bands at 762 cm^{-1}, 841 cm^{-1}, 961 cm^{-1}, 1022 cm^{-1}, 1139 cm^{-1}, 1285 cm^{-1}, 1524 cm^{-1} and 2173 cm^{-1} observed in PPy spectrum before the adsorption of metal ions have shifted to 782 cm^{-1}, 887 cm^{-1}, 964 cm^{-1}, 1037 cm^{-1}, 1158 cm^{-1}, 1301 cm^{-1}, 1537 cm^{-1} and 2293 cm^{-1}, respectively, after adsorption. All these shifts indicate the adsorption of arsenic ions by PPy conducting polymer fine powder. The bands at 1037 cm^{-1} due to N-H wagging and 1301 cm^{-1} due to C-N in-plane deformation shifted to 1022 cm^{-1} and 1285 cm^{-1}, respectively, after arsenic ions adsorption, suggests the metallic bond formation with nitrogen for PPy. Thus, from the FTIR spectrum, it can be evidenced that the fine powder from balanced oxidation polymerization shows maximum adsorption due to the complexation ionic adsorption affinity between the functionalized amine groups and arsenic ions.

The similar type of band shifts was observed for the adsorption of zinc and cadmium ions by the prepared polypyrrole adsorbent.

3.5. FE-SEM analysis

The morphology of prepared polypyrrole (Figure 11a and 11b) was studied by using a field emission scanning electron microscope. In Figure 11(a),flowery like compact surface indicates the polypyrrole formation. There was a very little visual changes observed in FE-SEM micrograph after the adsorption of metal ions (Figure 11.b). Similar micrographs have been observed for other heavy metal ions (zinc and cadmium). In this study, the surface area does not play the main role in adsorption by PPy like all other adsorbents but the nitrogen atoms in PPy are mainly responsible for arsenic ions adsorption. Similar results were found in another study of metal ions sorption [10].

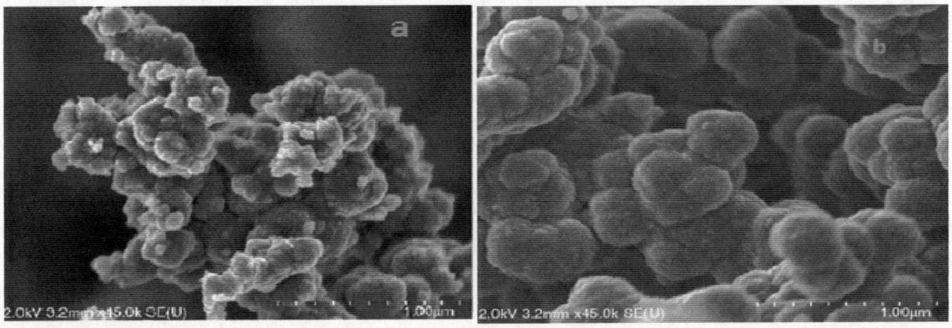

Figure 11. FE-SEM micrographs of PPy adsorbent (a) before and (b) after adsorption of metal ions (arsenic).

4. Conclusions

Polypyrrole conducting polymer prepared by chemical oxidation using iron (III) chloride hexahydrate has exhibited complete adsorption efficiency for the removal of arsenic ions from aqueous solution. The fine polypyrrole powder prepared from using the monomer to oxidant ratio of 1:1 appeared to be a very efficient (100%) conducting polymer-based adsorbent for arsenic ions while a lower efficiency was observed for cadmium and zinc ions. The complete removal of arsenic ions by Ppy from aqueous solution encourages to study further for it to be used to treat wastewater. The low preparation cost and ease of preparation appear attractive for its commercialization particularly for heavy metal ions removal from aqueous solutions.

Acknowledgment

The authors would like to acknowledge the financial support from University of Malaya Research Grant (UMRG) (grant no. RP014A-15SUS) under IPPP of University of Malaya.

References

[1] Nriagu J O and Pacyna J M 1988 *Nature* **333** 134.
[2] Bernardo G R and Rene R M 2009 *J. Hazard. Mater.***170** **(2)** 845.
[3] Javadian H 2014 *J. Ind. Eng. Chem.* **20** 4233.
[4] Deng S and Bai R 2004 *Water Res.* **38** 2424.
[5] Weidlich C Mangold K and Jüttner K 2001 *Electrochimica Acta* **47** 741.
[6] Ioanis A K and Zouboulis A I 2002 *Water Res.***36** 5141.
[7] Bhattacharya A K, Mandal S N and Das S K 2006 *Chem. Eng.* **123 (1)** 43.
[8] Mansour M S, Ossman M E and Farag H A 2011 *Desalination* **272 (1)** 301.
[9] Hasani T and Eisazadeh H 2013 *Synth. Metals* **175** 15.
[10] Jia Y F, Xiao, B and Thomas K M 2002 *Langmuir* **18** 470.